Lecture Notes in Artificial Intelligence 2801

Edited by J. G. Carbonell and J. Siekmann

Subseries of Lecture Notes in Computer Science

Springer
Berlin
Heidelberg
New York
Hong Kong
London
Milan
Paris
Tokyo

Wolfgang Banzhaf Thomas Christaller
Peter Dittrich Jan T. Kim
Jens Ziegler (Eds.)

Advances in Artificial Life

7th European Conference, ECAL 2003
Dortmund, Germany, September 14-17, 2003
Proceedings

Springer

Series Editors

Jaime G. Carbonell, Carnegie Mellon University, Pittsburgh, PA, USA
Jörg Siekmann, University of Saarland, Saarbrücken, Germany

Volume Editors

Wolfgang Banzhaf
Jens Ziegler
University of Dortmund, Department of Computer Science 11
Joseph-von-Fraunhofer-Str. 20, 44227 Dortmund, Germany
E-mail: {banzhaf, jens.ziegler}@cs.uni-dortmund.de

Thomas Christaller
Fraunhofer Institute for Autonomous Intelligent Systems (AIS)
Schloss Birlinghoven, 53754 Sankt Augustin, Germany
E-mail: thomas.christaller@ais.fraunhofer.de

Peter Dittrich
University of Jena, Institute of Computer Science
Bio Analysis Group, 07743 Jena, Germany
E-mail: dittrich@cs.uni-jena.de

Jan T. Kim
University of Lübeck, Institute of Neuro- and Bioinformatics
Seelandstr. 1a, 23569 Lübeck, Germany
E-mail: kim@inb.uni-luebeck.de

Cataloging-in-Publication Data applied for

A catalog record for this book is available from the Library of Congress

Bibliographic information published by Die Deutsche Bibliothek
Die Deutsche Bibliothek lists this publication in the Deutsche Nationalbibliographie; detailed bibliographic data is available in the Internet at <http://dnd.ddb.de>.

CR Subject Classification (1998): I.2, J.3, F.1.1-2, G.2, H.5, I.5, J.4, J.6

ISSN 0302-9743
ISBN 3-540-20057-6 Springer-Verlag Berlin Heidelberg New York

Springer-Verlag Berlin Heidelberg New York,
a member of BertelsmannSpringer Science+Business Media GmbH

http://www.springer.de

Typesetting: Camera-ready by author, data conversion by PTP-Berlin GmbH
Printed on acid-free paper SPIN: 10931943 06/3142 5 4 3 2 1 0

Preface

Artificial Life has come a long way. Since the inception of the field about 16 years ago with a seminal workshop at the Santa Fe Institute, the field has developed quickly. Its interdisciplinary roots have been both a blessing and a curse. Critical people would say that nothing was new in the ideas of Artificial Life, since many other disciplines had addressed the very same questions, though probably under different names. Other critics would state that the difficulty of interacting in an interdisciplinary way with colleagues from so many other and divergent fields would be so great that true progress could not come from such an enterprise, as those involved would be too busy understanding – or misunderstanding – each other. Admirers, on the other hand, would speak of a bold new attack on the most fascinating questions of science with this new approach. Others would say that new perspectives were opened by the questions the area of Artificial Life asked so pointedly. For those involved in this effort over some years, it has always been very interesting and fascinating to work on these questions.

From our discussions it also seems that Artificial Life is beginning to become mainstream. Evolutionary biology, computational and systems biology, and computational social science, to name a few, are disciplines benefitting from ideas hedged in Artificial Life. This, plus the success of open-ended evolutionary games in the entertainment industry, the sensibility achieved with decades of work behind us in artificial evolutionary approaches with fixed fitness measures, and the development of technology towards a networked, asynchronous, world of interacting entities, have all conspired to prepare the floor for Alife research coming into its own. Notably the concept of emergence of new qualities from the interaction of entities without this quality has been a huge success in recent years.

Artificial Life in Europe has always been a vivid activity. Beginning in 1991, a series of workshops and conferences have been held in close coordination with the emerging International Conference on Artificial Life. The present conference is the seventh in a series begun in Paris (France) in 1991, with its successors in Brussels (Belgium) in 1993, Granada (Spain) in 1995, Sussex (England) in 1997, Lausanne (Switzerland) in 1999, leading up to the preceding conference 2001 in Prague. They have left traces in the form of proceedings volumes published by Springer-Verlag or MIT Press. This, the 7th European Artificial Life Conference, held in Dortmund, Germany in September 2003, was no exception.

We would like to cordially thank Gundel Jankord for organizing the conference secretariat, Christina Lorenz and Philip Limbourg for their help in preparing these proceedings, our students at Dortmund University for their help in running the conference, and Alfred Hofmann from Springer-Verlag for accepting the conference into the LNCS/LNAI series. Dr. Holger Lange, Prof.Dr. Thomas Martinetz, and Dr. Frank Schweitzer, the other members of our organizing committee, did a wonderful job, and the members of this year's program committee are to be congratulated on their fine selection of papers. We would like to thank the additional reviewers who were really indispensable for getting the decisions right. Numerous others made themselves available when we needed them, and we would like to express our sincere gratitude to all who helped make ECAL 2003 in Dortmund possible.

More than 140 submissions are testimony to the fact that Alife is an active research area. The contributions selected for these proceedings, from various fields from artificial chemistries to robotics and autonomous agents, show the breadth and depth of thinking in our area. Origin of life, life-like behavior, and phenomena of the living are all central pieces of this endeavour. Evolution, development, and learning/adaptation are means to achieve survival in ever changing environments. These provide the themes in our community, and technology is both driving the progress and profiting from its results. With a number of invited lectures we tried to open our eyes to the development both within our field and beyond.

We think that the mystery of life and the living is still not solved. We hope that Alife research is one of the approaches that stands to answer particular questions in this regard, and stands to answer them decisively. Whether and when this happens is still an open question. May the meeting in Dortmund and these proceedings help us to move forward in this quest to solve one of the most fascinating problems of humankind.

June 2003

Wolfgang Banzhaf
Thomas Christaller
Peter Dittrich
Jan T. Kim
Jens Ziegler

Organization

ECAL 2003 was organized in collaboration with the Department of Computer Science, University of Dortmund and the Fraunhofer Institute for Autonomous Intelligent Systems (FhG-AIS), Sankt Augustin, Germany. It was held in the Zeche Zollern II/IV in Dortmund Bövinghausen, which is part of the Westfälische Industriemuseum.

Organization Committee

Program Co-chair:	Wolfgang Banzhaf University of Dortmund, Germany
Program Co-chair:	Thomas Christaller Fraunhofer AIS, Germany
Local Chair:	Jens Ziegler University of Dortmund, Germany
Tutorials:	Jan T. Kim Medical University of Lübeck, Germany
Workshops:	Peter Dittrich Friedrich-Schiller University Jena, Germany
Travel Grants:	Holger Lange Norwegian Forest Research Institute, Norway
Publicity:	Thomas Martinetz Medical University of Lübeck, Germany
Social Events:	Frank Schweitzer Fraunhofer AIS, Germany

Program Committee

A. Adamatzky
M. Asada
W. Banzhaf (Co-chair)
M. Bedau
H. Bersini
E. Bonabeau
S. Bullock
A. Cangelosi
T. Christaller (Co-chair)
K. Dautenhahn
P. Dittrich
K. Downing
M. Ebner
D. Floreano
I. Harvey
T. Hashimoto
P. Hogeweg
J. Kelemen
J. Kim
C. Lindgren
T. Martinetz
B. McMullin
J. Merelo
J.-A. Meyer
A. Moreno
C. Nehaniv

R. Pfeifer
D. Polani
S. Rasmussen
T. Ray
F. Schweitzer
M. Sipper
P. Stadler
C. Taylor
C. Teuscher
A. Thompson
C. Wilke
A. Wuensche
J. Ziegler

Additional Reviewers

C. Adami
D. Bisig
J. Blynel
J. Bongard
S. Bovet
C. Buckley
J. Cartlidge
F. Centler
D. Franks
T. Glotzmann
D. Harris
C. Hemelrijk
D. Koblitz
M. Krafft
C. Krause
H. Kunz
C.W.G. Lasarczyk
D. Marocco
C. Mattiussi
N. Matsumaru
N. Mitsunaga
Y. Nagai
M. Ogino
A. Perez-Uribe
C. Rawichote
F. Rochner
T. Schöler
T. Smaoui
P. Speroni di Fenizio
Y. Tada
T. Takuma
D. Thomas
I. Tsuda
J. Wantia
H. Williams
Y. Yoshikawa
K.-P. Zauner

Sponsoring Institutions

Deutsche Forschungsgemeinschaft, Germany
European Network of Excellence in Evolutionary Computation (EvoNet)
Gesellschaft der Freunde der Universität Dortmund e.V., Germany
Gesellschaft für Informatik e.V., Germany
Universität Dortmund, Germany
Westfälisches Industriemuseum

Table of Contents

Artificial Chemistries, Self-Organization, and Self-Replication

A Universal Framework for Self-Replication 1
Bryant Adams, Hod Lipson

Generic Properties of Chemical Networks: Artificial Chemistry Based on Graph Rewriting 10
Gil Benkö, Christoph Flamm, Peter F. Stadler

How to Program Artificial Chemistries 20
Jens Busch, Wolfgang Banzhaf

Artificial Life as an Aid to Astrobiology: Testing Life Seeking Techniques 31
Florian Centler, Peter Dittrich, Lawrence Ku, Naoki Matsumaru, Jeffrey Pfaffmann, Klaus-Peter Zauner

Molecularly Accessible Permutations 41
Dónall A. Mac Dónaill

Simulating Evolution's First Steps 51
Tim J. Hutton

Cellular Evolution in a 3D Lattice Artificial Chemistry 59
Duraid Madina, Naoaki Ono, Takashi Ikegami

Evolution of Rewriting Rule Sets Using String-Based Tierra 69
Komei Sugiura, Hideaki Suzuki, Takayuki Shiose, Hiroshi Kawakami, Osamu Katai

Models for the Conservation of Genetic Information with String-Based Artificial Chemistry 78
Hideaki Suzuki

Interaction Based Evolution of Self-Replicating Loop Structures 89
Keisuke Suzuki, Takashi Ikegami

Artificial Societies

Architectural Design for the Survival Optimization of Panicking Fleeing Victims 97
Rodrigo Escobar, Armando De La Rosa

Meta-evolutionary Game Dynamics for Mathematical Modelling of Rules Dynamics ... 107
Takashi Hashimoto, Yuya Kumagai

Preventing Bluff Agent Invasions in Honest Societies ... 118
Robert Lowe, Daniel Polani

Effects of Group Composition and Level of Selection in the Evolution of Cooperation in Artificial Ants ... 128
Andres Perez-Uribe, Dario Floreano, Laurent Keller

Effects of Learning to Interact on the Evolution of Social Behavior of Agents in Continuous Predators-Prey Pursuit Problem ... 138
Ivan Tanev, Katsunori Shimohara

War and Peace among Artificial Nations – A Model and Simulation Based on a Two-Layered Multi-agent System ... 146
Tatsuo Unemi, Yoshiaki Kaneko, Ichiro Takahashi

Cellular and Neural Systems

Developmental Neural Networks for Agents ... 154
Andy Balaam

Revisiting Idiotypic Immune Networks ... 164
Hugues Bersini

Production of Gliders by Collisions in Rule 110 ... 175
Genaro Juárez Martínez, Harold V. McIntosh, Juan Carlos Seck Tuoh Mora

A Computational Model of Neocortical-Hippocampal Cooperation and Its Application to Self-Localization ... 183
Michail Maniadakis, Panos Trahanias

Fascinating Rhythms by Chaotic Hopfield Networks ... 191
Colin Molter, Hugues Bersini

Solving a Delayed Response Task with Spiking and McCulloch-Pitts Agents ... 199
Keren Saggie, Alon Keinan, Eytan Ruppin

First Steps in Evolving Path Integration in Simulation ... 209
Robert Vickerstaff

Evolution and Development

On the Dynamics of an Artificial Regulatory Network ... 217
Wolfgang Banzhaf

Evolution and Growth of Virtual Plants 228
Marc Ebner

Evolvability of the Genotype-Phenotype Relation in Populations of Self-Replicating Digital Organisms in a Tierra-Like System 238
Attila Egri-Nagy, Chrystopher L. Nehaniv

An Evolutionary Approach to Damage Recovery of Robot Motion with Muscles 248
Siavash Haroun Mahdavi, Peter J. Bentley

Evolving Developmental Programs for Adaptation, Morphogenesis, and Self-Repair 256
Julian F. Miller

Evolving Embodied Genetic Regulatory Network-Driven Control Systems 266
Tom Quick, Chrystopher L. Nehaniv, Kerstin Dautenhahn, Graham Roberts

Evolution of Fault-Tolerant Self-Replicating Structures 278
Ludovic Righetti, Solaiman Shokur, Mathieu S. Capcarrere

Evolving the Ability of Limited Growth and Self-Repair for Artificial Embryos 289
Felix Streichert, Christian Spieth, Holger Ulmer, Andreas Zell

Evolutionary and Adaptive Dynamics

Caring versus Sharing: How to Maintain Engagement *and* Diversity in Coevolving Populations 299
John Cartlidge, Seth Bullock

Culture and the Baldwin Effect 309
Diego Federici

Evolutionary Network Minimization: Adaptive Implicit Pruning of Successful Agents 319
Zohar Ganon, Alon Keinan, Eytan Ruppin

Population Dynamics under Spatially and Temporally Heterogenous Resource Limitations in Multi-agent Networks 328
Thomas Glotzmann, Holger Lange, Michael Hauhs

Adaptive Coupling and Intersubjectivity in Simulated Turn-Taking Behaviour 336
Hiroyuki Iizuka, Takashi Ikegami

Distributed Genetic Algorithm: Learning by Direct Exchange of Chromosomes 346
Aleš Kubík

An Approach to Describe the Tierra Instruction Set Using Microoperations: The First Result 357
Shuichi Matsuzaki, Hideaki Suzuki, Minetada Osano

Critical Values in Asynchronous Random Boolean Networks 367
Bertrand Mesot, Christof Teuscher

Artificial Organisms That Sleep 377
Marco Mirolli, Domenico Parisi

Visualizing Evolutionary Dynamics of Self-Replicators Using Graph-Based Genealogy 387
Chris Salzberg, Antony Antony, Hiroki Sayama

The Baldwin Effect Revisited: Three Steps Characterized by the Quantitative Evolution of Phenotypic Plasticity 395
Reiji Suzuki, Takaya Arita

Does the Red Queen Reign in the Kingdom of Digital Organisms? 405
Claus O. Wilke

Languages and Communication

Conditions for Stable Vowel Systems in a Population 415
Bart de Boer

Piep Piep Piep – Ich Hab' Dich Lieb: Rhythm as an Indicator of Mate Quality 425
Eva van den Broek, Peter M. Todd

Evolving Agent Societies with VUScape 434
P.C. Buzing, A.E. Eiben, M.C. Schut

Why Synonymy Is Rare: Fitness Is in the Speaker 442
James R. Hurford

A Noisy Way to Evolve Signaling Behaviour 452
Tudor Jenkins

Language Games with Mixed Populations 462
Michael Lewin, Emmet Spier

Artificial Agents and Natural Determiners 472
Joris Van Looveren

Coevolution of Birdsong Grammar without Imitation 482
Kazutoshi Sasahara, Takashi Ikegami

Systemic Architecture for Audio Signal Processing 491
Rolf Schatten

Semantic Generalisation and the Inference of Meaning 499
Andrew D.M. Smith

Language Evolution in Populations: Extending the Iterated
Learning Model 507
Kenny Smith, James R. Hurford

Learning Biases for the Evolution of Linguistic Structure:
An Associative Network Model 517
Kenny Smith

The Learning and Emergence of Mildly Context Sensitive Languages 525
Edward P. Stabler, Travis C. Collier, Gregory M. Kobele,
Yoosook Lee, Ying Lin, Jason Riggle, Yuan Yao, Charles E. Taylor

THSim v3.2: The Talking Heads Simulation Tool 535
Paul Vogt

Grounded Lexicon Formation without Explicit Reference Transfer:
Who's Talking to Who? 545
Paul Vogt

Optimal Communication in a Noisy and Heterogeneous Environment 553
Willem Zuidema

Methodologies and Applications

A Clustering Algorithm Based on the Ants Self-Assembly Behavior 564
H. Azzag, N. Monmarché, M. Slimane, C. Guinot, G. Venturini

A Multi-agent Based Approach to Modelling and Rendering of 3D
Tree Bark Textures 572
Ban Tao, Zhang Changshui, Shu Wei

Measuring the Dynamics of Artificial Evolution 580
Mikhail S. Burtsev

Pattern Recognition in a Bucket 588
Chrisantha Fernando, Sampsa Sojakka

Discovering Clusters in Spatial Data Using Swarm Intelligence 598
Gianluigi Folino, Agostino Forestiero, Giandomenico Spezzano

When Can We Call a System Self-Organizing? 606
Carlos Gershenson, Francis Heylighen

Contextual Random Boolean Networks 615
Carlos Gershenson, Jan Broekaert, Diederik Aerts

Representation of Genotype and Phenotype in a Coherent Framework
Based on Extended L-Systems 625
Ole Kniemeyer, Gerhard H. Buck-Sorlin, Winfried Kurth

Integrated-Adaptive Genetic Algorithms 635
Henri Luchian, Ovidiu Gheorghieş

Simulating the Evolution of Ant Behaviour in Evaluating Nest Sites 643
James A.R. Marshall, Tim Kovacs, Anna R. Dornhaus,
Nigel R. Franks

Evolving Evolutionary Algorithms Using Multi Expression
Programming 651
Mihai Oltean, Crina Groşan

Modelling Artificial Ecosystem Selection: A Preliminary
Investigation 659
Alexandra Penn

Measuring Self-Organization via Observers 667
Daniel Polani

General Framework for Evolutionary Activity 676
Michael J. Raven, Mark A. Bedau

Developing and Testing Methods for Microarray Data Analysis
Using an Artificial Life Framework 686
Dirk Repsilber, Jan T. Kim

Approaching Virtual Organism by PheGe 696
Klaus Seidl

Robustness to Damage of Biological and Synthetic Networks 706
Roberto Serra, Marco Villani, Alessandro Semeria

An Agent-Based Approach to Routing in Communications Networks
with Swarm Intelligence 716
Hengwei Shen, Shoichiro Asano

Biomorphs Implemented as a Data and Signals Cellular Automaton 724
André Stauffer, Moshe Sipper

Robotics and Autonomous Agents

Analyzing the Performance of "Winner-Take-All" and "Voting-Based" Action Selection Policies within the Two-Resource Problem ... 733
Orlando Avila-García, Lola Cañamero, René te Boekhorst

Are There Representations in Embodied Evolved Agents? Taking Measures ... 743
Hezi Avraham, Gal Chechik, Eytan Ruppin

Evolving Fractal Gene Regulatory Networks for Robot Control ... 753
Peter J. Bentley

Explorations of Task-Dependent Visual Morphologies in Competitive Co-evolutionary Experiments ... 763
Gunnar Buason, Tom Ziemke

Optimal Morphology of a Biologically-Inspired Whisker Array on an Obstacle-Avoiding Robot ... 771
Miriam Fend, Hiroshi Yokoi, Rolf Pfeifer

Phase Transitions in Self-Organising Sensor Networks ... 781
Mark Foreman, Mikhail Prokopenko, Peter Wang

Artificial Metabolism: Towards True Energetic Autonomy in Artificial Life ... 792
Ioannis Ieropoulos, Chris Melhuish, John Greenman

An Imitation Game for Emerging Action Categories ... 800
Bart Jansen

Multi-agent Model of Biological Swarming ... 810
Robert Mach, Frank Schweitzer

Visually Guided Physically Simulated Agents with Evolved Morphologies ... 821
Ian Macinnes

The Robot in the Swarm: An Investigation into Agent Embodiment within Virtual Robotic Swarms ... 829
Chrystopher L. Nehaniv, Kerstin Dautenhahn, Adam Lee Newton

Building a Hybrid Society of Mind Using Components from Ten Different Authors ... 839
Ciarán O'Leary, Mark Humphrys

Requirements for Getting a Robot to Grow up ... 847
Peter Ross, Emma Hart, Alistair Lawson, Andrew Webb, Erich Prem, Patrick Poelz, Giovanna Morgavi

Coevolving Communication and Cooperation for Lattice Formation Tasks ... 857
Jekanthan Thangavelautham, Timothy D. Barfoot, Gabriele M.T. D'Eleuterio

Evolving Aggregation Behaviors in a Swarm of Robots ... 865
Vito Trianni, Roderich Groß, Thomas H. Labella, Erol Şahin, Marco Dorigo

Low-Level Visual Homing ... 875
Andrew Vardy, Franz Oppacher

Controlling a Simulated Khepera with an XCS Classifier System with Memory ... 885
Andrew Webb, Emma Hart, Peter Ross, Alistair Lawson

Collective Decision-Making and Behaviour Transitions in Distributed Ad Hoc Wireless Networks of Mobile Robots: Target-Hunting ... 893
Jan Wessnitzer, Chris Melhuish

Author Index ... 903

A Universal Framework for Self-Replication

Bryant Adams[1] and Hod Lipson[2]

[1] Department of Mathematics,
[2] Department of Mechanical and Aerospace Engineering,
Cornell University, Ithaca NY 14853, USA
badams@math.cornell.edu, hod.lipson@cornell.edu

Abstract. Self-replication is a fundamental property of many interesting physical, formal and biological systems, such as crystals, waves, automata, and especially forms of natural and artificial life. Despite its importance to many phenomena, self-replication has not been consistently defined or quantified in a rigorous, universal way. In this paper we propose a universal, continuously valued property of the interaction between a system and its environment. This property represents the effect of the presence of such a system upon the future presence of similar systems. We demonstrate both analytical and computational analysis of self-replicability factors for three distinct systems involving both discrete and continuous behaviors.

1 Overview and History

Self-replication is a fundamental property of many interesting physical, formal, and biological systems, such as crystals, waves, automata, and especially forms of natural and artificial life [1]. Despite its importance to many phenomena, self-replication has not been consistently defined or quantified in a rigorous, universal way. In this paper we propose a universal, continuous valued property of the interaction between a system and its environment. This property represents the effect of the presence of such a system upon the future presence of similar systems. Subsequently, we demonstrate both analytical and computational analysis of self-replicability factors for three distinct systems involving both discrete and continuous behaviors.

Two prominent issues arise in examining how self-replication has been handled when trying to extend the concept universally: how to deal with non-ideal systems and how to address so-called 'trivial' cases [2,3]. Moore [4] requires that in order for a configuration to be considered self-reproducing it must be capable of causing arbitrarily many offspring; this requirement extends poorly to finite environments. Lohn and Reggia [5] put forward several cellular-automata (CA) -specific definitions, and result in a binary criterion. A second issue that arose in the consideration of self-replicating automata was that some cases seemed too trivial for consideration, such as an 'all-on' CA, resulting in a requirement for Turing-universality [6].

W. Banzhaf et al. (Eds.): ECAL 2003, LNAI 2801, pp. 1–9, 2003.

The definition for self-replicability we propose here is motivated in part by (a) A desire to do more than look at self-replication as a binary property applicable only to certain automata, and, (b) The goal of encapsulating a general concept in a means not reliant upon (but compatible with) ideal conditions.

We wish to do this by putting self-replication on a scale that is algorithmically calculable, quantifiable, and continuous. Such a scale would allow for comparisons, both between the same system in different environments, determining ideal environments for a system's replication, as well as between different systems in the same environment, if optimizing replicability in a given environment is desired.

Rather than viewing self-replicability as a property purely of the system in question, we view it as a property of the interaction between a system and its environment. Self-Replication, as we present it, is a property embedded and based upon information, rather than a specific material framework. We construct replicability as a property relative to two different environments, which indicates the degree to which one environment yields a higher presence of the system over time. *Self*-replicability, then, is a comparison between an environment lacking the system and an environment in which the system is present. We will first introduce a number of definitions, and then give examples of replicability of three types of systems.

2 Definitions

Definition 1: ***Environment.*** *By Environment we denote a single state of a (presumably closed) system.*

For example, with a closed system such as a 5×5 grid, each of whose cells may be either 'on' or 'off', one environment, E_1, would be 'all 25 cells off', while another, E_2, might be 'every other cell on'.

Definition 2: ***Set of configurations.*** *Given an environment E, the set of configurations* $\overline{E}$ *is the set of all possible states of the system that includes state E. We call elements of* $\overline{E}$ *"E-configurations".*

We do not assume an arrow of time, and thus if, for some environments E_1, E_2, we have $E_1 \in \overline{E_2}$, then $E_2 \in \overline{E_1}$ and $\overline{E_1} = \overline{E_2}$. In the 5×5 grid, for example, the set of configurations would be the collection of all 2^{25} states the grid could be in.

Definition 3: ***Time development function.*** *A time development function is a map* $T : \overline{E} \times \Re^{+} \to \overline{E}$, *(respectively* $T : \overline{E} \times Z^{+} \to \overline{E}$ *for discrete time systems) constrained by* $T(E,0)=E$ *and* $T(T(E,y),x)=T(E,x+y)$.

With a time development function, we operate under the assumption that the progression between states as time passes is externally deterministic. We also assume there is no difference between a system that 'ages' five units and then ten units, and a system that first ages ten units, and then ages five units. When we write only *T(E)* rather than *T(E, n)*, it is assumed there is a natural increment of time and we are using *T(E, 1)*.

Definition 4: *Subsystem set.* *We denote by X* the collection of all subsystems of a given system X, and say X* is the subsystem set of X.*

Often, we are most interested in the subsystem set of an environment. For example, in the case where the system is a 5×5 grid and *E* is a state with all points being off, *E** would include, among its 2^{25} elements, four 4×4 binary grids with all points being off, five 1×5 binary grids with all points being off, and a number of L-shaped binary grids with all points being off. For further example, given a second system E_2, a 5×5 grid in which the top row of points were on, the rest off, E_2* would still have 2^{25} elements, each the same shape as a corresponding element in *E**, but now some elements would include 'on' points.

Definition 5: *Possible Subsystems.* *We define the possible subsystems* $\overline{E}^*$ *as the union of all F* such that* $F \in \overline{E}$.

In the case of the binary grid, the elements of $\overline{E}^*$ could be classified into 2^{25} distinct shapes, and each shape-class would have 2^n elements, with *n* being the number of cells in that shape.

Definition 6: *Dissimilarity pseudometric.* *To quantify the 'self' portion of self-replication, we assume that a dissimilarity metric* $d : \overline{E}^* \times \overline{E}^* \to \mathfrak{R}^+$ *is given. Recall a pseudometric* ***d*** *obeys* $d(x,y)+d(y,z) \geq d(x,z)$, $d(x,y) \geq 0$, *and* $d(x,x)=0$.

Note that *d* induces an equivalence relation on $\overline{E}^*$. That is, we say for $S_1, S_2 \in \overline{E}^*$, that $S_1 \equiv S_2$ exactly when $d(S_1,S_2)=0$. Presumably, the dissimilarity metric would be chosen such that it induced a natural equivalence relation, but this choice is not assumed.

Definition 7: *Presence.* *We define the presence* $P_\varepsilon(E,S)$ *of a subsystem S in an environment E within tolerance* ε *to be the measure E* normalized to 1. I.e, it is the probability that a randomly selected subsystem* $T \in E^*$ *will satisfy* $d(T,S) \leq \varepsilon$.

Essentially, the presence function measures 'how much' *S* is found in *E*. As a probability, *P* takes values in the interval [0, 1]. In the case of discrete environments, the presence function essentially reduces to counting, and in a continuous case we (temporarily) increase the measure of the set by using a nonzero tolerance.

Definition 8: *ε-Present, ε-Possible.* *When* $P_\varepsilon(E,S) \neq 0$, *we say that S is ε-present (in E). Also, we say that S is ε-possible (in E) when there is some time* $t \in R^+$ *such that S is ε-present in T(E,t).*

Definition 9: *Replicability, Momentary.* *Given a set of configurations* $\overline{E}$ *and two E-configurations* E_1, E_2, *we define the momentary relative replicability of a system S in* E_1 *relative to* E_2 *with tolerance* ε *at time t as*

$$R_M(S,E_1,E_2,\varepsilon,t) = \log\frac{P_\varepsilon(T(E_1,t),S)}{P_\varepsilon(T(E_2,t),S)} \tag{1}$$

The ratio in Eq. (1) serves to compare the probability at time t of finding S in the future of E_1 to the probability at the same time of finding S in the future of E_2. There are a few cases where Eq. (1) is undefined. If $P_\varepsilon(T(E_1,t),S)=P_\varepsilon(T(E_2,t),S)$ (including zero) we define R_M as 0. When $P_\varepsilon(T(E_1,t),S)=0$ but S is ε-present in $T(E_2,t)$, we define R_M as $-\infty$,and when $P_\varepsilon(T(E_2,t),S)=0$ but S is ε-present in $T(E_1,t)$, we define R_M as ∞.

In the case where S is not ε-possible in either or both of E_1 and E_2, we will not define replicability. In these cases, we would generate ratios comparing possibly nonzero quantities to zero quantities, which would fail to yield meaningful information. We explain the rationale for using the logarithm after all the definitions have been presented.

Definition 10: *Replicability, Over time.* *In the case where S is ε-possible in both E_1 and E_2, we also define the replicability over time τ_0 to τ_1 (in E_1 relative to E_2 with tolerance ε) as:*

$$R_T(S,E_1,E_2,\varepsilon,\tau_0,\tau_1) = \log\frac{\int_{t=\tau_0}^{\tau_1} P_\varepsilon(T(E_1,t),S)dt}{\int_{t=\tau_0}^{\tau_1} P_\varepsilon(T(E_2,t),S)dt} \tag{2}$$

Note that for discrete systems, integral reduces to a simple sum.

Definition 11: *Replicability, Overall.* *We define the Overall Replicability as the limiting case:* $R_O(S,E_1,E_2,\varepsilon) = \lim_{t\to\infty} R_T(S,E_1,E_2,\varepsilon,0,t)$

Note that, by using the logarithm of the fractions, we have an additive form for some basic relations. Letting $R(S, E_1, E_2)$ stand in for any of the defined types of replicability, letting S be a fixed system, and letting A, B, C ∈ $\overline{C}$, we have:

$$\begin{aligned} R(S,A,B)+R(S,B,C) &= R(S,A,C) \\ R(S,A,B) &= -R(S,B,A) \\ R(S,A,A) &= 0 \end{aligned} \tag{3}$$

The last relation highlights that replicability is being taken as a relative, rather than absolute, concept. When a replicability value is zero, we have two environments in which a system fares equally well. In order to relate the replicabilities of systems in different environments, we will specifically consider cases where S is present in E_1, is not present in E_2, and $d(E_1, E_2)$ is minimal. In these cases, we are making a comparison between a 'blank' environment and one that is minimally different, in order to reflect a minimally disturbing introduction of a system into the environment. For example, in modeling a crystal, we might consider a supersaturated solution as E_2, and a supersaturated solution with a tiny seed crystal as E_1, but we would not consider a fully crystallized configuration for E_1.

In particular, let an environment E and a system S be given, such that S is minimally ε-present in E. Let $E' = E - S$ denote some E'∈ $\overline{E}$ such that $d(E, E')$ is minimal, and S is not ε-present in E' but is ε-possible in E'. This leads to the particular case of self-replicability we wish to examine.

Definition 12: ***Self-Replicability, Overall.*** *With E, S, and E – S given as above, we define the self-replicability of a system S in an environment E as* $R_S(S,E,\varepsilon) = R_O(S,E,E-S,\varepsilon)$.

In essence, R_s looks at how present *S* becomes when it was at one point specifically in the environment, compared to how develops without prompting. If R_s is zero, then the inclusion of *S* neither adds nor detracts from the future presence of *S*. A positive value of R_s indicates that including the system results in a higher presence of the system in the future, i.e. the system appears to be self-replicating. A negative value indicates that including the system has a detrimental effect, and that the system is essentially self-defeating. Infinite values, both positive and negative, arise when the magnitude of the influence of the system increases at a more than linear rate as time goes on. This can happen in particular when the lifespan of a system is finite in one of *E* or *E–S* and infinite in the other.

3 Examples

3.1 Cellular Automaton

Our first system is a cellular automaton given by single-cycle graph with five nodes, each of which can take the state of 'on' or 'off', along with radius-1 evolution rules (Fig. la). Up to rotations and reflections of the graph, there are eight distinct states the system can take. There are two limit cycles under the evolution rule: First, the 'all on' state is sent to the 'all off' state, which is then sent to the 'all on' state again. Second, the remaining six states are transitive under the action of the evolution rule.

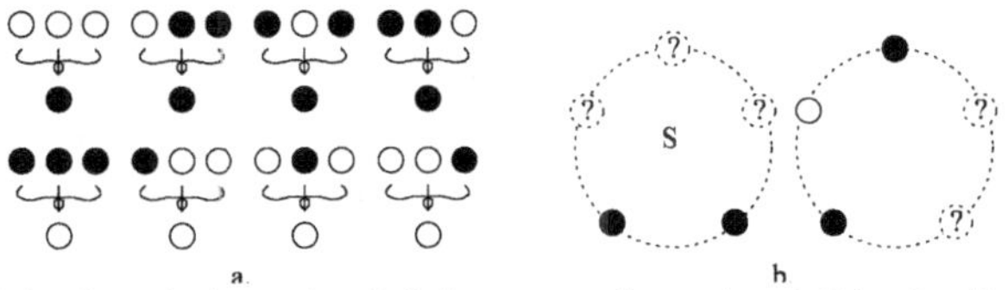

Fig. 1. (a) Table of the 8 evolution rules defining a one dimensional CA of radius I and (b) Two examples of subsystems. We will calculate the self-replicability of the one labeled *S*

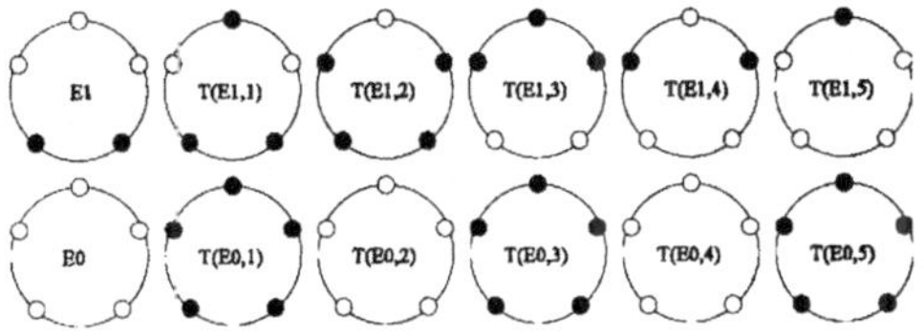

Fig. 2. Cellular Automaton system: Successive states of E_0 and E_1 under *T*

Examples of two subsystems are shown in Fig. lb. As this is a discrete system, we can move directly to the zero-tolerance case, and thus need to define only the preimage of zero under dissimilarity metric (i.e. establish the meaning of 'identical'.) There are 32 configurations, of which only eight are distinct, and each configuration has 32 subsystems. We define a dissimilarity metric that is zero when the subsystems being compared have, up to rotation and flipping, the same shape and pointwise parity. We now compute the values $T(E_0, n)$ and $T(E_0, n)$ by applying the evolution rule. The results are seen in Fig. 2. Note that $T(E_0, n) = T(E_0, n+2)$ and $T(E_1, n) = T(E_1, n+6)$. This cyclic nature of the environment under time development will allow us to find the (overall) self-replicability, rather than just the self-replicability over some constrained time. Next, we calculate the 'presence' of the subsystem S (see Fig. 1b) in each of $T(E_{0,1}, t)$, $t \in \{0,1,2,3,4,5\}$, by counting how many times it occurs in each time step. The results are presented in Table 1. We now take the log of the quotient of the sums, yielding log(7/15) = –0.762 for the self-replicability over $t = 0...5$.

Table 1. Cellular Automaton system: Time and Presence of S in E_0 and in E_1

Time (t)	**0**	**1**	**2**	**3**	**4**	**5**	**Totals**
$P(T(E_0,t),S)$	0	5	0	5	0	5	**15**
$P(T(E_1,t),S)$	1	1	3	2	0	0	**8**

3.2 Ring System

The second system gives an example where the amount of information in the environment grows as the time development function is applied. For conceptualization purposes, the system can be seen as an ideal closed fiber-optic ring (Fig. 3D), with a number of irregularities (Fig. 3B, Fig 3G) that act as beam splitters (Fig. 3B), into which a pattern of moving light, packets (Fig 3C) can be injected. We assume that light packets travel at a constant speed, clockwise or counterclockwise.

The problem becomes interesting when allowing for beam-splitting irregularities that divide the ring into irrational proportions. In this case, no steady state exists. While any initial distribution should become densely spread over the ring, the distribution is not necessarily uniform: thus, the presence of a given pattern is possibly nontrivial. For this example, a direct computer model was used, storing the initial conditions (initial light configuration and irregularity locations) and applying the time development rules to determine the location, direction, and intensity of all resulting light packets. The model was simulated and, in many cases, our hardware was not able to run enough time steps to suggest a long-term trend. However, all R_S values were no greater than 1 in absolute value. Those values that appeared to converge rapidly had a self-replicability factor between 0.5 and 0.9.

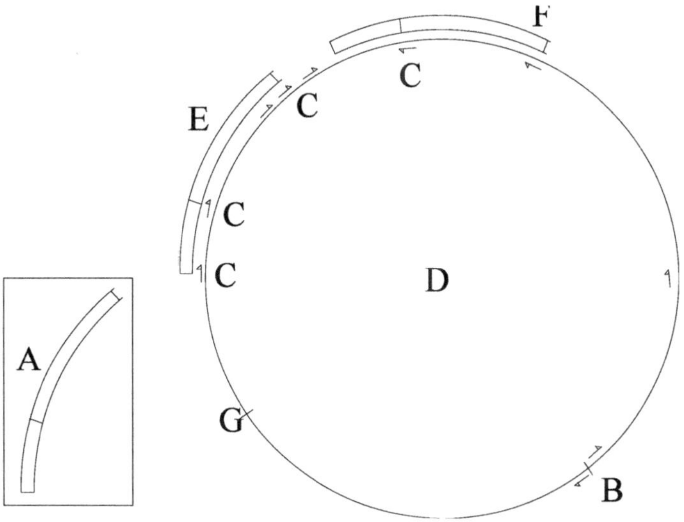

Fig. 3. Illustration of example Ring System. A: A system with potential replicability (a pattern to match) B: An irregularity with recently split packets C: optical packets D: center of the ring E: the system (A) acceptably matching a set of packets, F: the system (A) not acceptably matching a set of packets G: location of a second irregularity.

3.3 Crystal Growth System

The final example is a model of a physical system that is analyzed analytically, rather than explicitly simulated. The goal is to measure the replicability of a crystal satisfying two properties: (a) once a seed crystal is established, the presence of the crystal in the environment is given as a function of time. For this example, we specify an exponential function of time, and (b) there is a fixed nonzero probability rate $c \in (0,1]$ at any given time of a seed crystal spontaneously forming. This guarantees that a crystal S is always possible in an environment E, and the replicability thus is defined.

In this model, we require that the presence of crystal in the environment increase exponentially from the time a seed crystal is inserted, i.e. $P_\varepsilon(T(E_0,t),S)=e^t$. Thus, during the period while $e^t<1$ we compare an environment E_1, seeded with a crystal S, to an unseeded, homogeneous environment E_0. We specify that there is probability 'c' of a seed crystal spontaneously forming, and therefore a probability $(1-(1-c)^t)$ that a seed has formed by time t. Working out the self-replicability (details omitted due to editorial constraints) we obtain:

$$R_O(S,E_0,E_1,\varepsilon) = \lim_{x\to\infty} \log\left(\frac{e^x - 1}{\int_{t=0}^{x}\left(\int_{s=0}^{t} e^{t-s}(1-(1-c)^s)ds\right)dt} \right) \approx \log\left(1 - \frac{1}{\log(1-c)}\right) \quad (4)$$

So, we have a result that visibly converges in the long run (two examples see in Figs. 4,5), with the magnitude of the replicability being inversely proportional to the probability of seed formation. For example, if the probability of crystal formation c is at $c=0.1$, then $R_S = 2.35$, while when c is much smaller, such as $c=0.001$, we have $R_S=6.88$, and a high probability, like $c=0.999$, yields $R_S=0.13$.

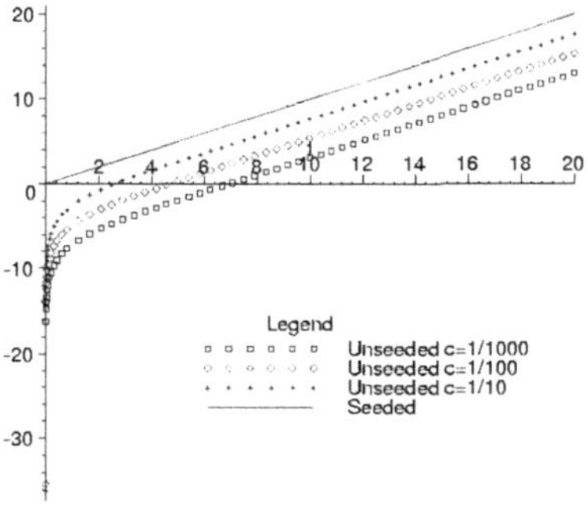

Fig. 4. Plot of the log of the crystal model presence, $log(P_\varepsilon(T(E_1,t),S))$ vs. t for different values of c, the probability of random crystal formation. We use the logarithm to make long-term behavior more visible.

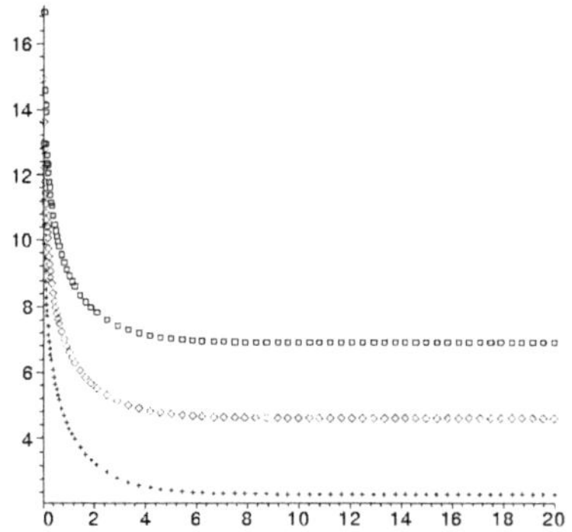

Fig. 5. Plot of $R_T(E_1,E_2,S,\varepsilon,0,t)$ vs. t for the crystal model , showing the same values of c as in Fig. 4. Note that R_T converges quickly (toward R_O) as t becomes large.

4 Conclusions

We have proposed a means of measuring the fundamental property of self-replication, seeking a graded measure that can be universally applied to systems from cellular automata to living systems. We provided three examples that involved various mixtures of discrete and continuous variables and were handled both through direct simulation and analytical modeling. This property is not simply seen as intrinsic to a system, but is found in the information that arises from the interaction between a system and its environment.

We currently see two branches of further immediate investigation. On the theoretical side, while a few properties such as additivity in Eq (3) are known, other properties such as continuity of the replicability function warrant further investigation. On the applied side, it would be desirable to find practical ways to calculate replicability in more intricate systems, such as simple biological auto-catalyzing enzymes and cell models, and artificial systems ranging from molecules, to blocks [7] and to machines [8]. Ultimately, we seek to get a better idea of the scale into which interesting replicabilities fall and the conditions under which self-replication is maximized.

Acknowledgment. This work was supported by the U.S. Department of Energy, grant DE-FG02-01ER45902.

References

1. Sipper, M., Reggia, J.A., *Go forth and replicate.* In: Scientific American 285/265(2), August 2001, 35–43
2. Nehaniv C., Dautenhahn K., *Self-Replication and Reproduction: Considerations and Obstacles for Rigorous Definitions.* Abstracting and Synthesizing the Principles of Life, Verlag Harri Deutsch, pp. 283–290, 1998.
3. McMullin, B, *John von Neumann and the Evolutionary Growth of Complexity: Looking Backward, Looking Forward* In: Artificial Life 6 (2000) 347–361 Sanchez,
4. Moore, E. F., *Machine Models of Self-Reproduction* in Burks, A. W., Essays on Cellular Automata (1970) 187–203
5. Lohn J.D. Reggia J.A. (1997). *Automatic discovery of self-replicating structures in cellular automata.* IEEE Trans. Evolutionary Computation, 1(3):165–178
6. Von Neumann, J, completed and edited by Burks, A. W., *Von Neumann's Self-Reproducing Automata* In: Burks, A. W., Essays on Cellular Automata (1970) 4–65
7. Penrose, L.S., *Self-reproducing machines.* In: Scientific American, 200 (6) (1959) 105–114
8. Chirikjian, G.S., Zhou, Y., Suthakorn, J., *Self-replicating Robots for Lunar Development,* In: IEEE/ASME Trans. on Mechatronics 7. 4 (2002) 462–472

Generic Properties of Chemical Networks: Artificial Chemistry Based on Graph Rewriting

Gil Benkö[1,2], Christoph Flamm[2], and Peter F. Stadler[1,2,3]

[1] Lehrstuhl für Bioinformatik, Institut für Informatik, Universität Leipzig, Kreuzstrasse 7b, D-04103 Leipzig, Germany.
peter.stadler@bioinf.uni-leipzig.de, http://www.bioinf.uni-leipzig.de/
[2] Insitut für Theoretische Chemie und Molekulare Strukturbiologie, Universität Wien, Währingerstrasse 17, A-1090 Wien, Austria
[3] Santa Fe Institute, Santa Fe, New Mexico

Abstract. We use a Toy Model of chemistry that represents molecules in terms of usual structural formulae to generate large chemical reaction networks. An extremely simplified quantum mechanical energy calculation and a straightforward implementation of reactions as graph rewritings ensure both transparency and closeness to chemical reality, both conditions that are necessary for the analysis of generic properties of large reaction networks. We show that some chemical networks graphs, e.g., repetitive Diels-Alder reactions, have the small-world property and exhibit a scale-free degree distribution. On the other hand, the Formose reaction does not fit well into this paradigm.

1 Introduction

Large-scale chemical reaction networks (CRN) appear as the metabolic networks of living cells [1], they describe the chemical processes in planetary atmospheres as well as in combustion and in combinatorial chemistry [2]. Surprisingly, their generic features have not been studied much in the past. For instance, it is unknown whether the small-world properties of metabolic networks [1,3] are characteristic for living systems, or whether they are common to most or all large chemical networks, as suggested by data reported in [4].

The systematic study of these questions requires a computational model for their generation. In a recent study we have introduced such a Toy Model in which generic properties of extensive chemical reaction networks can be explored in detail and that at the same time preserves the "look-and-feel" of chemistry.

In the following three section we briefly review the main ingredient of a computational implementation of a purely graph-based Toy Model [5] of artificial chemistry: molecules and the evaluation of their physical properties, reaction mechanisms and reactivities, and the formation of large-scale networks. In section 5 we investigate a few reaction networks in detail and provide first results towards identifying generic properties of chemical reaction networks.

W. Banzhaf et al. (Eds.): ECAL 2003, LNAI 2801, pp. 10–19, 2003.

2 Molecules

Much of theoretical chemistry is concerned with solving the time-independent Schrödinger equation

$$\widehat{H}\Psi = E\Psi\,, \tag{1}$$

in the Born-Oppenheimer approximation that describes the electrons in a fixed arrangement of nuclei [6]. For our purposes, however, this is too demanding in terms of computational resources. We therefore resort to an artificial chemistry that is derived from equ.(1) by a series of approximations detailed below. In [5] we have shown that such a model retains the "look-and-feel" of organic chemistry; it is hence much more suitable for exploring the generic features of large chemical networks than more abstract systems such as Walter Fontana's `AlChemy` [7], see also [8] for a recent review on AC models.

We use here a simplified version of Extended Hückel Theory (EHT). EHT [9] is obtained from the wave function Ψ, equ.(1), of the complete molecule, by first using the Born-Oppenheimer approximation to separate electron motion from nuclear motion. The *orbital approximation* then splits the electronic part into orthogonal functions Ψ_α called molecular orbitals (MOs). This brings the problem into a form tractable by the variational principle. The linear combination of atomic orbitals (AOs), the *LCAO approximation*, now assumes that one can write $\Psi_\alpha = \sum_i c_{\alpha,i}\chi_j$, where the AOs χ_j describe the spatial distribution of the individual electrons of an atom. The problem is thus reduced to starting with a basis set of atomic orbitals χ_i and a corresponding overlap matrix $S_{ij} = \int \chi_i\chi_j \,\mathrm{d}\tau$, and solving the generalized eigenvalue problem

$$\mathbf{H}\boldsymbol{c}_\alpha = E_\alpha\mathbf{S}\boldsymbol{c}_\alpha\,. \tag{2}$$

Here $\mathbf{H}$ is the one-electron Hamilton matrix, and $\boldsymbol{c}_\alpha$ denotes the vector of the AO coefficients belonging to the molecular orbital Ψ_α with orbital energy E_α. The EHT model is now obtained by parametrizing $\mathbf{H}$ in terms of the overlap integrals S_{ij}, and the *atomic valence state ionization potentials* I_i.

We further simplify the EHT approximation by parametrizing the values S_{ij} of the overlap matrix based only on the type of the AOs χ_i and χ_j at adjacent atoms instead of computing S_{ij} from orbital functions in 3D space. As a consequence we disregard the spatial embedding of the molecule and instead represent it uniquely by its *orbital graph* [10] which has the outer atom orbitals as its vertices while edges represent overlaps. The orbital graph in turn is uniquely determined by the chemical structure formula, the *chemical graph* of the molecule, by virtue of the VSEPR rules [11]. Our Toy Chemistry is thus implemented at the level at which (organic) chemistry is usually taught and described: the level of chemical structural formulae. By definition, any property of the molecule can now be computed at least in principle from its wave function, i.e., from the solutions $\boldsymbol{c}_\alpha$ of equ.(2). A detailed description of the model and its parametrization can be found in [5].

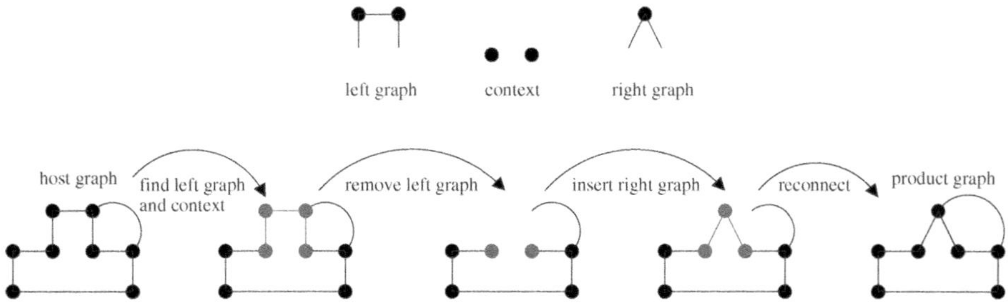

Fig. 1. Graph rewriting rule (*top*) and graph rewriting steps (*bottom*)

The main advantage of the Toy Model is that it incorporates a chemically meaningful energy function that is consistent with the framework of the simplified EHT theory: The total energy of atomization (TAE) of a molecule is given by

$$E = \sum_{\substack{\text{occupied} \\ \text{orbitals } \alpha}} n_\alpha E_\alpha \,. \tag{3}$$

From this quantity it is straightforward to compute the reaction energies as the difference of the TAEs of products and educts. The Toy Model hence has built-in chemical thermodynamics.

3 Reactions

With molecules represented as graphs it is natural to view chemical reactions as rewriting rules applied to molecular graphs. A graph grammar is a finite set of rules operating on edge and vertex labeled graphs. It typically consists of a left graph, a right graph, and a context graph defined as the parts of the graph that are removed, persist, and are newly introduced during a rewriting operation, Fig. 1 [12]. This graph rewriting formalism is very flexible and can be used to represent chemical reactions as well as chemically impossible yet strategically interesting reactions. A chemical reaction is the breaking, forming and changing of bonds. Thus the number and type of atoms must remain constant, which can be implemented by conservation of vertex labels. In analogy, the conservation of the number of valence electrons can be imposed on rewrite rules by ensuring conservation of total bond order. Both principles stem from the fact that chemical reactions are stoichiometric [13].

The simulation of unimolecular reactions is a straightforward application of rewrite rules to a molecule. A bimolecular reaction or any other similar rule is split by into one half reaction rule for each educt molecule, and a final reaction rule. The two half reaction rules do not modify existing bonds and atoms in the molecules, they just add flag nodes to the atoms that will be joined during the total reaction. They serve to identify those reaction sites for the reactivity

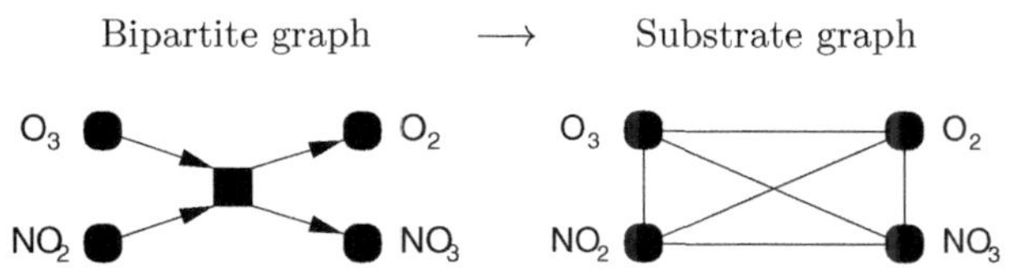

Fig. 2. One-mode projection from a bipartite graph to a substrate graph

evaluation described in the following. The evaluation determines which pairs of reaction sites from each of the two reactants are joined by a temporary edge. This temporary construct is then again submitted to the final reaction rule and transformed to the product.

The reaction rate for a specified reaction mechanism is at least in principle determined by the laws of quantum mechanics and can be expressed in terms of the wave function. It is natural therefore to use the Frontier Molecular Orbital (FMO) Theory [14] for this purpose. In the simplest version of FMO theory, the reactivity is inversely proportional to the difference of the HOMO and LUMO energies E_η and E_λ of the reactants. The regioselectivity is determined by the MO coeffcients at the reactive sites i, such that $\sum_i c_{\eta,i} c_{\lambda,i}$ is maximal. Again, we refer to [5] for a more detailed description.

4 Networks

The starting point for CRN generation is an initial set of molecules $\mathfrak{L}_0$. From this "seed set" we exhaustively generate the network in such a way that a reaction between the same molecules is computed only once, see [5]. While the generation of a large reaction network is rather straightforward, its representation is much less trivial. A plethora of graph-theoretical approaches have been discussed in the literature, see e.g. [15] for an overview. Following [16] a chemical reaction

$$aA + bB + \cdots \to vV + wW + \cdots \tag{4}$$

can be described as a weighted directed hyperedge in a directed hypergraph $\mathcal{H}(V, E)$ with vertex set V. A hyperedge $\boldsymbol{\rho} \in E$ is a pair of lists of vertices $(\{u_i\}, \{v_i\})$ where the u_i are the initial and the v_i are the terminal vertices of $\boldsymbol{\rho}$. In chemical terms, the u_i are the educts and the v_i are the products of the reaction $\boldsymbol{\rho}$. In addition we may have the *stoichiometric coefficients* (prefactors in the above reaction scheme) attached as weights to each u_i and v_i. The *stoichiometric matrix* $\mathbf{S}$, where $S_{x\boldsymbol{\rho}}$ is the coefficient of the chemical species x in reaction $\boldsymbol{\rho}$, is therefore the appropriate algebraic representation of a chemical reaction network.

Unfortunately, a well developed theory for the structural analysis of directed hypergraphs is not available. As an alternative, however, one can represent $\mathcal{H}(V, E)$ by an equivalent bipartite weighted digraph in which each hyperedge is "interrupted" by a special vertex representing the reaction itself. Incoming arcs

Table 1. Characteristics of the two example CRNs in [5] and the substrate graph of the *E. coli* energy and biosynthesis metabolism [1]

Network	n	$\langle k\rangle$	$\langle L\rangle$	$\langle L_{\mathrm{rand}}\rangle$	$\langle C\rangle$	$\langle C_{\mathrm{rand}}\rangle$
Formose	48	3.25	3.55	3.28	0.15	0.068
Diels-Alder	40	4.65	2.15	2.40	0.72	0.110
E.coli	282	7.35	2.9	3.04	0.32	0.026

connect this vertex with the educts, out-going arcs point to the products [17,18, sect. 2.3.2], Fig. 2. This bipartite network graph is still a faithful representation of the entire reaction network.

Reduced representations of chemical networks are useful in particular when studying connectivity and clustering properties where the distinction of two classes of nodes with completely different interpretations may lead to artifacts in the data. In [1], for example, the *substrate graph* is defined as having the chemical species as vertices and an edge connecting two species x and y if they take part in the same reaction. This amounts to replacing each directed hyperedge $\boldsymbol{\rho}$ by a clique. Alternatively, it is obtained from the bipartite representation by one-mode projection, Fig. 2, [19].

Let us denote the number of vertices by n, and let m be the number of edges of the substrate graph of a chemical reaction network. In the context of analyzing large networks it has turned out that the following measures yield useful characteristics:

The average node degree $\langle k\rangle = 2\,\frac{m}{n}$.
The average length of the shortest path between to nodes, $\langle L\rangle$, see e.g. [20].
Let d_x be the degree of a vertex x and let q_x be the number of edges that connect neighbors of a vertex x. The clustering at x can then be measured by the fraction of possible triangles, at x, i.e., $C_x = 2q_x/((d_x(d_x-1)))$. The clustering coefficient $\langle C\rangle$ then measures the overall cliquishness of the graph [21].
The distribution of detour length can be quantified in terms of the minimal cycle basis [4].

In the following section we will compare the characteristic quantities of chemical networks with two classes of randomly generated graphs: Erdös-Renyi (ER) graphs which feature statistically independent edges, and the AB model (see below). For later reference we note the expected mean path length and clustering coefficients for the ER model in terms of the average vertex degree $k = 2m/n$:

$$\langle L_{\mathrm{rand}}\rangle \approx \frac{\ln n}{\ln\langle k\rangle} \qquad \langle C_{\mathrm{rand}}\rangle = \frac{\langle k\rangle}{(n-1)} \tag{5}$$

Starting with the seminal paper by [21], it has been recognized that these real life networks differ qualitatively from the classical ER random graph models by the so-called *small-world* property: while the graphs are very sparse on average, the mutual distances between their vertices of both graph types are

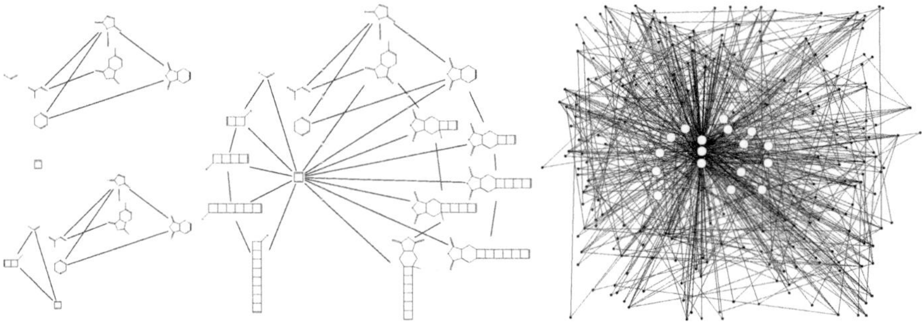

Fig. 3. Evolution of the network of a repetitive Diels-Alder network for different reaction rate thresholds. The CRN grows as the reactivity threshold $\ln k$ is lowered (from left to right: 108, 91.2, and 55.4). The simulation is started with 

approximately of the same size. The major difference lies in the degree of local clustering which is much higher in the real life networks. The characteristic parameters of scale-free networks scale such as those obtained by the preferential attachment procedure [22] (AB model) behave as follows [23]:

$$\langle L_{\mathrm{AB}} \rangle \sim n^{-0.75} \qquad\qquad \langle C_{\mathrm{AB}} \rangle \sim \frac{\ln n}{\ln \ln n} \,. \tag{6}$$

The cycles and edge-disjoint unions of cycles of a graph from a vector space (w.r.t. symmetric difference as addition) with dimension $\mu(G) = m - n + c(G)$, where $c(G)$ is the number of connected components. A *minimum cycle basis* (MCB) is a cycle basis that minimizes the total number of edges in the basis cycles. It can be shown that all MCBs have the same numbers n_ℓ of cycles of length ℓ [24]. We can therefore use the distribution of MCB cycle lengths to characterize the cycle structure of G, see also [4].

5 Results

The Diels-Alder reaction [25] has been extensively studied thanks to its importance in natural products synthesis and because it is easily tractable by simple semi-empirical methods. It is the typical test reaction for a semi-empirical quantum calculation methods such as ours, and furthermore for the numerous approaches of reaction description [18]. It involves the reaction between two linear π-systems of length 2 and 4, called *dienophiles* and *dienes*, and is thus called a [2+4]-cycloaddition. The product is again a dienophile and may react again in a Diels-Alder reaction. The reaction is used for the synthesis of polymers [25]. The generation of a repetitive Diels-Alder (DA) reaction network and a further network, the *Formose reaction*, is discussed in detail in [5]. Here their properties and those of DA networks with different reactivity thresholds are studied. The reactivity threshold determines which reactions are allowed, i.e. those above it. The series of CRN in Fig. 3 is obtained by repetitive Diels-Alder reactions of

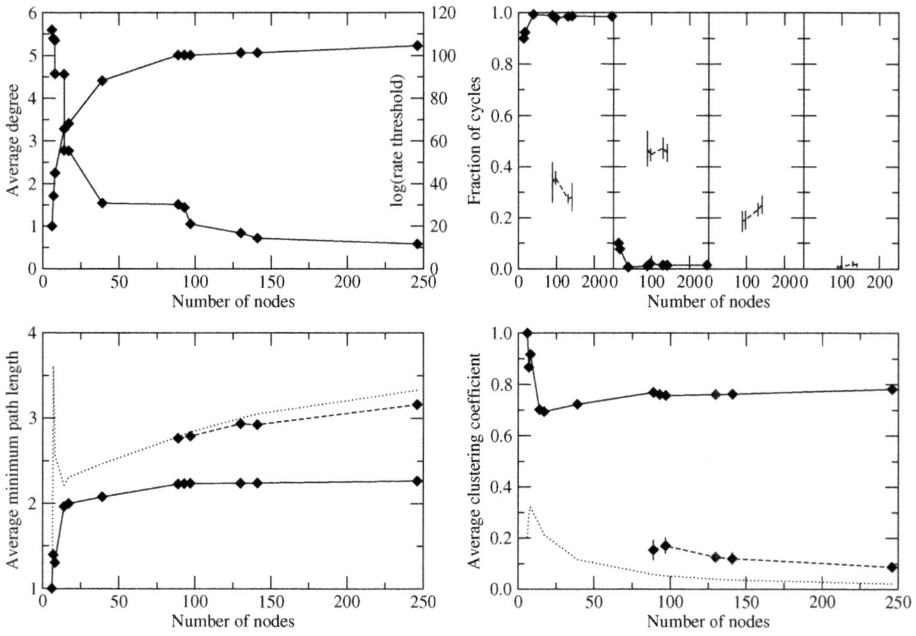

Fig. 4. Evolution of network properties for a repetitive Diels-Alder CRN with different reaction rate thresholds (*solid*) and the AB (*slashed*) and ER (*dotted lines*) model. The AB graphs were only generated in the region of $n \in [89, 246]$, where $\langle k \rangle \approx 5$. (a) Reaction rate threshold (*grey*) and $\langle k \rangle$ (*black*); (b) Cycle statistics of triangles and cycles of size 4, 5, and 6; (c) $\langle L \rangle$; and (d) $\langle C \rangle$ vs. networks size

a simple initial mixture of dienes and dienophiles with a decreasing threshold value.

Tab. 1 compares the network characteristics of the Diels-Alder, the Formose, and the *E. coli* metabolic network. They are all sparse graphs, i.e. they have much fewer edges than complete graphs, reflected by $m \ll \frac{n(n-1)}{2}$ or $\langle k \rangle \ll n$. Sparse networks are very common, ranging from the network of acquaintances to a neural network. In both cases, there are only few connections at each node. From the networks of Tab. 1, only Diels-Alder and *E. coli* fulfill the conditions $\langle C \rangle \gg \langle C_{rand} \rangle$ and $\langle L \rangle \leq \langle L_{rand} \rangle$ and thus are small-world networks in the strict sense.

The change of network characteristics with DA network size is shown in Fig. 4. Interestingly, both $\langle C \rangle$ and $\langle L \rangle$ stay constant for $n \in [89, 246]$. This property seems to be robust in chemical networks. This is not the case for ER and AB graphs as both $\langle C \rangle$ and $\langle L \rangle$ vary. Triangles dominate the minimum cycle basis of the substrate graph representation of a CRNs, see Fig. 4. For comparison we show the cycle distribution of scale-free graphs (AB model) with the same size and average degree. The number of triangles is $\Delta \approx \mu = m - n + 1 \approx n(\langle k \rangle - 2)/2$ for large n. So a linear rise of Δ with the network size is not surprising. As shown

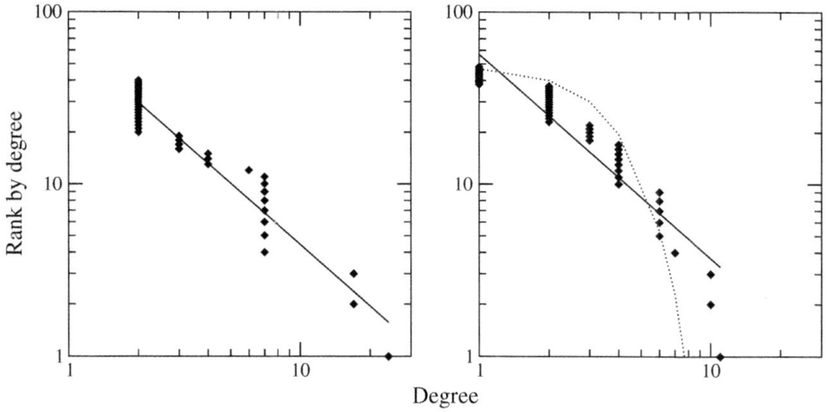

Fig. 5. Rank statistics of repetitive Diels-Alder (*l.h.s.*) and the Formose reaction network (*r.h.s.*). Datapoints are ranked by decreasing degree. The power-law regressions (solid lines) have slopes of -1.19 in both cases. The Poisson distribution expected for ER random graphs is shown for comparison as a dotted line

in Fig. 4, the fraction Δ/μ, however, also seems to be constant for increasing network size.

Finally, the degree distributions have been calculated (Fig. 5). Both networks are scale-free, i.e. their degree distributions follow a power law. The cumulative representation of Fig. 5 is equivalent to $\int_k^\infty P(x)\mathrm{d}x$ vs. k. The regression $\int_k^\infty P(x)\mathrm{d}x \sim k^{-1.19}$ is consistent with the values reported in [26]. The explanation of the origin of scale-freeness therein can be applied to the present examples. It relies on two generic mechanisms. First, the networks grows from an initial set of nodes by continuous addition of new nodes. Indeed, in the present case, there is an initial list of molecules, and the networks is built by adding molecules at every iteration (sect. 4). Second, the networks grows by *preferential attachment*, i.e. new nodes are preferably attached to nodes with high degree, or in the case of molecules, species who have already spawned many new other species are especially reactive and more likely to produce new molecules at each iteration. The power-law regression for the Formose reaction network fails for high k. The theoretical Poisson distribution for ER random graphs with high n, $P(k) = e^{-\langle k \rangle} \frac{\langle k \rangle^k}{k!}$, also fails in this range. A degree distribution following a truncated power law has been referred to as *broad-scale* [27]. Again, the exponent of the power law is constantly around 1.18 ± 0.05 for the repetitive Diels-Alder reaction network for sizes $n \in [39, 246]$.

The Formose reaction network, in contrast, does not fit the predictions of from scale-free models very well, Fig. 5. It includes many non-reactive species with respect to keto-enol condensation, which leads to small cliquishness and longer paths. Furthermore, it has particularly reactive species (formaldehyde) whose neighbors in the substrate graph do not react with each other: a keto-enol isomerization must toggle the "state" of a network species from non-reactive to reactive first. This leads in consequence to an elongation of paths and a reduction of cliquishness of the Formose reaction network.

6 Discussion

The study of Artificial Chemistries (AC) is closely related to the field of Artificial Life. The many models of AC can be categorized according to abstraction level and intended application. Very abstract models simulate AC in which strings or logical elements interact, such as in Fontana's λ-calculus. There, networks are built in a bottom-up approach to be studied phenomenologically, as in Artificial Life, for example for the emergence of structures or sustainability. On the other hand, models for practical applications tend to be top-down. They concentrate on describing interactions and try to predict, for example, the time and/or space evolution of concentrations.

The graph based Toy Chemistry Model possesses qualities from both approaches. The use of graph grammars to handle reactions makes our approach more bottom-up like, while the use of a simplified quantum mechanical energy calculation is clearly top-down. The graph based Toy Chemistry Model provides a very flexible and transparent framework for the exploration of the properties of chemical reaction networks. The built-in chemical thermodynamics makes the model self-consistent and helps to keep the resulting reaction networks close to chemical "reality". Since the used graph grammar (set of allowed reactions) can be chosen freely, a wide variety of reaction networks with different "chemical flavors" can easily be generated, which is an important prerequisite for the study of their network properties.

We found that the Diels-Alder reaction network exhibits small world properties similar to known cases e.g. the *E. coli* metabolic network. The Formose reaction network, on the other hand, does not fit the prediction for a small world network very well. We conclude that chemical networks do not fall into a single class of the small-world network classification scheme by Amaral *et al.* [27]. More detailed investigations will therefore be necessary.

References

1. Fell, D., Wagner, A.: The small world inside metabolic networks. Proc. R. Soc. Lond. Ser. B **268** (2001) 1803–1810
2. Höllering, R., Gasteiger, J., Steinhauer, L., Schulz, K., Herwig, A.: The simulation of organic reactions: From the degradation of chemicals to combinatorial synthesis. J. Chem. Inf. Comput. Sci. **40** (2000) 482–494
3. Jeong, H., Tombor, B., Albert, R., Oltvai, Z.N., Barabasi, A.: The large-scale organization of metabolic networks. Nature **407** (2000) 651–654
4. Gleiss, P.M., Stadler, P.F., Wagner, A., Fell, D.A.: Relevant cycles in chemical reaction network. Adv. Complex Syst. **4** (2001) 207–226
5. Benkö, G., Flamm, C., Stadler, P.F.: A graph-based toy model of chemistry. J. Chem. Inf. Comput. Sci. (2003) in press; presented at *MCC 2002*, Dubrovnik CRO, June 2002; SFI # 02-09-045.

6. Atkins, P., Friedman, R.: Molecular Quantum Mechanics. 3rd ed. edn. Oxford University Press (1997)
7. Fontana, W., Buss, L.W.: What would be conserved if 'the tape were played twice'? Proc. Natl. Acad. Sci. USA **91** (1994) 757–761
8. Dittrich, P., Ziegler, J., Banzhaf, W.: Artificial chemistries — a review. Artificial Life **7** (2001) 225–275
9. Hoffmann, R.: An Extended Hückel Theory. I. Hydrocarbons. J. Chem. Phys. **39** (1963) 1397–1412
10. Polansky, O.E.: Graphs in quantum chemistry. MATCH **1** (1975) 183–195
11. Gillespie, R.J., Nyholm, R.S.: Inorganic Stereochemistry. Quart. Rev. Chem. Soc. **11** (1957) 339–380
12. Nagl, M.: Graph-Grammatiken, Theorie, Implementierung, Anwendung. Vieweg, Braunschweig (1979)
13. Dugundji, J., Ugi, I.: Theory of the *be*- and *r*-matrices. Top. Curr. Chem. **39** (1973) 19–29
14. Fleming, I.: Frontier Orbitals and Organic Chemical Reactions. Wiley: New York (1976)
15. Bonchev, D., Mekenyan, O., eds.: Graph-Theoretical Approaches to Chemical Reactivity. Kluwer, Dordrecht, NL (1994)
16. Zeigarnik, A.V.: On hypercycles and hypercircuits in hypergraphs. In Hansen, P., Fowler, P.W., Zheng, M., eds.: Discrete Mathematical Chemistry. Volume 51 of DIMACS series in discrete mathematics and theoretical computer science., Providence, RI, American Mathematical Society (2000) 377–383
17. Balandin, A.A.: Multiplet theory of catalysis: Theory of hydrogenation. Classification of organic catalytic reactions. Algebra applied to strucutral chemistry. Volume 3. Moscow State Univ. (1970) in Russian.
18. Temkin, O.N., Zeigarnik, A.V., Bonchev, D.: Chemical Reaction Networks : a graph-theoretical approach. CRC Press Boca Raton, FL USA (1996)
19. Wasserman, S., Faust, K.: Social Network Analysis. Cambridge University Press (1994)
20. Newman, M.E.J., Moore, C., Watts, D.J.: Mean-field solution of the small-world network model. Phys. Rev. Lett. **84** (2000) 3201–3204
21. Watts, D.J., Strogatz, S.H.: Collective dynamics of 'small-world' networks. Nature **393** (1998) 440–442
22. Barabási, A.L., Albert, R., Jeong, H.: Mean-field theory for scale-free random networks. Physica A **173-187** (1999) 272
23. Bollobás, B., Riordan, O.: The diameter of a scalefree random graph. Technical report (2002) Preprint.
24. Chickering, D.M., Geiger, D., Heckerman, D.: On finding a cycle basis of with a shortest maximal cycle. Inform. Processing Let. **54** (1994) 55–58
25. Morgenroth, F., Müllen, K.: Dendritic and hyperbranched polyphenylenes via a simple Diels-Alder route. Tetrahedron **53** (1997) 15349–15366
26. Barabási, A.L., Albert, R.: Emergence of scaling in random networks. Science **286** (1999) 509–512
27. Amaral, L.A.N., Scala, A., Barthelemy, M., Stanley, H.E.: Classes of small-world networks. Proc. Nat. Acad. Sci. **97** (2000) 11149–11152

How to Program Artificial Chemistries

Jens Busch and Wolfgang Banzhaf

University of Dortmund,
Department of Computer Science, Chair of Systems Analysis (LS XI)
D-44221 Dortmund, Germany
{busch, banzhaf}@ls11.cs.uni-dortmund.de
http://ls11-www.cs.uni-dortmund.de

Abstract. Using the framework of artificial chemistries (ACs) an automated theorem prover (ATP) is constructed. Though it is an application of its own, in the context of ACs automated theorem proving can serve a second purpose. In this paper, we present a resolution-based AC named *RESAC*. Once converted to the first-order predicate calculus a problem straightly fits to this non-deterministic AC model. The calculus therefore provides a general and intuitive language for "programming" *RESAC*. The fixed implicit interaction scheme and predefined structure of the objects is advantageous and helps to predict the system's dynamics. Furthermore, the versatility of the methodology is demonstrated by implementing the Adleman problem. An analysis of the dynamic behavior is performed delivering insight into the synthesis of non-deterministic emerging processes. This analysis include a discussion of some general AC parameters.

1 Introduction

Inspired by real chemical processes, artificial chemistries (ACs) follow an intriguing computational paradigm. Abstract molecules (objects) interact in an autonomous, distributed and parallel way. Guided by reaction rules these objects act on one another, thereby proliferating or altering their information content or structure. Modified structure may imply a different function according to the imposed interaction scheme. Consequently, the role of a molecule can change over time and this way, AC provides a powerful framework to define constructive systems [10]. The actual state of the system is defined by concentration levels of molecules. A calculation step itself causes a change of such concentrations.

Formally, an AC is defined by the triple (S, R, A) [11]. S denotes a set of objects and R a set of interaction rules constituting the reaction schemes applied whenever molecules collide. The system's dynamics is controlled by an algorithm A describing how the rules are applied to a population of molecules. By this means a closed or an open reactor can be simulated, a well-stirred reaction vessel with no topology or for example a 3-D space like in [19].

On the one hand, ACs try to answer questions arising in evolution theory. Here, the emergence of global phenomena in restricted environments initiated

W. Banzhaf et al. (Eds.): ECAL 2003, LNAI 2801, pp. 20–30, 2003.

by a huge number of reactions among elementary entities is investigated. Results have been presented e.g. in [3,4,6,12,13,16]. On the other hand, researchers link theory with practical applications by taking advantage of the similarity between ACs and other complex dynamic systems (e.g. [2,7,8,15,20]).

A natural definition for many tasks is in terms of elementary interactions to be performed frequently. But what "language" should be used to encode a given task? Even more troublesome is the choice of a molecule representation and of an interaction scheme if one wants to deal with many problems from various domains instead of one specific task. In this paper, we present first-order predicate calculus (FOPC) as a general and intuitive molecule representation. Combined with resolution logics as an implicitly defined interaction rule we gain the dynamic system *RESAC* which can be interpreted in two different ways: (i) As implementation of an ATP (AC-based ATP). FOPC problem solving belongs to the class of hardest known problems in computer science, since the validity of FOPC-formulas is not decidable. In this area previous work was done and results have been presented in [9]. (ii) As object and reaction rule setup for the solution of arbitrary tasks (ATP-based AC) by defining both sets S and R.

Here, we are concerned with the second alternative. Regarded as a tool, the framework of ATP provides RESAC with a fixed setting (S, R), because data structure and interaction mechanism are defined independent from a specific task. This simplifies the AC design and the analysis of the system's dynamics considerably. However, the specification of an algorithm A is still necessary. In Sect. 3 we suggest such an algorithm, discuss several variants, and construct RESAC. As an example, the famous Adleman problem is run in RESAC in Sect. 4, and its dynamics is analyzed.

2 Basic Concepts of Logic and ATP

ATP might be considered one of the oldest branches of Artificial Intelligence. Given a *theory* $A = \{A_1, \ldots, A_n\}$ of expressions the task is to proof that another expression T is consistent with theory A ($A \models T$). Of course, checking all possible truth-assignments (*interpretations*) will solve the problem. Unfortunately systems of predicates have a potentially infinite number of interpretations, therefore this is not a practical approach. The combinatorial explosion of search space demands another, more sophisticated search strategy. Having an eye on the syntax instead on the semantics can be helpful to restrict the search space: A syntactical transformation mapping one expression onto a new expression consistent with the underlying theory is called *sound inference rule*. If expression T is producible by a sequence of inference rules operating on both, the theory A and every expression inferred from A so far, then $A \models T$ holds and T is called *theorem* of A. The protocol belonging to the inference sequence is said to be the proof of $A \models T$.

To construct RESAC, we will use concepts of FOPC which provides a representation for objects, their properties and relations. Note that this calculus is much more powerful than calculi restricted to Horn clauses used in PROLOG-

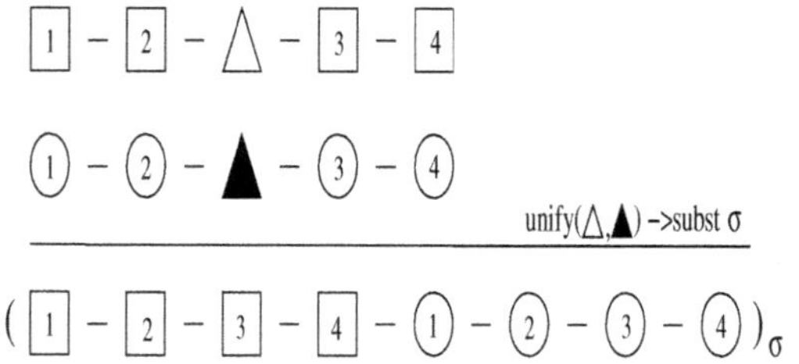

Fig. 1. Illustration of the resolution principle

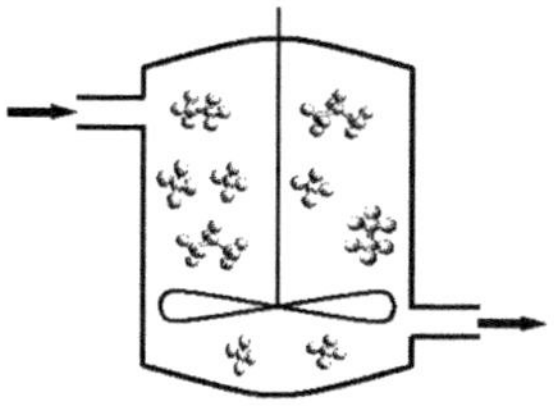

Fig. 2. Open, well-stirred reactor

based systems (e.g. [18]). In FOPC, logic expressions can be converted into an inferentially equivalent conjunction of *clauses* each being a disjunction of *literals*. These are either atomic expressions or negated atomic expressions. $|l|$ denotes the literal l without polarity.

Regarding RESAC, one inference rule is of special interest: binary resolution [17]. Given two clauses c_1 and c_2 with literals $l_1 \in c_1$ and $l_2 \in c_2$, a new clause is inferred by resolution if $|l_1|$,$|l_2|$ are unifiable[1] and of complementary polarity. The resolvent is obtained by joining c_1 and c_2 with the appropriate substitution applied and eliminating l_1 and l_2 (Fig. 1 outlines this scheme).

To be more effective, the *set-of-support strategy* may be applied: In this case, at least one of $\{c_1, c_2\}$ must have *support* in order to allow a resolution inference. Every result of an inference inherits the support. Initially, the support is given to a subset of A.

3 Programming ACs with RESAC

In this section, we shift focus to our resolution-based artificial chemistry, RESAC, which use concepts of FOPC to define the set of rules R and the set of objects S. We start the construction of RESAC, however, by defining the third component, the AC algorithm: The algorithm A driving the reactor M models a well-stirred reaction vessel (see Fig. 2). Thus arbitrary molecules can collide at any time. What seems to be an artificial simplification is, however, a concept of chemistry [14] if parameters like, for example, speed of stirring are chosen appropriately. Additionally, new start-clauses can feed the reactor—the so-called *inflow*—at a certain rate (*inflow rate*). The size of the vessel is kept constant, i.e. the inflow equals the dilution flux. An inflow rate $i, 0 \leq i \leq 1$,[2] indicates that $i * |M|$ start-clauses are inserted within one generation which, as usual, is defined to be $|M|$ collisions (independently of whether a collision causes a reaction or not). There are two special modes of inflow. First, an inflow rate 0 samples a closed reactor. Second, if a molecule is allowed to enter every time a reaction is not carried out (see below), we speak of an *elastic* inflow.

[1] Unification is the operation which is applied to terms in order to match them.
[2] synonymously referred to as $(i * 100)$%-inflow

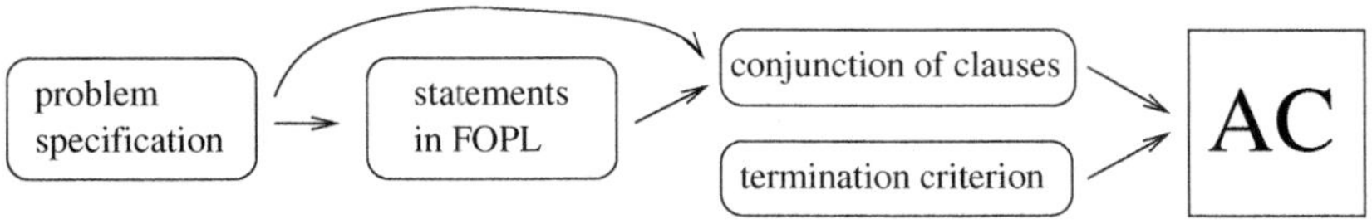

Fig. 3. Conversion steps from the original problem specification to AC

The set of objects S in the reactor covers molecules, each representing a clause of limited length[3]. These are arranged in a multiset-like structure [5] as implementation of the reactor model M. Thus all reactions performed map to set transformations. Input to RESAC are clauses $c_1, \ldots, c_n$, called *start-clauses*, which are derived from the problem specification (Fig. 3). Initially, the population holds η copies of each start-clause, and the size of the reactor $|M|$ equals $\eta * n$.

Concluding the description of RESAC, the remainder of this section is devoted to the reaction rule setup. The reaction scheme R between colliding molecules P and Q is given implicitly by the application of the resolution inference rule. If no resolution is feasible or permitted due to preconditions not met, the reaction is termed *elastic*, i.e. it does not take place at all. Otherwise, let V denote the inferred resolution result.This result is non-deterministic, because the pair of literals to be resolved is chosen randomly from the set of all resolvable literal pairs. Depending on a parameter *SELECT*, a reaction is then defined as follows:

$$\boxed{\textit{productive}\ \ \text{reaction:}}\ P + Q \Longrightarrow \begin{cases} U + V \text{ with } U \in \{P, Q\} & \text{if } \textit{SELECT}{=}\textit{educt} \\ P + Q + V & \text{if } \textit{SELECT}{=}\textit{free} \end{cases}$$

The parameter SELECT determines the implemented replacement policy. If *educt replacement* is applied the reaction scheme resembles an autocatalytic reaction with either molecule P or Q being preserved, whereas the other one is selected to be replaced by V. In contrast, *free replacement* lets V replace an arbitrary molecule.

The replacement policy matter is not discussed explicitly in most publications. Often free replacement is employed without comments on this issue. Although educt replacement may be considered closer to chemistry, it is rarely used (one exception is e.g. [7]). To reveal some of the differences between both replacement schemes, we investigate a basic setting with only three substances s_0, s_1, s_3, and simple reaction rule $s_1 + s_2 \Longrightarrow s_0$. Denoting the concentration of a molecule X at time step t with $|X|_t$, we set $(|s_1|_0, |s_2|_0, |s_0|_0) = (a, 1-a, 0)$, $0 \le a \le 1$. The following formulas are derived with $t \ge 0$, e stands for educt replacement, f stands for free replacement:

[3] In fact, the imposed restrictions are twofold: Molecules either too big or produced by too many reactions are not considered stable and are not allowed to be produced.

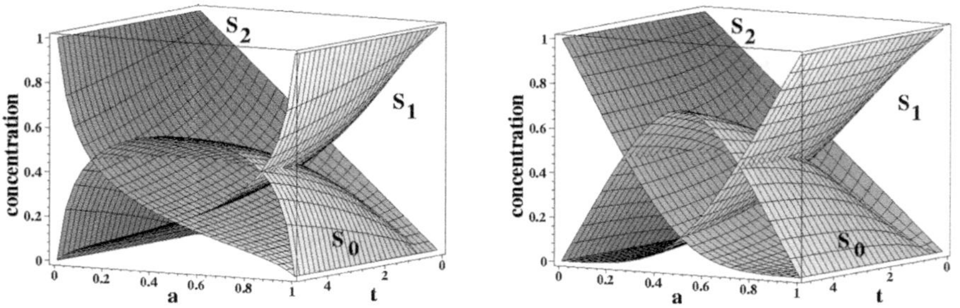

Fig. 4. Concentration transitions induced by $s_1 + s_2 \Longrightarrow s_0$ in a closed reactor. Explanations see text. *Left:* Free replacement. *Right:* Educt replacement

$$(|s_1|_t^f, |s_2|_t^f, |s_0|_t^f) = \left(\frac{a}{\sqrt{4\,(1-a)ta+1}}, \frac{1-a}{\sqrt{4\,(1-a)ta+1}}, \frac{\sqrt{4\,(1-a)ta+1}-1}{\sqrt{4\,(1-a)ta+1}}\right)$$

$$(|s_1|_t^e, |s_2|_t^e, |s_0|_t^e) = \begin{cases} \left(\frac{a(-1+2\,a)e^{t(-1+2\,a)}}{-1+a+e^{t(-1+2\,a)}a}, -\frac{(-1+2\,a)(-1+a)}{-1+a+e^{t(-1+2\,a)}a}, 1-\sum_{i=1}^{2}|s_i|_t^e\right) & a \neq \frac{1}{2} \\ \left(\frac{1}{t+2}, \frac{1}{t+2}, \frac{t}{t+2}\right) & a = \frac{1}{2} \end{cases}$$

In Fig. 4 the concentration development is plotted. When educt replacement is employed, s_1 and s_2 are consumed completely in order to produce s_0 iff $|s_1|_0^e = |s_2|_0^e$. With free replacement, however, $\lim_{t\to\infty} |s_0|_t^f = 1$ holds for $0 \neq a \neq 1$. As expected, the global effect of free replacement leads to remarkably different behavior when leaving the center of the simplex given by (s_1, s_2). If $a = \frac{1}{2}$, both variants have the same limits but convergence speed differs. On this issue, in the next section empirical results of an example are stated.

4 Setup and Analysis – An Example

To demonstrate and analyze RESAC, we reconsider Adleman's famous experiment in which he initiated the cooperation of computer science and molecular biology for solving the NP-complete directed Hamiltonian path problem (DHPP) using DNA parallel processing [1]. Though not complex in the given size, this problem is suitable for our investigations because of its simple semantical structure.

Given a directed graph with designated vertices v_{start} and v_{end} the DHPP consists of finding a path (a sequence of edges) starting at v_{start} and ending up in v_{end} going through each remaining vertex exactly once. The graph shown in Fig. 5 depicts the graph G Adleman solved at the molecular level. To represent G in FOPC we have to transfer it to a number of clauses in conjunctive normal form. Using the function *edge()* and predicates *Path()* and *Visited()* we chose a tripartite system of clauses (see again Fig. 5). The first functional group is formed by the specification of the graph. Second, a "`C onnector`" is introduced which

(0 → 1) Path(edge(0,1)), -Visited(0), -Visited(1)
(0 → 6) Path(edge(0,6)), -Visited(0), -Visited(6)

⋱

(5 → 2) Path(edge(5,2)), -Visited(5), -Visited(2)
(5 → 6) Path(edge(5,6)), -Visited(5), -Visited(6)

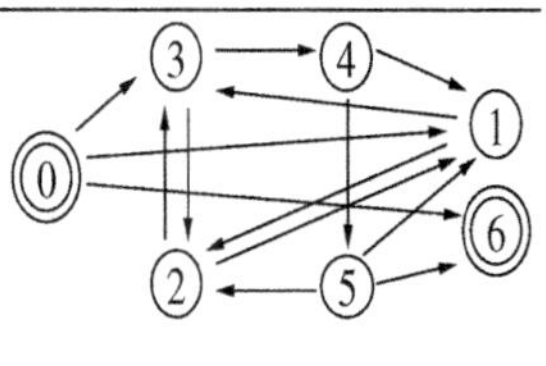

(`Connector`) -Path(edge(X,Y)), -Path(edge(Y,Z)), Path(edge(edge(X,Y),Z))
(`Solver`) -Path(edge(edge(edge(edge(edge(edge(0,X2),X3),X4),X5),X6),6))
(`Termination`) -Visited(0), -Visited(1), -Visited(2), ... , -Visited(5), -Visited(6)

(`Solver`$_{R1}$) -Path(edge(0,X2)),-Path(edge(X2,X3)), ... , -Path(edge(X6,6))
(`Solver`$_{R2}$) -Path(edge(edge(edge(edge(edge(edge(X1,X2),X3),X4),X5),X6),X7))

Fig. 5. Representation of the Adleman problem in FOPC

reflects the fact that if two paths $X \leadsto Y$ and $Y \leadsto Z$ are given also the path $X \leadsto Z$ exists. The solving clause "`Solver`" completes the definition of the set of start-clauses. Although the chosen representation is not optimal (see below) it is a somehow intuitive approach, and, moreover, it helps us to reveal some hidden properties of the different variants of RESAC. A simplified view of the solution process is as follows: The `Connector` joins pieces of paths to longer paths if possible. A path $v_{start} \leadsto v_{end}$ with 6 edges unifies and thereby reacts with the `Solver`. The *Path()*-literal is resolved, and a group of *Visited()*-literals remain. The terminating query criterion searches for such a group which comprises all nodes of G. If a reaction product matches this clause, a solution to this DHPP is found. By giving the support only to the `Solver` a kind of backward chaining is employed.

In our experiments we ran RESAC with 6 variants of inflow, 4 different sizes, and both replacement policies, each of these a 100 times. In Fig. 6 the percentage of solutions found (success rate), the average generation in which the solution was found (avg. success gen.), corresponding Std. Dev. and Std. Errors are listed. From the collected data we deduce 3 statements:

1. Free replacement outperforms or at least equals educt replacement in every experiment series with respect to a higher success rate, a lower avg. success gen., and a lower variability.
2. There seems to be an optimal inflow rate i_{opt}, such that for all inflow rates $i > i_{opt}$, it holds that the smaller i, the higher the success rate is, but, on the other hand, for inflow rates $i < i_{opt}$: the smaller i, the lower the success rate.
3. With increasing reactor size, the success rate is raising and solutions are found in earlier generations.

We now discuss these observations in indicated order:

(1): At first, we state the following two definitions which have been found useful for analyzing the system's dynamics: To measure the *productivity*, the number of productive reactions within one generation is divided by $|M|$. The *diversity* of M is the number of different molecules in M divided by $|M|$. In

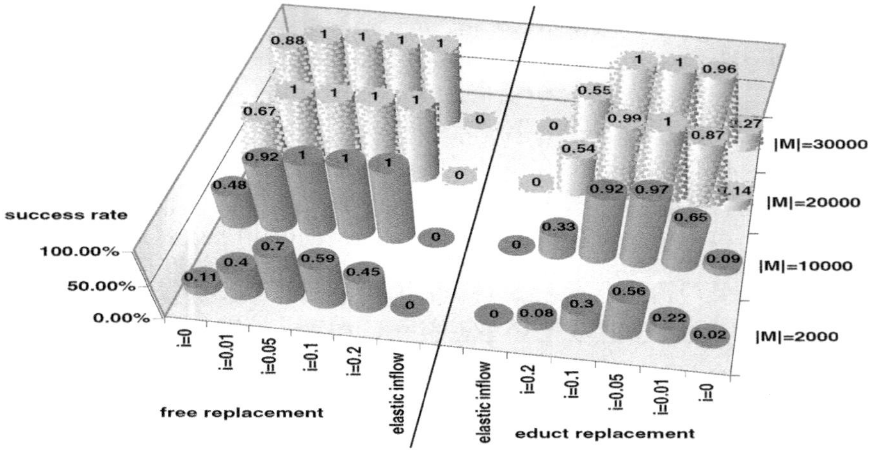

	inflow rate	$\|M\| = 2000$	$\|M\| = 10000$			$\|M\| = 20000$			$\|M\| = 30000$		
free replacement	0	-	-	-	-	-	-	-	-	-	-
	0.01	-	-	-	-	**41.2**	*14.9*	1.49	**37.3**	*15.4*	1.54
	0.05	-	**40**	*16.5*	1.65	**30.3**	*11*	1.1	**26.4**	*8.4*	0.84
	0.1	-	**35.3**	*12.9*	1.29	**27.2**	*9.4*	0.94	**25.7**	*7*	0.7
	0.2	-	**39.8**	*17.5*	1.75	**30.4**	*10.3*	1.03	**26.1**	*7.8*	0.78
	elastic	-	-	-	-	-	-	-	-	-	-
educt replacement	0	-	-	-	-	-	-	-	-	-	-
	0.01	-	-	-	-	-	-	-	-	-	-
	0.05	-	-	-	-	**42.7**	*13.6*	1.36	**36.8**	*10.5*	1.05
	0.1	-	-	-	-	-	-	-	**39.1**	*12.8*	1.28
	0.2	-	-	-	-	-	-	-	-	-	-
	elastic	-	-	-	-	-	-	-	-	-	-

Fig. 6. Adleman experiment in various ACs: i denotes the inflow rate, $|M|$ the reactor size. Each experiment was repeated a 100 times. Run time was limited to at most 100 generations. In the graph the success rate is compared. A 5%-inflow shows the best results. In the Table the average success generation (typed in **bold**), the Std. Dev. (typed in *italic*) and the Std. Errors are listed; "-" indicates that not all experiments were successful

fact, diversity plays an important role explaining the first observation. Molecule diversity in reactors employing free replacement is significantly higher than with educt replacement (Fig. 8 shows examples). This yields a better distribution of solution candidates (clauses) within the search space.

If we look for the reason we have to go into detail. Listing the concentration development of all start-clauses in a closed reactor with educt replacement (Fig. 7) reveals the extinction of the `C onnector` at an early stage of the experiment. As a consequence, productivity rate falls and approaches zero. What is the reason for this concise extinction of clauses which is a generally observed phenomenon?

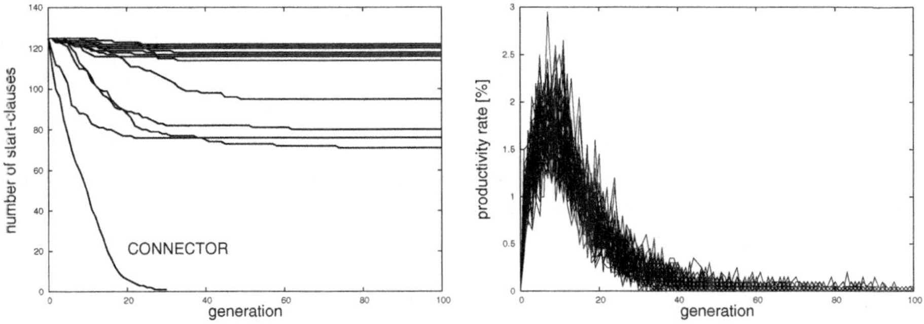

Fig. 7. Analysis of a closed reactor with size 2000 employing educt replacement. *Left:* Number of start-clauses. Note the development of the `Connector`. *Right:* Productivity in the same reactor[5]

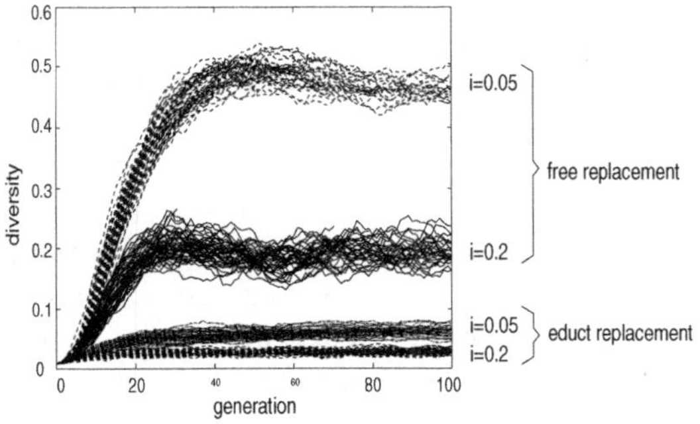

Fig. 8. Diversity in reactors with size 2000. Educt compared to free replacement. Diversity is significantly higher with free replacement. i denotes the inflow rate[4]

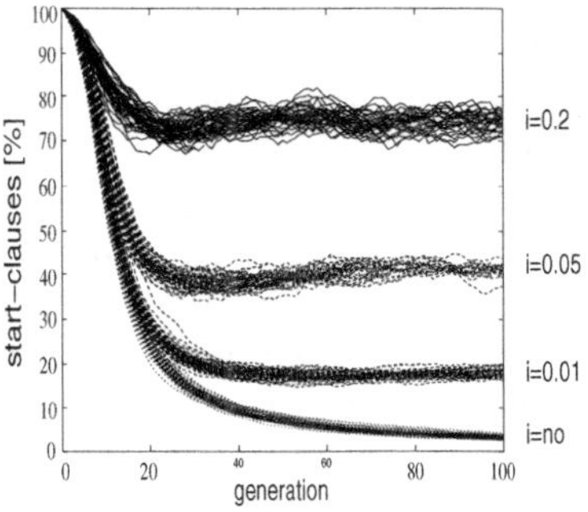

Fig. 9. A comparison of inflow variants i in a reactor with size 2000 employing free replacement[4]

With educt replacement one of the educts is replaced by the reaction product, thus, reactive molecules are more often replaced; a reaction pathway cannot be maintained if there is no inflow of the participating molecules. Moreover, diminishing molecules can prohibit the emergence of other pathways. On the other hand, non-reactive molecules obviously always are involved in elastic reactions and their concentration remains unchanged (niche building). Presumably useless most of the time, though, they use reactor space. As already seen in the basic experiment in the previous section free replacement policy, once again, has a global impact, thereby inducing a global pressure. But even if the `Connector` does not go extinct entirely, free replacement is superior because we observe a higher diversity of the free replacement reactor variants in open systems as well (Fig. 8).

(2): The higher the inflow, the higher is the fraction of start-clauses blocking the system by consuming valuable reactor space (Fig. 9). The elastic inflow

[4] A sample of 10 arbitrary runs is plotted simultaneously.

variant accompanied by immense inflow demonstrates this impressively (Fig. 10). In contrast, a low inflow rate leads to high diversity and a better covering of the search space. However, if the chosen inflow rate is too small, relevant start-clauses diminish which makes their interaction with other molecules a rare event, thus prolonging the solving process. To make, for example, a closed reactor successful, the run time and/or reactor size should be increased (see again Fig. 6). The optimal inflow rate i_{opt} is hard to find in the general case. In our experiments, a 5%-inflow has proven to be optimal.

(3): By enlarging the reactor, one can compensate for a smaller diversity caused, for example, by a high inflow rate. In generation 20 a reactor with 20000 molecules and 20%-inflow on average consists of nearly 2000 different molecules, approx. 3 times more than a 5%-inflow reactor with size 2000. In general, more molecules are (re)generated each generation (Fig. 11). Consequently, it is more likely to find a solution in an early stage of an experiment. Recalling the definition of a generation, however, this does not imply a faster execution. If measured in reactions, the large-sized reactor performs 10 times more reactions per generation than its small correspondent, such that a run is slower in terms of computational time.

Remark 1: Against the background of previous findings we may reconsider our representation of the Adleman problem. Basically, it has 2 drawbacks. The first difficulty is that "junk" clauses may emerge. Clauses considering more than 7 vertices are leading away from the goal. The second problem is the central focus on the `Connector`. Without restricting the solution space, one way to improve the situation is to remove this clause from the initial knowledge base and, instead, replace the `Solver` with $\texttt{Solver}_{R1}$ (Fig. 5). In fact, results improve remarkably: With educt replacement, 10%-inflow and a reactor with 20100 molecules all 100 runs succeeded. The avg. success gen. is 5.94 (Std. Dev. 2.2). However, due to the implemented simplifications this artificially designed problem does not show relevant complexity necessary for representing a general problem.

Remark 2: A slight modification of the `Solver` lets us find all paths of length 6 (see $\texttt{Solver}_{R2}$ in Fig. 5) . Every time the termination criteria is matched a directed Hamiltonian path is found. In this case, RESAC proceeds until a time limit is exceeded.

5 Conclusion

A computational system was set up whose overall behavior is characterized by emergent processes. With the resolution-based artificial chemistry *RESAC*, we intend to draw attention to the application-oriented branch of ACs, which probably is not as common as its use as a model ([10],p. 40). In previous publications the potential of resolution-based ACs was shown by implementing an automated theorem prover. In this paper, resolution logics enabled us to set up an AC based on a logic calculus. With this method, problems can be formulated within the AC framework by conversion to the first-order predicate calculus.

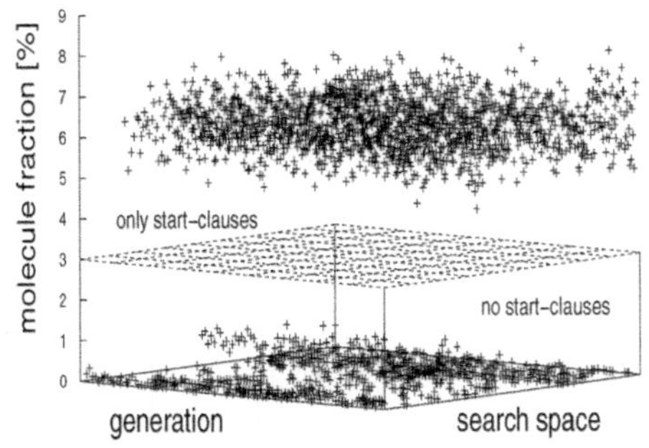

Fig. 10. Elastic inflow in a reactor with size 2000 (free replacement): The graph shows all clauses the system generates. All start-clauses have high concentrations—none of them is located below the plane. They block the system

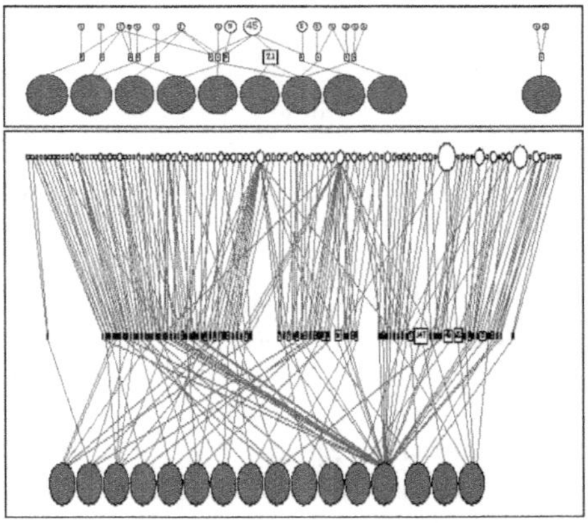

Fig. 11. Reaction network in generation 4, free replacement. Each line represents a reaction. *Top:* Reactor with 2000 molecules. *Bottom:* Reactor with 20000 molecules. Much more reaction pathways and molecules are built

We translated the directed Hamilton path problem presented by Adleman into FOPC. In 1994, Adleman implemented and solved the problem by means of DNA-computing, however, not much is revealed about the dynamics of such solving processes. Here, 48 variants of ACs were analyzed by investigating their dynamic behavior. Especially, the question how to proceed with the reaction product is addressed. Though these different setups cannot be very easily weighted against each other we identified some rules how performance of an application-oriented AC could be increased. These findings and the presented methods of dynamic system analysis may lead to a better understanding of the choices available at an early stage of AC design.

The presented system can easily be transferred to parallel computing platforms like multi-processor systems or distributed internet agents due to the inherent parallelism. Several realizations are currently under investigation.

Acknowledgments. We are grateful to Jens Ziegler for valuable comments and useful discussions.

References

1. L. M. Adleman. Molecular computation of solutions to combinatorial problems. *Science*, 266:1021–1024, 1994.
2. J. C. Astor and C. Adami. A developmental model for the evolution of artificial neural networks. *Artif. Life*, 6(3):189–218, 2000.
3. R. J. Bagley and J. D. Farmer. Spontaneous emergence of a metabolism. In C. G. Langton, editor, *Artificial Life II, Proceedings*, Cambridge, MA, 1992. MIT Press.

4. R. J. Bagley, J. D. Farmer, and W. Fontana. Evolution of a metabolism. In C. G. Langton, editor, *Artificial Life II, Proceedings*, Cambridge, MA, 1992. MIT Press.
5. J.P. Banatre and D. Le Metayer. Programming by multiset transformation. *Communications of the ACM*, 36(1):98–111, 1993.
6. W. Banzhaf. Self-organisation in a system of binary strings. In R. Brooks and P. Maes, editors, *Proceedings of Artificial Life IV*, pages 109–118, Cambridge, 1994. MIT Press.
7. W. Banzhaf, P. Dittrich, and H. Rauhe. Emergent computation by catalytic reactions. *Nanotechnology*, 7:307–314, 1996.
8. R.A. Brooks. Coherent behavior from many adaptive processes. In Dave Cliff, Philip Husbands, Jean-Arcady Meyer, and Steward Wilson, editors, *From animals to animats 3*, pages 22–29, Cambridge, MA, 1994. MIT Press.
9. J. Busch. Automated Theorem Proving for first order predicate calculus using Artificial Chemistries. In GI Gesellschaft für Informatik, editor, *Informatiktage 1999*, Bad Schussenried, November 1999. Konradin Verlag Robert Kohlhammer GmbH.
10. P. Dittrich. *On Artificial Chemistries.* PhD thesis, University of Dortmund, Department of Computer Science, D-44221 Dortmund, Germany, January 2001.
11. P. Dittrich, J. Ziegler, and W. Banzhaf. Artificial Chemistries - a Review. *Artificial Life*, 7(3):225–275, 2001.
12. W. Fontana. Algorithmic chemistry. In C. G. Langton, C. Taylor, J. D. Farmer, and S. Rasmussen, editors, *Artificial Life II: Proceedings of the Second Artificial Life Workshop*, pages 159–209. Addison-Wesley, Reading MA, 1991.
13. W. Fontana and L. W. Buss. The barrier of objects: From dynamical systems to bounded organizations. In J. Casti and A. Karlqvist, editors, *Boundaries and Barriers*, pages 56–116. Addison-Wesley, 1996.
14. C.G. Hill. *An introduction to chemical engineering kinetics and reactor design.* John Wiley & Sons, 1977.
15. Y. Kanada and M. Hirokawa. Stochastic problem solving by local computation based on self-organization paradigm. In *27th Hawaii International Conference on System Science*, pages 82–91, 1994.
16. S. A. Kauffman. *The Origins of Order.* Oxford University Press, New York, 1993.
17. J. A. Robinson. A Machine-Oriented Logic Based on the Resolution Principle. *Journal of the Association for Computing Machinery*, 12(1):23–41, January 1965.
18. T. Szuba and R. Stras. Parallel evolutionary computing with the random PROLOG processor. *Journal of Parallel and Distributed Computing*, 47(1):78–85, November 1997.
19. K.-P. Zauner and M. Conrad. Conformation-driven computing: Simulating the context-conformation-action loop. *Supramolecular Science*, 5:791–794, 1998.
20. J. Ziegler, P. Dittrich, and W. Banzhaf. Towards a metabolic robot control system. In M. Holcombe and R. Paton, editors, *Information Processing in Cells and Tissues.* Plenum Press, 1998.

Artificial Life as an Aid to Astrobiology: Testing Life Seeking Techniques

Florian Centler[1], Peter Dittrich[1], Lawrence Ku[2], Naoki Matsumaru[1], Jeffrey Pfaffmann[2], and Klaus-Peter Zauner[1]

[1] Jena Centre for Bioinformatics (JCB) and
Department of Mathematics and Computer Science
Friedrich-Schiller-University Jena
D-07743 Jena, Germany
`http://www.minet.uni-jena.de/csb/`
[2] Department of Computer Science
Wayne State University
Detroit, MI 48202, U.S.A.

Abstract. Searching for signatures of fossil or present life in our solar system requires autonomous devices capable of investigating remote locations with limited assistance from earth. Here, we use an artificial chemistry model to create spatially complex chemical environments. An autonomous experimentation technique based on evolutionary computation is then employed to explore these environments with the aim of discovering the chemical signature of small patches of biota present in the simulation space. In the highly abstracted environment considered, autonomous experimentation achieves fair to good predictions for locations with biological activity. We believe that artificially generated biospheres will be an important tool for developing the algorithms key to the search for life on Mars.

1 Life Detecting Machines

A spot of rust on a sheet of steel has much in common with a spot of mold on a sheet of fabric. Discerning alive matter among dead matter is a challenge even in our familiar surroundings. The ability to draw this line with some certainty is a rather recent achievement [1]. The U.S. National Aeronautics and Space Administration outlined a program to detect signs of fossil or present life within our solar system and beyond [2]. Crucial to this endeavor is the capability of recognizing life in whatever form it may take:

> "A strategy is needed for recognizing novel biosignatures. [...] For example, certain examples of our biosphere's specific molecular machinery, e.g., DNA and proteins might not necessarily be mimicked by other examples of life elsewhere in the cosmos. On the other hand, basic principles of biological evolution might indeed be universal." [2, p. 19]

This broad conception of biota was inherent in artificial life research from its very onset; in Langton's words:

W. Banzhaf et al. (Eds.): ECAL 2003, LNAI 2801, pp. 31–40, 2003.

> "[...] certainly, the dynamic processes that constitute life—in whatever material bases they might occur—must share certain universal features—features that will allow us to recognize life by its dynamic form, without reference to its matter." [3, p. 2]

He went on to point to the usefulness of synthetic life to provide context for the known forms of life. Thus artificial life would substitute for the lack of samples of life forms based on other materials. At the time, biologists may have dismissed the need for a context as broad as Langton envisioned. But when it comes to the quest of searching real life beyond the scope of our immediate environment, a narrow perspective on biology may in fact preclude the discovery of those life forms that would allow us to strengthen our as of yet weak notion of what really constitutes life. Accordingly, the abstraction artificial life aims at can make a significant contribution to the recognition of unknown life forms.

In the following, we report on the application of artificial chemistry modeling [4] to evaluate an autonomous experimentation technique with regard to its capability to detect chemical signatures of life.

1.1 Autonomous Experimentation

The search for signs of extraterrestrial biota is characterized by vast, hard to access areas and severe restrictions on communication bandwidth. Instrumentation with a high degree of automation is required. Autonomous experimentation utilizes computational discovery methods [5,6] to orchestrate the available sensor resources. In contrast to the majority of machine discovery algorithms that operate on a fixed dataset, autonomous experimentation goes beyond data analysis and include the experiments itself in the discovery process [7,8]. Kulkarni and Simon [9] demonstrated that an algorithm can successfully emulate the interplay of adjusting hypotheses and modifying experiments characteristic of human experimenters [10]. The integration of such algorithms into space probes [11] and planetary rovers [12] is key to the search for life beyond earth.

Here we employ an autonomous experimentation technique based on evolutionary computation. This technique, named *scouting*, has initially been developed to characterize protein response with regard to chemical signals [13]. Figure 1 illustrates its operation. The control computer simulates an evolutionary population. Each genome in the population represents a specification of experimental conditions (labeled $\mathbf{x}$ in Fig. 1). The first step toward the evaluation of the fitness of a genome is the generation of a prediction $\mathbf{r}'$ for the outcome of the experiment specified by the genome under consideration. This prediction is based on a database of experiences in which the observations from all experiments that have been conducted up to this point are stored. Subsequently the experiment specified by the genome is actually performed and provides an observation $\mathbf{r}$ for an experiment conducted under conditions $\mathbf{x}$. The more $\mathbf{r}$ deviates from what was expected for such an experiment (i.e., $\mathbf{r}'$), the larger the fitness value that is assigned to the genome that specified the experiment. In other words, specifying experiments that yield a lot of information is rewarded. As a consequence

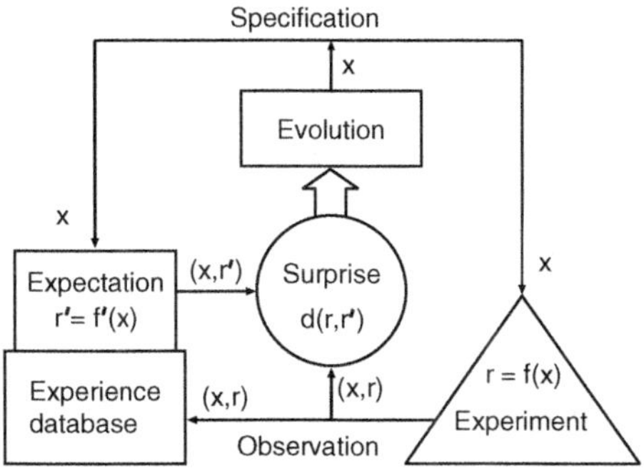

Fig. 1. Scouting combines the notion of information being equivalent to surprise value [14] with evolutionary computation [15] for autonomous exploration. See text for details

the next generation of the population in the evolutionary process will focus its attention on parameter combinations in the vicinity of the conditions that gave rise to the surprising experimental result.

The scouting method has been applied in wet-lab experiments in conjunction with a computer controlled fluidics system to autonomously investigate the response of the enzyme malate dehydrogenase to ion signals [16]. In this domain the scouting algorithm focuses measurements on areas of the parameter space that show unusual phenomena. However, the parameters set by the algorithm can also be spatial coordinates, in which case parameter space and real space coincide. The algorithm then probes the space for unusual observations.

The selection of measurement locations is of particular interest for instruments that can be applied locally only. Multi-touch surface sampling with altitude-controlled Montgolfiere balloons and air-born laser spectroscopy are examples. It is in this context that the artificial life simulation described in the next section provides a test scenario for autonomous experimentation.

2 Simulation of an Artificial Planetary Sphere

To create a test scenario for the study of the behavior of the scouting algorithm we simulated an abstract artificial planetary sphere with an inanimate chemistry that can be perturbed by locally introduced life forms. The assumption here is that life necessitates the synthesis of a larger variety of substances than commonly produced by inorganic reactions.

For simulating spatially inhomogeneous chemistry a two-dimensional cellular automata implementation of the chemical dynamics is convenient [17]. We employed an asynchronously updated, probabilistic variant in which the state of a cell represents a chemical component present at the location of the cell. All rules take the form $A + B \xrightarrow{p} C + D$. If component B is present within the set of four (von Neumann) neighbors of component A, then A may be substituted by C and B substituted by D. The probability for this event to take place is proportional to p. In principle the states of the cells could represent a complex local chemical composition; for simplicity we use the abstraction of a single substance. Note that the defined order of substitution in the rules enables diffusion to be expressed by sets of rules of the form $\mathrm{A} + \mathrm{B} \xrightarrow{p} \mathrm{B} + \mathrm{A}$.

2.1 The Chemical Model

As the basis for our model we selected a randomly created artificial chemistry for which all modes of organization (see next section) are known [18]. In this abstraction of molecular interaction, substances are created by the cooperative action of two catalytically acting substances. This was motivated by biopolymer synthesis and entails the assumption of an inexhaustible pool of building blocks being available. In contrast, the present model does take substrates explicitly into account, such that all reactions are of the form $\mathrm{X} \xrightarrow{(\mathrm{E},\mathrm{F})} \mathrm{Y}$, where X is a substrate molecule, E and F are co-acting catalysts and Y is the reaction product. To realize reactions in which a substrate, a product, and two catalysts participate with the binary reaction scheme of the cellular automata, intermediate agents have been introduced for each reaction. This is in agreement with real chemistry, where elementary steps are assumed to be predominantly binary and at most ternary. Thus, above reaction would be represented in the cellular automata by two rules:

$$\mathrm{E} + \mathrm{F} \xrightarrow{p_1} \mathrm{E} + \mathrm{I}, \tag{1}$$

$$\mathrm{X} + \mathrm{I} \xrightarrow{p_2} \mathrm{Y} + \mathrm{F}. \tag{2}$$

The intermediate agent I is assumed to be highly reactive and leads to a rapid transformation of substrate X to product Y, i.e., $p_1 \ll p_2$. Any one of a set of three arbitrarily selected special substances {0, 6, 7} can serve as substrate. Figure 2A shows the reaction network taken from [18]. Note that the reaction matrix is asymmetric, which effectively corresponds to the possibility of two products being catalyzed by the cooperative action of the two catalysts.

For the present purpose the reaction network shown in Fig. 2A was augmented by decay reactions that transform any substance other than intermediates into the substrate substances through rules of the form $\mathrm{A} + \mathrm{B} \xrightarrow{p} \mathrm{X} + \mathrm{B}$, where $A \in \{0, 1, \ldots, 9\}, \mathrm{X} \in \{0, 6, 7\}$ and B can take any value. Decay reactions for all but substance 1 have a probability of 0.075%. Substance 1 is instable and decays with the normal reaction probability of 0.75%. The probability for diffusion is 25%. Finally, an additional catalyst, number 10, was introduced as *biota*. It consumes the substrate molecules {0,6,7} and excretes the instable substance 1. All substances, but not the biota had the ability to diffuse. The resulting reaction network, illustrated in Fig. 2C, was simulated with the cellular automata.

2.2 Simulation of Spatial Chemical Composition

A cellular automata implementation of the model described above was run in a bounded simulation space of 200×200 cells. This space was initialized with a

A

	0	1	2	3	4	5	**6**	**7**	8	9
0		0						6		
1	2			5						
2				1						0
3	5			0						1
4	8		6	6	4		8	4	8	5
5				1		9				8
6		8	0					7	2	
7		5				5		9		
8	1	2	9	1		6	3			2
9	0	8		0	3		6		0	0

B

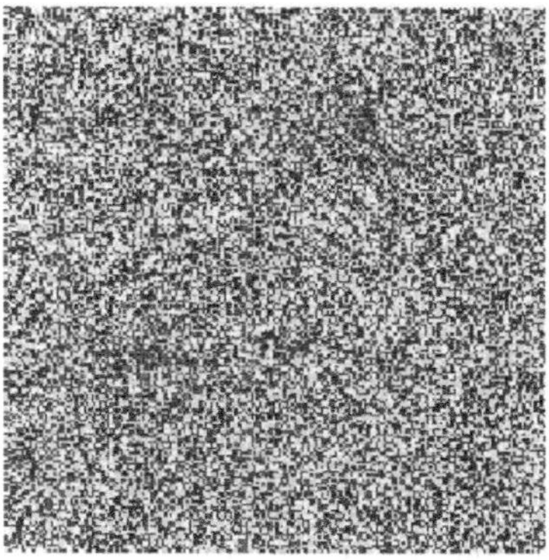

C

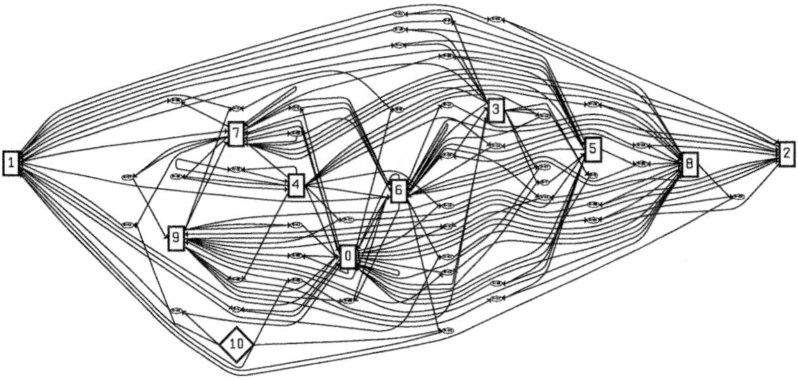

Fig. 2. (A) The network of potential reactions. Numbers in the leftmost column refer to substances that take the place of E in equation 1, numbers in the top row correspond to F in equations 1 and 2, and entries in the table correspond to Y in equation 2. Substances that can serve as substrates are marked with bold face. (B) Snapshot of the states of the 200×200 cell automaton after 10^8 individual updates. The different shades indicate cell states corresponding to different chemical compositions. All intermediate products are mapped to the same shade. (C) The reaction network used to model the planetary sphere. Numbered boxes represent the substances. The diamond shape (10) indicates the biota's interaction with its chemical environment. Reactions are shown as tiny nodes; for clarity the 123 fast intermediate reactions are not shown

pseudo-random distribution[1] of four of the ten substances, namely $\{0, 6, 7, 9\}$. These four substances form a closed and self-maintaining reaction system. In other words, they do not catalyze the production of any substance not present in the system (property of closure) and every substance is catalytically produced within the reaction system (property of self-maintenance). A reaction system that is closed and self-maintaining is called *organization* [19,20]. The chemical

[1] For our simulations, we initialized always: 12762 cells with substance 0, 12556 cells with substance 6, 13419 cells with substance 7, and 1263 with substance 9.

organization of the reaction system may change if it is perturbed by the introduction of new chemical species, such as substance 1 excreted by the biota. For all simulations in which the biota was present, exactly six cells where set to the biota state.

Diffusion rules (A+B $\xrightarrow{p_D}$ B+A with $p_D = 0.25\%$), here an abstraction of all transport phenomena, were implemented for all substances but not for the biota. The total number of rules, including diffusion rules for all substance combinations (inclusive intermediate substances) is 2873. To update the simulation space a cell and one of its von Neumann neighbors are pseudo-randomly selected. The subset of rules applicable to the two chemical components located in the two cells is determined and in accordance with their probabilities one of these rules may be applied to transform the content of the cells. Figure 2B indicates the distribution of chemical compounds in the simulation space after an average number of 2500 updates per cell. The grey level distribution is indicative of the complexity of spatial distribution of substances. The chemical composition of local areas in the simulation space shows considerable fluctuation. It provides a rich testbed for the scouting algorithm because in general no accurate prediction of the local chemical composition is possible. The dynamics of the artificial chemistry model was simulated and snapshots of the cellular automata state, such as depicted in Fig. 2B, were saved during the run to serve as input for the scouting algorithm.

3 Scouting Experiments

In the experiments described here the scouting algorithm is applied to detect unusual chemical signatures in a complex background chemistry [21]. It is assumed that the chemical composition can be sampled locally, but no a priori knowledge of the chemical effects of potentially present biota is available to the algorithm. For sampling the algorithm has to choose a location $\mathbf{x} = (x, y)$ in the simulation space. The measured data at this location is a vector $\mathbf{r} = (r_1, \ldots, r_n)$ containing for each of the n substances the fraction of cells in a ± 2 pixel vicinity that hold this substance. Correspondingly the entries $(\mathbf{x}, \mathbf{r})$ in the experience database and the expectations $(\mathbf{x}, \mathbf{r}')$ formed for a location prior to its sampling are such vectors (cf. Fig. 1). The surprise value d, which constitutes the fitness criterion, is computed as the Shannon entropy [14] of the difference between the actual response $\mathbf{r}$ and response that was expected $\mathbf{r}'$. The expectation $\mathbf{r}'$ for a location $\mathbf{x} = (x, y)$ is computed as distance-weighted average over the (up to) 25 measurements nearest to $\mathbf{x}$ available in the experience database. A population size of 10 genomes, all offspring of the single best parent was used for evolution (i.e., a $(1, 10)$-strategy [15]). We mutate an offspring by adding an equally distributed random number taken from within a radius of δ. The mutation strength δ depends on the previous surprise

$$d(\mathbf{r}, \mathbf{r}') = -\sum_{i=1}^{n} |r_i - r'_i| \ln |r_i - r'_i|, \qquad \delta = \begin{cases} 0.03 \text{ if } \sqrt{4.2} \le d(\mathbf{r},\mathbf{r}'), \\ 0.04 \text{ if } \quad 2.0 \le d(\mathbf{r},\mathbf{r}') < \sqrt{4.2}, \\ 0.1 \text{ if } \sqrt{3.5} \le d(\mathbf{r},\mathbf{r}') < 2.0, \\ 0.99 \text{ else }. \end{cases} \tag{3}$$

In the scouting algorithm applied for the experiments described below, these parameters have been established empirically and kept constant for all experiments. Methods to adapt the mutation strength dynamically during the sampling are under development. It should be noted that due to the highly dynamic fitness landscape, adaptation methods commonly used in evolutionary computation do not perform well in scouting (unpublished results).

Simulations were conducted for four situations, one patch with biota (6 cells), two (2×3 cells), and three patches with biota (3×2 cells). After 10^8 updates (2500 updates per cell) of the cellular automata a snapshot of the distribution of chemicals in the simulation space was saved.

Scouting experiments were run on these snapshots to sample 800 locations in the simulation space. Figure 3 shows the probed locations in the left column. Their density indicates areas of high interest to the scouting algorithm. We used hierarchical clustering with subsequent expectation maximization (performed with *mclust* in *R* [22,23]) to automatically identify clusters in the sampling positions. The number of clusters identified during expectation maximization can be seen in the middle column as peaks of the Bayesian information criterion. We used the samples allocated to a cluster (shown in Fig. 3 only for the case of three biota patches: panel J) and calculated their mean position as prediction for the location of biota in the simulation space. These predicted positions are marked with + in panels C, F and I. If the scouting is repeated with an alternate seed value for its random generator, the evolution of the sampling positions will take a different course; predictions from four additional runs are marked with · for comparison. The localization is fairly good, despite the complex chemical background (panel D, for example, shows the scouting of Fig. 2B).

Detection of the biota is possible with considerably less than the 800 samples used in Fig. 3, albeit with false positives and less accuracy. In Fig. 4 bars represent clusters, the solid bars correspond to clusters at biota patches, striped bars are clusters close to the correct location and white bars are false positives. The top graph (I) shows an experiment with a single biota patch that is detected with less than 100 samples and located with 300 samples. The middle graph (II) shows scouting with two biota patches, more samples are needed to find both locations. With three biota patches (graph III) one is detected early, but several false positives appear and one of the three patches is only approximately located (striped bar) with 800 samples.

4 Concluding Remarks

We have applied an autonomous experimentation technique for the search of unusual chemical signatures in a complex environment. Scouting was able to consistently detect biota signatures without any domain knowledge regarding the chemistry of the environment or the biota. This performance of course has to be seen in the light of the relative small search space and the highly abstracted environment applied here.

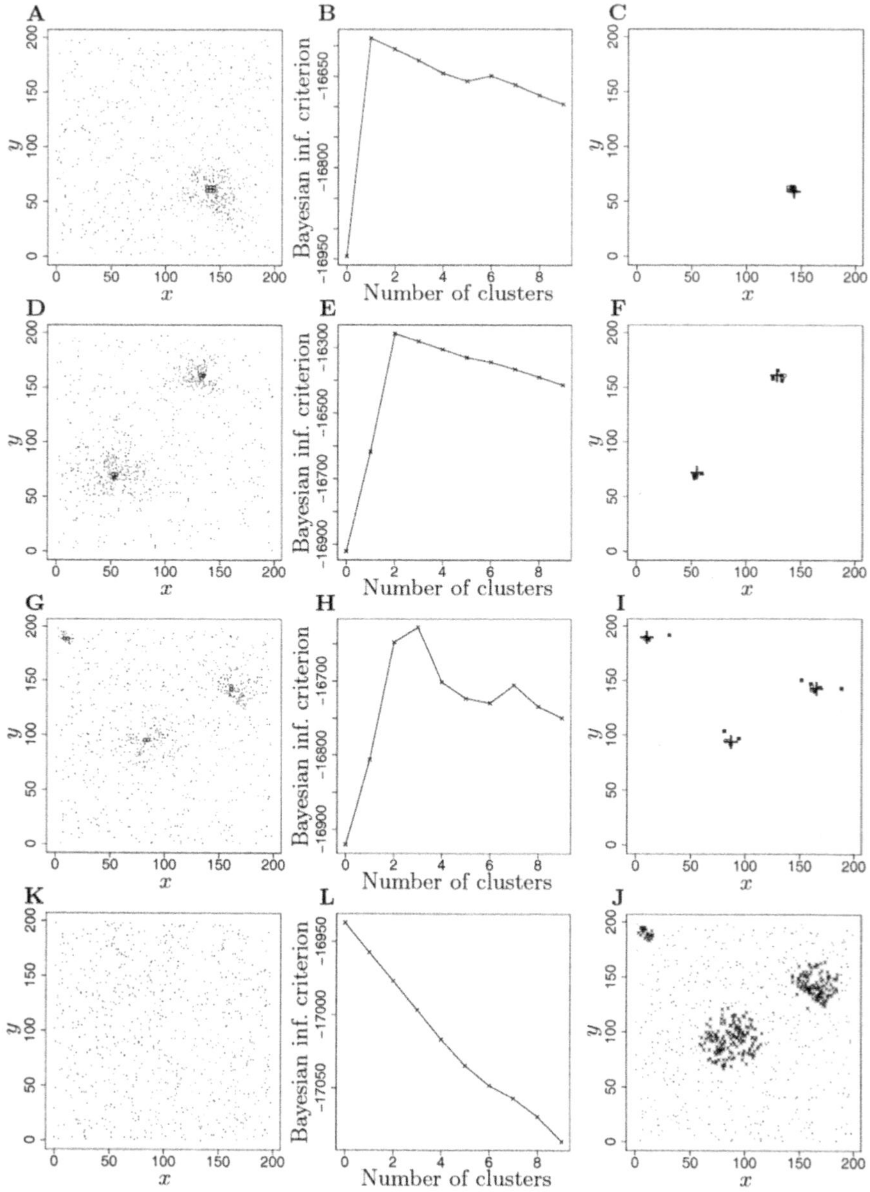

Fig. 3. Detection of an unusual chemical composition. Actual location of biota (◦) and the 800 locations (·) probed by scouting, are shown for 1, 2, 3, and 0 biota patches in (A, D, G, K). By clustering the biota can be localized, marked + (and ▪ for additional scouting runs) in C, F, I. In J the allocation of sample positions to clusters is shown for the case of three biota patches. A run without biota (K, L) does not lead to clusters. For clarity symbols for biota and cluster centers have been enlarged but refer to only one point in the plane. See text

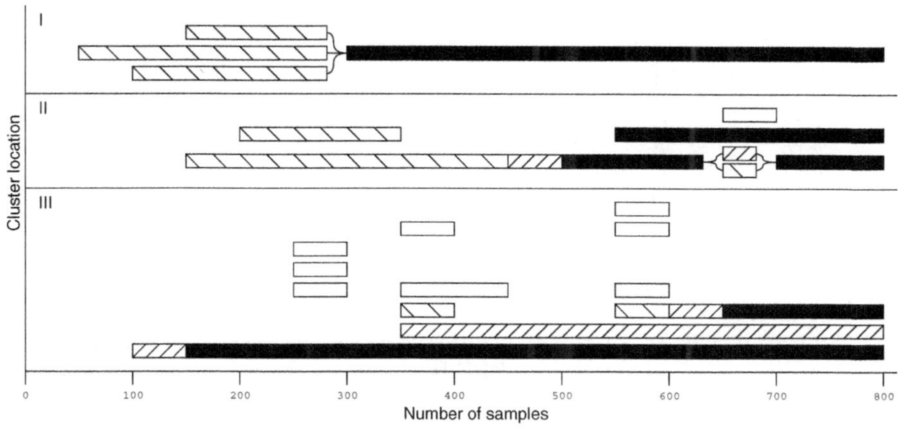

Fig. 4. Development of clusters for one, two and three biota patches (I, II, III). Clustering was performed every 50 samples, solid bars represent correctly localized biota, white bars false positives; see text for details

The path to algorithms that confidently can be intrusted with the search of life in remote locations of our solar system is still long. We belive that the broad variety of artificial life is an important resource to draw on for test scenarios during the development of such algorithms.

Acknowledgments. The authors thank E. G. Schukat-Talamazzini for helpful discussion. This article is based upon work supported by BMBF (Federal Ministry of Education and Research, Germany) under grant No. 0312704A to PD, and NASA under Grant No. NCC2-1189 to KPZ.

References

1. Toulmin, S., Goodfield, J.: The Architecture of Matter. Harper & Row, New York (1962)
2. Anonymous: Astrobiology roadmap. U.S. National Aeronautics and Space Administration (NASA) (November 2002) Available at: `http://astrobiology.arc.nasa.gov/roadmap/roadmap.pdf`
3. Langton, C.G.: Artificial life. In Langton, C.G., ed.: Artificial Life. SFI Studies in the Science of Complexity, Addison-Wesley, Redwood City, CA (1989) 1–45
4. Dittrich, P., Ziegler, J., Banzhaf, W.: Artificial chemistries—a review. Artificial Life **7** (2001) 225–275
5. Langley, P., Simon, H.A., Bradshaw, G.L., Żytkow, J.M.: Scientific discovery: Computational exploration of the creative processes. MIT Press, Cambridge, MA (1987)
6. de Jong, H., Rip, A.: The computer revolution in science: steps towards the realization of computer-supported discovery environments. Artificial Intelligence **91** (1997) 225–226

7. Rajamoney, S.A.: A computational approach to theory revision. In Shrager, J., Langley, P., eds.: Computational Models of Scientific Discovery and Theory Formation. Morgan Kaufmann, San Mateo, CA (1990) 225–253
8. Karp, P.D.: Hypothesis formation as design. In Shrager, J., Langley, P., eds.: Computational Models of Scientific Discovery and Theory Formation. Morgan Kaufmann, San Mateo, CA (1990) 275–317
9. Kulkarni, D., Simon, H.A.: Experimentation in machine discovery. In Shrager, J., Langley, P., eds.: Computational Models of Scientific Discovery and Theory Formation. Morgan Kaufmann, San Mateo, CA (1990) 255–273
10. Gooding, D.: Experiment and the Making of Meaning. Kluwer Academic, Dordrecht (1990)
11. Stolorz, P., Cheeseman, P.: Onboard science data analysis: Applying data mining to science-directed autonomy. IEEE Intelligent Systems & Their Applications **13** (1998) 62–68
12. Gilmore, M.S., Castano, R., Mann, T., Anderson, R.C., Mjolsness, E.D., Manduchi, R., Saunders, R.: Strategies for autonomous rovers at mars. J. Geo. Phys. Res. Planets **105** (2000) 29223–29237
13. Pfaffmann, J.O., Zauner, K.P.: Scouting context-sensitive components. In Keymeulen, D., Stoica, A., Lohn, J., Zebulum, R.S., eds.: The Third NASA/DoD Workshop on Evolvable Hardware, IEEE Comp. Soc., Los Alamitos (2001) 14–20
14. Cherry, C.: Chapter 5. In: On Human Communication: A Review, a Survey, and a Criticism. 2nd edn. MIT Press, Cambridge, MA (1966)
15. Beyer, H.G.: The theory of Evolution Strategies. Springer, Berlin (2001)
16. Matsumaru, N., Colombano, S., Zauner, K.P.: Scouting enzyme behavior. In Fogel, D.B. et al., eds: 2002 World Congress on Computational Intelligence, IEEE, Piscataway, NJ (2002) CEC 19–24
17. Adamatzky, A.: Identification of Cellular Automata. Taylor and Francis, London (1994)
18. Speroni di Fenizio, P., Dittrich, P.: Artificial chemistry's global dynamics. movement in the lattice of organisation. The Journal of Three Dimensional Images **16** (2002) 160–163
19. Fontana, W., Buss, L.W.: 'The arrival of the fittest': Toward a theory of biological organization. Bull. Math. Biol. **56** (1994) 1–64
20. Speroni di Fenizio, P., Dittrich, P., Ziegler, J., Banzhaf, W.: Towards a theory of organizations. In: German Workshop on Artificial Life, in print, Bayreuth, 5.-7. April, (2000)
21. Yung, Y.L., DeMore, W.B.: Photochemistry of Planetary Atmospheres. Oxford University Press, New York (1999)
22. Ihaka, R., Gentleman, R.: R: A language for data analysis and graphics. Journal of Computational and Graphical Statistics **5** (1996) 299–314
23. Fraley, C., Raftery, A.E.: MCLUST: Software for model-based clustering, discriminant analysis and density estimation. Technical Report 415, Department of Statistics, University of Washington, Seattle, WA (2002) Available at: `http://www.stat.washington.edu/www/research/reports/`

Molecularly Accessible Permutations

Dónall A. Mac Dónaill

Department of Chemistry, Trinity College, Dublin 2, Republic of Ireland
dmcdonll@tcd.ie

Abstract. The propagation of text expressed in an n-letter alphabet, A_n, may be effected by any of the set of associated permutations S_n, applied repeatedly or in suitable combinations until the original text is reproduced. Scaling as $n!$, the number of possible permutations rapidly becomes intractable, and individual permutations cannot be considered for all but the smallest alphabets. This paper explores how a molecular medium, in which replication must proceed by means of template propagation, might serve to limit the number of permutations which may be reasonably find a molecular expression. The analysis suggests that the number of molecularly realisable permutations is restricted to a limited variety of permutation types, scaling linearly with alphabet size. It is also suggested that alphabets with odd numbers of letters may be less accessible than alphabets with even numbers of letters.

1 Introduction

Our understanding of replicating systems seems to be advancing most rapidly in two related but largely independently developing fields. Much effort, in what has become to be known as 'artificial life', has been directed towards the study of fundamental principles of information processing in self-replication, independent of physical realisation [1], where the nature of the physical replicating machinery is often ignored [2]. On the other hand, supramolecular chemistry, exploring the processes of molecular recognition, self-assembly and template propagation, has largely focused on the physical realisation of replication and assembly, and has resulted in an enormously expanded view of this domain [3]. Notwithstanding these independent developments, it has however long been realized that the physical manifestation of the phenomenon of life lies at the intersection of these two fields, and indeed as far back as 1963 Stahl and Goheen proposed that some fundamental biological processes could be formally interpreted as molecular algorithms [4]. It certainly seems reasonable to take the view that many molecular biological processes, both as naturally manifest, and as might potentially be synthesized, should be subject to constraints arising from the nature of information itself, in addition to constraints relating to the molecular medium in which the information is encoded and propagated.

Replication of information in the molecular world is a particular manifestation of the more abstract problem of propagation based on n-letter alphabets, effecting repli-

W. Banzhaf et al. (Eds.): ECAL 2003, LNAI 2801, pp. 41–50, 2003.

cation through the realization of one or more of $n!$ available permutations. Permutations may be viewed as being effected by unspecified molecular machines, agents which both read and write molecular text, such as polymerase in the case of the nucleotide alphabet. Drawing from biochemical nomenclature we may term these hypothetical machines 'permutases'.

One immediately apparent problem is that the space of available permutations increases so rapidly with alphabet size that consideration of individual permutations is impossible for all but the smallest alphabets. However, template propagation in a molecular medium involves the flow of considerable amounts of matter, and this fundamentally distinguishes it from other sorts of information transfer. In virtual or digital electronic worlds, where transmission is not based on molecular flux [2], almost all permutations can be effected with more or less equivalent overhead. This is not self-evidently the case in molecular transmission where the overheads may be expected to vary with the complexity of the permutations considered. Accordingly, the purpose of this study is to explore potential constraints which might act to favour some permutases over others in the particular context of a molecular medium of expression. Our aim is to bridge the worlds of artificial life and supramolecular chemistry, marrying permutational algebra with molecular reality, and perhaps in the process learn something of the nature of molecular replication.

We begin in section 2 by briefly introducing the necessary aspects of permutational algebra, equivalence classes, and associated notation, and proceed in section 3 to consider the more probable limitations which might favour the physical realisation of some permutations over others.

2 Group Theory

The action of polymerase is to effect a permutation, reading a nucleotide and expressing its complement. Permutations do not change the information content, rather they disguise it. If the letters of an alphabet expressing a piece of text are subjected to a permutation, the information is retained in the transformed text. In cyclic notation[1] the function executed by polymerase may be considered as executing $f = (AT)(CG)$. Two-fold execution of polymerase yields the original text

$$f^2 = i;\; f = (AT)(CG),\; i = (A)(T)(C)(G). \tag{1}$$

where i is the identity operation. Similarly, if we permute the letters of the English alphabet, replacing each letter with that which follows two positions later, i.e. $f\colon\{a \to c,\, b \to d\ldots,\, x \to z,\, y \to a,\, z \to b\}$ we map the expression '*molecular biology*'

1 Cyclic notation provides a convenient method for describing permutations; subsets of permuted letters are bounded by parentheses, each letter being replaced in the permutation by the letter to its right, except for the final letter which is replaced by the first letter, completing the cycle. For example, in the six letter alphabet {a, b, c, d, e, f}, the permutation $f\colon\{a \to b,\, b \to c,\, c \to a,\, d \to d,\, e \to f,\, f \to e\}$ may be represented by (abc)(d)(ef).

to '*oqngewnct dkqnqia*'. Although the transformed text is apparently nonsense, the information is not destroyed, and all that is required for '*oqngewnct dkqnqia*' to become intelligible is the application of the inverse transformation or permutation f^{-1}, or alternatively, a combination of permutations equivalent to f^{-1}.

2.1 Alphabets, Permutations, and Complexity

Replication may proceed by permutation of a text into apparently nonsense forms, provided the latter are eventually permuted back into the sense form. Any permutation, or combination of permutations, can in principle achieve this, but with varying degrees of efficiency as judged by the number of required steps. We shall consider alphabets of arbitrary size, and define A_n as an alphabet of n letters; $A_n = \{a, b, c, \ldots\}$, and S_n as the associated set of $n!$ permutations; $S_n = \{\alpha, \beta, \gamma, \delta, \ldots\}$.

The increase in the size of S_n with n is worse than exponential, and although consideration of all permutations is straightforward for small n, it rapidly becomes intractable. For example, in an alphabet of size four (such as the nucleotide alphabet), S_n has a total of $4! = 24$ potential permutations, all of which can easily be examined. However, for an alphabet of size 10 the number of permutations rises to 3628800, while for the English alphabet we have an intractable $26!$ ($> 4.0 \times 10^{26}$) permutations. To place this in context, at a rate of one permutation per second it would take approximately 1.3×10^{19} years, roughly 10^{9} times the age of the universe to date ($\approx 1.3 \times 10^{10}$ years) [5] to consider all permutations. Clearly, it is not in general possible to consider each permutation in all but the smallest alphabets

Fortunately, closely related permutations, α and β, may be grouped into equivalence classes; provided there exists another permutation σ such that

$$\beta = \sigma\alpha\sigma^{-1}. \qquad (2)$$

For many purposes it is sufficient to consider a representative member of a class, and not each individual permutation. For S_3 the problem reduces from six permutations to three classes (Table 1). While the reduction is marginal for S_3, it rapidly becomes significant as n increases (Table 2). For S_{26} the problem reduces from approximately 4.0×10^{26} permutations to 2436 equivalence classes, and the time required to explore S_{26} at a rate of one permutation or equivalence class per second, falls from 1.3×10^{19} years to just over 40 minutes.

Table 1. Permutations and Equivalence Classes for S_3. (Equivalence class labels are those used in the C_{3v} point group in Chemical Group Theory)

Equivalence Class	Permutations	No. of permutations in class
E	(a)(b)(c)	1
C_3	(abc), (acb)	2
σ_v	(a)(bc), (c)(ac), (b)(ac)	3

2.2 Partitions, Types, and Classes of Permutations

Calculating the number of equivalence classes into which the permutations of S_n may be partitioned is equivalent to calculating the number of partitions $p(n)$ of integer n, where by partition we refer to the various combinations of integers, $m_i \leq n$, which sum to n. For example in the case of $n = 3$ the partitions are $1 + 1 + 1$, $2 + 1$, and 3; thus $p(3) = 3$. Comparison with the components of S_3 (Table 1) will show that the number of partitions of integer 3 equals the number of equivalence classes in S_3. Moreover, there is a one to one correspondence between integer partitions and permutation classes, with partitions '1 + 1 + 1', '2 + 1' and '3' corresponding to equivalence classes exemplified by (a)(b)(c), (ab)(c) and (abc) respectively. The number of partitions (or equivalence classes), $p(n)$, for an integer or permutation group may be determined by recursion [6]; partitions for $n = 2$ to $n = 20$ are given in table 2.

Table 2. Numbers of permutations and equivalence classes as a function of alphabet size from $n = 2$ to $n = 20$. S_n.is the set of $n!$ permutations of an alphabet of size n.

Permutation Group (Alphabet)	Number of Permutations = n!	$p(n)$ = no. of partitions or equivalence classes of n.
S_2	2	2
S_3	6	3
S_4	24	5
S_5	120	7
S_6	720	11
S_7	5 040	15
S_8	40 320	22
S_9	362 880	30
S_{10}	3 628 800	42
S_{11}	39 916 800	56
S_{12}	479 001 600	77
S_{13}	6 227 020 800	101
S_{14}	87 178 291 200	135
S_{15}	1 307 674 368 000	176
S_{16}	20 922 789 888 000	231
S_{17}	355 687 428 096 000	297
S_{18}	6 402 373 705 728 000	385
S_{19}	121 645 100 408 832 000	490
S_{20}	2 432 902 008 176 640 000	627

The standard notation describing partitions, or classes of permutations, requires counting the parts of each size [7]. Where there are α^i parts of size i, the 'type' of a partition may be written as

$$\text{type} = [1^{\alpha_1} 2^{\alpha_2} 3^{\alpha_3} \ldots n^{\alpha_n}] \tag{3}$$

where the number of permutations of a given permutation type or class is given by

$$\frac{n!}{1^{\alpha_1} 2^{\alpha_2} 3^{\alpha_3} \ldots n^{\alpha_n} \alpha_1! \alpha_2! \alpha_3! \ldots \alpha_n!} . \tag{4}$$

The representation of permutation types may be contracted by not explicitly describing instances for which $\alpha^i = 0$; the full and contracted notations are for S_3 are exemplified in Table 3.

Table 3. Permutations and Equivalence Class Types in S_3.

Equivalence Class	Permutation Type (Long form)	Permutation Type (Contracted form)
(a)(b)(c)	$[1^3 2^0 3^0]$	$[1^3]$
(a)(bc), (c)(ac), (b)(ac)	$[1^1 2^1 3^0]$	[12]
(abc), (acb)	$[1^0 2^0 3^1]$	[3]

Equivalence classes allow consideration of permutations in more tractable terms. We proceed by exploring how the choice of permutation or permutations with which to effect replication might be effected by the limitations of a molecular medium.

3 Molecularly Encoded Permutations

In general, replication of a text may be effected simply by executing permutation α = (a)(b)(c)(d)…(n), that is, the identity operation, or by a combination or repetition of permutations equivalent to α. An alphabet A_n of n letters has a total of $n!$ permutations available to it, any one of which could theoretically be used for the replication; The hypothetical molecular machines or enzymes giving effect to these permutations are termed permutases.

3.1 How Many Replicators?

The first question we approach is the number of permutations required. To illustrate this we consider A_4, containing letters a, b, c and d; table 4 gives some of the permutations in S_4. Permutation γ = (abcd), acting on the string 'aabbcdaa', yields the string 'bbccdabb', while the action of δ = (adcb) on the product string restores the original text; $\delta = \gamma^{-1}$. However, the repeated action of any single permutation can also restore the original text; thus while the initial action of γ yielded 'bbccdabb', γ^2 yields 'ccddabcc', γ^3 yields 'ddaabcdd', and finally, γ^4 gives 'aabbcdaa', restoring the original text. All permutations in S_4 could in principle regenerate the original text. Combinations such as $\delta\gamma$, and repetitions such as γ^4, and β^2, are equivalent to the identity permutation α generating perfect copies of an original string. However, notwithstanding their equivalent function, $\delta\gamma$, γ^4 and β^2 may be distinguished in one fundamentally important way. Whereas $\delta\gamma$ generates a copy of the original text through the agency of two distinct permutations, both γ^4 and β^2 involve the agency, albeit the repeated agency, of just a single permutation.

Table 4. Some Permutations in A_4.

Permutation Type	Permutations	Permutation
$[1^4]$	α	(a)(b)(c)(d)
$[1^2 2]$	β	(b)(d)(ac)
[4]	γ	(abcd)
[4]	δ	(adcb)

The emergence of polymerase is likely to have been one of the most difficult transitions in the origin of life. Whatever the probability of a single permutase emerging from a molecular soup, the possibility of two distinct such enzymes seems considerably remote, and three or more enzymes even more unlikely. It is much more probable that in early evolutionary history a single replicating/permuting enzyme emerged. Similar considerations should apply to artificial life, since the challenge of synthesizing a single replicating permutase is more tractable than that of synthesizing two or more distinct permutases which require integrated functionality. This realization enormously simplifies the problems, meaning that we need only consider the application of single permutases, albeit repeatedly, and combined permutases, corresponding to mixed permutations, can be set aside. The problem of molecularly realizable permutations reduces to

$$\lambda^N = \alpha . \tag{5}$$

where $\lambda \in S_n$, and N is an integer. All permutations satisfy this equation for some value of N.

3.2 Selection Based on Permutation Order

In the electronic world almost all permutations can be effected with equivalent overhead, and with an instruction cycle in the nanosecond range. Not so in the molecular world where replication by enzymes is comparatively slow. For example, in *Escherichia coli* RNA polymerase progresses along DNA at approximately 10 nucleotides per second (at 25°C) [8]. Replication fork propagation can be somewhat faster, running at up to 1000 nucleotides per second in DNA [9]. Assigning 2 bits/nucleotide this corresponds to a rate of 250 bytes per second, or a little over one hour per megabyte. There is little basis for expecting hypothetical synthetic replicators to differ significantly from polymerase in this respect. Consequently, as the cost of replications is incurred each and every time a permutation is executed, it would seem advantageous to miminise the number of repetitions necessary to achieve replication of the original text. For example, in A_6, permutases expressing permutations of type $[2^3]$, such as (ab)(cd)(ef), have a cycle length of two, and should be favoured over those expressing permutations of types $[1^2 4]$, $[3^2]$ or [15].

Table 5. Permutations and Equivalence Classes Types for S_6. $n! = 120$, $p(6) = 11$,

Permutation Type	Sample Permutation	Number of Permutations	Cycle Length N
$[1^6]$	(a)(b)(c)(d)(e)(f)	1	1
$[1^4 2]$	(a)(b)(c)(d)(ef)	15	2
$[1^2 2^2]$	(a)(b)(cd)(ef)	45	2
$[2^3]$	(ab)(cd)(ef)	15	2
$[1^3 3]$	(a)(b)(c)(def)	40	3
[123]	(a)(bc)(def)	120	6
$[3^2]$	(abc)(def)	40	3
$[1^2 4]$	(a)(b)(cdef)	90	4
[24]	(ab)(cdef)	90	4
[15]	(a)(bcdef)	144	5
[6]	(abcdef)	120	6

It is difficult to argue for a particular value for N since this will naturally depend on the individual nature of the chemical system in which the permutations are encoded. Nevertheless, all other factors being equal, N will be as small as possible as each expression increases the resources overhead. As a working hypothesis we set $N \leq 3$, rejecting permutations requiring larger N. Viable permutations are those which satisfy a modified form of equation (5)

$$\lambda^N = \alpha\,, \quad N \leq 3 \tag{6}$$

This eliminates the greater number of permutations; those satisfying (6) are of types

$$[1^{\alpha_1} 2^{\alpha_2}] \text{ or } [1^{\alpha_1} 3^{\alpha_3}] \tag{7}$$

The increase in permutations of types conforming to (7) is now linear in n, a critically reduced complexity from the $n!$ general permutations.

3.3 Chemical Manifestation of Cycles of Orders One, Two, and Three

Molecularly encoded life should be based on permutations drawn from this considerably reduced set. Cycles of order two reflect replication through the template directed generation of negatives, where twofold execution of permutations is sufficient to regenerate the original text. Nature's replication engine, polymerase, acting on nucleotides A, C, G and T is such a permutation, and specifically of type [2^2], see equation (1). Replication based on two-cycles in not in principle restricted to nucleotides, and alternatives to nucleotides have been proposed [10]

More generally, equation (7) informs us that cycles of up to size three are allowable. Fig. 1a gives a schematic representation of a triplex, from which template-based cycle of order three might emerge. The nucleotide assembly depicted in Fig. 1b is a rather simple manifestation of such a triplex [13], but serves to illustrate the possibility of three-cycles, although in the example given the G-aA interaction will show little discrimination. However, permutations involving three-cycles seem the least probable since they have the highest energetic and physicochemical overhead, requiring three executions of a permutase to replicate the original text.

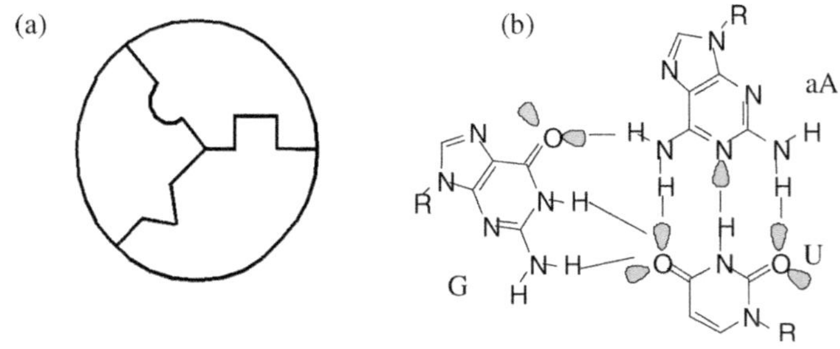

Fig. 1. (a) Schematic representation of triplex which from which replication based on a three-cycle may be developed. (b) simple triplex based on nucleotides aA, U and G.

Template propagation in cycles of order one requires self-recognition; this does not of itself pose a chemical difficulty [11], but does introduce some structural restrictions, since in a Watson-Crick (WC) presentation self-complementarity is not accessible (Fig. 2a). By contrast, in the anti-WC presentation, self-recognition is possible provided an even number of recognition features is employed (Fig. 2b). Should an odd number of recognition features be used, then self-association/self-recognition will be opposed in at least one position (Fig. 2c). However, if one of the features is made redundant, so as to restore an even number of active recognition features, then self-recognition can occur (Fig. 2d).

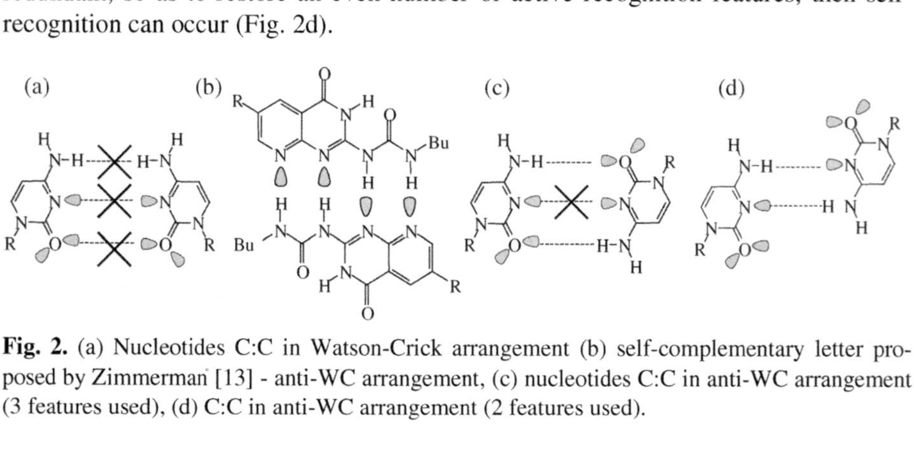

Fig. 2. (a) Nucleotides C:C in Watson-Crick arrangement (b) self-complementary letter proposed by Zimmerman [13] - anti-WC arrangement, (c) nucleotides C:C in anti-WC arrangement (3 features used), (d) C:C in anti-WC arrangement (2 features used).

The identity operation would have the least overhead, and would be by far the most efficient, but is only possible in permutations completely entirely of 1-cycles. For example, in A_4, the permutation (a)(b)(c)(d) requires that a, b, c and d are all self-complementary. However, the pool of self-complementary letters is quite restricted. To illustrate this we consider the 3-bit and 4-bit binary spaces, B^3 and B^4, using 0's and 1's to represent complementary molecular features. (See [14] for examples of the correspondence between molecular features and binary numbers.) Inspection will show that, while the elements of B^3 may be arranged into complementary pairs, there are no self-complements, even allowing order reversal. In B^4, however, of 16 available patterns, four are self-complementary, namely, 0011, 0101, 1010 and 1100, when presented in an anti-Watson-Crick arrangement; Figs. 2b and 3b show the molecular expression of patterns 1100 and 1010. Self-complementarity is therefore limited to a considerably reduced molecular suite, and only in anti-WC arrangements.

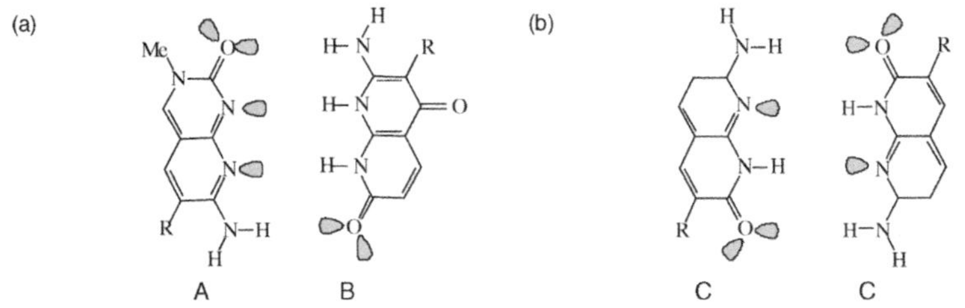

Fig. 3. A hypothetical three-letter alphabet composed of (a) a complementary pair, A and B, and (b) a self-complementary letter, C.

Space precludes a detailed consideration of the relationship between permutations and molecular alphabets, and we conclude with a number of observations. Firstly, we may note that replication in alphabets with an even number of letters require 2-cycles, or an even number of odd cycles (Table 5), e.g. permutations of type (ab)(cd)(ef) or (a)(bc)(def). As three-cycles are problematic, the more probable permutations will be composed of two-cycles with a limited number of one-cycles.

An alphabet with an odd number of letters must possess at least a single odd cycle (Table 3). For replication to occur in such an alphabet the replicating permutation must involve a one-cycle, requiring self–complementarity, or alternatively a three-cycle. Where the complexity of triplex replication eliminates three-cycles, an alphabet with an odd number of letters may only arise where some degree of self-complementarity is available. Given that an alphabet composed entirely of one-cycles seems improbable, odd-alphabets seem accessible only with a combination of complementary pairs and self-complementary letters. An example of such a potential alphabet is depicted in Fig. 3, where replication corresponds to two-fold execution of permutation (AB)(C), type [12]. Finally we note that the greater geometric restrictions associated with self-complementarity may make odd-alphabets less accessible than even alphabets.

4 Conclusion

The number of potential permutations associated with an alphabet of size n, increases rapidly with size so that consideration even for quite modest alphabets rapidly becomes intractable. However under the constraints of a molecular medium of expression the space of accessible permutations is a small subset of the entire space. For reasons of complexity replication should be effected by a single replicating machine, applied repeatedly. For reasons of resource efficiency the number of repeated applications should be minimized. The most efficient permutation would be the identity operation; however, molecular replication, proceeding by template propagation, has a limited capacity to express this. Subject to these pressures the most probable replicating machines are those composed of one- and two-cycles. This includes the permutation corresponding to polymerase, but also includes a variety of other hypothetical permutase types such as (AB)(C) acting on the three-letter alphabet depicted in Fig. 3.

References

1. Stauffer, A. Sipper, M: An Interactive Self-Replicator Implemented in Hardware. Artificial Life 8 (2002) 175–183
2. Hull, D.: Replication, The Stanford Encyclopedia of Philosophy (Winter 2001 Edition), Edward N. Zalta (ed.), URL = plato.stanford.edu/archives/win2001/entries/replication/
3. Lawrence, D. S., Jiang, T., Levett, M.: Self-Assembling Supramolecular Complexes. Chem. Reviews 95 (1995) 2229–2260
4. Stahl, W. R. Goheen, H. E.: Molecular Algorithms. J. Theor. Biol. 5 (1963) 266–287
5. Hu, W., Fukugita, M., Zaldarriaga, M., Tegmark M.: Astrophysical Journal, 549 (2001) 669–680
6. Biggs, N. L.: Discrete Mathematics, Oxford, 1989, at p. 423.
7. Ibid at p. 100
8. Wang, M. D., Schnitzer, M. J., Yin, H., Landick, R., Gelles, J., Block S. M.: Force and Velocity Measured for Single Molecules of RNA Polymerase. Science 282 (1998) 902–907
9. Cox, M. M., Goodman, M. F., Kreuzer, K. N., Sherratt, D. J., Sandler S. J., Marians, K. J.: The Importance of Repairing Stalled Replication Forks. Nature 404 (2000) 37–41
10. Szathmáry, E: What is the Optimum Size for the Genetic Alphabet? Proc. Nat. Acad. Sci. 89 (1992) 2614–2618
11. Peters, M., Rozas, I., Alkorta I., Elguero, J.: DNA Triplexes: A Study of Their Hydrogen Bonds. J. Phys. Chem. B 107 (2003) 323–330
12. Sijbesma, R.P., Beijer, F.H., Brunsveld, L., Folmer, B.J.B., Hirschberg, J.H.K.K.,. Lange, R.F.M., Lowe, J.K.L., Meijer, E.W.: Reversible Polymers Formed from Self-Complementary Monomers using Quadruple Hydrogen Bonding. Science 278 (1997) 1601–4
13. Zimmerman, S.: Supramolecular Polymer Chemistry and the Origin of Life, XIIth Internations Conference of Supramolecular Chemisrtry, Oct. 6–11, Israel, 2002
14. Mac Dónaill, D. A.: A Parity Code Interpretation of Nucleotide Alphabet Composition. Chem. Comm. (2002) 2062–2063

Simulating Evolution's First Steps

Tim J. Hutton

Biomedical Informatics Unit, Eastman Dental Institute for Oral Health Care Sciences, University College London, 256 Gray's Inn Road, London WC1X 8LD, UK.

Abstract. We demonstrate a simple artificial chemistry environment in which two small evolutionary transitions from the simplest self-replicators to larger ones are observed. The replicators adapt to increasingly harsh environments, where they must synthesise the components they need for replication. The evolution of a biosynthetic pathway of increasing length is thus achieved, through the use of simple chemical rules for catalytic action.

1 Introduction

The evolutionary growth of biological complexity [11] is an intriguing process but not one we are able to experiment with directly because of the time-scales involved. The creation of a virtual organism, one that replicates, competes for survival in a simulated environment and undergoes largescale evolutionary development is a primary goal of ALife research. Being able to tinker with such a creature inside a computer simulation would be a useful tool for exploring the requirements for life and for the growth of complexity to occur naturally.

Ideally, the constituent components of the creature would be closely related to those in nature, in order that any conclusions drawn about the simulated biology might be expected to hold for our actual biology. One way to achieve this is through direct simulation of our physics and chemistry but to accurately simulate even a single molecule stretches current computing capabilities. The IBM Blue Gene project aims to combine one million CPUs into a supercomputer in order to simulate the folding of proteins suspended in water. Even with an awesome 10^{15} floating-point operations per second it is estimated that simulating a system of 32,000 atoms for 100 microseconds would take three solid years of computing time [20].

Thus for the moment we are forced to abstract out some of the details, leaving a simulation that contains the features that we think are important but that requires far less computing power. There have been many different approaches taken to this problem over the years, with cellular automata (see [15]) and machine code systems (see [21]) perhaps the most popular. Biological processes can also be modelled using artificial chemistries [3], with a division between abstract chemistries that have no representation of the physical location of the components (eg. AlChemy [4], P-systems and membrane computing [14] and ARMS/ACS [16]) and more concrete artificial chemistries that do (eg. [12,10, 6]).

W. Banzhaf et al. (Eds.): ECAL 2003, LNAI 2801, pp. 51–58, 2003.

Modern Darwinian theory tells us that there are three fundamental requirements for the evolutionary growth of complexity. Firstly, there must exist entities that duplicate information (replicators). Secondly, there must exist an evolutionary path from the simplest replicator to the most complex, with only minimal changes at each step, achievable through random mutation. Thirdly, at every step, the minimally more complicated replicator must outperform (or at least survive in the presence of) all the surviving replicators. Without this third requirement there would exist an evolutionary path but there would be no drive to follow it. Of course this does not mean that *every* minimally more complicated replicator has to outperform all surviving replicators, merely that for an evolutionary path to have been followed this must have been true at every step.

In this paper we present a novel artificial chemistry (AC) environment that meets all three requirements and does indeed exhibit evolutionary growth, though in the current system only two small steps are demonstrated. Previous systems in which the evolutionary growth of complexity has been observed include Geb [2], an agent-based simulation in which neural networks evolve, and Avida [8], a machine code system in which replicators that perform certain prespecified operations are rewarded with the energy needed to execute their code. One advantage that concrete ACs have over such systems is that their representation is much closer to the substrate of our own biology - this should make the organisms more recognisable in their design solutions and any experimental findings should be more directly applicable to an understanding of natural evolution. Additionally, unlike the two systems mentioned, some ACs have the features that have been suggested as necessary for creative, open-ended evolution [19]: implicit reproduction, embeddedness of individuals, materiality and rich interactions. Of these four features, Geb arguably has two (implicit reproduction and rich interactions) while Avida, interestingly, has none.

The AC presented in [6] showed a strong survival pressure for smaller replicators, since these were able to copy themselves more rapidly. The same effect is observed in vitro with the Qβ replicase [13]. A fascinating question, and one we must answer if we are to understand why complex life evolved on Earth, is this: what are the features of a system that drive the evolutionary growth of complexity? Towards this question, though falling short of answering it, we can explore ways of extending the AC given in [6] to produce a growth in complexity. Note that achieving the equivalent of this in vitro is very difficult because we do not have the luxury of being able to change the rules of chemistry and cannot easily monitor the dynamics of the system but perhaps the same result might be achieved in other ways.

Satisfying the third requirement for evolution (that at least some more complex replicators should be naturally selected for) was found to be difficult in our AC. Without a membrane surrounding each replicator, frequently it was found that any phenotypic benefit of having certain bases in the molecule would be shared with other replicators nearby, thus conferring no survival advantage. One solution that was found was that the molecules should be able to catalyse the units needed for their replication ('food') from the surrounding primordial soup but that these food particles would soon be converted back into non-food particles if not utilised. The second part of this provision ensures that it is likely to be only the replicator that catalysed the food that gets the benefit. It was found that with limited flow between a sequence of environments in which it was possible for the replicators to catalyse the components they needed but increasingly difficult to do so, the replicators would adapt, increasing in length as they acquired more catalytic bases.

The process of synthesising required components is a major part of the activity of cells, and the development of this ability was an important step in evolution:

> "The mechanism for the evolution of long biosynthetic pathways was probably that envisioned by N. H. Horowitz [5]. Any organism that could convert some available compound to a compound required for cell reproduction could then survive in the absence of the formerly required compound. Organisms that developed pathways for the synthesis of required cell compounds had selective advantages over others." [9] (p. 8)

Firstly, in the next section we specify the chemical rules of the system and its starting configuration. In section 3 we show that by introducing mutation and occasional mixing between different compartments the molecules evolve upwards in length as they adapt to survive in the different environments.

2 System Description

As before (see [6]), we use a simple diffusion algorithm on a square grid in 2D, giving us the movement that brings our simulated atoms into contact with each other. (It should be noted that certain features of our atoms are not shared with actual atoms, and perhaps a closer analogy is with a small molecule such an amino acid.) Each atom moves at random to an empty square in its Moore neighbourhood, if there are any, with the additional constraint that it is also not allowed to move outside of the Moore neighbourhood of any atom it is bonded to.

Bonds between atoms are made and broken by reactions, which can also change the state of atoms (0,1,...) but not their type (a-f). One problem with the reactions in [6] was that the molecules had a tendency to become tangled with each other when replicating. While this gave them the ability to mutate we found that the rate of mutation was too high for the experiments here. A different set of reactions was found that gave much more robust replication, these are shown in Fig. 1a. By using three-way reactions we can ensure that in the normal replication sequence bonds only form between atoms in the same molecule, something which is very difficult to enforce when using only two-way reactions. Additionally we were able to reduce the number of states required from 10 to 7.

To introduce catalytic effects we also introduce reactions R9-R13 as shown in Fig. 2. R9 is responsible for converting the 'food' atoms (state 0) to non-food

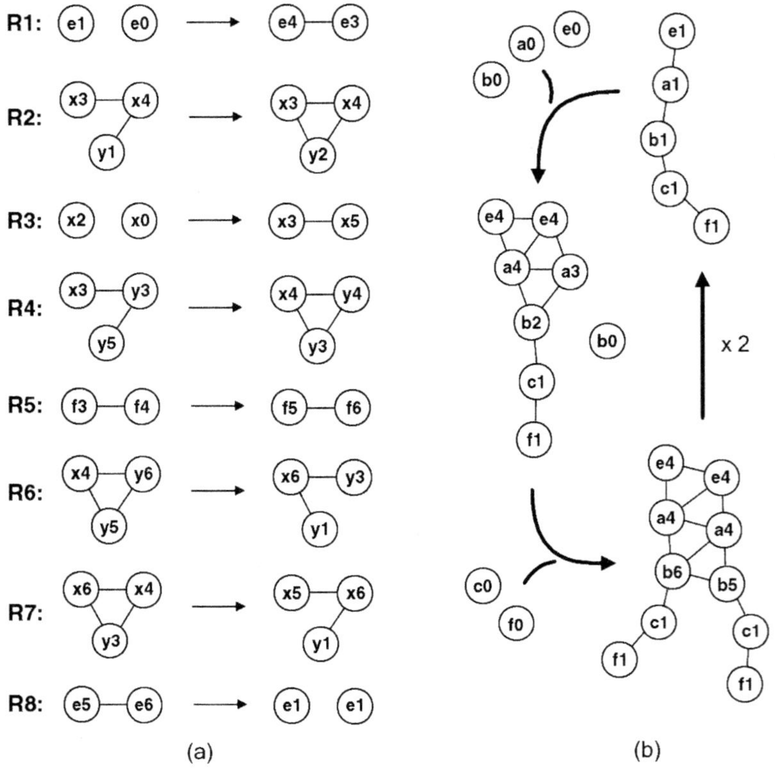

Fig. 1. *(a)* The reactions required for molecular replication in our artificial chemistry. Some act between just two atoms, while others involve three atoms. x and y are variables standing for any type (a-f). These reactions only specify the atoms that are directly involved, the atoms may be bonded to other atoms not shown. *(b)* Three stages in the replication sequence of an eabcf molecule. Any chain of atoms a-d in state 1 with an e1 at one end and an f1 at the other will replicate repeatedly using these reactions when immersed in a soup of free atoms in state 0.

atoms (state 7). One atom in state 7 in an environment containing atoms in state 0 will very rapidly convert them all to state 7 through a cascade of reactions. Reactions R10 and R11 introduce another non-food atom state, 8. Atoms in state 8 convert both state 0 and state 7 to state 8.

Reactions R12 and R13 provide the catalytic countermeasures to this removal of food. R12 says that an atom of type a with state i (where $i \in \{1, 2 \ldots 6\}$) will convert an atom in state 7 to state 0, thus rendering it available for use in replication. A molecule in the form e1a1f1 would therefore be able to replicate when only atoms in state 7 were available by first converting them to state 0. Similarly, atoms b1-b6 convert atoms in state 8 to state 7. Molecules with both a and b atoms in them would be able to replicate (albeit slowly) if only atoms in state 8 were available. These statements are verified in the next section.

Additionally, reaction R14 allows the spontaneous formation of the smallest possible replicator (e1f1), while R15 allows replicators to add or lose an atom (mutation). Unlike R1-R13, these reactions do not take place every time the right atoms come together, instead only happening with some low probability.

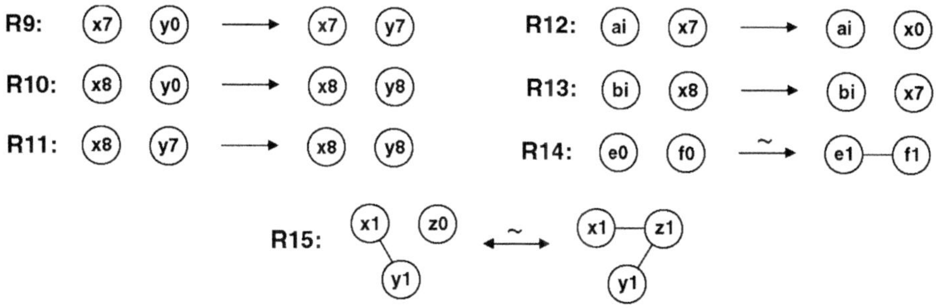

Fig. 2. R9-R13: Additional reactions that give replicating molecules some catalytic properties. x and y are variables standing for any type (a-f), i is a variable standing for any state in the range 1-6. R14-R15: Two additional reactions that allow evolutionary change. R14 permits the minimal replicator (e1f1) to appear spontaneously when an e0 and an f0 come into contact. R15 permits replicators to change length occasionally, either adding or losing an atom. Reactions R14 and R15 only happen with a low probability, P = 0.00001.

3 Observations

We initialize a virtual world with the above physics and chemistry and the following environment. The world is 152×50 and is divided into three zones by two vertical walls of static atoms in a non-reacting state (-1) at columns 51 and 102. The three zones are initialised with atoms in state 0, 7 and 8 respectively, at a density of 1 atom to every 6 squares.

As before [6], periodically the zones are flooded by removing all the atoms in one half and filling that half with atoms in the raw material state (in this case 0, 7 or 8 depending on the zone). This happens every $T_{\text{flood}} = 10,000$ timesteps. This repeated dilution ensures that only replicators can persist[1].

Every $10 \times T_{\text{flood}} = 100,000$ iterations we cause a degree of mixing between the zones. First we delete half of zone 3 and move half of zone 2 into it. Then, we move half of zone 1 into the empty half of zone 2 and refill the empty half of zone 1 with atoms of random type with state 0. Any replicators that get carried between zones will either be able to survive in the new chemical environment or will be removed with successive floods.

[1] Likewise, homeopaths should be careful about bacteria lest their dilutions inadvertently have a physical effect on their customers...

In this experiment we do not seed any of the zones with replicators but instead allow them to form and evolve naturally. We keep a record of each replication event (by waiting for R8 and then analysing the molecules that were involved) so that we can track the frequency of replication of each molecule in each zone.

Figure 3 shows the results of a typical run. In zone 1 (top) the molecule ef dominates, although various mutants are seen. In zone 2 there is no replication until the appearance of eaf at 90,000 iterations. The first eaf replicator could either have mutated in zone 1 and been transported over, or could have mutated from an ef molecule in zone 2 that had been transported previously. In zone 3 (bottom), there is no lasting replication until 290,000 iterations when ebaf molecules appear. In each zone there are occurances of mutated replicators that perform less well than the dominant replicators but these are quickly eradicated, indicating the selection pressures at work.

In zone 3, the ebaf molecule does not dominate forever. In this run, at approximately 2,200,000 iterations (not shown) the equivalent molecule eabf takes over, after a brief struggle.

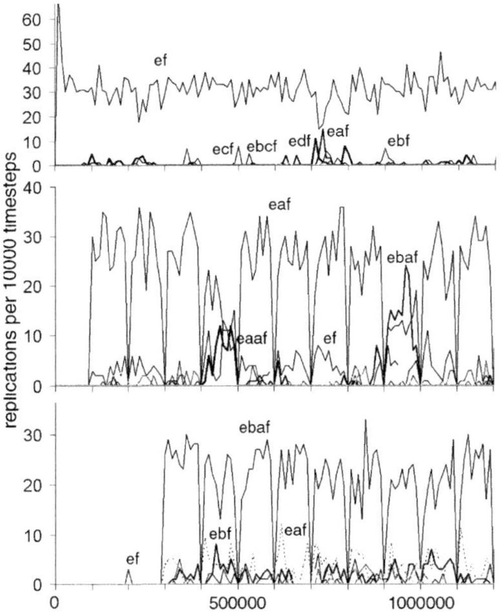

Fig. 3. The average replication rates of different replicators for the three different zones 1, 2 and 3 (top to bottom). While molcules in the form ef appear early on in zone 1, zones 2 and 3 remain without life until molecules eaf and ebaf respectively are evolved.

4 Conclusions

We have demonstrated a simple artificial chemistry environment in which the evolutionary growth in replicator length from 2 to 4 can be observed. While this is only a modest increase, the experiments show that adding simple catalytic properties to replicating molecules can create a survival differential. It has been suggested that the earliest evolution on Earth from the smallest replicating molecules to more complex ones must have involved some simple phenotypic properties of the molecule, since an explicit decoding mechanism could not yet have evolved [7,1,17]. While we may never know exactly which phenotypic properties were first acquired, the experiments in this paper confirm that evolutionary growth is possible with simple catalytic effects, and without an enclosing membrane.

Clearly a membrane is a desirable thing from a replicator's point of view. Being able to protect the genomic information from harmful reactions with unknown external agents, and additionally to keep the products of catalytic reactions to oneself are strong incentives to develop and maintain the necessary information sequence. It would be instructive to try to evolve this mechanism in an AC, perhaps by incorporating ideas from other ACs where membranes have been created [10,12]. It is also conceivable that we may one day be able to combine replicators and lipids in the laboratory to make synthetic proto-cells [18], this would also be exciting.

Perhaps the biggest drawback in the current system is that we are only getting as much out as we put in, since the evolutionary changes are in direct response to the environmental conditions and catalytic reactions that we concocted. A major goal of evolutionary systems is to find a way to provide the minimum of features after which evolution takes-off and develops complexity on its own. The only system to date that seems to have achieved this (other than biology on Earth) is Geb [2]. In our AC this would mean finding a set of reactions that allows the sequence of bases to affect the survival chances of the replicator. If such reaction-sets can be found then their common features should enable general conclusions to be drawn about the requirements for the evolutionary growth of complexity.

We have chosen to implement the artificial chemistry described on a 2D grid but it would work equally well with other physics, including continuous space in 2D or 3D. Some demonstrations of this and of the experiments in the paper are available at: `http://www.eastman.ucl.ac.uk/~thutton/Evolution/Squirm3`

References

1. A.G. Cairns-Smith. *Seven clues to the origin of life.* Cambridge University Press, Cambridge, 1985.
2. A. Channon. Improving and still passing the ALife test: Component-normalised activity statistics classify evolution in Geb as unbounded. In R. Standish, M.A. Bedau, and H.A. Abbass, editors, *Proc. Artificial Life VIII*, pages 173–181. MIT Press, 2002.

3. P. Dittrich, J. Ziegler, and W. Banzhaf. Artificial chemistries - a review. *Artificial Life*, 7(3):225–275, 2001.
4. W. Fontana and L.W. Buss. What would be conserved if "the tape were played twice"? *Proc. Nat. Acad. Sci. USA*, 91:757–761, 1994.
5. N.H. Horowitz. On the evolution of biochemical synthesis. *Proc. Nat. Acad. Sci. USA*, 31:153–157, 1945.
6. T.J. Hutton. Evolvable self-replicating molecules in an artificial chemistry. *Artificial Life*, 8(4):341–356, 2002.
7. G.F. Joyce and L. Orgel. Prospects for understanding the origin of the RNA world. In R.F. Gesteland, T.R. Cech, and J.F. Atkins, editors, *The RNA World*, pages 49–77, New York, 1999. Cold Spring Harbor Laboratory Press.
8. R.E. Lenski, C. Ofria, R.T. Pennock, and C. Adami. The evolutionary origin of complex features. *Nature*, 423:139–144, 2003.
9. L. Margulis. *Symbiosis in Cell Evolution.* Freeman, New York, 1981.
10. B. Mayer and S. Rasmussen. Dynamics and simulation of micellar self-reproduction. *International Journal of Modern Physics C*, 11(4):809–826, 2000.
11. B. McMullin. John von Neumann and the evolutionary growth of complexity: Looking backwards, looking forwards... *Artificial Life*, 6(4):347–361, 2000.
12. N. Ono and T. Ikegami. Artificial chemistry: Computational studies on the emergence of self-reproducing units. In J. Kelemen and P. Sosík, editors, *Proc. European Conference on Artificial Life*, pages 186–195. Springer, 2001.
13. L.E. Orgel. Selection in vitro. *Proceedings of the Royal Society B*, 205:435–442, 1979.
14. Gh. Paun. *Membrane Computing. An Introduction.* Springer-Verlag, Berlin, 2002.
15. H. Sayama. A new structurally dissolvable self-reproducing loop evolving in a simple cellular automata space. *Artificial Life*, 5(4):343–365, 1999.
16. Y. Suzuki and H. Tanaka. Chemical evolution among artificial proto-cells. In M.A. Bedau, J.S. McCaskill, N.H. Packard, and S. Rasmussen, editors, *Proc. Artificial Life VII*, pages 54–64. MIT Press, 2000.
17. E. Szathmáry and L. Demeter. Group selection of early replicators and the origin of life. *Journal of Theoretical Biology*, 128:463–486, 1987.
18. J.W. Szostak, D.P. Bartel, and P.L. Luisi. Synthesizing life. *Nature*, 409:387–390, 2001.
19. T. Taylor. Creativity in evolution: Individuals, interactions and environment. In P. Bentley and D. Corne, editors, *Proceedings of the AISB'99 Symposium on Creative Evolutionary Systems, The Society for the Study of Artificial Intelligence and Simulation of Behaviour.* Morgan Kaufman, 1999.
20. The IBM Blue Gene team. Blue gene: A vision for protein science using a petaflop supercomputer. *IBM Systems Journal*, 40(2), 2001.
21. C.O. Wilke and C. Adami. The biology of digital organisms. *Trends in Ecology and Evolution*, 17(11):528–532, 2002.

Cellular Evolution in a 3D Lattice Artificial Chemistry

Duraid Madina, Naoaki Ono, and Takashi Ikegami

The University of New South Wales, Sydney 2052, Australia
ATR Human Information Science Laboratories,
2-2-2 Hikaridai, "Keihanna Science City" Kyoto, 619-0288, Japan
Department of General Systems Sciences,
The Graduate School of Arts and Sciences,
University of Tokyo, 3-8-1 Komaba, Tokyo 153-8902, Japan
duraid@unsw.edu.au, nono@atr.co.jp and ikeg@sacral.c.u-tokyo.ac.jp

Abstract. We introduce a three-dimensional model of the formation of proto-cell structures. Our model is based on an artificial chemistry extended from an earlier two-dimensional realization by Ono and Ikegami. This model describes the chemical reactions of a primitive metabolic system and the spatial interactions of simple amphiphilic molecules which organize into membrane-like structures. The results demonstrate the emergence of dynamic three-dimensional cellular structures from a "primordial soup", and a variety of self-maintaining structures may be observed, depending on initial conditions.

1 Introduction

Theoretical studies of reaction-diffusion systems in non-equilibrium environments [1,2] have shown that self-organizing and self-reproducing chemical patterns can be demonstrated in quite simple chemical systems without relying on complex molecular devices. However, the dynamics of these patterns are such that it is impossible for different regions of a particular pattern to have any significant degree of individuality. To distinguish living organisms from mere dissipation structures, a new phenomenology of biological systems seems required.

Maturana and Varela [3] focussed on an essential feature of living organisms – "autopoiesis", their ability to produce and maintain their own boundaries. While they only considered a two-dimensional case, they stated that a three-dimensional extension to their model would be a simple one not involving any conceptual obstacles. While we indeed found this to be the case, in three dimensions there is a much greater diversity of structures to be found. Luisi[4] and others showed that some vesicles of amphiphilic lipids had a catalytic activity enabling them to assimilate resources so that they could reproduce themselves automatically.

Molecular simulations of the self-assembly of amphiphilic lipids have been studied thoroughly. However, because realistic simulation of lipid molecules is

W. Banzhaf et al. (Eds.): ECAL 2003, LNAI 2801, pp. 59–68, 2003.

a computationally difficult task, various models based on simpler discrete dynamics have been proposed [5,6,7]. Another issue is the metabolism of membrane molecules. Though there are various attempts to simulate realistic chemical pathways in the cell [8] we instead focus on abstract models of artificial chemistry [9] since our purpose is to understand primitive, minimal forms of life. In the spirit of Zeleny's and Varela's work, there have been studies of the organization and maintenance of proto-cell structures in computational models [10,11]. Along this line, Ono and Ikegami presented a model called "Lattice Artificial Chemistry" (LAC) [12,13] which simulates both spatial interactions and chemical reactions of abstract molecules within a simple consistent framework using a lattice method [14]. In the 2D version of the model, evolution of cellular structures from a random soup, maintenance and self-reproduction of cell structures, and evolution of catalysts through cellular selection were observed. In the present 3D case however, a vastly greater number of possible cellular morphologies is expected. Therefore, we expect the transition from 2D to 3D to produce qualitatively different behaviour.

One issue we wish to emphasize is that cell membranes are not merely vehicles which contain rich chemical networks. We argue that it may be the other way around: perhaps it is the membrane structures themselves that determine what kind of chemical reactions may occur within them. For example, we wonder if chemical reactions may proceed differently when constrained to occur in quasi-1D structures such as tube-like membranes. Such questions are themselves of interest, besides the problem of the origin of cells. In the next section, we first explain some details of our model. Essentially, we perform a natural extension of the repulsive interaction between hydrophilic and hydrophobic particles from 2D to 3D.

2 The Model: A 3D Lattice Artificial Chemistry

Here we extend Lattice Artificial Chemistry ([12,13]) to three-dimensional space.

Chemicals are represented by abstract particles which move on a cubic lattice. Note that any nonnegative integer number of particles are allowed to coexist on each lattice site. In the simulations reported in this paper the lattice size is 64^3 (with periodic boundaries) and there are 300 particles per site, on average. We consider here only the simplest autocatalytic chemical reactions which can produce membrane formation, and chemicals which can diffuse spatially.

i) Chemical reactions occur probabilistically according to the chemical potentials associated with them, though reactions on any particular site may be promoted by catalytic particles on that site. We use a system composed of two metabolic cycles of autocatalysts and membranes. An autocatalyst **A** catalyzes the reproduction of another such **A** and the production of membrane particles **M**. Resource particles are supplied from an external source at a constant rate, which is the energy supply driving the system. Further details of the model appear in [12,13].

ii) Diffusion of chemicals is biased according to the spatial gradient of the potential of particles in the local cubic neighbourhood (of size 27), which is computed by calculating the repulsive potential of all particles in the neighbourhood.

iii) In order to simulate the organization of membrane-like structures, we consider hydrophobic interactions. Specifically, we model the repulsion between two primary classes of particles, hydrophobic and hydrophilic, as being stronger than the repulsion between other particles. We also introduce neutral particles which are not strongly repelled by hydrophobic particles, so that they can diffuse through membranes.

Membrane molecules are represented as hydrophobic particles which have an anisotropic potential field. Figure 1) illustrates an example of the repulsion around a membrane particle. When phase separation occurs due the repulsion

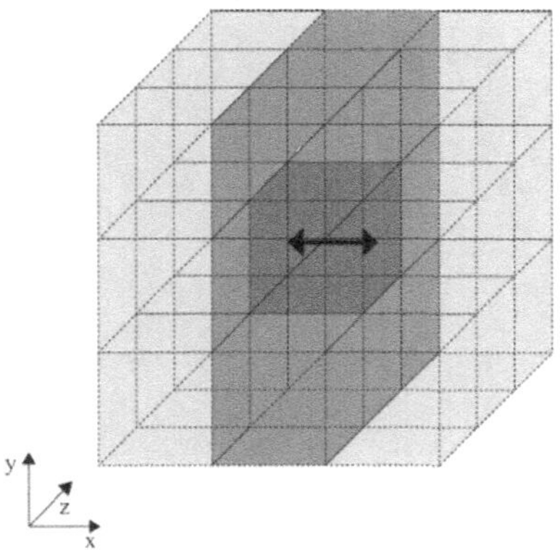

Fig. 1. The anisotropic repulsion field around an membrane particle. Deeper gray indicates stronger repulsion

between hydrophilic and hydrophobic particles, thin, membrane-like structures form as a result of this anisotropy. We assume that autocatalysts and resources are hydrophilic and neutral, respectively.

Below, we explain the details of the repulsive potential as this is a key part of the present model.

We model hydrophobic interaction using the following simple functions. First, the repulsion between hydrophilic and neutral particles depends only on the distance between the two particles

$$\phi_T(\mathbf{dr}) = R_T 2^{-dr^2} (T \in \{WW, WN, NN\}) \tag{1}$$

where R_T is the corresponding given constant of the repulsion coefficient for hydrophilic-hydrophilic, hydrophilic-neutral, and neutral-neutral interactions, and dr denotes the Manhattan distance between the particles in lattice units (i.e. $dr = 0, 1, 2$ or 3).

Second, elements of membranes (small clusters of membrane molecules) are represented by oriented particles as shown in Fig.2. Taking the symmetry of cubic lattice into account, we consider thirteen unique orientations. Thus, at each lattice site there are thirteen seperate populations of membrane particles, one for each orientation. In addition to diffusion, the rotation of membrane particles (i.e. the migration of membrane particles from one of the thirteen populations to another) is also biased according to the local potential field.

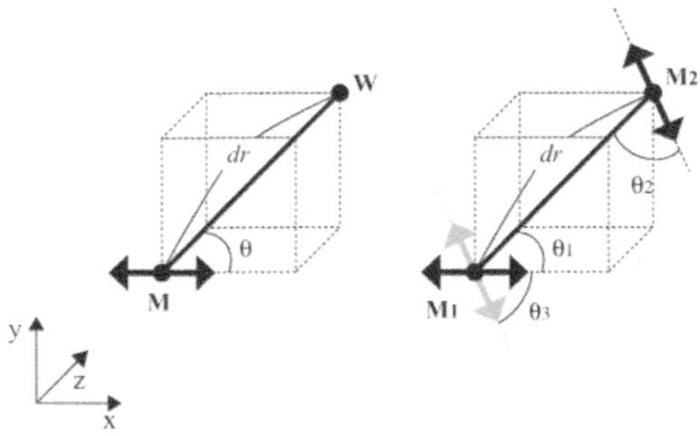

Fig. 2. The configuration of two particles.

The repulsive potential between hydrophobic and hydrophilic particles is calculated as follows:

$$\phi_T(\mathbf{dr}) = (R_{T_0} + R_{T_1}\sin^8(\theta))2^{-dr^2}(T \in \{WM, NM\}) \tag{2}$$

where R_{T_0} and R_{T_1} are given constant coefficients and θ denotes the angle between the orientation of the membrane particle and the position vector $\mathbf{dr}$ (see Fig2a).

Due to the second term, hydrophilic and neutral particles tends to avoid the "side" of membrane particles. The repulsion between membrane particles and hydrophilic particles is so strong that it causes phase separation between them. On the other hand, we assume that the repulsion to neutral particles is much weaker.

Finally, the repulsion between two membrane particles depends on both their orientations and their configuration. We calculate the repulsion between two membrane particles as follows:

$$\begin{aligned}\phi_T(\mathbf{dr}) = (R_{T_0} + R_{T_1}/2(\cos^4(\theta_1) + \cos^4(\theta_2)) \\ + R_{T_2}\sin^8(\theta_3)) * (2^{-|dr|^2})(T \in \{MM\})\end{aligned} \tag{3}$$

where R_{T_0}, R_{T_1} and R_{T_2} are given constants, θ_1, θ_2, are the angles between their orientations and the position vector, and θ_3 is the difference of their orientations. Due to the second term, membrane particles tend to avoid having their "heads" near each other and the third term makes them tend to align in the same direction. Note that the exponents of this equation were chosen empirically.

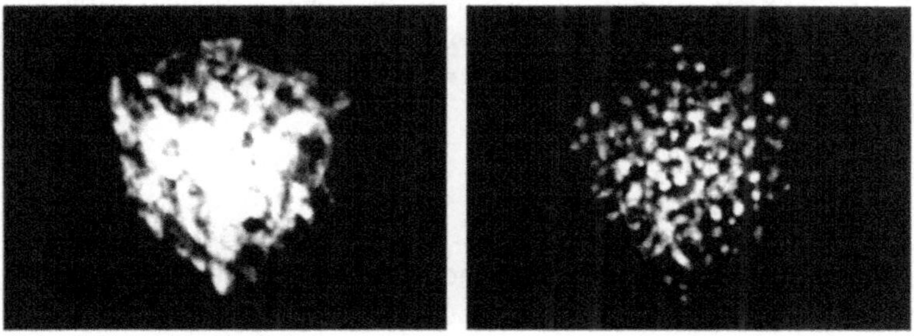

Fig. 3. Clouds of membrane particles (left) slowly relax into a low-energy configuration (right)

3 Results

As will be demonstrated in this section, this system exhibits a variety of qualitatively different behaviours. However, these may be classified into four categories, which arise from different levels of physical and chemical evolution:

	weak diffusion	strong diffusion
weak reaction	relaxation	simplification
strong reaction	cellular formation	cellular evolution

3.1 Relaxation

In this region, neither the chemical composition nor the physical structure of the system changes significantly over time. While random initial conditions demonstrate a slow phase separation-like behaviour between hydrophobic and hydrophilic particles, any features present in the initial conditions are preserved for long periods of time. Figure 3) shows the formation of dense "membrane clouds" – diffusive forces are weak enough that clusters of membrane particles are not encouraged to form a hollow core, and chemical reactions occur at a low, steady rate such that the clusters do not grow or decay. Over sufficiently long time scales, the system will relax into a low energy configuration such as the hexagonal packing of circles in 2D, or the face-centered cubic packing of spheres in 3D.

3.2 Simplification

In this region, while the populations of the different particle types remain steady, any membrane structures exhibit a strong tendency to degenerate or "simplify" into structures of lower genus or dimension. Stable structures such as a hollow sphere or flat sheet of membrane particles can survive for long periods of time, but introducing even minor defects into them leads to their rapid decay under the strong diffusive forces. For example, a punctured sphere flattens out into a sheet (see Figure 4), while two punctures in a flat sheet quickly coalesce into a single, larger defect.

3.3 Cellular Formation

When there are sufficient resource and catalytic particles available to allow the formation of membrane particles, a number of different cellular structures may be formed. We say that a cell has formed whenever a region of space is completely enclosed by a non-zero number of membrane particles. From random initial conditions, the evolution typically consists of the spontaneous formation of thin filaments of membrane particles, which begin to grow in diameter. Once they grow sufficiently, thin membrane surfaces may begin to form between nearby and similarly-aligned filaments (Figure 5). Since the diffusion term is proportional to the angle between membrane particles, any sharp corners in surfaces that are formed are smoothed out and the original thin filaments either become more rounded, or disappear. Figure 6) is an example of tube formation given specific initial conditions. While initial structure is often preserved, a variety of new structures also form. Instead of spherical cell structures, tubes grow to connect two parallel membranes. However, outside the regime of cellular evolution, the structures that are formed change slowly over time, and do not exhibit cell division.

3.4 Cellular Evolution

As soon as a number of filaments successfully transform into cells, these cells may begin to grow, compete for resources, and divide. A cell grows by absorbing resource particles from its neighborhood, which may or may not include other cells. As a cell grows, it is possible that it produces more membrane particles than it needs to maintain its structure. If this occurs, these excess particles can gather to seed the growth of a new membrane near the center of the cell. Eventually, this new membrane may grow to span the original cell completely, dividing it into two or more new, smaller cells. It is important for cells to maintain their structural integrity – if there is a defect in the membrane, catalysts may leave

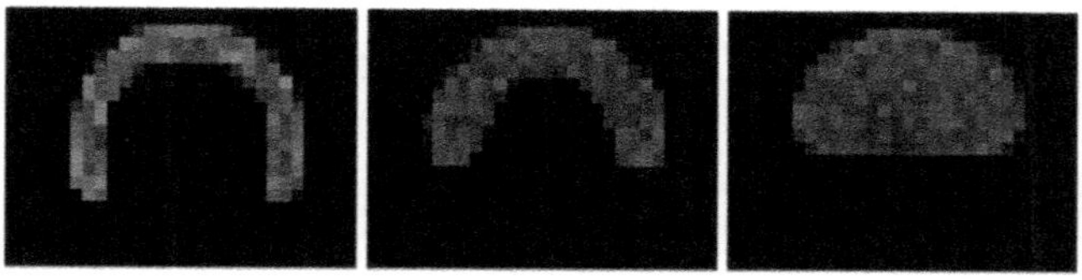

Fig. 4. Cross section: a punctured sphere rapidly loses its structure (800 timesteps)

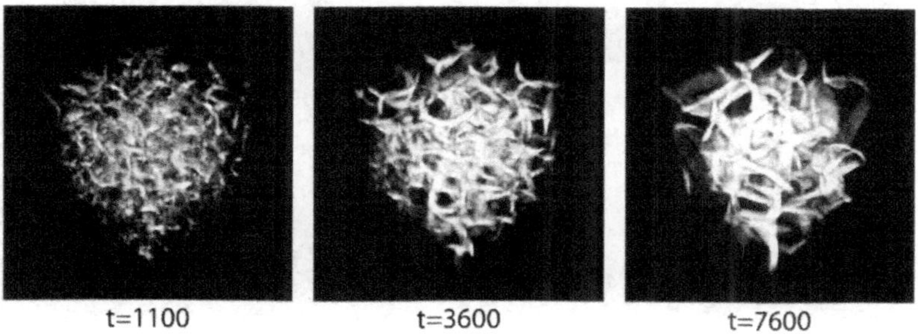

Fig. 5. From a random homogeneous initial configuration, filaments form and begin to organize

Fig. 6. From an initial configuration of two parallel membranes, tubular structures form to connect them

the cell by diffusing through the defect. If too many catalysts escape, those that remain will be insufficient to maintain the cell membrane, and eventually the cell will decay completely. Thus, in this region there is a crucial interaction between diffusion and reaction: the two can interact only as long as cellular structures exist. Figure 7) is an example of this behaviour. Figure 8) shows that while for a given set of model parameters the emergence of cells is robust, the detailed dynamics of cellular evolution may vary considerably. We hypothesize that this is due to our model having a sensitive dependence on initial conditions. While the initial state for every run in Figure 8) was uniformly random, in the cases where the standard deviation of catalytic densities was particularly low, this may have prevented the early formation of membranes in specific areas, which might otherwise have been able to grow by simply aggregating other membrane particles produced nearby. This would lead to a higher cell population of smaller, shorter-lived cells. Had the initial random distribution contained any significant deviations, these deviations would have seeded the early growth of a few membranes which could then grow to dominate the volume, as in Figure 5). One interesting aspect of cell division in our model is that cells do not have any mechanism with which to select an internal axis of division. Therefore, cell di-

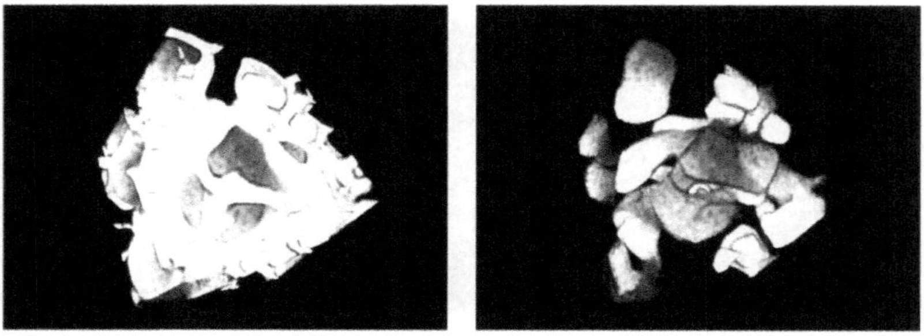

Fig. 7. Fully formed cellular structures (left) and the distribution of catalytic particles in the same structure (right). The effect of competition for resources is evident: A previously homogeneous distribution of resources has given way to cellular 'hoarding'.

vision does not resemble mitosis: when cells divide, they do not displace each other, but instead share a newly formed membrane surface. Furthermore, unlike real cell division, in our model there can often be a high degree of asymmetry between the "child" cells that are created. This leads to rapid disintegration of the smaller child cells, and the resources they would otherwise have consumed become available for another child cell to grow to closely resemble its parent. This phenomenon leads to the cyclic fluctuations in cell populations that can be observed in Figure 8).

4 Discussion

We have presented a three-dimensional model for the evolution of cellular structures. From random initial configurations, we have observed the emergence of proto-cells. These cells exhibit a significant degree of homeostasis in addition to the ability to spontaneously divide into smaller cells. The mechanism for this evolution consists of three stages: (1) local metabolism which produces membrane particles, (2) the formation of proto-cell structures, i.e. cellular structures which maintain their own membranes, in spite of external physical and chemical influences, including competition for resource particles from neighbouring cells, and (3) the division of those proto-cells successful enough to have an excess supply of membrane particles.

This behaviour is generally insensitive to the changing of model parameters, so long as the changes are not so severe as to completely prohibit the formation of membrane particles or to disallow the diffusion of other particles through membranes. In the present model, changing parameters may lead to different cell sizes, shapes and division rates, but the qualitative behaviour remains the same.

We note that while our model is at best an extremely primitive one, we feel that it is a promising approach to modelling proto-cells – that is, structures

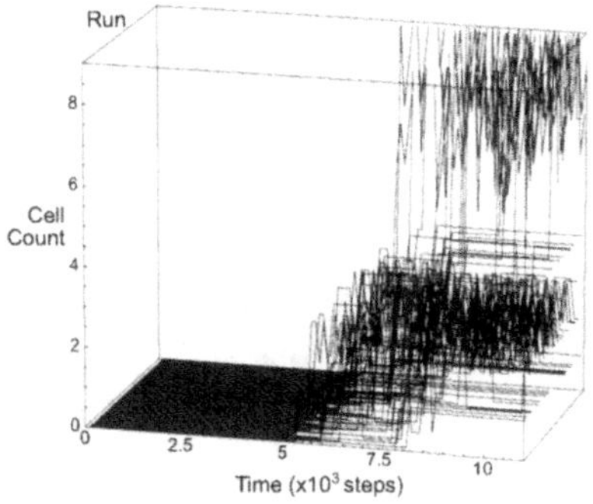

Fig. 8. The population dynamics of several runs from random initial conditions. For a significant length of time, the volume is devoid of cells. Eventually, cells may form, though only certain populations remain successful.

capable of self-maintenance and self-reproduction – within a non-equilibrium physico-chemical framework. In particular, we believe that it is the combination of membrane dynamics (driven by Gibbs energy) and physicochemical dynamics (supported by autocatalytic particles) which is key to any life-like behaviour observed.

In the simulations presented here, the supply of resource particles was homogeneous. If instead one enforces an inhomogeneous distribution of resources, one may observe cellular formation only in the region where there are sufficient resources. However, once cells are formed, they are able to migrate to and survive in regions of lower resource density.

Only the simplest reaction pathways exist in the present model. Moreover, cells are not generally formed in isolation, but adjacent to each other. This may limit the emergence of higher-order phenomena (such as cell differentiation and nontrivial interactions between groups of cells). Removing these limitations is one area of current work.

Lastly, the variety of types of membrane formation evident in the present 3D simulations tells us that possible proto-life forms may be radically different from the living structures we see today. In the 2D case, cellular reproduction of hexagonal shapes and some connected compartments is observed. In the 3D case, we observe stranger topologies. Taking Figure 6) as an example, two membrane sheets which are initially separate and parallel, if separated by an appropriate distance, will generate pipes between them. Catalytic particles can be transfered from one side to the other through these pipes, which reminds us of slime molds. With such connected membrane boundaries, preconceptions of self-reproduction become irrelevant. Instead of taking the concept of self-reproduction as primary, we should determine the necessary dynamics and environment that favour self-reproduction rather than connected units. Our message is that "vehicle takeover", not just the genetic takeover proposed by Cairns-Smith [15], deserves further investigation.

Acknowledgments. This work was partially supported by the COE project "complex systems theory of life", a Grant-in-aid (No. 13831003) from the Japanese Ministry of Education, Science, Sports and Culture and afiis project (Academic Frontier, Intelligent Information Science 2000-2004 Doshisha University). DM thanks TI for his generous support, and the UNSW HPCSU and APAC for their assistance.

References

1. Turing, A. M. The Chemical Basis of Morphogenesis, *Philosophical Transactions of the Royal Society (part B)*, **237** (1953) 37–72, (Reprinted in *Bulletin of Mathematical Biology* (1990) **52**, No.1/2 153–197).
2. Pearson, J. E. Complex Patterns in a Simple System, *Science*, **261** (1993) 189–192.
3. Varela, F. J., Maturana, H. R., Uribe, R. Autopoiesis: The Organization of Living Systems, its Characterization and a Model., *BioSystems*, **5** (1974) 187–196.
4. Luisi, P. L., Walde, P., Oberholzer, T. Lipid Vesicles as Possible Intermediates in the Origin of Life, *Curr. Opin. Coll.*, **4** (1999), 33–39.
5. Mayer, B., Köhler, G., Rasmussen, S. Simulation and Dynamics of Entropy-Driven, Molecular Self-Assembly Processes, *Physical Review E*, **55**, 4 (1997) 4489–4499.
6. Drefahla, A., Wahabb, M., Schillerb, P., Mögel, H.-J. A Monte Carlo Study of Bilayer Formation in a Lattice Model, *Thin Solid Films*, **327-329** (1998) 846–849.
7. Boghosian, B. M., Coveney, P. V., Love, P. J. A Three-Dimensional Lattice Gas Model for Amphiphilic Fluid Dynamics, *Proc. R. Soc. Longon A*, **456** (2000) 1431.
8. Kitano, H. Computational Systems Biology, *Nature*, **420** (2002) 206–210.
9. Dittrich, P., Ziegler, J., Banzhaf, W. Artificial Chemistries - A Review, *Artificial Life*, **7(3)** (2001) 225–275.
10. McMullin, B., Varela, F. J. Rediscovering Computational Autopoiesis, In proceedings of 4th European Conference on Artificial Life, Husbands, P., Harvey, I. (eds.) MIT press, Brighton, UK, (1997) 38–47.
11. Breyer, J., Ackermann, J., McCaskill, J. Evolving Reaction-Diffusion Ecosystems with Self-Assembling Structures in Thin Films, *Artificial Life*, **4** (1998) 25–40.
12. Ono, N., Ikegami, T. Model of Self-Replicating Cell Capable of Self-Maintenance, Proceedings of the 5th European Conference on Artificial Life (ECAL'99), Floreano, D., Nicoud, J. D., Mondada, F.(eds.), Springer, Lausanne, Switzerland (1999) 399–406.
13. Ono, N., Ikegami, T. Artificial Chemistry: Computational Studies on the Emergence of Self-Reproducing Units., Proceedings of the 6th European Conference on Artificial Life (ECAL'01) (eds.) Kelemen, J. and Sosik, S., Springer, Prague, Czech Republic (2001) 186–195.
14. Chen, S., Dawson, S. P., Doolen, G. D., Janecky, D. R., Lawniczak, A. Lattice Methods and Their Applications to Reacting Systems, *Computers. chem. Engng*, **19** (1995) 617–646.
15. Cairns-Smith, A.G., Genetic Takeover and the Mineral Origins of Life, Cambridge University Press (1982).

Evolution of Rewriting Rule Sets Using String-Based Tierra

Komei Sugiura[1], Hideaki Suzuki[2], Takayuki Shiose[1], Hiroshi Kawakami[1], and Osamu Katai[1]

[1] Graduate School of Informatics, Kyoto University
Yoshida-Honmachi, Sakyo-ku, Kyoto 606-8501, Japan
{sugiura, shiose, kawakami, katai}@sys.i.kyoto-u.ac.jp
[2] ATR Human Information Science Laboratories
2-2-2 Hikaridai, Seika-cho, Soraku-gun, Kyoto 619-0288, Japan
hsuzuki@atr.co.jp

Abstract. We have studied a string rewriting system to improve the basic design of Tierra. Our system has three features. First, the Tierra instruction set is converted into a set of string rewriting rules using regular expressions. Second, every agent is composed of a string as the genome and a set of string rewriting rules as its own instruction set. Third, a genetic operation and selection of rewriting rules are introduced to allow the agents to weed out the worst rules. We carried out experiments on how agents evolve through self-replication. The results have shown that our system can evolve not only the agents' genomes but also the rewriting rule sets.

1 Introduction

Artificial Life (ALife) research attempts to not only investigate complex phenomena of life but also to engineer artificial systems with life-like properties such as autonomy and evolvability. Accordingly, it is necessary to design the process for constructing the system rather than to design the system itself. Designers have to prepare two main components: a set of objects (symbols [1], character sequences [2, 3, 4], binary strings [5], λ-terms [6]) standing for agents and environments and a set of reaction rules (rewriting rules [5, 2, 3, 4, 1], λ-calculus [6]) defining the interaction among objects. The evolvability of an ALife system depends heavily on this basic design. Hence, these objects and reaction rules must be robust to changes and capable of open-ended evolution. Nevertheless, previous studies have focused mainly on the evolution of agents using fixed reaction rules. Man-made rules, however, do not always have the capability of open-ended evolution.

Suzuki has reported the optimization of string rewriting grammar [3], and Matsuzaki proposed a way to translate the Tierra [7] instruction set [8]. However, no previous research has realized a Tierra-like ALife system as a string rewriting system nor evolved a set of string rewriting rules.

W. Banzhaf et al. (Eds.): ECAL 2003, LNAI 2801, pp. 69–77, 2003.

In this paper, we propose an ALife system in which both the agents and the set of reaction rules can evolve together by providing the system with the ability to improve the rule set by itself. In order to realize this, we take three steps to develop a string rewriting system based on Tierra. First, we convert 32 Tierran instructions into a set of 140 independent rewriting rules (initial rule set). Each rule is represented as a sequence of the matching or substitution of regular expressions, which makes the rule robust against changes to it. Second, we equip every agent with both a string as the genome and a set of string rewriting rules and then make the agent self-replicate using its own rule set. Lastly, we introduce a genetic operation and selection of rewriting rules so that the agents can weed out the worst rules.

2 String-Based Tierra

2.1 Comparison with Tierra System

In this system, every Tierran instruction is denoted by a character, and the execution of an instruction is represented as the application of a string rewriting rule. Each agent is composed of a string as the genome and a rewriting rule set. An agent replicates its genome by using its own rewriting rules. In order to improve the rewriting rule set, we introduce a genetic operation and natural selection into the system. Table 1 compares our system with Tierra in terms of object, genome, and reaction rule.

Table 1. Comparison of proposed system and Tierra

	Proposed system	Tierra
Object	Character	Machine code
Genome	String	Sequence of machine code
Reaction rule	String rewriting rule	Execution of instruction

2.2 Set of Objects

In our system, the objects have three kinds of characters: instruction, register, and membrane.

Instruction Characters. Each instruction character corresponds to one of thirty-two Tierran instructions. For example:

$$\text{NOP0} = \text{‘}\texttt{f}\text{’}, \text{NOP1} = \text{‘}\texttt{t}\text{’}, \text{PUSHA} = \text{‘}\texttt{A}\text{’}, \ldots, \text{DIVIDE} = \text{‘}\texttt{z}\text{’}$$

Register Characters. Register characters represent an instruction pointer, registers (as, bx, cx, dx), and a ten-word stack contained by a Tierra CPU. Each agent has these characters in its genome as well as instruction characters. Instead of storing an address in RAM, a register character implies that it stores the number of instruction characters between the beginning of the genome and itself and also indicates the next instruction character. Containing these characters in a genome has two advantages:

- It makes it possible to represent the movement of a pointer, etc., as rewriting a string, and
- it enables agents to have multiple pointers, etc., which means the execution of the instruction can be made in parallel.

Membrane Characters. These characters are introduced to show the beginning and the end of a genome. The register characters of an agent are contained only inside its membrane.

Table 2 shows the characters used in our system.

Table 2. Characters used in proposed system

	Proposed system	Tierra
Instruction characters	`t, f, A, ..., Z, w, x, y, z`	Instruction words
Register characters	`p`	Instruction pointer (IP)
	`a, b, c, d`	Registers (ax, bx, cx, dx)
	`0, 1, ..., 9`	Ten-word stack
Membrane characters	`[, ]`	

2.3 Rewriting Rules Using Regular Expressions

We represent the execution of a Tierran instruction as string rewriting. Each Tierran instruction is converted into one or more string rewriting rules. This approach has the following three advantages over Matsuzaki's proposal [8]:

1. *Using regular expressions*
   We write string rewriting rules with Perl regular expressions [9], which enables us to carry out complicated rewriting. On the other hand, the rewriting rules proposed by Matsuzaki use their own form.
2. *Sequence of matching/substitution*
   A string rewriting rule is represented as a sequence of the matching or substitution of a regular expression combined by *AND* operators. This method simplifies the rules and provides them with robustness against changes.

3. *Independency among rules*
   We represent 32 Tierran instructions as 140 string rewriting rules, which are independent from each other. This rule set has an advantage over Matsuzaki's, where a priority order is needed owing to the non-independency among rewriting rules. In our system, rules are applied from top down sequentially, but the order is changeable. This means that our system has no priority order among rewriting rules.

The following two examples show how our rewriting rules are applied for 'INCA' and 'IFZ'.

Example 1. INCA (increment ax) is represented as follows:

`s/pN/Np/g && s/a($J)/$1a/g`

where:

`N` : instruction character corresponding to INCA

`$N` = `[`*register characters*`]`

`$J` = `$N*[`*instruction characters*`]$N*`

`$N` is the character class representing register characters, and `$N*` represents zero or more `$N`. Therefore, `$J` matches one arbitrary instruction character that has zero or more `$N` before and after.

Fig. 1 shows the execution of INCA as string rewriting. First, the former substitution makes `p` move right for one instruction character. Next, the latter substitution makes `a` move right for one instruction character as well. These operations correspond to the fact that the IP and register ax are incremented in Tierra.

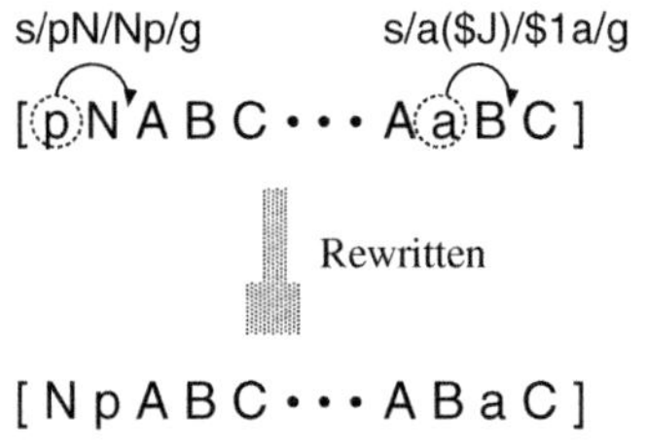

Fig. 1. Execution of INCA as string rewriting

Example 2. IFZ (if cx == 0 perform next instruction, otherwise skip it) needs multiple rules. IFZ is represented as follows:

```
  /\[$N*c/ && s/pU/Up/g
! /\[$N*c/ && s/pU($J)/U$1p/g
```

where '`U`' is the instruction character corresponding to IFZ.

Fig. 2 shows the execution of IFZ as string rewriting. The operation of "`/\[$N*c/`" returns *false*, i.e., the regular expression does not match the genome. Hence, the substitution of the upper rule is not executed, as shown in the left-hand figure. On the other hand, the operation of "`! /\[$N*c/`" returns *true*, so that the substitution of the lower rule is executed, as shown in the right-hand figure.

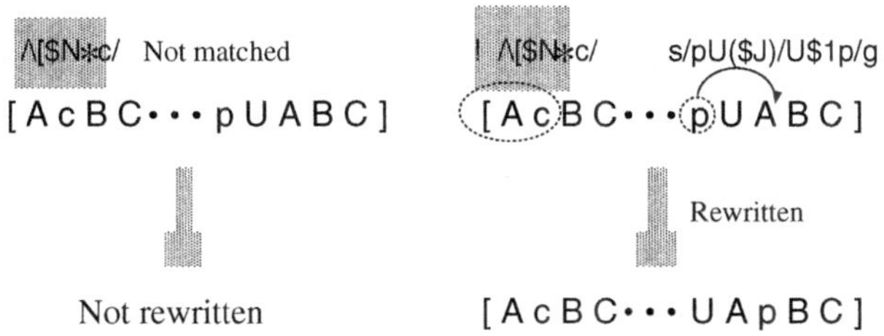

Fig. 2. Execution of IFZ as string rewriting

2.4 Selection of Rewriting Rules

In our system, each agent has its own rewriting rule set **R**, and it self-replicates using **R**. Here, every rewriting rule in **R** is executed from top down sequentially. The number of applied rules per iteration, therefore, depends on the order of the rules. Even if a couple of agents have the same genome, the more the order of its own rules is adapted to the genome, the fewer iterations the agents need to self-replicate.

We introduce the selection of rules in order to improve rewriting rule sets. Here, the fitness value of a rewriting rule is defined as the number of applied times. Fig. 3 shows schematically the selection of rewriting rules. Selection in **R** is executed as follows:

1. sort all of the rules according to the number of applied times,
2. delete l rules from the bottom,
3. generate new l rules by giving l elite rules genetic operation,
4. insert new rules in the middle of **R**.

This operation is executed with probability p per iteration. Through this operation process, rules with fewer applied times are removed and rules with many applied times are selected.

2.5 Selection of Individuals

In our system, a genome is represented as a string, so that its length is elastic. To avoid the infinite reproduction of agents, the maximum population N_{max} is

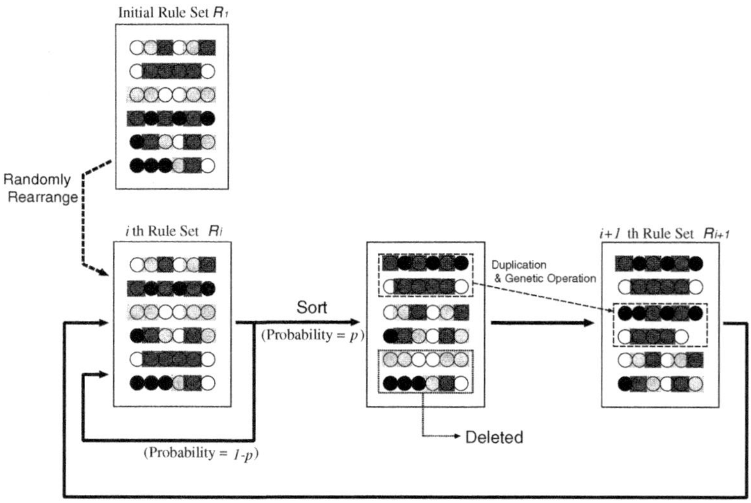

Fig. 3. Selection of rewriting rules

introduced. If the population of agents exceeds N_{max}, randomly chosen agents are deleted.

Our system also deletes agents that are not able to self-replicate. Suppose that an agent cannot apply any rule due to a disadvantageous mutation in its genome. To remove such agents, the system checks the number of applied rules per iteration and weeds out every agent without an applied rules in its rule set.

2.6 Mutation

Our system has the following two mutations for both genomes and rewriting rules, while the Tierra system has only mutation for genomes.

Genome Mutation. This string-based system has two other mutations, insertion and deletion, in addition to substitution, which is used in Tierra.

Rule Mutation. The rule mutation contains three kinds of operations: duplication, deletion, and modification. As an example, the modification of a rule R is represented as follows:

$$R = r_1 \ \& \ r_2 \ \& \cdots \& \ r_i \ \& \cdots \& \ r_n \Longrightarrow r_1 \ \& \ r_2 \ \& \cdots \& \ r_i' \ \& \cdots \& \ r_n$$

In order to change r_i into r_i', we introduce three kinds of operations as rule mutation: insertion, deletion, and substitution.

3 Results and Discussion

In our system, the ancestor self-replicates using the initial instruction set (140 rules), which is equivalent to the Tierran instruction set. The ancestor's genome in our system is a string converted from the genome of the Tierra ancestor. We started the experiment under the condition of that the number of ancestors is N_{max}. The parameters we use are as follows:

$$
\begin{aligned}
N_{max} &= \text{maximum number of agents} \\
m_g &= \text{genome mutation rate} \\
m_r &= \text{rule mutation rate} \\
p &= \text{probability of rule selection per iteration} \\
l &= \text{number of rules weeded out per rule selection}
\end{aligned}
$$

We obtained the most remarkable result under the condition of $N_{max} = 32$, $m_g = 0.01$, $m_r = 0.9$, $p = 0.02$, $l = 14$. It took about four hours for a server with a 2.8-GHz CPU to carry out 100,000 iterations.

In Fig. 4, the number of applied rules is plotted against the number of iterations. The number of applied rules is defined as the number of rewriting rules that are applied from the set of 140 rules within one individual. In the figure, the solid, dotted, and broken lines show the maximum number (best) of applied rules, the minimum number, and the average number, respectively. Fig. 4 shows that the maximum number of applied rules increases with the increase in iterations. Specifically, the maximum number is approximately 13 for iterations between 10,000 to 60,000 and increases to 25 for iterations of more than 80,000. This result means that the agents obtain important rules while weeding out unnecessary rules.

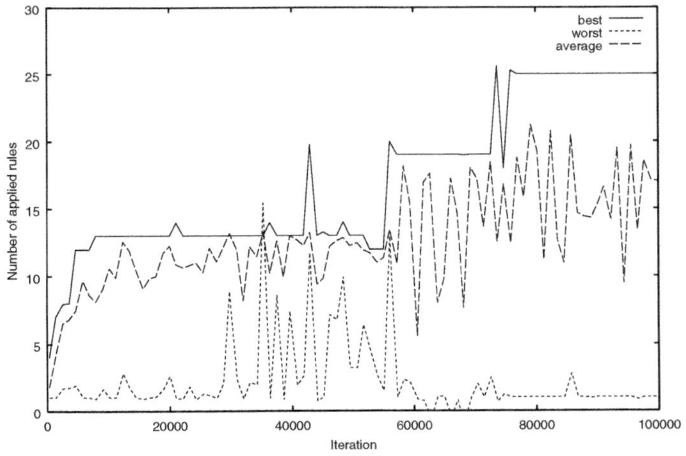

Fig. 4. Variation of applied rules

In Fig. 5, the interval of cell division is plotted against the number of iterations. The interval of cell division is defined as the number of iterations from the latest to the second latest cell division. Fig. 5 shows that the best interval, the average interval, and the worst interval, all drastically decrease with the increase in iterations. Specifically, the average interval is approximately 500 at 1000 iterations but decreases to 30 at 100,000 iterations. This result can be explained by the fact that the agents after 100,000 iterations have a better rule set than the ancestors.

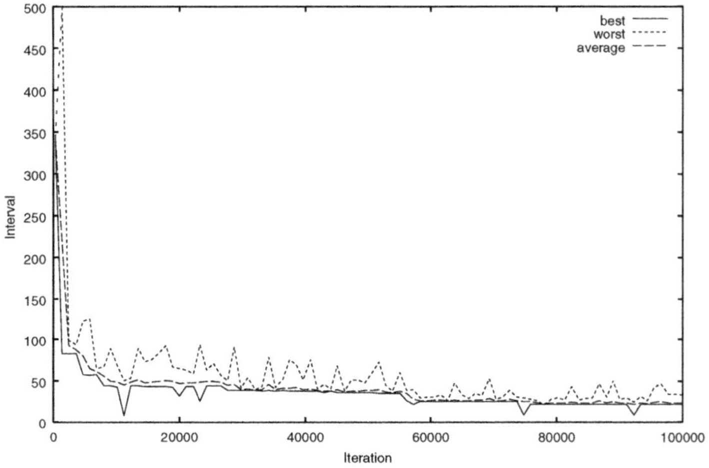

Fig. 5. Variation of interval of cell division

We also studied the variation of the genome length. The length of the ancestor was 87, while the length of the descendants after 100,000 iterations was reduced to 69. This indicates that the genome was improved through the iterations.

Thus we obtained three fundamental results: an increase in the number of applied rules per iteration, a decrease in the average interval, and a decrease in the genome length. These results clearly indicate that both agents' genomes and rewriting rules sets evolved in our system.

Finally, we briefly describe an example of newly generated rules. The ancestor replicates its genome by executing the reproduction loop dozens of times. The rewriting rules are performed 10 times in one loop. We have found that an agent after 40,000 iterations generates a new rule:

```
s/pfN/fNp/g && s/a($J)/$1a/g
```

This rule can execute the sequence of instructions '`fN`' (NOP0 and INCA) in one substitution. Namely, it reduces the number of rules necessary for the loop to 9. This result also supports the fact that both the agent's genomes and rewriting rule sets can evolve in our system.

4 Conclusions

In this paper, we studied the evolution of rewriting rule sets in order to improve the basic design of Tierra. In order to realize this, we took three steps to develop a string rewriting system based on Tierra. First, we converted 32 Tierran instructions into a set of 140 independent rewriting rules (initial rule set). Each rule is represented as a sequence of the matching or substitution of regular expressions combined by *AND* operators. Second, we equipped every agent with both a string as the genome and a set of string rewriting rules as its own instruction set, and we made the agents self-replicate by string rewriting. Lastly, we introduced a genetic operation and selection of rewriting rules so that the agents could weed out the worst rules. The ancestor of our system consists of both the genome equivalent to the Tierra ancestor and the initial rule set, and it self-replicates by string rewriting.

We carried out experiments and obtained three key results: an increase in the number of applied rules per iteration, a decrease in the average interval, and a decrease in the genome length. These results indicate that our system can evolve not only the agents' genomes but also the rewriting rule sets.

References

1. Suzuki, Y., Tanaka, H.: Chemical evolution among artificial proto-cells. In Bedau, M.A., McCaskill, J.S., Packard, N.H., Rasmussen, S., eds.: Artificial Life VII: Proceedings of the Seventh International Conference on Artificial Life, MIT Press (2000) 54–63
2. Suzuki, H.: Evolution of self-reproducing programs in a core propelled by parallel protein execution. Artificial Life **6** (2000) 103–108
3. Suzuki, H.: String rewriting grammar optimized using an evolvability measure. In Kelemen, J., Sosik, P., eds.: Advances in Artificial Life (6th European Conference on Artificial Life Proceedings), Springer-Verlag, Berlin (2001) 458–468
4. Suzuki, H., Ono, N.: Universal replication in a string rewriting system. In: Proceedings of the Fifth International Conference on Humans and Computers (HC-2002). (2002) 179–184
5. Dittrich, P., Banzhaf, W.: Self-evolution in a constructive binary string system. Artificial Life **4** (1998) 203–220
6. Fontana, W.: Algorithmic chemistry. In Langton, C.G., Taylor, C., Farmer, J.D., Rasmussen, S., eds.: Artificial Life II: Proceedings of the Second Artificial Life Workshop, Addison-Wesley (1992) 159–209
7. Ray, T.S.: An approach to the synthesis of life. In Langton, C.G., Taylor, C., Farmer, J.D., Rasmussen, S., eds.: Artificial Life II: Proceedings of the Second Artificial Life Workshop, Addison-Wesley (1992) 371–408
8. Matsuzaki, S., Suzuki, H., Osano, M.: Tierra instructions implemented using string rewriting rules. In: Proceedings of the Fifth International Conference on Humans and Computers (HC-2002). (2002) 167–172
9. Friedl, J.E.F.: Mastering Regular Expressions. O'Reilly (1997)

Models for the Conservation of Genetic Information with String-Based Artificial Chemistry

Hideaki Suzuki

ATR Human Information Science Laboratories
hsuzuki@atr.co.jp
http://www.atr.co.jp/his/~hsuzuki

Abstract. A model of artificial chemistry in which both genotypic and phenotypic strings mingle together and react with each other is presented. A cell that includes a string set for the replication and translation of genetic information is designed on this system, and the cells are reproduced by cell division cycles. From experiments that emulate cell selection, it is shown that a complete set of genes constituting the designed system can be stably transmitted to the offspring cells if they are put into a long single chromosome or the gene replication is strictly regulated and the replicated genes are transferred to the two daughter cells by a spindle.

1 Introduction

The problem of integrating and conserving genetic information has attracted a lot of attention among researchers of ancient biological evolution. In order to explain why and how living cells or proto-cells took the first step to store a large amount of information on genotypic molecules, several different models have been proposed. These include the hypercycle model [3,4], the stochastic corrector model [16,18,10], and the parabolic growth model [17,19,18,12]. Among them, the most relevant to this paper is the stochastic corrector model by Szathmáry and Demeter, which asserts that the symbiosis of two replicators with different growth rates is possible if the replicators are segregated into small subgroups and are subject to group selection. The diversity among subgroups is greater with a smaller subgroup size; hence, the model was able to effectively bring about and maintain the integration of two replicators when the subpopulation (cell) size was sufficiently small.

Another model addressed here is that for the origin of a chromosome, i.e., the linkage between genes. According to Maynard-Smith and Szathmáry [9], two replicators with different growth rates are more likely to be found in the same chromosome if the cell size is small, and the lack of either replicator could be fatal to the cell.

According to these models, a smaller cell size is favored for the integration/conservation of genetic information; however, a small cell has a serious

W. Banzhaf et al. (Eds.): ECAL 2003, LNAI 2801, pp. 78–88, 2003.

drawback for information conservation. This happens during the cell division cycle. If a genome is separated into a number of units and the mother cell distributes them equally to the two daughter cells, the possibility of both daughter cells inheriting a complete set of units decreases *geometrically* with the unit number. A smaller cell is always exposed to the danger of this possibility; hence, the stochastic corrector model, which favors a smaller cell size, cannot provide a clear answer to explain the conservation of genetic information coded in many DNA units. In order for a small offspring cell to have a complete set of genes, it might have to have machinery as sophisticated as that of the biological cells.

To study this problem, the paper presents a new model devised on string-based artificial chemistry (SAC) [11,14]. Although a variety of models have been proposed so far in the field of artificial chemistry [8,5,6,2], most of these models are unable to store or have difficulty in manipulating genetic information. (Only the α-universes by Holland [8] deal with the translation of coded information; nevertheless, the α-universes are still not satisfactory in that they use a fixed algorithm for replication and translation that is prepared outside.) SAC, which was devised using a strong resemblance to the real biological system, enables a modeler to design artificial genes for the replication and translation of genetic information and can provide a nice workbench for the problem of the stable transmission of genetic information. In the presented model, a set of strings (molecules) that constitutes a system for the gene replication/translation is prepared with the modified SAC, the string set is put into a cell, and cells are divided according to the multiplication of the inner strings. The replication and translation are accomplished by six or more artificial genes and the same number of proteins; hence, if these molecules are distributed equally to the daughter cells, the daughter cells cannot inherit a complete set of genes. The experimental results show that a simple cell with separated chromosomes cannot keep on reproducing, whereas the cells are able to stably and continuously divide themselves when the genes are linked to each other to make a long chromosome or the genes are replicated just once before the cell division and are forcibly transferred by an artificial spindle.

In the following, after the method of SAC and the cell selection scheme are introduced in Section 2, Section 3 describes the results of an experiment. Section 4 presents the conclusion and discussions.

2 The Model with SAC

String-based artificial chemistry (SAC; [11,14]) is a new methodology enabling the design of artificial molecules by using a strong comparison to biochemical reactions. Here, the author modifies the original SAC slightly and makes phenotypic molecules (proteins) directly represented by functional characters. SAC, in the revised form, takes the following steps to simulate a reaction. (1) A cell (or tank) containing a set of strings is prepared. There is neither spatial structure nor operator-operand discrimination between the strings. (2) To simulate chemical reaction processes, a pair of strings is randomly chosen from the cell/tank and made to react with each other. Though each string may be selected as an operator or as an operand by the reactor algorithm, the first string is regarded as an operator and the second string is regarded as an operand. (3) For each collision, the operator's characters are translated into string rewriting rules that are operated on the operand. If the rewriting rules do not match the operand, no modification is made on the operand (elastic collision). (4) After the rewriting (reaction) is finished, the operator and the (modified) operand are put back into the cell/tank.

2.1 Elementary Characters

The elementary characters used for the representation of every SAC string are listed in Table 1. The functions shown in this table are basically the same as those for the canonical regular expressions [7]. In addition, characters such as ., ‘, ", and / with novel functions are introduced to enable the operations of string separation, character deletion, character creation, and substring permutation, respectively. In this version of SAC, a single character can be freely deleted or created by ‘ and "; however, a simultaneous deletion/creation of two or more characters is not allowed (‘* and "* are prohibited).

Table 1. Elementary character set

Character	Function
.	Separation of the string
0,1,2,3	Non-functional
‘	Deletion of the next char
"	Creation of the next char
!,?,%	Wild-card char that matches an arbitrary single character
*	Wild-card char that matches an arbitrary length of (sub)string
/	Permutation of two substrings delimited by three /s
\	Suppression of the function of the next char
&	Logical AND operation between rewriting rules
$	Logical AND-NOT operation between rewriting rules
M,E	Ingredients for the membrane strings
L,R	Active transportation of the string starting with L/R (spindle)

2.2 Reaction

The reaction is begun by randomly choosing a pair of strings from the cell/tank. When the reaction occurs, the operator string is translated into a sequence of rewriting rules punctuated by &s or $s (if any), which are sequentially operated on the operand. This process can be illustrated as follows.

Example 1: Let an operator be

$$P_0 \equiv \texttt{\textbackslash 0\&\textbackslash !\$`0''1*0`1''2} \Rightarrow \texttt{[\textbackslash 0]\&[\textbackslash !]\$[00*01} \rightarrow \texttt{01*02]}, \tag{1}$$

where the operator string was first punctuated by & and \$ into substrings (\0, \! and 0‘0"1*0‘1"2), and then the left-hand side and right-hand side strings of the rewriting rules were created by deleting "x for the left-hand side and deleting ‘x for the right-hand side. When the operator reacts to the operand, the reaction goes, for example, as

$$\mathtt{003301} \xrightarrow{P_0} \mathtt{013302}. \tag{2}$$

Here, [\0] and [\!] are first operated and searched in the operand. Because there exists a 0 and there does not exist a ! in the operand (note that \! matches !), the next rule [00*01 → 01*02] is operated. (If a rule punctuated by & or \$ does not or does match the operand string, respectively, the reaction is stopped at that time.) From this operation, the header and tail characters of the operand are rewritten.

Example 2: Let an operator be

$$P_1 \equiv \mathtt{/2/\%/''\%} \Rightarrow \mathtt{[2\% \rightarrow \%2\%]}, \tag{3}$$

which exchanges 2 and the next character (that % matches), which are delimited by three /s, and at the same time inserts a character that matches % on the right of 2. This operator works, for example, as

$$\mathtt{002300} \xrightarrow{P_1} \mathtt{0032300}, \tag{4}$$

where % matches 3 in the operand so that 3 is inserted on the right of the 2.

After these operations, the operator and operand are put back into the cell/tank; however, if an operand includes .s (not \.) after the operation, the operand is separated at .s into plural strings, which are put back into the cell/tank. At every time step, this reaction procedure is repeated $R_{\mathrm{rct}} \cdot S^2$ times (where R_{rct} is a reaction rate and S is the string number in the cell/tank), and after that, ‘fragile’ strings, which start with two ‘s, are eliminated from the cell.

2.3 Replication and Translation System

As a replication and translation system, a set of strings that are logically equivalent to a universal replicator [20,14] is designed (Appendix A). The system is composed of three parts: genotypic strings (G_{c} and G_{s}), which encode genetic information, phenotypic strings (P_{c}), which copy the genotypic strings, and phenotypic strings (P_{s}), which construct phenotypic strings by translating information written on G_{c} and G_{s}. As shown in Appendix A, P_{s} is equipped with the characteristics for a ‘universal’ constructor because P_{s} can construct any operator string if an appropriate sequence is given in the genotypic sequence. In the following experiments, three different models are used: (i) a model with separated chromosomes without a spindle, (ii) a model with a single chromosome without a spindle, and (iii) a model with separated chromosomes with a spindle. Specifically, for Model (iii), the replication of the chromosomes is strictly regulated and a cell is controlled to start dividing after all chromosomes are replicated one time. See Appendix A for their detailed designs.

2.4 Cellular Evolution on SAC

After the string set described in Section 2.3 is put into an ancestor cell, the SAC reaction is begun in the cell. New genotypic and phenotypic strings are produced, and the size of the cell (number of contained strings) gradually grows. During this process, using a method similar to an artificial cell system in [15], a cell is divided into two daughters if the cell size (S) exceeds a particular threshold value (S_{max}) for Models (i) and (ii), or if S exceeds S_{max} and the cell includes a particular membrane string starting with ten consecutive Ms for Model (iii). The strings contained in the mother cell are distributed equally to the daughters, provided that for Model (iii), strings starting with a character L (spindle fiber) are forcibly put into the left daughter and those starting with R (spindle fiber) are forcibly put into the right daughter. After the cell division is finished, the header L and R characters and a membrane string MMMMMMMMMM··· are deleted by the reactor program.

All of the cells have ages. At every time step, a cell's age is incremented by one; however, when a cell divides itself, the two daughter's ages are reset to zero. The cells are arranged in a queue in the order of their ages, and when the total number of cells exceeds N_{max}, older cells are killed. This emulates cellular selection, which allows the survival of fitter cells able to multiply inner strings more swiftly and stably.

3 Experiments with SAC

To examine the reproductivity of the designed string sets, we put a string set comprised of only one copy of each DNA unit and protein into an ancestor cell and observe the effect of the string reaction and the cell division cycles. From preliminary experiments, the parameter values were determined as $N_{max} = 50$, $S_{max} = 100$, $R_{rct} = 1.0$ for Models (i) and (iii), and $R_{rct} = 10.0$ for Model (ii).

Figure 1 shows typical results. As time goes on, the cell number grows, and at the same time, the number of operators (phenotypic proteins) and operands (genotypic DNA units) swiftly increases for the three models. For Model (i), however, the number of operators and operands suddenly begins to decrease after 70 time steps. After that, this number gradually decreases, and when the time step number is about 150, active operands are almost completely missing from the cells. In order to improve this situation and make reproduction continue, a number of different parameter values (specifically, smaller S_{max}s) were tried in vain. When six genes are prepared apart and they are freely replicated by the collision with the operator strings, the disparity between the growth rates of the DNA units causes the loss of genes that take a longer time to replicate, and cells will finally lose a complete gene set and string reactivity. For Models (ii) and (iii), on the other hand, such loss of reactivity does not occur. Both S_{opt}

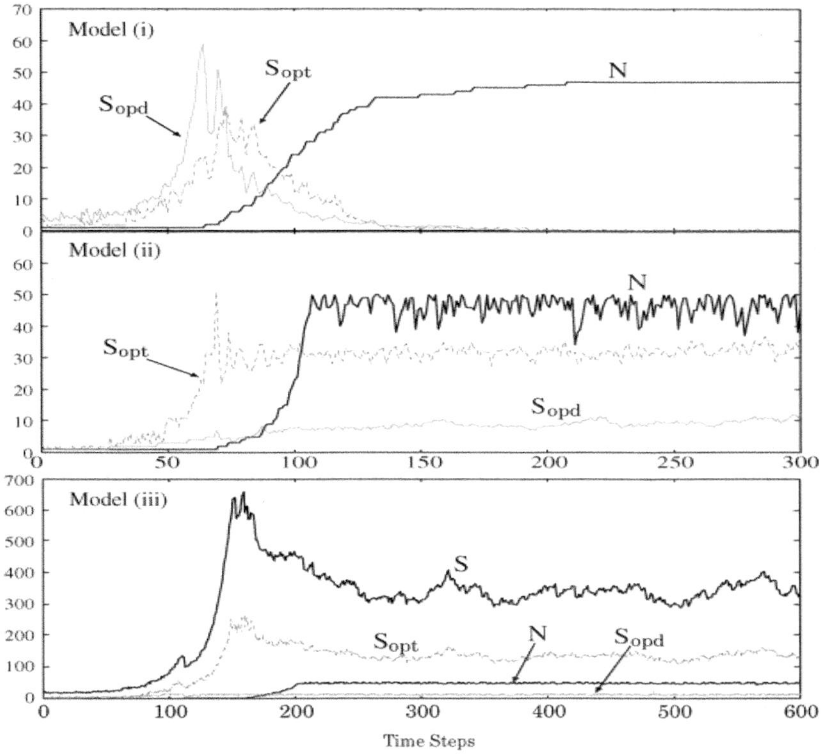

Fig. 1. Cell number (N), average cell size (S), average number of operators in a cell (S_{opt}), and average number of operands in a cell (S_{opd}) as a function of time step number. $S_{\text{opt}}/S_{\text{opd}}$ are the numbers of strings that work as operators/operands during the reaction processes. The longest job, which is an experiment for Model (iii), was performed by a two-and-a-half-day simulation run on a desktop computer with a Pentium III processor (900 MHz).

and S_{opd} basically increase until reaching particular numbers, and after that, the cells keep on reproducing themselves continuously. This result is achieved by bundling all genes into a single chromosome (Model (ii)) or regulating replication and distribution of multiple chromosomes (Model (iii)).

Although, as described in Appendix A, Model-(iii)'s string design is much more complicated than that of Model (ii), the presented design of Model-(iii) is considered to be one of the simplest models able to stably transmit genetic information that is coded in multiple chromosomes. Before obtaining the results for Model (iii), several different versions of models with a spindle were tested in the preliminary experiments, which revealed that a complete set of genes coded in separated chromosomes could not be conserved unless the chromosome replication were strictly regulated synchronously with the cell division cycle.

4 Conclusion and Discussions

Using replication and translation systems designed on SAC, the possibility of the conservation of genetic information during dynamic cell division cycles was studied. Genetic information for the perpetual multiplication of strings is composed of six to eight genes, and it was shown that the complete gene set was successfully transmitted from mother to daughters when they are linked on the same chromosome, or when they are replicated just once before the cell division and are forcibly transferred to the daughter cells by an artificial spindle. Due to the difference between replication rates of genes, a whole gene set could not be successfully transmitted without such mechanisms even when using the cell selection scheme.

Weak cell-selection pressure: The coexistence or symbiosis of replicators with different growth rates has been one of the most extensively studied topics in the study of the origin of life [3,4,16,9,10,18,12,13]. Instead of using replicator equations, the present paper addressed this problem by using a technique of artificial chemistry, which highlighted the difficulty of conserving genetic information and presented solutions. The model is striking for the following reasons. First, the designed translation system needs six or more replicators (genes) to maintain the metabolic reaction. This number is much larger than previous numbers (most of the time, two [16,18]), which raised a new problem for the distribution of a complete set of genes to the daughter cells. If the genes are on separated chromosomes, the cell must be huge to overcome this problem, which reduces the diversity between cells and weakens the cellular selection pressure. Second, because the presented models include not only genotypic replicators but also phenotypic strings (proteins), even if a replicator is lost from a cell by chance, the remaining proteins can propel reactions and the cells can grow for some time. This allows bad cells to reproduce for some time, which also weakens the cellular selection pressure. The cellular selection, which has been regarded as one of the primary mechanisms for information integration [16,10], is not effective for the present model. The genetic information was conserved only when all of the genes were tightly linked on the same chromosome (like procaryote) or they were actively transported to the daughter cells by a spindle while the cell is dividing (like eukaryote).

Acknowledgments. The author thanks Dr. Ono of ATR labs for his helpful comments on the analysis of simulation results. Dr. K. Shimohara of ATR labs actively encouraged the study. This study was supported by the Telecommunications Advancement Organization of Japan and Doshisha University's Research Promotion Funds.

References

1. Alberts, B., Bray, D., Lewis, J., Raff, M., Roberts, K., Watson, J.D.: Molecular Biology of the Cell, The Third Edition. Garland Publishing, New York (1994)

2. Dittrich, P., Ziegler, J., Banzhaf, W.: Artificial chemistries-a review. Artificial Life **7** (2001) 225–275
3. Eigen, M., Schuster, P.: The Hypercycle – A Principle of Natural Self-Organization. Springer-Verlag, Berlin (1979)
4. Eigen, M., Gardiner, W., Schuster, P., Winkler, R.: The origin of genetic information. Scientific American **244** (1981) 88–118
5. Fontana, W.: Algorithmic chemistry. In: Langton, C.G. et al. (eds.): Artificial Life II: Proceedings of an Interdisciplinary Workshop on the Synthesis and Simulation of Living Systems (Santa Fe Institute Studies in the Sciences of Complexity, Vol. 10), Addison-Wesley (1992) 159–209
6. Fontana, W., Buss, L.W.: The barrier of objects: From dynamical systems to bounded organizations. Santa Fe Working Paper #96-05-035 available at http://www.santafe.edu/sfi/publications/working-papers.html
7. Friedl, J.E.F.: Mastering regular expressions. O'Reilly (1997)
8. Holland, J.H.: Studies of the spontaneous emergence of self-replicating systems using cellular automata and formal grammars. In: Lindenmayer, A., Rozenberg, G. (eds.): Automata, Languages, Development. North-Holland, New York (1976) 385–404
9. Maynard-Smith, J., Szathmáry, E.: The origin of chromosomes. I. Selection for linkage. J. theor. Biol. **164** (1993) 437-446
10. Maynard-Smith, J., Szathmáry, E.: The major transitions in evolution. Springer-Verlag, Berlin (1995)
11. Ono, N., Suzuki, H.: String-based artificial chemistry that allows maintenance of different types of self-replicators. The Journal of Three Dimensional Images **16**(4) (2002) 148–153
12. Rasmussen, S., Chen, L., Stadler, B.M.R., Stadler, P.F.: Proto-organism kinetics: Evolutionary dynamics of lipid aggregates with genes and metabolism. Santa Fe Working Paper #02-10-054 available at http://www.santafe.edu/sfi/publications/working-papers.html
13. Rasmussen, S., Chen, L., Nilsson, M., Abe, S.: Bridging nonliving and living matter. Artificial Life. In press.
14. Suzuki, H., Ono, N.: Universal replication in a string-based artificial chemistry system. The Journal of Three Dimensional Images **16**(4) (2002) 154–159
15. Suzuki, Y., Tanaka, H.: Chemical evolution among artificial proto-cells. In: Bedau, M.A. et al. (eds.): Artificial Life VII: Proceedings of the Seventh International Conference on Artificial Life, MIT Press, Cambridge (2000) 54–63
16. Szathmáry, E., Demeter, L.: Group selection of early replicators and the origin of life. J. theor. Biol. **128** (1987) 463–486
17. Szathmáry, E., Gladkih, I.: Sub-exponential growth and coexistence of non-enzymatically replicating templates. J. theor. Biol. **138** (1989) 55–58
18. Szathmáry, E., Maynard-Smith, J.: From replicators to reproducers: the first major transitions leading to life. J. theor. Biol. **187** (1997) 555–571
19. von Kiedrowski, G.: Minimal replicator theory I: Parabolic versus exponential growth. In: Bioorganic Chemistry Frontiers **3** Berlin, Heidelberg (1993) 115–146
20. von Neumann, J.: Theory of self-reproducing automata. University of Illinois Press, Urbana. Edited and completed by A.W. Burks (1966)

A String Sets for the Replication and Translation System on SAC

A set of strings for Model (i) is described first. The set consists of twelve strings: three phenotypic strings (proteins) for P_c expressed as

00‘0‘0"1"1"0"2"0"2*0011"0"2"0"2 (A1a)

/\0\2\0\2/\\%/*"\\"%0202 (A1b)

00‘1‘1"0"0*‘0‘2‘0‘20011"\."0"0"0"0*‘0‘2‘0‘2"0"0"1"1 (A1c)

three proteins for P_s expressed as

00‘0‘0"1"1"0"2"2"0*0011"0"2"2"0 (A2a)

/\0\2\2\0/\\%/*"%0220 (A2b)

00‘1‘1"0"0*‘0‘2‘2‘00011"\.*‘0‘2‘2‘0 (A2c)

and six genotypic strings (DNA units) for G_c and G_s that are created from the above six proteins by inserting \ in front of every character (which suppresses the functionality of the protein characters) and inserting header and tail characters, 0000 and 0011, for each string. The functions of Strings (A1a)∼(A2c) are explained as follows:

String (A1a) ⇒ [0000*0011 → 00110202*00110202] : a copy starter that changes the header and tail characters and inserts a recognition sequence (0202).

String (A1b) ⇒ [0202\\%*0202 → \\%0202*\\%0202] : a copier that moves 0202 and appends a copy of characters at the end of the DNA unit.

String (A1c) ⇒ [0011*02020011*0202 → 0000*0011.0000*0011] : a copy terminator that deletes 0202, restores the header and tail characters, and inserts . between the original and copied strings.

String (A2a) ⇒ [0000*0011 → 00110220*00110220] : a translation starter that similarly changes the header and tail characters and inserts a recognition sequence (0220).

String (A2b) ⇒ [0220\\%*0220 → \\%0220*%0220] : a translater that appends a copy of characters while removing \s.

String (A2c) ⇒ [0011*02200011*0220 → 0000*0011.*] : a translation terminator that detaches the created phenotypic string.

The string set for Model (ii) is made by modifying the above set for Model (i). The primary difference between Model (i) and Model (ii) is that for Model (ii), all of the genotypic strings (DNA units) are concatenated to make a single very long genotypic string (of 604-character length). With this modification, the header and tail characters of genes are modified slightly (0110 for the header of a gene, 0101 for the tail of a gene, 0000 for the header of the first gene, and 3300 for the tail of the last gene), and the following two genes are added to replicate/translate the concatenated chromosome. The Model (ii) string set consists of eight proteins and one DNA unit.

/\0\2\0\2/!/*"!\0\2\0\2
⇒ [0202!*0202 → !0202*!0202] : a copier that replicates the header/tail characters between genes.

/\0\2\2\0/0101*0110/*3300/"\.*/\0\2\2\0/
⇒ [02200101*0110*3300*0220 → 0101*01100220*33000220.*] : a non-coding region skipper that detaches a created protein and moves the recognition sequence (0220) to the next gene.

The string set for Model (iii) is made by modifying the string set for Model (i) as follows. For Model (iii), we have to add functions for the creation/deletion of the spindle characters L and R and for the creation of a membrane string MM···. For the spindle characters, the copy terminator, String (A1c), is modified so that when it separates the operand string, it might add header characters, L and R, to the original and copied DNA units, respectively, and the copy starter, String (A1a), is modified so that it might be unable to work on an operand string with L or R. This also prevents DNA units from being replicated twice before the cell division is finished and Ls and Rs are eliminated by the reactor program. In order to create a membrane string for the division signal in a timely manner, on the other hand, we need a negative string signal that is created from a chromosome not yet replicated [1]. The signal string (‘‘M"\‘"\‘MMMM ⇒ [MMMMM → ‘‘MMMM]), which we call ‘membrane dissolver’, is a fragile string that makes a membrane string starting with five Ms fragile and is created by a particular protein expressed as

\L\0\0$\R\0\0$\0\0*\0\0\1\1"\."\‘"\‘"\M"\" "\\"\‘"\" "\\"\‘"\M"\M "\M"\M ⇒ [L00]$[R00]$[00*0011 → 00*0011.‘‘M"\‘"\‘MMMM] : a division suppressor that recognizes a DNA unit without the header character L or R and creates a membrane dissolver.

With Model (iii), a cell is divided when it includes a membrane string starting with ten Ms. To create this string, the copy terminator, String (A1c), is modified so that when it separates the operand string, it might create a new pair of strings, ‘membrane seed’ (EM) and ‘membrane breeder’ ("M\E\M ⇒ [EM → MEM]) that react with each other to polymerize the Ms in front of EM. If there is no chromosome not yet replicated, the breeder is able to polymerize ten Ms in front of the seed, whereas if there exists a chromosome whose replication is not yet finished, the division suppressor recognizes such a chromosome to create the membrane dissolver, which destroys the polymerized membrane. The copy starter, String (A1a), is also modified so that when it works, it might create a fragile string named ‘membrane breeder dissolver’ (‘‘\""\‘"\‘M\\\E\\\M ⇒ ["M\E\M → ‘‘M\E\M]) to destroy the membrane breeder, which is not a fragile string. Because the polymerization of five Ms takes several time steps, there is no possibility that MMMMMEM is destroyed by the membrane dissolver created by the division suppressor that recognizes the replicated chromosome itself, or that the membrane dissolver created from other non-replicated chromosomes cannot destroy a string MMMMM··· before it accumulates ten Ms in front of the seed. The Model (iii) string set consists of seven proteins (modified version of Strings (A1a) to (A2c) and the division suppressor) and seven DNA units.

Interaction Based Evolution of Self-Replicating Loop Structures

Keisuke Suzuki and Takashi Ikegami

General Systems Sciences,
The Graduate School of Arts and Sciences,
The University of Tokyo,
3-8-1 Komaba, Tokyo, 153-8902, Japan
{ksk, ikeg}@sacral.c.u-tokyo.ac.jp

Abstract. We propose an ecosystem of self-replicating cellular automaton loops by designing new rules for interactions between individual loops. The loops interact competitively with each other, from which a hypercycle-like network emerges. A spiral structure is seen to emerge from this model with five species. This spiral allows the formation of larger loops at the bounaries between different species. Unlike the single species case, our model allows larger loops to live for long periods of time, and they can replicate in spite of their necessarily lower replication speed.

1 Introduction

The study of self-replicators using 2-dimensional cellular automata (CA) originated in J.von. Neumann's universal constructor[1]. His self-replicator can construct structures that can be coded on a description tape, which is analogous to biological reproduction. By extending von. Neumann's original studies, we would like to understand how new replicators can come out and how different replicators can interact with each other. In particular, evolution isn't caused passively by random mutation but is caused actively by other replicators. The latter interactive and deterministic aspect of evolution is stressed by, e.g.[2,3]. Here, we consider about the evolution caused by only the interaction between replicators. As mentioned in [4], von. Neumann's model allows for increases in both functional and structural complexity. However, in our model we consider only structural complexity, i.e. the shapes of replicators. Without any explicit fitness function, replicators irreversibly change their shapes over time, which we call "evolution". Recently, Sayama has studied the evolution by interaction with his cellular automata replicating system[5]. In his model, while replicators can change their shape, no open-ended evolution appears. It is difficult for more complex replicators to emerge, because simpler self-replicators, which are smaller, can replicate faster than larger ones. This leads to the dominance of small replicators. In another of Sayama's models[6], which uses shape-encoding worms, more divergent forms can replicate but these become fragile against interactions.

W. Banzhaf et al. (Eds.): ECAL 2003, LNAI 2801, pp. 89–96, 2003.

In this paper we propose an improved self-replicating CA model which allows larger loops to evolve due to the spiral formation via the interaction between loops. In section 2, we introduce the present model. The rules governing interaction between loops are presented in section 3. In sections 4 and 5 we discuss our results. In particular, macroscopic spirals are seen to emerge, with larger loops found at the boundaries between spirals.

2 Shape-Encoding Loop

To allow the evolution of self-replicators, we use the shape-encoding loop model, which was designed by Morita and Imai[7]. Our model has a greater number of cell states, and new state transitions. It is defined on a two-dimensional, five-neightbor cellular space, and self-replicators are given by configurations of cell states on the space.

2.1 States of Cells

129 cell states are used in this model. There is a unique "background" (inactive)cell state. The remaining 128 states are used to compose self-replicating pattens as shown in Fig. 1

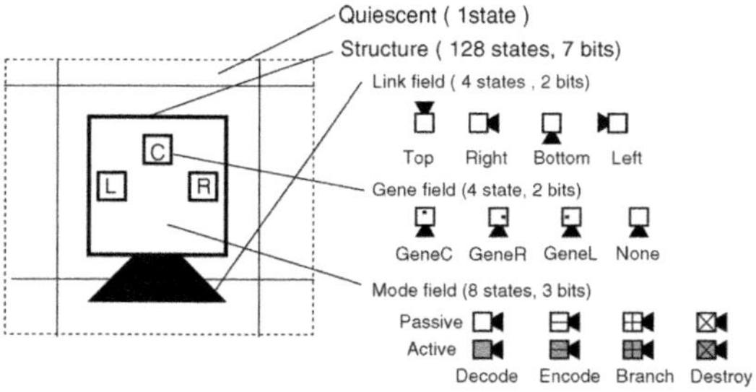

Fig. 1. An illustration of cell states. Three different fields are classified by function. The Link field (4-states) represents a cell direction, which is drawn as a filled trapezoid in the following figures. The Gene field (4-states) stores shape infomation for constructing new loop strutures, which are drawn as the three inner squares. The Mode field (8-states) represents special conditions used for self-replication, which are drawn as lines and colors in the outer squares.

2.2 Self-Replication Process

The shape infomation of replicators is coded in their gene fields. The construction of daughter loops follows this shape infomation. Unlike Langton's loops, each gene field is read only once during the construction of a daughter loop. As a result, the shape of self-replicators can be varied and complex.

Here, we demonstrate the self-replication process in four stages: “Finding corner”, “Encoding shape”, “Expanding arm” and “Withdrawing arm” (Fig.2). A loop of size N is in the finding corner stage when it has one ’branch-active’ (ba) state and N-1 ’decode-passive’ (dp) states. The ba state is transferred to its neighbouring cell in the anti-clockwise direction. When the ba cell reaches the first corner, it first turns into the ’decode active’ (da) state and then turns into the ’branch-passive’ (bp) state. Then the second ’encoding shape’ stage begins. Here, the ’encode’ state is transferred to all connected cells beginning with the cell connected to the first corner. As a result, the replicator’s shape information is encoded througout the loop, before returning to the corner bp state. In Fig.2, the shape information may be read as “LCLLCC”. In the third stage, which we call the expanding arm stage, new cells are successively added to the corner bp state. These new cells are called the expanding arm, and the shape information is sent from the parent cell, through the arm, to make a new (child) replicator. Finally, in the ’withdrawing arm’ stage, the ’destroy-passive’ state is created which removes the expanding arm cells until it reaches the first corner cell. When it reaches the corner, it turns the bp state already at the corner into the ba state, beginning the next replication cycle. In the next section, we define a transition rule for interactions between self-replicating worms.

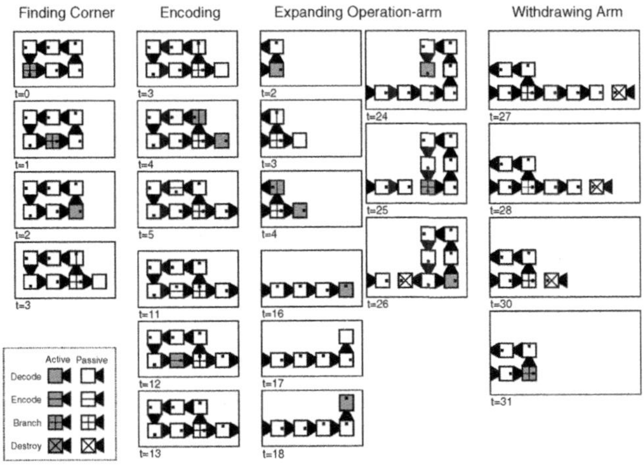

Fig. 2. Self-replicating processes of a 6-cell loop. An arbitrary loop structure can be replicated by the same process

3 Competitive Interaction

We define new rules for the interaction between self-replicating loops. Following the Evoloop model, we introduce dissolution cell states. Here, we use the destroy-active mode which is not utilized in normal self-replicating processes.

This destroy-active mode only emerges on a site where collision occurs between two loops and unlike the destroy-passive mode, it dissolves any type of cell, not just arm cells. Also, we can introduce a notion of "species" by adding a new index field. Species are defined as cells having the same value in this field.

We organize the competition between different loop species by introducing a hypercycle-like interaction. For example, the species S_2 is superior to S_1 but inferior to S_3(Fig3). The competitive interactions between the different species yield a variety of different self-replicators on the CA space, which have not been studied in previous self-replicating CA models. .

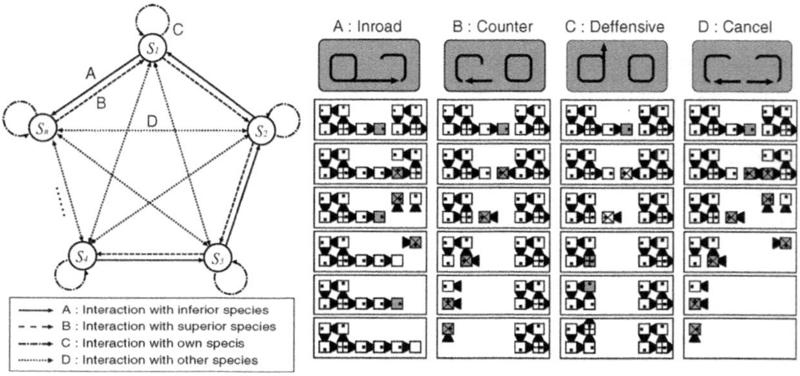

Fig. 3. Competitive interactions between species are illustrated in the left figure. Each species has exactly one superior and one inferior species. The strengths of interaction are given by the four collision styles depicted in the right figure. When the arm of a loop on the left side collides with a loop on the right, only the right-hand loop is destroyed by the 'Inroad' interaction, only the left loop is destroyed by 'Counter', both of them survive in 'Defensive', and both of them are destroyed in 'Cancel'.

4 Dynamics of Self-Replicating Loops

We now report on the dynamics of interacting self-replicating loops. Two main observations can be made. First, a spiral structure emerges from this system, which is not unexpected. Second, larger replicating loops with longer life times evolve at the boundaries between species.

4.1 Spiral Structures

On the boundary regions between species, loops of the superior species can clear space for their replication because they can succesfully invade loops of the inferior species. Consequently, the global dynamics of these self-replicators forms spiral structures, as shown in Fig.4. This kind of structure was often seen in the study of other spatial models where multiple elements interact(see e.g. [8]).

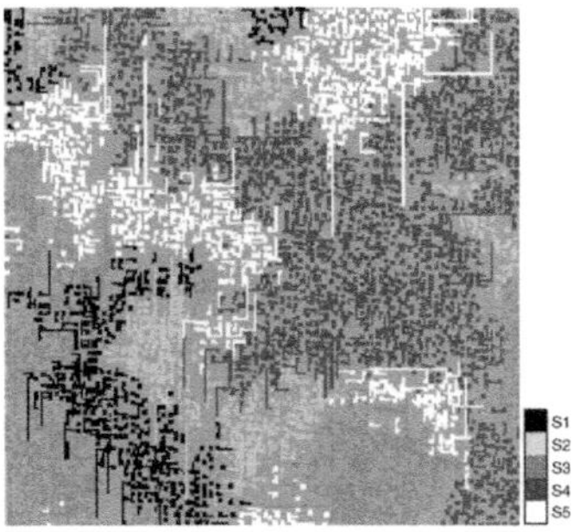

Fig. 4. A spiral structure produced by five interacting species in 200 ×200 cell space. The spiral rotates in a counter-clockwise direction

In the Evoloop model, larger replicators cannot survive for long periods of time, because smaller ones can replicate themselves faster, quickly spreading out in the CA space. Like Evoloop systems, most of the CA space is dominated by loops of minimal size, because they can replicate faster and spread out more quickly. However, we can also observe larger loops frequently emerging on the boundary between species as shown in Fig.5. Various kinds of shapes can exist in a region of inferior species. Although most loops cannot copy themselves, some loops can replicate correctly and breed locally.

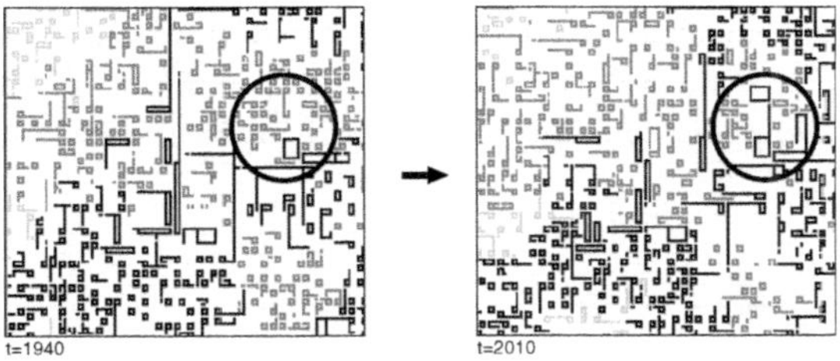

Fig. 5. Larger loops emerge on a boundary with an inferior species. The loops with a medium gray tone are of an inferior species to the loops in black. After 70 steps, the emergent larger loop (circled) sustains replication, invading the inferior species

4.2 Analysis of Self-Replicating Loop Structures

To understand how the spiral structure can generate larger loops, we compare the multi-species competition model with the single-species models. Fig.6 shows the frequency distribution of replicating loops for the competition model, and each of the 4 interaction types used individually. The initial condition for the competition case is a random arrangement of the five species over a 400 × 400 cell space. Only four-cell loops exist at the initial time step. For the other cases shown in Fig.6, only a single species which interacts with itself via the particular interaction is used. The results are accumulated over 10000 time steps. It is easy to see that the full competition model produces the largest loops(e.g. larger than 50 cell sizes). The competition model can provide enough space for replication at the boundary, but the single species models cannot do this.

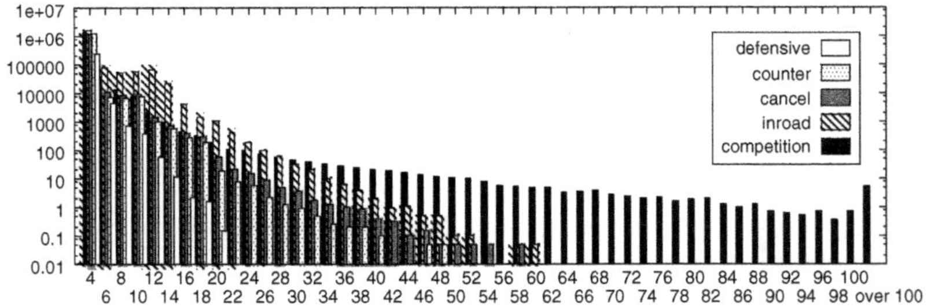

Fig. 6. Frequency histogram of loop sizes. A vertical bar represents the total number of loops of a particular size observed during 10000 steps. The initial state only includes loops of size 4. The CA space has 400 × 400 cell sites. Loops larger than size ∼50 only emerge in the competition model

Fig.7 shows that both the population of the loops and the boundary length between regions of different species oscillates globally. The length of the boundary to the inferior species rises soon after the loop population increases, and the boundary length to the superior species rises after that. The frequency of larger loops increases whenever a significant length of boundary to an inferior species arises.

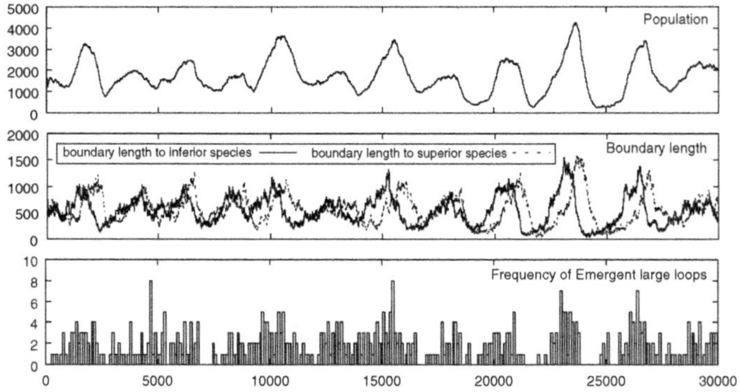

Fig. 7. Dynamics of one species in a five-species competition network. Fluctuations of the loop population (top) seem to be accompanied by changes in the boundary length (middle). The number of larger loops (size ≥ 20) loops increases whenever the length of the boundary to the inferior species increases(bottom)

Though larger loops can emerge, these loops vanish when they interact with other loops. Fig.8 shows the plot of lifetimes of loops according to their length. For the stability of loops on interaction, the 'Counter' and 'Defensive' interaction rules generate long lived loops, in contrast to the 'Inroad' and 'Cancel' rules. The 'Counter' and 'Inroad' rules are assymmetric interactions that generate relatively large loops. On the other hand, the 'Cancel' and 'Defensive' rules generate only small loops. The competition interaction can develop loops with both larger size and longer lifetimes.

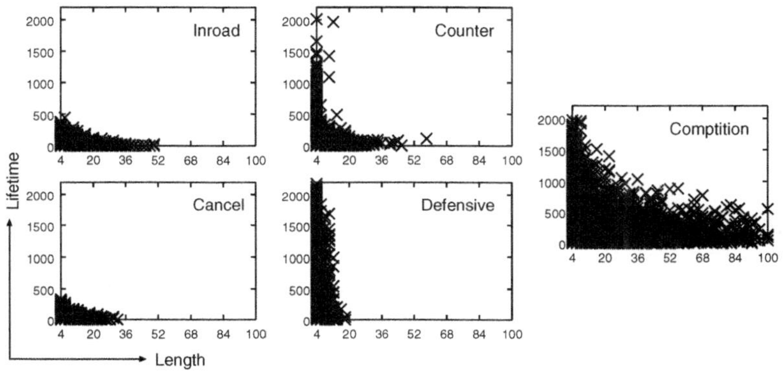

Fig. 8. Life time plots of loops interacting to each of the four interactions, and the full competition network, plotted as lifetime versus loop length. Because of the relative unstability of the inroad and cancel rules, they cannot generate long-lived loops. The distribution found in the competition model shows the existence of many large loops with long lifetimes.

5 Discussion

In this paper, we showed that introduction of species and competition between them enhances the evolvability of self-replicating loops. It is reported that hypercycle formation gives an evolutionary advantage as it leads to resistance against parasites[8]. We also insist that hypercycle formation indeed enhances evolvability. The ability for loops to clear sites is important for the existence of larger loops. How long these large loops can survive depends on the type of interaction between loops(e.g. the degree of loop destruction involved) Fig.8 reveals that more stable interactions (i.e. "Counter" and "Defensive) lead to many long lived loops. The coexistence of different interaction types leads to large loops with longer lifetimes.

There are many problems left to address. We have to allow both left and right interaction with loop arms, and also allow both clockwise and counter-clockwise loops. Then we expect that not only the size but the shape of loops can more easily evolve.

One of our future projects is to observe information flow from the microscopic to the macroscopic level, and vice versa. Flow from the microscopic to the macroscopic level is easier to observe. It is clear that properties of replicators (e.g. genetic information such as size and form) can affect the global behavior of the ecology (e.g. spiral waves). What is more interesting is to see how macrocsopic structure can affect the information at the microscopic level. For example, it would be interesting to see the emergence of different replicators reflecting the spatio-temporal macroscopic ecological pattern. Introducing a food-web structure may be another way to see information flow from the macroscopic to the microscopic level(see e.g. [9]).

Acknowledgements. This work is partially supported by Grant-in aid(No. 09640454 and No. 13-10950)from the Ministry of Education, Science, Sports and Culture. We'd like to thank Barry McMullin for his userful comments.

References

1. J. von Neumann. Theory of Self-Reproducing Automata. University of Illinois Press Urbana, Illinois, 1966. Edited and completed by A. W. Burks.
2. T. Ikegami, and T. Hashimoto. Active Mutation in Self-reproducing Networks of Machines and Tapes. Artificial Life 2(3), 305–318, 1995.
3. T. Ikegami, and T. Hashimoto. Replication and Diversity in Machine-Tape Coevolutionary Systems. Artificial Life V : Proceedings of the Fifth International Workshop on the Synthesis and Simulation of Living Systems, C. G. Langton and K. Shimohara, eds, 426–433. MIT press. 1997.
4. B. McMullin. John von Neumann and the Evolutionary Growth of Complexity: Looking Backward, Looking Forward... . Artificial Life 6(4), 347–361, 2000.
5. H. Sayama. A new structurally dissolvable self-reproducing loop evolving in a simple cellular automata space. Artificial Life 5(4):343–365, 1999.
6. H. Sayama. Self-replicating worms that increase structural complexity through gene transmission. Artificial Life VII : Proceedings of the Seventh International Conference on Artificial Life, M. A. Bedau, J. S. McCaskill, N. H. Packard and S. Rasmussen, eds, 21–30, MIT Press. 2000.
7. K. Morita, and K. Imai. A simple self-reproducing cellular automaton with shape-encoding mechanism. Artificial Life V : Proceedings of the Fifth International Workshop on the Synthesis and Simulation of Living Systems, C. G. Langton, and K. Shimohara, eds, 489–496, MIT Press. 1997.
8. P. Hogeweg. Multilevel evolution: replicators and the evolution of diversity. Physica D, 75:275–291, 1994.
9. H. Ito, T. Ikegami. Evolutionary Dynamics of a Food Web with Recursive Branching and Extinction. Artificial Life VIII : Proceedings of the Eighth International Conference on Artificial Life, K. S. Russell, M. A. Bedau, and A. A. Hussein, eds, 207–215, MIT Press. 2002.

Architectural Design for the Survival Optimization of Panicking Fleeing Victims

Rodrigo Escobar and Armando De La Rosa

Chalmers University of Technology, Complex Adaptive Systems Masters Program
412 96 Gothenburg, Sweden
{rodri_e,barbol_se}@hotmail.com
http://www.dd.chalmers.se/~armando/panic/report.htm

Abstract. Panic is a powerful and at times deadly force; hence the aim of the project is to provide initial modifications in architecture to a large room full of panicking individuals to optimize the evacuation. This was done using a model constructed by Helbing, Farkas, and Vicsek [1]. After reproducing the main results, this model was used to explore the behavior of a mixed crowd, in which only a percentage of the victims were panicking. The main part of the project is then focused on examining architectural adjustments to increase the victim flow out of the room. The outcome of these experiments yielded fruitful results, which at times almost doubled the victim flow velocity. This leaves great room for further more ambitious experiments like full-on evolution of architectural design.

1 Introduction

Panic is one of the few natural catastrophes caused by humans; by natural we mean that it is not deliberate. Collective hysteria may be set off by other disasters like fires, earthquakes, and such. At times it is prompted in less drastic scenarios such as concerts, religious gatherings, or sporting events, though with equally shocking consequences. Documentation exists as far back as the late XIX century and as recent as this year [2], proving and reminding us that we are still victims of this dilemma. Some characteristics of panic, according to Helbing et al., are the following:

- People try to move considerably faster than normal.
- Individuals start pushing, and interactions among people become physical.
- Moving, in particular passing thru a bottleneck, becomes uncoordinated.
- At exits, arching and clogging are observed. Jams build up.
- The interactions in the crowd add up and cause dangerous pressures up to 4,450 N/m^2, which can bend steel barriers or push down brick walls.
- Escape is further slowed by fallen or injured people acting as "obstacles".
- People show a tendency towards mass behavior.

As everyone increases their desired exit-speed in a panic situation, the overall outflow of the population is considerably slower than what it would be, had

W. Banzhaf et al. (Eds.): ECAL 2003, LNAI 2801, pp. 97–106, 2003.

everyone remained calmed. From a game theory perspective, it is easy to see how staying calm is not a stable strategy. The event, as suggested by Brown and Coleman, may be easily compared to the "prisoner's dilemma" or "common goods dilemma", where in a non-iterated game, it is reasonable for the players to take selfish actions rather than cooperate [3].

Due to the tragic nature of this phenomenon, it was with great interest that the search was set out to understand and to certain extent alleviate the problem. A solution to this predicament, considering that panicking people will not listen to the person yelling: "everybody stay calm", would be to redesign the architecture to maximize the output victim flow [4,5]. Ideally, this improvement would be incorporated from the initial design of a building, but for now, the focus is to offer more immediate solutions in a single large room, (15m x 15m), by small feasible alterations.

The model for the simulation is strongly based on the one by Helbing et al., [1], which belongs to the more recent type of models oriented towards self-driven many-particle system [5,6,7], as opposed to the ones oriented to fluid modeling [8,9]. Slight modifications and simplifications were made to adapt the system to the needs of the research, such as not taking into account injured pedestrians. The ratio of panic individuals vs. calm ones within the crowd while exiting a room was reviewed (a discussion not covered in [1]).

In the paper by Helbing et al., they describe their discovery that by placing a column near the exit of the room in an asymmetrical position, the victim flow is greatly improved. It is the aim of this project to expand on that kind of idea, making the world just a little safer.

2 Method

The model is a continuous one, both in time and space. Since solving the very large set of equations is impossible, time wise the simulation is done in a discrete way.

2.1 Equations

The equations are based on a simple physical force model. Each victim i of mass m_i is trying to move in the direction $\bar{e}_i^o$ with a speed v_i^o, and therefore adjusts its current velocity $\bar{v}_i$with characteristic time τ_i. At the same time, it is attempting to keep a prudent distance from other fellow victims j, walls W, and obstacles o; which are represented by the interaction forces: $\bar{f}_{ij}$, $\bar{f}_{iW}$, and $\bar{f}_{io}$, respectively. Thus, mathematically we have the following change in velocity:

$$m_i \frac{d\bar{v}}{dt} = m_i \frac{v_i^o(t)\,\bar{e}_i^o(t) - \bar{v}_i(t)}{\tau_i} + \sum_{j(\neq i)} \bar{f}_{ij} + \sum_{W} \bar{f}_{iW} + \sum_{o} \bar{f}_{io}. \qquad (1)$$

In the interaction forces in equations 2, 3, and 4, the exponential term denotes a repulsive force acting on victim i so that he/she will avoid physical contact with any other element of the workspace, (i.e. victim, wall, or obstacle), and because of the nature of an exponential term, this force will only be significant when the two objects are close enough. The term d is the distance from the pedestrian, and r is the minimum distance there should be between these two elements. This last term depends on the radius of the individual, which is randomly generated between the values of 0.25m and 0.35m; this randomness also helps to keep the system somewhat robust and immune to possible gridlocks by exactly balanced forces in symmetrical configurations. The normalized vector pointing from the element to the victim is $\bar{n}$.

In a panic situation the individuals no longer care if there is physical contact, and we then have that d is smaller than r. This situation gives rise to granular force interactions that are represented by the terms containing the function $g(x)$. This function is zero if pedestrians don't touch, and it takes the value of the argument x if they do. The term with k indicates the "body force" and the one with κ the "sliding friction" or tangential force. The other vectors and scalars in these terms, such as: $\bar{t} = (-n^2, n^1)$, $\Delta v_{ji}^t = (\bar{v}_j - \bar{v}_i) \cdot t_{ij}$, define these forces and are better illustrated in [1], from which this explanation is taken.

$$\bar{f}_{ij} = \left\{ A_i \exp\left[\frac{r_{ij} - d_{ij}}{B_i}\right] + kg\left(r_{ij} - d_{ij}\right) \right\} \bar{n}_{ij} + \kappa g\left(r_{ij} - d_{ij}\right) \Delta v_{ji}^t \bar{t}_{ij}. \quad (2)$$

$$\bar{f}_{iW} = \left\{ A_i \exp\left[\frac{r_i - d_{iW}}{B_i}\right] + kg\left(r_i - d_{iW}\right) \right\} \bar{n}_{iW} - \kappa g\left(r_i - d_{iW}\right) \left(\bar{v}_i \cdot \bar{t}_{iW}\right) \bar{t}_{iW}. \quad (3)$$

$$\bar{f}_{io} = \left\{ A_i \exp\left[\frac{r_{io} - d_{io}}{B_i}\right] + kg\left(r_{io} - d_{io}\right) \right\} \bar{n}_{io} + \kappa g\left(r_{io} - d_{io}\right) \Delta v_{oi}^t \bar{t}_{io}. \quad (4)$$

The parameters that tune the model to perform in a qualitative good way were directly taken from the original[1]. Since panic situations are dangerously unexpected, there's hardly any data to test it in a quantitative way.

2.2 Vector Field Solution

An important hurdle that was satisfactorily overcome, was defining in which direction each victim desired to move. In an empty room this was no problem. Since all agents know the coordinates of the exit location and their own coordinates, it's just a matter of normalizing a simple vector difference. The problem arises in rooms like the one shown in Fig. 1a, where if the victims apply the method just described, they would get stuck trying to go thru walls.

The solution was to create a vector field as shown in Fig. 1b. This vector field was created with a small enough grid-size and was smoothed out using a moving

[1] **Parameters:** A = 2 x 10^3 N. B = 0.08 m. τ = 0.5 s. k = 1.2 x 10^5 kg s^{-2}. κ = 2.4 x 10^5 kg m^{-1} s^{-1}. m = 80 kg. v^o_{calm} = 1 m s^{-1}. v^o_{panic} = 5 m s^{-1}.

average, so that it would produce a continuous field to this otherwise discrete method. The vectors point in the direction of the shortest Euclidian distance to the main exit, in similar way as done in [5,6].

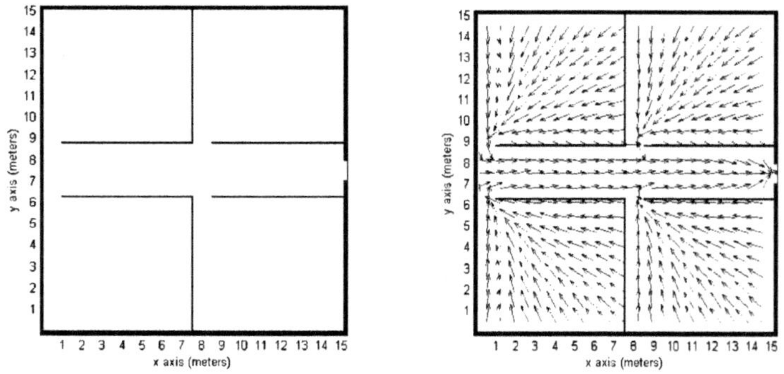

Fig. 1. a. Above, (room 102), is a perfect example of how if the victims headed for the main exit in a straight line, they would get stuck on corners and would never leave the room. **b.** The vector field created allows the agents to know in which direction to move so they will take the shortest Euclidian distance.

3 Experiments

The tests, which were run under the same conditions as specified in [1] (200 agents, an exit door of 1 m width, a room of 15 m x 15 m, and 45 seconds), resembled the calm and panic situations generated there. Once the model was implemented and similar results were achieved, the first experiment aimed to evaluate the flow of victims when the population is composed of panicking and non-panicking people. The percentage of individuals in panic was varied in increments of 10% from 0% to 100%, and the victim flow measured. This was done a total of 20 times each.

Finally, in the main part of the project, a few rooms were intuitively designed, adding some features to them like rounded obstacles (columns for example) and walls. The simulations were ran at least 20 times each to have an average flow of victims, measured in victims per minute (v/min), and a confidence interval that would allow us to compare our results to those obtained by Helbing et al. The corresponding videos were recorded to have the history of each of the tests, and to be able to alter or mix the first set of room designs in order to produce a new "generation" according to some clues and insights learned from their analysis. The rooms in the appendix labeled 01 to 09 correspond to the first set; and rooms 10 to 19 are variations and modifications made to this first one.

A third independent set was created that explores the idea of smaller rooms within a large one (rooms 100 to 106 in appendix), mimicking perhaps classrooms or conference halls. The same procedure was followed as before.

4 Results and Discussion

This simple and interesting model was suggested in a continuous time-space environment. The model includes repulsion forces in order to refrain the victims from coming in contact with walls, obstacles, and other people; this keeps them below their desired speed. The distance determines the forces that change the acceleration, and with it, the speed and position of each person.

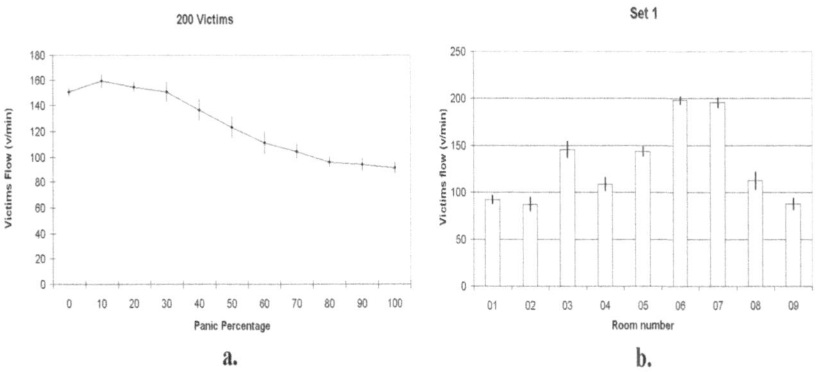

Fig. 2. a. Average victim flow for different panic/calm ratios with their confidence interval. With up to 30% of the individuals panicking the evacuation is still reasonable. Tests show that the need for 200 agents derives from the fact that for less than this, panic is actually favorable for the population. **b.** Victim flow charts for rooms 01 thru 09 with their confidence interval.

The first part analyzes the changes in victim flow when a mix of panicking and non-panicking victims composed the group of agents. Fig. 2a shows the flow for the whole range; the graph was plotted after 20 runs for each configuration. It shows an increment up to 10%, after which the flow decreases to reach its minimum at 100%. As Helbing et al. showed first, the clogging associated with panic is what makes the flow decrease. Thus, it's reasonable to think that this depends directly on the number of people used, and consequently in the percentage of them that are panicking, though it is also evident that some panic benefits the overall evacuation. We noticed in early stages of our work that more than 100 people were needed in the room to produce clogging, otherwise, with 100 people or less, the victim flow was actually enhanced with panic instead of being inhibited. Therefore, this study only has relevance when the number of people involved is high and thus, so is the risk of victims.

An important point of comparison is Room 01, an empty one. Helbing et al. did a lot of their experiments in a setting identical to this one and under the same conditions. It is therefore critical that we achieved similar results. For the authors, the victim flow out of this room was of 87 v/min. In our case, the average was of 92 v/min; though slightly higher than theirs, the confidence interval with a level of 95% was of (87, 96). The standard deviation was interestingly large, this due to the sensitivity of the duration of arching and clogging at the doors on the initial conditions of each run for the same room, which in this case is only the radius and position of each agent. With this in mind, the rest of the rooms that were designed (see appendix) were tested, and the goal of improving the victim flow was achieved. The average of the flow along with the confidence interval of the first set can be seen in Figure 2b.

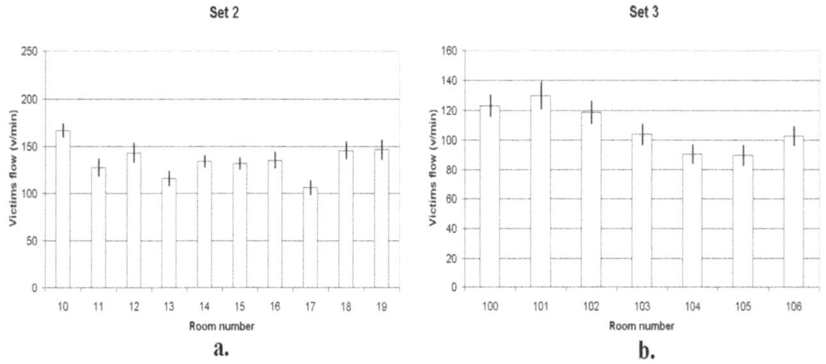

Fig. 3. Victim flow charts and confidence interval. **a.** Rooms 10 thru 19, which were a product of alterations and improvements done on Set 1. **b.** Rooms 100 thru 106. Set created independently from the two previous ones, designed to analyze the configuration of smaller rooms within a large room.

In the first set only two of the rooms gave the same results as Room 01, which serves as reference along with Room 03 since these are the rooms that Helbing et al. experimented with. In the bar graph (Fig. 2b) is shown that Room 02 and Room 09 worked similar to Room 01, and the modifications didn't promote the outflow of pedestrians. On the other hand Room 04 and Room 08 marginally improved the result of Room 01; Room 05 gave results almost as good as the one obtained with the single column suggested by [1]. However, the most interesting results were the ones obtained for Room 06 and Room 07, which show a noticeable gain even over the first improvement achieved with Room 03. Both rooms end up with an average flow of almost twice as much as the flow in Room 01, doubling the victim escape velocity of an otherwise empty room. Analyzing the population's behavior in the movie for Room 06, it was discovered that the space generated by the walls in front of the door, creates a region for a new -but smaller- "room", which, having less pedestrians in it, makes it easier for them to leave (Fig. 4a). The two new virtual doors at both sides of this "triangle" feed this new area, serving as "regulators" of the victim flow. The same explanation could be drawn over the success in Room 07. Here, the two columns allow three different paths and some free space (though not as much as in Room 06) to reach the exit (Fig. 4b). This single feature of a new "area", which is the main qualitative difference between these two rooms and the rest from its set, seems to be the clue that explains their superiority. Although other rooms (Room 04, 05 or 08) also provide various paths to the main exit, they didn't exhibit such positive results, (Fig. 4c).

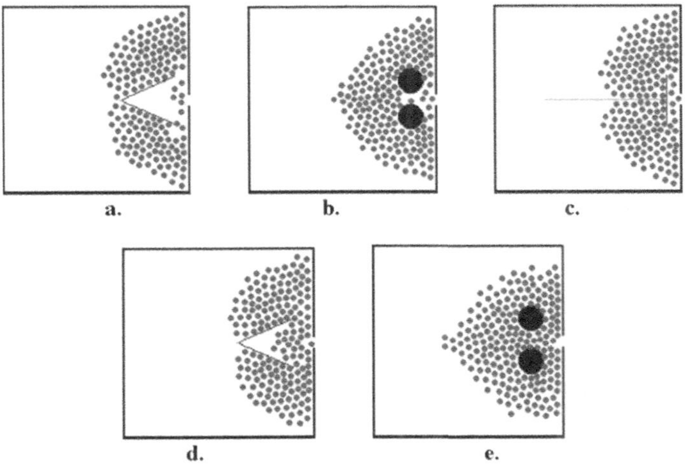

Fig. 4. Snap shots of various rooms. **a. b.** Room 06 and 07, showing the "waiting room" effect. **c.** Room 05 shows how only having "regulators" is not enough to reach high escape velocities. **d. e.** Room 16 and 17 demonstrate how slightly opening the "valves" causes overcrowding, this produces clogging which in turn slows down the victim flow.

Having arrived to this first conclusion, Set 2 was designed trying to exploit these features and analyze some variations to the rooms in Set 1, (see appendix). However, the second set did not display results as positive as those for rooms 06 or 07 in Set 1, though they did manage to exhibit improvement over an obstacle free room such as Room 01 (Fig. 3a). Here, it is interesting to see how a slight increase in the space available for pedestrians to get past the obstacles, (in Room 16 and 17 the triangle wall and the columns were moved 0.5m away from the wall in comparison with Room 06 and 07), made the flow decrease (Fig. 4d and 4e). It seems that as the obstacle is easier to avoid, the "flow control" mechanism mentioned before does not function as well, and the free space in front of the main exit gets rapidly crowded, enabling clogging and thus affecting the evacuation.

It is important to mention that in the second set of rooms, some of the designs were intended to mimic the ones in Set 1, but with the main door placed at a corner of the room rather then at the side. With this, those rooms were analyzed and compared to Room 02, which does not have any obstacles or walls, but does have the door placed in a corner. Again, the regulation of flow determined by different paths to a smaller "area" seems to be crucial, and those designs that did not manage to establish this could not obtain better results.

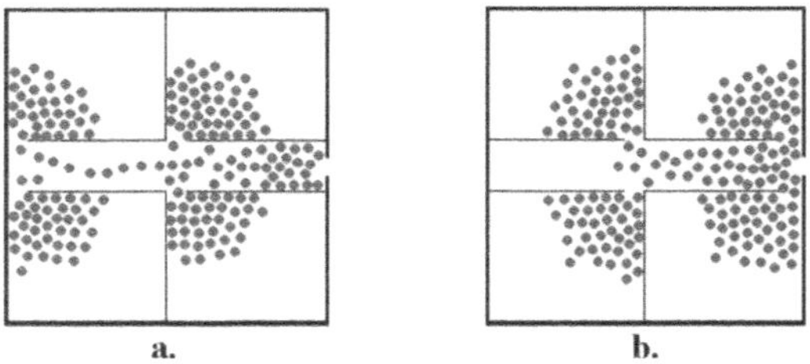

Fig. 5. a. Room 102 is an example of how having the doors to the corridor far from the main exit helps the crowd flow. **b.** Room 105 is the counterexample in which exits from the smaller rooms are too close to the main entrance.

The third set was an approach to a different view of the area, now containing four smaller rooms with their own door and equally distributed but keeping the goal to simulate 200 panicking pedestrians attempting to exit the entire area (see appendix). In this set the rooms differ from one other in the position where the doors for the smaller rooms are placed, the results of which can be seen in Fig 3b. The first three rooms, Room 100, 101 and 102, were slightly better than the rest, which on average behaved more or less the same. A closer look at these two groups suggests that when the doors of the small rooms are placed next to the main exit, this leads to clogging and to slower pedestrian flow, (Fig. 5).

The video simulation resultswere recorded for all rooms and offer a realistic dynamics of how actual individuals may be behaving in such a situation[2].

5 Conclusions

The first important conclusion is the direct effect the number of people participating in a panic evacuation has on the flow. If the number is sufficiently low (under 100), panic will actually make the evacuation faster. All other conclusions are based on this first one.

A second observation was the finding that in a mixed crowd, (some calmed, some panicked), as long as the majority stays calm, exiting the room will happen approximately as smooth as if all were calm. If however, the panicked individuals exceed the 30% threshold, the general flow will start to diminish.

[2] To see some video results please visit the web page or contact authors.

The last and most important discovery was the "waiting room" effect. In general, obstacles slow down the individual victim velocity, increasing the general victim flow. The "waiting room" effect is present when these obstacles that act as flow regulators, also create a small area right in front of the main exit. This "waiting room" can then be treated as a regular room but with very few victims, and as stated in the first conclusion, panic will boost the escape velocity in such conditions. However, if the inflow to the "waiting room" exceeds the outflow, overcrowding takes place and clogging arises, which is the main cause of victim flow decline.

Panic is still a serious problem that gives rise to injuries and deaths. This kind of simulation thus has tremendous potential to alleviate such grievances, whether it may be by panic-testing an architectonic design, by creating panic-safe architecture, or by merely making slight modifications to existing designs.

The model used is computationally expensive, especially since in this multi-agent based environment all interact with all, as suggested in [6]. To this, one still has to add the cost of discretely simulating a continuous model, where the time step has to be small enough to conserve its continuous characteristics.

6 Future Work

The next step will be to proceed on to non-interactive evolution by means of a genetic algorithm. With that, it would be viable to explore the benefits of asymmetry, which according to Helbing et al., proved so useful in placing a single column, as well as other unimaginable configurations. It is a well-known fact that evolution is a powerful thing.

In the simulations, the only sources of randomness, (an important element in any system that helps it become robust), were the diameter of the pedestrians and their initial location in the workspace. Other variables such as mass, level of panic (within individuals and not the entire population), and desired velocity, were kept constant for all agents. This was done with the purpose of making the original model easier to tune; since this has been satisfactorily accomplished it is now feasible to do simulations with a less homogeneous group of people. The idea would be to give the individuals different non-discrete values for the variables mentioned above or to include diffusion and decay properties of panic as modeled in [10].

Something not taken into account was injured people, this could arguably be a critical omission for the following reason: injured victims become obstacles and so affect the general flow of the other victims. Perhaps they do not act as stringent columns, (as proposed by Helbing et al.), but certainly carry a non-negligible aspect and are an important indicator of effectiveness. Hence, future models should include such considerations.

For now, the model is limited to flat surfaces. However, the real world contains non-flat elements that would play a decisive role, such as ramps or stairs. It is imperative then to incorporate these features in order to bring the model closer to reality helping understand and give better solutions to panic phenomenon.

Taking into consideration all the previous suggestions, it would then be possible to implement the model in more specific and realistic situations, hoping to supply designers with insight that would make architecture panic-safe.

Lastly, and on a different note, there is also interest in further pursuing the Game Theory analysis of these panic situations.

References

1. Helbing D., Farkas I., Vicsek T. Simulating dynamical features of escape panic. Nature 407. (2000) 487–490.
2. Helbing D., Farkas I., Vicsek T. (2000). Panic: a quantitative analysis. Web page supplement to reference 1. Available from: http://angel.elte.hu/~panic/
3. Axelrod R. The evolution of cooperation. U.S.A. Basic Books. (1984).
4. Stanton R.J.C., Wanless G.K. Pedestrian movement. Safety Science, Vol. 18, issue 4. (1995) 291–300.
5. Thompson P.A., Marchant E.W. Computer and fluid modeling of evacuation. Safety Science, Vol. 18, issue 4. (1995) 277–289.
6. Kirchner A., Schadschneider A. Simulation of evacuation processes using a bionics-inspired cellular automaton model for pedestrian dynamics. Physica A312. (2002). 260–276.
7. Still G.K. Crowd dynamics, simple solutions to complex problems. PhD Thesis, University of Warwick. (2000).
8. Smith R. A. Density, velocity and flow relationships for closely packed crowds. Safety Science, Vol. 18, issue 4. (1995). 321–327.
9. Riddel, J., Barr, I. In [5].
10. Volchenkov, D., Blanchard, P. A toy society under attack. How to make panic to subside. International Conference SocioPhysics. Bielefeld, Germany. (2002).

Appendix

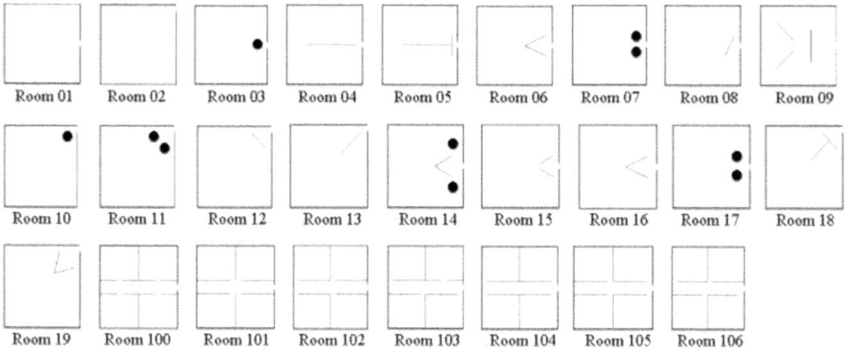

Meta-evolutionary Game Dynamics for Mathematical Modelling of Rules Dynamics

Takashi Hashimoto[1,2] and Yuya Kumagai[1]

[1] School of Knowledge Science,
Japan Advanced Institute of Science and Technology (JAIST)
1-1, Tatsunokuchi, Ishikawa, 923-1292, JAPAN
hash@jaist.ac.jp http://www.jaist.ac.jp/~hash/
[2] Language Evolution and Computation Research Unit,
School of Philosophy, Psychology and Language Sciences,
University of Edinburgh

Abstract. This paper proposes an evolutionary-game-theory model, called meta-evolutionary game dynamics, for studying the dynamics of rules and individual behaviour. Although there are two game theoretical views of rules, i.e., seeing rules as game forms and as equilibria of games, endogenous changes of rules are not modelled effectively by either of these two views. We introduce a model for integrating the two views, in which the interaction rules of replicator equations change dynamically. Computer simulations of an example of the model that include mutation and extinction of both strategies and games show (1) an intermittent change of strategy distributions, (2) a continual transition from a dominant strategy to another, and (3) metastable states distinct from the Nash equilibria. We discuss the notion of evolutionary stability of games and some natural examples showing rule dynamics. We conclude that meta-evolutionary game dynamics enables the study of the endogenous dynamic of rules. Our model contributes, therefore, the development of game theory approach to the dynamics of rules.

Keywords: Dynamics of rules; Mathematical modelling; Meta-evolutionary game dynamics; Replicator equations; Evolutionary stable games

1 Introduction

There are certain social "rules", such as institutions, norms, ethics, conventions, laws, and languages. Not all of these rules are given *a priori*, but many rules emerge spontaneously from our activities. These spontaneous rules constructed by human activities undergo change during the course of history. The changes of the rules are induced by interactions between the rules and the individual activities regulated by the rules.

W. Banzhaf et al. (Eds.): ECAL 2003, LNAI 2801, pp. 107–117, 2003.

1.1 Background: Game Theory Views of Rules

Conventions, norms, and institutions, which we refer to as "rules" in this paper, and their dynamics have been studied in terms of game theory and evolutionary game theory [1,2,3,4,5].

There are two views of rules in the framework of game theory [5].[1] One view is to see rules as game forms of games, and individuals' actions as moves or strategies of the game players [1,2]. In this view, individual behaviour is regulated via the payoffs in the games. Rules should be changed from the outside of the system. Rule changes are represented by, for instance, the change of payoff matrices of the existing games or the addition of new strategies.

The other view considers rules as equilibria of a game [3,4,5]. In this view, individuals following particular rules are players, and properties of the rules are strategies of the players at the equilibria. Exogenous features that are common to all players, such as environmental factors and laws, are specified as a game form including payoff functions. The actually playable equilibria are thought of as rules established and enforced.

This view may include the evolutionary game framework [6], in which the population or the proportion of individuals taking particular strategies changes as a result of games played. Although it seems in this framework that there is formation of a rule by means of transition from a non-equilibrium state to an equilibrium state, however, the process is actually a selection of one equilibrium out of several prescribed ones. Furthermore, the system no longer changes once it attains an equilibrium state, unless perturbations, like invasions of new strategies or changes of the payoff matrix, are given exogenously. Thus, the spontaneous change of rules cannot be treated.

1.2 Integration of Two Views

The common inclinations in both views are that established rules are static and do not change endogenously. As such, these views fail to grasp the fact that rules are dynamic in nature and undergo transition through individuals' activities. We need a new view and a mathematical tool to study the dynamic aspect of rules, one that recognizes that rules and the individuals' behaviour interact with each other and change concurrently. In this paper, we develop a mathematical model of the interaction between rules and individual activities, to expand evolutionary game theory to understand the endogenous dynamics of social structures.

The new view proposed here is an integrated version of the two views above. We basically express initial rules in a society by game forms. Multiple games, which are introduced explicitly, have their own weight factors representing the extent of importance of the rules. Each player plays all games with one strategy and gains weighted payoff from each game. The population of strategies varies in

[1] Aoki [5] lists three views of institutions, the third one of which is 'institutions as specific players of games'. Since this view is not pursued so much, we do not deal with it.

time through the payoffs. In addition, the weights of the games change, depending on the payoffs and the population of strategies. The change of the weights is governed by another rule, called a meta-rule.

The weighted sum of all games can be considered as indicating an entire rule in the society. Accordingly, this entire rule shifts continuously, rather than discretely as in transition among equilibria. The rules and configurations of strategies may also stay at an equilibrium, which is a stable rule in both senses described in the previous subsection.

The situation abstracted here may be acceptable when we suppose multinational firms engaged in business activities in multiple markets. Each market has its own scale and degree of importance in the world economy. The basic strategy of a firm is thought of as the same for all markets, even though the firm can adjust its tactics for each market. Firms change their shares according to the benefits obtained at all markets. The scale and the importance of each market also change in time with the activities of the firms.

1.3 Meta-rule and Hierarchy of Rules Plasticity

One might ask what a meta-rule represents. It is a principle regarded as more basic than an objective rule to be modelled, and therefore it is exogenous. In the above example, it is the basic principle of the market mechanism, such that a profitable market is more importance in the world economic system than an unprofitable market. The meta-rule should be appropriately set for an objective system. For instance, it may be the happiness of the majority for a utilitarian system, the diversity of individual opinions for a democratic system, or sportsmanship in sports.

Such a principle, of course, may change in time, or be replaced by a new principle. Thus, we might consider introducing a meta-meta-rule to capture the change of the principle. Logically speaking, the chain of meta-meta-$\cdots$-rules may cause a problem of infinite regression.

However, we allow a hierarchy of rules plasticity. We suppose that the rules are classified in terms of their variability. As exemplified above, a meta-rule are deemed more universal and invariable than rules at a focal level. For example, in the legal system of a country, the constitution hardly changes, criminal laws can be amended under the constitution, civil laws are more changeable, and so on. In this paper, we assume that the change of rules in the lower classes of the plasticity hierarchy can be neglected. By introducing an invariable meta-rule, we confine the dynamics within two levels, the rules and the behaviour of individuals.

2 Meta-evolutionary Game Dynamics

In this section, we formalize the views proposed in §1.2 to model dynamic change of rules. It is an extension of replicator dynamics [7] to include multiple games with variable weights. We call our formalization *meta-evolutionary game dynamics*.

Suppose N different strategies and the population share of individuals taking the i-th strategy denoted by x_i, where $\sum_{i=1}^{N} x_i = 1$. The replicator system is

$$\dot{x}_i = (u_i - \bar{u})x_i \ , \tag{1}$$

where u_i is the payoff of i-th strategy and $\bar{u} = \sum_{i=1}^{N} x_i u_i$ is the average of payoffs over all individuals. This equation implies that the share of the i-th strategy grows or shrinks in proportion to the difference between its payoff and the average payoff.

We introduce here M multiple games with weighting factors. The weight of the g-th game is expressed by w^g, where $\sum_{g=1}^{M} w^g = 1$. All individuals play all games simultaneously with their own strategies. Thus, the growth rate of the share of each strategy consists of the weighted sum of the payoffs at each game. Accordingly, the time evolution of the share is given by

$$\dot{x}_i = \sum_{g=1}^{M} w^g (u_i^g - \bar{u}^g)x_i \ , \tag{2}$$

where u_i^g and $\bar{u}^g = \sum_{i=1}^{N} x_i u_i^g$ are the payoff of the i-th strategy and the average payoff at the game g, respectively. We refer to this equation as the *weighted replicator equation.*

We assume that each game is evaluated in terms of certain principles or criteria, i.e., a meta-rule in our terminology. We also assume the weight changes according to the evaluation. It is further postulated that the higher a game is evaluated above average, the more the weight of the game grows. Thus, as with the dynamics of strategies, the dynamics of weights is described by replicator type equation (1), by introducing the evaluation λ^g of the game g,

$$\dot{w}^g = \frac{1}{\tau}(\lambda^g - \bar{\lambda})w^g \ , \tag{3}$$

where τ represents the time constant associated with the rate of change relative to the change of strategy shares, and $\bar{\lambda} = \sum_{g=1}^{M} w^g \lambda^g$ is the weighted average of the evaluation over all games. We refer to the whole system of equations (2) and (3), together with the definition of the evaluation function, which will be presented later, as the *meta-evolutionary game dynamics.*

A meta-rule regulating the dynamics of games is introduced as the form of the evaluation function. Basically, the meta-rule is considered as the function of the population shares and the payoffs of all strategies, namely, $\lambda^g = \lambda^g(\boldsymbol{x}, \boldsymbol{u}^g)$, where $\boldsymbol{x} = (x_1, x_2, \cdots, x_N)$ is a vector of population shares called the strategy profile and $\boldsymbol{u}^g(\boldsymbol{x}) = (u_1^g(\boldsymbol{x}), u_2^g(\boldsymbol{x}), \cdots, u_N^g(\boldsymbol{x}))$ is a vector of the payoffs at the game g.

We can define various evaluation schemes according to the objects to be modelled. For example, we can evaluate any single stock market based on the average profit it provides to all who invest in it. In the same way, we can evaluate any single game based on the average payoff it provides to all who play it. In this case, it is defined as $\lambda_{\mathrm{A}}^g(\boldsymbol{x}, \boldsymbol{u}^g) = \langle u_i^g(\boldsymbol{x})\rangle$, where $\langle\cdot\rangle$ means the average over all strategies. In this scheme, a game that gives large payoff to the players on average raises its weight. Another example is the inverse of variance, $\lambda_{\mathrm{IV}}^g(\boldsymbol{x}, \boldsymbol{u}^g) = (\langle u_i^g(\boldsymbol{x}) - \langle u_i^g(\boldsymbol{x})\rangle\rangle)^{-1}$. This evaluation function represents a situation in which equality or impartiality acquires importance as in a democratic society.[2]

We suppose that the state of a societal rule is characterized by the weighted average of the component rules. Then, the *total game* is defined as

$$G = \sum_{g=1}^{M} w^g A^g \ , \tag{4}$$

where A^g $(g = 1, \cdots, M)$ stands for the payoff matrices of a component game g. Note that this matrix is not constant, since each weight w^g changes in time according to Eq. (3). Using the total game G, the population dynamics, namely, the weighted replicator dynamics, Eq. (2), is reduced to the matrix form,

$$\dot{\boldsymbol{x}} = (G(t)\boldsymbol{x} - \boldsymbol{x}\cdot G(t)\boldsymbol{x})\,\boldsymbol{x} \ , \tag{5}$$

where the operator '·' is the inner product between vectors. This formalization clearly shows that meta-evolutionary game dynamics is a direct development of the replicator equation into one with a time-changing interaction matrix.[3]

3 Model

3.1 Specification of Simulation Model

We specify a particular model of meta-evolutionary game dynamics for 2×2 symmetric game matrices and individuals characterized by mixed strategies.[4] In this model, the basic equations (2) and (3) were reformalized into difference equations. In addition, mutation and extinction of both the strategies and the games were introduced as follows. The values in parentheses were used in the following simulations described in §4.

Games: All games are 2×2 symmetric matrices with elements $a_{ij} \in [-1, 1]$.

[2] The payoff matrices of all games should be appropriately normalized or limited for each evaluation function.

[3] The meta-evolutionary game dynamics may be represented as continuous-time infinite dynamic games or differential games [8]. The cost function to be maximized in that framework, however, is defined as a time integration of payoff at a point in time or at one-shot game. Furthermore, following the definition in [8], the payoff function at a point in time does not include payoff values at previous times.

[4] In the usual replicator dynamics, an individual is thought to be taking a pure strategy, and the strategy profile can be interpreted as a mixed strategy.

Strategies: We adopt a mixed strategy (s_1, s_2) for an individual, where $s_1, s_2 \in [0, 1]$, and $s_1 + s_2 = 1$. (We use 11 strategies $s_1 = (0.0, 0.1, \cdots, 1.0)$).
Mutation of strategies: The ratio $\mu_s (= 0.003)$ of the share of each strategy is transferred to the other strategies or to a new strategy at every step.
Mutation of games: The ratio $\mu_g (= 0.003)$ of the weight of a randomly selected game is transferred to a new game at every step. Each element of the new game is shifted by a random number of normal distribution with the average 1 and the variance 0 from the original game.
Extinction of strategies: The strategies with the share $x_i < \theta_s (= 10^{-7})$ are removed.
Extinction of games: The games with the weight $w^g < \theta_g (= 10^{-7})$ are removed.

3.2 Categorization of 2 × 2 Game Matrix

Followings are the normalization and the categorization of game matrices. Since the replicator dynamics is invariant under a local shift of payoff [6], given a 2×2 payoff matrix $A = \begin{pmatrix} a & b \\ c & d \end{pmatrix}$, then we can transform the payoff matrix without loss of generality to

$$A' = \begin{pmatrix} a - c & 0 \\ 0 & d - b \end{pmatrix} \equiv \begin{pmatrix} \alpha & 0 \\ 0 & \beta \end{pmatrix} . \tag{6}$$

Under this transformation, the set of Nash equilibria is invariant. We use the α-β space to view the motion of games. The α-β space is divided into four categories in terms of properties of the set Θ^{NE} of Nash equilibria and the convergence point $\hat{\boldsymbol{x}}$ of the replicator dynamics [6]. Let the strategies for 2×2 game be $\boldsymbol{e}_0 = (1, 0)$, $\boldsymbol{e}_1 = (0, 1)$, and $\boldsymbol{p} = \left(\frac{\beta}{\alpha+\beta} \frac{\alpha}{\alpha+\beta} \right)$. Then the four categories are defined as follows:

Category I For a game with $\alpha < 0$ and $\beta > 0$, $\Theta^{NE} = \{\boldsymbol{e}_0\}$ and $\hat{\boldsymbol{x}} = \boldsymbol{e}_0$.
Category II For $\alpha > 0$ and $\beta > 0$, $\Theta^{NE} = \{\boldsymbol{e}_0, \boldsymbol{p}, \boldsymbol{e}_1\}$. If the initial state $\boldsymbol{x}_0$ is $\boldsymbol{e}_0 < \boldsymbol{x}_0 < \boldsymbol{p}$, then $\hat{\boldsymbol{x}} = \boldsymbol{e}_0$. If $\boldsymbol{p} < \boldsymbol{x}_0 < \boldsymbol{e}_1$, then $\hat{\boldsymbol{x}} = \boldsymbol{e}_1$.
Category III For $\alpha < 0$ and $\beta < 0$, $\Theta^{NE} = \{\boldsymbol{p}\}$ and $\hat{\boldsymbol{x}} = \boldsymbol{p}$.
Category IV For $\alpha > 0$ and $\beta < 0$, $\Theta^{NE} = \{\boldsymbol{e}_1\}$ and $\hat{\boldsymbol{x}} = \boldsymbol{e}_1$.

4 Simulation Results of Average-Type Meta-rule

As a simple example of the meta-rule, we examined the average type meta-rule,

$$\lambda^g_{\mathrm{A}}(\boldsymbol{x}, \boldsymbol{u}^g(\boldsymbol{x})) = \frac{1}{N} \sum_{i=1}^{N} u^g_i(\boldsymbol{x}) . \tag{7}$$

This evaluation function formalizes a meta-rule that selects the rules that gives the maximum utility to individuals on average.

An example of the time evolution of population share for each strategy is shown in Fig.1. In this and the following results, only one game whose elements are all 0.0 exists and four strategies are selected using random numbers with uniform distribution at the initial state. The distribution of share intermittently changes in time. Moreover, there occasionally arise the transitions in a dominant strategy sometimes.

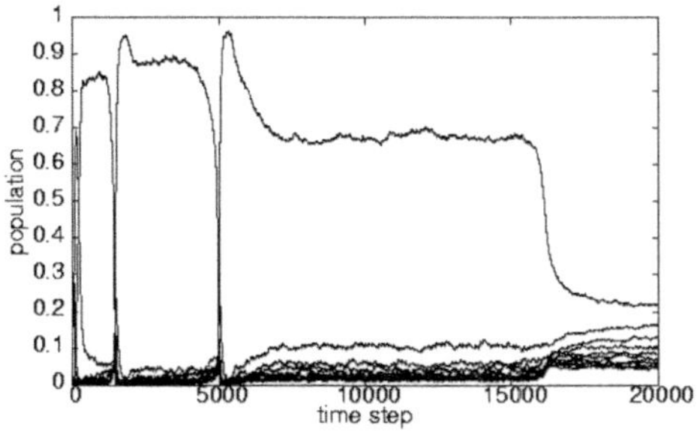

Fig. 1. The time evolution of the share of strategies for the average-type meta-rule. The x-axis is the time step and the y-axis is the share. The time constant $\tau = 4.0$. All 11 lines for the strategies are drawn. We can observe the changes of the distribution and transitions of the dominant strategies.

To see the movement of games, we depict in Fig.2 the change of the elements α and β of the normalized total game. The large changes in the population share correspond with the large changes in the total game. The drastic change of distribution of strategies and the change of dominant strategies are caused by the shift of the total game between or inside the categories. In the α-β plane, the total game starts from the origin, moves among the categories, and finally comes back toward the origin.

The usual replicator dynamics with a 2×2 matrix has a stable equilibrium at one of two pure strategies in each of the categories I, II, and IV. However, the actual distribution of the strategies in meta-evolutionary dynamics does not converge to these Nash equilibria of the total game, as shown in Fig.3. For categories I, II, and IV, this misconvergence is induced by the perturbation caused by strategy mutation. Therefore, the difference between the actual distribution and the equilibria is small.

For category III, the mutation also affects the misconveregence, but it seems to be more than a perturbation. The system presumably diverges in a disparate direction from the equilibrium under the appropriate mutation rate.[5] In the ideal (without perturbations) situation, the distribution of strategies would converge to the Nash equilibrium of the total game. With perturbations, however, the distribution of strategies converges to a metastable point, which is different from the equilibrium point. To understand the mechanism that causes the two points to be different, we should analyse each game not only the total game.

5 Discussion

[5] If the mutation rate is too large, the structure of the system is destroyed. Thus there is no remarkable dynamics as described here, and the difference between the equilibrium and the actual stationary point is just proportional to the mutation rate.

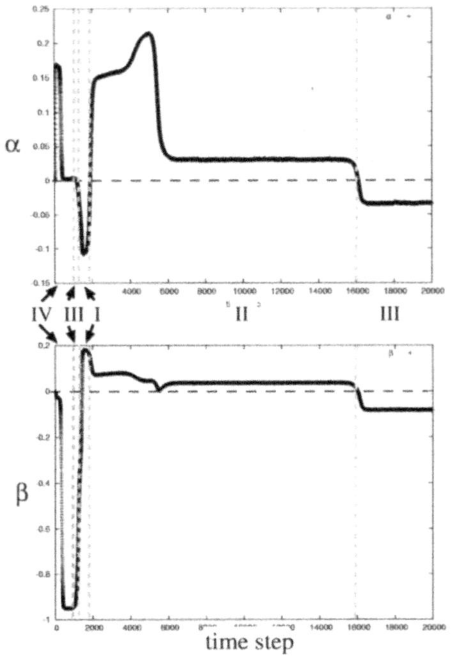

Fig. 2. The time evolution of α (upper) and β (lower) of the total game. The x-axis denotes the time step. Categories I $\sim$ IV (introduced in §3.2) and the line dividing them are indicated.

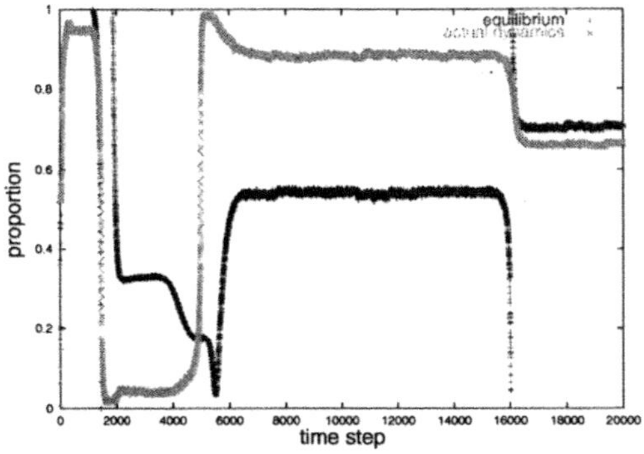

Fig. 3. The comparison of time evolutions of the actual dynamics ($\times$, gray) and the equilibrium point of the total game ($+$, black). The x-axis is the time step and the y-axis is the proportion of the first pure strategy. The symbol $\times$'s are for the actual dynamics, $\sum_i x_i s_1$, and $+$'s are that of the theoretically calculated equilibrium of the total game.

5.1 Evolutionary Stable Game

In meta-evolutionary game dynamics, the weights of games change according to Eq. (3). If a strategy profile is fixed, the games reach an equilibrium under the profile. Therefore, the evolutionary stability of games is formalized as the straightforward extension of the evolutionary stable strategy in evolutionary game theory [9].

Suppose a strategy profile $\boldsymbol{x}$. When a game g (possibly the total game) satisfies

$$\lambda^{g}(\boldsymbol{x}, \boldsymbol{u}^{g}) > \lambda^{g'}(\boldsymbol{x}, \boldsymbol{u}^{g'}) \tag{8}$$

for any g' under the profile $\boldsymbol{x}$, the game g is stable under the invasion of any game. Furthermore, if the strategy profile $\boldsymbol{x}$ is the evolutionary stable strategy, there is no incentive for both the game and the strategy to change. "Evolutionary stable strategy" satisfies

$$\boldsymbol{x} \cdot A^{g}(\epsilon \boldsymbol{y} + (1-\epsilon)\boldsymbol{x}) > \boldsymbol{y} \cdot A^{g}(\epsilon \boldsymbol{y} + (1-\epsilon)\boldsymbol{x}) \tag{9}$$

for sufficiently small $\epsilon > 0$ and for any strategy profile $\boldsymbol{y}$, where A^{g} is the payoff matrix of the game g. Thus, the system is at an equilibrium with respect to both the game and the strategy. We call the game g and the strategy profile $\boldsymbol{x}$ satisfying Eq. (8) and (9) an *evolutionary stable game* (ESG). The ESG of the average-type meta-rule under the specifications of §3 is $G^{ESG} = \begin{pmatrix} 1 & 1 \\ 1 & 1 \end{pmatrix}$. Therefore, the total game converges to the origin of the α-β space.

We found that, in the simulation of the average-type meta-rule with mutation and extinction, the meta-evolutionary game dynamics results in a metastable state distinct from the Nash equilibria. In general, for a system in which ESG is definable and mutation is introduced, since the strategy profile changes slowly around Nash equilibria, we infer the existence of a metastable state different from Nash equilibria.

We have also examined a system with the variance-type meta-rule

$$\lambda_{\mathrm{V}}^{g}(\boldsymbol{x}, \boldsymbol{u}^{g}(\boldsymbol{x})) = \frac{1}{N}\sum_{i=1}^{N}\left(u_i^{g}(\boldsymbol{x}) - \frac{1}{N}\sum_{i=1}^{N} u_i^{g}(\boldsymbol{x})\right) . \tag{10}$$

Since this evaluation function gives a high weighting factor to a game in which the variance of the payoff of strategies is large, the difference of payoffs between winners and losers increases and one strategy tends to dominate. In such a situation, the other strategies decrease their population, and sooner or later they become extinct. When they disappear, there remains only one strategy. In such a situation, the variance also disappears and, therefore, the evaluation of such a game decreases drastically. Accordingly, ESG is not defined for this meta-rule and the games and the strategies tend to be itinerant.

5.2 Various Systems with Dynamics of Rules

Natural examples in which rules of a system change with time are ubiquitous. Almost all adaptive systems show dynamics of rules. The brain has a hierarchical structure, i.e., upper levels regulate lower levels to some extent, the dynamics at the upper level induce change of the rule of dynamics at the lower level [10].

Cognitive development is a process of rewriting rules for representation [11]. Biological evolution is the transition of functions that determine characteristics of biological systems [12].

In this paper, we treat social systems such that the behaviour of elements at the micro or lower level not only is governed by a rule, the macro or upper level, but also can change the rule. In other words, the system has a dynamic interaction loop between the two levels. Language is another representative what having such dynamics. Grammars and lexicons of a language are kinds of rules that the users of the language have produced and followed. Such rules change temporally to a greater or lesser extent. Changes are brought about by the users' activities, especially writing and speaking the language. Unused, or dead, languages are indisputably invariable.

It is difficult to treat the dynamics of rules mathematically and generically. In the theory of dynamical systems, state changes of the system are governed by fixed functions. The change of functions is beyond the scope of the standard dynamical systems theory. Recently, some studies in iterated functional systems, skew product, and functional shift [13], have considered change of functions. To understand the dynamics of rules, we need to develop a mathematical framework for modelling the dynamics, as well as describing them empirically.

6 Conclusion

We conclude that the meta-evolutionary game dynamics proposed in the present paper is useful for studying the dynamic change of rules in which the interaction loop between individual behaviour and the state of rules matters. The endogenous change of rules cannot be studied effectively by either the game-form view or the equilibrium view. Such interesting dynamics, intermittent changes of the distribution of strategies, a continual drastic transition of the dominant strategies, and metastable states different from the Nash equilibria are not seen in the usual low-dimension replicator dynamics. Further study of meta-evolutionary game dynamics will contribute to an understanding of rule dynamics from the viewpoint of game theory. We must clarify foremost how such dynamics occur by looking closely at the dynamics of both the total game and of component games. Moreover, applicability of meta-evolutionary game dynamics to the empirical investigation of the formation of norms and institutions should be pursued.

Acknowledgement. The authors thank Prof. DiGiovanni for his critical reading of the manuscript. This work is partly supported by the Research Fellowship Program of Canon Foundation in Europe, and by a Grant-in-Aid for Scientific Research (No.12780269 and No.15700183) from the Ministry of Education, Culture, Sports, Science and Technology of Japan and by Japan Society for the Promotion of Science.

References

1. North, D. C.: *Institutions, Institutional Change and Economic Performance.* Cambridge University Press (1990)
2. Hurwicz, L.: Institutions as Families of Game Forms. *Japanese Economic Review,* **47** (1996) 113–32
3. Lewis, D.: *Convention.* Harvard University Press (1969)
4. Young, H. P.: *Individual Strategy and Social Structure – An Evolutionary Theory of Institutions.* Princeton University Press (1998)
5. Aoki, M.: *Toward a Comparative Institutional Analysis.* The MIT Press (1996)
6. Weibull, J. W.: *Evolutionary Game Theory.* MIT Press (1995)
7. Taylor, P., Jonker, L.,: Evolutionary Stable Strategies and Game Dynamics. *Mathematical Biosciences* **40**, (1978) 145–156.
8. Basar, T., Olsder, G. J.: *Dynamic Noncooperative Game Theory (2nd Edition).* Academic Press (1995)
9. Maynard-Smith, J.: *Evolution and the Theory of Games.* Cambridge University Press (1982)
10. Tsuda, I.: Toward an interpretation of dynamic neural activity in terms of chaotic dynamical systems. *Behavioral and Brain Sciences* **24** (2001) 793–847
11. Karmiloff-Smith, A.: *Beyond Modularity – A Developmental Perspective on Cognitive Science.* The MIT Press (1992)
12. Maynard-Smith, J., Szathmary, E., *The Major Transitions in Evolution*, Oxford University Press (1995)
13. Namikawa, J., Hashimoto, T.: Dynamics and computation in functional shifts. Submitted to *Nonlinearity*

Preventing Bluff Agent Invasions in Honest Societies

Robert Lowe and Daniel Polani

Adaptive Systems Research Group
Faculty of Engineering and Information Sciences
University of Hertfordshire
College Lane, Hatfield Herts AL10 9AB
United Kingdom
R.Lowe@herts.ac.uk

Abstract. Frequently debated issues in the domain of game theory involve the issue of signalling strategies used in order to resolve conflicts between agents over indivisible resources and to reduce the costly outcomes associated with fighting. Signalling behaviour, used by agents of different strengths, to aid resource acquisition was modelled using an artificial life simulation environment. Honest signalling and the bluff strategy based on Enquist/Hurd's adapted pay-off matrix (1997) were evaluated relative to different proportions of resident strong agents capable of imposing a 'punishment' cost on bluffer agents. We found that in order for honest signalling to be immune to invasion by a bluff strategy, the number of punishment enforcers in the society must be high. Additionally, the number of punishment enforcers is more influential in preventing bluff agent invasions than the severity of punishment.

1 Introduction

A contentious issue in the area of signalling theory involves the extent to which animals claiming exaggerated strength ('bluffing'), via the transmission of signals normally associated with animals of superior strength, can successfully ward off rivals for indivisible resources. Game theory and models of strategic signalling are inextricably linked regarding agent competition for indivisible resources. Two famous models; "Hawk v Dove" (Maynard Smith and Price, 1973, 1982) and the "War of Attrition" model (Canning and Maynard Smith, 1978) have contributed much to our early understanding of how signalling strategies may be mathematically modelled though they have failed to account for the asymmetries of strength that occur in nature. Models that cater for such asymmetries have looked at the effects of production costs and costly consequences. Zahavi (1975) and Grafen (1990) respectively established and mathematically validated the "Handicap principle" pertaining to nature's enforcement of honesty amongst the weak. Stronger agents are better able to demonstrate their prowess and thus procure the female resource via the emission of a costly to produce signal that weaker agents cannot afford to produce. Enquist (1985) and Hurd (1996, 1997) collaborated on a pay-off matrix pertaining to costly consequences as opposed to costs of production. They suggested that honest signalling of strength is the "Evolutionary Stable Strategy" (ESS[1]) given that the cost of

[1] The ESS is the optimal strategy and cannot be invaded by other strategies over evolutionary time.

W. Banzhaf et al. (Eds.): ECAL 2003, LNAI 2801, pp. 118–127, 2003.

being punished for bluffing by stronger agents is sufficiently high. Caryl (1987) however refuted Enquist's claims and argued that weak agents may choose to bluff via claims of exaggerated strength provided that there exists the possibility of escaping the costs of retribution, a notion empirically corroborated by Noble (2000) using an artificial life simulation environment. Noble found that the ease to bluffers of escaping determined the likelihood of their adopting dishonesty as a resource procuring strategy. Maynard Smith (1982), Mesterton-Gibbons (1995) also support the notion of successful bluffing strategies and point to evidence of their existence in the wild.

In terms of general tendencies towards cheating or dishonest behaviour in human society there is evidence that a reduced probability of an agent escaping punishment for dishonesty is a greater deterrent than is the threat of an elevated punishment or cost. Statistics from the "British Crime Survey 2000" and "Criminal Statistics England and Wales 2000" indicate that severe sentencing is less of a deterrent than the possibility of being caught. In America the notorious '3 strikes' threat - a life sentence for committing a third imprisonable offence - has not even affected crime rates suggesting that increased severity of punishment might not deter those who would try to circumvent the system. Bluffing and even dishonest behaviour in general is thought to be most advantageous if used rarely. After all signalling within a society must be predominantly honest if it is to be informative. However, such purely communicative signalling of strengths and needs will always be susceptible to cheats and parasites.

Using the later adaptations to the original Enquist pay-off paradigm (1985) made in 1996 and 1997 by Hurd and Enquist/Hurd respectively we implemented an artificial life simulation environment and ran experiments in order to ascertain under which conditions honest or purely communicative signalling may be corrupted by the bluff strategy. The artificial life approach was adopted since it allows for the possibility of controlling the various parameters being tested under various conditions and offers the benefits of computational power to gauge the trends of the strategies over generational time. Of particular relevance is Enquist's hypothesis that the punishment cost for bluffing must be sufficiently high in order that the bluff strategy does not invade the honest signalling ESS:

$$E(-1) > (C(0) + \frac{1}{2} \cdot V) \qquad (1)$$

That is the cost to the bluffer $(E(-1))$ must be greater than the cost of fighting an equally matched opponent added to half the value of the resource. Two key parameter values were manipulated in the various experiments in order to gauge when this equation can be satisfied if at all; 1) *Cost* of the punishment to a bluff agent administered by a strong agent (the enforcer). 2) *Number* of strong punishment enforcing agents relative to their weaker counterparts. Enquist/Hurd's paradigm refers to strong and weak agents where bluffing among the weak pays relative to being honest only to the extent that it is not sufficiently punished by the strong honest agents. In order to maintain the proportion of strong enforcers to weak agents it was necessary to artificially 'clamp' the strong agents' ability to reproduce. Consequently, the finding of any evolutionary stable strategy or ratio regarding weak bluff and honest agents will be conditional upon this point. The paper is structured as follows; Section 2 describes the key areas of the Enquist/Hurd pay-off matrix, Section 3 describes the artificial life architecture used and how Enquist/Hurd's pay-off matrix was implemented, Section 4 provides the results section, Section 5 provides the discussion and offers concluding remarks.

2 The Enquist/Hurd Model

The Enquist/Hurd Model has an abundance of literature pertaining to it; Johnstone (1998), Bergstrom, Lachmann and Szamado (2000, 2001). For this reason it was chosen as a valid, contemporary and mathematically quantifiable means of evaluating the conditions necessary to prevent bluffers from prevailing. The model describes pay-offs associated with different signalling strategies and consequent behaviours used by two agents when attempting to obtain an indivisible resource. The disputing agents simultaneously exchange signals that by convention are associated with strength i.e. the agents assume their opponents' signals are honest indications of strength though the signals themselves have no associated performance cost or inherent value. On receipt of the transmitted signals the agents must then choose the pay-off optimizing behavioural response according to their strength and the perceived strength of their opponent. The Enquist pay-off matrix thus models two phases; the signal and response phase which are often central to resolving disputes between two agents over indivisible resources. The signalling phase involves the simultaneous exchange of one of two arbitrarily selected signals that may then become conventionally associated with discrete strengths (weak and strong). The response phase relates to the behaviour of the agents once the opponent's signal has been received. Agents can choose among 'Attack', 'Pause-Attack' (allowing opponents the chance to escape if they are perceived to be weaker) and 'Give-Up' responses which depends on the perception of their opponent's strength as well as knowledge of their own. The two agent 'strategists' under investigation in this study are 'honest' and 'bluff' agents. Honest agents adhere to the signalling convention that highlight their weakness whilst bluffers use the strategy of adopting the signal normally associated with agents of superior strength. Honest agents will attack agents of perceived same strength whilst weak honest agents will flee stronger agents without incurring a cost and stronger agents will pause-attack perceived weak agents allowing them time to escape to the benefit of both. Bluffers will flee stronger attacking agents but will be met with some inescapable cost and will mimic the pause-attacking behaviour of the strong agent when confronted by a weak opponent thus enabling the costless procurement of the resource. In this sense bluff agents are 'conditional strategists' since they depend on the gullibility of the weak honest agents and the existence of the strong agents, whose behaviour they mimic, in order to survive. Finally a bluff agent encountering a fellow bluffer will perceive his opponent as being of superior strength and should thus flee, Enquist and Hurd in such cases however randomly allocate the reward of the resource to one or other of the agents.

3 Experimental Design and Methodology

3.1 ALife Environment Configuration

Run on a JBuilder3 Applet the Artificial Life Simulation Environment is initialised with fourteen resource cells and sixty agents who move in a continuously spaced, discretely timed environment. The resources were spatially distributed in such a way that effects of 'clustering' were minimised. The agents themselves have 360 degree vision, simple collision detection, a metabolic rate associated with energy reduction, maintain constant speed and make simple if-then based decisions according to the rules that Enquist/Hurd's pay-off matrix outlines regarding disputing behaviour. Additionally hunger state was recorded. If energy dips below the hunger threshold then the agent will actively seek a resource with which to 'refuel'. Agents must stay at the resource for a constant time period before any of the value of the resource is 'consumed'. This allows the possibility of a number of independent disputes over individual resources to emerge when another 'hungry' agent is in the vicinity of the same occupied resource. All disputes are between two agents; the agent at the resource and the first selected challenger. Agents are capable of producing clones if their energy reaches a certain level and can die of starvation, old age or due to the energy costs incurred during a dispute. All such instances of death ultimately relate to energy levels dropping below zero however.

3.2 Modelling Costs

The costly consequences associated with producing a given signal faithfully model Enquist/Hurd's pay-off matrix. The following critical constant parameter values were chosen in accordance to the specifications of the Enquist/Hurd pay-off matrix:

$V = 1000$. (Value of resource) $C(0) = 430$. (Cost of fighting to equally matched agents)

One specification of the Enquist model is that the indivisible resource must be worth fighting over with cost of fighting an equally matched opponent exceeded by half the value of the resource (note: $\frac{1}{2} \cdot V = 500$ which is greater than $C(0) = 430$). The total combined costs of fighting to the two agents $C(0) + C(0)$ must be slightly less than the value of the resource. Two equally matched opponents fighting here; $C(0)+C(0) = 860$, constitutes a high combined cost, slightly less than the resource thus satisfying the aforementioned criteria. We used parameters $E(-1)$ representing the 'punishment' cost of bluffing a stronger agent and R representing ratio of strong agents to weak agents.

As mentioned earlier Enquist and Hurd (1997) proposed that the value of $E(-1)$ should be high; more precisely (1) should be satisified to enable Honesty to be the unique evolutionary stable strategy. In this case $E(-1)$ would have to be greater than 930 i.e. $E(-1) > (430 + 500)$. Any value less than or equal to this would lead to honest signalling being successfully invaded by bluffers and therefore signalling would cease to be useful as a method of pure communication. $E(-1)$ bluff punishment cost was set to values of 300, 400, 500, 600, 700, 800, 900 and also 930 (the threshold, in this experiment, beyond which Enquist and Hurd predict honesty is the ESS) in the first set of experiments mentioned directly below.

Two sets of experiments were run with the following ratios for strong to weak agents;

1. Agent society where Strong to Weak ratio R = 1:1 (high 'punishment enforcer' presence)

2. Agent society where Strong to Weak ratio R = 1:2 (low punishment enforcer presence)

Whilst Ratio R for the combination of weak and strong agents is changed in these experiments from R = 1:1 to R = 1:2 the weak agent ratio of bluffer/honest agents was initialized at 1:1 for all experiments. Enquist/Hurd's pay-off matrix related to the pay-offs of two agents contesting a resource whereas in our experiments such contests occured in the context of a group or 'society' of agents. As such the initial expected dispute pay-off was held constant for bluffer agents existing in $R = 1:1$ and $R = 1:2$ and did not change over time. In this way it was possible to see whether bluffers are more or less successful if they are more likely to avoid encountering and thus being punished by the strong punish enforcers even if $E(-1)$ is proportionately higher.

Thus, where R = 1:1 the initial expected cost to a bluffer agent will be:

$$(V + \frac{1}{2} \cdot V - (E(-1) \cdot 2))/4. \tag{2}$$

That is the pay-off for the possibility of meeting another bluff agent plus a weak honest agent plus two strong honest agents (note: the ratio is one strong for every weak agent) all divided by four. So, given an initial starting value therefore of $V = 1,000$ and $E(-1) = 500$, a bluffer's initial average pay-off would be $(1500 - 1000)/4 = 125$. The effective pay-off will change over time with the proportion of static strategies. However, we did not study strategic adaption here, rather fixed 'innate' strategies. To convert the $E(-1)$ bluffer cost value from R = 1:1 to R = 1:2 the equation:

$$(V + \frac{1}{2} \cdot V - E(-1))/3 \tag{3}$$

must be used. In order to maintain the same average initial pay-off we feed the known value of the resource (V) into the equation: $(1500/3 - (E(-1)/3)) = 125$. Therefore $500 - 125 = (E(-1))/3$ and so $E(-1) = 1125$. This method can be used to convert all $E(-1)$ values for given strong to weak agent ratios in a society. The converted values used in experiment set 1 are thus; 825, 975, 1125, 1275, 1425, 1575, 1725, 1770 respectively. Parameters such as; initial energy, hunger threshold and fertility level were held constant. Space however, precludes an in-depth description here.

All individual experiments consisted of 100 simulation runs with 20,000 time steps analysed after every 2,500 time steps. Agent strong/weak ratios were maintained via externally imposing a rule that if either strong or weak agents exceed their ratio at any point then no agent of this strength may reproduce until the ratio has returned to its experimental value. This was not intended to be biologically realistic, however it was imperative to maintain these ratios to a high degree since it was not the success of the strong agents that was being observed rather the relative success of the bluff strategy when agents are weak. This is the 'clamped' parameter referred to earlier and the reason why any suggestion of evolutionary stability will be conditional. In this model strengths are discrete and non-evolvable since focus is placed upon weak agent strategies. There is consequently a mild fluctuation of ratios during the experiments particularly when agents are removed from the experiment through death. Such fluctuations however, were not considered to be of great significance. Motivated by the findings on punishment evasion of Caryl(1987) and Noble(2000) we hypothesize that societies where punishment enforcers are relatively abundant (50% of society) bluffers will be less successful than in societies where punishment enforcers are relatively scarce (33% of society) but that administer a proportionately higher punishment cost.

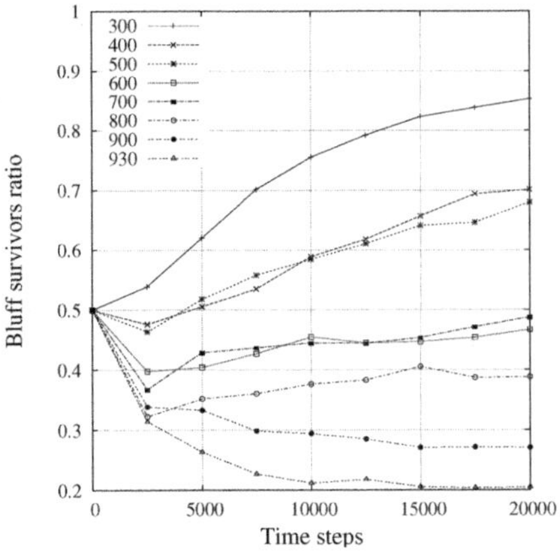

Fig. 1. Graph of mean values showing the proportion of bluff survivors out of all weak agents. Ratio of strong to weak is 1:1

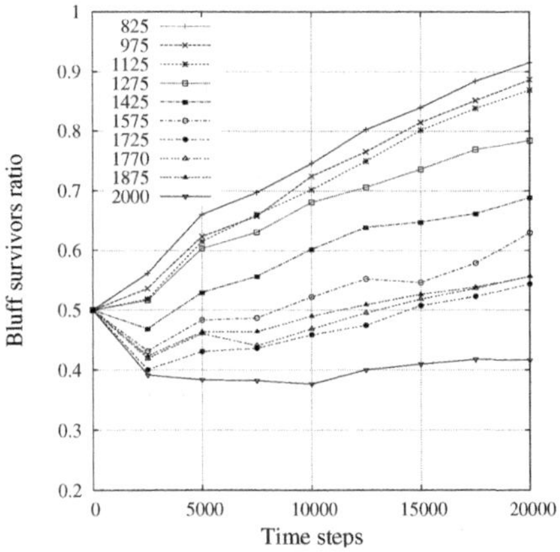

Fig. 2. Graph of mean values showing the proportion of bluff survivors out of all weak agents. Ratio of strong to weak is 1:2

4 Results and Analysis

Fig.1 and Fig.2 show the proportion of bluffer agents changing over time. All the plots represent the values averaged over 100 simulation runs. Fig.2 shows a general tendency for bluffers to have an increasingly high survival rate over time relative to honest agents while Fig.1 shows that bluffers will begin to dominate at $E(-1)$ values of less than 600. Comparing Fig.1 with Fig.2, the results indicate that bluff agents are less successful in terms of mortality rate relative to their weak honest agent counterparts when $R =$ 1:1 than when $R =$ 1:2. The difference can be seen to be greatest where the $E(-1)$ is greatest i.e. 930 compared to the converted equivalent 1770 and this difference was found to be highly statistically significant (95%) using a t-test. This is in accordance with our hypothesis. It would seem that given the relatively low likelihood of being 'caught' bluffing by a strong agent, there is a sufficiently high number of bluffers avoiding any form of punishment which are then able to proliferate at the expense of the weak honest agents. The proportionately increased punishment cost administered when $R =$ 1:2 doesn't appear to be as inhibiting to bluff agent success in this respect. Conceivably, bluff agents receiving high $E(-1)$ costs are relatively better able to reproduce and/or avoid a relatively extreme punishment when there are fewer enforcers ready to mete out this punishment.

There is a general tendency for bluff agents, and honest agents for that matter, to either dominate the set of weak agents or be dominated depending on the parameter value $E(-1)$. The higher is this value, unsurprisingly the greater is the likelihood that the conditional ESS will be honesty. Previously it had been stipulated that according to Enquist and Hurd, $E(-1)$ should not exceed 930 in these experiments in order that the bluff strategy successfully invade the honest approach. In experiments where weak to strong ratio equals one to one (Fig.2) we note that bluff strategists predominate at

$E(-1) \leq 500$. A punishment cost value greater than this leads to a notable decline in bluffer survival rate.

Interestingly in Fig.1 we note that despite the increasing punishment to bluffers for being caught, as high a bluff cost of 1725 (the converted equivalent of $E(-1) = 930$) and up to 1875 does not appear to represent an increased deterrent to bluffers. Only by 2000, the maximum punishment cost that is equivalent to certain death do we see a drop in bluffer performance where mean and median values tend to hover around 0.4 at the various time steps analysed albeit with a slight tendency to increase over time. This suggests that even the long term effects of the severest punishment here won't necessarily impede bluffers indefinitely.

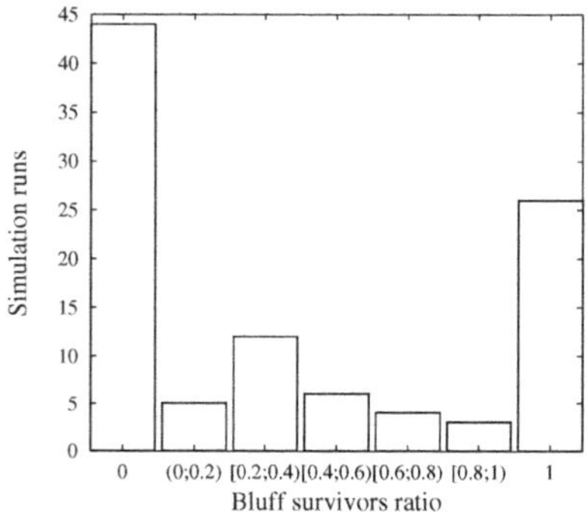

Fig. 3. Column Chart showing proportion of Bluff survivors by 20,000 time steps for strong to weak ratio 1:1

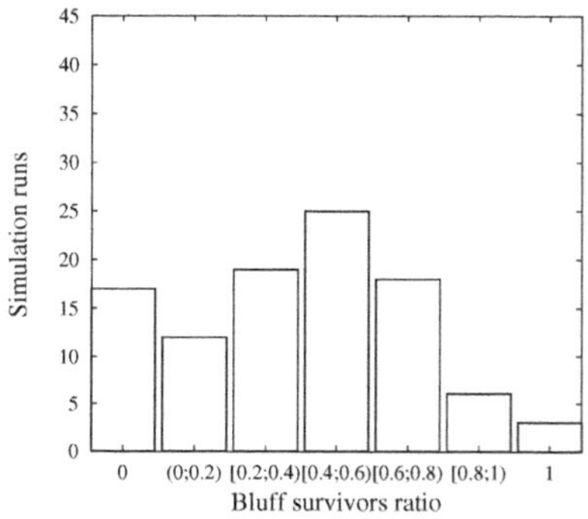

Fig. 4. Column Chart showing proportion of Bluff survivors by 20,000 time steps for strong to weak ratio 1:2

Fig.3 with Fig.4 show the distributions of proportions of bluff survivors by the final time step analysed i.e 20,000. Fig.3 shows the distribution of R = 1:1 when $E(-1) =$ 800. Fig.4 shows the distribution of R = 1:2 when $E(-1) = 2000$. The $E(-1)$ values compared in the two figures were chosen since they both had means of under 0.5 (around 0.4) and therefore bluffers were in the minority as is the empirical norm. The bluff survival rates interestingly, also showed signs of stabilizing around this point over generational time hinting at possible stable ratios of honest and bluff signalling agents. Fig.3 displays a tendency for either bluffers or weak honest agents to be the pure conditional ESS. 100% dominance of either strategy happens on 70% of the occasions. This might suggest that a high enforcer presence provokes an extreme dominance of bluffers or weak honest agents. Alternatively this may be symptomatic of a drift convergence effect whereby on occasion one strategy, by chance, might be successful to the extent that it may have many agents reproducing at the same time and this can then perpetuate. Fig.3 is fairly typical of the distributions found in societies with 1:1 weak to strong ratio particularly where proportion of bluff and weak honest survivors were similar. Fig.4 alternatively shows something approaching a normal distribution. Strategic saturation by bluffers or weak honest agents only accounts for 20% of agent survivor distributions at time step 20,000. The most common proportion of agents was found where bluff survivor rate was between 0.4 and 0.6 which is in accordance with the median at 0.409 and the mean at 0.417 and similar to the average value over all time steps for $E(-1) = 2000$ of 0.39 to 2 d.p and 0.40 to 2 d.p for median and mean values respectively. This might hint at a mixed conditional evolutionary stable ratio of bluffers to weak honest agents occurring at around 2:3. An honest society therefore may not be completely invaded by a bluffing strategy at this, slightly higher level than the 930 threshold needed to be exceeded in order to satisfy Enquist/Hurd's equation. The differences in the two sets of experiments might be explained by the fact that with a greater number per se of weak agents in the 1:2 strong to weak societies there is greater immunity to the random effects of drift convergence. Experiments with very high numbers of agents might gauge better therefore the possibility of stable ratios existing here.

5 Discussion

The prevalence of bluffing and 'bluffer strategists' existing in the wild and even in human society is undeniable. In general, the likelihood of the bluff strategy proliferating is dependent on; the frequency of its being punished, the severity of the punishment and how such average costs are perceived relative to the benefits of the bluff. If an agent bluffs in the wild, can this agent escape unharmed from a stronger opponent having attempted to fool him via an exaggerated signal or display of strength? If so what is to stop the bluff strategy from propagating throughout the agent population? If this happens then the honest signalling 'strategy' could not be stable in pure form and bluffers would begin to dominate. Ultimately though, bluffing would serve little purpose if it became too successful since attention to signals would cease. Therefore bluffers must be in the minority if signalling is to function as a communicative strategy and there must exist a genuine threat of punishment for bluffing to ensure this minority. Maynard Smith (1982) suggested that gains from bluffing have to be exceeded by the costs of such dishonesty in order for honest signalling strategies to be stabilized and indeed the experimental results found here show that different ratios of strong to weak agents suggest honest signalling to be the ESS above certain values of bluff cost.

The main finding of the experiments run was that with a comparable, expected average pay-off to disputing bluff agents in agent societies comprising a different proportion of punishment enforcers the number of enforcers has more influence than the severity of the punishment cost. Bluff agents were able to flourish when ratios of strong to weak agents were 1:2 with the equation (1) proving decisive in this case. Bluff agents receiving costs above this equation i.e at $E(-1) = 2000$ (the converted equivalent of 1163) were only partially successful at invading the honest approach. An extreme punishment i.e. death was needed therefore to prevent the bluff strategists from saturating the weak agent population when strong punish enforcers were significantly reduced. This was however, still less effective than having higher numbers of punishment enforcers.

5.1 Evaluation of Main Findings and Further Research

Enquist and Hurd proposed that bluff punishment cost must be high in order that honest signalling remain stable over evolutionary time. The equation (1) was to be adhered to in order to meet this requirement. It was found that this equation was better satisfied when proportions of strong (effective cost enforcers)/weak agent ratios were at 1:2 than when R = 1:1. Given that bluffing proved less successful at a strong to weak ratio of 1:1 it is anticipated that Enquist/Hurd's equation would be satisfied at a strong to weak ratio of around 2:3. The experimental findings suggests that bluffers are more likely to excel when the chances of them being caught for the 'crime' is reduced, even allowing for proportionately higher punishment severity.

The Enquist/Hurd pay-off matrix as a suitable paradigm for evaluating the potential for bluffers to invade agent societies of pure honest signallers can be criticized on the grounds that its adherence to discrete as opposed to continuous strengths is not ethologically realistic. It could be argued that for the purposes of simplicity and comprehension however a binary signalling paradigm is more expressive regarding advertisement of strength and related resource procuring strategies. Another possible criticism is that the simultaneous exchange of signals between agents vying for an indivisible resource is not what is found in nature. Asynchronous exchange of conventional signals is the norm and strategies regarding not only type but also amount of information are of greater relevance here. Agents might gain from at least revealing some information that would hint at a possible range of strengths to which they belong but conversely may only be willing to reveal all if they are sure they have higher strength. A ternary as opposed to a binary signalling system with asynchronous signalling of strength ranges has been implemented adapting the Enquist/Hurd pay-off matrix. Early results have been promising regarding its validity as a more empirically relevant and yet simple signalling model. An in-depth study as well as an information theoretic approach to analyse the communication is planned in the near future.

One limitation of this paper is the stated effect of the artificial enforcement of strong/weak ratios on the possibility of measuring valid evolutionary stable strategies. Long-term evolutionary mechanisms for adapting and maintaining relative proportions of 'strong' and 'weak' agents and the effect this has upon strategies in itself could be an important area of research but may be more applicable to similarly evolving, rather than static, strategies. In this paper the notion 'conditional evolutionary stable strategy' has been coined as a short-term measure of long-term dependencies. The origins and evolution of emotional expression with possible implications for antagonistic signalling strategies is a current area of research.

5.2 Concluding Comments

The implications of the findings in this study are that to ensure that honest signalling prevails there needs to be a high number of agents in the society capable of inflicting costly but not necessarily severe, instant punishment on the perpetrators of the societal 'cheats'. Bluffers in nature have been observed to excel up to a point where the pay-off has permitted or where punishment evasion has been possible. It may even be relevant to extrapolate to human societies and individuals' motivation to 'cheat the system'. In this context it is interesting to note the "Youth Survey 2002" carried out by MORI on behalf of the Youth Justice Board which discovered that school students and excluded youths interviewed in the UK rated probability of being caught as the greatest deterrent to the carrying out of a criminal and therefore dishonest activity. Mechanisms in nature that prevent cheats and bluffers from invading honest societies are selected and stabilized over evolutionary time. Such mechanisms usually involve instant punishment for attempting to 'parasitize' the honest members of the society. In general, similar instant punishments are impractical and even ethically dubious in human societies. There are indications that swiftness and consistency of punishment are however, equally important mechanisms for keeping dishonesty in check.

References

1. Adams, E.S and Mesterton-Gibbons, M (1995). The cost of threat displays and the stability of deceptive communication, *Journal of theoretical biology*, 175, 405–421.
2. Lachmann, M., Szamado, S., and Bergstrom, C. T. (2001). Cost and conflict in animal signals and human language. *Proceedings of the National Academy of Sciences of the United States of America*, 98(23):13189–13194.
3. British Crime Survey (2000), Criminal Statistics England and Wales 2000. Payback-Crime and Punishment, `http://www.payback.org.uk/candp_key-issues.html`, 05/03/03.
4. Canning and Maynard Smith (1978) cited in Dawkins R (1999), *The Selfish Gene*, 75–76, 130.
5. Caryl, P.G (1987) Acquisition of information in contests: The gulf between theory and biology. ESS Workshop on Animal Conflicts, Sheffield, UK.
6. Enquist, Magnus (1985), Communication during aggressive interaction with particular reference to variation in choice of behaviour, *Animal Behaviour (1985)*, 33, 1152–1161.
7. Enquist, Magnus and Hurd, P.L (1997) Conventional Signalling in Aggressive Interactions: the Importance of Temporal Structure, *Journal of Theoretical Biology (1998)*, 192, 197–211, Department of Zoology, University of Stockholm.
8. Gale J.S and Revd Eaves L.J (1975) cited in Dawkins R (1999), *The Selfish Gene*, 283.
9. Grafen, A. (1990). Biological Signals as handicaps, *Journal of Theoretical Biology*, 144, 517–546.
10. Hurd, P.L (1996) Is Signalling of Fighting Ability Costlier for Weaker Individuals?, *Journal of Theoretical Biology (1997)*, 184, 83–88, Department of Zoology, University of Stockholm.
11. Johnstone R. A (1998) Game Theory and communication. *Game Theory and Animal Behaviour*, pp 94–117. Oxford University Press, New York.
12. Lachmann M, Bergstrom C.T, and Szamado S(2000). The death of costly signalling? `http://octavia.zoology.washington.edu/signalling/death.html`, 27/02/03.
13. Maynard Smith and Price (1973, 1982) cited in Dawkins R (1999), *The Selfish Gene*, 69–79.
14. Noble, J (2000) Talk is cheap: Evolved strategies for communication and action in asymmetrical animal contests. In J.-A. Meyer, A. Berthoz, D. Floreano, H. Roitblat, and S. Wilson, editors, SAB00, pages 481–490. Honolulu, Hawaii: MIT Press.
15. Zahavi A (1975) Mate selection – a selection for a handicap, *Journal of Theoretical Biology*, 53, 205–214.
16. Youth Survey (2002) Summary of the MORI 2002 Youth Survey, `http://www.youth-justice-board.gov.uk/policy/Mori_exc_summary.pdf`, 05/03/03.

Effects of Group Composition and Level of Selection in the Evolution of Cooperation in Artificial Ants

Andres Perez-Uribe[1], Dario Floreano[1], and Laurent Keller[2]

[1] Autonomous Systems Lab., Swiss Federal Institute of Technology, Lausanne, Switzerland, http://asl.epfl.ch

[2] Institute of Ecology, Laboratory for Zoology, University of Lausanne, Switzerland, http://www.unil.ch/izea

Abstract. Since ants and other social insects have long generation time, it is very difficult for biologists to study the origin of complex social organization by guided evolution (a process where the evolution of a trait can be followed during experimental evolution). Here we use colonies of artificial ants implemented as small mobile robots with simple vision and communication abilities to explore these issues. In this paper, we present results concerning the role of relatedness (genetic similarity) and levels of selection (individual and colony-level selection) on the evolution of cooperation and division of labor in simulated ant colonies. In order to ensure thorough statistical analysis, the evolutionary experiments, herein reported, have been carried out using "minimalist" simulations of the collective robotics evolutionary setup. The results show that altruistic behaviors have low probability of emerging in heterogeneous colonies evolving under individual-level selection and that colonies with high genetic relatedness display better performance.

Keywords: Evolution, cooperation, division of labor, altruism, social insects.

1 Introduction

It has been estimated that ants compose about 15% of the animal biomass in most terrestrial environments, and up to one-third of the animal biomass of the Amazon rain forest consists of ants and termites [1]. Their success might come from the fact that social interactions can compensate for individual limitations, both in terms of physical and cognitive capabilities. Indeed, herds and packs allow animals to attack larger prey and increase their chances for survival and mating [2], while organizations and teams facilitate information sharing and problem solving. The complexity of any society results from the local interactions among its members. Synthesizing and analyzing coherent collective behavior from individual interactions is one of great challenges in both ethology and artificial intelligence.

W. Banzhaf et al. (Eds.): ECAL 2003, LNAI 2801, pp. 128–137, 2003.

Social insects not only exhibit highly organized collective behaviors. They provide some of the most remarkable examples of altruistic behavior with their worker caste, whose individuals forego their own reproduction to enhance reproduction of the queen. These and other examples of group harmony and cooperation have given rise to the concept that colonies are harmonious fortress-factories in which individual-level selection is muted, with the result that colony-level selection reigns. In other words, the colony often appears to behave as a "super-organism" operating as a functionally integrated unit [3]. However, the concept of a super-organism as being the only unit at which natural selection operates has been challenged both on theoretical grounds and by the observation that life within the colony is not always as harmonious as it may first appear. Social life may involve conflicts of genetic self-interest, resulting in tactics of coercion, manipulation and even deadly aggression between colony members in the name of genetic self-interest. These conflicts arise because colony members may favor individuals that are more closely related (share more genes identical by common ancestry) to maximize their inclusive fitness [4]. These conflicts, in turn, have negative effects at the colony-level because they may decrease the overall productivity. These costs at the level of the colony are expected to lead to counter-strategies to suppress selfish behaviors [5]. In other words, the actual conflict should generally be lower than the potential conflict [6]. Understanding exactly how and to what degree actual conflict is suppressed and how this increases overall group productivity is the key to understanding the extent to which social insect colonies can be viewed as adaptively organized group-level units. In our ongoing project, we intend to investigate this issues by evolving colonies of artificial ants implemented as simulated robots endowed with simple behaviors. In order to ensure thorough statistical analysis, the evolutionary experiments have been carried out using genetic algorithms coupled to fast "minimalistic" simulations of the robotic evolutionary setup.

In this paper, we describe a set of four experiments, which have been used to test the effects of genetic relatedness and levels of selection in the evolution of cooperation and labor division. Relatedness is known to have played a major role in favoring the evolution of altruism in social insects [7,8,5] and other animals and we would like to determine whether the role of relatedness can be experimentally demonstrated with artificial ants.

These explorations will provide hypotheses for the evolution of cooperation and division of labor in biological ants and guidelines for the design of autonomous robots capable of cooperation and task self-allocation, because most of the work done so far with real robots make use of pre-designed and fixed rules [9]. Accomplishing tasks with a multi-robot systems is appealing because of its analogous relationship with populations of social insects. Researchers argue that by organizing simple robots into cooperating teams, useful tasks may be accomplished otherwise impossible using a single robot [10].

In the next section we describe the model we have used to study the evolution of cooperation and the division of labor. In section 3, we describe the set of

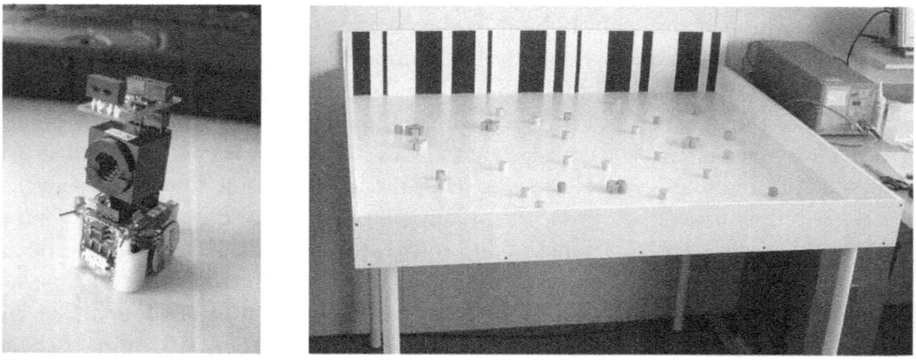

(a) (b)

Fig. 1. Experimental setup. (a) Alice robot with proximity (infrared sensors), vision and radio modules. ALICE is a 21 x 21 x 20 mm micro-robot weighing less than 10g. It has an autonomy of up to 10 hours. (b) Real setup for the foraging task (the small and large objects represent food items, the wall with black and white stripes represents the nest).

experiments we have performed. In section 4, we discuss our current results, and finally in section 5, we present our concluding remarks.

2 The Model

In theoretical biology, the study of the evolution of cooperation has been mainly undertaken using formal mathematical models and by using evolutionary game models [11]. A very well known example is the Prisoner's Dilemma game model [12] which has been particularly useful for studying the evolution of cooperation via reciprocity. The agent-based modeling approach is an alternative to the equation-based methods. These models offer the possibility to implement complex interactions between important parameters of the models which are sometimes impossible to incorporate into equation-based models, thus, the simplifications necessary to make the model tractable are less likely to bias the results [13].

2.1 Foraging Task

We use an agent-based model of a colony of artificial ants performing a foraging task. The agents or artificial ants (e.g., robots or simulated robots) are supposed to look for food items randomly scattered in a foraging area. There are two kinds of food items, small food items which can be transported by single agents to the nest, and large food items, which can only be transported if two ants cooperate. When a cooperative foraging ant happens to find a large food item, it sends a local message asking for help. Given the local nature of the help message, another cooperative individual will only be able to help the first one, provided that it happens to be close to it and hear its message. For the sake of simplicity large food items can only be transported by a pair of ants and we have not included a pheromone-like communication among ants.

In a preliminary set of experiments, we tested the technical feasibility of the experiments using an enhanced version of the ALICE micro-robot [14] including five infra-red sensors and a linear vision system (see Figure 1a). Interestingly, the ALICE micro-robot have limitations somehow similar to real ants due to their small size. First, the power available with the on-board batteries and the properties of the micro-motors limit the maximal force for pushing food items. Second, the complexity of the control software that can be implemented on the on-board micro-controllers is severely constrained by local-range sensors and relatively low computational power. Both these aspects, which can also be found in real ants, could benefit from cooperative strategies of a group of robots.

In Figure 1b) we show the real experimental setup where other members of our project tested the ALICE micro-robots and obtained robots capable of locating small objects (using infrared sensors) and pushing them toward a wall (i.e., the nest).

In the first phase of the project we use a "minimalistic" simulator to carry out a large number of experiments. The minimalistic simulator includes stochastic effects and time-dependent dynamics modeled upon the constraints of the physical setup.

In our experimental setup, each ant is endowed with a set of three 5-bit genes (g_0, g_1, and g_2) which encode three threshold values (T_0, T_1, and T_2) that are used to determine if a given basic behavior (b_0, b_1 or b_2) is activated or not during a step of a foraging trial.

Table 1.

b_0	b_1	b_2	**Description of the resulting strategy**
0	0	0	do nothing
1	0	0	if a small food item is found, bring it to the nest, ignore large food items, and do not help other ants
0	1	0	if a large food item is found, stay and ask for help, ignore small food items, and do not help other ants
0	0	1	if a help message is perceived, go and help, ignore small and large food items
1	1	0	if a small food item is found, bring it to the nest, if a large food item is found ask for help, but do not help other ants
1	0	1	if a small food item is found, bring it to the nest, help other ants, but ignore large food items
0	1	1	if a large food item is found, stay and ask for help, ignore small food items, and help other ants
1	1	1	if a small food item is found, bring it to the nest, if a large food item is found, stay and ask for help, and help other ants

By varying the energetic value of the food items, one can put more or less pressure on the advantage of cooperative behaviors. For example, if a large food item is valued ten times more than a small one, it may pay off for ants to recruit other ants and transport the large food item toward the nest area even if this means that those ants will have to share it with the whole colony.

2.2 Behavior Activation

The values of the binary variables b_0, b_1, and b_2 indicate whether a particular behavior, or a combination of behaviors has been activated or not (see Table 1). The members of a colony activate their behaviors in a random order.

The expression of a given behavior b_i depends on the number of foragers already engaged in that behavior and is mediated by an individual threshold T_i whose value is genetically encoded (i.e., by g_i). More precisely, if the proportion of members of the colony having activated a given behavior j is smaller than the corresponding threshold T_j^k of ant k, behavior b_j^k is set to '1' (i.e., it is activated). This threshold mechanism was motivated by a model of division of labor in insect societies proposed by Bonabeau and colleagues [15]. In our model, high threshold values indicate that behaviors are more likely to be activated when several individuals have already activated that behavior than low threshold values, and low threshold values indicate that behaviors have less probability of being activated when several individuals have already activated that behavior.

```
Algorithm : Evolutionary experiment()

for run ← 1 to 10
  Initialize population genotypes
  for generations ← 1 to 100
    Generate 20 random groups of 20 individuals
    for group ← 1 to 20
      for trial ← 1 to 20
        for steps ← 1 to 5
          Perform behavior activation
          Perform foraging task
      Compute individual and group performance
    Perform reproduction (crossover and mutation) and selection
```

Fig. 2. General loop of an evolutionary experiment.

2.3 Trial Description

A trial consists of a series of steps and every step is divided into two phases: in the first phase, each ant activates one or more basic behaviors (see section

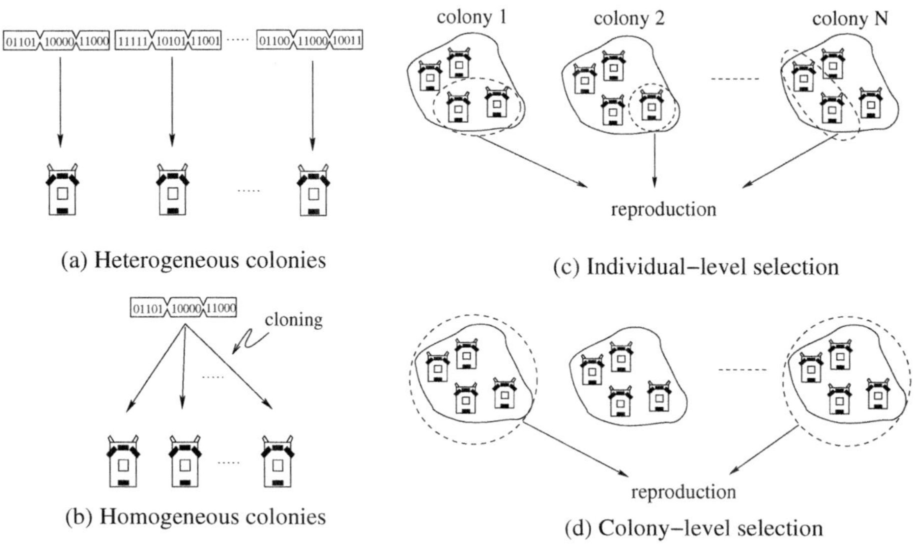

Fig. 3. Generation of (a) heterogeneous and (b) homogeneous colonies, and implementation of (c) individual and (d) colony-level selection. See Section 3 for more details.

2.2), while in the second phase, the group of 20 ants is tested on the foraging task. In the reported experiments there are 4 small and 4 large food items in the arena. To simulate the spatial distribution of the food items, an ant has a probability of $1 - NOPP$ of finding a food item. In these experiments, the no-operation parameter $NOPP$ was set to 0.2, which means the ants happen to find an object with a probability of 80% during the first step of a trial. When the number of available food items diminishes (i.e., they are "transported to the nest") the probability for an ant to find an object also diminishes. Each colony has 5 time steps to achieve the foraging task. The performance of the colonies is then measured using the average score obtained during 20 different trials. In Figure 2 we show the whole sequence of steps of an experiment. See our web site (http://asl.epfl.ch) for a more detailed description of the 'minimalist" probabilistic simulator of the foraging task.

In our experiments, the small food items provided a score of 1.0 to the single ant having transported it to the nest, while the large food items provided a total score of 16.0. However, the large food items are shared with the whole colony, thus each individual gets a score of 0.8 when a couple of cooperating ants "managed to transport" a large food item to the nest. According to these payoffs, the individuals that do not cooperate can get 0.8 points for every large food item transported by other individuals of the colony, whereas the individuals that cooperate and share food, have a cost of 0.2 points compared to the score they would made if they concentrated on the small food items. Nevertheless, the total performance of the colony is maximized if the individuals prefer to transport large food items than acting "selfishly".

3 Experiments

Four experiments have been designed to determine to what extent the level of cooperation among ants is influenced by the level of genetic relatedness (heterogeneous or homogeneous colonies) and the level of selection: individual or colony-level selection (See Figure 3). The four experiments have been carried out using a form of artificial evolution where only the "fittest" individuals or colonies are allowed to reproduce [16]. The offspring are generated by exchanging genes between the parents with a probability of 20% (crossover probability) and by adding a random value to each gene.

A population of 400 individuals is initialized by setting up their genotypes as follows: g_0 = rnd('00000', '11111'), that is, a 5-bit random string, g_1 = '00000', and g_2 = '00000'. This means that the whole population is initially composed of individuals that do not cooperate. The individuals are organized into 20 colonies of 20 ants each. Individuals of a colony can be all clones (homogeneous) or have different genes (heterogeneous). In the case of individual-level selection, individuals are selected from different colonies on the basis of their performance, and reproduced to form 20 new colonies. In the case of colony-level selection, all individuals of the colonies with the highest fitness are reproduced to form 20 new colonies.

4 Results and Discussion

In Figure 4 we show the evolution (during 100 generations) of the frequency of altruistic individuals in the simulated ant populations according to the four experimental setups, previously described. Herein, we consider an individual to be "altruistic" when it does not "pay attention" to the small food items and concentrate only on the large food items, whether looking for large food items or helping other individuals to transport large food items.

Figure 4 shows the evolution of the mean frequency of individuals in the population having activated the strategies: $[b_0 = 0, b_1 = 1, b_2 = 0]$, $[b_0 = 0, b_1 = 0, b_2 = 1]$, and $[b_0 = 0, b_1 = 1, b_2 = 1]$. Notice that the percentage of altruistic individuals within the population of heterogeneous colonies evolved using individual-level selection remains below 10%. However, in the other three setups there is a gradual dominance of altruistic individuals in the population. In particular, the resulting number of altruistic individuals is higher when using *colony-level selection* (Figure 4b and Figure 4d). This is understandable because colony-level selection favors the individuals that work for the colony and not the ones that specialize in the foraging of small food items for their own benefit.

In Figure 5 we show the evolution of the mean performance of the homogeneous and heterogeneous colonies evolved using individual and colony-level selection. The performance differences appear to be related to genetic relatedness. Indeed, the homogeneous colonies display higher mean fitness than the heterogeneous colonies after the 100-generation runs. In order to validate this, we performed 10-run evolutionary experiments during 2000 generations and measured the mean performance from generation 1000 to 2000. Then, we performed

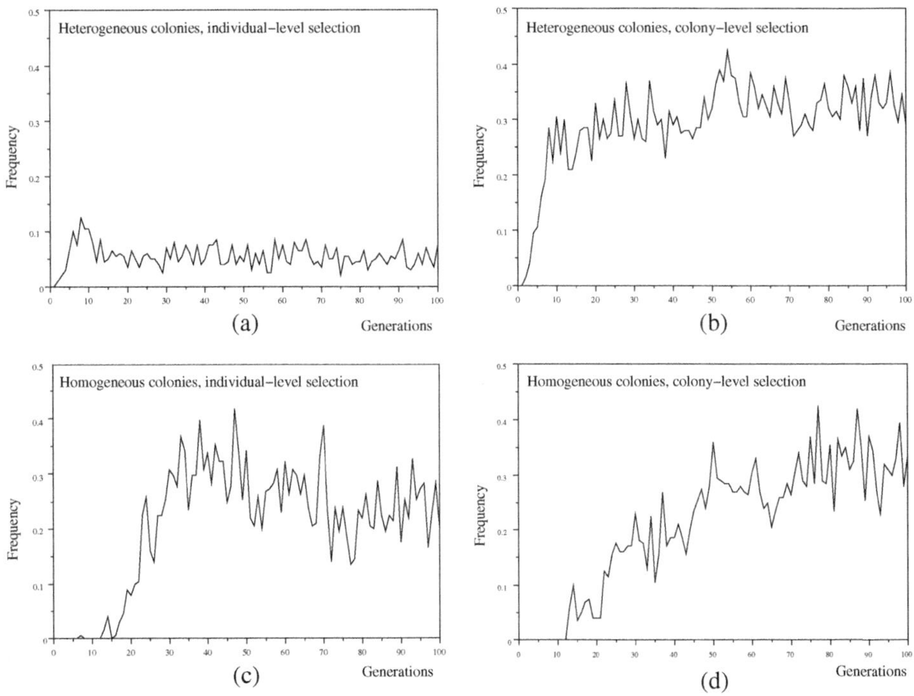

Fig. 4. Evolution of the frequency of altruistic individuals in the simulated ant populations (average of 10 runs) given the following experimental setups: (a) Heterogeneous colonies, individual-level selection, (b) Heterogeneous colonies, colony-level selection, (c) Homogeneous colonies, individual-level selection, and (d) Homogeneous colonies, colony-level selection.

a Wilcoxon test to determine if the difference between the mean performance of the colonies were "statistically significant". These results show that: (1) the mean performance of the homogeneous colonies evolved using colony-level selection P_{HmC} is higher than the mean performance of the heterogeneous colonies evolved using colony-level selection P_{HtC}, that is, $P_{HmC} > P_{HtC}$ [$p < 0.0001$]; (2) the mean performance of the homogeneous colonies evolved using individual-level selection P_{HmI} is higher than the mean performance of the heterogeneous colonies evolved using individual-level selection P_{HtI}, that is, $P_{HmI} > P_{HtI}$ [$p < 0.0001$]; (3) the mean performance of the heterogeneous colonies evolved using colony-level selection P_{HtC} is higher than the mean performance of the heterogeneous colonies evolved using individual-level selection P_{HtI}, that is, $P_{HtC} > P_{HtI}$ [$p < 0.0001$]. However, there is no significant difference between the mean performance of the homogeneous colonies evolved using colony-level selection P_{HmC} and the mean performance of the homogeneous colonies evolved using individual-level selection P_{HmI}.

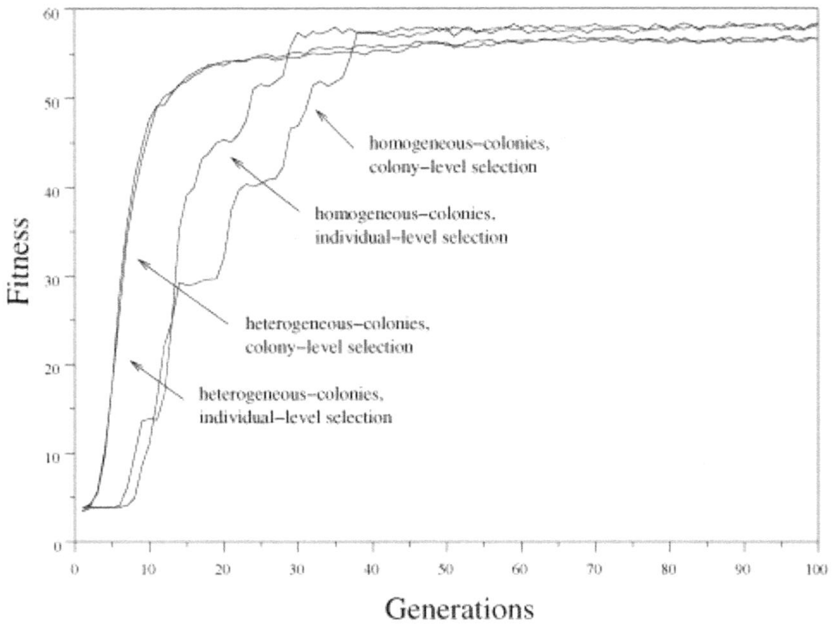

Fig. 5. Evolution of the mean performance of homogeneous and heterogeneous colonies under individual and colony-level selection (each curve is the average of 10 different runs).

5 Concluding Remarks

In this paper we present a set of experiments, where minimalistic simulations are used to study the effects of group composition and levels of selection in the evolution of cooperation in artificial ants. (1) We show that altruistic behaviors are favored by colony-level selection and that altruistic behaviors have low probability of emerging in heterogeneous colonies evolving under individual-level selection. (2) Concerning the prediction, made by some biologists [17,18], that groups should be more efficient when selection acts at the level of the colony and when there is higher relatedness within groups, we have so far found that homogeneous colonies, indeed, performed better than heterogeneous ones. However, we did not find a clear superiority in the performance of colonies evolved using colony-level selection. This may be due to the fact that individuals can simultaneously activate self-interest and altruistic behaviors. We are currently pursuing our research performing further exploration of these issues and the evolution of division of labor. Our next step is to validate these experiments with the physical robots, where interaction dynamics may significantly affect costs and benefits.

Acknowledgments. The authors gratefully acknowledge the support from the Fonds UNIL-EPFL 2002, the Rectorate of the University of Lausanne, and the

Swiss National Science Foundation. Moreover, we thank A. Colot and G. Caprari for their help with the ALICE robot, the implementation of the real arena, the development and test of the new ALICE robots, and the pictures.

References

1. Hölldobler, B., Wilson, E.: The Ants. Springer-Verlag, Berlin (1990)
2. Gadagkar, R.: Survival strategies: cooperation and conflict in animal societies. Harvard University Press (1997)
3. Wheeler, W.: The social insects: their origin and evolution. Kegan Paul, London (1928)
4. Hamilton, W.: The genetical evolution of social behavior 2. Journal of Theoretical Biology **7** (1964) 17–52
5. Keller, L., Reeve, H.: Dynamics of conflicts within insect societies. In: Levels of Selection in Evolution. Princeton University Press (1999) 153–175
6. Ratnieks, F., Reeve, H.: Conflict in single-queen hymenopteran societies: the structure of conflict and processes that reduce conflict in advanced eusocial species. Journal of Theoretical Biology **158** (1992) 33–65
7. Bourke, A., Franks, N.: Social evolution in ants. Princeton University Press (1995)
8. Sundström, L., Chapuisat, M., Keller, L.: Manipulation of sex ratios by ant workers: A test of kin selection theory. Science **274** (1996) 993–995
9. Krieger, M., Billeter, J., Keller, L.: Ant-like task allocation and recruitment in co-operative robots. Nature **406** (2000) 992–995
10. Brooks, R.: Challenges for complete creature architectures. In Meyer, J.A., Wilson, S., eds.: From animals to animats. First International Conference on Simulation of Adaptive Behavior, The MIT Press (1991) 434–443
11. Dugatkin, L., Reeve, H., eds.: Game Theory and Animal Behavior. Oxford University Press (1998)
12. Axelrod, R., Hamilton, W.: The evolution of cooperation. Science **211** (1981) 1390–1396
13. Pepper, J., Smuts, B.: The evolution of cooperation in an ecological context: an agent-based model. In Kohler, T.A., Gumerman, G.J., eds.: Dynamics in Human and Primate Societies: Agent-Based Modeling of Social and Spatial Processes, Oxford University Press (2000) 45–76
14. Caprari, G., Estier, T., Siegwart, R.: Fascination of down scaling - alice the sugar cube robot. Journal of Micro-Mechatronics **1** (2002) 177–189
15. Bonabeau, E., Theraulaz, G., Schatz, B., Deneubourg, J.L.: Response threshold model of division of labour in a ponerine ant. Proc. Roy. Soc. London B **265** (1998) 327–335
16. Nolfi, S., Floreano, D.: Evolutionary Robotics. The Biology, Intelligence, and Technology of Self-organizing Machines. The MIT Press, Cambridge, Massachusetts (1999)
17. Keller, L., Ross, K.: Selfish genes: a green beard in the red fire ant. Nature **394** (1998) 573–575
18. Keller, L., Chapuisat, M.: Cooperation among selfish individuals in insect colonies. BioScience **49** (1999) 899–909

Effects of Learning to Interact on the Evolution of Social Behavior of Agents in Continuous Predators-Prey Pursuit Problem

Ivan Tanev[1] and Katsunori Shimohara[1, 2]

[1] ATR Human Information Science Laboratories,
2-2-2 Hikaridai, "Keihanna Science City", Kyoto 619-0288, Japan
[2] Graduate School of Informatics, Kyoto University,
Yoshida-Honmachi, Sakyo-ku, Kyoto 606-8501, Japan
{i_tanev, katsu}@atr.co.jp

Abstract. We present the results of our work on the effect of learning to interact on the evolution of social behavior of agents situated in inherently cooperative environment. Using continuous predators-prey pursuit problem we verified our hypothesis that relatively complex social behavior may emerge from simple, implicit, locally defined, and therefore – robust and highly-scalable interactions between the predator agents. We argue that the ability of agents to learn to perform simple, atomic acts of implicit interaction facilitates the performance of evolution of more complex, social behavior. The empirical results show about two-fold decrease of computational effort of proposed strongly typed genetic programming (STGP), used as an algorithmic paradigm to evolve the social behavior of the agents, when STGP is combined with learning of agents to implicitly interact with each other.

1 Introduction

Over the past few years, multi-agent systems (MAS) have become more and more important in many aspects of computer science such as distributed artificial intelligence, distributed computing systems, robotics, artificial life, etc. MAS introduce the issue of collective intelligence and of the emergence of behavior through interactions between the agents. Currently, the main application areas of MAS are problem solving, simulation, collective robotics, software engineering, and construction of synthetic worlds [3]. Considering the latter application area and focusing on the autonomy of agents and the interactions that link them together [12], the following important issues can be raised: What is minimum amount of perception information needed to agents in order to perceive the world? How can agents cooperate? What are the lower bounds of communications, required for them to coordinate their actions? What approaches can be applied to automatically construct the agents' functionality, with the quality of such a design being competitive to the design handcrafted by human? These issues are of special interest, since the aim is to create MAS, which is scalable, robust, flexible, and able to adapt to changes.

W. Banzhaf et al. (Eds.): ECAL 2003, LNAI 2801, pp. 138–145, 2003.

Within this context, the *objective* of our research is an automatic design of autonomous agents which situated in inherently cooperative environment are capable of accomplishing complex tasks through interaction. We adhere to the methodological holism based on the belief that any complex system or society [10][11], and multi-agent society in particular [7] is more than the sum of its individual entities that compose it. From such a holistic viewpoint, we investigate the following aspects:

- The way in which the social behavior, needed to accomplish the complex task might emerge in MAS from relatively simply defined interactions between the agents. We are interested in the ultimate case of such simplicity, which we regard as Occam's razor in interactions – local, implicit, proximity defined, and therefore, robust, flexible and highly scalable interactions, and
- The effect of the ability of entities in MAS to learn simplest actions on the efficiency of evolution of more complex, social behavior.

We consider the first of the abovementioned issues as beyond the scope of this document. The interested reader is suggested to refer to [13] for details related to it. Instead, focusing on the second aspects, we intend to highlight the issues of applying genetic programming paradigm for evolving the social behavior of agents and investigating the effects on ontogenetic learning on the performance of such evolution.

The remaining of the document is organized as follows. Section 2 introduces the task, which we use to test our hypotheses. Same section discusses the architecture of the agents and briefly elaborates on the strongly typed genetic programming (STGP), developed as an algorithmic paradigm to evolve the functionality of agents. The proposed mechanism of learning to interact is introduced in Section 3. The same section presents empirical results of speedup effects of the learning on the performance characteristics of STGP. The conclusion is drawn in Section 4.

2 Background

2.1 Instance of Predator Prey Pursuit Problem

The general, well-defined and well-studied yet difficult to solve predator-prey pursuit problem [1] is used to verify our hypothesis that ability of entities to learn simple actions in MAS might facilitate the evolution of relatively complex social behavior. The evolved social, surrounding behavior emerges from simple, local, implicit, proximity-defined, and therefore – robust and highly scalable interactions between the predator agents [13]. The problem comprises four predator agents whose goals are to capture a prey by surrounding it on all sides in a world. In our work we consider an instance of the problem, which is more realistic and probably more difficult for predators than in commonly considered in the previous work [5][6]. The world is a simulated two-dimensional continuous torus. The moving abilities of four predator agents are continuous. We introduce proximity perception model for predator agents in that they can see the prey and only the closest predator agent, and only when they are within the limited range of visibility of their simulated sensors. The prey employs random wandering if there is no predator in sight and a priori handcrafted optimal escaping strategy as soon as predator(s) become "visible". The maximum speed of prey is higher than the maximum speed of predators. In order to allow predators to stalk and collectively approach the prey the range of visibility of predators is more than the range of visibility of the prey. We consider this case in order to create inherently cooperative environment in that the mission of predators is nearly impossible unless they collaborate with each other.

2.2 Architecture of the Agents

We adopted the subsumption architecture of the agents [2] comprising of functional modules distributed in three levels, corresponding to the three different aspects ("levels of competence") of agent's behavior: wandering/exploring, greedy chasing and surrounding (social behavior). We focus our attention on the issues of evolving the top-level, highest priority module – surrounding the prey, assuming that wandering/exploring and greedy chase are a priori handcrafted and incorporated into the agent's controller. As documented in [13] such an approach allows to simultaneously evolve (i) the capability of agents to resolve social dilemmas, determined by the way social behavior overrides greedy chasing when prey is in sight, and (ii) the capability to resolve the exploration-exploitation dilemma, determined by the ability of surrounding to override wandering/exploring when prey is invisible.

2.3 Algorithmic Paradigm: Strongly-Typed Genetic Programming

Limiting the Search Space of Genetic Programming. We consider a set of stimulus-response rules as a natural way to model the reactive behavior of predator agents [7], which in general can be evolved using artificial neural networks, genetic algorithms, and genetic programming (GP). GP is a domain-independent problem solving approach in which a population of computer programs (individuals) is evolved to solve problems [8]. The simulated evolution in GP is based on the Darwinian principle of reproduction and survival of the fittest. Figure 1 illustrates sample stimulus-response rule, expressing a reactive behavior of turning to the relative bearing of the peer agent (`Peer_a`) plus `10` (e.g. degrees) as a result of stimulus of its own speed being less than `20` (e.g. mm/s).

```
IF (Speed<20) THEN Turn(Peer_a+10)
```

Fig. 1. Sample stimulus-response rule

The strength of GP to automatically evolve a set of stimulus-response rules featuring arbitrary complexity without the need to a priori specify the extent of such complexity might imply an enormous computational effort caused by the need to discover a huge search space while looking for the potential solution to the problem. Agreeing with [11] that for huge and multidimensional search spaces the introduction of "pruning algorithms" is a critical step towards efficient solution search, we impose a restriction on the syntax of evolved genetic programs based on some a priori known semantics [13]. The approach is known as strongly typed genetic programming [9]. The grammar of proposed strongly typed genetic programming (STGP) establishes generic data types of visible angle, distance, speed, and Boolean with the corresponding allowed ranges of the values for their respective instances (variables and ephemeral constants). The allowed ranges of these generic types correspond to the physically reasonable limits of perceptions and moving abilities of agents. In addition, the grammar of STGP explicitly stipulates the data type of the results of arithmetical and logical expressions, and the allowed data type of operands (perception variables and ephemeral constants) involved in these expressions. Since the proposed approach is not based on domain-specific knowledge, STGP cannot be regarded as a "stronger" approach compromising the domain-neutrality of the very GP paradigm itself. The limitations imposed to the syntax of genetic programs are based (i) on the natural presumption that the predator agents are aware of the physically reasonable limits of their perception and moving abilities; and (ii) on the common rule in strongly-typed algorithmic languages that all the operands in addition, subtraction and comparison operations should have the same data types.

Main Attributes of STGP. The function set of STGP [13] comprises IF-THEN statement, arithmetical operations and comparison operators. Terminal set features local, proximity defined perception variables, continuous moving abilities and ephemeral constants. The representation of genetic programs is based on widely adopted document object model (DOM) and extensible markup language (XML): genetic programs are represented as a DOM-parsing trees featuring corresponding flat XML text. Both the genetic operations and the evaluation of individuals are performed on their respective DOM-parsing trees using off-the shelf, platform- and language neutral DOM-parsers, and XML-text representation is employed as a flat format, feasible for migration of genetic programs among the computational nodes in eventual distributed implementation of STGP. Selection mechanism is binary tournament selection. Crossover operation is defined in a strongly typed way in that only the nodes (and corresponding subtrees) of the same data type can be swapped. The sub-tree mutation is also allowed in strongly typed way in that a random node in genetic program is replaced by syntactically correct sub-tree. Breeding strategy is homogeneous: the performance of single genetic program, cloned to all the agents is evaluated. We consider such a strategy as adequate to the symmetrical nature of the world which populated with identical predator agents is unlikely to promote any behavioral specialization among them. The fitness of the genetic program is evaluated as average of the fitness measured over 10 different, randomly created initial situations. The fitness measured during the trial starting with particular initial situation is evaluated as a length of the radius vector of the derived agents' behavior in the virtual energy-distance-time space accounting for (i) the average energy loss of the agents during the trial, (ii) the average distance of the agents to the prey by the end of the trial, and (iii) the elapsed time of the trial. The energy loss estimation takes into account both the basal metabolic rate of the agents and the energy loss for moving activities. Smaller values of fitness function correspond to better performing predator agents.

3 The Effect of Ability to Learn to Interact on the Evolution of Social Behavior of Predator Agents

3.1 Ability to Learn to Interact

In the proposed representation of genetic programs the same data structures (DOM parsing trees) are maintained during both genetic manipulations and evaluation of the behavior of the agents, and consequently, no gene expression mechanism is employed. However explicitly referring to the genetic representation of agents accessed during genetic manipulations and evaluation as genotype and phenotype respectively, we view the process of learning as a modification to the phenotype yielding improved performance of the agents. Motivated by the intention to mimic the biologically plausible examples of learning, we consider a pure Darwinian phenotype plasticity (a case of Baldwinian evolution) which (in contrast to, for example, Lamarckian learning) implies that the ontogenetic changes to the phenotype are not inherited through the germ line. We are interested in the indirect influence of ability of agents to learn simple actions on the performance characteristics of evolution of more complex, social behavior in MAS. We consider the following aspects of ontogenetic learning via phenotype plasticity: (i) the objective of learning (i.e. learning task), (ii) the way of selecting the phenotypic fragment to be modified, and (iii) the modification algorithm.

Adhering to the heuristic view that ability of agents to interact with each other might contribute to more quickly attain well-performing social behavior, we define the objective of learning as simple increase of amount of atomic actions related to interactions between the agents. We regard referring to relative position of the peer agent (i.e. accessing corresponding sensory variables) as an implicit interaction between the agents. The overall amount of references of the sensory variables related to the peer agents are counted during the trial and the difference between the values of the counter for both after- and prior modification of the phenotype is considered as an estimation of the ability to achieve the learning objective. The overall fitness is calculated as a combination of the fitness, obtained from the (i) behavior of phenotype of agents before learning and the (ii) degree of achieving the learning objective. The learning algorithm, incorporated into the fitness evaluation routine is shown in Figure 2. Function `dF` (Figure 2, Line 19) maps the potentially huge difference in amount of interactions in both of considered cases into desirable, limited range of values used to alter the originally obtained fitness value. The implementation of `dF` is based on natural logarithmic function `ln`, the nonlinearity of which provides the desired sensitivity for small differences of amount of interactions while limiting resulting value when these differences are relatively big.

Three different ways of selecting the modified fragments in the phenotype have been implemented: (i) modification of random tree node (RM), (ii) modification of tree node selected from the set of nodes that correspond to the most recently introduced (via mutation) nodes in genotypic representation of the agent (MRM), and (iii) modification of tree node selected from the set of nodes that correspond to the least active nodes in genotypic representation of the agent (LAM). The case (ii) is based on the empirical observation that new structures in genotype, introduced by random mutations most likely yield a inferior phenotypic structures, while (iii) relies on the hypothesis that the amount of interactions can be increased by altering the less active, or ultimately, the inactive fragments in phenotypic representations of the agents (caused by introns, e.g. mutually exclusive sequences of stimulus response rules).

The algorithm of modification of selected phenotypic fragment is subtree random mutation. Both the selection and modification mechanism are implemented in routine `Modify(OriginalGP)` shown in Figure 2, Line 14.

```
 1. Procedure EvalWithLearning(      OriginalGP: TypeGP;
 2.                                   LearningCycles,
 3.                             [out] Fitness,
 4.                             [out] SuccessfulSituations:integer);
 5. var
 6.   GP: TypeGP;
 7.   i, OriginalAmountOfInteractions, AmountOfInteractions: integer;
 8. begin
 9.   Clone_GP_To_All_Agents(OriginalGP);
10.   Evaluate_Behavior([out] Fitness,
11.                     [out] SuccessfulSituations,
12.                     [out] OriginalAmountOfInteractions);
13.   for i:=0 to LearningCycles-1 do begin
14.     GP := Modify(OriginalGP);
15.     Clone_GP_To_All_Agents(GP);
16.     Evaluate_Behavior([out] AmountOfInteractions);
17.     if AmountOfInteractions > OriginalAmountOfInteractions
18.       then Fitness :=   Fitness
19.                   - dF(AmountOfInteractions-OriginalAmountOfInteractions);
20.   end;
21. end;
```

Fig. 2. The learning algorithm. Line numbers and output specifiers of function parameters are explicitly given for better readability

3.2 Empirical Results

Parameter Values of STGP. The parameters of STGP used in our experiments are as follows: the population size is 400 genetic programs, the selection ratio is 0.1, including 0.01 elitism, and the mutation ratio is 0.02, equally divided between sub-tree mutation and transposition. The termination criterion is defined as fitness of the best genetic program being less than 300. This value corresponds to the successful team of predator agents, which captures the prey in all 10 initial situations, and the average time of capture is around the middle of trial interval of 600 steps. A single learning iteration is performed.

The Effect of Learning on Computational Effort of STGP. Computational effort (the amount of individuals needed to be evaluated in order to obtain the solution with specified probability, e.g. 0.95) for four cases is obtained from the probability of success p(t) by each of 20 independent runs in a way as suggested in [8]. The four cases correspond to the (i) evolution without learning, evolution influenced by learning via (ii) RM, (iii) MRM, and (iv) LAM respectively. The results, shown in Figure 3 indicate p(t)=0.95 for about 32,000 evaluated individuals for evolution without learning. The result verifies the very concept that relatively complex social behavior is evolvable and it emerges from simple, implicit, locally defined, and therefore – robust and highly scalable interactions between the predator agents. The best computational effort of about 16,000 individuals (including the computational cost of learning – the amount of individuals, additionally evaluated during the learning cycle as shown in Figure 2, Line 16), achieved in case of evolution influenced by learning via LAM (denoted as E+LAM), indicating about two-fold improvement of computational effort of STGP. Figure 4 illustrates the effect of learning to interact on the performance of evolution of social behavior. The individual, initially situated at position (1) on the fitness landscape, during the learning climbs the learning landscape up to position (2). Without accounting for interactions, the fitness of individual after learning might be better or worse (as shown in Figure 4) than fitness before learning (position (1)). However, the ability to climb the learning landscape (which, due to presumed dynamic correlation of the learning landscape [4] with the probability of being closer to the optimal solution is associated with advance towards the proximity of the optimality (3)), is rewarded with fitness adjustment giving competitive advantage for the survival of the individual.

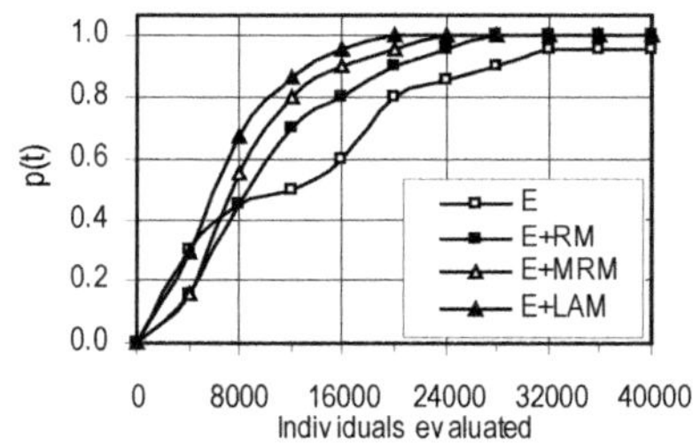

Fig. 3. Probability of success of evolving social behavior of the agents

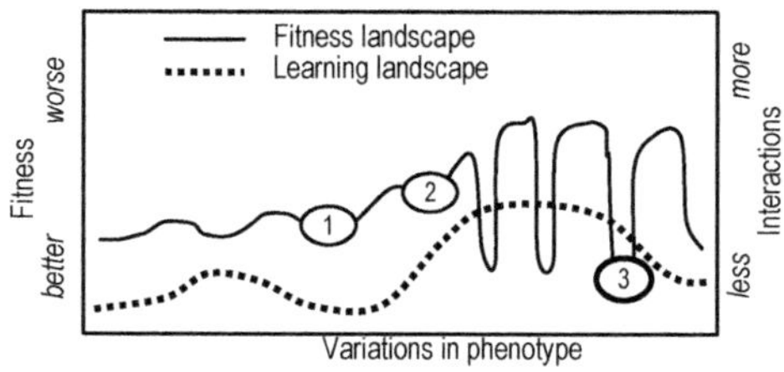

Fig. 4. Fitness and learning landscapes. Individual, initially situated at position (1) on the fitness landscape, during learning climbs the learning landscape (position (2)), which presumably correlates with advancing towards the areas of optimal solutions (3).

4 Conclusion

We presented the result of our work on the effect of learning to interact on the evolution of social behavior of agents situated in inherently cooperative environment. Using continuous predators-prey pursuit problem we verified our hypothesis that relatively complex social behavior may emerge from simple, implicit, locally defined, and therefore – robust and highly scalable interactions between the predator agents. We argued that the ability of agents to learn to perform simple, atomic acts of implicit interaction with each other facilitates the performance of evolution of more complex, social behavior. The empirical results show about two-fold improvement of computational effort of proposed strongly typed genetic programming (STGP), used as an algorithmic paradigm to evolve the social behavior of the predator agents, when STGP is embeds ontogenetic learning to implicitly interact with each other.

Acknowledgements. This research was conducted as part of "Research on Human Communication" with funding from the Telecommunications Advancement Organization of Japan.

References

1. Benda, M., Jagannathan, B., Dodhiawala, R.: On Optimal Cooperation of Knowledge Sources. Technical Report BCS-G2010-28, Boeing AI Center, Boeing Computer Services, Bellevue, WA (1986)
2. Brooks, R.A.: A Robust Layered Control System for a Mobile Robot. IEEE Journal of Robotics and Automation Vol.2(1) (1986) 14–23
3. Ferber, J.: Multi-Agent Systems: An Introduction to Distributed Artificial Intelligence, Harlow: Addison Wesley Longman (1999)
4. Floreano, D., Nolfi, S., Mondada, F.: Co-evolution and ontogenetic change in competing robots,In J.P.Mukesh, V.Honavar & K. Balakrishan (Eds.), Advances in Evolutionary Synthesis of Neural Networks. Cambridge, MA: MIT Press (2001) 273–306.
5. Haynes, T., Sen, S.: Evolving Behavioral Strategies in Predators and Prey, in Gerard Weiss and Sandip Sen, eds., Adaptation and Learning in Multi-Agent Systems (1996) 113–126
6. Haynes, T., Wainwright, R., Sen, S., Schoenefeld, D.: Strongly Typed Genetic Programming in Evolving Cooperation Strategies, in L. Eshelman, ed., Genetic Algorithms: Proceedings of the Sixth International Conference (ICGA95), M. Kaufmann (1995) 271–278
7. Holand, J.H.: Emergence: From Chaos to Order, Cambridge, Perseus Books (1999)
8. Koza, J.R.: Genetic Programming: On the Programming of Computers by Means of Natural Selection, Cambridge, MA, MIT Press (1992)
9. Montana, D.: Strongly Typed Genetic Programming, Evolutionary Computation, Vol.3, No.2 (1995) 199–230
10. Morgan, C.: Emergent Evolution, New York (1923)
11. Morowitz, H.,J.: The emergence of Everything: How the World Became Complex, Oxford University Press, New York (2002)
12. Parunak, H. Van D., Brueckner, S., Fleischer, M., Odell, J.: Co-X: Defining what Agents Do Together, Proceedings of the AAMAS 2002 Workshop on Teamwork and Coalition Formation, Onn Shehory, Thomas R. Ioerger, Julita Vassileva, John Yen, eds., Bologna, Italy (2002)
13. Tanev, I., and Shimohara, K., On Role of Implicit Interaction and Explicit Communications in Emergence of Social Behavior in Continuous Predators-prey Pursuit Problem, Proceedings of the Genetic and Evolutionary Computation Conference (GECCO-2003), Chicago, IL, USA (2003) (to appear)

War and Peace among Artificial Nations – A Model and Simulation Based on a Two-Layered Multi-agent System

Tatsuo Unemi[1], Yoshiaki Kaneko[1], and Ichiro Takahashi[2]

[1] Department of Information Systems Science,
[2] Department of Economics,
Soka University, Tangi-machi, Hachiōji, Tokyo 192-8577, JAPAN,
unemi@t.soka.ac.jp, ykaneko@intlab.soka.ac.jp, itak@soka.ac.jp

Abstract. This paper attempts to model the dynamics of war and peace in a virtual world to deepen our understanding why wars continue arising in spite of many peace keeping efforts. Land is modelled as a toroidal plain divided into square grid. Each nation consists of several villages that own a limited number of adjoining tiles of grids. A nation sometimes invades an adjacent village of another nation and occupies it. Inhabitants in each village consists of farmers, officers, and soldiers. Village can enlarge their territory by developing an empty tile, and can be independent from the nation they currently belong to. Each nation and village have their own *meme* for decision making that mutates when their successors inherit the strategies. In order to investigate what occurs in the virtual world we built a simulation system with a GUI that facilitates parameter set-up. The results displayed a typical phenomenon: repeated rise and fall of big nations. With other parameter settings, we also observed a period of equilibrium by a number of nations followed by a period of domination by one large ruling nation.

1 Introduction

War is barbarous and inhuman. Nothing is more cruel, nothing more tragic. ... *Daisaku Ikeda*

War should be avoided in any situation. It destroys many things we wish to protect and keep. One of the naive questions asked is why we cannot prevent the foolish activities of war. Many persons have approached this issue from various standpoints such as social movement, religious organisation, statement propagation, international negotiation, contract and law, *et al.* The motivation behind building of a computational model of international relationship based on a multi-agent system, was to contribute in solving this serious problem and in understanding why the problem is difficult to solve. Recent researches on Artificial Life [1], Simulating Societies [2] and Artificial Societies [3] including authors' researches [4] on evolutionary ecology give important clues to our study. The current theory of evolution seems to deem this problem as insolvable since people

W. Banzhaf et al. (Eds.): ECAL 2003, LNAI 2801, pp. 146–153, 2003.

who employ peaceful strategies always suffers the domination of an aggressive nation with strong military power.

The following sections describe the model, some scenarios of simulation, suggestions for further studies, and concluding remarks.

2 Simplified World Model

Our simple model consists of a world involving two layers of organisations, *villages* and *nations*. We assumed a simple economy of production, storage, redistribution, and consumption of food, but we did not consider any other factors such as money, trading, technology development, and so on.

Land is fixed as the size of the toroidal surface divided into square grid. Each *tile* grid is either occupied by a village or is empty. Each village occupies one or more adjoining tiles. Each nation consists of one or more adjoining villages. More details are as follows.

2.1 Villages

Village is the unit of habitation in this model which includes its own population, land and food stock.

Population of each village increases with the growth rate, $\rho - 1$ every step if sufficient amount of food is available. Specifically, the population size of village i at step t is

$$P_{i,t} = \min(\lambda_p \rho P_{i,t-1}, F_a/\zeta), \tag{1}$$

where F_a is the amount of food available, ζ is the amount of food that is needed by one person per step, and λ_p is a random real number of fluctuation. People living in the villages have a choice of working as: a farmer, an officer, or a soldier. The farmers produce food, the officers work in public services such as food delivery, and soldiers fight against foreign nations according to decisions made by the government. The government of a nation decides the proportion of each type of worker as will be described later. Some portion of soldiers are killed in battle when the nation invades or is invaded by another. A village is ruined when its inhabitants disappear because of war or starvation.

The amount of food produced at a village in one step is given by

$$F_p = \lambda_f \min(\alpha P_f, \beta L), \tag{2}$$

where α is food production per farmer, P_f is the number of farmers, β is food production per tile of land, and L is the size of village in unit of tiles. λ_f denotes a random real number of fluctuation. A village stores the food produced for people in the next step. A fixed portion γ of food stock decays. Thus, the amount of food stock in village i at step t is

$$S_{i,t} = (1 - \gamma) S_{i,t-1} - \zeta P_{i,t-1} + F_{p,i,t} + F_{d,i,t} \tag{3}$$

where the second term in the right hand side is the amount consumed by the people, and F_d is the net amount of food delivered by national government as will be described later. The available food F_a in equation (1) is $\theta_a S$ where θ_a is a real value in the village *meme* of which range is $(0, 1]$. If $\theta_a = 0$, nobody can live in the village. If $\theta_a = 1$, the inhabitants soon disappear due to starvation because they consume the whole amount of food at once.

Each village attempts to expand its territory by developing an adjacent empty tile if the condition $\alpha P_f / \beta L > \theta_d$ is satisfied, where α, P_f, β and L are as used in equation (2), and θ_d is a threshold value in the village *meme*. The above equation implies that a village tries to expand itself when it has too many farmers, considering the size of its territory. When there is no empty tile around it, the trial has failed. The maximum number of tiles for one village was limited because a village was considered to be a unit of community where people can manage his or her resources effectively. Here, we denote the maximum number as $L_{\max}$. When the size L becomes greater than $L_{\max}$ after an expansion, a village is split into two of half sizes. Both of them inherit the meme from the original village, but one of the offspring is randomly chosen to be mutated.

2.2 Nations

Nation manages the food redistribution, and determines labour allocation into three types of workers. It is also the unit of military action, but its features are described later since it concerns the relationship between villages and nations.

Since each village is exposed to an independent random shock in food production. A nation can help stabilise food supply for villagers by delivering food from villages with good harvest to ones with bad harvest. The proportion of delivery from a village in nation k is

$$p_{d,k} = \phi_k \min(F_{r,k}^- / F_{r,k}^+, 1) \tag{4}$$

where ϕ_k is a real number obtained from the *meme* of the nation k with range $(0, 1]$, $F_{r,k}^+$ is the sum of surplus food over the nation k, and $F_{r,k}^-$ is the sum of deficiency. The amount of delivery F_d in equation (3) is

$$F_d = \begin{cases} p_{d,k} f_{r,i} & \text{if } f_{r,i} > 0 \\ p_{d,k} (F_{r,k}^+ / F_{r,k}^-) f_{r,i} & \text{otherwise} \end{cases} \tag{5}$$

where $f_{r,i}$ is the surplus in village i, that is, $f_{r,i} = S_i - \zeta P_i$. If there is not enough surplus to cover the deficiency, the villages with surplus provide ϕ_k times of the whole of surplus, and the government delivers them to poor villages according to their deficiency. Rich villages are forced to behave altruistically, but this form of alliance is beneficial for them as well if ever they fall into deficiency.

A nation determines the allocation of labour into the three types of workers and the same decided-proportion is applied to all the villages it rules. Each nation employs officers and soldiers from villages. The total number of officers of a nation k is $P_{o,k} = aP_k + b$, where a and b are constant values and P_k is the population size of the nation k. The number of soldiers determines the military power of the nation. In this model, each national government prefers keeping the number of soldiers as follows.

$$\hat{P}_{s,k} = \frac{w_1 \sigma_1 \max_{j \in N_k} P_{s,j} + w_2 \sigma_2 \min_{j \in N_k} P_{s,j} + w_3 \sigma_3 \bar{P}_{s,k}}{w_1 + w_2 + w_3} \quad (6)$$

$$\bar{P}_{s,k} = P_k - P_{o,k} - \frac{\zeta P_k - S_k}{\alpha} \quad (7)$$

where N_k is the set of neighbour nations of k, $w.$ and $\sigma.$ are real values in the nation *meme*. Three factors are considered in deciding on the scale of military power: defence against its strongest neighbour, invasion of its weakest neighbour, and capacity of support. The terms from left to right of the numerator in equation (6) corresponds to these factors accordingly. σ.s are biases and w.s are weights. Their ranges are $\sigma_1 \in [0, 1.5], \sigma_2 \in [0, 2.5], \sigma_3 \in [0, 1]$, and $w. \in [0, 1]$. In each step, national government employs officers from the villages, aims at keeping the ideal number of soldiers, and then allocates the rest of the population as farmers. If $\hat{P}_{s,k} > P_k - P_{o,k}$, that is, if it fails to keep the target number of soldiers, all of the people other than officers are employed as the soldiers, and the nation has no farmers.

2.3 Independence and Move

A village can choose either to become independent or change the nation it belongs to. The index of *richness*, defined as the amount of food stock per inhabitant $R_k = S_k/P_k$, is assumed to be used as the criterion in deciding which nation is better to belong to. In each step, each village computes the richness indices of its adjacent neighbouring nations and compares them with the nation it currently belongs to. Taking the expected richness after independence into consideration, it decides the choice of its association state. It chooses independence or moves to an adjacent nation if it can improve the current situation. This action taken by a village will not be interfered by the nations unless the village is under a target of invasion. The meme of the newly independent nation is a mutant of the original one.

2.4 Invasion and Occupation

When the nation has military power, it tries with probability ψ_k to find a victim from adjacent villages beyond the national border. ψ_k is a real number in $[0, 1]$ in the nation *meme*. The nation invades the richest village of a weak nation to occupy it. The criteria of weakness of victim nation is defined as $P_{s,v}/P_{s,k} < \mu_k$, where $P_{s,v}$ is the number of soldiers of victim nation, $P_{s,k}$ is the number of soldiers of invading nation, and μ_k is a real number in the range $[0.2, 0.8]$ of the nation *meme*. The winning probability for the offending nation is

$$P(o) = x^2/(1 + x^2), \quad (x = \nu P_{s,k}/P_{s,v}), \quad (8)$$

where ν is a constant in the range $(0, 1]$ representing how advantageous an invader is. The above equation guarantees that the more soldiers are provided, the more there is of the probability to win. Moreover, the side defending the target village is more likely to succeed when both sides have the same scale of military power. After a battle, some portion of soldiers are killed. In this model, the percentage of war casualties in each nation is calculated independently as the probability distribution of triangle shape, where its peak is achieved at the centre value of the number of soldiers of a weaker nation.

2.5 Geographical Constraint

Because we assume all the villages belonging to a nation adjoin to each other, a nation can invade only an adjacent village, and a village can choose to belong only to the adjacent nations. This geographical constraint also implies that a nation is sometimes separated into multiple nations after a village becomes independent, moves to another nation, or falls into ruins. All of these separated nations inherit the same meme from the original nation, but they mutate independently from each other after separation.

2.6 Inheritance and Mutation of Meme

As described above, some parameters for decision making are encoded in the memes in villages and nations. The values of all of these parameters are real numbers, though their ranges are different. In the current implementation, each parameter is encoded by eight bits byte representing an unsigned integer, which is linearly transformed onto the specific range. The mutation is done by adding a random integer with Gaussian distribution $N(0, \sigma_m)$ to the original byte and adjusting the value range from 0 to 255. It is forced to be 0 or 255 if the value is less than 0 or greater than 255, respectively. In addition to the split cases of independence and separation, the memes of any villages and nations mutate at an interval of various steps. The number of steps in one interval is also encoded on the meme. In this model, the interval could be interpreted as the duration of a King's reign or President's term.

3 Some Scenarios

We examined several combinations of parameter settings and initial conditions in the simulation of a numerous number of steps and trials, under 256×256 tiles torus world. Starting from 256 randomly arranged nations, each of which owns four villages of ten tiles, we observed the following two types of typical scenarios.

3.1 Rise, Division, Then Ruin

One typical scenario was repetition of rise, division, and then ruin of relatively big nations. Figure 1 shows a typical process of division and ruin. The parameter

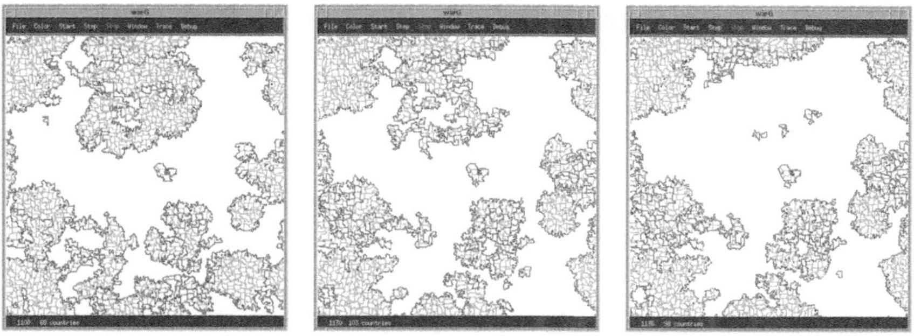

Step 1100 Step 1130 Step 1135

Fig. 1. A typical process of division and ruin of a big nation. The small area surrounded by grey border is a village, and the large area surrounded by black border is a nation. A relatively big nation in Step 1100 at upper middle position was separated into small nations before Step 1130, and then fell in ruin in Step 1135.

settings are $\rho = 1.1, \alpha = 2, \beta = 100, \gamma = 0.05, \zeta = 1, L_{\max} = 50\text{tiles}, P_{o,k} = 0.2P_k + 70, \nu = 0.5, \sigma_m = 25.5$, and $\lambda_p, \lambda_f = N(1, 0.05)$.

Once the government makes an occasional mistake because of mutation or neighbour nations' change, some relatively rich villages becomes independent from the nation and/or some poor villages disappear. This type of event causes the separation of the nation into a number of smaller nations. In this model, it is difficult for a small nation adjacent to a big one to keep being independent because it has to have military power strong enough to defend against the adjacent big nation. Inevitably, a small nation is invaded and occupied by another big nation, or falls under starvation because it allocated too many of its inhabitants as soldiers, sacrificing food production.

3.2 Equilibrium and Domination

Making the scale of mutation smaller, we can observe a more stable domination of land by a number of nations. If there were no mutations and no fluctuations, the world would be consisting of a number of stable nations and would never change. Figure 2 shows an example pattern of the case of $\sigma_m = 12$. The left figure is a type of equilibrium. But this situation does not remain the same for a long time because of small but nonetheless existing mutation. The right side of the figure shows a pattern of the later step. The nations in the previous world falls into ruin and one super power dominates almost the entire world.

The left plots of figure 3 shows how the number of wars in each step changes over time in the case of Figure 2. Between approximately 100th and 500th steps, a large number of small nations aggressively fight against each other. After this "war age," the world shifts to relatively stable state of equilibrium with a number of rivalling nations, in which wars are relatively suppressed from 600th to 1300th step. But this situation breaks up before 1400th step.

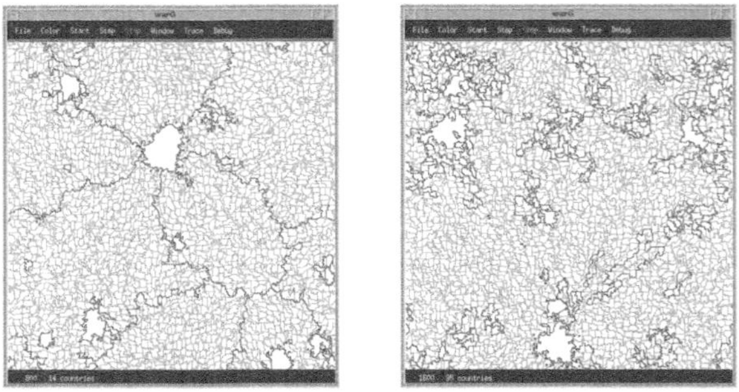

Step 800 Step 1600

Fig. 2. Patterns of equilibrium and domination under the lower rate of mutation.

In terms of the frequencies of war, it seems better to be in equilibrium than under one nation's domination. But the right plots of figure 3 suggests that it is not true from another point of view. It shows the proportion of war casualties for population in each step. The scale of war casualties during the equilibrium is not smaller than those during the earlier "war age." It once shrinks at around 1100th step, but literally exploded at 1300th step, when more than 22% of people were killed by war. We should notice that the number of casualties becomes very small after this massacre. This low casualties is resulted from the small portion of soldiers in the dominant nation facing practically no rivalry.

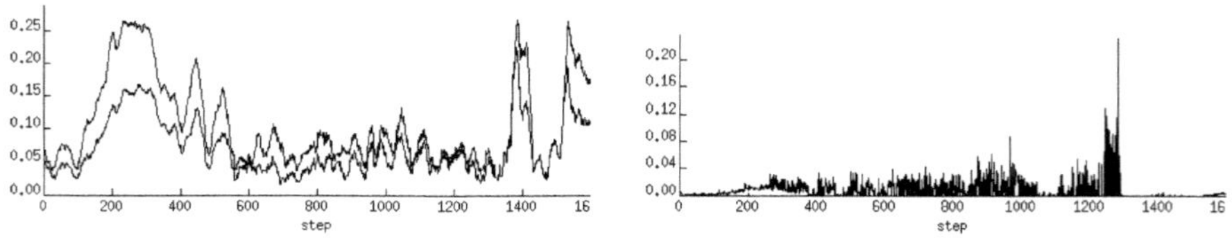

Fig. 3. Left: The moving average (window size = 20 steps) of the times of wars per nation in each step. The lower lines indicate the times of successful occupation by invaders. Right: The proportion of war casualties in each step.

4 Next Step

The model described above lacks many features, and we can improve it by having more appropriate similarities with real history. The items listed below are the candidates we will implement in the near future.

1. *Upper layer over nations*: Any type of coalition or alliance among nations can provide stronger influence in both peace keeping and domination. More flexible organisation of layers should be considered.

2. *Lower layer under villages*: Social movements in democratic society against war are mainly motivated by the emotion of hating inhuman behaviour. It is important to consider the dynamics in the bottom layer, organised by people, to consider whether democracy is effective to suppress wars.
3. *Distribution of troops*: The reason why a small nation hardly remains in this model is that a big nation can fight with all of soldiers it employed from widely distributed villages. This is not realistic. The cost of deploying troops and the strategy of distributed dispatch of troops should be considered.
4. *Nonuniform distribution of fertility*: Conflict among nations sometimes arises because of limited natural resources. Nonuniform distribution of fertility of land tiles can capture this situation.
5. *Similarity and possession norms*: Riolo *et al* [5] showed that similarity among agents can lead to the emergence of co-operation. Flentge *et al* [6] demonstrated that conflicts become less likely to occur as the possession norms emerges. We will incorporate these important features into the model.

5 Conclusion

We constructed a world model of nations, in a form of two-layered multi-agent system, which endogenously choose the strength of military forces. We demonstrated two typical processes of the virtual world: repetition of rise, division and ruin of nations; a type of equilibrium among a number of nations followed by a period of domination by one huge nation. We examined these different periods of the world in terms of the frequency of wars and the scale of casualties.

Through the repeated experimental simulation, we obtained expectation for this approach to be promising and useful to understand the dynamics of international relationships, though it is necessary to extend to a more realistic model by adding features including ones listed in the previous section.

By modifying some features such as geographical constraints, this approach seems to be applicable not only to international issues but also to economic battles among companies in a market.

The authors hope the usage of this particular system can contribute to realising a more peaceful world.

References

1. Langton, C. G., ed.: Artificial life. Addison-Wesley (1989)
2. Gilbert, N., Doran, J., ed: Simulating societies, UCL Press, London (1994)
3. Epstein, J. M., Axtell, R.: Growing artificial societies. MIT Press (1996)
4. Unemi, T.: Should seeds fly or not? Proc. of the Seventh International Conference on Artificial Life. (2000) 253–259
5. Riolo, R. L., Cohen, M. D., Axelrod, R.: Evolution of cooperation without reciprocity, Nature **414** (2001) 441–443
6. Flentge, F., Polani, D., Uthmann, T.: Modelling the emergence of possession norms using memes, Journal of Artificial Societies and Social Simulation **3** (2001) http://www.soc.surrey.ac.uk/JASSS/4/4/3.html

Developmental Neural Networks for Agents

Andy Balaam

CCNR, University of Sussex

Abstract. A system for generating neural networks to control simulated agents is described. The networks develop during the lifetime of the agents in a process guided by the genotype and affected by the agent's experience. Evolution was used to generate effective controllers of this kind for orientation and discrimination tasks as introduced by Beer. This scheme allows these behaviours to be generated quickly and effectively and may offer insights into the effects of developmental processes on cognition. For example, development may allow environmental regularities to be recognised without genetic prespecification. Possible future research into the abilities of these controllers to adapt to radical changes and to undertake widely varying tasks with a single genotype is described.

1 Introduction

Much of the agent-based modelling in Adaptive Systems research involves using evolution to find genetic 'blueprints' for controllers which perform a given task. A natural extension to this idea is to incorporate a developmental process that allows one to use evolution to find genetic 'recipes' for generating useful controllers.

The use of a developmental process to generate phenotype and behaviour from a genotype may have several advantages over the direct specification of the phenotype in the genotype. It may allow some of the work required to perform a task to be done during the development process (responding to the regularities found in a given environment, perhaps) rather than requiring evolution to solve these problems. It may also offer the ability to produce complex behaviours later in the lifetime of the agent, building upon simpler behaviours acquired earlier. In the longer term there are rich possibilities for this work to feed into biological and psychological studies of development and cognition.

Thus it is interesting to investigate developmental processes in the design of controllers. This paper presents the first steps in the design of a developmental controller which seeks to build upon the work already done in agent-based evolutionary modelling in two ways: first, it is based upon continuous-time recurrent neural networks [1] which have been studied extensively, and second it deals with 'minimally cognitive' tasks as defined by Beer [1,19]. By taking this approach we are able to understand these new systems using our understanding of existing systems.

Since systematic study is essential in exploring a new area, so far these controllers have been evolved to perform extremely simple tasks. However, already

W. Banzhaf et al. (Eds.): ECAL 2003, LNAI 2801, pp. 154–163, 2003.

some very interesting consequences of the developmental process have been observed. In section 6 it will be argued that the abilities and scientific interest of these controllers may be expected to increase as the complexity of the tasks they are required to perform increases.

2 Background

2.1 The Importance of Development in Living Systems

Recent work in neuroscience has uncovered a much greater incidence of structural change (including significant growth and death of neurons) in brains than previously expected. This change occurs even after brain development is complete. It is increasingly believed that this kind of structural change is vital for cognition and memory.

Recent studies have shown that newly generated neurons not only appear in the brain, but that they become involved in its functional activity. For example, van Praag et al [20] characterised new neurons in adult mice over time after their appearance and found that they developed morphologies similar to those of mature neurons. They were able to show strong evidence that these new cells received synaptic input from their neighbours and were functionally incorporated into the hippocampal network.

Meanwhile, momentum for the general view that change is vital and pervasive in adult brains is continuing to grow [14]. For example, Ivanco and Greenough [11] argue that the changes brought about by learning and experience are exhibited as physical changes indicating functional reorganisation, and that the mechanisms that bring these changes about may be the same mechanisms that repair tissue after damage to the brain.

The use of neural network models by neuroscientists to investigate the role of structural change in real brains is becoming common ([4], [13],[16],[17], [18]).

If cell growth and death and synaptic plasticity are important to the cognitive capabilities of animals, it is reasonable to assume that in order to produce cognitive capabilities in artificial agents we will need to allow for analogous processes of structural change.

2.2 Adaptive Systems Approaches

In the early 1990s several researchers approached the problem of developmental neural networks. Most of the work done at this time used string and graph rewriting grammars to model the development process. Perhaps the most successful of these models was that of Gruau [7,8,9], who used a graph rewriting grammar coupled with a tree-like genotype to create modular neural networks that were grown in a development phase before the lifetime of an agent. Thus the development was decoupled from the experience and sensory capabilities of the agent and encapsulated in a very abstract mathematical model.

In 1995 Jakobi [12] designed a developmental neural network scheme involving models of DNA molecules, protein transcription and diffusion and a genomic regulatory network. A controller was seeded at the beginning of an agent's lifetime with a single unit and others grew according to the rules of the genotype and protein interactions. This development occurred within a two-dimensional controller space, and took place before the agent's lifetime began.

After several years with very little activity, momentum has been growing recently behind research into developmental controllers [15].

Elliot and Shadbolt [3] have successfully shown that obstacle avoidance behaviour can be generated in a real robot by the use of a hand-designed (as opposed to evolved) developmental system that models some of the processes found in neuroscience.

There is a growing level of interest in plasticity in neural networks, especially in terms of their synaptic connections. Several pieces of work have been done using Floreano and Mondada's plastic neural networks [2,5,6], to investigate the potential benefits of this limited form of structural change in controllers.

3 Method

The environments and tasks modelled are designed to be identical to some of those used by Beer [1] in his research into minimally cognitive behaviours. In addition, the controllers used are closely linked with Beer's continuous-time recurrent neural networks (CTRNNs).

In parallel with the developmental experiments, replications of Beer's experiments (using CTRNN controllers identical to those used in [1]) on the tasks used were undertaken. These may serve as controls to quantify the performance and the behaviour of the controllers.

3.1 Continuous-Time Recurrent Neural Networks

Neurons in both the developmental networks and the standard CTRNNs are governed by the following equation: [1]

$$y_{i,t} = y_{i,t-\Delta} + \left(\frac{\Delta}{\tau_i}\right)\left(-y_{i,t-\Delta} + \sum_j w_{j,i} z_{j,t-\Delta} + S_i\right)$$

where $\sum_j$ denotes the sum over all neurons j connected to neuron i, and

$$z_{i,t} = \frac{1}{1+\exp(-(y_{i,t}+b_i))}$$

and $y_{i,t}$ is the cell potential of neuron i at time t, Δ is the time step size being used (0.1 in this work), τ_i is the time constant of neuron i, $z_{i,t}$ is the firing rate of neuron i at time t, $w_{j,i}$ is the weight of the connection from neuron j to neuron i, S_i is the amount of sensory input to neuron i and b_i is the bias of neuron i.

3.2 Developmental Neural Networks

The design of the developmental controllers is intended to strike a balance between biological inspiration and pragmatic assumptions about what kind of controllers will generate useful behaviours in a reasonable time. Development takes place during the lifetime of the agent.

The concepts inspired by biological development are: growth from smaller towards larger networks, an identical genotype for each unit and the use of simulated chemical gradients that specify which part of the genome is relevant to a given unit.

The general scheme is as follows: the controller is located within a two-dimensional space. Certain areas of the space correspond to the agent's sensors (neurons within those areas receive input from the corresponding sensor) and certain areas correspond to motors (the firing rates of neurons in these areas affect the action of the corresponding motor) as illustrated in figure 1. The controller is seeded at the beginning of the agent's life with one neuron per sensor on the agent.

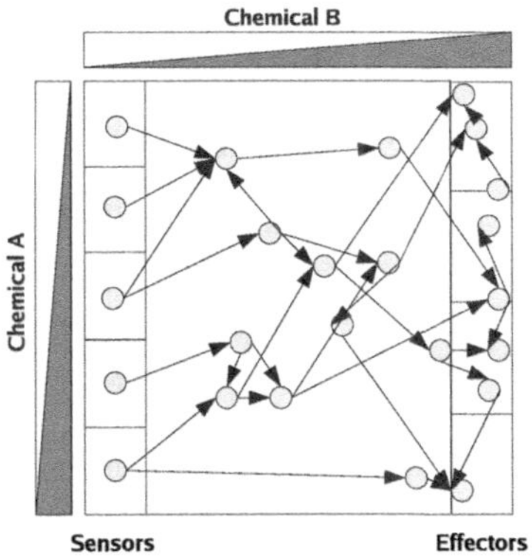

Fig. 1. The controller is a two-dimensional space with two chemical gradients over it, and specific sensor and effector regions.

There are two chemical gradients across the space, running horizontally and vertically. When a neuron is created, the levels of chemicals locally and the parent neuron's cell potential (if it has a parent) serve as inputs to functions the output of which provide the new neuron's time constant, bias, and energy. The shape of those functions is defined by the genotype.

At every time step the neuron's cell potential and the local chemical levels are fed into functions defined by the genotype to give the amount by which that neuron's 'growth sum' will change. When the growth sum goes beyond a threshold - if the neuron has any remaining energy - the neuron grows a new 'child' neuron, using functions defined by the genotype to determine the angle and distance at which to grow the new neuron. A connection is created linking from the parent to the child, and its weight is found using a function defined by the genotype. The energy of the parent neuron is decreased by 1 whenever it grows a child. If a neuron already exists at the point in the space where a new one is to be grown, a link is formed to the existing neuron instead.

The functions mentioned above are designed to approximate an arbitrary function using the following form:

$$p_n = \sum_{i=1}^{N_c} \sum_{j=1}^{N_s} \left(a_{i,j,1} \sin\left(a_{i,j,2} x_i + a_{i,j,3}\right)\right)$$

where p_n is a property of the neuron ($n = 1, \ldots, 7$), $a_{i,j,k}$ is a genetically-determined value, x_i is the local concentration of chemical i or the cell potential, N_c is the number of chemicals + 1 for cell potential (= 3 in this work), N_s is a constant (10 in this work).

The genotype is a list of 630 real values, since $7 \times N_c \times N_s \times 3 = 630$. The 7 occurs because there are 7 properties: bias, time constant, energy, growth increment, direction of growth, distance to grow and new connection weight. So the genotype may be thought of as a vector function, taking 3 inputs (chemical 1, chemical 2 and cell potential) and having 7 outputs.

3.3 Minimally Cognitive Behaviours

The environment and agent are designed to be exact replicas of those used in [1] including relative distances and speeds.

The environments in which the agents behave are extremely simplified situations designed to encapsulate certain fundamental elements of cognition. Agents themselves are circular and able to move only left and right at the bottom of a two-dimensional environment. Objects fall from above the agents and tasks require either 'catching' (matching horizontal positions when vertical positions are matched) or 'avoiding' objects in different circumstances.

Agents have distance-sensitive ray sensors (analogous to infra-red sensors on real robots) which produce large input when an object crosses a ray close to the agent and small input when it crosses far away. These sensors are arranged in a fan shape pointing out from the top of the agent over an angular range of $\pi/6$ (figure 2).

3.4 Genetic Algorithm

A generational, asexual genetic algorithm using rank selection with elitism (4% elitist fraction) on a population size of 50 is used, with real-valued genotypes. In CTRNNs mutation is performed by adding a random displacement vector whose direction is uniformly distributed on the M-dimensional hypersphere and whose magnitude is a Gaussian random variable with mean 0 and variance 10. In developmental networks mutation involves randomly selecting 5 of the 630 real values on the genotype and randomising them within their allowed ranges.

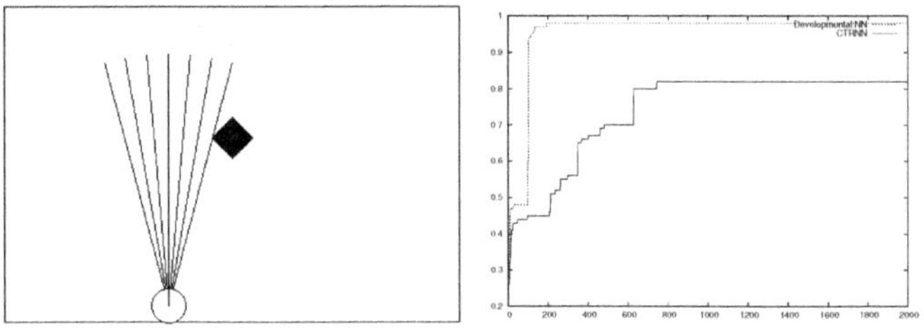

Fig. 2. Left: The agent has ray distance sensors and may move left and right in the environment. Right: Fitness vs. generation. The best fitness achieved with a developmental controller in the orientation experiment far out-performed the best achieved by a CTRNN.

The fitness of an agent over several presentations is calculated as follows:

$$F = 0.45f_m + 0.45f_w + 0.1f_c$$

where f_m is the mean fitness over the presentations, f_w is the fitness achieved in the worst (lowest fitness) presentations, and f_c is the fitness awarded for attributes of the controller. f_m, f_w and f_c are $\in [0:1]$ and thus $F \in [0:1]$.

In CTRNN controllers f_c is always equal to 1, and in developmental controllers f_c is equal to the mean (over the presentations) of the quantity x/W where x is the horizontal position of the rightmost neuron in the controller at the end of the agent's lifetime and W is the width of the controller space. So agents are rewarded for growing from the seed neurons on the left towards the motor region on the right.

4 Experiments and Results

The simulated environment was 400 wide by 275 high (the units are arbitrary) and contained the agent (which was circular with radius 15), and objects: circles of radius 13 and squares (called 'diamonds' since they were rotated by 45°) of side 26.

Orientation. In the first experiments circles were dropped at varying speeds and directions from the top of the environment and agents were required simply to orient themselves to the circles, catching them. Fitness for a single presentation was awarded as $F_p = 1 - \frac{d}{W}$ where d was the distance between the agent and the circle at the end of the presentation and W was the width of the environment (i.e. the maximum possible distance). The final fitness of an agent was found by combining fitnesses for each presentation as specified in section 3.4.

The developmental controllers performed at least as well at this task as standard CTRNN controllers. In fact, over the course of 20 evolutionary runs of each type (2000 generations in each), the maximum fitness acquired for developmental controllers (0.98 - the maximum possible was 1) was significantly higher than that acquired for CTRNNs (0.82), and 8 of the 20 developmental runs achieved a fitness higher than 0.82.

Discrimination. The second set of experiments involved the agents discriminating between objects of different shapes. Circular or diamond-shaped objects were dropped vertically from the top of the environment at different horizontal positions and fitness was awarded as follows:

$$F_p = \begin{cases} 1 - \frac{d}{D} & \text{when the object was a circle} \\ \frac{d}{D} & \text{when the object was a diamond} \end{cases}$$

where d was the distance between the agent and the object at the end of the presentation (capped to be $\leq$ D) and D was a maximum distance less than the width of the environment.

In these experiments the developmental controllers performed reasonably well, with the best agents, when examined manually, clearly correctly discriminating in 15 of 20 presentations. However, CTRNN controllers perform better at this task, with the best controllers correctly discriminating in 18 of 20 presentations.

5 Analysis

A number of interesting interactions and dynamics arise in the evolved solutions to the tasks described above. These fall into several categories:

Specialisation. A potential advantage of the use of developmental controllers over standard CTRNNs is that development allows the agent to specialise according to the environment in which it finds itself. Of course, in the tasks so far studied the need for specialisation does not appear pressing to an observer, but specialisation did occur in some cases. For example, as shown in figure 3 the same agent (in this case in the discrimination task) developed different controller structures depending on the environmental conditions. The behaviours exhibited by this agent are shown in figure 4.

While the tendency to specialise may have negatively impacted the performance of agents in this simple task (due to problems with consistency), it seems likely that this feature will be extremely useful when tackling more difficult and complex tasks (see section 6).

Indirect genotype-phenotype mapping. The mapping between genotype and phenotype in the simulations described is indirect, in contrast with much of the work being done in agent-based robotics today. The advantages of this, beyond its biological inspiration, are that it allows smooth changes of interesting kinds (such as rotations of whole sub-trees of neurons due to the change in angle of a single growth), it allows for modularity and genotype reuse (for example similar structures may be developed at the top and bottom of the controller if the vertical chemical gradient is insignificant in a certain area), and most of

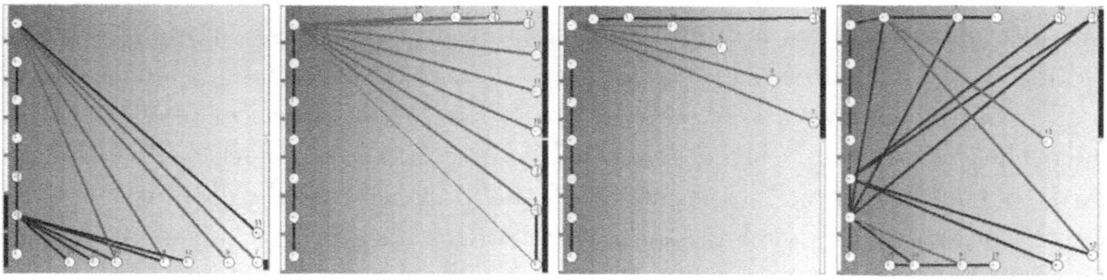

Fig. 3. The controller structures grown by a single agent under different environmental conditions. The two structures on the left show the controller when circles are dropped at different horizontal positions. The two on the right show what happens when diamonds are dropped. Circles must be caught and diamonds avoided. Sensor neurons (with which the controller is seeded at the beginning of a lifetime) appear on the left, and effector neurons on the right.

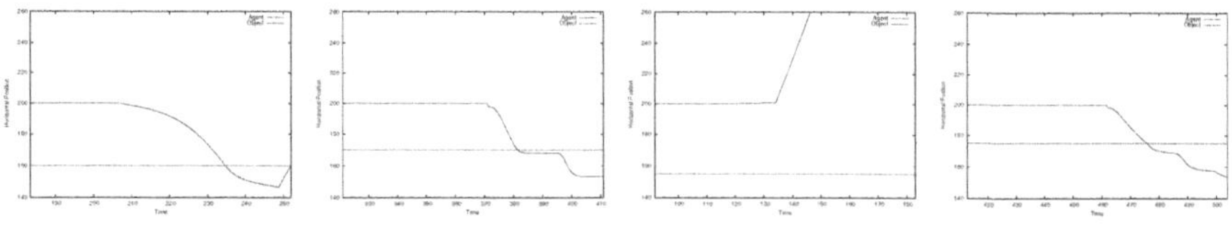

Fig. 4. The behaviour of the agent with the controller illustrated in figure 3. The horizontal line shows the horizontal position of the falling object over time. The other line shows the position of the agent.

all it is scalable to large numbers of neurons without a corresponding increase in genotype length. This allows the evolutionary process to find solutions with appropriate numbers of neurons without this constraint being imposed from outside.

Neutrality and rich fitness landscapes. The fitness landscape of these controllers contains a great deal of neutrality; for example, the growth of a neuron into an area of the space which does not affect the motors has no effect on behaviour.

Moreover, the fact that solutions were found to the problems studied in reasonable timescales implies that the landscape is quite rich with effective solutions, despite the relatively vast size of the genotype (630 values as compared with 12 and 47 values in the CTRNN versions of the orientation and discrimination tasks respectively). This suggests that the space of all developmental controllers is quite densely packed with interesting behaviours.

6 Conclusions and Future Work

The experiments described here represent the first steps in the design of developmental controllers that satisfy two key criteria: that they should encapsulate some of the properties of development that are useful for producing complex and interesting behaviours, and that they should be simple enough to allow for evolution of such interesting behaviours in a reasonable time. This early work goes towards justifying the assertion that the latter criterion may be satisfied, since the controllers evolved respectably in comparison with CTRNNs. In regard to the former criterion, some of the potential of development has been shown, notably the development of controllers specifically suited to particular environmental conditions, but future work will be aimed at showing further that development is useful in producing more interesting behaviours.

Initial directions will be towards understanding the capabilities of these controllers as they are by testing them with more complex environments and tasks, and then working towards increasing the power of the controllers by discovering how to maximise the advantages of developmental processes and minimise the disadvantages. Adjustments are likely to be required, especially to the nature of the functions described in section 3.2. Some of the drawbacks of the current design are becoming clear, including a high chance of catastrophic mutations, and local fitness maxima caused by the unidirectional nature of the development process.

Development in biological organisms can allow the regularities in the environment to be automatically incorporated during the organism's lifetime into its cognitive faculties, rather than predicted and prespecified in the genotype. This effect may allow artificial developmental controllers to lift some of the cognitive burden away from direct evolutionary control, into the developmental process. Thus environments with existing but varying regularities could be used to test whether developmental controllers can more easily take advantage of these regularities than conventional controllers. An extreme case of this would be asking a controller to perform two different behaviours in two completely different environments, distinguishable only by the sensory input the agent receives in each environment.

Another way in which developmental processes are of benefit is in adaptation to radical changes to an agent or its environment. For example as an organism grows larger over its lifetime developmental processes allow it to cope with this change. Interesting experiments may be imagined involving predictable but radical changes to an agent's environment. For example, an agent that first has to navigate a corridor before emerging into an open space where it has to perform some other task (approaching a certain area, say) may well benefit from being fitted with a developmental controller that can change in structure at the crucial moment to perform a different behaviour. Developmental processes may also benefit agents that need to adapt to unpredictable changes as well.

Along with the goal of producing controllers capable of complex and interesting behaviours, a long term goal of this work is to shed light on the relationship between development and cognition. If the results of experiments such as Held's [10] demonstration of links between behaving and learning in kittens could be reproduced in an artificial system, the benefits of artificial systems could be put to good use in understanding the nature of such systems in general terms.

References

1. Beer, R.D. (1996) 'Toward the evolution of dynamical neural networks for minimally cognitive behavior' In *SAB4*
2. Di Paolo, E.A. (2000) 'Homeostatic adaptation to inversion of the visual field and other sensorimotor disruptions' *Proc. of SAB 2000.*
3. Elliot, T. and Shadbolt, N. (2001) 'Growth and repair: Instantiating a biologically inspired model of neuronal development on the Khepera robot' *Robotics and Autonomous Systems* 36 pp 149–169.
4. Elman, J. (1993) 'Learning and development in neural networks: The importance of starting small' *Cognition* 48 pp 71–99.
5. Floreano, D. and Mondada, F. (1996) 'Evolution of plastic neurocontrollers for situated agents' In *SAB4*
6. Floreano, D. and Mondada, F. (1998) 'Evolutionary neurocontrollers for autonomous mobile robots'. *Neural Networks* 11, pp 1461–1478.
7. Gruau, F. and Whitley, D. (1993) 'Adding learning to the cellular development of neural networks: Evolution and the Baldwin effect' *Evolutionary Computation* 1, 3:213–234.
8. Gruau, F. (1994) 'Neural Network Synthesis Using Cellular Encoding and the Genetic Algorithm' PhD. Thesis, Ecole Normale Superieure de Lyon, LPIMAG
9. Gruau, F.(1995) 'Automatic Definition of Modular Neural Networks' *Adaptive Behaviour* 3.2 pp 151–183.
10. Held, R. (1965) 'Plasticity in sensory-motor systems' *Scientific American*, 213(5), 84–94.
11. Ivanco, T., and Greenough, W. (2000) 'Physiological consequences of morphologically detectable synaptic plasticity: potential uses for examining recovery following damage.' *Neuropharmacology* 39, pp 765–776.
12. Jakobi, N. (1995) 'Harnessing Morphogenesis' Technical Report School of Cognitive and Computing Sciences, University of Sussex.
13. Karmiloff-Smith, A. (1992) *Beyond Modularity: A Developmental Perspective on Cognitive Science*, MIT Press.
14. Kolb, B., Forgie, M., Gibb, R., Gorny, G. and Rowntree, S. (1998) 'Age, Experience and the Changing Brain' *Neuroscience and Biobehavioural Reviews* 22.2 pp 143–159.
15. Nolfi S., Miglino O. and Parisi D. (1994) Phenotypic plasticity in evolving neural networks *Proceedings of the International Conference From Perception to Action* pp 146–157
16. Quartz, S. and Sejnowski, T. (1997) The neural basis of cognitive development: a constructivist manifesto' *Behavioural and Brain Sciences* 20 pp 537–596.
17. Quartz, S. (1999) 'The constructivist brain' *Trends in Cognitive Sciences* 3.2 pp 48–57.
18. Quinlan, P. (1998) 'Structural change and development in real and artificial neural networks' *Neural Networks* 11 pp 577–599.
19. Slocum, A., Downey, D., Beer, R. (2000) 'Further Experiments in the Evolution of Minimally Cognitive Behaviour' in *SAB6.*
20. van Praag, H. et al (2002) 'Functional neurogenesis in the adult hippocampus' *Nature* 415, pp 1030–1034.

Revisiting Idiotypic Immune Networks

Hugues Bersini

IRIDIA – Université Libre de Bruxelles
CP 194/6
50, av. Franklin Roosevelt
1050 Bruxelles – Belgium
bersini@ulb.ac.be
http://iridia.ulb.ac.be/

Abstract. This paper will summarize the computational abstractions lying behind the simulation of immune idiotypic networks and present the most relevant emergent outcomes of the simulation, the same outcomes that give support to Varela, Coutinho and Stewart, when they see in this network one privileged way to address conflicting immunological questions. These more recent years, we have attended a resurgent interest for evolving networks showing interesting connectivity structure. The idiotypic network turns out to be such a kind of network and, for a well-tuned probability of matching two antibodies, the so called "percolation value", it exhibits a particular connectivity structure. For this value, the network is able to demonstrate the most effective memorisation capacity and to usefully keep separated its tolerant from its immune responses. Would it be possible that such a kind of biological network, by deliberately tuning the matching probability of its cells to a given value, improving the communication among surrounding cells, exhibit as a secondary effect both the emergence of a dichotomous behaviour, tolerant and immune, and the best memorisation of previous encounters.

1 Introduction

Since Oudin's experimental finding and Jerne's enthusiastic emphasis [4,5] on the existence of idiotypic network, that antibodies can mutually stimulate themselves in a way very similar to the stimulation antigen exerts on antibody has been convincingly revealed by a large set of experiments. Roughly, in an idiotypic network, there is no intrinsic difference between an antigen and an antibody, and any node of the network can bind and be bound by any other. At least, the difference is not of a recognition type but rather at the production level. Apart from the way they get into the system and the way their concentration changes by interacting with this system, nothing really distinguishes an antigen from an antibody. In the absence of any particular external interaction, attributed to antigenic impacts, the network maintains a normal background autonomous activity.

W. Banzhaf et al. (Eds.): ECAL 2003, LNAI 2801, pp. 164–174, 2003.

Varela and Coutinho are responsible for a radical shift by substituting a self-recognition system with a self-assertion one [8]. In the later, the frontier is progressively constructed as an outcome of the network evolution and of its sustaining autonomous activity. What is self is not frozen and determined by whatever exogenous characteristics of the impacts to be recognized by the system, but instead produced by the autonomous activity of the network. Gradually some impacts, without pre-labelling as self or non-self, will either smoothly integrate the system (then consequently be assessed as self) or be rejected from this same system (and consequently assessed as non-self). Self will only be defined by what the system accepts as a whole to blend into itself. No pre-categorization exists outside and the environmental post-categorization into tolerance and reactivity is under the sole responsibility of the network. What they rather found in this network is the necessary qualities to provide some sort of ideal cellular communication pathway, which allows the system to preserve a perfect integrity and to maintain a homeostasis, despite the extraordinary variety of the cells to control and the unpredictable and constant fire of impacts targeting these cells. Stewart [7] pushed the most radically in that direction, by suggesting that the defensive role of the immune system is secondary and that it was evolutionary selected as a consequence of its first and primordial function, namely to provide this ideal communication pathway.

These recent years, we have attended a resurgent interest for evolving networks showing interesting scale-free connectivity structure [1,3]. Although the most representative of these networks have been spotted in the human and social worlds (like the Internet or epidemic networks), the fact that such a connectivity structure allow the nodes to optimally connect (these networks exhibit the small-world property allowing a fast and reliable communication), has encouraged some authors to believe that some particular type of connectivity, like a scale-free one, should hopefully be shared by biological networks. Some cellular and genetic networks seem, as a matter of fact, to structure their connectivity in such a way. A scale-free connectivity, not only allow the nodes to nearly optimally communicate, but also make the network more robust since the large majority of nodes, being weakly connected, can be deleted from the network with minor consequence.

As we have shown for many years, idiotypic network is indeed an evolving network whose connectivity might be far from random. The paper will show that, for a well-tuned probability of matching two antibodies, called the "percolation value", for which the network springs into existence, the network could exhibit a specific connectivity structure. Also, for this particular tuning, the network could demonstrate the most effective memorisation capacity and to usefully keep separated its tolerant from its immune responses. The structural robustness, freely coming out from such a heterogeneous connectivity structure, will be questioned to supply the immune system with the functional homeostasis necessary to mutually regulate the complex web of surrounding cells. Would it be possible that such a kind of biological network, by deliberately tuning the matching probability of its clones to a given value, improving the inter-cells communication, will exhibit as a secondary effect both the emergence of a dual behaviour, tolerant and immune, and the most effective memorisation of previous encounters. Roughly said, the functions generally taken as the key ones of the immune system could be retrogressed to the consequences of a network simply targeting a better integrative function.

2 Simulation of the Idiotypic Network

We suppose each antibody to be identified by a binary string of N bits and the affinity an antibody i exerts on an antibody j defined in the following way:

$$affinity(i,j) = C_i(t)*(DHamming(i,j) - L) \qquad 2.1$$

with DHamming(i,j) being the Hamming distance between the binary strings i and j, $C_i(t)$ the concentration of cell i at time t and L the affinity threshold (the key parameter to be varied in the simulations to come). The affinity can only be positive or null.

All antibodies i exert on any antibody j a field of affinity aff_i obtained by summing this affinity for all the cells currently present in the system:

$$aff_j = \sum_i affinity(i, j) \qquad 2.2$$

At every time step, the concentration of the antibody j evolves in the following way:

$$\begin{gathered} if\,(T_{low} < aff_j < T_{high})\ C_j(t+1) = C_j(t) + 1 \qquad 2.3 \\ else \\ C_j(t+1) = C_j(t) - 1 \end{gathered}$$

if $C_j(t) <= 0$ *the cell j disappears from the system*

Important here is the presence of the two bounding thresholds: $\mathbf{T_{low}}$ and $\mathbf{T_{high}}$. These two bounds reproduce in a much simpler way the bell-shaped form of the activation functions present in more sophisticated simulations of the dynamics of the model. These two bounds clearly justify why the simulation of the system do not explode i.e. why the whole network together with the concentration of the cells don't grow to infinity.

Two types of simulation will be presented, without network and with network. To begin with the first one, the classical Burnetian clonal selection view, we will assume that only the antigens are subject to this field of affinity and that, reciprocally, only the antigens exerts affinity on the cells. aff_j is computed just by summing the field exerted by the antigens for the cells and by the cells for the antigens. For instance, for the cells:

$$aff_j = \sum_{i=antigens} affinity(i, j)$$

To differentiate the exogenous mode of production of the antigens from the endogenous mode of productions of the antibodies, we will suppose that the antigen concentration just decreases if the immune cells bind them enough and decreases in proportion of the received affinity. Take an antigen j, if the affinity it receives is above the minimum threshold, it will decrease in concentration according to:

$$if\ (aff_j > T_{low})\ C_j(t+1) = C_j(t) - k^*(affj/T_{low})\quad with\ k\ a\ given\ temporal\ rate \qquad 2.4$$

$$if\ C_j(t)\ <= 0,\ the\ antigen\ j\ disappears\ from\ the\ system$$

In the less classical view, the "network view", all cells bind to all cells. The way we will compute the aff_j received by any cell is as follows:

$$aff_j = \sum_{i=cells} affinity(i,j) + \sum_{i=antigens} affinity(i,j) \qquad 2.5$$

This time, the affinity received by any cell is a sum of the exogenous stimulation of the antigens and the endogenous stimulation of the cells themselves. Suppress the first term of the sum, and you are back to the previous case.

2.1 Percolation and Repertoire Self-Selection

In the simulations to come, no external antigen is present, just the network of antibodies and the way it appears and evolves is examined. Every time step one new cell is recruited with concentration 20. The concentration evolves as indicated in 2.3. T_{low} = 100 and T_{high}=1000. The simulation is run for 6000 time steps.

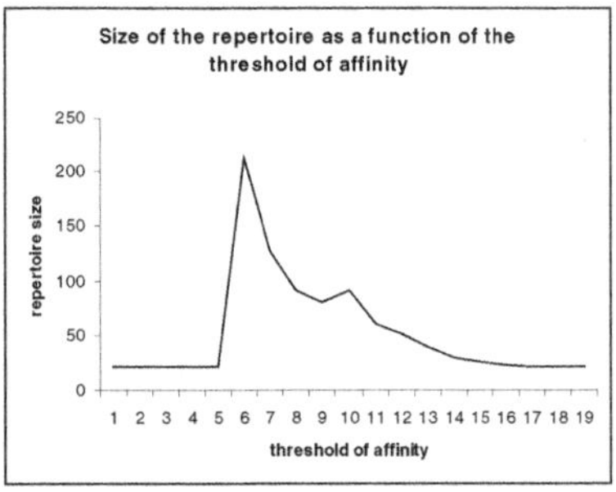

Fig. 1. Size of the repertoire as a function of the affinity threshold from 19 to 0. The percolation takes place at 14

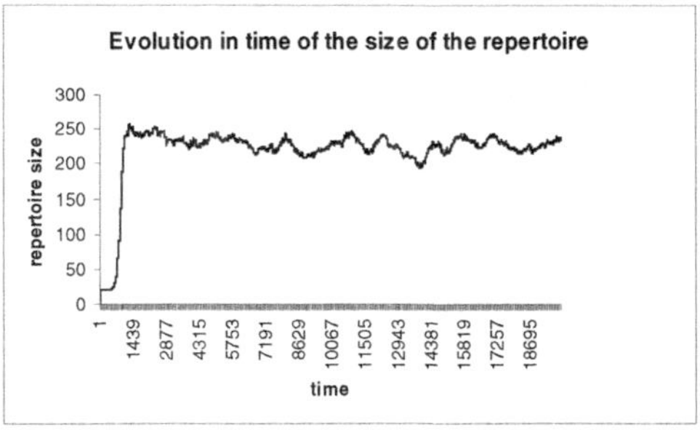

Fig. 2. Size of repertoire in time for 20000 simulation time steps

In Figure 1, the size of the repertoire (number of cells) at the end of the simulation is plotted as a function of L varying from 19 to 0 (the value 19 implies a very specific type of matching since one cell will exert affinity only on its perfect complement, while a value 0 implies that all cells exert affinity on all of them, the abscise should be read as 20 - L). Two phenomena deserve attention. The first one is that nothing happens until L reaches 14. The network does not pop up since the probability of two antibodies matching is too small. The construction of the network can only be triggered of above this threshold. So, for an idiotypic network to pop up and to turn its virtual existence into reality, a minimum probability of match is required. There is a balance to be found between three values: the initial concentration Ci(0), the number of recruited cell at each time step and the affinity threshold. If the probability to recruit two complementary cells is not high enough for compensating the time for the cells to be recruited and to die (if not stimulated), the network will never spring into existence. Notice that this percolation phenomenon extends to all of the three values just discussed.

The second interesting outcome of the simulation shown in figure 1 is, just following the percolation, the slow decrease of the stabilized size of the repertoire as a function of the threshold increase. As also verified in similar simulations achieved by De Boer and Perelson [2] who discuss it as the most interesting outcome of their simulations, the higher the probability of match between two cells the smaller the size of the repertoire. We attend a self-regulation of the repertoire size as a function of the matching probability, the higher the probability for one cell to match a second one, the smaller the number of necessary candidates to indeed succeed this match. This size is very small as compared to the potential number of diverse cells (here 250 out of one million). However it can increase as a function of various parameters of the simulation, like as a function of T_{high}.

Figure 2 shows the evolution in time of the size of the repertoire for one simulation. It is easy to see the percolation of the network beginning at around 1500 time steps, followed by a very rapid expansion, to then lightly fluctuate around a stabilized value. Despite the constant turnover of the cells (new cells are constantly recruited in the network), this size reaches a fixed point the value of which is inversely proportional to the probability of match between two cells.

2.2 Interaction with External Antigens

We will now present results obtained by a new original set of simulations focusing on the interaction of the system with external antigens, both in the presence and in the absence of the network effect. In the simulations to come, every 200 time steps, an antigen is recruited in the system with a initial concentration of 100. Since the simulation will last for 10000 time steps, 50 antigens are integrated in the system. The interaction of the antigen with any antibody, how its concentration decreases as a result of this interaction, is like indicated in 2.4 and 2.5. The simulations are done again with L varying from 19 to 0, from a very specific matching on the left of the figure to a very non-specific on the right. The simulations are performed first without interaction among the antibodies (without network) and then with interaction among the antibodies (with network). The two results are shown in the same figure 3.

What is shown in this figure is the number of tolerated antigens at the end of the simulation as a function of the affinity threshold. First, without the network, the results are quite easy to interpret. For a low probability of match, until L=16, all antigens survive, at least during the 10000 time steps. The probability to recruit an antibody capable of binding any antigen is null. Despite the fact that these antibodies virtually exist, the rate of recruitment of new cells together with the huge variety of cells make statistically impossible the generation of the right candidates. From L=16 on, the probability to match the antigens gradually increase until L=9, for which no antigen is able to remain in the system for the whole duration of the simulation.

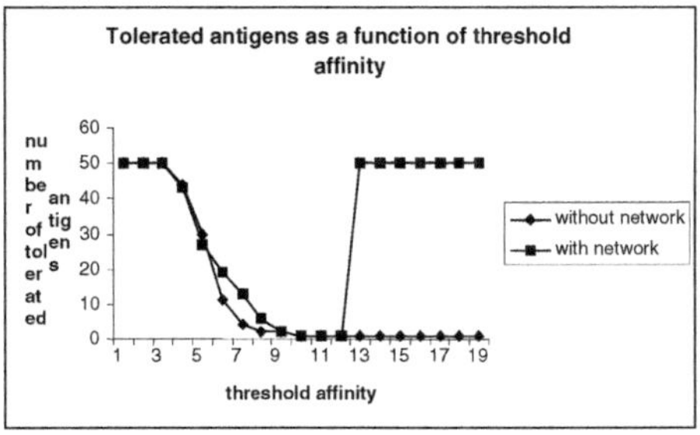

Fig. 3. Number of tolerated antigens as a function of the affinity threshold with and without network.

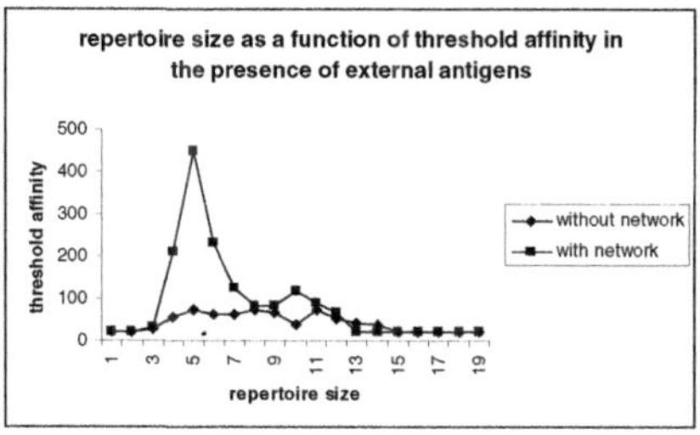

Fig. 4. Repertoire size as a function of the affinity threshold with and without network

Much more intriguing are the results obtained in the presence of the idiotypic network, so with antibodies mutually stimulating and sustaining themselves. Until L=8, the results are pretty much the same as without the network despite the fact that, as shown in figure 4, the size of the repertoire with and without the network is quite different. With the network effect, for L=16, the percolation takes place and the repertoire substantially grows. The percolation occurs at L=16 instead of 14 in the previous simulations, since the presence of external antigens in higher concentration than the cells makes the network to spring into existence much more easily.

The idiotypic network autonomously divides the binary space into two zones: a tolerant and a reactive one despite the completeness of the repertoire. Paradoxically, since the causes are different, with respect to the external antigens, the network forces the whole system to behave as if such a network did not exist. The network inhibits the antibodies that could match the antigens and eliminate them. Indeed, for L<8, all antigens become tolerated because the network inhibit all cells and maintain the size of the repertoire to a very low value. The most surprising conclusion turns out to be that, as regards the instantaneous reactivity to external intrusions, and within the most interesting range of L (8 < L < 16), the network does not propose any innovative effect. Its self-assertion is nothing more than the replica, when no network is at play, of the difficulty to match and eliminate in due time any external impact. Therefore, if the network does not play any interesting role to improve the response of the system to external intruders, what is it useful for? The next section will try to answer this question by investigating the memory capacity of such a network.

2.3 Network Memory Capacity

In order to test the memory capacity of the network, the following experiment has been done. At the end of the 10000 time steps, allowing the network to develop and to stabilize in the presence of the 50 external antigens (some being tolerated others being rejected), a set of 50 new antigens are submitted to the network two times. The first time, the antigens are randomly shaped, while the second time the antigens are selected to be very similar to the ones encountered during the network development (a

Hamming distance of one with respect to them). The simulation is released again for 2000 extra time steps. Following this additional run, the number of tolerated antigens among these two new sets is plotted in figure 5 as a function of the affinity threshold. The results indicate that, for L=15 and L=14, the antigens similar to the ones encountered during the development of the network are more likely to be rejected than the random ones (around 10 additional antigens are rejected). Roughly, the network memorizes its antigenic encounters in a way reminiscent of vaccination since it reinforces its defensive capacities against antigens similar to previously encountered ones.

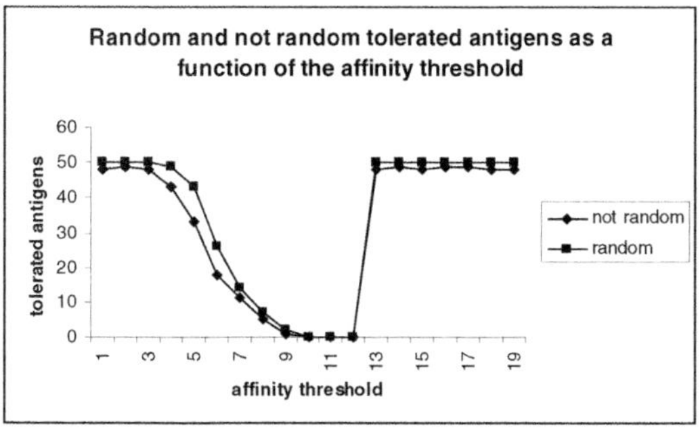

Fig. 5. The number of tolerated antigens following the development of the network. The first set of antigens is randomly shaped. In the second set, the antigens have a Hamming distance of 1 with respect to the ones encountered during the development of the network

However, the essential result here is that, as a matter of fact, the network can show a memory capacity and tend to react differently in time (here more defensively) when presented with antigens similar to previously encountered ones. This memory capacity seems to be more important for intermediate values of the affinity threshold closed to the percolation value (L=14). For this same value of L, when the network is presented with random antigens, nearly half of them are tolerated while the other half is rejected, a kind of homogenous splitting of the somatic environment.

One classical way to explain how the immune system tolerates self is to admit that early in time, during the embryonary development, the immune system behaves in a reverse way to later in time. Namely, during its early development the system eliminates instead of amplifies cells that react with external antigens (the "clonal deletion" theory). Since these antigens are more likely to represent the inside than the outside of the body, they serve a kind of learning purpose. They teach the immune system about the self and suppress any attempt to aggress this same self. One basic problem with such a vision is to resolve the mystery of this dual behaviour: how a system that eliminates any reactive cells during its early life can later in time reverse its behaviour and positively reinforces this reactivity. Again the idiotypic network can shed some new lights on this question like the simulations to come will show.

In these simulations, two sets of 50 antigens are presented to the network, one at time 0 of the simulation and the second set at time 6000. The simulation again lasts for 10000 time steps. At the end of this simulation, the network is now presented with four sets of 50 antigens. The two first sets are random while the third one contains antigens at Hamming distance 1 from the 50 antigens presented at time 0 and the fourth set from the 50 antigens presented at time 6000. The question is the following: will the network behave differently in response to antigens similar to the very early ones and antigens similar to the very late ones. So can this dual behaviour in time be just obtained by a network reacting in one singular way but to stimuli occurring at successive times? The figures 6 and 7 seem to give a positive answer to this question. In figure 6 the affinity threshold is varied from 17 (for which all antigens are tolerated) to 11 (for which none are tolerated). It is easy to see that the most important temporal memory effect is obtained for L=14 where 22 antigens similar to the early ones are tolerated, while this number falls to 16 for antigens similar to late ones (these values have been averaged on a huge number of simulations). Roughly the network tends to be slightly more tolerant for antigens similar to the very early encounters, in accordance with the classical immunological vision, but with no need for any mysterious switch in behaviour.

This temporal memory effect turns out to be even sharper for higher value of the concentration of the antigens presented at time 0. For instance in figure 7, the number of tolerated antigens for the four sets as a function of the affinity threshold is again shown. However this time, the concentration of the initial antigens is set to 1000 instead of 100.

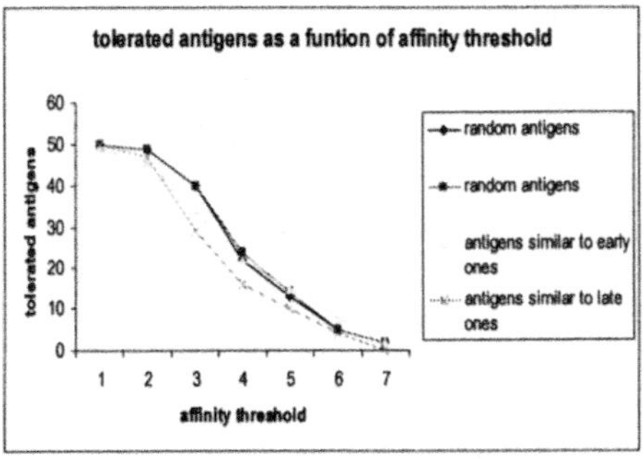

Fig. 6. Number of tolerated antigens among 4 sets of 50 antigens. The two first sets are random while the other two contain antigens similar to early ones and then to late ones. The affinity threshold is varied from 17 (for which all antigens are tolerated) to 11 (for which none are tolerated).

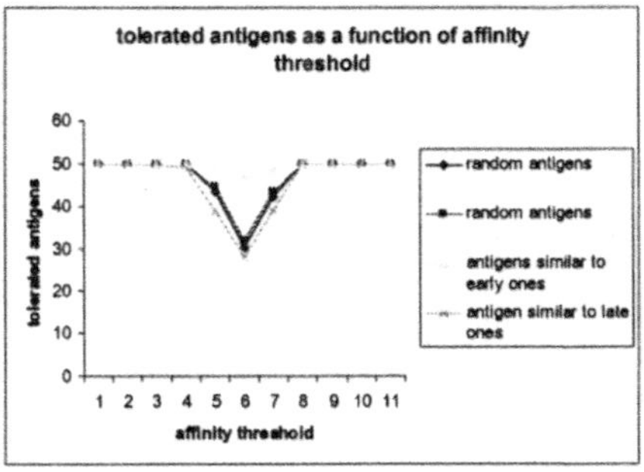

Fig. 7. Number of tolerated antigens among 4 sets of 50 antigens. The two first sets are random while the other two contain antigens similar to early ones and then to late ones. The affinity threshold is varied from19 (for which all antigens are tolerated) to 9 (for which none are tolerated). The concentration of the initial antigens is 1000.

Again for L=14, the difference is the greatest between the number of tolerated antigens among the first set (48) and the number of tolerated antigens among the second set (28).

2.4 Connectivity in Idiotypic Networks

Recently, there has been a new wave of interest for networks of various types, to be found in nature, in human and animal society or in technology, and which have the capacity to grow and to exhibit, due to a particular way of growing, a scale-free connectivity structure [1]. For such a structure, the number of nodes showing a given degree of connectivity (the number of partners they connect with) plotted as a function of this degree follows a power law: the higher this degree of connectivity the smaller the number of nodes it concerns. This connectivity endows the network with interesting properties: the first one is to reduce the number of nodes to cross in order to connect any pair of them (the so-called small-world property) and to allow this reduction in path at a low structural price (maintaining the number of connections to a small value), the second one is to make this network more robust, since its core nodes, to which the network functionality is the most sensitive, are also in minority. Such a connectivity pattern has been observed in various nets and has been suggested as the outcome of an implicit optimisation process targeting a reliable and fast communication at a low structural price [3].

The idiotypic network being one singular growing and evolving network, it is attempting to analyse its connectivity structure, first as a function of the affinity threshold. Four of these connectivity patterns, obtained for L taking values 14, 13, 12 and 11, at the end of 10000 time steps, with no external antigen, are shown in figure 8 (plotting the number of cells as a function of the number of their partners). Here a partner of a cell is any other cell with a positive affinity (having Hamming distance > L). The number of partners varies from 0 to n (and not from 1 to n like shown in the figure). Therefore, for all four values of L, some antibodies are totally disconnected. For lower values of L, so in presence of less specific cells, the obtained connectivity structure is more reminiscent of a homogeneous random graph. The figures show that the connectivity structure changes as a function of the probability of match between two antibodies.

Additionally, if plotting the connectivity in time, the earliest situation is characteristic of a random homogeneous graph with high average connectivity per cell. Later in time, while the network self-stabilizes the size of its repertoire, the average connectivity decreases and its structure becomes heterogeneous, with very few highly connected nodes and many weakly connected ones. Interestingly enough, these results are in agreement with the rare experimental data obtained on real antibodies, and which indicate first that the connectivity in the embrionnary state is higher than in the adult state and, second, that this connectivity does not seem to be homogeneous, presenting a small set of very sticky antibodies on the one hand and a large amount of poorly connected antibodies on the other hand. This heterogeneous connectivity goes in line with the Varela and Coutinho's separation of the immune system into a strongly connected central part and a weakly connected peripheral one [8].

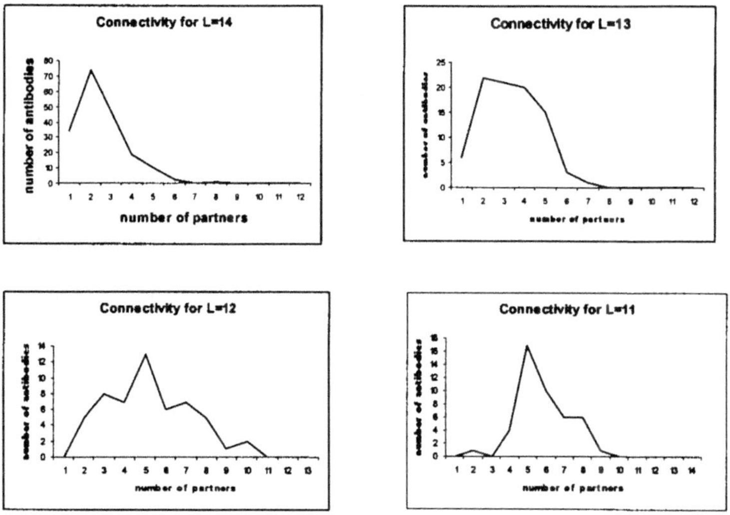

Fig. 8. The four connectivity structures obtained for the following values of L: 14, 13, 12 and 11. The value at the extreme left always gives the number of nodes disconnected from the network (i.e. with number of partners = 0).

3 Conclusions

Although this immune network is initially a growing one, every new node and new link of the network is the result of a random process. Consequently, this singular heterogeneous connectivity structure is more likely to result from a percolation threshold phenomenon than from any preferential attachment [1] during the growing in size of the network. It is well known that percolation phenomena tend to be associated with a singular connectivity structure [6]. However, in addition to shaping the connectivity in such a way, the previous sections have shown that this value of L improves the spontaneous tendency of the network to split its responses into tolerant and reactive ones and to memorize the previous encounters. For this threshold probability of match, it seems that a lot of unexpected properties could appear for free, like the dual behaviour and the memory. Consequently and in step with Stewart hypothesis [7], what we view as the primary functions of the immune system: recognition, tolerance, defence and memory, could turn out to be secondary with respect to the quest for an optimal connectivity structure favouring a fast and reliable communication pathway. Evolution could have privilege this particular matching probability so as to improve the integrative properties of the network with, as side-effects, this reinforced selectivity for entering the network and this capacity to more effectively reproduce a behaviour simply adopted during previous encounters with external antigens.

References

1. Barabasi, L-A. and R. Albert. 1999: Emergence of scaling in random networks. *Science* 286, pp. 509–512.
2 De Boer, R.J., Hogeweg, P. and A.S. Perelson. 1992: Growth and Recruitment in the immune network. In A.S. Perelson and G. Weishbuch (Eds.) – *Theoretical and Experimental Insights into Immunology*, pp. 223–247
3. Ferrer Cancho, R. and R.V. Solé. 2001: Optimization in complex networks. *Cond-mat/0111222*.Appendix: Springer-Author Discount
4. Jerne, N. 1974: Towards a network theory of the immune system – *Annals of Institute Pasteur/Immunology (Paris)* – 125C: 373–389.
5. Oudin, J., and M. Michel – 1963 *C.R. Acad. Sci.* Paris, 257, 805.
6. Stauffer, D. and A.Aharony. 1994: *Introduction to Percolation Theory*, 2nd ed, Taylor and Francis, London.
7. Stewart, J. 1994: *The Primordial VRM System and the Evolution of Vertebrate Immunity.* Austin, TX: R.G. Landes.
8. Varela, F. and A. Coutinho. 1991: Second Generation Immune Network – in *Immunology Today*, Vol. 12 No5 – pp. 159–166.

Production of Gliders by Collisions in Rule 110

Genaro Juárez Martínez[1], Harold V. McIntosh[2], and Juan Carlos Seck Tuoh Mora[3]

[1] Departamento de Ingeniería Eléctrica, Sección Computación, CINVESTAV-IPN, Av. IPN 2508, San Pedro Zacatenco, Apartado Postal 14-740, 07360, México D.F.
genarojm@correo.unam.mx

[2] Departamento de Aplicación de Microcomputadoras, Instituto de Ciencias, Universidad Autónoma de Puebla, Apartado Postal 461, 72000, Puebla, Puebla México
http://delta.cs.cinvestav.mx/~mcintosh

[3] Centro de Investigación Avanzada en Ingeniería Industrial, Universidad Autónoma del Estado de Hidalgo, Carr. Pachuca-Tulancingo Km. 4.5, Pachuca Hidalgo 42184, México
jseck@uaeh.reduaeh.mx

Abstract. We investigate the construction of all the periodic structures or "gliders" up to now known in the evolution space of the one-dimensional cellular automaton Rule 110. The production of these periodic structures is developed and presented by means of glider collisions. We provide a methodology based on the phases of each glider to establish the necessary conditions for controlling and displaying the collisions of gliders from the initial configuration.

1 Introduction

The interest in the study of Rule 110 begins with the investigations by Stephen Wolfram in one-dimensional cellular automata. Wolfram detects that this automaton can support complex behaviors identifying the existence of well-defined periodic structures in the evolution space, as Douglas Lind describes in the appendix of [21]. In the cellular automata environment, a periodic structure moving through time is called a glider.

Cellular automata are discrete dynamical systems which evolve through time, these systems may support complex and self-reproducing behaviors, as it is described by the precursor of this theory John von Neumann in [20].

The most famous cellular automaton is The Game of Life developed by John Horton Conway [6], it is a binary two-dimensional automaton and it has been used to implement artificial life, for instance, state zero represents a dead cell and state one a live one. In this context the initial configuration has a number of live beings who vary from generation to generation applying the evolution rule, reproducing a set of very interesting behaviors.

W. Banzhaf et al. (Eds.): ECAL 2003, LNAI 2801, pp. 175–182, 2003.

The wide variety of behaviors presented by this type of automata is the reason to investigate and simulating certain biologic, chemical, mathematics, physics and computing processes [1]. In the computing theory environment, Conway demonstrates that The Game of Life is universal on simulating a registry machine, constructing logics gates with gliders [2].

An important result in cellular automata theory in the last twenty years is developed by Matthew Cook in the middle of the 90's. Cook proves that 110 is universal by means of simulating a cyclic tag system [3], [22] y [12]. This demonstration and the one corresponding with The Game of Life use gliders to represent data and operations.

Cook establishes a classification of the gliders in Rule 110, the list was available in [4][2] and one part of the same appears in [22]. The list of gliders proposed by Cook is more complete than the one presented by Lind, because it includes quite a rare extensions of gliders, gliders of complex construction and the existence of a glider Gun.

In The Game of Life the glider Gun is very important to construct the registry machine, but in Rule 110 the glider Gun does not have an relevant paper in this sense. In this case blocks of well-defined gliders are used to simulate each one of the parts of the cyclic tag system. This is a direct application of the gliders, although the existence of gliders is an interesting subject by its own sake. For instance, the number of collisions of gliders in Rule 110 is unlimited because gliders can be grouped for traveling together in the evolution space and some of them have extensions, this is important because several complex processes may be implemented [1].

One of the questions is to know if Rule 110 can reproduce by itself each one of its periodic structures, we solve this problem for gliders without extensions.

In this paper we show that each glider (without extensions) proposed by Cook is obtained by collisions, using glider phases aligned by ether. In order to reproduce each collision we use the phases [8] [12] to control gliders by means of establishing a horizontal measurement in the initial configuration. We take the binary productions presented in [10] for classifying all the binary collisions producing a particular glider, this detailed examination demonstrates that D_2, $Bbar_n$, $Bbar8_n$ and H gliders and the glider Gun cannot be obtained through binary collisions, but they are product of multiple collisions. Throughout this paper we use the classification proposed by Cook to identify each glider.

The collisions illustrated in this work were yielded by the OSXLCAU21 system which may be freely obtained in [24]. This system applies the glider phases to construct suitable initial configurations in Rule 110, and it also allows to filter ether using an adequate selection of colors.

---

[1] In the web site of Tim Tyler there are several examples including attractive applets and other directions with excellent works in these subjects, http://timtyler.org/

[2] By some legal problems, the list and some other information developed by Cook about Rule 110 were retired [7]

2 Representation of Rule 110 by Means of Phases

A relevant feature of Rule 110 is that it is a binary one-dimensional cellular automaton with a very simple local dynamics, but the global behaviors that the rule is able to construct are highly complex. In this way, we shall provide the basic terms used in cellular automata theory.

A one-dimensional cellular automaton is composed by a linear array of cells where each cell takes one value from 0 or 1, this is the initial configuration of the system. The evolution rule is defined by the transformations of the neighborhoods, a neighborhood is formed by a central cell, a neighbor to the right and another to the left. In order to calculate the following configuration the evolution rule is applied to all neighborhood in the array at the same time. This process is repeated a number of times producing the global evolution of the automaton; an example of the global evolution of Rule 110 using a random initial configuration is illustrated in Figure 1.

The number 110 represents the rule in decimal notation which originally is the binary number 01110110, in this way the neighborhoods 000, 100 and 111 evolve into 0 in the following generation and the neighborhoods 001, 010, 011, 101 and 110 evolve into. Notice that the ending cells of the ring at both sides specify incomplete neighborhoods, in order to solve this problem we concatenate the initial cell with last one to have complete neighborhoods in each position of the evolution space.

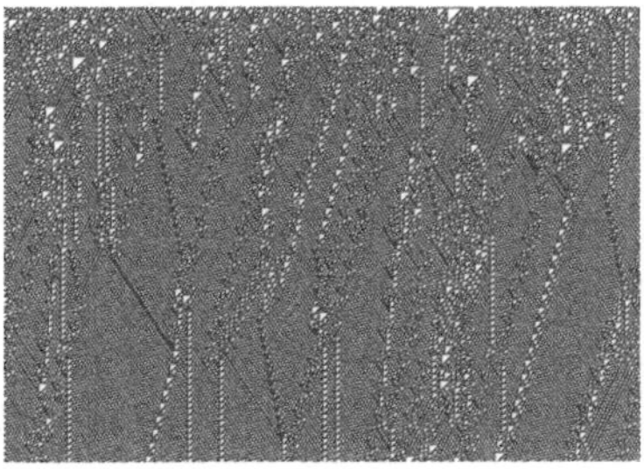

Fig. 1. Random evolution in Rule 110.

In Figure 1 we have regions with stable behaviors represented by the periodic background called ether by Cook, the periodic regions are determined by gliders and the chaotic regions may be generated from the initial configuration or as product of collisions with a short or long duration. In this figure we can also see all the possible gliders which arise in a natural way as Lind describes in the appendix of [21].

The evolution of Rule 100 can be seen as a covering of the evolution space by means of triangles formed by the cells of the automaton, this is defined by Harold V. McIntosh in [15]. T_n defines a triangle where very side has n cells for

$n \in \mathbb{Z}^+$; an interesting question is to determine the largest triangle produced by a collision in Rule 110 [16].

Ether is represented by a periodic sequence which moves 14 cells to the right in 7 generations. A triangle T_3 represents ether and it establishes our horizontal measurement aligning a pair of them to obtain a periodic sequence. We shall first represent in a systematic way each one of gliders classified by Cook and then we shall control each production using the glider phases. For this reason we shall establish a horizontal measurement f_{i}_i for $1 \leq i \leq 4$ as Figure 2 illustrates.

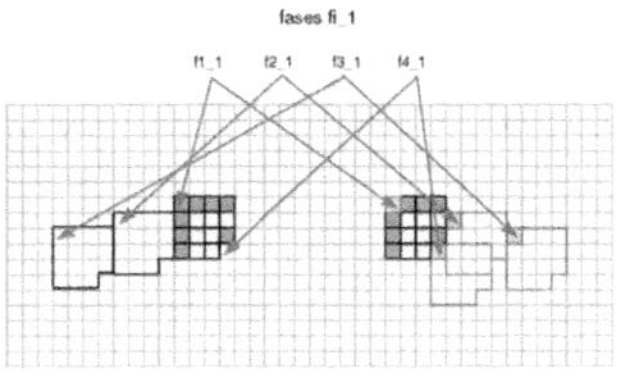

Fig. 2. Phases in Rule 110.

For instance, the sequence 11111000100110 represents the phase f_1_1 of ether and we describe it as e(f_1_1). This alignment allows to identify four different periodic sequences f_i to represent a given particular glider. In the case of gliders with more than one alignment with ether, they have several phases to initiate from the initial configuration. The second subindex i indicates the phase of the T_3 triangle; the other phases are a permutation of the first one. Therefore the f_i_1 phases are enough to establish the horizontal measurement.

Gliders are codified in the following way: $\#_1(\#_2,f_i\_1)$, where $\#_1$ represents a glider according to the classification of Cook and $\#_2$ is the phase of the glider if it is greater than one.

3 Producing Gliders by Means of Collisions

Gliders arising in a natural way in the evolution space are not difficult to obtain analyzing all the binary collisions; but Cook mentions that there are gliders which cannot be produced in the evolution of the automaton and they can only be specified from the initial configuration. However, although the Bbar8 and H gliders and the glider Gun are indeed complicated structures which do not arise commonly in the evolution space, they can be generated by means of collisions. In this sense specialized computing searches were developed, finding several interesting results.

Figure 3 depicts the production of each glider and the lower part of each figure describes the sequence codified in phases to produce each collision. In the case of the Bbar8 glider the collision is complicated and the synchronization of each one of its parts is unique; the change of a single cell disturbs completely

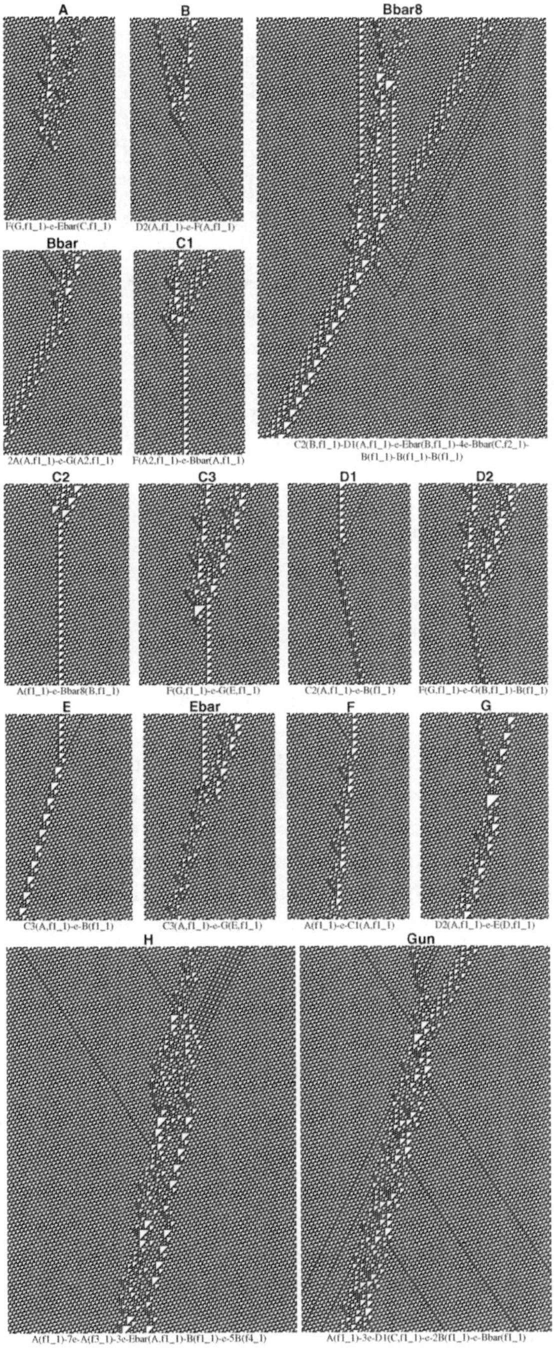

Fig. 3. Gliders of Rule 110 reproduced by collisions.

the final result. The same phenomenon is presented on synchronizing the gliders to simulate a cyclic tag system [17].

An important point is that the succession of collisions must be defined taking into account the order of the glider speeds. For instance in order to generate the Bbar8 glider, the C_2 and D_1 gliders have speed of 0 and of 1/5 respectively. [3]. Thus the existence of the D_1 glider preceded by the C_2 glider may be questioned whether it is product of a collision or not.

The A glider is generated by the Ebar and F gliders, the most of the collisions between these gliders have a soliton-like behavior. [4] Solitons have their own interest in cellular automata theory as Kenneth Steiglitz describes in [18] and [13]. However it does not imply that they are the only gliders in Rule 110 able to simulate solitons [9].

The G glider is produced by an isolated triangle T_{13}, an relevant question is to know if each glider may be yielded by an isolated triangle. For instance the T_{10} triangle induces an Ebar glider, the T_8 defines an E and the T_1 generates the A and B gliders.

H glider is produced by internal collisions among several gliders and small chaotic regions interacting at the same time; in this sense we can suppose the existence of more complex gliders. An important restriction is that every glider must advance with increments of 2/3 and going back with decrements of -1/2 in each phase. On the other hand although we have a complex glider in its construction, it can cover the evolution space and sometimes without ether [11]; for instance the H glider has two ways of covering the evolution space yielding a really exotic result.

Finally, the existence of a glider Gun is important in two aspects: the first is the straightforward representation of the unlimited growth of the automaton, and the second is the possibility of constructing a self-reproducing system, a relevant result developed by von Neumann in cellular automata theory. Glider Gun arises more frequently than Bbar8 and H gliders; nevertheless the quick interaction with other structures or chaotic regions avoids to form and conserve the glider because in a similar way with H glider, its period is very large.

4 Concluding Remarks

The list of gliders without extensions proposed by Cook is fully reproduced by means of collisions, the initial configurations used for obtaining this result have been constructed through the phases of each glider. In several paragraphs we have described the similarities between Rule 110 and The Game of Life, something interesting is that although Rule 110 has a one-dimensional evolution and its evolution rule is defined by eight neighborhoods, the global behavior is

[3] The glider speed is determined by the displacement of its cells between its period

[4] Soliton is a solitary wave with a nonlinear behavior which interacts with other waves conserving its form and speed, and suffering just small displacements in each collision [19]

very difficult to analyze. Another open question is to demonstrate that there are not more gliders in Rule 110.

The glider phases specify the horizontal measurement in periodic sequences aligned by ether, in fact they are sequences of the extended de Bruijn diagram [14], this diagram is useful to calculate all the periodic sequences of a given cellular automaton. In this way the de Bruijn diagrams may show the whole set of gliders in Rule 110, the problem is that these diagrams have an exponential growth for large gliders.

An interesting point of Rule 110 is the existence of complex behaviors in a periodic background, something that does not happen in The Game of Life which has a stable background. Rule 110 also defines several periodic backgrounds with other combinations of triangles. In these backgrounds the existence of others gliders can be discussed, although they must be carefully established because a small irregularity destroys these gliders and forms natural ether.

There is an unlimited number of interactions in Rule 110 and in the same way as The Game of Life, a time must pass to find new devices like blinkers, flip-flop configurations, structures eating other gliders, or large still-life regions. In this sense a complete and interesting study is presented in [5].

The analysis of automata with complex behaviors produced by the existence of gliders is an interesting area as Andrew Wuensche describes in [23]. Wuensche defines a very practical process to filter gliders in periodic backgrounds of distinct cellular automata. A further work is to project Rule 110 in two and three dimensions for detecting other properties which have not been observed in one dimension. In the tridimensional case, the evolution space covered by tetrahedrons defining gliders and a periodic background must be spectacular.

Acknowledgements. In special to Matthew Cook, to Departamento de Aplicación de Microcomputadoras in UAP, the support of CONACyT with registry number 139509 and to Centro de Investigación Avanzada en Ingeniería Industrial in UAEH.

References

1. Andrew Adamatzky (Ed.), *Collision-Based Computing*, Springer, 2002 (ISBN 1-85233-540-8).
2. Elwyn R. Berlekamp, John H. Conway and Richard K. Guy, *Winning Ways for your Mathematical Plays*, Academic Press, 1982 (ISBN 0-12-091152-3) Vol. 2, chapter 25.
3. Matthew Cook, "Universality in Elementary Cellular Automata," personal communication.
4. Matthew Cook, "Introduction to the activity of rule 110" (copyright 1994-1998 Matthew Cook), http://w3.datanet.hu/~cook/Workshop/CellAut/Elementary/Rule110/110pics.html, January 1999.

5. Matthew Cook, "Still Life Theory," in *New Constructions in Cellular Automata* p. 93, (Santa Fe Institute Studies on the Sciences of Complexity) Oxford University Press, April 2003 (ISBN 0-1951-3717-5).
6. Martin Gardner, "Mathematical Games – The fantastic combinations of John H. Conway's new solitaire game Life," *Scientific American* **223**, pp 120–123, 1970.
7. Jim Giles, "What kind of science is this?" *Nature*, **417** 216–218 (16 May 2002).
8. Genaro Juárez Martínez, "Fases $f_{i\_}i$ en la Regla 110," *en preparación.*
9. Genaro Juárez Martínez, "Solitones en el autómata celular unidimensional Regla 110," April 8, 2002 http://delta.cs.cinvestav.mx/~mcintosh/oldweb/working.html.
10. Genaro Juárez Martínez and Harold V. McIntosh, "ATLAS: Collisions of gliders like phases of ether in rule 110," http://delta.cs.cinvestav.mx/~mcintosh/comun/s2001/s2001.html, August 2001.
11. Genaro Juárez Martínez, Harold V. McIntosh and Juan Carlos Seck Tuoh Mora, "Estructuras periódicas cubriendo el espacio de evoluciones en el autómata celular unidimensional Regla 110," http://delta.cs.cinvestav.mx/~mcintosh/oldweb/working.html, Abril 8, 2002.
12. Genaro Juárez Martínez, Juan Carlos Seck Tuoh Mora and Harold V. McIntosh, "Reproducing the cyclic tag systems developed by Matthew Cook with Rule 110 using the phases $f_{i\_}i$," *in preparation.*
13. Mariusz H. Jakubowski, Ken Steiglitz and Richard Squier, "Computing with Solitons: A Review and Prospectus," *Multiple-Valued Logic*, Special Issue on Collision-Based Computing, vol. 6, Numbers 5–6, 2001 (ISSN 1023-6627).
14. Harold V. McIntosh, "Linear cellular automata via de Bruijn diagrams," http://delta.cs.cinvestav.mx/~mcintosh/oldweb/pautomata.html, 1991.
15. Harold V. McIntosh, "Rule 110 as it relates to the presence of gliders," http://delta.cs.cinvestav.mx/~mcintosh/oldweb/pautomata.html, January 1999.
16. Harold V. McIntosh, "A Concordance for Rule 110," http://delta.cs.cinvestav.mx/~mcintosh/oldweb/pautomata.html, April 2000.
17. Harold V. McIntosh, "Rule 110 Is Universal!," http://delta.cs.cinvestav.mx/~mcintosh/oldweb/pautomata.html, June 30, 2002.
18. James K. Park, Kenneth Steiglitz and William P. Thurston, "Soliton-like behavior in automata," *Physica D* **19**, 423–432, 1986
19. John Scott Rusell, "Report of Waves," *Re. Brit. Assoc. for the Advancement of Science*, pp. 311–390, 1844.
20. John von Neumann, *Theory of Self-reproducing Automata*, University of Illinois Press, Urbana and London 1966.
21. Stephen Wolfram, *Theory and Applications of Cellular Automata*, World Scientific Press, Singapore 1986.
22. Stephen Wolfram, *A New Kind of Science*, Wolfram Media, Inc., Champaign, Illinois, 2002 (ISBN 1-57955-008-8).
23. Andrew Wuensche, "Classifying Cellular Automata Automatically," *Complexity*, Vol. 4, no. 3, 47–66, 1999.
24. "OSXLCAU21" system is available for OpenStep and Mac OS X operating systems in http://delta.cs.cinvestav.mx/~mcintosh/comun/s2001/s2001.html, August 2001.

A Computational Model of Neocortical-Hippocampal Cooperation and Its Application to Self-Localization

Michail Maniadakis and Panos Trahanias

Institute of Computer Science
Foundation for Research and Technology-Hellas (FORTH)
71110 Heraklion, Crete, Greece
and
Department of Computer Science
University of Crete
71409 Heraklion, Crete, Greece
{mmaniada,trahania}@ics.forth.gr

Abstract. Recently many computational modules of hippocampal system have been proposed, investigating mainly the development of place cells, similar to mammals, but without employed by other structures for further use. We propose a biologically plausible computational model of neocortical-hippocampal cooperation, which is based on familiarity recognition by neocortex, followed by a recall process in the hippocampus. Our model is implemented and tested in a simulated robotic platform, which shows that neocortex is able to interact with hippocampus for the development of a self-localization behaviour.

1 Introduction

The hippocampus is one of the most studied areas of the mammalian cortex because of its prominent role in the memorization of spatial information. Different groups of cells have been detected in the mammalian hippocampus, which preferably fire when the animal is in a particular portion of its environment, but they are largely independent of its orientation and actual view [1]. These cells are usually termed *place cells*. Following recent trends in the area, we focus our study in the investigation of the entorinal cortex (EC) from parahippocampal region and dentate gyrus (DG) and Amon's horn structures CA3, CA1 from hippocampal formation. Lately, place cells have been detected in all these structures.

Early approaches in hippocampal comutational models consist of an arrangement of appropriately connected neurons on a planar map. For example in [2] allothetic visual information is used to perform quantization of the environment, while idiothetic motor information is integrated in a feed-forward neural model to connect locations. A similar idea is used by [3] but with a much more sparse representation of the environment. However, according to [4], [5], the existence of a topographical relation between environmental location and hippocampal cells seems not valid.

W. Banzhaf et al. (Eds.): ECAL 2003, LNAI 2801, pp. 183–190, 2003.

Since the anatomical structure of CA3 consists of a high number of synapses within pyramidal cells, it is usually assumed to perform relational computations. This is taken into consideration by recent hippocampal models. For example in [6] a neural model with recurrent connections similar to CA3 is proposed, which operates in two modes (learning-recall). A combination of planar map with recurrent connections which use attractor dynamics is presented in [7].

The projection from CA1 to EC is usually omitted in many proposed models. This is a very critical design decision, since a recurrent cellular structure is computationally represented by a feed forward one. A computational model with re-entrant projections from CA1 to EC is presented in [12], but it is not tested for the development of place cells.

Certain hippocampal models are oriented towards the development of navigation abilities [8], [2]. These models assume that information about goal location is given in the hippocampus, even though there is no experimental evidence of its existence [9]. Our approach complies with the belief that hippocampal system is not directly involved in navigation, since it has been experimentally proved that rats with hippocampal lesions are able to navigate to visible goals [10],[11]. Thus, in the present study we don't investigate navigational abilities of hippocampus. In contrast, we focus on the cooperation of neocortical and hippocampal structures to infer self-location, which can be further used by the neocortical structures (prefrontal, motor cortex) responsible for action.

A similar work investigating neocortical-hippocampal cooperation [12] uses separate input and output EC structures to mediate interaction. However, it has only been tested in the memorization of static relational information but not the development of place cells.

In summary, existing hippocampal models exhibit shortcomings, with more important ones the lack of neocortical structures with efferent projection to hippocampus, the lack of detailed EC computational model, the lack of re-entrant projection from CA1 to EC, and the non-cooperative performance of hippocampal and neocortical areas.

In this work, we present a detailed hippocampal model with separate, biologically plausible, computational modules for each hippocampal area (EC, DG, CA3, CA1) and we test its interaction with neocortex. Similar to other models we use a large number of recurrent connections in CA3 to perform relational computations. Instead of representing EC by a poor input layer of sensory allothetic information, we provide a model of neocortical sensory association area (AC) with afferent and efferent EC connectivity. Our model is able to develop place cells in all hippocampal areas, similar to mammalian hippocampus. Even though our approach is currently tested in a spatial problem, we believe that it is also able to encode individual events within episodes, whether spatial or not [5], by encoding the relation within event features. The existence of a detailed computational model for AC, the single model for EC, together with the recurrent connectivity within AC-EC and EC-CA1 modules, constitute the main contributions of our approach.

In the following section we present the details of our approach for the hippocampal processing of sensory stimuli and the cooperation with neocortex. The

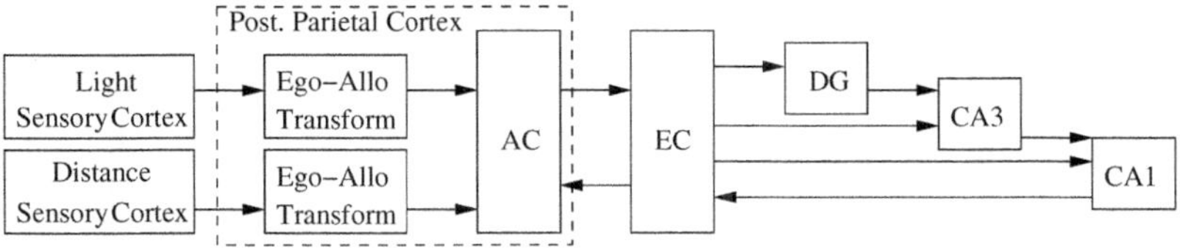

Fig. 1. The flow of information in the proposed neocortical-hippocampal model

results from the application of our model in a spatial learning task of a simulated robot are presented in section 3. Finally, conclusions and suggestions for further work are drawn in the last section.

2 Methodology

Our approach is in accordance with the flow of information in the mammalian central nervous system. The general layout is shown in Fig 1. Environmental information reaches the somatotopically organized sensory cortex. Posterior Parietal Cortex receives afferent projection from sensory cortices to perform high order processing. This part of the cortex undertakes egocentric to allocentric transformation of sensory information, and also relates different stimuli in the Association Cortex (AC). Experimental evidence for ego-allo transformation in Posterior Parietal Cortex is given in [9]. AC projects to EC which is the gate of hippocampus, and has efferent projection to all hippocampal formation structures. DG examines current information to emphasize novel features and projects to CA3 which perform temporal and spatial relational processing. The results of CA3 together with the detailed input from EC are passed to CA1 to perform fine tuning. The integration of the results with new sensory information takes place in EC which receives afferent projection from CA1. Efferent projections from EC to AC are used to store the statistical regularities of environmental stimuli in neocortex for further use (e.g. action planning).

2.1 Egocentric-Allocentric Transform of Sensory Information

Animals receive egocentric information from their sensors which is modulated by their orientation in the environment. However, it has been experimentally shown that hippocampal system process allocentric (orientation invariant) information. Thus, allocentric information is fed in the hippocampus, which is a common hypothesis for all computational models.

The existing computational models assume a hard-wired transformation from egocentric to allocentric information. Instead, we have implemented a simple computational module to perform this transformation, given the current orientation ϕ of the animal. For the sake of simplicity we assume that the number of head-direction (HD) neurons is equal to the number of light or distance sensors; let this number be M. Each HD neuron has a preferred direction θ of maximal activation and follows the gaussian model, similar to real HD cells [13]. Let us assume that the information of the $i - th$ egocentric distance sensor is given by h_i. The allocentric distance measure is achieved by the following summation over all HD neurons $f_i = \frac{\sum_{j=0...M-1} h_{(i+j)modM} e^{-(\phi-\theta_{(M-j)modM})^2}}{\sum_{j=0...M-1} e^{-(\phi-\theta_{(M-j)modM})^2}}$ where f_i is the new orientation invariant measure. It is interesting to observe that this formula can be directly used to combine our approach with other computational models that develop HD cells (e.g. [14]).

2.2 The Cortical Computational Module

We have implemented a general computational module to represent cortical areas. This model bears similarities with other works in the definition of synapses [14], and the computational model of the neurons [6]. In contrast to the majority of hippocampal models, a separate module is employed to represent each area of hippocampus and neocortex.

Synapse Definition. Each cortical computational module consists of a population of excitatory and inhibitory neurons with a predefined position. A rectangular plane with both sets of neurons uniformly distributed simulates the cortical area. In order to achieve common spatial properties for neurons in the middle and neurons in the borders of the plane, we assume that opposite planar sides are met and the neurons near by can be connected. Interconnectivity of neurons follows the general rule of locality [14] for both intracortical and cortico-cortical connectivity

The local intracortical connectivity is expressed by the rule that close neurons are more likely to be connected. It is simulated with a stochastic gaussian linking. If d is the distance of neurons a,b then $t = e^{\frac{-d^2}{\sigma}}$ defines the probability of synaptic connection for those neurons, (σ determines the sharpness of connectivity). Using a random $r \in [0, 1]$ a synapse from a to b is defined if $r < t$. The synapse is assigned with a random weight $w_{ab} \in [0, 1]$ to represent its strength. The same process is repeated for all pairs of neurons (excitatory to excitatory, excitatory to inhibitory, inhibitory to excitatory) to define the local connectivity within a cortical module. We assume that inhibitory neurons are used to enforce separability within local planar areas, thus inhibitory-inhibitory synapses are not used, and only three different sets of synapses are defined W^{EE}, W^{EI}, W^{IE}.

Long range cortical connectivity follows the general rule that neighbouring cells project to neighbouring areas. Thus, different cortical modules are connected using their spatial properties inherited by the planar model with one-way synapses. For each cortical module B, which receives afferent projection from a module A, a set of input neurons Inp is defined, equal to the number of excitatory neurons in module A. They are located in the same positions as their respective excitatory neurons in module A. The connectivity from input neuron a to excitatory neuron b is defined by employing the same gaussian stochastic linking described above, with σ_{InpE} connectivity variance. Thus a set of input synapses V with random synaptic weights is defined in module B.

Neuron Model. All neurons of the cortical module, follow a modified version of the Wilson-Cowan neuronal model, similar to [6]. Let p represent the potential, and q the activation of a neuron. The potential of each neuron is updated based on the afferent input information, and the excitatory and inhibitory signals accepted by neighbouring neurons. This is expressed mathematically, in a single form for both excitatory and inhibitory neurons, by: $\frac{1}{\mu}\Delta p_b = -p_b + \sum_{v_{ab}\in V} v_{ab}inp_a + \sum_{w_{ab}\in W^{EE}\cup W^{EI}} w_{ab}q_a - \sum_{w_{ab}\in W^{IE}} w_{ab}q_a$ where μ presents the membrane time constant. Then, the activation of the neuron is defined using the non-linear sigmoid function $q_b = \frac{1}{1+e^{-\alpha(p_b-\beta)}}$ where β stands for the threshold, and α is the slope of the activation function.

Learning Rules. A learning process adjusts the initially random strength of synapses to encode the spatial properties of the environment. We have used four biologically plausible Hebbian-like rules to train cortical modules. Principal Component Analysis rule is used to maximize information flow within neurons, while Anti-Hebbian learning facilitates the development of a novelty detection mechanism. We have also used Postsynaptic and Presynaptic rules, which simulate the biological heterosynaptic and homosynaptic learning. Assuming that there is a synapse with strength z_{ab} from neuron a with activation q_a to neuron b with activation q_b, then learning rules are described bellow.

PCA Rule [15]: $\Delta z_{ab} = q_b(q_a - q_b z_{ab})$.
AntiHebbian Rule [16]: $\Delta z_{ab} = k + \frac{-2q_a q_b}{{q_b}^2+1}$, where $k > 0$.
PostSynaptic Rule [17]: $\Delta z_{ab} = z_{ab}(q_a - 1.0)q_b + (1.0 - z_{ab})q_a q_b$.
PreSynaptic Rule [17]: $\Delta z_{ab} = z_{ab}(q_b - 1.0)q_a + (1.0 - z_{ab})q_a q_b$.

3 Experimental Results

In order to evaluate our neocortical-hippocampal model we have tested it on a simulated Khepera robot. The goal of our experiments was to implement a self-localization process of the robot in the environment. In the following we present a detailed sample experiment. We specify the actual experimental setup and then report on the obtained results.

3.1 Experimental Setup

Our model (Fig. 1) receives information from 8 distance and 8 light sensors with similar range of view, which are uniformly distributed in a circle around the robot. This somatotopic relation is preserved in the Sensory Cortex, similar to mammals. Each sensor is assigned a position in a circle of the Distance or Light Sensory cortex. Both circles have diameter of 60, and a random centre. Then, environmental information is transformed from egocentric to allocentric measures using the process described in section 2.1. Both the allocentric Distance and Light Sensory Cortex are projected to AC and then to hippocampus via EC. After hippocampal processing, recalled memory is projected back to AC, again via EC.

We use planes of common size 100×100 distance units for all cortical modules. The parameters used for cortical module construction are shown in Table 1. All modules have a small number of recurrent interconnections to simulate synapses within cortical layers. The activation of excitatory and inhibitory neurons have a local and global effect, respectively. This model of local excitation (through excitatory neurons) and global inhibition (through inhibitory neurons), imposes competition within different areas of the cortical plane. However, in CA3 a large value of σ_{EE} is used to simulate high recurrent connectivity within CA3 pyramidal cells.

The neurons in most modules have relatively high values of potential change μ to capture the fast environmental changes, together with a high slope α for sharp representation. Only CA3 neurons have a large range of variance in potential change to explore temporal relations (some neurons are updated rapidly and some others very slowly). This fact, in conjunction with the large number of synapses in CA3, allows a spatio-temporal relational processing in this module. The thresholds in each module have been set experimentally and do not have a special biological meaning.

Table 1. The first 2 columns show the number of excitatory and inhibitory neurons in each module. The following 7 columns show the variance σ_{InpE} used for cortico-cortical connectivity (Dsc and Lsc stands for Distance and Light Sensory Cortex). The next three columns show the variance for intracortical connectivity, and the last 3 the parameters of neurons.

	EN	IN	Dsc	Lsc	AC	EC	DG	CA3	CA1	σ_{EE}	σ_{EI}	σ_{IE}	μ	α	β
AC	64	36	15	15	-	10	-	-	-	10	20	30	[0.8-0.9]	2.5	1.8
EC	49	25	-	-	10	-	-	-	10	12	15	20	[0.7-0.8]	3.0	1.2
DG	49	25	-	-	-	10	-	-	-	12	15	20	[0.7-0.9]	3.0	0.5
CA3	49	25	-	-	-	12	10	-	-	20	25	30	[0.2-0.9]	2.0	0.8
CA1	49	25	-	-	-	12	-	10	-	10	15	25	[0.8-0.9]	3.0	0.4

Table 2. Each cortical computational module uses a different set of learning rules to adjust synaptic weight.

	Input-Exc.	Exc.-Exc.	Exc.-Inh.	Inh.-Exc.
AC	PCA	PreSyn.	PostSyn.	PreSyn.
EC	PCA	PostSyn.	PCA	PreSyn.
DG	PCA	AntiHeb.	PCA	PCA
CA3	PostSyn.	PostSyn.	PostSyn.	PreSyn.
CA1	PCA	PostSyn.	PCA	PCA

Different learning rules were used to adjust the various types of synapses in each module. Similar to the learning functionality of neocortex and hippocampus, we assume that hippocampus is specialized for rapid memorization, while neocortex is specialized for slowly learning about statistical regularities in the environment. Thus, the learning rate for AC was set to 0.01, while hippocampal structures learned with a three times faster rate. Learning rules applied in each module are shown in Table 2.

3.2 Findings

In our experiment, we set the simulated robot to move around in the environment in a wonder mode for 10000 steps. The range of view of the robot was limited to a maximum of two sides at a time (those which are vertically connected). We have used an environmental shape with many angles to enforce the locality of robot view. The simulated workspace employed is depicted in the middle of Fig 2; each light source is depicted on it by an arrow.

The results of place cell development in CA1 for different locations (those marked with the numbers 1 through 9) are shown in Fig 2(a). Evidently, these results indicate the effective development of place cells in CA1 and compare promisingly to other results from the literature [2],[6]. Furthermore, similar neural firing is present in the other areas of hippocampal system (EC,DG,CA3).

The recalled locations are also transfered to AC (Fig 2(b)) and are readily available for further use (e.g. motor control). We believe that this novel result, i.e. transfer of recalled location to AC, confirms the validity and suitability of the proposed model. Additionally, in order to test the stability of our model we let the robot run for 50000 steps and we got very similar results.

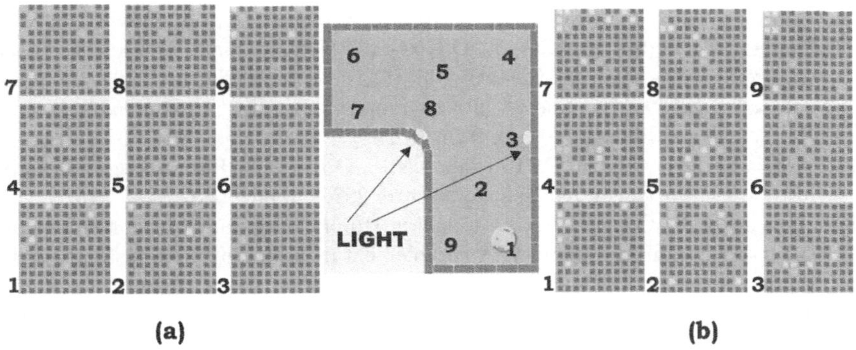

Fig. 2. Results of Self-localization via place-cell development in (a) CA1 and, (b) AC

4 Conclusions

We have presented a model of neocortical-hippocampal cooperation and its application to self-localization. Our approach complies with the main structural, functioning and learning properties of the mammalian cortex, constituting it a biologically plausible model. The introduced model has been successfully tested in the development of place cells for localization, similarly to the mammalian hippocampus. Moreover, this knowledge is also projected to neocortex for further use. Since AC in mammals is one of the areas which relate sensory and motor processing, our future work aims at the use of spatial knowledge for action planning in the motor cortex.

References

1. O'Keefe J.: Place units in the hippocampus of the freely moving rat. Exp. Neurol., 51, pp. 78–109, 1976
2. Arleo A., Gerstner W.: Spatial cognition and neuro-mimetic navigation: A model of hippocampal place cell activity. Biological Cybernetics, 83, pp. 287–299, 2000.
3. Hafner V.V.: Learning places in Newly Explored Environments. Proc. SAB2000.
4. O'Keefe J., Burgess N., Donnett J., Jeffery K., Maguire E.: Place cells, navigational accuracy, and the human hippocampus. Ph. Tran. R. Soc. 353, pp. 1333–1340, 1998.
5. Eichenbaum H., Dudchenko P., Wood E., Shapiro M., Tanila H.: The hippocampus, memory and place cells: Is it a spatial memory or a memory space? Neuron, vol. 23, pp. 209–226, 1999.
6. S. Kali, P. Dayan: The involvement of recurrent connections in area CA3 in establishing the properties of place fields: a model. J. Neurosc. 20(19) pp. 7463–77 2000.
7. Samsonovich A. McNaughton B.L.: Path integration and Cognitive Mapping in a Continuous Attractor Neural Network Model. J. Neurosc., 17(15), pp. 5900–20, 1997.
8. Hasselmo M. E., Hay J., Ilyn M., Gorchetchnikov A.: Neuromodulation, theta rhythm and rat spatial navigation. Neural Networks, 15, pp. 689–707, 2002.
9. Burgess N., Becker S., King J., O'Keefe J.: Memory for events and their spatial context: models and experiments. Ph. Trans. R. Soc., 356, pp. 1493–1503, 2001.
10. Jarrard L.E.: On the role of hippocampus in learning and memory in the rat. Behav. Neural Biology, 60, pp. 9–26, 1993.
11. Morris R.G.M., Garrud P., Rawlins J.N.P., O'Keefe J.: Place navigation impaired in rats with hippocampal lesions. Nature, 297, pp. 681–683, 1982.
12. Norman K.A., O'Reilly R.C.: Modelling Hippocampal and Neocortical Contributions to Recognition Memory: A complementary Learning Systems Approach University of Colorado, Boulder, ICS, Technical Report 01-02, 2001.
13. Taube J.S.: Head direction cells and the neuropsychological basis for a sense of direction. Progress in Neurobiology, 55, pp. 225–256, 1998.
14. Redish A.D., Elga A.N., Touretzky D.S.: A Coupled attractor model of the rodent head direction system. Comp. in Neural Systems, 7, pp. 671–685, 1996.
15. Oja E.: A simplified neuron model as a principal component analyzer. Journal of Mathematical Biology, 15, pp. 267–273, 1982.
16. Schraudolph N.N., Sejnowski T.J.: Competitive Anti-Hebbian Learning of Invariants Advances in Neural Information Processing Systems, 4, pp. 1017–1024, 1992.
17. Urzelai J., Floreano D.: Evolutionary Robotics: Coping with Environmental change. Proc. GECCO2000.

Fascinating Rhythms by Chaotic Hopfield Networks

Colin Molter and Hugues Bersini

Laboratory of Artificial Intelligence
CP 194/6 University of Brussels, Brussels, Belgium
cmolter@iridia.ulb.ac.be

Abstract. This papers aims to introduce a new way to store inputs in the network. By this way, it appears that the spontaneous dynamics of the network will increase in complexity by increasing the size of the learning set. This experimental work might give additional support to the Skarda and Freeman strong intuition that chaos should play an important role in the storage and the search capacities of our brains. An analysis of the type of chaos exploited to code these inputs will be related with the "frustrated chaos" described in previous papers. A live demonstration can be shown where rhythms and melodies associated with the dynamics can be heard.

1 Introduction

If one of the 2 main types of artificial neural networks, the feed-forward neural network, has been successfully used in different kind of applications, from classification to control, the other promising one, the fully connected recurrent neural network (RNN), has still not found useful applications despite many trials in distinctive domains.

Work has been done in order to implement supervised learning methods on RNN (Backpropagation through time [1], Extended Kalman-filtering approaches [2], Echo state network [3], ...) but no ones seems to justify an application for a concrete problem. The drawbacks of these methods may vary, but generally comes from slow convergence (if the convergence has been guaranteed), expensive calculations (NP-hard problem [4]), some crucial parameters which have to be finely tuned, demanding a lot of expertise to the user.

Work has also been done in order to use RNN for storing data. The successful hebbian learning algorithm on Hopfield network has been found. But unfortunately, its capacity is very poor (less than 14 % of the network's dimension).

Despite these failures, since the seminal paper of Skarda and Freeman dedicated to chaos in real brains [5], many authors have shared the idea that chaos is the perfect regime to store and efficiently retrieve information in neural networks [6], [7], [8], [9], [10], [11], [12], [13], [14]. Indeed this complex dynamical regime, although very simply produced, inherently possesses an infinite amount of cyclic regimes to be exploited for coding memories. Moreover, the dynamics

W. Banzhaf et al. (Eds.): ECAL 2003, LNAI 2801, pp. 191–198, 2003.

randomly wanders around these unstable regimes in an autonomous way, thus rapidly proposing alternative responses to external stimuli, and able to easily switch from one to another of these potential attractors in response to a new coming stimulus.

Exhaustive experimental descriptions of bifurcations and their resulting dynamics occurring are still missing. However it has been suggested by Bersini et al [15], [16] that chaotic dynamics occurring in these networks are different from other ones. This new kind of chaos has been described as *frustrated*. It belongs to the family of intermittency type of chaos, first described in [17], [18]. It is a dynamical regime that appears in a network when the global structure is such that local connectivity patterns responsible for stable oscillatory behaviors are intertwined, leading to mutually competing attractors and unpredictable itinerancy among brief appearance of these attractors. These successive appearances of almost stable regime are separated by chaotic bursts. The relative duration of the dynamics into one of these temporary attractors depends on the respective size of its basin of attraction. All basins are intertwined in a fractal way.

Some extended analyzes of small RNN (from 2 to 10) are performed here. For this purpose, a platform has been developed enabling to trace different kinds of diagrams with different tools (bifurcation diagrams, Lyapunov exponents diagrams, FFT, Return Map,...).

An attempt to develop a new usage of RNN has also been investigated with promising applications. By introducing a symbolic discretization of the outputs, it is shown how these networks could be practically exploited to code information. For each input to store, the only rule will be to have a periodic output. Said differently, each stored input has to generate a distinguished and dynamically stable symbolic output. We'll see that by increasing the amount of information to be stored, it entails the network to spontaneously reinforce the "chaoticity" of its autonomous dynamics. No learning rule will be developed to find "good" weight matrix : it has appeared that a random search biased by an heuristic gave more convincing and faster results (the supervised learning algorithm ESN is also based on random matrix [3]. To increase the size of the learning set, instead of increasing the size of the network, or the period of the cycle in output, we chose to join multiple neuro-modules together, the input feeding one of By this way drawbacks of resolution of an NP-hard problem are avoided.

In the next section, the experimental set up is described. Once the network has learned, section three shows the results. Our main two expectations will be verified. First of all, by plotting the cardinality of the learning set of these new regimes as a function of the Lyapunov exponents, a clear monotonous relationship will appear. The more input to be stored in the network, the more chaotic regimes are favored, as Skarda and Freeman anticipated, to indeed make sense of the world. It gives sense to the suggestion that chaotic behavior is required by the brain in order to perform efficient non-deterministic symbolic computation during cognitive tasks [19]. Secondly, once the network has learned and is subsequently feed with a new input, in-between learned ones (so to some extent mixing some features of them), the resulting dynamics reflects this ambiguity by adopting the type of frustrated regimes presented in the previous publications with, as bounding cycles, the ones that the learning process associated with the input to be stored.

In order to illustrate these results, the networks have been used to store rhythms and melodic phrases. By modifying the inputs, chaotic music appears during phase transition. Bigger melodic phrases are generated by coupling these networks as explained above.

2 Description of the System

The Analogic Hopfield Networks to be tested here are kept small (from 2 to 10 neurons) and fully connected. Neurons activation is continuous, ranging from 0 to 1, and the evolution in time of the activation is discrete. Moreover, each neuron receives a constant input. The mathematical description of the Hopfield Network is the very classical one. The evolution equation of a neuron x_i is given by the equation 1.

$$x_i^k(t) = \frac{1}{2}\left(1 + tanh\left(g\sum_{j=1}^{n} w_{ij}x_j(t) + i_i^k\right)\right) \tag{1}$$

with as parameters:

- n : the number of neurons of the network (from 2 to 10);
- g : the slope of the sigmoid. The value has been set arbitrarily to 9, not too small to avoid global stability,nor to big which would be equivalent to a triggering function;
- w_{ij} : the weight between the neuron j and i (from -1 to 1). The matrix will be noted W ;
- i_i^k : the constant input k of the neuron i (from -1 to 1);

The network has to learn a set of k different inputs where every input is a point in a n-dimensional space, given or generated randomly. Simply said, the network will be stimulated by a constant input to learn, and its stabilized resulting dynamics will be the way to "learn" or to "store" this input. To avoid transitory effect 1000 time steps are first computed before observing the output. For sake of simplicity, the activations of all neurons are averaged, and it is this average value whose dynamical behavior needs to be associated with each input to be stored. Let's call this average dynamical output $x^k(t)$. Its value is discretized in a certain number of equal intervals, n_{dis}, a further parameter of the simulation. For instance, for $n_{dis} = 7$, the $x^k(t)$ takes only seven discrete symbolic values: a, b, c, d, e, f, g. Such kind of symbolic dynamics is being more and more proposed to extract invariant grammatical and statistical characteristics from time series [20]. The resulting dynamics generates words in the following way : at each time step, the letter to add is the one associated with the interval containing $x^k(t)$.

Each input of the learning set triggers and thus becomes associated with a fixed point or a cyclic output. The maximum period of the cycle, p_c, is a further parameter of the simulation. For instance tuning p_c to 1 forces the network to only deliver symbolic fixed points in response to any input, while tuning it to 2, one possible output turns out to be *abababab*.... The maximum number of possible outputs to be triggered by the input is proportional to ${n_{dis}}^{p_c}$, exponentially increasing with the degree of discretization, and boosting, as compared with the original fixed point Hopfield networks, the storing capacity. For example, if $k = 5$, $n_{dis} = 10$, and $p_c = 3$, for a randomly chosen input set, after some iterations, the program could find a weight matrix, which associates to the 5 inputs the stabilized attractors $((aaa...), (aagaag...), (ghcghc...), (ggg...), (gcdgcd...))$.

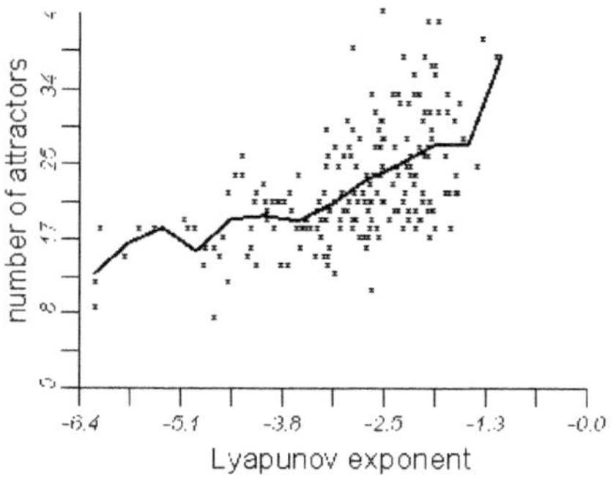

Fig. 1. Plot of the number of robust dynamic attractor from an input set of 200 inputs in function of the averaged Lyapunov exponent (obtained from the computation of the lyapunov exponent of 200 inputs randomly chosen). 200 randomly chosen matrix of 3 neurons. It shows that "chaotic" network can have bigger input learning set.

Finally, in order to guarantee the robustness of the learning, in a way reminiscent of the original use of Hopfield Nets as associative memory, the same dynamical output must be associated with a variety of inputs, all selected in a small hypersphere of diameter d_{pert} around the original input to learn.

The way the network is used differs slightly from usual. We can distinguish 2 cases. Firstly, a matrix is chosen and we just look to which inputs this matrix could be relevant. In the second case, we look for a matrix which could associate a cyclic output to any given input of a learning set. In any cases, there is a learning process, the matrix is randomly chosen with the use of some heuristics: it has been proved statistically that the weight average has to be small, and slightly negative.

This brute force learning process is justified by the very irregular and fractal bifurcation diagrams (like can be seen in figure 2) obtained by continuously changing one of the synaptic value. Since the successive attractors appear in the network in a critical way, it becomes difficult to imagine a smooth learning process (like a gradient-based method) that slightly modifies the weight as a feedback function of the output of the network. However, because of the exponential explosion of the time required to learn an increasing number of inputs, a more clever learning process has to be found. Construction of bigger network by joining with a synaptic weight neuro-modules having strong autonomous chaotic behavior seems to give good results.

Finally, in order to show how, by increasing the number of input to learn, chaos more and more prevails as the spontaneous regime of the network, we tested the dynamics of the network when presenting it with 100 new random inputs. For each associated resulting dynamics, the biggest Lyapunov exponent is computed and averaged over the 100 inputs. The computation of the Lyapunov exponent is done following the method described in [21].

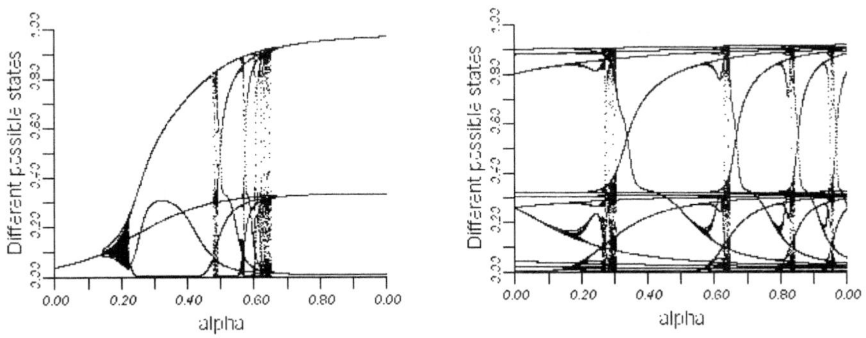

Fig. 2. Bifurcation diagrams shown as a function of the input i varying linearly between two of the input to learn i^j, i^k. Chaos appears in a critical way and randomly merge the dynamical attractors associated with the learned inputs. The right figure shows a zoom of the chaotic attractor.

3 Numerical Results

Two kinds of results are shown in this section. For all the results, the attractor in the output will have a period of maximum 4.

The main result is shown in the figure 1. Two things appears, firstly, that a small number of neurons with a well chosen weight matrix can be used to store a big set of inputs (30 inputs for 3 neurons). Secondly that the size of the learning set increases monotonously in function of the averaged Lyapunov exponent. The others results illustrate the type of chaos appearing in these networks. Practically, in order to find a network having good guess to generate chaotic dynamics, we used the previous results and we choose a network having a big input learning set. It connects this work with the works presented in previous publications [16] analyzing the nature of the frustrated chaotic attractor in small Hopfield Networks. One can see the presence of chaos and how this chaos seems to be built upon the cyclic responses associated with the two inputs. When presenting with ambiguous inputs, the resulting dynamics reflects this ambiguity by exhibiting a frustrated regime merging the dynamics that have been previously associated with the input to learn.

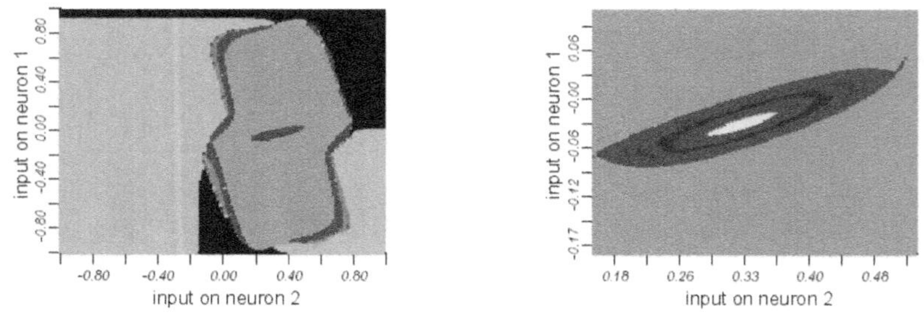

Fig. 3. The dynamic of the output is plotted in function of the variation of the input to 2 neurons. The left figure is an enlargement of the right one. The fractal structure appears.

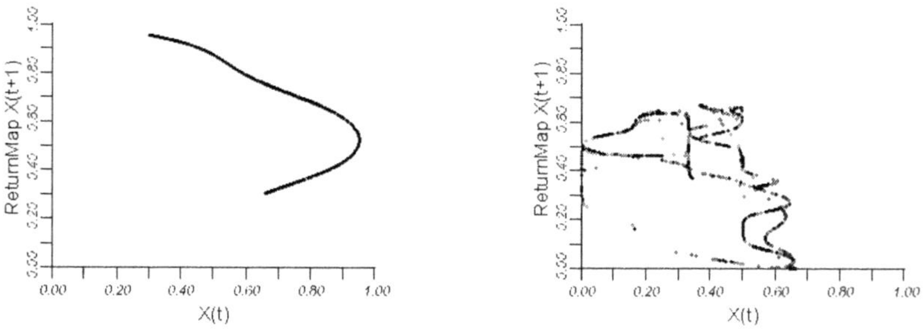

(a) RNN of 2 neurons

(b) RNN of 6 neurons

Fig. 4. Return map coming from 2 networks of different size and, by consequent, from different dynamical domain. For each, it has been verified by viewing the graph of the FFT, that it corresponds to a strange attractor. We see that by increasing the size of the network, the structure becomes more complex.

Figure 3 shows the fractal structure occurring often during the transition between 2 attractors In order to have more information from these diagrams, chaotic attractors have been analyzed (and confirmed) with the help of the Lyapunov exponent, bifurcation diagrams, the FFT and the return map.

In the figure 4, 2 return map from strange attractors are plotted, each one coming from a different dynamical domain and from a network of different size. All chaotic dynamics have given a structured return map (like the return map of the logistic map). Two constatations have been made, firstly that all points coming from the same chaotic dynamical domain have similar return maps. Secondly, when the size of the network is increased, the return map of these strange attractors becomes more complex.

4 Fascinating Rhythms

In order to have a better representation of the attractors of the network, the output of the network is feeding a rhythm or a melody generator. In the first case, at each letter associated to the output, a rhythm in 4 times is associated, while in the second case, each neuron is connected to a selected instrument, specifying its frequency. The platform enables to modify the input. So by varying the sliders, it is possible to shift from one attractor to another one, and to hear the fascinating dynamics occurring between the two. It's also possible to create multiple small neuro-modules and to join them by a variable synaptic weight. More complex melodies can be heard by this way.

5 Conclusion

This paper aims at fulfilling two experimental missions. The first one is to show what several authors claim in more intuitive and qualitative ways namely how the harder the learning process is for an unconstrained Neural Network the more chaos prevails as a natural and spontaneous dynamical regime.

The second one is to show how the frustrated chaos in Hopfield Networks can be construed as a dynamical version of the "hesitation" of the network between possible responses to associate with the current situation. Ambiguous input provokes ambiguous dynamics [22].

References

1. Werbos, P.: Backpropagation through time: what it does and how to do it. Proceedings of the IEEE 78 **10** (1990) 1550–1560
2. Feldkamp, L., Prokhorov, D., Eagen, C., Yuan, F.: Enhanced multi-stream kalman filter training for recurrent networks. Nonlinear modeling: advanced black-box techniques (1998) 29–53
3. Jaeger, H.: The "echo state" approach to analyzing and training recurrent neural networks. GMD - German National Research Institute for Computer Science **GMD Report 148** (2001)
4. Orponen, P., Šíma, J.: On the computational complexity of binary and analog symetric Hopfields Nets. Neural Computation **12** (2000) 2965–2989

5. Skarda, C., Freeman, W.: How brains make chaos in order to make sense of the world. Behavioral and Brain Sciences **10** (1987) 161–195
6. Babloyantz, A., Lourenço: Computation with chaos: A paradigm for cortical activity. Proceedings of National Academy of Sciences **91** (1994) 9027–9031
7. Freeman, W.: Simulation of chaotic eeg patterns with a dynamic model of the olfactory system. Biological Cybernetics **56** (1987) 139–150
8. Gelder, T.V.: The dynamical hypothesis in cognitive science. Behavioral and Brain Sciences (1997)
9. Gray, C., Köning, P., Engel, A., Singer, W. Nature **338** (1989) 334–337
10. Ishii, S., Fukumizuet, K., Watanabe, S.: A network of chaotic elements for information processing. Neural Networks **9** (1996) 25–40
11. Kozma, R., Freeman, W.: Chaotic resonance - methods and applications for robust classification of noisy and variable pattern. International Journal of Bifurcation and Chaos **11** (2001) 1607–1629
12. Pasemann, F.: Neuromodules: A dynamical systems approach to brain modelling. Supercomputing in brain Research, World Scientific, Singapore (1995) 331–348
13. Tsuda, I.: Towards an interpretation of dynamic neural activity in terms of chaotic dynamical systems. Behavioral and Brain Sciences **24** (2001)
14. Rodriguez, E., Jamel, B., Lachaux, J., Lutz, A., Martinerie, J., Reunault, B., Varela, F.: She's got bette davis eyes: Multi-band brain synchrony during a complex visual task. Nature **397** (1999) 430–433
15. Bersini, H., Calenbuhr, V.: Frustrated chaos in biological networks. J. theor. Biol **177** (1997) 199–213
16. Bersini, H.: The frustrated and Compositionnal Nature of Chaos in Small Hopfield Networks. Neural Networks 11 (1998) 1017–1025
17. Pomeau, Y., Manneville, P.: Intermittent transitions to turbulence in dissipative dynamical systems. Comm. Math. Phys. **74** (1980) 189–197
18. Ott, E.: Chaos in Dynamical Systems. Cambridge University Press (1993)
19. Kentridge, R.: Computation, chaos and non-deterministic symbolic computation : The chinese room problem solved ? Psycoloquy (**12**) 17
20. Hao, B.: Symbolic dynamics and characterization of complexity. Physica D **51** (1991) 161–176
21. Dechert, W., Sprott, J., Albers, D.: On the probability of Chaos in large Dynamical Systems: A monte carlo study. Journal of Economic Dynamics & Control **23** (1999) 1197–1206
22. Kelso, J., Case, P., Holroyd, T., Horvath, E., Raçzaszek, J., Tuller, B., Ding, M.: Multistability and metastability in perceptual and brain dynamics, Ambiguity in Mind and Nature : multistable cognitive phenomena. Volume 64. Springer Series in Synergetic (1995)

Solving a Delayed Response Task with Spiking and McCulloch-Pitts Agents

Keren Saggie[1], Alon Keinan[1], and Eytan Ruppin[1,2]

[1] School of Computer Sciences,
Tel-Aviv University, Tel-Aviv 69978, Israel,
{keren,keinan}@cns.tau.ac.il, ruppin@post.tau.ac.il
[2] School of Medicine,
Tel-Aviv University, Tel-Aviv 69978, Israel

Abstract. This paper investigates the evolution of evolved autonomous agents that solve a memory-dependent delayed response task. Two types of neurocontrollers are evolved: networks of McCulloch-Pitts neurons, and spiky networks, evolving also the parameterization of the spiking dynamics. We show how the ability of a spiky neuron to accumulate voltage is utilized for the delayed response processing. We further confront new questions about the nature of "spikiness", showing that the presence of spiking dynamics does not necessarily transcribe to actual spikiness in the network, and identify two distinct properties of spiking dynamics in embedded agents. Our main result is that in tasks possessing memory-dependent dynamics, neurocontrollers with spiking neurons can be less complex and easier to evolve than neurocontrollers employing McCulloch-Pitts neurons. Additionally the combined utilization of spiking dynamics with incremental evolution can lead to the successful evolution of response behavior over very long delay periods.

1 Introduction

The evolution of artificial neural networks of embedded autonomous agents constitutes a powerful tool for creating agents with complex behavioral abilities, and studying properties of the emergent networks dynamics. In this paper we investigate the evolution of Evolved Autonomous Agents (EAAs) that solve a delayed response task. Delayed response tasks are challenging since they are characterized by a significant delay between the stimulus and the corresponding appropriate response, which make them impossible to solve by a simple sensory-motor mapping. Such tasks are very common in real-life situations, and are thought to be solved by working memory. In the traditional delayed-response *match-to-sample* animal behavior task, an animal is presented with a cue, which determines the action that it should perform, but the animal's action should be taken only after a delay of a few seconds. Here, an autonomous agent is navigating in an arena occupied with food and poison, and should consume as much food as possible, while avoiding poison. In order to eat, the agent has to remain still on a food item for exactly K steps, after which it can consume the food.

W. Banzhaf et al. (Eds.): ECAL 2003, LNAI 2801, pp. 199–208, 2003.

In order to be able to wait precisely K steps facing fixed, invariant inputs during the waiting time, the agent has to "remember" how many time steps it has already waited. Similarly to the match-to-sample task in animals, the cue (the presence of food) determines the corresponding action that should be performed after a substantial delay.

Two types of neurocontrollers are evolved: networks of McCulloch-Pitts (MP) neurons, and spiky networks of *Integrate-And-Fire* neurons. Models of spiking neurons have been extensively studied in the neuroscience literature. Spiky networks have a greater computational power than networks of sigmoidal and McCulloch-Pitts neurons [1], and are able to model the ability of biological neurons to convey information by the exact timing of an individual pulse, and not only by the frequency of the pulses [2,3]. It is appealing to use spiky neural networks in EAAs studies since they are biologically more plausible: Biological neurons perform integration over their pre-synaptic inputs such that a neuron accumulates voltage over time, and fires if its voltage exceeds a threshold. After firing its voltage is returned to a resting voltage, having a refractory period that inhibits its firing. Spiking dynamics may be useful in solving delayed response tasks, since such tasks require memory. Recent studies that combine evolutionary computation with spiky neural networks have analyzed properties of the spiking dynamics in the evolved networks, for example, whether the spiking dynamics result in a time-dependent or a rate-dependent computation, and the effect of noise on the emerging network [4,5].

This study is the first to address the questions of **whether an evolved network with spiking neurons is truly spiky, and how can we define and measure the spikiness level of each neuron.** In this work we try to answer these questions, by presenting two new fundamental ways by which we define and quantify the spikiness functional contribution, and the level of integration of inputs over time of a spiky neuron. Evolving MP and spiky neurocontrollers **enables us to compare and analyze the resulting network solutions** in terms of the difficulty of the evolution, the delayed response processing, and other network properties. The rest of this paper is organized as follows: Section 2 describes the network architectures and the evolutionary procedure. Section 3 analyzes the evolved neurocontrollers and their dynamics. In Sect. 4 we present and quantify two basic properties of spikiness in embedded agents, with which we analyze the evolved spiky agents. These results and their implications are discussed in Sect. 5.

2 The Model

2.1 The EAA Environment

The EAA environment is described in detail in [6]. The agents live in a discrete 2D grid "world" surrounded by walls. Poison items are scattered all around the world, while food items are scattered only in a "food zone" in one corner. The agent's goal is to find and eat as many food items as possible during its life, while avoiding the poison items. The fitness of the agent is proportional to the number of food items minus the number of poison items it consumes. The agent is equipped with a set of sensors, motors, and a fully recurrent neurocontroller of binary neurons.

Four sensors encode the presence of a wall, a resource (food or poison, without distinction between the two), or a vacancy in the cell the agent occupies and in the three cells directly in front of it . A fifth sensor is a "smell" sensor which can differentiate between food and poison underneath the agent, but gives a random reading if the agent is in an empty cell. The four motor neurons dictate movement forward (neuron 1), a turn left (neuron 2) or right (neuron 3), and control the state of the mouth (open or closed, neuron 4).

In previous studies [6], eating occurs if the agent stands on a grid cell containing a resource for one step. Here, we have modified this task to include a delayed-response challenge: In order to eat, the agent has to stand on a grid-cell containing a resource for a precise number of steps K, without moving or turning, and then consume it, by closing its mouth on the last waiting step. Closing its mouth after standing on a food item for more or less than K steps does not increase its fitness. Hence, in essence, the agent has to learn to precisely count to K. The agent's lifespan, defined by the number of sensorimotor steps available to it, is limited. Waiting steps are not counted as part of the lifespan in order to facilitate the evolution of the delayed-response task.

2.2 The Neurocontrollers

All neurocontrollers are fully-recurrent with self-connections, containing 10 binary neurons (out of which 4 are the motor neurons), such that the 5 sensors are connected to all network neurons. We compare between neurocontrollers with McCulloch-Pitts (MP) neurons, employed conventionally in most EAA studies, and ones with spiky *Integrate-And-Fire* neurons. In both types of networks, a neuron fires if its voltage exceeds a threshold. The spiking dynamics of an Integrate-And-Fire neuron i are defined by

$$V_i(t) = \lambda_{\rm i}(V_i(t-1) - V_{rest}) + V_{rest} + \frac{1}{N}\sum_{j=1}^{N} A_j(t)W(j,i), \tag{1}$$

where $V_i(t)$ is the voltage of neuron i at time t, $\lambda_{\rm i}$ is a **memory factor** of neuron i (which stands for its membrane time-constant), $A_j(t)$ is the activation (firing) of neuron j at time t, $W(j,i)$ is the synaptic weight from neuron j to neuron i, N is the number of neurons including the input sensory neurons, and V_{rest} stands for the resting voltage (set to zero in all simulations). After firing, the voltage of a spiky neuron is reset to the resting voltage, with no refractory period.

The voltage of a spiky neuron results from an interplay between the history of its inputs and the current input field. The memory factor, which ranges between 0 and 1, determines the amount of integration over time that the neuron performs: The higher the memory factor, the more important is the neuron's history ($V_i(t-1)$), compared with the current input field (the last summand in (1)). A spiky neuron with a zero memory factor is exactly identical to an MP neuron, in which only the current input field determines the voltage. The memory factor is different for each neuron, as different neurons may have different roles, each demanding a different amount of integration over time.

A *genetic algorithm* is used to evolve the synaptic weights $W(j, i)$ and, for spiky neurocontrollers, the memory factor parameters, both directly encoded in the genome as real valued numbers. Evolution is conducted over a population of 100 agents for 30000 generations, starting from random neurocontrollers, using a mutation rate of 0.02 and uniform point-crossover with rate of 0.35.

3 Delayed Response Processing

In this section we present the basic evolutionary results for both types of agents (Sect. 3.1), afterwhich the successfully evolved spiky and MP networks are compared in terms of localization of processing (Sect. 3.2), and in the way they solve the counting task (Sect. 3.3). This section is concluded with the results of incremental evolutions of the task (Sect. 3.4).

3.1 Performance Evaluation

Successful agents that solve the delayed-response task were evolved with both MP and spiky networks. The evolution of the delayed-response task is fairly difficult, and many evolutionary runs converged without yielding successful agents. We measure the difficulty of each task as the fraction of successful runs (Fig. 1A). Evidently, the task is harder as the agent has to wait for a longer delay period. More important, successful spiky neurocontrollers evolve more easily than MP networks.

3.2 Delayed Response Processing Localization

It is interesting to compare between the spiky and the non-spiky networks, in terms of localization of processing. The localization score [7] is a measure of how distributed is the agent's task in the network. The lower this measure is, the more the processing is evenly spread in the network across many neurons. The localization score is obtained from lesioning experiments [7], where numerous multiple lesions are afflicted upon a network, and the corresponding performance score of the agent is recorded. Based on these data, the causal importance (contribution) of each element to the behavioral task is assessed by the Multi-lesion Shapley value Analysis (MSA) [8]. By lesioning the network's neurons we compute the general neuronal contributions, from which the localization score is calculated.

We have shown that the difficulty of evolving the counting-task increases with the delayed response period (Sect. 3.1). Is there a correlation between the difficulty of evolving a network that solves a task, and the localization level of the resulting network? Figure 1B plots the average localization score of successful agents as a function of the delayed response period. Since the tasks differ

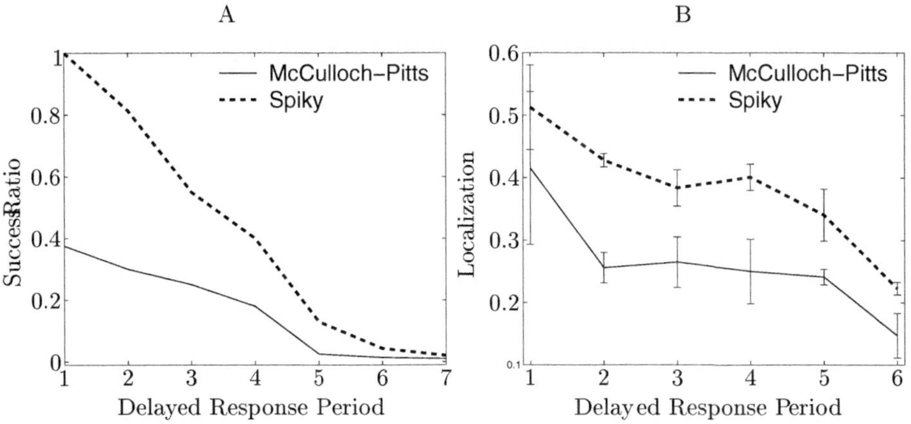

Fig. 1. (A). Task difficulty vs. delayed response period: Success ratio for a task is the fraction of evolutionary runs, out of 30, in which the best agent achieved a behavioral fitness above a pre-defined threshold. (B). Localization score (mean and standard deviation across several successfully evolved agents) vs. delayed response period

only by the duration of this counting period, the localization differences can be attributed to the counting process. In both types of agents, the network's localization decreases with the length of the delayed response period, and MP agents are less localized than the spiky ones. Interestingly, the correlation coefficient between the average localization score and the evolutionary success fraction is high: 0.87 for the spiky networks, and 0.79 for the MP ones.

3.3 Analysis of the Delayed Response Processing

We now turn to investigate the actual counting process in the successful neurocontrollers, studied by a receptive-field analysis, during which the activation patterns of an agent's neurons are analyzed in view of its behavior. For a specific agent, the firing sequence during the delayed response period usually repeats itself precisely in all counting incidents. Figure 2A presents the activation pattern sequence during the delayed response period of agent MP5, an MP agent with a delay period of 5 time-steps. Since the agent does not move during the delayed response period, the sensory inputs are constant, inhibiting the first 3 motor neurons. On each time-step, a different sub-group of neurons is active, such that at the last waiting step the agent closes its mouth and eats, and a step after the forward neuron is reactivated, and the agent moves. In general, since the sensory inputs stay exactly the same throughout the counting period, for an MP agent to count to K, the network has to pass through K different activation states. Thus an MP neurocontroller with N neurons can theoretically count to 2^N.

Figure 2B shows the firing pattern of the counting neurons of S7, a spiky agent with a delay period of 7 waiting time-steps. The food inputs activate

A

	1	2	3	4	5	6	7	8	9	10
	1	0	0	0	0	1	1	0	1	0
1	0	0	0	0	0	1	1	0	1	0
2	0	0	0	0	0	1	0	0	0	1
3	0	0	0	0	0	0	0	0	1	0
4	0	0	0	1	0	1	1	0	0	0
5	**0**	**0**	**0**	**0**	**0**	**0**	**1**	**0**	**1**	**1**
6	1	0	0	0	0	0	0	0	1	0

B

	1	4	7
	1	0	0
1	0	0	0
2	0	1	0
3	0	1	0
4	0	1	0
5	0	1	0
6	0	1	1
7	**0**	**0**	**1**
8	1	0	0

C

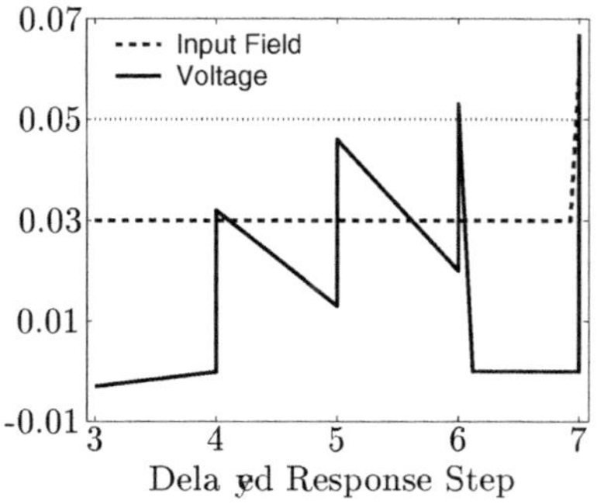

Fig. 2. *Counting processing.* (A). The activation pattern of the neurons of agent MP5 throughout the delayed response period. An activation value of 1 states that the neuron is firing. Each column represents a neuron, when the leftmost neurons are the forward, left, right and open-mouth motor neurons, respectively. Each row represents a different time-step, in an ascending order. Eating occurs at the fifth time-step (marked in bold) when the agent closes its mouth, which is open in the previous step. After eating the forward motor is activated and the agent moves. (B). The activation pattern of neurons 1, 4 and 7 of agent S7 during the delayed response period (all other neurons do not fire throughout this period). Columns and rows as in A, eating occurs at the seventh time-step. (C). The input field and the voltage of neuron 7 of agent S7 during time steps 3 to 7 of the delayed response period. The memory factor of neuron 7 is 0.43, the threshold above which the neuron fires is 0.05 (dotted line). The voltage decays between input ticks, and is recalculated with each new input field. This decay is exponential (1), the linear decay portrayed is only for illustration purposes. After firing at the sixth time step the voltage is returned to V_{rest}, and the neuron fires again at the seventh step without integration

neuron 4, which has a positive synapse to neuron 7. Neuron 7, a spiky neuron with a memory factor of 0.43, gradually accumulates voltage during steps 3 to 5, till it passes the threshold and fires at the sixth time step. This firing inhibits the forth neuron on the seventh step, closing the agent's mouth to consume the food. After eating, the forward neuron is activated by neuron 7, and the agent moves to another grid-cell. As shown in Fig. 2C, during steps 3 to 6 of the delayed response period the input field of neuron 7 is constant, bellow the threshold, hence the neuron's history is the one causing the increase in its voltage. In a spiky network, the same network activation pattern can be repeated several times during the counting process, since the state of a neuron consists also of its voltage. Thus, theoretically, under spiky dynamics, a network can count with a single neuron, that accumulates voltage over $K-1$ steps, and fires on the Kth step. The evolved spiky networks do not posses such efficient counting, but usually one or more neurons use their spikiness to accumulate voltage and "count" for a few steps, and as a result, less neurons participate in the counting process compared with MP networks, explaining the higher localization levels of the spiky networks (Sect. 3.2).

3.4 Incremental Evolution

Since evolving agents with a large delay period is a difficult evolutionary task (Sect. 3.1), we use the incremental evolution technique: Each evolutionary process starts with a population of 100 mutated copies of an already evolved agent with a delay period of K, and develops an agent with a delay period of $K+1$. Incremental evolutions were easier and shorter than regular evolutions for both spiky and MP agents, and spiky incremental evolutions were easier and shorter than incremental MP ones (around 90% success ratio, 6000 generations needed in typical spiky incremental evolutions, and around 70% success ratio, 15000 generations needed in typical MP incremental evolutions). For MP networks, we have succeeded in developing agents that count to 15. Examining the activation patterns of the agents developed via this incremental evolution revealed that agents counting to 3 use two neurons in the counting process, agents that count to 4-5 do it with four neurons, agent that count to 6-7 use five neurons, and an additional neuron is used for counting to 8 and above, demonstrating that under MP dynamics more neurons are needed to count to a higher number.

For spiky networks, we have evolved agents that count up to 35! Out of these, the spiky networks with a delayed response period of 3 to 9 use two neurons for counting: one is spiky, and the other has a vanishing memory factor, and de facto behaves like an MP neuron. In the networks with a delayed response period of 10 and higher this MP-like neuron was transformed into a real spiky neuron with a memory factor of 0.38, such that the two spiky neurons count together. The difference between the agents is only the amount of time-steps that each spiky neuron "counts", influenced mainly by the memory factors. Correspondingly the networks of the agents that count to 10 and higher have a constant localization value, irrespective of K. This result demonstrates that a spiky network can indeed employ a very efficient localized counting method (as described in Sect. 3.3). The evolutionary process does not always find such efficient solutions, since there is no evolutionary pressure towards this direction. Spiky agents with even larger delay periods can be easily constructed by manipulating the memory factors and the synaptic weights, as well as by continuing with the incremental evolutionary runs.

4 The Different Faces of "Spikiness"

4.1 Two Measurements of Spikiness

Are the evolved agents with spiking dynamics really spiky? First, having encoded the neuronal memory factors in the genome gives rise to the possibility that the evolution will come out with non-spiky solutions. Second, **even if the memory factor is high (1), it does not ensure that the neuron indeed utilizes its "integration potential" in its firing**. For example, a neuron may receive a large excitatory input field in every time step and fire in a very high frequency, without performing any integration over its past input fields. That is, given its input field, its pattern of firing would be indistinguishable from an MP neuron.

Third, **even if the spikiness is utilized for firing, it does not necessarily contribute to the agent's performance**. Essentially, we aim to distinguish between the observation that a given neuron has been assigned *spiking dynamics* by evolution, i.e. obtained a non-vanishing memory factor, and the true level of its *spikiness*, i.e., the amount by which it really "utilizes" its spiking dynamics. In this section we present two methods for measuring the spikiness level of a neuron, based on two fundamentally different perspectives.

Relevant Spikiness (RS). The first measurement for spikiness answers the following question: **how much are the spiking dynamics of a neuron needed for good performance of the agent?** Intuitively, if while abolishing the spiking dynamics of a neuron, the agent's performance deteriorates considerably, then its spiking dynamics contribute to the agent's behavior. If, in contrast, the fitness of the agent is maintained, this neuron's spiking dynamics are functionally insignificant.

To quantify this type of spikiness we define a new lesioning method: This **λ-lesioning** method *lesions only the memory factor of the neuron*, leaving the rest of its dynamics unaltered. Lesioning the memory factor is done by clamping it to zero, which turns the neuron into an MP one. Fitness scores are measured for all multi-lesion configurations with the λ-lesioning method, and the MSA is utilized to quantify **the contribution of the memory factor of each neuron to successful behavior**. .

Evident Spikiness (ES). The second index of spikiness measures **how much do the spiking dynamics of a neuron influence its firing**: If the firing pattern of a neuron stays the same regardless of whether it possesses spiking dynamics or not, then we can consider it as non-spiky. The ES index is calculated by comparing the firing of a spiky neuron to that of an MP neuron receiving an identical input field on each time step (last summand in (1)). The fraction of time steps in which there is a difference between the binary activations of the spiky neuron and a corresponding "benchmark" MP neuron, quantifies **the average percentage of lifetime steps in which the spiking dynamics "made a difference" in the firing of the neuron examined**. The RS and ES measures are normalized such that the sum over all neurons equals one.

4.2 Analysis of Spikiness

We examine the spikiness level of the neurocontrollers evolved with spiking neurons, focusing on two agents: S5 and S7, with a delay period of 5 and 7 waiting time steps, respectively. For S5, Figure 3A compares the general contributions of the neurons (yielded by the MSA, Sect. 3.2) with the RS values, indicating how much the spikiness of each neuron contributes to successful behavior. Notably, neurons 1, 4, 9 and 10 contribute significantly to the agent's behavior, as shown by their general MSA contributions, while the spikiness of only neurons 1 and 10 has a significant contribution, according to their RS values. Figure 3A also presents the ES values. Clearly, the two methods for measuring "spikiness" yield different results: Neuron 5 receives a very high ES score, and a near-zero RS score. A more pronounced difference is apparent in Fig. 3B, which shows both measures for agent S7, whose delayed response processing was analyzed in Sect. 3.3. In this case, the seventh neuron gets the highest RS value, but receives a pretty low ES score. Neurons 5 and 8 are the most spiky ones according to the ES measure, but receive low RS values.

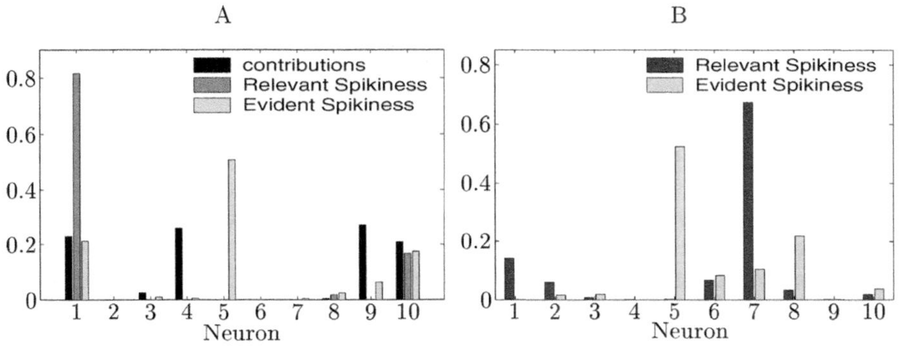

Fig. 3. *Comparison between spikiness measurements.* (A). RS and ES scores for the neurons of agent S5, along with their general neuronal contributions. (B). RS and ES scores for agent S7

The difference between the results of the two spikiness measurements originates in their different nature: The ES method measures the role that spiking dynamics play in determining the firing of a neuron *irrespectively of that neuron's functional and behavioral role.* The RS method is a functional measurement, which considers as spiky only neurons whose spikiness contributes to behavior. Usually such neurons also have high neuronal contributions, and are generally important to behavior. Additionally, these neurons must have a non-zero ES score, since influencing the firing pattern of the neuron is a prerequisite of having a behavioral contribution. In the case of agent S7, as shown in Sect. 3.3, neuron 7's spikiness plays a pivotal role in the agent's counting ability. When this neuron is λ-lesioned, the agent does not eat, explaining the very high RS value assigned to it. However, since the fraction of steps in which the spiking dynamics influence the activation of neuron 7 is only about 3% of its total lifetime steps (the counting steps), this neuron receives a low ES score. In almost all evolved spiky agents, most of the network's neurons have low spikiness functional contributions (RS), and usually, neurons with high RS scores are involved in the counting process. The two spikiness measurements show that the evolved agents with spiking dynamics are truly spiky, both in using past history to determine the activation patterns of some neurons, and in utilizing the spiking dynamics for successful behavior.

5 Conclusions

In this work we have succeeded in evolving agents solving a non-trivial, memory-dependent delayed response task, without using any special structure for counting, or giving an external reinforcement to the agent while waiting. Although incremental evolutions were easier and shorter than regular evolutions, even by a regular evolutionary process we evolved an agent that stands still on a food item for precisely 7 time-steps and then consumes it. By analyzing the counting process, we have shown how the ability of a spiky neuron to accumulate voltage is utilized for counting. We have further illustrated that in a delayed-response task, as may be the case for other memory-dependent tasks, networks of spiking neurons can be less complex and easier to evolve, compared with MP networks. These results demonstrate the importance of choosing a network architecture with characteristics and abilities that match the requirements of the given task.

Additionally, the study of spiky neural network in the context of EAAs brings forward basic questions regarding spiking dynamics that have not yet been raised. We have shown that the presence of spiking dynamics does not necessarily transcribe to actual spikiness in the network, and that the spikiness level can be defined and quantified in several functionally different ways. Specifically, the spikiness functional contribution (RS) and the spikiness evident influence (ES) each point to different neurons in the studied neurocontrollers.

References

1. Maass, W.: Networks of Spiking Neurons: the third generation of neural network models. Neural Networks 10, 1656–1671, (1997)
2. Bugmann, G.: Biologically Plausible Neural Computation. Biosystems 40, 11–19, (1997)
3. Maass, W., Ruf, B.: On Computation with pulses. Information and Computation 148(2), 202–218 (1999)
4. Floreano, D., Mattiussi, C.: Evolution of Spiking Neural Controllers for Autonomous Vision-based Robots, Evolutionary Robotics IV, Berlin, Springer-Verlag, (2001)
5. Paolo, E.A.D.: Spike-timing dependent plasticity for evolved robot control: neural noise, synchronization and robustness. To appear in Adaptive Behavior **10** (2003)
6. Aharonov-Barki, R., Beker, T., Ruppin, E.: Emergence of memory-driven command neurons in evolved artificial agents. Neural Computation **13** (2001) 691–716
7. Aharonov, R., Segev, L., Meilijson, I., Ruppin, E.: Localization of Function Via Lesion Analysis. Neural Computation **15** (2003)
8. Keinan, A., Hilgetag, C.C., Meilijson, I., E. Ruppin: Fair attribution of contribution: Shapley value analysis of neurocontrollers. preprint (2003)

First Steps in Evolving Path Integration in Simulation

Robert Vickerstaff

Centre for Computational Neuroscience and Robotics, School of Biological Sciences, University of Sussex, Brighton, BN1 9QG, UK

Abstract. Path integration is a widely used method of navigation in nature whereby an animal continuously tracks its location by integrating its motion over the course of a journey. Many mathematical models of this process exist, as do at least two hand designed neural network models. Two one dimensional distance measuring tasks are here presented as a simplified analogy of path integration and as a first step towards producing a neuron-based model of full path integration constructed entirely by artificial evolution. Simulated agents are evolved capable of measuring the distance they have travelled along a one dimensional space. The resulting neural mechanisms are analysed and discussed, along with the prospects of producing a full model using the same methodology.

1 Introduction

Path integration (PI) is a class of navigation strategy whereby an animal tracks its location by integrating its velocity (or the equivalent such as double integrating its acceleration, counting steps taken etc.) over the course of a journey. It is known to be used by many animal species, being particularly well studied in insects (see e.g. [1]). For some species it may be the only available method of navigation, as seems to be the case for the desert ant *Cataglyphis fortis* (see e.g. [2]), for others it is used in addition to or closely coupled with other methods such as land mark recognition [3]. A series of mathematical models of PI have been produced describing different solutions to the problem, usually based on either polar or Cartesian two dimensional coordinate systems. Two hand designed neural network models have also provided existence proofs of a system based on neuron-like elements.

In this paper we describe work aimed at producing another neuron-based PI model, but one constructed automatically using a genetic algorithm (GA) - to our knowledge an approach that has not been employed for modelling this problem before. It is hoped that a PI system produced by an evolutionary search method will complement the existing models, in that, while the resulting network may be harder to analyse, it will not be so directly influenced by the biases of the designer. In fact an evolved PI network may be more simple than a hand designed one, since a wider range of network styles and architectures can be considered using a GA than by hand design, making a wider range of neural mechanisms available for the solution.

W. Banzhaf et al. (Eds.): ECAL 2003, LNAI 2801, pp. 209–216, 2003.

Presented here are the results of a first step towards such a PI system, based on two simple distance measuring problems (used as a one dimensional analogy of PI in higher dimensions). Existing PI models are presented in section 2; the tasks and neural networks used for the present work are described in section 3; the evolved neural networks are described and tested in section 4 and the prospects for modelling PI in two dimensions using the same approach are discussed in section 5.

2 Existing Path Integration Models

Most existing PI models are mathematical descriptions of the calculations required to update a two dimensional position vector of the nest's location relative to the agent or vice versa. The equations are usually iterative formulae describing the vector's components at time $t+1$ given their values at time t and the agent's motion between t and $t+1$ written using either polar or Cartesian coordinate systems. The animal is assumed to be moving on a flat two dimensional plain. For reviews see [5,4].

At least two neuron based models also exist ([7,6]), whereby the calculations are carried out using networks of simple model neurons. Wittmann and Schwegler [7] solve the problem using a nest centred polar coordinate system which performs accurate PI. The vector is elegantly encoded using a sinusoidal array - an array of neurons whose activity pattern describes a sine wave in space. The wave's amplitude encodes distance from the nest, while the phase encodes bearing. Updating the vector is achieved by simple element-to-element addition of two such arrays. Hartmann and Wehner [6] also solve the problem using nest centred polar coordinates, but their system uses an approximate method designed to mimic the systematic errors seen in ant PI. The vector is represented separately as a distance and a bearing: the distance is encoded as an activity pattern extending and receding along a chain of neurons, while the bearing is encoded as a patch of activity moving clockwise and anti-clockwise around a circle of neurons.

The work presented here is not yet at a stage where direct comparison with either of the above classes of model is possible, for example the issue of what type of coordinate system to use does not arise since only a scalar value is needed. However, the method used to encode this scalar value can be compared to the two neural models. Is the neuron firing rate used or some other property of the dynamics? Is the value encoded using a single neuron or several? Also, the effect of modelling sensing errors or noise within the neural network can be studied as was the case with at least one of the mathematical models (Benhamou *et al.* (1990) cited in [5]).

3 Methods

A continuous time recurrent neural network (CTRNN, [8]) was used to control the agent. The neural network was defined by a genotype under the control of a genetic algorithm. The network received input from a speed sensor and a food sensor (which signaled when the agent had reached its goal and should return to the nest), and controlled three actuators: a forward motor, a reverse motor and a signal (which the network must activate when the agent has returned to the nest). The agent moved as a particle in a one dimensional space controlled by simple physical modelling. Key characteristics of each agent's behaviour were recorded during trials and used to assign a fitness to the neural controller at the end of each trial. Two tasks were set for the agent in separate experiments: (i) the two-way task: move forwards until reaching a randomly placed food item (indicated by the food sensor input changing from 0 to 1), return to the starting point and signal (defined as the firing rate of the signal neuron changing from a supra-threshold to a super-threshold level); (ii) the one-way task: move forward until reaching a randomly placed food item, then continue past the item for a distance equal to that already travelled before signalling. In both cases the location the agent is required to signal at is referred to as the nest for simplicity.

The CTRNNs used model a neural network using a continuous time description of neural firing rates, using the differential equation (adapted from [8]):

$$\tau_i \frac{dv_i}{dt} = -v_i + \sum_{j=1}^{N_{inputs}} w_{ji}\sigma_j \tag{1}$$

where $i \in \{1, \ldots N_{neurons}\}$, τ_i is a time constant, v_i is the cell membrane potential, w_{ji} is the weighting for input j of neuron i ($N_{inputs} = 10$ for all results presented here) and σ_j is the current value of input j. The input may be from a neuron or sensor. For a neuron $\sigma_j = \frac{1}{1+e^{-(g_k(v_k+b_k))}}$ where k is the inputting neuron and g and b are gain and bias parameters respectively. For a sensor the input is the current sensor activation (i.e. the speed or food sensor) linearly mapped such that $(\sigma_j = I_k) \in [0,1]$ under all possible conditions, where k is the inputting sensor.

τ_i, w_{ji}, g_i, b_i and the type (neuron or sensor) and subscript (k) of the input are all evolvable properties controlled by the genotype. τ_i and g_i are encoded in log_{10} form and raised to the power of ten before being used. Cell potentials are initialised at the beginning of each simulation run to values, v_{0i}, also encoded by the agent's genotype. A neuron is assigned to control each of the three outputs (forward motor, reverse motor, signal), control can be switched to other neurons by mutation.

Three global parameters control the amount of noise in the simulation. η_v controls uniform noise in the initialisation of the cell potentials such that $v_i \in [v_{0i} - \eta_v, v_{0i} + \eta_v]$. η_I controls uniform sensory noise whereby $I'_k \in [I_k - \eta_I, I_k + \eta_I]$ each time step. η_σ controls uniform neural noise whereby $\sigma'_k \in [\sigma_k - \eta_\sigma, \sigma_k + \eta_\sigma]$ each time step. The equations were numerically integrated using Euler's method with a step size of 0.2 while the genetic algorithm was running, evolved agents were tested with a step size of 0.2 and 0.01 to check for artifacts due to numerical integration (all subsequent analysis was done with step size 0.2 since the level of noise is dependent on the step size).

The agent's location in the one dimensional arena is specified as a single point with zero volume, x. It's motion is influenced by the forces exerted by its two motors and by *wind* which can exert a drag force on the agent. The wind speed is a further source of random noise in the simulation, being set to a uniform random value from the interval $[-1, 1]$ for a uniform random length of time $\in [0.2, 20]$ time units (not integration steps). Wind speeds are positive in the same direction as positive agent speeds. The agent's motion is controlled by Newtonian dynamics assuming a linear drag model:

$$F = F_{max}(\sigma_f - \sigma_r) - C_{drag}(\frac{dx}{dt} - w), \quad \frac{d^2x}{dt^2} = F/(M_{agent} + M_{food})$$

where F is the resultant force, F_{max} is the maximum force a motor can exert (set to 1), σ_f, σ_r are the firing rates of the forward and reverse motor control neurons respectively, C_{drag} is the linear drag coefficient (set to 1), $\frac{dx}{dt}$ is the agent's speed, w is the wind speed, $\frac{d^2x}{dt^2}$ is the agent's acceleration, M_{agent} is the agent's mass (set to 1) and M_{food} is the mass of food carried by the agent, equal to 0 before it has reached to food and 1 after that. This equation was integrated using Euler's method to give the speed, then integrated again exactly (i.e. assuming only a constant acceleration) to give the agent's new location.

A generational genetic algorithm (GA), [9], population size 15, was used to evolve the agent's neural network. The fittest 5 genotypes were copied, mutated and used to replace the worst 5 genotypes each generation. Each agent was evaluated using 5 trials and assigned the fitness of the worst trial.

Fitness was calculated as a value $\in [0, 1]$ using a series of formulae reflecting key data collected during the trial: nearest approach to food, furthest distance from food, time reached food, nearest approach to nest, furthest distance from nest, time reached nest, minimum value of signal neuron, maximum value of signal neuron, trial status when signalled, location when signalled. The fitness rewards were calculated for three criteria: (i)if and when it reached the food, (ii)if and when it reached the nest (after first reaching the food) and (iii)if and when it signalled the nest location. For (i) high values for nearest-to-food and furthest-from-food were penalised, high values for time-to-food were penalised but always gave a higher reward than not reaching the food at all. Likewise for nearest-to-nest and furthest-from-nest in (ii), likewise reaching the nest was always better than having only reached the food. For (iii) if the agent didn't signal rewards were given for higher maximum and lower minimum values of the signal neuron, but signalling anywhere was always rewarded more than not signalling at all; signalling after reaching the food was better than signalling before it. All rewards were calculated such that they were scaled to a well defined interval, and scaled relative to each other so that 94% of the maximum fitness (1.0) was accounted for by the accuracy of signalling the nest location after first reaching the food, the other criteria being present to encourage the earlier stages of behaviour to evolve first. Signalling accuracy was assessed by an absolute rather than relative error measure, rewards were therefore normalised relative to the size of the food placement zone rather than the actual distance between the food and nest in a given trial. The location of the food for each of the five trials was set using a stratified random scheme, such that the defined food placement zone was divided into five equal sections, each being sampled randomly in one of the trials. Agent's were initially evolved to perform the task with the food zone set between 5 and 10 distance units. Once satisfactory results were obtained the food zone was doubled in length successively until a zone from 5 to 400 distance units was reached.

The GA used was capable of varying the number of neurons in the network using neuron duplication and deletion operators; this feature was used in the experiments where the network size was allowed to vary. Other neuron level mutation operators were used in all experiments: homologous recombination (replacement by the most similar neuron in another genotype), overwriting (deletion followed by duplication of a remaining neuron, to allow duplication in a fixed length genotype) and randomisation (deletion followed by replacement by a randomly generated neuron). Neuron parameter mutations were performed using creep mutations for floating point values, where the new value was the old value plus a Gaussian random value with zero mean whose standard deviation was scaled by the size of the valid range of the parameter and a global mutation rate parameter.

4 Results and Analysis

Initially two sets of experiments were performed for each of the two tasks, for both of which neural network size was fixed to 3 (one to control each of the actuators). These were: (i) with wind, but no other sources of noise ($\eta_v = 0, \eta_I = 0, \eta_\sigma = 0$); (ii) as (i) but $\eta_v = 1, \eta_I = 0.25, \eta_\sigma = 0.025$. GA runs were also performed where the number of neurons was allowed to change by mutation, but for both tasks these runs did not produce any significant improvement in agent fitness or novel mechanisms compared to the agents with only three neurons.

The Two-way Task. Here the agent must move forward to the food, then return to the starting point and signal. In the experiments without noise agents evolved able to signal the nest location accurately, in spite of the buffeting effects of the wind. Five duplicate GA runs were performed, all of which evolved the same method of solving the task. For the best solutions the fitness was greater than 0.99 no matter where the food was within the placement zone, with the worst ones still consistently greater then 0.95. The forward motor neuron flipped from high to low when the food was reached, likewise the reverse motor neuron flipped from low to high, bringing about the agent's change of direction. This can be achieved by a simple reactive mechanism since the food sensor input remains low until the food is reached, then becomes high until the end of the trial[1]. The signal neuron acted as an integrator of the speed sensor input. This is possible

[1] Earlier runs (data not shown) had been tried where the food sensor only stayed high within a short distance of the food; agents evolved successfully to perform under this scheme; it was changed to the current scheme to simplify subsequent analysis slightly

for a single neuron since, if the $-v_i$ term in equation (1) is small, then the neuron simply integrates the sum of its weighted inputs. The best agents used a very small range of v to encode the agent's location, to minimise the decay and therefore perform more accurate integration. This came at the price of being very sensitive to noise if introduced, especially where $\eta_v > 0$, since this acted to effectively randomise the neuron's state at the start of the trial. For example setting $\eta_v = 0.1$ resulted in fitnesses varying between 0.05 and 1.0 for trips up to 150 distance units and between 0.6 and 1.0 for trips between 150 and 400 units. The best agents were also able to flip the two motor states rapidly once the food was reached. The less efficient solutions entered a short phase where only one of the motor controllers had flipped states, resulting in no net force being produced by the motors. For the best solutions, the agent's location varied linearly with the integrator neuron's state, see Fig. (1,left) for the network of the fittest agent.

To test the generality of the integration mechanism, trials were performed where the agent was held still once it had reached the food for 500 time units before being allowed to return as normal. This reduced the fitness of the agent only marginally to around 0.9 for short journeys, and had even less effect on longer runs.

The experiments with noise produced similar results, except that the signal neuron encoded the agent's location using a cell potential range approximately 50 times larger, (approximately 3.3 v units per 100 distance units compared to approximately 0.065 per 100). This reduced the effects of initial neuron potential noise. Agents scored greater than 0.9 for most trials, dropping to 0.85 for the longest trips. Plots of agent location against signal potential were still noticeably linear, but included substantial wobbles (data not shown). For short journeys the agents were prone to signal very early before reaching the food due to noise pushing the signal neuron above threshold.

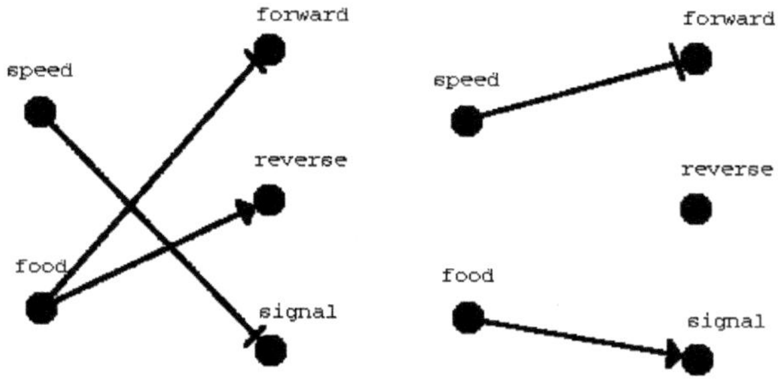

Fig. 1. Evolved networks which solve (left) the two-way and (right) the one-way tasks. All redundant connections have be pruned. Excitatory connections are shown with arrowheads, inhibitory ones with bars. Sensors are on the left of each network, actuator control neurons are on the right.

The One-way Task. This task involved simply travelling forward from the starting point, passing the food and signalling when an equal distance past the food had been travelled. This makes the simple one neuron integrator solution impossible since the two legs of the journey do not result in different speed sensor inputs. The experiment without noise evolved the same solution in all five runs. It varied its forward motor output in a way which damped down the effects of the wind on the agent's speed whereby the motor output dropped rapidly once the speed sensor got above approximately 0.8, and rose once it was below approximately 0.55. The signal neuron did not receive input from the speed sensor, only from the food sensor. Initially the signal neuron's v decreased very slowly due to a large τ and low negative $\frac{dv}{dt}$, at an approximately linear rate. Upon reaching the food the food sensor input became high, and the v increased at an approximately linear rate due to this change, see Fig.(1, right) for the network of the fittest agent. The agent was therefore only measuring time and not integrating the speed sensor input. By reducing speed variations with compensatory motor output this process approximates integration of the speed. The accuracy of this agent is approximately the same as the three neuron two-way task agent evolved without noise, with fitnesses of greater than 0.98 regardless of the location of the food. This mechanism simply 'assumes' the first and second legs of the journey will take the same amount of time to complete. In trials where the wind speed was held at 0.5 for the first leg and at -0.5 for the second leg the agent consistently signalled too early.

The experiment with noise did not evolve the wind damping strategy seen in the above experiment, but otherwise used the same time measuring mechanism. Fitness was markedly reduced, scoring between 1.0 and 0.85, except for short journeys where the signal was once again prone to be triggered by noise very early in the trial, resulting in very low fitnesses.

5 Discussion

The two distance measuring tasks were each solved by a different strategy. Of these the solution to the first task solves the problem in the most general way, in that it can truly integrate the speed sensor input. This relies on using a single neuron as an integrator, which is not entirely without precedence in the neuroscience literature [10]. Alternatively, each CTRNN neuron could be assumed to model a group of similar neurons. This result is markedly different from and more simple than the use of a chain of neurons to encode a scalar value [6], and comes quite naturally once neurons are described using a continuous time non-reactive model. This simple mechanism could be extended to perform PI in two dimensions by using two integrator neurons to record the agent's location as, for example, a nest-centred Cartesian coordinate, although it remains to be seen if such a mechanism could be produced by a GA. The agents performing the one-way task did not evolve a mechanism capable of generalised integration, but instead simplified the problem by using a negative feedback loop to travel at a more constant speed. This could of course have been modelled more abstractly without concern for the dynamics of the agent's motion, by simply assuming a constant speed. In nature, however, an animal's motion and PI mechanism have evolved together. When testing the hypothesis that an animal's nervous system might only need to be capable of performing approximate PI to navigate successfully, explicit modelling of the agent's motion may prove beneficial for this reason. These results have shown two simple strategies for performing a simple analogy of path integration. The methods used here will now act as the basis for further work aimed at producing a two dimensional path integration model by evolutionary search.

References

1. Collett, M., Collett, T.S.: How do insects use path integration for their navigation? Biol. Cybern. **83** (2000) 245–259
2. Wehner, R., Gallizzi, K., Frei, C., Vesely, M.: Calibration processes in desert ant navigation: vector courses and systematic search. J. Comp. Physiol. A **188** (2002) 683–693
3. Wehner, R., Michel, B., Antonsen, P.: Visual navigation in insects: coupling of egocentric and geocentric information. The Journal of Experimental Biology **199** (1996) 129–140
4. Benhamou, S., Seguinot, V.: How to Find one's Way in the Labyrinth of Path Integration Models. J. theor. Biol. **174** (1995) 463–466
5. Maurer, R., Seguinot, V.: What is Modelling For? A Critical Review of the Models of Path Integration. J. theor. Biol. **175** (1995) 457–475
6. Hartmann, G., Wehner, R.: The ant's path integration system: a neural architecture. Biol. Cybern. **73** (1995) 482–497
7. Wittmann, T., Schwegler, H.: Path integration - a network model. Biol. Cybern. **73** (1995) 569–575
8. Beer, R.D.: Toward the evolution of dynamical neural networks for minimally cognitive behavior. In: Maes, P., Mataric, M., Meyer, J., Pollack, J., Wilson, S.(eds.): From animals to animats 4: Proceedings of the Fourth International Conference on Simulation of Adaptive Behavior. MIT Press (1996) 421–429
9. Goldberg, D.E.: Genetic algorithms in search, optimization, and machine learning. Addison-Wesley, Reading, Mass. (1989)
10. Egrov, A.V., Hamam, B.N., Fransen, E., Hassel, M.E., Alonso, A.A.: Graded persistent activity in entorhinal cortex neurons. Nature **420** (2002) 173–178

On the Dynamics of an Artificial Regulatory Network

Wolfgang Banzhaf

Department of Computer Science, University of Dortmund, D-44221 Dortmund
banzhaf@cs.uni-dortmund.de
http://ls11-www.informatik.uni-dortmund.de/people/banzhaf/
Tel: +49 231 9700-953 Fax: +49 231 9700-959

Abstract. We investigate a simple artificial regulatory networks (ARNs) able to reproduce phenomena found in natural genetic regulatory networks. Notably heterochrony, a variation in timing of expression, is easily achievable by simple mutations of bits in the genome. It is argued that ARNs are useful and important new genetic representations for artificial evolution.

1 Introduction

Regulatory networks are a fascinating new area of research in biology [4,5]. With the advent of whole genome information and the realization that - in higher organisms - but a tiny fraction of DNA is translated into protein, the question what the rest of DNA is doing becomes all the more pressing. Regulation seems to be a very reasonable answer for a functional role of unexpressed DNA. Even in single-celled organisms, regulation takes up a substantial part of their highly compressed genomes. According to Neidthardt et al. [18], 88 % of the genome of the bacterium *E.Coli* is expressed, 11 % is suspected to contain regulatory information (see also Thomas [22]).

It is also recognized that the information on DNA-strings controlling the expression of genes is key to understanding the differences between species and thus to evolution [12]. How evolution actually managed to evolve different structures of multicellular organisms from more or less the same proteins seems to be a question of the sequence and the intensity of events during development. Even at the level of biochemical reactions necessary for, e.g. metabolism, regulation plays an important role, thus spanning reaction times from milliseconds (physiology) to Megayears (evolution).

There are three major genetic mechanisms, all tied to regulation [5], which allow such a variety of reactions of living organisms to the pressure for survival:

1. Interactions between the products of genes
2. Shifts in the timing of gene expression (heterochrony)
3. Shifts in the location of gene expression (spatial patterning)

W. Banzhaf et al. (Eds.): ECAL 2003, LNAI 2801, pp. 217–227, 2003.

Regulatory networks are used by Nature to set up and control mechanisms of evolution, development and physiology. Regulatory networks unfold the patterns and shapes of organism morphologies and of their behavior. In addition, they mediate between development and evolution, since many evolutionary effects can be followed through their regulatory causes.

It is therefore natural to ask whether we can learn from this type of controlling the organization of matter in the area of artificial evolution. Over the last decades, simple approaches to artificial evolution have proven useful in optimization problems of engineering and in combinatorial problems of computer science [19,21,11,9,15,2,8,7]. Most of these approaches, however, use a primitive genotype-phenotype mapping without implementing any dynamics into this process of mapping information into behavior. Only recently it was realized that it may the dynamical (and possibly complex) mapping of genotypes to phenotypes, that allows natural systems to evolve with ease (see, for example, [13]).

How can we make use of these insights in artificial evolutionary systems? Previous work in the area is scattered. Eggenberger [6] has studied the patterning of artificial 3D-morphologies. Reil [20] has set up an artificial genome and studied some consequences for artificial ontogeny. Kennedy [14] examined a model of gene expression and regulation in an artificial cellular organism. Bongard and Pfeifer have considered the relation between evolving artificial organisms and behavior [3].

In this contribution we shall present a recently conceived model of a regulatory networks which should be useful for artificial evolutionary systems. The model is a simplification and abstraction of what the author perceives as key elements of the protein-genome interaction in regulatory networks. It is not yet connected to a semantics of structures or behavior, but shows a rich behavioral dynamics. Thus it seems most appropriate to study this artificial regulatory network (ARN) in the context of Artificial Life.

The paper is organized as follows: Section 2 explains the overall view of the artificial regulatory network model, section 3 views it from the static and dynamic perspective. Section 4 explains the concept of heterochronic control. Section 5 exemplifies the plasticity of such systems to evolutionary pressure, section 6 summarizes the discussion and outlines future steps.

2 The Artificial Regulatory Network

Our ARN consists of a bit string with direction (like $5' \rightarrow 3'$ in DNA), the 'genome', and mobile information-carrying molecules, 'proteins', which are equipped with bit patterns for interaction with the genome. Together, they represent a theoretically closed world with a network of interactions between genome and proteins, and a dynamics of protein concentration development determined by this network.

More technically, a mechanism for reading off genes and for producing proteins of particular bit-patterns is given in the form of a 'genotype-phenotype mapping'. Proteins are able to wander about and to interact with any pattern

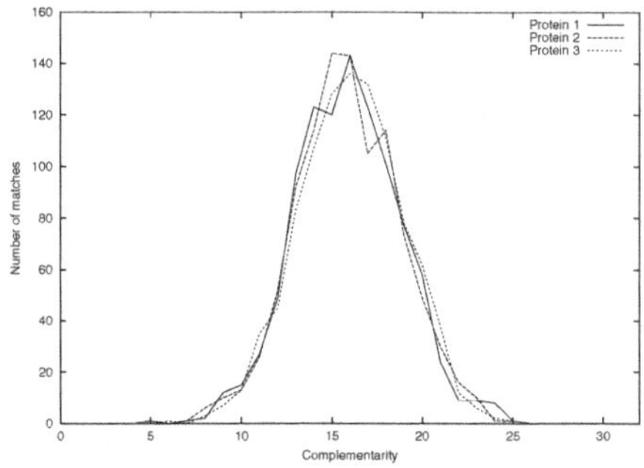

Fig. 1. Distribution of matches for sample proteins with their genome. The distribution is roughly Gaussian, as should be expected from random genomes.

on the genome, notably with 'regulatory sites' located upstream from genes. By attaching to these special sites, they can positively (by enhancing) or negatively (by inhibiting or silencing) influence the production of (other) proteins. We observe the production of proteins and the dynamics of their concentration changes which are a result of the interplay between all the interactions taking place simultaneously. At this time, there is no energy or raw material considered in the system.

The genome is implemented as a sequence of 32-bit (integer) numbers. The length of the sequence, l_G, determines the length of the genome and is frequently used as a parameter. A particular START pattern, the 'promoter', is used to signal the beginning of a gene on the bit string (analogous to an open reading frame (ORF) on DNA), starting at the next integer. The signal used is arbitrary and was chosen as '01010101', a one-byte pattern which in a genome generated by randomly choosing '0's and '1's will appear with a probability of $2^{-8} \approx 0.0039 = 0.39\%$. Genes have a fixed length of $l_g = 5$ 32-bit integers resulting in a bit pattern of 160 bits for each gene (this could be changed later into a STOP signal).

Upstream from the promoter site two special sites are located, one enhancer site and one inhibitor site, both of length 32 bits. Attachment of proteins to these sites will result in changes in protein production of the corresponding gene. It is assumed that an equally low production of proteins takes place if both sides are unoccupied. Usually, however, there will be proteins around to influence the expression rate of a particular gene, and we shall look at that in more detail later. In this simple model, we restrict ourselves to just one regulatory site for expression and suppression of proteins, a radical simplification with regard to natural genomes, where 5-10 regulatory sites are the rule that might even be occupied by complexes of proteins.

Proteins are produced from genes by feeding their bit patterns into a genotype-phenotype mapping function and producing mobile elements carrying other bit patterns, the proteins. In this model, therefore, we disregard the transcription process completely. Further, there are no introns, no RNAs and no translation procedure resulting in a different alphabet for proteins. Instead, proteins consist of bit patterns of a particular type: Each protein is a 32-bit number resulting from a many-to-one mapping of its gene: On each bit position in the gene's integers the majority rule is applied so as to arrive at one bit for the protein. In the case of a tie (not possible with an odd number for l_g), this is resolved by chance.

Proteins can now be examined as to how they 'match' the genome: Each bit pattern of a protein can be compared to the genome pattern with the overlap being the number of bits set in an XOR operation. Thus, complementarity between genome and protein bit patterns determines their match. In general, it can be expected that a Gaussian match distribution results when shifting proteins over all the sequence of a random genome. Notably, there are a few high-matching and a few low-matching positions and many average-matching ones on the genome.

3 Static and Dynamic View

Let us first look at examples from the static perspective. Table 1 gives three examples of genomes with increasing size. We list the number of genes which roughly follows the 0.39% rule, the maximum match between resulting proteins and their genome at any location, and the number of times such a maximum match has been found.

Table 1. Sample genomes of increasing size. As can be seen the number of proteins with maximum match remains about the same, but their specificity increases.

Genome length l_G	Number of genes	Maximum match	Frequency of max. match
1000	3	25	3
10000	37	28	4
100000	409	30	3

If we look at the distribution of matches the picture according to Figure 1 emerges for a sample genome with 3 proteins: Roughly a Gaussian distribution of matches is found for each of these genes.

Let's change perspective and look from the genome's point of view, and more specifically, from the point of view of regulation sites. A number of proteins are produced and floating by, with some providing better matches to the site, other proteins providing worse matches. In principle, each protein has the potential to interact with each regulatory site, and the degree of matching will determine the probability of occupation of a certain site with a certain protein.

Because proteins are competing for attachment to regulatory sites, the probability of occupation with a particular protein is dependent on the degree of matching of all other proteins to this site.

Under the simplifying assumption that occupation of two regulatory sites per gene modulates the expression of corresponding protein, a network of interac-

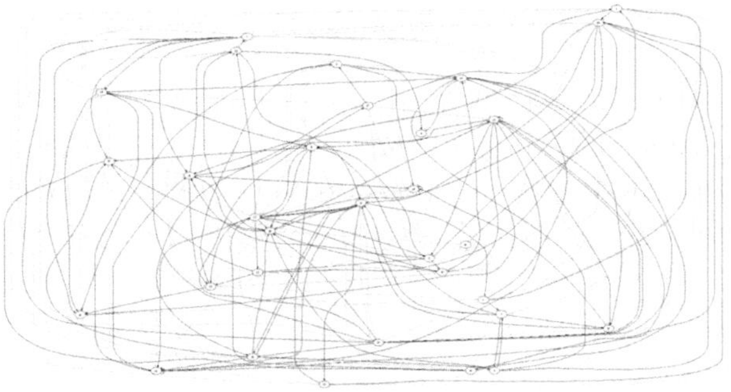

Fig. 2. Distribution of matches of proteins of a larger example. 9 out of 32 proteins are depicted and their matches with the enhancer (black) and inhibitor (gray) regulatory sites of all 32 genes of this example ($l_G = 10,000$).

tions between genes and proteins can be deduced, which can be parametrized by strength of match. Figure 2 shows a sample network, again taken from the example genome with 32 genes / proteins.

No evolution has taken place (recall these genomes are generated by randomly drawing bits), and the network of interactions shows a highly random view of the resulting interactions. These networks must be considered very complex (in terms of layers vs. participating nodes) and a deep hierarchy of interactions is visible[1].

In the rest of this section we shall concentrate on the dynamics of the interaction network. A match between a protein and regulatory site of a gene leads to activation or inhibition of protein production of the corresponding gene. Generally, the influence of a protein i with $i = 1, ..., n_p$ on an enhancer/inhibitor site is exponential in the number of matching bits, $exp(\beta(u_i - u_{max}))$ where u_{max} is the maximum match achievable.

The concentration of protein molecules c_j of protein j modulates this strength to produce the following excitatory / inhibitory signals for the production of protein i:

$$e_i = \frac{1}{N} \sum_j c_j e^{\beta(u_j^+ - \bar{u}_{max}^+)} \tag{1}$$

$$in_i = \frac{1}{N} \sum_j c_j e^{\beta(u_j^- - \bar{u}_{max}^-)} \tag{2}$$

where a scaling was done as to have a maximum match for the best matching protein, both in excitatory and inhibitory signals.

[1] As has been observed in natural genome organization, shallow hierarchies, up to the point of modularity, are a hallmark of biological organisms. It is interesting to note that a simple process of duplication and divergence suffices to reach a similar state, even from a random genome [1].

Given these signals, protein i is produced via the following differential rate equation

$$\frac{dc_i}{dt} = \delta(e_i - in_i)c_i - \Phi \tag{3}$$

A flow term assures that concentrations remain in the simplex, $\sum_i c_i = 1$, resulting in competition between sites for proteins.

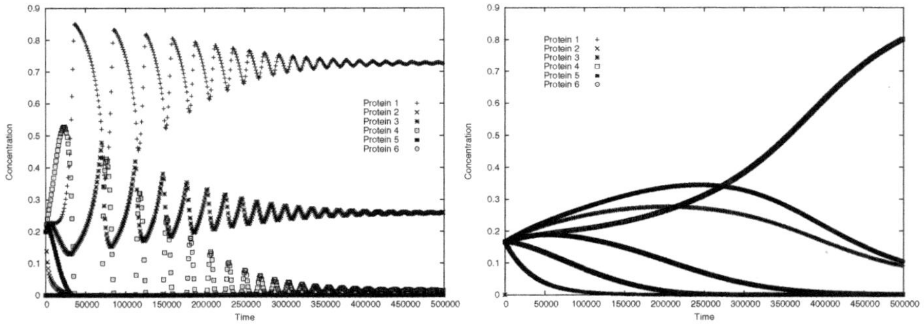

Fig. 3. Two different dynamical systems realized by two different genomes. Left: A dampened (nonlinear) oscillator type of dynamics. Right: Slow and smooth development of concentrations.

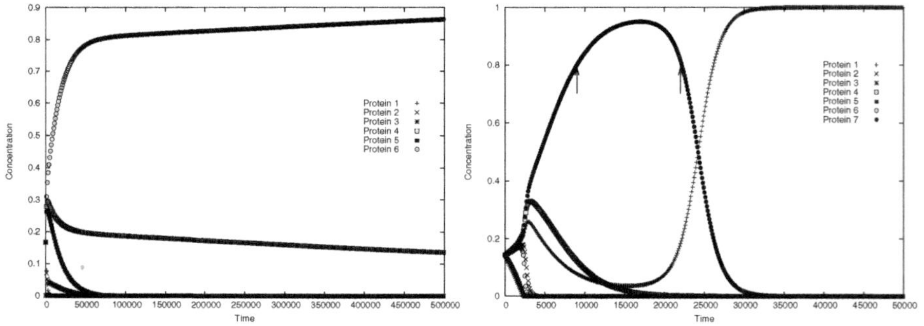

Fig. 4. Two other dynamics. Left: Settlement into a point attractor is well under way. Right: Extended transition phase with one protein achieving high values of concentration then switching the state to expression of another gene.

If we look at the dynamics of concentration changes of proteins, starting from a state of equal concentration that reflects the native low-level expression of all genes, we can observe that some proteins increase their level of concentration, then fall again, with usually one being left over. Thus, a typical dynamic system behavior can be seen, well known as 'point attractor' in dynamical systems theory [10].

For different random genomes (different number of genes, matching etc) the dynamics is remarkably different. There are cases of longer and shorter time scales, there are complicated and simple dynamics. Figures 3 - 4 show four different dynamics resulting form four different genomes.

It should be noted that this richness of dynamics is merely a result of different genomes of the same length, with different patterns for proteins resulting in different matching and regulation results. No development has yet been put in place. There are three types of competition effective simultaneously, (i) competition of proteins for binding sites (only regulatory sites considered), (ii) competition of binding sites for proteins, and (iii) competition of genes for raw material for production of proteins. This latter competition is implemented by normalizing the strength of the inhibition / enhancement signals through division by N.

4 Heterochronic Control

If we look at this from the perspective of how many proteins are above or below a certain production threshold we can observe the turning on or turning off of genes (on/off could be set equal to x 2 or x 1/2 of initial production, or it could be based, as here, on an absolute concentration value, 0.8 here). This translates into a timing of onset/termination of protein production. Figure 4, right, for instance, shows the timing of onset and termination of concentrations above 0.8 for protein 7 (arrows), $t_{on} = 9,000, t_{off} = 22,000$.

Changing the degree of matching between regulatory sites and proteins by one or two bits can result in dramatic changes in the dynamics, but it must not. Sometimes there are no changes at all and we have a neutral variation. Sample changes that actually varied the expression are shown in Figure 5. It is interesting to note that variations in patterns are translated by the ARN into time variations, similar to what was observed in natural GRNs [5]

Heterochrony, i.e. a variation in the timing of onset or offset of certain genes are heavily used in development for generating particular structural effects [17, 16]. As we can see by comparing Figures 5, left and right, and 4, right, small changes cause small effects. The same principle could be also of use in physiological reactions, for instance under the control of external factors exceeding certain threshold values.

Interestingly, the range of possible changes is partitioned logarithmically, due to the change of occupation probability, that is depending on an exponentiated matching difference between proteins and DNA bit-patterns. This can be seen most easily, if we put all concentration curves of protein 1 and 7 into one plot, see Figure 6. We can clearly see the range of changes expanding with further additions of bit flips.

5 Evolution

The most important question to be addressed with such a model is whether it would be possible to define arbitrary target states and evolve the genome / pro-

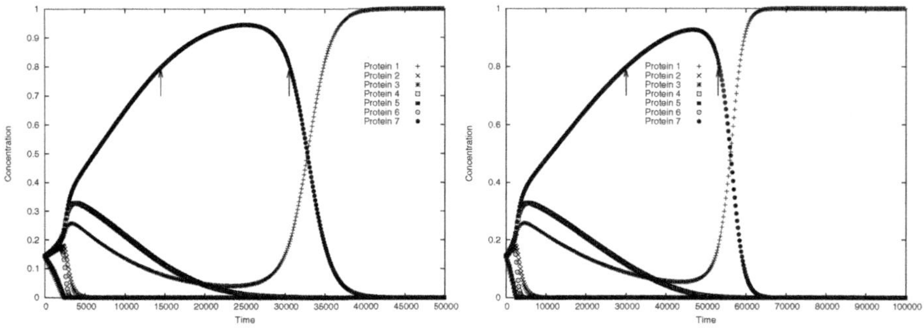

Fig. 5. Genome of Figure 4. Left: Degree of matching between protein 7 and inhibitory site to gene 4 changed by one bit. Timing of expression of protein 7 changes substantially: $t_{on} = 14,500, t_{off} = 30,500$. Right: Degree of matching between protein 7 and inhibitory site to gene 4 increased by another bit, timing changes even further $t_{on} = 30,000, t_{off} = 53,000$.

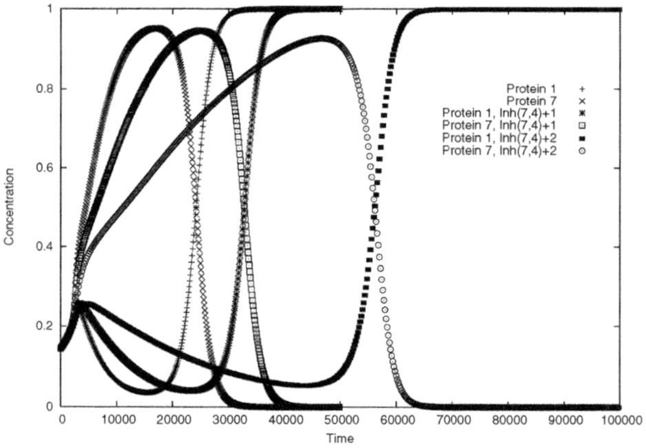

Fig. 6. Genome of Figure 4. Degree of matching between protein 7 and inhibitory site to gene 4 changed progressively. Timing of expression of protein 7 changes in increasing step sizes.

tein network toward this target state. Our first results in a typical simulation are shown in Figure 7. It shows the progress of a network in approaching the target concentration of a particular protein, here protein 6. As we can see, the evolutionary process quickly converges towards this target state. It must be emphasized, that the very simplest way of doing evolution was used here, a $(1+\lambda)$ evolution strategy, with $\lambda = 1$ [19]. Various experiments were performed with the same genome (not shown here), allowing evolution of other concentration levels for other proteins. We can see from the figure, that steep declines in the deviation (error) curve are followed by stagnation periods. These stagnation periods, are, however, accompanied by continued changes in the genome under evolution. It

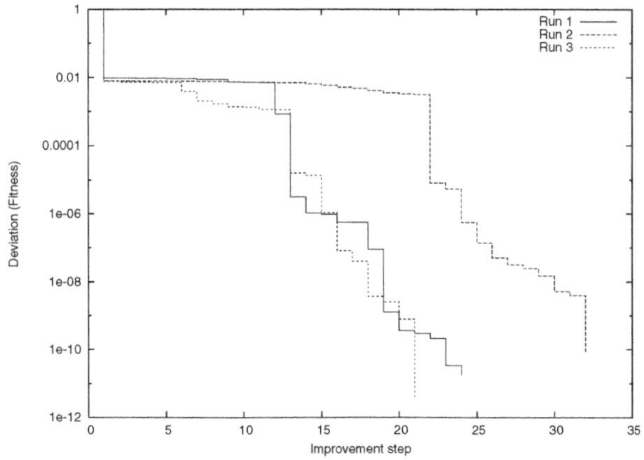

Fig. 7. Evolution at work: 3 different runs of a (1+λ) strategy to arrive at a prespecified concentration of one particular protein: $c_6 = 0.085$ at time $t = 100$.

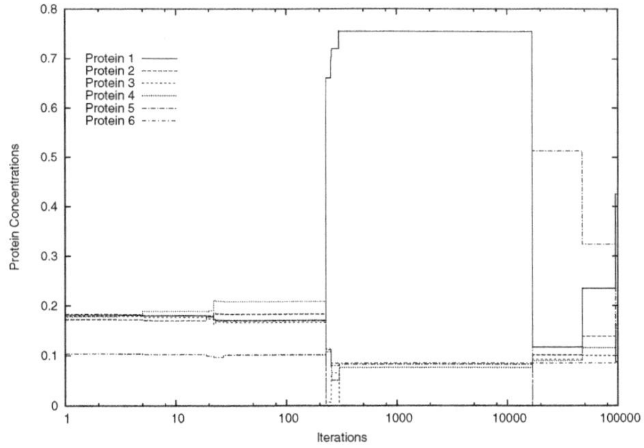

Fig. 8. Evolution at work: Same run as in Figure 7, with all protein conentrations protocolled at $t = 100$. As can be seen, protein concentration of selected protein 6 meanders towards goal state $c_6 = 0.085$, whereas other protein concentrations pass through huge swings.

is merely the mapping of the genome that does not show many consequences of these variations. By construction we designed a system with many neutral pathways. During some periods it does not look as if there is evolutionary progress, but nevertheless changes happen in genomes due to neutral steps.

This can be seen if we consider the changes in concentration levels of all proteins at $t = 100$ in Figure 8. Here we can discover that all protein concentrations change over time, with many stagnation periods for all proteins. Huge steps are sometimes shown by certain proteins, which are not reflected in the fitness of an individual, due to the focus on measuring only the deviation from $c_6 = 0.085$ for fitness.

6 Summary

In this contribution we have shown that a simple model for artificial regulatory networks can be formulated which captures essential features of natural genetic regulatory networks. Although our investigation is preliminary in that it is only qualitative in results, the different behavior of these networks from usual genetic represenations can be seen already from the few examples shown here.

With this contribution we have just started on a path that relates changes in time and intensity to tiny pattern changes on bit strings. As such, the network picture of a genome might be a very fruitful approach and could possibly provide the algorithmic "missing link" between genotypes under constant evolutionary changes and remarkably stable phenotypes that we find in the real world.

Acknowledgement. The author gratefully acknowledges a sabbatical stay at the Institute for Genomics and Bioinformatics at UC Irvine, where part of the ideas that lead to this work were born. Especially he wants to acknowledge the hospitality of its director, Prof. Pierre Baldi, and of its manager, Mrs. Ann Marie Walker.

References

1. W. Banzhaf. Artificial Regulatory Networks and Genetic Programming. In R.L. Riolo et al., editor, *Genetic Programming – Theory and Applications*, 2003, to appear.
2. W. Banzhaf, P. Nordin, R. Keller, and F. Francone. *Genetic Programming – An Introduction.* Morgan Kaufmann, San Francisco, CA, 1998.
3. J. Bongard and R. Pfeifer. Behavioral selection pressure generates hierarchical genetic regulatory networks. In W. B. Langdon et al., editor, *Proceedings of the Genetic and Evolutionary Computation Conference*, page 132. Morgan Kaufmann, 2002.
4. J.M. Bower and H. Bolouri (Eds). *Computational Modeling of Genetic and Biochemical Networks.* MIT Press, Cambridge, MA, 2001.
5. E.H. Davidson. *Genomic Regulatory Systems.* Academic Press, San Diego, CA, 2001.
6. P. Eggenberger. Evolving morphologies of simulated 3d organisms based on differential gene expression. In Inman Harvey and Phil Husbands, editors, *Proceedings of the 4th European Conference on Artificial Life*, pages 205–213. Springer, 1997.
7. D.B. Fogel. *Evolutionary Computation.* IEEE Press, New York, 1995.
8. L.J. Fogel, A.J. Owens, and M.J. Walsh. Artificial intelligence through a simulation of evolution. In M. Maxfield, A. Callahan, and L.J. Fogel, editors, *Biophysics and Cybernetic Systems*, pages 131–155, 1965.
9. D.E. Goldberg. *Genetic Algorithms in Search, Optimization and Machine Learning.* Addison-Wesley, Reading/Mass., 1989.

10. M. Hirsch and S. Smale. *Differential Equations, Dynamical Systems and Linear Algebra.* Academic Press, Reading, MA, 1997.
11. J.H. Holland. *Adaptation in natural and artificial system.* MIT Press, Cambridge, MA, 1992.
12. L. Hood and D. Galas. The digital code of dna. *Nature*, 421:444–448, 2003.
13. H. Kargupta. Editorial: Special Issue on Computation in Gene Expression. *Genetic Programming and Evolvable Machines*, 3:111–112, 2002.
14. P.J. Kennedy and T.R. Osborn. A model of gene expression and regulation in an artificial cellular organism. *Complex Systems*, 13, 2001.
15. J.R. Koza. *Genetic Programming: On the Programming of Computers by Natural Selection.* MIT Press, Cambridge, MA, USA, 1992.
16. M. McKinney. Heterochrony: Beyond words. *Paleobiology*, 25:149–153, 1999.
17. M. McKinney and K. McNamara. *Heterochrony: The Evolution of Ontogeny.* Plenum Press, New York, NY, 1991.
18. F.C. Neidhardt. *Escherichia Coli and Salmonella typhimurium.* ASM Press, Washington, DC, 1996.
19. I. Rechenberg. *Evolutionsstrategie "93.* Frommann Verlag, Stuttgart, 1994.
20. T. Reil. Dynamics of gene expression in an artificial genome - implications for biological and artificial ontogeny. In D. Floreano et al., editor, *Proceedings of the 5th European Conference on Artificial Life*, pages 457–466. Springer, 1999.
21. H.-P. Schwefel. *Evolution and Optimum Seeking.* Sixth-Generation Computer Technology Series. John Wiley & Sons, Inc., New York, 1995.
22. G.H. Thomas. Completing the e. coli proteome: a database of gene products characterised since completion of the genome sequence. *Bioinformatics*, 7:860–861, 1999.

Evolution and Growth of Virtual Plants

Marc Ebner

Universität Würzburg, Lehrstuhl für Informatik II
Am Hubland, 97074 Würzburg, Germany
ebner@informatik.uni-wuerzburg.de
http://www2.informatik.uni-wuerzburg.de/staff/ebner/welcome.html

Abstract. According to the Red Queen hypothesis, an evolving population may be improving some trait, even though its fitness remains constant. We have created such a scenario with a population of coevolving plants. Plants are modeled using Lindenmayer systems and rendered with OpenGL. The plants consist of branches and leaves. Their reproductive success depends on their ability to catch sunlight as well as their structural complexity. All plants are evaluated inside the same environment, which means that one plant is able to cover other plants leaves. Leaves which are placed in the shadow of other plants do not catch any sunlight. The shape of the plant also determines the area where offspring can be placed. Offspring can only be placed in the vicinity of a plant. A number of experiments were performed in different environments. The Red Queen effect was seen in all cases.

1 Motivation

The fitness of coevolving species may remain at the same level over time even though each individual may be continually improving some specific trait. This has been called the Red Queen hypothesis [20], after the Red Queen chess piece in Lewis Carroll's "Through the Looking Glass". The queen has to keep running just to stay in the same place. The classic example of the Red Queen effect is the chase between cheetahs and gazelles. If the population of prey becomes faster, some of the predators are unable to catch their prey. The slow ones will not be able to reproduce. After some generations, mutations might lead to predators which are a little faster than their ancestors. These individuals will again be able to catch their prey easily and the status quo is maintained. In the course of evolution both populations will become better at achieving their task. We have visualized the Red Queen effect in a population of artificial plants.

Plants are usually represented as Lindenmayer systems or L-systems for short [19]. Prusinkiewicz and Lindenmayer [19] have shown how complex, photo-realistically looking plants can be created from a relatively small number of rules. See Deussen et al. [3] for an introduction to the realistic modeling and rendering of plant ecosystems. Computer-simulated evolution of virtual plants was pioneered by Niklas [16]. He performed an adaptive walk through a space of branching patterns. Jacob [5,6,7,8,9] used a variant of genetic programming

W. Banzhaf et al. (Eds.): ECAL 2003, LNAI 2801, pp. 228–237, 2003.

to evolve context-free and context-sensitive L-Systems representing plants. He used a combination of the number of blossoms, the number of leaves and the volume of the plant as a fitness function. Two-dimensional plant morphologies were evolved by Ochoa [17]. Kokai et al. [11,12] also evolved L-Systems. They tried to generate rules which, when executed and viewed, look identical to a given image. Mock [15] developed a system where the user could take the role of a virtual gardener selecting plants for reproduction. He also worked with an explicit fitness function that rewarded plants which are short but wide. Artificial models for natural plant evolution were developed by Kim [10].

In contrast to most previous work on artificial plant evolution we allow interactions between the plant and its environment. All plants are placed in the same environment and evaluated together. Plants need to catch as much virtual sunlight as possible in order to reproduce successfully. Because all plants are evaluated inside the same environment, one plant may shadow the leaves of other plants or it may even shadow its own lower leaves. If a plant reproduces twice into the next generation, it will have to compete with a copy of itself. We will see that after a short number of generations, maximum fitness starts to fluctuate around a constant level. This contrasts sharply with a more continuous evolution when plants are evaluated in isolation. However, even though fitness is no longer rising, plants are still evolving and adapting to their environment. At this point an evolutionary arms race [2] sets in. The plants need to grow higher and higher in an attempt to gain more sunlight.

2 Plant Representation

Plants are modeled as deterministic, context-free L-Systems. A L-System is basically a string rewrite system where the individual letters of a word are transformed according to a set of rules. The set of rules is applied to all letters in parallel. A L-System is defined by an alphabet V, a starting word ω and a set of rules P [19]. The starting word ω is a non-empty string $\omega \in V^+$. This word is transformed using the set of rules. The set of rules is defined as a subset of $V \times V^+$. Each rule consists of a predecessor a and a successor χ. Thus, a single rule can be written as $a \to \chi$. The letter a will be replaced by the string χ wherever it appears in the word. In case no successor is defined for a predecessor a, it is assumed that the rule $a \to a$ also belongs to the set of rules P. This rule leaves the letter a unchanged. By replacing all predecessors with their successors, a new word is derived from the original one. This process is repeated 5 times (for the experiments described in this paper). The final word is interpreted as a sequence of commands for a three-dimensional drawing device.

Our alphabet consists of the symbols:

$$V = \{\texttt{f}, \texttt{l}, +, - <, >, /, [,], \texttt{A}, ..., \texttt{Z}\}. \quad (1)$$

The letter `f` produces a branch segment. It draws a cylinder and moves the drawing device forward by a distance equal to the length of the cylinder. The letter `l` produces a leaf. Here, the position of the drawing device does not change.

initial word: $\texttt{f}$
rules:

$$\texttt{f} \rightarrow \texttt{f}\backslash\backslash < \texttt{l}[\texttt{Afl}]\backslash\backslash\backslash - \texttt{f}$$
$$\texttt{A} \rightarrow \texttt{B} - \backslash\texttt{f}$$
$$\texttt{B} \rightarrow \texttt{CA}$$
$$\texttt{C} \rightarrow\ >$$

Fig. 1. Sample grammar of an evolved plant. The plant was evolved using coevolution on a flat landscape.

All leaves of the plant have the same shape and size. These are the building blocks which are used to build the plant. The symbols `+`, `-`, `<`, `>`, `/`, and `\` can be used to change the orientation of the drawing device. The orientation can by changed in discrete steps of 22.5 degrees in both directions around any of the three axis of the coordinate system. Thus, the sequence `f+f` creates a bent branch segment. Branching structures can be created with the symbols `[` and `]`. Whenever the symbol `[` is encountered during the interpretation of the drawing commands, the current state, i.e. the three-dimensional position and orientation, of the drawing device is pushed on a stack. Whenever the symbol `]` is encountered the topmost state is popped from the stack. This restores the position and orientation of the drawing device to a previously saved state. Thus, the sequence `f[-f]+f` creates a Y-shaped branching structure. Additional symbols `A` through `Z` can be used to define substructures. These symbols only play a role during development. When interpreting the final word, they cause no operation. A sample grammar of an evolved plant is shown in Figure 1.

3 Plant Evaluation

All individuals of a population are evaluated at the same time. Each individual has a specific position (x, y) and orientation α inside an rectangular evaluation area. Fitness is a function of the plant's ability to collect sunlight as well as the plant's structural complexity. We assume that the sun is positioned directly above the plant. The leaves of the plant may be used to collect virtual sunlight. Figure 2 shows a plant viewed from above. The plant was rendered with OpenGL using orthographic projection into a 512×512 image. Shading calculations were turned off. Cylinders were drawn in black and leaves in green. OpenGL uses a technique called Z-buffer for hidden surface removal. Thus, the number of green pixels in the image is a measure of the plant's ability to collect sunlight. Use of the Z-Buffer to estimate the amount of sunlight hitting a plant was proposed by Beneš [1].

Fig. 2. The image on the left shows an image of the plant shown in Figure 1. The plant is viewed from above, and rendered using orthographic projection. Leaves are drawn with green color. The number of green pixels is used to calculate the plant's fitness. The image on the right shows the shadow of the plant. Offspring may grow anywhere on this footprint of the plant. Thus a plant's offspring are always placed in the parent's vicinity.

In a natural environment a plant faces competition from many sources. One prime source of evolution may be the presence of other individuals. We have modeled this by evaluating all individuals of the population at the same time. All plants are rendered in a single image using orthographic projection. Each plant is assigned a unique green color to draw the plant's leaves. Cylinders are again drawn in black. A sample population of 7 plants is shown in Figure 3. The two images on the left show the plant population and the image on the right shows the image which is used to determine the fitness values of the plants. If a plant grows very tall it may shadow other plants. Thus, if coevolution is used, there is an incentive for the plants to grow higher and higher. Note that this type of coevolution does not use two separate populations. We only work with a single population. All individuals of the population are evaluated in a single environment.

The amount of light received is only one part of the fitness calculation. The second part is due to the structural complexity of the plant. In our model a branch segment costs 1 point and a leaf costs 3 points. This basic cost is multiplied with a factor which depends on the distance to the root of the plant. Branches and leaves which are far away from the root are more costly than branches and leaves which are close to the root. The structural complexity of a plant is defined as

$$\text{complexity} = \sum_{b \in B} \text{cost}_{\text{branch}} \cdot \text{factor}^{\text{height}(b)} + \sum_{l \in L} \text{cost}_{\text{leaf}} \cdot \text{factor}^{\text{height}(l)} \quad (2)$$

where B is the set of branch segments, L is the set of leaves and height is the number of branch segments between the current position and the root of the

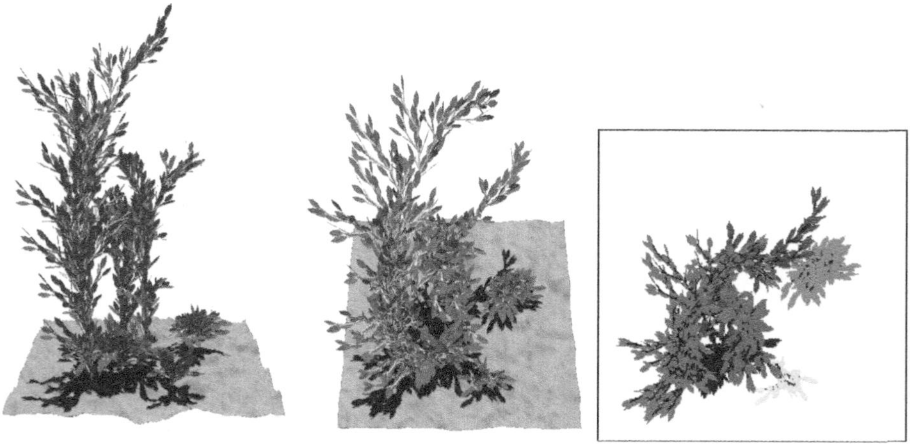

Fig. 3. The two images on the left show 7 different evolved plants. Large plants shadow smaller plants. The image on the right shows the image used to evaluate the plant's fitness values. Each plant has a unique color which is used to render the plant's leaves.

plant. The following parameters were used during the experiments: factor = 1.1, $\text{cost}_{\text{branch}} = 1$, and $\text{cost}_{\text{leaf}} = 3$. Fitness is calculated by subtracting the structural complexity from the amount of light received:

$$\text{fitness} = 10 \cdot \text{pixels} - \text{complexity} \tag{3}$$

where pixels is the number of pixels covered by the plant's leaves. The number of green pixels is weighted with a factor of 10 which was determined experimentally. If this value is too small, plants will not be able to grow very high as the number of green pixels per plant is fixed when distributed evenly among all plants. On the other hand, if this value is very large, plants have the potential to grow very high. Negative fitness values are not allowed. If this occurs, the fitness is set to zero.

4 Evolution of Artificial Plants

We use an evolutionary algorithm, a variant of genetic programming [13,14] to automatically generate new populations of plants [4]. To create the individuals of the next generation we first chose a genetic operator at random. Depending on the type of genetic operator one or two individuals are selected from the original population. Parents are selected using tournament selection. This selection process does not depend on the spatial location of the plants. The following genetic operators were used: permutation, mutation, insertion, add-branch, deletion, one-point-crossover, sub-tree-crossover, delete branch, add-rule, delete-rule. For a description of these operators see Ebner et al. [4]. Offspring are basically

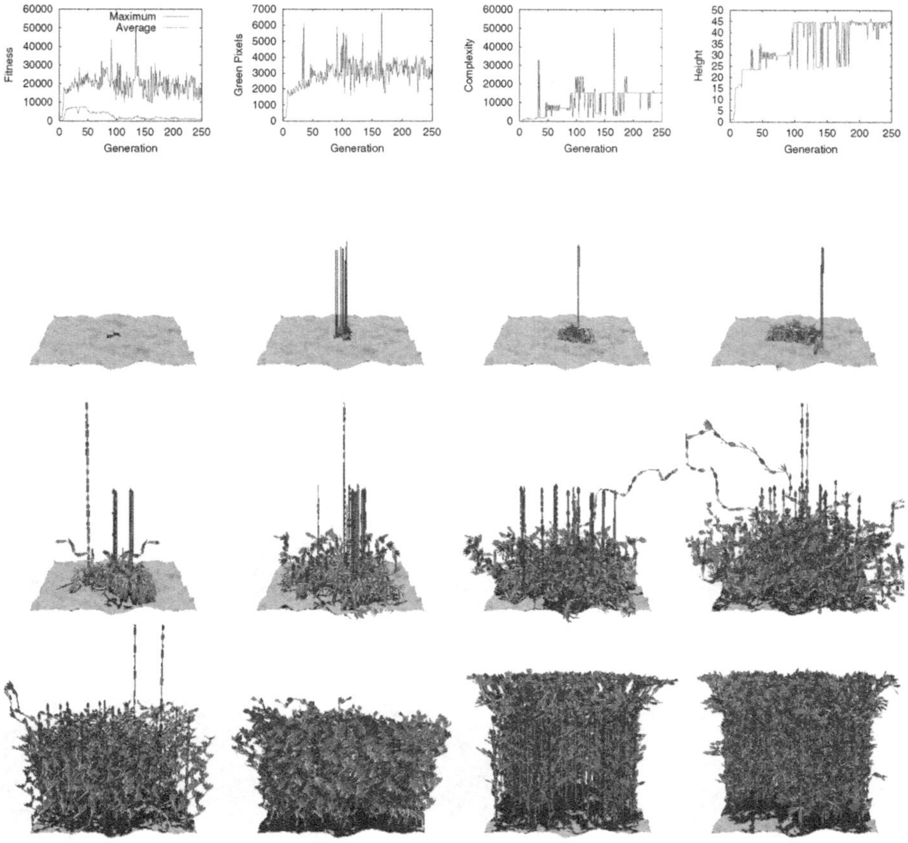

Fig. 4. Results for the experiment on a flat landscape. The first graph shows maximum and average fitness over time. The second graph shows the number of green pixels, the third graph shows plant complexity and the fourth graph shows the height of the best plant of each generation. The images below show the population at generation 0, 4, 6, 7, 8, 9, 10, 14, 32, 64, 128, and 250. The population spreads from a small area in the middle and eventually populates the whole area.

created by making small changes to the selected individual (mutation) or by combining segments of two selected individuals (crossover). Offspring are inserted into the next generation until the generation is filled. Whenever an offspring is created, we have to determine the position where it will grow. This position is determined by rendering the plant using OpenGL with orthographic projection. Cylinders and leaves are both drawn in black. A sample footprint of an evolved plant is shown in Figure 2. From this footprint, we randomly select a position to place the offspring. The orientation of the plant is chosen randomly. Thus, offspring always grow in the vicinity of their parents.

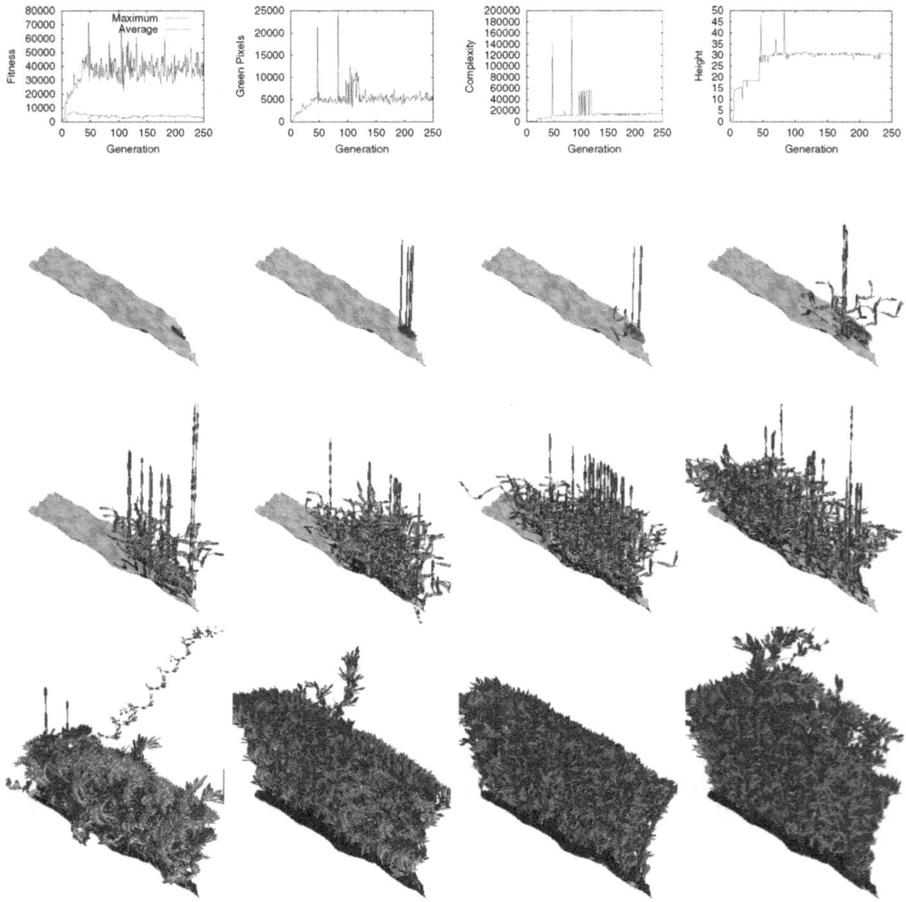

Fig. 5. Results for the experiment on a landscape with a large slope. The images below show the population at generation 0, 4, 6, 7, 8, 9, 10, 14, 32, 64, 128, and 250. The population spreads from a small area on the right side, climbs up the slope and eventually populates the whole area.

5 Experiments

We ran experiments on three different landscapes. The landscapes were generated using Perlin noise [18]. The first landscape is essentially flat with small height differences. Individuals are initially placed randomly within a small circular area in the center of the evaluation area. A population of 200 individuals with tournament selection and a tournament size of 7 was used. The first generation contains only individuals with the single rule $\mathtt{f} \to \mathtt{f}$. The starting word is `f`. Therefore, all individuals of the first generation have a fitness of zero. Each genetic operator was applied with a probability of 0.1 to generate an offspring.

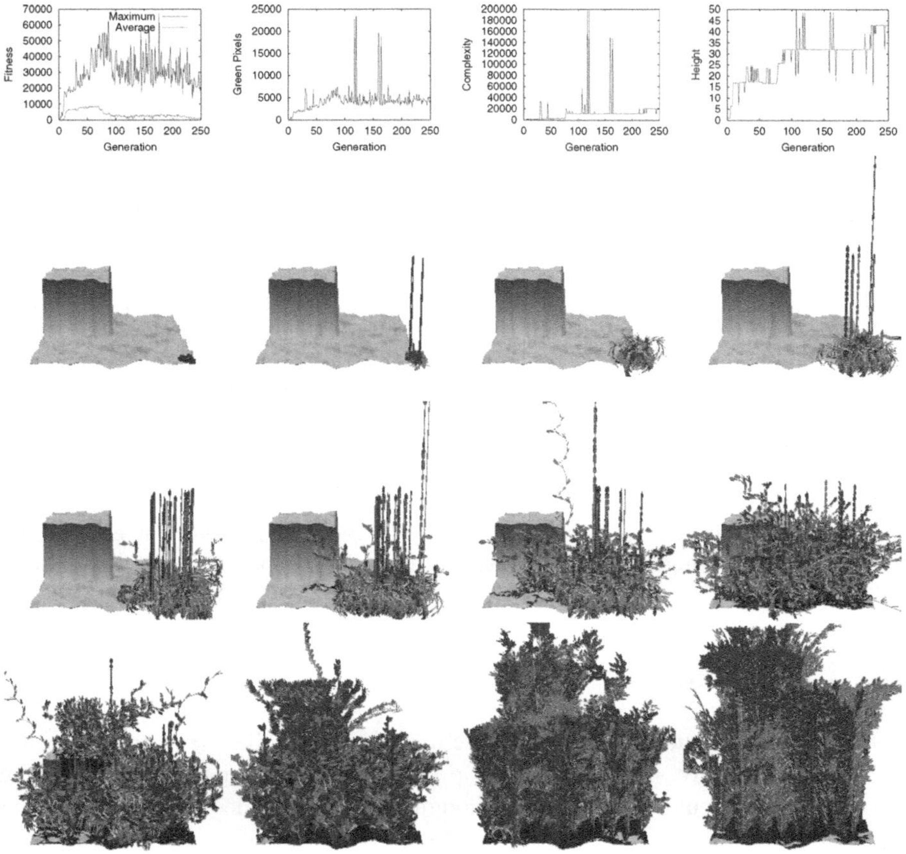

Fig. 6. Results for the experiment on a landscape with two different height levels. The images below show the population at generation 0, 4, 7, 8, 9, 10, 11, 15, 32, 64, 128, and 250. The population spreads from a small area in the lower right corner and eventually populates the whole area.

Results for the flat landscape are shown in Figure 4. The second experiment was carried out on a landscape with a large vertical slope. Results for this experiment are shown in Figure 5. The third experiment was carried out on a landscape where one fourth of the landscape was elevated. Results for this experiment are shown in Figure 6.

In all three cases, maximum fitness stops increasing after generation 50. However, the evolutionary process has not come to a halt yet. Plants are still adapting to their environment and changing their shape. Note, that we did not impose a maximum height for the plants. However, the number of green pixels per plant is fixed when distributed evenly among plants. Thus, this also limits the maximum height attainable. Maximum complexity can be no larger than the first term

Fig. 7. Best individuals of experiments 1, 2, and 3. The fourth image shows a plant which was evolved by evaluating the individuals in isolation. This mode of evaluation produces a bush-shaped plant. It contrasts sharply with the plants evolved using coevolution.

of the fitness function. Plants evolve to the point where no further increase in structural complexity is possible. Plants all over the environment converged to very similar looking plants. The best individuals of generation 250 are shown in Figure 7.

6 Conclusion

Due to the Red Queen effect, evolution may occur even if fitness seems to remain constant. We have shown this for a population of artificial plants. The plants adapt to their environment even though maximum fitness starts to fluctuate around a constant level. Plants were modeled using Lindenmayer systems and evaluated using OpenGL. They need to catch virtual sunlight in order to reproduce. OpenGL proved to be very effective to evaluate the amount of sunlight each plant receives. A similar mechanism was used to determine the position where offspring could be placed. Offspring were always placed in the vicinity of their parent. Three types of environments were used to visualize the Red Queen effect. An essentially flat landscape, a landscape with a large slope and a landscape with two height levels. In all three cases, instances of the plants quickly populated the environment. After all of the environment was populated, an arms race set in, which further shaped the plants. Plants need to outgrow their competitors in order to gain sunlight. This lead to the evolution of higher and higher plants.

References

1. B. Beneš. An efficient estimation of light in simulation of plant development. In R. Boulic and G. Hegron, eds., *Computer Animation and Simulation 96*, pp. 153–165, Springer-Verlag, Berlin, 1996.
2. R. Dawkins and J. R. Krebs. Arms races between and within species. *Proc. R. Soc. Lond. B*, 205:489–511, 1979.

3. O. Deussen, P. Hanrahan, B. Lintermann, R. Měch, M. Pharr, and P. Prusinkiewicz. Realistic modeling and rendering of plant ecosystems. In *SIGGRAPH '98 Conf. Proceedings, Comp. Graphics, Orlando, FL*, pp. 275–286. ACM Press, 1998.
4. M. Ebner, A. Grigore, A. Heffner, and J. Albert. Coevolution produces an arms race among virtual plants. In J. A. Foster, E. Lutton, J. Miller, C. Ryan, and A. G. B. Tettamanzi, eds., *Genetic Programming: Proc. of the Fifth Europ. Conf., EuroGP 2002, Kinsale, Ireland*, Springer-Verlag, Berlin, 2002.
5. C. Jacob. Genetic L-system programming. In Y. Davudor, H.-P. Schwefel, and R. Männer, eds., *Parallel Problem Solving from Nature – PPSN III. The Third Int. Conf. on Evolutionary Computation. Jerusalem, Israel*, pp. 334–343, Springer-Verlag, Berlin, 1994.
6. C. Jacob. Evolution programs evolved. In H.-M. Voigt, W. Ebeling, I. Rechenberg, and H.-P. Schwefel, eds., *Parallel Problem Solving from Nature – PPSN IV. The Fourth Int. Conf. on Evolutionary Computation. Berlin, Germany*, pp. 42–51, Springer-Verlag, Berlin, 1996.
7. C. Jacob. Evolving evolution programs: Genetic programming and L-systems. In J. R. Koza, D. E. Goldberg, D. B. Fogel, and R. L. Riolo, eds., *Proc. of the First Annual Conf. on Genetic Programming*, pp. 107–115, The MIT Press, Cambridge, MA, 1996.
8. C. Jacob. Evolution and coevolution of developmental programs. *Computer Physics Communications*, pp. 46–50, 1999.
9. C. Jacob. *Illustrating Evolutionary Computation with Mathematica.* Morgan Kaufmann Publishers, San Francisco, CA, 2001.
10. J. T. Kim. Lindevol: Artificial models for natural plant evolution. *Künstliche Intelligenz*, 1:26–32, 2000.
11. G. Kókai, Z. Tóth, and R. Ványi. Application of genetic algorithms with more populations for Lindenmayer systems. In E. Alpaydin and C. Fyfe, eds., *Int. ICSC Symposium on Engineering of Int. Systems EIS '98, University of La Laguna, Tenerife, Spain*, pp. 324–331, ICSC Academic Press, Canada/Switzerland, 1998.
12. G. Kókai, Z. Tóth, and R. Ványi. Evolving artificial trees described by parametric L-systems. In *Proc. of the 1999 IEEE Canadian Conf. on Electrical and Computer Engineering, Shaw Conference Center, Edmonton, Alberta, Canada*, pp. 1722–1727. IEEE Press, 1999.
13. J. R. Koza. *Genetic Programming. On the Programming of Computers by Means of Natural Selection.* The MIT Press, Cambridge, MA, 1992.
14. J. R. Koza. *Genetic Programming II. Automatic Discovery of Reusable Programs.* The MIT Press, Cambridge, MA, 1994.
15. K. J. Mock. Wildwood: The evolution of L-system plants for virtual environments. In *Int. Conf. on Evolutionary Computation, Anchorage, Alaska*, pp. 476–480, 1998.
16. K. J. Niklas. Computer-simulated plant evolution. *Scientific American*, 254(3):68–75, 1986.
17. G. Ochoa. On genetic algorithms and Lindenmayer systems. In *Parallel Problem Solving from Nature – PPSN V*, pp. 335–344, Springer-Verlag, Berlin, 1998.
18. K. Perlin. Noise, hypertexture, antialiasing and gesture. In D. S. Ebert, F. K. Musgrave, D. Peachey, K. Perlin, and S. Worley, eds., *Texturing and Modeling: A Procedural Approach. 2nd Ed.*, pp. 209–274, AP Professional, Cambridge, 1998.
19. P. Prusinkiewicz and A. Lindenmayer. *The Algorithmic Beauty of Plants.* Springer Verlag, New York, 1990.
20. L. Van Valen. A new evolutionary law. *Evolutionary Theory*, 1:1–30, July 1973.

Evolvability of the Genotype-Phenotype Relation in Populations of Self-Replicating Digital Organisms in a Tierra-Like System

Attila Egri-Nagy and Chrystopher L. Nehaniv

University of Hertfordshire
Department of Computer Science, Faculty of Engineering and Information Sciences,
College Lane, Hatfield, Hertfordshire AL10 9AB, United Kingdom
{A.Nagy, C.L.Nehaniv}@herts.ac.uk

Abstract. In other Tierra-like systems the genotype is a sequence of instructions and the phenotype is the corresponding executed algorithm. This way the genotype-phenotype mapping is constrained by the structure of a creature's processor, and this structure was fixed for an evolutionary scenario in previous systems. Our approach here is to put the mapping under evolutionary control. We use a universal processor (analogous to a universal Turing-machine) and put the structural description of the creature's processor as well as the instruction set of the actual processor into the organism's genome. The life-cycle of an organism begins with building its actual processor, then the organism can start executing instructions in the rest of its genome with the newly built processor. Since the definitions of the processors and instruction sets are in the genome, they are subject to mutations and heritable variation enabling their evolution. In this work we investigate the evolutionary development of the processor structures. In evolving populations, changes in the components (registers, stacks, queues), variations in instruction-set size and the redefinition of the instructions can be observed during experiments.

1 Introduction

There is a trend in artificial life research that researchers try to put more freedom in their models. One of the big leaps was to implement ecological natural selection instead of artificial selection (i.e. optimizing an objective function). Using artificial selection in genetic algorithms may yield powerful optimization methods for very different problems, but the evolution of the system does not resemble biological evolution. In the best case the system converges to a solution satisfying externally imposed criteria. The evolutionary development is not open-ended. However, in Tierra-like systems [11,1,5,2], digital organisms self-replicate *in silico*, and there is no global aim to reach; those entities which do not survive and replicate fast enough simply vanish from the population. But still the possible results are highly determined by the construction of the system: the evolved programs for example are written in the same assembly language as their progenitor. This language was designed by a human, not evolved. In this

W. Banzhaf et al. (Eds.): ECAL 2003, LNAI 2801, pp. 238–247, 2003.

paper we present a further improvement of the original idea of digital evolution. We give freedom to evolving populations of digital organisms (here called *archeans*) to change their 'genetic language' namely their instruction sets and the structures of their processors. This is done by putting the description of the processor and instruction set into the genome itself, thus the 'semantics' of the program is in the program itself (cf. [15]), and hence under evolutionary control.

2 Evolvability and Universal Processors

We handle the heritable information in digital organisms – a sequence of executable instructions or data cells or more simply a sequence of integers – as strings. Moreover descriptions of their processors and the instruction sets that run on them are parts of these strings. Variation in these structures thus becomes heritable. This results in new kinds of heritable variability and augments evolvability of the genotype-phenotype mapping [8,17,13,6].

Among processors smooth evolvability means that in most cases we can apply transformations (delete, insert, replace) on these sequences without drastically changing their semantics with the hope that some of these minor changes will be favored by natural selection. This property does not appear in real processors where just a simple bit-flipping can cause the abnormal termination of execution. Therefore the design principles for processors will be different from the current industrial standards when the main aim is to enable open-ended evolution of machine-code programs driven by natural selection.

What are these design principles? By what means can we discover them? Should we try out various processors and check how they perform under evolution [10]? There are already significant achievements implemented in the previous systems (Tierra [11], Avida [1], Primordial Soup [5]). The most prominent examples are as:

- Labels and pattern-matching can be used instead of hardcoded addresses in transfer of flow of control, subroutine calls, etc. [11].
- Fault tolerance: there is no combination of instructions which makes the processor crash. Even in the worst cases violations result in nothing happening (like performing a `nop` operation) [11].
- Using different instruction sets or subsets of a bigger and redundant instruction set in order to test their evolvability [12,1].

In these previous works of using a processor with fixed structure it simply executes the instructions and during execution a copy of the executed program is made. This is the replication process. Of course different fixed processors have been used but not in the same evolutionary run [12,10]. One of our original goals of designing `Physis`, yet another digital evolution system, was to provide possibilities to implement several processor architectures (including Tierra, Avida, Primordial Soup) within the same experimental tool ensuring common metrics for evaluation [2]. But this has been superseded by the general idea of using evolvable architectures, so our approach here is now radically different. The idea is if we don't know how to construct a processor which supports evolvability of self-replicating programs then we should *let it be evolved!* If we provide freedom for the evolutionary process in tinkering with the processor structure, evolvable processors may arise in the course of evolution. The guiding principle behind this: the evolutionary potential of a processor itself is an important property at lineage level and thus may be favoured by natural selection.

According to this we can outline a 'meta' design principle suggesting dynamic structures for processors. The internal structure (registers, stacks, queues and their sizes) and the instruction set may be varied over evolutionary time between different creatures in the population.

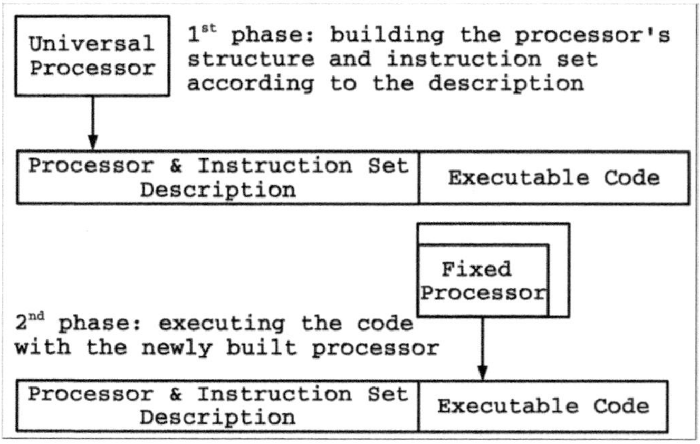

Fig. 1. Execution phases of the universal processor.

In the case of the universal processor the execution consists of 2 phases:

1. First the structure of the processor is built according to a description contained in the code of the digital organism itself. This structure remains constant for the whole lifespan of the organism.
2. Then the actual code is executed on the newly built processor. During replication the processor description is also copied and thus it is subject to mutation. So it may be changed.

The underlying theoretical construction is the universal Turing-machine [16]. Something similar happened in the hardware industry by introducing the `Crusoe` processors [9].

This idea is also motivated by biological analogies. In real life the decoding system, the mechanisms inside the cell, such as genetic code controlling transcription, translation, protein biosynthesis and genetic regulatory control (see, e.g. [13, Ch. 6]) evolved together with the genetic information, with DNA. How the genetic code for protein biosynthesis evolved is one of the biggest questions in today's evolutionary biology. *In silico* experiments may shed light on this mystery.

3 Processor Description Language

Obviously we need a genetic language to describe the inner structure of the processor and its instruction set. As mentioned before we treat a digital organism as a string of integers comprising a circular genome (as in prokaryotes or Avida). Unlike in other Tierra-like systems, the context of decoding determines whether an integer is an executable instruction or a component of the structural description: the processor treats an integer as an instruction in the instruction fetch phase while the same integer references a structural element (such as a register identifier) if an operand is needed. This is accomplished by using not directly integers $\mathbb{Z}$ but $\mathbb{Z}_m$, integers modulo m, where m is the number of components of the required type. This mapping of integers into either structural elements or instruction identifiers or positions in the genome constitutes an abstract analogue of DNA to RNA transcription (often said to be lacking in such systems, cf. [4]).

3.1 Genetic Description of the Processor Structure

Part of an archean's genome specifies its processor architecture. There are 3 basic building blocks – we call them structural elements or primitives representing basic data structures in the organism's specification of the processor that its code will run on. Each structural primitive has a corresponding symbol and we need a special mark `B` for separating structural elements:

`R` register
`S` stack
`Q` queue
`B` blank

For example, a `RRSSSBSS` string in the genome represents two registers and two stacks with sizes of 3 and 2. (A stack or queue of size 1 acts as a register, so it would be an equivalent possible choice to use `SB` instead of `R`.) For every structural element (stack, register or queue) a unique integer is assigned, by which it can then be referenced by executable instructions in the archean's genome. The structural element referenced by a simple zero is the instruction pointer by default.

3.2 Genetic Description of the Instruction Set

The instruction set of an archean is defined by using the universal processor's *basic instruction set*. Those instructions can be used as the smallest building blocks of a newly defined instructions. A digital organism cannot execute basic instructions directly but executes instructions defined in the genome itself with the syntax below. The descriptive part of the creature's genome is separated from the executable part by a special `SEPARATOR` instruction which marks where execution should begin. Instruction definition in an archean's genome specifies instructions that are comprised of basic instructions of the universal processor.

The basic instruction set was designed to fulfill the following requirements:

- each instruction should be as simple as possible:
  - an instruction as elementary building block represents a single action not a compound one
  - an instruction is independent from any addressing mode
- according to RISC philosophy only dedicated `load/store` instructions can access the memory [14]
- the instruction set should be complete (any more complex operation should be easily definable)

The basic instructions can be categorized in the following way (see [2] for more details and exact specifications):

Data transfer. There are 3 types of data transfer: between structural elements, between a structural element and the environment, between structural elements and memory.
`in` reads data from environment,
`out` writes data into the environment,
`load` copies from memory to a structural element,
`store` copies from a structural elements to memory,
`move` general move between structural elements.

Control-flow. Two instructions are enough to build any control structure.
`jump` unconditional jump,
`ifzero` skips the next instruction if operand not zero.

Arithmetic and Logic. These are the usual instructions for mathematical and logical operations. The operands are structural elements and the result is stored in a structural element. These can be mixed arbitrarily. A defined instruction can even use one structural elements for operands and results, if it is for example a stack.
`compare, add, sub, neg, div, mod, shift-l, shift-r, and, or, xor, not, is_sep ...`

Biological. These instructions are required for replication.
`cinc,cdec` cyclic increment/decrement for circular genomes,
`allocate` allocates memory,
`divide` splits the child and the parent program.

In order to enable inter-organism communication or parallelly executing organisms this basic instruction set could be extended by the appropriate additional basic operations.

The description of an instruction begins with an `I` symbol after which comes the code of the newly defined instruction – just like defining a subroutine in a higher level programming language. This technique enables evolution of modularization. The semantics of the instruction is defined by a microprogram written in the assembly language of the universal processor[1]. In short:

```
I basic_instr [operand...] [basic_instr [operand...]]...
```

[1] It can be said that a CISC processor is defined on a RISC architecture [14].

For example `I move 0 1` is an instruction that copies the content of the instruction pointer to the first structural element. The operands point to structural elements which are indexed by a nonnegative number (modulo the total number of structural elements). Also a number uniquely identifies an instruction (modulo the size of instruction set). Again it depends only on the context of execution whether an integer denotes a structural element or executable instruction or simply a data cell.

4 The Original Replicators

For starting the evolutionary process we need an original replicator. For the time being this should be written by a human. As such it consists of 3 distinct parts. (Fig. 2). In spite of this clear structure this organism is far from being optimal in terms of replication speed.

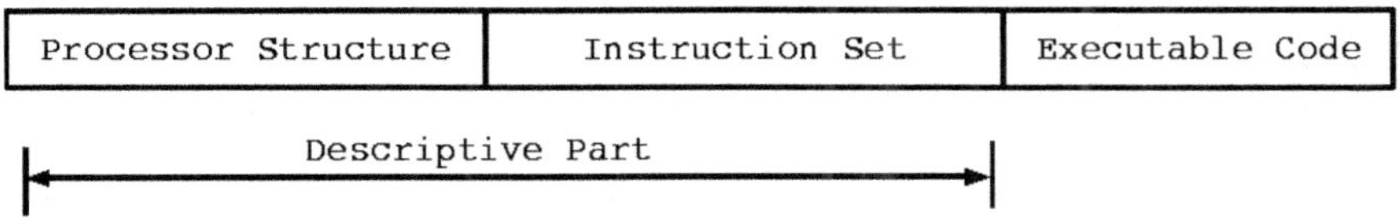

Fig. 2. The structure of an original archean replicator. The lengths of the different sections are not proportional to the number of cells. (length of structure description 8, definition of the instructionset 52, executable code for self-replication 18 instructions)

Two original replicators were designed having different processor architectures. One uses 4 registers (Fig. 3), and the other uses 2 registers and a stack of size 2. The instruction set reflects the difference but otherwise both have the same replication algorithm (count the instructions cyclically forward to the `SEPARATOR` in one loop, allocate memory, copy the instructions in another loop, divide).

5 Experimental Set-ups and Results

We are interested in changes of the processor structures over evolutionary time and in checking whether universal processors have at least as much evolutionary potential as their fixed counterparts or not. To assess the evolutionary potential of our system, we implemented two scenarios. In the first, no external or environmental rewards were provided and the archeans evolved only subject to the constraints of their biotic environment. In the second, archeans were rewarded with additional clock cycles for exploiting external environmental resources by performing certain operations.

R	I	I	I	0
B	move	load	load	1
R	0	2	1	2
B	2	4	4	3
R	I	is_sep	I	4
B	clear	4	rel-store	5
R	1	4	1	6
	I	ifzero	2	3
	move	4	4	7
	0	I	I	8
	3	jump 3	dec	2
	I	I	1	11
	inc	allocate	I	9
	1	1	ifnotzero	10
	I	I	1	12
	cinc	move	I	6
	2	1	divide	13
		2	SEPARATOR	

Fig. 3. The genome of the original archean replicator (read top-down, left-right). The first column is the structrual description (4 registers), the last column is the archean's actual executable code for self-replication and the three columns in the middle constitute the definition of the instruction set. For example the number 13 in the executable part refers to the last defined instruction containing only the `divide` basic instruction.

We carried out 5 different experiments for each setup. The maximum population size was 20000, the system runs for 200000 update cycles. This results in roughly a few thousands of generations of archeans.

We use an observational *fitness* f for measuring (but not guiding!) the evolutionary performance, defined as in Avida [3]:

$$f = \frac{m}{\gamma} \tag{1}$$

where m is merit, defined as the organism's effective length (the number of executed instructions in the genome) possibly multiplied by the bonus values earned by performing computational tasks, and γ is the gestation time (i.e. the number of processor cycles needed for self-replication).

5.1 The Simple Scenario

In this case fast replication is rewarded via natural selection: each organism receives constant size timeslice of 11 cycles and no computational task is rewarded. Evidently in the presence of these constraints the organisms try to decrease their gestation time either by shortening the genome or optimizing the replication algorithm. Most of the evolved organisms have less than half of the original replication time (Fig. 4). The handwritten progenitor is inefficient: it executes extra instructions when measuring its size by scanning its genome from SEPARATOR cyclically, but the evolutionary process was able to improve this design in each run by one of the following mechanisms: instead of measuring its size in a loop the archean starts to execute the descriptive part of the genome resulting in faster determination of size, or it just produces its size by using the cdec instruction, which is decrementation modulo the length of the genome.

Organism	Genome length	Gestation times (1st & 2nd offspring)	Effective length
4regs original	78	857, 857	17
evolved	76 ± 3.7	498 ± 84.6, 414 ± 52.2	59.2 ± 26.2
2regs1stack	81	1369, 1369	17
evolved	80.4 ± 2.3	732.2 ± 43.7, 662.2 ± 34.9	49.6 ± 20.8

Fig. 4. Comparing the two different ancestral archeans and their evolved descendents.

As general trends the instruction set usually gets shorter by losing one or two instructions (but retains its size by defining empty instructions), and the use of operands read from the genome instead of hardcoded in the instruction set appears. The structure of the processors remains generally constant in each run although variants appear but never dominate in the population.

5.2 The Evolutionary Learning Scenario

In this case we would like to test whether universal processors have some drawbacks beyond being slower as compared to static processors. This scenario uses a task-handler subsystem, like in Avida [3], to model the exploitation of external environmental resources (other than time and space). In addition to replicating themselves digital organisms can get longer execution time by performing simple calculations i.e. reading integers from the environment, transforming and writing the result back. All activities of the organisms are monitored. The average timeslice given to an archean for one update is 11 processor cycles.

In this case changing the instruction set is more vital, as the instructions for communicating with the environment are not defined in the progenitor. The execution of the descriptive part had to evolve to provide the instructions for I/O operations necessary to perform the tasks.

The results of these experiments support the hypothesis that digital organisms with universal processors are able to learn the same set of computational tasks as the ones with fixed processors. There were no observed signs of the different evolutionary potential so far.

Fig. 5 shows the dynamics of the maximum and the average fitness through time. Leaps indicate the evolutionary innovations (i.e. the learning of some tasks). The evolutionary learning process shows the dynamics described by the theory of punctuated equilibrium [7].

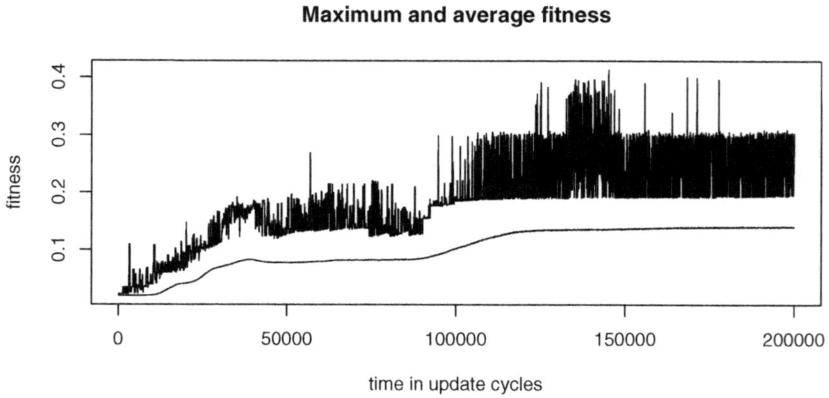

Fig. 5. The dynamics of average (below) and maximum fitness (above).

5.3 Evolution of Instruction Set and Processor Structure

The observations in terms of the evolution of processor structure can be summarized in two categories. One is for those attributes which did not change:

- Evolution is conservative in changing the processor's inner structure. Although there were several variants (e.g. with fewer or more registers, stacks or queues instead of a register, etc.), they could not gain domination over the population. The ratio of the original processor structure did not fall below 90% in the course of evolution observed.
- The number of defined instructions was also preserved during the evolutionary process similarly to the processor's structure. The rigidity of the mapping of integers to instructions might be a possible explanation for this observation.

Persistent change in the processor structure and instruction set size may require more time or some kind of macro-evolutionary steps in order to appear. Alternatively we may consider the problem of redesigning of the universal processor's structure in terms of a more flexible mapping of instruction codes to instructions.

The second category comprises attributes which were observed to vary:

- The instruction set was highly modified, especially when evolutionary learning was involved: instruction definitions were extended with new basic instructions or even completely changed.
- The sectioning present in the original handwritten organism vanished, i.e. the descriptive part and the executable code became interwoven. This way some parts of the organism's code were reused and had two or more different meanings depending on the context.

6 Conclusions

We have demonstrated that the evolutionary potential of universal processors is at least high as for the fixed processors in previous systems. The observed evolutionary changes in processor structures justify the need for further extending these experiments. While the number of instructions remained unchanged in our experiments, the instruction set itself did change radically during the course of evolution. The observed conservatism in number and type of processor components and instruction set size indicates a rigidity against changing the interpretation of instructions analogous to conservatism of DNA to RNA transcription in living organisms [13, Ch. 5-6]. By evolving the genotype-phenotype mapping by puting processor design and instruction set definition under evolutionary control, we may realize digital organisms with greater evolutionary potential.

References

1. Avida. `http://dllab.caltech.edu`. Digital Life Laboratory.
2. Physis. `http://physis.sourceforge.net`.
3. Chris Adami and C. Titus Brown. Evolutionary learning in the 2D artificial life system "Avida". *Proc. Artificial Life IV*, pages 377–381, 1994. MIT Press.
4. Christoph Adami and Claus O. Wilke. The biology of digital organisms. *Trends in Ecology & Evolution*, 17(11):528–532, November 2002.
5. Marc de Groot. Primordial soup. unpublished.
6. C.L. Nehaniv (ed.). *BioSystems*, special issue on evolvability. 69(2-3), 2003.
7. Stephen Jay Gould and Niles Eldredge. Punctuated equilibrium comes of age. *Nature*, 366:223–227, November 1993.
8. Marc Kirschner and John Gerhart. Evolvability. *PNAS*, 95:8420–8427, 1998.
9. Alexander Klaiber. The technology behind the Crusoe processors, 2000. `http://www.transmeta.com`.
10. Charles Ofria, Christoph Adami, and Travis C. Collier. Design of evolvable computer languages. *IEEE Transactions on Evolutionary Computation*, 6(4):420–424, 2002.
11. Thomas S. Ray. An approach to the synthesis of life. *Artificial Life II., Studies in the Sciences of Complexity*, IX:371–408, 1992. Addison Wesley.
12. Thomas S. Ray. Evolution, complexity, entropy, and artificial reality. *Physica D*, 75:239–263, 1994.
13. J. Maynard Smith and E. Szathmáry. *The Major Transitions in Evolution.* W.H. Freeman, 1995.
14. Daniel Tabak. *RISC systems and applications.* Research Studies Press, 1996.
15. Tim Taylor. Some representational and ecological aspects of evolvability. In Carlo C. Maley and Eilis Boudreau, editors, *Artificial Life 7 Workshop Proceedings*, pages 35–38. 2000. Available online at: `http://homepages.feis.herts.ac.uk/~nehaniv/al7ev/cnts.html`.
16. Alan Turing. On computable numbers, with an application to the Entscheidungsproblem. *Proceedings of the London Mathematical Society*, 42:230–265, 1936.
17. G.P. Wagner and L. Altenberg. Complex adaptations and the evolution of evolvability. *Evolution*, 50(3):967–976, June 1996.

An Evolutionary Approach to Damage Recovery of Robot Motion with Muscles

Siavash Haroun Mahdavi and Peter J. Bentley

Dept. Computer Science, University College London, Gower St. London WC1E 6BT, UK
{mahdavi, p.bentley}@cs.ucl.ac.uk

Abstract. Robots that can recover from damage did not exist outside science fiction. Here we describe a self-adaptive snake robot that uses shape memory alloy as muscles and an evolutionary algorithm as a method of adaptive control. Experiments demonstrate that if some of the robot's muscles are deliberately damaged, evolution is able to find new sequences of muscle activations that compensate, thus enabling the robot to recover its ability to move.

1 Introduction

It has long been the dream of science fiction to have robots capable of self-repair and recovery from damage. While many of the ideas seem plausible (for example, R2D2 repairing other robots in the movie *Starwars*), most rely on conventional methods of replacing damaged mechanisms or switching to redundant systems. Consequently, they require the overheads of spare components or systems and a sufficiently dexterous and knowledgeable repair robot (which in reality is always a human being).

However, these may not be the most efficient approaches of coping with damage. Natural systems recover from damage in very different ways: they either regrow damaged parts of themselves or they adapt their behaviour to compensate for the loss of functionality. A dog may not be designed to walk on three legs, but it can learn to do so very effectively if it must.

In this work we focus on such biologically-inspired ideas. A self-adapting snake (SAS) robot is created, using shape memory alloy as its "muscles". An evolutionary algorithm is then used to evolve a sequence of muscle activations in order to enable the robot to propel itself along. By deliberately damaging some of the muscles of the SAS, we can then assess the capability of evolution to adapt the motion of the robot and use the remaining muscles in a more efficient manner.

2 Background

2.1 Self-Repairing and Shape Memory Alloy Robots

Although evolutionary design [1] is common, little work is evident in the field of self-repairing robotics. Most current research seems to be limited to theory and future predictions. Bererton and Khosla propose that there will be large self-sufficient robot

W. Banzhaf et al. (Eds.): ECAL 2003, LNAI 2801, pp. 248–255, 2003.

colonies operating on distant planets. They describe methods where teams of robots could repair each other, and they have done experiments on visual docking systems for this purpose [2]. Michael claims that in the future 'fractal robots' will emerge which can completely change their shape and so if damaged, can reassemble them selves to recover their previous capabilities [10]. Kamimura has built robots that can reconfigure themselves to perform different tasks (as have Dittrich et al [3]), though he has not yet looked into the idea of self-repair [7]. Perhaps the work most similar to that described here was performed by Støy [12], in which a chain robot made of nine modules is robust to signal loss and able to continue effective locomotion. However, unlike the system described here, the controllers are preprogrammed and incapable of learning novel movement strategies after damage to the robot.

The use of smart materials in robotics has already been investigated. For example, Kárník looked at the possible applications of walking robots, which use artificial muscles with smart materials as a drive [8]. Mills has written a book aimed at beginners which teaches them how to make simple eight-legged robots using smart materials as actuators [11]. However, no one has used smart materials and evolutionary algorithms together in a robot before. This work uses nitinol wires as muscles within a robot snake, with the muscle activation sequences evolved using genetic algorithms and finite state machines [5].

2.2 Smart Material

Nitinol, an alloy made of **Ni**ckel and **Ti**tanium, was developed by the **N**aval **O**rdinance **L**aboratory. When current runs through it, thus heating it to its activation temperature, it changes shape to the shape that it has been 'trained' to remember. The wires used in this project simply reduce in length, (conserving their volume and thus getting thicker), by about 5-8 % [6].

Shape Memory Alloys, when cooled from the stronger, high temperature form (Austenite), undergo a phase transformation in their crystal structure to the weaker, low temperature form (Martensite). This phase transformation allows these alloys to be super elastic and have shape memory [6].

The phase transformation occurs over a narrow range of temperatures, although the beginning and end of the transformation actually spread over a much larger range of temperatures. Hysteresis occurs, as the temperature curves do not overlap during heating and cooling [6]. With thick wires, this could bring about problems for the SAS as the NiTi wires would take some time before returning to their original lengths, however, due to the very small diameter of the NiTi wires used (~0.15mm), the hysteresis was almost negligible as the cooling to below the Martensite temperature, (Mf), was almost instantaneous [4].

3 Building the Self Adaptive Snake (SAS)

The main bulk of the robot snake is made of foam. This provides a restoring force great enough to restore the wires to their original lengths after each activation. The SAS used in the experiments uses twelve NiTi wires (diameter=0.176mm, activation (Austenite) temperature=70°C, recommended current 200mA). The body of the robot snake is split into four segments (this is an extension from the first SAS prototype described in [5] which used only a single segment). These segments are readily detachable from each other and the whole structure can be expanded or segments replaced. Each segment has three NiTi wires running down its length and a central copper wire that runs through the foam and supplies the power, much like a spinal chord carries nerve impulses to muscles through the body see Fig. 1. These four segments are connected together and the ends of the 'spinal chord' are connected together to create a continuous connection along the length of the robot snake see Fig. 2. The total weight of the robot snake is approximately 150g.

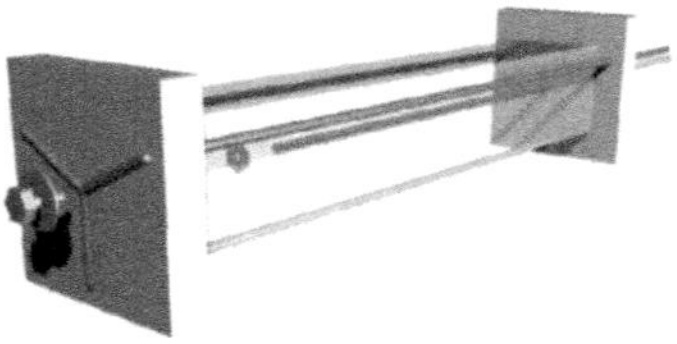

Fig. 1. A single segment of the robot snake showing the three NiTi wires and the central copper wire.

Fig. 2. A photograph of the SAS showing all four segments of the robot snake connected together.

Finite State Machine (FSM)

The NiTi wire activations are determined by a FSM, this is evolved by a special genetic algorithm designed especially for this task. The initial state of the FSM is constructed randomly and an array of FSMs was constructed representing a population of solutions.

Each member of the population is made up of a string of 0s and 1s. Each string consists of two segments, 'sequence' and 'next time', see Fig. 3. The 'sequence' is the part that is sent to the SAS, and determines which wires are to be activated at that particular time. The 'next time' is the part that tells the program to which time slot in the current row it should then jump. For the SAS, the 'sequence' length is twelve bits, and the 'next time' length is six bits, therefore the total length of each string is $2^6 \times (12 + 6) = 1152$ bits.

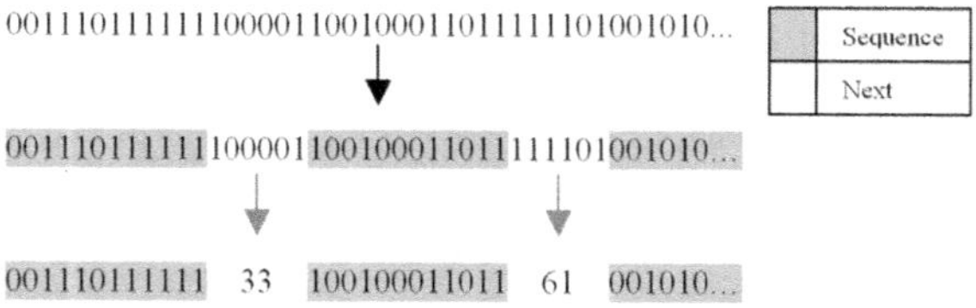

Fig. 3. Binary string is split into 'sequence' and 'next time' segments.

The program is allowed to send 30 sequences to the SAS before stopping, lasting a total of approximately 40 seconds. Finite state machines that act as repeating patterns are easily created if the jump points point to each other in a loop of some sort. Indeed, very rarely is there a string that did not loop at all.

These patterns are then sent to the SAS for evaluation. The fitness given to each individual in the population is directly proportional to how far the SAS travels after the sequences have been sent.

The computer interfacing hardware, though quite complex, has two simple tasks to perform. The first is to supply enough power to each NiTi wire (~200mA). This is achieved with the use of some Darlington amplifiers. The second task to perform is the ability to activate muscles in parallel. For this, the microcontroller used is the Motorola MC68HC908JK1 [9]. It has 20 pins and with the structure of the circuit board is capable of activating up to 12 pins in parallel (PTBx and PTDx excluding PTB0 and PTB3). [4,5]

Genetic Algorithm

After the program has gone through the whole population, the genetic algorithm is used to evolve the next generation of solutions. This method is described below using a sample string containing 6 'sequence' bits. Two strings are chosen from the population using roulette wheel selection. See Fig. 4

10110 **3**	111000 **4**	101111 **5**	110000 **1**	101011 **7**	001010 **3**	111001 **6**
010110 **2**	100100 **4**	100001 **5**	110111 **3**	001000 **6**	110010 **2**	100101 **3**

Fig. 4. Example of two strings chosen from the population of current solutions.

The repeating pattern that each individual defines is illustrated below see Fig. 5 & 6. These patterns can be extracted and better observed as simple loops. These simple loops are now taken as the chromosomes of the solutions and so only these parts of the complete string are crossed over. As can be observed in Figs 5 & 6, these chromosomes can be of varying lengths. In order to keep the length of the complete string constant, the shorter of the two chromosomes is taken, and a crossover section within its length is randomly cut. A section of the same size as that of the first chromosome is also cut at a random point along the second chromosome. These two sections are then swapped. Note that in order for the loops to remain functional, only the 'sequence' sections of the chromosomes are in fact crossed over leaving the 'next time' sections in unison, see Fig. 7.

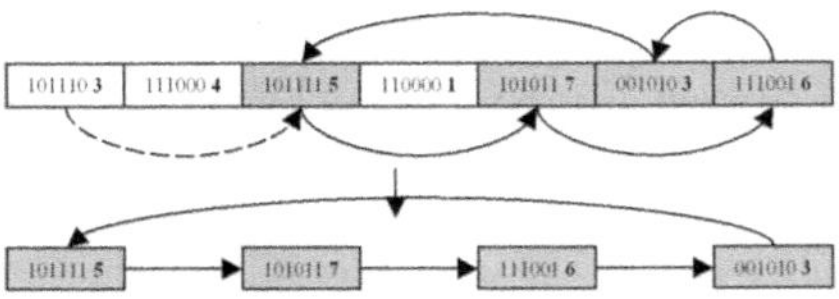

Fig. 5. The first string is converted into the loop that it represents.

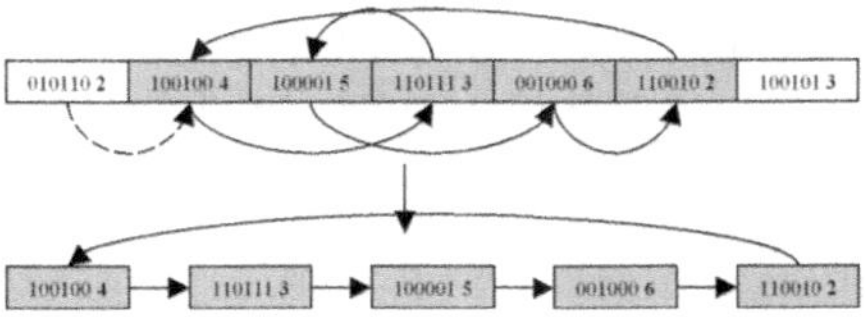

Fig. 6. The second string is converted into the loop that it represents.

Once this is done, mutation is carried out, where each bit of the string has an equal chance of being mutated, Fig. 8. These chromosomes are then placed back into their original slots in the complete string.

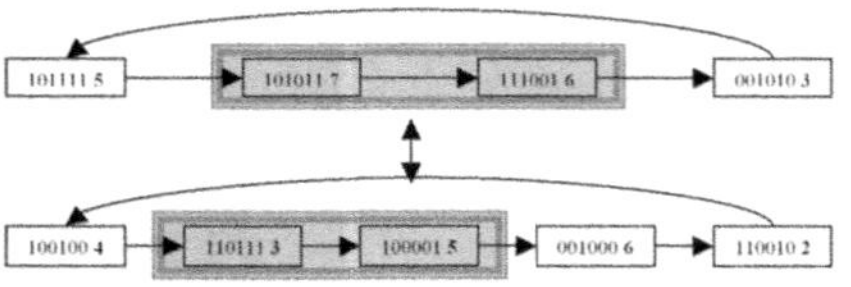

Fig. 7. Sections of the same size are swapped leaving the 'next time' sections intact.

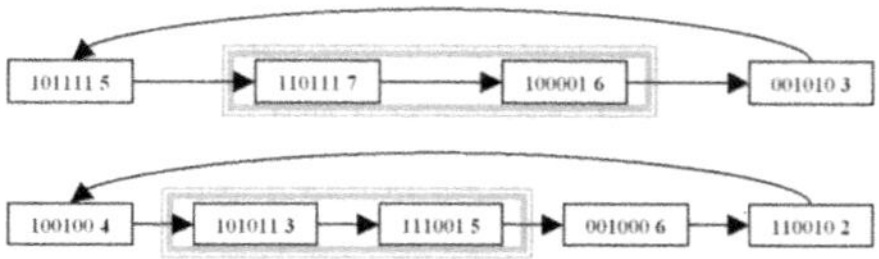

Fig. 8. Mutation is carried out only in the sections crossed over.

Only these chromosome fragments within the entire string are mutated, this ensures that these loops are accessed. If the rest of the sequence was mutated, then there is a good chance that a completely different loop would be accessed, thus the offspring would in no way resemble its parent and its fitness would be completely independent to that of the parent.

Though the sizes of both loops are conserved, mutation has a chance of completely redefining every loop, and so completely new solutions are still given a chance to be discovered which could have very different loop sizes. The genetic algorithm is elitist and so all but two of the solutions in the new population of solutions are constructed in the way described above. The last two solutions in the new population are a direct copy of the best two solutions of the previous population. This ensures that the maximum fitness in each generation never decreases.

4 Experimentation

4.1 Objectives

The objectives of this experiment are to investigate whether the genetic algorithm can find new methods of locomotion to recover from damage caused to the robot snake. In order to test this, a series of experiments are performed to investigate the evolution of movement for the SAS. The SAS's motion is evolved until the maximum fitness in the population of solutions remained unchanged for seven generations (chosen because of feasibility and time constraints). This was taken as a (somewhat local) maximum for the current configuration of the snake and its environment. Two wires that are widely used by the most fit individuals of the previous generations are then damaged. This was done by blocking access to them via software. The genetic algorithm is then allowed to continue evolving motion until another maximum is reached. Again two widely used NiTi wires were damaged and evolution was again allowed to continue.

4.2 Set up

The microcontroller board is attached to the top of the robot snake and all the NiTi wires are connected to the IO ports of the board, see Fig. 3. The only wires leaving the snake are the power supply and RS232 (both designed to minimise weight and resistance to movement). Another important factor that was realised in previous experiments [4,5] and taken into consideration is that the resistance of the wires remains more or less constant no matter how far the SAS moves.

The distance travelled by the robot snake can be measured to the nearest millimetre. Experimentation was done to observe the true accuracy by which the distance travelled should be stated. The results show that the distance travelled was always within the nearest millimetre, and so the distances travelled by the SAS can be stated to the nearest millimetre without being over accurate [4].

As the program starts, an initial population of randomly generated finite state machines is created. Each individual finite state machine is then sent to the SAS for evaluation. The fitness of any member is determined solely on how far the snake travels (in mm) in the forward direction. The total time taken to complete each generation is roughly twenty minutes (size of population (20) × time taken for each evaluation and reinitialisation (~1 minute)). The total time taken for the whole experiment to complete was 33 hours.

5 Results

Figure 9 shows the distance travelled at each generation. The greatest distance travelled by the SAS before any wires were damaged was 92mm. Immediately after two commonly used wires were damaged, the maximum distance travelled by the SAS in the next generation was 27mm. This is equivalent to a 71% drop in performance.

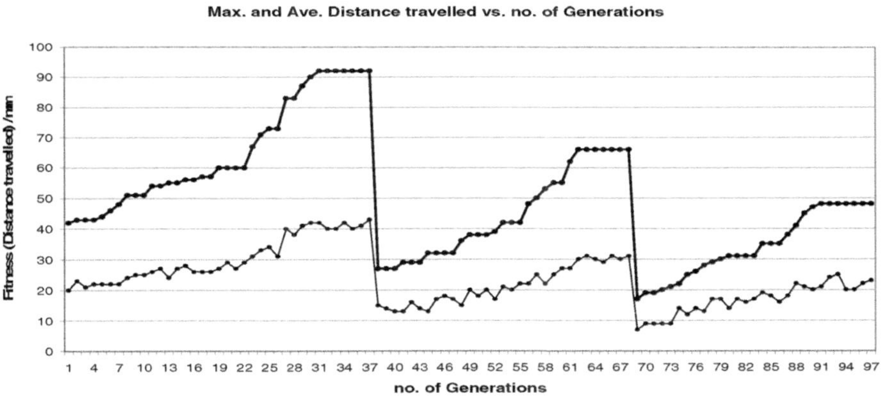

Fig. 9. Maximum fitness (bold) and average fitness plotted at each generation.

The SAS was then allowed to evolve for another 20 generations. This improved the distance travelled to a maximum of 66mm. This is an improvement in performance of 244%, and motion has been recovered to within 72% of the original undamaged SAS. The SAS was then once again damaged, removing another two commonly used wires from action, resulting in a 74% drop in performance. Finally the SAS was then allowed to evolve for a final 18 generations. This improved the distance travelled to a maximum of 48mm. This is an improvement in performance of 282%, and motion has been recovered to within 52% of the original undamaged SAS.

6 Analysis

During the evolution of the SAS, numerous interesting methods of locomotion were carried out. This section seeks to analyse the control strategies of the solutions at the three maximums of the graph above. These methods of locomotion correspond to the best sequence found before any wires were damaged, the best sequence found after the first pair of wires were damaged, and the best sequence found after the second pair of wires were damaged.

Though the sequence lengths could vary from a length of one to a maximum of sixty-four, the sequence lengths that travelled the furthest seemed to average a length of ten. The evolved sequences at these points are very complex. By way of illustration, the two most commonly used wires out of twelve in those sequences are shown below, see Fig. 10, 11 & 12. In each of the sequences, the two wires highlighted were simultaneously activated and deactivated in nearly every step, while a complicated sequence of activations was sent to the rest of the wires. The two wires that are highlighted in Fig. 10 are also the wires that were subsequently damaged, resulting in a new sequence (and thus an entirely new movement plan) that relied on two different wires being activated and deactivated as illustrated in Fig.11. Once again, the two wires highlighted were the most commonly activated in this second solution and so were subsequently damaged resulting in the final solution. This relied on two different wires being activated and deactivated more often than any other wire in order to move effectively (and thus a third entirely new movement plan), as illustrated in Fig. 12.

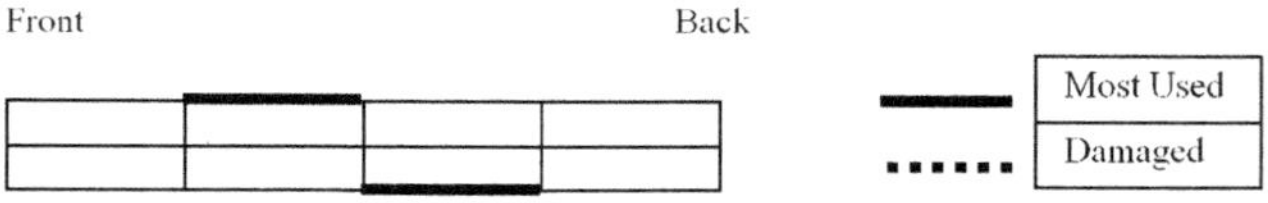

Fig. 10. Plan view of muscle wires in the four segments of the SAS (horizontal lines). The two most commonly used wires when there is no damage are shown in bold.

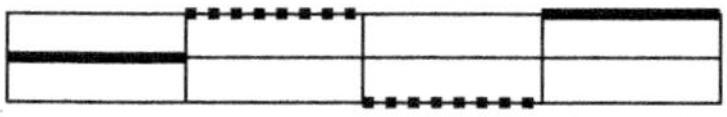

Fig. 11. The two most commonly used wires after two wires are damaged (bold), damaged wires are shown as dotted lines.

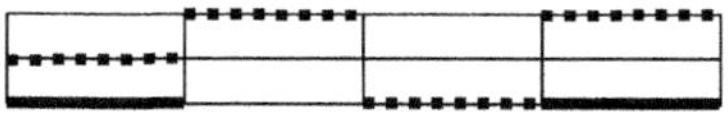

Fig. 12. The two most commonly used wires after four wires are damaged.

In previous experiments performed with fewer wires, the motion exhibited by the SAS could be compared to various types of snake undulation [4,5]. This has not been the case with twelve wires. There are two possible reasons for this. Firstly that the genetic algorithm had not had enough time to evolve elegant snakelike motion. The second reason could be that the genetic algorithm was exploiting asymmetries in the construction of the robot snake, the difference in the tautness of the NiTi wires, and the individual properties of these wires. This being the case, these more asymmetrical sequences would prove to be more effective for creating motion than if the SAS had explicitly tried to mimic snakelike motion.

7 Conclusion

This work has shown that it is possible to use a combination of a self-adaptive snake robot (using shape memory alloys to provide flexibility) and evolutionary control to produce a robot that is capable of recovering from damage by learning new movement strategies. The experiment demonstrated that a high percentage of locomotion can be recovered more than once, even when muscle wires crucial to a locomotion strategy are disabled. Further work will investigate exactly how the movement of the robot snake is achieved, through analysis of the evolved control strategies. A testbed will also be created, allowing robots and their locomotion to be assessed automatically, enabling faster and longer evolution runs and increased robot capabilities.

Acknowledgements. Thanks to Sara and Saba Haroun Mahdavi for their invaluable assistance in the experiments. This work is being performed in collaboration with BAE Systems.

References

1. Bentley, P. J. (Ed.) (1999) *EvolutionaryDesign by Computers.* Morgan Kaufmann Pub.
2. Bererton, C and Khosla, P. A Team of Robots with Reconfiguration and Repair Capabilities. In proceedings International Conference on Robotics and Automation, 2000.
3. Dittrich, P., Skusa, A., Kantschik, W., and Banzhaf, W. (1999) Dynamical Properties of the Fitness Landscape of a GP Controlled Random Morphology Robot. Proc. of GECCO '99, p. 1002-1008, Morgan Kaufmann, San Francisco, CA, 1999
4. Haroun Mahdavi, S. (2002) Evolving Motion Master's Dissertation, MSc IS, University College London, Dept. Computer Science.
5. Haroun Mahdavi, S and Bentley, P. J. (2002) Evolving Motion of robot with muscles. To appear in Proc. of *EvoROB2003, the 2nd European Workshop on Evolutionary Robotics.*.
6. Hodgson,D.E., Wu,M. H. and Biermann, R. J. (1999) Shape Memory Alloys. Shape Memory Applications, Inc. http://www.sma-inc.com/SMAPaper.html
7. Kamimura, A et al. (2001) Self-Reconfigurable Modular Robot – Experiments on Reconfiguration and Locomotion. Proc. of IEEE Int. Conf. on Intelligent Robots and Systems, 590–597, Hawaii, USA.
8. Kárník, L. (1999) The possible use of artificial muscles in biorobotic mechanism. In ROBTEP'99, Kosice, SF TU Kosice, 1999, pp. 109–114. ISBN 80-7099-453-3.
9. MC68HC908JK1 Microcontroller User's Manual (2002). Motorola Literature Distribution: P.O. Box 5405, Denver, Colorado 80217
10. Michael, J. *Robodyne Cybernetics Ltd.* www.fractal-robots.com
11. Mills, Jonathan W. (1999) Stiquito for beginners. ISBN 0-8186-7514-4, 1999.
12. Støy, K (2002) Using Role Based Control to Produce Locomotion in Chain-Type Self-Reconfigurable Robots. IEEE Transactions on Mechatronics, 7(4), pages 410–417, 2002.

Evolving Developmental Programs for Adaptation, Morphogenesis, and Self-Repair

Julian F. Miller

School of Computer Science, University of Birmingham, Birmingham, UK, B15 2TT
j.miller@cs.bham.ac.uk
http://www.cs.bham.ac.uk/~jfm

Abstract. A method for evolving a developmental program inside a cell to create multicellular organisms of arbitrary size and characteristics is described. The cell genotype is evolved so that the organism will organize itself into well defined patterns of differentiated cell types (e.g. the French Flag). In addition the cell genotypes are evolved to respond appropriately to environmental signals that cause metamorphosis of the whole organism. A number of experiments are described that show that the organisms exhibit emergent properties of self-repair and adaptation.

1 Introduction

The mechanisms whereby mature multicellular biological organisms are constructed from the original fertilized egg are extraordinary and complex. Living organisms are perpetually changing yet outwardly they often give the appearance of stasis. This is accomplished by cells constantly reproducing and dying. There is no central control behind this. Such a process leads naturally to the self-repairing properties of organisms. At present humans design systems using a top down design process. Until now, this methodology of design has served humanity well. However when one looks at the problem of constructing software and hardware systems with of the order of 10^{13} or more components (roughly the number of cells in the human body) it appears that we face a design crisis. As the number of interacting components grows the complexity of the system grows also. This can lead to many problems. It becomes difficult to formally verify such systems due to the combinatorial explosion of configurations or states. The cost of maintenance grows alarmingly also and their unreliability increases. Electronic circuits are now routinely constructed at sub-micron level. This is leading to immense wiring problems. As the trend for increasing miniaturization of circuits continues, it will become ever more difficult to supply the huge amounts of data required to define or program these devices. Living systems demonstrate very clever solutions to these problems. The information defining the organism is contained within each, and every, part. There is no central design. Consequently, it is likely that designers will have to increasingly turn to methods of constructing systems that mimic the

W. Banzhaf et al. (Eds.): ECAL 2003, LNAI 2801, pp. 256–265, 2003.

developmental process that occur in living systems. To some extent, such an agenda has been embraced in the nascent field of amorphous computing [1].

Genetic programming is a paradigm in which computer program are subjected to process that mimics Darwinian evolution [19]. The dominant representation of programs chosen is that of a tree in which there is no distinction between genotype and phenotype. This is a drawback if one is interested in problems that involve phenotypes of arbitrary size and complexity. Natural evolution works on the level of the genotype that is formed when an ovum is fertilized to form a zygote. This undergoes an extraordinary process of cell replication and differentiation to construct an entire organism. If higher level organisms were really colonies of cells with different genotypes it would have been much harder for evolution to evolve organisms of the complexity and sophistication of many living creatures. Thus, it seems imperative that the conventional GP paradigm must be extended to encompass development. The poor scalability of directly encoded systems (i.e. a one-to-one mapping from genotype to phenotype) is particularly evident in the evolution of neural networks, where each link requires a floating-point weight that must be determined.

The work presented in this paper is motivated by an number of questions: (1) How can we define programs that run inside cells that construct complex structures of arbitrary size, when each cell runs an identical program? (2) Is it possible to evolve organisms that can adapt to environmental signals? (3) Can organisms be evolved that are capable of self-repair? It is not the aim of this work to model closely natural developmental processes, but rather, to explore a simple idealization of biological development in the hope that it will exhibit some of the advantages of biological systems. It is hoped that computer science might benefit from such studies.

The plan for the paper is as follows: A review of relevant related work is given in section 2. Section 3 describes how the cells and their environment are represented, and the cell program's inputs and outputs. Section 4 describes the form of genetic programming used to evolve the cell program. Section 5 describes the experiments and the results obtained. In section 6 some conclusions and ideas for future work are given.

2 Related Work

Fleischer and Barr created a sophisticated multi-cellular developmental test bed and included realistic models of chemical diffusion, cell collision, adhesion and recognition [9]. Their purpose was to investigate cell pattern generation. They noted that the design of an artificial genotype that develops into a specific pattern is very difficult. They also noted that size regulation is critical and non-trivial and that developmental models tend to be robust to perturbations.

A number of researchers have studied the potential of Lindenmeyer systems [20] for developing artificial neural networks (ANNs) and generative design. Boers and Kuiper have adapted L-systems to develop the architecture of artificial neural networks (ANNs) (numbers of neurons and their connections) [4]. They used an evolutionary algorithm to evolve the rules of a L-system that generated feed-forward neural networks. Backpropagation was used, and the accuracy of the neural networks on test data was assigned to the fitness of the encoded rules. They found that this method produced more modular neural networks that performed better than networks with a predefined structure. Kitano developed another method for evolving the architecture of an artificial neural network [17] using a matrix re-writing system that manipulated adjacency matrices. Although Kitano claimed that his method produced superior results to direct methods (i.e. a fixed architecture, directly encoded and evolved), it was later shown in a more careful study that the two approaches were of equal quality [23]. Gruau devised an elegant graph re-writing method called cellular encoding [11]. Cellular encoding is a language for local graph transformations that controls the division of cells that grow into artificial neural networks. This was shown to be an effective method for optimizing both the architecture and weights at the same time, and they found that, to achieve as good performance with direct encoding, required the testing of many candidate architectures [12]. Others have successfully employed this approach in the evolution of recurrent neural networks that control the behaviour of simulated insects [18]. Recently Hornby and Pollack have also evolved context free L-systems to define three dimensional objects (table designs) [15]. They found that their generative system could produce designs with higher fitness and faster, than direct methods. Jacobi created an impressive artificial genomic regulatory network, where genes code for proteins and proteins activate (or suppress) genes [16]. He used the proteins to define neurons with excitatory or inhibitory dendrites. This allowed him to define a recurrent ANN that was used to control a simulated Khepera robot for obstacle avoidance and corridor following. Nolfi and Parisi evolved encoded neuron position and branching properties of axonal trees that would spread out from the neurons and connect to other neurons [22] and in later work introduced cell division using a grammar [6]. Astor and Adami have created a developmental model of the evolution of an ANN that utilizes an artificial chemistry [2]. Eggenberger suggests that the complex genotype-phenotype mappings typically employed in developmental models allow the reduction of genetic information without losing the complex behaviour. He stresses the importance of the fact that the genotype will not necessarily grow as the number of cells, thus he feels that developmental approaches will scale better on complex problems [8]. Sims evolved the morphology and behaviour of simulated agents [24]. Bongard and Pfeifer have evolved genotypes that encode a gene expression method to develop the morphology and neural control of multi-articulated simulated agents [5]. Bentley and Kumar examined a number of genotype-phenotype mappings on a problem of creating a tessellating tile pattern [3]. They found that the indirect developmental mapping (that they refer to as an implicit embryogeny) could evolve the tiling patterns much quicker, and further, that they could be subsequently grown to (iterated) much larger sized patterns. One drawback that they reported, was that the implicit embryogeny tended to produce the same types of patterns. Other researchers are more motivated by fundamental biological aspects of cell behaviour. Furusawa and Kaneko are modeled cell internal dynamics and its relationship to the emergence of cell multi-cellularity[10]. Hogeweg has carried out impressive work in computer models of development and constructed a sophisticated model of cells (biotic) by modeling their internal dynamics by groups of cells in a cellular automaton that are subject to energy minimization [13][14]. The energy minimization leads to cell movement and sorting by differential cell adhesion. The cell genome was modeled as 24 node Boolean network that defined cell signaling and adhesion. She used a fitness criterion that was related to the difference in the gene expression in all the cells. She evolved organisms that exhibited many behaviours that are observed in living systems: cell migration and engulfing, budding and elongation, and cell death and re-differentiation.

3 Cell and Chemical Representation

The cell's genotype is a representation of a feed-forward Boolean circuit (that implements the cell program). This maps the cell's input conditions to output behaviour. A cell is a square in a non-toroidal two-dimensional cellular automaton. Each live cell sees its own state and the states of its eight immediate neighbours. It also sees the amount of chemical (at present a single chemical) at the location of each of its eight neighbours. Using this information, the cell's program decides on the amount of chemical that it will produce, whether it will live, die, or change to a different cell type at the next time step, and how it will grow.

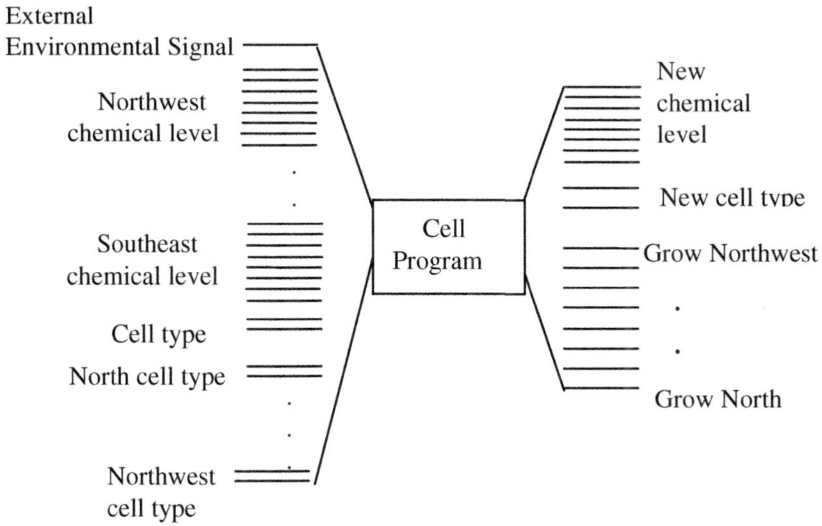

Fig. 1. If an external environmental signal is (not) defined a cell's program reads 83 (82) binary bits: 64 bits represent the amount of chemical at the eight immediate neighbours and 18 bits represent all the cell types in the Moore neighbourhood. There are 18 output bits in the program: eight for the amount of chemical that will be placed at the locations where the replicated cell exists at the next time step, two for the new cell type, and eight bits that define the places where the new cell can grow

Unlike real biology, when a cell replicates itself, it is allowed to grow in any or all of the eight neighbouring cells simultaneously (this is done to speed up growth, mainly for reasons of efficiency). In all the experiments reported in this paper there are three cell types (blue, red, white) and the amount of chemical is represented by an eight-bit binary number. The cell types are represented by two-bit binary codes: 00 = dead or absence of a cell, 01 = blue cell, 10 = red cell, 11 = white cell. Only live cells have their programs executed. Initially a single cell is placed in the grid (the zygote). In the experiments reported in this paper, the zygote has been chosen to be white In addition an amount of chemical equal to 255 (the maximum) is placed at the zygote's location. If two or more cells decide to grow into the same location at the next time step, the last such cell in the scan path overwrites all previous growths. This was chosen as it greatly simplified the process of constructing the newly grown organism. The two dimensional grid is scanned from the top-left corner to the bottom right. The process of constructing the new organism at time t+1 from the organism at time t is the following: Every live cell from the top-left to the bottom-right has its program run (all cells run the same program). A new map (initially empty) is created and filled with cells that have either grown, or not died, in the map at time t. After all the programs inside the living cells have been run, the map at time t+1 replaces the map at time t. The chemical map is updated in a similar manner. A depiction of the cell's inputs and outputs is shown in Fig. 1. The chemicals obey the diffusion rule (1).

$$(c_{ij})_{new} = 1/2(c_{ij})_{old} + \frac{1}{16}\sum_{k,l \in N}(c_{kl})_{old} \,. \tag{1}$$

This ensures that over time, chemicals diffuse away from their point of origin. The rule was designed so that diffusing chemical would be conserved (apart from the loss when the level falls below a level of one). Note that, since cell's can determine their own new level of chemical there is no strict conservation of chemical in the entire system. Let N denote the neighbourhood with neighbouring position k,l, the chemical at position i,j at the new time step is given by (1).

4 Cartesian Genetic Programming and the Cell Program

Cartesian Genetic Programming was developed from methods developed for the automatic evolution of digital circuits [21]. CGP represents a program or circuit as a list of integers that encode the connections and functions. The representation is readily understood from a small example. Consider the one bit binary adder circuit (Fig. 2). This has three inputs that represent the two bits to be summed and the carry-in bit. It has two outputs: sum and carry-out. CGP employs an indexed list of functions that represent in this example, various two input logic gates and three input multiplexers. Suppose that in a function lookup table AND is function 6, XOR is function 10 and MUX is function 16. The three inputs A, B, Cin are labeled 0, 1, 2. The output of the left (right) XOR gate is labeled 3 (5). The output of the MUX gate is labeled 6. The AND output is labeled 4. In Fig. 2 a genotype is shown and how it is decoded to a phenotype (the one-bit binary adder). The integers in italics represent the functions, and the others represent the connections between gates, however, if it happens to be a two input gate then the third input is ignored. It is assumed that the circuit outputs are taken from the last two nodes. The second group of four integers (shown in grey) represent an AND gate that is not part of the circuit phenotype. Since only feed-forward circuits are being considered, it is important to note that the connections to any gate can only refer to gates that appear on its left.

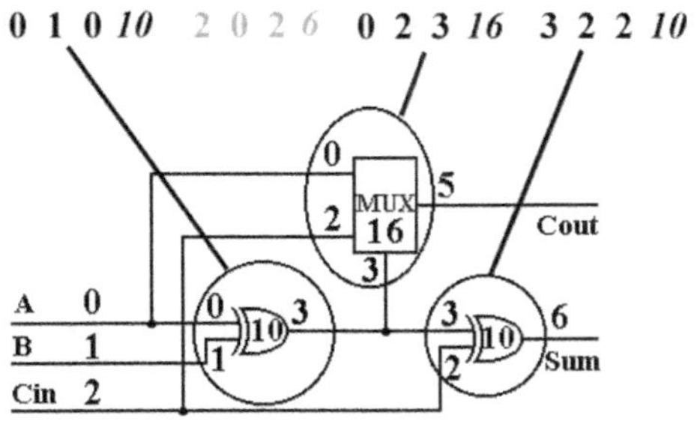

Fig. 2. One-bit adder circuit.

Typically CGP uses point mutation (that is constrained to respect the feed-forward nature of the circuit). Suppose that the first input of the MUX gate (0) was changed to 4. This would connect the AND gate into the circuit (defined by the four grey genes). Similarly, a point mutation might disconnect gates. Thus, CGP uses a many to one genotype-phenotype mapping, as redundant nodes may be changed in any way and the genotypes would still be decoded to the same phenotype. The (1+4)-ES evolutionary algorithm uses characteristics of this genotype-phenotype mapping to great advantage (i.e. genetic drift). This is given below:

1. Generate 5 chromosomes randomly to form the population
2. Evaluate the fitness of all the chromosomes in the population
3. Determine the best chromosome (called it *current_best*)
4. Generate 4 more chromosomes (offspring) by mutating the *current_best*
5. The *current_best* and the four offspring become the new population
6. Unless stopping criterion reached return to 2

Step 3 is a crucial step in this algorithm: if more than one chromosome is equally good then the algorithm always chooses the chromosome that is not the *current_best* (i.e. equally fit but genetically different). This step allows a genetic drift process that turns out be of great benefit [25][27]. The mutation rate is defined to be the percentage of each chromosome that is mutated in step 4. In all the experiments described in this paper only four kinds of MUX logic gates were employed defined by the expression f(A,B,C)=AND(A, NOT(C)) OR AND(B, C). The four types correspond to cases where inputs *A* and *B* are either inverted or not. Since in this paper 18 output bits are defined, the rightmost 18 multiplexers are used. The first of these defines the most significant bit of the new chemical level.

5 Evolutionary Experiments and Results

The French Flag model of Lewis Wolpert [26] was the inspiration for the task the maps of cells were to achieve. The following experiments were carried out.

1. Produce a growing and recognizable French flag
2. Produce a growing pattern of blue spots
3. Produce an adaptive growing pattern of blue or red spots

In the first experiment two organisms representing French flags were defined. These were used to compare the evolved organism at particular times in the development of the organism. The evolved organism and target organism were compared cell by cell and a cumulative score of correctness was calculated. This was the fitness of the cell genotype used in the evolutionary algorithm.

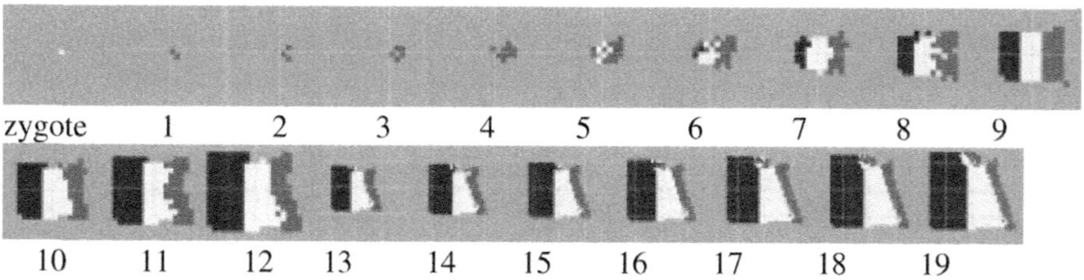

Fig. 3. Best solution for task 1 (fitness test points at time = 7 and 9)

When the organism produced in task 1 is iterated beyond iteration 9 it begins to deform a little (Fig. 3.). However it still remains recognizable as the French flag. In a further post-evolution experiment the organism in task 1 was subjected to severe damage. Firstly, a large rectangular section of the organism at iteration 9 was removed and the organism allowed to develop. Secondly, the white and red sections were removed. Fig. 4 shows how the organism is able to reconstruct itself, though it retains some scarring. Note the images changes scale at iteration 13 (Figs. 3 and 4).

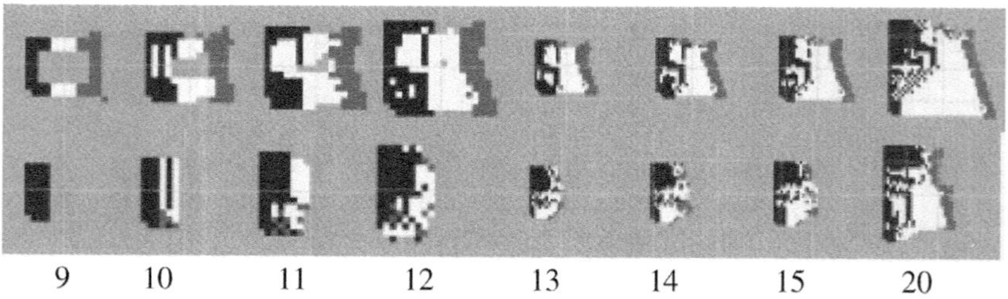

Fig. 4. The organism for task one is damaged but repairs itself

The second experiment required the evolved organism to produce a growing (and always faithful) pattern of blue rectangular spots surrounded by white cells.

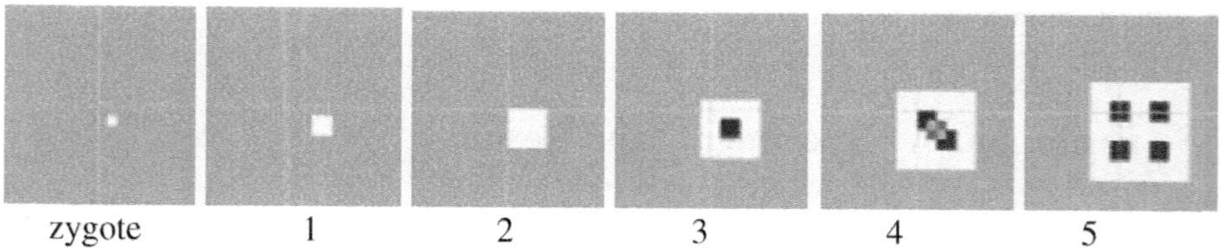

Fig. 5. A perfect solution for task 2 (single fitness test point at time = 5). The temporal pattern alternates indefinitely between blue spots and an elegant tiling pattern.

The third experiment utilized the external environmental signal and required the evolved organism to be a growing blue spots pattern when the signal was zero and a growing red spots pattern when the signal was one. The fitness of the evolved organism was compared against the desired blue and red spots pattern and a fitness computed in the usual manner. The experimental parameters were: population = 5, generations = 30,000, runs = 10, mutation rate = 1%, max number of multiplexers = 200.

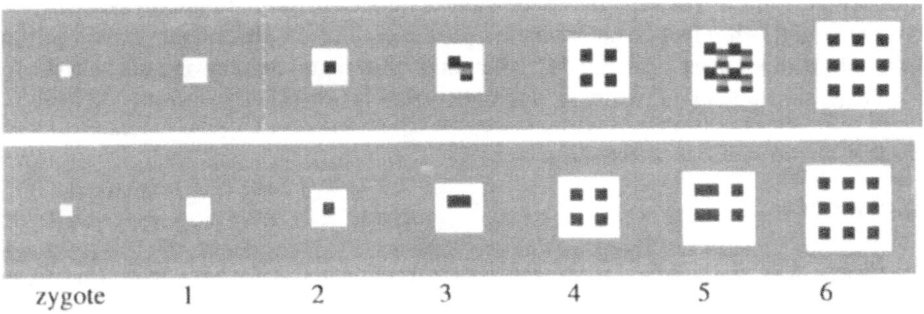

Fig. 6. A perfect solutions for task 3 (fitness test points at time = 4 and 6)

The behaviour of the adaptive organism in task 3 changes when exposed to new environmental signal that it hadn't encountered during evolution. Fig. 7 shows that for an alternating environmental signal (0 then 1 alternately) the organism adopts a stable but very different behaviour. If the environmental signal takes other temporal sequences of 0 and 1 much more chaotic behaviour is obtained. If the chemical is changed, or the chemical program bits ignored, the organisms behaviour can be radically affected. In the case of task 3 the chemical map behaved in a very different way depending on the environmental signal. Organisms evolved without chemicals were of much lower fitness and lacked stability. The 18 programs (see Fig. 1) that produce each of these behaviours have been examined and are difficult to interpret. Given the 83 binary input bits it would be fiendishly difficult for a human designer to create programs that meet the defined requirements. Detailed analysis of these will be published in due course.

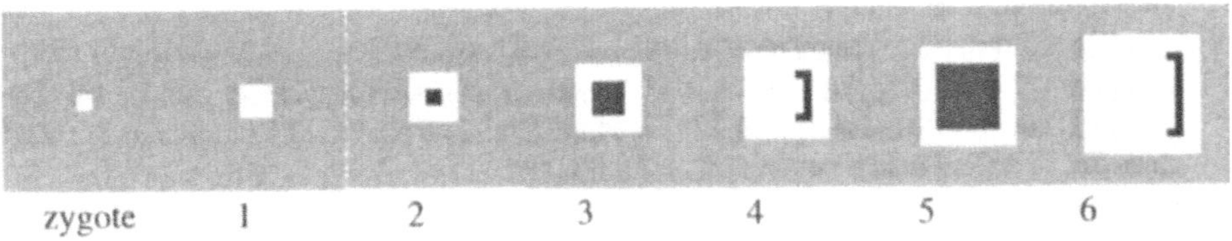

Fig. 7. Behaviour of task 3 organism with alternating environmental signal

6 Conclusions and Further Work

A developmental genotype has been described that can under evolution, and solve tasks with phenotypes of arbitrary size capable of self-repair and adaptation under global environmental change. In the future, coordinated cell growth and movement will be investigated. Careful work remains to be undertaken to examine the role of chemicals. Future investigations will consider the chemical as energy so that cell growth and death might become an emergent phenomenon, this might also be helpful in establishing a more open ended evolution, where multi-cellular organisms fight for survival, thus linking morphology with behaviour.

References

1. Abelson H., Allen D., Coore D., Hanson C., Homsy G., Knight Jr. T., Nagpal R., Rauch E., Sussman G. J., Weiss R. (1999), "Amorphous Computing", MIT Tech. report, AI Memo 1665.
2. Astor J. C., and Adami C. (2000), "A Development Model for the Evolution of Artificial Neural Networks", Artificial Life, Vol. 6, pp. 189–218.
3. Bentley P., and Kumar S. (1999), "Three ways to grow designs: A comparison of embryogenies for an Evolutionary Design Problem", in Proceedings of the Congress on Evolutionary Computation, IEEE Press, pp. 35–43.
4. Boers, E. J. W., and Kuiper, H. (1992), "Biological metaphors and the design of modular neural networks", Masters thesis, Department of Computer Science and Department of Experimental and Theoretical Psychology, Leiden University.
5. Bongard J. C. and Pfeifer R. (2001), "Repeated Structure and Dissociation of Genotypic and Phenotypic Complexity in Artificial Ontogeny", in Spector L. et al. (eds.) Proceedings of the Genetic and Evol. Comput. Conference, Morgan-Kaufmann, pp. 829–836.
6. Cangelosi, A., Parisi, D., and Nolfi, S. (1993), "Cell Division and Migration in a 'Genotype' for Neural Networks", Tech. report PCIA-93, Inst. of Psych., CNR, Rome.
7. Dellaert, F. (1995), "Toward a Biologically Defensible Model of Development", Masters thesis, Dept. of Computer Eng. and Science, Case Western Reserve University.
8. Eggenberger P. (1997), "Evolving morphologies of simulated 3D organisms based on differential gene expression", Proc. Of European Conf. on Artificial Life, pp. 205–213.
9. Fleischer, K., and Barr, A. H. (1992), "A simulation testbed for the study of multicellular development: The multiple mechanisms of morphogenesis", in Langton C. G (ed.) Proceedings of the 3rd Workshop on Artificial Life, Addison-Wesley, pp. 389–416.
10. Furusawa C., and Kaneko, K. (1998), "Emergence of Multicellular Organisms with Dynamic Differentiation and Spatial Pattern", in Adami C. et al. (eds.) Proceedings of the 6th International Conference on Artificial Life, MIT Press.
11. Gruau, F. (1994), "Neural Network Synthesis using Cellular Encoding and the Genetic Algorithm", PhD thesis, Ecole Normale Supérieure de Lyon.
12. Gruau, F., Whitley, D., and Pyeatt, L. (1996) "A Comparison between Cellular Encoding and Direct Encoding for Genetic Neural Networks", in Proc. of the 1st Annual Conference on Genetic Programming, Stanford.
13. Hogeweg, P. (2000), "Evolving Mechanisms of Morphogenesis: on the Interplay between Differential Adhesion and Cell Differentiation", J. Theor. Biol., Vol. 203, pp. 317–333.

14. Hogeweg, P. (2000), "Shapes in the Shadow: Evolutionary Dynamics of Morphogenesis", Artificial Life, Vol. 6, pp. 85–101.
15. Hornby G. S., and Pollack J. B. (2001), "The Advantages of Generative Grammatical Encodings for Physical Design", in Proceedings of the Congress on Evolutionary Computation, IEEE Press, pp. 600–607.
16. Jacobi, N. (1995), "Harnessing Morphogenesis", Cognitive Science Research Paper 423, COGS, University of Sussex.
17. Kitano, H. (1990), "Designing neural networks using genetic algorithms with graph generation system", Complex Systems, Vol. 4, pp. 461–476.
18. Kodjabachian, J. and Meyer, J-A. (1998), "Evolution and Development of Neural Controllers for Locomotion, Gradient-Following and Obstacle-Avoidance in Artificial Insects", IEEE Transactions on Neural Networks, Vol. 9, pp. 796–812.
19. Koza, J. R.: Genetic Programming: On the programming of computers by means of natural selection. MIT Press (1992), Genetic Programming II: Automatic Discovery of Reusable Subprograms. MIT Press (1994)
20. Lindenmeyer, A. (1968), "Mathematical models for cellular interaction in development, parts I and II", Journal of Theoretical Biology, Vol. 18, pp. 280–315.
21. Miller, J. F. and Thomson, P. (2000), "Cartesian genetic programming", in Proceedings of the 3rd European Conf. on Genetic Programming. LNCS, Vol. 1802, pp.121–132.
22. Nolfi, S., and Parisi, D. (1991) "Growing neural networks", Technical report PCIA-91-15, Institute of Psychology, CNR, Rome.
23. Siddiqi, A. A., and Lucas S. M. (1998), "A comparison of matrix rewriting versus direct encoding for evolving neural networks", in Proceedings of the 1998 IEEE International Conference on Evolutionary Computation, IEEE Press, pp. 392–397.
24. Sims K. (1994), "Evolving 3D morphology and behaviour by competition", in Proceedings of Artificial Life IV, pp. 28–39.
25. Vassilev V. K., and Miller J. F. (2000), "The Advantages of Landscape Neutrality in Digital Circuit Evolution", 3rd Int. Conf. on Evolvable Systems: From Biology to Hardware, LNCS, Vol. 1801, Springer-Verlag, pp. 252–263.
26. Wolpert, L. (1998), "Principles of Development", Oxford University Press.
27. Yu, T. and Miller, J. (2001), "Neutrality and the evolvability of Boolean function landscape", in Proceedings of the 4th European Conference on Genetic Programming, Springer-Verlag, pp. 204–217.

Evolving Embodied Genetic Regulatory Network-Driven Control Systems

Tom Quick[1], Chrystopher L. Nehaniv[2], Kerstin Dautenhahn[2], and Graham Roberts[1]

[1] Department of Computer Science
University College London
Gower Street, London WC1E 6BT, U.K.
{t.quick,g.roberts}@cs.ucl.ac.uk
[2] Adaptive Systems Research Group, University of Hertfordshire
Hatfield, Herts AL10 9AB, U.K.,
{c.l.nehaniv,k.dautenhahn}@herts.ac.uk

Abstract. We demonstrate the evolution of simple embodied Genetic Regulatory Networks (GRNs) as real-time control systems for robotic and software-based embodied Artificial Organisms, and present results from two experimental test-beds: homeostatic temperature regulation in an abstract software environment, and phototactic robot behaviour maximising exposure to light. The GRN controllers are continually coupled to the organisms' environments throughout their lifetimes, and constitute the primary basis for the organisms' behaviour from moment to moment. The environment in which the organisms are embodied is shown to play a significant role in the dynamics of the GRNs, and the behaviour of the organisms.

1 Introduction: Genetic Regulatory Networks and Environmental Coupling

In this paper we present initial results from a novel experimental system, *Biosys*, in which Genetic Regulatory Networks (GRNs) play a critical and active role throughout the lifetimes of evolved artificial organisms, continually driving the interplay between organism and environment, giving rise to coherent observable emergent behaviours.

The role of GRNs in the development of living organisms is well-known [1, 2]. Moreover, they play a key role in cellular metabolism throughout organisms' lifetimes [3]. For both prokaryotes and eukaryotes, there is continual, tightly woven interaction between an organism's genes, the protein machinery of the cell or cells in which the genes exist, and the organism's environment. Our model can be applied to achieving a basis for such control in artificial organisms. In this work we focus on the evolution of heritable regulatory mechanisms based on model GRNs in populations of artificial organisms, and show how these mechanisms can drive behaviour via continual coupling to an environment.

W. Banzhaf et al. (Eds.): ECAL 2003, LNAI 2801, pp. 266–277, 2003.

In Artificial Life (ALife) research, GRNs typically appear in one of two broad research contexts. The first is to study them as dynamical systems in isolation from environmental stimulus, for example, identifying basins of attraction and other dynamical systems features [4,5,6]. The second is to recognise them as an emulatable feature of biological developmental processes, contributing to an at least partially realistic mapping between genotype and phenotype. This approach is outlined by Dellaert and Beer [7], and has been utilised to grow neural networks [8], as well as body plans and nervous systems for multi-cellular artificial agents [9]. However, a common feature of these models that stands in stark contrast to biological organisms, and which we seek to address, is that once a phenotype has been produced from a genotype, the genome becomes totally redundant, having no further role to play in the structure or behaviour of the resultant artificial organism.

A similar developmental approach, though with greater attention to biological detail is employed by Eggenberger [10] and Kumar [11], who focus on growth and morphogenesis rather than agent control. In this context, issues regarding the post-development role of the genome are not directly relevant; however even during development, the role of the environment is minimal in these models, in contrast to the 'eco-devo' approach to development [12,13].

Artificial chemistry-based control systems modelled on cellular signal transduction have been developed that do exploit moment to moment interaction between an artificial organism and an environment as a basis for producing emergent behaviours [14]. However, genes only factor as part of the Evolutionary Algorithms used to produce the chemistries involved. Cellular components such as proteins and regulatory networks do not feature at all.

In contrast, real-time GRN-driven control does feature in wet-ware synthetic gene network engineering research [15]. Cross-fertilisation of principles and theory with ALife seems likely, particularly as the level of biological detail in ALife models increases to a level that makes the two programmes commensurate.

In section 2, we discuss our experimental methods, focusing on how *Biosys* works, how it is embodied, and describe the Genetic Algorithm (GA) used. We then present results from experiments evolving GRN-driven controllers embodied in two different environments, as follows. First, section 3 presents homeostatic behaviour in an abstract thermal environment, illustrating the workings of our model and some key principles. Second, section 4 presents phototactic, light-maximising robot behaviour in a simulated physical environment. Finally, in section 5 we present our conclusions.

2 Methodology: Embodiment, GRNs, and Evolution

2.1 Embodied Systems

In previous papers we have defined 'embodiment' in terms of the conditions necessary for structural coupling to occur between a system (for example an organism or a software agent) and an environment:

> A system X is embodied in an environment E if perturbatory channels exist between the two. That is, X is embodied in E if for every time t at which both X and E exist, some subset of E's possible states have the capacity to perturb X's state, and some subset of X's possible states have the capacity to perturb E's state. [16]

We constructed the model presented here with this definition in mind. One of the roles played by proteins in our model is to provide the basis for cellular embodiment, by conveying 'bridging' input and output signals between the GRN and the environment (see fig. 1). Sensory input from the environment is manifest as an increase in protein concentrations in the cell, and effector output is determined by protein concentrations. Hence the GRN is *continually coupled* to the organism's environment, perturbing and being perturbed, acting as a real-time control system. The concept of behaviour emerging from such coupling is central to autopoietic theory [17], and is analysed from a dynamical systems perspective by Beer [18], in the context of coupling between a neural network and an environment.

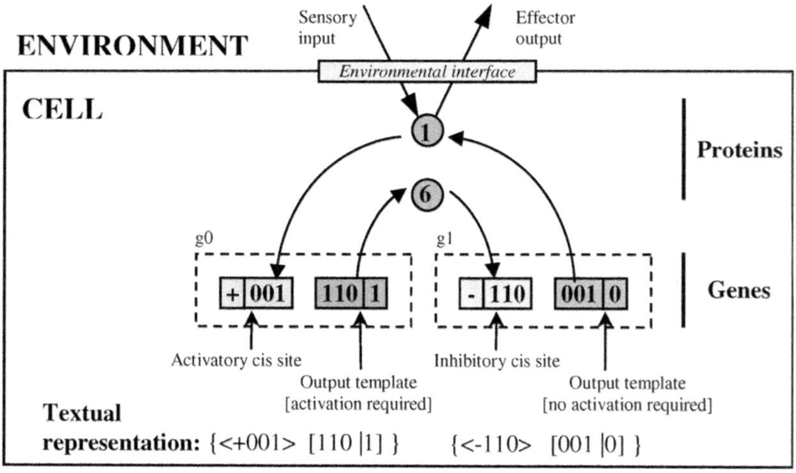

Fig. 1. Schematic gene–protein–environment interaction, with a textual representation of genes illustrated

2.2 The *Biosys* Model

In the simple single-celled non-spatial biological model currently used in *Biosys*, a cell consists of a genome and proteins (fig. 1). Genes produce proteins, which in turn both regulate the activity of other genes and control effector output. Sensor input is manifest via proteins introduced into the cell.

We use the term 'cell' loosely, to refer to this collection of genes and proteins. When we later refer to a cell responding to an environment in some way, this is very close to saying that the cell's genome's regulatory dynamics respond in some way, as the only other constituents of the cell are the proteins, which do relatively little other than convey signals between genes, and to and from the environment.

The number of genes in the model does not vary between evolutionary runs. There is also a fixed alphabet of *protein types*—eight in each experiment, represented as a binary integer between 0 and 7 (000–111). Each protein type has an evolvable *decay rate*, and is present at a concentration ranged between 0 and an evolvable *saturation value*. Proteins are produced at a variable rate within the cell via gene expression, and do not directly interact with one another. We refer to protein types with the notation Pn, where n is an integer between 0 and 7.

In a simplification of biological GRN control (cf. [1,19]), each gene on the genome has the same number of 'cis' *regulatory sites*, which act as *activators* and *inhibitors*, illustrated in fig. 1. Each site has a binary value between 0 and 7, representing a protein type. Identity-based template matching is used to *bind* proteins to these sites. For example, if a regulatory site has the value 5 (101), then instances of P5 (101) will bind to it. An evolvable global *binding proportion* parameter determines how many protein instances out of all available matching instances bind. Where a given value appears on more than one regulatory site on the genome, available proteins are distributed evenly amongst them.

The level of protein binding at each gene's cis regulatory sites is used to calculate the gene's *activation level*. Starting from zero, this is derived by adding the number of protein instances bound to activatory sites and subtracting the number bound to inhibitory sites.

Each gene produces a single protein type as its *output*. The quantity output depends on the gene's activation level, which is passed through a sigmoid *output function* (derived from tanh, the hyperbolic tangent), to calculate an output value between zero and a ceiling. Each gene has one of two *activity types* (also under evolutionary control), corresponding to one of two possible output functions. One type produces output as the activation level rises above zero, and hence is 'off' by default. The other produces output from just below zero, and hence is 'on' by default.

This model bears formal similarities to those used by Kauffman [4] and Reil [5]. Of the differences between those and the research presented here, the most significant are environmental coupling—which we focus on in the experimental sections below, and variable gene activation and output. Variable activation and output are illustrated in fig. 2, which shows output from the simple regulatory network illustrated in fig. 1, but isolated from environmental coupling. The first gene (g0), which is 'off' by default, produces P6 (110), and is up-regulated by the product of the second gene (g1), P1 (001), which is 'on' by default. g1 is in turn down-regulated by g0's P6 output. This creates an oscillating negative feedback loop controlling g1—as its P1 output increases, this increases g0's P6 output, which in turn inhibits g1. As g1's P1 output subsequently falls, so does g0's P6 output, reducing the inhibitory pressure on g1, which becomes active once more.

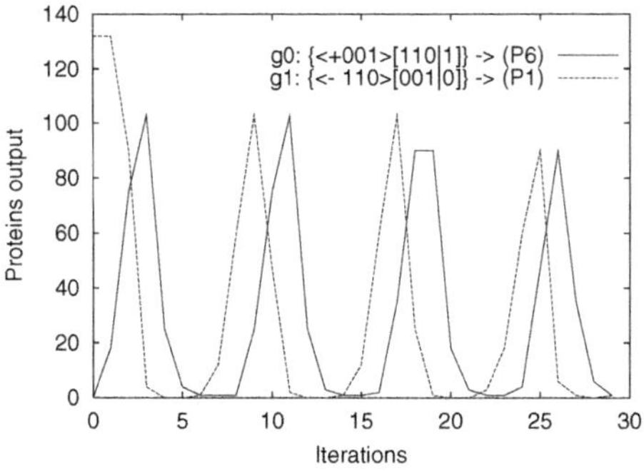

Fig. 2. Oscillating output in a simple two-gene network. *g0*, which is off unless activated, inhibits *g1*; *g1*, which is on by default, activates *g0*

2.3 Genetic Regulatory Control during the Life-Cycle

The model is run as a series of iterations, where an iteration is one step in an individual cell's 'lifetime'. Each iteration comprises the following sequence. First, unbound proteins are bound to matching regulatory sites. Second, transcription and translation occurs. For each gene, an activation level is calculated based on the level of binding at each cis regulatory site, then passed through the gene's output function to determine how many protein instances to produce. Third, proteins are decayed using each type's decay rate. Finally, an environmental interface performs effector output based on unbound protein concentrations, and adds proteins to the cell based on sensory readings (cf. fig. 1).

2.4 Evolution

We use a binary genome to encode individuals. Each gene is encoded with 8 bits (see fig. 1). There are four bits per cis regulatory site: one bit to indicate whether the site has an inhibitory or activatory effect, and three to encode the integer value of a matching protein type. The output region uses three bits to encode an output integer value, and one bit to indicate whether an activation level greater than zero is needed to produce an output.

The genome also encodes a number of global parameters not illustrated in fig. 1. There are nine four-bit parameters that each represent a proportion value in the range 0.0 to 1.0, via a lookup table. Eight of these parameters specify a rate of decay for each of the eight protein types (*decay rate*). The ninth describes what proportion of available proteins to bind to matching cis regulatory sites (*binding proportion*). An additional three-bit parameter specifies the maximum number of proteins of each type that can be present in the cell (*saturation value*), again via a look-up table.

Individuals were evolved using a basic fixed-length haploid GA, with single point cross-over, tournament selection and weak elitism (a single fittest individual is retained from each population into the next). Cross-over occurred with a probability of 0.9, and mutation with a probability of 0.01 per bit. Fitness functions were specified in terms of high-level phenotype behaviour in relation to an environment.

3 Homeostatic Behaviour

In this section we describe results from evolving GRN-controlled cells in an abstract software environment. The environment consisted of a single variable, which we arbitrarily termed 'temperature' (this makes the target behaviour easy to visualise, as being like a heating unit attached to a thermostat). Individual cells' fitnesses were assessed in terms of their homeostatic ability with regard to this variable over 100 iterations. Our intention was to observe coupling between regulatory dynamics and environment in as simple a context as possible.

3.1 Experimental Setup

Two protein types were used to couple cells to the environment. For every 'degree' the temperature was below the target on a given iteration, the same number of instances of P0 were introduced into the cell, representing input from sensors. For effector output, the number of unbound instances of P7 in the cell was summed, and the temperature increased by that amount (always greater than or equal to 0).

For each cell, the temperature was reset to a target figure of 50, with a possible range of 0–100. The temperature was then altered by 5 'degrees' at each iteration, initially falling, then switching direction at iterations 25, 75 and 90.

The GA was configured to maximise fitness, normalised to the range 0.0–1.0, calculated by summing deviations from the target, not including environmental increases. Individual fitness scores were inversely proportionate to this deviation measure. Individuals were immediately terminated with a fitness of 0.0 if the temperature fell to zero. Penalty points were accumulated and used to scale fitness downwards for failing to output heat when the temperature was below the target, and for outputting heat when above the target.

Each evolutionary run used a population size of 200 over 250 generations. We performed three evolutionary runs for each of five different specification cells—increasing the number of genes (N) from one to five. In all cases, each gene had a single cis regulatory site ($K = 1$). Note that unlike Kauffman's models, if $N > 1$, $K = 1$ does not imply that each gene is only affected by one other gene, as many genes may output proteins of a specific type, and hence exert a regulatory effect on a single gene.

3.2 Results

Independently of the value of N, individuals' fitnesses fall into three broad bands, most visible in the $N = 1$ runs (fig. 3). At the lowest level of fitness are cells whose genomes do not code for the output protein P7: they do not output any heat,

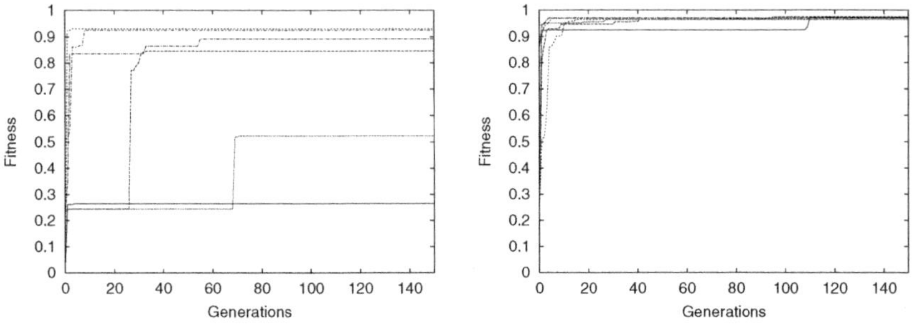

Fig. 3. Fitness over 6 evolutionary runs, showing the first 150 of 250 generations (pre-convergence), population size 200, for temperature controller cells with $N = 1, K = 1$ (*left*) and $N = 5, K = 1$ (*right*)

and quickly 'die'. At a low to mid range are cells that do produce P7, but whose genomes have no cis regulatory sites matching P0, which effectively provides information about the environment. Their heat output is therefore unresponsive to the environment's ambient temperature. Finally, with fitnesses around 0.8–1.0 are cells that are fully coupled to the environment, sensitive to P0 and outputting P7 in response. For $N = 1$ genomes, this requires a gene of the form: $\{\langle +000 \rangle [111|1]\}$ (see textual gene representation in fig. 1). This gene is inactive until activated by P0, producing P7. Hence the gene is activated when the cell senses a fall in temperature below the target, causing its effector surface to output heat in response. Additional minor increases in fitness can be achieved through alterations to the global parameters that are also encoded on the cells' genomes, such as protein decay rates.

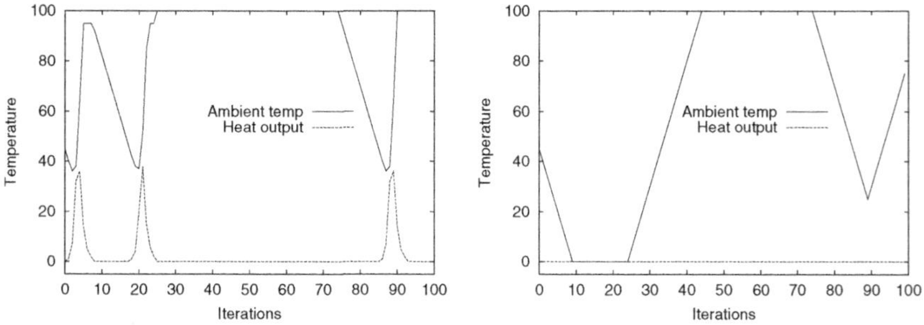

Fig. 4. Ambient temperature and cellular heat output with (*left*) and without (*right*) environmental coupling. Cell and environment are otherwise identical. Temperature changes visible in non-coupled environment (*right*) are autonomous

The left plot in fig. 4 shows the behaviour of the fittest cell from generation 34 of the second $N = 1$ run (fitness: 0.84). As the temperature falls below 50 degrees, the cell's output quickly rises, raising the temperature, causing in turn a drop in output, due to falling levels of P0, which reduces the gene's activation level. For contrast, the right plot shows data for the same environment and cell with no interaction between them—the cell outputs nothing, whilst the environment's temperature drifts up and down. The marked difference between the two illustrates clearly the impact of the coupling relationship between the cell's genome and the environment. It is from this tightly coupled relationship that the emergent behaviour exhibited by the organism arises.

This is an important point. Simulated in isolation, a GRN with $N = 1, K = 1$ cannot hope to display much variety of behaviour. However, coupling to an environment creates a situation functionally equivalent to the GRN being a sub-set of a larger, effectively invisible regulatory network—with a potentially infinite non-repeating sequence of states therefore available to the sub-network, depending on the nature of the environment. The relevance of this phenomenon becomes apparent in the next section, in which we show phototactic, light-maximising robot controller behaviour emerging from the interplay between an environment and a rudimentary two-gene network.

4 Phototactic, Light-Maximising Behaviour

This section describes results from evolving GRNs in a slightly more complex and realistic environment—a bounded flat physical surface with a light-source suspended in the middle, simulated using Cyberbotics' 'Webots' version 2. Individual cells' fitnesses were assessed in terms of their ability to maximise the amount of light encountered during their lifetimes.

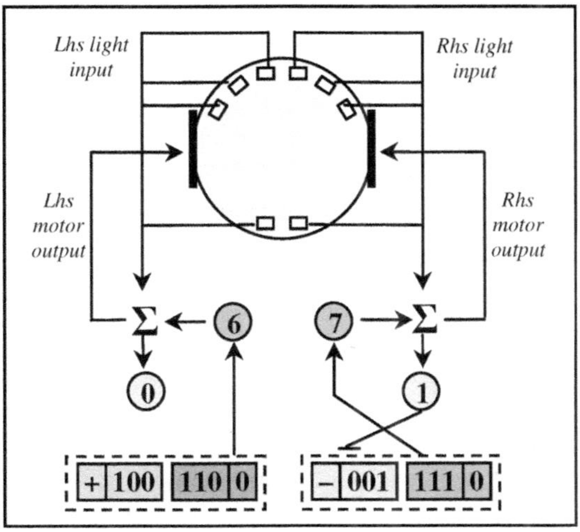

Fig. 5. Interface between Khepera robot and cell. Lhs light input generates instances of P0; rhs light instances of P1. Instances of P6 set the left wheel's speed; P7 the right wheel's speed. The genome shown has evolved to constantly drive both wheels, with rhs light input inhibiting right wheel speed. This produces the behaviour shown on the third plot in fig. 7

4.1 Experimental Setup

In this experiment we created an environmental interface for the cells, illustrated in fig. 5, inspired by Braitenberg's 'vehicles' [20]. The Khepera (K-Team) robot has eight infrared sensors capable of detecting visible light, spaced at intervals around a cylindrical body, with a pair of motor-driven wheels on opposite sides. At each iteration the environmental interface sums light readings from the left hand side (lhs) sensors, scales this figure by a hard-coded gain value, and inputs the equivalent number of instances of P0 into the cell. The same process is repeated on the right hand side (rhs), and input as instances of P1. For motor output, the interface sets the speed of the left wheel to the number of instances of P6, and the speed of the right wheel to the number of instances of P7. No other pre- or post-processing was performed.

The mapping from sensor readings to proteins, and proteins to motor output is arbitrary, and remained fixed throughout evolutionary runs. It is the task of the evolutionary process to produce genomes that respond to and produce these proteins in such a way that behaviours with high fitness values emerge.

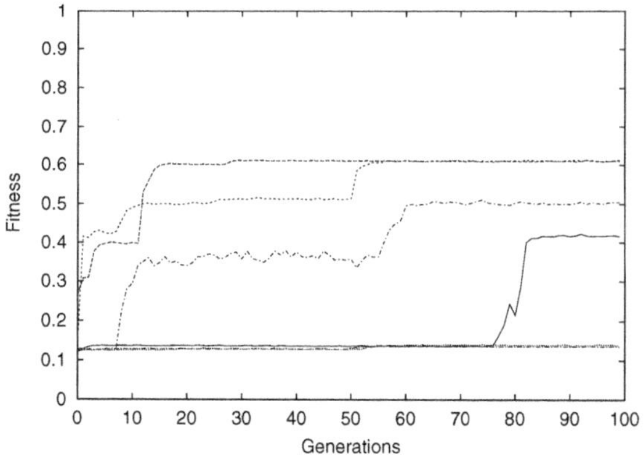

Fig. 6. Fitness over 6 evolutionary runs of 100 generations, population size 150, for phototactic robot controller cells with $N = 2, K = 1$

The GA was configured to maximise fitness, measured as: the amount of light encountered during the cell's lifetime as a proportion of a maximum value. Optimal behaviour would be to move rapidly to the region where light-intensity is greatest (a circle of some radius r from the light-source), and then to remain in that region with as many sensors as possible facing the light.

Individuals that exhibited no movement in the first 10 iterations of their lives were terminated, retaining their fitness scores. Six evolutionary runs were performed, each using a population of 150 over 100 generations. Each individual had two genes with one cis regulatory site per gene ($N = 2, K = 1$). Each individual began its lifetime in the same location.

4.2 Results

Over evolutionary time (fig. 6), the populations took a similar strategic route to that seen in the single variable homeostatic experiments: exploiting increased coupling to the environment to achieve greater fitness. Fig. 7 illustrates typical behaviours.

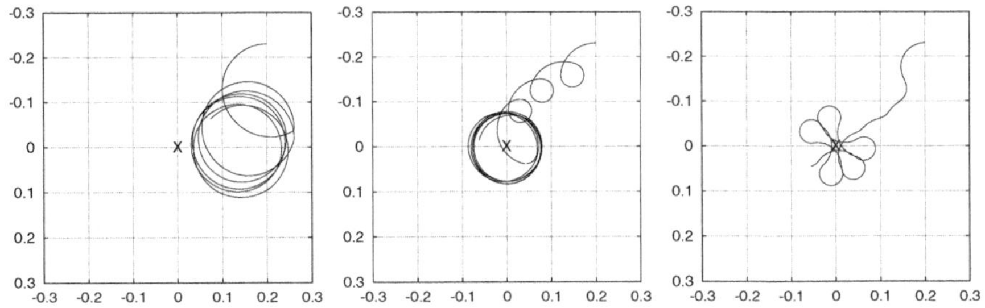

Fig. 7. GRN-driven phototactic robot controller trajectories (N=1,K=1), on a 2D surface. Axes show distance in metres, central 'X' marks light-source. From left to right, plots show fittest individuals (*looper*, *biased-looper* and *weaver*) from run 3 at generations 8, 41 and 83

Initial fitness improvements were quickly achieved through outbound coupling: producing P6 and P7, and hence motor output and movement. The first trajectory plot in fig. 7, generated by individual we will call *looper*, from generation 8 of run 3 (fitness: 0.42), is an example of this. Its genome (cf. fig. 1):

$$\{\langle +100\rangle[111|0]\}, \{\langle -101\rangle[110|0]\}$$

is not sensitive to environmental input (P0 or P1), but continually produces P6 and P7. Its genome also encodes a lower decay rate parameter (not shown) for P7 than P6, so the cell outputs more of the former than the latter, and hence the right wheel turns more quickly than the left, resulting in a continual anti-clockwise looping motion.

Inbound coupling appeared next, 'reading' the changes in input arising from motor output, with cis regulatory sites matching P0 or P1. The second trajectory plot in fig. 7, from generation 41 of run 3, is from a cell (*biased-looper*; fitness: 0.51) with the same genes as *looper*, but with an inhibitory cis site on the second gene matching P1 ($\langle -001\rangle$), corresponding to light from the right hand side. This produces a biased looping behaviour. As the robot loops away from the light source, light levels increase on the rhs, resulting in increased levels of P1 in the cell. This inhibits the production of P6 by the second gene, slowing the left hand wheel, tightening the loop. As the robot turns to face the light again, the rhs faces away from the light, and P6 production increases. This results in a net movement towards the light source, until the robot is circling it. At this point, interaction between cell and environment is stabalised, with the rhs continually facing away from the light.

In the transition between the second and third plots in fig. 7 (from *biased-looper* to *weaver*), the relationship between inputs and outputs were re-arranged, producing a genome as shown, from generation 83 of run 3 (fitness: 0.61):

$$\{\langle +100\rangle[110|0]\}, \{\langle -001\rangle[111|0]\}$$

This causes rhs sensory input (P1) to inhibit rhs effector output (P7). The first gene's output is changed to P6, which it produces constantly. This relationship is shown in fig. 5. The net effect is that whilst the controller maintains a tendency to loop to the left seen in *looper* and *biased-looper*, as the robot turns away from the light, increasing rhs input (P1), production of P7 is inhibited. This slows the right wheel, causing a loop to the right. This control strategy allows for a self-correcting, more or less direct route towards the light-source, around which the robot then orbits, weaving a series of broad left and tight right turns.

It is important to recognise the role the environment plays, through the cells' embodiment, in generating the behaviour observed, selecting from the cell's range of available 'next' states. As the space of possible states of system and environment increases and the scope for mutual perturbation between them becomes greater, so the space of potential behaviour (observable interaction) increases.

5 Conclusion

A key conclusion from a biological perspective, and certainly for other ALife models, is that the significance of the environment as a source of input external to the model should not be overlooked. Particularly from a dynamical systems perspective, the consequences of unpredictable input constantly perturbing the system are profound.

We have demonstrated the potential of environmentally coupled systems based on a GRN model to act as control systems, producing simple coherent behaviours. The simplicity of the GRN controllers is surprising given the behaviours exhibited during their evolution (see fig. 7). In a sense this illustrates the power of adaptive structural coupling that resides in even the most 'basic' of real biological systems—which combine GRN dynamics, sophisticated protein-based signal transduction pathways and many other features besides, coupled to the rich environment of the physical world, all under the aegis of the laws of physics.

Many interesting questions remain to be addressed. Given a more feature-rich, structurally plastic environment and a larger space of possible behaviours, what level of control system sophistication can be achieved with GRNs with higher values of N and K? How would this impact on evolvability? What consequences follow from increasing the sophistication and realism of the model, for example by adding protein–protein interactions and support for development involving differentiated multi-cellularity? Given a definition of embodiment in terms of structural coupling, and given the apparent relationship between system–environment coupling and potential behavioural complexity, what is the precise relationship between embodiment and adaptive behaviour, across different types of environment?

Acknowledgments. The first author wishes to thank Sanjeev Kumar for his advice and suggestions regarding biological processes and the binary representation and implementation of cis regulatory sites.

References

1. Davidson, E.H.: Genomic Regulatory Systems: Development and Evolution. Academic Press (2001)
2. Arthur, W.: The Origin of Animal Body Plans: A Study in Evolutionary Developmental Biology. Cambridge (2000)
3. Alberts, B., et al.: Molecular Biology of the Cell. 4th edn. Garland (2002)
4. Kauffman, S.: Metabolic stability and epigenesis in randomly constructed genetic nets. Journal of Theoretical Biology **22** (1969) 437–467
5. Reil, T.: Dynamics of gene expression in an artificial genome—implications for biological and artificial ontogeny. [21] 457–466
6. de Jong, H., et al.: Hybrid modeling and simulation of genetic regulatory networks: A qualitative approach. In Pnueli, A., Maler, O., eds.: Hybrid Systems: Computation and Control. Volume 2623 of LNCS., Springer-Verlag (2003) 267–282
7. Dellaert, F., Beer, R.: A developmental model for the evolution of complete autonomous agents. In Maes, P., et al., eds.: From Animals to Animats 4: Proc. 4th Int. Conf. on Simulation of Adaptive Behaviour, MIT Press (1996) 393–401
8. Astor, J.C., Adami, C.: A developmental model for the evolution of artificial neural networks. Artificial Life **6** (2000) 189–218
9. Bongard, J.C.: Evolving modular genetic regulatory networks. In: Proc. IEEE 2002 Congress on Evo. Comp., IEEE Press (2002) 1872–1877
10. Eggenberger, P.: Evolving morphologies of simulated 3D organisms based on differential gene expression. In Husbands, P., Harvey, I., eds.: Proc. 4th European Conf. on Artificial Life, MIT Press (1997) 205–213
11. Kumar, S., Bentley, P.J.: Biologically inspired evolutionary development. In: Proc. 5th Int. Conf. on Evolvable Systems: From Biology to Hardware. LNAI, Springer (To Appear)
12. Gilbert, S.: Ecological developmental biology: Developmental biology meets the real world. Developmental Biology **233** (2001) 1–12
13. Bolker, J.A.: From genotype to phenotype: Looking into the black box. In Kumar, S., Bentley, P., eds.: On growth, form and computers. Academic Press (To appear)
14. Ziegler, J., Banzhaf, W.: Evolving control metabolisms for a robot. Artificial Life **7** (2001) 171–190
15. Weiss, R., et al.: Genetic circuit building blocks for cellular computation, communications, and signal processing. Natural Computing: An Int. Journal **2** (2003)
16. Quick, T., Dautenhahn, K., Nehaniv, C.L., Roberts, G.: On bots and bacteria: Ontology independent embodiment. [21] 339–343
17. Maturana, H.R., Varela, F.J.: Autopoiesis and Cognition. D. Reidel (1980)
18. Beer, R.D.: A dynamical systems perspective on agent-environment interaction. Artificial Intelligence **72** (1995) 173–215
19. Ptashne, M.: A Genetic Switch: Phage Lambda and Higher Organisms. Blackwell Science (1992)
20. Braitenberg, V.: Vehicles: Experiments in Synthetic Psychology. MIT Press (1984)
21. Floreano, D., Nicoud, J.D., Mondada, F., eds.: Advances in Artificial Life: 5th Euro. Conf. on Artificial Life. Volume 1674 of LNAI, Springer (1999)

Evolution of Fault-Tolerant Self-Replicating Structures

Ludovic Righetti[1], Solaiman Shokur[1], and Mathieu S. Capcarrere[2,3]

[1] Logic Systems Laboratory
Lausanne Swiss Federal Institute of Technology, CH-1015 Lausanne
Name.Surname@epfl.ch, http://lslwww.epfl.ch
[2] Computer Science Institute
University of Lausanne, CH-1015 Lausanne
[3] Computing Laboratory
University of Kent, Canterbury CT2 7NF
mathieu@capcarrere.org, http://www.cs.ukc.ac.uk

Abstract. Designed and evolved self-replicating structures in cellular automata have been extensively studied in the past as models of Artificial Life. However, CAs, unlike their biological counterpart, are very brittle: any faulty cell usually leads to the complete destruction of any emerging structures, let alone self-replicating structures. A way to design fault-tolerant structures based on error-correcting-code has been presented recently[1], but it required a cumbersome work to be put into practice. In this paper, we get back to the original inspiration for these works, nature, and propose a way to *evolve* self-replicating structures, faults here being only an idiosyncracy of the environment.

1 Introduction

Self-replication is at the very core of the cellular automata inception. The original study of von Neumann that lead to the development of CA was it not centered around the question of self-reproduction[1] ? This birthmark lead to many studies of self-replication by means of CA, some famous landmarks being Langton's loop [6], Tempesti's loop [14] or the simplest self-replicating loop to date, Chou and Reggia's [12].

Self-replicating structures in cellular automata have thus generated a large quantity of papers [13]. However, most of these have been devoted to the study of the self-replicating process in itself, as a model of one of life primary properties. The underlying motivation of such researches could be expressed as a question: What minimal information processing is required for self-replication to happen?

[1] Barry McMullin in [10] argued very interestingly that in fact self-reproduction is a paraphernalia rather than the center of von Neumann's work. We would not quite disagree with him but think that this "means" of getting complexity has turned out to be an interesting problem in itself, as demonstrated by the large subsequent literature.

W. Banzhaf et al. (Eds.): ECAL 2003, LNAI 2801, pp. 278–288, 2003.

Simplification was thus a prime constraint in all these studies, unfortunately, not only as a guide to extract the quintessential ideas behind self-replication, but also for obvious practical reasons. We believe that for the latter reasons rather than the former principle, an always present quality in natural systems was omitted: *robustness*.

The usual model of cellular automata is discrete in time and space and all cells update synchronously according to a deterministic rule. No fault ever occurs in the transition rule nor on the cell current state. Nevertheless if we are to look at real biology, it appears at all levels that either insensitiveness to faults or self-repair is a prime condition of survival, the environment being extremely noisy and uncertain. Most fault resistant cellular automata designed in the past[3, 4,5] are based on a hierarchical constructions, i.e., an alteration of the inner architecture of the CA. In this paper, we rather propose to develop fault-tolerant transition rules, thereby keeping the classical CA struture unchanged. We evolve CA that are fault-tolerant in their functioning itself, at a minimal cost in terms of space, time or memory.

In section 2 of this paper, we present the main previous work on the evolution of self-replicating structures which served as the base for this work. In section 3, we then accurately define what we mean by a faulty environment. Section 4, presents our evolutionary model, notably detailing the algorithm and parameters we used to successfully evolve fault-tolerant self-replicating objects. Finally in section 5, we show the results we obtained with a straightforward approach and then present refinements that lead to the evolution of (relatively) very robust self-replicating structure. We conclude in section 6.

2 Previous Work on the Evolution of Self-Replicating Structures

If *hand-designed* self-replication in CA has attracted its chunk of interest, very few have tackled *the evolution* of self-replication, a more arduous but also a more "natural" approach[2]. The main works on the evolution of self-replication were lead by Jason Lohn and James Reggia [8]. Let us now briefly present the framework of this previous study thereby introducing ours as it was used as the general grounds for our work.

2.1 Framework of Previous Evolution of Self-Replication

The three important aspects in an evolutionary algorithm are: the encoding, the fitness function, and finally the structure of the algorithm itself.

The encoding of the transition rules of the CA is rather straightforward. As von Neumann neighborhood was used, the transition rule is a function $t : q^5 \rightarrow q$,

[2] Beyond the scope of CAs, there are some very interesting studies of evolution of self-replication, to cite but one, Pargellis [11]. Talking of this work, we should underline here that we evolve worlds where robust self-replication emerges, rather than setting-up a world where self-replication evolves.

where q is the set of all possible states of each cell. If an order on q^5 is defined, then a string of symbols of q of length q^5 is enough to characterize a transition rule. In other words, as the set of all possible neighborhoods is ordered, the first symbol of this string is the future state matching the first neighborhood, the second symbol is the future state matching the second neighborhood, and so on and so forth. Then the chromosome is just that characterizing string.

The fitness function is the core of the evolutionary algorithm. In the original population no candidate solution is even close to self-replication, the quality of a fitness function will thus lie in its ability to detect elements with great potentials. Lohn's fitness function is decomposed into three parts F_g, F_p, and F_r: – F_g measures the tendency to grow for each element (each state). It is bounded between 0 and 1 and is equal to 0 if all elements disappear and to 1 if all elements present in the seed configuration increases in numbers.– F_p measures the quality of the relative position of each element. This is the most important fitness bit as it will make the difference between chaotic growth and self-replication. It is bounded between 0 and 1, 0 meaning that the neighborhood of an element has no connection to its neighborhood at the previous time step and 1 meaning the exact opposite. Replication would imply high values for both F_g and F_p but not necessarily vice-versa. – F_r, hence, measures directly the number of replicants. More precisely it is a function of the number of replicants. It is also bounded between 0 and 1. For each evaluation the CA is initialized with a seeding configuration and is evaluated on T time steps. Finally the three fitness components are weighted out against each other. Lohn used $F = 0.05F_g + 0.75F_p + 0.2F_r$. As we will see these weights are of prime importance.

The algorithm itself is a simple straightforward genetic algorithm using a fitness proportionate selection, a cross-over rate of 0.8 and a mutation rate of 0.1. The only specific operator is the crossover operator which is multi-point: the order of the characterizing string is such that the states of the center cell of the first q^4 neighborhoods matching the first q^4 characters are identical, the same for the following q^4 characters, and so on and so forth; simply a crossover point is chosen for each q^4 long chunk, for each "gene"; the crossover happens in parallel on all genes.

2.2 Previous Results

The complexity of the task depends on the size of the structure that should replicate. This is determined by the configuration used as the seed on which the evolution takes place. The results published by Lohn and Reggia [7,8,12] show a success rate of 93% for structures of size 2 (linear), 22% for structures of size 3 (right angle) and only 2% for structures of size 4. A run is succesful if the CA replicates at least once. We can note that the latter result is statistically not significant and we will thus concentrate on 3 element structures.

3 A Non-deterministic Cellular Automata Space

The idea of fault-tolerance is rather vague. We propose here to delimit the scope of validity of our work.

Formally cellular automata are d-dimensional discrete space, in which each point, called a cell, can take a finite number of state q. Eevery cells update synchronously its own state according to the states of a fixed neighborhood of size n following a deterministic rule. In CAs, faults may basically occur at two levels: the synchronization and the cell transition rule. We do not cater specifically for the former problem in this paper as the latter encompasses it. Though as demonstrated in [2], limiting ourself to the former would have been simpler as it is possible to simulate any q-state synchronous CA with a $3.q^2$ state asynchronous CA. As for the latter problem, which could be further divided into reading and writing errors, we model it as a non-deterministic rule. More precisely, there is a probability of faults p_f, such that any given cell will follow the transition rule with probability $(1 - p_f)$, and take any state with probability p_f. This model, though apparently catering only for writing errors, does simulate reading errors[3]. One may object that we did not mention the major fault problems when a cell simply stops functioning at all. This is not our purpose to treat that kind of 'hardware' problem here, but we may say that our model could be implemented over an embryonics tissue [9] which deals with this kind of malfunction. Besides, one may note that such a permanent failure is not, in itself, fatal to our system, but only weakens all intersecting neighborhoods. As we have shown in [1] this noisy environment is more than enough to irrevocably disturb and destroy any of the most famous self-replicating structures.

In the scope of the experiments presented in this paper, we evolve a perfectly standard CA which works in a faulty environment. However, unlike in our preceding paper [1], cells which are not updated, i.e. cells which are quiescent surrounded only by quiescent cells, are *not* subject to faults. Matter is not created out of nothing.

4 The Evolutionary Framework

We will now briefly overview the parameters and the algorithm we used, thereby allowing for future replication of our results.

The algorithm used is a straightforward classical Genetic Algorithm. Our initial population size for all experiments was 200 and the mutation and crossover rate were respectively 0.1% and 80%.

The selection method chosen is a peculiar version of rank selection. Every individuals are sorted, the worst ranked first and the best ranked last. Then the probability of choosing an individual with rank i for reproduction in a population of size N is $\frac{i}{\sum_{k=1}^{N} k}$.

[3] Strictly speaking, it is not exactly equivalent as a writing error induces a reading error for *all* surrounding cells.

The encoding is exactly the same as for Lohn's work. We use the set of states q to be $\{0, 1, 2, 3\}$, where 0 is the quiescent state. However there is a subtlety in the sense that each state, except the quiescent state, has a direction component which may be one of the four directions $\leftarrow, \uparrow, \rightarrow, \downarrow$. This direction component is ignored by the transition rule. More explicitely, $3 \leftarrow, 3 \uparrow, 3 \rightarrow$ or $3 \downarrow$ will be taken simply as 3. The future state, on the other hand, may be any of the $4 * 3 + 1$ possible states, thereby leading to a redundant encoding. Thus the length of our encoding string is 4^5. This choice was made as it yielded better results.

The crossover used is the multi-point crossover described earlier.

The fitness function was at first exactly the one described by Lohn which we talked about earlier. This choice was motivated by the fact that fault-tolerance should naturally evolve from the environmental pressure: if it is not robust, it won't replicate. It turned out, however, that Lohn's fitness lead to very poor result, *even in non-faulty environment.* We thus adapted the fitness as outlined now.

The fitness F is decomposed in three subfitnesses F_r, F_g and F_p. F_r, the fitness that evaluates the number of replicants is defined as in Lohn's work, that is: $F_r = (1 + exp^{-(max(r_t)-\theta)})^{-1}$, where r_t is the number of replicants at time t and θ is the number of replicants from which the fitness importance of this measure decreases, here θ is 4. F_p the fitness that measures the quality of the position of the elements is also left unchanged. Let S_v be the average number of elements adjacent to element v. Let M_v^t be the number of elements v at time t. Finally, let $m_v(t)$ be the number of elements adjacent to v at time t that were of the same type and position relative to v at time 0. Then we define $o_v(t)$ to be 0 if $M_v^t \leq 1$ and $\frac{m_v(t)}{M_v^t . S_v}$ otherwise. Then $Fp = \frac{1}{T \Sigma_v S_v} \Sigma_v \Sigma_{t=1}^T S_v o_v(t)$. If F_r and F_p were left unchanged, we found necessary to modify F_g. The first runs highlighted a high tendency of the structure to grow as a compact pack, which in conjunction with a good relative position of elements, lead to very high fitness and premature convergence. Following the observation that interesting patterns of growth exhibited a growth rate of 1.7, we modified Lohn's definition of F_g as follows. Let $p_v(t)$ be 1 if $M_v^t > M_v^{t-1}$ and $M_v^t < 2 * M_v^{t-1}$, be 0.5 if $M_v^t = M_v^{t-1}$ and be 0 otherwise.

Finally we weighted quite differently from Lohn's work the fitnesses, here $F = 7.5F_g + 72.5F_p + 20F_r$.

The seeding configuration chosen contained a right angle three-element structure as shown in Figure 1. As said earlier two-element structures were devised as too easy and four-element results were deemed as too incidental. The grid is toroidal and of global size 50x50.

0	0	0	0
0	1	0	0
0	2	3	0
0	0	0	0

Fig. 1. The seeding configuration for all runs.

5 Results

In this last section we present the fault-tolerant self-replicating CAs obtained through evolution. As we will see these results, though structurally simple, are as complex as their non-robust counterpart found by evolution. In section 5.1, we present a decisive improvement in the encoding. In section 5.2, we detail the good results obtained through a "straightforward" evolution. Finally, in section 5.3, we show how a more refined evolutionary process can lead to great improvements of the results.

5.1 Preliminary Experiments: On the Importance of Destruction

As said earlier, the fitness was modified to reduce the "attractiveness" of the compact growth strategy. This improved things but compact growth remained a strong basin of attraction and further fitness manipulation did not seem to do the trick. However it appears, if one thinks of what the main problem is with this behavior of the CA, that in fact growth is equivalent to a lack of cell destruction. In other words, not enough rules lead to the quiescent state. The main reason for this is quite obvious given our orientation insensitiveness. That property implies that there are 4 possible encodings for all states except for the quiescent state which has only one encoding. Thereby in the system as defined earlier there is only 1 chance in 13 (7.7%) of "choosing" cell destruction as its transition rule.

The main idea here is thus to introduce a redundant encoding for the quiescent state. As we found out from a batch of 50 experiments, in a safe environment, the optimal number of way of coding the quiescent state was 5. This is slightly more than 4, which is the number of redundant encoding for all the other states. Thereby the quiescent state must be over-represented to create better condition of success. To sum up the results, while the success rate was only 8% with no redundant encoding, it jumped to 60% with the optimal number of 5 quiescent states. We can note in passing that this is way better than Lohn's results [12].

5.2 First Experiments

The main idea behind that paper was that robustness in living organisms evolved as a consequence of their environmental pressure. Environment here is considered in its largest sense, that is the external environment pressure such as hardship, starvation, attack by predators etc. but also the system internal pressure such as defects in the cellular replication process, in the accuracy of the sensor, etc. and obviously anything in-between such as virus or parasites. Thereby, after the fine tuning of the fitness exposed before, the experiments run were intended as "spontaneous" evolution of robust self-replicating structures.

To do this we just changed the normal CA environment with the faulty one described in section 3, introducing a failure rate p_f. This failure rate means that, statistically, p_f faults occur every 100 transitions. This implies that the probability of one fault occurring on a structure of size n is $1-(1-p_f)^n$, where the size n of the structure encompasses all the surrounding quiescent cells of the non-quiescent structure. Thereby even the seeding three-element structure involves 15 cells, and thus a p_f of 1% implies a global failure rate of 14%; with a size of 100, which is quickly reached, the global failure rate is 63%. Using this environment, we are not facing a deterministic automata anymore. Running the same CA on the same seeding configuration may not lead to the same results. Therefore we introduce a new parameter, n_i, which denotes the number of repeated runs of the same CA on the seed configuration used for evaluation. The n_i fitnesses then obtained are averaged to minimize as much as possible the stochastic bias.

The error rate tested out, p_f, were 0.5, 1, 1.5 and 2. As could have been expected from the simple global failure equation above, high error rates of 1.5 and 2% lead very quickly to complete chaos. On the other hand, evolving with $p_f = 0.5$ leads to too little selection pressure and, while a high success rate[4] is attained, the solutions found turn out to be more brittle than those found with a rate of 1%, which thus seems to be the most appropriate.

As for the number of iterations, on a theoretical point of view the higher the better. However, practically, evaluating the CA is the most time consuming part of the algorithm. Thereby one should aim at diminishing as much as possible that value.

Table 1. Different success rates for different values of n_i and p_f. 'runs' is simply the number of runs on which the statistics were established, 'successful runs' are runs for which at least one replication occurred, 'interesting CAs' are those truly replicating, and finally 'robust CAs' are those at least moderately fault-tolerant.

n_i	p_f	runs	successful runs	interesting CAs	robust CAs
1	0.5	50	34	8	1
1	1	50	29	7	0
5	0.5	50	27	11	5
5	1	50	34	12	4
10	0.5	50	25	10	6
10	1	50	33	13	7

From table 1 it is hard to draw definitive conclusions but one may say: – First and foremost, there is a saturation to the improvement that may be brought by an increase in n_i. Quite clearly a few repeats are enough to discriminate between different CAs. It may be concluded that a n_i of 5 is a good trade-off between the time required and the number of robust CAs found. Nevertheless, if we refine the analysis further by visual testings of the solution found, it turns out that they are less resistant than the ones found with a n_i of 10. The same remark is true for the lower p_f. While the number of robust CAs found is somewhat equivalent to those found with a p_f of 1, they appear visually to be brittler. Therefore the

[4] An evolution is rated as successful if the best CA replicates at least once.

most favorable conditions seems to be a n_i of 10 and a p_f of 1, while the more reasonable ones would be $n_i = 5$ and $p_f = 1$.

The batch of experiments with the optimal settings for n_i and p_f (10 and 1) is studied here in more details. Figure 2 illustrates the fitness evolution through time of a typical successful run. Three kinds of successful strategies were found by evolution. The first strategy, which may be designated as opportunist, replicates along a diagonal in a safe environment, but uses error to create other lines of replication. Figure 3 illustrates a typical run of this strategy. The limit of this approach is that it does not correct error as such and given the toroidal nature of the grid error accumulates to the point where the CA collapses into chaos.

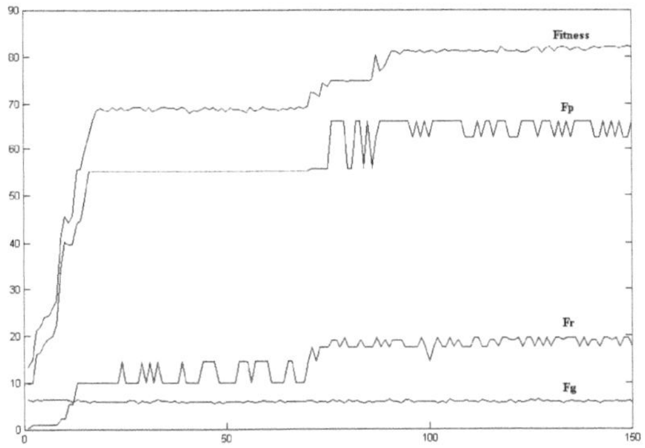

Fig. 2. Typical run fitness evolution through time. Here the population size is 200, the error rate, p_f, is 1%, the redundancy of quiescent state is implemented and n_i is 10. As one may see F_r and F_p are exhibiting exactly the same behavior after generation 8. However before F_r is on a plateau while F_p is continuously increasing. Quite obviously the relative position of the element increases before any real replication occurs. F_g remains almost constant as from generation 2 onwards. The selected individuals always tend to overgrow, thereby encountering the cut-off introduced in the fitness.

The second strategy, which one could call the gnat strategy, is simply based on fast exponential replications to compensate for the losses due to errors. Though the replication process in a safe environment of such automata is usually truly impressive, this strategy fails here and it eventually ends up in total chaos in a closed environment. Finally the third strategy, the no-error-zone strategy, simply consists at maintaining some sort of gap between the chaos generated by errors and the single diagonal line of replication. This last strategy, though not visually very interesting, is quite efficient at surviving. However like the other two strategies, it fails eventually.

As we see, evolving robust self-replication by adding pressure from the environment is possible. However, even if faults are handled with, it does not satisfy

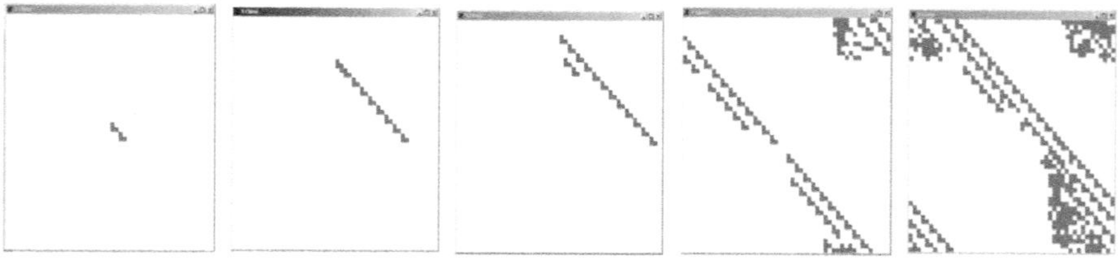

iteration 3 iteration 17 iteration 22 iteration 48 iteration 60

Fig. 3. Example of a "successful" strategy. An error appears at t=17 and is used to create a new front of replication at t=22, but then error accumulates and while the replication process still goes on, one may see at t=60 the seeds of the chaos to come. Though evolved with a p_f of 1%, it is demonstrated here with a p_f of 0.5%.

t=50 t=60

Fig. 4. A selected individual from the previous runs is shown on the left at t=50. On the right, the result from the evolution of the population generated from that individual. As one may see the garbage has disappeared and with a p_f of 0.5 as shown here the CA on the left does never collapse into chaos.

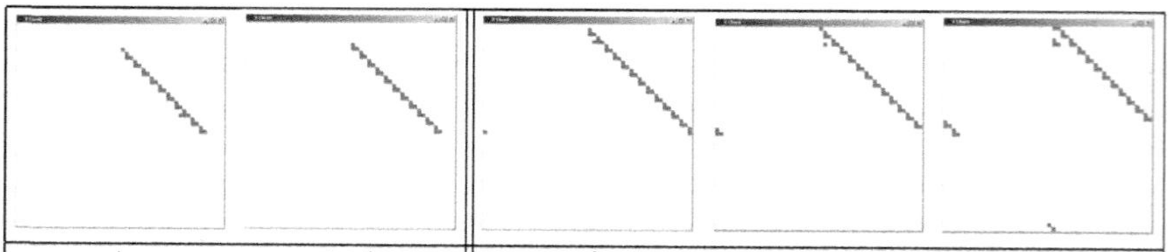

t t+1 t' t'+1 t'+2

Fig. 5. The improved rule of figure 4 is shown on the left as correcting an error, while on the right it expels the error at time t'+1 and uses this to create a new self-replication from at t'+2.

our original aim of correcting errors. In consequence, errors always submerge the CAs found through evolution. These mitigated results lead us to develop a more complex evolutionary process which is described in the next section.

5.3 Improving the Evolutionary Process

The results found until here were not completely satisfying. What was lacking in most of the solutions found was the ability to repair. To concentrate on this, we fixed the genes that were involved in the replication process under a deterministic environment and only evolved the genes left unused. This choice is not totally unreasonable if one thinks about short timescale evolution. In that context the more important genes of a population are fixed in the sense that every individuals have the same alleles at the same loci. To do this, a promising individual that has been found in the preceeding runs, is selected. It is then evaluated for a dozen time steps in a safe environment. The rules used during that evaluation are marked. It is then easy to generate a population of 200 individuals, random on all the unmarked rules, and identical on the marked ones. Except for the first population, the rest of the algorithm is identical as before.

As one can see in figure 4, this two-steps evolutionary process improves dramatically the results of the evolution. It gives rise to an almost perfectly fault-tolerant, self-replicating CA, correcting errors as expected (see figures 4 & 5).

6 Concluding Remarks

In this paper we have shown that it was possible to evolve fault-tolerant self-replicating structures using a straightforward evolutionary approach under faulty conditions pressure. Moreover, by implementing a two step evolutionary process, we have shown that error-correcting self-replicating CAs could be evolved.

Obviously the structures found were of small size, but not smaller than the one found before in deterministic space and the results are even better in terms of success rate. This indicates that research efforts should now concentrate first on refining the evolutionary process under deterministic condition. Indeed, the improvement we brought to the original algorithm may lead in the future to the evolution of more complex structures.

References

1. Daniel C. Bünzli and Mathieu S. Capcarrere. Fault-tolerant structures: Towards robust self-replication in a probabilistic environment. In J. Kelemen *et al* (eds), *European Conference on Artificial Life 2001*, pages 90–99. Springer Verlag, 2001.
2. Mathieu S. Capcarrere. *Cellular Automata and Other Cellular Systems: Design & Evolution.* Phd No 2541, Swiss Federal Institute of Technology, Lausanne, 2002.
3. Peter Gács. Self-correcting two-dimensionnal arrays. In Silvio Micali (ed), *Randomness in computation*, vol. 5 of *Advances in Computing Research*, pages 223–326, Greenwich, Conn, 1989. JAI Press.
4. Peter Gács. Reliable cellular automata with self-organization. In *Proceedings of the 38th IEEE Symp. on the Foundation of Computer Science*, pages 90–99, 1997.
5. Masaretu Harao and Shoici Noguchi. Fault tolerant cellular automata. *Journal of computer and system sciences*, 11:171–185, 1975.
6. Christopher G. Langton. Self-reproduction in cellular automata. *Physica D*, 10:135–144, 1984.
7. Jason D. Lohn and James A. Reggia. Discovery of self-replicating structures using a genetic algorithm. In *Proceedings of the 1995 IEEE International Conference on Evolutionary Computing (Perth)*, pages 678–683. Springer-Verlag, 1995.

8. Jason D. Lohn and James A. Reggia. Automatic discovery of self-replicating structures in cellular automata. *IEEE Trans. on Evol. Computation*, 1:165–178, 1997.
9. Daniel Mange, Moshe Sipper, Andre Stauffer, and Gianluca Tempesti. Towards robust integrated circuits: The embryonics approach. *Proc. of the IEEE*, 88(4):516–541, April 2000.
10. Barry McMullin. John von neumann and the evolutionary growth of complexity: Looking backwards, looking forwards... In Mark A. Bedau *et al*, (eds), *Artificial Life VII: Proceedings of the seventh international conference*, pages 467–476, Cambridge, MA., 2000. MIT Press.
11. A. N. Pargellis. Digital life behavior in the amoeba world. *Artificial Life*, 7(1–2):63–75, Winter 2001.
12. James A. Reggia, Hui-Sien Chou, and Jason D. Lohn. Self-replicating structures: Evolution, emergence and computation. *Artificial Life*, 4:283–302, 1998.
13. Moshe Sipper. Fifty years of research on self-replication: An overview. *Artificial Life*, 4:237–257, 1998.
14. Gianluca Tempesti. A new self-reproducing cellular automaton capable of construction and computation. In F. Morán *et al* (eds), *Proc. 3rd European Conf. on Artificial Life 1995*, vol. 929 of *LNAI*, pages 555–563. Springer-Verlag, 1995.

Evolving the Ability of Limited Growth and Self-Repair for Artificial Embryos

Felix Streichert, Christian Spieth, Holger Ulmer, and Andreas Zell

Center for Bioinformatics Tübingen (ZBIT), University of Tübingen,
Sand 1, 72074 Tübingen, Germany,
streiche@informatik.uni-tuebingen.de,
http://www-ra.informatik.uni-tuebingen.de/welcome.html

Abstract. In this paper we address the problem of limited growth and the difficulty of self-repairing in the field of Artificial Embryology. For this purpose, we developed a topological simulation environment for multiple cells which is continuous and structure-oriented, with a dynamically connected network of growing cells and endogenous communication between these cells. The cell behavior is simulated based on models of gene regulatory networks like Random Boolean Networks and S-Systems. Evolutionary Algorithms are used to evolve and optimize the parameters of the models of the gene networks. We compare the performance of Random Boolean Networks and S-Systems when optimized by Evolutionary Algorithms on the problem of limited growth and two types of cell death with and without signaling of cell death on the problem of self-repair.

1 Introduction

The research field of Artificial Embryology (AE) is rather new and only few papers have been published on this topic. Most researchers regard AE mainly as a simulation environment for early developmental processes, to test or to develop biologically plausible theories of pattern formation. Others apply the generative grammatical encodings of AE together with Evolutionary Algorithms (EA) to shape and structure optimization problems like growing neural network topologies or Evolutionary Engineering (EE). For both topics the generative grammatical encoding of AE offers the advantage of better scaling behavior over direct encoding schemes.

One application for AE is using it in multi-cellular simulation environments in the field of theoretical biology. An example for such a multi-cellular simulation environment is the *Cell Programming Language* (CPL) by Agarwal [1]. Here the cell behavior can be programmed with a number of high level instructions like move(direction) or differentiateto(tissuetype). Another AE environment was introduced by Fleischer [13,12] which uses cells with attributes like position, shape, and concentrations of biochemicals. The behavior of a cell is based on programmable differential equations determining the concentration of biochemicals. Another example for cell behavior based on differential equations is the *Cellerator* by Shapiro and Mjolsness [22].

W. Banzhaf et al. (Eds.): ECAL 2003, LNAI 2801, pp. 289–298, 2003.

Regarding the two applications of AE as coding scheme for growing neural nets and general shapes in EE, growing neural nets seems to be more popular and there are many publications in this research area [3,7,10]. EE on the other hand seems to lack a suitable representation scheme for general shapes. There are only few successful examples of EE in the literature based on specialized EA representations, for instance, Funes and Pollak [14] were able to evolve *Lego*® bridges and cranes using a tree based EA representation. The general problem of shape representation in EE was discussed by Schoenauer on a topological optimum design problem in [21]. Where he compared several parametric representations in respect of performance. In his conclusions he indicates that generative grammatical encodings could yield major advantages for EE. Actually Hornby and Pollak describe the advantages of generative grammatical encodings as they occur in AE by comparing a parametric, context-free L-System, generating LOGO style GP code, and a standard non-generative GP to build a model of a desk [16].
In this paper we address two fundamental problems for AE, first the problem of limited growth and second the problem of self-repair for artificial embryos, because without the ability of limited growth AE becomes useless as dynamic simulation environment in theoretical biology. The same holds true for EE. We solve these problems not by creating suitable rules by ourself but by optimizing the cell behavior, i.e. the parameters of the cell behavior with an EA. We compare the performance and the properties of two models of cell behavior on the problems of limited growth and self-repair.
The next section gives an overview over some of the related work in the field of AE. In section 3 we describe our implementations of a simulator for a multi-cellular environment and of the EA. Results on the problems of limited growth and self-repair are discussed in section 4. Finally, conclusions on the achieved results and an outlook is given in section 5.

2 Related Work

In one of the earliest work by Hugo de Garis [5] the cell behavior was similar to neighborhood interaction rules of cellular automata and determined whether a cell is to remain inactive, to die or to split as long as a maximum number of cell division was not exceeded. By using Genetic Algorithms (GA) he was able to grow simple convex shapes but failed to evolve non-convex shapes. Therefore, de Garis extended the cell model with sensors for gradients of biochemicals and added iteration counters [4]. Additionally, the cell behavior was enhanced by 'operons' imitating the dynamics of gene regulatory networks that allowed rules to be switched on or off. With additional sources of biochemicals placed in the environment he was able to grow simple non-convex shapes.
Frank Dellaert examined a biologically defensible model of development and evolved the shape and behavior of simple agents using a GA [6]. To emulate the behavior of gene regulatory networks he used Random Boolean Networks (RBN) and took the binary states of RBNs to detect cell differentiation. Delleart was able to evolve differentiated agents with sensors and actuators, but

to do so he used asymmetrical division for the very first division to establish the anterior/posterior orientation of the organism. He gave cells adjacent to the horizontal midline of the organism a special input signal. As an additional restriction, only 64 cells were allowed to grow, after which the simulation stopped. Peter Eggenberger also used the concept of gene regulation to control the behavior of developing cells [9,11]. The cell behavior was based on structural genes, turned on or off by regulating genes depending on a local concentration of biochemicals. Activated structural genes produced biochemicals, cell adhesion molecules, receptors for biochemicals and caused cell division and death. The activation states of genes were used for cell differentiation. Although Eggenberger used EA to optimize shapes, the results he presented relied on additionally created exogenous sources of biochemicals which indicated additional positional information to the organism.

A topological model was introduced by Duvdevani-Bar and Segel [8]. The behavior of the cells was based on the reaction-diffusion model devised by Gierer and Meinhardt [15]. They created examples for animal/vegetal region differentiation and examples for cell migration and neural differentiation in the visual system of Drosophila.

Only Fleischer and Barr discussed the topic of limited growth in [12]. They gave examples how to create rules that result in limited growth by:

- using a threshold of a biochemical concentration to disable cell division.
- exhausting a limited, not regenerating factor necessary for cell division.

All other papers did not explicitly address the problem of limited growth, neither did they mention if the organism were able to maintain the cell differentiation over an extended period of time. And although most researchers were able to produce more or less fancy shapes, we first want to concentrate on the most basic problems of AE and then start to evolve complex shapes with EA.

3 Artificial Embryology

For discussing our multi-cellular simulation environment we will distinguish between cell behavior, cell mechanics, and the cell model that merges both. The cell behavior computes the upcoming cell states given by concentrations of n biochemicals $\boldsymbol{B} \in [0;1]^n$. The cell model translates the current cell state into cell actions like cell differentiation, growth (mitosis), cell death, etc.. The cell mechanics simulates the effect of cell actions on the neighborhood topology and spatial distribution of cells in the organism.

Using the classification scheme suggested by Prusinkiewicz [20] our cell model is continuous structure-oriented, synchronous, but time discrete, with a network topology and dynamic neighborhood relations. The communication between cells is based on lineage and endogenous interaction.

To simulate the development of an organism we start with a single cell c_0 with $\boldsymbol{B}_{t=0} = [0]^n$, compare Fig. 1 at $t = 0$. For each discrete time step we compute the cell model for each cell c_j of the organism. The cell model determines if cell actions take place and then simulates the cell behavior and the cell mechanics.

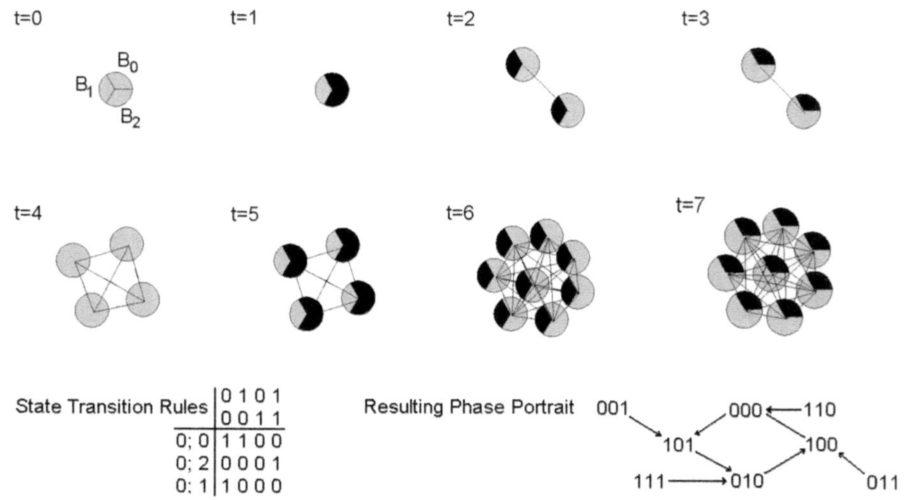

Fig. 1. Simulated organism with RBN ($n = 3; k = 2$), $\boldsymbol{T} = [0.5]^3$, $\boldsymbol{R}_{prod} = [1]^3$, $\boldsymbol{R}_{dgr} = [1]^3$, $\boldsymbol{D} = [0]^3$ and $\boldsymbol{B}_{t=0} = [0]^3$. Here the cells state equals the state of RBN.

3.1 Cell Model

All possible actions performed by cell c_j are based on the current state of the cell's biochemicals $\boldsymbol{B}$ or on external causes, as for example in the case of cell death or diffusion. Currently, only a small number of cell actions are possible:
Cell growth (mitosis): In the case of $B_0 = 1$ cell mitosis can happen if and only ifenough space is available. To prevent proliferation a cell is only allowed to split, if no neighboring cell is closer than $\frac{t_{dist}}{2}$. During mitosis the daughter cell inherits all properties of c_j. For both cells the values of $\boldsymbol{B}$ are halved, assuming that both cells instantaneously reach the original volume without generating additional biochemicals. Compared with Fig. 1 in this example mitosis occurs at $t = 1$, $t = 3$, $t = 5$ etc..
Cell death: Currently, death can only be caused by external events, see Chap. 4.2. When dying the cell removes all connections to neighboring cells and is removed from the environment. In an alternative implementation the dying cell injects an additional biochemical into neighboring cells indicating it's death.
Diffusion of biochemicals: Endogenous interaction between cells is simulated by exchanging biochemicals $\boldsymbol{B}$ between neighboring cells depending on a concentration gradient and the diffusion rate $\boldsymbol{D}$.

3.2 Cell Behavior

The behavior of a cell resembles a gene regulatory network of n genes which computes the concentration of biochemicals $\boldsymbol{B_{t+1}} = f(\boldsymbol{B_t})$. There are a number of alternatives to model gene regulatory networks: Random Boolean Networks

(RBNs) [18], Qualitative Network models [2], Weight Matrices [24], Dynamic Bayesian Networks [19], S-Systems [17], general differential equations [15] and many more.
We decided to compare RBNs and S-Systems since they are the most different and are both well documented and examined.

Random Boolean Networks. (RBN) are one example for a model of gene regulatory networks [18,25]. They consist of a state vector of n booleans, each representing the state (on/off) of a single gene. In our cell model each one is associated directly with a biochemical B_i. The state transition rule S_i for each state is defined by n boolean functions with k biochemicals as input I_i. An example for state transition rules and the resulting phase portrait, describing the successive states of an RBN is given in Fig. 1.
Since this behavior is not very realistic we extended the binary RBN to a real-valued RBN. In this, the boolean input states for the RBN are calculated from B and each one is set *true* if the associated biochemical B_i exceeds a given threshold T_i. If the subsequent boolean state is *true*, the activated gene produces biochemicals proportional to the production rate $\boldsymbol{R}_{prod}$ or the deactivated gene degrades the biochemical proportional to the degradation rate $\boldsymbol{R}_{dgr}$.

S-Systems. (*synergistic* and *saturable* systems) have been suggested by Irvine and Savageau [17]. An S-System is given by a set of nonlinear differential equations:

$$\frac{dB_i(t)}{dt} = \alpha_i \prod_{j=1}^{n} B_j(t)^{\mathcal{G}_{i,j}} - \beta_i \prod_{j=1}^{n} B_j(t)^{\mathcal{H}_{i,j}} \tag{1}$$

For each simulation time step the equations (1) are integrated using a Runge-Kutta algorithm with fixed step size of $t_{step} = 0.02$. To prevent the S-System of running into a fixed state with $\boldsymbol{B} = [0]^n$ we demand $\boldsymbol{B}_i \geq 0.0001 \; \forall i$.

3.3 Cell Mechanics

Our cell mechanics are based on a 'winged vertex' structure which is given by the cell's position (vertex) and the cell's neighborhood relations (wings). Both are updated by the cell mechanics to achieve evenly distributed cells with a smooth neighborhood topology. In the case of a two-dimensional world we used $r_{cell} = 1$ for the cell radius, $t_{dist} = 2 \cdot r_{cell}$ as a target distance between cells and $t_{neigh} = 7$ as a maximum number of neighborhood relations.
Regarding the neighborhood relations each cell c_j adds the t_{neigh} most closest cells as neighbors. An alternative method to calculate the neighborhood topology are Voronoi diagrams [8].
To update the position of c_j, we use the neighborhood topology to simulate the forces of each neighbor exerted on c_j. This is simulated repeatedly by a system of springs to reach a state of equilibrium. To introduce an element of chance, brownian motion is added to the cells position. An example distribution

of cells can be seen in Fig. 1. A broken symmetry in this example can occur due to limited space available for cell mitosis and due to asymmetrical diffusion of biochemicals because of local differences in the neighborhood topology.

Although the example in Fig. 1 connotes reproducibility, the brownian motion of cells has a major impact on the results. Compare Fig. 2 for four examples of the same real-valued RBN based cell behavior. Only in three simulations the organism developed the upper left convexity. This can be explained by properties of the discrete RBN behavior. Only slight changes in the neighborhood topology can cause variations in the concentration of biochemicals. When using boolean discretization these variations can lead to major changes in cell behavior and finally in different shapes of the organism.

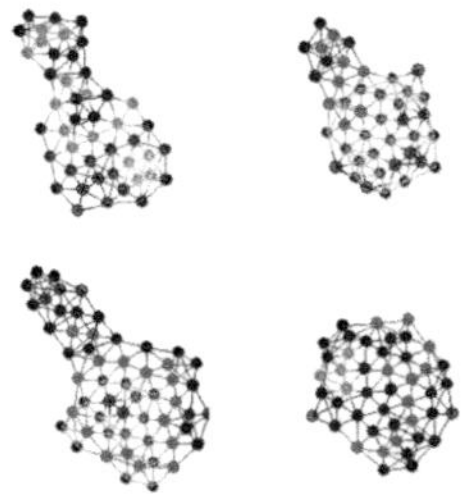

Fig. 2. Example organisms

4 Results

We used an EA-hybrid to optimize the parameters of the real-valued RBN and the S-System to evolve suitable cell behavior for organisms with the ability of limited growth and self-repair formulated as maximization problem. The EA-hybrid allows us to use different EA genotypes and mutation/crossover operators on the level of individuals at the same time. We used an $(\mu = 50, \lambda = 100)$ EA population strategy, tournament selection with $t_{group} = 8$ and a mutation and crossover probability of $p_{mut} = p_{cross} = 0.7$ for each selected EA operator. The mutation on the GA genotype alters one bit per mutation and uses one-point crossover. The GA genotype codes the state transition rules for the real-valued RBN and a BitMask for Structure Skeletalizing for the S-System, as introduced in [23]. On the ES genotype we used global ES mutation and discrete recombination. The ES genotype is used for encoding $\boldsymbol{T} \in [0;1]^n$, $\boldsymbol{R}_{prod} \in [0;1]^n$, $\boldsymbol{R}_{dgr} \in [0;1]^n$ when the real-valued RBN was used, for $\boldsymbol{\alpha}$, $\boldsymbol{\beta}$, $\mathcal{G}$ and $\mathcal{H}$ while optimizing S-Systems and for the diffusion rate $\boldsymbol{D} \in [0;0.5]^n$ in both cases.
To evaluate the fitness each individual is simulated for $t_{max} = 100$ time steps. For each experiment we performed 10 independent runs over 100 generations.

4.1 Limited Growth

The fitness function for the problem of limited growth is defined as the inverse of minimizing $\Theta - \|O(x_i)_t\|$, weighted with t to favor fast growing organisms:

$$\Phi(x_i) = \frac{1}{1 + \sum_{t=0}^{t_{max}} \left(\frac{t \cdot (\Theta - \|O(x_i)_t\|)^2}{t_{max}} \right)} \tag{2}$$

where $\Theta = 30$ gives the desired size of the resulting organism and $\|O(x_i)_t\|$ is the size of the organism at time t.

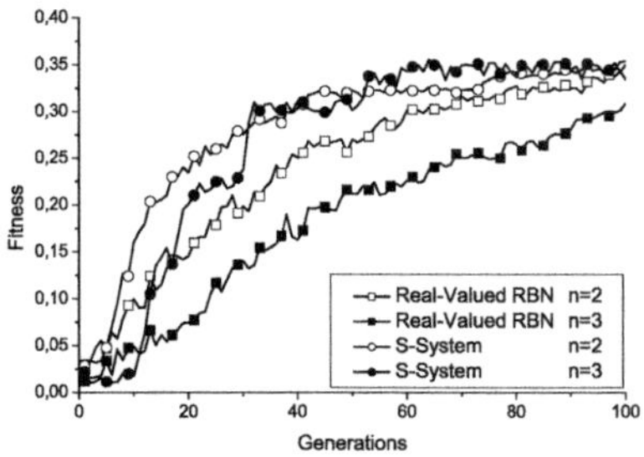

Fig. 3. Avg. of best solutions

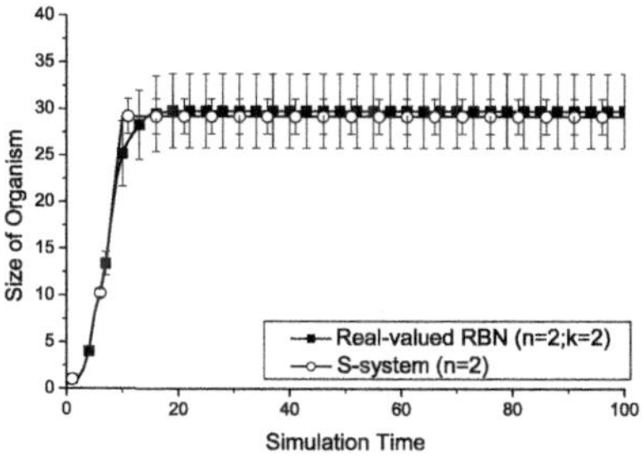

Fig. 4. Avg. size of best samples

We examined the performance of the real-valued RBN and the S-System with either two or three genes on this problem. Fig. 3 shows that rules for limited growth can be evolved in both cases, i.e. using real-valued RBN and S-System. It also demonstrates that only two genes are sufficient and that rules for two genes evolve slightly faster. Although the real-valued RBN and the S-System reach the same level of fitness, the EA optimizing the S-System converges faster and the rules evolved are better suited regarding initial speed of growth and show smaller deviations regarding the size of the organism. Fig. 4 shows the behavior of sample solutions averaged over ten simulations. The corresponding parameters for the examples are given in Tab. 1.

Table 1. Model parameter for given examples

Two genes RBN Fig. 4						Two genes S-System Fig. 4						
I	S	T	R_{prod}	R_{dgr}	D	α	$\mathcal{G}$		β	$\mathcal{H}$		D
1;1	1010	0.00	1.00	1.00	0.18	6.35	1.06	1.33	0.0	0.0	-0.57	0.33
1;1	1111	0.50	0.32	0.00	0.15	2.31	-2.93	2.85	0.0	2.49	0.00	0.00

Three genes RBN Fig. 7 exp. 2						Three genes RBN Fig. 8 exp. 2					
I	S	T	R_{prod}	R_{dgr}	D	I	S	T	R_{prod}	R_{dgr}	D
0;1	1001	0.58	0.52	1.97	0.45	0;2	1101	0.97	1.00	0.71	0.04
0;1	0010	0.92	0.77	1.00	0.06	0;3	0101	0.94	0.56	0.59	0.49
1;1	0100	0.21	0.49	0.28	0.32	0;3	1000	0.73	0.98	0.48	0.13

The behavior of the RBN example can be easily interpreted: 1. Rule for B_0 states, turn gene on in case of $B_1 \leq 0.5$ and turn off if $B_1 > 0.5$, 2. Rule for B_1 says always on. This will cause the cells to split ($B_0 = 1$) until the threshold value for B_1 is exceeded. Considering that B_i is halved each time a cell splits, this set of rules gives the organism enough time to grow to the target size.

4.2 Self-Repair

Next we wanted the EA to evolve limited growth with the additional ability of self-repair, i.e. if the organism is partially destroyed or wounded, it is able to

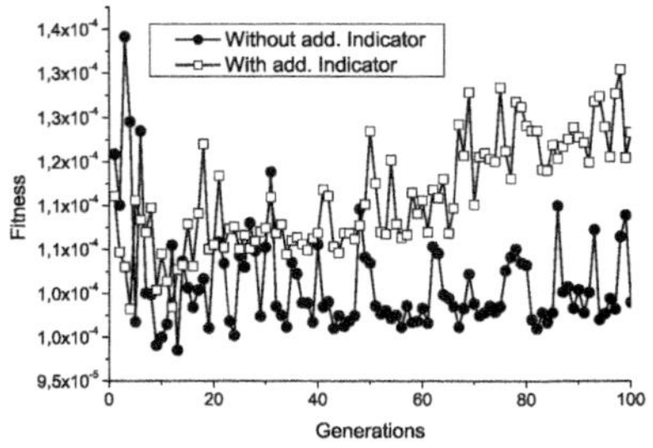

Fig. 5. Averaged best fitness

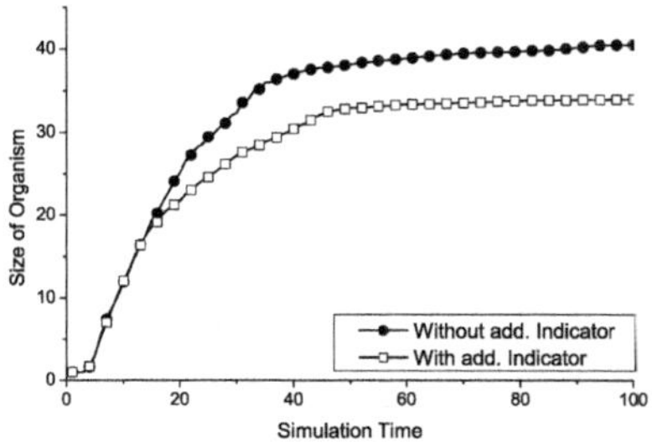

Fig. 6. Av. size on test case 1

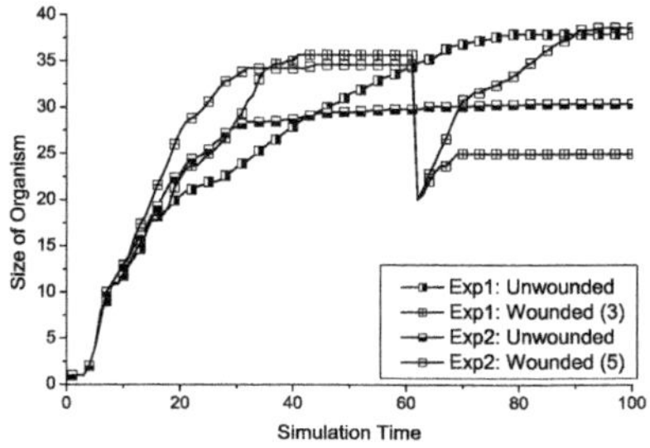

Fig. 7. Sample solutions - indicator

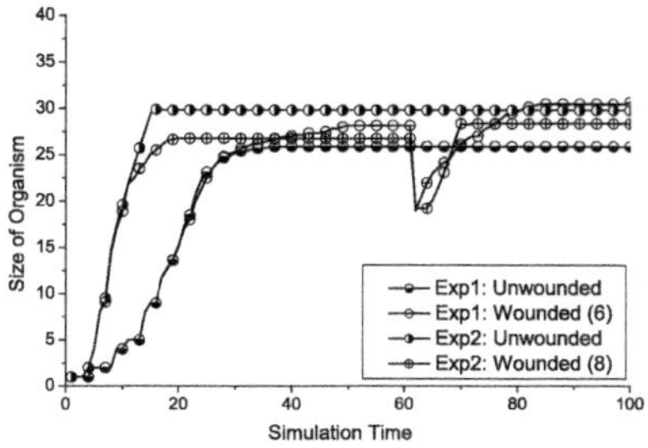

Fig. 8. Sample solutions + indicator

regrow to the former size. To prevent the EA to exploit loopholes, we used two test cases to evaluate the fitness. First the organism was tested whether it is able to grow to a limited size and remain stable in this configuration, otherwise continuous or time dependent growth could feign the ability of self-repair. Then, an additional simulation run was performed as a second test case where the organism was wounded at $t = 60$ by externally killing a group of cells. For both cases the fitness function (eq. 2) was slightly adapted by using two sums in the denominator, one for each test case. This allows lazy evaluation.
Only RBNs with $(n = 3; k = 2)$ were used on this problem and we compared the standard cell model with the alternative version with death signaling. It is shown in Fig. 5 that organisms with the additional indicator for wounds perform better than the organisms without. Fig. 6 shows the averaged size of organisms on the first test case without wounding. Here, the organisms without the additional indicator fail to produce limited growth, but tend to use unlimited growth to cope with wounds. Fig. 7 and Fig. 8 show examples of the evolved cell behaviors with the number of successful simulations of the second test case given in parenthesis. In case of example 2 in Fig. 8, B_0 and B_2 keep a fragile balance by inhibiting and activating each other in order to reach and keep the target size of the organism. In case of wounding the additional B_3 interrupts this cycle by inhibiting B_2. The missing B_2 keeps the organism growing until B_3 is totally degraded. After exhausting B_3 the balance between B_2 and B_0 can be reestablished.

5 Conclusions and Future Research

We have shown that Evolutionary Algorithms can be used to optimize models of gene regulatory networks for solving the AE problem of limited growth and astonishingly only two genes seem to be necessary to achieve limited growth. When comparing RBNs and S-Systems as models for gene regulatory networks, the S-Systems have shown a more reliable behavior. RBNs on the other hand are more prone to stochastic events due to the discrete nature of the state transition rules. But RBN are much faster to simulate.
Regarding the ability of self-repair we were able to show that an additional biochemical indicating the death of neighboring cells, enables the Evolutionary Algorithm to solve this problem more easily and more reliably.
In future experiments we want to apply the S-System to the problem of self-repair. Further on, we will focus on evolving organisms that differentiate into multiple cell types. Using RBN the cell differentiation can be determined by the attractor cycles of RBN. Another issue of further interest are extended cells mechanics to allow cell migration and tissue evagination.

References

1. P. Agarwal. The cell programming language. 2(1):37–77, 1994.
2. T. Akutsu, S. Miyano, and S. Kuhara. Algorithms for identifying boolean networks and related biological networks based on matrix multiplication and fingerprint function. *Journal of Computational Biology*, 7(3):331–343, 2000.
3. H. de Garis. Brain building: The genetic programming of artificial nervous systems and artificial embryos. in book 'Progress in Neural Networks, Vol. 4', ed. O.M. Omidvar, Ablex Publ. Corp., 1991.
4. H. de Garis. Artificial embryology – the genetic programming of cellular differentiation. In *Artificial Life III Workshop*, Santa Fe, New Mexico, USA, June 1992.
5. H. de Garis. Artificial embryology: The genetic programming of an artificial embryo. In B. Souček, editor, *Dynamic, Genetic, and Chaotic Programming*, pages 373–393. John Wiley, New York, 1992.
6. F. Dellaert. Toward a biologically defensible model of development. Masters Thesis, Case Western Reserve University, Dept. of Computer Engineering and Science, 1995.
7. F. Dellaert and R. Beer. A developmental model for the evolution of complete autonomous agents. In *From Animals to Animats IV, 4th International Conference on Simulation of Adaptive Behavior*, pages 393–401, Cambridge, MA., 1996. The MIT Press/Bradford Books.
8. S. Duvdevani-Bar and L. Segel. On topological simulations in developmental biology. *Journal of Theoretical Biology*, 131:33–42, 1994.
9. P. Eggenberger. Cell interactions as a control tool of developmental processes for evolutionary robotics. In *From Animals to Animats IV, 4th International Conference on Simulation of Adaptive Behavior*, Cambridge, MA., 1996. The MIT Press/Bradford Books.
10. P. Eggenberger. Creation of neural networks based on developmental and evolutionary principles. In *Proceedings of the International Conference on Artificial Neural Networks*, Lausanne, Switzerland, October 8–10 1997.

11. P. Eggenberger. Evolving morphologies of simulated 3d organisms based on differential gene expression. In P. Husbands and I. Harvey, editors, *Proceedings of the 4th European Conference on Artificial Life*, Cambridge, MA., 1997. The MIT Press.
12. K. Fleischer. Investigations with a multicellular developmental model. In C. Langton and K. Shimohara, editors, *Artificial Life V: Proceedings of the 5th International Workshop on the Synthesis and Simulation of Living Systems.* MIT Press, 1996.
13. K. Fleischer and A. H. Barr. A simulation testbed for the study of multicellular development: Multiple mechanisms of morphogenesis. In *Artificial Life III.* Addison-Wesley, 1993.
14. P. Funes and J. B. Pollack. Computer evolution of buildable objects. In P. Husbands and I. Harvey, editors, *Proceedings of the 4th European Conference on Artificial Life*, pages 358–367, Cambridge, MA, 1997. MIT Press.
15. A. Gierer and M. Meinhardt. A theory of biological pattern formation. Kybernetik 12, 30–39, 1972.
16. G. S. Hornby and J. B. Pollack. The advantages of generative grammatical encodings for physical design. In *Proceedings of the 2001 Congress on Evolutionary Computation (CEC2001)*, pages 600–607, COEX, World Trade Center, 159 Samseong-dong, Gangnam-gu, Seoul, Korea, 27–30 2001. IEEE Press.
17. D. Irving and M. Savageau. Efficient solution of nonlinear ordinary differntial equations expressed in s-systems canonical form. *SIAM Journal of Numerical Analysis*, 27:704–735, 1990.
18. S. Kauffman. *The Origins of Order.* Oxford University Press, New York, 1993.
19. K. Murphy and S. Mian. Modeling gene expression data using dynamic bayesian networks. Technical report, Computer Science Division, University of California, Berkeley, CA., 1999.
20. P. Prusinkiewicz. Modeling and visualization of biological structures. In *Proceedings of Graphics Interface*, pages 128–137, Toronto, Ontario, 19-21 May 1993.
21. M. Schoenauer. Shape representations and evolution schemes. In L. Fogel, P. Angeline, and T. Back, editors, *Proceedings of the 5th Annual Conference on Evolutionary Programming*, pages 121–129, Cambridge, MA, 1996. MIT Press.
22. B. E. Shapiro and E. D. Mjolsness. Developmental simulations with cellerator. In T.-M. Yi, M. Hucka, M. Morohashi, and H. Kitano, editors, *Proceedings of the 2nd International Conference on Systems Biology*, pages 342–351, California Institute of Technology, Pasadena, CA, USA, November 4–7 2001. Omnipress.
23. D. Tominaga, N. Kog, and M. Okamoto. Efficient numeral optimization technique based on genetic algorithm for inverse problem. In *Proceedings of German Conference on Bioinformatics*, pages 127–140, 1999.
24. D. C. Weaver, C. T. Workman, and G. D. Stormo. Modeling regulatory networks with weight matrices. In R. B. Altman, A. K. Dunker, L. Hunter, and T. E. Klein, editors, *Pacific Symposium on Biocomputing*, volume 4, pages 112–123, Singapore, 1999. World Scientific Press.
25. A. Wuensche. Genomic regulation modeled as a network with basins of attraction. In R. Altman, A. Dunker, L. Hunter, and T. Klein, editors, *Pacific Symposium on Biocomputing*, volume 3, pages 89–102, Singapore, 1998. World Scientific Press.

Caring versus Sharing: How to Maintain Engagement *and* Diversity in Coevolving Populations

John Cartlidge and Seth Bullock

School of Computing, University of Leeds, Leeds, LS2 9JT, UK

Abstract. Coevolutionary optimisation suffers from a series of problems that interfere with the progressive escalating arms races that are hoped might solve difficult classes of optimisation problem. Here we explore the extent to which encouraging moderation in one coevolving population (termed parasites) can alleviate the problem of coevolutionary disengagement. Results suggest that, under these conditions, disengagement is avoided through maintaining variation in relative fitness scores. In order to explore whether standard diversity maintenance techniques such as resource sharing could achieve the same effects, we compare moderating virulence with resource sharing in a simple matching game. We demonstrate that moderating parasite virulence differs significantly from resource sharing, and that its tendency to prevent disengagement can also reduce the likelihood of coevolutionary optimisation halting at mediocre stable states.

1 Introduction

Offering an attractive alternative to standard evolutionary approaches—by removing the difficult but necessary task of defining an adequate fitness function—competitive coevolution has been successfully utilised for optimisation in several domains (for e.g., [1,2]) However, as an optimisation technique, competitive coevolution suffers from some difficult problems stemming from the relative nature of fitness assessment—individuals receive a fitness based upon their success against contemporary opponents.

In general, coevolutionary systems are difficult to *direct*. Individuals may *over-fit* their contemporary competitors, resulting in potentially *brittle* solutions [3]. Rather than enter a progressive *arms-race*, competing populations may stabilise into a sub-optimal equilibrium, or *mediocre stable state* [2,4] As individuals are only rewarded for outperforming their *contemporary* opponents, it is possible for earlier adaptations to lost, potentially leading to cycling [4,5]. Finally, if one population outperforms the other to the extent that every opponent is beaten, the gradient for selection disappears and the populations *disengage* and drift [3,6]. As drift during disengagement is *random* rather than *neutral*, near-optimal populations are likely to degenerate.

Although there are methods for counter-acting particular coevolutionary problems, (e.g., fitness sharing and the "hall of fame" [7]), few of these address the problem of disengagement (one exception is the domain-specific approach presented in [4]). In this paper we present a biologically inspired method for combating coevolutionary disengagement by *moderating parasite virulence*. Continuing upon previous work [6], we demonstrate that reducing parasite virulence can reduce the effects of disengagement in

W. Banzhaf et al. (Eds.): ECAL 2003, LNAI 2801, pp. 299–308, 2003.

the *Counting Ones* domain. The possibility that disengagement could also be combated via existing techniques for maintaining population diversity is considered. In order to explore this, a simple *Matching Game* is used to compare the effects of moderating virulence with those of *resource sharing*, a common diversity maintenance technique. We demonstrate that, whilst fitness sharing encourages phenotypic diversity via niching, which is prone to mediocre stability, moderating virulence tackles disengagement through increasing diversity in *relative fitness scores*, which actively resists mediocre stability in the scenarios that we consider. Fundamentally, moderating parasite virulence is *not* just an apparatus for maintaining population diversity.

2 Moderating Parasite Virulence

Artificial coevolutionary systems are often described as analogous to natural predator-prey or host-parasite systems. Given that most coevolutionary algorithms employ only two populations, the host-parasite analogy is probably closer [8]—one population is typically considered to pose problems for the other resulting in a series of adaptations and counter-adaptations that may result in an escalating arms-race.

Coevolutionary algorithms typically differ from natural systems in the way that they deal with parasite virulence. In order to ensure survival long enough to reproduce, it is not always in the best interests of a natural parasite to be as virulent as possible [9]. As a result, virulence varies dramatically between natural host-parasite systems (compare, for instance, cholera and the common cold), and over time within a particular system (e.g., the history of the myxoma virus in Australian rabbit populations [10]). However, when parasites are used in artificial coevolution, they are generally encoded to be maximally virulent—their fitness varies inversely with the success of the hosts that they compete against. Might coevolutionary algorithms benefit from treating parasite virulence more naturally?

Coevolutionary disengagement occurs when one population secures a significant advantage over the other, such that each competitor from the advantaged population beats each of their opponents in competition. In this way, individuals in both populations become indiscriminable in terms of fitness, all scoring maximally in one population and minimally in the other. Without intra-population fitness diversity there can be no selective forces—each individual is *as likely* to reproduce as any other—resulting in coevolutionary drift. Disengagement continues until the populations re-engage by chance, by which time there may have been a dramatic reduction in the objective quality (*absolute* fitness) of both populations—disengagement hinders optimisation [6].

Often, coevolutionary systems are asymmetric—hosts and parasites may differ genetically (in terms of encoding) or behaviourally (in terms of goal strategy). Such asymmetry may result in an inherent advantage for one population. When coevolving pursuers and evaders, for example, it is often much easier, at least initially, to be a successful evader [5]. Given that disengagement results from one population out-performing the other, it is intuitive that an inherent asymmetrical advantage toward a particular population will encourage the likelihood of coevolutionary disengagement.

However, if parasite virulence were moderated—favouring parasites that achieve less than 100% success against opponents—this trend could be reversed. By rewarding parasites capable of discriminating hosts—those that occasionally lose—with *more* offspring, the asymmetrical advantage will be reduced. Moderating virulence may thus reduce the likelihood of disengagement. Critically, preventing disengagement will improve coevolutionary optimisation if a reduction in periods of degrading coevolutionary drift can be achieved without sacrificing the selection pressure that ensures progress.

Implementing Moderated Virulence. Canonically, parasites receive fitness proportionally to their ability to defeat the hosts they compete against. In order to moderate parasite virulence it is necessary to change this relationship. Throughout this paper we use the term *score* to refer to the ability of a parasite to defeat the hosts it is pitted against. Parasite scores are normalised with respect to the maximum score achieved that generation such that the best current parasite always achieves a score of 1. We define parasite fitness as a function of score, x, and virulence, λ ($0.5 \leq \lambda \leq 1.0$), such that: $f(x,\lambda) = \frac{2x}{\lambda} - \frac{x^2}{\lambda^2}$. Thus, a parasite achieves optimum fitness by winning a proportion of contests equal to a fraction λ of that achieved by the best parasite. By varying λ, parasites can be encouraged to be more, or less, virulent. Although there is a continuum of possible curves, throughout this paper, we use only three values of λ. These are labelled as *maximum* virulence ($\lambda = 1.0$) where parasites are encouraged to beat as many hosts as possible, *moderate* virulence ($\lambda = 0.75$) where parasites are encouraged to achieve a win-rate three-quarters that of the highest scoring current parasite, and *null* virulence ($\lambda = 0.5$), where the fittest parasites achieve half the win-rate of their highest-scoring conspecifics. Notice that a value of lambda lower than 0.5 would encourage cooperation between parasites and hosts.

3 Study One: Counting Ones

In order to introduce the concept of coevolutionary disengagement, Watson and Pollack [3] used a minimal substrate to highlight the effects of disengagement in the *Counting Ones* domain. In this section we utilise an adaptation of the Counting Ones domain to demonstrate the effect that moderating virulence has upon disengagement.

Two reproductively isolated populations of size 25 are coevolved. Individuals in each population consist of binary strings containing 100 bits, with each bit initialised to 0 in generation 0. The *aim* of the Counting Ones problem is to evolve strings containing as many ones as possible. Of course, in this *toy* example, as observers we can assess the absolute fitness or objective quality of each individual by counting its 1-alleles. This allows us, as experimenters, a useful way of measuring progress. However, the coevolutionary algorithm does not make use of this absolute measure, only having access to the relative fitness measure described below.

Members of one population are selected to play a set of *pair-wise* contests against a random sample of 5 opponents from the competing population. During each contest, the individual with the genotype containing the greatest number of 1-alleles receives a fitness point. Each opponent receives half a fitness point for contests resulting in a draw. Individuals in both populations reproduce asexually with parents chosen through tournament selection (tournament size 5; winner reproduces). Offspring have a small probability of mutation, m. Unless specified otherwise, $m = 0.03$.

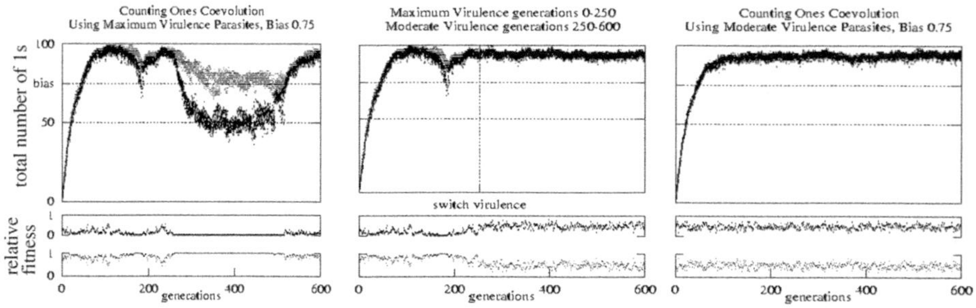

Fig. 1. Typical results from the Counting Ones domain ($B_p = 0.75$; same random seed each run).

An asymmetry was introduced by varying mutation bias B_p ($0 \leq B_p \leq 1$) in favor of one of the two coevolving populations, henceforth classified as the *parasite* population. Given mutation at a particular parasite locus, the substitution of a 1 or 0 occurs with probability B_p and $(1 - B_p)$, respectively. In contrast, the coevolving *host* population substitutes a 0 or 1 with *equal likelihood* whenever mutation occurs. We thus see that if $B_p > 0.5$, then there is a bias in favour of evolving parasites with more ones—an asymmetry that favours the parasite population.

This type of problem asymmetry is common in coevolutionary problems. It is often the case that one side of the coevolutionary contest enjoys some (perhaps temporary) advantage over the other in terms of the ease with which successful counter-adaptations are discovered. For example, at the outset of coevolving list-sorting algorithms it is easier to find a list that is difficult to sort than an algorithm that is difficult to defeat [1].

Several parasite mutation bias values and three parasite virulence levels were tested over a series of runs; $B_p = [0.50, 0.99]$ and $\lambda = 0.50, 0.75, 1.0$. Unless otherwise stated, the value of λ remained constant throughout each run.

Results. Fig. 1 displays three stereotypical runs, each using a parasite mutation bias, B_p, of 0.75. When using maximally virulent parasites (Fig. 1, left) the populations have a tendency to disengage. This can be observed between generations $150 - 175$ and again between $250 - 500$. During these periods of disengagement the populations drift back to their relative baseline performance, equal to the mutation bias, $B_p = 0.75$ and $B_{host} = 0.5$. Only once the populations re-engage by chance is there an improvement in absolute fitness. Repeating the run with the same random seed, the second period of disengagement depicted in Fig. 1 is prevented if moderate virulence is introduced at generation 250 (middle). Notice that the left and middle graphs are identical until generation 250—the point at which parasite virulence is changed to moderate. In contrast to maximum virulence, when moderate parasites are used (right), the populations remain engaged throughout the entire run, achieving a continuously high level of performance.

These results are sensitive to variation in both population size and the number of opponents played by each individual. As either parameter increases, the probability of disengagement decreases due to the increased frequency of meeting varied opponents. However, the results observed in this section are qualitatively robust to mutation rate ($m = [0.005, 0.05]$) and tournament size ($tourney = [2, 15]$).

Fig. 1 clearly demonstrates that moderating parasite virulence in asymmetric coevolution can reduce the effects of disengagement. Further results (not shown[1]) also suggest that the greater the inherent asymmetry, the greater the effect moderating virulence has upon results—i.e., in order to reduce disengagement, virulence should be reduced as engagement increases. The asymmetry imposed in this model gave the biased parasite population a great advantage over the coevolving host population. Purely by stochastic effects one would expect all the individuals from parasite populations to contain more ones than individuals in host populations. This is observed in Fig. 1. The mutation bias alone pushes the populations toward the expected ratio of ones to zeros, i.e., 0.50 for hosts and 0.75 for the parasites. The difference in fitness acceleration between populations, forced by the mutation bias, ensures that once disengagement occurs, the populations quickly diverge to different equilibrium levels. Both populations will remain disengaged until a very large, and thus very *unlikely*, mutation occurs allowing the gap to be, at least temporarily, bridged, for e.g., Fig. 1, left, generation 500.

The first non-zero parasite generation will on average contain many more ones than the host population. However, under moderated virulence, any parasites that beat all opponents are *less* fit than those parasites that lose a small percentage of contests. In this way, acceleration is decreased as the parasites resist their mutation bias. Moderate virulence parasites appear to actively prevent disengagement. Using the continuous selection pressure hosts evolve to a greater level than would otherwise be possible. It should not be overlooked, however, that moderate parasites gain from this relationship too, as both populations evolve to a greater standard than either would alone (Fig. 1, right). However, as parasite virulence is decreased there is a tendency for coevolution to stagnate at a sub-optimal but highly engaged fluid local optimum. In order to push populations to optimal solutions, stronger selection pressure is required (see below).

Diversity Maintenance. Disengagement occurs when intra-population fitness diversity reduces to zero. Moderating virulence counter-acts disengagement by selecting for reproduction parasites that are occasionally beaten. This preserves a selection gradient for hosts which, in turn, maintains relative fitness diversity in both populations.

A tendency toward reduced population diversity (and associated premature convergence) has long been a major concern of the evolutionary computation research community. As such, a suite of diversity maintenance techniques have been proposed, including for e.g., competitive fitness sharing [7], resource sharing [4], and spatial embedding [1]. These approaches are attempts to maintain genetic diversity on the assumption that a loss of diversity can be harmful to optimisation as it may restrict search to local optima.

Resource (or competitive fitness) sharing maintains genetic diversity in a population by encouraging niching—individuals are rewarded for being able to beat opponents that few others can. This idea has been extended to coevolutionary scenarios where opponents are treated as a commodity or resource. Rather than gain a fitness point for each victory against an opponent (simple fitness), one fitness point is shared among the competitors that beat a particular individual. Thus, individuals are rewarded less for *how many* opponents they beat and more for *who* they beat, rewarding genetic novelty and maintaining diversity.

[1] see [6], also http://www.comp.leeds.ac.uk/johnc/publications/ecal03long.pdf

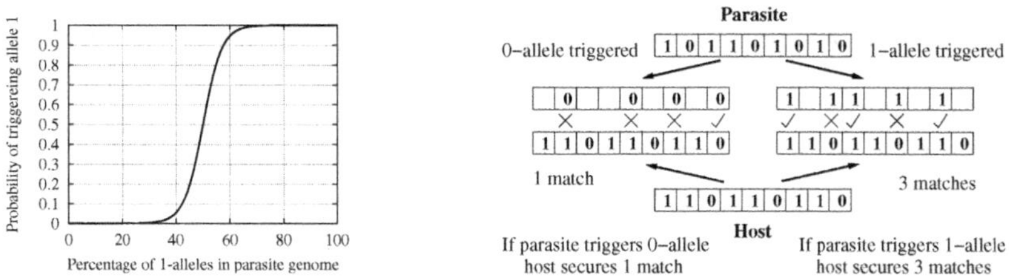

Fig. 2. The probability, $p = \frac{1}{2}\left(1 + tanh\left(\frac{x-50}{7}\right)\right)$, of parasites playing the 1-allele half of the matching game (left). Depending upon which half of the matching game the parasite plays, hosts try to match *either* 0– or 1-alleles (right).

Since disengagement is associated with a loss of diversity, could it be prevented by simple diversity maintenance approaches? Perhaps moderating virulence is only preventing disengagement by mimicking these existing techniques? If so, it is largely superfluous. The next study contrasts moderating virulence with resource sharing in order to explore whether they are effectively the same or different in some fundamental sense.

4 Study Two: Matching Game

In order to compare the influence of parasite (and host) resource sharing with that of moderating parasite virulence, we need to choose an appropriate and simple problem domain. Here we develop a simple matching game, in which hosts are rewarded for matching parasites, but parasites are punished. Games with this type of dynamic often suffer from coevolutionary cycling, as hosts chase parasites through the strategy space. Although a global optimum strategy exists, populations are easily diverted from it as they exploit the temporary idiosyncrasies of their opponents. Resource sharing is one way of discouraging this type of short-termist behaviour. By maintaining a diverse strategy-base in each population, the value of exploiting idiosyncrasies is reduced, encouraging generalists. Unfortunately, an alternative mediocre stable scenario is possible in which populations "speciate" such that they exhibit a number of different sub-optimal strategies that together form a stable combination. In this sense, the game is similar to any number of scenarios in which a generalist strategy is globally optimum, but difficult to evolve in practice—e.g., scissors-paper-stone, immune systems, etc.

Two genetically isolated populations of size 50 are coevolved—hosts and parasites. Individuals in each population consist of binary strings containing 100 bits, initialised *randomly* in generation 0. Each generation, members of the host population are selected to play a set of pair-wise contests against a random sample of 10 opponents from the parasite population. The aim for hosts is to match as many parasite alleles as possible. Antagonistically, parasites aim to mis-match host alleles. Both populations breed asexually, with each individual having a small probability of unbiased mutation per locus, $m = 0.03$ Tournament selection was used (tournament size 5) with the winner of each tournament always chosen to reproduce.

Not all loci are involved in this matching game. For parasites with many 1-alleles, the matching game tends to involve only those loci at which the parasite possesses 1-alleles.

For parasites with many 0-alleles, the game tends to involve only those loci at which the parasite possesses 0-alleles. Whether 1-allele loci or 0-allele loci are involved is determined probabilistically. The probability, p, of a game involving matching 1-alleles increases with the total number of 1-alleles. Once the game has been decided, a host wins by matching alleles in *at least* $T = 30$ loci, else the parasite wins (see Fig. 2).

Having several antagonistic points of attraction, the Matching Game domain is designed to exhibit interesting coevolutionary dynamics. Mutation bias attracts both populations toward genotypes containing 50% 1-alleles and 50% 0-alleles. However, given a host plays a parasite at the 1-allele (0-allele) half of the matching game, it is advantageous for the host to have *as many* 1s (or 0s) as possible. Thus, the host population is attracted toward homogeneous genotypes (all 1s or all 0s). The direction of attraction for hosts (toward either 100% 1- or 0-alleles) depends upon the frequency with which the parasite population plays either the 1-allele or 0-allele halves of the game. This occurs with increasing frequency the further parasite genotypes vary from 50% 1s. Thus, parasites are also attracted away from 50% 1s, but in the *opposite* direction to hosts. Parasites deviating too far from 50% 1s, however, become *too predictable*. In general, the most difficult parasites to match are those having approximately 50% 1s.

This matching game resembles the density classification task for 1-D cellular automata. The aim of the density classification task is to evolve a set of cellular automata (CA) rules which correctly classify the density of an initial condition (IC)—a binary string—as having less than, or more than, 50% 1s. Whilst coevolving CA rules, Juillé and Pollack found it necessary to moderate the virulence of ICs in order to stop disengagement, despite the use of resource sharing [4]. Although very successful, their approach is domain dependent, relying heavily upon domain-specific knowledge. As such, it is not transferable to arbitrary coevolutionary optimisation scenarios. Here we tease apart the contribution of two domain general approaches to improving coevolutionary optimisation, resource sharing and moderating parasite virulence.

As before, three λ values were tested over a series of runs; $\lambda = 1.0$ (maximum), $\lambda = 0.75$ (moderate) and $\lambda = 0.5$ (null). The value of λ remained constant throughout each run. Runs were performed under four conditions: maximum virulence without resource sharing (i.e., standard coevolution); maximum virulence with resource sharing; moderated virulence without resource sharing; both moderated virulence and resource sharing. Under each condition, the degree of niching or genotypic diversity within each population was calculated using a *linkage disequilibrium* measure, particularly sensitive to the effects of resource sharing.

Results. Fig. 3 displays four stereotypical graphs from the Matching Game domain, resulting from the four test conditions. Both resource sharing and moderating parasite virulence have clear effects on coevolutionary dynamics.

Under condition one—maximum virulence with no resource sharing, i.e., typical coevolutionary optimisation (top-left)—the system exhibits cycling. After the initial generations, hosts may begin to recruit more 1-alleles in order to defeat parasites playing the 1-allele half of the game. However, as parasites counter-adapt, by recruiting more 0-alleles, they increase the likelihood of playing the 0-allele half of the game. In response, hosts appear with a greater proportion of 0s, with the entire population

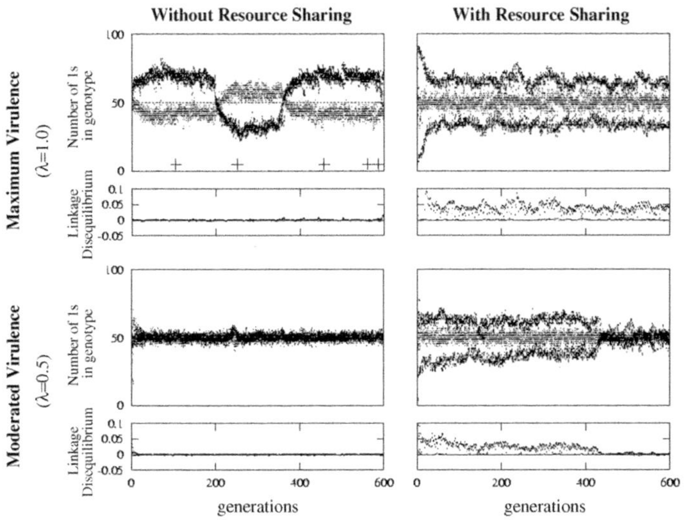

Fig. 3. Typical coevolution in the Matching Game domain.

eventually switching strategy, in order to concentrate on winning the 0-half of the game. Subsequently, parasites again regain the *upper-hand* by recruiting 1-alleles, and so on. Under these conditions, the Matching Game is inherently easier for the parasite population. Hosts find it difficult to be successful generalists—incapable of matching parasites along *both* dimensions—and are encouraged to become brittle *specialists*. As a result, maximally virulent parasites win the majority of competitions and occasionally win *all* competitions, resulting in disengagement (indicated by crosses).

Under condition two—maximum virulence with resource sharing (top-right)—the system reaches mediocre stability. At the beginning of the run, hosts immediately niche into two groups, each specialising on one half of the matching game. In order to be as unpredictable as possible, parasites tend toward 50% 1s—any deviation from this distribution will be punished by one of the specialist host niches. At this mediocre equilibrium the host population as a whole achieves roughly 50% victories over parasites, but each individual host is extremely vulnerable to parasites playing the opposite half of the game. In contrast, parasites tend to become maximally unpredictable and play either half of the matching game with roughly equal probability. Although a generalist strategy exists for hosts, they are unable to discover it.

Under condition three—moderated virulence ($\lambda = 0.5$) without resource sharing (bottom-left)—the system stabilises with generalist hosts. After the initial generations, the host population settles into generalist strategies, capable of matching some parasites whichever allele is triggered. Moderate virulence ensures that parasites are rewarded when occasionally matched, thus allowing hosts to succeed without having to concentrate on winning one half of the matching game. It should be noticed that moderating virulence *does not* result in host-parasite collusion, which would tend to result in homogeneous parasites—the simplest to match. Rather, parasites remain challenging and unpredictable. Any deviation from 50% 1s is quickly punished by the generalist hosts. As such, both hosts and parasites engage in competition in the most *difficult* regions of space. This is equivalent to discovering the "play random" strategy in scissors-paper-stone, or a generalist immune system capable of defeating a wide range of intruders.

Under condition four—moderated virulence with resource sharing (bottom-right)—the system initially achieves mediocre stability, before encouraging hosts to become generalists, strongly engaged with parasites. Early in the run, resource sharing encourages the host population into two niches, each concentrating on one half of the matching game. In this way, the system reaches mediocre stability with hosts and parasites sharing victories. However, unlike condition two, mediocre stability does not persist. Recall that null virulence encourages parasites to achieve a win-rate half that of the highest scoring parasite. This scheme lures parasites away from the mediocre equilibrium at which they achieve a 50% win-rate. As parasites become more easily matched, they reduce the pressure upon hosts to concentrate on one half of the matching game. In this way hosts are steered toward a more generalist strategy of 50% 1s. Hosts engage parasites in a difficult region of space, unattainable without moderated parasite virulence.

Results clearly demonstrate that imposing moderate virulence on parasites alters coevolutionary dynamics in a fundamentally different way to that achieved by resource sharing. Whilst resource sharing encourages *within-population* genetic (and phenotypic) diversity, observable as niching in the host population, moderate virulence encourages diversity in *relative fitness*.

5 Conclusions

We have demonstrated the effects that resource sharing and moderating virulence have upon coevolutionary dynamics. Though both tools increase diversity each affects coevolution in a fundamentally different manner. Resource sharing encourages a population to diversify into separate niches, thus reducing the likelihood of *over-focusing* and coevolutionary cycling. However, niching may produce mediocre stability—reaching a sub-optimal equilibrium—whereby niches *share* success. In contrast, moderating virulence does not encourage intra-population diversity and rather encourages engagement—the extent to which coevolving populations *interact*.

Resource sharing adds a second layer of coupling between conspecifics. In addition to the standard competition that conspecifics experience—striving to beat more opponents than each other—they are forced to *share* their success with one another. This encourages individuals to beat different opponents—i.e., to be different from one another. Niching results from this additional intra-population coupling. In contrast, moderating parasite virulence increases *inter-population* coupling—it ensures that individuals in one population *care* about the success of individuals in the other. In particular, through attempting to achieve moderate success, parasites care about the variation in relative fitness achieved by their opponents—they are selected to cause a range of scores in their opponents. However, this is not achieved through niching, or genetic diversity *per se*. Rather, it is a direct consequence of the moderation that maintains engagement.

It is true that increased genetic diversity has some relationship with coevolutionary engagement. If genetic diversity reduces to zero, populations will disengage (individuals will achieve equivalent scores). However, the converse is not true. Genetic diversity *does not* ensure engagement. Both populations may feature a diverse array of phenotypes, yet still suffer disengagement if each and every phenotype in one population defeats each and every phenotype in the other. Indeed, periods of disengagement often increase genetic diversity through random drift without necessarily increasing engagement. While this coevolutionary coupling (engagedness) is affected by genetic diversity (and noise, sampling error, etc.), it is not determined by it. These considerations ensure that moderating virulence and resource sharing are *complimentary*, rather than *exclusive*, tools. It is not necessary to choose one over another. Indeed, the greatest success may result from using both [4].

Finally, it is important to re-iterate the fact that moderating virulence involves a *trade-off* between engagedness and optimisation. Whilst reducing λ increases engagement, optimisation may suffer as a result, since parasites are being encouraged to present a less-than perfect opponent. Balancing this trade-off—maintaining engagement while encouraging optimisation—proves to be quite difficult. In general, a parameter such as λ must be constantly altered over the course of coevolution to achieve this optimal balance. Currently we have no successful way of adaptively altering λ. Future work involving an experimental interface for *manually steering* virulence during coevolution [11], will aim to discover insights into disengagement that lead to a technique for automatically adapting virulence in the necessary manner.

In conclusion, this paper has demonstrated that moderating virulence is *not* an alternative diversity maintenance technique but a complementary, and novel domain-independent tool, capable of improving coevolutionary optimisation through reducing coevolutionary disengagement.

References

1. Hillis, W.D.: Co-evolving parasites improve simulated evolution as an optimization procedure. Physica *D* **42** (1990) 228–234
2. Pollack, J., Blair, A., Land, M.: Coevolution of a backgammon player. In Langton, C.G., Shimohara, T., eds.: Artificial Life V, MIT Press (1996)
3. Watson, R.A., Pollack, J.B.: Coevolutionary dynamics in a minimal substrate. In: Proc. of the Genetic and Evolutionary Computation Conference, IEEE Press (2001) 702–709
4. Juillé, H., Pollack, J.: Coevolutionary learning: a case study. In Shavlik, J.W., ed.: Proc. of the Fifteenth International Conference on Machine Learning, Morgan Kaufmann (1998)
5. Cliff, D., Miller, G.F.: Tracking the Red Queen: Measurements of adaptive progress in coevolutionary simulations. In Morán, F., Moreno, A., Merelo, J.J., Chacón, P., eds.: Third European Conference on Artificial Life, Springer (1995) 200–218
6. Cartlidge, J., Bullock, S.: Learning lessons from the common cold: How reducing parasite virulence improves coevolutionary optimization. In Fogel, D., ed.: Congress on Evolutionary Computation, IEEE Press (2002) 1420–1425
7. Rosin, C.D., Belew, R.K.: New methods for competitive coevolution. Evolutionary Computation **5** (1997) 1–29
8. Janzen, D.H.: When is it co-evolution? Evolution **34** (1980) 611–612
9. Futuyma, D.J., Slatkin, M.: Coevolution. Sinauer Associates, Sunderland, Mass. (1983)
10. Fenner, F., Ratcliffe, F.N.: Myxomatosis. Cambridge University Press, Cambridge (1965)
11. Bullock, S., Cartlidge, J., Thompson, M.: Prospects for computational steering of evolutionary computation. In Bilotta, E., ed.: Artificial Life 8, Workshops, MIT Press (2002) 131–137

Culture and the Baldwin Effect

Diego Federici

Norwegian University of Science and Technology
Department of computer and information science
N-7491 Trondheim, Norway
federici@idi.ntnu.no,
http://www.idi.ntnu.no/~federici

Abstract. It is believed that the second phase of the Baldwin effect is basically governed by the cost of learning. In this paper we argue that when learning takes place the fitness landscape undergoes a modification that might block the Baldwin effect even if the cost of learning is high. The argument is that learning strategies will bias the evolutionary process towards individuals that genetically acquire better compared to individuals that genetically behave better. Once this process starts the probability of experiencing the Baldwin effect decreases dramatically, whatever the learning cost. A simulation with evolving learning individuals capable of communication is set to show this effect. The set of acquired behaviors (culture) competes with the instinctive one (genes) giving rise to a co-evolutionary effect.

1 Introduction

1.1 The Baldwin Effect

In the context of the debate between Darwinism and Lamarckism, James Mark Baldwin (1896) proposed that phenotypic plasticity might be regarded as "a new factor in evolution" [1]. Phenotypic plasticity allowing adaptation, would smooth the fitness landscape increasing the efficiency of the evolutionary process [2,3]. However, phenotypic plasticity has inherent costs associated with the training phase in terms of energy, time and eventual mistakes. For these reasons, in a second phase, evolution may find a way to achieve the same successful behaviors without plasticity.

Thus the Baldwin effect has two phases. During the first phase, adapting individuals can, in same cases, acquire behaviors that help them achieving higher fitness scores. But because of the costs of adaptation, there is an evolutionary advantage towards the discovery of equivalent instinctive behaviors. Thus in this second phase, a behavior that was once learned may eventually become instinctive (see also below and [1,2,3,4,5]) In computer science, the phenotypic plasticity is analog to a local search strategy. The evolutionary process and the local search may be used in combination, often achieving higher efficiency than either of the methods alone [5,6].

W. Banzhaf et al. (Eds.): ECAL 2003, LNAI 2801, pp. 309–318, 2003.

There are three basic requirements for the second phase to take place. First there must be a cost for the local search. In this way, the evolutionary process will have a reason (in terms of inclusive fitness) for the genetic assimilation to take place. This also means that in some settings, i.e. in a fast changing environment, genetic assimilation will never take place. With those setting, plasticity would be *the* optimal strategy. We will refer to this characteristic by *Assimilation Advantage*.

Also, genetic assimilation requires for the optimal strategy, acquired first through local search, to be expressible by the genotype. This might be impossible under some genotype-phenotype mapping strategies in which, the phenotype plasticity is required as part of the developmental process [1]. We will refer to this characteristic as *Genotypic Expressibility*.

In addition, the probability of the assimilation depends on the distance between the genotype using plasticity and the one not using it. The distance would be measured using the metric imposed by the genetic operators. A small distance is possible if there is a strong neighborhood correlation in the transformation from genotypic to phenotypic space. Where the distance is too high, the probability of genetic assimilation could be so little to be considered actually impossible. We will refer to this as *Genotypic-Phenotypic Correlation*

1.2 How Learning Effects Evolution

In a now famous paper, Hinton and Nowlan [6] proved that with the help a local search mechanism it is possible to speed up evolution in a hard fitness landscape.

In the Hinton and Nowlan example, the adaptive solution is ideally placed in the middle between the lowest fitness solutions and the single high one, hence smoothing the fitness landscape. Adaptation in this case is a step towards the discovery of the best non-adaptive solution. The same considerations apply to other examples such as [2,8] among others.

In these cases adaptation success is not affected by any genetically coded learning strategy. We argue that the search for good learning strategies might distract the evolutionary process from the discovery of fit non-adaptive behaviors. In other words, co-evolution of learning and the non-learning strategies modifies the fitness landscape. The quality of these modifications is an other factor that governs the second phase Baldwin effect.

2 Learning, Culture, and Fitness

Learning can be seen as the process of acquiring behaviors. The difficulty and time lost acquiring behaviors constitute a cost of learning. We have to introduce a clear distinction between instinctive and acquired behaviors. Instinctive are those behaviors that emerge steadily and directly from the genotype, while acquired ones are those that emerge through the interaction with the environment.

[1] Like in the development of the retina [7]

If we consider individuals belonging to a population sharing the same genotype, their individual fitness can be considered the sum of shared population fitness (PopFit), fitness change due to local environmental characteristics (LEFit) and fitness change due to individual specific behavior (IFit): $Fitness = PFit + LEFit + IFit$.

The LEFit can be considered as noise and could be absent in ideal experimental settings (all individuals having the exact same initial conditions or long fitness tests). Fitness deriving from acquired behaviors (IFit) constitutes the value of the learning process and incorporates the cost of learning.

When PFit $\gg$ IFit, the advantage for plasticity is negligible. Otherwise acquired behaviors may provide an advantage to genotypically similar individuals (see figure 1). In this case, there is strong evolutionary pressure towards the discovery of better acquisition mechanisms.

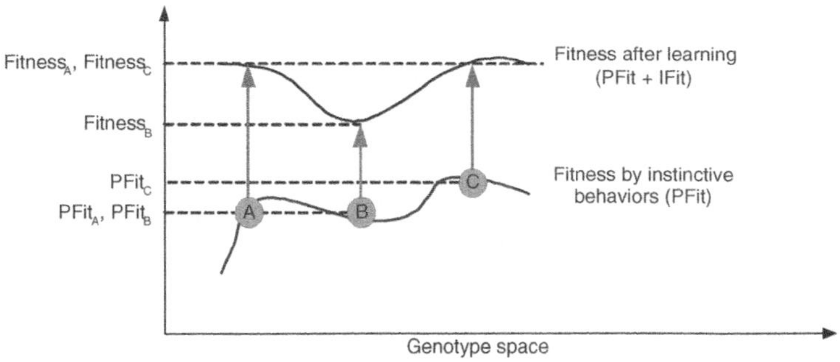

Fig. 1. Two similar individuals (A and B), share similar genotype and PFit values. By acquiring different behaviors they achieve different fitness scores. On the other side, two genotypically different individuals (A and C) reach the same fitness values because the same acquired behavior shadows the instinctive ones.

2.1 Memes

As genes form the transmission medium of biological systems, memes [9] do for acquired behaviors. Memes will be considered behavioral information blocks[2]. Basically memes are those things that "leap from brain to brain" [9] carrying a behavioral content. To strike a comparison to human society, we will call the set of transmittable behaviors *Culture*.

3 Simulation Details

We set up a population of learning individuals. Each individual/agent is equipped with a single layer neural network (NN) subject to an evolutionary process and a

[2] Memes usually have a wider definition, but considering only the behavioral ones, the discussion is simplified

classifier-like system (memes), see Figure 3. Agents perceive resources and other bots from all tiles in a hamming distance of 2 (see figure 2), this constitutes the input vector.

The NN produces an output vector with the expected reinforcement for each of the possible actions: don't move, go north, west, south and east.

Memes remind the agent the reinforcement experienced in the past. They are constituted by an input pattern P, an action a and an expected reinforcement R. If the pattern P matches the present input vector, then the meme replaces the output of the genetically evolved NN with R for the given action a. Basically the meme can recognize a particular sensory context (P) and reminds the agent that in the past he had performed a certain action (a) and the action yielded a given reinforcement (R).

The four expected reinforcements, generated by the NN and eventually modified by the memes, are used to stochastically select the action performed by the agent.

Agents score fitness by collecting resources spread at random in a toroidal map. Each resource type gives a fixed amount of reinforcement. If an agent does not move or collides with another agent it receives a small negative reinforcement[3]. Fitness is the sum of all reinforcements received over a fixed number of iterations. Agents undergo a steady state selection with a replacement fraction of 25%. Surviving individuals keep their memes. The offspring is placed in the proximity of a parent.

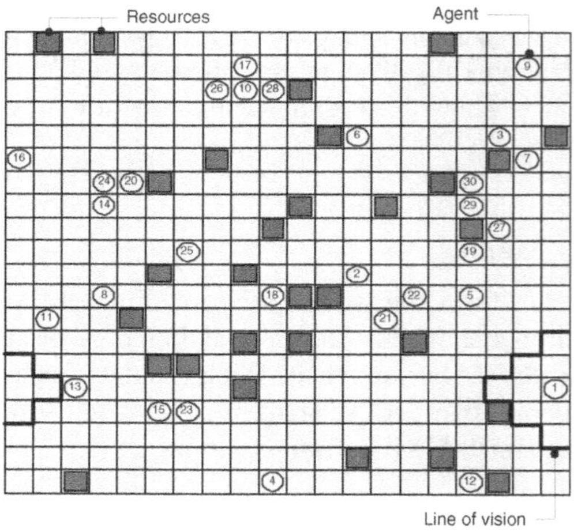

Fig. 2. Simulated Environment. The vision range of agent 1 is printed in a thick line. Two different types of resources are present. The resources represented by a darker color give a fixed negative reinforcement and fitness value, while the others give a fixed positive value. Resource types never change value and when consumed are regenerated on a random tile. Newly generated agents are placed in proximity of a parent.

[3] These penalties were added to speed up the evolutionary process

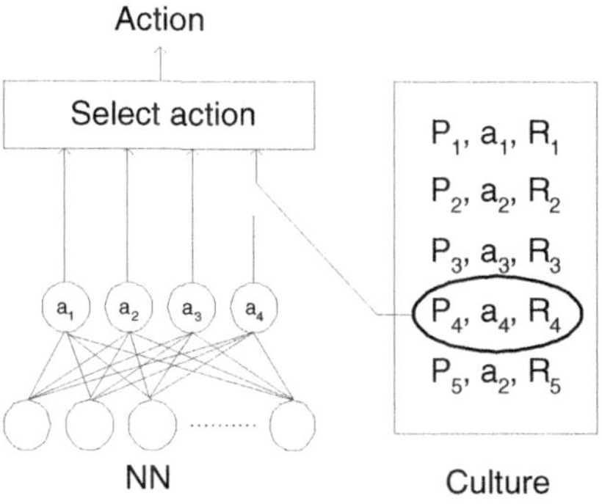

Fig. 3. Agent controller. The genetically evolved NN and the acquired culture are activated in parallel. When a pattern P matches the current input, the corresponding meme is activated (encircled in the figure). Its expected reinforcement R_4 replaces the NN output for action a_4. The performed action is selected stochastically giving a higher probability to higher expected reinforcements.

3.1 Cultural Evolution

An agent culture is built by a certain number of memes (20 at maximum in these simulations). Memes can be acquired in two different ways.

First by transmission, whenever two individuals are next to each other. Every iteration a fixed number of memes can be transmitted. These memes are probabilistically selected from the transmitter culture according to their estimated value. The number of memes transmitted when two agents are in contact (communication speed, CS) is varied from 0 (no transmission) to 20 (all the meme pool is transmitted in a single iteration). As the CS increases, the agents can acquire fit behaviors earlier during the fitness evaluation, hence the *Assimilation Advantage* is reduced.

The second way is through operant conditioning.

Operant conditioning is a learning mechanism that has been observed in a variety of animals. When an animal experiences a reinforcement, its brain tries to explain what caused the reward. The effect is that the behaviors that are thought to be responsible of the reinforcement are rewarded. Learning appears to build a relationship between behavior and reinforcement based on two general assumptions: the behavior that steadily is followed by reinforcement is held responsible for it, behavior and reward must fall into a certain time window. These assumptions have strong biological and psychological support [10,11,12].

Whenever an agent experiences an unexpected reinforcement a meme is generated from this situation. The value of the meme changes as it is used, increasing when it helps predict the expected reinforcement. This inhibition/enhancement is an explicit measure of the memes fitness, and is used to drive the memetic evolutionary process. Unfit memes can be explicitly identified and dropped, fit ones will proliferate through transmission and new variants will eventually be generated.

Memes variants are generated by merging, a stochastic generalization mechanism. Merging is the memetic equivalent of crossover and mutation in genes. It can occur if two memes code the same action and expected reinforcement. In this case, the merging probability is proportional to the hamming distance between the memes input matching patterns. Those parts that are different in the two input matching patterns are replaced by *don't care* symbols.

Merging is a weak simplification of a boolean function:
given $(P_1 \wedge a \mapsto R)$ and $(P_2 \wedge a \mapsto R)$ then with probability $\sim d_H(P_1, P_2)$ replace them with $((P_1 \circledast P_2) \wedge a \mapsto R)$; where $P_i \in$ pattern, $a \in$ action, $R \in$ reinforcement, d_H is the Hamming distance, and $\circledast$ is a bitwise operator $\circledast(b_i, b_j) = \{ b_i$ if $b_i = b_j$, *don't care* if $b_i \neq b_j \}$.

If it does not merge, a meme can be added only if the meme pool size does not exceed the maximum. If the maximum is exceeded a meme is dropped, the less general being dropped with higher probability. Because merging of memes can sometimes produce unfit memes, if the expected reward does not match the one experienced, the meme responsible for the error is removed.

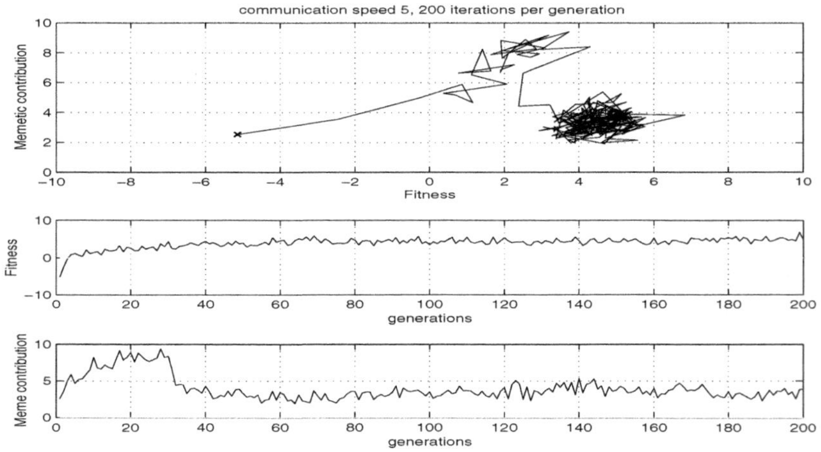

Fig. 4. Run showing the Baldwin effect, convergence to the $\mathbb{MG}$ attractor. The memetic contribution to the fitness is at first high. As it decreases, the memetic behavior is partially assimilated in the genes.

4 Results

Agents can transmit some of their memes whenever their are next to each other. The number of memes transmitted when two agents are in contact (communication speed, CS) is varied from 0 (no transmission) to 20 (all the meme pool is transmitted in a single iteration).

As the CS increases, the agents can acquire fit behaviors earlier during the fitness evaluation, hence the *Assimilation Advantage* is reduced.

It is possible to evaluate the amount of fitness generated through acquired behaviors by stopping the simulation every generation and recording how much

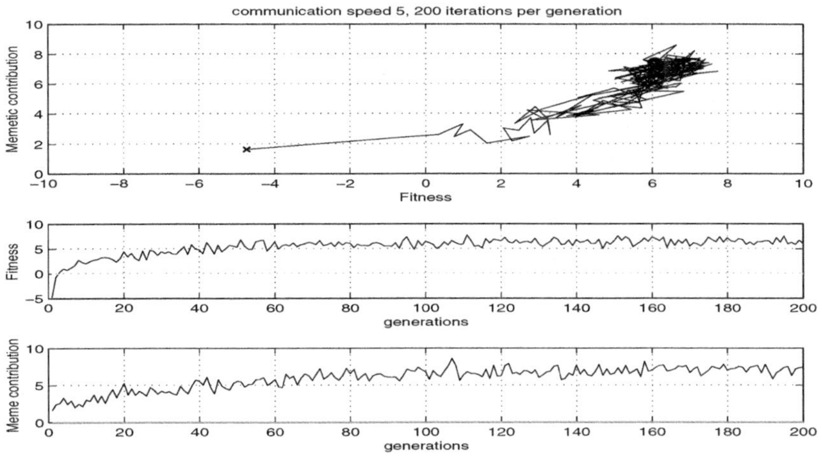

Fig. 5. Run not showing the Baldwin effect, convergence to the $\mathbb{M}$ attractor. Genetic assimilation does not take place.

fitness the agents score with and without the help of memes. It is so possible to plot how much the memes contribute to the total fitness score (memetic contribution). Figures 4 and 5 show three different plots. The first one is a state diagram, memetic contribution vs. fitness, showing the trajectory that a population undergoes during evolution. The second one displays the amount of average fitness scored by the population, and the third the quantity of memetic contribution.

The state diagram is particularly useful because it shows the attractors of the evolutionary process. Figure 4 shows the attractor for a typical population that went through the second phase of the Baldwin effect, figure 5 one that doesn't show it.

In almost every setting two attractors, such as those in figure 4 and 5 are present. The two attractors show the convergence basin of different strategies. The first relies both on memes and genes ($\mathbb{MG}$) with the instinctive behaviors capable of scoring some fitness. The second relies on memes only ($\mathbb{M}$).

It is then understandable that while the dynamic towards $\mathbb{MG}$ is associated to the second phase of the Baldwin effect, the path to $\mathbb{M}$ is not.

Figure 6 shows with which frequency a population falls in the $\mathbb{MG}$ attractor.

One would expect to experience a decreasing number of populations falling in the $\mathbb{MG}$ attractor as CS is being increased. In fact, increasing CS reduces the cost of learning and the advantage for genetic assimilation. Instead the frequency decreases to a minimum and then increases again. This proves that the Baldwin effect is influenced by other factors apart the cost of learning.

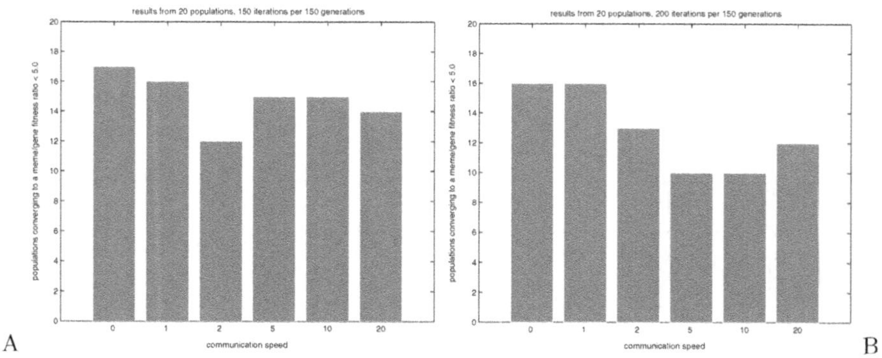

Fig. 6. Populations falling into the $\mathbb{MG}$ attractor with each CS setting

4.1 Different Strategies

An agent can acquire fit behaviors in three possible ways: evolving the proper NN, building memes by operant conditioning, and acquiring memes from other agents. CS can affect only the latter.

Being born next to its parent without any meme, an agent is faced with a dilemma. It can either try:

1. first receive as many memes as possible from other agents, and then exploit them to score fitness
2. start scoring fitness immediately and acquire memes by itself

Either choice requires a genetically coded strategy, a social one in the first case, and an asocial one in the latter. The social strategy, scoring fitness mainly by memes, is the one that converges to $\mathbb{M}$. $\mathbb{MG}$ is instead the attractor of the asocial strategy.

CS does not only change the cost of learning, modulating the acquisition speed. It also changes the nature of the two strategies $\mathbb{M}$ and $\mathbb{MG}$.

As CS decreases the social strategy becomes more difficult. In other words it requires a more committing strategy and a more specialized genotype. In fact, it must strongly avoid any asocial behavior, otherwise it will risk to interrupt the acquisition phase. On the other side when CS is maximal, the parent's culture is acquired in a single shot. The social strategy is achieved simply by being born. In this case $\mathbb{M}$ and $\mathbb{MG}$ will be maximally overlapping because an asocial individual will receive the same amount of culture as a social one.

The difference between the two strategies determines the probability of moving from one to the other. The higher the difference the lower the probability.

A second aspect is that the quality of culture (the amount of fitness it can guarantee) is not fixed. As genes evolve every generation, memes do every iteration. If CS is zero, agents cannot share their memes and the culture quality improves in a given generation but not across generations. If CS is high, even if an agent dies some of its memes might survive and continue evolving on other hosts.

The fitness advantage of a strategy or the other, is subjected by the actual level of cultural evolution. When many agents do not socialize, culture improves slowly and the social strategy is less attractive. With many social agents, cultural evolution is faster and offers a greater prize for adaptivity.

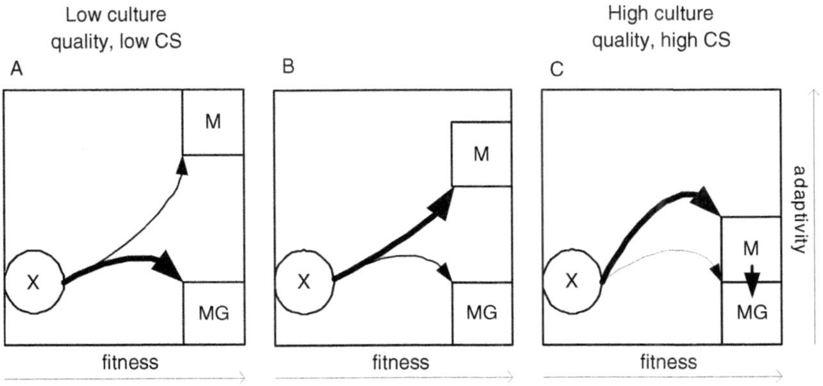

Fig. 7. Effects of CS to evolvability

As the cultural quality evolves or fails to evolve, the fitness landscape changes. If cultural evolution proceeds steadily, the $\mathbb{M}$ strategies will be able to dig into their attractor increasing their stability. Still, if $\mathbb{MG}$ lies too close to the $\mathbb{M}$ attractor, this is insufficient to prevent the second phase of the Baldwin effect. Figure 7 summarize these concepts.

Figure 7A. With low CS, the quality of culture is not enough so that the asocial strategy $\mathbb{MG}$ is more convenient (the thicker line indicates a higher transition probability).

Figure 7B. CS increases and the $\mathbb{M}$ attractor becomes stronger. As the attractors are far away in genotypic space, the probability to move from $\mathbb{M}$ to $\mathbb{MG}$ is very low. The frequency of the second phase of the Baldwin effect is minimal.

Figure 7C. CS continues to increase, the social and asocial strategy are very similar and passage from $\mathbb{M}$ to $\mathbb{MG}$ is more probable.

5 Conclusions

We have provided an example in which genetic assimilation cannot be explained by the Baldwin effect alone.

It is suggested that the Baldwin effect considers cases in which the genotype cannot modify directly the adaptation mechanism. Under these conditions, the genetic search can evolve only the non-adaptive part of the phenotype, and only that can provide a continuous fitness improvement. Examples for this case can be found in [6,2,8,13].

We argue that the adaptive behavior must be somehow expressed in the genotype, so that evolution affects also the adaptation mechanism and its quality.

Evolution can than proceed in at least two directions, one towards the discovery of better adaptive strategies (in this case the social behavior), the second towards the discovery of fitter instinctive behaviors (the asocial behavior).

The state of the evolution of adaptivity affects both the cost of learning and the correlation between genotype and phenotype. This can cause that both the cost of learning and the correlation decrease, so that the probability of observing the Baldwin effect is not a monotonic function.

Even in a static environment[4] as the one provided in this paper, the fitness landscape will undergo a dynamic. Under these circumstances, genetic assimilation is ruled more by the quality of the fitness landscape dynamic than by the assimilation advantage.

The simulations presented in this paper are very computational expensive and have been run on ClustIS[5] cluster.

References

1. J.M. Baldwin: A new factor in evolution. American Naturalist **30** 441–451 (1896)
2. K. Downing: Reinforced Genetic Programming. Genetic Programming & Evolvable Machines. Kluwer Publishers 259–288 (2001)
3. P. Turney: Myths and legends of the Baldwin Effect. Proceedings of the ICML-96 (13th International Conference on Machine Learning) 135–142 (1996)
4. P. Turney, D. Whitley, R. Anderson: Evolution, learning and instict: 100 years of the Baldwin effect. Evolutionary Computation **4:3** 206–212 (1996)
5. R.K. Belew, M. Mitchell: Adaptive individuals in evolving populations: models and algorithms. addison-wesley (1996)
6. G.E. Hinton, S.J. Nowlan: How learning can guide evolution. Complex Systems **1** 495–502 (1987)
7. L.M. Chalupa, B. L. Finlay: Development and organization of the retina : from molecules to function. Plenum Press, New York (1998)
8. R. French, A. Messinger: Genes, phenes and the Baldwin effect. In Proceedings of Artificial Life IV 277–282 (1994)
9. R. Dawkins: The selfish gene. Oxford university press (1976)
10. E.R. Kandel, J.S. Schwartz, T.M. Jessell: Principles of neural science. Elsevier Science Publishing Co., New York (2000) 4th edition
11. H. Markram, J. L'bke, M. Frotscher, B. Sakmann: Regulation of Synaptic Efficacy by Coincidence of Post-synaptic APs and EPSPs. Science, **275** 213–215 (1997)
12. T.J. Carew Behavioral Neurobiology: The Cellular Organization of Natural Behavior Sinauer Associates (2000)
13. T. Arita, R. Suzuki Interactions between learning and evolution: The outstanding strategy generated by the Baldwin effect. In Proceedings of Artificial Life VII, 196–205 (2000)

[4] where the best strategy does not require adaptation

[5] information can be obtained from http://ClustIS.idi.ntnu.no/

Evolutionary Network Minimization: Adaptive Implicit Pruning of Successful Agents

Zohar Ganon[1], Alon Keinan[1], and Eytan Ruppin[1,2]

[1] School of Computer Sciences, Tel-Aviv University,
Tel-Aviv 69978, Israel
{zganon,keinan}@cns.tau.ac.il, ruppin@post.tau.ac.il
[2] School of Medicine, Tel-Aviv University,
Tel-Aviv 69978, Israel

Abstract. Neurocontroller minimization is beneficial for constructing small parsimonious networks that permit a better understanding of their workings. This paper presents a novel, *Evolutionary Network Minimization (ENM)* algorithm which is applied to fully recurrent neurocontrollers. ENM is a simple, standard genetic algorithm with an additional step in which small weights are irreversibly eliminated. ENM has a unique combination of features which distinguish it from previous evolutionary minimization algorithms: 1. An explicit penalty term is not added to the fitness function. 2. Minimization begins after functional neurocontrollers have been successfully evolved. 3. Successful minimization relies solely on the workings of a drift that removes unimportant weights and, importantly, on continuing adaptive modifications of the magnitudes of the remaining weights. Our results testify that ENM is successful in extensively minimizing recurrent evolved neurocontrollers while keeping their fitness intact and maintaining their principal functional characteristics.

1 Introduction

Pruning a neural network is a standard approach in the connectionist literature by which unimportant weights are removed, usually in order to enhance the network's generalization performance [1]. In this paper we focus on pruning and its applications in the context of *Evolved Autonomous Agents (EAA)*. To demonstrate this we propose a new network minimization algorithm, *Evolutionary Network Minimization (ENM)*. ENM prunes successful networks while maintaining their fitness and principal functional characteristics intact. It is essentially a standard genetic algorithm with an additional step during reproduction in which weights are irreversibly eliminated.

Minimization of successful agents is motivated by several goals. First, it enables better understanding of the way an agent works: Finding a small underlying skeleton of the agent enables a direct inspection of its functioning. Second, the minimized agent is more efficient computationally, hence, if the agent is implemented in hardware, fewer components will be needed. Third, the size and complexity of the minimized agent may stand as an estimate for the complexity of the task it solves (while the size of the original agent is arbitrary).

W. Banzhaf et al. (Eds.): ECAL 2003, LNAI 2801, pp. 319–327, 2003.

Most pruning algorithms in the neural networks literature, such as pruning by the weights magnitude, Optimal Brain Damage (OBD) [2], Optimal Brain Surgeon (OBS) [3] and contribution based pruning [4,5], are non evolutionary methods. OBD and OBS analytically predict the effect of a weight removal on the error function and all algorithms allow for retraining after the elimination of weights[1]. However, in the case of EAA no explicit error function is available for such an analysis and for retraining. Furthermore, the neurocontrollers may be fully recurrent, as in the agents presented throughout this paper. Pruning algorithms incorporating an explicit network complexity term in the fitness function and performing an evolutionary search have also been suggested [6,7]. ENM does not rely on such an arbitrarily defined explicit complexity term in the fitness function. It is more akin to variable length encodings algorithms, an example of which is a genetic algorithm with an evolved mask representing the eliminated weights [8]. Within ENM, the pruning is based on accumulation of eliminated weights.

The rest of this paper is organized as follows : Sect. 2 describes the EAA environment. Sect. 3 presents the ENM algorithm, and in Sect. 4 we describe and analyze the results of minimizing neurocontrollers of autonomous agents. These results testify to the effectiveness and consistency of ENM, and illustrate that the minimized agents preserve the essential functional characteristics of the original ones. They further demonstrate the usage of a minimized network for understanding the agent's functional characteristics. We summarize in Sect. 5.

2 The Evolved Autonomous Agents Environment

The EAA model used in this paper is described in detail in [9]. A population of agents performing a task of navigation and foraging is evolved in a discrete grid arena surrounded by walls. Poison items are scattered in the whole arena, while food items are scattered only in a "food zone" in one corner. The agent's goal is to find and eat as many food items as possible during its life, while avoiding poison. Its fitness is the number of food items minus the number of poison items it has consumed, normalized by the total number of food items that were available giving a maximal value of 1. Due to the short lifetime, a fitness score of 1 is practically unattainable.

The sensory system of the agent is made of 5 sensors: *front*, *left*, *right* and *here* provide an indication whether there is a resource (food or poison, without a distinction between the two), a wall or an empty cell in front, left-front, right-front and underneath the agent, respectively. The *smell* sensor gives an indication of food or poison if they are directly underneath the agent and a random reading otherwise. In several simulations the agent was also equipped with a position sensor indicating its x and y grid coordinates, otherwise it had only local information from its 5 sensors. Four motor neurons dictate movement forward, a turn left or right, and control the state of the mouth (open or closed). Eating takes

[1] OBS suggests a formula for recalculating the magnitude of the weights, rather than simple retraining.

place if the agent opens its mouth and is neither moving nor turning. The agent is controlled by a fully recurrent neurocontroller of binary McCulloch-Pitts neurons, with the exception of the non-binary sensor neurons which have no input from other neurons.

Previous analysis [9] revealed that successful agents possess one or more *command neurons* that determine the agent's behavioural strategy. Artificially clamping these command neurons to either constant firing activity or to complete quiescence causes the agent to constantly maintain one of the two behavioural modes it exhibits, regardless of its sensory input. These two behavioural modes are *exploration* and *grazing*. Exploration, which takes place when the agent is outside of the food zone, consists of moving in straight lines, ignoring resources in the sensory field that are not directly under the agent, and turning at walls. Grazing, which takes place when the agent is in the food zone, consists of turning towards resources to examine them, turning at walls and maintaining the agent's location on the grid in a relatively small region.

In this paper we apply the ENM to three successful evolved agents: Z10, S22 and SP10 with 10, 22 and 10 neurons (including the motors), respectively. SP10 is also equipped with a position sensor.

3 The ENM Algorithm

The ENM algorithm receives as input an evolved neurocontroller and outputs a pruned neurocontroller evolved to achieve a high fitness on the original task. ENM is a genetic algorithm which evolves a population of genomes, directly encoding the networks' weights. The algorithm starts by cloning the input network's genome to form the first generation. Then, a standard genetic algorithm is used, with an additional important elimination step after the selection, crossover and mutation steps. In this step, weights are eliminated (zeroed) according to an *elimination criterion*. In order to create pressure for the accumulation of eliminated weights, they cannot be revived by mutation. The fitness function used for evaluating the new generation genomes is simply the fitness function of the original task. The evolution continues until both the average fitness and the average number of non-pruned weights in the population converge to stable values. The result of the evolutionary process is then a population of networks from which ENM selects and outputs the network with the best fitness.

The elimination criterion used throughout this paper is as follows: Weights in the range $[-0.2, 0.2]$ are eliminated, while keeping all weights in the range $[-10, 10]$ by truncating oversized weights in order to keep them within a reasonable distance from the elimination range. Before running ENM, the input network is scaled, without changing the network's operation, such that the magnitude of all weights is smaller than the maximal value (10). The genetic operators are as in the original evolution creating the successful agents: Roulette wheel selection is used with uniform point crossover (probability of 0.35). Mutation is applied to all non zero weights, adding a uniformly distributed random value between -0.2 and 0.2.

4 Results

4.1 Effectiveness and Consistency

We applied ENM for minimizing agent S22 (described in Sect. 2). Figure 1 shows the average fitness and average fraction of non zero weights throughout the minimization process. Evidently, **ENM markedly reduces the number of network connections while preserving high fitness values.** The minimized neurocontroller consists of only 26 weights out of the original 594. The average fitness declines in the beginning of the evolutionary process, then is gradually restored. This initial decline is due to the elimination of small but important weights that were mutated to magnitudes below the elimination threshold in some of the agents during the first generations. As the evolution continues, the important weights are "pushed" farther away from the elimination threshold, allowing for the regain of the average fitness.

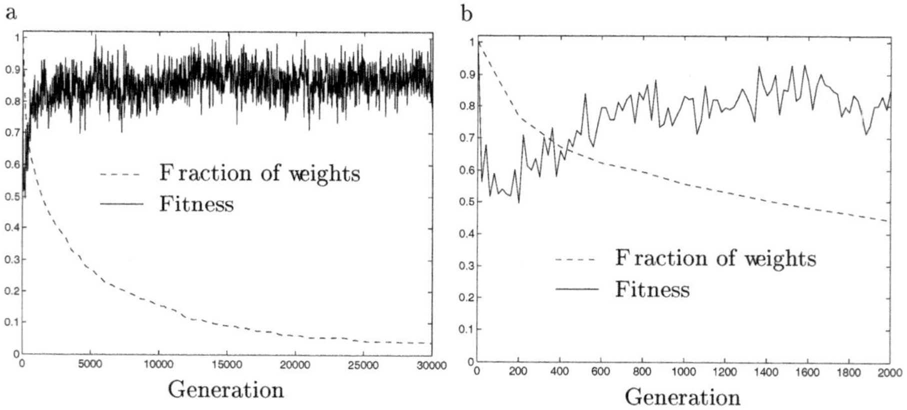

Fig. 1. Average fitness and average fraction of non zero weights across generations during ENM of S22. The fitness is normalized such that the fitness of the original agent equals 1. The averaging is over a population of 100 agents. (a) ENM during 30000 generations. (b) Focusing on the first 2000 generations, where the high fitness is restored after a decrease in the first generations

A comparison of the ENM with random pruning and magnitude based pruning is presented in Fig. 2 . Magnitude based pruning removes in each iteration the weight with the smallest absolute size[2]. ENM prunes 95% of the weights while maintaining the original fitness almost intact, whereas the other methods result in a large decrease in the fitness already at much smaller levels of pruning.

In order to test the consistency of the ENM, we repeatedly applied it 10 times for minimizing each of the evolved agents S22, Z10 and SP10. Fig. 3a

[2] This is done without retraining since the agents are evolved, rather than trained.

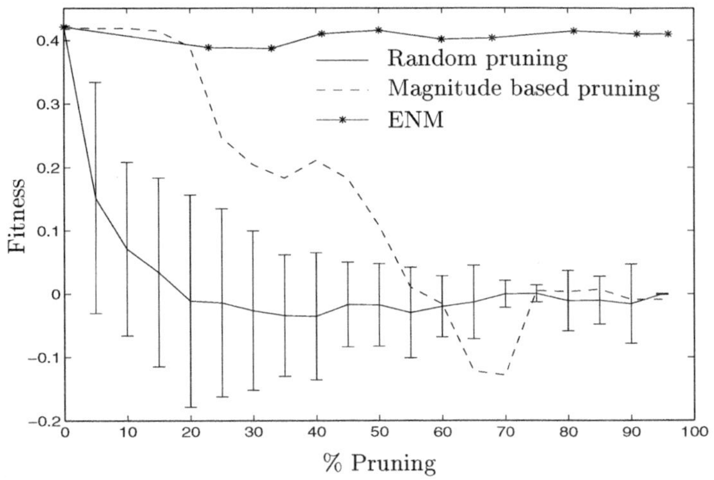

Fig. 2. Comparison of different pruning methods for minimizing S22. The agent's fitness is plotted against the percentage of pruned weights, using three different pruning methods: ENM, magnitude based pruning and pruning the weights in random order (mean and standard deviation across 25 runs). ENM converged to a final network architecture after 95% of the weights were pruned

presents the fitness values of the resulting minimized agents. In all three cases the fitness is very close to the fitness of the original intact agent, with small standard deviations, testifying to the consistency of ENM. Figure 3b presents the number of weights of the minimized agents. Evidently, the number of weights is very small compared with the original intact agents and is consistent across different ENM runs.

4.2 Preservation of Functional Characteristics

Several findings lead to the conclusion that **the ENM, although pruning most of the weights, preserves the essential functional characteristics of the network.** Observing the agents reveals similar patterns of behavior in the original and minimized agents. For example, Z10 has a unique characteristic, not found in other successful agents evolved in the same environment: When walking along a wall, the agent keeps a one cell distance from it. The ENM minimized agent of Z10 exhibits this exact same unique characteristic.

Repeating the ENM many times and considering the surviving weights in each, we note that about half the weights of a typical minimized network (9 out of 21) are weights that survive in all the runs (Fig. 4). This observation indicates that the ENM strives to preserve important weights: If the surviving weights were to be selected randomly then the number of surviving weights should have been strictly monotonously decreasing with the frequency of survival. Figure 4 shows that this is not the case.

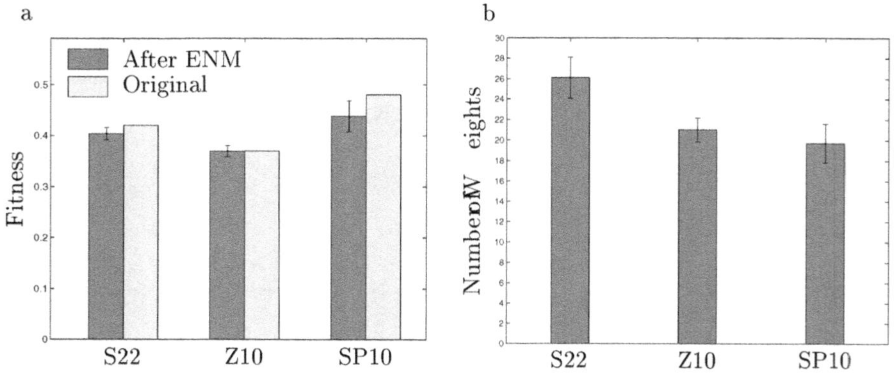

Fig. 3. Fitness and number of weights of ENM minimized agents. Mean and standard deviation across 10 ENM runs. (a) Fitness after minimization portrayed along with the fitness of the original agents. (b) Number of weights after minimization. The original S22, Z10 and SP10 agents have 594, 150 and 170 weights, respectively

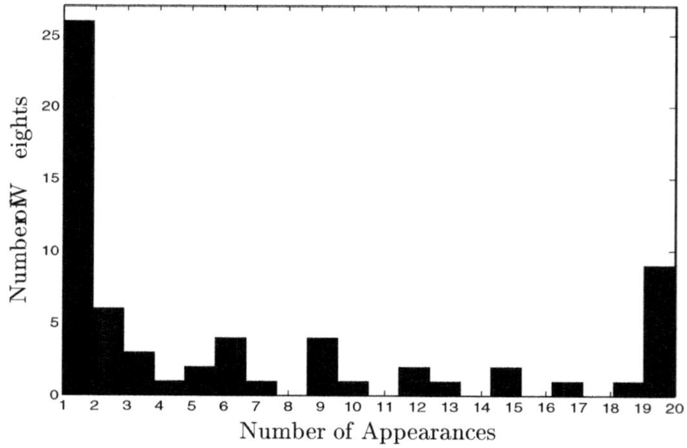

Fig. 4. Appearances of weights across different ENM runs of Z10. The histogram plots the number of weights (y-axes) surviving the pruning x times (x-axes) out of 20 ENM runs. For instance, 26 different weights survived only 1 run and 9 weights survived all 20 runs

We turn to further investigate the hypothesis that the more important weights of the input network stand a better chance of surviving minimization. To quantify the importance of each weight, we use a lesioning paradigm [4], afflicting numerous multiple lesions upon the weights, while recording the corresponding performance score of the agent after each such lesion. Based on these data, the causal importance (contribution) of each weight to the behavioral task is assessed by the Multi-lesion Shapley value Analysis (MSA) [5]. Formalizing the hypothesis in statistical terms, the question is whether the fraction of times the weights survive the minimization, out of the 20 runs (Fig. 4), is correlated with their contributions. We perform a non-parametric statistical test (Spearman rank correlation test) to find that this correlation is indeed significant (one sided p-value of 0.006). We further test whether the survival fraction is correlated with the magnitude of the weights and find this correlation to be non-significant (one sided p-value of 0.256). This analysis testifies that **the functional importance of a weight, rather than its magnitude, is significantly correlated with the probability of the weight to survive an ENM minimization. This ENM tendency to remove less important weights results in preservation of the agent's functional characteristics**. The latter is obviously a much desired property if one wants to use ENM to obtain insights regarding the workings of the original, non-pruned agent.

4.3 Analyzing the Minimized Neurocontroller

One of the major problems in the field of EAA is understanding the way neurocontrollers operate. An important application of pruning which preserves the agent's functional characteristics is that it helps in understanding a neurocontroller's operation by enabling direct inspection of the small core neurocontroller that it yields. The conclusions derived from such an analysis of the minimized neurocontroller may then be further tested on the functionally similar original agent. To demonstrate this, we applied ENM to the fully recurrent network of agent Z10, consisting of 150 weights. The original network is complex and is non-amenable for any visual inspection of its workings. The resulting pruned network consists of 20 weights (Fig. 5) and preserves the original high fitness of 0.36. **The minimized network is simple enough to be fully understood by observing its structure**. In the following we discuss the network structures underlying three behavioral features of the agent. The rest of the behavioral features may be explained in a similar, meticulous manner.

1. **When reaching a cell with food, the agent stops and eats it**. Standing on a food item, both the *here* and the *smell* neurons are active. Together, they inhibit the *move* motor, causing the agent to stop. Furthermore, the *smell* neuron activates the *eat* motor. The inhibition of the *move* motor is ceased after eating.
2. **When not in a food cell, the agent always moves**. Interneuron 1 is always active (a bias) due to a feedback from the *move* motor. When there is no movement (when the agent stands on a food item), interneuron 1 is active due to activation from the *eat* neuron. Whenever the agent is not in a food cell, interneuron 1 causes the firing of the *move* motor.
3. **Switching between exploration and grazing behavioral strategies**. When interneuron 3 is silent, interneuron 1 activates the *right* motor which then activates the *left* motor. Both stay active, initiating no turn, which constitutes the exploration (Sect. 2). When interneuron 3 is active, triggered by food consumption causing the silencing of the *move* motor, it inhibits the

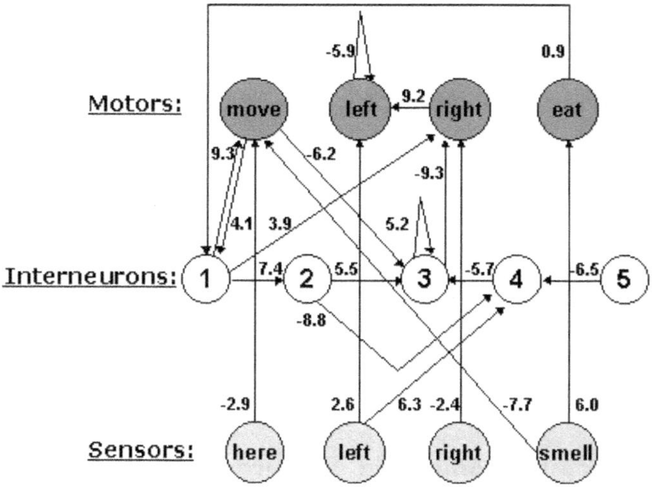

Fig. 5. Z10 pruned. The neurocontroller of a pruned descendant of Z10, consisting of 20 weights. Two neurons are not shown in the figure since all their weights were eliminated and they were pruned

right motor. This causes the agent to turn left whenever the *left* sensor is active, constituting the grazing (Sect. 2). Thus, interneuron 3 is a command neuron controlling the behavioral strategy. Switching from grazing back to exploration occurs when the agent stands in front of a wall, since the *left* sensor activates interneuron 4 which inhibits the command neuron.

4.4 Comparing Agents

Using such an analysis on several successful agents solving the same task, one may differentiate between *general principles* and *specific principles.* General principles are ones common to many successful agents. In this task, the existence of a command neuron (or several command neurons) is a general principle, as well as the loop causing the constant activation of the *move* motor. Another general principle is the mechanism which causes the agent to stop on food, based on the inhibition of the *move* motor by the *here* and the *smell* sensors. On the other hand, a specific principle is only exhibited by a specific agent. In this task, the exact mechanism for toggling the state of the command neuron is a specific principle, as well as the way the command neuron controls the behavior of the agent.

5 Summary

We demonstrated pruning and its applications in the context of Artificial Life using a new evolutionary network minimization algorithm, ENM. Given a neurocontroller, ENM consistently finds smaller neurocontrollers, while preserving its fitness and main functional characteristics. ENM is a very simple evolutionary process and hence can be easily implemented in existing systems. Such pruning allows for the analysis of an agent workings, as well as for finding general and specific principles underlying different agents solving a specific task. It also allows for the creation of more efficient implementations. We aim to further investigate the usage of pruning for estimation of task complexity. Our preliminary results indicate that the number of weights in the minimized networks may stand as an estimate for the minimum network size capable of successfully solving the task, which in turn may be used to approximate the task's complexity.

Acknowledgments. We acknowledge the valuable help provided by Ranit Aharonov and Isaac Meilijson and the technical help provided by Oran Singer. This research has been supported by the Adams Super Center for Brain Studies in Tel Aviv University and by the Israel Science Foundation founded by the Israel Academy of Sciences and Humanities.

References

1. Reed, R.: Pruning algorithms – a survey. IEEE Transactions on Neural Networks **4** (1993) 740–747
2. LeCun, Y., Denker, J., Solla, S., Howard, R.E., Jackel, L.D.: Optimal brain damage. In Touretzky, D.S., ed.: Advances in Neural Information Processing Systems. Volume 2., San Mateo, CA, Morgan Kauffman (1990) 598–605
3. Hassibi, B., Stork, D.G.: Second order derivatives for network pruning: Optimal brain surgeon. In Hanson, S.J., Cowan, J.D., Giles, C.L., eds.: Advances in Neural Information Processing Systems. Volume 5., San Mateo, CA, Morgan Kaufmann (1993) 164–171
4. Aharonov, R., Segev, L., Meilijson, I., Ruppin, E.: Localization of function via lesion analysis. Neural Computation **15** (2003) 885–913
5. Keinan, A., Hilgetag, C.C., Meilijson, I., Ruppin, E.: Fair attribution of functional contribution in artificial and biological networks. submitted (2003)
6. Zhang, B.T., Mühlenbein, H.: Genetic programming of minimal neural nets using Occam's razor. In Forrest, S., ed.: Proceedings of the 5th International Conference on Genetic Algorithms, ICGA-93, University of Illinois at Urbana-Champaign, Morgan Kaufmann (1993) 342–349
7. Kalganova, T., Miller, J.: Evolving more efficient digital circuits by allowing circuit layout evolution and multi-objective fitness. In Stoica, A., Lohn, J., Keymeulen, D., eds.: The First NASA/DoD Workshop on Evolvable Hardware, Pasadena, California, IEEE Computer Society (1999) 54–63
8. Whitley, D., Starkweather, T., Bogart, C.: Genetic algorithms and neural networks: Optimizing connections and connectivity. In: Parallel Computing. Volume 14. (1990) 347–361
9. Aharonov-Barki, R., Beker, T., Ruppin, E.: Emergence of memory-driven command neurons in evolved artificial agents. Neural Computation **13** (2001) 691–716

Population Dynamics under Spatially and Temporally Heterogenous Resource Limitations in Multi-Agent Networks

Thomas Glotzmann[1], Holger Lange[2], and Michael Hauhs[1]

[1] Bayreuther Institut für Terrestrische Ökosystemforschung (BITÖK), University of Bayreuth, Dr.-Hans-Frisch-Straße 1-3, 95448 Bayreuth, Germany
{thomas.glotzmann, michael.hauhs}@bitoek.uni-bayreuth.de
http://www.bitoek.uni-bayreuth.de

[2] Norwegian Forest Research Institute (Skogforsk), Høgskoleveien 12, N-1432 Ås, Norway
holger.lange@skogforsk.no
http://www.skogforsk.no

Abstract. An individual-based agent model is presented which resembles aspects of natural evolution in ecosystems under selective pressure due to limited resources. The environmental conditions are determined by spatial and temporal variability of resource abundances. The agents have to choose between three different types of resources; the one consumed most during lifetime solely counts for the fitness of the individual agent. Simulation runs show that populations specialized in different resource types are mutually influencing each other under temporal variation of a single resource type. Mobility of agents in a locally heterogenous world enables recolonization after a population has starved to death. Wavelet analysis of the population time series reveals that some observed population dynamics show phenomena such as localized periodicities which cannot be explained by linear dependencies on the resource input dynamics.

1 Introduction

The autonomy of agents is expressed in the ability of an agent to decide at every instance of time whether it attempts to reproduce or continues to acquire and compete for (maybe scarce) resources. Theoretical computer science introduces the notion of the behavior of (non-ending) processes. For such systems one can seek a set of primitive (but interactive) behaviors and compare these with the phenomenology of living systems.

We investigate a Java-based simulation tool called the POOL-WORLD system (in German: URSUPPE), distributed over several computers. The general setting of this multi-agent simulation is along the lines of classification given in [3]. There are passive entities (resources), which are just objects in the world, and active entities performing actions with behavioral rules. However, in the setup

W. Banzhaf et al. (Eds.): ECAL 2003, LNAI 2801, pp. 328–335, 2003.

given here, we do not provide our agents with memory, thus social interactions and group forming cannot occur.

Pool-World realizes a collection of habitats (so-called places) for agents [4]. In the places, there is a collection of different symbols (an alphabet). The agents are capable of reading these symbols with specific probabilities for access, and must do so to acquire the required resources. Within the setup introduced here, the quantity of the most frequently read symbol is rewarded. If the amount of every read symbol types falls below a certain threshold, the agent goes extinct.

The model structure is similar to Holland's Echo [7] as well as the so-called Eco-Grammar-Systems [2,9]. Contrary to those, however, our approach is completely asynchronous and places no restrictions on grammar or language structure. The elements (symbols) are considered just as resources necessary for maintenance of agents. The symbol distribution depends on time as well as spatial position.

Agents compete for consumption of symbols; the pools are replenished at a variable rate, eventually limiting the agent population. Under unfavourable circumstances (insufficient adaptation), an agent tends to migrate to another place. Successful (according to an adjustable fitness threshold, see below) agents are capable of asexual reproduction at the end of their lifetime.

An agent is equipped with a set of parameters constituting its genome [5]. One set of parameters reflects the sensitivity of the interfaces, i.e. defines the average probability with which each symbol is taken from the environment. An additional parameter regulates the migration tendency according to its past performance in the current environment.

A fitness value is assigned to each agent which is simply the amount of the resource acquired most often. If the agent's lifetime, which is equal for all agents, is exhausted, the agent reproduces, dies, and is removed from the population.

2 Example Applications

2.1 Model Description

We exemplify population dynamics with three possible types of specialists with alternating periods of coexistence and transition. Two settings with a single-place-environment and one with a triangular network of three places are shown. The resources, i.e. the model's alphabet, consist of three different symbols (tagged "R","G" and "B", resembling the colors red, green and blue). Each agent has to acquire as much as possible of one of these three resources, which are considered equally effective for the agent's metabolism. Therefore the agent is provided with three distinct input-interfaces, one for each symbol to be taken from the environment (the agent's current place). The symbols taken by the agent are removed, reducing the place's stock of resources. In each place the symbol stocks are refilled with a rate determined by a certain function of time for each symbol. Thus an agent "sees" a certain concentration for each resource in its place at a time. Limitation occurs when the number of agents increases.

There are weights assigned to the agent's input-interfaces specifying the probability with which each performs at a time. They are encoded in the genome and are subject to mutation. Assuming constant refill rates it seems obvious that agents which take a certain symbol exclusively, disregarding the other two, are best off. In such a scenario the ratios of the three specialists' populations should reflect the ratios of the resources' refill rates. On the other hand those extreme specialists are probably vulnerable under temporally varying refill rates.

The fitness value assigned to the agents is equal to the number of symbols ot the resource acquired most within their lifetime. The total sum of acquired symbols does not count at all. As a consequence an agent's fitness depends on the places it visits as well as on the population around. There is no other means for evaluating an agent's ability than testing it in a run under these circumstances. It has to be pointed out that, since the agents show stochastical behavior, the evaluation duration, i.e. their lifetime, is crucial. The longer the evaluation takes, the more likely random extremes in performance (random gamblers) do not occur. In the present example the lifetime of the agents is set to 1000 symbol-access operations.

At the end of an agent's life it may place children with mutated genomes if its fitness exceeds a certain value which is set to ten in the example. If not, the agent has no offspring. The reproduction rate is set to 1.01, which means that an agent has one offspring plus a second one with a probability of 1 %. The agent itself is removed in every case.

Additionally the agents have got a parameter determining the probability to move from one place to another called the emigration threshold. Whenever an agent does an operation, it evaluates the time elapsed since his last success in acquiring a symbol (idle time). With this information the probability for emigration to an adjacent place is calculated (see [5]).

At its destination the resource symbol concentrations may be more or less advantageous with respect to the agent's specialization. The emigration threshold parameter is also subject to mutation. The move requires energy, thus a fixed value is subtracted from the agent's lifetime (in this example: 10 each time). So an agent could move around up to 100 times and pick a symbol where it's most abundant without specializing the weights of its input-rules.

Three examples are given subsequently with three different types of resource refill dynamics: a complex, a sine and a noisy signal.

2.2 Results

A Complex Input Signal. Fig. 1 shows the evolutionary development of the agents' input interfaces along with a complex inflow dynamics of the "green" resource (called the "COMPLEX" run). In the figure resource inflow is shown as aggregates of 100 cycles. The inflow of "R" and "B" is set up to about 50 and 5 symbols, respectively, per 100 cycles.

It turns out that the rare "B"-resource doesn't play a significant role. Almost no specialists in the "blue" resource occur during the entire run. The abundance of the "green" resource is temporally clearly dominating but repeatedly drops to

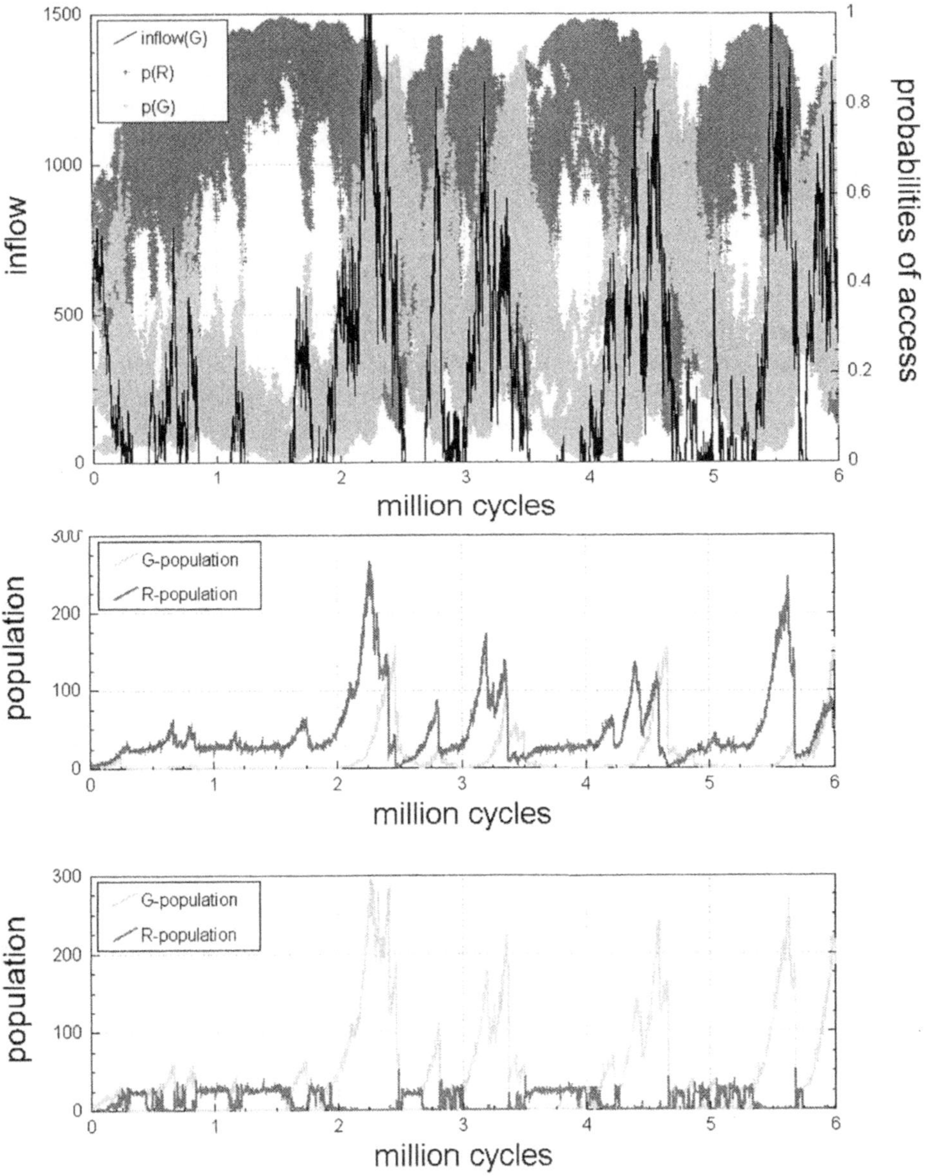

Fig. 1. Adaptation of the agent's input interfaces, resource inflow and population dynamics. Top: The "G"-inflow is a complex signal. The values give the number of resource input per 100 cycles. Specialists for "R" and "G" dominate alternately. Center: Populations of agents specialized in "R" and "G", respectively. Bottom: Populations of agents who actually acquired mostly "R" and "G", respectively.

zero as well. As figure 1 shows the population dynamics as well as the specialisation of the agents follow that dynamics with a short delay. The "R"-specialisation behaves reciprocal. Although the inflow of the "red" resource remains constant during the entire run, the "red" specialists show a significant sensitivity on the "G"-variation. This phenomenon can be explained by the occurence of the respective resource types in relative concentrations, i.e. the "visibility" from the agents' point of view varies. Thus, the single components of the ecosystem are tightly related comparable to communicating vessels.

The two population plots of figure 1 display the population dynamics of the specialists which have the highest weight on the respective interface ("R" or "G") (upper plot) and those who actually acquire the respective resource mostly. During times of rapid increase of the "green" resource even "red" specialists catch more "green" than "red" symbols because of the enormous disequilibrium of their abundances. These populations then gradually evolve towards "green" specialisation. The contrary behaviour is seen in times with a lack of "G".

A Triangular World. In contrast to the run introduced above the three resource types are assigned to three places, each of which provides a single resource type, the other two are set to zero. Top: $Place_1$ provides a "B"-inflow of about 50 units per 100 cycles. The refill rate of "G" in $Place_2$ is set to a sine function with a period of ten hours of simulation time (about 15 million cycles). Finally $Place_3$ provids a constantly high "R"-refill rate at about 500 units per 100 cycles. The emigration threshold of the agents is set to 1000 and the move costs are 10 life-cycles for each move.

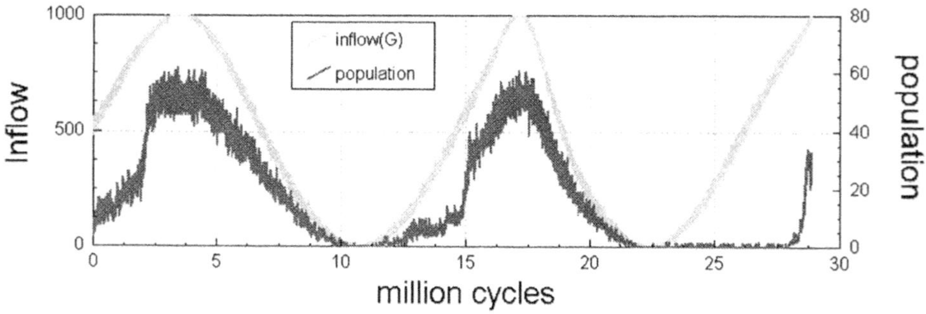

Fig. 2. Resource inflow and population dynamics of $Place_2$. The values give the number of resource inputs per 100 cycles. The "G"-inflow follows a sine curve between 0 and 1000 units. The agent population follows this trail and is repeatedly freshed up from the other places after deadly starvation.

$Place_1$ is able to support agent populations of between one and five individuals only, $Place_3$ between 20 and 30. As shown in figure 2 in $Place_2$ an exponential population growth is seen up to a maximum due to resource limitation indicated by the maxima of the sine curve of the "green" resource. There the population periodically starves out at the minima of this curve. Then a new seed is provided by immigrants from the other two places. Additionally the population growth is intensified by arriving immigrants which are able to live on the "G"-resource because of its (temporally) great abundance even if not yet adapted. The populations on $Place_1$ and $Place_2$ however are not significantly influenced by emigrants from $Place_2$, for they are permanently at their limits determined by the constant resource abundance.

Noisy inflow. The subsequent run labelled "RND" is set up analogous to "COMPLEX", but the inflow of the "green" resource follows a noisy signal varying from zero to 1000 per 100 cycles (fig. 3). Each of these random values is held constant for 10 seconds (approximately 1080 cycles). The plot, however, shows several peaks and dips in population seemingly following different periods, which cannot be explained by the inflow dynamics. Standard windowed Fourier analysis (not shown here) of the population time series reveals different periods present only locally.

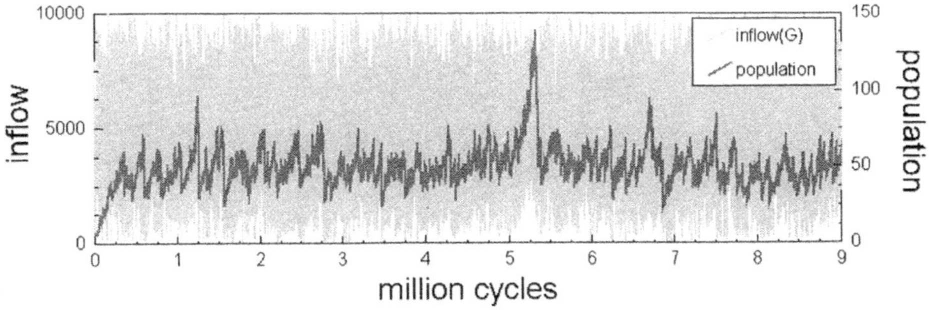

Fig. 3. Resource-inflow and population of run "RND"

Time series with transient long-term-structures cannot adequately be described by global power spectra. In such cases a *wavelet analysis* is recommended to decompose a time-series in the time-frequency-space, which is a common method in geophysical studies since a couple of years. A detailed description is given in [10].

The wavelet power-spectrum is usually visualized as a contour-plot, the coefficients of which are assigned to a colour-scale. In the region of long periods marginal phenomena caused by extrapolation of the original time series (extended periodically or with zero padding) are present. Therefore a so called *cone of influence* is drawn in the diagram marking the region free from those marginal effects. Finally a significance test for the consistency of the global power-spectrum with a chosen background spectrum is performed, revealing structures in the local wavelet spectrum which cannot completely be described by this background process.

Fig. 4 illustrates the wavelet spectrum for the population time series based on a *Morlet-wavelet* (see [10]). For the scaling period lenghts of 8 up to 4096 where chosen, subdivided into sections of integer powers of two, which again

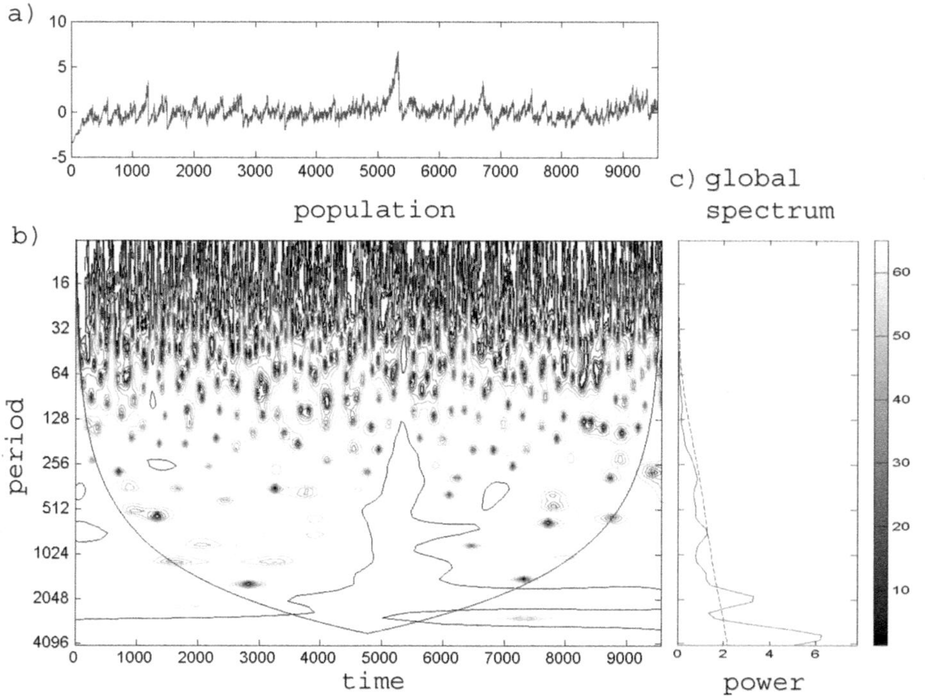

Fig. 4. a) Population time series of run "RND" (mean value subtracted). The simulation time is shown as 1000-aggregates (i.e. generations). b) Local wavelet spectrum with cone of influence. c) Global spectrum (integral of the local wavelet spectrum) (*solid line*), 5 %-significance level in comparison to an AR(1)-process (red noise) (*dotted line*)

are subdivided in eight equidistant intervals each. For the background spectrum an AR(1)-process (red noise, see [6]) with a variance and a lag-1 autocorrelation equal to that of the original time series was chosen. The regions lying beyond (to the right of) the 5 %-significance line (dotted line in fig. 4c) deviate significantly from red noise. Significant local structures are surrounded by an iso-line.

Primarily there is a significant wedge at about 5200 (5.2 million cycles) destroying the long-term structure. It resembles the major peak in the population series. Especially the region of periods greater than 128 is scarcely dotted by large islands, revealing transient long-term structures, which can neither be found in the original time series nor the global power spectrum. Among others there are ones at a period of 512 at the range of 1000–1800 (1–1.8 million cycles) containing the population peak at 1200 and the period of ~1600 dominating at a range from 2500–3500 as well as from 7000–7500.

In a couple of runs with the same parameter settings it turned out that each run seems to develop its own history based on the interaction of agents and environment together with the stochastical process of resource inflow on one hand and evolution on the other hand.

3 Conclusions

We could demonstrate that a model of relatively low complexity is able to exhibit behaviors analogous to those found in natural evolutionary systems. The origin of distinct behavioral categories lies in the stochasticity of resource consumption on one hand, and in the simple spatiotemporal structuring of the environment on the other. The model is asynchronous, rendering individual agents autonomous in time and behavior. Migration as well as evolution seem to bring structure-building power into systems with input of arbitrary complexity. As seen in run "COMPLEX" a highly complex signal puts its stamp on the whole system's history causing secondary effects as observed in the dynamics of the "R"-specialists' population. To a noisy input (low complexity) however the system answers with a more complex structure which cannot easily be explained by linear dependencies and global periodicity. The cause for this phenomenon we see in the power of interaction which occurs mainly at two levels: First at the level of individuals interacting with the environment and with one another. Secondly evolution itself is an interactive process on the level of whole populations, which can be seen as kind of intelligence.

References

1. Bedau, M. A., Snyder, E., Packard, N.H.: A Classification of Long-Term Evolutionary Dynamics. In: Adami, C., Belew, R.K., Kitano, H. and Taylor, C.E. (eds.): Artificial Life VI. MIT Press, Cambridge (1998)
2. Csuhaj-Varjú, E., Kelemen, J., Kelemenová, A., Păun, G.: Eco-Grammar Systems: A Grammatical Framework for Studying Lifelike Interactions. Artificial Life 3 (1997) 1–28
3. Ferber, J.: Multi-agent systems. An Introduction to Distributed Artificial Intelligence. Addison-Wesley (1999)
4. Glotzmann, T., Lange, H., Hauhs, M., Lamm, A.: Evolving Multi-Agent Networks in Structured Environments. In: Kelemen, J. und Sosik, P. (eds.): Advances in Artificial Life. Springer-Verlag, Berlin Heidelberg New York (2001) 110–119
5. Glotzmann, T.: Ein agentenbasiertes Simulationssystem zur Entwicklung ökosystemarer Szenarien in strukturierten Umgebungen. To appear in: Bayreuther Forum Ökologie, Bayreuth (2003)
6. Hipel, K., McLeod, A.: Time series modelling of water resources and environmental systems. Elsevier, Amsterdam London New York Tokyo (1994)
7. Hraber, P.T., Jones, T., Forrest S.: The Ecology of Echo. Artificial Life 3 (1997) 165–190
8. Schmid, T.: Evolutionäre Anpassungen in einem verteilten Multi-Agenten-System. Diplom Thesis, Univ. Bayreuth (2001)
9. Suzuki, Y., Tsumoto, S., Tanaka, H.: Analysis of Cycles in Symbolic Chemical System based on Abstract Rewriting System on Multisets. Artificial Life V. MIT press (1997) 522–528
10. Torrence C., Compo G.P.: A Practical Guide to Wavelet Analysis. Bulletin of the American Meteorological Society (1998) 61–78

Adaptive Coupling and Intersubjectivity in Simulated Turn-Taking Behaviour

Hiroyuki Iizuka and Takashi Ikegami

Department of General Systems Sciences,
The Graduate School of Arts and Sciences, University of Tokyo,
3-8-1 Komaba, Tokyo 153-8902, Japan
{ezca, ikeg}@sacral.c.u-tokyo.ac.jp

Abstract. Turn-taking behaviour is simulated with a coupled agents system. Each agent is modelled as a mobile robot with two wheels. A recurrent neural network is used to produce the motor outputs and to hold the internal dynamics. Agents are developed to take turns on a two dimensional arena by causing the network structures to evolve.
Turn-taking is established using either regular or chaotic behaviour of the agents. It is found that chaotic turn-takers are more sensitive to the adaptive inputs from the other agent. On the other hand, regular turn-takers are comparatively insensitive to noisy inputs due to their restricted dynamics. From various observations, including turn-taking with virtual agents, we claim that the chaotic turn-taking agents become less robust when coping with virtual agents but at the same time, those agents are more adaptable to each other than the regular turn-taking agents. All these findings are discussed and compared with Trevarthen's double monitor experiments.

1 Introduction

Intersubjectivity is a term used in psychology and philosophy to describe the sharing of mental states and intentions with others. Trevarthen was the first person to notice its importance [11]. This intersubjectivity is strongly connected to social behaviour between two or more entities. Interacting socially with others requires more than mere interaction and synchronization of actions but coordinated behaviour of entities that have rich dynamics. There are many ways to understand psychological phenomena by computer simulations and robot experiments rather than by studying human behaviour directly [2,10].

By conceiving couplings between agents with rich internal dynamics, we should develop new ways of understanding their dynamics [4,5,6]. We generalize from turn-taking behaviour to autonomous role-changing, such as games of tag among children, and investigate the generic underlying mechanisms using the dynamical systems method. Therefore this study focuses on different perspectives to the fixed role-playing games (e.g. a pursuit-evasion game [1]). Here we take turn-taking as the simplest example that shows diversity of dynamics. It is necessary for turn-taking behaviour to autonomously exchange roles along

W. Banzhaf et al. (Eds.): ECAL 2003, LNAI 2801, pp. 336–345, 2003.

with the context constructed by the entities' behaviours, e.g., chaser-evader and speaker-listener. When taking turns in a two-person conversation people usually avoid overlapping or interrupting each other's speech without setting any explicit cue to switch speakers. Some cues for this include eye contact and the detection of intonation changes. It is considered that turn-taking is established by coordination between predictions and the internal neural net that computes the output from inputs. Thus, the coupling between agents means a coupling of anticipatory systems with intrinsic dynamics.

By introducing the agent architecture, evolutionary algorithm and the turn-taking environments, we explore three topics in our simulations. They are dynamics repertoire, noise-driven turn-taking behaviour and turn-taking with virtual agents. The present paper is a continuation of the work of the previous study [8]. The basic model set-up is the same, but here we have a greater variety of turn-taking behaviours, which enables us to perform experiments described in this paper.

2 The Model

We modelled a playing tag game in which the role of chaser, or evader, is not given to players in advance. There are some game models in which the roles are not predefined [9,3]. Reynolds showed that the abilities of chasing and evading also evolve simultaneously by genetic programming in a game of tag, which is a symmetrical pursuit-evasion game. The variety of the behaviour of agents adapting to their environments is worth noting. In Reynolds' game, switching between evader and chaser is predefined to happen when both agents come into physical contact. The difference between Reynolds' model and ours is the spontaneous emergence of behaviour. Whether an agent plays the role of a chaser or an evader will be dynamically determined in our model. On the other hand, Di Paolo modelled and studied social coordination with agents that interact acoustically. To avoid misperceiving the acoustical signals, their emission timings were entrained in an anti-phase state; the resulting behaviour resembles a turn-taking process.

There is a difference between Di Paolo's turn-taking and ours. Both turn-taking behaviours are established by the coordination of agents through the history of their interactions. Di Paolo modelled turn-taking as a result of anti-phase signals to avoid signal interference; however, we modelled turn-taking behaviour as a result of coupling between richer internal dynamics. Therefore, in this paper we pay more attention to the diversity of behaviour patterns.

2.1 Game and Environment

Here each agent has a circular body of radius R, with two diametrically opposed motors (Fig. 1). The motors can move the agent backwards and forward in a two-dimensional unstructured and unlimited arena. The motion is described by the following equation of motion of an agent's heading angle (θ) and the velocity (v) in that direction.

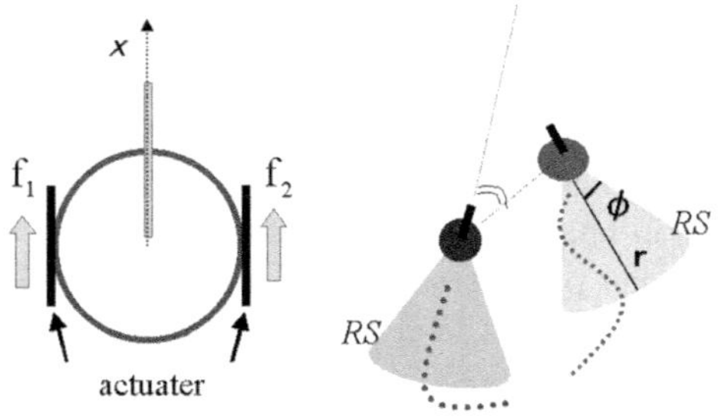

Fig. 1. Left: a schematic view of the mobile robot with two wheels (actuators). It computes the forward force vector and the torque strength from the force vector (f_1, f_2) on each actuator. Right: Two mobile robots interact to perform turn-taking behaviour by sensing each other's position, relative distance and heading angle. It is robot A's turn when A enters B's rear side (RS) position. The shape of this RS is parameterized by r and ϕ.

$$M\dot{v} + D_1 v + f_1 + f_2 = 0, \quad (1)$$
$$I\ddot{\theta} + D_2\dot{\theta} + \tau(f_1, f_2) = 0, \quad (2)$$

where f_1 and f_2 are the forward driving force, and τ denotes the torque. Each agent has a heading angle, which is denoted by θ. D_1 and D_2 express the resistance coefficients, and the agents have mass (M) and inertia (I). We solve the equations iteratively using the Runge-Kutta method. At each time step, agents compute the forces from the inputs using the internal neural nets described below.

We assume there is no collision between agents because we focus on the internal states of the agents that generate turn-taking. Two agents try to coordinate the turn-taking behaviour; each trying to get behind the other. Because they cannot get behind each other simultaneously the turn-taking cannot be achieved if both agents play chaser. Naturally, if both agents play evader, mutual turn-taking cannot be achieved, either. Therefore, it is necessary to have spontaneous symmetry break-down so that one plays the role of chaser and the other plays the role of evader. However, mere symmetry-breaking is not sufficient; temporal role-changing is also required. By using recurrent neural networks, we focus on how the turn-taking dynamics are self-organized.

2.2 Agents

We designed the agents to have recurrent neural networks (Fig. 2). Inputs to an agent are the other agent's position, distance and heading angle, relative to the agent. They move freely in the arena using two motors, the outputs of which are computed at every game time-step. The agent predicts the other's next relative position, assigned three output neurons. The dynamics of the recurrent neural network are expressed by the following equations at each time-step t,

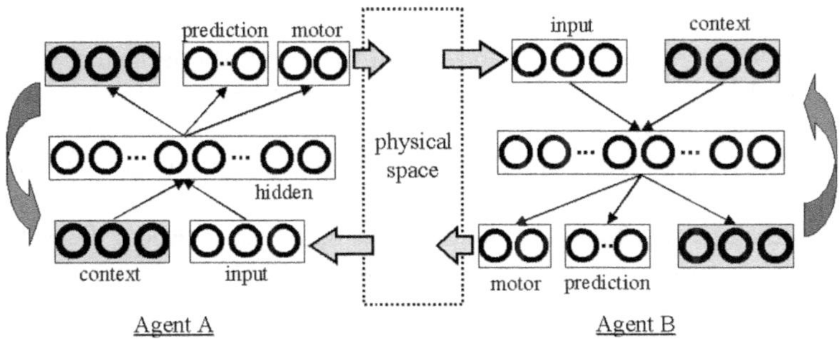

Fig. 2. Recurrent neural networks with three layers. Input nodes receive the other agent's relative position. The last layer consists of three types of nodes: context, prediction and motor output. Context nodes feed back to the input layer. Prediction nodes output the other's relative position in the next time-step. Motor nodes output the force vector, f_1 and f_2.

$$h_j(t) = g(\sum_i w_{ij} y_i(t) + \sum_l w'_{lj} c_l(t-1) + b_{j1}), \tag{3}$$

$$z_k(t) = g(\sum_j u_{jk} h_j(t) + b_{j2}), \tag{4}$$

$$c_l(t) = g(\sum_l u'_{jl} h_j(t) + b_{j3}), \tag{5}$$

$$g(x) = 1/(1+\exp^{-x}), \tag{6}$$

where y_i, z_k, h_j and c_l represent input, output, hidden and context nodes, respectively. The respective number of nodes in these layers is set to $(I, K, J, L) = (3, 5, 10, 3)$ throughout this paper. The symbols w_{ij}, u_{jk}, w'_{lj} and u'_{jl} denote the weights from input to hidden, hidden to output, context to hidden, and hidden to context neurons, respectively, while the parameter b gives a bias node. In this paper, we do not treat the results of predictions, which are discussed in [7]. This network architecture evolves using a genetic algorithm, which is explained in the following section.

3 Evolutionary Design of Neural Architecture

We update the weights according to turn-taking performance. In practice, the weight set of the neural networks has a vector representation of the real weight values, which evolve using a genetic algorithm (GA).

We use a GA to evolve two separate populations, to avoid agents of a single genotype from dominating, in which case turn-taking is played among genetically close agents. As a result, a player has to play against itself, which we want to avoid. Each population contains P individuals. The performance of all P^2 paired agents from the separated populations are evaluated at each generation. Agents that can equally exchange turns are evaluated to have greater fitness. At first, individuals in each population are initialized with random weight values. Then we calculate the fitness of each individual, based on its performance.

The highest value is given when both agents take their turn alternately and the agents can predict each other's behaviour. A one-sided (i.e. role-fixed) behaviour is associated with the lower fitness values. Practically, the fitness of an agent a from a population (A) against an agent b from the other population (B) is calculated as follows. Below, we define a total fitness F as the sum of two fitnesses associated with prediction and turn-taking, respectively. When an agent gets behind, the other agent has, by definition, its turn and the rear scope is specified as RS, which is parameterized by two parameters r and ϕ (see Fig. 1). The agent in this scope is said to be having its turn and is being rewarded. A spatial position of the b-th agent at time-step t is represented by $Pos_b(t)$. This is compared with the a-th agent's prediction value $Pos_{a\to b}$. Therefore the squared difference (Eq.(11)) evaluates the precision of the a-th agent's prediction.

$$F_a = s_1 \times F_a^{turn} + s_2 \times F_a^{predict}, \tag{7}$$

$$F_a^{turn} = \frac{1}{P}\sum^{P}\left(\sum_t^{T} g_a(t) \times \sum_t^{T} g_b(t)\right), \tag{8}$$

$$g_a(t) = \begin{Bmatrix} 1 \; Pos_a(t) \in RS_b(t) \\ 0 \; Pos_a(t) \notin RS_b(t) \end{Bmatrix}, \tag{9}$$

$$F_a^{predict} = -\frac{1}{P}\sum^{P}\left(\sum_t^{T} P_a(t) \times \sum_t^{T} P_b(t)\right), \tag{10}$$

$$P_a(t) = (Pos_b(t) - Pos_{a\to b}(t))^2, \tag{11}$$

The performance of turn-taking is evaluated for different lengths of time ($T = 500, 1000$ and 1500), so that agents cannot tell when the evaluation time is over. Evaluating the turn-taking performance at each GA generation, we leave the best E individuals in each population and let them reproduce with certain mutation rates. The GA proceeds by repeating this procedure and the recurrent neural networks evolve.

Noise: Note that sensory noises are added to the input neurons during each run. Therefore agents have to take turns under a noisy environment.

4 Simulation Results

4.1 Dynamics Repertoire

Simulation was done with a GA using 15 individuals ($P = 15, E = 4$). Figure 3 shows examples of the spatial trails of an agent from different GA generations with different initial population structures.

We can approximately classify these trails into two patterns based on their appearance. When spatial trails consist of regular curves and the turns are exchanged almost periodically (which corresponds to an abrupt turning point),

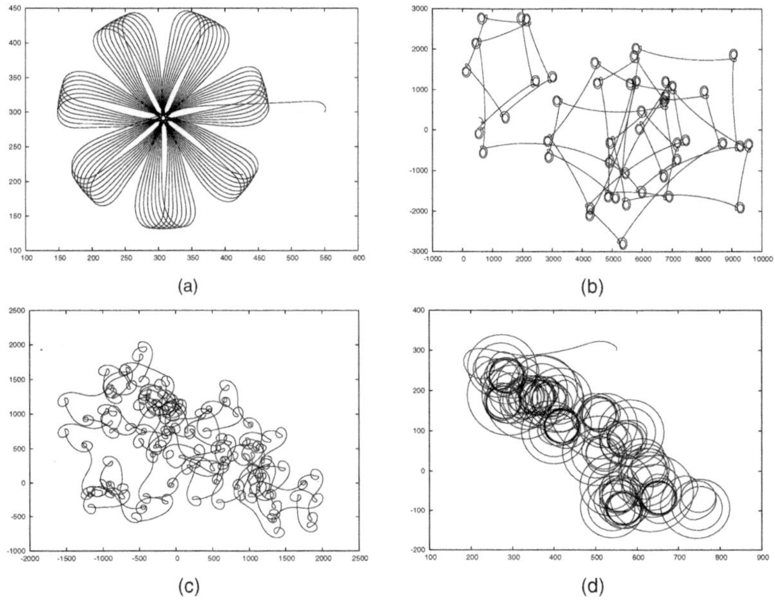

Fig. 3. Spatial trails of turn-taking behaviour observed in the simulations. In order to clarify the qualitative difference of turn-taking structures, a spatial trail of only one of two agents is shown. The other agent moves around these trails generating almost similar trails. All games in these graphs are started from (550, 300). (a) is a example of geometric turn-taking. (b),(c) and (d) are examples of chaotic turn-taking behaviour.

we call them *geometrical* turn-taking. On the other hand, if spatial trails have irregular curves with non-periodic turn-taking, we call them *chaotic* turn-taking.

In the earlier GA generations, the agents with geometric turn-taking have a higher performance (Fig. 3(a)). The behaviour structure is as follows: one agent follows the other and passes it; then it slows as does the other agent; then both agents simultaneously turn around quickly. This returns the agents to the original pattern. A series of behaviour patterns repeats almost periodically, and in this way the context nodes are periodically activated.

In the later GA generations, more chaotic patterns emerge (Fig. 3 (b), (c) and (d)). In contrast to the geometrical patterns, the turns are exchanged in different places with irregular time intervals. Therefore, the spatio-temporal pattern becomes chaotic. The corresponding context space plots show some tangled continuous-line formations.

The evolution of geometrical turn-takings to chaotic turn-taking is explained as follows: The evolutionary pressure of GA at first makes agents behave in a stable way in the noisy environment. We believe that robustness against noise prefers geometrical turn-taking.

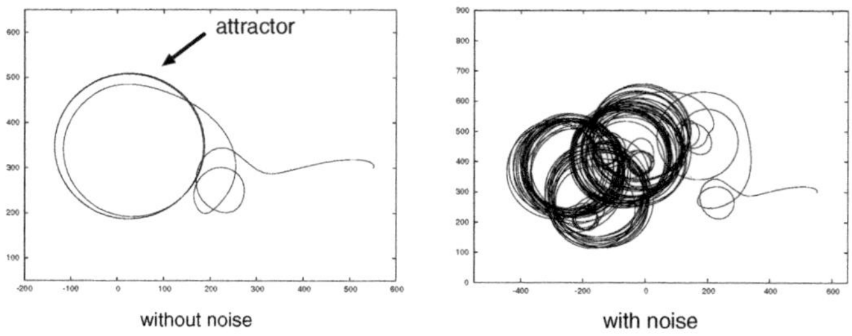

Fig. 4. Noise-driven turn-taking behaviour. There is an attractor of role-fixed behaviour. By adding noise to the agents, an agent can slip out of the attractor and successfully perform turn-taking.

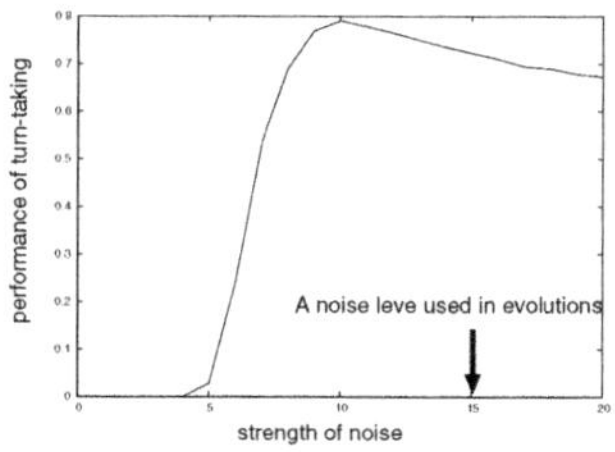

Fig. 5. The performance of turn-taking behaviour as a function of noise strength. Under the lower level of noise, agents cannot perform turn-taking. Beyond a certain noise level, agents take advantage of noise to perform turn-taking. This critical noise level is lower than the one used in evolution.

4.2 Noise-Driven Turn-Taking

Some turn-taking behaviours are established by taking advantage of the noise (Fig. 4). As shown in the figure, there is a strong attractor of a circular pattern of role-fixed chasing and evading without exchanging turns. Turn-taking emerges beyond a certain noise level (Fig. 5). Below that level, the attractor of the fixed role is too stable to escape. In another case, there are three attractors without noise. One is that agent A chases the rear side of agent B closely. Another one is the opposite, and the last one is that in which both agents chase each other. Every three attractors consist of circular orbits. The transition between attractors is caused by noise. Without noise, agents are trapped by one of the attractors.

Compared with these noise-driven behaviours, chaotic turn-takers can establish turn-taking behaviour without noise. Even if noise is introduced into the system, chaotic turn-takers can establish turn-taking behaviours independent of small noises. Namely, they do not utilize noise but suppress the effect of noise to perform turn-taking. On the other hand, noise-driven turn-takers need noise to perfrom turn-taking.

In the next section, we discuss this dynamic adaptability.

4.3 Turn-Taking with Virtual Agents

A difference between mere oscillator entrainments and dynamic coupling is found in the 'adaptability' of the coupling between the agents that is established by evolution. In order to clarify the nature of the dynamic interactions more concretely, we compare the behaviour of "live interaction" with "recoded interaction". The "Live interaction" is normal interaction between evolved agents, and the "recorded interaction" is that between an agent and a virtual agent, defined below.

First, we select the two best agents, A and B, from each population. Turn-taking between these agents is studied without introducing noise. This is what we term 'live interaction'. The trails of the agents are recorded during the run. Then, turn-taking between agent A and the recorded trail of agent B (i.e. a virtual agent) was conducted. This is what we term a 'recorded interaction'. We perturb the recorded trail and simulate the changes in the turn-taking dynamics.

Figure 6 shows the growth of a discrepancy between A-virtual B and A-perturbed virtual B (chaotic turn-takers). During the initial few hundred steps, no discrepancy is observed. The behaviours are similar as is shown in the figure. However, a small noise is amplified and the orbit drastically changes from the original orbit around 800 time-steps. In terms of the turn-taking behaviours, the adaptive agent cannot recover harmonization with the perturbed virtual agent any longer. The agent approaches the trail and tries to dynamically resume the original turn-taking behaviour.

Another example (the agents at 3,000 generations) is shown in Fig. 6 (b). These agents establish geometric turn-taking. In this case, agents can adequately cope with the perturbed virtual agent. Note that agents constructing geometric turn-taking behaviour do not always, but frequently do, have a tendency to cope with a perturbed virtual agent. It depends on the timing and strength of the perturbation. Sometimes turn-taking behaviour breaks down when additional noise is added to the recorded trail. However, there are some examples in which turn-taking recovers after a certain period of discrepancy.

5 Discussion

It is found in the experiments of turn-taking against virtual agents, that chaotic turn-takers are much more sensitive to the difference between the live and recorded inputs. In other words, turn-taking is driven by the ongoing interaction. On the other hand, geometric turn-takers are robust against the difference due to the restricted number of dynamics (may be only one or two). Therefore, turn-taking is driven by the stiffness of the individual dynamics. This is also confirmed by the experiments with different noise structures (Fig. 7) and also by the fact that chaotic turn-taking takes over from the geometrical turn-taking in the evolutionary context of our GA simulations. We speculate that geometric turn-takers can take turns only with fewer agents than the chaotic turn-takers. In summary, we claim that the chaotic turn-taking is less robust against noise but has more adaptability, compared with the geometric turn-taking.

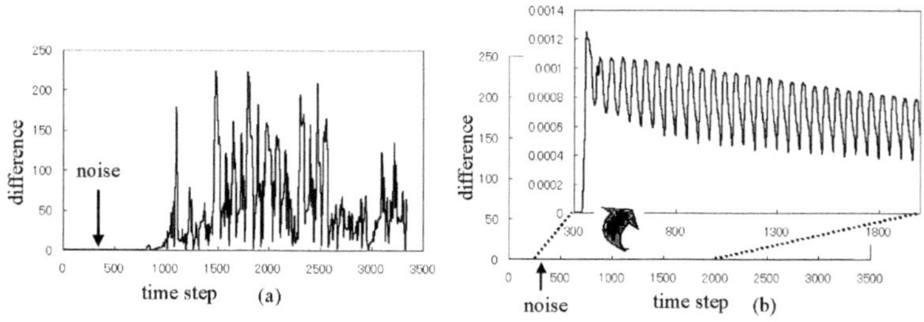

Fig. 6. Differences of orbits between agent's trails in a game with an adaptive agent and with the recorded trail. A small noise is introduced at 340 time-steps. If there is no noise, no difference is observed. Agents used in (a) and (b) correspond to those in Fig. 3 (c) and (a), respectively. The difference is amplified if agents fail to establish turn-taking.

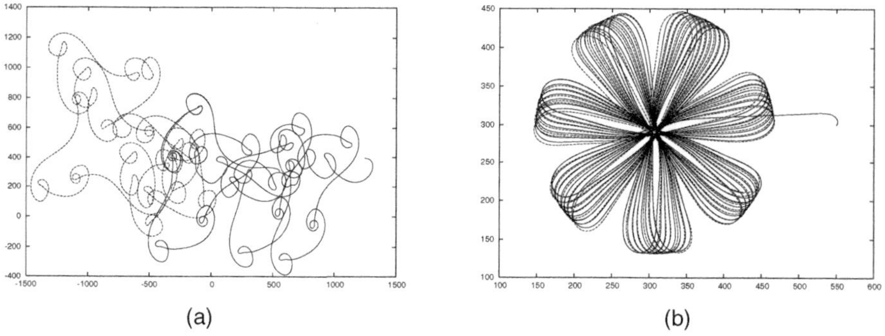

Fig. 7. Differences of spatial trails between adaptive agents without noise (solid) and with noise (dotted) are plotted. They start from the same initial points, (550, 300). (a) chaotic turn-taking (b) geometric turn-taking

We compare this simulation result with Trevarthen's double-monitor experiments between a baby-infant and its mother [12]. Mother and baby-infant only communicate through videos that display their faces to each other. It is reported that for the baby-infant to engage with the mother, the correct style and timing are required. If the recorded video of the mother is displayed to the baby-infant, the baby-infant becomes withdrawn and depressed. This is also true for the mother when she watches the recorded video of the baby-infant.

Trevarthen's experiments show that it is not necessarily important for the baby-infant that the mother is displayed on the monitor. It can be assumed that the most important clue under interactions is the ongoing anticipation of a partner. The baby-infant performs some actions and anticipates the mother's reactions reflecting the baby-infant's actions, and this is also true with respect to the mother's anticipation of the baby-infant. Interactions in social behaviour, including turn-taking, can be established when these anticipations are mutually formed dynamically. In our simulations, when an agent calculates outputs, this calculation simultaneously affects the internal dynamics. That is, the actions performed form its internal dynamics much as actions form anticipations in the statement above. The agent receives inputs as a partner's actions reflecting the agent's own actions. We maintain that turn-taking is established when these structures are mutually organized. Turn-taking is thus broken in the simulation with virtual agents. We therefore claim that this mutual adaptive coupling of actions and internal dynamics between agents is related to intersubjectivity.

Acknowledgments. This work is partially supported by Grant-in aid (No. 09640454 and No. 13-10950) from the Ministry of Education, Science, Sports and Culture.

References

1. Cliff, D., Miller, G.F.: Co-evolution of Pursuit and Evasion II: Simulation Methods and Results. From Animals to Animats 4:Proceedings of the Fourth International Conference on Simulation of Adaptive Behavior (SAB96), MIT Press Bradford Books (1996) 506–515
2. Dautenhahn, K.: Embodiment and Interaction in Socially Intelligent Life-Like Agents. Springer Lecture Notes in Artificial Intelligence Volume 1562 Springer (1999) 102–142
3. Di Paolo, E.A.: Behavioral coordination, structural congruence and entrainment in a simulation of acoustically coupled agents. Adaptive Behavior 8:1 (2000) 25–46
4. Ikegami, T., Taiji, M.: Structures of Possible Worlds in a Game of Players with Internal Models. Acta Polytechnica Scandinavica Ma. 91 (1998) 283–292
5. Ikegami, T., Taiji, M.: Imitation and Cooperation in Coupled Dynamical Recognizers. Advances in Artificial Life, Springer-Verlag (1999) 545–554
6. Ikegami, T., Morimoto, G.: Chaotic Itinerancy in Coupled Dynamical Recognizers. (submitted to Journal of American Institute of Physics)
7. Ikegami, T., Iizuka, H.: Joint Attention and Dynamics Repertoire in Coupled Dynamical Recognizers. The Proceedings of Imitation in Animals and Artifacts II (2003)
8. Iizuka, H., Ikegami, T.: Simulating Turn-taking Behaviours with Coupled Dynamical Recognizers. The Proceedings of Artificial Life 8, Standish et al.(eds), MIT Press (2002) 319–328
9. Reynolds, C.W.: Competition, Co-evolution and the Game of Tag. Artificial Life IV, Brooks & Maes (eds), MIT Press (1995) 59–69
10. Scassellati, B.: Imitation and Mechanisms of Joint Attention: A Developmental Structure for Building Social Skills on a Humanoid Robot. Computation for Metaphors, Analogy and Agents, Vol. 1562 of Springer Lecture Notes in Artificial Intelligence, Springer-Verlag (1999)
11. Trevarthen, C.: Descriptive Analyses of Infant Communicative Behaviour. In: Studies in Mother-Infant Interaction, H.R. Schaffer (ed.), London: Academic Press (1977)
12. Trevarthen, C.: The Self Born in Intersubjectivity: The Psychology of an Infant Communicating. The Perceived Self, U. Neisser(ed.), Cambridge University Press (1993) 121–173

Distributed Genetic Algorithm: Learning by Direct Exchange of Chromosomes

Aleš Kubík

Institute of Computer Science, Silesian University,
Bezručovo nám. 13, 746 01 Opava, Czech Republic
ales.kubik@fpf.slu.cz

Abstract. Genetic algorithms is a technique widely used to evolve controllers of agents or robots in dynamic environments. In this paper we describe a modification to a single-robot-based evolution of a controller - a distributed parallel genetic algorithm where the pool of chromosomes is dispersed over a multi-robot society. Robots share their experience in solving the task by direct exchange of individually evolved successful strategies coded by chromosomes.

1 Introduction

Evolutionary and genetic algorithms are suitable tool to evolve controllers from initially pseudo-randomly designed population of controllers evaluated by fitness function [4], [3].

Either the robot on its own runs a genetic algorithm to produce a controller with desired behavior (see e.g. [2])[1], or the evolution shapes individual robot behavior but that is influenced by couplings with other robots in a multiagent scenario (described in [9], [10], [8]).

In this paper we describe the experiments with the evolution of controllers using chromosomes dispersed over the whole society of robots. Robots create random sets of chromosomes and exchange successful population members among each other. In this case we can go along without a central process but the robots have to communicate with each other and when more robots are present even coordinate the exchange of chromosomes. As a testbed of our ideas we have chosen collision avoidance learning representing almost benchmark behavior in the experimental robotics.

In this place it should be cited the work of [11] where the authors implement a distributed genetic algorithm in a collective robotics scenario. As for the design and methodology our approach is nevertheless closer to standard genetic algorithm solution with combined individual and parallel evolution, selection, reproduction, fitness evaluation of genome etc.

[1] A good overview of evolutionary robotics focusing on single robot scenarios represents the work of [7].

W. Banzhaf et al. (Eds.): ECAL 2003, LNAI 2801, pp. 346–356, 2003.

2 Experimental Robots

In our experiments we used two Khepera robots ([6]) that are well suited for educational as well as experimental purposes (see [5]). The robot body has about 5cm in diameter and is equipped with two servomotors driving two wheels and eight infrared sensors (six in the front and two at the rear side of the robot - see figure 1) that serve to avoid obstacles.

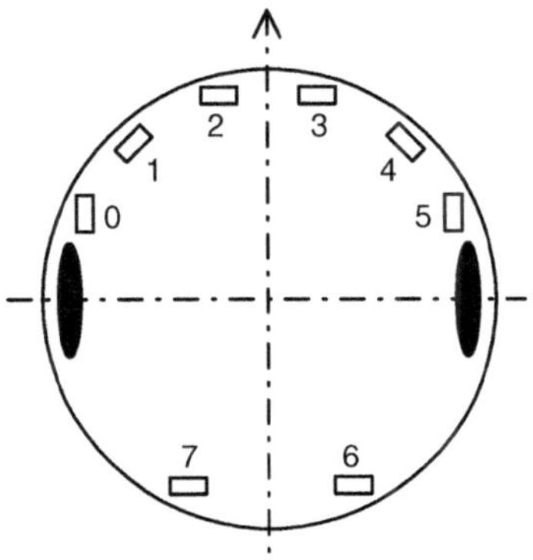

Fig. 1. Schema of a Khepera robot with two wheels and eight IR proximity sensors. The arrow indicates the front movement of a robot.

Programs written in C language run on the on-board processors of the robots. Robots thus conserve autonomy in relation to other robots and human intervention. It is possible that the program runs on the host computer and with the robot communicates through the serial communication line. Such type of a communication was used only for the purpose of collecting data from the experiments. Communication routines as well as program execution manages Motorola 68331 controller with 256 KB RAM and 512 KB ROM.

We used an additional component - a radio extension turret that is mounted on top of a robot in a plug and play fashion. It enables a peer-to-peer communication between robots. Robots communicate via sending and reading buffers each of which holds 16 bytes of data. Messages are sent directly to a robot with a specified identification number. There is a built-in mechanism of checking the correctness of messages. The sender considers a message to be sent correctly only if it receives the acknowledgement from the receiver. It repeats the sending process after timeout when it receives no acknowledgement. The message is considered lost after 10 unsuccessful trials.

3 The Learning Task

Robots learn to avoid dynamic and static obstacles. The behavior is coded by a very simple neural network resembling the perceptron. It controls the movements of a robot in a straightforward fashion. If the sensors of either side (left or right) of a robot perceive any kind of an obstacle, the motor on this side will speed up the wheel it drives. The algorithm is inspired by the work done in [1]. Genetic algorithm is used to evolve weight vector of input-output neuron connections. These are coded as chromosomes. Each robot evolves its own population of controllers. One of them sends (this will be called sender) the successful candidates to the second robot (denoted as receiver). The receiver uses both its own population of chromosomes as well as received chromosomes from the sender. In this way the receiver learns also from the experience of a sender. We expect that the receiver will learn better and faster than the sender.

3.1 Basic Processes

In this section we describe the algorithms of employed processes of both robots.

The program body of a sender is composed of two main processes that are running interchangeably. One is responsible for running evolution of weight vectors and testing the resulting neural controllers. The second process is sending 1/4 of most successful population chromosomes to the receiver after each generation finishes its learning. Pseudocode of the sender's processes can be expressed as follows:

process 1:

```
1. create weight vectors with rand. values;
2. for each generation do {
3.  for each weight vector do {
     run the neural controller for 800 cycles;
     compute fitness value of the controller;
     change speed randomly for 100 cycles;
    }
4.  sort the population according to fitness;
5.  transform the chromosomes into an array
      of buffers of 16 bytes each;
6.  suspend this process;
7.  wait for a signal from the process 2;
8.  choose the best 1/2 of a population
     and place it in the next generation;
9.  crossover the parent members to create
     offspring weight vectors;
10. mutate the new population;
11. go to step 3;
   }
```

process 2:

```
1. in endless loop do {
2.   wait for a signal from the process 1;
3.   for each message buffer in the array do {
       send the buffer to the receiver;
       suspend the task for 1000 ms;
     }
4.   send a signal to the process 1;
   }
```

After each generation first process transforms the best-valued weight vectors into an array of buffers size of which depends on the size of population. The size of each message is 16 bytes. Then it gives control to the second process waiting for response in the background. This is done via its suspension that switches the control to other processes. After the process 2 flushes all of the buffers from the array it sends a signal to the first process and awaits a signal too. Processes of a sender don't run in a parallel manner because of problems with interference between the radio transmission and communication with host computer that these processes maintain.

The receiver consists of two main processes described by following pseudocode:

process 1:

```
1. create weight vectors with rand. values;
2. for each generation do {
3.  for each weight vector do {
      run the neural controller for 800 cycles;
      compute fitness value of the controller;
      change speed randomly for 100 cycles;
    }
4.  sort the population according to fitness;
5.  replace the second 1/4 of a population
      members with received weight vectors;
6.  choose the best 1/2 of a population
      and place it in the next generation;
7.  crossover the parent members to create
      offspring weight vectors;
8.  mutate the new population;
9.  go to step 3;
   }
```

process 2:

```
1. in endless loop do {
2.   receive an array of message buffers
3.   if the new array differs
       from a previous one do {
         for each message buffer in the array
         do {
```

```
            transform the message to
            a part of a new weight vector;
            make a copy of a message array;
          }
        }
    }
```

The receiver continually checks for new messages from the sender. The buffers are not flushed instantly which means that if no new message arrives the received buffer still holds old data. That is why the robot must remember the old array of buffers which it always compares with newly received ones. If new array of buffers arrives, the receiver rewrites the received weight vectors and uses it in its own genetic algorithm. Processes of the receiver run concurrently and there is no need of synchronization.

3.2 Neuro-controller

A neural controller with primitive architecture (see figure 2) controls the robot collision avoidance.[2] There are 8 input neurons fully connected with two output neurons that compute the speed of two motors. These connections are labeled by integer-valued weights. Each of the input neurons has the information about the value read by one of the proximity sensors.

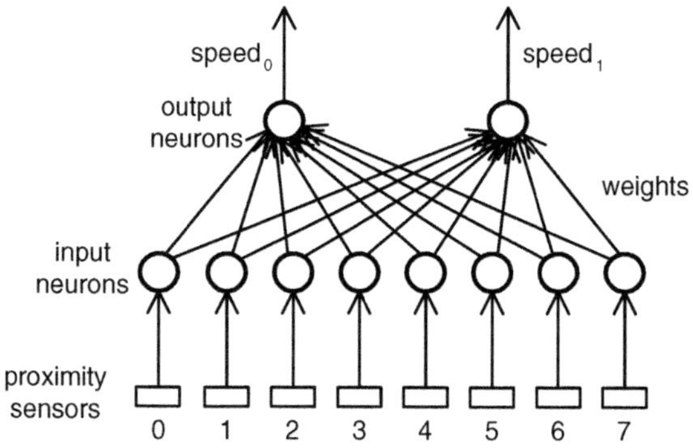

Fig. 2. Schematic view of a neural network architecture used in the experiments.

The speed of each motor is computed according to the Eq. (1),

$$speed_i = potential_i/400 + 10; \; i = \{0, 1\}, \tag{1}$$

where

$$potential_i = \sum_{j=0}^{7} (weight_{ij} * irvalue_j), \tag{2}$$

[2] It was a purpose to design an artificial neural network as simple as possible.

where $weight_{ij}$, $i = \{0, 1\}$, $j = \{0, 1, \ldots 7\}$ is a vector of weights represented as a two-dimensional array of integer values from the open interval $(-30, 30)$, and $irvalue_j$, $j = \{0, 1, \ldots 7\}$ is a vector of values read by proximity sensors. Speed of motors must be normalized and set to some non-negative value if the value of $potential_i$ is close to 0.

3.3 Parameters of a Genetic Algorithm

The genetic algorithm was used to evolve the weight vector $weight_{ij}$, $i = \{0, 1\}$, $j = \{0, 1, \ldots 7\}$. In each run we evaluated the population of neural controllers with the fitness function consisting of three components taken from [2] (see Eq. (3)).

$$\Phi = V(1 - \sqrt{|speed_0 - speed_1|})(1 - i), \tag{3}$$

where V is an average speed of two wheels, and i is a sensory value recorded by a proximity sensor with highest activation. These three components of a fitness function measure an average speed of a robot (it should be maximized), speed difference of rotating wheels (it should be minimized), and the highest activation of proximity sensors (it should be minimized).

The sender learns as follows. From each population it picks 1/2 of chromosomes with the highest fitness values and places it in the next generation. Then pairs of parental chromosomes are taken sequentially and the crossover at the random spot is performed to produce offspring chromosomes. Mutation of chromosomes was done with 5% probability.

The receiver robot combines 1/4 of its own chromosomes with the best 1/4 of the sender's best population members to produce the offspring generation as explained in the section 3.1. We experimented with various size of population - 12, 20, and 40 chromosomes. The smallest size of population was chosen so that it almost always finds "acceptable" solution for at least one of the robots - a controller that performs well judged not by the fitness value but by looking at the moving robot performance. We were mainly interested in the best controller, not the average performance that can be determined only computationally.

4 Results

In this section we present results of our experiments. Each experiment consists of 5 independent runs that were evaluated for the highest and average values. Each run consisted of 20 generation of evolving neural controllers.

4.1 Experiment 1

In the first experiment we used 12 chromosomes in one population. The performance of the best controller of a receiver is much better and it also learns faster in comparison with a sender robot (see figure 3a)[3]. Due to a small population

[3] Capital S and R in subsequent figures denote sender and receiver, resp.

size the learning peaks in the 14th generation. Its performance is even better on average even if not so markedly than the performance of best controllers. To measure how sharing the chromosome pool improved the performance of a

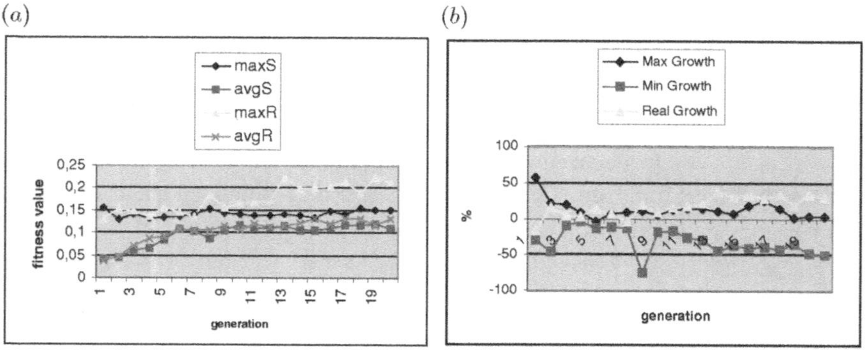

Fig. 3. Experiment with the population of 12 chromosomes. (a) Max. and avg. fitness values for both robots. (b) Expected vs. real fitness growth in receiver's chromosomes.

receiver we compared in each run best evolved controllers of the sender and the receiver. The difference (that could be called maximum expected fitness growth) measured in percentage shows how big the improvement could be in case of best-performing receiver's controllers. Figure 3b shows also the lowest expected fitness growth[4] along with the real improvement. In the first experiment real growth in fitness is higher than we would expect.

4.2 Experiment 2

In this experiment the population of controllers consisted of 20 members. Again the performance of a receiver was better in highest fitness values but approximately the same on average (see figure 4). In this experiment the receiver learned even faster than in previous case (the highest fitness value in the 5th generation), but from this point on the learning started slowly to degrade.

The expected versus real growth in fitness of receiver's best controllers can be seen in figure 5a.

The learning of a receiver is much more unstable with big differences between the highest and the smallest fitness values in each generation illustrated in figure 5b. There are mainly three reasons to this fact:

- Controllers acquired from the sender with worse performance in comparison with the receiver's own population pool can perturb learning and destabilize

[4] This denotes to what degree the received controllers could negatively influence the performance of a receiver.

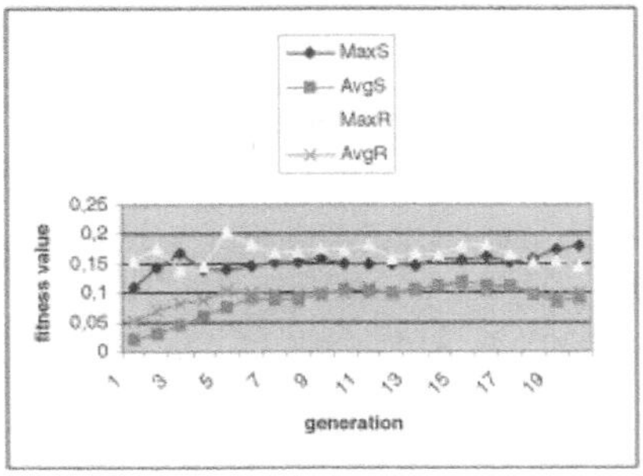

Fig. 4. Experiment with the population of 20 chromosomes. Max. and avg. fitness values for both robots.

(*a*)

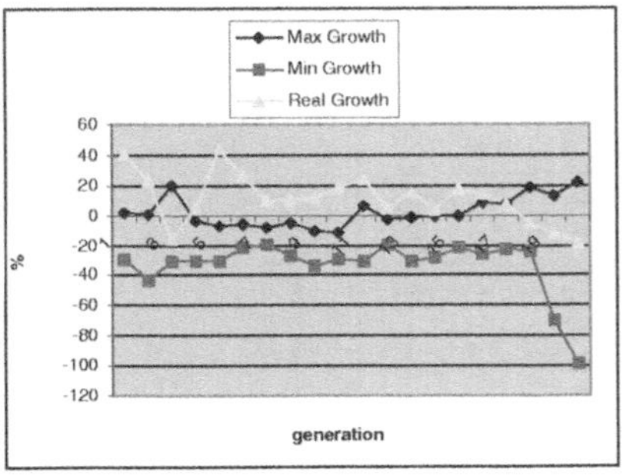

(*b*)

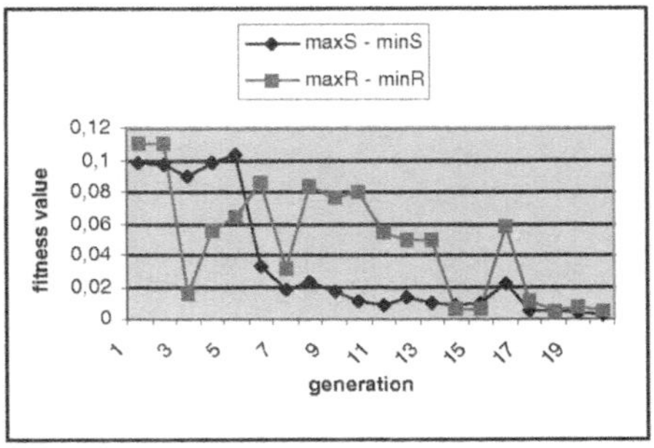

Fig. 5. Experiment with the population of 20 chromosomes. (a) Expected vs. real fitness growth in receiver's chromosomes. (b) Max. - min. fitness values taken from the run with the smallest difference.

it. Even if it should not influence the best chromosomes it has an influence on creating suboptimal solutions.

- Communication errors. The communication among Khepera robots is not very reliable. There appears loss of data during the transmission which we roughly estimate to be 5 to 15 %. If the robot fails to correctly determine when new array of message buffers arrives it uses old values instead which deteriorates the performance of a given generation. The bigger the population of chromosomes is, the more fatal the consequences appear to be. Firstly, the robots must exchange number of messages that equals half of a number of chromosomes in the population.[5] The more the messages robots exchange, the more probable is the occurrence of transmission errors.

[5] Radio turrets of Khepera robots use a protocol that enables to transfer only messages of size 16 bytes (more precisely unsigned bytes), it means only small positive integers. One weight vector that contains 16 weights must therefore be split into 2 messages. Each message holds 1/2 of the weight vector values plus the indication of a sign of a particular weight value. E.g. message containing data {... 1 7 1 3 0 19 0 21 0 19 0 2 1 16 1 15} is a half of a chromosome with values {7 3 -19 -21 -19 -2 16 15}.

– Time delay in sending and testing chromosomes. Because of non-concurrent nature of sender's processes it takes more time for the sender to run all the generations when compared to the receiver because the sender stops the genetic algorithm while sending data to the other robot. That is when there can occur transmission lags in that the sender can send the receiver chromosomes from generations that are older than the currently executed generations of the receiver. The learning of the receiver can suffer from this fact.

The consequences of the facts explained above will be more visible in the last experiment.

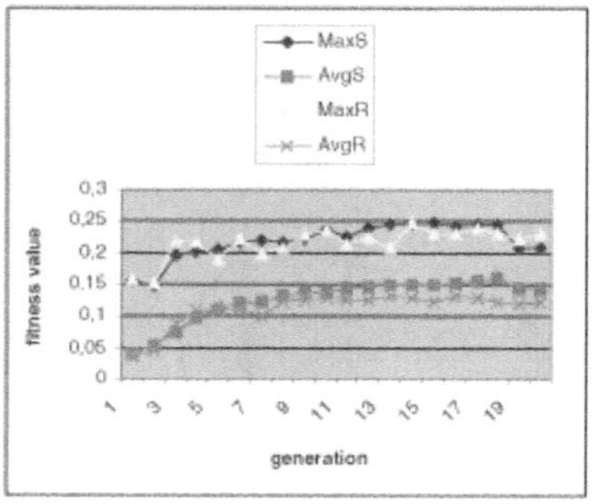

Fig. 6. Experiment with the population of 40 chromosomes. Max. and avg. fitness values for both robots.

4.3 Experiment 3

Even if both robots achieved the highest fitness values in this experiment (with 40 chromosomes in population), the receiver robot didn't perform better than the sender. Its best controllers were better in comparison with best-performing sender's controllers only in a couple of generations and it completely failed when we take average performance into account (see figure 6). In the last experiment the improvement in receiver's fitness got hardly positive values as depicted in figure 7a.

Figure 7b shows instability of the receiver's evolving populations that is on average higher than in previous experiment.

5 Conclusions

In this paper we proposed distributed genetic algorithm in a multi-robot society. We described the experiments with learning to avoid obstacles and compared the performance of a robot that learns only by individual evolution of controllers with the one that learns also by the experience of another robot through direct

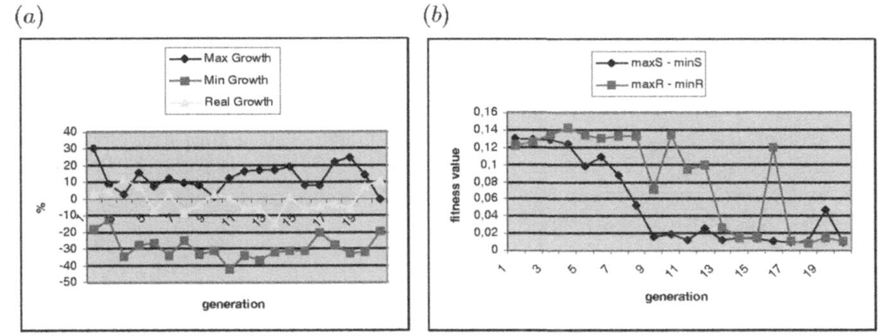

Fig. 7. Experiment with the population of 40 chromosomes. (a) Expected vs. real fitness growth in receiver's chromosomes. (b) Max. – min. fitness values taken from the run with the smallest difference.

exchange of chromosomes coding weight vectors of neural controllers. We discussed stability and the speed of learning in both cases as well as measured expected vs. observed growth in fitness values of chromosomes in the population of a second robot.

References

1. V. Braitenberg, *Vehicles. Experiments in Synthetic Psychology*, MIT Press, Cambridge, MA, 1984.
2. D. Floreano, and F. Mondada, Automatic Creation of an Autonomous Agent: Genetic Evolution of a Neural Network Driven Robot, From Animals to Animats III: Proceedings of the Third International Conference on Simulation of Adaptive Behavior (D. Cliff, P. Husbands, J. Meyer, and S. Wilson, eds.), MIT Press-Bradford Books, Cambridge, MA, pp. 402–410, 1994.
3. D. E. Goldberg, *Genetic algorithms in search, optimization and machine learning*, Addison-Wesley, Redwood City, CA, 1989.
4. J. H. Holland, *Adaptation in natural and artificial systems*, The University of Michigan Press, Ann Arbor, 1975.
5. J. Kelemen, and A. Kubík, RADIUS: Looking for Robot's Help in Computer Science Research and Education, *ERCIM News* (51), pp. 48–49, 2002.
6. F. Mondada, E. Franzi, and P. Ienne, Mobile Robot Miniaturization: A Tool for Investigation in Control Algorithms, *Proceedings of the Third International Symposium on Experimental Robotics*, Kyoto, Japan, 1993.
7. S. Nolfi, and D. Floreano, *Evolutionary Robotics: The Biology, Intelligence and Technology of Self-Organizing Machines*, MIT Press, Cambridge, MA, 2000.
8. S. Nolfi, and D. Floreano, Co-evolving Predator and Prey Robots: Do 'Arm Races' Arise in Artificial Evolution?, *Artificial Life*, Vol. 4(4), pp. 311–335, 1998.
9. M. Quinn, Evolving Communication Without Dedicated Communication Channels, *Advances in Artificial Life: Proceedings of the. 6th European Conference on Artificial Life* (J. Kelemen, and P. Sosík, eds.), Springer Verlag, Berlin, pp. 357–366, 2001.

10. M. Quinn, L. Smith, G. Mayley, and P. Husbands, Evolving Teamwork and Role-Allocation with Real Robots, *Proceedings of the Eighth International Conference on Artificial Life* (R. K. Standish, M. A. Bedau, and H. A. Abbass, eds.), MIT Press, Cambridge, MA, pp. 302–311, 2002.
11. Watson, Richard A., Ficici, Sevan G. and Pollack, Jordan B., Embodied Evolution: Embodying an Evolutionary Algorithm in a Population of Robots. *1999 Congress on Evolutionary Computation* (Angeline, Michalewicz, Schoenauer, Yao, and Zalzala, eds.), IEEE Press, pp. 335–342, 1999.

An Approach to Describe the Tierra Instruction Set Using Microoperations: The First Result

Shuichi Matsuzaki[1,2], Hideaki Suzuki[2], and Minetada Osano[1]

[1] Graduate School of Computer Science and Engineering,
Aizu University
Aizu-Wakamatsu City, Fukushima 965-8580 Japan
d8022201@u-aizu.ac.jp
[2] ATR Human Information Science Labs
2-2-2 Hikaridai Seika-cho, Soraku-gun, Kyoto 619-0288 Japan

Abstract. In constructing an Artificial Life system with a high potential for evolution, one of the elements that we should design most carefully is the machine structure, that is, the architecture of the system controlling the operations of a digital creature. This paper proposes an approach to modify the machine structure of Tierra in order to obtain higher evolvability. The structure of the Tierran digital creature that we propose is stimulated by the self-reproducing model proposed by von Neumann, which enables self-reproduction in a simple architecture. Verification of a digital creature's structure was carried out through computer simulation. As a result, the capabilities of self-reproduction and parasitism were demonstrated.

1 Introduction

In the development of an Artificial Life (Alife) system, it is important that the system be designed so that 'emergence' (the appearance of higher functions among elements) may occur. It is therefore necessary for us to prepare a good artificial environment during system construction, since the possibility of emergence happening in the system usually depends on the quality of the environment prepared by the designer. The actual biological system provides a good example for deciding what is needed in an Alife system. Today, it is widely accepted that an appropriate state of ancient earth permitted life to emerge and a variety of complex life forms to evolve. If we could identify the conditions necessary for this evolution [4], it would greatly assist in designing of a 'good' Alife system. Artificial environments prepared for an Alife system in a computer usually have a two-layered structure: the information space (hardware layer) and the reaction rule set (software layer), which cannot be changed during evolution.
Alife is an approach that uses a policy of "minimum implementation" in the design. In order to have a system with high evolvability, we should prepare the artificial environments so that the software layer may be changed/optimized by evolution as much as possible. From this point of view, we propose optimizing the instruction set of "Tierra", which was designed by Tom Ray [1]. In the

W. Banzhaf et al. (Eds.): ECAL 2003, LNAI 2801, pp. 357–366, 2003.

original experiment of Tierra, [1,2], the instruction set, a set of machine codes that specifies the operation was pre-programmed and thus could not able to be modified. If we were able to equip creatures with different instruction sets, the evolution of the instruction set in addition to the usual evolution of Tierra might take place. The model we propose in this paper aims to achieve this kind of evolution. As a result, the constructed architecture of the digital creature is considerably modified from that of the original Tierran creature.

In Section 2, we discuss past research on Tierra and the self-reproducing machine proposed by von Neumann, which formed the basis our self-reproducing model. We show our model in Section 3 by describing the architecture of the components. Experiments and results of a simulation are reported in Section 4. Our concluding remarks and future work are discussed in Section 5.

2 Previous Self-Reproducing Models

Since Ray proposed Tierra in 1992, it has been recognized as the most typical and representative Alife model. The Tierra simulation experiment produced stimulating results on the behaviors of virtual creatures. However, we believe that Tierra has not yet reached a state of completion, at least from the viewpoint of self-reproduction. Consequently, we propose another Alife model by following the excellent suggestions proposed by von Neumann (Fig. 1). He designed a self-reproducing machine comprised of a machine with the function of a universal constructor and copier and a tape consisting of descriptions of both the constructor and copier (conceptually, these objects consist of descriptions that might be called a description in the following discussion). The constructor manufactures another machine called a replicant, and the copier makes a copy of the descriptions for the replicant. Both the constructor and copier use the data of the description in each working process. Although von Neumann introduced the tape as simply a storage device to keep the coded information for the entire body of the machine during the self-reproduction process, we regard this self-reproducing machine as a model enabling evolution of genotype-to-phenotype association. We take this view because the constructer, whose codes themselves are also written on the description tape, translates the genotypic information on the tape into the phenotypic machines.

Here, we return to our discussion of Tierra. Figure 2 shows the machine structure of Tierra. In Tierra, the tape in von Neumann's model can be regarded as a core memory and the copier as a mixture of the functions of a program, a central processing unit (CPU), and an operating system (OS). Although the role of the constructor is achieved by the functions of the CPU and OS, these structures are implicitly programmed in the system. Therefore, a creature in Tierra can create its replicant's CPU by simply executing one instruction ('divide'), but almost all of the procedures necessary for the construction are executed by the operating system. In the simulation, the programs embedded in the operating system cannot be changed by any operation. This implies the impossibility of

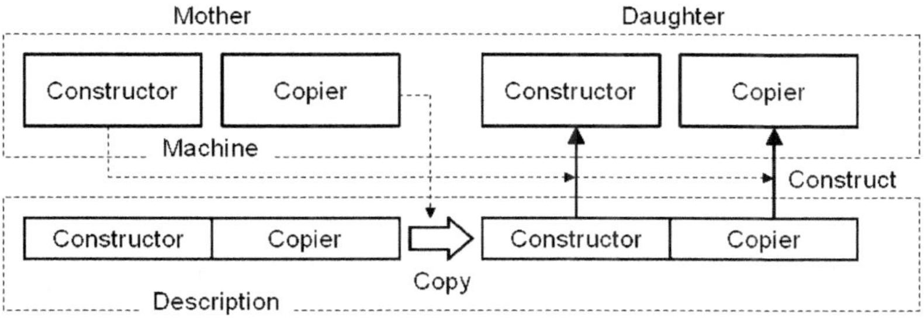

Fig. 1. John von Neumann's Self-reproducing Machine's structure

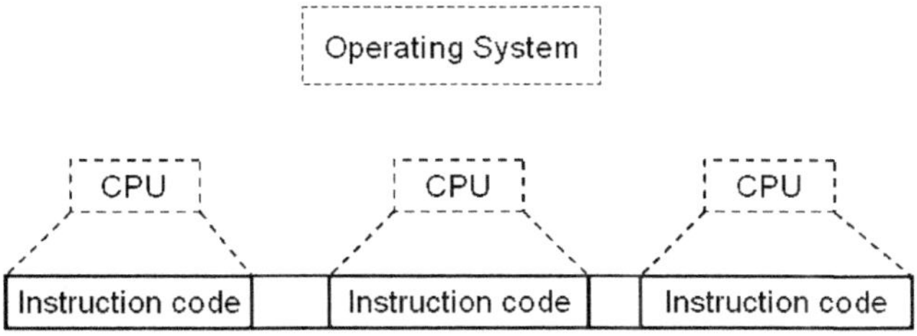

Fig. 2. Machine structure of Tierra

optimizing the instruction set in the current Tierra structure.
Noteworthy features of the Tierra structure are as follows.

- The instruction code plays mostly the same role as the copier, but there is no explicit constructor (it is programmed in the OS).
- In the simulation of Tierra, only the instruction code evolves, and the other functions are embedded in the operating system.

3 Proposed Model

3.1 Architecture

The architecture of individuals in this model is shown in Fig. 3. Since the entire model itself was designed based on Tierra, of course the individual of our model and that of Tierra still have a lot of common machineries. Nevertheless, the architecture is more strongly influenced by von Neumann's machine architecture rather than by Tierra.

The structure is divided into two parts: 'Machine' and 'Description'. In the machine, there is a control memory and a set of registers (Table. 1). The control

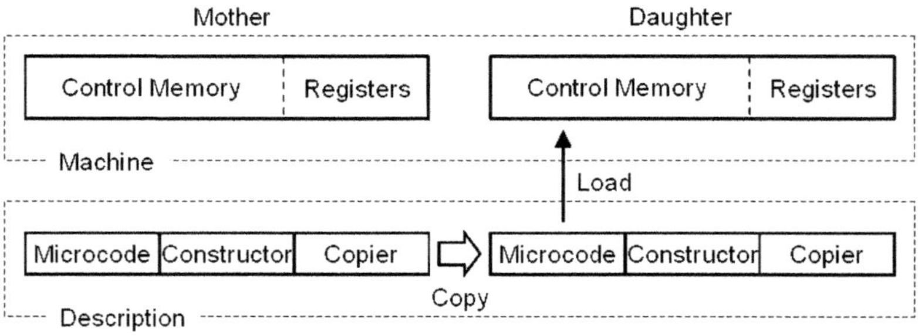

Fig. 3. Machine structure of Tierra redesigned

memory is the only device that is recognized as a driving device from the operating system, and all operations are propelled by the interpretation and execution of 'microcodes' in the control memory. As in the canonical technique of microprogramming [7], the microcodes are expressed as register-transfer operations (Table 2). In the process of self-replication, the codes are initially written in the core memory as a part of the description tape (i.e., the codes are copied from a mother individual to a daughter individual), and then they are copied (loaded) into the control memory. This process of loading the microcodes into the control memory is coded in the 'constructor'.

The code of the 'copier', which is also written in the core memory, is almost the same as the original Tierran program. The code of the copier is represented by a sequence of approximately eighty instructions of the original Tierra.

During the self-reproduction process of original Tierra, a creature, after allocating memory space for the daughter, executes 'divide' instruction to create the daughter CPU. In our model, the divide is eliminated, and instead a sequence of instructions in the constructor (Table 3) is added to the copier and executed. This additional routine, which is a sequence eighteen instructions, includes the function of the 'load' instruction and accomplishes the job of making machine architecture for the daughter CPU.

The description of the ancestor creature by the proposed model consists of 18 instruction words for the constructor, 86 instruction words for the copier, and 1,477 microcodes, which occupy 1581 data words in the core..

The following points give an outline of this model's features.

- The model is Tierra-based, but the structure of individuals is largely influenced by von Neumann's self-replicating machine structure.
- The individuals have two components: 'machine' and 'description'. The machine replicates (make copies) of both the individual's machine and description by interpreting microcodes.

- The entire code, that is, microcodes, constructor and copier, is explicitly coded on the core memory. Therefore, all code elements are changeable by mutation.

3.2 The Model vs. Tierra

In the previous subsection, we explained the architecture of our model. Here, we discuss the advantages of our model by comparing it with Tierra.

Each individual of Tierra has the instructions of Tierran language, that is, a virtual machine language (e.g. pushax, jmp, mal). The instructions are interpreted by the operating system and executed from the top during the self-replication. We should note that the instructions aren't a variable function, in other words, they are pre-programmed in the operating system. Therefore, the mapping from the code (in the core memory) to the instruction set (programed in OS) and the program that codes the Tierran instruction are never modified. With respect to the former problem, McMullin et al. detected new instructions with mutated genotype-phenotype mappings by using a variable lookup table that is explicitly coded in the core memory [9]. In our model, we aim to make not only the genotype-phenotype mapping but also the program itself mutable. As we discussed, the instructions of this model are interpreted by micro operations in the control memory 2. The important point here is that the microcode and the machine constructor are both coded in the core memory as a part of the genotype.

We expect the following changes to individuals by mutation in this model.

- By the mutation in the copier's code, the program (instruction) can be changed.
- By the mutation in the microcode, the micro operation loaded into the control memory can be changed.
- By the mutation in the constructor, the arrangement of the micro operations can be changed.

4 Computer Simulation

The experiment carried out to investigate this model were simulations using the system proposed in Section 3. We conducted two kinds of experiments. One was a simulation started with an ancestor in the initial condition. The other simulation started with an ancestor and a parasite individual that we designed. For both simulations, the rate of the mutations were set under the condition of $M_c = 0.0001$, $M_r = 0.000001$, where M_c represents the mutation occuring during the replication of core data words by movii instruction and M_r represents the mutation randomly occuring at some point in the core memory [1]. We should note here that these mutation rates are set much lower than those of Tierra because the time step of this model is counted by one execution of a micro operation, while the time step of Tierra is counted by one execution of an instruction.

Table 1. List of the names of registers and memory used to represent microoperations. Twelve registers are added to the original eight.

cm[ ]	control memory
cma	memory address of cm
ax	address register (original)
bx	address register (original)
cx	numerical register (original)
dx	numerical register (original)
dcm[ ]	control memory of daughter
dcma	address pointer of dcm[ ]
j	numerical register used in jmp and mal
sb	address register for jmp (backward)
sf	address register for jmp (forward)
st[ ]	stack (original)
pst[ ]	stack of searching template
sp	stack pointer of ST (original)
ma	memory address of m (or iginal)
psp	stack pointer of pst[ ]
m	value of the memory pointed by ma
state	state of the current memory
ef	error flag (original)
load	load flag

Table 2. Examples of microoperations. There are 157 sets of micooperations for translating thirty-two instruction sets. Once the instruction word written in the constructor or copier is interpreted, the corresponding microoperation is executed in sequence.

Original instruction word	Microoperation
fetch(if cma == 0)	cma ← m
nop0	cma ← 0
pushax	st[ ] ← ax, sp++, cma ← 0
popax	sp–, ax ← st[ ], cma ← 0
movcd	dx ← cx, cma ← 0
movab	bx ← ax, cma ← 0
movii	st[ ] ← ma, sp++, ma ← bx, st[ ] ← m, ma ← ax, m ← st[ ], sp–, ma ← st[ ], cma ← 0
sub_ab	cx ← ax - bx, cma ← 0
sub_ac	ax ← ax - cx, cma ← 0
inc_a	ax++, cma ← 0
zero	cx ← 0, cma ← 0
ifz	cma ← 0, if(c != 0) ma++
load	st[ ] ← ma, ma ← bx, dcm ← m, cx–, if(c == 0) ma ← st[ ], cma ← 0

Table 3. Description of the constructor. These are the codes for loading the replicant's microcode into the control memory.

```
pushbx /* push bx onto stack */
pushcx /* push cx onto stack */
adrf   /* search forward for template, put address in ax, template size in cx */
nop1   /* instruction for template */
nop0   /* instruction for template */
nop0
nop1
mov_ab /* bx = ax */
adrf
nop0
nop0
nop0
nop1
sub_ac /* ax = ax - cx */
sub_ab /* cx = ax - bx */
load   /* load microcodes into control memory */
popcx  /* pop from stack into cx */
popbx  /* pop from stack into bx */
```

In addition, the number of execution steps required for self-reproduction is much larger than that of the original Tierra because our ancestor creature (explained below) has a total of 1581 codes (i.e., the number of microcodes, copier and constructor), whereas the original ancestor creature has 80 instructions (usually, the ancestor of our model requires over 400,000 steps). Both simulations were run with the RAM space of 200,000 bits, which is the area in which approximately 100 ancestor-size individuals can exist at the same time.

In the first experiment, we used the ancestor individual we designed, which can self-replicate in the same manner as the Tierran individual (see Section 3). In the initial state, the simulation is started from one ancestor individual. Figure 4 shows the results of this simulation. The graph shows the appearance of offspring individuals that are able to self-replicate (therefore, so-called 'sterile' offsprings aren't included in the graph). The noteworthy result in this experiment is the appearance of the new individual, which has a far smaller individual size (97 data words) in approximately 34,000,000th step, and individuals of the same size subsequently appeared more frequently. Actually, these individuals are organized by only the copier, and they use the microcode and constructor of the nearest individual when they self-replicate. Interestingly, this type of parasite is recurrently created from a creature that has the same size as the ancestor. After the birth of these creatures, they can reproduce many offsprings; howver, the daughters of the parasites (the second generation parasites) cannot self-reproduce. Nonethless, we regard these individuals as a new kind of parasite individual.

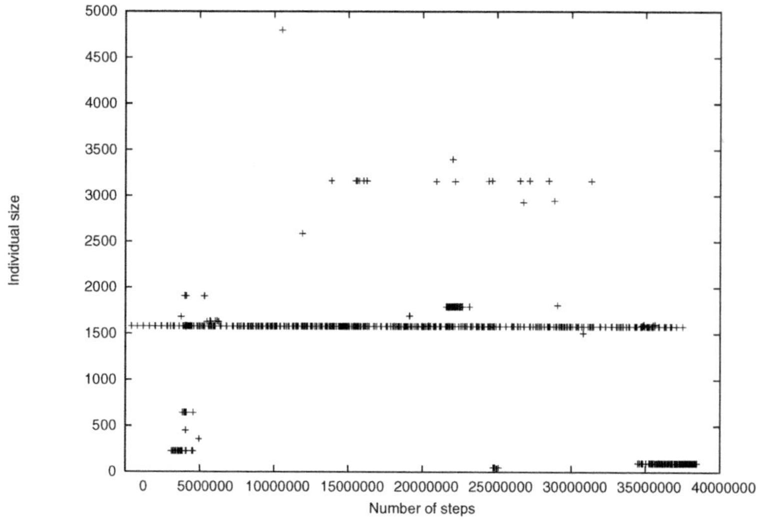

Fig. 4. The simulation starting with the ancestor individual

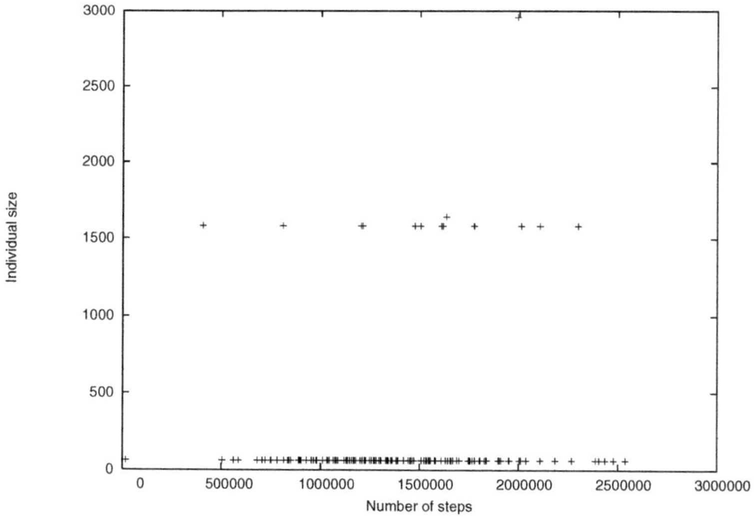

Fig. 5. The simulation starting with the ancestor and the parasite individual

Next, we designed a parasite individual by hand that has the same instructions as the Tierran parasite individual. The individual became smaller (62 data words) than the parasite individual that emerged in the first experiment. This individual certainly can self-replicate in the same way as the Tierran parasite individual. The second experiment was carried out with one ancestor and one parasite individual. Figure 5 shows the results of this simulation. In this graph, we can see that the number of parasite individuals increases rapidly compared to the number of ancestor individuals. This result largely comes from the difference in the necessary steps for self-replication, in other words, the steps of the parasite individual are much shorter than those of the ancestor individual. In addition, it appeared that the necessary steps for self-replication of the parasite individual depend on the distance from the host individual, while the necessary steps for self-replication of the Tierran parasite individual never depended on this.

5 Conclusions and Future Works

In this paper, we proposed a new Alife model based on Tierra. However, the individual in this model is designed based on von Neumann's self-replication model, which has advantages over the individual designed in Tierra. In our experiment, we found the appearance of parasite individuals that self-replicate by using the other individuals' code. Although the experiments conducted this time found a mutant caused by a mutation of the copier, mutants caused by the other parts of description, constructor and microcode could not be detected. Mutation in the constructor and microcode is expected to cause the appearance of new species in our futher experiments.

Acknowledgements. The authors thank Dr. Hart, Dr. Lee, and Mr. Sugiura of ATR labs for fruitful discussions. The authors also thank Dr. Shimohara of ATR labs for his sincere encouragement. The second author's study was supported by the Telecommunications Advancement Organization of Japan and Doshisha University's Research Promotion Funds.

References

1. Ray, T. S. " An approach to the synthesis of life" In C. G. Langton, C. Taylor, J. D. Farmer, and S. Rasmussen, Eds., Artificial Life II. Redwood City, CA: Addison-Wesley (1992) pp. 371–408
2. Ray, T.S. and Hart, J.: Evolution of differentiated multi-threaded digital organisms. In: Adami, C., Belew, R.K., Kitano, H., and Taylor, C.E. (eds.): Artificial Life VI: Proceedings of the Sixth International Conference on Artificial Life, MIT Press, Cambridge, MA (1998) pp. 295–304
3. Suzuki, H. An example of design optimization for high evolvability: string rewriting grammar. Biosystems V.69, N.2-3 (2003) 211–221
4. Suzuki, H., Ono, N., and Yuta, K.: Several necessary conditions for the evolution of complex forms of life in an artificial environment. Artificial Life V.9, N.2 (2003) 537–558
5. von Neumann, J.: Theory of self-reproducing automata. University of Illinois Press, Urbana. Edited and completed by A.W. Burks (1966)
6. Sipper, M.: Fifty Years of Research on Self-Replication: An Overview. Artificial Life 4(3): 237–257 (1998)

7. Mano, M.M.: Digital logic and computer design. Prentice-Hall, Inc., NJ (1979)
8. McMullin, B.: John von Neumann and the Evolutionary Grawth of Complexity: Looking Backwards, Looking Forwards...: Artificial Life 6: pp. 347–361 (2000)
9. McMullin, B., Taylor, T., and von Kamp, A.: Who Needs Genomes?: Atlantic Symposium on Computational Biology and Genome Information Systems and Technology (2001)
10. Matsuzaki, S., and Suzuki, H.: Tierra instructions implemented using string rewriting rules: Proceedings of the Fifth International Conference on Human and Computer (HC-2002) pp. 167–172
11. Suzuki, H., and Ono, N.: Universal Replication in a String Rewriting System: Proceedings of the Fifth International Conference on Human and Computer (HC-2002) pp. 179–184

Critical Values in Asynchronous Random Boolean Networks

Bertrand Mesot and Christof Teuscher

Logic Systems Laboratory, Swiss Federal Institute of Technology (EPFL), CH-1015 Lausanne, Switzerland
bertrand.mesot@epfl.ch, christof@teuscher.ch

Abstract. Wherever we see life, we see different kinds of complex networks, reason why they are studied across various fields of science. Random Boolean Networks (RBNs) form a special class in which the links between the nodes and the boolean functions are specified at random. Whereas synchronous RBNs were investigated in detail, there has little been done around their asynchronous counterpart, although there is evidence that most living systems are governed by asynchronous updating. Derrida's annealed approximation predicts a critical connectivity value of $K = 2$ for synchronous RBNs. We present a similar and original approach for asynchronous RBNs and show that they do not possess such a critical connectivity value. The asynchronous and nondeterministic updating scheme introduces perturbations that reach about 25% of the nodes and thus prevents the networks to become stable. Further, our numerical simulations show that asynchronous RBN tend to amplify small and to reduce big perturbations.

1 Introduction

Wherever we see life, we see different kinds of complex networks, reason why they are studied across various fields of science. Many of the natural networks such as ecological food webs, genetic networks, social networks, neural networks, and even the World Wide Web share common and global statistical features and motifs [20,18,14].

Among the different kinds of networks, *Random Boolean Networks (RBNs)* (sometimes also called *Kauffman nets or model*) form a special class in which the links between the nodes and the node's boolean transfer functions are specified at random. They are often specified by two parameters: N, the number of nodes and K, the number of incoming links per node (sometimes, K indicates the average number of links). Synchronous RBNs have been seriously investigated by Kauffman [12,13], Weisbuch [22], and many others as models for biological phenomena such as genetic regulatory networks and embryonic development.

Indeed, randomly connected networks with various kinds of nodes have been analyzed much earlier. The first persons to mention randomly connected networks were Ashby [3] and Allanson [1] who investigated in 1956 networks of dynamical systems. Rozonoér [16] analyzed the properties of networks consisting of elements whose properties depend on parameters chosen at random. In

W. Banzhaf et al. (Eds.): ECAL 2003, LNAI 2801, pp. 367–376, 2003.

1971, Amari [2] published a paper on the characteristics of randomly connected threshold-element networks with the intention of understanding some aspects of information processing in nervous systems. He showed that two statistical parameters are sufficient to determine the characteristics of such networks.

Whereas synchronous RBNs as abstract models of specific biological systems were investigated in detail, there has little been done around their asynchronous counterpart, although there is evidence that most living systems are governed by asynchronous updating. Harvey and Bossomaier note, that "[...] for many biological phenomena asynchronous versions are more plausible" [9]. Observed global synchronous behavior in Nature usually simply arises from the local asynchronous behavior.

This paper principally addresses the question whether asynchronous RBNs have a similar phase transition and critical value for K like synchronous RBNs. In particular, we use a similar approach to asynchronous RBNs as Derrida used to find the critical connectivity for synchronous RBNs. Our results are then verified by numerical simulations.

The reminder of the paper is as following: Section 2 introduces the principal characteristics of synchronous and asynchronous RBNs. Section 3 gives an overview on Derrida's annealed approximation that predicts $K = 2$ for synchronous RBNs. Our approach for asynchronous RBNs is presented in Section 4 and the numerical results in Section 5. Section 6 concludes the paper.

2 Synchronous versus Asynchronous Random Boolean Networks

Random boolean networks (RBNs) are usually considered as a more general case of classical cellular automata (CA). In the synchronous version, both are examples of discrete deterministic dynamical systems made up from simple components that process data in parallel. The RBN architecture is in many ways similar to weightless neural networks [21]. Kauffman's studies [13] have revealed surprisingly ordered structures in randomly constructed networks. In particular, the most highly organized behavior appeared to occur in networks where each node receives inputs from two other nodes ($K = 2$). It turned out that the networks exhibit three major regimes of behavior: *ordered* ("solid"), *complex* ("liquid"), and *chaotic* ("gas"). The most complex and interesting dynamics correspond to the liquid interface, the boundary between order and chaos. In the ordered regime, little computation can occur. In the chaotic phase, dynamics are too disordered to be useful. The most important and dominant results of Kauffman's numerical simulations can be summarized as follows [13]: (1) The expected median state cycle length is about $\sqrt{N}$. (2) Most networks have short state cycles, while a few have very long ones. (3) The number of state cycle attractors is about $\sqrt{N}$. (4) The most interesting dynamics appear with an average connectivity of $K = 2$ (the boundary between order and chaos).

Very few work has been done around asynchronous random boolean networks (ARBN), although the updating scheme in discrete systems plays a crucial role for its properties. Moreover, for many physical and biological phenomena, the assumption of asynchrony seems more plausible. Harvey and Bossomaier [9] have shown that ARBNs behave radically different from the deterministic synchronous version. Di Paolo [15] provided further analysis and mainly investigated rhythmic and non-rhythmic attractors. Although ARBNs cannot exhibit strictly cyclic behavior (due to their random updating scheme), he has shown that they can all the same model rhythmic phenomenon. Recently, Gershenson [7] provided a first classification of the different types of RBNs. The study also revealed that the RBNs point attractors are independent of the updating scheme and that they are more different depending on their determinism rather than depending on their synchronicity.

An *attractor* in a dynamical system is an *equilibrium state*. Following Hopefield [10], the attractors of networks represent a sort of content addressable memory. Each attractor is encompassed by a *basin (domain) of attraction*. A deterministic complex dynamical system with a finite number of states ultimately "settles down" in an attractor after a finite time. If the state vector comes to rest completely, it is called a *fixed point*. If the state vector settles into a periodic motion, it is called a *limited cycle*. Due to their indeterminism, asynchronous RBNs do not have cyclic attractors but only point and loose attractors. Similar to the synchronous version, the number of point attractors is independent of K [9].

Note that there is also a growing interest in asynchronous cellular automata in various problem domains (see for example [4,5,17,19]).

3 Derrida's Annealed Approach to Synchronous RBNs

As stated above, the average connectivity of $K = 2$ presents a critical connectivity for classical synchronous RBNs. This value is obtained by numerical simulations as well as by several theoretical methods. In 1986, Derrida and Pomeau [6] proposed the *annealed approximation* which allowed to predict $K = 2$ as the critical value of K. This section shall briefly recall the basic ideas of Derrida's approach that we then apply in a similar manner in Section 4 to asynchronous RBNs.

Assume that we have a network made up of N nodes, each being randomly connected to K other nodes. Each node can be in one out of two possible states, 0 or 1, and the node's state after the next update is defined by a randomly chosen boolean function that takes the K incoming links as inputs. The *network state* at time t is defined as the vector of node states at time t, which we write as $\mathbf{s}^t$.

The network states $\mathbf{s}^t$ for $t \geqslant 1$ are correlated to the network wiring and the node's boolean functions. Derrida noticed that this is somehow difficult to formalize, which lead him to the correct assumption that, since everything (i.e., wiring, transfer functions) is random in RBNs, randomly generating a new network after each update should not fundamentally affect the overall dynamics of the system. In order to find out the critical value for K, he compared the dynamics of two identically connected networks with state vectors $\mathbf{s}_1^t$ and $\mathbf{s}_2^t$. Let

$\mathbf{s}_2^t$ be a perturbed (i.e., changing the state of a random number of nodes) copy of $\mathbf{s}_1^t$ and let us define d_t as being the normalized *Hamming distance* between $\mathbf{s}_1^t$ and $\mathbf{s}_2^t$. The main question is then as follows: which value of K allows $d_t \to 0$ when $t \to +\infty$?

To answer this question, let a_t be the probability for a node to have the same value in both networks at time t. Let us then make two subsets of nodes called A and B. A contains all nodes that have equal states in $\mathbf{s}_1^t$ and in $\mathbf{s}_2^t$, B contains nodes with unequal states. Obviously, the probability for a node to belong to A at time t is a_t. Moreover, for each node two possibilities arise: (1) all of its input nodes belong to A, and (2) at least one of the input nodes belongs to B. Hence, in the first case, the node's input will be the same in both networks since each input node belongs to A. This does not hold in the second case as at least one of the input nodes belongs to B. For a given node, the probability of having all of its inputs in A is therefore $(a_t)^K$, and $\left(1 - (a_t)^K\right)$ otherwise. Consequently, the node's probability of being in the same state in $\mathbf{s}_1^t$ as in $\mathbf{s}_2^t$ at time $t+1$ is 1 in the first case and 1/2 in the second as we suppose that states 0 and 1 are equally distributed. Indeed, if a node has the same input in both networks, it will surely have the same output and consequently be in the same state in both networks at the next time step. In the second case, inputs are different and outputs will thus be equal with probability 1/2.

We can therefore describe the evolution of a_t as a function of t by means of the following recursive equation:

$$a_{t+1} = (a_t)^K + \frac{1}{2}\left(1 - (a_t)^K\right) = \frac{1 + (a_t)^K}{2} .$$

By taking into account that $d_t = 1 - a_t$, we get:

$$d_{t+1} = \frac{1 - (1 - d_t)^K}{2}$$

which describes the evolution of d_t as a function of time t. Figure 1 plots d_{t+1} as a function of d_t for $K = 1, 2, 3, 4, 10$. One can easily see that $K \in \{1, 2\}$ are the two only values that allow $d_t \to 0$ when $t \to +\infty$. Geometrically speaking, the plots for $K = 1$ and $K = 2$ lie below the identity function $d_t = d_{t+1}$ which implies that d_{t+1} tends toward 0 as time increases.

The results suggest that synchronous RBNs with a connectivity of $K = 2$ (the "edge between order and chaos") are particularly resistant to perturbations. This is mainly due to a phase transition in the number of frozen components within a network (see also [13]). Naturally, the question arises whether such a critical value exists in asynchronous RBNs. To the best of our knowledge, our attempt is the first one to investigate this question.

4 Our Approach to Asynchronous RBNs

As mentioned in Section 1, Harvey and Bossomaier claimed that asynchronous systems are biologically more plausible for many phenomena than their synchronous counterpart. However, studying asynchronous RBNs is often all but a

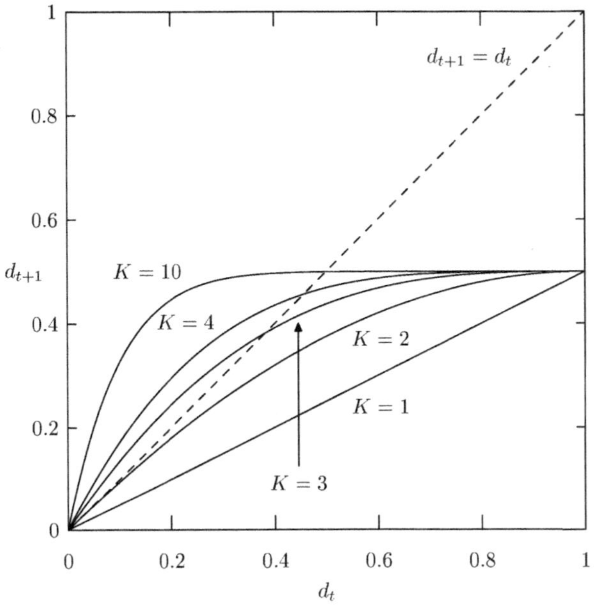

Fig. 1. d_{t+1} in function of d_t for $K = 1, 2, 3, 4, 10$. For $K = 1$ and $K = 2$, d_{t+1} tends toward 0 as time increases. For more information, see also Kauffman [13].

trivial task, mainly due to the nondeterminism introduced by the asynchronous node updating scheme.

In this paper, we will only consider the following case of asynchrony: at each time step, 1 to N nodes are randomly selected and synchronously updated. As shown in the previous section, we consider two identically connected networks with different state vectors $\mathbf{s}_1^t$ and $\mathbf{s}_2^t$. Again, consider $\mathbf{s}_2^t$ as being a perturbed copy of $\mathbf{s}_1^t$. In the asynchronous case, three possibilities arise for each node: (1) the node is updated in both networks, similarly to synchronous RBNs; (2) the node is updated in only one of the networks; and (3) the node is not updated at all. We therefore need to calculate the node's probability of being updated. Let m be the number of nodes updated at a given time t. Hence, the node's probability of being updated knowing m is m/N. As $m \in [1, N]$ and as we suppose that all values are equally probable, the probability of being updated becomes:

$$P(\text{being updated}) = \frac{1}{N} \sum_{m=1}^{N} \frac{m}{N} = \frac{N+1}{2N}\,.$$

Given this, we can now describe the probability of the three above-mentioned situations:

$$p_2 = \left(\frac{N+1}{2N}\right)^2, \quad p_1 = 2\left(\frac{N+1}{2N}\right)\left(1 - \frac{N+1}{2N}\right) \quad \text{and} \quad p_0 = \left(1 - \frac{N+1}{2N}\right)^2$$

where p_i describes the probability that the node is updated in i networks. Note that p_1 is counted twice because there are two possibilities of updating a node in only one of both networks.

Nodes can now again be separated into two subsets A and B with the same meaning as described in Section 3. Hence, if a node has all of its inputs in A and is updated in both networks, it will surely hold the same value in $\mathbf{s}_1^{t+1}$ as in $\mathbf{s}_2^{t+1}$. If only one of both networks updates the node, there will be one chance out of two to be in a different state. Finally, if none of the networks updates the node, it will keep its current state at time $t+1$ and thus hold the same value in both networks if and only if the node belongs to subset A. After adding together all these probabilities we get:

$$(a_t)^K \left[\frac{N+1}{2N} + \left(\frac{N-1}{2N} \right)^2 a_t \right] . \tag{1}$$

This represents the contribution of the nodes having all of their input nodes in A to the overall probability a_{t+1} that a node has the same state in both networks at time $t+1$. To this term we must add the contribution of the nodes that have at least one connection in B. Note that when a node is updated in both networks, the probability of being in the same state at time $t+1$ is not 1 but 1/2, similarly to the synchronous case. We therefore obtain:

$$\left(1 - (a_t)^K\right) \left[\frac{3/2 \cdot N^2 + N - 1/2}{4N^2} + \left(\frac{N-1}{2N} \right)^2 a_t \right] . \tag{2}$$

And finally, the probability a_{t+1} that a node has the same state in both networks at time $t+1$ is obtained by adding 1 and 2 when $N \to +\infty$. Note that $N \to +\infty$ simplifies the writing of the forthcoming equations and is justified by the fact that, as long as we consider probabilities over nodes, a "sufficient" number of them is required. The resulting recursive equation for asynchronous RBNs is thus as follows:

$$a_{t+1} = \frac{1}{8} (a_t)^K + \frac{1}{4} a_t + \frac{3}{8} .$$

As for synchronous RBNs, this leads to the equation that defines the evolution of perturbations in asynchronous RBNs:

$$d_{t+1} = \frac{5}{8} - \frac{1}{8} (1 - d_t)^K - \frac{1}{4} (1 - d_t). \tag{3}$$

Figure 2 shows the plots of Equation 3 for $K = 1, 2, 3, 4, 10$. The following observations can be made:

1. Critical values for K do not seem to exist as no plot is strictly below the identity function $d_{t+1} = d_t$.
2. Each plot starts from one and the same point, situated at $d_{t+1} = 0.25$. This means that two networks with the same initial state will become different on 25% of their nodes at the next time step. Hence, $d_t = 0 \Rightarrow d_{t+1} = 0$ is no longer valid.

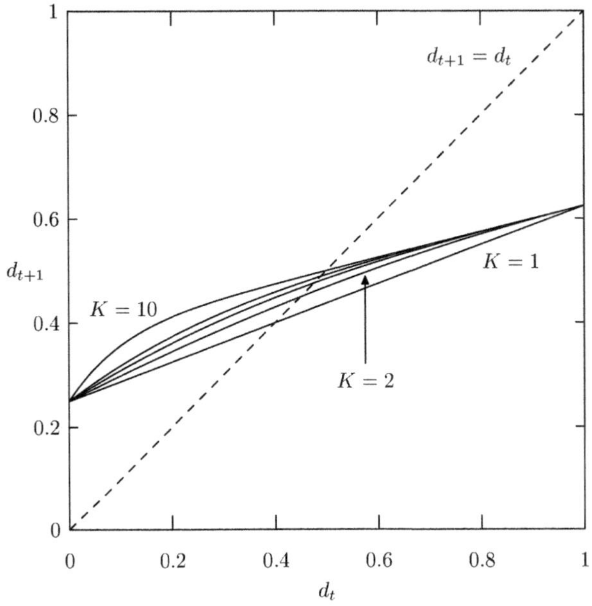

Fig. 2. d_{t+1} in function of d_t for $K = 1, 2, 3, 4, 10$. Note that no critical value of K seems to exist.

From the point of view of the above presented theoretical analysis, it follows that asynchronous RBNs do not seem to be as tolerant to perturbations as synchronous RBNs. In fact when d_t is lower than 0.5, d_{t+1} is always bigger than d_t, independently of K. This implies that, instead of reducing perturbations, asynchronous RBNs create new ones.

Remember that Derrida allowed his networks to change at each time step. This simplification has proved to be correct for synchronous networks, i.e., not affecting the overall network dynamics, however, one might ask whether this hypothesis is correct for asynchronous RBNs too. The next section shall address this question by means of numerical simulations.

5 Numerical Results

Figure 3 shows numerical results for $K = 1, 2, 3, 4, 10$ obtained with networks of $N = 200$ nodes. For each value of d_t between 1 and 200, 200 randomly generated pairs of network states were tested during $t = 600$ time steps. Then, for each value of d_t, the mean value of d_{t+1} was computed.

Compared to Figure 2, Figure 3 shows mainly two differences. The plot obtained for $K = 1$ lies very close to the identity function $d_{t+1} = d_t$. Hence, the theoretical results do not correspond to the numerical simulations for that case. A possible explanation is that $K = 1$ networks are highly ordered ("solid") and that they possess a large number of short attractors. Indeed, two identical net-

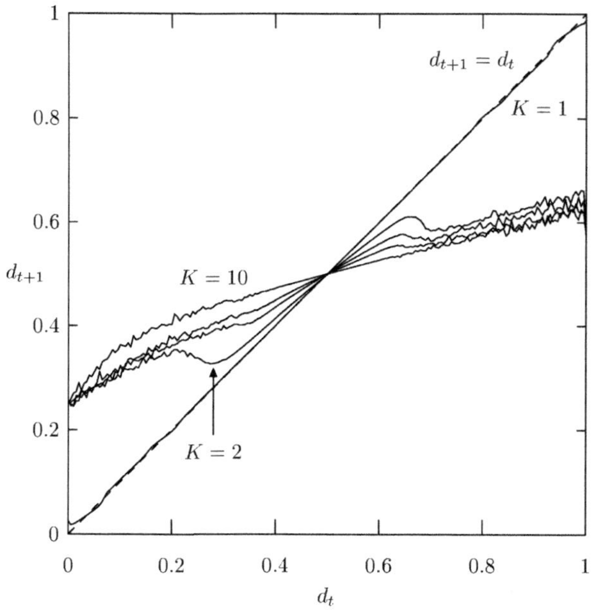

Fig. 3. Numerical results obtained for $K = 1, 2, 3, 4, 10$ with asynchronous RBNs made up of 200 nodes.

works with their state vectors at distance d_t will very quickly end up into different attractors that are separated approximately by the same distance. Remember, however, that our theoretical model is rewired constantly and that therefore no attractors are possible, hence the difference between the simulations and reality.

The second important difference is that our theoretical model also partly fails to predict the behavior of the $K = 2$ plot for $d_t \in [0.2, 0.7]$. The numerical results show that networks with two incoming connections per node can also be very stable within this interval as the plot lies rather close to the identity function.

Finally, the mean error between the theoretical model and the numerical simulations is about 3.3% for $K = 2$, 1.6% for $K = 3$ and falls below 1% for $K > 3$. Hence, the bigger K becomes, the better the model fits to reality.

6 Conclusion and Future Work

We have investigated the dynamic behavior of asynchronous RBNs by means of a method inspired by Derrida's annealed approximation. The main question was whether there is a similar phase transition and critical value for K in asynchronous as in synchronous RBNs.

Our original theoretical approach and numerical simulations revealed that asynchronous RBNs do not have a critical connectivity value similar to synchronous RBNs for K. The asynchronous and nondeterministic updating scheme introduces perturbations that reach about 25% of the nodes and thus prevents the networks to become stable. Although there were some small discrepancies between the theoretical model and the numerical simulations, we can say that our approach to asynchronous RBNs predicts the most important characteristics of their overall dynamics. From the numerical simulations we can conclude that asynchronous RBNs tend to amplify small perturbations, to reduce big ones and to keep them constant when they are located in a region around $d_t = 0.5$. Furthermore, our findings confirm what Harvey and Bossomaier [9] have shown: asynchronous RBNs behave radically different from the deterministic synchronous version.

Synchronous RBNs have mainly been used as abstract models of specific biological systems, however, some other applications exist. In an interesting attempt, for example, Hurford [11] used synchronous RBNs to cast several essential properties of natural languages. He modeled a language as an attractor of a boolean network. To the best of our knowledge no useful and practical applications seems to exist for asynchronous RBNs and one might question whether they are really biologically plausible models. Synchronous updating as used in synchronous RBNs is not usually seen in Nature, although synchronous behavior might arise from the synchronization of asynchronous elements (e.g., [8]). Strict synchrony as well as nondeterministic asynchrony present two extremes—neither is usually observed in Nature. Biological networks do not make use of any global clock (i.e., synchronizing) signal, but usually immediately react to perturbations. In genetic regulatory networks, for example, a change in a gene's activation state may instantly imply other state changes in a deterministic way elsewhere in the network. From this point of view, synchronous RBNs are closer to reality than the asynchronous RBNs studied in this paper. Based on our results, we seriously doubt whether purely asynchronous RBNs are of any interest in modeling biological systems. There exist, however, many variants of asynchronous, deterministic, and lossless information transfer methods (e.g., asynchronous logic based on Muller-C elements) that do not even make use of local clock signals.

Acknowledgments. This work was supported in part by the Swiss National Science Foundation under grant 20-63711.00, by the Leenaards Foundation, Lausanne, Switzerland, and by the Villa Reuge, Ste-Croix, Switzerland.

References

1. J. T. Allanson. Some properties of randomly connected neural nets. In C. Cherry, editor, *Proceedings of the 3rd London Symposium on Information Theory*, pages 303–313, Butterworths, London, 1956.
2. S. I. Amari. Characteristics of randomly connected threshold-element networks and network systems. *Proceedings of the IEEE*, 59(1):35–47, January 1971.
3. W. R. Ashby. *Design for a Brain.* Wiley, New York, 1956.
4. H. Bersini and V. Detours. Asynchrony induces stability in cellular automata based models. In R. A. Brooks and P. Maes, editors, *Proceedings of the Artificial Life IV conference*, pages 382–387, Cambridge, MA, 1994. MIT Press.

5. M. S. Capcarrere. Evolution of asynchronous cellular automata. In J. J. Merelo Guervós, A. Panagiotis, and H.-G. Beyer, editors, *Parallel Problem Solving from Nature*, volume 2439 of *Lecture Notes in Computer Science*, pages 903–912, Berlin, Heidelberg, 2002. Springer-Verlag.
6. B. Derrida and Y. Pomeau. Random networks of automata: A simple annealed approximation. *Europhysics Letters*, 1(2):45–49, 1986.
7. C. Gershenson. Classification of random boolean networks. In R. K. Standish, M. A. Bedau, and H. A. Abbass, editors, *Artificial Life VIII. Proceedings of the Eight International Conference on Artificial Life*, Complex Adaptive Systems Series, pages 1–8. A Bradford Book, MIT Press, Cambridge, MA, 2003.
8. L. Glass and M. C. Mackey. *From Clocks to Chaos: The Rhythms of Life*. Princeton University Press, 1988.
9. I. Harvey and T. Bossomaier. Time out of joint: Attractors in asynchronous random boolean networks. In P. Husbands and I. Harvey, editors, *Proceedings of the Fourth European Conference on Artificial Life*, pages 67–75. MIT Press, Cambridge, MA, 1997.
10. J. J. Hopfield. Neural networks and physical systems with emergent collective computational abilities. In *Proceedings of the National Academy of Sciences of the United States of America (PNAS)*, volume 79, pages 2554–2558, April 1982.
11. J. R. Hurford. Random boolean nets and features of language. *IEEE Transactions on Evolutionary Computation*, 5(2):111–116, April 2001.
12. S. A. Kauffman. Metabolic stability and epigenesis in randomly connected genetic nets. *Journal of Theoretical Biology*, 22:437–467, 1968.
13. S. A. Kauffman. *The Origins of Order: Self-Organization and Selection in Evolution.* Oxford University Press, New York; Oxford, 1993.
14. R. Milo, S. She-Orr, S. Itzkovitz, N. Kashtan, D. Chklovskii, and U. Alon. Network motifs: Simple building blocks of complex networks. *Science*, 298:824–827, October 25 2002.
15. E. A. Di Paolo. Rhythmic and non-rhythmic attractors in asynchronous random boolean networks. *Biosystems*, 59(3):185–195, 2001.
16. L. I. Rozonoér. Random logical nets I. *Automation and Remote Control*, 5:773–781, 1969. Translation of Avtomatika i Telemekhanika.
17. B. Schönfisch and A. de Roos. Synchronous and asynchronous updating in cellular automata. *BioSystems*, 51(3):123–143, September 1999.
18. O. Sporns and G. Tononi. Classes of network connectivity and dynamics. *Complexity*, 7(1):28–38, 2002.
19. W. R. Stark and W. H. Hughes. Asynchronous, irregular automata nets: The path not taken. *BioSystems*, 55(1-3):107–117, February 2000.
20. S. H. Strogatz. Exploring complex networks. *Nature*, 410:268–276, March 8 2001.
21. C. Teuscher. *Turing's Connectionism. An Investigation of Neural Network Architectures.* Springer-Verlag, London, September 2001.
22. G. Weisbuch. *Complex Systems Dynamics: An Introduction to Automata Networks*, volume 2 of *Lecture Notes, Santa Fe Institute, Studies in the Sciences of Complexity.* Addison-Wesley, Redwood City, CA, 1991.

Artificial Organisms That Sleep

Marco Mirolli[1,2] and Domenico Parisi[1]

[1] Institute of Cognitive Sciences and Technologies,
National Research Council
Viale Marx 15, 00137, Rome, Italy
parisi@ip.rm.cnr.it
[2] Philosophy and Social Sciences Department,
University of Siena
Via Roma 47, 53100, Siena, Italy
mirolli2@unisi.it

Abstract. Populations of artificial organisms live in an environment in which light is cyclically present (day) or absent (night). Since being active during night is non-adaptive (activity consumes energy which is not compensated by the food found at night) the organisms evolve a sleep/wake behavioral pattern of being active during daytime and sleeping during nighttime. When the population moves to a different environment that contains "caves", they have to get out of a cave although the dark conditions of the cave may tend to induce sleep. We study various solutions to these problems: evolving a light sensor, evolving a biological clock, evolving both a light sensor and a biological clock. The best solution appears to be evolving a light sensor that modulates a biological clock, a solution which may also be appropriate to solve other problems such as adapting to seasonal changes in daytime length.

1 Introduction

A population of organisms lives in an environment that contains randomly distributed food elements. The reproductive chances of each individual depend on the individual's ability to eat. The sensory receptors inform the individual about the location of the food elements and the individual responds to this information by approaching the food elements and eating them. Individuals that eat more are more likely to leave offspring than those that eat less. In fact, it is the selective reproductive rate that originally creates the ability to find food in these organisms. Notice that moving in the environment has a cost in terms of energy. For each movement an individual's energy is reduced by some amount. However, the energy acquired by eating a single food element is considerably greater than that consumed at each time step by moving around in the environment. Therefore, the organisms tend to move all the time.

Consider now a slightly different scenario. Everything is identical with the exception that periodically the light which illuminates the environment goes away. Cyclically, a number of time steps with light (day) are followed by a number of time steps with no light (night), and then light returns, and so on. When it is dark the organisms cannot see well so that they can make mistakes in identifying the location of food elements. As a consequence, it is uneconomical for them to move around in search of

W. Banzhaf et al. (Eds.): ECAL 2003, LNAI 2801, pp. 377–386, 2003.

food at night in that the energy spent by moving around may not be compensated by the energy acquired by eating. For the organisms it would be adaptive to develop a somewhat more complex behavior: sleep (don't move and therefore don't consume energy) at night, and be active (move around looking for food) during daytime. The individuals that develop this more complex behavioral pattern, which includes not only a component of food finding ability but also a component of appropriate sleep/wake behavior, are more likely to reproduce than the individuals that perhaps are very good at finding food but don't sleep or stay awake appropriately.

What kind of organisms can develop this more complex behavior? How should the nervous system that controls their behavior be organized in order to make this more complex behavior possible? While the sleep/wake behavioral pattern is a classical research topic with real organisms [3][6], little work has been done on artificial organisms or using simulations. (For neural network models of possible functions of sleep, see [4][10]. For non-agent-based mathematical models of sleep regulation, cf. [1][2]). In this paper we study the evolutionary emergence of the sleep/wake pattern by describing various simulations in which we manipulate the nervous system that controls the behavior of the organisms and explore the consequences of possessing different types of nervous systems. (For a general description of the approach, cf. [8].) In Section 2 we describe what is common to all the simulations. In Section 3 we describe what makes the various simulations different and the results of the simulations. In Section 4 we discuss the results.

2 Night and Day

The environment is a square of 15x15=225 cells. Each individual organism lives alone in its own copy of the environment. At any given time the environment contains 3 food elements, randomly distributed. When the organism happens to step on a food element, it eats the food element and a new food element appears in a randomly selected position. Each food element contains 30 energy units. One unit of energy is consumed at each time step when the organism is moving while no energy is consumed when the organism is not moving (it sleeps).

The entire lifetime for all individuals lasts 9 days. One day is 24 hours of 60 minutes each, where 1 minute corresponds to a single input/output cycle of the organism's neural network. Hence an individual lives a total of 12960 minutes. One day includes 12 hours of light (daytime) and 12 hours of dark (night). Both light and dark do not appear suddenly but they set in gradually.

In all simulations the organisms' behavior is controlled by a core neural network with two sensory input units encoding the position of the nearest food element, three hidden units, and two motor output units encoding the organism's movements (Fig. 1a). Among the output options there is a "do nothing" option, and when this option is chosen the organism does not move and we say that it "sleeps". When the other output options are chosen the organism is "awake" and it turns or moves around in the environment searching for food.

The activation level of the 2 input units encodes the location of the nearest food according to the following schema:

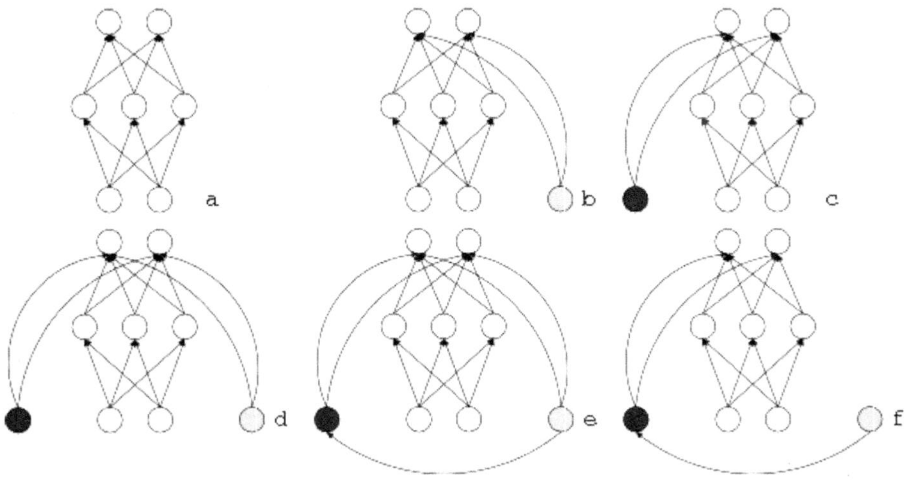

Fig. 1. Core neural network common to all organisms (a). Network with light sensor (b). Network with biological clock (c). Network with both light sensor and biological which independently influence behavior (d). Network with light sensor both independently influencing the organism's behavior and modulating the biological clock (e). Network with light sensor modulating the biological clock (f). (The network architectures were chosen to be as simple as possible.)

Food element exactly in front of the organism = 1;1
Food element exactly to the left of the organism = 1;0
Food element exactly to the right of the organism = 0;1
Food element somewhere in the front left quadrant = 1; 0,5
Food element somewhere in the front right quadrant = 0,5;1
Food element in the back of the organism = 0;0

The reliability of the organism's perceptual system depends on the light conditions of the environment. With probability = 1- L, where L is how much light is currently present in the environment, the correct input is replaced by a randomly selected input.

The 2 output units encode the organism's response in the following way:

1;1 = go one cell forward
1;0 = turn 90° to the left
0;1 = turn 90° to the right
0;0 = don't move (sleep).

The initial population is made up of 200 individuals whose neural network's weights are randomly selected in the range between –2 and +2. At the beginning of its life, each individual is positioned in a randomly selected location in the environment and at the end of life the individual is assigned a fitness value which increases with the number of food elements eaten and decreases with the number of time steps in which the individual has been active, i.e., in which its output has been different from 00. The fitness formula is the following:

F = (30 x number of food elements eaten – number of active time steps)

The 20 individuals with highest fitness are selected for reproduction. They generate 10 offspring each and the new 20x10=200 individuals constitute the second generation (reproduction is asexual). An offspring inherits the same connection weights of its parent except that each weight has 25% probability to have its value changed by adding or subtracting a quantity randomly selected between –1 and +1 to the current weight value. (This high mutation rate has been chosen to avoid convergence on sub-optimal behavior of always sleeping.)

The population lives for 1000 generations in the environment we have described which contains only the food elements. Then, for the next 1000 generations, i.e., from generation 1001 to generation 2000, the population moves to a different environment in which 85 out of the total 225 cells are "caves". In a "cave" it is deep dark and therefore when an individual enters a cave it perceives the food elements very badly. Caves represent a problem for our organisms. If the organisms have developed a behavioral pattern according to which they are active when there is light and they sleep when it is dark, they may simply go to sleep when entering a cave and never awake again since light never returns in a cave. How can they solve this problem?

We describe the main results of our simulations (20 replications for each simulation) using two measures: (a) total quantity of energy (= fitness), (b) total amount of time spent sleeping vs. being awake.

3 Simulations

3.1 Light Sensor

If an organism has absolutely no way to know when there is light in the environment (day) and when it is dark (night), it cannot behave differently during daytime and during nighttime. The best strategy in these circumstances is to be always active looking for food but this strategy is less than optimal given that being active at night consumes energy which is not compensated by the food found. However, if the organisms are informed by their senses about the light conditions of the environment, they can exploit this information by developing a better adaptive pattern: they can be active during the day and sleep at night. In the first simulation the organisms' neural network includes an additional input unit (light sensor) which is directly connected with the motor output units (Fig. 1b). The activation function of this unit is 1- L (light). Therefore the unit reaches its maximum activation value at midnight (deep dark).

The population develops the desired more complex behavior in few generations and in all 20 runs of the simulation. The organisms tend to sleep half of their life, i.e., a total of around 6000 cycles (Fig. 2a), they sleep at night and are awake during daytime, and they tend to sleep in a single long sleep episode with perhaps 6-7 short sleep episodes both at dusk and at dawn, that is, just before and after the nightly long sleep episode. Their energy at the end of the first 1000 generations is quite high: it is around 15500 energy units for the best individual and around 14000 units for the average individual (Fig. 2b).

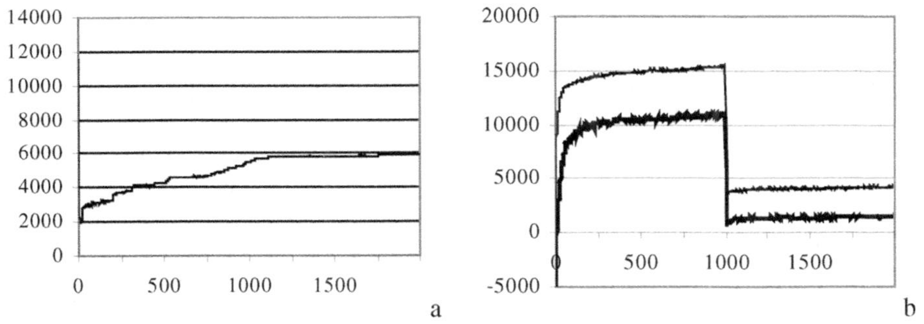

Fig. 2. Organisms with light sensor. Total sleeping time for average individuals (a) and total fitness (energy) for the best (thin line) and average (thick line) individuals (b). Generations 1-1000: environment without caves (average of 20 replications). Generations 1001-2000: environment with caves (average of 15 replications).

The possession of a light sensor provides a very efficient solution for the problem of sleeping at night in the environment in which the population lives for the first 1000 generations, an environment in which there are no caves and therefore "dark" univocally means "night". However, when the population moves to the environment with caves in which "dark" does not univocally means "night", the population tends to crash. In the new environment the information provided by the light sensor becomes ambiguous. "Dark" can mean "night", and in these circumstances the organism should respond by sleeping. But it can also mean "cave", and in these circumstances the organism should respond not by sleeping but by getting out of the cave. Hence, if the organism's sleep-wake behavior relies uniquely on the light sensor, the organisms will be disadapted to the new environment. In fact, in 15 out of 20 replications of the simulation, when an organism enters a cave it falls asleep and it never wakes up again since light never returns in a cave. The organisms sleep all the time and their energy goes to zero. In only 5 out of 20 replications (not included in the data of Fig. 2: Generations 1001-2000) the population shows some re-adaptation to the new environment. In these replications the organisms evolve new connection weights that allow them not to rely only on the light sensor in deciding whether to sleep or be active (consider that their motor output depends also on the changing input encoding food location) so that when they enter a cave they may occasionally respond by moving ahead and leaving the cave. However, their sleeping behavior is seriously disturbed (they have an average of 200 sleep episodes in a day) since with this re-adaptation they may happen to sleep even during daytime.

At any rate, given most initial conditions (replications of the simulation), the population with the light sensor simply would become extinct in the new environment. How can the problem be solved? How can organisms sleep during the night and be active during daytime if sensory information about the light conditions of the environment is ambiguous, sometimes meaning "night" and sometimes "cave" - which require two different responses. One solution is to develop a biological clock.

3.2 Biological Clock

In this second simulation there is no light sensor but the organisms evolve a biological clock that controls their sleep-wake behavior. The biological clock determines the activation level of an internal unit (biological clock unit) which is directly connected with the motor output units (Fig. 1c). The genetic information contained in the unit (i.e., in the nucleus of the neuron which is simulated by the unit) imposes a cyclical activation level to the unit [5][7][9]. The cyclical activation depends on two parameters which are encoded in an individual's inherited genotype together with the connection weights of the individual's neural network: the total length of the clock's cycle and the cycle's time of onset, that is, its synchronization with the day/night cycle. The best adapted individuals will have a biological clock's cycle of 24 hours and the cycle will be synchronized with the day/night cycle.

Evolving a biological clock that can appropriately regulate the sleep-wake behavioral pattern requires that two different things evolve at the same time: (1) the parameters of the biological clock, and (2) the synaptic weights of the connections linking the biological clock unit to the motor output units. In the simulation with the light sensor evolution's only task was to evolve the appropriate connection weights linking the light sensor unit to the motor output units since the activation level of the light sensor unit itself was determined by the external physical environment and it did not originate from inside the organism's body. In this new simulation, the activation level of the biological clock unit is determined internally and evolution has to take care of both aspects of the biological clock mechanism. Hence, its task is more complex.

The results for the simulation with the biological clock are shown in Fig. 3.

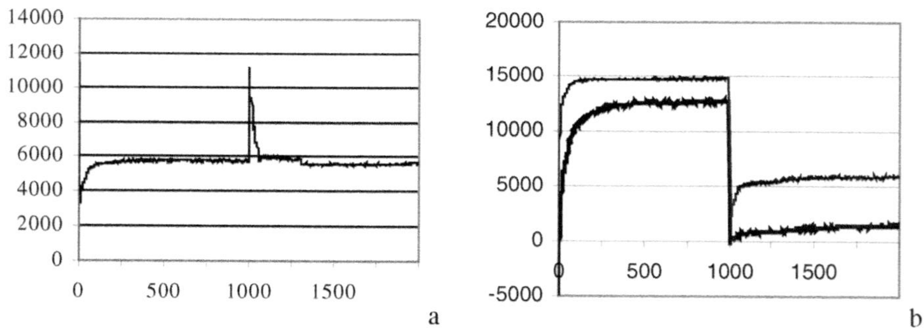

Fig. 3. Organisms with biological clock. Total sleeping time for average individuals (a) and total fitness (energy) for the best (thin line) and average (thick line) individuals (b). Generations 1-1000: environment without caves (average of 18 out of 20 successful replications). Generations 1001-2000: environment with caves (average of 18 out of 18 successful replications).

For the first 1000 generations, when the organisms live in the environment with no caves, the results are similar to those of the simulation with the light sensor except that (1) the population with the biological clock evolves more slowly (this is clear with respect to total sleeping time but is also true of fitness), and (2) the average fitness remains somewhat lower (11000 energy units vs. 14000). Furthermore, while the simulation with the light sensor was successful for the first 1000 generations in all 20 replications, the simulation with the biological clock succeeds in only 18 out of 20 replications. All these differences appear to be due to the fact that for the reasons discussed above the biological clock mechanism is more complex to evolve than the light sensor.

However, what we are interested in are the results of the second part of the simulation, from generation 1001 to generation 2000. In all 18 successful replications of the simulation, the organisms with the biological clock are still able to acquire enough energy when they move to the environment with caves. In fact, their sleep/wake pattern is not affected at all: even in the environment with caves the organisms sleep only half of their time and they sleep at night. Their total energy is lower than in the environment without caves but this is because the environment containing the caves is an intrinsically more difficult environment, independent of the sleep-wake behavior of the organisms. In the environment with caves 85 of the 225 total cells of the environment, the "cave" cells, are always dark, which means that when an individual enters one of these cells food cannot be seen well even during daytime. Hence, the environment with caves has a lower "carrying capacity" than the environment with no caves. This is reflected in the lower energy of the population when it moves to the environment with caves.

The possession of a biological clock has the important consequence that the organisms have no special problems and can remain adapted when they move to the new environment containing the caves. The organisms of the simulation with the light sensor fell permanently asleep when they entered a cave since for them "night" and "cave" were the same thing: no perceived light. The new organisms "know", because their biological clock tells them, when it is night and when it is day, and they have this information both when they are in the open environment and when they are in a cave. Since they do not depend on sensory access to light conditions for their knowledge of night and day, if the parameters of their biological clock are appropriately chosen they sleep when it is night and they are active when it is daytime independent of whether they are in the open environment or in a cave. Furthermore, their sleeping behavior does not appear to be disturbed as it was in those few replications of the simulation with the light sensor in which there was some re-adaptation to the new environment.

3.3 Both Biological Clock and Light Sensor

A third possibility is to have a neural network with both a light sensor and the biological clock. We have explored three different arrangements: (1) the light sensor and the biological both independently have an influence on the organism's behavior (the light sensor unit and the biological clock unit are directly connected with the motor output units: Fig. 1d); (2) the light sensor and the biological clock directly influence the organism's behavior but the light sensor also modulates the biological clock (the light sensor unit is connected with both the motor output units and the biological clock unit: Fig. 1e); (3) the light sensor has the only role to modulate the biological clock (Fig. 1f).

The results indicate that solution (1) gives the worst results from all points of view. It may be useful to have both a light sensor and a biological clock but the two mechanisms for regulating sleep should not function independently from each other but they should interact and be coordinated together within the brain. This is what happens in solutions (2) and (3), where the activation level of the biological clock unit depends both on the biological clock (an internally generated input) and on the activation level of the light sensor (an input from the external environment). Solution (2), where the light sensor both directly influences behavior and modulates the biological clock, gives the best results in terms of fitness but it is very difficult to evolve because of its complexity so that in almost half of the replications of the simulation the population crashes.

Solution (3), where the light sensor has the only task to modulate the biological clock, seems to give the best overall results (Fig. 4).

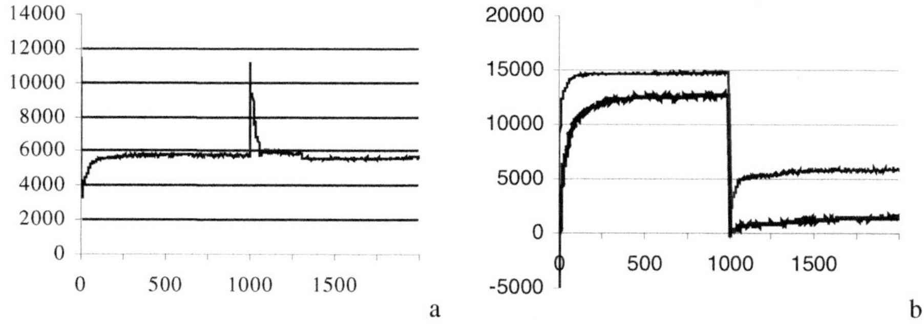

Fig. 4. Organisms with light sensor which modulates the biological clock. Total sleeping time for average individuals (a) and total fitness (energy) for the best (thin line) and average (thick line) individuals (b). Generations 1-1000: environment without caves (average of 20 replications). Generations 1001-2000: environment with caves (average of 20 replications).

First, all the 20 replications of the simulation succeed both in the first 1000 generations (environment without caves) and in the subsequent 1000 generations (environment with caves). Second, the neural network in which the light sensor modulates the biological clock has the advantage, with respect to the neural network with the biological clock but no light sensor, that the appropriate sleep/wake behavioral pattern evolves earlier, and the advantage, with respect to the neural network with the light sensor but no biological clock, that the correct sleep/wake behavioral pattern is maintained in the environment with caves (after a few generations of re-adaptation), at least in terms of total sleeping time, although the number of daily sleep episodes is greater than in the condition with only the biological clock. Finally, the neural network of Solution (3) reaches the highest fitness, at least for the best individual, in the environment with caves.

4 Discussion

If for a population of organisms it is adaptive to be active and look for food during daytime and to sleep at night the population is able to evolve the appropriate sleep/wake behavioral pattern provided that the organisms either directly sense the light conditions of the environment or can evolve an internal biological clock that tells them when to be active and when to sleep. In our evolutionary scenario the reason why it is adaptive to sleep at night is that when it is dark food cannot be seen well and therefore the little food which can be found does not compensate for the energy spent by being active at night. Future work might involve using our simulation scenario to investigate other possible reasons why it is adaptive to sleep. For example, one might need to sleep in order to eliminate dangerous substances accumulated when one is awake [3][6] or one might need to sleep in order to consolidate learning [4][10].

We have explored two different mechanisms that may underlie an organism's sleep/wake behavioral pattern: a light sensor and a biological clock. The light sensor is easier to evolve because the population has only to evolve the appropriate neural connection weights for utilizing the information directly provided by the senses concerning the light conditions of the environment. However, the information provided by the light sensor becomes misleading if it is possible for the organism to enter a local environment where it is dark during daytime, e.g., a cave. In these circumstances the organism would fall for ever asleep since light never returns in a cave.

Compared to the light sensor the biological clock is a more complex mechanism to evolve. The organism's body must evolve a cyclical mechanism, the clock, with the appropriate cycle length (24 hours) and the appropriate synchronization with the light/dark environmental cycle, and it must evolve the appropriate neural structure (connection weights) for utilizing the biological clock to generate the sleep/wake behavioral pattern. The advantage of the biological clock is that an organism won't fall asleep when entering a cave during daytime because its biological clock tells the organism not to do so.

Organisms can have both a light sensor and the biological clock. (In the real brain the light sensor may correspond to information in the retina directly reaching the neurons of the suprachiasmatic nuclei in the anterior hypotalamus, a subcortical structure. These neurons, influenced by the biological clock and by this input, trigger the pineal gland to produce melatonin that causes the brain to become asleep [5][7][9]) This double mechanism may not only produce better results than each mechanism working alone, as indicated by our simulations, but it might solve other adaptive problems not considered in our simulations. For example, we have assumed an environment in which the length of daytime and nighttime is always the same. This might apply tropical environments but it becomes progressively less true for environments at higher or lower latitudes. In the more northern regions of the Earth, for example, nights are shorter and days longer during the winter and the opposite is true in the summer. Humans tend to respond to these seasonal changes by sleeping more during the winter and less during the summer. The biological clock is an evolved mechanism and by itself it may not be able to keep trace of these relatively short-term environmental changes. The possession of a light sensor, in addition to the biological clock, might solve the problem. If the sleep/wake behavior of the organism uses information from both the biological clock units and from the light sensor unit, the information provided by the light sensor concerning the current light conditions of the environment can modulate the action of the biological clock and induce the observed seasonal changes in the sleep/wake behavior, re-synchronizing this behavior with the seasonally changing environmental conditions [3][6]. This might be another direction of future research.

Simulations of sleep/wake behavior using neural networks that live in and interact with an environment are interesting because they demonstrate an interesting aspect of the complexity of behavior of real organisms. In our organisms the reproductive chances of an individual depend on a complex interaction between the individual's ability to approach food and its sleep/wake behavioral pattern. The two "abilities" are not simply additive. An individual of course must be able to approach a food element efficiently when the food element is perfectly perceived in full daytime and it must also be able to sleep at night and be awake during the day. But if the environment becomes dark during daytime, e.g., when the organism enters a cave, the organism must respond to the input by getting out of the cave and in order to do so it must *both* avoid to fall asleep *and* ignore food.

Another interesting aspect of our simulations is that they make it clear that while organisms are not responsible for the inputs that arrive from the external environment to their neural network and they must only possess (evolve) an ability to respond appropriately to these inputs, the situation is different for the inputs that arrive to an organism's neural network from inside the organism's own body. The organism is responsible for these inputs. It must possess (evolve) not only an ability to respond appropriately to the inputs originating within its own body but also an ability to generate the appropriate inputs inside its body. In our simulations the biological clock mechanism includes both an ability to generate the appropriate inputs for the biological clock unit and the appropriate synaptic weights for the connections linking this unit to the motor output units. Evolution must evolve both aspects - one neural, the other one non-neural – of the biological clock.

References

1. Beersma, D.G.M.: Models of human sleep regulation. Sleep Medicine Review, Vol. 2 (2000) 31–43
2. Borbely, A.A., Achermann, P.: Sleep homeostasis and models of sleep regulation. Journal of Biological Rhythms, Vol. 14 (1999) 557–568
3. Dement, W.C., Vaughan, C.: The Promise of Sleep. Delacorte Press, New York (1999)
4. Destexhe, A., Sejnowski, T.J.: Sleep Oscillations. In: Arbib, M.A. (ed.): The Handbook of Brain Theory and Neural Networks. MIT Press, Cambridge (Mass.) (2003) 1049–1053
5. Hastings, M.: The Brain, Circadian Rhythms, and Clock Genes. British Medical Journal, Vol. 317 (1998) 1704–1707
6. Kleitman, N.: Sleep and Wakefulness. University of Chicago Press, Chicago (1963)
7. Lavie, P.: Sleep-Wake as a Biological Rhythm. Annual Review of Psychology, Vol. 52. (2001) 277–303
8. Parisi, D., Cecconi, F., Nolfi, S.: Econets: neural networks that learn in an environment. Network, Vol. 1 (1990) 149–168
9. Reppert, S.M., Weaver, D.R.: Coordination of Circadian Rhythms in Mammals. Nature, Vol. 418 (2002) 935–941
10. Robins, A., McCallum, S.: The Consolidation of Learning during Sleep: Comparing the Pseudorehearsal and Unlearning Accounts. Neural Networks Vol. 12 (1999) 1191–1206

Visualizing Evolutionary Dynamics of Self-Replicators Using Graph-Based Genealogy

Chris Salzberg[1], Antony Antony[1], and Hiroki Sayama[2]

[1] Section Computational Science, Universiteit van Amsterdam, The Netherlands
[2] Dept. of Human Communication, University of Electro-Communications, Japan
{chris,antony}@phenome.org, sayama@hc.uec.ac.jp

Abstract. We present a general method for evaluating and visualizing evolutionary dynamics of self-replicators using a graph-based representation for genealogy. Through a transformation from the space of species and mutations to the space of nodes and links, evolutionary dynamics are understood as a flow in graph space. Mapping functions are introduced to translate graph nodes to points in an n-dimensional visualization space for interpretation and analysis. Using this scheme, we evaluate the effect of a dynamic environment on a population of self-reproducing loops. Resulting images visually reveal the critical role played by genealogical graph space partitioning in the evolutionary process.

1 Introduction

Research on artificial self-replication has resulted in a variety of systems exhibiting complex evolutionary dynamics[11]. Of crucial interest in the analysis of these systems are the localized events (interaction, mutation) that collectively decide the path of global trends in evolution. Few attempts have been made to visualize the topology of this transition-space in a general way. For example, the method proposed by Bedau and Brown[2] circumvents this problem and purports to characterize the evolutionary activity of genotypes through their relative concentration in a population. While useful as a global indicator, this approach overlooks the very fluctuations which enable evolution to occur: these are the genealogical *links* relating distinct species through their ancestry. Without these links, a gap remains between the global dynamics we observe and the localized interactions which trigger their emergence.

In this paper we attempt to bridge this gap. We do so by introducing a method to transform the space of self-replicator species and their mutations to an abstract graph space where nodes and links represent species and mutations, respectively. Within this new space, temporal evolution of populations and their genealogical connectivity is conceptualized as a *flow*, with individual replicator species classified according to their reproductive capability on a scale between *source* and *sink*. We then show an example of graph-based genealogy visualization applied to a simple self-replicating cellular automaton, the "Evoloop"[9], to demonstrate the effectiveness of our method in capturing the critical role played by genealogical graph space partitioning in the evolutionary process.

W. Banzhaf et al. (Eds.): ECAL 2003, LNAI 2801, pp. 387–394, 2003.

2 Graph-Based Genealogy Analysis

We derive a general, graph-based picture of evolution from a set of basic definitions. In what follows we limit ourselves within asexually reproducing systems where reproduction occurs via binary fission.

2.1 Definitions

Our method assumes an arbitrary system of self-replicators evolving over time in a spatial domain $\mathbf{C}$ which we call *configuration space.* This may be a cellular automata space[5,9], the memory and CPU of a computer[7,13], or any other well-defined domain. The intrinsic structure of individual self-replicators in this domain is described uniquely by a sequence of digits from an arbitrary alphabet which we call an *identifier.* Identifiers may be gene sequence, program code, or any other well-defined modular form that specifies the replicator's evolutionary identity. We call the space of these identifiers Θ.

To each replicator in configuration space we associate a position $r \in \mathbf{C}$ and identifier $\theta \in \Theta$; the pair (r, θ) uniquely describes an *instance.* We group replicators with the same identifier together as a *species.* To keep track of different species we assign an arbitrary but unique index k to each, which is used for this purpose only. We define $\theta(k)$ as a mapping from species index k to its actual identifier.

The creation of a new replicator is defined as a *birth* and is described by the *parent* species θ_p, *child* species θ_c, time of birth t_b, and location of birth r_b. A birth for which $\theta_p \neq \theta_c$ indicates mutation has occurred. Borrowing the idea by Bedau and Brown[2], we summarize the time evolution of birth events over the domain $\mathbf{C}$ by the birth "trigger" function δ_b:

$$\delta_b(\theta_p, \theta_c, t_b, r_b) = \begin{cases} 1 & \text{if a new replicator of } \theta_c \text{ is born from a parent} \\ & \quad \text{of } \theta_p \text{ at time } t_b \text{ at position } r_b, \\ 0 & \text{otherwise.} \end{cases} \quad (1)$$

In what follows, we use the function δ_b as our fundamental quantity. A more complete description would also include an analogous death trigger function δ_d identifying the event of replicator death; for simplicity we omit this function here. Note that most existing analysis of artificial models implicitly assume tree-based genealogy [7,9,13]; equation (1) makes no such assumption. We do however assume that we may identify parent and child in a unique and unambiguous manner, and that births may be precisely tracked in both time and space. While practically impossible in real biological systems, such complete data collection can be done for artificial evolutionary models.

2.2 Genealogy on a Graph

The basis for our genealogy analysis is a transformation from configuration space $\mathbf{C}$ to a directed graph $\mathbf{G}$ which we call a "genealogy graph". Although not strictly necessary, we assume time and space to be discrete hereafter. Definitions describing this graph cover a time window from t_i to t_f, which is written as $\mathbf{T} = (t_i, t_f)$. We assign a node in the graph for each species, and associate it with a unique index k and an initial population $\mathrm{P}(k, t_i)$. Directed edges in this graph represent ancestral links created with the birth of replicators: following detection and identification, a parent node k is assigned to the newborn node l. The pair of species $(\theta(k), \theta(l))$ are henceforth distinguished as parent and child relative to the directed link between them.

Note that nodes in **G** represent *groups*, not *instances*, of self-replicators. Here we aim at systems of what Szathmáry calls "limited heredity replicators" [12], for which the number of possible different types is about equal to or smaller than their population so that the same type may be realized many times during evolutionary exploration processes. Nodes may thus have multiple incoming links corresponding to mutations undergone by instances of several distinct species. For different systems where each birth always produces a novel, almost unique type, each node has one (and only one) incoming link so that conventional tree-based genealogy becomes more relevant.

To track the evolution of graph-based genealogy, we introduce a traversal frequency function $\mathrm{F}(k, l, \mathbf{T})$ describing the number of link traversals (births) from node k (parent) to node l (child) in the interval $\mathbf{T}$, derived from δ_b as:

$$\mathrm{F}(k, l, \mathbf{T}) = \sum_{t'=t_i}^{t_f} \sum_{r \in \mathbf{C}} \delta_b(\theta(k), \theta(l), t', r). \tag{2}$$

From the traversal frequency, we derive three important quantities: the number of incoming link traversals $\mathrm{I}(k, \mathbf{T})$, outgoing link traversals $\mathrm{O}(k, \mathbf{T})$ and buckle (self-link) traversals $\mathrm{B}(k, \mathbf{T})$ in the interval $\mathbf{T}$:

$$\mathrm{I}(k, \mathbf{T}) = \sum_{l \neq k} \mathrm{F}(l, k, \mathbf{T}), \tag{3}$$

$$\mathrm{O}(k, \mathbf{T}) = \sum_{l \neq k} \mathrm{F}(k, l, \mathbf{T}), \tag{4}$$

$$\mathrm{B}(k, \mathbf{T}) = \mathrm{F}(k, k, \mathbf{T}). \tag{5}$$

In transforming from δ_b to $\mathrm{F}(k, l, \mathbf{T})$, information about spatial distribution and genetic details of replicators have been lost. However, population dynamics and genealogy for our system — the quantities of interest for our analysis — are well-described. Moreover, the emphasis in this formalism is on the dynamics of evolutionary *connectivity* rather than on global cumulative trends. The advantage to such an approach is that we retain statistical properties of individual species within the context of their ancestral links; these links play a critical role in the emergent phenomena which we observe.

2.3 Evolution as Flow

The framework defined above inspires a change in the way we understand the dynamics of evolution. The alternative we propose here is the idea of *evolution as flow*. In this picture, an evolutionary system is composed of a subset of nodes in genealogical graph space, each of which represents a distinct type of replicator to which a population and collection of active incoming and outgoing links are attributed. Self-replication and variation correspond to the traversals of self-links and outgoing links in graph space, respectively, which induce a collective motion of the global population.

To reflect the above ideas in terms of a *flow*, we derive a quantity which we call the *production*:

$$\mathrm{Prod}(k, \mathbf{T}) = \mathrm{O}(k, \mathbf{T}) + \mathrm{B}(k, \mathbf{T}) - \mathrm{I}(k, \mathbf{T}) \tag{6}$$

According to this definition, replicator species which frequently construct other species as well as replicators of their own species will have a high production; those which are frequently constructed by other species but fail to self-replicate will have negative production. Note that the production is not the same as the fitness of a species; in addition to self-reproduction (B), it also considers the existence of outgoing and incoming links (O and I) and thus emphasizes genealogical connectivity. Borrowing terminology from physics and network theory, we associate a node with highly positive production to an evolutionary *source*, whereas a node with highly negative production we call an evolutionary *sink*. Ranking nodes in this way quantifies the role species play in the evolutionary process. The balance in equation (6) is hence between the capability of a species to *produce* versus the tendency to *be produced*.

To visualize this evolutionary flow in a genealogy graph, we introduce a vector mapping function $\mathbf{M} : \Theta \mapsto \mathbf{R}^n$ from the space of species identifiers to an n-dimensional possibility space. This vector mapping function consists of n real-valued functions $\{\mathrm{M}_j : \Theta \mapsto \mathbf{R};\ \ j = 1, \ldots, n\}$, which, as a whole, map a species $\theta(k)$ to a point $\mathbf{x}^k$ in an n-dimensional space. The point $\mathbf{x}^k = (x_1^k, \ldots, x_n^k)$ represents the co-ordinates of the kth species in this new visualization space.

Note that we have put no requirements on the nature of the functions M_j, hence they can be many-to-one and thus the point $\mathbf{x}^k$ is not necessarily unique to species $\theta(k)$. The optimal choice of functions will map highly connected nodes — groups of species related by strong ancestral links — close to each other in the visualization space. The transformation from graph links to lines connecting the points $\mathbf{x}^k$ will then be more easily understood when viewed in $\mathbf{R}^n$. Note that we have chosen this graph-space mapping for its simplicity and generality; many others exist and could also be used.

3 Example: Evoloop with Dynamic Environment

In this section we apply the proposed method to a simple self-replicating cellular automaton, the "Evoloop" [9]. Definitions discussed in Section 2.1 are linked to model-specific events governing self-replication in the Evoloop CA. From these definitions we construct a mapping from CA space to a 2D visualization space in the manner outlined in Section 2.3. The introduction of a dynamic environment [1] is used to demonstrate the insight gained by this technique.

3.1 Defining Trigger Functions

As our chosen model, the Evoloop satisfies the three conditions necessary for evolution to occur: replication, variation (mutation) and differential fitness (competition)[3]. This model is also simple and scalable, taking the form of 9-state cellular automata with a von Neumann neighbourhood. Structurally, the Evoloop is divisible into two basic components: an inner and outer sheath of square or rectangular shape and a gene sequence of moving signal states. Coordination of the duplication process is controlled via the sequence of genes within the external sheath; mutations occur through extrinsic interaction, leading to a change in the gene sequence of offspring loops. This results in a uniquely emergent process of evolution, one which — due to their robustness and high replication rate — generally favours smaller-sized loops[9].

Our analysis begins with a model-specific definition for *birth*. We use for this purpose the umbilical cord dissolver, a state which appears upon closure of the arm and then retracts towards the parent loop. The local configuration highlighted in the second frame of Fig. 1 signals that such an event has occurred. At birth, loops are assigned a *genotype* $g \in \Gamma_g$ corresponding to the configuration of genes in their gene sequence and a *phenotype* $p \in \Gamma_p$ describing their size (length and width). According to earlier definitions, the space of identifiers Θ is the space of all possible combinations of genotype and phenotype, hence $\Theta = \Gamma_g \times \Gamma_p$ (with a constraint applied so that phenotype must be large enough to contain accompanying genotype). Each $\theta \in \Theta$ contains the necessary information to reconstruct a loop in the exact configuration as it was when it was born. Given the time t_b of the middle frame of Fig. 1 and the location r_b of the umbilical cord dissolver, the birth trigger function $\delta_b(\theta_p, \theta_c, t_b, r_b)$ now has a precise meaning for the Evoloop model. With δ_b defined, graph-based parameters derived in Section 2.2 are now described.

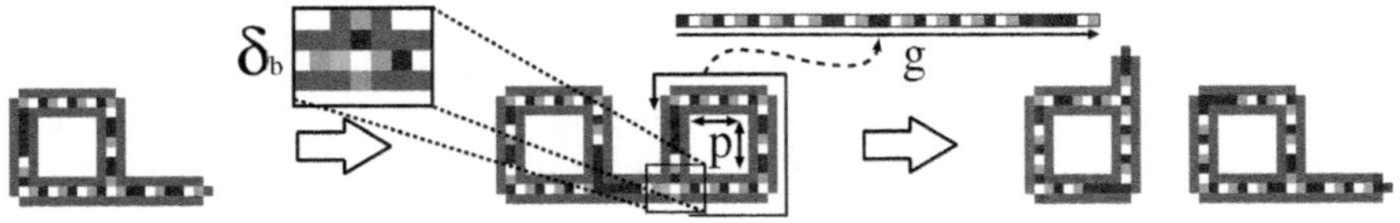

Fig. 1. Self-replication of the Evoloop. g is the genotype, p the phenotype (length and width), and δ_b (middle frame) the configuration which triggers a birth event.

3.2 Mapping to a 2D Space

The fact that $\Theta = \Gamma_{\mathrm{g}} \times \Gamma_{\mathrm{p}}$ presents a convenient separation for constructing mapping functions onto a 2D visualization space. We associate x and y axes with genotype g_k and phenotype p_k for identifier $\theta(k)$, respectively, so that

$$\mathbf{x}^k = \mathbf{M}(\theta(k)) = \begin{pmatrix} \mathrm{M}_x(\theta(k)) \\ \mathrm{M}_y(\theta(k)) \end{pmatrix} = \begin{pmatrix} \mathrm{M}_x(g_k) \\ \mathrm{M}_y(p_k) \end{pmatrix} . \tag{7}$$

For the phenotype-map, we apply a simple transformation $M_y(p = l \times w) = \sqrt{lw}$. Details of the genotype-map are omitted here as they involve model-specific parameters and weighting factors. We instead point to an important feature: that this function is heavily size-dependent yet also highly affected by changes in genotypical configuration. Points in 2D space hence line up near the diagonal, with mutation-induced changes in gene positioning leading to spatial separation in the x-y plane between distinct species. For more details we refer to [8].

For visualization of species according to eq. (7) we use the following scheme:

- Evolutionary *sources* ($\text{Prod}(k, \mathbf{T}) > 0$) are mapped to circles.
- Evolutionary *sinks* ($\text{Prod}(k, \mathbf{T}) < 0$) are mapped to triangles.
- Sizes of sources and sinks are scaled relative to the magnitude $|\text{Prod}(k, \mathbf{T})|$.
- Links between species are represented by lines whose thickness is determined by the cumulative transversal frequency over the time window $\mathbf{T}$. Arrow heads are omitted here to make the plots concise. In the case of bidirectional links, thickness of the higher-frequency direction is used.

3.3 Results

Figure 2 demonstrates results on an 800 × 800 grid, beginning with a single loop of species 9f4945/8×8 (genotype/phenotype) using the compressed hexadecimal notation presented in [1]. Data is binned over periods of $\mathbf{T}$ = 10K iterations. The observed evolution towards small-sized species is expected, however Fig. 2 also reveals the paths through which smaller species achieve domination. In particular, the second frame shows that the majority of well-traversed paths (drawn in black) lead to species of a smaller phenotype. The many kinks in this path indicate the existence of species which are neither source nor sink. These "transient" species [1] replicate a different species than their own, playing the role of intermediary between stable species in the process of evolution; this can be understood as the active mutation discussed by Ikegami[4]. Due to their low numbers, many traditional methods of population analysis would miss these species, yet they play a crucial role in shaping the evolution of populations.

For comparison, we contrast the above result with a different case using the dynamic environment made of "persistent dissolver", another kind of dissolving

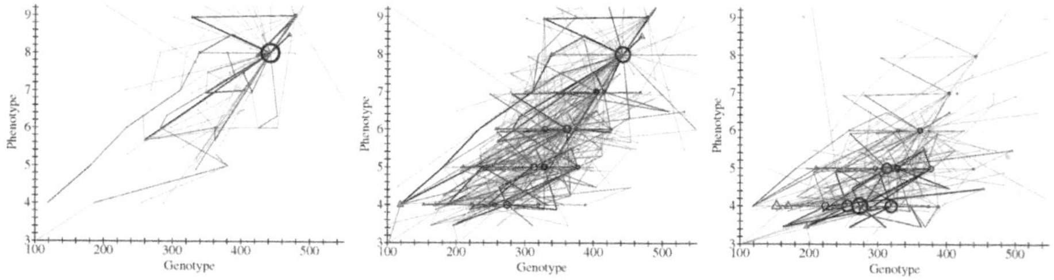

Fig. 2. Evolution in genealogy graph space. Each image covers 10K iterations. Sequence runs from $\mathbf{T} = (0\text{K}, 10\text{K})$ to $\mathbf{T} = (20\text{K}, 30\text{K})$. Visualization scheme described in text.

state that lasts for a substantially longer timescale, introduced in [1]. The effects of this new environment have been found to encourage diversity leading to speciation, punctuated equilibrium, and evolutionary "bottlenecking". Fig. 3 depicts evolution of the same species, now coupled with the dynamic environment. Loop species are observed to collectively explore a broader portion of genealogical graph-space, roaming to both smaller and larger sizes. We notice for instance that the original size-8 loop evolves into a larger size-9 loop, then to another, different size-8 loop (identified as k, l and m in Fig. 3). A strong direct link and many lower frequency links form the path between these species. Hence lateral motion in graph-space (evolution to other species of the same size) has been achieved via loops of other sizes. The fact that this graph-space exploration occurs can be understood in terms of the partitioning of graph-space. A comparison between Fig. 2 and 3 reveals that a number of high-frequency (black) links leading to smaller loops have been cut due to local extinction caused by the dynamic environment. The system has thus explored a much more diverse subset of graph-space. While previously observed[1] via direct data analysis, the visualization of Fig. 3 offers a more complete representation of this behaviour.

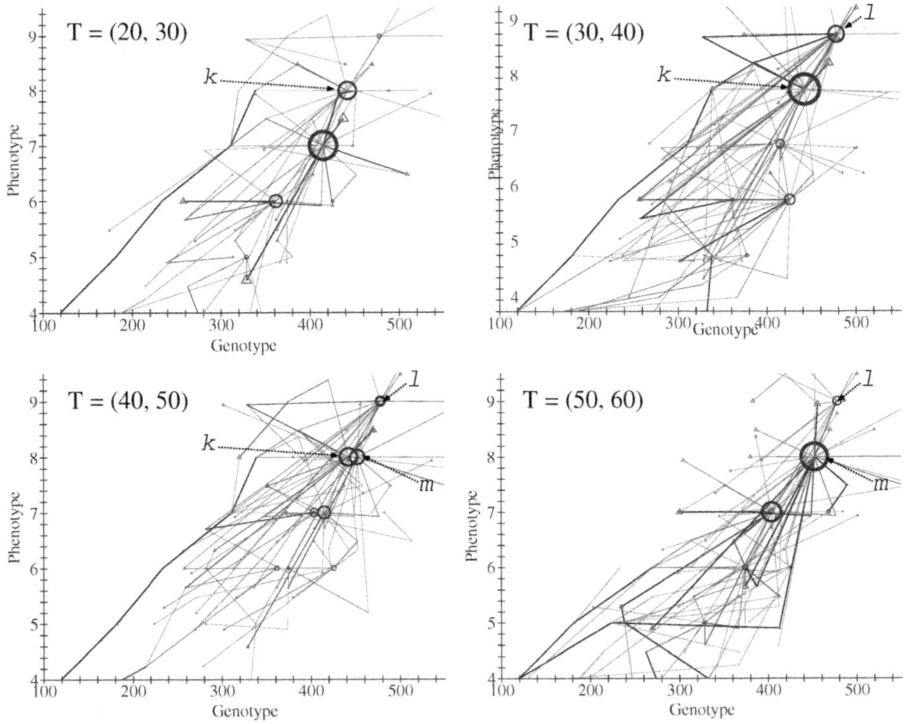

Fig. 3. Evolution in genealogy graph space with persistent dissolver. Each image covers 10K iterations. Sequence reads row-wise left to right, from $\mathbf{T} = (20\text{K}, 30\text{K})$ to $\mathbf{T} = (50\text{K}, 60\text{K})$. Species $k = 9\text{f}4945/8\times 8$, $l = 9\text{fa}4\text{a}29/9\times 9$, $m = 13\text{e}9251/8\times 8$.

4 Conclusion

Our approach in this paper was to emphasize the importance of genealogical connectivity in visualizing the evolutionary dynamics of self-replicators. Results obtained with the Evoloop demonstrated the potential for genealogical complexity in even a simple CA model. Applied to this model, the proposed method was shown to highlight critical evolutionary paths in genealogical graph space.

Related work has been performed by a number of groups. Schuster and Fontana [10] focus on adaptive trajectories of RNA sequences through RNA "shape"-space using the concept of a "relay series". A key distinction is that the relay series is *target*-oriented; gene sequences which do not form part of the relay series are not visualized. This is not true of the method presented here. Lenski et. al. [6] explore the evolutionary origin of complex features. Genealogical analysis in this work is also exact; there are no "missing links" in evolutionary paths. Yet the focus in [6] is on functional genomics for the Avida model rather than evolutionary dynamics for a general system; here the latter target is emphasized.

Current work focuses on applying the visualization presented here to a new class of Evoloop species with multiple graph-space attractors. For more information on this project, see: `http://artis.phenome.org`.

References

1. A. Antony, C. Salzberg, and H. Sayama. A closer look at the evolutionary dynamics of self-reproducing cellular automata. Accepted pending revisions.
2. M. Bedau and C. Brown. Visualizing evolutionary activity of genotypes. *Artificial Life*, 5:17–35, 1999.
3. D. Dennett. *Encyclopedia of Evolution*, pages E83–E92. Oxford University Press, New York, 2002. ed. Pagel, M.
4. T. Ikegami. Evolvability of machines and tapes. *Artificial Life and Robotics*, 3:242–245, 1999.
5. C. Langton. Self-reproduction in cellular automata. *Physica D*, 10:135–144, 1984.
6. R. Lenski, C. Ofria, R. Pennock, and C. Adami. The evolutionary origin of complex features. *Nature*, 423:139–144, 2003.
7. T. Ray. An approach to the synthesis of life. In *Artificial Life II*, volume XI of *SFI Studies on the Sciences of Complexity*, pages 371–408. Addison-Wesley Publishing Company, Redwood City, California, 1991.
8. C. Salzberg, A. Antony, and H. Sayama. in preparation.
9. H. Sayama. A new structurally dissolvable self-reproducing loop evolving in a simple cellular automata space. *Artificial Life*, 5:343–365, 1999.
10. P. Schuster and W. Fontana. Chance and necessity in evolution: lessons from RNA. *Physica D*, 133:427–452, 1999.
11. M. Sipper. Fifty years of research on self-replication: An overview. *Artificial Life*, 4:237–257, 1998.
12. E. Szathmary. A classification of replicators and lambda-calculus models of biological organization. In *Proceedings: Biological Sciences*, volume 260, pages 279–286, 1995.
13. G. Yedid and G. Bell. Macroevolution simulated with autonomously replicating computer programs. *Nature*, 420:810–812, 2002.

The Baldwin Effect Revisited: Three Steps Characterized by the Quantitative Evolution of Phenotypic Plasticity

Reiji Suzuki and Takaya Arita

Graduate School of Human Informatics, Nagoya University
Furo-cho, Chikusa-ku, Nagoya 464-8601, Japan
{reiji, ari}@info.human.nagoya-u.ac.jp
http://www2.create.human.nagoya-u.ac.jp/~{reiji, ari}/

Abstract. An interaction between evolution and learning called the Baldwin effect has been known for a century, but it is still poorly appreciated. This paper reports on a computational approach focusing on the quantitative evolution of phenotypic plasticity in complex environment so as to investigate its benefit and cost. For this purpose, we investigate the evolution of connection weights in a neural network under the assumption of epistatic interactions. Phenotypic plasticity is introduced into our model, in which whether each connection weight is plastic or not is genetically defined and connection weights with plasticity can be adjusted by learning. The simulation results have clearly shown that the evolutionary scenario consists of three steps characterized by transitions of the phenotypic plasticity and phenotypic variation, in contrast with the standard interpretation of the Baldwin effect that consists of two steps. We also conceptualize this evolutionary scenario by using a hill-climbing image of a population on a fitness landscape.

1 Introduction

The Baldwin effect[1] is known as one of the interactions between evolution and learning, which suggests that individual lifetime learning (phenotypic plasticity) can influence the course of evolution without the Lamarckian mechanism. This effect explains these interactions by paying attention to balances between benefit and cost of learning through the following two steps[2]. In the first step, lifetime learning gives individual agents chances to change their phenotypes. If the learned traits are useful for agents and make their fitness increase, they will spread in the next population. The learning acts as a benefit in this step. In the second step, if the environment is sufficiently stable, the evolutionary path finds innate traits that can replace learned traits, because of the cost of learning. This step is known as genetic assimilation[3]. Through these steps, learning can guide the genetic acquisition of learned traits without the Lamarckian mechanism in general. Figure 1 roughly shows the concept of the Baldwin effect which consists of two steps described above.

W. Banzhaf et al. (Eds.): ECAL 2003, LNAI 2801, pp. 395–404, 2003.

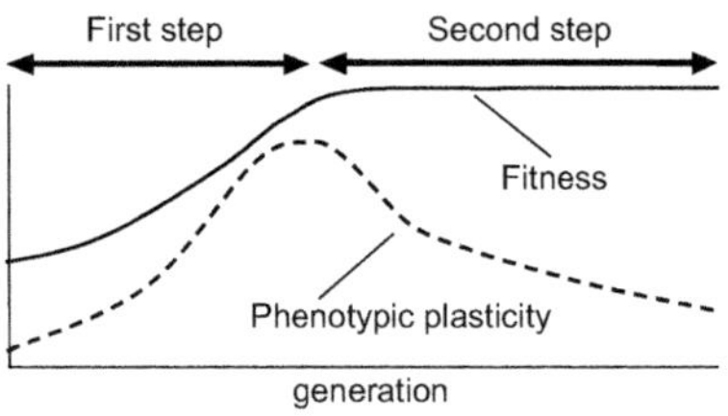

Fig. 1. Two steps of the Baldwin effect

Hinton and Nowlan constructed the first computational model of the Baldwin Effect and conducted an evolutionary simulation[4]. This had a large impact not only on biologists but also on computer scientists. The essential point of this study is that they introduced the quantitative evolution of phenotypic plasticity into their model, in other words, they allowed a population to adjust how much it depends on these two adaptive mechanisms through evolution. They clearly revealed the Baldwin effect by showing the transition of phenotypic plasticity as typically conceptualized in Figure 1, especially in the first step.

Since Hinton and Nowlan, many studies had discussed the interactions between evolution and learning based on various versions of Hinton and Nowlan's model. Harvey observed the existence of the second step in their model by conducting long term experiments and discussed the effect of genetic drift on this step[5]. Turney have discussed the Baldwin effect in view of bias shifting algorithms in machine learning by introducing bias strength (a probability which decides whether each phenotype is determined by learning or genetic information) into a simple model along the line of Hinton and Nowlan's[12]. Watson and Pollack extended their model so as to discuss the evolution of symbiotic interactions by replacing learning (random search) with lifetime interactions among organisms[6]. They demonstrated how symbiotic relationships can guide the genetic make-up of adaptive organisms. Arita and Suzuki discussed the Baldwin effect in dynamic environments by focusing on benefit and cost of learning caused by interactions among agents[7]. They adopted the evolution of strategies for iterated Prisoner's Dilemma, and then introduced a relatively simple learning rule termed as Meta-Pavlov learning into strategies as phenotypic plasticity which is quantitatively allowed to evolve. They found that the population evolved to be cooperative and stable through the two steps of the Baldwin effect.

There have also been many studies which have discussed interactions between evolution and learning with more sophisticated learning mechanism or in more complex environments. Especially, they often adopted the evolution of neural networks. Ackley and Littman's work is another pioneering example of such cases[8]. In their model, each individual seek for food avoiding carnivores in an artificial eco-system using the neural network which maps from the environmental input to the actual behavior. It also has another network which evaluates the environmental conditions and reinforces the connection weights in the former network. They successfully showed that reinforcement learning and evolution of initial connection weights together were more successful than either alone in producing adaptive population that survived to the end of their simulation. Parisi and Nolfi discussed the effect of the correlation between the learning task and the evolutionary task by evolving a population of neural networks which forage distributed foods in the environment[9]. They pointed out that even if the learning task is not so correlated with the evolutionary task, it can facilitate the increase in the fitness of the evolutionary task. Sasaki and Tokoro also studied the relationship between learning and evolution using a model, where individuals learn to distinguish poison and food by modifying the connection weights in the neural network[10]. They compared the Darwinian and Lamarckian evolutionary mechanism in dynamic environments, and then concluded that the Darwinian evolution is more robust against environmental changes.

In the latter cases of studies, they assumed epistatic interactions among loci and enabled continuous change of them based on learning. However, their discussions were mainly based on two extreme cases, that is, experiments with or without learning process because the quantitative evolution of phenotypic plasticity was not introduced. Thus, the two steps of the Baldwin effect were not clearly discussed in comparison with Hinton and Nowlan's experiment. But it seems essential to clarify how the effect of the quantitative evolution of phenotypic plasticity on evolutionary scenario in such complex environments as discussed in the former cases of studies.

Our purpose is to give a valuable insight into the interaction between evolution and learning by focusing on the quantitative evolution of phenotypic plasticity in complex environments. We adopt the evolution of connection weights in a neural network as a situation where there are epistatic interactions among loci. We introduced the phenotypic plasticity into our model, in which whether each connection weight is plastic or not is genetically defined and plastic weights can be adjusted by back-propagation through learning processes.

The rest of the paper is organized as follows: Section 2 describes a model for investigating the interaction between evolution and learning by evolving connection weights and their plasticity in neural networks. In Section 3, based on the results of experiments, we show that the evolutionary scenario consists of three steps unlike the standard interpretation of the Baldwin effect. Furthermore, we investigate the transition of the benefit and cost of learning in each step by considering the effect of difference in evolutionary state of fitness. Section 4 summarizes the paper and conceptualizes this evolutionary scenario by using a hill-climbing image of a population on a fitness landscape.

2 Model

2.1 Genetic Description of Neural Network

We investigate the evolution of connection weights in a neural network as a situation where there are epistatic interactions among loci. There are P individuals in a population and each individual has a feed-forward multi-layer neural network which consists of $N+1$ neurons in the input layer, $M+1$ neurons in

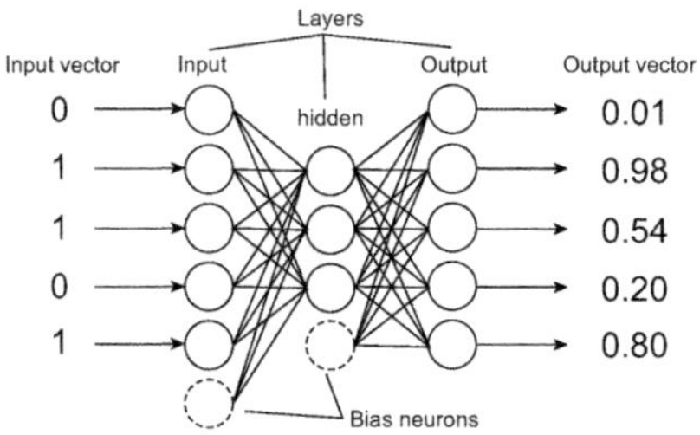

Fig. 2. The topology of neural network in case of $N = 5$, $M = 3$

the hidden layer and N neurons in the output layer as shown in Figure 2. We represent the bias of each neuron by including extra "bias" neurons into input and hidden layers. Each neuron has a standard sigmoidal activation function as its mapping from a weighted sum of input values to the output value.

Each agent has a set of $(N + 1)M + (M + 1)N$ genes termed GW. Each gene in GW decides initial real value of each connection weight as its phenotype (phenotypic value). Here, we introduce the phenotypic plasticity into our model. Each agent also has another set of $(N + 1)M + (M + 1)N$ binary genes GP which decides whether each corresponding phenotype of GW is plastic ("1") or not ("0"). The connection weights whose corresponding bit in GP equals to "1" can be adjusted based on the learning process.

2.2 N-M-N Encoder Decoder Problem

We adopt a version of the N-M-N encoder-decoder problem so as to evaluate the fitness of each agent. This problem is well known as a benchmark problem of learning algorithms. The objective is to achieve a mapping of N input units to M hidden units (encoding) and a mapping of M hidden units to N output units (decoding), in which each N-length input vector consists of "0" or "1" and its desirable output vector is identical to the input vector itself. As we assume a complete encoder-decoder, all possible 2^N combinations of input and output vectors are evaluated in this model.

2.3 Learning

We use a batch-type back-propagation algorithm which includes the effect of plasticity of connection weights as a learning mechanism of each agent[11]. Each individual updates each connection weight $w^{(t)}$ at time t to learn a correct mapping described before for L times by using the following equation:

$$\Delta w^{(t+1)} = -\eta \cdot p \cdot \sum_{v \in V} \left. \frac{\partial E_v(w)}{\partial w} \right|_{w=w^{(t)}} + \alpha \cdot \Delta w^{(t)}, \tag{1}$$

where $\Delta w^{(t+1)}$ represents the difference between $w^{(t+1)}$ and $w^{(t)}$, η is the learning rate, p denotes the phenotypic plasticity ("0" or "1") of w, V is a set of all possible 2^N input vectors, and α denotes the momentum parameter. $\partial E_v(w)/\partial w|_{w=w^{(t)}}$ is the local gradient function of the squared error with respect to w when we compared the input vector v (that is the same as the training vector) with the output vector using connection weights at time t. It can be derived mathematically from back-propagation. The essential point of this equation is that there is an additional value p in the first argument on the right side. If $p = 1$, the weight is updated based on a standard back-propagation mechanism, otherwise it is not updated at all. Thus, we can regard GP as the parameter which decides the learning rate of each connection weight respectively.

2.4 Evolution

After all individuals have finished learning processes, we calculate the fitness of each individual by using the mean square error of all pairs of input and output vectors based on the following equation:

$$fitness = 1.0 - \frac{1}{N \cdot 2^N} \sum_{v \in V} \sum_{i=0}^{N-1} (Out_{v,i} - In_{v,i})^2, \tag{2}$$

where $In_{v,i}$ denotes the input value into the ith neuron in the input layer and $Out_{v,i}$ denotes the output value of ith neuron in the output layer when we inputted the vector v. Note that we do not introduce the explicit cost of learning process itself into the fitness evaluation.

Finally, the population in the next generation is generated by a simple genetic operation as follows: First, the worst individual's sets of genes (GW and GP) are replaced by copies of the best individual's. Then, every gene of all individuals is mutated with a probability p_m, respectively. A mutation in GW adds a randomly generated value within the range $[-d, d]$ to the current weight and a mutation in GP flips the current binary value.

The GP in our model corresponds to a binary case of bias strength in Turney's model[12] as described before. But our model is clearly different from his in the following sense: 1) there are epistatic interactions among loci in our model, 2) we focus on the continuous changes of phenotypic value through learning process, 3) we assume no effects of noise on learning.

3 Experimental Results

3.1 Interaction between Evolution and Learning

We have conducted evolutionary experiments using the following parameters: $P = 20$, $N = 5$, $M = 3$, $L = 10$, $\eta = 0.2$, $\alpha = 0.5$, $p_m = 0.005$ and $d = 1.0$. The initial population was generated on condition that initial connection weights were randomly decided within the range $[-d, d]$ and the proportion of

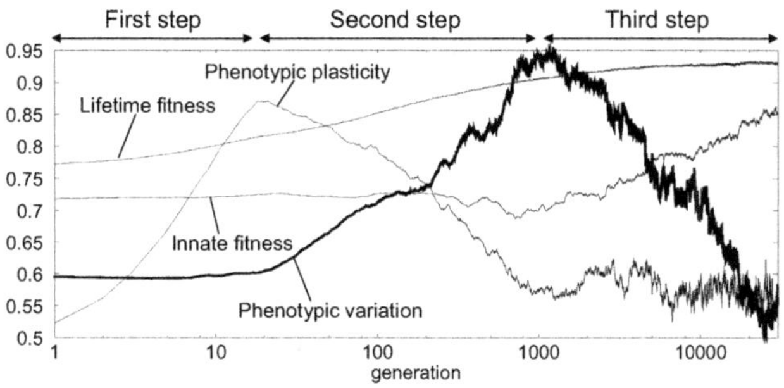

Fig. 3. Evolutionary dynamics of fitness and phenotypic plasticity

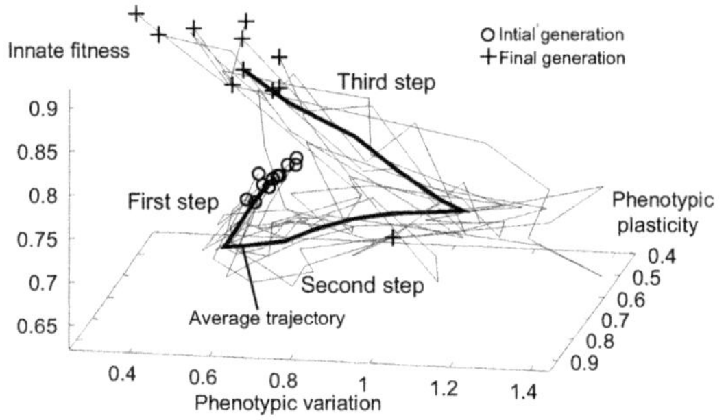

Fig. 4. The evolutionary trajectories of 10 trials

"1" in GP of each individual was uniformly distributed. Figure 3 shows the course of evolution over 30000 generations. The results are averages over 10 trials. The horizontal axis represents the generation in logarithmic scale. The *lifetime fitness* denotes the average actual fitness among all individuals that is calculated after learning process in each generation, and the *innate fitness* is the average potential fitness among all individuals calculated before learning process using initial (genetically defined) connection weights. The *phenotypic plasticity* represents the average proportion of "1" in all GP s and the *phenotypic variation* is the average absolute difference between the initial weight and the resultant weight adjusted by learning process among all plastic phenotypes. Figure 4 shows the evolutionary trajectories of these 10 trials drawn in the space of phenotypic plasticity, phenotypic variation and innate fitness.

By focusing on the transitions of these indices, we have found that the evolutionary scenario consists of three steps unlike the standard interpretation of the Baldwin effect as shown in Table 1, although there is some degree of variation in the actual transitions of these indices among trials as illustrated in Figure 4. Starting from the initial population, we observe an increase in both lifetime fitness and phenotypic plasticity while the innate fitness remained steady, then the phenotypic plasticity rose to the peak value 0.87 at around 20th generation. This means that more plastic individuals could obtain higher fitness and could

Table 1. Transitions among three steps

step (generation)	1st (-20)	2nd (20-1000)	3rd (1000-)
lifetime fitness	increasing	increasing	slightly increasing
innate fitness	steady	steady	increasing
phenotypic plasticity	increasing	decreasing	steady
phenotypic variation	steady	increasing	decreasing
the standard interpretation	1st	1st and 2nd	2nd

occupy the population because of the benefit of learning. Thus, we can regard that the population was in the first step of the Baldwin effect.

Subsequently, while the lifetime fitness still gradually increased, the phenotypic plasticity decreased to about 0.58 until around 1000th generation. This step has both properties of the first and second steps in the standard interpretation of the Baldwin effect in the following two different points of view. When we focus on the transition of the phenotypic plasticity, we can say that the population was in the second step in the sense that the increased fitness by learning was getting dependent of fewer plastic phenotypes. At the same time, in view of the transition of the innate fitness, it still remained flat and the phenotypic variation grew larger. Thus, the population was getting more strongly dependent of remaining plastic phenotypes, then we can also say that the population was still in the first step in this point of view.

Finally, the lifetime fitness still slightly increased and the phenotypic plasticity was approximately kept steady. The innate fitness eventually began to increase, however in contrast, the phenotypic variation decreased. Obviously, the genetic assimilation[3] was occurring in the remaining plastic phenotypes because these initial phenotypic values were getting closer to resultant phenotypic values after learning. Thus, the population entered into the second step in the standard interpretation of the Baldwin effect completely.

3.2 Evolutionary Benefit and Cost of Learning

Here we discuss the benefit and cost of learning in each step by considering the effects of difference in the phenotypic plasticity on fitness so as to clarify the meanings of the three steps. Figure 5 illustrates the effect of difference in evolutionary state on fitness in 10th, 500th and 30000th generations in a trial. For every individual in each generation, we generated copies of the individual whose relative differences in the number of plastic phenotypes between themselves and the original one were ± 1 or ± 2 by flipping the binary values at a randomly selected locus in their GP s for several times, and then evaluated their lifetime fitness. The horizontal axis represents the relative difference in the number of plastic phenotypes, and the vertical axis represents the fitness of individuals after learning process. For example, an individual whose relative difference equals to "+1" is produced by changing randomly selected a "0" to "1" in a copy's GP. The results are averages over 5 operations described above. We can grasp how

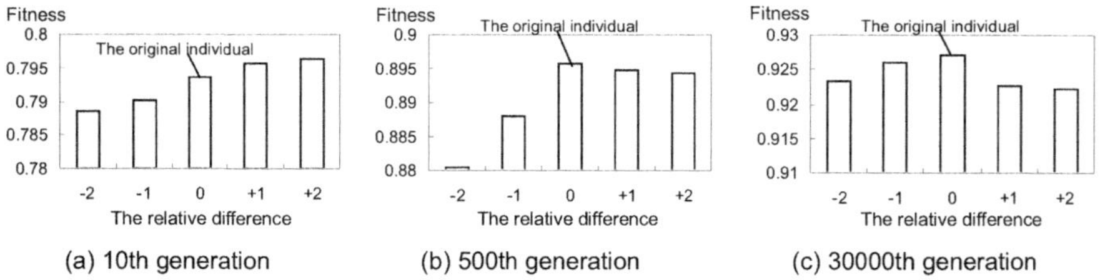

Fig. 5. The effect of difference in evolutionary state on fitness

the selective pressure was working on the evolution of phenotypic plasticity in each step by comparing these figures.

Figure 5(a) shows that the fitness always increased as the number of plastic phenotypes increased, thus learning worked as a benefit in the first step.

In the second step, we observe the peak of the fitness at the number of the plastic phenotypes of the original individual as shown in Figure 5(b). It is noteworthy that there is a negative effect on the fitness increase as the number of plastic phenotypes increases from original individual, although there is no explicit cost of learning as mentioned before. The reason is supposed to be the implicit cost of learning caused by epistatic interactions. In contrast with Hinton and Nowlan's model, a contribution of each phenotypic value to the individual's fitness strongly depends on the other phenotypic values because each value is one of connection weights in the neural network. Similarly, the learning of a value of plastic phenotype based on back-propagation also affects learning processes of the other values of plastic phenotypes, and such an affection caused by additional plastic phenotypes does not always yield an effective increase in the whole fitness. Thus there is the implicit cost of learning which corresponds to the *genetic costs* in DeWitt's classification of costs and limits of phenotypic plasticity[13]. However, such a detrimental effect could be reduced if more learning iterations were conducted, while we adopted small number of learning iterations $L = 10$ in this experiment. This limitation enlarges the difference of fitness caused by the effect because the learning process is required to achieve higher fitness in the limited number of iterations. In fact, the phenotypic plasticity decreased more gradually on condition that L was larger in our experiment.

Also, the decrease in the number of plastic phenotypes still yielded the decrease in the fitness because the basic benefit of learning obtained through the first step have to be maintained. Thus, we can say that the population evolved to have necessary and sufficient amount of phenotypic plasticity through the second step. In addition, we can also say that the decrease in the phenotypic plasticity was not merely due to the random drift in GP.

When the population was in the last step, the fitness tended to decrease more rapidly as the number of plastic phenotypes increased in comparison with the second step as shown in Figure 5(c). It means that the implicit cost of learning got larger in this step. Mayley pointed out that there should be a neighborhood correlation between genotype and phenotype space to guarantee a genetic

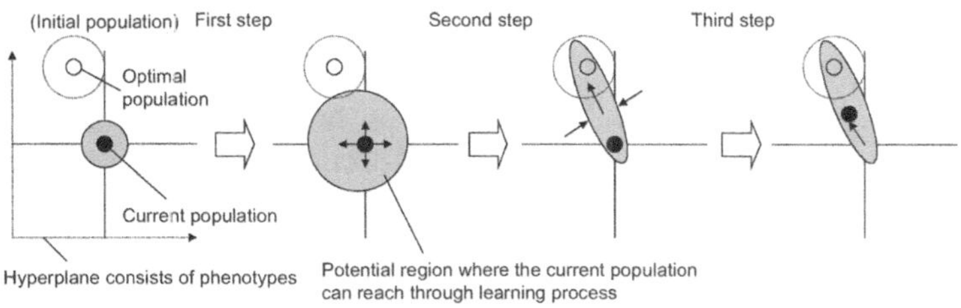

Fig. 6. Three steps of the evolutionary scenario illustrated by a hill-climbing image of a population on a fitness landscape

assimilation to occur[14]. In this model, as the phenotypic plasticity decreases, resultant values of plastic phenotypes obtained by learning tend to correlate more with the genetically specified (initial) values because they are affected by other fewer phenotypes during learning process. Therefore, the decrease in the phenotypic plasticity with the consequent result of genetic assimilation in the third step supports his claim, and suggests that the evolution of phenotypic plasticity automatically enables the population to enter into the second step in the standard interpretation of the Baldwin effect. Besides, if there are no correlations among phenotypes, the evolutionary scenario is supposed to consist of the standard two step process of the Baldwin effect which is approximately similar to that of Hinton and Nowlan's experiment.

4 Conclusion

We have discussed how evolution and learning mutually interact by focusing on the quantitative evolution of phenotypic plasticity based on the evolution of connection weights in a neural network. We observed the Baldwin effect and have found that the evolutionary scenario consists of three steps.

Finally, we conceptualize this evolutionary scenario by using a hill-climbing image of a population on a fitness landscape, although there is another discussion on how to draw an actual landscape of population[15]. As shown in Figure 6, each two dimensional space represents a hyperplane which consists of possible phenotypic values (sets of connection weights). The population can be represented as a particular point in this plane and its height corresponds to the fitness of the population. Let us assume that there was a peak of fitness on the white filled circle and initial population existed on the black filled circle. The gray region represents the potential area where the current population can reach through learning process. In the first step, the population expands its potential region toward every direction by increasing the phenotypic plasticity of population so as to search for the peak of fitness.

However, the implicit cost of learning limits the maximum area of its potential region. Then, the population transforms the shape of its potential region by whittling away excessive phenotypic plasticity and expands it toward the peak.

These correspond to the decrease in the phenotypic plasticity and increase in the phenotypic variation in the second step. Finally, when the potential region reaches the peak of the fitness, the population begins to approach the peak at last, that is, the occurrance of the genetic assimilation in the third step.

References

1. Baldwin, J. M.: A New Factor in Evolution, *American Naturalist*, Vol. 30, pp. 441–451 (1896).
2. Turney, P., Whitley, D. and Anderson, R. W.: Evolution, Learning, and Instinct: 100 Years of the Baldwin Effect, *Evolutionary Computation*, Vol. 4, No. 3, pp. 4–8 (1996).
3. Waddington, C. H.: Canalization of Development and the Inheritance of Acquired Characters, *Nature*, Vol. 150, pp. 563-565 (1942).
4. Hinton, G. E. and Nowlan, S. J.: How Learning Can Guide Evolution, *Complex Systems*, Vol. 1, pp. 495–502 (1987).
5. Harvey, I.: The Puzzle of the Persistent Question Marks: A Case Study of Genetic Drift, *Proceedings of the Fifth International Conference on Genetic Algorithms*, pp. 15–22 (1993).
6. Watson, R. A. and Pollack, J. B.: How Symbiosis Can Guide Evolution, *Fifth European Conference on Artificial Life*, pp. 29–38 (1999).
7. Arita, T. and Suzuki, R.: Interactions between Learning and Evolution: The Outstanding Strategy Generated by the Baldwin Effect, *Proceedings of Artificial Life VII*, pp. 196–205 (2000).
8. Ackley, D. and Littman, M.: Interaction between Learning and Evolution, *Proceedings of Artificial Life II*, pp. 487–509 (1991).
9. Parisi, D. and Nolfi, S.: The Influence of Learning on Evolution, *Adaptive Individuals in Evolving Populations*, pp. 419–428 (1996).
10. Sasaki, T. and Tokoro, M.: Evolving Learnable Neural Networks under Changing Environments with Various Rates of Inheritance of Acquired Characters: Comparison between Darwinian and Lamarckian Evolution, *Artificial Life*, Vol. 5, No. 3, pp. 203–223 (1999).
11. Rumelhart, D. E., Hinton, G. E. and Williams, R. J.: Learning Representations by Back-Propagating Errors, *Nature*, Vol. 323, pp. 533–536 (1986).
12. Turney, P.: How to Shift Bias: Lessons from the Baldwin Effect, *Evolutionary Computation*, Vol. 4, No. 3, pp. 271–295 (1996).
13. DeWitt, T. J., Sih, A. and Wilson, D. S.: Costs and Limits of Phenotypic Plasticity, *Trends in Ecology and Evolution*, Vol. 13, pp. 77–81 (1998).
14. Mayley, G.: Landscapes, Learning Costs and Genetic Assimilation, *Evolutionary Computation*, Vol. 4, No. 3, pp. 213–234 (1996).
15. Provine, W. B.: *Sewall Wright and Evolutionary Biology*, The University of Chicago Press (1986).

Does the Red Queen Reign in the Kingdom of Digital Organisms?

Claus O. Wilke

Digital Life Laboratory 136-93, California Institute of Technology
Pasadena, CA 91125, USA
wilke@caltech.edu
http://dllab.caltech.edu/~wilke

Abstract. I investigate the competition dynamics between two identical clones of digital organisms, for three sets of clones taken from different locations in the fitness landscape. When the clones are taken from the base of a fitness peak, such that beneficial mutations are abundant, then both gain in fitness during the competition (Red Queen effect), until eventually one clone drives the other to extinction. When beneficial mutations are rare or completely absent, on the other hand, then either the clone that finds a beneficial mutation first wins, or the clone that loses the highest-fitness mutant first loses the competition. The time until one of the two strains dies out is in general shorter in the Red Queen case than in the other cases. I discuss the relevance of my findings for competition studies with RNA viruses.

1 Introduction

Two major principles of evolutionary biology have been observed in competition experiments between variants of RNA viruses with identical or almost identical fitness: competitive exclusion and Red Queen dynamic [1,2]. The competitive exclusion principle refers to the ultimate outcome of these competition experiments, and states that when two or more species live on the same resource, eventually all but one will die out [3]. The Red Queen dynamic refers to the initial phase of these competition experiments, where the two competing virus variants both increase in fitness while they remain in roughly equal concentrations. Van Valen [4] had originally proposed the Red Queen dynamic as a metaphor for the struggle for existence of species in complex ecosystems. These species, just like the Red Queen in Lewis Carroll's *Through the Looking Glass*, would constantly have to run (that is, to adapt to changing conditions) just to remain where they were (see also Ref. [5]). The term Red Queen dynamic is most commonly applied to coevolutionary races, for example in host-parasite competition, but in the RNA virus studies and in the present study, Red Queen dynamic refers to the simultaneous adaptation of two clones to the same niche.

Solé et al. [6] studied the competition of neutral virus variants in a simple bitstring model with additive fitness, related to Kauffman's NK model [7], and could confirm both competitive exclusion and Red Queen dynamic in their

W. Banzhaf et al. (Eds.): ECAL 2003, LNAI 2801, pp. 405–414, 2003.

model. Moreover, Solé et al. showed that the competitive exclusion principle follows immediately from the quasispecies equations [8,9] that describe virus evolution.

Unlike competitive exclusion, the Red Queen effect is not a necessary consequence of the quasispecies dynamic, as we can see from a simple thought experiment: Assume that we let compete two viruses that are both situated on top of the highest peak in the fitness landscape. None of the two viruses can accumulate any further beneficial mutations, and the outcome of a competition between them will be determined by genetic drift. Clearly, the Red Queen dynamic can occur only if beneficial mutations are sufficiently abundant, so that adaptation happens on a faster time scale than genetic drift. The additive model of Solé et al. has a fairly high rate of positive mutations for all but the very best sequences in the fitness landscape, which explains why Solé et al. observed the Red Queen dynamic. Simple additive or multiplicative fitness landscapes lead to a smooth increase in average fitness over time [10,11], and such increase has been reported for RNA viruses [12]. However, in many cases beneficial mutations are rare, which leads to the frequently-observed stepwise increase in average fitness [13, 14,15,16].

Here, I present a study of the influence of the immediate neighborhood in fitness space on the competition dynamics of two identical clones of digital organisms (self-replicating computer programs) [17]. I carried out this study using the Avida platform. Avida is similar to Tom Ray's Tierra [18], and has been described extensively in the literature [19,20,21,22]. In Avida, the self-replicating computer programs are rewarded with increased CPU speed when they perform certain logical computations. By changing the bonus structure, that is, by changing which computations are rewarded with what bonuses, the researcher can shape the fitness landscape in which the digital organisms evolve. I studied the intra-strain competition dynamic for three different strains of digital organisms that were located on the top (A), at the base (B), and some distance away from the nearest fitness peak (C) (Fig. 1).

2 Materials and Methods

2.1 Computer Experiments

I used Avida version 1.99, which is available from `http://sourceforge.net/projects/avida`. I used the default setup, apart from the following modifications: I switched off length changes, and set insertion and deletion mutations to zero. As the original ancestor of all organisms in this study, I used the handwritten organism `organism.heads.100`, which comes with the Avida distribution. I used two different environment files (the environment determines the logical operations for which organisms are rewarded, and thus defines the fitness landscape). The first one, environment 1, did not reward any logical computations, so that fitness was a direct function of an organism's replication efficiency. The second one, environment 2, rewarded all possible one-, two-, and three-input logical operations. In this environment, replication efficiency was only a minor

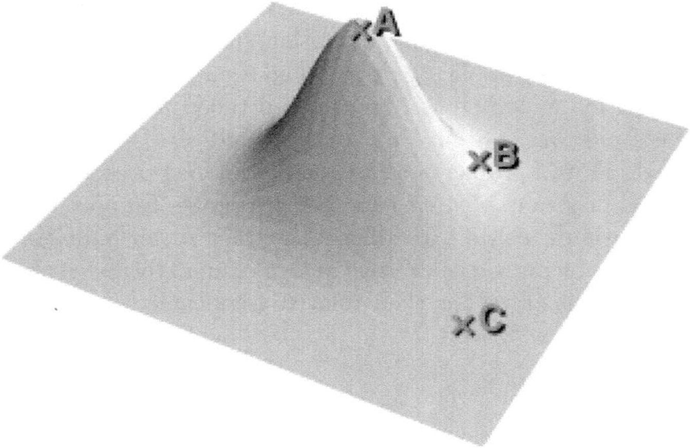

Fig. 1. Schematic illustration of the location in fitness space of the three strains used in this study. Strain A was taken from the top of a fitness peak, strain B from the base of a fitness peak, and strain C from some distance away from the nearest fitness peak. For simplicity of illustration, this figure shows only one fitness peak, whereas in reality the fitness peak of strain A was different from the one of strains B and C.

component of fitness, whereas the successful completion of as many logical computations as possible resulted in high fitness values. The mutation rate in all experiments was 0.75 miscopied instructions per offspring organism, unless explicitly stated otherwise.

I created the strains A, B, and C (as illustrated in Fig. 1) as follows. For strain A, I inocculated a population of $N = 5000$ organisms with the handwritten ancestor, and propagated this population for 14,000 generations in environment 1. Then, I extracted the most abundant organism, which replicated roughly two times faster than the original ancestor. By generation 14,000, the population had not experienced any significant fitness improvements for over 10,000 generations, which showed that it had indeed reached the top of a fitness peak. For strain B, I first evolved from the handwritten ancestor an organism that could perform all possible one- and two-input logical computations. Then, I inocculated a population of size $N = 200$ with this evolved organism, in order to let the organism accumulate mutations. After 50 generations, I extracted from the population a variant with drastically reduced (by a factor of fifteen) fitness. This variant differed in six instructions (out of 100) from its evolved ancestor, and had lost five logical operations. It was therefore at the base of a fitness peak in environment 2, where many additional logical computations could be learned. For strain C, I simply used the handwritten ancestor (also in environment 2). The handwritten ancestor did not perform any logical computations. In Avida, the evolution of the first logical computation is harder than the evolution of additional ones, which guaranteed that C was further away from the fitness peak than B.

I carried out all competition experiments with the following protocol. For each strain, I used three different population sizes, $N = 100$, $N = 1000$, and $N = 10,000$. I did 100 replicates per population size and competing strain. In each competition experiment, I inocculated the population with N identical copies of the strain under study, and marked half of them with an inheritable neutral marker. Then, I propagated the population in the respective environment (environment 1 for strain A, environment 2 for strains B and C) at fixed size N until the population consisted of either all marked or all unmarked organisms. Each generation, I recorded the average and maximum fitness of the marked and unmarked organisms, as well as their relative frequencies.

2.2 Coalescent Theory

For the competition of two clones in a completely flat landscape, we can calculate the distribution of extinction times from coalescent theory [23]. Let $p_j(t)$ be the probability that after t generations, each of the N organisms in the population is a direct descendant of one of a group of only j organisms that were present at $t = 0$. Further, let $m_j = [1 - (1/2)^{j-1}]$ be the probability that this group of j contained both marked and unmarked organisms. Then, the probability that neither the marked nor the unmarked clone is extinct after t generations is

$$P(t_{\text{extinct}} > t) = \sum_{j=2}^{N} m_j p_j(t) \,. \tag{1}$$

The quantity $p_j(t)$ has been given in various forms in the literature [24,25,26]. The following version is based on the result by Tavaré [26]:

$$p_j(t) = \sum_{k=j}^{N} \rho_k(t)(-1)^{k-j}(2k-1)C_{jk} \,, \tag{2}$$

where $\rho_k(t) = \exp[-k(k-1)t\sigma^2/(2N)]$, and C_{jk} satisfies the recursion (for $k \geq j$):

$$C_{11} = 1 \,, \tag{3a}$$

$$C_{jk+1} = \frac{(N-k)(k+j-1)}{(N+k)(k-j+1)} C_{jk} \,, \tag{3b}$$

$$C_{j+1k} = \frac{(k+j-1)(k-j)}{j(j+1)} C_{jk} \,. \tag{3c}$$

The quantity σ^2 is the variance of the offspring distribution. We obtain σ^2 as follows. The organisms in Avida replicate by splitting into two daughter organisms. Therefore, the maximum number of offspring per generation is two (fitness influences the generation time, but not the number of offspring per generation). Random removal of organisms reduces this number to either one or zero in each one-third of the times. Therefore, $\sigma^2 = 2/3$ (if all organisms have equal fitness).

3 Results and Discussion

I measured the time to extinction of the losing clone in all competition experiments and derived the distributions of extinction times. The nine distributions (three strains at three different population sizes) are shown in Fig. 2. The extinction times are the longest for competitions of strain A, intermediate for competitions of strain C, and shortest for competitions of strain B. For strains A and B, the extinction times grow systematically with increasing population size, whereas for strain C, the extinction times grow from $N = 100$ to $N = 1000$, but then decrease again for $N = 10,000$. (Notice that the line with solid diamonds lies on the left of the line with solid squares in Fig. 2b.)

Typical competition dynamics for the three strains are displayed in Figs. 3-5. In all three figures, I show the average and maximum fitness of the winning and the losing clone as a function of time, as well as the relative concentration of the winning clone as a function of time.

For strain A, average and maximum fitness of the two competing clones remained typically almost unchanged until the losing clone disappeared (Fig. 3). Changes in relative clone size were mostly due to genetic drift. However, the theoretical prediction Eq. (1) works only moderately well for the smallest population size, and overestimates the extinction times for the larger population sizes substantially (Fig. 2). I verified that this effect is not a shortcomming of the theoretical prediction by carrying out additional competition experiments at

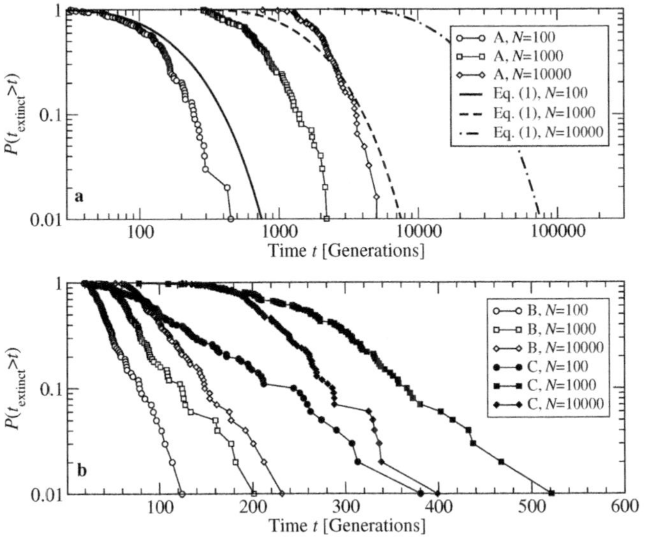

Fig. 2. Distributions of extinction times for all three strains and all three population sizes. I plot the cummulative frequency of all extinction times larger than or equal to the given generation, $P(t_{\text{extinct}} > t)$. Part a shows also the prediction for neutral evolution according to Eq. (1).

zero mutation rate (data not shown). For these experiments, the extinction data were in very good agreement with the theoretical prediction, which implies that the reduced extinction times in the competitions with mutations must be due to the accumulation of deleterious mutations. Indeed, in Fig. 3, we see that the maximum fitness of the losing clone is—after approximately 500 generations—consistently lower than the maximum fitness of the winning clone. Even though the difference in maximum fitness between the two clones is very small, it is sufficient to accelerate the extinction process. And the smaller the losing clone becomes, the more likely it is to experience even further reductions in its maximum fitness. The final stage of the competition is mutational meltdown [27]: Decreasing clone size accelerates loss of the highest fitness mutants, which in turn results in even further reduction of clone size. The clone decreases in size and loses fitness at an ever accelerating pace, until it has disappeared.

Fig. 3 clearly shows that mutational meltdown takes place towards the end of the competition and leads to a reduced extinction time. At first glance it is surprising that mutational meltdown should be responsible for the increasing deviations between theory and measured extinction times as the population size increases. After all, mutational meltdown is commonly associated with small population sizes. The reason why this effect here becomes more pronounced at larger population sizes is the following: When the relative difference in fitness between two mutants is smaller than the inverse of the population size, then these two mutants are effectively neutral, in the sense that they are equally af-

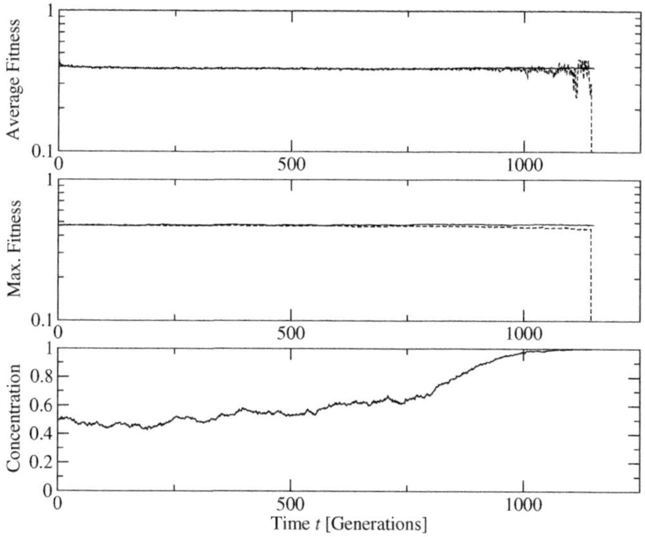

Fig. 3. Typical dynamic of a competition experiment for strain A ($N = 10,000$). The top two graphs show the average and the maximum fitness of the winning (solid lines) and the losing (dashed lines) clones as a function of time. The bottom graph shows the relative concentration of the winning clone as a function of time.

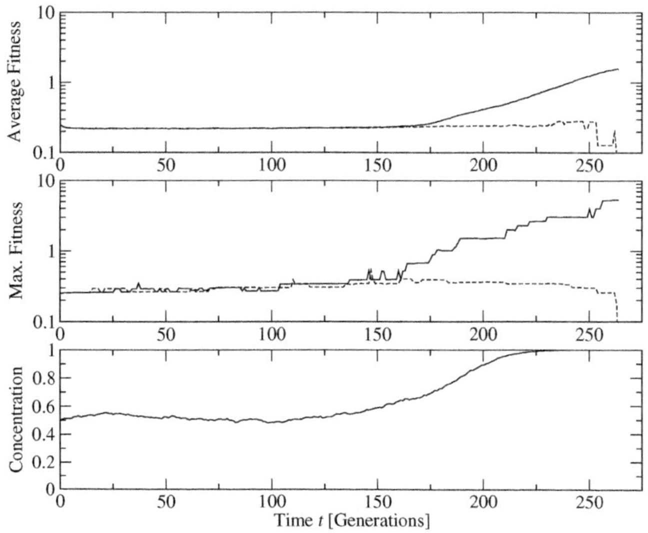

Fig. 4. Typical dynamic of a competition experiment for strain C ($N = 10,000$). The top two graphs show the average and the maximum fitness of the winning (solid lines) and the losing (dashed lines) clones as a function of time. The bottom graph shows the relative concentration of the winning clone as a function of time.

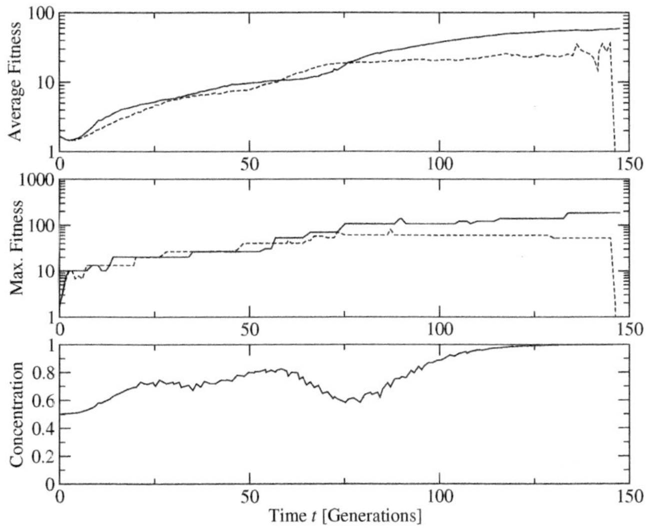

Fig. 5. Typical dynamic of a competition experiment for strain B ($N = 10,000$). The top two graphs show the average and the maximum fitness of the winning (solid lines) and the losing (dashed lines) clones as a function of time. The bottom graph shows the relative concentration of the winning clone as a function of time.

fected by genetic drift [28]. Therefore, larger population sizes can resolve finer fitness differences between mutants. In the case of strain A, the fitness difference between the winning and the losing clone is miniscule, so that at small population sizes, drift is the dominant factor. Once the population size is sufficiently large, however, this small fitness difference renders the population dynamic deterministic, and the clone that loses the fittest mutant first will slowly but surely disappear.

A competition between two clones of strain C typically started out just like one between two clones of strain A. However, often one of the two clones managed eventually to acquire a beneficial mutation with substantial selective advantage, and would then quickly exclude the other clone (Fig. 4). Since beneficial mutations were fairly rare for strain A, the second clone did not have the chance to pick up an even better mutation in the short time that remained before the clone died out. The time to the first beneficial mutation therefore determined the distribution of extinction times, unless it was much larger than the typical time to extinction by genetic drift. In general, the time to the first beneficial mutation grows with decreasing population size, while the time to extinction by drift grows with increasing population size. For $N = 100$, beneficial mutations were very rare, and the extinction times were dominated by the effects of drift. For $N = 1000$, beneficial mutations were still rare, but nevertheless so abundant that the extinction times were clearly shorter than the ones for drift alone. Finally, for $N = 10,000$, beneficial mutations were so frequent that the time to extinction was on average even shorter than it was for $N = 1000$.

For strain B, we observed a competition dynamic very similar to the one described by Solé et al. [6]. Both the marked and the unmarked clone gained substantially in fitness during the competition, and both clones would alternatingly take the lead in fitness gains (Fig. 5). However, this Red-Queen dynamic came at a price: The time to extinction of either clone was consistently shorter than for strains A or C. Apparently, the constantly changing relative growth rates of the two competing clones introduced increased fluctuations in the clone sizes, so that one of the clones was quickly reduced to a size at which it became prone to mutational meltdown, or was at least substantially impaired in its ability to acquire further beneficial mutations.

4 Conclusions

The location in fitness space from where a strain is taken has a strong influence on its competition dynamic. A clear arms race of mutants with ever increasing fitness can only be observed when beneficial mutations are abundant. When beneficial mutations are rare or completely absent, then either the clone that finds a beneficial mutation first wins or the clone that loses the highest-fitness mutant first loses. In general, it seems that a positive mutation rate will always reduce the competition time, so that the loser dies out earlier than it would in the absence of mutations. The Red-Queen dynamic, where both clones acquire mutants of ever increasing fitness, is particularly unstable. In this case, competitions last the shortest.

The results that I have presented here were obtained in computer experiments with digital organisms. Therefore, it is not a priori clear that my conclusions apply directly to RNA viruses. Nevertheless, I believe it is very likely that they do. In particular, the way in which RNA virus strains are typically prepared for competition experiments (frequent transfers at small population sizes, which leads to accumulation of deleterious mutations, and transfers to new environments, where many advantageous mutations can be acquired) are similar to my preparation of strain B. The fact that they show competition dynamics very similar to that of strain B is therefore reassuring. To test my other predictions in a virus system, one would need strains similar to A or C. In principle, it should be possible to prepare a virus strain which is located at or near the top of a fitness peak, by propagating the virus for a long time in a constant environment at a large effective population size. With such a strain, the competition dynamic should be more similar to my strain A, or maybe C, than B. If this turns out to be true, then competition experiments at various population sizes can be used as a reliable tool to map out the neighborhood in fitness space of a particular virus strain.

Acknowledgments. This work was supported by the NSF under contract No. DEB-9981397. I would like to thank D. A. Drummond for interesting discussions on coalescent theory, and C. Adami for helpful comments on the manuscript.

References

1. D. K. Clarke, E. A. Duarte, S. F. Elena, A. Moya, E. Domingo, and J. Holland. The red queen reigns in the kingdom of RNA viruses. *Proc. Natl. Acad. Sci. USA*, 91:4821–4824, 1994.
2. J. Quer, R. Huerta, I. S. Novella, L. Tsimring, E. Domingo, and J. J. Holland. Reproducible nonlinear population dynamics and critical points during replicative competitions of RNA virus quasispecies. *J. Mol. Biol.*, 264:465–471, 1996.
3. G. Hardin. The competitive exclusion principle. *Science*, 131:1292–1297, 1960.
4. L. van Valen. A new evolutionary law. *Evol. Theory*, 1:1–30, 1973.
5. M. Ridley. *The Red Queen: Sex and the Evolution of Human Nature*. MacMillan, 1994.
6. R. V. Solé, R. Ferrer, I. González-García, J. Quer, and E. Domingo. Red queen dynamics, competition and critical points in a model of RNA virus quasispecies. *J. theor. Biol.*, 198:47–59, 1999.
7. S. Kauffman. *Origins of Order*. Oxford University Press, Oxford, 1990.
8. M. Eigen and P. Schuster. *The Hypercycle—A Principle of Natural Self-Organization*. Springer-Verlag, Berlin, 1979.
9. M. Eigen, J. McCaskill, and P. Schuster. Molecular quasi-species. *J. Phys. Chem.*, 92:6881–6891, 1988.
10. L. S. Tsimring, H. Levine, and D. A. Kessler. RNA virus evolution via a fitness-space model. *Phys. Rev. Lett.*, 76:4440–4443, 1996.

11. I. M. Rouzine, J. Wakeley, and J. M. Coffin. The solitary wave of asexual evolution. *Proc. Natl. Acad. Sci. USA*, 100:587–592, 2003.
12. I. S. Novella, E. A. Duarte, S. F. Elena, A. Moya, E. Domingo, and J. J. Holland. Exponential increases of RNA virus fitness during large population transmissions. *Proc. Natl. Acad. Sci. USA*, 92:5841–5844, 1995.
13. R. E. Lenski and M. Travisano. Dynamics of adaptation and diversification: a 10,000-generation experiment with bacterial populations. *Proc. Nat. Acad. Sci. USA*, 91:6808–6814, 1994.
14. S. F. Elena, V. S. Cooper, and R. E. Lenski. Punctuated evolution caused by selection of rare beneficial mutations. *Science*, 272:1802–1804, 1996.
15. W. Fontana and P. Schuster. Continuity in evolution: On the nature of transitions. *Nature*, 280:1451–1455, 1998.
16. E. van Nimwegen, J. P. Crutchfield, and M. Mitchell. Statistical dynamics of the royal road genetic algorithm. *Theoretical Computer Science*, 229:41–102, 1999.
17. C. O. Wilke and C. Adami. The biology of digital organisms. *Trends. Ecol. Evol.*, 17:528–532, 2002.
18. T. Ray. An approach to the synthesis of life. In C. G. Langton, C. Taylor, J. D. Farmer, and S. Rasmussen, editors, *Artificial Life II*, pages 371–408. Addison-Wesley, 1991.
19. C. Adami. *Introduction to Artificial Life*. Springer, New York, 1998.
20. R. E. Lenski, C. Ofria, T. C. Collier, and C. Adami. Genome complexity, robustness and genetic interactions in digital organisms. *Nature*, 400:661–664, 1999.
21. C. Adami, C. Ofria, and T. C. Collier. Evolution of biological complexity. *Proc. Natl. Acad. Sci. USA*, 97:4463–4468, 2000.
22. C. O. Wilke, J. L. Wang, C. Ofria, R. E. Lenski, and C. Adami. Evolution of digital organisms at high mutation rate leads to survival of the flattest. *Nature*, 412:331–333, 2001.
23. J. F. C. Kingman. On the genealogy of large populations. *J. Appl. Prob.*, 19A:27–43, 1982.
24. R. C. Griffiths. Lines of descent in the diffusion-approximation of neutral Wright-Fisher models. *Theor. Pop. Biol.*, 17:37–50, 1980.
25. P. Donnelly. The transient behaviour of the Moran model in population genetics. *Math. Proc. Cambridge Philos. Soc.*, 95:349–358, 1984.
26. S. Tavaré. Line-of-descent and genealogical processes, and their applications in population genetics models. *Theor. Pop. Biol.*, 26:119–164, 1984.
27. M. Lynch and W. Gabriel. Mutation load and the survival of small populations. *Evolution*, 44:1725–1737, 1990.
28. J. F. Crow and M. Kimura. *An Introduction to Population Genetics Theory*. Harper & Row, New York, 1970.

Conditions for Stable Vowel Systems in a Population

Bart de Boer

AI-lab, Vrije Universiteit Brussel, Pleinlaan 2, 1050 Brussel, Belgium
bartb@arti.vub.ac.be

Abstract. This paper describes an investigation of two computer models of how vowel systems can be transferred from one generation to the next. Humans tend to reduce the articulation of the vowels (and other speech sounds) they produce. If infants would learn on the basis of these reduced signals, vowel systems would collapse rapidly over time. As this is not observed in practice, some mechanism must be present to counter it. Two candidate mechanisms are investigated in this paper: compensatory expansion of articulations learned on the basis of reduced speech sounds and learning on the basis of more carefully articulated speech. It turns out that larger vowel systems with central vowels can only remain stable when learning is based on carefully articulated speech.

1 Introduction

This paper investigates under what circumstances vowel systems are stable when they are transferred from generation to generation in a population of language users. It is obvious that stable transfer occurs. This can be deduced from the observation that children learn to reproduce the vowel systems of their parents as closely as possible in many ways. Hence, for example, the subtle distinctions in pronunciation between closely related dialects, which can be identified readily by speakers. Such stable transfer is not directly obvious. After all, in rapid, casual speech (the kind people use most often) articulation of speech sounds is reduced. This means that the acoustic distance between the different speech sounds becomes less. In vowels this is especially noticeable. The simplest assumption one can make about learning of speech by infants is that they make a statistical analysis of the speech sounds they hear. However, in that case one would expect systems of speech sounds to become slightly more reduced in every new generation, as the most frequently occurring speech is reduced. This would lead to a relatively rapid collapse of the system of speech sounds.

Such collapses are not generally observed in the way languages change over time. Sound systems of languages sometimes change rather rapidly over time (e. g. [8]), but such change is different in nature from the collapse due to reduction in articulation. In the case of the collapse of a vowel system, for example, one would expect vowels to move more and more to a central articulation (such as the vowel in the English word *the*) and different vowels to merge when they get too close together acoustically. In reality one observes complex shifts, mergers and splits of vowels (e.g. [4]). However, these changes preserve the general positions of- and distances between the different vowels in the vowel system.

W. Banzhaf et al. (Eds.): ECAL 2003, LNAI 2801, pp. 415–424, 2003.

Therefore the assumption of purely statistical learning must be fallacious. There must be a mechanism that prevents collapse of speech sounds when infants learn them. Two possible mechanisms will be compared in this paper. In order to keep the simulation tractable, only transfer of vowel systems was studied. An extra advantage of vowel systems is that their acquisition and transfer has been relatively well studied.

Both learning mechanisms use the statistical distribution of vowels in the input to determine the position of the vowels that are learned. The first mechanism (the *compensation mechanism*) uses all the available input, but uses the knowledge that the input consists of a reduced version of the original vowel system. It undoes the contraction of the vowel system by applying an expansion that approximates the inverse of the contraction.

The second mechanism (the *infant-directed speech mechanism*) assumes that the input the infant really uses to learn is not as reduced as rapid casual speech. Instead of using all available data for a statistical analysis, the infant only learns on the basis of speech that is articulated more carefully than rapid, casual speech. Such input can be recognized from characteristics of the signal itself (slower speed, clearer intonation, higher volume) or from the setting in which the sound is perceived (one-to-one interactions between the infant and the caretaker, for example).

In a sense, these mechanisms are minimal implementations of learning bias. Bias can be caused by a tendency of a learning mechanism to find representations that are different from the statistical distribution of the input data. It can also be caused by the selection of input data, such that the distribution of the data on which learning is based is different from the distribution of the complete data set. The first possibility is represented by the compensation mechanism; while the second possibility is represented by selective attention to better articulated input data. The algorithms presented below are minimal. The compensation mechanism is the *exact* inverse (except for noise) of the reduction while the selection mechanism only pays attention to the *best* articulations. Any other implementation of these mechanisms must perform less well.

There is *a priori* evidence that both processes could play a role. In order for infants to imitate adults, they must be able to match the sounds they produce with the sounds they perceive from adults. The same vowel produced by an infant and by an adult is quite different acoustically, due to the infant's shorter vocal tract. Therefore infants have to do a normalization of the signals they perceive. If infants are able to do this, it is not unlikely that they are able to do a similar normalization of reduced speech sounds to more expanded versions of the same speech sounds.

Also, there is ample evidence that the kind of speech that is addressed to infants is more carefully articulated than ordinary, adult-directed speech [7]. Such speech is distinguished by slower speed, exaggerated intonation and higher volume. It has been found that infants tend to prefer such speech to ordinary speech [2]. It can therefore be inferred that the kind of input that children use to base their vowel systems on does not necessarily consist of the rapid, casual speech that is used most often between adults. Combined with the fact that this special infant-directed register is found almost universally in different cultures, it would appear that such input plays an important role in the way speech sound systems are learned by children.

This paper uses a computer model to investigate the dynamics of both mechanisms. The computer model is inspired in part by the work on language games [13,14] and in part by the work on the iterated learning model [5,6]. The computer model consists of a population of agents. All of these agents can produce and perceive speech sounds in a human-like way. Some of these agents model infants and others model adults. Adults talk to infants in a more or less reduced register, and infants learn the adults' vowel system on the basis of the distribution of the input signals they receive. The imperfectness of articulation is modeled by adding noise to the articulatory positions that agents try to achieve. After a while, infants change into adults, old adults are removed from the population and new infants are added to it. It can be monitored over time how the vowel systems in the population change.

2 The Simulation

The simulation is based on a population of agents that can interact with each other. Agents exist in two modes: adult mode and infant mode. Adult agents have a fixed repertoire of vowels, while infant agents are busy acquiring a repertoire. Interactions consist of an adult agent producing a vowel from its repertoire, and an infant agent listening to that sound and updating its repertoire of speech sounds. Each agent participating in the interaction is chosen at random. After a certain number of interactions, the adult agents are removed from the population, infant agents become adult agents and a new batch of infant agents is added to the population. The agent's vowel systems are logged and it can be investigated how they change over time.

The agents can produce and perceive speech sounds in a human-like way. For this purpose they are equipped with a vowel synthesizer and a means to calculate distances between vowels. The vowel synthesizer is the same as the one used in [1]. It has three articulatory inputs: the position of the tongue in the front-back dimension, the height of the tongue and the amount of rounding of the lips. These correspond to the parameters that phoneticians usually employ to describe vowels [9, ch. 9]. In the model each input can have values between 0 and 1, where 0 corresponds to most front, lowest and least rounded, while 1 correspond to most back, highest and most rounded. Thus the vowel [a] can be described by the parameter values (0, 0, 0) while the vowel [u] can be described by the parameter values (1, 1, 1). Its outputs consist of the frequencies of the first four formants (resonances of the vocal tract). For example, for the inputs (0, 0, 0), the output would be (708, 1517, 2427, 3678) and for the inputs (1, 1, 1) the output would be (276, 740, 2177, 3506). With this articulatory model all ordinary vowels can be generated

The perception function is very similar to the one used in [1] but it follows the original perception model by Schwartz *et al.* [12] more closely. It is based on the observation that most vowel signals can be simplified with what is called an effective second formant. Vowel signals generally have multiple peaks in their frequency spectrum, each peak corresponding with a resonance frequency of the vocal tract. However, one can generate an artificial signal that sounds very similar to the original vowel using a frequency spectrum that has only two peaks. The position of the first peak in such a spectrum is equal to the position of the first peak in the original spectrum, but the position of the second peak is a non-linearly weighted sum of the positions of the second, third and fourth peaks in the original spectrum. This second peak is called the *effective second formant.*

The position of the effective second formant can be calculated using a formula due to Mantakas *et al.* [11] that was adapted by [12]. In the perception model, frequencies in Hertz are converted to the more perceptually inspired Bark frequency scale. Equal distances in the Bark scale correspond to equal perceived differences in pitch.

The conversion is performed with the following formula:

$$Bark = 7 \cdot \sinh^{-1}(Hertz/650)$$

Given the first formant and the effective second formant in Barks, the perceptual distance between two vowels *a* and *b* can then be calculated as follows:

$$D = \sqrt{(F_{1,a} - F_{1,b})^2 + \lambda^2 (F'_{2,a} - F'_{2,b})^2}$$

where F_1 is the first formant and F'_2 the effective second formant (of vowels *a* and *b*, respectively). *D* is the perceptual distance and λ is a constant that regulates the importance of the effective second formant relative to the first formant. It has a value of 0.3 in all experiments presented here, a value which has been found to generate perceptually realistic distances [12].

Agents engage in interactions. For each interaction an adult agent and an infant agent are chosen randomly from the population. In all experiments presented here, the population contains 20 adult agents and 20 infant agents. An adult chooses a random vowel from its repertoire and produces this. The infant perceives this signal and uses it to derive the possible vowels that are used in the population. In all the experiments, 10 000 interactions are performed, after which all the adult agents are removed from the population, and the infant agents turn into adults.

As has been mentioned above, adults can produce utterances from their repertoire of vowels. For each of the agent's vowels the articulatory parameters are stored. However, humans cannot produce utterances perfectly. In order to model this, noise is added to the articulatory parameters before they are synthesised. For each of the articulatory parameters, this noise is taken from the normal distribution with mean 0 and standard deviation 0.05. Also, the sloppiness of agents' articulations can be modelled by adding a bias for articulations to become more centralized. This is done in the following way:

$$x_{sloppy} \leftarrow x + \alpha(0.5 - x) \tag{1}$$

where x is any of the articulators (position, height or rounding) and α is a constant that determines how much the articulations are attracted towards the centre. If α is 0, no reduction takes place, and if α is 1 all articulations are reduced to 0.5. A realistic value for α is relatively hard to determine for rapid, casual speech, as it is not known where the original articulations are supposed to lie, but a realistic value would be at least 20% (0.2). This can be deduced from the fact that the surface of the acoustic space (F_1-F'_2 space) that is used for infant-directed speech, which can be considered carefully articulated, is at least 1.4 times the surface that is used for ordinary adult-directed speech (Hiu Mei Liu, *personal communication*). This amounts to a linear reduction of about 20%.

While the adult agents produce sounds, the infant agents perceive and learn speech sounds. In reality, learning must occur incrementally in infants. This means that each sound an infant perceives must slightly update the representations in its brain until the different vowel categories are represented. This could in principle be modelled with a neural network, but for simplicity of implementation, it was decided to use a batch-learning algorithm. Thus infant agents participate in a number of interactions and store all the signals that were used in those interactions. Then, when they are converted into adults, they derive the vowel categories from these stored signals.

Vowel categories are derived with an unsupervised classification algorithm called iterative valley seeking [3]. Iterative valley seeking is a parameter-free classification algorithm. This means that it makes no assumptions about the distribution that underlies the observed data points. It assumes that each peak in a distribution corresponds with a category, and it tries to locate these peaks by using the local density of data points. In order to estimate the local density, the method uses one parameter: the radius R of each point's local neighbourhood. This parameter determines how much the distribution of data points is smoothed when determining the peaks.

This process works well in practice, and is not extremely sensitive to the value of R. A value of 0.3 was used here. However, in the simulations there existed a tendency for outliers to be interpreted as real vowels. These vowels were then produced when the infant became an adult, and spread rapidly through the population. In order to prevent such spurious vowels to spread through the population, a requirement was added that classes needed to contain a minimum number of data points before they were accepted as real vowels. In order to determine whether to accept a class as representing a genuine vowel or not, first the average number of data points per class was calculated. Every class that had more than one third of the average number of data points was accepted as a genuine vowel.

Iterative valley seeking and the selection criterion result in a number of classes with example data points, but not yet in an acoustic or articulatory representation of a vowel. The acoustic representation was determined directly from the data points of the class. It was decided that the best way to determine the acoustic prototype of a class was to take the point at which the density of the class was highest. As no assumptions could be made about the distribution of the data points, the K-nearest neighbour method was used, with $K = 3$. This means that the densest part of the class was assumed to lie at the point that had the nearest third neighbour. This simple method turns out to work well for the kinds of distributions and the number of data points used here.

The articulatory prototype corresponding to this acoustic data point was then found by the agent talking to itself. It started with a vowel at articulatory position (0.5, 0.5, 0.5). Small modifications were made to this, these were articulated and it was measured whether they moved the acoustic signal closer to the target signal. This process was iterated until no more improvement could be achieved. The size for each modification step was a random value from the normal distribution with mean 0.1 and standard deviation 0.01 .

In the implementation of the compensation mechanism, agents re-expand their vowel repertoire in order to compensate for the reduced articulation of the adult agents. This expansion is the opposite of the reduction described in equation (1):

$$x_{expanded} \leftarrow x - \beta(0.5 - x)$$

where β is a constant that should have the value $\alpha/(1-\alpha)$ in order to compensate exactly for a reduction of size α. In all experiments presented here that use expansion, the value of β is chosen to compensate exactly for the value of α.

These methods result in reliable learning of a vowel system by a population of infant agents. However, the intention of the research was to check whether vowel systems would be stable when transfer was repeated over many generations.

3 Experiments

In the experiments, the stability of vowel systems over time was investigated. First, it had to be established that vowel systems do indeed collapse when learning is based on reduced speech. Therefore, the evolution over time of a five-vowel system (consisting of /i/, /e/, /a/, /o/ and /u/) was investigated when a minimally realistic amount of 20% reduction was present. The first generation of adults was initialized with the original vowel system, and this was transferred to successive populations of infants. As can be seen in figure **1** the system collapsed to a one-vowel system within 25 generations. In this figure, the vowel system of one agent from the population is plotted for each generation. The first- and effective second formant are plotted in such a way that the position of the vowels correspond to the way phoneticians usually plot them. The open carets represent the original vowels, the black dots vowels in subsequent generations and the open square the final vowel. This result establishes the necessity of an alternative mechanism to preserve vowel systems.

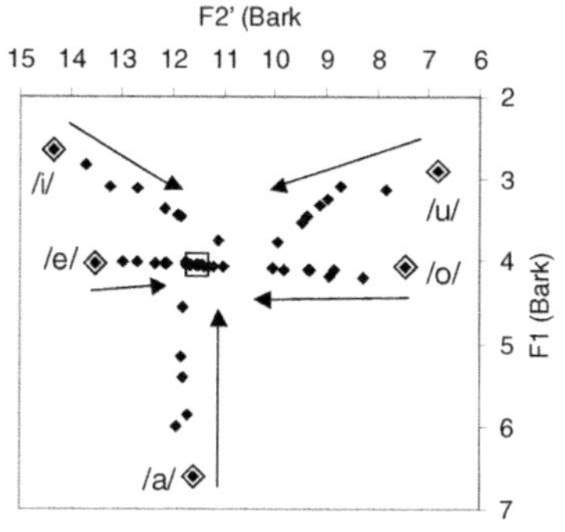

Fig. 1. Collapse of a five-vowel system under 20% reduction

The first experiments were done with the same five-vowel system. The compensation mechanism was implemented with the same 20% reduction, but now with a 25% expansion that, in the ideal case, would exactly compensate the reduction. The infant-directed speech mechanism was implemented as a reduction of only 2%. Some reduction was used, as it cannot be assumed that infant-directed speech is perfect. However, no expansion was assumed. It turned out that no significant difference could be observed between the two mechanisms and that vowel systems were almost perfectly preserved over runs of 100 generations. An example of vowel system preservation in both cases is presented in figure 2. Again starting positions are indicated with open carets, intermediate ones with black dots and final ones with open squares.

The case of seven-vowel systems was more interesting. Three different seven-vowel systems were investigated. All these systems contained the vowels /i/, /e/, /a/, /o/ and /u/. One system (type 1) contained /↔/ (as in "bed") and /⅃/ (as in "pot")Type 2 contained /y/ and /ø/ (front rounded vowels, as occur in French end German), while type 3 contained the vowels /}/ (Russian "ы" or Turkish "ı") and /∴/ (the vowel in English "the").

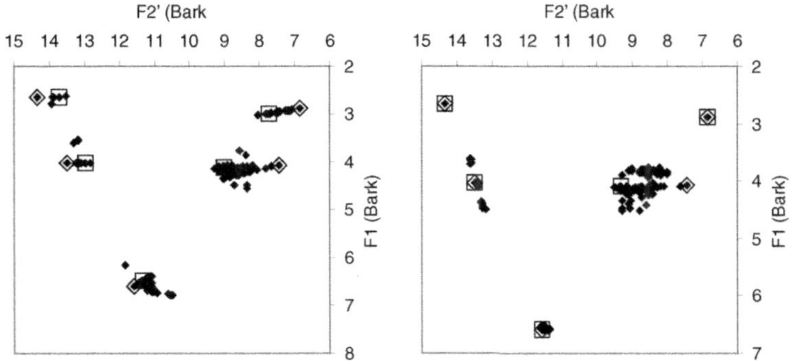

Fig. 2. Evolution of five vowel systems over time. The left frame shows the effect of the compensation mechanism and the right frame shows the effect of the infant-directed speech mechanism

In order to obtain a statistically significant comparison of the two mechanisms when transferring these vowel systems, the number of vowels per agent for each generation was monitored. This turned out to be a good measure of vowel system preservation. The evolution of the different types of vowel systems over 250 generations is shown in figure 3. Black lines indicate performance of the infant-directed speech mechanism while gray (orange) lines indicate the performance of the compensation mechanism. Neither of these mechanisms is better than the other in all cases. Also, for both mechanisms, significant reduction of the vowel systems is observed. Apparently, neither of the mechanisms is sufficient for preserving seven-vowel systems.

The combination of both mechanisms was therefore tested. This was implemented by using a reduction of 2% and a corresponding expansion of 100/49 ≈ 2.04%. As can be observed in figure 4 this resulted in much better preservation of vowel systems over time. Although vowel systems are still reduced over time, this happens so slowly that it could be realistic. It should also be kept in mind that the learning mechanism was constructed to avoid learning vowels that occurred infrequently. Once a vowel is lost, it is therefore not possible to relearn it, so that the system is biased towards shrinking vowel systems over time.

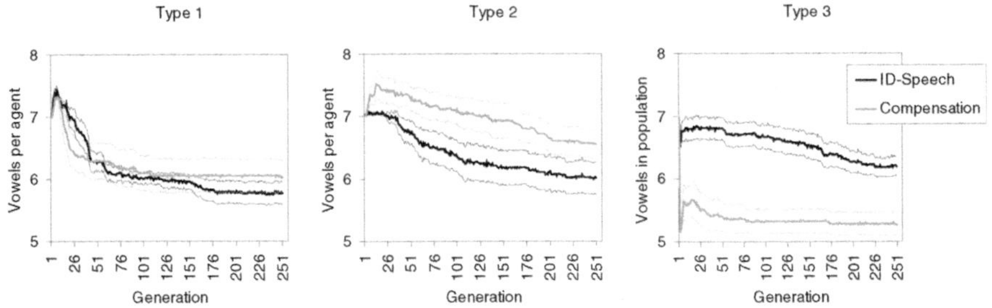

Fig. 3. Change of seven-vowel systems over time. Thick lines indicate averages, thin lines indicate 95% confidence intervals

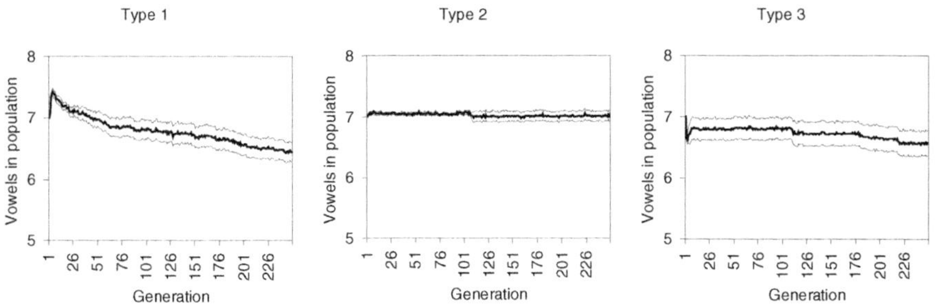

Fig. 4. Evolution of average vowel system size when both compensation and infant-directed speech are used. Dotted lines indicate 95% confidence intervals

4 Conclusion and Discussion

Infants learn vowel systems on the basis of the sounds that their parents produce. However, the most frequent register of adult speech, rapid casual speech, is often very much reduced. It is therefore clear that infants cannot learn the categories of speech sounds on the basis of an unbiased statistical analysis of the speech signals they perceive. This would result in a rapid collapse of the language's vowel system, as was shown in the first experiments presented above.

In the other experiments, two minimally sufficient models of preserving structure in vowel systems were compared. Although the exact behaviour of the models is of course dependent on the details of the learning algortihms, they were kept as simple as possible, and in a sense "minimal"; any other learning method would give less good learning results. Also, the simulation study was a qualitative study, and only measured the stability of vowel systems over time. No attempt was made to model the exact duration and trajectory of collapse of real human vowel systems. Therefore it is probably safe to draw qualitative conclusions on the basis of the models.

Both methods were based on statistical learning of the input data. In the first method, infant agents learned the number and the position of vowel categories on the basis of speech that was reduced by about 20%. However, the infant agents compensated for this by pushing the learned vowel categories away from the center of the articulatory space. In the second method, infant agents learned the position of vowel categories on the basis of speech that was only slightly reduced (2%). The infant agents did not compensate for the reduced articulation.

In the absence of experimental evidence, the method by which infants compensate is to be preferred, as infants have to perform renormalization of speech sounds in any case. The infant vocal tract produces signals that are quite different from those produced by the adult vocal tract for similar articulatory gestures. In order to learn to imitate speech sounds, infants need to be able to compensate for this. If such compensation needs to take place, an extra compensation to account for reduced articulation can be assumed to take place as well.

The criterion for comparison of the different methods is the number of generations over which vowel systems are preserved in the population. Although vowel systems of human languages can change rather rapidly [8] complex vowel systems of languages can remain stable for longer periods of time as well. Such stability must be present before other historic processes (influence of phonetic context, for example) can change vowel systems. A reasonable threshold for stability is preservation of a vowel system over approximately 50 generations, which corresponds to 800–1000 years in a (prehistoric) human population. After such a period of time, different historic processes generally have changed human vowel systems.

It was shown in the experiments that both methods were perfectly able to preserve five-vowel systems without central vowels. Such systems remained stable over at least 100 generations. On the basis of this data, no conclusion can therefore be drawn about which method best explains learning by infant. When learning of seven-vowel systems was tested, it was found that neither mechanism alone was able to assure stable transfer in all cases. However, a combination of both mechanisms was able to achieve stable transfer of vowel systems.

Either mechanism can account for learning of small vowel systems. As has been outlined above, the compensation mechanism is to be preferred in such a case. Although a more sophisticated compensation mechanism could probably be designed to preserve stability of more complex vowel systems, each automatic compensation mechanism must have vowel combinations that it cannot successfully reconstruct from the reduced input. After all, information does get lost in the reduction. Possibly compensatory mechanisms for seven vowel systems can still be designed, but human vowel systems can have up to at least 17 different vowels (Norwegian [15]), and given the experimental results, it is unlikely that such complex systems can be stably reconstructed from reduced articulations.

Therefore, for successful transmission of complex vowel systems, learning needs to be based on more carefully articulated examples as well. This would indicate that special infant-directed speech registers (motherese) are more important when a language has more vowels. Whether this really is the case has not been experimentally investigated so far, but data from a study of infant-directed speech in different languages [7] seems to indicate that the more vowels a language has, the more carefully articulated special infant-directed speech is. From the data in [7], it can be measured how much more area of the 2-D acoustic space is used for infant-directed (ID) speech than for adult-directed (AD) speech. It turns out that Russian, with a small (six-) vowel system has an ID/AD ratio of 1.73, English, with an intermediate size vowel system has a ratio of 1.85 and Swedish, with a large vowel system has a ratio of 1.96. Additional data [10] shows that Mandarin Chinese with a vowel system that is slightly smaller than that of Russian has a ratio of about 1.4. These data seem to indicate that languages with large vowel systems use more carefully articulated infant-directed speech. This data corresponds well with the finding of this paper that vowel systems with larger number of vowels need carefully articulated examples to be transferred from one generation to the next.

The conclusion that can be drawn from the experiments presented here as well as from the infant-directed and adult-directed speech data is that carefully articulated examples are not necessary as long as small and simple vowel systems need to be learned. Compensation for reduced articulation can be performed by simple expansion of the learned vowel prototypes. However, more complex vowel systems can only be learned successfully when more carefully articulated examples are available. Such carefully articulated examples can be found in infant-directed speech, and it is found that this speech register is more pronounced in languages with larger vowel systems.

References

1. de Boer, B.: Self-organization in vowel systems, Journal of Phonetics 28(4), (2000) 441–465
2. Fernald, A.: Four month-old infants prefer to listen to motherese. Infant Behavior and Development 8, (1985) 181–195
3. Fukunaga, K.: Introduction to statistical pattern recognition, Boston: Academic Press (1990).
4. Hock, H. H.: Principles of Historical Linguistics, second edition, Berlin: Mouton de Gruyter. (1991)
5. Kirby, S.: Function, Selection and Innateness: The emergence of language universals, Oxford: Oxford University Press. (1999)
6. Kirby, S.: Natural Language from Artificial Life, Artificial Life 8 (2002) 185–215
7. Kuhl, P. K., Andruski J. E., Chistovich, I. A., Chistovich, L. A. Kozhevikova, E. V., Rysinka, V. L., Stolyarova, E. I., Sundberg, U. & Lacerda, F.: Cross-Language Analysis of Phonetic Units in Language Addressed to Infants, Science 277 (1997) 684–686
8. Labov, W.: Principles of Linguistic Change: internal factors, Oxford:Blackwell (1994)
9. Ladefoged, P. & Maddieson, I.: The Sounds of the World's Languages, Oxford: Blackwell (1996).
10. Liu, H.-M., Tsao, F.-M. & Kuhl, P. K. Support for an expanded vowel triangle in Mandarin motherese. International Journal of Psychology, 35(3–4) (2000) 337
11. Mantakas, M, J.L. Schwartz & P. Escudier Modèle de prédiction du 'deuxiéme formant effectif' F2'—application à l'étude de la labialité des voyelles avant du français. In Proceesings of the 15th journées d'étude sur la parole. Société Française d'Acoustique, (1986) 157–161.
12. Schwartz, J.-L., Boë, L.-J., Vallée, N. & Abry, C.: The Dispersion-Focalization Theory of vowel systems. Journal of Phonetics 25, (1997) 255–286.
13. Steels, L.: A Self-Organizing Spatial Vocabulary. Artificial Life 2(3) (1995) 319–332.
14. Steels, L.: The Synthetic Modelling of Language Origins, Evolution of Communication 1(1) (1997) 1–34.
15. Vanvik, A.: A phonetic-phonemic analysis of Standard Eastern Norwegian. Norwegian Journal of Linguistics 26 (1972) 119–64.

Piep Piep Piep – Ich Hab' Dich Lieb[1]: Rhythm as an Indicator of Mate Quality

Eva van den Broek and Peter M. Todd

Centre for Adaptive Behavior and Cognition
Max Planck Institute for Human Development
Lentzeallee 94, 14195 Berlin, Germany
evdbroek@phil.uu.nl, ptodd@mpib-berlin.mpg.de

Abstract. Rhythm is common in courtship signals of many species. Here we explore whether regularly repeating rhythmic patterns can serve as indicators of underlying mate quality. We find through simulation that rhythmic signals allow the greatest discrimination between high- and low-quality males when low quality is associated with timing errors in artificial songs. However, rhythmic signals are difficult to evolve in our framework, leading to the conclusion that other pressures may have been involved in their appearance.

1 Introduction

Rhythmically repeated behaviors are common in nature: locomotion, breathing, chewing, and the like all rely on regular repetition for their effectiveness, and special neural circuits (central pattern generators, or CPGs) evolved early on to ensure that such behaviors would be performed with the proper rhythmic timing [1]. In addition to these important life functions, animal signals and displays often take a rhythmic form, from the regular flashing of fireflies or stridulations of crickets to the alarm calls of squirrels or the songs of birds and humans. Indeed, barring perhaps the performances of some Eurovision Song Contest entrants, rhythm is one of the most distinctive hallmarks of human music.

Given their prevalence, it is natural to ask whether the rhythmically repeated nature of these signals has some adaptive function. Perhaps rhythm is attention-grabbing; on the other hand, arrhythmic displays, by contrasting with commonly-seen rhythmic motor patterns, could be more surprising and draw more attention. More plausibly, rhythm may be used as a signal of an individual's underlying traits, and in particular may indicate factors that are important in mate choice. Karl Grammer and his colleagues have been studying the traits that human rhythmic behavior may convey. In their studies, people were brought to the lab and asked to "dance to their own rhythm". The participants' movements were filmed and analyzed using neural networks which came up with surprisingly accurate estimates of the dancers' personality

---

[1] "Peep peep peep, I love you", from the song "Guildo hat euch lieb", by Guildo Horn And The Orthopaedic Stockings, Germany's entry in the Eurovision Song Contest 1998: http://willow.dyndns.org/rachel/doh/jukebox/guildo.htm

W. Banzhaf et al. (Eds.): ECAL 2003, LNAI 2801, pp. 425–433, 2003.

traits according to the standard "Big Five" dimensions[2]—showing that traits that are often important in mate choice can be elicited from rhythmic motion patterns.

Further hints that rhythm may have the function of signaling mate quality can be found in the static visual domain. Regularly repeated forms that are easy to compare with each other can be used as quality indicators: Stripes and other regular patterns on fish, insects, snakes, and other animals might make errors stemming from developmental noise more salient. As Geoffrey Miller [2] put it, "From the viewpoint of signaling theory, repetitions across space (bilateral symmetry, radial symmetry, stripes) and across time (rhythm, repetition) are efficient ways to indicate developmental stability, a major component of fitness" (p. 67). This line of reasoning fits with the great amount of research in the past fifteen years devoted to finding the correlation between fluctuating asymmetry and developmental stress or genetic imperfection [3].

Besides revealing developmental noise, rhythmic displays in the temporal domain could heighten the salience of neural noise or disorders (leading to disruptions in motor control and in the generation of behavioral patterns), or show off positive aspects of quality such as respiratory fitness. In this paper, we test this function of rhythmic behavior as a proximate cue of underlying mate quality or fitness. We do this by constructing both coevolutionary and optimization models of populations of interacting artificial birds that produce and evaluate songs—temporal signals—for mate choice.

We use birdsong as our domain for testing these more broadly applicable ideas about signal rhythmicity for two main reasons. First, birdsong is used as a courtship signal (among other functions), indicating its possible quality-revealing function. In many bird species, sexual selection has resulted in complex, elaborated songs with rhythmic elements [4]. Second, there is evidence that rhythm in birdsong is disrupted by low-quality aspects of an individual singer. Birdsong production depends on features of the brain that easily break down under developmental stress or poor nutrition [5] (making song a revealing handicap). Consistent variation exists among individuals with respect to the temporal aspects of song delivery [6], and differences in rhythm are not only perceived but also important in inducing responses [7]. Rhythm may also be disrupted by noise at the neural level, leading to unwanted song timing variations. Courting male zebra finches seem to attempt to overcome this by holding down their level of neural noise when their songs are directed towards a female [8]. Thus the amount of rhythmicity in birdsong may serve as an indicator of developmental noise, neural noise, or current condition (e.g., energy reserves), all of which may be useful for a discriminating female to assess in a singing male.

2 Modeling the Function of Rhythmic Signals

Models of signal design have typically focused on aspects other than regularity or rhythmicity, with a few exceptions. Johnstone [9] used neural networks to investigate a universal sensory bias for symmetry. Enquist and Arak [10] proposed that regularity arose as a by-product of the need to recognize objects irrespective of their position (but see [11] for problems with this approach). In addition to the (pre-biased) sensory mechanisms of the receiver, the environment is a factor influencing signal design. In

---

[2] *http://evolution.anthro.univie.ac.at/ishe/multimedia/alysis.html*

noisy environments, signals might evolve to be more redundant (which can be instantiated as rhythmicity) and therefore easier to discriminate from the background [12].

A coevolutionary modeling approach similar to that taken here was used by Werner and Todd to explore signal design in birds in terms of novelty rather than rhythm [13]. They emphasized neophilia and an evolutionary pressure towards constantly changing signals, as opposed to a female preference for regularity. In their model, females used inherited song preferences along with a "desire for surprise" to select males who evolved over time to perform a wide and ever-changing diversity of songs. This result is complementary to the approach taken here, because the pressure for novelty can only operate after a pressure for regularity has first given rise to specific (possibly rhythmic) patterns and hence expectations that can then be violated.

2.1 Performing Songs with Evolved Templates

To explore whether regular repeated (rhythmic) signals could evolve as useful indicators of quality in mate choice, we constructed a set of models with two types of individuals: male and female "birds". (In our models, the birds do not have a specific predetermined sex, but can be drawn from the population to serve either as a mother or a father.) Both types are born with a basic song template, which they inherit from their parents. We leave out learning and fix the template at birth (thus making the model more similar to non-passerines than passerine songbirds).

The lives of our artificial birds are simple: They merely seek mates and reproduce. As in nature, the *in silico* males have to advertise themselves to potentially interested females by singing. In fact, in the model this is *all* that males do: males are reduced to a song template performed by a singing mechanism. The way a template is expressed (i.e., the behavioral phenotype that is produced) depends on the quality of the male, as described below. Females are the choosy sex, because the number of offspring they can have is limited (they can only mate once per generation, whereas a particularly tuneful or lucky male could mate many times). They consist of a judging device that uses an inherited song template of the same form as used by the singing males.

The song template encodes a temporal sequence of a very simple sort, representing what is happening in the song at a sequence of equally-spaced points in time. Here we cut the notion of a "song" down to its simplest binary form: At each instant, a note can either be sung or not, so that the song template entry can be either on or off (1 or 0). Thus, an example of a template with ten time-steps (the length that we typically use here) specifying the sequence of notes from left to right could look like "0010110010". Every individual has two templates, one of which is used depending on the individual's current sex role: the male template that is expressed as a song, and the female template that is used to judge male songs.

Like most sexually selected traits, the expression of the song has both a genetically transmitted and a condition-dependent component. Whereas the template is inherited from the parents, the condition or quality (in range 0.0-1.0) is randomly assigned to males at birth (females do not have condition-dependent traits). Quality has two effects for males. First, it impacts on the song he sings as he tries to attract a female and be chosen by her for mating. The male attempts to copy his inborn template perfectly in his song, but he may make mistakes in relation to his quality. More specifically, the probability that the male bird makes a mistake on any single note while singing his template is inversely related to his quality (noisiness = (1-quality)/5, so that the "average" male with quality .5 has a .1 chance of making a mistake at each note).

Second, quality also determines the relative number of offspring a male can have, if a female chooses him. The function mapping male quality to the number of offspring controls the speed of evolutionary convergence. There must be enough difference between the fertility of low- and high-quality males to ensure that evolution can proceed appropriately, but without resulting in premature convergence. To achieve this balance, we let each mated pair containing a male of quality Q produce 6·Q eggs. These eggs get put into the "egg pool" from which a fixed number (typically 200) are drawn at random to "hatch" and yield the next (non-overlapping) generation. Because the father's randomly assigned quality only influences the number of eggs produced, quality should be regarded in this model as male investment and is not a heritable feature (as opposed to "good genes" models).

What kind of errors can a male make in his song? Given that he can only produce one kind of note or a pause at each time-step (1 or 0), we only need to consider a small set of possibilities. Specifically, a male could introduce a gap at some point and shift the rest of his song one time step forward (a note *insertion*), or he might forget a note, thereby shifting his song one time step backward (a *deletion*) [14]. We assume every insertion to be a pause, putting a zero on the actual time step and shifting the rest of the template one position to the right, thereby dropping the last note. A deletion is implemented by shifting the rest of the template to the left and inserting a zero at the end. These two ways of making mistakes are combined by defining an equal chance for either mistake to be made at each position in the template.

2.2 Evaluating Songs with Templates and Preference Tables

Each female's goal (evolutionarily speaking) is to select a high-quality mate with whom she will have many offspring. However, she cannot assess quality directly, and so must rely on some aspect of the male songs to make her choice. Whether or not rhythm is one of the aspects that will evolve to aid in this choice is what we want to find out. All of the females get to select a mate, but they do not listen to the songs of all the males in the population before choosing; instead, a "choir" of a fixed number of singing males is selected randomly for each female to pick from. This limited sample reflects the time pressure to mate that real birds face. The number of males that a female can choose from regulates the force of selection. The larger her choir, the more strongly she can express her preferences (because she is more likely to find a male that she strongly prefers), and the more quickly the population may converge.

We build in a psychological mechanism by which the female makes her choice. The female follows a best-of-N rule [15], picking as her mate the male in her choir whose performed song (phenotype) is closest—in some sense, defined below—to her own template (genotype). Given the possible noise in the male's singing, similarity between his (hidden) genotype and his (heard) phenotype is not guaranteed; but when noise rates are low, each female will on average pick the male with the song template genotype that is closest (again, in some defined sense) to her own template genotype.

How does the female judge the similarity between the male song she hears and her own internal song template? The calculated similarity or distance depends on her preferences, for instance whether she is annoyed by missing notes, or by new notes that she was not expecting, or both. Exactly how a female judges is captured in her preference table (see Table 1), which defines how she rates what she hears (in the male song) at each moment against what she expected (in her template). These preferences are then summed over all time-steps in the song to yield the female's judgment of that male.

Table 1. Female preference table for judging male songs

	Expected	0	1
Heard	0	X1	X2
	1	X3	X4

As an example, a simple symmetrical preference table, one which is not biased towards a prevalence of either ones or zeros, is expressed by X1...X4 = {1, 0, 0, 1}. A female using this table effectively just calculates the Hamming distance between her template and the male song. Another psychologically possible preference table is an asymmetrical one in which there is no reward for producing a pause at the right time (X1), but there is a big punishment for being off the beat by producing a note in the wrong place (X3), so X1...X4 = {0, 0, -1, 1}. There are a number of other plausible preference tables that a female could use, so to keep from biasing our results with an arbitrary choice we have explored the space more widely. Assuming that the four values in the preference table come from the set {-1, -0.5, 0, 0.5, 1}, we limit ourselves to considering those preference tables where X4 = 1, and X4 >= X1 > X2 > = X3. In other words, the value placed on an expected note (X4) is always greater than or equal to the value for an expected pause (X1). Furthermore, expected events are always valued more than unexpected events, and an unexpected pause is penalized less than an unexpected note (i.e., a female's expectations will be violated more by an unexpected note than by a pause—see [16]). There are 20 such preference tables in all, shown in Table 2.

Once every female (that is, every individual in the population in their female roles) has chosen a mate, offspring are made from combinations of the templates of the parents. Crossover happens (with low probability) only between the two female- or male-associated templates, so that the templates used for judging cannot be mixed with those for singing. (This excludes a Fisherian runaway process, but correlations between the two templates within individuals can still arise). The mutation probability at each time-step is 0.01. Finally, as mentioned earlier, some number (depending on the father's quality) of the mated pair's eggs hatch and are placed in the offspring pool from which the next generation is randomly selected. Males in the new generation again are assigned a random quality.

3 Results: High Expectations and Poor Performances

Given this framework, we now ask what types of male songs females can use to judge male quality and make good mate choices. Next, because good mate choices would lead to more offspring, we ask whether appropriate female templates will evolve to allow such choices, and whether male songs will coevolve to the same patterns, as is necessary for female choice to work. Our simulations aimed to find out if the female expectations and male performances could evolve to interact in this way.

3.1 Template Discriminability

Some songs that males could sing may make their quality evident—that is, any errors they produce would be obvious—while other songs would hide their quality. For instance, a totally silent "song" (0 at every time-step) would not allow any timing errors to be perceivable, because there are no notes to shift about in time. Females will gain an advantage if they (and the males) use song templates that allow them to judge male quality most effectively, discriminating accurately between high-quality low-error males and low-quality high-error ones. Which templates are the most discriminable for our female birds, given particular preference tables? We addressed this question using an optimization approach, looking for highly-discriminable templates for each of the plausible preference tables (in Table 2) in the following manner.

For each of the 1024 possible templates with length 10, we created a series of "noisy songs" by using an error rate that on average gives each song one insertion or deletion error. We computed the female's preference for each noisy song given the current preference table, and then normalized each of these by dividing by the maximum possible preference rating for this template, which is the score, from 0-10, that the female gives if she hears an exact copy of this template. (Otherwise in some cases, for instance for an asymmetric score table, there could be an undesired difference between the scores for the templates with different numbers of notes.)

The gap between the best normalized score possible for each template (given a certain fixed preference table), which is 1.0, and the mean over all noisy performances of that template, is a measure of how well a female with this template can discriminate between low quality (one mistake on average) and high quality (no mistakes) males. For each of the 20 preference tables, we compared all of the templates to find the one with the largest such difference. This most discriminating template is then optimal for a female with this preference table, in the sense that it maximizes her quality-assessing abilities which determine how well her genes will be passed on.

As shown in Table 2, across the 20 psychologically plausible preference tables the most discriminating song template was rhythmically alternating (1010101010 or 0101010101) in all but four cases. (Two of the 16 rhythmic templates added an additional 1 at the end, but in these cases the fully rhythmic template had very similar discriminability). Only the four non-rhythmic templates had $X1 \leq 0$. Thus, placing a positive value on producing a pause when it is expected (X1) seems essential for rhythmic templates to be the most discriminating. Given this caveat, rhythmic song templates appear to be the most useful type of signal, for our simulated birds at least, for discriminating high quality males from low quality ones.

Table 2. The 20 psychologically plausible preference tables where X4 >= X1 > X2 > = X3, along with the most-discriminating song template found for each. Only four preference tables have a non-rhythmic pattern as most discriminating.

X1	X2	X3	X4	Template
-0.5	-1	-1	1	0010101001
0	-1	-1	1	0000000001
0.5	-1	-1	1	0101010101
1	-1	-1	1	0101010101
0	-1	-0.5	1	0000000100
0.5	-1	-0.5	1	0101010101
1	-1	-0.5	1	0101010101
0.5	-1	0	1	0101010101
1	-1	0	1	0101010101
1	-1	0.5	1	0101010101
0	-0.5	-0.5	1	0000000001
0.5	-0.5	-0.5	1	0101010101
1	-0.5	-0.5	1	1010101011
0.5	-0.5	0	1	0101010101
1	-0.5	0	1	0101010101
1	-0.5	0.5	1	0101010101
0.5	0	0	1	0101010101
1	0	0	1	0101010101
1	0	0.5	1	0101010101
1	0.5	0.5	1	1010101011

3.2 What Will Evolve?

Now that we have found that rhythmic templates allow the greatest discrimination of error-prone males, the next question is whether or not they will arise during evolution. Will the rhythmic templates be of sufficient help to their female owners in enabling them to pick more males with a higher quality, thereby producing more offspring that inherit the same adaptive template? It all depends on how the male songs evolve at the same time. To find out, we built a co-evolutionary model with a population of 200 birds seeking mates and reproducing to form the next generation. In this model, it is assumed that the female judging system is not perfect with regard to the information it receives. To increase the need for using highly discriminable templates, we added Gaussian noise (sd = 1.5) to each female's appraisal of a male song.

We began by exploring what would evolve when we initialized male and female templates at random, using the basic symmetric preference table (X1…X4 = {1, 0, 0, 1}) in 100 simulation runs with a choir size of 20. We measured the average number of alternations between 0 and 1 notes in all evolved templates in each generation (where the maximum, for the fully rhythmic length-10 template, is 9, and the randomly expected mean is 4.5). The average number of alternations after 1000 generations is only slightly higher than the randomly expected mean, indicating that the population tends to converge to essentially random templates over time. On the other hand, we did find some indication that females with more alternating (rhythmic) templates are able to choose higher-quality males, but this effect does not seem strong enough to drive the evolution of rhythmic templates very far.

While rhythmic templates do not seem to "pop up" in our simulations, we can proceed to ask whether they would be stable over time if they arose in the population for some other reason. To answer this question, we started evolutionary runs from particular initial populations. First, we set all templates in the initial generation, both male and female, to the rhythmic template with nine alternations (1010101010). As the population evolves, the mean number of alternations per template starts diminishing after 300 generations. The resulting number of alternations after 1000 generations is 7.5, averaged over 10 runs. Conversely, when the population is initiated with a constant template (1111111111), it evolves at a higher rate toward random levels of alternation. Thus, the rhythmic template does appear to be evolutionarily more stable than the nonrhythmic template, but not entirely impervious to drift.

4 Discussion

Our model shows that rhythmic songs are more discriminating with regard to the sorts of mistakes that male birds make. However, rhythmic song templates are not easy to evolve in our binary-encoding framework, nor particularly stable against change once they have predominated in a population. Given that rhythmic signals do seem to be useful quality indicators, but are difficult to evolve in our model, the question arises whether other selective pressures or historical contingencies favoring rhythm may have been present in the evolution of those rhythmic signals we find in the world today. One plausible source of such historical contingencies exists at the neural level, where preexisting central pattern generators originally used for other purposes such as locomotion may have been exapted for new use in creating rhythmic signals. Such rhythmic circuits may also have led to sensory biases in the receivers to prefer regularly repeating displays. A promising direction in which to extend our models here would therefore be to build song templates in both males and females on top of a neural substrate in which rhythmic circuits could naturally arise, for example Jordan- or Elman-style sequential connectionist models or continuous-time recurrent neural networks (some of which have already been explored in the context of music production and perception—see [16]).

Further evidence for or against the mate-quality-indicator hypothesis for rhythm may come from uncovering the neural structures leading to rhythmic behavior and the genetic and environmental factors that can impact on the production of rhythmic signals. If a lack of noise (whether on a developmental, neural, or condition level) is indeed correlated with a fitness advantage for a bird, a rhythmic song might be the sonic equivalent of the peacock's tail through which a male can reveal his mate quality.[3]

[3] Whether or not this can help explain why Guildo Horn only placed seventh in the 1998 Eurovision Song Competition remains to be examined.

Acknowledgments. We wish to thank John Hutchinson, Miguel Rodriguez-Gironés, Carel ten Cate, Seth Bullock, and our anonymous reviewers for valuable input and comments.

References

1. Mathayomchan, B., Beer, R.D.: Center-crossing recurrent neural networks for the evolution of rhythmic behavior. Neural Computation 14 (9) (2002) 2043–2051
2. Miller, G.: Mental traits as fitness indicators: Expanding evolutionary psychology's adaptationism. In: Evolutionary perspectives on human reproductive behavior, Ann. of the NY Acad.of Sc. 907 (2000) 62–74
3. Møller, A.P., Swaddle, J.P.: Asymmetry, developmental stability, and evolution. Oxford University Press, Oxford (1997)
4. Searcy, W.A., Yasukawa, K.: Song and female choice. In: Kroodsma, D.E., Miller, E.H. (eds): Ecology and evolution of acoustic communication in birds. Cornell U. Press (1996)
5. Nowicki, S., Searcy, W.A., Peters, S.: Brain development, song learning and mate choice in birds: a review and experimental test of the "nutritional stress hypothesis" J. Comp. Physiol. A 188 (11-2) (2002) 1003–1014
6. Riebel, K., Slater, P.J.B.: Temporal variation in male chaffinchsong depends on the singer and the song type. Behaviour 140 (2003) 269–288
7. Slabbekoorn, H., Ten Cate, C.: Collared dove responses to playback: Slaves to the rhythm. Ethology 105 (1999) 377–391
8. Hessler, N.A., Doupe, A.J.: Social context modulates singing-related neural activity in the songbird forebrain. Nat. Neurosc. 2 (3) (1999) 209–211
9. Johnstone, R.A.: Female preference for symmetrical males as a by-product of selection for mate recognition. Nature 372 (6502) (1994) 172–175
10. Enquist M., Arak A.: Symmetry, beauty and evolution. Nature 372 (6502) (1994) 169–172
11. Bullock, S., Cliff, D.: The role of 'hidden preferences' in the artificial co-evolution of symmetrical signals. P. Roy. Soc. Lond. B Biol. 264 (1381) (1997) 505–511
12. Enquist, M., Arak, A.: Neural representation and the evolution of signal form. In: Dukas, R. (ed): Cognitive ecology: The evolutionary ecology of information processing and decision making. University of Chicago Press, Chicago (1998) 21–87
13. Werner, G.M., Todd, P.M.: Too many love songs: Sexual selection and the evolution of communication. Com. Adap. Systems (1997) 434–443
14. Podos, J.: Motor constraints on vocal development in a songbird. Anim. Behav. 51 (1996) 1061–1070
15. Real, L.: Search theory and mate choice. I. Models of single-sex discrimination. American Naturalist 136 (1990) 376–404
16. Helekar, S.A., Marsh, S., Viswanath N.S., Rosenfield, D.B.: Acoustic pattern variations in the female-directed birdsongs of a colony of laboratory-bred zebra finches. Behav. Proc. 49 (2000) 99–110
17. Griffith, N., Todd, P.M. (Eds.): Musical networks: Parallel distributed perception and performance. MIT Press/Bradford Books, Cambridge, MA (1999)

Evolving Agent Societies with VUScape

P.C. Buzing, A.E. Eiben, and M.C. Schut

Department of Artificial Intelligence Vrije Universiteit Amsterdam 1081 HV Amsterdam, The Netherlands {pcbuzing,gusz,schut}@cs.vu.nl

Abstract. The main contribution of this paper is twofold. Firstly, it presents a new system for empirical investigations of evolving agent societies in SugarScapelike environments, which improves existing Sugarscape testbeds. Secondly, we introduce a framework for modelling communication and cooperation in an animal society. In this framework the environmental pressure to communicate and cooperate is controllable by a single parameter. We perform several experiments with different values for this parameter and observe some surprising outcomes.

1 Introduction

There are quite a few systems enabling experimental investigations with artificial societies based on the SugarScape framework [5], for example, [8, 2]. The added value of our new system VUSCAPE comes at two levels: on the conceptual level, it incorporates significant extensions of the SugarScape framework itself, and on the technical level it allows quick and easy specification of system variants together with an extensive monitoring of system behavior[1].

Our implementation offers practical advantages over existing social simulation software like Ascape [8], Repast [2] and Swarm [3]. Firstly, VUSCAPE gives the user more flexibility because all settings can be specified either in configuration files or by command line arguments. This enables the user to automate experiments, which substantially speeds up the time needed for, for example, investigating effects of varying experimental parameters (often combinatorial number of sessions). Data is automatically saved at specified locations, enabling detailed experimental logging. Secondly, VUSCAPE is programmed and documented such that it is very easy to add, replace or delete code when changes or extensions to the model need to be implemented. Finally, there is no direct necessary connection between the actual program and the graphical user interface. Initial exploration can be easily done by using the graphical user interface; for automated experimentation, the system can be run solely by configuration files or command line. In this paper, we use VUSCAPE to perform experiments on the effects of communication and cooperation (inspired by [7, 1]).

This paper is organised as follows. In Section 2 we present the artificial world VUSCAPE, that is used in the empirical study described in this paper. The built-up of individual agents is discussed in Section 3; our agents exhibit specific behaviours in this study, which we briefly summarise in this Section. Section 4 presents the empirical results as obtained in the VUSCAPE investigations. Finally, in Section 5 we summarise our findings and present issues for further research.

[1] The source code can be found at `http://www.cs.vu.nl/ci/VUScape/`.

W. Banzhaf et al. (Eds.): ECAL 2003, LNAI 2801, pp. 434–441, 2003.

2 VUScape

The artificial world used in this paper is VUSCAPE, inherently based on the well known SugarScape world, as introduced by Epstein and Axtel [5] as a generic testbed for social simulation. For the purpose of the study described in this paper, we extended the SugarScape world in a number of ways, thereby introducing the possibility to research the specific emergent phenomena of our interest. Additionally, these adaptations can be considered to extend the SugarScape domain in an interesting generic way, opening up possibilities to investigate SugarScape worlds in wider perspectives.

Like SugarScape, the VUSCAPE world is a two dimensional grid, wrapped around the edges. Each position corresponds with an area which can contain multiple agents and an amount of sugar.

We made the following adaptations to the SugarScape world:

- **Cooperation** – Cooperative behaviour of agents is stimulated by means of setting a maximum amount of sugar that an agent can eat on its own. This maximum amount of sugar can be set by the cooperation threshold parameter described in Section 3.
- **Communication** – Agents receive information (listen) from other agents about amounts of sugar at their locations and broadcast (talk) the amount of sugar at their own locations.
- **Explorative Behaviour** – Agents in VUSCAPE have explorative behaviour as such that an agent randomly moves around in case it does not know of any sugar to move to or to eat. Traditionally, in SugarScape, agents do not exhibit such explorative behaviour but stay at their location in such situations.
- **Increased grid-point inhabitance** – We allow for multiple agents to be at a single grid-point in VUSCAPE, whereas this is not allowed in SugarScape.
- **Randomised sugar distribution** – The conventional sugar distribution in SugarScape is based on two sugar-hills that are located in the world at appropriate distances from each other. In this world sugar is concentrated and grows back at the given location making passive (i.e., not too mobile) agents viable. As we are interested in explorative behavior, we initially distribute so called sugar seeds randomly over the grid and move a seed to another random location if its sugar has been eaten.
- **Randomised age initialisation** – As in SugarScape models, the age parameter of agents is initialised uniformly (at 0), this brings with it some surprising phenomena that are not necessarily realistic. Additionally, the effects of such a setup echoes through the whole simulation until finished.We initialise the ages of agents randomly (between fixed minimum and maximum values), thereby preventing the mentioned methodological errors resulting from initialisation at 0.

We can catagorise these changes over two dimensions. Firstly, we can divide them according to whether a change is *methodological* or *experimental*. For example, the moving sugar seeds and randomised age initialisation are methodological, i.e., we do not explore their effects, but only investigate worlds that work according to these modifications. On the other hand, the cooperation threshold and communication are experimental extensions, as such that we describe empirical findings on their effect in Section 4. Secondly, some changes relate to the world, while other changes relate to the agents.

2.1 Execution Cycles

The VUSCAPE world evolves with discrete time steps, called cycles. In one such an execution cycle, the world (including agents) is updated. For the agents, this means that they can generate and execute actions; the world then processes these actions and generates outcomes, feeding these back to the agents. More precisely, the following stages take place in chronological order within a single execution cycle. During a single cycle, all stages are executed for each agent in parallel[2].

1. An agent *gathers information* about the presence of sugar in the world. This is done by means of listening (from other agents along the axes) and looking (by looking at the directly surrounding locations and the current location). Upon completion of this stage, the agent has at its disposal an array of locations and amounts of sugar on these locations.
2. Based on this array, the agent picks out the location with most sugar and moves to this location. In case there are multiple locations with the most amount of sugar, the agent chooses a random one from these locations and moves there[3].
3. Having arrived at the sugar, this sugar is *harvested* in case the amount is under the cooperation threshold. If the amount is above the cooperation threshold, the agent cooperates immediately if there are more agents at the location. Otherwise, it *broadcasts* (with some probability) to the other agents among the same axes that it needs help.
4. If possible, the agent *reproduces* and generates offspring. For this, it is (at least) necessary that there is another agent of the opposite sex at the location. (Other conditions to reproduction are applicable, and are discussed in detail below.)

In Figure 1, the agent control as described above is shown. Additionally, it shows the way communication in VUSCAPE is currently implemented.

2.2 World Features

A world in VUSCAPE has specific features that can be set as parameters in empirical investigations performed in the testbed. We describe the features below.

Traditionally, the initial *population size* can be set. Typically, in SUGARSCAPE this is set to 400. An initial population is usually half female, half male. The *width* and *height* of the world can be set, typically to 50×50. Finally, the *sugar grow-back rate* determines at which speed sugar grows back in the world.

As mentioned above, sugar is randomly distributed over the world during initialisation in VUSCAPE. The amount of sugar to be distributed can be set by the *sugar richness* parameter: this is the average amount of sugar that is on each location. Sugar richness of

[2] Conceptually, all agents execute these stages in parallel. However, technically, the stages are partially executed in sequence. Therefore, the order of agents, performing their control loops, is randomised over the execution cycles to prevent order effects.

[3] Choosing the closest by (and breaking ties randomly) is the method traditionally used in SUGARSCAPE. We have the option in VUSCAPE to either choose a random one or the closest by. Since a move action does not have a cost associated with it, we do not consider choosing the closest by having a reasonable rationale.

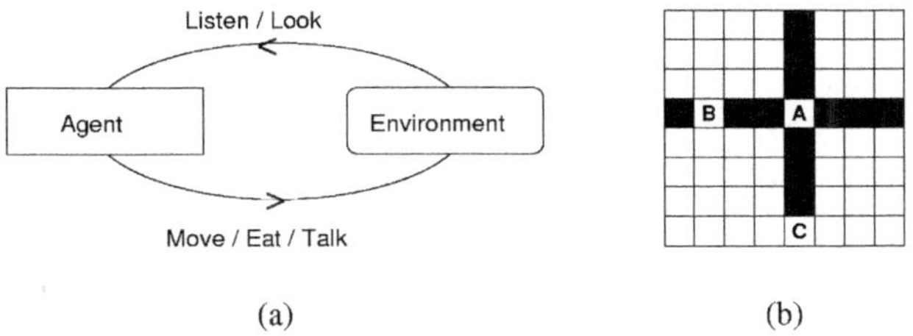

Fig. 1. (a) The agent control loop in VUSCAPE. (b) Communication in VUSCAPE over the axes. In this example, agent A broadcasts information, which is received by agents B and C.

1.0 means that on a 50×50 world, 2,500 units of sugar are distributed. Possibly, there are more seeds per location. The actual sugar that can be consumed by agents grows from sugar seeds. Each sugar seed has a maximum amount of sugar it can grow back to after consumption. For example, when a sugar seed with maximum 4 is consumed, it grows back to 4 until consumed again. In VUSCAPE, this can be set by a parameter; usually, one allows for sugar seeds with maximum grow-backs 1, 2, 3 and 4. The initial sugar distribution determines what seeds are on which location. Each seed is initialised with its maximum sugar, achieving the initial sugar richness as desired. Although VUSCAPE allows for other possibilities here we initialize in such a way that the number of sugar seeds is equal for all maximum grow-back values. For 2,500 sugar units, this means that there are 250 seeds with maximum 1, 250 with maximum 2, etc. thus amounting to 2,500 in total.

In order not to depend too much on the initial sugar distribution, we allow for the sugar to be redistributed during the existence of the world. Therefore, in VUSCAPE, seeds can be moved to another location after consumption. A parameter allows for this option to be on or off. By doing this, we stimulate agents to travel more to other locations. For example, it is not possible anymore for an agent to stay put at a sugar seed is that sufficiently feeds the agent; instead, it encourages the agent to travel to find even better locations.

Monitors. In VUSCAPE it is possible to monitor a variety of variables as the basis of obtained empirical findings. Although a number of monitors have been predefined, it is possible to put a monitor on every numeric variable in the source code. If a variable is monitored, this means that every execution cycle, the value[4] of the variable recorded. In Table 1 we briefly enumerate the predefined monitors here. Generally, monitors refer either to variables within the *agent* or within the *world*[5].

[4] Each variable can monitored as average, minimum, maximum, sum, variance, standard deviation, frequency, or any combination of these.

[5] The world is called a scape in VUSCAPE.

Table 1. An overview of VUSCAPE monitors.

Type	Name	Denotes	Domain
Agent	age	age of the agent	[0:100]
Agent	listenPref	whether agent listens	[0:1]
Agent	talkPref	whether agent talks	[0:1]
Agent	sugarAmount	sugar contained by an agent	[0: ∞]
Agent	inNeedOfHelp	percentage of agents on sugar > coopTresh	[0:1]
Agent	cooperating	percentage of agents that cooperates	[0:1]
Agent	exploreCell	percentage of agents that moved to a new cell	[0:1]
Agent	hasEaten	amount of food that agent has eaten	[0:4]
Agent	numberOfAgents	number of agents	[0:∞]
Agent	numberOfBirths	number of just born agents	[0:∞]
Agent	numberOfDeaths	number of just died agents	[0:∞]

3 Agents

Agents have a specific genetic make-up in VUSCAPE. In this Section we describe the particular *features* of an agent and the specific *behaviours* that the combination of features brings forth.

3.1 Agent Features

An agent consists of and possesses some particular *features* that determine the make-up of a particular agent. In VUSCAPE, a number of these features have directly been taken from the original SUGARSCAPE model. Additionally, we extended these features by including a cooperation threshold, reproduction threshold and initial amount of sugar.

From the traditional SUGARSCAPE model [5], we took the basic agent features which make up an agent. These features include metabolism, gender, child bearing, death, vision, allow sex and replacement.

Each agent has at its disposal a maximum amount of sugar that it can harvest on its own. As mentioned previously, this amount is called the *cooperation threshold* (or: *maximum sugar harvest* MSH). If an agent is at a location at which the amount of sugar is over this threshold, it needs other agents to harvest the sugar. If there are more agents at such a location, these agents harvest the sugar together and the sugar is evenly distributed over these agents. In the empirical investigations described below, the cooperation threshold is the same for all agents.

As agents realistically need energy (here: sugar) to reproduce, VUSCAPE offers the possibility to set the amount of sugar needed for mating by setting the *reproduction threshold*. If the amount of sugar contained in an agent is over this threshold, then (in prevailing circumstances) this is agent is able to reproduce. The offspring of two parents receives half of the amount of sugar from each parent at birth. Whereas in SUGARSCAPE, the reproduction threshold (called *endowment* in SUGARSCAPE) is implemented as being the same value as the initial amount of sugar an agent received, the VUSCAPE implementation enables one to set this parameter independently of the initial amount of sugar.

Table 2. Experimental settings.

Parameter	Value	Parameter	Value
Height of the world	50	Minimum death age	60
Width of the world	50	Maximum death age	100
Initial number of agents	400	Minimum begin child bearing age	12
Sugar richness	0.5	Maximum begin child bearing age	15
Sugar growth rate	3.0	Minimum end child bearing age male	50
Minimum metabolism	1.0	Maximum end child bearing age male	60
Maximum metabolism	1.0	Minimum end child bearing age female	40
Minimum vision	1.0	Maximum end child bearing age female	50
Maximum vision	1.0	Reproduction threshold	5
Minimum initial sugar	50.0	Mutation sigma	0.01
Maximum initial sugar	50.0	Sex recovery	0

3.2 Agent Behaviours

For agents, we distinguish four types of behaviour: mating, consumptional, communicative and cooperative behaviour, similarly to [4]. With the exception of communicative behaviour, these behaviours are hardwired into the agents and, as such, do not evolve during a run.

Mating Behaviour. Agents reproduce like in the original SUGARSCAPE [5]. Mutation can be applied to the talk and listen genes: after each value has been inherited, a random value, drawn from a Gaussian distribution with zero mean, is added to the inherited value (similar to [6]). In cross-over, the preferences for talking and listening are both inherited as the values for these characteristic of the wealthiest parent (the one with most sugar).

Consumptional Behaviour. The consumptional behaviour prescribes an agent to always go to the location with the most sugar out of all perceived locations. If there are multiple such locations, the agent picks a random one from these locations.If an agent has arrived at a location with sugar, it eats the complete amount of sugar if it can, i.e., if the amount is under the cooperation threshold. If the amount is over this threshold, the agent cannot consume, but might communicate to other agents that it needs help.

Communicative Behaviour. Two features of an agent determine its communicative behaviour, namely the features to *talk* and to *listen*. The listen feature is used in the observation process of the agent. By listening, the agent receives information from other agents about amounts of sugar at the locations of those agents. The talk feature determines whether the agent performs a communicative action itself, namely broadcasting to other agents: 1) the amount of sugar that is on its location, and 2) the coordinates of its location. After initialisation, the average talk preference over all agents is 0.5. With a preference p, an agent broadcasts the amount of sugar at its location with probability p in case it needs help to harvest the sugar at its location.

Cooperative Behaviour. The maximum amount of sugar at one location that an agent can eat alone is called the cooperation threshold. This parameter allows for tuning the necessity to cooperate. If it is zero, all pieces of food must be eaten by at least two cooperating agents. Gradually raising this value the necessity to cooperate diminishes.

Explorative Behaviour. Agents in VUSCAPE have explorative behaviour as such that an agent randomly moves around in case it does not know of any sugar to move to or to eat.

4 Experiments

In this section we describe our experiments with different values for the cooperation threshold, or maximum sugar harvest, MSH. Its value is gradually varied from 0 (every piece of food must be eaten by at least two agents) to 4 (cooperation is never necessary). The other parameters of the system are summarized in Table 2.

Additionally to varying the MSH we also test two options for the communication facility. Runs with communication use the talk and listen features as described in Section 3, runs without communication have them disabled. In our present setup agents in a communicating society always listen, only their "talkativeness" is determined by an individual gene.

The system's behavior is illustrated here by two measures, the population size and the number of cooperation acts (when two or more agents effectively eat a pile of sugar together). Figure 2 shows the outcomes of 10 independent runs overlaid[6].

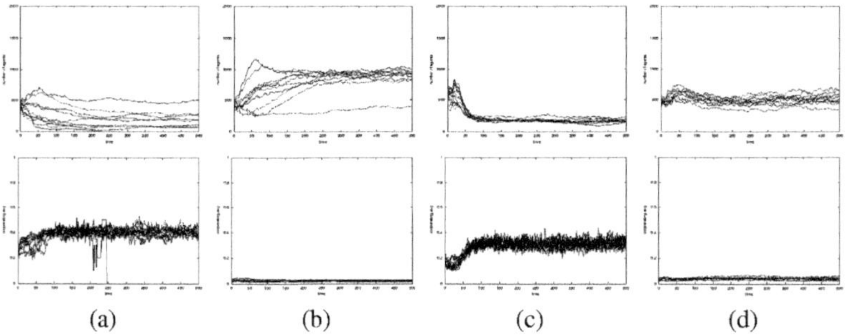

Fig. 2. Population size (above, from 0 to 2000 on y-axis) and percentage of cooperations (below, from 0 to 1 on y-axis) over 500 execution cycles (on x-axis). Results are shown (from left to right) for MSH = 0 (a) and (b) MSH = 1 *without* communication and for MSH = 0 (c) and MSH = 1 (d) *with* communication.

5 Discussion

Somewhat to our surprise the most visible effect of communication is not that it implies larger surviving populations, but that it makes the behaviour of the system more stable

[6] To save space, the graphs are plotted small, only clearly revealing the trends as mentioned in the text. The reader can find these and the omitted Figures in readable format on http://www.cs.vu.nl/ci/VUScape/graphs/.

with much less variation than without communication. Using the communication feature all runs have terminated, while without it some ended in extinction by starvation. Yet, our experimental data does not supply hard evidence for the superiority of a communicating species. At the moment we are investigating other communication schemes (e.g. spherical instead of axis-wise) to check if this is an implementation artefact or has a conceptual reason.

As for the number of cooperations, we can observe that for a hard world (with low MSH) it is greater: more agents cooperate in such worlds. This is in line with our expectations. We observe that in these worlds, the difference between the cooperation curves with and without cooperation is relatively small. It can be hypothesised that the random exploratory behaviour of the agents results in sufficient random encounters to support large populations. Additionally, we have checked the evolution of the distribution of the "talkativeness gene". Preliminary results indicate a tendency to become more talkative over time, which suggests an evolutionary advantage of talking and thereby initiating cooperation. However, this issue needs further investigation.

References

1. E. Avdis and K. Dautenhahn. Self-organisation of communicating agents – linguistic diversity in populations of autonomous agents. In *Proceedings of the Eighth Symposium on Intelligent Robotic Systems*, 2000.
2. N. Collier. Repast: An extensible framework for agent simulation. Technical report, Social Science Research Computing, University of Chicago, Chicago, Illinois, 2000.
3. M. Daniels. Integrating simulation technologies with swarm. In *Proceedings of the Workshop on Agent Simulation: Applications, Models, and Tools*, University of Chicago, Illinois, 1999.
4. A. E. Eiben, D. Elia, and J. I. van Hemert. Population dynamics and emerging mental features in AEGIS. In W. Banzhaf, J. Daida, A. Eiben, M. Garzon, V. Honavar, M. Jakiela, and R. Smith, editors, *Proceedings of the Genetic and Evolutionary Computation Conference*, volume 2, pages 1257–1264, Orlando, Florida, USA, 13–17 July 1999. Morgan Kaufmann.
5. J. Epstein and R. Axtell. *Growing Artificial Societies: Social Science From The Bottom Up*. Brookings Institute Press, 1996.
6. D. Hales. Evolving specialisation, altruism and group-level optimisation using tags. In J. Sichman, F. Bousquet, and P. Davidsson, editors, *Multi-Agent-Based Simulation II. Lecture Notes in Artificial Intelligence 2581*, pages 26–35. Berlin: Springer-Verlag, 2002.
7. J. Noble. Cooperation, conflict and the evolution of communication. *Journal of Adaptive Behaviour*, 7(3/4):349–370, 1999.
8. M. Parker. Ascape: Abstracting complexity. Technical report, Center on Social and Economic Dynamics, The Brookings Institution, Washington D.C., Washington, USA, 2000.

Why Synonymy Is Rare: Fitness Is in the Speaker

James R. Hurford

Language Evolution and Computation Research Unit,
School of Philosophy, Psychology and Language Sciences,
University of Edinburgh, Edinburgh EH8 9LL, Scotland, UK,
jim@ling.ed.ac.uk,
http://www.ling.ed.ac.uk/~jim/

Abstract. Pure synonymy is rare. By contrast, homonymy is common in languages. Human avoidance of synonymy is plausibly innate, as theorists of differing persuasions have claimed. Innate dispositions to synonymy and homonymy are modelled here, in relation to alternative roles of speaking and hearing in determining fitness.
In the computer model, linguistic signs are acquired via different genetically determined strategies, variously (in)tolerant to synonymy or homonymy. The model defines communicative success as the probability of a speaker getting a message across to a hearer; interpretive success is the probability of a hearer correctly interpreting a speaker's signal. Communicative and interpretive success are compared as bases for reproductive fitness. When communicative success is the basis for fitness, a genotype evolves which is averse to synonymy, while tolerating homonymy. Conversely, when interpretive success is the basis for fitness, a genotype evolves which is averse to homonymy, while tolerating synonymy.

1 Synonymy and Homonymy in Languages

Synonymy, at least of the pure variety, is rare. If found at all, it is usually in

- contact between languages (*napkin/serviette, eggplant/aubergine*) or dialects (*flat/apartment, bonnet/hood, boot/trunk, mobile/cellphone*),
- handy abbreviation (*microphone/mike, bicycle/bike, hippopotamus/hippo*),
- specialized domains where euphemism is rife, such as sex (*fuck/shag/ ...*), death (*croak/expire/ ...*) and bodily functions (*shit/crap/ ...*).

Human avoidance of synonymy is plausibly innate; theorists of starkly differing persuasions have claimed so. In the functionalist/empiricist tradition, Clark's Principle of Contrast [2] attributes synonymy-avoidance to a (presumably inborn) "pragmatic" tendency of humans to seek and/or create new meanings, rather than accept one meaning for several different forms. Markman [7] notes a disposition in children to assume, initially at least, that no two words may overlap in meaning. The formalist Wexler has proposed an innate Uniqueness

W. Banzhaf et al. (Eds.): ECAL 2003, LNAI 2801, pp. 442–451, 2003.

Principle [16], which prevents the child from internalizing more than one form per meaning; Pinker [9] also subscribes to this Uniqueness Principle.

By contrast, homonymy is common in languages. Consider the many and unrelated senses of such words as *bank, list, file, toast* and *port*. A priori, this is puzzling, as one would suppose a well-designed language to be able to tolerate synonymy without threat to its communicative efficiency, but to outlaw homonymy, as it potentially gives rise to interpretations by a hearer differing from the speaker's intended message. Computer languages and command systems, for example, frequently allow aliases (alias synonyms), but often cannot handle different intended uses of the same term in the same syntactic context (homonymy).

This paper suggests one reason why synonymy-avoidance is in some sense innate, and why there is no parallel innate tendency to avoid homonymy.

2 The Model

2.1 Initial Vocabularies

Each simulation begins with an initial vocabulary of 50 signs. A sign is a form-meaning pair. To compose a single sign, a form is drawn randomly from a set of 100 possible forms ($\mathbf{f}_1 \ldots \mathbf{f}_{100}$), and a meaning is drawn randomly from a set of 100 possible meanings ($\mathbf{m}_1 \ldots \mathbf{m}_{100}$). This method of composing an initial vocabulary gives rise to accidental instances of both synonymy and homonymy, as illustrated below.

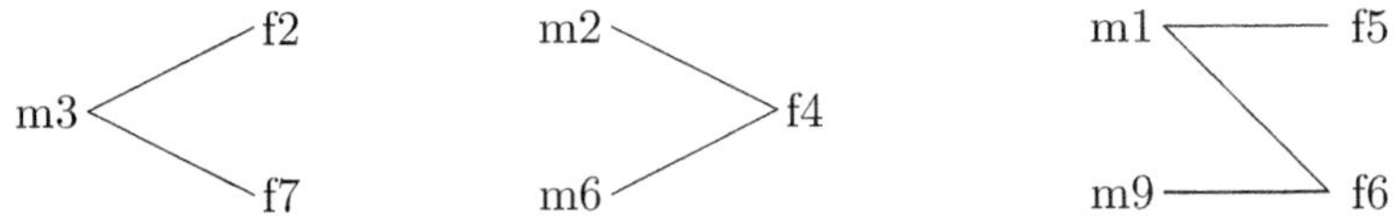

Synonymy (two signs) — Homonymy (two signs) — Synonymy and homonymy (three signs)

In the initial 50-sign vocabularies there were on average between eight and nine such instances of synonymy and of homonymy.

2.2 Population of Learners

All simulations started with populations of 30 agents, composed of 10 innate 'synonym-rejectors', 10 innate 'homonym-rejectors', and 10 innate 'allowers'. These distinct genotypes competed over many generations, until one genotype had completely eliminated the others. The distinct genotypes are defined by their learning behaviour when exposed to a vocabulary of signs. Learning proceeded by each learner being exposed to each sign in the vocabulary of the preceding generation. As the name suggests, when a synonym-rejector was exposed to several distinct signs involving the same meaning (but different forms), it only acquired the first, ignoring all the others. Likewise, when a homonym-rejector was exposed to two or more signs involving the same form (but different meanings), it only acquired the first, ignoring the others. An agent with the 'allower' strategy was more permissive, acquiring all the signs to which it was exposed.

After all the initial population had acquired their individual vocabularies, each individual's fitness was assessed on the basis of how well it could transmit messages to, or interpret signals from, the rest of the population.

Where an agent has several (f) forms for one meaning (i.e. synonyms), it chooses randomly between those forms when attempting to convey that meaning. Thus with f forms for one meaning, the probability of using any particular form for that meaning is $1/f$.

An agent's interpretation of a signal was modelled under three 'context' conditions, 'zero-context', 'full-context' and 'half-context'. In the zero-context condition, where a hearer agent has several (m) meanings associated with one form (i.e. in a case of homonymy), it chooses randomly between those meanings when it hears that form. Thus if a hearer has m different meanings for one form, the probability of his interpreting that form as any particular meaning is $1/m$. In the full-context condition, a hearer agent is always empowered to know which meaning of a form the speaker agent intended; so in this condition the probability of interpreting a form as the meaning intended by the speaker is 1, and the probability of interpreting it as any other meaning is 0. In the half-context condition, a hearer agent is partly biased toward the meaning of a form which a speaker agent intended. This is precisely defined so that the probability of the hearer interpreting a form as the meaning intended by the speaker is $(1+1/m)/2$, where the form has m meanings in the hearer's vocabulary at the time.

Two measures of fitness were defined, Communicative Potential (CP) and Interpretive Potential (IP), as follows. For any pair of agents, S and H, S's communicative potential relative to H and relative to a given meaning M is expressed as $CP(S,H,M)$ and defined thus:

$$CP(S,H,M) = \sum_{i=1}^{n} [P(S \text{ transmits form } i \text{ for } M) \times P(H \text{ interprets } i \text{ as } M)]$$

where n is the number of forms associated with M in S's vocabulary. Thus $CP(S,H,M)$ is the probability, given that S intends to convey M, of S successfully communicating M in an encounter with H. The converse of CP is IP (interpretive potential). $IP(H,S,M) = CP(S,H,M)$. $IP(H,S,M)$ is the probability, given that S is thinking of M, and transmits a form for it, of H interpreting that form as meaning M. A more general measure $CP(S,H)$ can be defined, not relative to any particular meaning, but averaging over all meanings in the system.

$$CP(S,H) = \frac{\sum_{i=1}^{n} CP(S,H,i)}{n}$$

where n is the number of meanings in the system. $CP(S,H)$ is the probability, given that S intends any arbitrary meaning, of S successfully conveying that meaning in an encounter with H. Conversely, $IP(H,S) = CP(S,H)$. $IP(H,S)$ is the probability of H successfully interpreting any form emitted by S. An individual's CP (or IP) relative to the whole population is obtained by averaging over its CP (or IP) relative to all other individuals in the whole population. (These are exactly the same fitness measures as used in [3] and [8].)

2.3 The Simulation Cycle

Each simulation went repeatedly through the following cycle:

1. Establish CP (and/or IP) for each individual relative to the whole population,
2. Nominate parents, on weighted basis of CP (and/or IP),
3. Breed new population of 30 individuals from nominated parents, passing on a parent's innate acquisition strategy ('allow', 'syn-rej', or 'hom-rej'),
4. Each new individual acquires a sign-set on the basis of its inherited innate acquisition strategy and exposure to the complete sign-set of the previous generation of the population,
5. If there is still variation in the population, go back to step 1 and re-cycle.

Two orthogonal sets of conditions were investigated. One set of conditions was the 'context effect', mentioned above; here there were three conditions, zero, half, and full. The other set of conditions was the fitness measure; here again there were three conditions. In one condition, an individual's fitness was proportional to its CP, in another, to its IP, and in the third condition to the average of its CP and IP. Thus fitness was variously identified with success as a speaker, success as a hearer, and overall success as both a speaker and hearer. The two intersecting sets of conditions gave a 3×3 experimental design. For each of the 9 combinations of conditions, 100 simulation runs were carried out. A run terminated when the whole population was homogeneously of one innate type of sign-acquirer. Each of the 900 runs started with a different randomly generated initial vocabulary, as described in section 2.1.

2.4 Results

Figures in cells in the table below show how many simulations, out of 100, each genotype 'won'.

	CONTEXT EFFECT					
	zero		*half*		*full*	
$fitness = CP$	*allow*	6	*allow*	20	*allow*	48
	syn − rej	94	*syn − rej*	80	*syn − rej*	52
	hom − rej	0	*hom − rej*	0	*hom − rej*	0
$fitness = IP$	*allow*	0	*allow*	12	*allow*	32
	syn − rej	4	*syn − rej*	8	*syn − rej*	39 ∗
	hom − rej	96	*hom − rej*	80	*hom − rej*	28
$fitness = \frac{CP+IP}{2}$	*allow*	10	*allow*	29	*allow*	47
	syn − rej	41	*syn − rej*	59	*syn − rej*	52
	hom − rej	49	*hom − rej*	12	*hom − rej*	1

Dissection of the Basic Result. The most striking results appear in the top two cells of the left-hand column. These directly compare selection for communicative potential with selection for interpretive potential, in the zero-context condition. In the zero-context condition, no help is given to hearers in disambiguating ambiguous signals; they choose at random from the possible meanings of a form. These boxes show clearly that, where fitness is associated with success in getting ones meaning across (i.e. with CP), selection favours individuals who resist acquiring synonyms, while agents who resist acquiring homonyms are never favoured. Conversely, where fitness is associated with success in correctly identifying another agent's meaning (i.e. with IP), selection favours individuals who resist acquiring homonyms, while agents who resist acquiring synonyms are very seldom favoured.

If this result is puzzling, it is helpful to consider a single small example as a microcosm of what happens in the simulation. Consider the case of learners exposed to two signs, exhibiting synonymy. Exposed to such data, synonym rejectors acquire a single sign, whereas homonym rejectors and allowers acquire both signs, as diagrammed below.

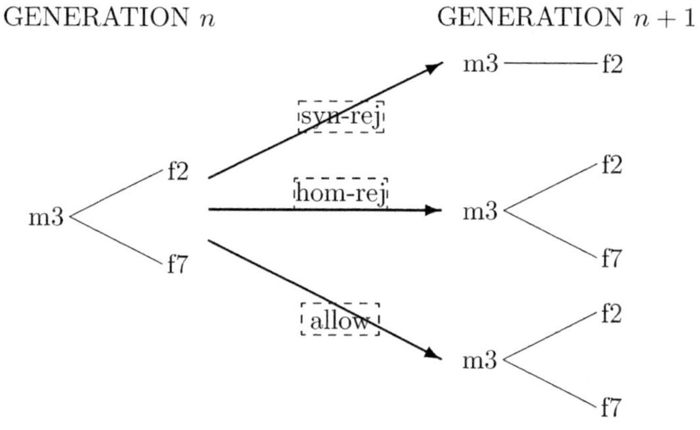

Now consider a mini-population consisting of just three agents, one of each type, with minimal vocabularies as in this diagram. That is, the synonym rejector has a single sign relating meaning $\mathbf{m}_3$ to form $\mathbf{f}_2$, whereas the homonym rejector and the allower both have two signs involving this meaning, relating it to forms $\mathbf{f}_2$ and $\mathbf{f}_7$. If these three agents now attempt to communicate with each other about meaning $\mathbf{m}_3$, their rates of success (i.e. their CP and IP relative to meaning $\mathbf{m}_3$) are as summarized in the following table.

	syn-rej	hom-rej	allow	CP
syn-rej	1.0	1.0	1.0	1.0
hom-rej	0.5	1.0	1.0	0.833
allow	0.5	1.0	1.0	0.833
IP	0.667	1.0	1.0	

The marginal figures here are averages[1]. It can be seen that, in this case, synonym rejectors make themselves harder for the other types to communicate with, thus disadvantaging them, while keeping their own perfect communicative potential (CP). But the synonym rejector's IP is less than that of the other two types.

Next consider a converse case of learners exposed to two signs exhibiting homonymy. Exposed to such data, homonym rejectors acquire one sign, whereas synonym rejectors and allowers acquire two, as diagrammed below.

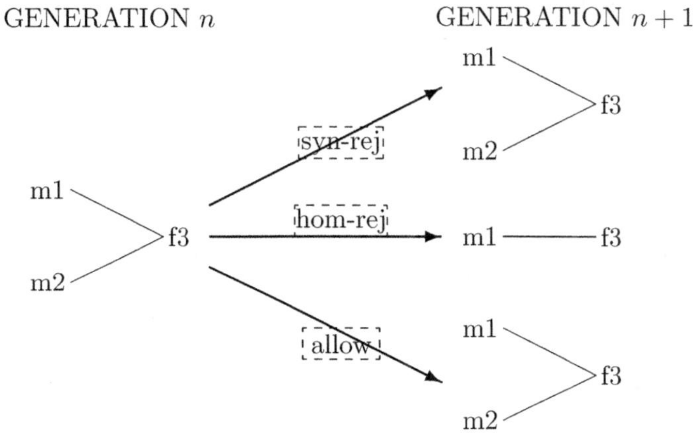

Again, we can imagine a mini-population of just three agents, one of each type, with the mini-vocabularies as on the right of this diagram. If these agents try to communicate with each other about meanings $\mathbf{m}_1$ and $\mathbf{m}_2$, they will have varying success, as shown in the following table. (Here, there are two numbers in the main cells, the first relating to communication about meaning $\mathbf{m}_1$ and the second to communication about meaning $\mathbf{m}_2$. Again, the marginal CP and IP figures are averages, and the earlier footnote still applies.)

	syn-rej	hom-rej	allow	CP
syn-rej	0.5 0.5	1.0 0.0	0.5 0.5	0.5
hom-rej	0.5 0.0	1.0 0.0	0.5 0.0	0.333
allow	0.5 0.5	1.0 0.0	0.5 0.5	0.5
IP	0.417	0.5	0.417	

It seems reasonable to interpret the homonym rejector as being unable either to transmit or receive messages involving meaning $\mathbf{m}_2$, thus getting a zero score for both CP and IP relative to that meaning. It can be seen that all types have a problem with homonyms; the marginal averages are lower in this table than in the preceding table. But, on IP score, the homonym rejector does better than the other types. Conversely, on CP score, the synonym rejector and allower outperform the homonym rejector.

[1] These are averages over all figures in the rows and columns, thus including the case where an agent talks to one of its own type, as well as to agents of the other two types. If, alternatively, the figures in the main diagonal of the table are ignored, the resulting averages make the central point of this subsection even more strongly.

Putting the two tables together, thus combining the two possible basic scenarios (synonymy and homonymy), it is clear that the synonym rejector outperforms the other two types on CP, and the homonym rejector outperforms the other two types on IP. This dissection of the typical elementary interactions in the simulations shows why, in the overall results at the beginning of this section, in the zero-context condition, synonym rejectors were by far the biggest winners when fitness was correlated with CP, and conversely why homonym rejectors were easily the biggest winners when fitness was correlated with IP.

Other Results

The Context Effect. The context effect modelled the influence of context in helping to disambiguate homonyms. The full context condition effectively banished the problem of homonymy, and the half context condition partially banished it. The homonym rejection strategy is an innate solution to the homonymy problem, shown to be effective by the results in the middle box of the left-hand column of the main results table. Banish the problem, and there is no need for a solution. Thus, as one moves rightward across the table, in the middle row, homonym rejectors fare progressively worse. (In the top row, they could not get any worse, starting and staying at zero wins against the other strategies.)

The asterisk in the middle right-hand cell of the main results table above indicates that the numbers in this cell only add up to 99, as one of the runs did not terminate in the time allotted. The numbers in this cell are more finely balanced between all three competing strategies than in any of the other cells. In this combination of conditions, 'full-context' gives all agents a form of mind-reading to disambiguate ambiguous signals, and fitness is based on IP. Thus these conditions give no sign-acquisition strategy a clear advantage.

Fitness Averaged between CP and IP. The bottom row of the main results table shows numbers of winners of each type when fitness was based on the average of CP and IP (with various context effects). Here, not surprisingly, with zero context effect (left-hand box), synonym rejectors and homonym rejectors were about evenly balanced, but both had a clear advantage over allowers. Now look rightward in the table along the bottom row, first through half-context diminishing the problem of homonymy, and then to full-context eliminating the problem entirely; homonym rejectors become progressively and markedly less successful, with their niche being taken over largely by allowers and to a lesser extent by synonym rejectors.

3 Discussion

3.1 Relation to Other Work

The model used in this study bears obvious similarities to several other models, but differs in various ways from all of them.

In the present model, the signs are learned, and not innate like those in the models of [15,1].

The present model is not a model of how populations, starting with no shared conventional meaning-form pairings, evolve socially coordinated signalling systems. In the present model, the same initial (randomly generated) vocabulary is given to each member of the population in the initial generation; agents of different innate types subsequently modify these vocabularies, but only by eliminating some signs. In this way, the present model differs significantly from those in [8,10,11,13,14].

In K.Smith's model [10], various innate learning biases compete over many generations, as in the present model. He found that suitably biased (homonymy- and synonymy-avoiding) innate learning rules did not evolve in the absence of a pre-existing shared vocabulary in the population. In another model [11], K.Smith showed that only certain innate learning biases can give rise to the cultural evolution, over many generations, of a communally shared set of meaning-form pairings. Combined with the results obtained in the present model, these studies give us a chicken-and-egg problem. This model shows that innate anti-synonymy and anti-homonymy biases may evolve naturally, given a pre-existing shared code; K.Smith's studies show that a common code cannot arise unless the innate biases are already in place. In a further study, [12] he argues that the required biases may have arisen for reasons independent of communication. It would be worthwhile, in future work, to try to solve the chicken-and-egg problem here with a more direct approach, involving 'mutual bootstrapping' of the cultural evolution of a code and the biological evolution of the learning biases. This might be possible by starting with very small sets of meanings and available signals, possibly as small as 2. All the studies reported here started with larger sets of available meanings and forms, e.g. 10 meanings in K.Smith's case, and 50 meaning-form pairs in the present study.

The present model is technically an instance of the 'Iterated Learning Model' (ILM) [4,6,11]. Such models explore the evolution of signalling systems under conditions of repeated cultural transmission over many generations. In the basic form of the ILM, there is no biological evolution; all the interesting changes are in the culturally transmitted abstract system symbiotically 'inhabiting' the population. An extended version of the ILM is the 'Evolutionary Iterated Learning Model' (EILM) in which the coevolutionary interactions between biological and cultural evolution can be explored; an instance is [5]. In the present model, there is indeed both cultural transmission (of signs) across generations and biological transmission (of sign-acquisition strategies), and so the model is technically an instance of an EILM. But in the current model, the dynamics on which I have focussed exhibit neither interesting changes in the culturally transmitted vocabularies nor coevolutionary interaction between cultural and biological mechanisms. The interesting conclusions relate to the evolved genotypes, under the various conditions. In this sense, the present model is more like a genetic algorithm in spirit, with experimental variation of the fitness function.

There is a rather obvious, but trivial, sense in which the culturally transmitted vocabularies evolve in the present simulations. When a population is totally taken over by innate synonym-rejectors, the population's shared vocabulary, by definition, contains no synonyms. Likewise, the vocabulary of a population taken over by homonym-rejectors contains homonyms. Further experiments, not reported in detail here, show that, if random mutation between the three genotypes is permitted in a population which has culturally converged on a synonymy-free vocabulary (with no subsequent cultural innovation in the vocabulary), the population's gene-pool can drift to reincorporate synonym-rejectors. In such a situation, the fixed cultural environment presents no problem arising from synonymy, and thus there is no selection of genotypes to cope with such problems. Here synonymy is analogous to a culturally eradicated disease; where there is no threat from the disease, genes can be reintroduced which lessen the resistance to it, with no ill effects (until the disease somehow resurfaces).

3.2 Conclusion

There may be several reasons why synonymy is rare in human languages. This model has explored the dynamics of communication involving a many-to-many mapping between a fixed set of meanings and a set of forms. The model reveals a systematic relationship between synonymy/homonymy and the transmission or reception of messages. It would be interesting to explore the possibility of a changing set of available meanings; new, more specialized, or perhaps metaphorical, meanings could be constructed out of a basic set. Such meaning-creation is certainly active in the dynamics of human languages; it is not incompatible with the present model. Economy of storage might be another factor in synonymy avoidance – why remember more than one word for the same meaning?

Computers are ideally adapted as receivers of instructions. That is why computer languages, unlike human languages, abhor homonymy. There may be various factors contributing to the rarity of synonyms in human languages. This model builds one piece of evidence that humans evolved to be well adapted as senders of messages; accurate reception of messages was less important in our prehistory. We may be primarily speakers, and secondarily listeners.

References

1. Ackley, D., Littman, M.: Altruism in the Evolution of Communication. *Artificial Life*, **4** (1994) 40–48.
2. Clark, E. V.: On the Pragmatics of Contrast. *Journal of Child Language*, **17** (1990) 417–431.
3. Hurford, J. R.: Biological Evolution of the Saussurean Sign as a Component of the Language Acquisition Device. *Lingua*, **77** (1989) 187–222.
4. Kirby, S.: Spontaneous evolution of linguistic structure: an iterated learning model of the emergence of regularity and irregularity. *IEEE Journal of Evolutionary Computation* **5(2)** (2001) 102–110.

5. Kirby, S., Hurford, J.R.: Learning, culture and evolution in the origin of linguistic constraints. In *Fourth European Conference on Artificial Life*, edited by P. Husbands and I. Harvey. MIT Press, Cambridge, MA. (1997)
6. Kirby, S., Hurford, J.R.: The Emergence of Linguistic Structure: An overview of the Iterated Learning Model. In *Simulating the Evolution of Language*, edited by A. Cangelosi and D. Parisi, Springer Verlag. (2002)
7. Markman, E. M.: *Categorization and Naming in Children: Problems of Induction.* MIT Press, Cambridge MA. [esp. Chs 8 & 9] (1989)
8. Oliphant, M., Batali, J.: Learning and the Emergence of Coordinated Communication. *Center for Research on Language Newsletter*, **11(1)**. (University of California at San Diego.) (1997)
9. Pinker, S.: *Language Learnability and Language Development*, Harvard University Press, Cambridge MA. (1984)
10. Smith, K.: The Importance of Rapid Cultural Convergence in the Evolution of Learned Symbolic Communication. *Advances in Artificial Life: Proceedings of the 6th European Conference on Artificial Life*, (2001) 637–640.
11. Smith, K.: The Cultural Evolution of Communication in a Population of Neural Networks. *Connection Science*, **14(1)** (2002) 65–84.
12. Smith, K.: *The Transmission of Language: models of biological and cultural evolution*, PhD thesis, University of Edinburgh. (2003)
13. Steels, L.: *The Talking Heads Experiment*, Volume 1. Words and Meanings. Antwerpen: Laboratorium. (1999) Special pre-edition.
14. Steels, L., Kaplan, F., McIntyre, A., van Looveren, J.: Crucial Factors in the Origins of Word-Meaning. In *The Transition to Language*, edited by A. Wray. Oxford University Press, Oxford. (2002)
15. Werner, G., Dyer, M.: Evolution of Communication in Artificial Organisms. *Artificial Life*, **2** (1992) 659–687.
16. Wexler, P., Culicover, P.: *Formal Principles of Language Acquisition*, MIT Press, Cambridge MA. (1980)

A Noisy Way to Evolve Signaling Behaviour

Tudor Jenkins

School of Cognitive and Computer Sciences,
University of Sussex, Brighton, BN1 9QH, U.K.
tudor@wontfail.com
http://www.wontfail.com

Abstract. This paper looks at the way signaling behaviour can arise within a population of evolving agents involved in complex task domains where problem-solving behaviours need to be developed and integrated with appropriate signaling strategies. A method is proposed to overcome the difficulties of evolving separate yet compatible parts required by transmitters and receivers that serve no function but communication. The validity of this method is supported by a series of experiments. These not only succeed in evolving agents capable of controlling and enhancing complex behaviours through signaling but also demonstrate how bigger search spaces with more signal channels than might be needed can lead to faster adaptation.

1 Introduction

A greater understanding of how signaling can evolve within an artificial population serves a dual purpose. This is to assist in our understanding of natural communication systems and also to enable the development of artificial life capable of grounded symbolic communication which, it has been argued [1,2], is a prerequisite for many higher level aspects of intelligent behaviour. In this paper we consider why there are still great difficulties in generating signaling behaviour within an adaptive population of simulated robots even when conditions to favour cooperative communication prevail. Quinn [3] has shown how iconic signaling can arise under these conditions but we shall consider the more general case where agents can use arbitrary signals on dedicated signaling channels. Although signaling behaviour exploiting these arbitrary signals might be harder to acquire [4], enhancing our understanding of how this might be achieved has great value as there are many problem domains where iconic signals will not suffice.

The difficulty stems from the number of different and possibly independent features that have to arise and be coordinated in order for any benefits to accrue from signaling behaviour. Consider we have two agents involved in an interaction from which both will benefit if one (T) conveys some information to the other (R) and this second agent acts appropriately to the value of the information. The state of this information only agent T can potentially retrieve from either the environment or its inner state. In order for these two to benefit

W. Banzhaf et al. (Eds.): ECAL 2003, LNAI 2801, pp. 452–461, 2003.

1. T must detect and discriminate this situation that warrants a signal and then,
2. activate its signal generating device to create a signal unique to the situation.
3. R must detect the signal and discriminate it from others and finally,
4. R needs to modify its behaviour accordingly to take advantage of this new information.

Where there are a small number of possible world states that agents are able to sense and a restricted number of possible behavioural repertoires and signals they can generate relating to an interaction, then the search space of suitable control systems is not too large. Most existing work within this domain has applied this approach with a view to understanding higher level characteristics of signaling systems [5,6]. However, it seems intuitive that more general communicative abilities with the possibility of increased complexity to cover non-finite problem domains (including most real world problems) must be integrated at a very low level of control capable of generating an infinite range of behaviours. But it is just this approach where the four separate parts mentioned above will each present a very big search space. The time taken for the four separate stages to develop in their own way and coincidentally match could take very long, so a way of speeding up the process is needed. One approach whereby this can be achieved occurs when the production of signals arises directly from a behavioural trait caused from the transmitting agent's own reaction to the information. For example, T might detect a resource to its right and turn in that direction. If this change in direction could possibly be detected by R then it could act as a signal conveying information about the direction of the resource. The transmitting behaviour would arise as a consequence of an adaptation to some fitness enhancement other than signaling. Only the receiving side of the signaling interaction would then have to develop for rewards to be gained from the signaling interaction. Another way that this four stage problem can be overcome for communication to develop is by reception behaviour developing to take advantage of information provided by the environment. With one half of the signaling behaviour in place, it is then a far simpler matter for potential transmitters to adapt to imitate the environmental signal and thus influence the receivers in a similar way to the environment. Krebs and Dawkins [7] suggest that signals developed in this way are a means for transmitters to exploit the receiving strategies of others. Both of these approaches provide iconic signaling that we previously referred to. Effective as these approaches are for a large class of signaling behaviours, many tasks do not offer the opportunity of rewards for the partial development of signaling behaviour.

2 A Noisy Approach

What is needed is a more general approach, not dependent on the possible presence of environmental props, but still providing a means to ease the generation and integration of the four stages discussed in the last section. Such a method will now be proposed and details of an experiment to assess one aspect of its performance follow. As an agent interacts with its environment at any stage there is a chance that one of its sensors or an inner state might reflect the value of a particular state of the environment or the agent itself that could be useful information to another agent. Under conditions of cooperation, conveying this information to the other agent would be advantageous to both. By providing agents with some device which enables them to transmit information about all their sensor and inner states it would be possible to ensure that any other agent capable of detecting the full range of these signals would be privy to all the information available to the transmitting agent. Thus, this receiving agent could base its behaviour on this information as well as the state of the environment directly amenable to its other sensors. Unfortunately this is not a realizably approach due the vast number of different signals that would have to be transmitted at any time. Restricting the range to some fixed subset would not offer a suitable solution either as the very information required might be outside of this and where agents adapt, the range of possible information available changes.

A solution would be to restrict the number of signals whilst allowing a change over time[1] to what bits of information they refer. Where there is a correlation between any particular signal and some information of value to receiver agents then whilst that particular signal persists, there is an opportunity for the receivers to adapt to use that information. Prior to any receiving strategies developing in this way, there would be no selective pressure for the maintenance of any particular signal to state correspondence, allowing a drift and thus exploration of new combinations. However, once a cooperative receiving strategy does arise to use some signal then the fitness landscape changes for the transmitters and that particular signaling strategy gets selected for. This approach relies on the production of a lot of signals most of which are worthless, which is why it is referred to as the *Noisy Approach*[2]. There are several important factors that will effect the success of any population relying on this method to generate a communication system. Particularly, the number of possible signals available and the rate at which these signals modify what information they are correlated to. In order to increase the rate at which the transmission of different information is tried, the rate of modification needs to be as high as possible. However, if this rate is too high there will not be enough opportunity for the receiving strategies to develop and spread sufficiently to provide a selective pressure to maintain the signal.

As the number of simultaneous signals available to a population of potential transmitters increases so too does the speed with which one of these will evolve to produce a signal conveying information required for successful communication behaviour. Having as many signaling channels as possible does not offer the best approach though. For members of this population evolving their behaviour so that they can take advantage of information provided by a particular signal then

[1] A genotypic time scale for evolution and phenotypic for learning.

[2] In nature this would restrict its application to conditions free from enhanced risk of predation.

it is better if this signal is one of as few as possible. Clearly there is a conflict of interests about the optimum number of signal channels between the transmission side of populations' control and their receiving side. The rest of this paper details an experiment designed to confirm the validity of the noisy approach and test the hypothesis that there will be a finite non-zero optimum number of signal channels for the fastest development of communication.

3 Task

A population of agents were evolved using a genetic algorithm where the fitness was established from the ability of the agents to perform a particular task enhanced by communication. This process was repeated a number of times with all parameters remaining the same between runs except the number of signaling channels available to the agents. By measuring the time taken for the various populations to converge on a particular behavioural strategy dependent on signaling it was possible to establish how the number of signal channels influenced the evolution of signaling. Within these experiments only a single signal was necessary for optimum behaviour but in the later discussion issues relating to the development of multiple signals are considered.

A typical problem that has been successfully attempted within the field of evolutionary robotics is the task of navigation within T-junction mazes to locate resources placed at the end of one of the two corridors leading off at right angles from the base corridor. A variation on this problem is easily amenable to a signaling task. Two agents are placed in the base corridor at the far end from the junction and one of them (T) is free to explore the maze which contains a variety of features to assist the agents with navigation. Once it encounters an area of resources placed randomly at the end of one of the other two corridors the second agent (R) is free to move. Though a small reward is offered to the initial agent (T) based on the speed with which it finds the resources, a substantially greater reward is available, equally to both agents, that is related to the speed with which the second agent reaches the resources. As the agents have no way of individually determining the location of the resources without actually being over them, the second agent can enhance its performance if the initial explorer provides information about the location through a signal.

4 Experimental Details

The environment in which agents operate is a simple T-shaped maze with impenetrable walls. Only two agents can be placed in this arena at any one time. The corridors have a width of 10 units, where one unit represents the width of an agent, and the base corridor's length is 20 units whilst the length of the two other corridors leading off from the junction are 30 units each. Agents are placed in the arena within 5 units from the end of the base far from the junction with a randomly assigned orientation. At the beginning of any individual trial with two agents the arena is cleared of any residual resources and an area occupying all space within 10 units from the end of one of the non-base corridors (goal zones) selected at random are filled with resources. Agents are able to travel over resources unencumbered.

A series of features are also placed in the maze. Whilst occupying a fixed position they take up no physical space. The reason for these features is to give agents an ability to identify their position within the maze environment both as a guide to assist in navigation and also to act as a possible reference for signals. The features allow an agent to navigate the maze without the need to develop some complex representational mapping. In these experiments 3 different features are used, one placed at the junction and one in each of the goal zones.

Signals produced by the agents are propagated through the environment. These signals are analogous with sound signals over short distances. They travel very fast throughout the maze, covering every area in one time step with no decay of amplitude. Furthermore, it is impossible to determine the direction of a signals origin from its physical properties. This is to force the signals to take on a symbolic form. The signals have only one value associated with them and that is the channel on which they were generated. This is equivalent to the pitch. No two signals can interfere with each other.

The agents are simplified models of Khepera robots. Collisions with other agents and corridor walls are non-elastic and frictionless. Movement is achieved by two aligned high friction wheels which also provide steering by variation in wheel speed. The power to each wheel motor is determined by the output of the actuators associated with it. One actuator applies forward torque and another, reverse torque. The agents are restricted to a top speed equal to their diameter per time step.

Agents are equipped with a minimal number of sensors in order to achieve their navigational goals. *Base sensors* are placed around the agent, typically 4, pointing in different directions. These sensors act effectively as a compass allowing an agent to identify its orientation. *Feature sensors* correspond to the various features placed within the environment of which there are 3.They are omnidirectional and are activated by the proximity of a feature. A *Resource sensor* is also available. This is activated if the agent is in contact with a resource. One final set of sensors is for detecting signals. The number of separate signal sensors associated with predetermined channels is set for each experimental run. If a signal is present on a particular channel then the sensor corresponding to that channel is activated for a time step with a value proportional to the strength of the signal. Actuators corresponding to each of these channels are also present in each agent. When these receive activation they produce a signal with strength proportional to the activation at the frequency corresponding to the channel of the actuator, though there is a 10 percent chance of an adjacent channel being used. The introduction of this noise is hoped to be useful in future experiments where several signals are to be derived from existing ones.

The agents are each controlled by a discrete time recurrent neural network, many of the parameters of which are specified by a genetic algorithm described later. The overall architecture of the network is shown in figure 1.

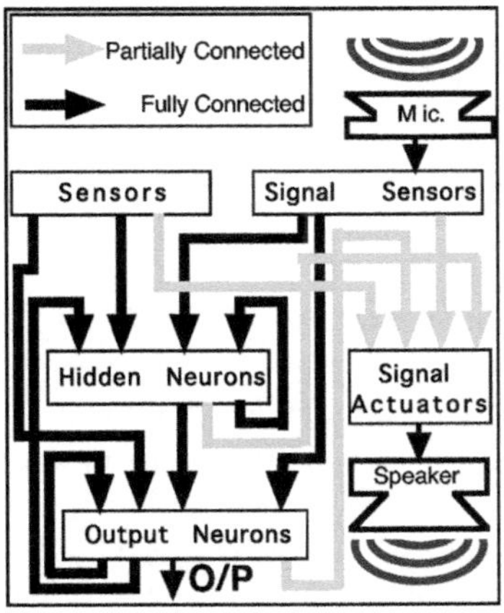

Fig. 1. Control architecture.

Neurons are grouped into four main categories, sensors, hidden neurons, outputs and signal actuators. The characteristics of the neurons in the second and third of these categories are the same, it is the connections from them that divides them into separate categories. Signal sensors form a subset of the total sensors available to the agents and are only separated in 1 to draw awareness to them.

The value of each of the actuators (used to power wheels or produce noise) associated with an output neuron is simply the value of the output of that particular neuron restricted to within the limits of zero and one. Input neurons take their value from sensors associated with each of them. The output of these input neurons is a direct copy of the sensor value.

The inputs of both the output and hidden neurons are fully connected by weighted connections within the network to the outputs of all other neurons except the signal actuators. The inputs of the signal actuator neurons are connected in a similar fashion to the outputs of other neurons within a controller, however each of these is not connected to all other neurons but simply a restricted number of them (ten). The neurons to which these signal actuator neurons are connected is genetically determined.

The functionality of the various neurons differ. The sensor neurons have already been accounted for. All other neurons besides the signal actuator neurons work as follows. At each time step the output value of a neuron is established in two stages. Firstly, the input to the neuron is worked out by the following equation:

$$\dot{A}_j = \frac{-A_j + \sum_{i=1}^{k} w_{ij} O_i + I_j}{\tau_j} \tag{1}$$

where A_j is the activation of the jth neuron, w_{ij} is the weight of the connection between the ith and j^{th} neuron, O_i is the output of the ith neuron, I_j is a constant input value to the jth neuron, and τ_j is the time constant of the jth neuron. The output of a neuron is calculated by applying this input to a standard sigmoid function:

$$O_j = \frac{100}{1 - e^{(t_j - A_j)}} \tag{2}$$

where t_j is the threshold value of that particular neuron. These values, w_{ij}, I_j, t_j, and τ_j are genetically specified for each neuron and their possible values fall within the following ranges: $-500 \leq w_{ij} \leq 500$; $-500 \leq I_j \leq 500$; $100 \leq \tau_j \leq +500$; $-500 \leq tj \leq 500$.

A simpler method is employed for the activations of the signal actuator neurons. The products of all of the outputs of those neurons connected to a signal actuator neuron and the weights of the connections are summed. If this results is a positive value then this is given as the output of the neuron up to a fully activated value of 100 otherwise the neuron is switched off.

A simple generational evolutionary algorithm was employed. The population was stored in a one dimensional string and at the end of the trials for each generation, the genotype of every agent was replaced by recombination and mutation of the genotypes of the two highest scoring local agents where local was defined for these experiments to be 2 places either side. There was a direct mapping between genotypes and phenotypes.

Each agent performed 2 trials with both its immediate neighbours in the role of potential transmitter and receiver. After each of these trials a score attributed to the agents was added to their overall fitness. A trial would last for 500 time steps at which point it would terminate automatically. If however the second agent reached a resource before this time then the trial would end prematurely. This fitness was used as the score for breeder selection. A fitness function was needed that not only rewarded behaviour arising from signaling but also some of the sub behaviours required of the agents to be in a position to take advantage of signaling. These were initially moving around the maze and then effective strategies for exploring the maze. For this, a score was allocated to each of the agents scaled to a maximum of 50 for the number of different walls that they both came into contact with away from the base corridor. A second part to the fitness function provided a significantly higher score of up to 500 points for the time that the second agent could take to get to the resource. A best possible performance was calculated. This yielded a score of 500 and any time longer than this was scaled linearly down as the agents took longer to a minimum of zero if the task was not completed within the timeframe allocated. Though this was sufficient to generate communicative behaviour, a final bonus score was added to make any communication easy to notice from the scores. This bonus was 2000 if the agents managed to perform their two trials, as either transmitter or receiver, with one other individual with near optimal performance (90%). An efficient signaling strategy would ensure that any agent could collect this bonus 4 times as well as a further score of 500 x 8 giving a total score of 12,000. Agents not capable of signaling could expect to gain the bonus only one in every four times thus expecting a score of approximately 5000.

5 Results

5.1 Signaling Behaviour

Trials were conducted with a range of different signal channels available to the agents. The trials were executed on a G3 300Mhz processor and each took between a day to four days to complete. In order to test the system and gain some control of what possible behaviours could be expected without communication, tests were conducted with no signaling channels available. From these tests the agents were restricted to limited scores, however they did manage to develop a very efficient way of exploring the maze but could expect to get no further than that. The two agents in a trial would both behave in similar ways as this control provided no difference between their tasks. An agent would initially rotate just short of facing directly towards the junction then head off in a straight line. On reaching the junction the agent would always be to one side and it would rotate towards the goal corridor that was on this side. This rotation was precipitated by the detection of a feature present at the junction. Once the agent reached the end of this corridor relying on a local feature or a resource to establish this there would be two possible courses of action.

With no resource there, the agent would rotate on the spot and head directly back across the junction to the end of the other possible goal corridor. If the corridor contained a resource then its behaviour could take on any form, normally just moving to a corner and remaining there. On contact with a resource the second agent would be released and perform similar behaviour. Even agents with signaling capabilities nearly always settled on a similar strategy for the transmitter agent performing the initial exploration. A maximum fitness of approximately 4000 was achieved within 15,000 generations. The populations were left to evolve for 200,000 generations but no significant changes occurred.

A range of different numbers of channels were now employed between 1 and 20. An attempt was made to try more than twenty but this resulted in trials taking far too long on account of the severely reduced speed of execution of the agents control. On each of these trials significantly higher scores were obtained than in the zero signal channel control. The populations were obtaining average fitnesses of between 9,000 and 11,000 and in nearly every case this occurred before 60,000 generations. As well as being indicated by the scores, observations of the evolved agents revealed clearly that a communication system was in place. Two dominant signaling tactics evolved.

When the first agent encountered a goal corridor without any resources, it would do either one of two things: (1) Continue, as in the case with no signaling, to change direction and head for the second goal corridor. On reaching this, normally only one of the number of signal channels would change state always from inactive to active and thus produce a novel signal also, the agent would rather than move around randomly, stay fixed in the same spot. (2) Generate a signal, turn and proceed towards the other goal corridor where it would drive into a corner, forcing it to stop yet continue with the signal. This would have exactly the same effect on the second agent as the first approach. In both of these cases the second agent's behaviour is nearly always very similar to the first agent on account of them being very closely related. Thus, without a particular signal the environment it faces is the same as the first one as it set out so it will behave in the same way as the first agent. This is why the first agent very rarely generates a signal when a resource is encountered in the first direction that it looks. When the second agent responds to a signal it would proceed as before but the initial rotation to get it heading in the direction of the junction would carry on a bit longer and the agent would head towards the junction pointing slightly towards the turning towards the resources. At the junction a slight modification in direction would ensure the agent would head directly towards the resources.

Though channels have an equal likelihood of starting in an on or off state and switching to the other to indicate a signal, all observations have indicated a preference for switching from off to on. A possible reason for this might be that as both agents in an interaction will be very closely related, if the neutral position for the channel to convey a signal is on and it is switched to off by the initial agent to indicate the position of a resource then the second agent, not in contact with a resource will not notice the switch as it would be still producing a signal on the channel.

In the trials where agents generated a signal on not finding a resource and then headed for the other possible resource location it was interesting to see whether the agents were using the signal for their guidance. By taking one of these agents and placing it in the maze but disabling the particular channel it was possible to test this idea. Of 10 agents taken from different populations 7 of them failed to find the resource corridor once the channel was disabled. By using the signal channel, the agents were able to activate their own signal sensors and use this as part of their own control mechanism. Clearly signaling has a better chance of arising when the control of some information is processed by the agents "thinking aloud". Future designs should perhaps ensure the selection of signaling sensors and actuators suited to encourage this.

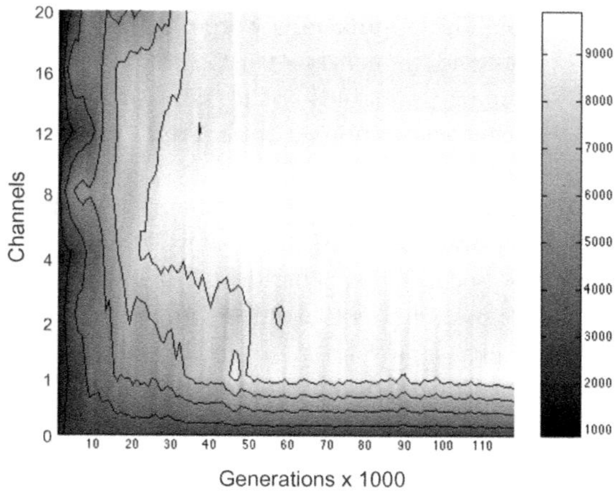

Fig. 2. How signal acquisition varies with number of available channels.

5.2 Measuring the Effects of Channel Variation

A series of experiments were conducted to establish how the number of channels available to agents effects their ability to evolve a specific signaling behaviour. In all 5 trials each were carried out on populations of agents with 0, 1, 2, 4, 8, 12, 16 and 20 signal channels. The trials were run until a score of over 9000 was achieved continuously for 10,000 generations, though terminating after 120,000 where this was not achieved. After considerable time, the 5 sets of scores for each number of channels were obtained and then combined to form an average performance. These were then plotted intermediate values were interpolated in order for contours of fitness to be drawn figure(2). In the graph the space between the first and second contour from the top right (excluding the closed contours) represents the fitness at which some signaling is providing an enhancement to fitness. From this it is clear that even though only a single signal is required for this interaction, the smaller search spaces where only one and two signal channels are available do not provide the optimum conditions for the signaling to arise. There is also a slight decrease in performance as the number of signal channels exceeds 16, however this could be accounted for by the relatively small sample size.

6 Conclusions and Outlook

The results demonstrate that it was possible to evolve signaling using the multiple signal channels without having to resort to providing a task where either the reception or transmission behaviour are stably developed independently of the communication dependent task. It was clearly demonstrated that by using a greater than minimum number of signal channels required, signaling behaviour could evolve more rapidly, though there was no conclusive evidence that further increases in the number of channels lead to a reduction in the rate at which the signaling could evolve. Signaling did arise with just a single channel but it took a considerably greater number of generations.

References

1. Clark, A.: Being There. MIT Press, Cambridge (1997)
2. Vygotsky, L.,S.: Thought and Language. MIT Press, Cambridge (1986)
3. Quinn, M.: Evolving communication without dedicated communication channels. In Kelemen, J., Sosik, P. (Eds), Proceedings of Sixth European Conference Artificial Life (ECAL2001). Springer (2001)
4. Deacon, T.: The Symbolic Species. Allen Lane, London (1997)
5. Hashimoto, T., Ikegami, T.: Emergence of net-grammar in communicating agents. Biosystems **38** (1996)
6. Steels, L.: The origins of syntax in visually grounded robotic agents. In Pollack, M. (Ed.) (1997) Proceedings of IJCAI 97. Morgan Kauffman Publishers, Los Angeles.
7. Krebs, J., Dawkins, R.: Animal signals: Mind reading and manipulation. In Krebs, J., Davies, N. (Eds) Behavioral Ecology: An Evolutionary Approach (2nd Edition) Blackwell, Oxford (1984)

Language Games with Mixed Populations

Michael Lewin and Emmet Spier

School of Cognitive and Computing Sciences,
University of Sussex, Brighton, UK
mlewin@bigfoot.com

Abstract. This paper presents an adaptation of Luc Steels's model of Category Formation and Language Sharing. The simple competitive learning algorithm is proposed as a more general means of creating categories from real-world perception. The model is shown to achieve high levels of coherence and to be very robust when two distinct populations are mixed together, with both populations learning each other's words.

1 Introduction

Luc Steels has established a model of concept sharing in artificial agents using two mechanisms: firstly, the partitioning of a perceptual Input Space as a means of *Category Formation*, and secondly, the use of Language Games amongst agents as a means of *Language Sharing*. The result is a population of agents with a high degree of coherence in their use of language. See [15] for an overview of Steels's "Talking Heads" experiment, which utilises these two mechanisms.

This paper is an extension of Steels's work (especially [9], [10] and [11]) which attempts to create a more general model of category learning. The protocol of the Language Games [9] is largely unchanged, but the nature of the perceptual input and the subsequent method of Category Formation is very different. *Simple competitive learning* [4] is used to partition the Input Spaces, which are vector spaces and can be of any dimension. This increased generality should enable Steels's model to be applied successfully to a wide variety of practical situations.

The use of *symbols* has been a source of much debate in Cognitive Science. Newell and Simon's *physical symbol system hypothesis* [6] – that physical symbol systems are necessary and sufficient conditions for intelligence – has been challenged, in particular by Searle's Chinese Room argument [8] and behavioural-based AI approaches (e.g. Brooks [1]). In response to Searle, Harnad [3] suggested that the only way to escape the "symbol/symbol merry-go-round" of empty syntax is for a set of elementary symbols to be grounded through perception in the real world. All other symbols can then be derived from this elementary set. This *symbol grounding problem* is an important theme in Steels's work, which addresses the need to bridge the gap between perception and his concepts by partitioning a continuous perceptual Input Space into a finite set of discrete categories.

Brooks [1] proposed that we can do away with symbols altogether, but his claim can only be justified by empirical evidence. Thus far, non-symbolic approaches have been effective in models of low-level tasks, but intuition suggests

W. Banzhaf et al. (Eds.): ECAL 2003, LNAI 2801, pp. 462–471, 2003.

that symbols are still useful in talking about – and perhaps even essential in dealing with – higher level tasks, in particular language.

As a means of compromise in the symbol debate, and possibly of bridging the gap between low- and high-level tasks, Vogt [16] suggests that Pierce's definition [7] of symbol is adopted[1]. A *semiotic symbol* is defined as the relationship between a referent (e.g. an object), its meaning, and an arbitrary or conventionalised form (e.g. a word). It is difficult to describe exactly what "meaning" is, but Vogt says it "can be viewed as a functional relation between a form and a referent".

Category Formation is central to our model, so how does this relate to the semiotic symbol? According to Vogt, "although a category should not be equated with a meaning, it is labeled as such when used in communication, because it forms the memorized representation of a semiotic symbol's meaning". The similarity between "meaning" and "concept" is exposed here – a "category" can be seen as an elementary form of "concept", where the concept is defined only in terms of its members rather than by some kind of meta-description of the category.

The paper is laid out as follows: Section 2 gives an introduction to Steels's framework and explains how our model differs from Steels's. Details of the model are given in Sections 3 and 4. Steels's work is simulated in Section 5, and the importance of a *forgetting mechanism* is highlighted. The robustness of the model when two different populations are mixed together is demonstrated in Section 6.

2 Extending Steels's Model

The key difference between the proposed model and Steels's is the nature of the Input Spaces and the categories formed within them. In Steels's model, an Input Space is taken to be the one-dimensional bounded real segment [0,1]. Initially the entire space constitutes one category. New categories are formed as a consequence of an unsuccessful Language Game; an existing category is bisected to form two new ones in the hope of creating a distinction between two previously indistinguishable objects.

In our model, an Input Space is taken to be the n-dimensional vector space $\mathbb{R}^n$. In contrast to Steels's model, these Input Spaces may be of any dimension and take arbitrarily large positive or negative real values. Categories are formed using a simple competitive learning algorithm [4], an example of unsupervised learning. It partitions the space according to the naturally existing clusters of vectors[2]. The input vectors could be defined by a statistical distribution or sampled from a real-world situation.

Steels's model can account for higher dimensional spaces by combining several one-dimensional spaces which are considered to be orthogonal. This leads to

[1] Steels also advocates Pierce's terminology in [15].

[2] It must be assumed that the input is presented in such a way that members of the same category are indeed clustered together. This is what Harnad calls the "Categorical Perception (CP) Effect" [2], but it will not be discussed further here.

a partitioning of the space that is less natural and more restricted than the competitive learning algorithm's, as illustrated in Fig. 1.

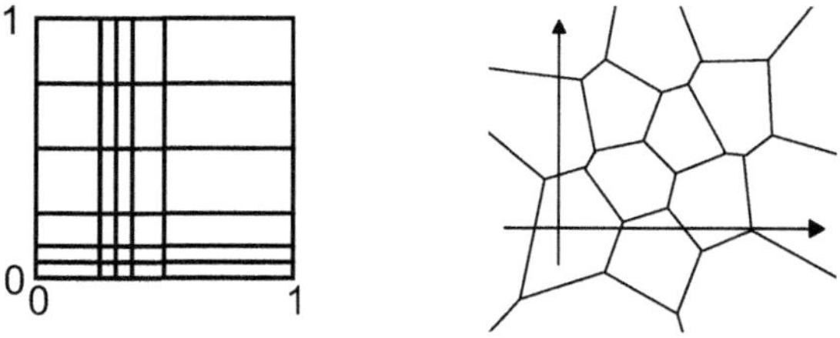

Fig. 1. The difference between Steels's model (left) and the new model (right) is the nature of the Input Spaces and the categories formed within them.

In Steels's model, the agents associate words with *feature sets*, where a *feature* is defined as an *attribute-value pair*. For example, a one-dimensional attribute such as "horizontal position" could be divided into values such as "far left", "left", "centre" and "right". In our model, each *dimension* of the Input Space will usually correspond to an *attribute*. When the Input Space is one-dimensional, the result is similar to Steels's model – a feature is a region in a single dimension.

In some cases, one may wish to model an attribute as multi-dimensional. For example, colour can be thought of as a three-dimensional space whereby every colour is defined by its red, green and blue components. In such a case it is natural to treat the Input Space as three-dimensional and give names to the regions of this space, without allowing the agent direct access to the three underlying dimensions. The categories will correspond not to regions in a single dimension but to regions in the higher dimensional space. They can still be thought of as *attribute-value* pairs, for example "the colour purple", where "colour" is the attribute and "purple" its value (a region in $\mathbb{R}^3$).

It is also possible, however, to take a single high-dimensional Input Space to be the entire space of possibilities of objects in the world. Thus agents would create names for objects (e.g. "tree") as opposed to attributes (e.g. "big") and the lower-lying attributes of an object (as defined by the dimensions of the space) would not be directly accessible to the agent. It is a question of design whether the agents will create words which correspond to objects, attributes or both.

In this paper, the agents are simulated and their perceptional stimuli are abstract (clusters of points in a vector space). There would be no difficulty, however, in transferring the model to a physical situation such as that which Steels has presented [15].

3 Simple Competitive Learning

Our model uses the standard winner-takes-all algorithm with leaky learning. The network consists of a fixed number of *nodes* and with each node is associated a vector of the same dimension as the Input Space. Initially the vectors are randomly distributed according to the uniform distribution U[-2,2]. Whenever the network is presented with an input vector, the *winning node* is the one whose corresponding vector is closest (in Euclidean distance) to it. The winning vector is then updated according to the following rule:

$$d\boldsymbol{w_i} = \eta(\boldsymbol{I} - \boldsymbol{w_i}) \tag{1}$$

where $\boldsymbol{I}$ is the input vector and $\boldsymbol{w_i}$ is the weight vector corresponding to the winning node i. The learning rate η was always set to 0.1.

The winning vector, which was already closest to the input vector, moves closer still. This is the positive feedback mechanism which enables the prototype vectors to identify clusters in the Input Space. This alone, however, will not always be successful in identifying the clusters. There will often be *dead nodes* which are never the winner and hence never move. To remedy this, it is common practice to add *leaky learning* to the model; the winner is updated according to (1) and all the other vectors are updated by the same rule but with a smaller learning rate. We used a rate of $\frac{\eta}{100}$.

This leads to a very stable situation. The prototype vectors are able to locate the clusters quite quickly and, more importantly, once the clusters have been located the prototype vectors stay in the region of that cluster. Thus there is no need to reduce the learning rate after training, as is common with other neural network algorithms.

At any time the Input Space can be thought of as being partitioned by the *"Voronoi Sets"* [4] defined by the prototype vectors. Each such set, or *proximity neighbourhood*, contains precisely one prototype vector and is defined as the set of points which are closer to that prototype vector than any other.

4 Language Games

Once the necessary categories have been formed, the question arises of how a whole society of agents can share their categories using language, i.e. a set of words with commonly agreed meanings. In a realistic setting, it is important to consider language in the context of some *selective pressure*: language will only arise because it is beneficial in some way to either the individual or the species. Steels [12] [13], however, argues that the contrived notion of a *Language Game* is a useful tool for rigorously analysing the process of language acquisition. Here, a Game is defined as "a routinised sequence of interactions between two agents involving a shared situation in the world". In this case it involves a speaker trying to describe an object to a hearer by uttering words which refer to the object's features. An overview of our model is shown in Fig. 2.

The protocol of the Language Games is almost identical to that used by Steels. Two agents are selected at random to play the roles of *speaker* and *hearer* respectively. A fixed number of objects are created[3] to form the *context* and one

[3] In Steels's model the objects are chosen from a finite (and small) collection. In our model, they are created afresh each time, allowing for an infinite variety of objects.

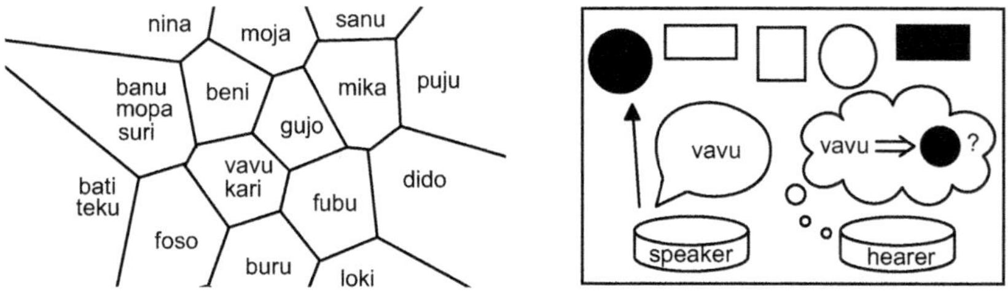

Fig. 2. Overview of the model. Words can be associated with the Voronoi Sets formed by competitive learning (left). The agents achieve a consensus through interacting in Language Games (right).

of them is designated (as if the speaker had pointed at it) to be the *subject* which the speaker will try to describe.

The speaker then generates a *distinctive feature set (DFS)* – a set of features which the topic possesses and which none of the other objects in the context has. If more than one DFS exists, the smallest is chosen. When more than one minimal DFS exists, preference is given to feature sets for which the agent already has a word. To choose between two such feature sets, the entry's *"score"* $(successes - failures)$ is used as a tie-breaker – the entry with the highest score is selected. This is the crucial positive feedback mechanism which drives the system towards coherence – agents prefer words which are already established and successfully used in communication. If no DFS exists, the Language Game is aborted.

The speaker converts its DFS to a word[4] using the *cover* function.[5] This searches the lexicon for an occurrence of the feature set and returns the corresponding word. If more than one such occurrence exists, the entry's *score* $(successes - failures)$ is again used as a tie-breaker. If no such occurrence exists, it is possible (with probability $p_w = 0.05$) for a new word corresponding to the DFS to be created and stored in the lexicon. If this does not occur, the game is aborted. The parameter p_w affects how many different words enter the population, for the lower it is, the more chance there is for a single word to be spread around the population instead of many words with the same meaning being invented by different agents.

The hearer then converts the word back to a feature set using the *uncover* function. This simply searches the lexicon for that word. If the word does not appear in the hearer's lexicon, its meaning is guessed – a new entry is created comprising the uttered word and a DFS selected by the hearer. It is possible that this is not the same DFS selected by the speaker, so ambiguities can arise.

[4] all "words" have four letters of the form consonant, vowel, consonant, vowel. Of course any arbitrary symbols could be used as "words".

[5] The *cover* and *uncover* functions defined here are simpler than Steels's [9]. In particular, *utterances* (sets of more than one word) were deemed unnecessary.

If the uttered word was not new to either the speaker or hearer then the game is evaluated (as if the hearer had nodded or shaken its head in response). The hearer interpreted the word as a feature set. If this is a DFS for the subject, the game is recorded as a success. If not, it is recorded as a failure. This feedback will affect the scores of the word/category pairs in the agents' lexicons.

5 Simulation of Steels's Model

Initially, a situation similar to that proposed by Steels in [10] was considered. A population of 30 agents was trained for 10,000 input steps before the Language Games commence to allow the competitive learning algorithm ample time to establish stable categories. A series of 200,000 training Language Games was then carried out. Each Language Game involved 5 objects. Every object comprised 5 parts, each drawn from a one-dimensional Input Space. The five Input Spaces all had the same distribution: just two clusters distributed normally with means at 0 and 1 respectively and identical variance σ^2. This is illustrated in Fig. 3a.

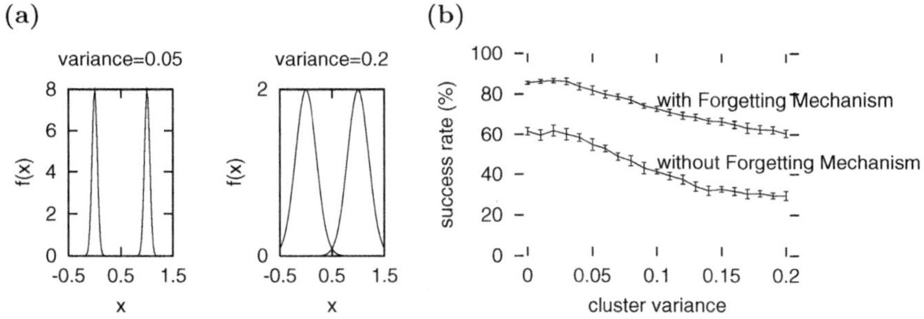

Fig. 3. (a) Points in the Input Space are defined by two normal distributions centred at 0 and 1. The variance σ^2 affects how much overlap there is between the two clusters. **(b)** Long-Term Success Rate against cluster variance σ^2, with and without forgetting mechanism. Each bar shows the mean and one standard deviation over 10 runs.

When σ^2 is very small, the agents' categories are almost fixed – only the "leaky learning" mechanism causes a slight plasticity in the prototype vectors' positions. This is comparable to Steels's model. As σ^2 increases, agents' categories become less stable and so different agents will have different perceptual categories, leading to a decrease in success of the Language Games.

A useful way to quantify this is to define the *Long-Term Success Rate* ρ as

$$\rho = \left(100 \times \frac{successes}{successes + failures}\right) \qquad (2)$$

where the successes and failures, as defined in Section 4, are recorded over 1,000 test Language Games after the 200,000 training Language Games have been carried out.

In later work [15] Steels adds a *forgetting mechanism* to his model. This allows word/category pairs to be removed from an agent's lexicon if they have not been used for a long time (20,000 rounds), or if their score $successes - failures$ falls below a given threshold (0).

The forgetting mechanism significantly increases the communicative success of the population. Without it, any word/category pair recorded in an agent's lexicon remains fixed for the duration of the experiment. Thus if there is a single case of misunderstanding between two agents, the erroneous word/category association will lead to repeated failures in the Language Games.

Results

Figure 3b shows the Long-Term Success Rate ρ with and without the forgetting mechanism. It is clear that ρ is significantly higher when the forgetting mechanism is used, achieving highs of almost 90% for small σ^2. By comparison, without the forgetting mechanism, ρ is never more than 65%. In both cases, ρ decreases as σ^2 increases, as expected[6]. When $\sigma^2 = 0$ these results replicate Steels's.

6 Mixing Two Populations

An agreed set of linguistic conventions can be tested by allowing two populations to develop independently before mixing and engaging in Language Games together. The agents will already have established their own linguistic conventions, but they should also be able to learn the new words of the foreign population. This can be considered a crude model of the meeting of two different nationalities with entirely different words for the same meanings.

Steels [14] demonstrated that under his model, a linguistically naive agent can be added to the population and will successfully learn the linguistic conventions present in the society. Here, two mature populations are mixed together – a much sterner test of stability.

Two populations of equal size (30 agents)[7] were trained for 10,000 rounds and then engaged in training Language Games for 200,000 rounds using identical parameter values to those in Section 5. Then the agents in the second population were pooled with a set of *visitors* from the first population to create a mixed population. A further series of training Language Games was then carried out in which the speaker and hearer were drawn at random and with equal probability from this mixed population. Thus it was possible that the speaker and hearer came from different populations, but there would still be interactions between

[6] The model was also effective with higher dimensional Input Spaces. Increasing the number of dimensions does not directly affect ρ, but other related parameters, such as the number of categories formed by the agents and the degree to which data is clustered together, will. Details can be found in [5].

[7] If these two populations were not of equal size then the agents from the smaller population would engage in more Language Games. Thus their words would have a higher score than those of the other population, causing them to dominate.

agents from the same population as well. This phase was continued for another 200,000 rounds in order to investigate the long-term effect of mixing populations. The effect of varying the number ν of visitors from 1 to 30 was investigated.

Results

Figure 4 shows two different measurements of success plotted against ν. Each bar shows the mean and one standard deviation over 40 recorded values of the success rates. The upper set shows the Long-Term Success Rate ρ (as defined in Section 5) measured over 1,000 test Language Games after the mixed population has engaged in 200,000 training Language Games. The lower set shows a Short-Term Success Rate ρ_* measured over the first 1,000 training Language Games immediately after mixing.

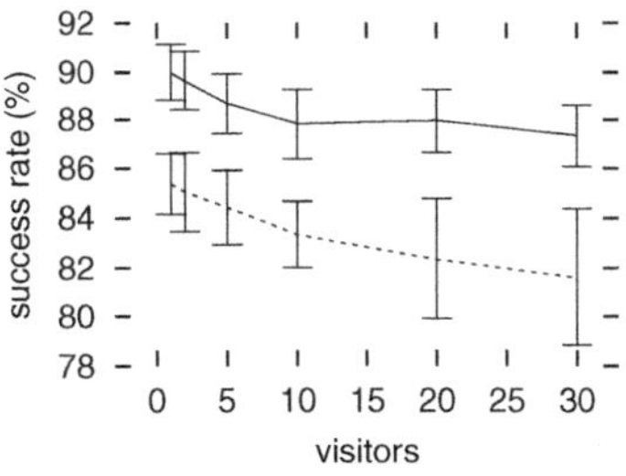

Fig. 4. Long-Term (solid) and Short-Term (dashed) Success Rates after mixing populations for varying number ν of visitors. Cluster variance $\sigma^2 = 0.001$.

The graph demonstrates that the model is very robust when two populations are mixed. Even the Short-Term Success Rate is very high, around 85% for small ν, indicating that the agents are able to take on the new words very quickly without a major drop in communicative success. For a fixed value of ν, there is an improvement of about 5% from the Short- to the Long-Term Success Rate. The mixed population is able to recover from the initial drop in communicative success - in fact, it can attain success rates as high as would have been seen had the mixing not occurred (not shown). Both rates appear to fall with ν, but there is quite a large variance over the 40 runs. For a given value of ν, the variance of ρ_* is greater than that of ρ.

In order to quantify the amount of lexical exchange taking place, it is useful to define the *Foreign Word Rate* – the proportion of times that a word spoken or heard by an agent comes from the population to which it does not belong. Figure 5 shows this rate in four different situations – spoken or heard, and by an agent from the visiting or host population – for different values of ν.

Words are foreign to the hearer much more often than they are to the speaker because agents tend to prefer to utter their own words. That foreign words are ever spoken, however, implies a successful integration of the two populations.

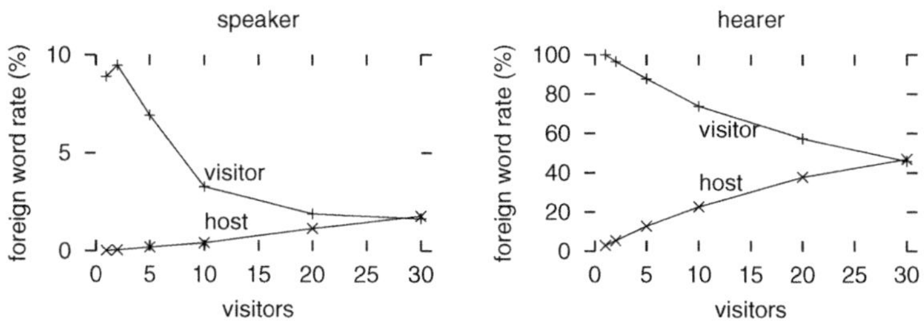

Fig. 5. Foreign Word Rates after 200,000 rounds for mixed populations. Each population comprised 30 agents but the number of visitors is varied. On the left, visiting and host speakers are compared. On the right, visiting and host hearers are compared.

As ν increases, the hosts speak or hear foreign words more frequently while the visitors do so less. In fact the ratio between the visitors' foreign word rate and the hosts' is the inverse of the ratio between the number ν of visitors and the 30 hosts. This can be explained as follows: the population ratio governs the proportion of times a hearer will encounter a foreign speaker and (probably) hear a foreign word. Successful interactions of this kind will increase the score of the foreign word and, in time, the foreign words can become the preferred word for a particular feature set and will be uttered when that agent is the speaker.

7 Discussion

This paper has demonstrated that an unsupervised learning technique, simple competitive learning, can be successfully coupled with Luc Steels's model of Language Games. High levels of coherence were achieved, although the presence of a "forgetting mechanism" was crucial for this.

The new technique used in the Category Formation stage has some important similarities to that used in the Language Sharing stage. Both achieve coherence through a *positive feedback* mechanism, namely the winner-takes-all rule and the preference for successful words respectively. Also, although an initial training period is necessary for the agents' networks to stabilise, there is no discontinuity between this stage and the subsequent stage of Language Games. At all times, the agent's perceptual apparatus is presented with objects from the environment, and the agent's internal network is updated according to the competitive learning algorithm. The result is a more unified overall framework.

Steels believes that the interaction between language and conceptualisation is very important, and his model achieves this since new categories were formed as an outcome of the Language Games. In our model, there is much less interaction because the number of categories is a fixed parameter of the agent's network, but the model is stable nonetheless. It is not practical to create arbitrarily fine distinctions between objects as Steels proposes; sometimes, two objects really are perceptually identical. Furthermore, the generalisation through prototypes is a more robust model in a noisy environment because in Steels's model fine distinctions would be created between identical objects as an artifact of the noise.

The lexicons' robustness has been demonstrated through the agents' ability to adapt to new linguistic conventions without sacrificing their existing ones. Even when the visiting population is small, foreign words can be adopted and spread around a population of agents without any significant fall in coherence, something commonly observed in human language.

References

1. Brooks R. A.: Intelligence without representation. Artificial Intelligence **47** (1991) 139–159
2. Harnad, S.: Category induction and representation. In: Harnad, S. (Ed), Categorical perception: The groundwork of cognition (1987)
3. Harnad, S.: The symbol grounding problem. Physica D **42** (1990) 335–346
4. Hertz, J., Krogh, A., Palmer, R.: An Introduction to the Theory of Neural Computing. Pub. Addison-Wesley CA (1991)
5. Lewin, M.: Concept Formation and Language Sharing: Combining Steels' Language Games with Simple Competitive Learning. Masters thesis, School of Cognitive and Computing Sciences, University of Sussex (2002). Also available at `http://www.cogs.susx.ac.uk/lab/adapt/EASy_MSc_abs_02.html`
6. Newell, A., Simon, H. A.: Computer Science as empirical enquiry: Symbols and search. Communications of the ACM **19** (1976) 113–126
7. Pierce, C. S. Collected papers, vol I–VIII. Cambridge MA: Harvard University Press (1931–1958)
8. Searle, J.: Minds, Brains, and Programs. Behavioral & Brain Sciences **3** (1980) 417–458
9. Steels, L.: Emergent Adaptive Lexicons. In: Maes, P., Mataric, M.J., Meyer, J.-A., Pollack, J., Wilson, S.W. (Eds), From animals to animats 4: proceedings of the Fourth International Conference on Simulation of Adaptive Behavior (1996) 562–567
10. Steels, L.: Constructing and Sharing Perceptual Distinctions. In: van Someren, M.W., Widmer, G., 9th European Conference on Machine Learning (1997) 4–13
11. Steels, L.: Synthesising the Origins of Language and Meaning Using Co-Evolution, Self-Organisation and Level Formation. In: Hurford, J.R, Studdert-Kennedy, M., Knight, C. (Eds), Approaches to the Evolution of Language: social and cognitive bases (1998) 384–404
12. Steels, L.: Language Games for Autonomous Robots. IEEE Intelligent Systems **16(5)** (2001) 16–22
13. Steels, L., Kaplan, F.: AIBO's first words. The social learning of language and meaning. Evolution of Communication **4(1)** (2001)
14. Steels, L., Kaplan, F., McIntyre, A., Van Looveren, J.: Crucial Factors in the Origins of Word-Meaning. In: Wray, A. (Ed), The Transition to Language (2002) 214–217
15. Steels, L.: Grounding Symbols Through Evolutionary Language Games. In: Cangelosi, A., Parisi, D. (Eds), Simulating the Evolution of Language (2002) 211–226
16. Vogt P.: The physical symbol grounding problem. Cognitive Systems Research **3** (2002) 429–457

Artificial Agents and Natural Determiners

Joris Van Looveren

AI-lab, Vrije Universiteit Brussel
Pleinlaan 2, 1050 Brussels, Belgium
joris@arti.vub.ac.be

Abstract. Language is a complex phenomenon. Utterances arise from complex interactions between semantics and grammar. Usually, semantics and grammar are studied separately from each other. This paper introduces a model that makes it possible to study the interaction between these two parts of language. The model comprises a population of agents that feature a relatively complex semantic module, and a lexicalisation module that can produce utterances from semantic representations. The language the agents use to communicate is developed through their interactions, without central control. The concrete focus of the model is on determination: whether or not the referents of an utterance are definite.

1 Introduction

Language is regarded by everyone as one of the characteristics that identify our species. It is an important part of our identity, and therefore everybody is curious as to how it works. The sciences of language and the mind have already worked for a long time on this problem, and have developed into several subfields with their own specialisations. This shows that language is a very complex phenomenon that is not easily explained.

There are many ways in which language is studied. Linguistics studies language directly, but other, more indirect ways have been developed to find out how language and the processes that produce it in the brain work. Aside from relatively standard reaction time experiments, there is a growing body of neurological research that seeks to visualize the processes that go on in the brain, to support functional hypotheses that have been made, or in order to develop models that are more biologically plausible (see e.g. [BH99]). However, because the brain's language functionality is tightly interwoven with its other functionalities (if only because the neocortex is folded into a very small volume) it is very difficult to separate the language-related activation of the brain from the activation for other functions that go on at the same time, and that are presumably not steady.

Evidence from language produced by people with brain lesions provides yet another peek at what goes on inside, but again, because it is not known whether language uses separate brain modules or shares its modules with other functionality, it is hard to derive any firm conclusions from this evidence. (Pinker provides examples of these works in [Pin94].)

W. Banzhaf et al. (Eds.): ECAL 2003, LNAI 2801, pp. 472–481, 2003.

Because it is so hard to draw conclusions from any type of evidence by itself, it is important to gather evidence of as many types as possible in order to make the picture as complete as possible.

In recent years, a new method of language research has emerged that takes a completely different approach. Instead of trying to unravel language producers themselves, the goal in this method is to try to create systems that reproduce phenomena encountered in natural language, both in terms of interactions between language users as within a single language user. The design and implementation of such systems, especially if they succeed in reproducing the phenomena they try to model, may lead to new insights about their natural counterparts.

In this paper, we will look at such a system, and how it may be used to study determination. In the following sections, we will first look at determination and its behaviour in natural language. Section 3 explains the model and the preliminary experiments that have been done with it. Finally, section 4 concludes.

2 Determiners

Just like several other phenomena in linguistics, and probably science in general, determiners got their name based on initial, fairly superficial knowledge of their functions and behaviours: "the apple" vs. "an apple". The former indicates a specific apple, for example one that can be seen while looking around in the room, while the latter refers to any apple.

Of course, despite the fact that the name stuck, determination turned out to be much more subtle than mere identification of referents: the term *identifiability* usually refers to referents that can be found in the immediate (physical) context of the conversation. The case when a referent is not immediately available is called *familiarity*. Another case is *uniqueness*, where a definite form indicates uniqueness of the referent in the context. There are still other subcases that deal e.g. with mass nouns ("drinking water is good for you"). I refer e.g. to [Haw78, Lyo99] for a more complete treatment of the possible functions determiners can have.

Not only are there many differences in the semantic functions that determiners embody; they have also different grammatical modes of appearance: in some languages, such as English, they are function words, while in other languages they are affixes; table 1 shows the determiner paradigm in Swedish, which is a mixture of both. Historically, determiners usually seem to have derived from words that have the strongly pragmatic function of drawing attention to a referent in the outside world (demonstratives, or even the verb "see" in one instance [HK02]).

As a practical matter for my own work, I chose one of the many functions that determiners may have, and started to work from there: reference to previously mentioned objects in a conversation. At the moment, the work is concentrated on implementing the semantic basis for this function, which is based on concepts from computational linguistics (see later). In the next phase of the project, the semantic part will be linked to a grammaticalization mechanism that will

Table 1. The determiner paradigm in Swedish

	Simple		With adjective	
Indefinite	*en* bok *ett* hus	a book a house	*en* bra bok *ett* stort hus	a good book a big house
Definite	bok*en* hus*et*	the book the house	*den* bra bok*en* *det* stora hus*et*	the good book the big house

allow to study the grammatical realisation of semantic functions such as tracking references in an ongoing conversation, and the impact this has on communicative efficiency and "mental" load.

3 Model and Experiments

3.1 Paradigm

The model presented in this paper uses the *language game* paradigm: a population of agents resides in an environment, and plays controlled interactions about a topic in this environment. The agents in the simulation are capable of perceiving the environment, and can produce and interpret utterances. The main components of such an experiment are:

World. The agents in the population need an environment that contains objects to talk about. The complexity of this environment is one of the factors that determines the complexity of the language that will eventually emerge.
In its simplest form, the environment might contain only the agents; the agents then describe themselves; either by giving names to each other, or by describing their features.
On the other end of the spectrum we find worlds of arbitrary complexity: the world could be a complex 3D simulator, or it could actually be the real world, where the agents are embodied in actual robots that have sensors to capture information about the world and actuators to influence the world.

Agents. The agents process the information they get from their sensors and use this information to produce language. The information usually passes through several modules: perception, semantics/pragmatics, and grammar. These modules transform the information according to the task that the agents should perform, which can also be arbitrarily simple or complex: it could range from simply describing objects to survival tasks. The task is closely linked to the protocol of the language games.

Protocol. The core of the experimental framework is the fact that agents interact with each other. The protocol for these interactions is fixed and reflects the the agents' task. In most experiments up to now the task consisted simply of describing objects in the environment (speaker) and recognizing/guessing the topic (hearer). The interactions are usually called *language games*.

A number of previous experiments have shown that the language game paradigm is useful for studying language; notably experiments on lexicon emergence and evolution have shown how lexical items propagate through a population, and how a population can arrive at a coherent set of lexical conventions without central control [Ste96]. A more involved lexicon experiment called the Talking Heads experiment showed that the results from the simple experiment mentioned before also works in a more complex setting, where agents appear and disappear, and the environment changed regularly [Ste99,BSvL98].

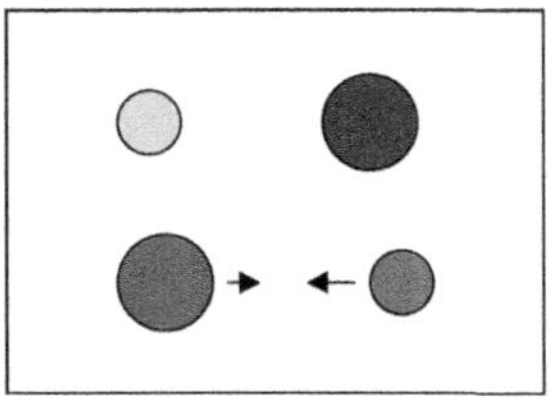

Fig. 1. An example of a scene in the world. The arrows indicate movement.

3.2 World

In our model, the world is a constantly changing environment in which a predefined number of objects can move around. At regular intervals, a randomly selected agent is asked to describe a certain object or event in its environment. The object or event that has to be described is called the *topic* of the interaction; the other ones constitute the *context*.

Whenever an agent engages in such an interaction, it is provided with the current state of the environment; fig. 1 gives a visual representation of what the environment can look like.

The agents' actual perceptions are descriptions in the form of feature-value pairs, as in the following list:

```
(ITEM 5  ((HPOS 0.25) (VPOS 0.75) (AREA 0.33) (COLOR RED)))
(ITEM 10 ((HPOS 0.75) (VPOS 0.75) (AREA 0.66) (COLOR BLUE)))
(ITEM 8  ((HPOS 0.75) (VPOS 0.25) (AREA 0.66) (COLOR RED)))
(ITEM 12 ((HPOS 0.75) (VPOS 0.75) (AREA 0.33) (COLOR RED)))
(DERIVED-ITEM 329 ((O1 5) (O2 10) (APPROACH)))
```

The objects in this list are selected from a pre-generated set of objects, while the events (of type `derived-item`) are generated randomly based on the selected set of objects.

3.3 Agent

The goal of the agents in communication is to distinguish the topic from the other objects in the context. It is composed of four different modules, as shown in fig. 2:

a memory, which stores the perception from the world, a semantic module which generates and interprets meanings about the world, a lexicalisation module that is capable of constructing utterances for meanings, and a discourse stack.

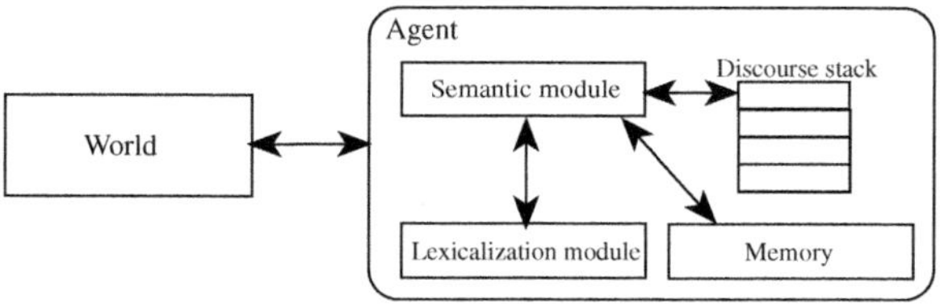

Fig. 2. A schematic representation of the model

Memory. The agent's memory is a simple database that holds the agent's perceptions.

Semantic module. The semantic module creates descriptions about the world using a type of procedural semantics based on predicate logic. Procedural semantics means that semantic descriptions are not just predicate logic clauses describing the state of an object, but the clauses can actually be evaluated by an interpreter within the context of the agent's perceptions. For example, a predicate `blue(x)` may be implemented such that it evaluates to *true* whenever its argument `x` has the feature `(color blue)`, such as item 10 in the example list shown above.

Concretely, there are a number of basic predicates like `blue(x)` that can interact with a database that stores an agent's perceptions. These predicates can be combined into small programs, where the interactions between the predicates are represented by variables that link the different predicates' arguments together. Such combinations of predicates can describe complex aspects of the environment, such as "*the blue square approaches the circle.*" An example of a semantic description the following:

```
(x | approach(x) ^ agent(y,x) ^ blue(y) ^
     square(y) ^ patient(z,x) ^ circle(z))
```

This sequence of predicates says that `x` must be an "approach" event that has an agent `y` which is blue and square and a patient `z` which is a circle. The agent has an evaluator that is able to process these sequences of predicates within the set of perceptions that the agent acquired from the current context, and construct a set of bindings for `x`, `y` and `z` such that the entire predicate is true if and only if such a set exists. Variable `x` is the head of the meaning; this is indicated by the explicit mention to the left of the description. In this case, the "approach" event is the head. The same sequence of predicates can also be used to express the head `y`, which could be paraphrased in natural language as "*the blue square approached by the circle.*"

The mechanism that allows an agent to construct meanings of arbitrary complexity is *abstraction* or naming. The agent is not only capable of constructing complex semantic descriptions, as shown above, but it is also capable of giving new descriptions names and incorporating them in its repertoire of predicates that it uses to construct semantic descriptions. A description added in this way can subsequently be used in new descriptions in the same way that primitive predicates are used to construct lower-level semantic descriptions. For example, the description above with topic `y`, abstracted as `operation1(y)` could be used as the topic in a description for "*the blue square approaching the circle is large*:"
`(w | operation1(w) ^ large(w))`

This mechanism also completes the compositionality of the semantic system: descriptions are functions of their elements, which in their turn can be composed of several other descriptions, and so on.

Meaning construction essentially proceeds in two phases. The first phase is to check whether the agent already knows a description that works. This is done by checking the topic against the descriptions that are already in the agent's set of known meanings. If a working description is found there, the agent stops searching and uses that meaning. If no pre-existing valid meaning is found, the agent begins a heuristic search procedure, in which the generator tries out combinations of (primitive and compound) predicates that could produce valid meanings, until it finds one or more descriptions that actually work. When such a description is found, it is stored in the agent's so that it is available for retrieval immediately later on. In order to keep the meaning cache size at a reasonable level, each description has a score. Whenever a description is used successfully, its score increases; whenever it fails, its score is lowered. By periodically removing descriptions with low scores from the memory, only the most useful ones are kept.

Lexicalisation. Lexicalisation is currently handled by a module that is capable of using utterances composed of several words, as opposed to single words in previous experiments. However, there is no grammar involved; agents use utterances like *big red* to refer to two features of the referent.

The fact that there is no grammar involved, means that words cannot be linked to other words. There is no possibility of signaling beforehand that two words represent different aspects of the same referent. The hearer agent can establish this fact only by referring to the context when it reconstructs an utterance's meaning. Also, when an utterance refers to several objects (e.g. when expressing a relation such as *near* between two objects), there is no way to infer which feature description refers to which object.

Discourse stack. In computational linguistics, discourse is usually handled using a stack [GS86]. New referents are put on the stack, and old referents disappear from the stack as they become less and less relevant in the discourse. The model in this paper makes use of this idea, and equips its agents with a *discourse stack*. Whenever an interaction was successful, the topic of that interaction is put on top of the stack. The stack has a fixed depth, so old referents fall out of the stack after a number of successful interactions.

The agents can refer to this stack using a primitive semantic predicate that can bind its argument to a referent on the stack, or test if a bound variable corresponds to a referent on the stack.

3.4 Experiments

Three types of experiments have been done. The difference between the three types are the primitive operations that the semantic module uses to build meanings. The results shown below are tentative, as they are based on a very small sample of experiments. All experiments used a population of 5 agents.

Type 1. Agents in type 1 experiments only have primitive predicates that discriminate features from objects themselves, such as color or size. These are the basic experiments that will be the yardstick for the results of the more complex types of experiments.

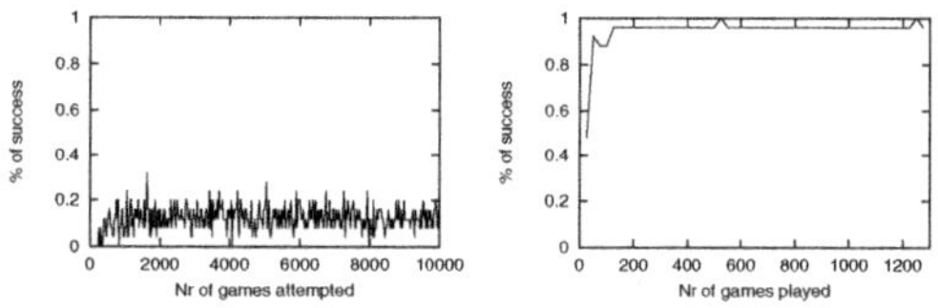

(a) Success (b) Played success

Fig. 3. Success and played success in Type 1 experiments

The utterances created in the experiments describe meanings such as `red(x)` or `blue(x) ^ big(x)`. As such, they correspond to earlier experiments, such as [Ste96], and should perform likewise.

The graph in fig. 3 shows the overall communicative success in the population, as it was measured in previous experiments. An average value of 15% is not very impressive. It turns out that in a large number of games, the speaker is not even capable of constructing an utterance. This means that, in all these games, there is actually no communication; this arises from a mismatch between the structure of the world and the agents' semantic capabilities. Naturally, these games fail. If, for now, we discount these games, we get the graph in fig. 3 (b). This graph looks indeed like the success graphs from previous experiments, so that at least the base expectation (comparable performance to previous experiments) is met.

In the other experiments, we will look only at the games in which there actually is communication.

Type 2. The agents in these experiments can not only discriminate features from objects, but can also discriminate relations between objects, such as `near(x)`. It should be said that the played success (fig. 4 (a)) in this experiment seems to descend again towards the end; it is not known yet if this is really the case or if it is an artefact. Subsequent experiments will have to be run for a longer time to decide this.

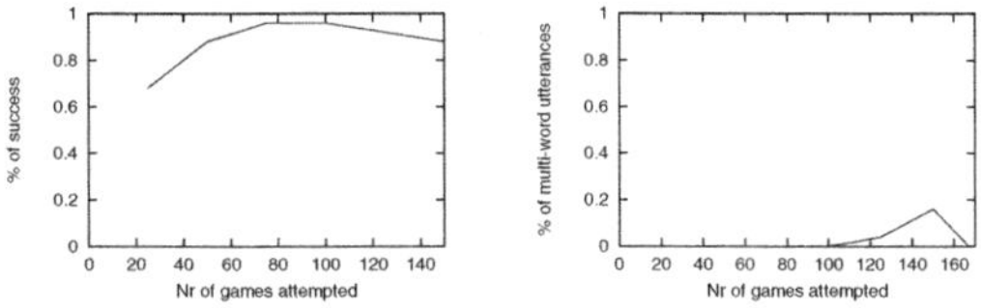

(a) Played success (b) Use of multiple words

Fig. 4. Played success and use of new features in Type 2 experiments

The graph in fig. 4 (b) shows that the agents do indeed use the capacity to use more complex utterances, but only to a very limited extent. This is due to the fact that when a speaker can find a meaning to describe a referent, it usually is a simple one. This means that the world (in combination with the primitive predicates) allows either very simple descriptions, i.e. contexts where the topic is very easily distinguishable from the other objects, vs. contexts where the topic is hardly distinguishable at all. This points again to the mismatch between the world and the primitive operations used by the agents.

Type 3. This is the most complex type of experiments. The agents in these experiments have the ability to refer back to objects previously mentioned in the conversation. As such, these experiments implement the semantics that can be recruited for the expression of determination in the grammar.

Currently, the agents know how to use several words in an utterance (fig. 5 (c)), but utterances are only compositional, not grammatical. Also, as can be seen in fig. 5 (b), the agents do make use of the new possibility to refer to an object that was mentioned previously in about 20% of the cases. However, with only a few short experiments done, it is hard to interpret this number.

3.5 Discussion

The results up to now seem to be relatively trivial, but are nevertheless crucial: the agents in the population actually make use of the new capabilities they are given. Obviously, if this were not the case, the system would have to be redesigned. The results also establish a base line on which subsequent experiments must improve. The current experiments do not use grammar, and because grammar increases the complexity of the individual, there has to be a trade-off in

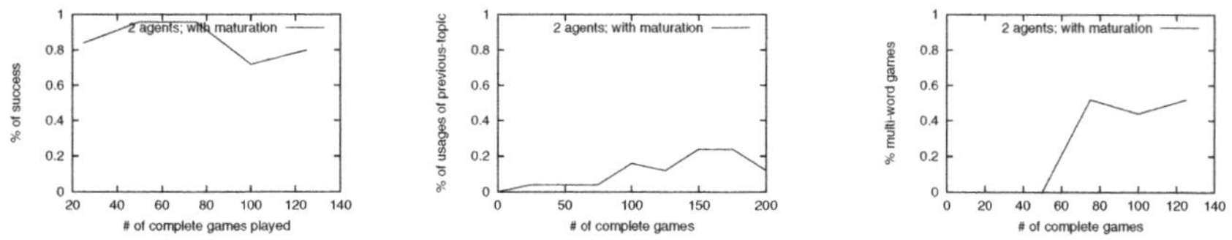

(a) Played Success (b) Use of previous topic (c) Use of multiple words

Fig. 5. Played success, and use of new features in Type 3 experiments

terms of performance. If grammar-enabled agents do not do better than agents with pure lexicon-based communication, why bother with the added complexity? Thus, in future experiments we should see a significant benefit from using grammar in at least one way, such as expressiveness, communicative efficiency, resistance against noise, etc.

When increasing the complexity of the agents, it was observed initially that the performance of the system suffered: the agents were not capable any more of building a reliable communication system among them. The reason for this turned out to be that, although semantically they were perfectly capable of constructing meanings for topics like they were before, these meanings were much less lexicalizable. In principle, the agents were able to lexicalize parts of a meaning using lexicon entries that cover those parts, but because meanings were complex from the beginning, there never were any applicable lexicon entries. Thus, in absence of a mechanism that could extract common subparts from complex meanings, the agents kept creating new words even for meanings that were very similar to other meanings that they already knew.

Two solutions for this problem were contemplated: an extraction mechanism, which would be a complex and cumbersome solution, or, simply making sure that the agents would start out using simple, concrete primitive operations only. After a certain period they would be allowed to use their full arsenal of primitive operations, including the more complex ones. The second method was tried, and proved succesful at this level of the experiment: when the agents had an opportunity to learn "simple" words first, these words could be reused later in more complex utterances. This two-stage learning process mimics a maturation process, and could indicate that this type of bootstrapping might allow simpler learning mechanisms to learn what otherwise would require more complex mechanisms, or could even be untractable. Of course, it remains to be seen if this approach is still successful when the experiment becomes more complex.

4 Conclusion

Determination is a grammatical manifestation of one or more semantic features. The work in this paper focuses on identification or recognition of a given referent in the course of a conversation. Theories within linguistics and computational linguistics usually posit some kind of discourse stack that keeps track of the referents mentioned in a conversation, and an amount of shared knowledge in which a hearer can search for referents. The model implements such a stack.

The model builds on a concept borrowed from game theory, in which individuals in a population interact with each other. The individuals in the model have implementations of several abilities they need in order to communicate: perception, semantics, and lexicalisation. At present, the perception and lexicalisation of the model are only very rudimentary, but the semantic module is relatively well developed: the agents can not only detect features of referents, but also relations between referents, and are able to refer to objects that have been referred to previously in a conversation.

The experiments done up to now indicate that the performance of the model at least equals that of previous, less complex experiments. The agents also make use of their new capabilities of using more complex utterances (as opposed to using only single words) and referring back to previous topics. However, much work remains to be done, both in terms of the architecture of the system (e.g. improve the match between the world and the agents' semantic predicates), and in terms of measuring the performance of the model and its indviduals (e.g. lexical coherence between the agents, performance in inferencing and how grammar influences the performance).

Acknowledgements. The research presented in this paper is sponsored by the Flemish institute for the advancement of innovation in science and industry (IWT).

References

[BH99] Colin M. Brown and Peter Hagoort. *The Neurocognition of Language.* Oxford University Press, Oxford, UK, 1999.

[BSvL98] T. Belpaeme, L. Steels, and J. van Looveren. The construction and acquisition of visual categories. In A. Birk and J. Demiris, editors, *Proceedings of the 6th European Workshop on Learning Robots, Lecture Notes on Artificial Intelligence*, Lecture Notes on Artificial Intelligence, Berlin, 1998. Springer.

[GS86] Barbara J. Grosz and Candace L. Sidner. Attention, intentions, and the structure of discourse. *Computational Linguistics*, 12(3), 1986.

[Haw78] John A. Hawkins. *Definiteness and indefiniteness: a study in reference and grammaticality prediction.* Croom Helm, London, UK, 1978.

[HK02] Bernd Heine and Tania Kuteva. *World Lexicon of Grammaticalization.* Cambridge University Press, Cambridge, MA, 2002.

[Lyo99] Christopher Lyons. *Definiteness.* Cambridge University Press, UK, 1999.

[Pin94] Steven Pinker. *The language instinct, how the mind creates language.* Harper Perennial, 1994.

[Ste96] L. Steels. Emergent adaptive lexicons. In P. Maes, editor, *Proc. of the Simulation of Adaptive Behaviour Conference.* MIT Press, Cambridge, Ma., 1996.

[Ste99] Luc Steels. The talking heads experiment. Available from the VUB Artificial Intelligence Laboratory, Brussels, Belgium, 1999.

Coevolution of Birdsong Grammar without Imitation

Kazutoshi Sasahara and Takashi Ikegami

Department of General Systems Studies, Graduate School of Arts and Sciences,
University of Tokyo,
3-8-1, Komaba, Meguro-ku, Tokyo 153-8902, Japan
{sasahara,ikeg}@sacral.c.u-tokyo.ac.jp
http://sacral.c.u-tokyo.ac.jp

Abstract. The mating song of the male Bengalese finch can be described by a finite-state grammar and has the feature that more complex songs are preferred by females [1]-[3]. These facts suggest that complex song grammars may have evolved via sexual selection. How, then, do the female birds gauge a song's complexity? Assuming that they can measure the complexity of a song while communicating with a male, but without making a model of the song, we studied the evolution of song grammars. In our simulation, it was demonstrated that song grammars became more complex through communication between coevolving males and females. Furthermore, when singing and listening were subject to fluctuations, peculiar features were observed in communication and evolution.

1 Introduction

From the standpoint of Chomsky's theory, the most unique function of human language is its "recursiveness", which enables potentially infinite expressions from finite elements [4]. This plays a fundamental role in grammar. Since "linguistic behavior does not fossilize", the questions of how mankind attained this function and how language has been complicated in modern times are difficult to deal with scientifically and they remain significant open questions.

In addition to mankind, songbirds and whales have a similar capability. They can combine finite sound elements recursively to have vocal communication [5]. Therefore, their song can be regarded as analogous to human language. The study of such language-like behavior is one way to understand the evolution of language.

We focus on the Bengalese finch (*Lonchura striata var. domestica*), in particular the song of the male as it courts a female. This song consists of a combination of chunks, each of which is a sequence of sound elements. It can be described by a finite-state grammar, unlike alarm calls and threats. In addition, it was shown in Okanoya's experiments that more complex courtship songs were preferred over monotonous ones, and that they promoted the reproductive behavior of females [1][2]. These facts suggest that males with complex songs have been chosen by females and that song grammars have evolved as a result of sexual selection [3].

W. Banzhaf et al. (Eds.): ECAL 2003, LNAI 2801, pp. 482–490, 2003.

The handicap principle [6], which says that the complex song of a male may be a genuine indicator of his superiority, supports this scenario.

How do female birds perceive the complexity of a song? One possibility proposed by Okanoya is that she can judge it through communication, without making a model of the male song. Based on the observation that the female Zebra finch chirps (interjects) in synchrony with the male, we supposed that the female Bengalese finch also interjects in this way, measuring how many interjections succeed according to her preference to evaluate the song quality. What is significant in this hypothesis is that without a model of the song in the female brain, she can still judge its quality according to her preference and then complex song grammars may evolve. To explore this, we attempted to model the evolution of song grammar through communication. Furthermore, we introduced the effects of mis-singing and mis-listening and studied how such fluctuations influenced communication and evolution. In this paper, we verify our hypothesis, examine the behaviors observed, and argue that song grammars become more complex as a result of coevolution.

2 Modeling

Here we model the coevolution of males and females as a communication game. The song grammar of a male bird is expressed as a sequential machine[8], $M_i = (Q, \Sigma, \Delta, \delta, \lambda, q_0)$, which is a finite automaton(FA) with output function, where Q is a finite set of states, q_0 is an initial state, Σ is a finite set of input symbols, Δ is a finite set of output symbols, δ is a state transition function, $Q \times \Sigma \to Q$, λ is an output function, $Q \times \Sigma \to \Delta$. In this model, $\Delta = \{blank, A, B, C, D, E, F, G, H\}$, each letter represents a song chunk and "*blank*" represents a silent interval between chunks. The whole output sequence expresses a courtship song. Furthermore, two types of dummy male are introduced as probes for investigating evolutionary features of the system. One is a "random bird", which has a random number generator instead of a FA and sings a random song. The other is a "monotonous bird", which has a clock generator that outputs chunks and blanks alternately. He sings a simple periodic song.

Female birds, on the other hand, have another type of FA, $M_i' = (Q, \Sigma, \delta, q_0, F)$, which represents her preference, where Q, Σ, δ, and q_0 are the same as above, F is a set of accepting states, which is a subset of Q. She changes her internal state by listening to the song and interjects if she is in an accepting state. If she is able to interject when the male bird sings *blank*, we say that the interjection is successful. Examples of male and female FAs are shown in Fig.1.

Moreover, the song fluctuation effects, mis-singing and mis-listening, are taken into account. Some chunks are easily mistaken (in both singing and hearing) while others are not. Specifically, in the pair (A,B), "A" tends to be mistaken for "B" once every three times and vice versa. Similarly, the mistake rate is 1/5 for (C,D), 1/10 for (E,F) and 1/20 for (G,H). Hence, the male does not always sing his grammar exactly and the female does not always accept his song exactly, even if it is perfect.

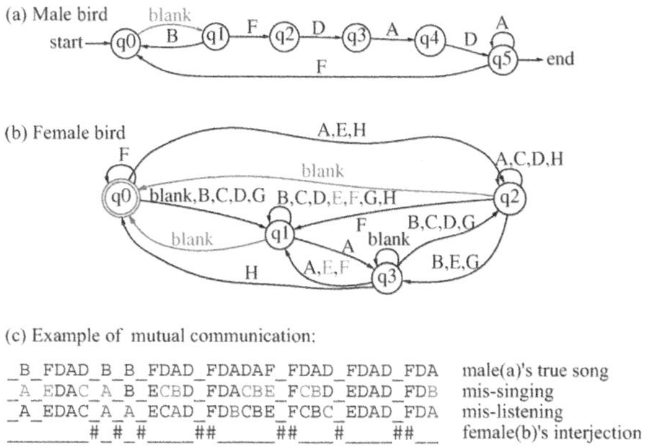

Fig. 1. Examples of finite automata(FA) and communication. (a) Shows a male FA and (b) shows a female. The double circle represents an accepting state. (a) and (b) are a well-suited pair that achieved good communication at $t = 5,000$. (c) Illustrates a communication process including song generation, interjection and fluctuations in singing and listening. "_" denotes "*blank*" and "♯" denotes interjection.

Given n types of males and females, communication is performed as follows. Every time step, each normal male with a FA sings for length L_i^{song} chunks, according to his grammar, M_i. Dummy birds sing according to their own mechanisms. Note that they make mistakes in singing at the rates given above. Subsequently, each female interjects to the songs in accordance with her preference, M_i'. Females also make mistakes in hearing. An example of communication is illustrated in Fig.1.

In communication between the populations of males of the i-th type and females of the j-th type, their scores are calculated as follows.

$$a_{ij} = \frac{N_j^{success}}{N_i^{blank}} - C \cdot N_i^{node} \tag{1}$$

$$b_{ji} = \frac{N_j^{success}}{N_j^{interj}} - C \cdot N'^{node}_j \tag{2}$$

where a_{ij} is the score of the i-th male population, b_{ji} is the score of the j-th female population, $N^{success}$ is the number of successful interjections, N^{blank} is the total number of blanks, N^{interj} is the total number of interjections and N^{node} is the number of nodes in M_i, N'^{node} is the number of nodes in M_i'. The first terms denote the success rate of interjection and the second terms denote the cost (in nodes) they pay for their ability to sing complex songs. Note that the dummy birds have no node-cost because they have no FA.

In following cases, however, it is considered that the communication ends in failure, and both male and female are given zero score: (i) Every element of the song is *blank* (inappropriate song), (ii) every reaction of the female is interjection (inappropriate interjection), (iii) the female interjects exactly when the male sings *blank* and never otherwise (no novelty). We consider (iii) so that we can take into account the importance of novelty of songs for females' preferences [7].

Let x_i be the relative population of the i-th male bird, and y_i be that of the i-th female bird. Supposing that they produce offspring relative to their scores as defined previously, the relative populations are calculated by the replicator equations [9], with the discrete-time Runge-Kutta method:

$$\frac{dx_i}{dt} = x_i \left(\sum_{j=1}^{n} a_{ij} y_j - \sum_{k=1}^{n} \sum_{l=1}^{n} x_k a_{kl} y_l \right) \tag{3}$$

$$\frac{dy_i}{dt} = y_i \left(\sum_{j=1}^{n} b_{ij} x_j - \sum_{k=1}^{n} \sum_{l=1}^{n} y_k b_{kl} x_l \right) \tag{4}$$

where, $0 \leq a_{ij} \leq 1$, $0 \leq b_{ij} \leq 1$, and $\sum_i x_i = \sum_i y_i = 1$.

We define the linearity of a FA, LI, as $LI \equiv N^{node}/N^{arrow}$, where N^{arrow} is the number of arrows leaving from a node. If $N^{node} = N$, this value ranges between $1/N \leq LI \leq 1$ as N^{arrow} varies from N^2 to N. More complex FAs have lower values of LI.

The following mutations occur with the same probability R^{mutat} to males and females at each time step:

(a) **Arrow Mutation**: Change the transitions of the FA with the number of nodes remaining fixed.
(b) **Node Mutation**: Change the number of nodes(± 1) and then add or remove arrows as required.
(c) **Length Mutation**: Change $L^{song}(\pm 1)$.
(d) **Selection1**: Remove the males in proportion to LI or remove a randomly selected females.

Note that sexual selection doesn't involve any relation between complex sexual display and male survival superiority. In selection1, however, we hypothesize that having a complex grammar is correlated to male survival ability. Therefore, males with higher values of LI are regarded as less able to survive. In addition, the following selection is performed:

(d) **Selection2**: Remove the population with a relative population lower than the threshold, V^{thresh}.

The dummy males only mutate according to (c) and (d). The females are prone to every mutation except (c). After the selections, a normal bird type is randomly chosen and first mutated according to (a) or (b), and then mutated according to (c), before being reintroduced into the system as a new bird type with relative population $1/n$. In this way, the number of populations is kept constant.

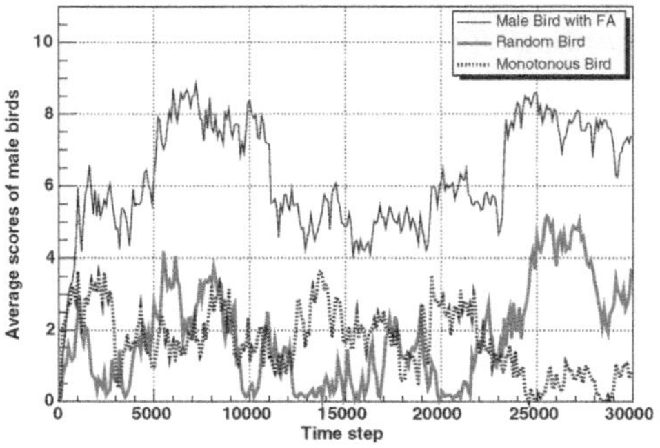

Fig. 2. Average scores of males. The solid line shows the average score of normal males, the thick line shows that of the random bird, and the dotted line shows the monotonous bird.

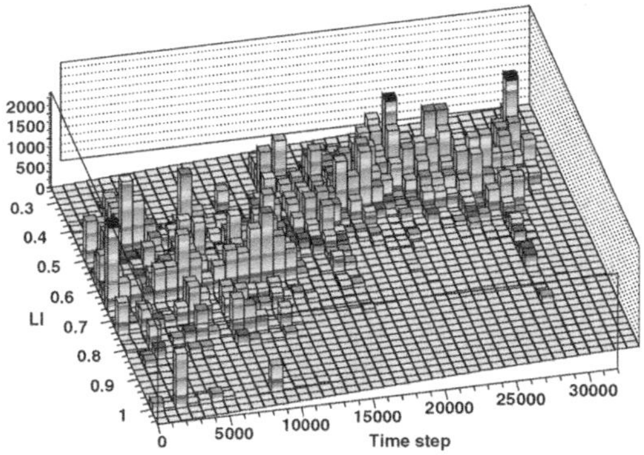

Fig. 3. Time evolution of linearity(LI). We observe that evolution stagnates around $t = 10,000$ and that a rebound phenomenon occurs between $t = 23,000$ and $27,000$.

3 Simulation Results

Our simulations consisted of ten types ($n = 10$) each of both male and female populations, where one of the male populations was comprised of random birds, and another of monotonous birds. The other eight were normal birds with FAs. The initial relative populations were $x_i = y_i = 0.1$. Every FA was constructed randomly with the initial conditions $N^{node} = 2$ and $L^{song} = 5$. Other significant parameters were $C = 0.001$, $R^{mutat} = 0.06$, $V^{thresh} = 0.001$, max L^{song}, max N^{node} and N'^{node} were 50.

The average scores of the males as a function of time appear in Fig.2. We see from Fig.2 that the average score of the normal males increased rapidly until about $t = 2,000$ and then their score varied between around 4 and 8. Very similar evolution was observed for the females. On the other hand, the dummy birds couldn't maintain high scores. The normal males' scores are much larger than those of dummy birds. This clearly indicates that the females could interject properly to the output sequences from the song grammars, and could react to behavior between periodic and random. Note that the monotonous birds did obtain non-zero score. Without fluctuations, the monotonous males would suffer complete interjection and could obtain low scores only when they were matched with monotonous females. Mis-singing and mis-listening, however, enabled them to acquire higher (but still low) scores. On the other hand, some proper complex songs might yield lower scores because of fluctuations. Song fluctuation has an effect of obscuring the complexity of messages which expresses "honest signal of male superiority". Therefore, the handicap principle does not work perfectly in a noisy environment.

The time evolution of linearity(LI) appears in Fig.3. From this, it can be seen that the males' grammars were generally having smaller values of LI, i.e. they were becoming more complex over time. This shows clearly that the song grammars evolved to become more complex as a result of communication. Furthermore, the "rebound phenomenon" - that complexity of grammars returned to being relatively simple after being somewhat complex - was observed. In Fig.3, it can be seen between $t = 23,000$ and $27,000$ and it is consistent with the period in which males' scores were rising again in Fig.2. This rebound phenomenon is due to males' grammars becoming so complex that females couldn't successfully interject, so that they began to choose slightly less complex grammars. Thus, we conclude that song grammars cannot become arbitrarily complex and instead they attain complexities within the range that females are able to interject appropriately. In a sense, communication itself seems to be the trigger that causes the complexity of grammars to increase, at the same time determining its maximum. Subsequently, we may observe "stagnated evolution" - that simple males were selected and dominated the system around $t = 5,000$. In this period, communications seemed to be successful, as shown in Fig.2. However, since the females became too complex (by increasing their number of nodes) at this time, after that the simple males suffered perfect interjection and disappeared from the system. Without selection1 (i.e. without considering the relation between complex grammar and survival superiority), we did not observe any clear contrast between the rebound phenomenon and the stagnated evolution, even though the song grammars gradually became complex and saturated at the higher LI comparing with the case of selection1. This suggests that with only sexual selection the complexity cannot develop enough. We hypothesize that the higher complex grammars can only attain at the expense of the lower complex grammars.

Fig.4 shows the change of the average values characterizing the communication: the song length, the number of kinds of chunks and the number of nodes in males and females. It was found that L^{song} had almost reached a maximum after $t = 6,000$. This indicates that females were apt to succeed in interjection with the longest songs, even taking into account the risks of mis-singing, mis-listening and mis-interjection. Moreover, it turned out that the number of nodes of female FAs decreased at around $t = 10,000$ which was the stagnation period, and around $t = 23,000$ when the rebound phenomenon occurred. Females with FAs of only a few nodes are not sensitive to the order of chunks and expresses preferences to the chunks themselves. On the other hand, if the number of nodes

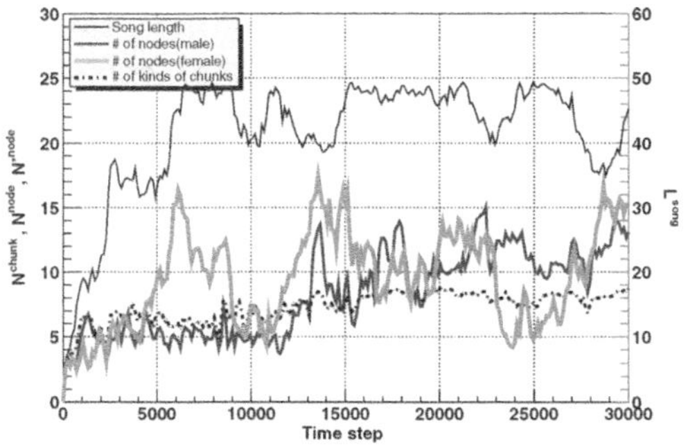

Fig. 4. Dynamics of communication features. Each line shows a value averaged over all birds. The normal line shows the song length(L^{song}), the medium line shows the number of nodes in males FA(N^{node}), the thickest line is the number of nodes in females FA(N'^{node}), and dotted line shows the number of kinds of chunks(N^{chunk}), used in songs.

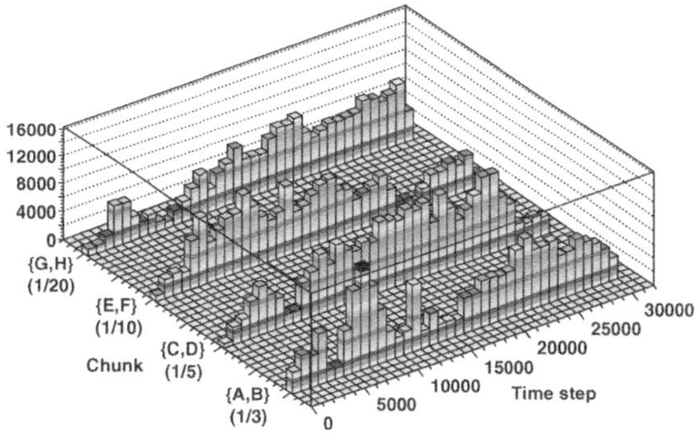

Fig. 5. Trend of songs favored by females. Chunks used in songs oscillate over time.

increases, the females become more complex and sensitive to phrasing. Such a correlation between the number of nodes in females and the linearity of the males they prefer is a reasonable trait.

Finally, Fig.5 shows the number alternations of chunks used in the songs over time, which includes the pairs {A,B}, {C,D}, {E,F}, {G,H}. Even though these pairs have different mistake rates, we didn't find one to be evolutionarily superior. Since the females did not care about the mistake rates, the popularity of different chunks was free to oscillate over time. Indeed, novelty in songs (due to different chunks changing their popularity) can be regarded as a feature of songs favored by females. At t=5,000, {E, F} was the most popular chunk and we could find the fluctuation-absorbing structure in a female FA in Fig.1. That is, since E and F were on the same arrow of her FA, the interjection could be successful even if she mistook E for F. Such structures were commonly found in simulations with song fluctuation.

4 Conclusion

In this paper, we did not require the driving force behind grammar evolution to be a diverse outside environment, but only to be communication itself [10]-[14]. We focused on whether song grammars could evolve to become complex through relatively brief, song-interjection communication. Consequently, we could clearly demonstrate that song grammars could evolve to become complex as a result of females' interjection. This supports Okanoya's hypothesis. In addition, it turned out that the song grammars could not become arbitrarily complex, and instead they evolved towards a boundary where interjection is successful, but not perfect. For example, the rebound phenomenon, which showed that overly complex grammars would revert to simpler ones, was observed.(see Fig.2 and Fig.3) That is, the complexity evolved by sexual selection must be understandable by the cognitive system of partner.

Subsequently, we investigated how song fluctuation in singing and listening affected communication and evolution. In such an environment, it was possible that the simple males might be selected and the complex males might not be. The song grammars, however, generally have a tendency to become complex, although not only excellent grammars are inherited in the presence of fluctuation. Moreover, particular fluctuation-absorbing structures were found. In this model, such fluctuation functioned as a mechanism which influenced evolution in a form distinct from mutation.

Acknowledgements. This work was partially supported by Grant-in aid(No. 09640454) from the Ministry of Education, Science, Sports and Culture. KS would like to thank Okanoya for his helpful suggestions.

References

1. Honda, E., and Okanoya, K.(1999), Zoological Sciecne, **16**, pp. 319–326.
2. Hosino, T., and Okanoya, K.(2000), NeuroReport, **11**, pp. 2091–2095.
3. Okanoya,K.(2002), The Transition to Language, pp. 46–63, Oxford University Press.
4. Hauser, M.D., Chomsky, N. and Fitch, W.T.(2002), Science, **298**, pp. 1569–1579.
5. Aitchison, J.(2000), The Seeds of Speech: Language Origin and Evolution, Cambridge University Press
6. Zahavi, A. and Zahavi, A.(1997), The Handicap Principle, Princeton University Press.
7. Werner, G.M. and Todd, P.M.(1997), Fourth European Conference on Artificial Life, pp. 434–443.
8. Hopcroft, J.E. and Ullman, J.D.(1979), Introduction to Automata Theory, Languages and Computation, Addison Wesley.
9. Hofbauer, J., Sigmund, K.(1998), Evolutionary Games and Population Dynamics, Cambridge University Press.
10. Suzuki, J. and Kaneko, K.(1994), Physica D, **75**, pp. 328–342.
11. Hashimoto, T. and Ikegami, T.(1996), Biosystems **38**, pp. 1–14.

12. Steels, L.(1999), The Talking Heads Experiment. Volume 1. Words and Meanings, Antwerpen.
13. Nowak, M., Komarova, N.L. and Niyogi, P.(2001), Science, **291**, p. 114–118.
14. Nowak, M., Komarova, N.L. and Niyogi, P.(2002), Nature, **417**, pp. 611–617.

Systemic Architecture for Audio Signal Processing

Rolf Schatten

Division of Neuroinformatics
Department of Computer Science
University of Bonn
Roemerstr. 164, D-53117 Bonn, Germany
schatten@nero.uni-bonn.de
http://www.nero.uni-bonn.de

Abstract. This paper proposes a layered *systemic architecture* for audio signal processing. The described systems consists of several building blocks connected in different ways and thus enabling different behaviour. Three different systems are proposed constructed with almost the same building blocks fulfilling three different tasks of audio processing: learning to hear, learning to reproduce and learning to associate. The systemic architecture facilitates the connection of all three proposed subsystems to get one big system fulfilling all three proposed tasks of audio signal processing.

1 Introduction

The behaviour of most animals is not only addicted to inherited attitudes but additionally learned knowledge is used to grow up and survive the environment. *Growing up* here means acquiring new abilities on already adopted knowledge.

1.1 Systemic Architecture

As an architecture with the capability to grow up the *systemic architecture* is proposed in [Goe01]. This architecture consists of simple so called *building blocks* each connected in a layered, well ordered manner with a sensory upstream and an actuatory downstream. Every building block consists of two elements, the *classificator* and the *action generator* (see fig. 1). Once a block has finished learning the acquired knowledge can be used as reinforcement signal to train the actuator to reproduce the data the classifier learned to distinguish. The unsupervised learning in the classification modules and the supervised learning in the actuator-modules is organised within a "syllabus" which schedules the learning of each module. The first implementation of the systemic architecture used to control a growing up robot is described in [KS02]. On the lowest level of the architecture the classificator gets sensory values. On higher levels it gets the output of the lower levels to process. The higher the information is processed

W. Banzhaf et al. (Eds.): ECAL 2003, LNAI 2801, pp. 491–498, 2003.

within the architecture the higher the information is structured. Using the systemic architecture for audio signal processing the lowest level has to detect the basic elements within the stream of raw audio data. The next layer should group these basic elements to something like words. The third layer consists of an associator to learn associations between these learned words and other environmental sensory input.

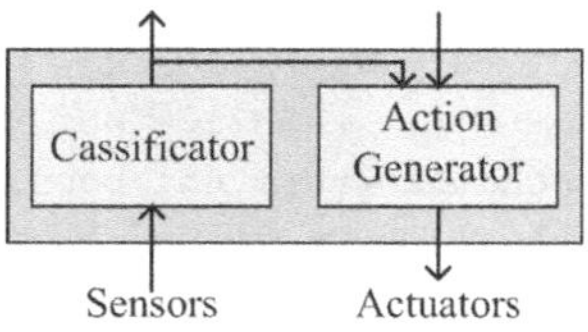

Fig. 1. Every *building block* consists of two elements, one classificator in the sensory upstream and one action generator in the actuatory downstream. Within the *systemic architecture* several building blocks are connected in a layered, well ordered manner. The figure presents the special case of a building block on the lowest level of the architecture connected directly to the sensors and actuators.

1.2 Classificators

Classificators are used to extract features from the sensory- respectively the data-stream. On the one hand this is done to reduce information within the data-stream. But the main reason for classifying the data-stream is to abstract the information. Basically all kinds of unsupervised learning algorithms like LVQ, SOMs, and so on are adequate for classification. Within the audio signal processing system the k-means clustering showed satisfying results.

1.3 Action Generators

With the help of *action generators* the system is able to act within its environment. The action generators in the lowest layer of the architecture are connected directly with the actuators to manipulate the environment while the action generators on higher levels send their output as input to the lower action generators. Like the classificators the action generators have to learn to generate their output. But while classificators learn unsupervised the action generators may learn supervised getting a feedback signal from the classificator.

1.4 Associators

Associators are used to connect at least two systems build up with systemic building blocks. Each system deals with different kind of sensor information and the associators function is to build associations between these different up-streams. The associative building block consists of several memory blocks, one for each connected system, where information from the different sensory upstreams are received. These memory blocks are as well input and return value buffer for a neural associator who is implemented to learn to associate between the different memory blocks (see fig. 2). Since the information is stored in memory blocks and since the associator is able to manipulate these memory blocks the two connected systems may manipulate each other. If the associator connects a moving robot on the one side and a language system on the other side it is able to associate the different kinds of sensory input. To learn associations the linear associator as suggested in [Roj93] is used. It consists of two layers of neurons, x and y, fully connected by weights w to store the associations. In the system for audio signal processing the weights in the associator are adapted with an adjusted learning rule $\Delta w_{ij} = \eta \cdot (x_i - x_i') \cdot (y_j - y_j')$ where x_i' is the value of neuron x_i after processing the input y and vice versa and where η is the learning rate.

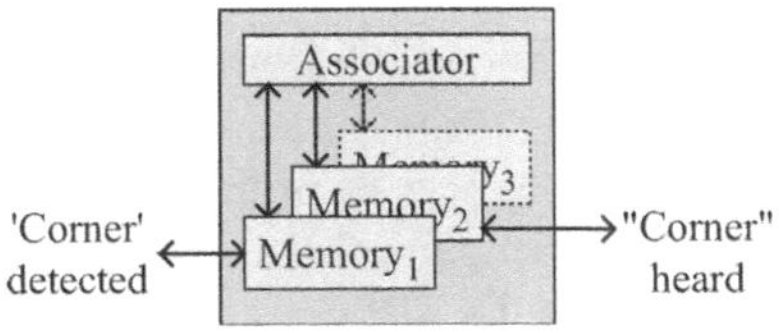

Fig. 2. Associators connect at least two systems and build associations between these sensory inputs.

2 Learning a Language Using Systemic Architecture

The systemic architecture is designed for growing up structures which will be able to learn to classify sensory data and learn to detect the elements within these data streams. This approach shows how to use systemic architecture to learn an artificial language. Such a system grows up by stepwise learning to classify, group, associate, and reproduce the elements of the language. These subtasks are realised within building blocks of identical type.

2.1 The Task of Language Learning

To reduce the complexity of the language and keep it computable we envisage an artificial language which is used to show the feasibility of the approach. It is possible to apply the presented approach to natural spoken language. The suggested language is based on 3 to 7 basic elements (e.g. characters, phonemes, ...) forming up to 100 words (e.g. symbols, icons, ...). A grammar combining these words to sentences creates characteristic word correlations.

2.2 Learn to Hear

To fulfil the task of language learning using the systemic architecture a time schedule is needed to keep a strict order of learning within the building blocks. Three subsequent levels of learning and levels of capability are envisaged:

1. classify basic elements,
2. classify words,
3. build word correlations.

The presented systemic architecture is up to the mark of a language learning system, it fulfils the described requirements. To perform the different subtasks of learning basic elements, words or word correlations several individual building blocks are used. According to the levels of learning the building blocks are arranged in a layered way. The layers are connected in a solely 'upstream' manner to pass the information to the next layer. Additional to the connections a time schedule to coordinate the three learning phases is implemented.

Layer 1: Classify Basic Elements: The first layer within the system gets mono audio data, e.g. directly recorded from a microphone or artificially generated. The sample frequency is adjustable. Motivated by language processing systems a sample frequency of 8 kHz is preferred which showed very good results. Audio signals of 480 ms length which exceed a minimum amplitude will be Fourier-transformed (FFT) with 512 points as a pre-processing similar to the cochlear spectral processing. Since the FFT-packages have a 256 points overlap the audio-patterns are 30 time units long. To reduce data the resulting frequencies will be divided in 24 critical bands as suggested in [Zwi61]. Thus we get a 720 dimensional input patterns for the classificator. To find clusters in these patterns the k-means clustering as described in [ARS98] is used. Since the k-means algorithm is susceptible to bad initialisations of the start centres and is not able to auto detect the parameter k, multiple initialisations are run in parallel and the 'best one' is taken for classification. To identify the best clustering the silhouette coefficient

$$s(o) = \frac{b(o) - a(o)}{\max\{a(o), b(o)\}}$$

as described in [Ber02] with $a(o)$ as the distance of pattern o to the next centre and $b(o)$ as the distance of pattern o to the next but one centre showed to work very well (see fig. 5). After having classified the basic elements {#, _, o} of the language the building block sends the stream of basic elements (#o#_ooo_#o##_##o#...) to the next layer.

Layer 2: Classify Words: Inside the building block used in the second layer (see fig. 4) a further classifier is used. Within our artificial language the words are separated by a pause (except the noise). Since the distinction between signal and gaps is done in layer 1 this layer only has to store the identified words (e.g. code book).

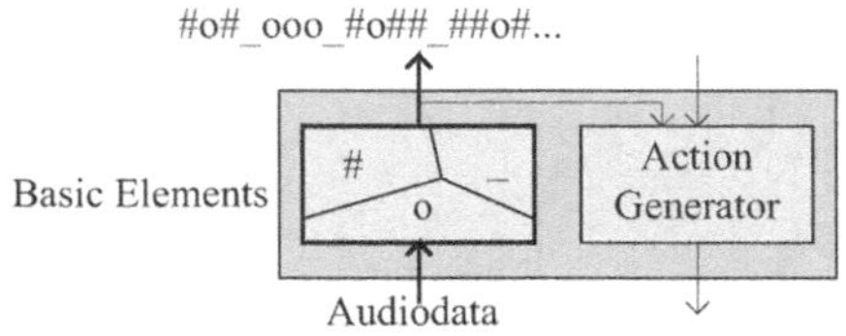

Fig. 3. This picture illustrates the first level of the language learning system. This level contains one systemic architecture building block with a classifier {#, _ , o} (left), and an action generator (right). Within this scenario only the classifier from the building block is used. Audio data is fed into this classifier to categorise this data into a stream of basic elements (#o#_ooo_#o##_##o#. . .).

Layer 3: Building Word Correlations: The building block at the third level of the language learning system builds correlations within the stream of words. Different methods are useable to get a representation of the word correlations within the language (e.g. N-Gram statistics, Hidden Markov Models, finite state machines, regular expressions, learnable finite state machines, . . .)

2.3 Learn to Understand

Learning to understand what you hear is an other aspect of language learning and audio processing. Understanding means to learn to associate between different kinds of internal representations of one and the same object, behaviour, ... Thus the associator described in section 1.4 is connected to the audio processing system (see fig. 4). This associator gets further inputs from a second system e.g. implemented on a robot working on ultrasonic or infrared sensory streams. The associator learns to associate the words of the language with the detected objects from the environment. If both branches provide their information synchronously the associator is trained to combine words with real world events. Thus the system gets an internal representation of spoken words. Not only sensory inputs but as well actuatory behaviour can be associated to the words sensed and classified by the audio processing system. Since the memory building blocks of the several systems within the associator are connected, the speech system is primed to produce a word via the downstream branch of action generators when the robot detects the associated object within its environment. Thus the robot is able to give us a 'status report' of what it is doing while performing a task.

2.4 Learn to Speak What You Hear

A further approach for an audio processing system is to reproduce the sound it heard and classified. Therefore the same building blocks as described before are used but now the action generator produces some sound which is passed to the classifier. Thus the building block gets a reinforcement signal which may be used to train the action generator supervised to reproduce sound it heard before.

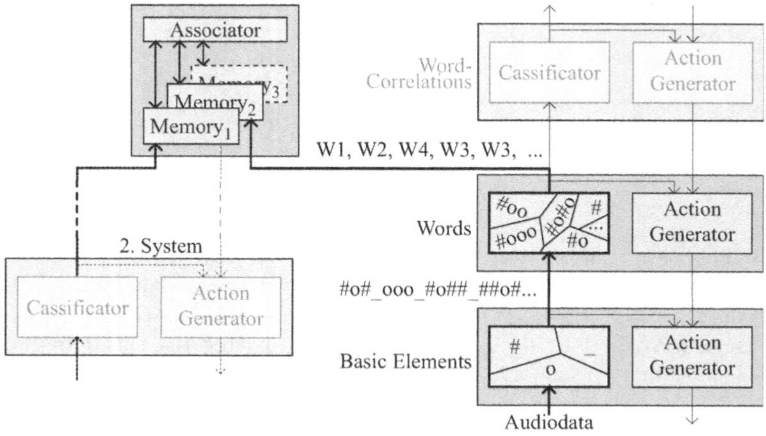

Fig. 4. This picture shows a word extracting system connected to an associator building block. A second detection system, e.g. a second system implemented on a robot working on sensor streams is connected to the same associator building block. The three layers on the right comply exactly to the system presented in section 2.2.

The system contains up to two layers of building blocks with the loop from the sensory branch closed to the actuatory branch. After learning to extract basic elements the system is able to learn how to reproduce these basic elements. To reproduce words the system has to learn within the second layer to reproduce the adequate sequences of basic elements. Once learned to reproduce the basic elements of the language upon request, the second building block can use the learned capability of classifying element streams into words for producing those words. Closing the loop between the classifier to the action generators the system is enabled to reproduce what it has classified before.

3 Results

Very important for the efficiency of the proposed system for audio signal processing is the operativeness of the k-means clustering and especially the significance of the silhouette coefficient. To get some auditive test data the words *start*, *stop*, *go*, *kurt*, *left*, *right*, and *back* were spoken and recorded with a microphone. Each word was spoken 35 times. Every pattern extracted by the preprocessing algorithm in layer 1 of the audio processing system was tagged manually. While classifying this tag was not used. After classifying the tag was used to automatically get the number of wrong classified patterns. The k-means clustering algorithm was applied with values of $k = 2 - 40$. For each clustering 300 initialisations has been done which results in 11700 runs altogether. After each run the silhouette coefficient was calculated. Figure 5 shows the misclassification versus the silhouette coefficient. It can be seen that a large silhouette coefficient is correlated with a small number of wrong classified patterns. This demonstrates

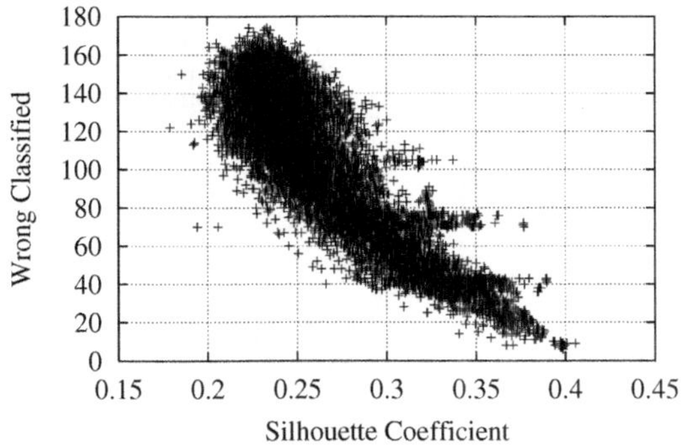

Fig. 5. The silhouette coefficients and the number of wrong classified patterns of the seven natural spoken words *start*, *stop*, *go*, *kurt*, *left*, *right*, and *back* are shown. Every word was spoken 35 times. The clustering algorithm tried to find the best clustering for k=2-40. Each value of k was initialised 300 times. The figure shows the higher the silhouette coefficient is the better the words are clustered.

that the silhouette coefficient is a good measure for getting the parameter k and finding a good clustering not affected by bad initialisations of the k centres. Since the linear associator used within the system has to deal with noise it was tested with 10 neurons in both layers where $x_i = y_i$. This means, neuron i on the left side is associated to neuron i on the right side. To measure the firmness of the linear associator up to 0, 3, and 7 neurons were added on both sides during the training phase. Figure 6 shows the number of wrong associated neurons after every learning step. After learning has finished the associative building block learned to assign correct the symbols of the different connected subsystems.

4 Discussion

Different approaches for audio processing systems using the systemic architecture are presented. Each system uses the same systemic building blocks. According to the different tasks the connections between and within the building blocks are different. All presented approaches may be used for several tasks within the wide range of audio processing systems. In favour the task of language learning was described as a possible application. Therefore the audio processing system may be connected with a mobile robot and learns to associate spoken words with sensory input or actuatory behaviour of the robot. Thus the system gets an internal representation which may be used to command the robot or to describe objects within the robots environment. Another possible task for the audio processing system is to serve as an additional sensory input for the mobile robot. Therefore the system is not only connected to, but also mounted on the robot. The system uses the microphone to record sounds while the robot moves around or while it bumps in an object. The system is able to learn the different kinds of sound and

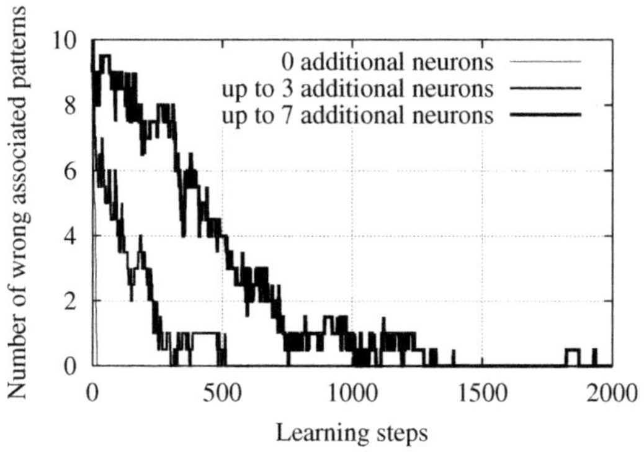

Fig. 6. The linear associator as proposed in [Roj93] was tested with 10 neurons on both sides where neuron x_i should be associated with neuron y_i and vice versa. To measure the firmness of the associator against noise up to 0, 3, and 7 neurons were additionally activated on both sides. The figure shows, that the associator associates correct even with a high level up to 70% of additive noise.

to classify them. Thus the robot gets an additional information about the kind of floor it is passing and about the kind of object it hits.

Acknowledgement. Part of this work is supported by the European Commission's Information Society Technologies Programme, project SIGNAL, IST-2000-29225. Partners in this project are Napier University, National Research Council Genoa, Austrian Research Institute OFAI and the University of Bonn.

References

[ARS98] Alsabti, Ranka, and Singh. An Efficient Parallel Algorithm for High Dimensional Similarity Join. In *IPPS: 11th International Parallel Processing Symposium*. IEEE Computer Society Press, 1998.

[Ber02] Pavel Berkhin. Survey of clustering data mining techniques. Technical report, Accrue Software, San Jose, CA, 2002.

[Goe01] Nils Goerke. Perspectives for The Next Decade of Neural Computation. In *Proc. of NATO Advanced Research Workshop on: Limitations and Future Trends in Neural Computation*, LFTNC'01, Siena, 2001.

[KS02] Florian Kintzler and Rolf Schatten. Systemic Architecture for Growing up Animats. In *On Growing up Artifacts that Live – Basic Principles and Future Trends*, SAB'02-Workshop, 2002.

[Roj93] R. Rojas. *Theorie der neuronalen Netze*. Springer-Verlag, 1993.

[Zwi61] E. Zwicker. Subdivision of the Audible Frequency Range into Critical Bands (Frequenzgruppen). *JASA*, 33(2):248, 1961.

Semantic Generalisation and the Inference of Meaning

Andrew D.M. Smith

Language Evolution and Computation Research Unit,
School of Philosophy, Psychology and Language Sciences, University of Edinburgh
andrew@ling.ed.ac.uk

Abstract. In this paper, a computational model of a successful negotiated communication system is presented, in which language agents develop their own meanings in response to their environment and attempt to infer the meanings of others' utterances. The inherent uncertainty in the process of meaning inference in the system leads to variation in the agents' internal semantic representations, which then itself drives language change in the form of semantic generalisation.

1 Introduction

Modern evolutionary linguistics is primarily occupied with understanding the apparently paradoxical universality and massive diversity found among human languages; much of the recent work within this paradigm uses A-life computational simulation techniques to explore these questions. It is important to note that the evolution of language should not just be considered in a genetic framework, but also in a cultural one; children acquire their language based on the linguistic information produced by those around them. Recently, researchers have shown that the cultural transmission of language itself leads to adaptive pressures which can explain some of the characteristic properties of language [4,7,15]. Cultural explanations of the structure of language mean, importantly, that there may well be less need to fall back on the existence of a somewhat ethereal language acquisition device [11].

Many A-life models of language evolution focus on the emergence of syntactic structure in their agents' language, but fail to explain the origin of the semantics on which their 'emergent' syntax is built. One of the most intriguing universal features of language is the ceaseless and inevitable nature of language change, driven by language variation within speech communities [18]; in this paper I present a model of conceptual development and negotiated communication between agents, in which agents can create meanings individually and still co-ordinate a joint language. I show, moreover, that the inherent uncertainty in the process of meaning inference in such a usage-based model of language [5], leads to variation in the agents' conceptual structures, which itself drives language change in the form of semantic generalisation.

2 Meaning Representation and Creation

A large proportion of recent A-life research into language evolution has been focused on the emergence of syntactic structure, or more specifically compositionality [1,4,7].

W. Banzhaf et al. (Eds.): ECAL 2003, LNAI 2801, pp. 499–506, 2003.

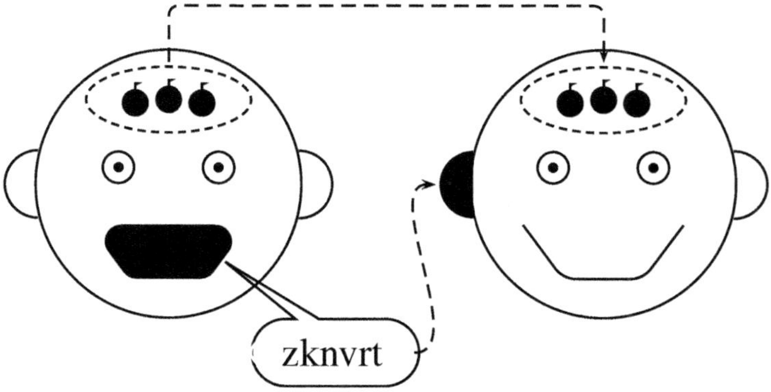

Fig. 1. A communicative episode which consists of the explicit transfer of both a signal 'zknvrt' and a meaning 'three apples' from speaker to hearer.

Although the precise implementations vary to some extent, in all compositionality models structure arises in the signal space as a direct consequence of the agents recognising and coding regularities between parts of signals and parts of meanings. There are a number of problems with models like these, particularly with respect to their semantic representations, which I will discuss here briefly.

Firstly, the agents are always provided with a structured meaning space, which is explicitly linked to an unstructured signal space. As Nehaniv [9] has pointed out, structure develops in the signal space precisely because the agents use signals coupled with already structured meanings; the signals are essentially parasitic on the meanings, and structure only emerges as the agents decipher the pre-defined semantic coding system.

Secondly, no treatment of meaning can avoid addressing the fundamental concepts of both *sense* and *reference* [6]. A basic model of sense relations would include at least, for instance, some notion of antonymy, the relationship between a word and its opposite[1], which is "one of the most important principles governing the structure of languages" [8, p.271], and might be expected also to include other notions such as *hyponymy*, which describes the relationship between a subset and a superset, as for instance between *cat* and *animal*. Reference, on the other hand, is defined in terms of the objects or actions in the external world to which a word refers in a particular context. It is very difficult to relate the 'semantic' structure in such compositionality models to sense relationships in the semantics of real languages in any way, and, even more problematically, there is no reference at all in the meaning systems, either because there is no external world in the experiment at all [1], or because the world is inaccessible to the agents [4,7]. Without reference, and with only a very tenuous link to sense, it is clear that such representations of meaning are actually semantic in name only.

[1] There are many types of opposition in language, including gradable antonyms, such as WET/DRY, expressing meanings on a relative scale; ungradable binary antonyms, such as ALIVE/DEAD, expressing complementary propositions; and converses, such as ABOVE/BELOW, BUT , which refer to the same relationship from opposite points of view.

Thirdly, as I have argued elsewhere [13,14], the lack of reference in the meaning representations of these models leads to an implementation of communication which is seriously flawed, because of its necessary reliance on *explicit meaning transfer* [13]. Figure 1 shows an example of such a communication episode, with the speaker on the left uttering a word *zknvrt*, which is received by the hearer on the right. Simultaneously, however, the meaning THREE APPLES, is transferred directly from speaker to hearer. During communication, the hearer is given explicitly not only the signal, but also both the meaning, and the information that it should make the appropriate association between the particular signal and meaning. Such a model not only sidesteps the very important Quinean [12] problem of how hearers interpret the meanings of unfamiliar words from a set which is in principle infinite, but also actually undermines the need for language at all: if meanings can be transferred accurately by telepathy, then the signals are not being used to convey meaning. There is little motivation, therefore, for the emergence of a system in which the agents spend their time and energy developing a communication system which encodes exactly the same information as another system which they are born with and which already works perfectly.

The semantic model in this paper, on the other hand, tries to avoid these pitfalls; the basic procedure of agent-based grounded meaning creation through discrimination games was originally presented by Steels [16], and has been used by others both in simulated worlds [2,14], and on robots interacting with objects in a real environment [17,19]. An agent is situated in a world made up of abstract objects, with which it interacts by playing discrimination games, which consist of four parts:

Scene-setting: a set of objects, called the context, is chosen from the world and presented to the agent; one of these objects is chosen to be the target of discrimination.
Categorisation: the agent cycles through the objects in the context, returning, for each, an association with one or more of its semantic representations.
discrimination: the agent tries to define a distinctive category for the target object, i.e. a category which both represents the target and does not represent any other object in the context.
Adaptation: the agent modifies its internal conceptual structure in some way.

In the model used in these experiments, the objects in the world have a number of abstract, meaningless features, the values of which are normalised to lie in the range 0.0... 1.0, and the agents have a specific sensory channel corresponding to each feature, on which they build a hierarchical semantic representation, or a discrimination tree [16]. Each node on the discrimination tree is a discrete category, corresponding to a particular contiguous segment of the continuous feature value space. Categories are created by splitting the sensitivity range of a node into two discrete, equally sized segments, each of which is therefore sensitive to half the range of the previous node. The trigger for the creation of new meanings is failure in the discrimination game; it is important to note, however, that there is no pre-definition of which categories should be created, nor is there any guarantee that the newly created categories will turn out to be useful in future discrimination games.

Because of this, agents develop different, though equally valid, representations of the same world. I quantify the similarity of two agents' meaning structures by averaging the similarity of the particular discrimination trees built on each of their sensory channels.

If $k(t,u)$ is the number of nodes which trees t and u have in common, and $n(t)$ is the total number of nodes on tree t, then the similarity between any two trees t and u is:

$$\tau(t,u) = \frac{1}{2}\left(\frac{k(t,u)}{n(t)} + \frac{k(t,u)}{n(u)}\right) \tag{1}$$

I then obtain an overall measure of *meaning similarity* σ between two agents, by averaging τ over all their sensory channels. If a_{ij} identifies channel number j on agent i, and each agent has c sensory channels, then the meaning similarity σ between agents a_1 and a_2 is:

$$\sigma(a_1,a_2) = \frac{1}{c}\left(\sum_{j=0}^{c-1} \tau(a_{1j},a_{2j})\right) \tag{2}$$

If two agents a_1 and a_2 have identical conceptual structures, where $\sigma(a_1,a_2) = 1$, then I refer to their meanings as being *synchronised* [14]. Meanings in this model are expressed using the notation *j-r*, where j identifies the number of the sensory channel, and r is a sequence of 0s and 1s representing the path from the tree's root to the node in question; if the tree's root is on the left, and it branches to the right, then 0 signifies a traversing of the lower branch, and 1 the upper branch.

Importantly, the meanings are grounded in the world, created in response to the environment, and encode both sense and reference relations: the hierarchical nature of the tree means that meanings nearer the root of trees (and therefore with a shorter route r) are more general than those nearer the leaves, and that a node and its daughter nodes can be related by hyponymy; while the meanings also refer to the properties of objects in the world.

3 Communication and the Inference of Meaning

Avoiding the problems with previous models discussed above, the speaker's meaning is *not* transferred to the hearer, in contrast to associative learning models [4,3], nor does the hearer know which object in the context is being referred to, in contrast to Steelsian guessing games [17]; the communication process is made up of three separate sections:

Production: the speaker, having found a distinctive category in a discrimination game, chooses a signal to represent this meaning.
Transfer: the signal is transferred to the hearer.
Interpretation: the hearer interprets the signal from the context in which it is heard.

Each agent maintains a dynamic lexicon of associations between signals and meanings, for use both in production and in interpretation. Each entry in this lexicon contains a signal s, a meaning m, a count u of the pair $< s, m >$'s mutual association, and a representation p of the agent's confidence in the association between the pair $< s, m >$. A signal-meaning pair can be used both by being uttered by the speaker and by being understood by the hearer, so that u is the total number of communicative episodes in which the agent either uttered s to represent m, or interpreted s as representing m. An agent's confidence in a signal-meaning pair is based solely on the relative co-occurrence of signals and meanings, or the proportion of times in which s has been used that it has been associated with m. More formally, $p(s, m)$ can be expressed as:

$$p(s, m) = \frac{u(s, m)}{\sum_{i=1}^{l} u(s, i)}, \tag{3}$$

where l is the number of entries in the lexicon. Communicative success is based on referent identity; speaker and hearer communicate successfully by referring to the same object, but they are not obliged to use the same meaning to do so.

Because meanings are not explicitly transferred between agents, the hearer must infer the meaning of the signal from the context. It is important to note that the hearer does *not* know which object is the target object, and so it must try to come up with descriptions for all the objects in the context. The hearer plays a series of discrimination games and creates a list of possible meanings to consider, each of which describe only one of the objects in the context. The hearer has *no* other information, so all these meanings are equally plausible; it therefore associates each of them with the signal in the lexicon, modifying the values of u and p in the lexical entries accordingly. Having modified its lexicon, it chooses, from the list of possible meanings, the meaning in which it has the highest confidence p. This modification of the lexicon in context is the only way in which the agents learn; in contrast to similar language game models [17], they receive no feedback about the success of either their communication games or their lexicon development. I have shown previously [13] that communication is successful under these circumstances if the speaker chooses a signal that it thinks the hearer will understand. Of course, the speaker does not have access to the hearer's lexicon, as this would defeat the object of ruling out mind-reading, so it bases its decision on what *it itself* would understand, if it heard the signal in this context, without knowing the target object. This technique for choosing signals is a version of the obverter mechanism [10], modified so that the only lexicon an agent can investigate is its own, and which I therefore call *introspective obverter*.

4 Meaning Similarity and Meaning Variation

The introspective obverter algorithm allows the agents to develop successful communication systems; I have shown elsewhere that to achieve optimal communication, the agents should have synchronised meaning structures [14]. One effect of an optimal communication system, however, is that the agents quickly settle on a common language, which is consistently reinforced through continued use and is completely stable, in contrast to the fluidity of human language.

The development of perfectly synchronised meaning structures is, however, very unlikely, given the inherent randomness in the agents' meaning creation algorithms; so what happens to the agents' language when their meanings are not synchronised and each is trying to communicate their language to the other? I explore this by tabulating detailed extracts from their lexicons through the progress of a simulation. The agents firstly develop most of the meaning structure which enables them to succeed in the discrimination games, and only then do they begin to communicate with each other. Extracts from two sample lexicons after four hundred communicative episodes, showing

Table 1. An extract from two lexicons, showing the high degree of coordination between the signal-meaning pairs.

Agent 1				Agent 2			
Signal	Meaning	Usage	Confidence	Signal	Meaning	Usage	Confidence
bc	3-1	2	0.5	egla	1-11	9	0.45
yq	1-01	8	0.32	bc	3-1	2	0.4
tjop	3-01	7	0.32	tnip	3-10	6	0.35
...	...	...	...	...	...	...	...
egla	1-11	8	0.23	yq	1-01	7	0.25
tnip	3-10	6	0.17	tjop	3-00	6	0.2

Table 2. The meaning exists in both semantic structures, and so context-driven disambiguation leads to an agreement over the meaning of the signal.

Agent 1				Agent 2			
Signal	Meaning	Usage	Confidence	Signal	Meaning	Usage	Confidence
jch	2-000	9	0.69	jch	2-000	9	0.17

the three words in which each agent has the highest confidence, is given in table 1. Although the agents have different levels of confidence in the signal-meaning pairs, they have broadly settled on a common language, with most words referring to the same meaning for both agents.

Later on, the first agent uses a meaning 2-000 in a discrimination game, but has no word for this meaning in its lexicon. These are exactly the circumstances under which we allow lexical innovation to occur, and so it coins a new word jch and utters this to the other agent. The development of the new word jch in both agents' lexicons depends on a number of parameters, including naturally the number of times it is chosen to be uttered, but also crucially whether the meaning to which it is linked is present in both agents' semantic representations, and it is this semantic development of the word jch which I will now discuss.

If the meaning 2-000 does exist in both agents' semantic representations, then we find, after rolling the simulation on a few hundred episodes, that the agents' lexicons contain the extracts shown in table 2. The second agent has associated jch with many different meanings, having heard it in a number of different contexts. The meaning 2-000, however, has occurred in each of these contexts, and so the agent's confidence in this particular signal-meaning pair is higher than in any other; repeated exposure in different contexts has disambiguated the agent's set of possible semantic hypotheses.

If the meaning 2-000 does not exist in both agents' representations, however, then the process of coordination is not so smooth, as we see in table 3. The second agent has again associated jch with a large set of possible meanings, but this time meaning 2-000 cannot be among them, as it is not in this agent's semantic repertoire, and so it is instead most confident in the meaning $4 - 01$. The agents now each have a different meaning associated with the word, and this has interesting consequences. The second

Table 3. The meaning does not exist in both agents' semantic structure, and no agreement over the meaning of the signal is reached. As a consequence, both agents' confidence in their respective lexical association falls.

Agent 1				Agent 2			
Signal	Meaning	Usage	Confidence	Signal	Meaning	Usage	Confidence
jch	2-000	9	0.47	jch	4-01	6	0.11

agent now uses jch to represent meaning 4-01, and utters this in a context in which the first agent's meaning 2-000 is not a possible meaning. This leads the first agent to become less confident in its original association; when agents are using the word for different meanings over time, both agents' confidence in their respective original associations for the word fall. There is now a conflict between the agents' use of the word jch, which must in the end be resolved by one agent losing so much confidence in its preferred association that it chooses another meaning for the word.

Because of the hierarchical nature of the meaning creation process, it is more likely that the meanings which the agents have in common, and on which they can agree word associations, are the more general meanings, which are nearer the roots of the discrimination tree. When the conflict over the use of the word jch is resolved, therefore, it is likely that the eventual meaning to which it is attached is, other things being equal, a relatively more general meaning than it was originally coined to serve; its use in a variety of contexts by agents who have different semantic structures has led to its meaning becoming more generalised.

The agents' lexicons are of course dynamic, and do not stop developing because the conflict over one word has been temporarily resolved. If the first agent comes across a context in which the original meaning of jch, namely 2-000, is needed, it now has no word which it can use and so must coin another one; the generalisation process itself has led to the loss of words from some part of the lexicon, which, if they are needed again, leads inevitably to more innovation. Of course, if meaning 2-000 still does not exist in the other agent's semantic structure, then the same process of conflict over meaning, generalisation and innovation is likely to happen all over again. We find, therefore, that meaning uncertainty, which is inevitable when meanings must be inferred instead of given, leads to the development of a continuous cycle of language innovation and semantic generalisation and extension.

5 Conclusions

I have presented a model in which agents individually create meanings based on their interactions with their external environment, and develop a co-ordinated communication system without either feedback or the explicit transfer of meaning. If the agents have very similar conceptual structures, then they are able to develop optimal communication systems by inferring the meanings of unfamiliar words through their exposure in different contexts. On the other hand, variation in conceptual structure itself creates pressure leading to a cycle of innovation and semantic generalisation. Work is currently under way to explore this trade-off between semantic uniformity which helps communication and semantic variability which drives language change.

References

[1] J. Batali. The negotiation and acquisition of recursive grammars as a result of competition among exemplars. In E. Briscoe, editor, *Linguistic Evolution through Language Acquisition: Formal and Computational Models*, pages 111–172. Cambridge University Press, Cambridge, 2002.

[2] T. Belpaeme. *Factors influencing the origins of colour categories*. PhD thesis, Artificial Intelligence Lab, Vrije Universiteit Brussel, 2002.

[3] A. Billard and K. Dautenhahn. Experiments in learning by imitation – grounding and use of communication in robotic agents. *Adaptive Behavior*, 7(3/4):415–438, 1999.

[4] H. Brighton. Compositional syntax from cultural transmission. *Artificial Life*, 8(1):25–54, 2002.

[5] W. Croft. *Explaining Language Change: an evolutionary approach*. Longman Linguistics Library. Pearson, 2000.

[6] G. Frege. Über Sinn und Bedeutung. *Zeitschrift für Philosophie und philosophische Kritik*, 100:25–50, 1892.

[7] S. Kirby. Spontaneous evolution of linguistic structure: an iterated learning model of the emergence of regularity and irregularity. *IEEE Journal of Evolutionary Computation*, 5(2):102–110, 2001.

[8] J. Lyons. *Semantics (2 vols.)*. Cambridge University Press, Cambridge, 1977.

[9] C. L. Nehaniv. The making of meaning in societies: Semiotic and information-theoretic background to the evolution of communication. In *Proceedings of the AISB Symposium: Starting from Society — the application of social analogies to computational systems, 19–20 April 2000*, pages 73–84, 2000.

[10] M. Oliphant and J. Batali. Learning and the emergence of coordinated communication. *Center for Research on Language Newsletter*, 11(1), 1997.

[11] S. Pinker and P. Bloom. Natural language and natural selection. *Behavioral and Brain Sciences*, 13:707–784, 1990.

[12] W. v. O. Quine. *Word and Object*. MIT Press, 1960.

[13] A. D. M. Smith. Establishing communication systems without explicit meaning transmission. In Jozef Kelemen and Petr Sosík, editors, *Advances in Artificial Life: Proceedings of the* 6th *European Conference on Artificial Life*, number 2159 in Lecture Notes in Artificial Intelligence, pages 381–390. Springer-Verlag, 2001.

[14] A. D. M. Smith. Intelligent meaning creation in a clumpy world helps communication. *Artificial Life*, 9(2):559–574, 2003. in press.

[15] K. Smith. Compositionality from culture: the role of environment structure and learning bias. Paper given at the 4th International Conference on the Evolution of Language, Harvard University, 2002.

[16] L. Steels. Perceptually grounded meaning creation. In M. Tokoro, editor, *Proceedings of the international conference on multi-agent systems*. MIT Press, 1996.

[17] L. Steels and F. Kaplan. Bootstrapping grounded word semantics. In E. Briscoe, editor, *Linguistic Evolution through Language Acquisition: formal and computational models*, pages 53–73. Cambridge University Press, Cambridge, 2002.

[18] R. L. Trask. *Historical Linguistics*. Arnold, 1996.

[19] P. Vogt. *Lexicon Grounding on Mobile Robots*. PhD thesis, Vrije Universiteit Brussel, 2000.

Language Evolution in Populations: Extending the Iterated Learning Model

Kenny Smith and James R. Hurford

Language Evolution and Computation Research Unit, School of Philosophy, Psychology and Language Sciences, The University of Edinburgh,
Adam Ferguson Building, 40 George Square, Edinburgh EH8 9LL
kenny@ling.ed.ac.uk

Abstract. Models of the cultural evolution of language typically assume a very simplified population dynamic. In the most common modelling framework (the Iterated Learning Model) populations are modelled as consisting of a series of non-overlapping generations, with each generation consisting of a single agent. However, the literature on language birth and language change suggests that population dynamics play an important role in real-world linguistic evolution. We aim to develop computational models to investigate this interaction between population factors and language evolution. Here we present results of extending a well-known Iterated Learning Model to a population model which involves multiple individuals. This extension reveals problems with the model of grammar induction, but also shows that the fundamental results of Iterated Learning experiments still hold when we consider an extended population model.

1 Introduction

Language is culturally transmitted — children learn their language on the basis of the observed linguistic behaviour of others. A recent trend has been to explain the structural properties of language in terms of adaptation, by language, to pressures arising during its cultural transmission. Using this approach, properties of language such as recursion [10], generalised phrase structure [4] and compositionality [2,8] have been shown to be adaptations which help language survive the repeated cycle of production and learning. A particular feature of this work has been its foundations in 1) linguistic theory and 2) experimentation based on computational models in the A-Life tradition.

These models suffer from an impoverished treatment of population dynamics. As discussed in Section 2.1, populations are modelled as a series of discrete, non-overlapping generations (with each generation typically consisting of a single agent), or a mono-generational collection of multiple agents. Both of these treatments of population dynamics are rather unsatisfactory, particularly given (as discussed in Section 2.2) the apparent importance of factors such as population structure and demography in language evolution in the real world. This paper has three main aims:

1. to highlight this problem and the desirability of addressing it.
2. to describe initial investigations into the impact of population dynamics on language evolution (see Section 3), highlight the problems encountered (Section 4) and present results (Section 5).
3. to outline future directions to take in addressing these issues (Section 6).

W. Banzhaf et al. (Eds.): ECAL 2003, LNAI 2801, pp. 507–516, 2003.

2 Population Dynamics and Linguistic Evolution

2.1 Population Dynamics in Models

Computational models of the cultural transmission of language are based around the Expression/Induction (E/I) cycle [5]. In such models, the internal knowledge of individuals determines the observable behaviour that those individuals produce (the Expression part of the process). This behaviour then forms the basis for the formation of internal knowledge in other individuals, via learning (the Induction process). This in itself is a fairly general model of cultural transmission.

Within the E/I framework, two distinct basic models have emerged which I will term the Iterated Learning Model (ILM, a term introduced by Brighton [2]) and the Negotiation Model (NM, a term based on Batali's work [1]). In both these models, agents acquire their linguistic competence based on observations of the behaviour of other agents. The difference between these two basic models is in their treatment of population dynamics.

In the NM, a population consists of a mono-generational collection of agents. At each time-step, members of the population produce linguistic behaviour, which is observed and learned from by other members of the population. There is no population turnover in the NM — new agents do not enter the population and no agents are removed.

Unlike in the NM, there is population turnover in the ILM. This turnover is typically generational, where the entire population is replaced at each time-step, although gradual turnover variants do exist. In either case, transmission is exclusively from fully enculturated individuals to naive individuals. Furthermore, the population at each generation is typically taken to consist of a single agent.[1]

2.2 Population Dynamics in Language

This simplification of population dynamics is a reasonable idealisation in tackling the complex problem of modelling the cultural evolution of linguistic structure. Indeed, it would be entirely acceptable were it not for the fact that the real-world data shows that issues of population structure and demography play an important role in shaping language during language birth and language change.

Based on a review of language birth events, Sonia Ragir has suggested that "the emergence of new languages, both signed and spoken, depends on: (1) a *critical mass* of individuals generating shared patterns of linguistic practices; (2) historical continuity maintained by a continuous influx of new participants into the language pool" [13]. Similar factors have been highlighted in studies of language change driven by dialect contact. In their list of factors influencing the outcome of dialect contact, Kerswill & Williams emphasise the importance of "[t]he proportion of children to adults in the immediate post-settlement years" [7, p75] and "[t]he presence of the possibility of forming new social networks among children and younger people: These possibilities are influenced by demographic factors such as high density of population, a 'critical mass' of population, and the presence of universal schooling" [7, p75].

[1] There are implementations of the ILM which move away from the single agent population model [4,9]. However, these models either suffer from a very rigid, tightly spatially organised population model (as in [9]) or a strongly biased model of a learner (as in [4]).

3 The Basic Model

The linguistic facts outlined in the previous section highlight the importance of population dynamics in the cultural evolution of linguistic form. E/I models potentially constitute a powerful tool for the investigation of such social and population dynamics and their impact on the structure of emergent or extant languages. However, as they currently stand, the ILM and NM implementations of the E/I framework do not allow us to address such issues directly. It is our goal to develop models, building on existing models as much as possible, which are explicitly designed to allow the investigation of population dynamics and their influence on linguistic structure.

The first step in this process is to verify that the elementary results obtained through Iterated Learning experiments pertain when we move to a non-trivial model of populations — in other words, do structured languages still emerge when we move away from extremely small population models? In order to investigate this question, we have adopted the model of grammars and grammar induction developed by Simon Kirby [8, 10].[2] These elements of the model are briefly described below in Section 3.1. We slot this model of language and language learning into a generational turnover ILM, designed to examine the impact of learning from multiple individuals (described in Section 3.2).

3.1 Grammars and Grammar Induction

Representation. An individual's linguistic competence consists of a definite-clause grammar with attached semantic arguments. These definite-clause grammars consist of a set of rules, where the left hand side consists of a non-terminal category and the semantic label for that category and the right hand side consists of zero or more non-terminal categories, with semantic labels, and zero or more strings of characters,[3] which correspond to phonetically-realised components of a signal. Semantic representations are predicate-argument structures.[4] Two example grammars[5] might be:

Grammar 1

$S\,/\,p(x,y) \rightarrow N/x\ V/p\ N/y$
$V\,/\,\text{love}' \rightarrow$ `loves`
$N\,/\,\text{lynne}' \rightarrow$ `lynne`
$N\,/\,\text{garry}' \rightarrow$ `garry`
plus various other lexical rules
⋮

[2] The SICStus Prolog source code for Kirby's model is available at http://www.ling.ed.ac.uk/lec/software.html

[3] We consider the case where there are 26 possible characters, corresponding to the 26 characters of the Roman alphabet.

[4] We consider the limited semantics where there are five two-place predicates and five possible arguments. As in Kirby's models, we rule out the case where the first and second arguments are identical. This yields $5 \times 5 \times 4 = 100$ possible meanings.

[5] Atomic semantic elements are marked with primes, characters are represented in `typewriter font`, upper case italics represent non-terminal categories and lower case italics represent variables over semantic elements. *S* is the privileged, top-level non-terminal category.

Grammar 2:

S / love′(lynne′,garry′) $\rightarrow$ `lynnelovesgarry`
S / love′(lynne′,beppe′) $\rightarrow$ `igajojopop`
S / kick′(trevor′,barry′) $\rightarrow$ `mo`
plus various other sentence-level rules
⋮

Both grammars would produce the string `lynnelovesgarry` meaning love′(lynne′, garry′), but clearly do so in rather different ways — Grammar 1 would do so in a compositional manner (each subpart of the meaning corresponds to a subpart of the signal), whereas Grammar 2 would do so in a rote-learned, holistic manner.

Learning. Learners in Kirby's model are presented with a set of utterances, where utterances consist of meaning-signal pairs, and induce a grammar based on these utterances. Grammar induction consists of two main processes — rule incorporation and grammar compression. In an incorporation event a learner is presented with a meaning-signal pair $\langle m, s\rangle$ and forms a rule: $S/m \rightarrow s$. This amounts to simply memorising an observed utterance.

After every incorporation event, learners attempt to extract regularities and compress their grammars. This involves two main sub-processes, chunking and merging. During chunking, pairs of rules are examined in the search for meaningful chunks, which are then separated out into new syntactic categories. This leads to an *increase* in grammar size. Grammar compression is then achieved by merging similar rules. Understanding these processes is not essential for the understanding of this paper, and we refer the reader to previous papers [8,10] for a full explanation. The key point is that learners exposed to utterances containing repeated meaningful subparts of signal will tend to arrive at grammars like Grammar 1 above, which will in turn lead to them generating utterances with regular meaning-signal correspondences. In contrast, learners exposed to an idiosyncratic set of utterances will arrive at grammars more like Grammar 2 above, and will in turn produce irregular, unprincipled sets of utterances.

Production and Reception. When called upon to produce an utterance for a meaning, or to interpret a received signal, agents search, depth-first, for a combination of rewrite rules which will cover the given meaning or signal. If such a set of rules cannot be found during production then the producer applies an invention procedure, with parts of the meaning which are not expressible using the grammar being expressed with random strings. The inventor learns from its own invention, via the process described above.

3.2 The Population Model

We will begin by verifying that results obtained using this model extend to the case where populations consist of multiple individuals. We will consider a generational turnover ILM — the population consists of a set of non-overlapping generations, where each generation consists of N agents. This model of population turnover was chosen as a starting point as it most closely resembles the population model used in the majority of simulations of the emergence of linguistic structure, and can be straightforwardly extended to a more complex model involving overlapping generations.

Each individual in the population at generation n receives e exposures to the linguistic behaviour produced by agents at generation $n-1$. A single exposure consists of a meaning-signal pair. In previous models, the set of meaning-signal pairs which an individual learns from is typically drawn from the behaviour of a single individual. In this model each learner has p cultural parents. Those cultural parents are selected at random, with replacement, from the N agents in the previous generation. Each of the e meaning-signal pairs the learner observes is then generated by a randomly selected member of this set of cultural parents.

4 Problems and Solutions

4.1 Problem: Signal Growth

When $p = 1$ the simulation effectively reduces to the simple case where each generation consists of a single agent. For the case where $p > 1$ (individuals potentially have more than one cultural parent), this simple extension to the generational turnover ILM runs into problems. The length of the right hand side of rules rapidly increases over generations, due to the addition of strings of meaningless terminal characters. This can result in the right hand side of rules expanding to include several hundred terminal characters. This radical string growth is a consequence of three factors:

1. the emergence of multiple, overlapping but non-identical grammars in the population.
2. inconsistent training data presented to learners (as a consequence of point 1).
3. the greedy induction algorithm.

Radical right hand side growth only occurs when you consider multiple cultural parents because the process requires variability in the input to the language learner, which does not occur in the typical single-agent ILM. The "spare" terminal characters are motivated by a small number of observed utterances. The initial addition of the spare characters is due to the greedy nature of the induction algorithm — learners do not consider multiple possible grammars at a time, nor are they capable of backtracking from the kind of over-generalisation that introduces spare characters.

4.2 Solutions to the Problem

This is clearly an undesirable artifact of the model of grammar induction, which is only exposed by moving away from the simplest possible population model. How can this problem be resolved? The first obvious solution is to move from greedy incremental to batch grammar induction — rather than compressing the grammar as much as possible after each observation, the learner should only compress its grammar after a reasonable number of utterances have been observed. This approach has been adopted previously [15].

While batch induction may be suitable for strictly generational models of population turnover, it is unsuitable for a more realistic model of population turnover. A batch learning procedure draws a strict distinction between learners and non-learners. This kind of clear-cut distinction is undesirable in a model involving a less restrictive population dynamic — ideally, we want learners to produce observable behaviour which can be learned from by other learners. Incremental induction allows maximal flexibility in this regard.

The second possible alternative is to allow learners to entertain multiple hypotheses at any one time — rather than maintaining a single grammar, and compressing it as much as possible, a learner can maintain multiple competing grammars, compressing them differently and therefore having multiple ways of expressing certain meanings. In this scenario, we allow learners to retreat from incorrect generalisations of the type which occur in the more greedy induction process.

This multiple-competing-hypothesis approach has been used in implementations of the NM [1]. There are two main problems with this approach. The first is that there has to be pruning of grammars from time to time — maintaining all possible grammars rapidly becomes intractable. Identifying which grammars can be safely pruned is in itself a non-trivial issue. The second problem is that there must be a way of evaluating the competing grammars, in order to decide which is best. Batali implements this in his costing system. However, this costing system is rather ad-hoc. We would like the evaluation metric to be well-motivated, in terms of justifiable compression, the types of grammar evaluations that language learners make, or some other general consideration, such as the nature of the communicative task. We are working towards just such a grammar induction model, in conjunction with Simon Kirby and Willem Zuidema at the LEC.

A more immediate solution can be obtained by including justifiable production biases in the model. In other words, we allow the inducer to arrive at the wrong hypothesis, but filter out the more outrageous consequences of this hypothesis. In the case of this model, this is achievable by building in a production preference in favour of short utterances — production proceeds in a depth-first manner, as before, but at each step the producer chooses the rule with the shortest right hand side. This production bias plausibly applies in humans, as a preference for *signal simplicity* [12], or as a consequence of more general principles of least effort [14]. Including this production bias eliminates the string growth problem.

5 Results: Language Evolution in Populations

Runs of the ILM were carried out to verify that convergence on a shared, expressive grammar was possible in the context of a population-level ILM, and to ascertain whether the number of cultural parents each individual learns from (p) has any impact on cultural evolution. For the results presented here, we use a population size of ten ($N = 10$), and vary the number of cultural parents used in each run ($1 \leq p \leq 10$). Each learner observes $e = 50$ meaning-signal pairs, each of which is produced by one of its p cultural parents. Runs were allowed to proceed for 1000 generations. Ten runs of the ILM were carried out for each condition.

We evaluate three aspects of the populations' grammar at each generation:

Grammar Size: The number of grammar rules each agent has after learning.
Coverage: The proportion of meanings each agent can express after learning, without recourse to invention.
Communicative Accuracy: The proportion of meanings an agent is able to successfully communicate to two randomly selected communicative partners.[6]

Grammar size and coverage give an indication of the degree of structural regularity in an agent's grammar. A grammar size of approximately 50 (equal to the number of exposures each individual receives) and coverage of around 0.5 indicates an idiosyncratic, non-compositional grammar, where each meaning is associated with an unstructured signal in a holistic fashion. A grammar size of around 11 and coverage of 1 indicates a highly compressed, highly expressive compositional grammar (one sentence-level rule, plus five lexical rules for predicates and five lexical rules for arguments). Coverage of 1 also indicates stability — if all meanings can be expressed, then the underlying grammar must exhibit regularities, which subsequent learners can extract. Communicative accuracy indicates the degree of within-generation coherence in the population. A communicative accuracy of 1 indicates that agents agree on the signals which should be used to express each possible meaning.

It should first be noted that, for all values of p (numbers of cultural parents), the majority of runs converged on compressed, expressive grammars which resulted in high intra-generational communicative accuracy. In other words, fundamental results from the ILM extend to the case where the population dynamics are non-trivial — compressed, compositional, communicatively-functional grammars can evolve through Iterated Learning in populations where learners learn from multiple cultural parents.

p does appear to have some effects on the structure of the emergent grammars. Fig. 1 plots mean grammar size, coverage and communicative accuracy against p. The mean values of the three measurements were calculated by averaging over the final ten generations of each of the ten runs for each condition. As can be seen from this figure, there appears to be a "sweet spot" for p, at around $p = 4$ or 5. For these values of p, coverage is at a maximum, grammars are highly compressed and intra-generational communicative accuracy is high. Away from this value of p, coverage is (fractionally) lower, grammars are larger and communicative accuracy is lower. However, this analysis is complicated somewhat by the fact that there were non-convergent runs for $p = 9$ and 10 (as suggested by the lower average coverage for these values of p). Excluding these non-convergent runs, the overall trend is weakened, although still visible.

Number of cultural parents also impacts on the speed with which populations converged on a shared, stable grammar. Measuring stability in these simulations is a somewhat fraught task — the stochastic sampling of the language of the previous generation can always introduce change into an apparently stable system. The best approximation of stability is coverage — a system which can express a high proportion of meanings without recourse to invention tends to be stable. Fig. 2 plots p against time to convergence, according to two measures of stability based on coverage. As can be seen from

[6] In order to evaluate communicative accuracy between a speaker and a hearer, the speaker produces a signal for each possible meaning, and the hearer parses that signal to arrive at a meaning. If the speaker's meaning matches the hearer's interpreted meaning, the interaction is a success.

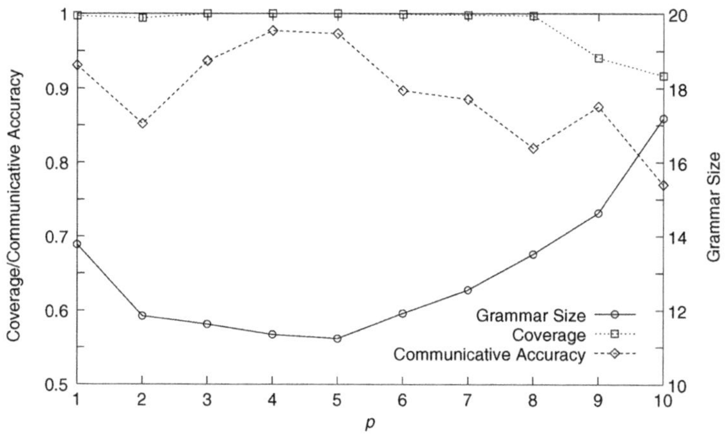

Fig. 1. The relationship between p and language structure, as evaluated by grammar size, coverage and communicative accuracy. For $p = 4$ or 5, cultural evolution in populations leads to smaller grammars and higher intra-generational communicative accuracy.

this figure, low or high values of p tend to result in longer times to convergence, although this is perhaps less clear with respect to the stricter measure of convergence. The main point is that the values of p which were identified as the "sweet spot" in Fig. 1 tend to lead to more rapid convergence — there are certain values of p which optimise grammar compression and functionality, and lead to more rapid convergence within a population. When p is too low, learners view few alternative possible grammars and have little opportunity to preferentially acquire more compressible grammars. When p is too high, learners observe too many competing grammars, resulting in instability and difficulty in achieving consensus. At the optimal value of p, these factors are better balanced — learners witness enough variability to allow their biases to come into play, while at the same time being able to achieve stability and consensus.

6 Future Directions

We aim to expand upon this initial research in two ways. Firstly, we will develop a model of grammar induction which does not suffer from the type of problem outlined in Section 4 above. This inducer will maintain multiple potential candidate grammars, and therefore be able to retreat, to a degree, from inappropriate generalisations.

Secondly, we will develop more sophisticated models of population dynamics. As discussed in this paper, we have verified that it is possible for a population of agents to converge on a compressed, compositional, shared grammar through Iterated Learning. We will move away from the strictly generational turnover approach, to a situation where there is gradual population replacement, and unrestricted learning interactions. We are particularly interested in the effects of high degrees of learner-learner contact, which the literature on language birth and change suggests is a key factor.

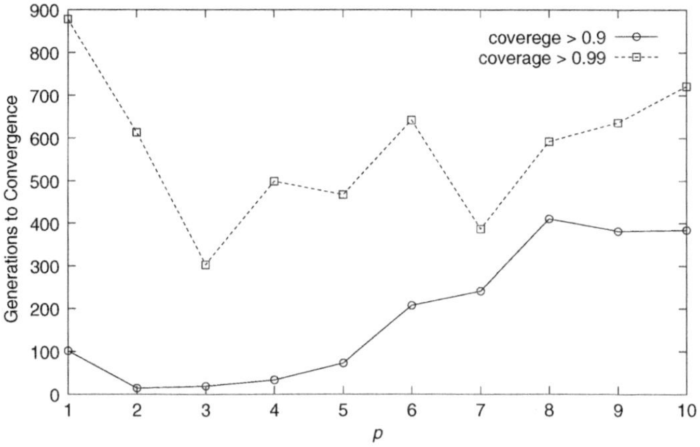

Fig. 2. Time to convergence as a function of p. A population was considered to have converged if it arrived at grammars with a certain level of coverage (either 0.9 or 0.99), and stayed above that level of coverage for the remainder of the simulation run. At least 80% of all runs reached convergence according to the weaker definition, and 70% according to the stronger definition. Once again, there appears to be a sweet spot for p, which leads to more rapid convergence.

7 Conclusions

A-Life techniques can be applied to the investigation of the cultural evolution of linguistic structure. Implementations of the Expression/Induction framework (both Iterated Learning and Negotiation Models) typically suffer from an impoverished treatment of population dynamics. This is unfortunate, given that factors such as population structure and demography appear to play an important role in the processes of language birth and language change. It is our aim to develop this modelling approach to allow the impact of population dynamics on linguistic evolution to be investigated and quantified.

As a first step in this process, we must verify that previous models can be extended to a treatment of language evolution in populations, and that the key results derived from these models pertain in the new situation. The model outlined in this paper demonstrates that this is the case — general, compositional grammars can evolve culturally within a population. Furthermore, the number of cultural parents each learner has (one aspect of population dynamics) has some impact on the structure of the emergent languages and the speed with which they evolve. However, in addition to this positive result, extension to a population model also reveals some flaws in a well-known Iterated Learning Model, which are not exposed in a minimal population model.

We are developing our approach along two lines. Firstly, we are involved in tackling the problems arising from incremental induction and greedy compression, by working on a new, theoretically well-motivated model of grammar induction. Secondly, we are extending our model to a wider range of population dynamics. This kind of experimentation, drawing on A-Life techniques, promises to be profitable in investigating language evolution in general, and, in particular, in identifying important factors in language evolution in populations.

References

1. J. Batali. The negotiation and acquisition of recursive grammars as a result of competition among exemplars. In Briscoe [3], pages 111–172.
2. H. Brighton and S. Kirby. The survival of the smallest: Stability conditions for the cultural evolution of compositional language. In Kelemen and Sosík [6], pages 592–601.
3. E. Briscoe, editor. *Linguistic Evolution through Language Acquisition: Formal and Computational Models*. Cambridge University Press, Cambridge, 2002.
4. J. R. Hurford. Social transmission favours linguistic generalization. In Knight et al. [11], pages 324–352.
5. J. R. Hurford. Expression/induction models of language evolution: dimensions and issues. In Briscoe [3], pages 301–344.
6. J. Kelemen and P. Sosík, editors. *Advances in Artificial Life: Proceedings of the 6th European Conference on Artificial Life*. Springer-Verlag, Berlin, 2001.
7. P. Kerswill and A. Williams. Creating a new town koine: Children and language change in Milton Keynes. *Language in Society*, 29:65–115, 2000.
8. S. Kirby. Syntax out of learning: the cultural evolution of structured communication in a population of induction algorithms. In D. Floreano, J. D. Nicoud, and F. Mondada, editors, *Advances in Artificial Life: Proceedings of the 5th European Conference on Artificial Life*. Springer, Berlin, 1999.
9. S. Kirby. Syntax without natural selection: how compositionality emerges from vocabulary in a population of learners. In Knight et al. [11], pages 303–323.
10. S. Kirby. Learning, bottlenecks and the evolution of recursive syntax. In Briscoe [3], pages 173–203.
11. C. Knight, M. Studdert-Kennedy, and J.R. Hurford, editors. *The Evolutionary Emergence of Language: Social Functions and the Origins of Linguistic Form*. Cambridge University Press, Cambridge, 2000.
12. R. W. Langacker. Syntactic reanalysis. In C. N. Li, editor, *Mechanisms of Syntactic Change*, pages 57–139. University of Texas Press, Austin, TX, 1977.
13. S. Ragir. Constraints on communities with indigenous sign languages: clues to the dynamics of language origins. In A. Wray, editor, *The Transition to Language*, pages 272–294. Oxford University Press, Oxford, 2002.
14. G. K. Zipf. *Human behaviour and the principle of least effort : an introduction to human ecology*. Addison-Wesley, Cambridge, MA, 1949.
15. W. H. Zuidema. Emergent syntax: The unremitting value of computational modeling for understanding the origins of complex language. In Kelemen and Sosík [6], pages 641–644.

Learning Biases for the Evolution of Linguistic Structure: An Associative Network Model

Kenny Smith

Language Evolution and Computation Research Unit,
School of Philosophy, Psychology and Language Sciences,
The University of Edinburgh,
Adam Ferguson Building, 40 George Square, Edinburgh EH8 9LL
kenny@ling.ed.ac.uk

Abstract. Structural hallmarks of language can be explained in terms of adaptation, by language, to pressures arising during its cultural transmission. Here I present a model which explains the compositional structure of language as an adaptation in response to pressures arising from the poverty of the stimulus available to language learners and the biases of language learners themselves.

1 Introduction

The goal of evolutionary linguistics is to explain the origins and development of human language — how did language come to be structured as it is? Recent research, much of which uses A-Life techniques to develop its argument, attempts to answer this question by appealing to cultural evolution (see [4] for review). Language is culturally transmitted to the extent that language learners acquire their linguistic competence on the basis of the observed linguistic behaviour of others. A key contribution of those working within the cultural framework is to show that the cultural transmission of language leads to an adaptive dynamic — the adaptation, by language itself, to pressures acting during its cultural transmission. This cultural evolution can lead to the emergence of at least some of the characteristic structure of language.

This process of cultural adaptation must be dependent on some biological endowment. What is not clear is what form this endowment takes, or to what extent it is language-specific. In this paper I will present a series of experiments, using a computational model of the cultural transmission of language, which allow us to refine our understanding of the necessary biological basis for a particular structural characteristic of language, compositionality. In this model a learner's biological endowment consists of a particular way of learning, with an associated learning bias.

2 Elements of the Model

I will present an Iterated Learning Model (ILM) which allows us to investigate the role of stimulus poverty and learning bias in the evolution of compositional language. The ILM is based around a simple treatment of languages as a mapping between meanings and signals (see Section 2.1). Linguistic agents are modelled using associative networks

W. Banzhaf et al. (Eds.): ECAL 2003, LNAI 2801, pp. 517–524, 2003.

(see Section 2.2). These agents are slotted into a minimal population model to yield the ILM. For the purposes of this paper, we will consider an ILM with the simplest possible population dynamic[1]— the population consists of a set of discrete generations, with each generation consisting of a single agent. The agent at generation n produces some observable behaviour (in this model a set of meaning-signal pairs), which is then learned from by the agent at generation $n + 1$.

2.1 Compositionality and a Model of Languages

Compositionality relates semantic structure to signal structure — in a compositional system the meaning of an utterance is dependent on the meaning of its parts. For example, the utterance "John walked" consists of two words, a noun ("John") and a verb ("walked"), which further consists of a stem ("walk-") and a suffix (-"ed"). The meaning of the utterance as a whole is dependent on the meaning of these individual parts. In contrast, in a non-compositional or *holistic* system the signal as a whole stands for the meaning as a whole. For example, the meaning of the English idiom "bought the farm" (meaning died) is not a function of the meaning of its parts.

The simplest way to capture this is to treat a language as a mapping between a space of meanings and a space of signals. In a compositional language, this mapping will be neighbourhood-preserving. Neighbouring meanings will share structure, and this shared structure will result in shared signal structure — neighbouring meanings in the meaning space will map to neighbouring signals in signal space. Holistic mappings are not neighbourhood-preserving — since the signal associated with a meaning does not depend on the structure of that meaning, shared structure in meaning space will not map to shared signal structure, unless by chance.

For the purposes of this model, meanings are treated as vectors and signals are strings of characters. Meanings are vectors in some F-dimensional space, where each dimension takes V possible values. F and V therefore define a meaning space $\mathcal{M}$.[2] The world, which provides communicatively relevant situations for agents in the model, consists of a set of objects, where each object is labelled with a meaning drawn from the meaning space $\mathcal{M}$.[3] Signals are strings of characters of length 1 to l_{max}, where characters are drawn from the character alphabet Σ.[4] l_{max} and Σ therefore define a signal space $\mathcal{S}$.

Given these representations of meanings and signals, we can now define a measure of compositionality. This measure is designed to capture the notion given above, that compositional languages are neighbourhood-preserving mappings between meanings and signals, and is based on a measure introduced in [1]. Compositionality (c) is based on

[1] More complex population dynamics have consequences for the cultural evolution of communication systems, as discussed in Smith & Hurford (this volume).

[2] The structure of this meaning space has been shown to have consequences for the cultural evolution of compositional structure [2]. However, I will not vary this parameter. All results reported here are for the case where $F = 3$ and $V = 5$.

[3] All results presented here are for the case where the world contains 31 objects, each object is labelled with a distinct meaning, and those meanings are drawn from a hypercube subspace of the space of possible meanings.

[4] For the results reported here, $l_{max} = 3$ and $\Sigma = \{a, b, c, d, e, f, g, h, i, j\}$.

the meaning-signal pairs that an agent produces, and is the Pearson's Product-Moment correlation coefficient of the pairwise distances between all the meanings and the pairwise distances between their corresponding signals.[5] $c = 1$ for a perfectly compositional system and $c \approx 0$ for a holistic system.

2.2 A Model of a Linguistic Agent

We now require a model of a linguistic agent capable of manipulating such systems of meaning-signal mappings. I will describe an associative network model of a linguistic agent. This model is based upon a simpler model of a linguistic agent, used to investigate the cultural evolution of vocabulary systems [6]. The main advantage of this model is that it allows the biases of language learners to be manipulated and investigated. For full details of the network model, the reader is referred to [5].[6]

Representation. Agents are modelled using networks consisting of two sets of nodes $\mathcal{N}_M$ and $\mathcal{N}_S$ and a set of bidirectional connections $\mathcal{W}$ connecting every node in $\mathcal{N}_M$ with every node in $\mathcal{N}_S$. Nodes in $\mathcal{N}_M$ represent meanings and partial specifications of meanings, while nodes in $\mathcal{N}_S$ represent partial and complete specifications of signals.

As summarised above, each meaning is a vector in F-dimensional space where each dimension has V values. *Components* of a meaning are (possibly partially specified) vectors, with each feature of the component either having the same value as the given meaning, or a wildcard. Similarly, components of a signal of length l are (possibly partially specified) strings of length l. Each node in $\mathcal{N}_M$ represents a component of a meaning, and there is a single node in $\mathcal{N}_M$ for each component of every possible meaning. Similarly, each node in $\mathcal{N}_S$ represents a component of a signal and there is a single node in $\mathcal{N}_S$ for each component of every possible signal.

Learning. During a learning event, a learner observes a meaning-signal pair $\langle m, s \rangle$. The activations of the nodes corresponding to all possible components of m and all possible components of s are set to 1. The activations of all other nodes are set to 0. The weights of the connections in $\mathcal{W}$ are adjusted according to some weight-update rule. In Section 4 this weight-update procedure will be a parameter of variation. However, initially, we will consider the rule

$$\Delta W_{xy} = \begin{cases} +1 \text{ if } a_x = a_y = 1 \\ -1 \text{ if } a_x \neq a_y \\ \;\;0 \;\text{ if } a_x = a_y = 0 \end{cases} \qquad (1)$$

where $W_{xy} \in \mathcal{W}$ gives the weight of the connection between nodes x and y and a_x gives the activation of node x. The learning procedure is illustrated in Fig. 1 (a).

Production. An *analysis* of a meaning or signal is an ordered set of components which fully specifies that meaning or signal. During the process of producing utterances, agents are prompted with a meaning and required to produce a meaning-signal pair. Production

[5] Distance in the meaning space is measured using Hamming distance. Distance in the signal space is measured using Levenstein (string edit) distance.

[6] Available for download at http://www.ling.ed.ac.uk/~kenny/publications.html

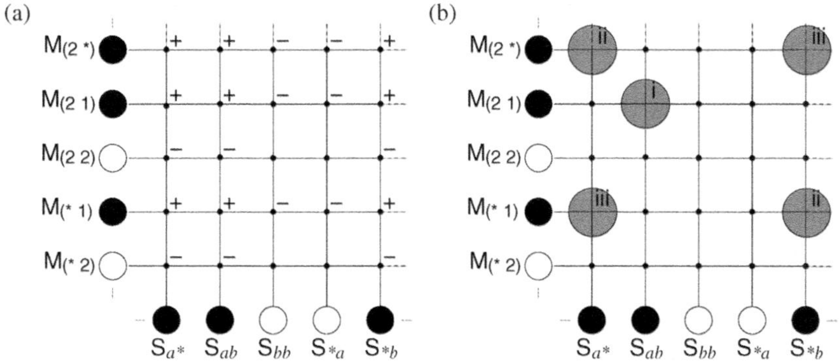

Fig. 1. (a) Storage of the meaning-signal pair $\langle (2\ 1), ab \rangle$. Nodes are represented by large circles and are labelled with the component they represent. For example, $M_{(2\ *)}$ is the node which represents the meaning component $(2\ *)$, where $*$ is an unspecified feature value. Nodes with an activation of 1 are represented by large filled circles. Small filled circles represent weighted connections. During the learning process, nodes representing components of $(2\ 1)$ and ab have their activations set to 1. Connection weights are then either incremented $(+)$, decremented $(-)$ or left unchanged. (b) Retrieval of three possible analyses of $\langle (2\ 1), ab \rangle$. The relevant connection weights are highlighted in grey. The weight for the one-component analysis $\langle \{(2\ 1)\}, \{ab\} \rangle$ depends on the weight of the connection between the nodes representing the components $(2\ 1)$ and ab, marked as i. The weight for the two-component analysis $\langle \{(2\ *), (*\ 1)\}, \{a*, *b\} \rangle$ depends on the weighted sum of two connections, marked as ii. The weight of the alternative two-component analysis $\langle \{(2\ *), (*\ 1)\}, \{*b, a*\} \rangle$ is given by the weighted sum of the two connections marked iii.

proceeds via a winner-take-all process. In order to produce a signal based on a given meaning $m_i \in \mathcal{M}$, every possible signal $s_j \in \mathcal{S}$ is evaluated with respect to m_i. For each of these possible meaning-signal pairs $\langle m_i, s_j \rangle$, every possible analysis of m_i is evaluated with respect to every possible analysis of s_j. The evaluation of a meaning analysis-signal analysis pair depends on the weighted sum of the connections between the relevant nodes. The meaning-signal pair which yields the analysis pair with the highest weighted sum is returned as the network's production for the given meaning. The production process is illustrated in Fig. 1 (b).

3 A Familiar Result

I will begin by replicating a familiar result: the emergence of compositional structure through cultural processes depends on the presence of a *transmission bottleneck* [2, 3]. Recall that a learner in the model acquires their linguistic competence on the basis of a set of observed meaning-signal pairs. That set of meaning-signal pairs is drawn from the linguistic behaviour of some other individual, which is a consequence of that individual's linguistic competence. I will investigate two possible conditions. In the *no transmission bottleneck* condition, this set of meaning-signal pairs contains examples of the signal associated with every possible meaning, and each learner is therefore presented with the complete language of the agent at the previous generation. In the *transmission*

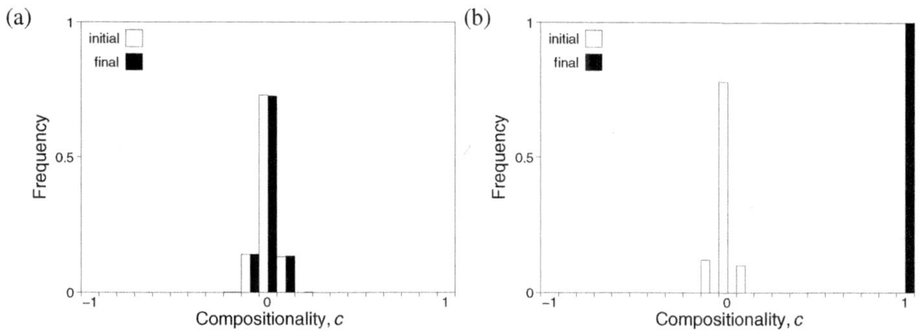

Fig. 2. The impact of the transmission bottleneck. (a) gives frequency by compositionality for runs in the no bottleneck condition. (b) gives frequency by compositionality for runs where there is a bottleneck on transmission.

bottleneck condition, the set of observed behaviour does not contain examples of the signal associated with every meaning, therefore each learner is presented with a subset of the language of the agent at the previous generation.[7] The transmission bottleneck constitutes one aspect of the poverty of the stimulus problem faced by language learners — they must acquire knowledge of a large (or infinite) language on the basis of exposure to a subset of that language.

In both conditions, the initial agents in each simulation run have all their connections weights set to 0, and therefore produce every meaning-signal pair with equal probability. Subsequent agents have connection weights of 0 prior to learning. Runs were allowed to progress for a fixed number of generations (200). Figs. 2 (a) and (b) plot compositionality by frequency for the initial and final languages, for the no bottleneck and bottleneck conditions respectively.[8]

As can be seen from the figure, in the absence of a bottleneck on transmission, there is no cultural evolution and compositional languages do not emerge. In contrast, in the bottleneck condition highly compositional systems emerge with high frequency — cultural evolution leads to the emergence of compositional language from initially holistic systems. This confirms, using a rather different model of a language learner, previously established results [2,3].

In the absence of a transmission bottleneck, the initial, random assignment of signals to meanings can simply be memorised. Consequently, there is no pressure for compositionality and the holistic mapping embodied in the initial system persists. However, holistic systems cannot survive in the presence of a bottleneck. The meaning-signal pairs of a holistic language have to be observed to be reproduced. If a learner only observes a subset of the holistic language of the previous generation then certain meaning-signal pairs will not be preserved — the learner, when called upon to produce, will produce some other signal for that meaning, resulting in a change in the language. In contrast,

[7] For all simulations involving a transmission bottleneck described in this paper, learners observed approximately 60% of the language of the previous agent.

[8] The results for the no bottleneck condition are based on 1000 independent runs of the ILM. The results for the bottleneck condition are based on 100 runs — fewer runs are required as there is less sensitivity to initial conditions.

compositional languages are generalisable, due to their structure, and remain relatively stable even when the learner observes a subset of the language of the previous generation. Over time, language adapts to this pressure to be generalisable. Eventually, the language becomes highly compositional, highly generalisable and consequently highly stable.

4 Exploring Learning Biases

To what extent is this fundamental result, that the transmission bottleneck leads to a pressure for compositional language, dependent on the model of a language learner? There is indirect evidence that this result is to some extent independent of the model of a language learner — a wide range of learning models all produce this fundamental result (see [7] for review). However, do these models share a common element? Is there some *learner bias* common across all these models which is required for compositional language to evolve culturally?

In order to investigate this question, further experiments were carried out, in which the parameter of interest is the weight-update rule used to adjust network connection weights during learning. The general form of the weight-update rule is as follows:

$$\Delta W_{xy} = \begin{cases} \alpha \text{ if } a_x = a_y = 1 \\ \beta \text{ if } a_x = 1 \wedge a_y = 0 \\ \gamma \text{ if } a_x = 0 \wedge a_y = 1 \\ \delta \text{ if } a_x = a_y = 0 \end{cases} \tag{2}$$

For the results described in the previous Section, $\alpha = 1, \beta = \gamma = -1, \delta = 0$. I will now consider a wider range of weight-update rules, restricting myself to rules where $\alpha, \beta, \gamma, \delta \in \{-1, 0, 1\}$. This yields a set of $3^4 = 81$ possible weight-update rules. In order to ascertain the biases of the different weight-update rules, each weight-update rule is subjected to three tests:[9]

Acquisition test: Can an isolated agent using the weight-update rule acquire a perfectly compositional language, based on full exposure to that language? To evaluate this, an agent using the weight-update rule was trained on a predefined perfectly compositional ($c = 1$) language, being exposed once to every meaning-signal pair included in that language. The agent was judged to have successfully acquired that language if it could reproduce the meaning-signal pairs of the language in production and reception.

Maintenance test: Can a population of agents using the weight-update rule maintain a perfectly compositional language over time in an ILM, when there is a bottleneck on transmission? To evaluate this, 10 runs of the ILM were carried out for the weight-update rule, with the agent in the initial generation having their initial connection weights set so as to produce a perfectly compositional language. Populations were defined as having maintained a compositional system if c remained above 0.95 for every generation of ten 200 generation runs.

[9] A similar technique has been applied to the investigation of learning biases required for the cultural transmission of vocabulary systems [6].

Table 1. Summary of the results of the three tests. The table gives the three types of performance exhibited, and the number (out of 81) of weight-update rules fitting that pattern of performance.

Test result			Number of rules
Acquire?	Maintain?	Construct?	
no	no	no	63
yes	no	no	16
yes	yes	yes	2

Construction test: Can a population of agents using the weight-update rule construct a highly compositional language from an initially random language, when there is a bottleneck on transmission (as happened in the results outlined in the previous Section)? To evaluate this, 10 runs of the ILM were carried out for the weight-update rule, with the agent in the initial generation having initial connection weights of 0 and therefore producing a random set of meaning-signal pairs. Populations were defined as having constructed a compositional system if c rose above 0.95 in each of ten 200 generation runs.

The results of these sets of experiments are summarised in Table 1. Only a limited number of weight-update rules (two of 81) support the evolution of compositional language through cultural processes. Why? What is it about the assignment of values to the variables α, β, γ and δ in these rules that make them capable of acquiring, maintaining and constructing a compositional system?

A full analysis is somewhat involved, and I will simply summarise the key point here — for full details the reader is referred to [5]. The two weight-update rules which pass the acquisition, maintenance and construction tests satisfy three conditions: 1) $\alpha > \beta$; 2) $\delta > \gamma$; 3) $\alpha > \delta$. These two rules[10] are the only weight-update rules which satisfy these conditions. By returning to the network and examining the way in which connection weights change on the basis of exposure to individual meaning-signal pairs, we can identify the consequences of these restrictions.

1. $\alpha > \beta$ ensures that, if an agent is exposed to the meaning-signal pair $\langle m_i, s_j \rangle$, they will in future tend to prefer produce s_j when presented with m_i, rather than $s_{k \neq j}$.
2. $\delta > \gamma$ ensures that, if an agent is exposed to $\langle m_i, s_j \rangle$, they will prefer *not* to produce s_j when presented with $m_{k \neq i}$.
3. $\alpha > \delta$ ensures that, if an agent is exposed to $\langle m_i, s_j \rangle$, they will tend to reproduce this meaning-signal pair in a manner which involves the maximum number of components.

Points 1 and 2 in combination lead to a preference for *one-to-one* mappings between meanings and signals — agents with the appropriate weight-update rules are biased in favour of learning languages which map each meaning to a constant signal (one-to-many mappings are avoided, see Point 1), and which map each distinct meaning onto a

[10] To be explicit, the two rules are:
$\alpha = 1,\ \beta = -1,\ \gamma = -1,\ \delta = 0$
and $\alpha = 1,\ \beta = 0,\ \gamma = -1,\ \delta = 0$.

distinct signal (many-to-one mappings from meanings to signals are avoided, see Point 2). Point 3 corresponds to a bias in favour of memorising associations between elements of meaning and elements of signal, rather than between whole meanings and whole signals.

5 Conclusions

I have presented an Iterated Learning Model of the cultural evolution of compositional structure. This model has been used to replicate a familiar result — the poverty of the stimulus available to language learners (as imposed by the transmission bottleneck) leads to the emergence of compositional structure. However, novelly, this cultural evolution has been shown to be dependent on language learners possessing two biases:

1. a bias in favour of one-to-one mappings between meanings and signals.
2. a bias in favour of exploiting regularities in the input data, by acquiring associations between parts of meanings and parts of signals.

Both these biases are present in most computational models of the evolution of linguistic structure and, significantly, there is also evidence to suggest that human language learners bring these biases to the language acquisition task [7]. Compositionality, a fundamental structural property of language, can therefore be explained in terms of cultural evolution in response to two pressures — a pressure arising from the poverty of the stimulus, and a pressure arising from the biases of language learners. The source of this learning bias in humans is a topic for further research — is the bias a consequence of some general cognitive strategy, or a specific biological adaptation for the acquisition of language?

References

1. H. Brighton. Experiments in iterated instance-based learning. Technical report, Language Evolution and Computation Research Unit, 2000.
2. H. Brighton. Compositional syntax from cultural transmission. *Artificial Life*, 8(1):25–54, 2002.
3. S. Kirby. Syntax out of learning: the cultural evolution of structured communication in a population of induction algorithms. In D. Floreano, J. D. Nicoud, and F. Mondada, editors, *Advances in Artificial Life: Proceedings of the 5th European Conference on Artificial Life*. Springer, Berlin, 1999.
4. S. Kirby. Natural Language from Artificial Life. *Artificial Life*, 8(2):185–215, 2002.
5. K. Smith. Compositionality from culture: the role of environment structure and learning bias. Technical report, Language Evolution and Computation Research Unit, 2002.
6. K. Smith. The cultural evolution of communication in a population of neural networks. *Connection Science*, 14(1):65–84, 2002.
7. K. Smith. Learning biases and language evolution. In *Proceedings of the 15th European Summer School on Logic, Language and Information*. forthcoming.

The Learning and Emergence of Mildly Context Sensitive Languages

Edward P. Stabler[1], Travis C. Collier[2], Gregory M. Kobele[1], Yoosook Lee[2], Ying Lin[1], Jason Riggle[1], Yuan Yao[2], and Charles E. Taylor[2]

[1] UCLA Linguistics
Box 951543, Los Angeles, CA 90095-1542 USA
[2] UCLA Organismic Biology, Ecology and Evolution
Box 951606, Los Angeles, CA 90095-1606 USA
http://taylor0.biology.ucla.edu/al/

Abstract. This paper describes a framework for studies of the adaptive acquisition and evolution of language, with the following components: language learning begins by associating words with cognitively salient representations ("grounding"); the sentences of each language are determined by properties of lexical items, and so only these need to be transmitted by learning; the learnable languages allow multiple agreements, multiple crossing agreements, and reduplication, as mildly context sensitive and human languages do; infinitely many different languages are learnable; many of the learnable languages include infinitely many sentences; in each language, inferential processes can be defined over succinct representations of the derivations themselves; the languages can be extended by innovative responses to communicative demands. Preliminary analytic results and a robotic implementation are described.

1 Introduction

Human language is learned in context, apparently easily, and not only by the most highly intelligent among us. In fact, human language recognition is 'mandatory' and 'automatic' in a certain sense, like human vision: when you look at a typical visual scene, you cannot help seeing it as having 3D objects in it, objects that can remain constant through significant variations in lighting and orientation. The large array of color and light intensities is interpreted as a small array of objects arranged in 3D space. Similarly, when you hear a sentence like (1), you cannot help hearing it as a linguistic structure of a certain kind with familiar objects in it that appear in various forms.

(1) The economy, Bush said, has been improving

There is another important similarity between visual and language perception: just as we cannot help noticing complete objects when parts of them are occluded and discontinuous in the visual image. points out that there are apparently similarly discontinuous presentations of units in human language. In (1), for example, not only is the subject *the economy* separated from the predicate *has*

W. Banzhaf et al. (Eds.): ECAL 2003, LNAI 2801, pp. 525–534, 2003.

been improving by another subject and predicate, but also *have* and *-en* plausibly form a unit, and so do *be* and *-ing*. We see these same familiar elements in examples like this:

(2) The economy improves

(3) The indexes have fall-en

(4) Expectations are ris-ing

(5) The economy Bush said has be-en improv-ing

Recognizing familiar structures like this – through their discontinuous presentations – is apparently extremely easy for speakers of English, and presents no particular challenge for any normal language learner.

Notice also that while there is noise in any image (e.g., surfaces are not perfectly smooth because of rendering and print limitations), even if there were no noise at all (e.g. images given by functions that deliver arbitrarily precise results about light and color placement), there would still be a challenge in saying how a 3D-like description is computed from the 2D one. There is also noise in language perception (visual or auditory or tactile), but even in a noiseless setting, the challenge of computing the conceptual representation from the percept is non-trivial.

Are there models of language users that account for these analogies between language and non-linguistic perception? Recent linguistic and learning results provide a setting in which this and related questions can be studied. The range of patterns and discontinuities found in human languages goes beyond the power of context free formalisms, but not by much: the "mildly context sensitive" (MCS) grammars seem capable of capturing them elegantly, and there are positive learning results that can be extended to them. This paper uses a particular MCS formalism called 'minimalist grammar' (MG) to support a model of language emergence and transmission among agents who already have some (perceptual, maybe even "non-cognitive") abilities for recognizing simple objects from discontinuous presentations. The model has these properties:

i. Dependencies among words can be inferred and grounded in certain settings.

ii. MG languages are specified by their lexicons: i.e. by the association of words with features that specify dependencies.
So the plasticity of language comes from the finitely many learned lexical associations only, not from variations in the basic (possibly "non-cognitive") mechanisms for recognizing composites of these elements.

iii. A certain (infinite) class of MG languages, the "rigid" MG languages, are (provably) learnable from dependency structures.

iv. Recursive language structure emerges quickly and easily in this class.

v. Each MG derivation is unambiguously specified by the sequence of lexical items in it. This provides a natural conceptual representation for reasoning. So on this view, rather than translating linguistic structures into quite different conceptual representations for reasoning, language takes the shape it does partly because that is the shape of objects we reason with.

vi. Language learning can be complete and perfect in principle (provably), but in many actual settings it will be imperfect, and these transmission errors may propagate in a population.

After briefly introducing this model, we describe a robotic realization.

1.1 Grounding

Humans learn language in context. It is natural to assume that learning begins with the identification of correspondences between linguistic tokens and cognitively salient features of the context, where the cognitively salient features will tend to be similar across conspecific participants in a common situation. Many studies have considered this problem (Niyogi, 2002; Vogt, 2000; Kirby, 1999; Steels, 1996). In the present context, we have framed the grounding problem as a learning problem, so that the methods of formal learning theory can be brought to bear. Following (Siskind, 1996), we can regard the evidence available to the learner as a sequence of paired linguistic signals and conceptual representations, where the linguistic signals come from speakers with a certain unambiguous "target" pairing between morphemes and conceptual elements ("sememes"). Even in the ideal setting where all possible elements appear in all possible utterances somewhere in the data – "positive text" – Siskind's algorithm will not always succeed in identifying the target pairing, but (Kobele et al., 2003) provides an algorithm that is guaranteed to succeed in a broad (and defined) range of conditions. Single word and short utterances can be especially valuable in the early stages, while in later stages the appearance of a new word will often correspond to the appearance of a new sememe, and the association can be confidently established by a single exposure.

Again following (Siskind, 1996), notice that once a unary sememe (a property) has been paired with a symbol P, and a constant sememe (a name) with a symbol a, it is not difficult to guess (and sometimes confirm by observation) that the utterance aP signifies that the thing associated with a by grounding has the property associated with P, and in this case we also obtain the information that the language allows the order Subject-Predicate, and information about how the meaning of this structure is related to the meanings of its parts. In the case where the agent observes two things associated with morphemes a, b in a binary relation associated with some morpheme R, the utterance aRb conveys similar information. In general, with richer assumptions about the syntax of the language, more sophisticated grounding strategies can be explored.

2 Minimalist Grammars

The syntax of the present study is inspired by linguistic proposals (Chomsky, 1995; Brody, 1995) that have aggressively stripped redundancy from linguistic structures. (Stabler and Keenan, 2003) provides a formal model of some of these proposals, and many variants have now been considered (Kobele, 2002; Michaelis, 1998; Michaelis, 2001; Frey and Gärtner, 2002;

Kobele et al., 2002; Niyogi, 2001). Most of these systems are weakly equivalent (and strongly similar) to the simplest variant, and also to many other independently proposed "mildly context sensitive" grammatical systems (Weir, 1988). These systems define an MCS strict superset of the context free languages they are efficiently parsable, and an infinite class of these languages is learnable from semantic structures of a certain kind, as described below.

A 'minimalist grammar' (MG) has two parts: (i) a finite lexicon and (ii) two fixed structure building rules. The **lexicon** is finite set of lexical items, each of which associates a string with an arbitrary finite sequence of features of one of these 4 kinds:

	(examples)
category features	N,V,A,P,C,D,T,...
selector features	=N,=V,=A,=P,=C,=D,=T,...
licensor features	+wh,+k,...
licensee features	-wh,-k,...

So for example, a grammar could have a set of 4 lexical items like this:

G0={ read::=D =D V, Hamlet::D, she::D -k, will::=V +k T }

The features of the first item in G0 indicate that it is pronounced "read," and that it must select two expressions with category D ("determiner") to form a complex (a "verb phrase") with category V. The second item "Hamlet" has category D. The third, "she," has category D and -k indicates that it must move to a "case" position. And the fourth item, "will," selects a verb phrase and triggers a case movement to form a T ("tensed") phrase. The notions of selection and movement are defined by the structure building operations.

There are two structure building operations which build trees, and these operations are constant across all grammars: *merge* and *move*. Merge is a binary partial function whose first argument has =F (for some F) as its first syntactic feature, indicating that it wants to select a constituent whose category is F, an operation defined as follows. If the first argument, the selector, is a simple lexical item, then the selected element forms a tree by attaching to the right. If the selector is already a tree, then the second argument attaches to the left:

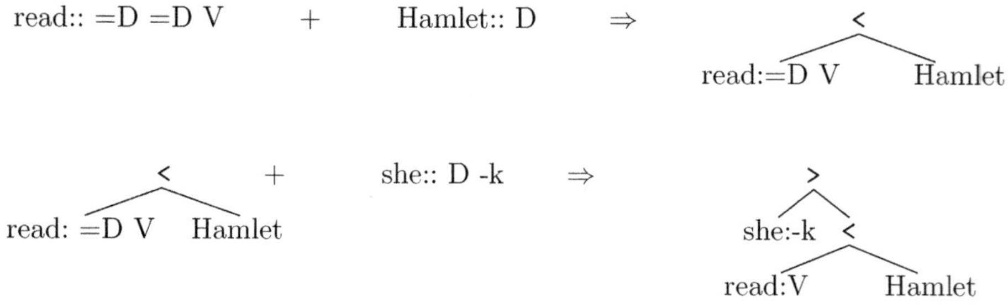

When merge applies it deletes the pair of features that trigger its action (here, =D and D), building binary tree structures like those shown, with an order symbol at each internal node that "points" towards the "head" of the complex. The features of a tree or subtree are always just the features of its head. And in any tree, any subtree is "maximal" iff no larger subtree has the same head. Notice also that the operations delete features in order, from left to right. The only features affected by merge are those on the heads of the two arguments.

The second structure building operation, move, is a partial unary function on trees. It applies to a tree whose head has a +X feature, if there is exactly one maximal subtree whose head has -X. In that case, that maximal tree with the -X head is moved up to attach at the left edge of the complex as shown. Like merge, this operation deletes a pair of features (here, +k and -k) and creates a discontinuous dependency:

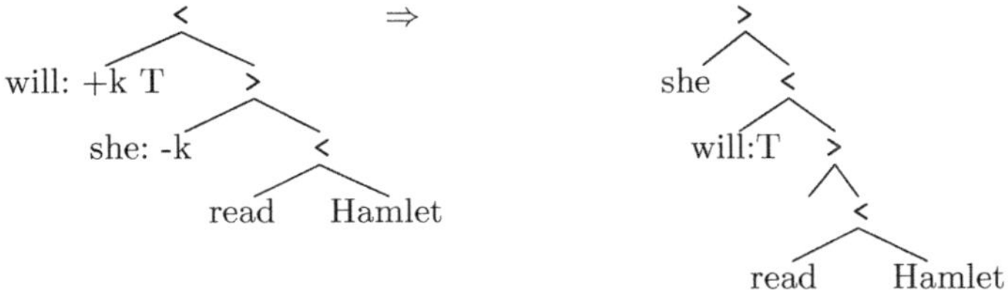

If T is the "start category" of the grammar, then the tree on the right is a completed derivation of the sentence *she will read Hamlet*, since T is the only remaining syntactic feature. Our syntactic representations are sparser than usual, but it is a simple matter to translate derivations in this formalism into the representations that are common in linguistic theory.

Since merge and move do not vary, a minimalist grammar (MG) is completely specified by its lexicon, like G0 above. A derivation from any such grammar simply involves the application of operations that check features until the only remaining feature is the "start category" T. Numbering the lexical items of G0

G0={ 1.read::=D =D V, 2.Hamlet::D, 3.she::D -k, 4.will::=V +k T }

it is easy to see that

$$\text{move}(\text{merge}(4, \text{merge}(\text{merge}(1, 2), 3)))$$

is the tree for *she will read Hamlet* shown just above. That derivation can also be completely specified by saying which feature checking relations obtain among the lexical items, as depicted by the following graph:

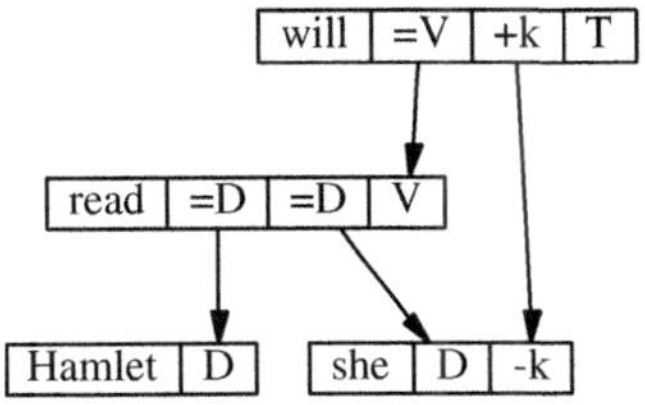

Notice that in this graph, each arc is a feature checking relation established in the derivation of the string, but not only that, each arc corresponds to a prominent semantic relation of the sort mentioned in the previous section, except for the "movement" arc that connects +k to -k, and this latter arc is directly evidenced in a change from canonical string position, moving one of the verb's arguments away from the verb. This sets the stage for a dependency-based approach to learning, where the syntactic dependencies either mirror semantic ones or are evidenced by string position. Clearly the learner does not have prior information about the identity of the syntactic features involved in the utterance, but, stripping away these features, we have a dependency structure, a "d-structure", which the learner can plausibly identify (given reasonable assumptions about the nature of the observed utterances):

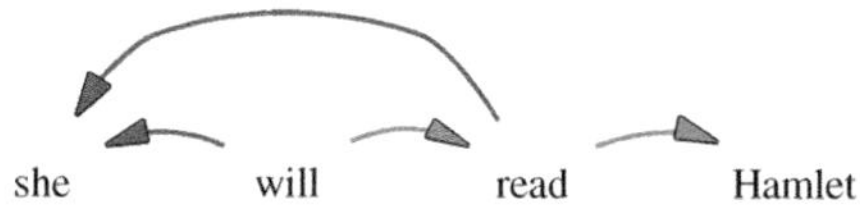

3 Learning

In the simple 'Gold paradigm' model of learning mentioned above, generous assumptions about the data and available computational resources assure that finite languages can be learned, but when the targets come from an infinite class of infinite languages, most of the results are negative. For example, no strict superset of the finite languages is learnable. However, we do not want to learn all of the finite languages, and the languages we want to learn have some structure in common. There is a tradition of positive results of this kind. (Angluin, 1982) shows that an infinite subset of the finite state languages can be learned from example sentences, in the Gold sense of perfect identification. And (Kanazawa, 1998) shows that from some basic dependency structures among the words of the language of a reduced classical categorial grammar, it is possible to identify the target grammar itself (up to alphabetic variants), if the language is *rigid* in the sense that no word has more than one syntactic category.

Kanazawa's results can be extended to MG languages that are rigid in the sense that no two different grounded morphemes (i.e. morpheme-sememe pairs) have the same syntactic roles (Retoré and Bonato, 2001; Stabler, 2002). This rigidity assumption might seem to be a very un-human-like requirement, since many words are ambiguous: *bank* is a noun and a verb. But the rigidity requirement is much less of an imposition when the lexicon is grounded. Then there are multiple elements *bank*: one grounded in financial institutions that is a noun; one grounded in river edges that is a noun; one grounded in a certain financial activity that is a verb. And as we have seen, the learner can begin grounding the language before calculating the syntactic dependencies of its elements. This is the assumption adopted here, and this assumption allows the learner to gather information about a word from multiple utterances, with a straightforward extension of the well-understood process of grammar "unification." We describe how the learner proceeds with a simple example.

The learner begins by identifying semantic relations among morphemes (described above) and shifts in the positions of semantically related elements, like the position of subject of in *she will read Hamlet*, (as indicated in the d-structure above). The d-structure is a labeled acyclic graph with a root. Checking the d-structure above, it is easy to see that the root is *will*, so we assign this root the start category T. Then we have various different arcs to other elements which we assume to be ranked in salience, or priority: internal argument<external argument<oblique arguments. Assigning arbitrary categories $A, B, C, \ldots$ to each semantic relation, labeling the end of each arc with a new category and the origin with the corresponding selection feature, and then labeling the end of each string-shifting arc with a new licensee feature and the origin with the corresponding licensor, we obtain a grammar that generates exactly the string given, with the d-structure given, and no other. In this case, the grammar is (up to alphabetic variance):

G1={ read::=A =B C, Hamlet::A, she::B -D, will::=C +D T }

Suppose that the learner now hears the sentence *she will read Hamlet and MacBeth* with the d-structure:

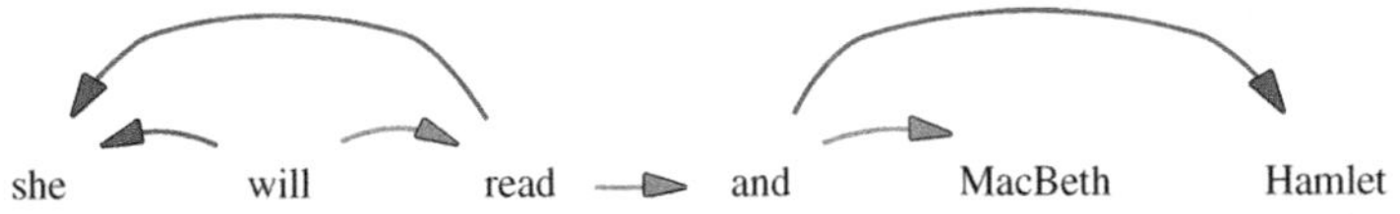

The learner computes a grammar that generates exactly this string, using all new features:

G2={ read::=E =F G, and::=H =I E, MacBeth::I Hamlet::H
she::F -J, will::=G +J T }

Now, if the common words of the two utterances are grounded to the same semantic values, the learner unifies the grammars to obtain a single rigid grammar that can generate both sentences:

G3={ read::=A =B C, and::=I =A A, MacBeth::I Hamlet::A
she::B -D, will::=C +D T }

At this point, the grammar is already recursive in the lexical entry for *and*, and generates an infinite language. This happens because the learner noticed first that the object of *read* has some category A, and then noticed that the object of *read* can be a complex that properly includes another constituent of that same category A. So in this framework, recursion comes from the assumption that two expressions with the same sound and same meaning will play the same role in each utterance, and this can happen with just a few, simple, grounded utterances. (This is a kind of "bottleneck," but it is a very narrow, semantically-based one – cf. Kirby 1999.)

4 Conceptual Representation and Language Generation

We observed above that $move(merge(4, merge(merge(1,2),3)))$ is the derivation of *she will read Hamlet* from grammar G0. It is not hard to show that the sequence of lexical items in this derivation, 4123, has at most one derivation, and this holds regardless of the grammar. (A given sequence of words, on the other hand, like *she will read Hamlet* can be multiply, even infinitely ambiguous, as in context free languages.) In fact, the sequence 4123 is a Polish notation logic in which the functors of specific arity always precede their arguments. In standard propositional logic, the ambiguous $\neg p \vee q$ corresponds to two unambiguous Polish expressions: $\vee \neg pq$ and $\neg \vee pq$. In the same way, the sequence of lexical items used in a MG derivation, in order, is always unambiguous, compact notation suitable for conceptual representation and inference. Human languages seem to be designed well for monotonicity-based inference methods. rather than the long sequences of modus ponens steps which can achieve the same results, but a careful discussion is beyond the scope of this paper.

The function from lexical sequences (like 4123) to pronouncable strings (like *she will read Hamlet*) can be computed in linear time, and this provides the basis for a theory of language generation ("vocalization"): when a conceptual representation is formed of elements that are grounded to linguistic expressions, the step to the well-formed string is linear; and when some element of a conceptual representation is not associated with a linguistic element, the speaker can choose a semantically related element, or innovate a new expression for the audience to learn. We are currently extending this simple approach to allow generation to be influenced by the speaker's model of the audience.

Previous work on language generation has tackled much more difficult problems, based in part on the idea that the conceptual to linguistic map must be very highly many-to-one. We believe that this component of the problem is overestimated: by and large, different linguistic expressions correspond to different conceptual representations. Even with this assumption though, many open problems remain.

5 Robotic Realization and Future Work

The previous sections have briefly introduced the assembly of results that was given in (i-vi) on page 526. Human perceptual and conceptual abilities are quite impressive and still largely unknown, so it is appealing to study simpler artificial systems in an environment that is well understood. We are studying several robotic environments with roughly the configuration diagrammed below. In one study, the "external events" being monitored were network communications, which can be described in a simple English-like language. An early version of this system was reported in (Wee et al., 2001), and the basic linguistic components of the current system are publically available from the project web page (given with the addresses on the first page). In a networked environment now under study, different machines describe their network communications to each other,

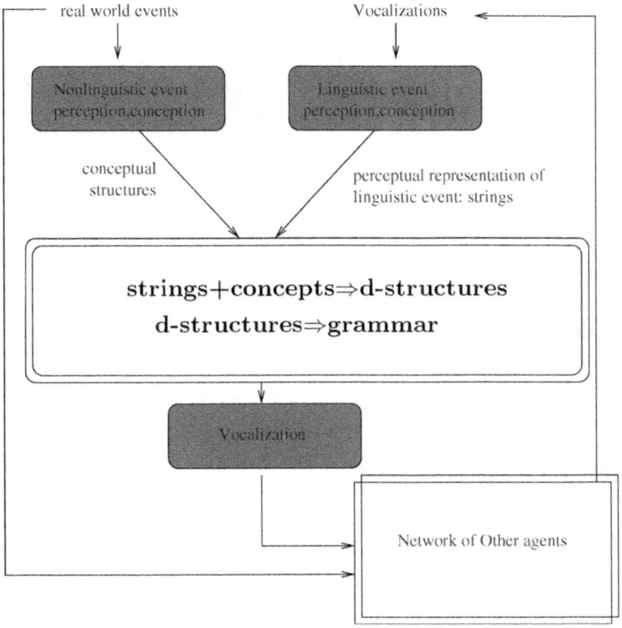

and since many of these will be common, the machines can, in certain settings, successfully ground their utterances. In this setting, we are exploring several kinds of language innovation: shortening common expressions ('hypocoristics'), and innovation of new morphemes. The learning model extends immediately to guarantee that other members of the language community will be able to learn these new elements, but the dynamics of these interactions in real settings have not yet been studied. We are also studying extensions of the system to more challenging perceptual environments with more severe grounding problems.

This work was supported by the UCLA Center for Embedded Network Sensors, the Defense Advance Research Projects Agency (DARPA), administered by the Army Research Office under Emergent Surveillance Plexus MURI Award No. DAAD19-01-1-0504, and DARPA MURI award administered by the US Airforce No. F49620-01-1-0361. Any opinions, findings, and conclusions or recommendations expressed in this publication are those of the authors and do not necessarily reflect the views of the sponsoring agencies.

References

Angluin, Dana. 1982. Inference of reversible languages. *Journal of the Association for Computing Machinery*, 29:741–765.

Brody, Michael. 1995. *Lexico-Logical Form: A Radically Minimalist Theory.* MIT Press, Cambridge, Massachusetts.

Chomsky, Noam. 1956. Three models for the description of language. *IRE Transactions on Information Theory*, IT-2:113–124.

Chomsky, Noam. 1995. *The Minimalist Program*. MIT Press, Cambridge, MA.

Frey, Werner and Hans-Martin Gärtner. 2002. On the treatment of scrambling and adjunction in minimalist grammars. In *Formal Grammar'02*.

Kanazawa, Makoto. 1998. *Learnable Classes of Categorial Grammars*. CSLI Publications/FOLLI, Stanford, California.

Kirby, Simon. 1999. Learning, bottlenecks, and the evolution of recursive syntax. In E.J. Briscoe, editor, *Linguistic Evolution Through Language Acquisition: Formal and Computational Models*, Cambridge. Cambridge University Press.

Kobele, Gregory M. 2002. Formalizing mirror theory. *Grammars*, 5:177–221.

Kobele, Gregory M., Travis Collier, Charles Taylor, and Edward Stabler. 2002. Learning mirror theory. In *6th International Workshop on Tree Adjoining Grammars and Related Frameworks*.

Kobele, Gregory M., Jason Riggle, Travis Collier, Yoosook Lee, Ying Lin, Yuan Yao, Charles Taylor, and Edward Stabler. 2003. Grounding as learning. In *Language Evolution and Computation Workshop, ESSLLI'03*.

Michaelis, Jens. 1998. Derivational minimalism is mildly context-sensitive. In *Logical Aspects of Computational Linguistics, LACL'98*, NY. Springer.

Michaelis, Jens. 2001. *On Formal Properties of Minimalist Grammars*. Ph.D. thesis, Universität Potsdam. *Linguistics in Potsdam 13*.

Niyogi, Sourabh. 2001. A minimalist implementation of verb subcategorization. In *Seventh International Workshop on Parsing Technologies, IWPT'2001*.

Niyogi, Sourabh. 2002. Bayesian learning at the syntax-semantics interface. In *Procs., 24th Annual Meeting of the Cognitive Science Society*, pages 697–702.

Retoré, Christian and Roberto Bonato. 2001. Learning rigid Lambek grammars and minimalist grammars from structured sentences. In L. Popelínský and M. Nepil, editors, *Proceedings of the Third Learning Language in Logic Workshop, LLL3*, pages 23–34.

Siskind, Jeffrey Mark. 1996. A computational study of cross-situational techniques for learning word-to-meaning mappings. *Cognition*, 61:39–91.

Stabler, Edward P. 2002. Structures for learning. In *CoLogNet Lecture, ESSLLI'02*, Trento. Publication forthcoming.

Stabler, Edward P. and Edward L. Keenan. 2003. Structural similarity. *Theoretical Computer Science 293*: 345–363.

Steels, Luc. 1996. Synthesizing the origins of language and meaning using coevolution, self-organisation and level formation. In J. Hurford, C. Knight, and M. Studdert-Kennedy, editors, *Evolution of Human Language*. Edinburgh University Press, Edinburgh, pages 161–165.

Vogt, Paul. 2000. *Lexicon grounding on mobile robots*. Ph.D. thesis, Vrije Universiteit Brussel, Laboratorium voor Artificiële Intelligentie.

Wee, K., T. Collier, G. Kobele, E.P. Stabler, and C. Taylor. 2001. Natural language interface to an intrusion detection system. In *International Conference on Control, Automation and Systems, ICASE'01*.

Weir, David. 1988. *Characterizing mildly context-sensitive grammar formalisms*. Ph.D. thesis, University of Pennsylvania, Philadelphia.

THSim v3.2: The Talking Heads Simulation Tool

Paul Vogt

Induction of Linguistic Knowledge, Tilburg University
P.O. Box 90153, 5000 LE Tilburg, The Netherlands.
Language Evolution and Computation Research Unit
University of Edinburgh, UK.
paulv@ling.ed.ac.uk

Abstract. The field of language evolution and computation may benefit from using efficient and robust simulation tools that are based on widely exploited principles within the field. The tool presented in this paper is one that could fulfil such needs. The paper presents an overview of the tool – THSim v3.2 – and discusses some research questions that can be investigated with it.

1 Introduction

The field of language evolution and computation has become increasingly popular in the alife community, see [4,10] for recent overviews. Up to now, research has focused on the origins and evolution of simple symbolic communication systems (or *lexicons*), e.g., [3,7,11,13,14,16,18] and on the evolution of syntax, e.g., [1,2, 9]. The bulk of the research so far have assumed a predefined meaning space [1,2, 9,7,11,14,18]; an assumption that inevitably leads to the *symbol grounding problem* [6], which relates to the question how agents interpret symbols meaningfully. Relatively few experiments are known that tackles the symbol grounding problem in relation to lexicon formation, both in simulations [3,12,16] and robotic platforms [14,15]. The simulation tool *THSim v3.2* presented in this paper has been designed to study various aspects of grounded lexicon formation.

One of the most familiar experiments on lexicon grounding is the *Talking Heads* experiment [14]. In this experiment a population of agents developed a lexicon from scratch by engaging in language games. The agents were embodied as pan-tilt cameras looking at a white-board on which geometrical figures were pasted. The experiment was set up such that human users could interact with the experiment through the Internet by launching agents, controlling their whereabouts and altering their lexical entries. Although this was an interesting feature that showed how the system could deal with the resultant open dynamics, its disadvantage was that the experiments were uncontrolled and therefore few scientifically valid experiments could be done. Nevertheless, some interesting controlled experiments were done with the Talking Heads, see, e.g., [8,14]. However, many open questions remain that could be studied with the Talking Heads.

W. Banzhaf et al. (Eds.): ECAL 2003, LNAI 2801, pp. 535–544, 2003.

To study these questions in a cheap, efficient and open manner, a new Talking Heads simulation tool has been designed. The development of this tool was part of a Dutch project to develop on-line education programmes in Knowledge Technology and has been made accessible for students and teachers associated with the project.[1] Currently, version 3.2 is available on the web for everyone who wishes to use it.[2]

This paper presents the main the functionalities of the tool. More details are available in the tool's manual. The next section presents a detailed overview of the tool's functions. Section 3 presents some open questions that can be studied using THSim.

2 The Talking Heads Simulation Tool

The Talking Heads simulation tool (THSim) has been designed as a tool to investigate various aspects of language evolution in a controlled and robust manner. In addition, it is a tool that allows visualisation of the ongoing processes, which is useful for a researcher using the tool or for users who are interested in learning about simulating language evolution, such as (under)graduate students. Although written in Java, which makes the simulation platform independent, the tool is sufficiently fast. The software is designed in an object-oriented fashion to make changes in the program relatively straightforward. In this section, an overview of the tool is presented.

2.1 The User Interface

When THSim is started with the user-interface, a canvas is displayed on the computer screen with four different windows, see Figure 1. Clock-wise from the top-left these windows are called 'GEOM world', 'Control', 'Statistics' and 'Language games'. The 'GEOM world' [14] is the environment of the population. This environment contains 10 different geometrical figures (such as rectangles, circles, triangles and various regular and irregular polygons) of randomly generated, but distinctive colours that can be displayed on the screen. When a language game is played, a given number of objects are generated randomly and displayed on the screen. These objects constitute the context of the language game. Features relating to these objects are handed to the agents that play the language game. These features are generated such that the information they contain could be extracted from a camera image in a similar way as in the original Talking Heads experiment.

The 'Control' panel allows the user to set various parameters and to control the simulation. Some of these parameters will be discussed briefly in Section 3.

---

[1] The project is called LOK (*national education web knowledge technology*), and its web-pages – which are only available in Dutch – can be found on http://www.ou.nl/lok/.

[2] http://www.ling.ed.ac.uk/~paulv/thsim.html

Fig. 1. A screen shot of THSim with default parameter settings shown in the control panel.

The 'Statistics' window shows a graph displaying some relevant statistics of the simulation that is being played. The statistics that are shown relate to the communicative success and the discriminative success of the simulation. Communicative success measures the effectiveness of inter-agent communication, while discriminative success measures the effectiveness of the conceptualisation or meaning formation.

The 'Language games' window displays some information about each language game that is being played. This information include the agents that participate, the features relating to the topic of the language game (open boxes), the categories that the agents use to categorise the topic (coloured boxes), the way the agents name the topic and the outcome of the language game.

2.2 Basic Functionality: Language Games

THSim's functionality is based around the language game model (see, e.g., [13, 14]). A language game is played by two agents who are selected from the population. Both agents jointly attend to a scene that is generate by the GEOM world and which constitutes the context of the game. The speaker of the language game selects one object from the context as the topic and tries to produce an utterance to name this topic. Before producing an utterance, however, the speaker has to categorise the object. This is done using a variant of the *discrimination game* (see, e.g., [14,15]). After having produced an utterance, the hearer of the game tries to interpret the utterance. It first has to play one or more discrimination games to construct a meaning on which it can match the utterance. After the hearer interprets the utterance, both agents adapt their lexicons depending on certain constraints. One constraint is the type of language game that the agents play. Currently three types of language games have been implemented in THSim: the observational game, the guessing game and the selfish game. Below follows a detailed description of these three games; for more details consult [16,18]. But before explaining the language games, the discrimination game is explained.

Discrimination game. The objective of the discrimination game is for each individual agent to find a categorisation of the topic that distinguishes this topic from all other objects in the context. By playing a number of discrimination games, each agent a gradually constructs its ontology $\mathcal{O}_a$. An ontology is a set of categories $\mathcal{O}_a = \{c_0, \ldots, c_p\}$. The categories c_i are represented as prototypes $\mathbf{c}_i$, which are points in an n-dimensional meaning space. At the start of an agent's lifetime, the ontology is empty. The game basically consists of four steps: feature extraction, categorisation, discrimination and adaptation.

Feature extraction: When the agents attend to the scenery of the GEOM world, they detect a number of features $f_k \in [0,1]$ for each object in the context. Hence, each object can be described by an n-dimensional feature vector $\mathbf{f}_i = (f_1, \ldots, f_n)$, where n equals the number of features an agent extracts and which is equal to the dimension of the meaning space. Currently, six features are implemented: the Red, Green and Blue components of the RGB colour space, the position on the X-axis, on the Y-axis and the shape feature (A). Shape feature A is calculated by the function $2 \cdot (\frac{A_o}{A_{bb}} - \frac{1}{2})$, where $\frac{A_o}{A_{bb}}$ is the filling-ratio of the object's area A_o and the area of the object's bounding-box A_{bb} – which is the smallest rectangle that can be drawn around the object. This feature produces a value that indicates the shape of the object; e.g., all rectangles have values of 1.0, circles have values of 0.57 and triangles have values of 0.0.

Categorisation: In the categorisation phase, a category $c_j \in \mathcal{O}_a$ is searched for each feature vector $\mathbf{f}_i$ such that the Euclidean distance $||\mathbf{f}_i - \mathbf{c}_j||$ is smallest. This is the *1-nearest neighbourhood search* [5].

Discrimination: In this phase it is verified whether the category for the topic is distinctive from the categories relating to the other objects in the context. If no such category exists, the discrimination game fails and the ontology has to be expanded (see Adaptation). Otherwise, the discrimination game is a success and the resulting *distinctive category*, c_d, is forwarded to the production or interpretation phase of the language game.

Adaptation: At the end of the discrimination game, the ontology of the agent is adapted according to the outcome of the game. There are two possibilities:

1. **Failure:** If the game is a failure, the ontology is expanded with a new category for which the feature vector $\mathbf{f}_t$ of the topic is used as an exemplar.
2. **Success:** In case of success, the distinctive category c_d is adjusted to make it a more representative sample of the objects it categorised. There are currently three implementations of this adaptation and the default calculates the centre-of-mass of the feature vectors it represents and is defined as follows:

$$\mathbf{c}_d = \frac{U(c_d) \cdot \mathbf{c}_d + \mathbf{f}_t}{U(c_d) + 1} \tag{1}$$

In this equation $U(c_d)$ is the frequency with which category c_d was used as a distinctive category.

Three different language games. In order to investigate the impact of non-verbal social interactions on lexicon formation, three different language games have been designed: the observational game (OG), guessing game (GG) and selfish game (SG).

In a language game, a speaker agent tries to verbalise the meaning of the topic, which a hearer agent tries to interpret. Depending on their success, both agents adapt their lexicons. New lexical elements may be constructed by invention or adoption, and association scores indicating the effectiveness of elements are increased or decreased. The way the agents evaluate the effectiveness of the game and the way scores are adapted depend on the type of language game they play.

OG: The speaker informs the hearer which topic it selected before the verbal interaction. Scores are adapted following Hebbian learning.

GG: The speaker does not inform the hearer about the topic prior to the verbal communication, but verifies if the hearer guessed the right topic. The scores are then adapted similar to reinforcement learning.

SG: The hearer is not informed about the topic, nor is the success of the game evaluated. Both agents adapt their scores in a way that relates to Bayesian learning.

Each agent $a \in \mathcal{A}$ has a private lexicon $\mathcal{L}_a = \{l_0, \ldots, l_q\}$. A lexical element l_i is defined as a triplet containing a word w_i, a meaning m_i and an association score σ_i, i.e. $l_i = \langle w_i, m_i, \sigma_i \rangle$. A word is constructed from 1 to 3 consonant-vowel pairs, where the consonants and vowels are taken from a given alphabet.[3] The meanings are categories that were used in a language game at least once, which means they must have been used distinctively at least once. The association score is a real value between 0 and 1, indicating the effectiveness of the lexical element. There are two different implementations of the association scores: one is

[3] The words that are constructed can look like "pi", "wilo" and "wateve" for example.

Table 1. A score-base description of the three language games. The first column numbers the steps taken, the second column indicates for which type of game the steps are applied, and the final column describes the step.

St.	Game	Description
1.	all	Two agents are randomly selected from the population $\mathcal{A}$, one becomes the speaker S, the other becomes hearer H.
2.	all	The agents 'observe' the GEOM world where a context $C = \{o_0, \ldots, o_r\}$ of geometrical figures o_i is constructed. In addition, a focus of attention $F \subseteq C$ is established by selecting a number of objects randomly from the context. F and C are shared by both agents.
3.	all	S selects an $o_i \in F$ as the topic t_S of the language game.
4.	**OG**	S informs H what object it selected as topic.
5.	all	S plays a discrimination game to find a distinctive category c_d for the topic. If the discrimination game fails, the language game fails too and stops here. Note that discrimination games are always played relative to the context C.
6.	all	S produces an utterance $u = w_i$, where the word w_i is from a lexical element $l_i \in \mathcal{L}_S$ for which $m_i = c_d$ and $\sigma_i \geq \sigma_j$ for all other elements $l_j \in \mathcal{L}_S$ for which $m_j = c_d$. If no such element is found, a new element $l = \langle w, c_d, 0.01 \rangle$ is added to the lexicon $\mathcal{L}_S$, where w is a newly invented word. This construction occurs only with a certain *word-creation probability* pWC. If no utterance is produced, the language game fails and stops here.
7.	all	When, and if, H receives the utterance u, it plays one or more discrimination games. If it does not know the topic, as is the case in the GG and SG, it does so for every $o_i \in F$. The results of the discrimination game(s) is stored in the distinctive category set D. If $D = \emptyset$, the language game fails and stops here.
8.	all	If $D \neq \emptyset$, H tries to interpret the utterance by searching an element $l_i \in \mathcal{L}_H$ for which $w_i = u$, $m_i \in D$ and $\sigma_i \geq \sigma_j$ for other elements $l_j \in \mathcal{L}_H$ with $w_j = u$ and $m_j \in D$. The object that was categorised with m_i becomes H's topic t_H. (If H was informed about the topic as in point 4, H can only find one matching element and the above still holds.)
9.	**GG**	If H guessed a topic, it informs S which object it guessed. S verifies whether this is the same topic and provides H with *corrective feedback*. If $t_S = t_H$, the GG was successful. If H's topic was different from S's, there was a mismatch in reference and S presents H the topic.
10.		At this point the language game finishes and both agents adapt their lexicons. The adaptations differ for the three different games:
	OG	If H found a lexical element to cover u, the OG was successful and both agents increase the association score of the used element l_i by $\sigma_i = \eta \cdot \sigma_i + 1 - \eta$, where $\eta \in \langle 0, 1 \rangle$ is a learning parameter (at default $\eta = 0.9$). In addition, the association scores of competing elements l_j are laterally inhibited by $\sigma_j = \eta \cdot \sigma_j$. An element l_j is competing if $(w_j = u) \wedge (m_j \neq m_i)$ or if $(w_j \neq u) \wedge (m_j = m_i)$. The OG fails if H does not know the word in relation to the distinctive category of the topic. In that case, H adopts u and adds the element $l = \langle u, c_d, 0.01 \rangle$ to $\mathcal{L}_H$. In turn, S lowers the association score of the used element l_i by $\sigma_i = \eta \cdot \sigma_i$.

St.	Game	Description
	GG	The adaptation for the GG is basically the same as in the OG: In case of success the score of the used element is increased while competing elements are inhibited. In case of failure, S lowers the association score and, if necessary, H adopts u and associates it with the topic, which has then
10.	**GG**	been indicated by S. Note that the GG fails if $t_S \neq t_H$, which is the case when H does not know u or when it misinterprets u. In case H misinterprets u, it also decreases the association score of the used element.
	SG	In the SG no success or failure is evaluated. S increases and the association score of the used element as in the case of a successful observational game and laterally inhibits competing elements. H increases the association scores for elements that are in the focus of attention F (i.e. elements l_i for which $w_i = u$ and $m_i \in D$). If there is a meaning $m \in D$ that is not yet lexicalised, the association $l = \rangle u, m, 0.01\langle$ is added to $\mathcal{L}_H$ first. Competing elements that fall outside the scope of F are laterally inhibited as before.

referred to as *score-based*, the other as *usage-based*. The difference will be made clear shortly. At the start of an agent's lifetime $\mathcal{L}_a = \emptyset$.

Given the discrimination game and the lexicon, the *score-based* language games are implemented as outlined in Table 1.

In the *usage-based* language games the association scores σ_i are calculated by $\sigma_i = \frac{U(w \wedge m_i)}{U(w)}$, where $U(w \wedge m_i)$ is the usage frequency of the co-occurrence of word w and meaning m_i. The denominator of this equation $U(w)$ is the usage frequency of word w disregarding which meaning it was used. The demoninator may be omitted by the hearer H as it only tries to interpret one utterance. The speaker S, however, considers different words when trying to produce an utterance.

The way the usage frequencies $U(w \wedge m_i)$ are updated at the end of a language game differs for the different games. For the OG and GG, $U(w \wedge m_i)$ is increased by 1 only if the association was used in a successful game, while $U(w)$ is increased by 1 everytime the word w is used, disregarding the game's success. For the SG, the update is more complicated: S increases both $U(w \wedge m_i)$ and $U(w)$ by 1 for the produced utterance u, and H increases $U(w \wedge m_i)$ and $U(w)$ by 1 for every element with $w = u$ and $m_i \in D$. As in the score-based SG, the association between w and $m_i \in D$ is added to the lexicon first it it does not yet exists, where initially $U(w \wedge m_i) = 0$. (Note that for the SG, the equation for calculating σ_i can be reformulated in terms of the conditional probability $P(m|w)$ that, given the occurrence of w, meaning m occurs, see [18].)

Iterated learning model. One interesting aspect of studying language evolution is investigating how the language of one generation may be transmitted to the next generation. To study this aspect, the Iterated Learning Model (ILM) has been proposed [2,9]. A population in the ILM contains a group of N adult agents and a group of N learner agents. The adults have passed the stage of learners and are supposed to have mastered the language. The learners learn from the adults by interacting with them using language games. The ILM has been adapted for THSim and iterates the following two steps:

1. K language games are played.
2. The adults are removed from the population and replaced by the learners. N new agents are placed in the learners group.

As a default setting, the speaker of a language game is always selected from the adult population and the hearer from the learner group, except in the first iteration where each agent is equally likely to be selected as speaker or hearer. It is possible to vary this setting as in the simulations reported in another paper in this volume [16].

3 Research Questions

In this section, a few research questions will be discussed that can be investigated using THSim v3.2 as it is. The discussions will pose a research question, discuss briefly why the question is interesting and indicate what parameters should be set in THSim, apart from the default settings as shown in Fig. 1. When using THSim to investigate computational expensive settings, it is wise to start THSim without the user interface.

What Is the Influence of Perceptual Noise on Lexicon Formation?

In the default setting, the agents detect the features of the objects in GEOM world without noise and both agents in one game thus detect the same features. In a real world setting, such as the real Talking Heads, the perception of the world includes noise caused by physical factors such as varying lighting conditions, changes in temperature and the different locations of the agents participating in the game.

To investigate the effect of noise on the lexicon formation, it is possible to vary *pNoise* between 0 and 1. When this is done, each agent distorts the originally generated features f_i by $f_i' = f_i + \text{pNoise} \cdot (0.5 - X)$, where $X \in [0, 1]$ is a randomly generated real value.

Up to which Population Size Can the Simulations Be Scaled with or without Using an Incremental Population Growth?

As language societies are typically large and most simulations reported so far only have relatively low populations, it is interesting to investigate to what extent a population can increase without too much loss of performance.

It is possible to investigate this by changing the parameter *nAgents*, which effects the population size from the beginning of the simulation. Another realistic scenario would be to have language spread incrementally over a society. This can be investigated by setting the parameter *incrPop=true* in combination with *ILM=true* and *nIter>1*. The population then will grow incrementally in each new iteration by the parameter *growth*, until the maximum population size *maxAgents* is reached.

How Should Prototypes Move through the Meaning Space?
One property of the discrimination game is that when it is played successfully, the prototype of the distinctive category is moved. The default implementation calculates the centre-of-mass of the feature vectors for prototypes which have been used distinctively, see Eq. 1. It is interesting to investigate the effect of varying the methods with which the prototypes are moved.

THSim allows the experimenter to alter the update type by varying the parameter *uPType*. The default is 'centre-of-mass', others are 'simulated annealing', 'walk' and 'none'. In simulated annealing, the prototype $\mathbf{c}$ is moved toward the feature vector $\mathbf{f}$ by $\Delta\mathbf{c} = (\mathbf{f} - \mathbf{c}) \cdot e^{-\frac{|\mathbf{f}-\mathbf{c}|}{T}}$, where $T = 1.0$ is the initial temperature which decays with $T = 0.9 \cdot T$ after each update of $\mathbf{c}$. In walk the prototype is moved by $\Delta\mathbf{c} = \epsilon \cdot (\mathbf{f} - \mathbf{c})$, where $\epsilon = 0.01$ is a constant step-size. In future releases of THSim, it will be possible to vary the constants T and ϵ.

What Is the Effect of Varying Game Types?
As in [18], it is possible to compare the effect of applying the different language game types, while keeping the type fixed during one simulation. Additionally, it is possible to investigate the effect of varying the game type during one simulation. This is interesting as it is likely that human language users use various strategies to learn the meaning of words.

If one wishes to play either an OG, GG or SG during a simulation, one can select the game using the parameter *gameType*. If one wishes to investigate the effect of allowing agents to vary the games between OG, GG and SG in one simulation, one must set the parameter *varGames=true*. Then one can vary the probabilities with which an OG or GG is played by setting *pOG* and *pGG*. The probability with which the SG is played is automatically calculated by *pSG* = 1 − *pOG* − *pGG*. In order to investigate the added effect of learning in the SG with respect to the two other games, one can turn off the hearer's adaptation in the SG by deselecting *adaptSG*.

4 Conclusion

This paper presents an overview of the recently released simulation toolkit THSim. This toolkit, which is a simulation of the Talking Heads experiment, can be used by investigators who are interested in grounded lexicon formation. It must be stressed that every simulation starts with a population of agents that have no linguistic knowledge whatsoever, including knowledge of how to categorise their world. All this knowledge is bootstrapped during a simulation. Future releases will add other interesting functionalities, which include the formation of grammatical structures that is currently under construction [17].

Acknowledgements. This paper was written during a Visiting Research Fellowship at the Language Evolution and Computation Research Unit of the University of Edinburgh, awarded by the Royal Society of Edinburgh and the Caledonian Science Foundation. The members of LEC are thanked for their valuable comments and suggestions on this work.

References

1. J. Batali. Computational simulations of the emergence of grammar. In J. R. Hurford, M. Studdert-Kennedy, and C. Knight, editors, *Approaches to the Evolution of Language*, Cambridge, UK, 1998. Cambridge University Press.
2. H. Brighton. Compositional syntax from cultural transmission. *Artificial Life*, 8(1):25–54, 2002.
3. A. Cangelosi and D. Parisi. The emergence of "language" in an evolving population of neural networks. *Connection Science*, 10:83–93, 1998.
4. A. Cangelosi and D. Parisi, editors. *Simulating the Evolution of Language.* Springer, London, 2002.
5. T.M. Cover and P.E. Hart. Nearest neighbour pattern classification. *Institute of Electrical and Electronics Engineers Transactions on Information Theory*, 13:21–27, 1967.
6. S. Harnad. The symbol grounding problem. *Physica D*, 42:335–346, 1990.
7. J. R. Hurford. Biological evolution of the saussurean sign as a component of the language acquisition device. *Lingua*, 77,2:187–222, 1989.
8. F. Kaplan. *L'émergence d'un lexique dans une population d'agent autonomes.* PhD thesis, Laboratoire d'infromatique de Paris 6, 2000.
9. S. Kirby. Spontaneous evolution of linguistic structure: an iterated learning model of the emergence of regularity and irregularity. *IEEE Transactions on Evolutionary Computation*, 5(2):102–110, 2001.
10. S. Kirby. Natural language from artificial life. *Artificial Life*, 8(3), 2002.
11. M. Oliphant. The learning barrier: Moving from innate to learned systems of communication. *Adaptive Behavior*, 7 (3-4):371–384, 1999.
12. A. D. M. Smith. Establishing communication systems without explicit meaning transmission. In J. Kelemen and P. Sosík, editors, *Proceedings of the 6th European Conference on Artificial Life, ECAL 2001*, LNAI 2159, pages 381–390, Berlin Heidelberg, 2001. Springer-Verlag.
13. L. Steels. Emergent adaptive lexicons. In P. Maes, editor, *From Animals to Animats 4: Proceedings of the Fourth International Conference On Simulating Adaptive Behavior*, Cambridge Ma., 1996. The MIT Press.
14. L. Steels. *The Talking Heads Experiments. Volume 1. Words and Meanings.* Special pre-edition for LABORATORIUM, Antwerpen, 1999.
15. P. Vogt. Bootstrapping grounded symbols by minimal autonomous robots. *Evolution of Communication*, 4(1):89–118, 2000.
16. P. Vogt. Grounded lexicon formation without explicit meaning transfer: who's talking to who? In *Proceedings of ECAL*, 2003. this volume.
17. P. Vogt. Iterated learning and grounding: from holistic to compositional languages. In *Proceedings of the European Summer School in Logic, Language and Information*, 2003.
18. P. Vogt and H. Coumans. Investigating social interaction strategies for bootstrapping lexicon development. *Journal for Artificial Societies and Social Simulation*, 6(1), 2003. http://jasss.soc.surrey.ac.uk.

Grounded Lexicon Formation without Explicit Reference Transfer: Who's Talking to Who?

Paul Vogt

Induction of Linguistic Knowledge, Tilburg University
P.O. Box 90153, 5000 LE Tilburg, The Netherlands
Language Evolution and Computation Research Unit
University of Edinburgh, UK.
paulv@ling.ed.ac.uk

Abstract. This paper presents a first investigation regarding lexicon grounding and evolution under an iterated learning regime without an explicit transfer of reference. In the original iterated learning framework, a population contains adult speakers and learning hearers. In this paper I investigate the effects of allowing both adults and learners to take up the role of speakers and hearers with varying probabilities. The results indicate that when adults and learners can be selected as speakers and hearers, their lexicons become more similar but at the cost of reduced success in communication.

1 Introduction

In the past decade, an increasing number of alife researchers have investigated the evolution of language, see [3,5] for recent overviews. One area of this field concentrates on the emergence of simple symbolic communication systems or *lexicons*. The focus of this paper will be on *lexicon grounding*, such as previously studied in [6,7,10]. This means that the evolved lexicons acquire their meanings through their interaction with their environment.

An example of a grounded experiment where agents develop a simple communication system is the *Talking Heads* [7]. In this experiment, a population of robotic agents developed a lexicon from scratch by interacting with each other by means of *guessing games*. In a guessing games a speaker attempts to produce an utterance to name an observed object, and then the hearer tries to guess the reference of this utterance. If the hearer fails to identify the correct reference, it receives corrective feedback from the speaker, which is used to learn the meaning of words. One problem with using *corrective feedback* is that it is not often observed in children's language acquisition [1].

In other approaches toward lexicon formation, the topic of communication was given to hearers prior to the verbal communication, see, e.g. [10,12]. Recently an alternative language game has been developed independently by Smith [6] and Vogt [10]. In this approach, which I call *selfish games* (SG), agents do not give each other the reference of communication in a non-verbal manner, neither before nor after the verbal interaction. In [12] we have shown that a lexicon is learnable

W. Banzhaf et al. (Eds.): ECAL 2003, LNAI 2801, pp. 545–552, 2003.

using the selfish game model, provided that language learners can learn from more experienced speakers. This result was shown using a simulation where we used iterated learning to model a population dynamics; if no population dynamics was used, the selfish game was only learnable for populations of 2 agents [6,12]. The problem with this work was that the meanings were predefined and hence the simulations were subject to the symbol grounding problem [4].

The *Iterated Learning Model* (ILM) [2], has been designed to model a culturally transmitted evolution of language. In the ILM the population contains two groups of agents: adults and learners. The adults have passed the stage of learners and are supposed to have mastered the language, while the learners are novices who acquire the language by observing the linguistic behaviour of adults. After a given period of time the adults are replaced by the learners and novel learners are added to the population. In this model, adults take up the role of speakers in a language game, while learners play the role of hearers. In human societies, however, adults and learners both take the role of speaker and hearer. In the implementations of Brighton [2], this is not very important as the populations only contain one adult and one learner. However, when the population is larger, this has the effect that emerging languages may not converge, because the adults do not learn from each other, although they might invent new parts of the language [12].

In this paper, the selfish game and he ILM are combined to form a model of language evolution based on the idea that language evolved as a complex dynamical system [7] by means of cultural evolution [8]. I investigate some aspects of modelling language evolution and acquisition. In particular, I investigate the effects of both grounding and the way communication lines are distributed on a population's ability to form a shared lexicon. Although the model is intended to study aspects of human language evolution and acquisition, the purpose of this paper is to show the effect of these aspects on lexicon formation without concerning too much about its relations with actual human behaviour.

2 The Method

The simulations presented in this paper were all carried out with the Talking Heads simulation tool *THSim v3.0* [11].[1] THSim can be used to simulate various aspects of language evolution and is based on the Talking Heads experiment [7].

As the (usage-based) selfish game model is explained elsewhere in this volume [11], this section only gives a brief summary. At the start of the selfish game, the agents observe a context containing 4 objects. In the current implementation, the agents only extract features relating to the RGB colours of the objects, so the lexicon they develop are colour names. The colours are generated with random values between 0 and 1 for each RGB component. Thus one might also say that the agents receive an arbitrary 3 dimensional feature vector.

The speaker selects one object as the topic of the game and the hearer has to guess what object the speaker selected. The speaker forms a meaning of the topic

[1] Downloadable from http://www.ling.ed.ac.uk/~paulv/thsim.html.

by categorising the extracted feature vector of the topic. Given a meaning, the speaker produces an utterance, which the hearer tries to interpret in relation to a meaning that refers to an object in the context. If the speaker fails to produce an utterance based on its lexicon, it expands its lexicon by inventing a new word. If the hearer does not know the utterance in relation to the meanings of the context objects, the utterance is associated with each meaning it does not know the word for yet. The speaker increases the usage of the used word-meaning association, the hearer increases the usage of each word-meaning association present in the context. The usage is used as a selection criterion for production and interpretation.

The selfish game is embedded in the iterated learning model, so after a given number of selfish games, the adults are replaced by the learners and new learners enter the population. This process then repeats for a given number of iterations (or generations).

3 Results

Given the models of the selfish game and iterated learning, a number of simulations were carried out, which will be presented in this section. The first simulation investigated the emergence of a lexicon in a population that used the SGs in combination with the ILM. In this simulation the speakers were all selected from the adult group and the hearers from the learner group, except during the first iteration where each agent is equally likely to be selected as speaker and hearer. This approach during the first iteration is taken, because it was found that otherwise no shared lexicon could emerge at all. From the second iteration onward the adults do not communicate among each other, nor do the learners. This is the control condition that reflects the original description of the ILM [2] and which was also applied in our previous (ungrounded) simulations [12].

In the second set of simulations, the probability with which the speakers and hearers were selected from the adult group was varied, thus allowing more lines in the communication. The probabilities are not selected to reflect any actual distribution of communication lines in human societies, although it should be clear that adults do not only communicate with learners as the ILM proposes. The purpose of this study is to investigate what effect the various probabilities have on lexicon formation.

In all simulations presented the population size N was set to 10. For each condition, 10 simulation runs were done for 10 iterations, where each iteration involved 50,000 SGs.

In order to analyse the effectiveness of the simulation two measures were used: *communicative success* and *similarity*. Communicative success measures, after each SG, the proportion of successful SGs over the past 100 games. The (lexical) similarity measure gives the proportion of agents $n(w)$ that use the same words w, weighted by the frequencies $U(w)$ with which the agents used the words, and averaged over the whole population of N. Similarity is calculated as follows:

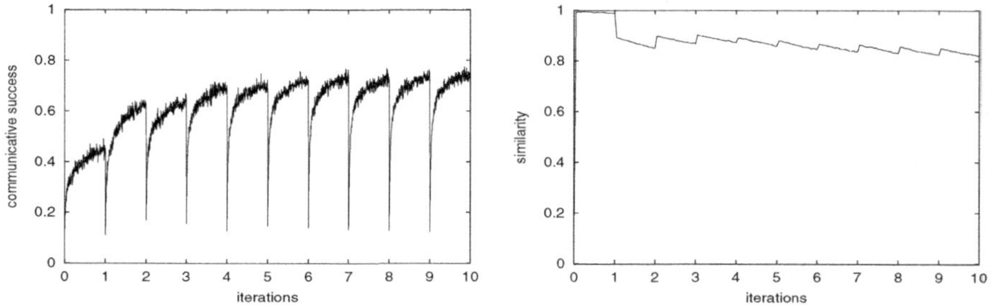

Fig. 1. The results of the simulations with only adult-learner interactions. The figures show (a) the communicative success and (b) the lexical similarity.

$$\text{sim} = \frac{1}{N} \sum_{w} \frac{n(w) \cdot U(w)}{\sum_w U(w)} \tag{1}$$

Similarity was calculated every 2500 games over the past 2500 games.

Adult-learner communication only. Figure 1 shows the results of the first simulation, averaged over 10 runs with different random seeds. Figure 1 (a) shows that the communicative success rose over iterations up to a value of around 75% in the final iteration and was still rising at the end of the iterations. It is noteworthy that the rise in success between the first and second iteration was much larger than between the other iterations. This has to do with the fact that, in the first iteration, all agents were equally likely to be selected as speaker or hearer. In between two successive iterations, the communicative success drops. This is because at those points new learners enter the population, and it takes a while before these new agents have acquired the lexicon from the new adults.

Although communicative success rose in subsequent iterations, this increase was small. More obvious is that the communicative success rose faster *within* succeeding iterations and was still rising at the end of the iterations. Because running all simulations of this paper is time consuming and running them for 50,000 games per iteration sufficiently illustrates the tendencies of the model, all simulations were run for 50,000 selfish games.

The lexical similarity revealed a different evolution, see Fig. 1 (b). This measure, which should ideally be 1.0, decreased from the second iteration onward to values slightly above 0.8. In addition, the trend within each iteration was decreasing. These findings reveal that the emerged internal lexicons (I-lexicons) of the agents was more similar in the first iteration than in following ones. This result should not be so surprising as from the second iteration onward the adults did no longer communicate with each other. As a consequence, the adults were not able to learn from each other and the learners learnt the different lexicons from these five different adults, causing the I-lexicons to diverge.

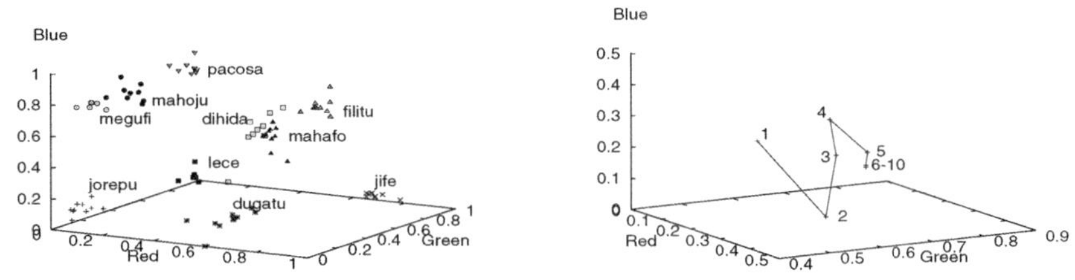

Fig. 2. Figure (a) shows a part of the lexicon in the RGB meaning space that emerged during the final iteration of one run of the simulation with only adult-learner interactions. Figure (b) shows the movement of the word "lece" over the different iterations. See the text for more details.

One of the main differences with the work presented here and that published in [12] is the lexicon grounding. Figure 2 shows some aspects of the emergent lexicons plotted in the meaning space. Fig. 2 (a) shows a part of the population's lexicons that emerged in the final iteration of one simulation run. The points in the RGB meaning space are the centre-of-mass prototypes that an agent used in association with a word in successful SGs. The prototypes shown belong to the 10 most frequently used words (the lexicons contained an average of about 110 elements). The plot shows that the prototypes used for a word are nicely clustered in the meaning space. This means that the distance between prototypes used in association with a particular word is relatively small. The plot also shows a more or less even distribution of the meaning space, which nicely corresponds to the distribution of the randomly generated colours.

Fig. 2 (b) shows the movement over the different iterations of the prototype for the word "lece" as used by the most dominant agent within an iteration. The movement starts in the first iteration around point $(0.4, 0.4, 0.3)$ and is $(0.1, 0.8, 0.1)$ from the sixth iteration onward. So after a relatively long walk through the meaning space, the meaning for "lece" becomes a rather stable point. This is because "lece" has become a word referring to the colours that are around this point, a property that is transmitted culturally to subsequent generations.

Inter-group interactions. In the second series of simulations the probability for selecting a speaker and hearer from the adult population (pS and pH) was varied. pS was varied between 0.5 and 1.0 and pH was varied between 0.0 and 0.5, both with intermediate steps of 0.1. The results of the simulations are shown in Figure 3.

The results are interesting. Fig. 3 (a) shows that when pS is 1.0 the communicative success is highest for all values of pH. When $pS = 0.9$, the communicative success is still fair when $pH \leq 0.2$, but drops significantly for higher values of pH. The results are worst for $pS = \{0.6, 0.7, 0.8\}$, while the success for

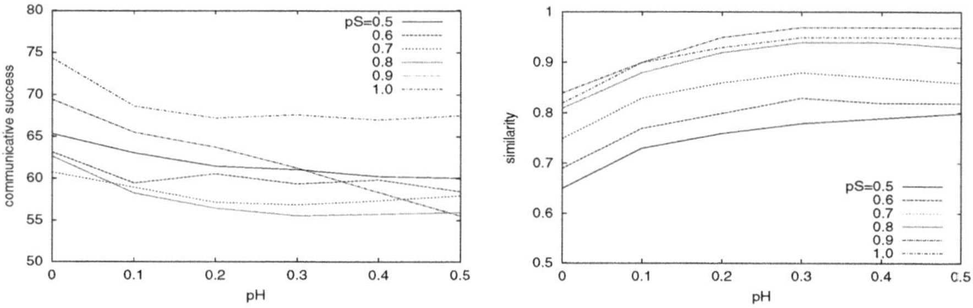

Fig. 3. The results of varying the inter-adult interaction probabilities. Figure (a) shows the communicative success and (b) the similarity both measured over the final 2500 games of the simulations. The y-axes show the different measures and the x-axes the different values for pH. The different lines in each graph relate to the different values of pS.

pS = 0.5 is higher than expected. It is also interesting to see that in all cases the communicative success more or less stabilised for pH ≥ 0.3, except for pS = 0.9.

The similarity reveals more structured results, see Fig. 3 (b). The similarity increases for higher values of pH, which is to some extent also true for increasing values of pS. The similarities for pS = 0.9 exceed the values obtained for pS = 1.0 for pH ≥ 0.2, but statistical tests revealed that this difference is not significant.

Extending each iteration to 100,000 instead of 50,000 SGs leads to an increase in communicative success of approximately 5%, while the similarity remained more or less the same. These results were obtained with four simulations for which $pS = \{0.5, 1.0\}$ and $pH = \{0.0, 0.5\}$.

4 Discussion

The simulation without inter-adult and inter-learner communication gives results that are, to a certain extent, similar to the results published in [12]. The main difference is that the level of communicative success is about 25% lower than in [12]. The reason for this lower performance is that, in the current simulation, the lexicon is grounded, whereas in [12] the meanings were predefined and shared by all agents. As a consequence, each agent develops its own ontology of meanings, which differ from each other. This, in turn, makes the learning problem in the grounded setting much harder. 75% communicative success is around the same level of success obtained for the observational games and guessing games in experiments using mobile robots [10]. Furthermore, the success is higher than the overall success of the Talking Heads experiment, which was on the average around 60% with large fluctuations [9]. This difference is partly explainable in the real world setting and partly in the uncontrolled population dynamics of the Talking Heads experiment. It is to be expected that – through the absence of corrective feedback in the SG – the Talking Heads experiment, which used guessing games, would outperform the current experiment [12]. Considering that the context size was fixed at 4, the results are significantly higher than chance, which is 25%. In addition, lexical similarity reveals that the lexicons are shared to a high degree. Furthermore, the results reveal an emergence of clustering of I-lexicons across agents. Hence, although the results are lower than those obtained in [12], the SG combined with iterated learning and grounding performs rather well.

In this paper the quality of the emerging I-lexicons was analysed using the newly designed similarity measure. The simulations show that similarity tends to decrease over different iterations when adults do not interact with each other. Starting from a shared lexicon in the first iteration – where all agents communicate with each other – the lexicons tend to diverge in following iterations. This differs from [12], where the opposite was observed. However, in [12] the distinction between adults and learners was made from the first iteration causing distinct I-lexicons between adults in the first iteration. Further research is required to explore if the opposing differences are indeed caused by the different conditions in the first iteration.

One interesting finding is that for fixed values of pS communicative success tends to decrease while similarity increases, which occurs when adults may be selected as hearers, i.e. when $pH > 0$. When adults communicate with each other, they learn from each other, causing their I-lexicons to become more similar. This similarity is in turn passed on to the learners of a population, which in the next iteration become adults themselves and pass on even more similar lexicons. The decrease in communicative success can be explained by considering two effects: First, when adults can be selected as hearers, the learners will receive less instances to learn from and consequently it takes more SGs to learn the lexicon. This is confirmed by the increase of communicative success in the extended simulations. The lack of change in similarity in these extended simulations further indicate that there is no causal relation between communicative success and similarity. Second, when learners are selected as speakers ($pS < 1.0$), they may attempt to invent parts of the lexicon themselves, which hampers the effectiveness of the communication, thus leading to a slower rise in communicative success.

5 Conclusions

In this paper the effects of grounding and varying communication lines on lexicon formation were studied using a simulation of the Talking Heads. In this simulation, the selfish game – that models communication without explicit reference transfer – and the ILM – that models cultural transmission over successive generations – were combined in a grounded setting.

The effect of grounding is that lexicon formation is harder than when the meanings are predefined; a result that is well known. In addition, the simulations show that the selfish game offers a method for constructing lexicons grounded from the population's interaction with the environment; a result that has not been shown before for populations larger than 2.

Varying the distribution with which adults and learners communicate among each other revealed that when speakers communicate more often with each other, the emerging lexicons become more similar. However, this is at the cost of a slower learning process among the learners.

Future investigations should concentrate on scaling up the system and perhaps allowing more realistic interaction strategies, which should reflect the psycholinguistic research.

Acknowledgements. The author is sponsored by a Visiting Research Fellowship awarded by the Royal Society of Edinburgh and the Caledonian Science Foundation. The author wishes to thank Andrew and Kenny Smith for the valuable comments made on earlier versions of this paper.

References

1. P. Bloom. *How Children Learn the Meanings of Words.* The MIT Press, Cambridge, MA. and London, UK., 2000.
2. H. Brighton. Compositional syntax from cultural transmission. *Artificial Life*, 8(1):25–54, 2002.
3. A. Cangelosi and D. Parisi, editors. *Simulating the Evolution of Language.* Springer, London, 2002.
4. S. Harnad. The symbol grounding problem. *Physica D*, 42:335–346, 1990.
5. S. Kirby. Natural language from artificial life. *Artificial Life*, 8(3), 2002.
6. A. D. M. Smith. Establishing communication systems without explicit meaning transmission. In J. Kelemen and P. Sosík, editors, *Proceedings of the 6th European Conference on Artificial Life, ECAL 2001*, LNAI 2159, pages 381–390, Berlin Heidelberg, 2001. Springer-Verlag.
7. L. Steels, F. Kaplan, A. McIntyre, and J. Van Looveren. Crucial factors in the origins of word-meaning. In A. Wray, editor, *The Transition to Language*, Oxford, UK, 2002. Oxford University Press.
8. M. Tomasello. *The cultural origins of human cognition.* Harvard University Press, 1999.
9. J. Van Looveren. Robotic experiments on the emergence of a lexicon. In B. Kröse, M. de Rijke, G. Schreiber, and M. van Someren, editors, *Proceedings of the 13th Belgian/Netherlands Artificial Intelligence Conference, BNAIC'01*, 2001.
10. P. Vogt. Bootstrapping grounded symbols by minimal autonomous robots. *Evolution of Communication*, 4(1):89–118, 2000.
11. P. Vogt. THSim v3.2: The Talking Heads simulation tool. In *Proceedings of ECAL*, 2003. this volume.
12. P. Vogt and H. Coumans. Investigating social interaction strategies for bootstrapping lexicon development. *Journal for Artificial Societies and Social Simulation*, 6(1), 2003. http://jasss.soc.surrey.ac.uk.

Optimal Communication in a Noisy and Heterogeneous Environment

Willem Zuidema

Language Evolution and Computation Research Unit
School of Philosophy, Psychology and Language Sciences
and Institute for Cell, Animal and Population Biology
University of Edinburgh
40 George Square, Edinburgh EH8 9LL
Scotland, United Kingdom
jelle@ling.ed.ac.uk
http://www.ling.ed.ac.uk/~jelle

Abstract. Compositionality is a fundamental property of natural language. Explaining its evolution remains a challenging problem because existing explanations require a structured language to be present before compositionality can spread in the population. In this paper, I study whether a communication system can evolve that shows the preservation of topology between meaning-space and signal-space, without assuming that individuals have any prior processing mechanism for compositionality. I present a formalism to describe a communication system where there is noise in signaling and variation in the values of meanings. In contrast to previous models, both the noise and values depend on the topology of the signal- and meaning spaces. I consider a population of agents that each try to optimize their communicative success. The results show that the preservation of topology follows naturally from the assumptions on noise, values and individual-based optimization.

1 Major Transitions in the Evolution of Language

Human languages are unique communication systems in nature because of their enormous expressiveness and flexibility. They accomplish this by using combinatorial principles in phonology, morphology and syntax [8], which impose important requirements on the cognitive abilities of language users. Explaining the origins of the structure of language and the human abilities to process it is a challenging problem for linguistics, cognitive science and evolutionary biology. Mathematical and computational models have been invaluable tools for getting a grip on this problem [11].

Jackendoff [8] has laid out a scenario for the various stages in the evolution of human language from primate-like communication, that reflects a growing consensus and can be summarized with the following "major transitions":

1. From situation-specific signals (e.g. alarm calls), to signals that are non-situation-specific but from a closed class;

W. Banzhaf et al. (Eds.): ECAL 2003, LNAI 2801, pp. 553–563, 2003.

2. From (1) to an open, unlimited (learned) class of signals and, subsequently, a phonological combinatorial system;
3. From (1) to the concatenation of signals and, subsequently, the use of ordering of signals to convey semantic relations ("compositionality");
4. From (2) and (3), which constitute the ingredients of a protolanguage, to hierarchical phrase structure and recursion,
5. From (4) to modern language, with a vocabulary for abstract semantic relations, grammatical categories, grammatical functions and a complex morphology.

Presumably, all transitions have greatly increased the number of distinct "signs" (signal–meaning pairs) that can be expressed, transmitted, memorized and learned. Jackendoff argues convincingly that modern languages contain "fossils" of each of the intermediate stages. E.g. the compound noun construction in English can be viewed as a fossil of stage (3): the meaning of words like "doghouse" and "housedog" is deducible (but not completely specified) from the meaning of the component words and the order in which they are put.

Less consensus exists on how the transition from each stage to another could have happened. Some have argued for extensive innate, language-specific cognitive specializations that have evolved under natural selection (e.g. [1,8]. This is an appealing position, in line with dominant "nativist" theories in linguistics and evolutionary biology. Unfortunately, explanations of this type have generally remained much *underspecified*. Jackendoff admits: *"I will not inquire as to the details of how increased expressive power came to spread through a population [...]. Accepted practice in evolutionary psychology [...] generally finds it convenient to ignore these problems."([8], p. 237)*

Ignoring this problem is an unfortunate tradition. Understanding how innovations can spread in a population is the essence of any evolutionary explanation, and a better end-result is neither a sufficient nor a necessary condition for the spread of innovations. Specifically in the case of language, the spread of innovations is not at all obvious, even if the end-result – when the whole population has adopted an innovation – is demonstrably better, because of two important difficulties that arise from the *frequency-dependency* of language evolution : (i) if only the hearers benefit from communication, it is not clear why speakers would evolve as to give away – altruistically – more and more information [15,21]; (ii) even if both speakers and hearers benefit, it is not clear how there can be a positive selection pressure on a linguistic innovation if that innovation appears in a population that uses a language without it, and, moreover, how that pressure can be strong enough to prevent it from being lost by drift [6,3,21].

A number of researchers have explored the possibilities of general learning and cognitive abilities and cultural evolution explaining the transitions instead (see [20,11] for reviews and references), or, of cultural evolution facilitating the genetic evolution of linguistic innovations [5,9]. These models are useful in clarifying the conditions for the "major transitions", but face some new difficulties themselves as well: (i) in many cases, the assumed cognitive abilities are much more language-specific than one would like; (ii) cultural evolution, such as the progressively better structured languages in the "Iterated Learning Model" [10, 2], only takes off when there is already some initial, random structure in the language.

Explaining the evolution of aspects of natural language like combinatorial phonology and compositionality thus remain challenging problems because both the genetic and the cultural evolution explanation require a structured language to be already present in the population before the linguistic innovations can successfully spread in a population. In this paper, I focus on compositionality: the property that the meaning of the whole (e.g. a sentence) is a function of the meaning of the parts (e.g. the words) and the way they are put together. I do not study the evolution of compositionality itself, but explore a possible route for a structured language to emerge without the capacity for compositionality present in the population. That structure is *topology preservation* between meaning-space and signal-space, i.e. similar meanings are expressed with similar signals.

In the next section I present a formalism to describe a communication system where there is noise in signaling and variation in the values of meanings. In contrast to previous models, both the noise and values depend on the topology of the signal- and meaning spaces. In section 3 I present a model of a population of agents that each try to optimize their communicative success under these circumstances. The results, in section 4, show that the preservation of topology between meaning-space and signal-space follows naturally from the assumptions on noise, values and individual-based optimization.

2 A Formalism for Communication under Noisy Conditions

Assume that there are M different meanings that an individual might want to express, and F different signals (forms) that it can use for this task. The communication system of an individual is represented with a *production matrix* $\mathbf{S}$ and an *interpretation matrix* $\mathbf{R}$. $\mathbf{S}$ gives for every meaning m and every signal f, the probability that the individual chooses f to convey m. Conversely, $\mathbf{R}$ gives for every signal f and meaning m, the probability that f will be interpreted as m. $\mathbf{S}$ is thus a $M \times F$ matrix, and $\mathbf{R}$ a $F \times M$ matrix. Variants of this notation are used by [7,14] and other researchers.

In addition, following [13], I assume that signals can be more or less similar to each other and that there is noise on the transmission of signals, which depends on these similarities. Further, I assume that meanings can be more or less similar to each other, and that the value of a certain *interpretation* depends on how close it is to the *intention*. These aspects are modeled with a *confusion matrix* $\mathbf{U}$ (of dimension $F \times F$) and a *value matrix* $\mathbf{V}$ (of dimension $M \times M$). This notation is an extension of the notation in [13], and was introduced in [22].

These four matrices together can describe the most important aspects of a communication system: which signals are used for which meanings by hearers and by speakers, how likely it is that signals get confused in the transmission, and what the consequences of a particular successful or unsuccessful interpretation are. Interestingly, they combine elegantly in one simple expression for the expected payoff w_{ij} of communication between a hearer i and a speaker j [22]:

$$w_{ij} = \mathbf{V} \cdot (\mathbf{S}^i \times (\mathbf{U} \times \mathbf{R}^j)) \tag{1}$$

In this formula, "×" represents the usual matrix multiplication and "·" represents dot-multiplication (the sum of all multiplications of corresponding elements in both matrices; the result of dot-multiplication is not a matrix, but a scalar).

A hypothetical example, loosely based on the famous Vervet monkey alarm calls [17], might make the use of this formalism and measure clear. Imagine an alarm call system of a monkey species for three different types of predators: from the air (eagles), from the ground (leopards) and from the trees (snakes). Imagine further that the monkeys are capable of producing a number (say 5) of different sounds that range on one axis (e.g. pitch, from high to low) and tahte these are more easily confused if they are closer together. Thus, the confusion matrix **U** might look like in the left matrix of figure 1.

$U =$

	received signal				
sent signal ↓	*1kHz*	*2kHz*	*3kHz*	*4kHz*	*5kHz*
1kHz	0.7	0.2	0.1	0.0	0.0
2kHz	0.2	0.6	0.2	0.0	0.0
3kHz	0.0	0.2	0.6	0.2	0.0
4kHz	0.0	0.0	0.2	0.6	0.2
5kHz	0.0	0.0	0.1	0.2	0.7

$V =$

	intentions		
interpretations ↓	*eagle*	*snake*	*leopard*
eagle	0.9	0.5	0.1
snake	0.2	0.9	0.2
leopard	0.1	0.5	0.9

Fig. 1. Confusion and value matrices for the monkeys in the example, describing the noise in signaling and the value of intention–interpretation pairs in their environment.

Further, although it is obviously best to interpret a signal correctly, if one makes a mistake, typically not every mistake is equally bad. For example, if a leopard alarm is given, the leopard response (run into a tree) is best, but a snake response (search surrounding area) is better than an eagle response (run into a bush, where leopards typically hide) [17]. Thus the value matrix **V** might look something like the right matrix in figure 1.

For any given production and interpretation matrix, we can through equation (1) calculate the expected payoff from communication. Assume a speaker i with its $\mathbf{S}^i$ as the left matrix in fig. 2, and a hearer j with its $\mathbf{R}^j$ as the right matrix in that figure. The expected payoff of the interaction between i and j

$S =$

	sent signal				
intention ↓	$1kHz$	$2kHz$	$3kHz$	$4kHz$	$5kHz$
eagle	1.0	0.0	0.0	0.0	0.0
snake	0.0	0.0	1.0	0.0	0.0
leopard	0.0	0.0	0.0	0.0	1.0

$R =$

	interpretation		
received signal ↓	*eagle*	*snake*	*leopard*
$1kHz$	1.0	0.0	0.0
$2kHz$	1.0	0.0	0.0
$3kHz$	0.0	1.0	0.0
$4kHz$	0.0	0.0	1.0
$5kHz$	0.0	0.0	1.0

Fig. 2. Production and interpretation matrices for the monkeys in the example, describing which signals they use for which situations.

if the constraints on communications are as in **U** and **V** in fig. 1 is, by proper application of equation (1): $w_{ij} = 0.7 \times 0.9 + 0.2 \times 0.5 + 0.2 \times 0.5 + 0.6 \times 0.9 + 0.2 \times 0.5 + 0.1 \times 0.5 + 0.2 \times 0.9 + 0.7 \times 0.9 = 2.33$

In this simple example, the matrices **U** and **V** are very small, and reflect only a 1-dimensional topology in both signal and meaning space. The matrices **S** and **R** are set by hand to arbitrarily chosen values. In contrast, in the simulations of this paper we will consider larger and more complex choices for **U** and **V**, and we will use a hill-climbing algorithm to find the appropriate (near-) optimal settings for **S** and **R**.

3 Distributed Hill-Climbing

Based on the measure of equation (1), I use a hill-climbing algorithm to improve the communication. To speed up the simulations, I make the simplification that the values in the **S** and **R** matrices are all either 1 or 0, i.e. they are deterministic encoders and decoders, which can be shown to always perform better than their stochastic versions [18,16]. Hill-climbing in the simulations reported here is *distributed*, i.e. I simulate a population (size 400) of agents that each try to optimize their success in communicating with a randomly selected other agent (see the author's website for details). Experiments (not reported here) suggest that distributed hill-climbing, although orders of magnitude faster, leads to very similar results as global hill-climbing. In some of the simulations, agents are placed on a grid (size 20×20) and interact only with their direct neighbors (8, except for agents at the edge which have less neighbors), but also in this condition very similar results are obtained.

The motivation for this style of optimization is (i) that it is fast and straightforward to implement; (ii) that it works well, and gives, if not the optimum, a good insight on characteristics of the optimal communication system; and (iii) that it shows possible *routes* to (near-) optimal communication systems, and in a sense forms an abstraction for both learning and evolution.

The **V** and **U** matrices can be chosen to reflect all kinds of assumptions about the signal and meaning space. In this paper I vary whether all meanings are equally valuable ($v = 1.0$, labeled "homogeneous"), or get assigned a random value ($0.0 < v \leq 1.0$, labeled "heterogeneous"). I further vary whether or not there is a topology, and if so, of which dimensionality. The diagonal elements of **V** are always v and of **U** always 1.0. Without a topology ("0d"), the off-diagonal elements in **U** or **V** are 0. With a topology, the off-diagonal elements are given by $\mathbf{V}(p,q) = v/(1 + D(p,q))$ and $\mathbf{U}(p,q) = 1/(1 + D(p,q))$, where $D(p,q)$ gives the squared Euclidean distance between the positions of the two meanings or signals i and j. In the 1-dimensional condition ("1d"), the position of a meaning or signal is simply defined as its index. In the 2d condition, the meaning and signal spaces are 2-dimensional surfaces of size $(\sqrt{M} \times \sqrt{M})$ or $(\sqrt{F} \times \sqrt{F})$. The x-coordinate is then given by the largest integer smaller than the root of the index: $x = int(\sqrt{i})$. The y-coordinate by: $y = i \; modulo \; x$. After these values are set, the rows of both **U** and **V** matrices are normalized.

I monitor the behavior of the model with two measures. The first is the average payoff, as given by equation (1), averaged over all individuals interacting with all other individuals, both as speaker and as hearer. The second is a measure for the degree of topology preservation between the meaning space and the signal space in the emerging languages. Following [2], I use the correlation ("Pearson's r") between the distance between each pair of meanings and the distance between the corresponding signals:

$$r = \underset{m,m' \in M}{correlation} \left(D\left(m, m'\right), D\left(S[m], S[m']\right) \right), \tag{2}$$

where $S[m]$ gives the most likely signal used to express m according to **S**.

For 2-dimensional meaning spaces I also visualize the topology preservation by plotting all meanings as nodes in a meaning space, and connecting those nodes where the corresponding signals are (one of maximal 4) neighbors in signal-space.

4 Results

Figure 3 shows the average payoff and topology preservation for simulations under 3 different conditions: (i) homogeneous and no topology in the meaning space ("**V**:0d"); (ii) homogeneous and **V**:1d; (iii) heterogeneous and **V**:0d. The results are plotted with a logarithmic x-axis. They show that convergence is more than 10 times faster if there is a topology in the meaning space. Recall that in the topology condition, interpretations with a meaning close to the intention are also rewarded. That fact facilitates establishing conventions regarding which signals to use for which meanings, because it creates more possibilities to break the initial symmetry (when no convention is established, every signal–meaning pair is equally good or bad).

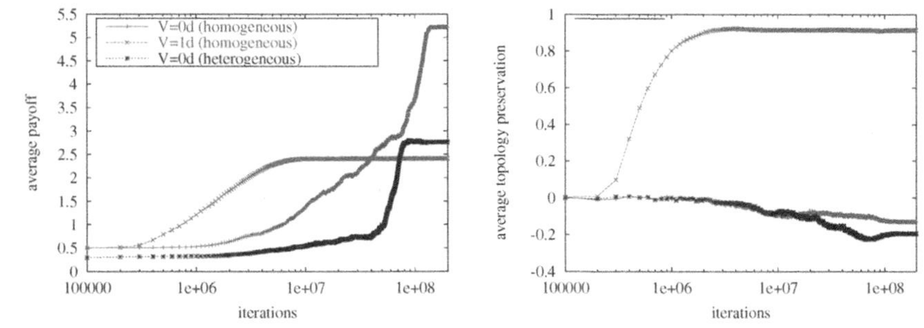

(a) Payoff ($\overline{w}$)

(b) Topology Preservation ($\overline{r}$)

Fig. 3. Average payoff (a) and degree of topology preservation (b) for 2×10^8 iterations under 3 conditions: (1) **V**:0d homogeneous, (2) **V**:1d homogeneous; (3) **V**:0d heterogeneous. The maximum average payoffs that are reached depend on the arbitrary chosen values of the **V** matrices; hence, only the shapes of the curves are important. Common parameters are P=400, M=16, F=49, **U**:1d.

Figure 4 shows the average payoff and topology preservation for 60 simulations where the dimensionality of the signal space is varied, and where hearers are selected randomly from either the whole population ("dis"), or from one of the speaker's 8 neighbors ("spatial"). In all cases, the payoff reaches high levels (when the signal space is 1d) or intermediate levels (when the signal space is 2d and the overall noise-level is consequently higher because each signal has more neighbors). Also, in all cases the topology preservation reaches high levels (when the dimensionalities of meaning and signal space match) or intermediate levels (when the dimensionalities mismatch).

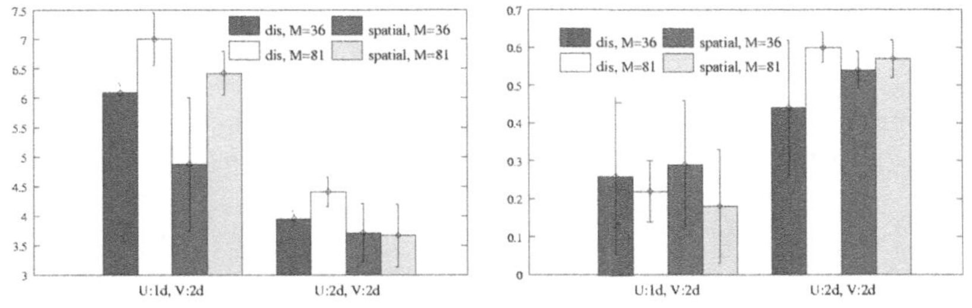

(a) Payoff (w)

(b) Topology preservation (r)

Fig. 4. Average payoff (a) and degree of topology preservation (b) after 5×10^7 iterations for different parameters. Error-bars indicate standard deviations. Common parameters are P=400, M=36 and **V**:2d heterogeneous.

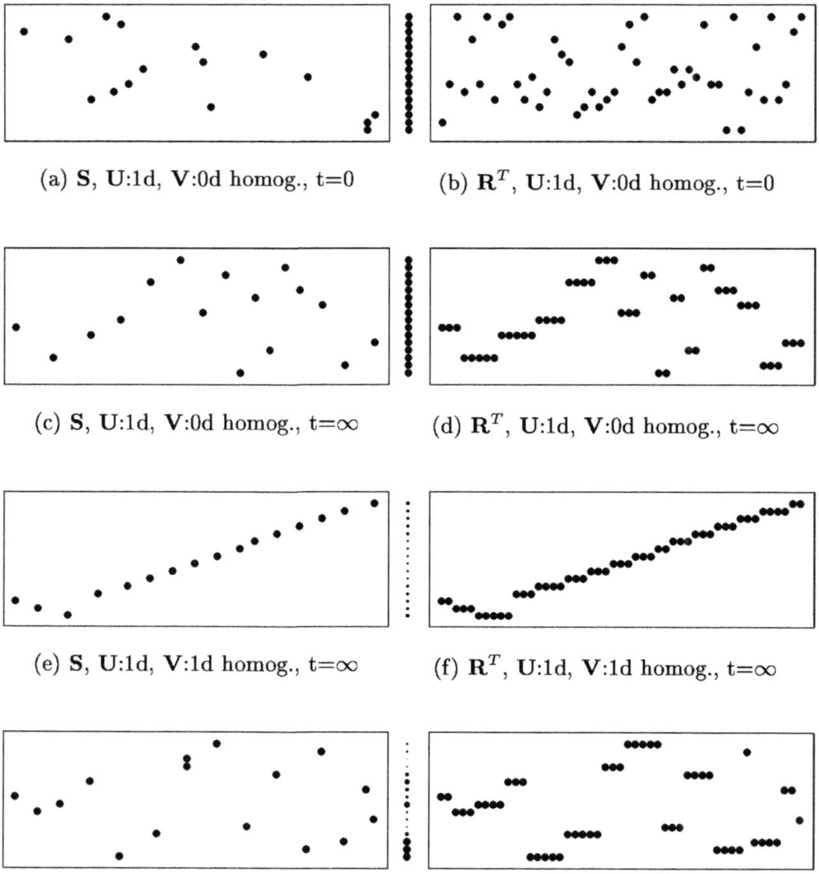

(a) **S**, **U**:1d, **V**:0d homog., t=0 (b) $\mathbf{R}^T$, **U**:1d, **V**:0d homog., t=0

(c) **S**, **U**:1d, **V**:0d homog., t=∞ (d) $\mathbf{R}^T$, **U**:1d, **V**:0d homog., t=∞

(e) **S**, **U**:1d, **V**:1d homog., t=∞ (f) $\mathbf{R}^T$, **U**:1d, **V**:1d homog., t=∞

(g) **S**, **U**:1d, **V**:0d heterog., t=∞ (h) $\mathbf{R}^T$, **U**:1d, **V**:0d heterog., t=∞

Fig. 5. (a)-(h) Examples of **S** and **R** matrices from the simulations of figure 3. For easy comparison, the **R** matrices are transposed so that in all matrices meanings differ on the vertical axis, and signals on the horizontal axis. Between the matrices the diagonal values of the **V** matrix are plotted, where the diameter of a circle corresponds to value of the corresponding meaning. Common parameters are P=400, M=16, F=49.

The emerging communication systems are visualized in fig. 5 and 6 and can be summarized with the following properties:

Specificity: every meaning has exactly one signal to express it and vice versa (i.e. no homonyms, and no real synonyms: if different signals have the same meaning they are very similar to each other).

Coherence: all agents agree on which signals to use for which meanings, and vice versa. Specificity and coherence are also found in "language game" models where there is no noise on signaling (e.g. [14,19]).

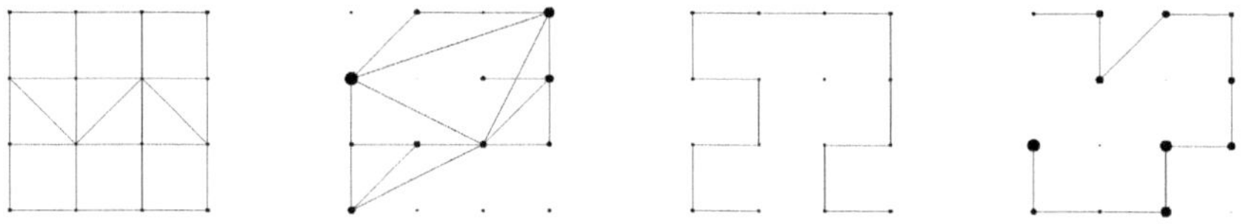

(a) **U**:2d, **V**:2d ho., $\overline{r} = 0.67$ (b) **U**:2d, **V**:2d he., $\overline{r} = 0.65$ (c) **U**:1d, **V**:2d ho., $\overline{r} = 0.46$ (d) **U**:1d, **V**:2d he., $\overline{r} = 0.07$

Fig. 6. Topology preservation at equilibrium in 4 simulations with 1d and 2d **U** matrices, and homogeneous and heterogeneous 2d **V** matrices. Nodes are meanings (diameters correspond to value), edges connect neighbors in signal space (several signals can map to a single meaning, such that nodes can have many neighbors; some meanings are not expressed, and the corresponding nodes are not connected). Common parameters are P=400, M=16, F=49.

Distinctiveness: in the **S** matrices, the used signals are maximally dissimilar to each other, so that they can be easily distinguished (compare figure 5a, at the start of the simulation, with 5c, at equilibrium). In the **R** matrices, clusters of neighboring signals all are interpreted as the same meaning. Typically, the most central signal (except at the edges) in such a cluster is the one that is actually used by the **S** matrix (compare figure 5c with 5d). Distinctiveness is also found in the "imitation game" [4], where no meanings are modeled.

Topology preservation: if there is a topology in both the meaning- and signal-space (as determined by **V** and **U**), similar signals tend to have similar meanings [22]. This preservation is not perfect (there is one major irregularity and several minor ones in the signal–meaning mapping of figure 5e and f. The topology preservation, according to equation (2), is $\overline{r} = 0.915$), but in all simulations performed it is surprisingly high. "Bad" solutions, such as the **S** and **R** of figures 5c and d ($\overline{r} = -0.073$), are stable once established in the population, but have a much smaller basin of attraction. In the case of a two-dimensional meaning space, we can draw plots like figures 6a-d, which show that the topology is almost perfectly preserved if the dimensionalities of the meaning- and signal-spaces match (6a), although it is skewed if different meanings receive very different values (6b). But even if the dimensionalities do not match, there is a strong tendency to preserve topology as well as possible (6c and d).

Valuable meanings first: When one analyzes the intermediate stages between the random initialization and the equilibrium solutions (not shown here; see author's website), it becomes clear that with a heterogeneous **V** valuable meaning-signal pairs get established first, and change little afterward.

Meanings sacrificed: Finally, when the **V** matrix is heterogeneous (figure 6b and d), or there is a dimensionality-mismatch (figure 6c and d), one can observe that meanings with very low value are sacrificed for the benefit of robust recognition of more valuable meanings (a similar observation was made in [13]).

These sacrificed meanings "deliberately" get expressed with a signal that will be interpreted with a meaning that is very close.

5 Conclusions

In this paper, I have shown that from simple assumptions about topologies in meaning- and signal-space, and individual-based optimization, communication systems can arise that show a structured mapping from meanings to signals. In a population where such a language is spoken, the fundamental new phenomenon of compositionality can presumably much more easily evolve.

There is no space here to explore the many connection between these simulations and the fields of Information Theory [16] and Evolutionary Game Theory [12]. In a sense, the matrices of figure 5 and 6 describe evolutionary stable strategies, under the constraints of communication over a noisy channel. These connections, and the analytic proofs that can be worked out in these frameworks, will be the topic of future work.

Acknowledgments. Funding from the Prins Bernhard Cultuurfonds and a Marie Curie fellowship of the European Commission is gratefully acknowledged. I thank Gert Westermann and Andy Gardner for their remarks.

References

[1] D. Bickerton. *Language and Species.* University of Chicago Press, 1990.
[2] H. Brighton. *Simplicity as a Driving Force in Linguistic Evolution.* PhD-thesis, Theoretical and Applied Linguistics, University of Edinburgh, 2003.
[3] L.L. Cavalli-Sforza and M.W. Feldman. Paradox of the evolution of communication and of social interactivity. *Proc. Nat. Acad. Sci. USA*, 80:2017–2021, 1983.
[4] B. de Boer. Self organization in vowel systems. *J. of Phonetics*, 28:441–465, 2000.
[5] T. Deacon. *Symbolic species, the co-evolution of language and the human brain.* The Penguin Press, 1997.
[6] R.A. Fisher. On the dominance ratio. *Proc Roy Soc Edin*, 42:321–431, 1922.
[7] J. Hurford. Biological evolution of the Saussurean sign as a component of the language acquisition device. *Lingua*, 77(2):187–222, 1989.
[8] R. Jackendoff. *Foundations of Language.* Oxford University Press, 2002.
[9] S. Kirby and J. Hurford. Learning, culture and evolution in the origin of linguistic constraints. In P. Husbands and I. Harvey, editors, *Proceedings 4th European Conference on Artificial Life*, pages 493–502. MIT Press, Cambridge, MA, 1997.
[10] S. Kirby. Syntax without natural selection. In C. Knight et al., editors, *The Evolutionary Emergence of Language.* Cambridge University Press, 2000.
[11] S. Kirby. Natural Language from Artificial Life. *Artificial Life*, 8(2):185–215, 2002.
[12] J. Maynard Smith. *Evolution and the Theory of Games.* Cambridge University Press, Cambridge, UK, 1982.
[13] M.A. Nowak and D.C. Krakauer. The evolution of language. *Proc. Nat. Acad. Sci. USA*, 96:8028–8033, 1999.

[14] M. Oliphant and J. Batali. Learning and the emergence of coordinated communication. *Center for research on language newsletter*, 11(1), 1996.
[15] M. Oliphant. The dilemma of Saussurean communication. *BioSystems*, 37(1-2):31–38, 1994.
[16] J.B. Plotkin and M.A. Nowak. Language evolution and information theory. *Journal of Theoretical Biology*, pages 147–159, 2000.
[17] R.M. Seyfarth and D.L. Cheney. Some general features of vocal development in nonhuman primates. In C.T. Snowdon and M. Hausberger, editors, *Social influences on vocal development*, pages 249–273. Cambridge University Press, 1997.
[18] C.E. Shannon. A mathematical theory of communication. *The Bell Systems Technical Journal*, 27:379–423 and 623–656, 1948.
[19] L. Steels, F. Kaplan, A. McIntyre, and J. Van Looveren. Crucial factors in the origins of word-meaning. In A. Wray, editor, *The Transition to Language*. Oxford University Press, Oxford, UK, 2002.
[20] L. Steels. The synthetic modeling of language origins. *Evolution of Communication*, 1:1–35, 1997.
[21] W. Zuidema and P. Hogeweg. Selective advantages of syntactic language: a model study. In Gleitman and Joshi, editors, *Proceedings of the 22nd Annual Meeting of the Cognitive Science Society*, pages 577–582. Lawrence Erlbaum, 2000.
[22] W. Zuidema and G. Westermann. Evolution of an Optimal Lexicon under Constraints from Embodiment. *Artificial Life*, 2003. (accepted).

A Clustering Algorithm Based on the Ants Self-Assembly Behavior

H. Azzag[1], N. Monmarché[1], M. Slimane[1], C. Guinot[2], and G. Venturini[1]

[1] Laboratoire d'Informatique de l'Université de Tours,
École Polytechnique de l'Université de Tours - Département Informatique
64, Avenue Jean Portalis, 37200 Tours, France.
Phone: +33 2 47 36 14 14, Fax: +33 2 47 36 14 22
hanene.azzag@etu.univ-tours.fr
{monmarche,slimane,venturini}@univ-tours.fr
http://www.antsearch.univ-tours.fr/webrtic
[2] C.E.R.I.E.S.,
20 rue Victor Noir, 92521 Neuilly sur Seine Cedex.
christiane.guinot@ceries-lab.com

Abstract. We have presented in this paper an ants based clustering algorithm which is inspired from the self-assembling behavior observed in real ants. These ants progressively become connected to an initial point called the support and then successively to other connected ants. The artificial ants that we have defined similarly build a tree where each ant represents a node/data. Ants use the similarities between the data in order to decide where to connect. We have tested our method on numerical databases (either artificial, real, and from the CE.R.I.E.S.). We show that AntTree improves the clustering process compared to the Kmeans algorithm and to AntClass, a previous approach for data clustering with artificial ants.

1 Introduction

Natural systems have evolved in order to solve many problems that can be related to the data clustering problem. For instance, different species have developed social behaviors to tackle the problem of gathering objects or individuals, like in brood sorting or cemetery organization in real ants [3]. Therefore, several studies involve ants behavior and data clustering [9] [7][4] [11]. We are interested in this paper in showing how to adapt a new biological model to this clustering problem where the data must be hierarchically organized in a tree. This model is based on the ability of ants to build live structures with their bodies [8]. Each ant represents a data and is initially placed on a fixed point, i.e. the root of the tree that we will call the support in the following. The behavior of an ant consists in moving on already connected ants and in connecting itself at a convenient location in the tree. This behavior is directed by the local structure of the tree and by the similarity between data represented by ants. When all ants are connected, the resulting tree can be interpreted as a partitioning of the data in order to solve a clustering problem [6].

W. Banzhaf et al. (Eds.): ECAL 2003, LNAI 2801, pp. 564–571, 2003.

The remainder of this paper is organized as follows: in section 2, we give an overview of the underlying biological model. In section 3, we present the details of the AntTree algorithm. Its properties and comparative results are given in section 4. In section 5, we finally draw some conclusions and describe future evolutions of this method, like its use in Web portal automatic construction.

2 Self-Assembly Behavior in Real Ants

Ants are able to build mechanical structures by a self-assembling behavior. Biologists observe for instance the formation of drops of ants [12], or the building of chains of ants [8]. These types of self-assembly behaviors have been observed with *Linepithema humiles* Argentina ants and African ants of gender *Oecophylla longinoda*. The goal of drop structures built by *L. humiles* is today still obscure. This ability has been recently experimentally demonstrated [12]: ants fix themselves by mean of their tarsus. The drop can sometimes fall down. For *Oecophylla longinoda* ants, it can be observed that two types of chains are built: on the one hand chains of ants fixed with their tarsus are used to cross an empty space, and on the other hand, chains of ants hung by their mandibles and their petioles to build their nest [8]. In both cases, these structures disaggregate after a given time.

Our computer model especially uses the following principles of the ants self-assembly behavior: ants start building the structure from a fixed support (stem, leaf,...) and they may move on the structure in order to become connected. All positions can be reached but for instance in the case of chains, ants preferably fix themselves at the end of the chain because they are attracted by gravity or by the object to reach. Most of the ants which are connected to the structure can not move anymore. In the case of a chain of ants, this corresponds to ants placed in the middle of the chain. A small proportion of connected ants may easily become disconnected from the structure, like for instance the ants placed at the end of a chain. Therefore one may observe a growth but also a decreasing of the structure.

In general, the motivation for using bio-inspired clustering techniques is twofold: they can avoid local minima thanks to their probabilistic behavior and they may produce high quality results without any prior knowledge of data structure (such as the number of clusters or an initial partition). In this work, in addition to these motivations, we are especially interested in showing that this new biological model may be a promising technique for achieving tree-based clustering. One should also notice that ants have been used to build specific kind of trees in [1] but with an ACO algorithm.

3 The Proposed Algorithm: AntTree

3.1 Global Principles

To obtain a partitioning of the data[6], we build a tree where nodes represent data and where edges remain to be discovered. Each data can be described

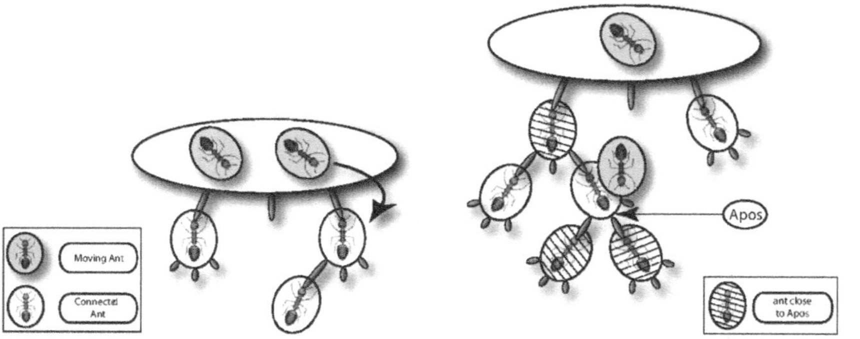

(a) (b)

Fig. 1. Representing connected and moving ants (a), and an ant's neighborhood (b)

by any representation language provided that there exists a similarity measure $\text{Sim}(i, j)$ between two data (d_i, d_j), $i \in [1, N]$, $j \in [1, N]$. This function must return values in $[0, 1]$ where 0 means that d_i and d_j are totally different and 1 means that they are identical. The main principles of our algorithm called AntTree are the following (see figure 1(a)): the root of the tree (the support) is represented by a node a_0. Ants gradually become connected to this initial node, and then successively to the ants fixed to this node, and so on until all ants are attached to the structure (AntTree stopping criterion). Ants move over the other connected ants and decide where to connect to the structure according to the value returned by $\text{Sim}(i, j)$ and according to the local neighborhood of the moving ants. We consider that each ant $a_i, i \in [1, N]$ has one outgoing link and no more than $L_{\max}$ incoming links (the structure is a tree with maximum degree of $L_{\max}$). We define for each ant a_i a similarity threshold $T_{\text{Sim}}(a_i)$ and a dissimilarity threshold $T_{\text{Dissim}}(a_i)$ that will be locally updated by a_i. These thresholds will be used to determine whether the data d_i represented by a_i is sufficiently similar or sufficiently dissimilar with an other data represented by an other ant.

During the building of the structure, each ant a_i will be either moving on the tree or will be connected to the tree. In the first case, we denote by a_{pos} the ant (or the support) over which a_i is located (and moving). a_i is completely free to move on the support or toward another ant within the neighborhood of a_{pos}. This neighborhood is defined by all ants connected to a_{pos}, considering that edges are undirected (see figure 1(b)). At each step, an ant a_i is selected in a sorted list of ants (we will explain how this list is sorted in section 4.3) and will connect itself or move according to the similarity with its neighborhood.

3.2 Ants Local Behavior

The first ant is straight connected to the support a_0. Next, for each ant a_i, two cases need to be considered. The first case is when a_i is on the support. Let a^+ denotes the ant which is the most similar to a_i among the ants already connected to the support. If a_i is similar enough to a^+ according to its similarity threshold (i.e. $\text{Sim}(a_i, a^+) \geq T_{\text{Sim}}(a_i)$), then a_i is moved toward a^+ in order to be possibly clustered in the same sub-tree, i.e. the same cluster. Else (i.e. a_i is not similar enough to a^+), if a_i is dissimilar enough to a^+ (i.e. $\text{Sim}(a_i, a^+) < T_{\text{Dissim}}(a_i)$), then it is connected to the support. This means that we create a new sub-tree, where ants will be as much dissimilar as possible to the ants in the other sub-trees connected to a_0. If no incoming links are available for a_0, then a_i is moved toward a^+. Finally, if a_i is not similar or dissimilar enough to a^+, we update its thresholds with $T_{\text{Sim}}(a_i) \leftarrow T_{\text{Sim}}(a_i) * 0.9$ and $T_{\text{Dissim}}(a_i) \leftarrow T_{\text{Dissim}}(a_i) + 0.01$. a_i becomes more tolerant and increases its probability to be connected the next time it will be considered (in the meantime, other ants will have changed the tree). We have experimentally chosen the values 0.9 and 0.01 because they provide good results. The similarity threshold must be decreased with a higher rate than the dissimilarity threshold is increased because of the distribution of similarities.

The second case is when a_i is on an other ant denoted by a_{pos} (a^+ also denotes the ant which is the most similar to a_i among the ants connected to a_{pos}). If there is a free incoming link for a_{pos} and if a_i is similar enough to a_{pos} (i.e. $\text{Sim}(a_i, a_{pos}) \geq T_{\text{Sim}}(a_i)$) and dissimilar enough to ants connected to a_{pos} (i.e. $\text{Sim}(a_i, a^+) < T_{\text{Dissim}}(a_i)$), then a_i is connected to a_{pos}. In this case, a_i represents the root of a new sub-tree/sub-cluster below a_{pos}. Its dissimilarity with other ants directly connected to a_{pos} is such that sub-clusters of a_{pos} will be well "separated" from each others (while being similar to a_{pos}). Else, a_i is randomly moved toward a neighbor node of a_{pos} and its thresholds are updated in the same way as in the previous case. So, a_i will move in the tree to find a better location where to be connected. The algorithm ends when all ants are connected.

4 Results

4.1 Testing Methodology

We have used databases in which data are represented with numerical attributes and one class label. Some databases are artificial and have been generated with gaussian and uniform laws (Art1,..., Art8). Other databases (Iris, Wine, Glass, Pima, Soybean and Thyroid) are taken from the the Machine Learning Repository [2] and correspond to standard benchmarks. Finally, we have used the data from the CE.R.I.E.S. that concern the healthy human skin domain [5]. For all these data, there exits a class label which is not given to the clustering methods but which we use for evaluation purposes. This is done by comparing the real partitioning of the data with the obtained one. We use a clustering error measure Ec which evaluates the proportion of misclassified couples of data: for all couples

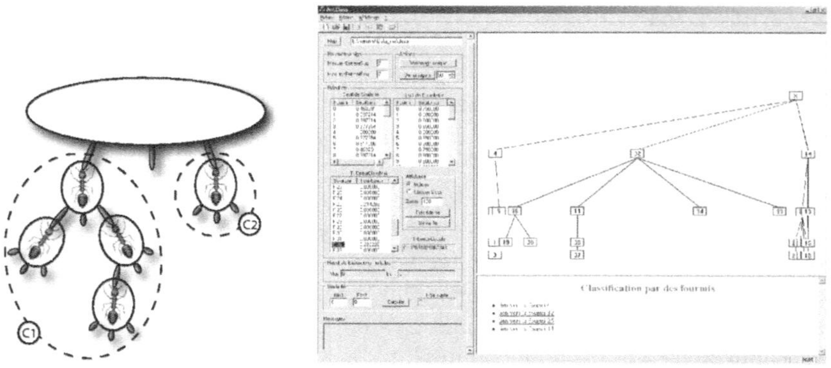

(a) (b)

Fig. 2. Interpreting the tree as "flat" clusters (a) and tree visual and interactive exploration using for instance hyperbolic zooming (b)

(d_i, d_j), increase the error Ec if d_i and d_j have been clustered in the same class (while being in different real classes) or if the two data have been separated (while they belong to the same real class). We use an Euclidean distance as a similarity measure (data are previously normalized between 0 and 1).

4.2 Interpreting the Results

Results can be interpreted in different and complementary ways. First, the obtained tree-structured partitioning can be interpreted as a "flat partitioning", with the aim for instance of comparing our approach with other "flat clustering" methods (like the Kmeans in the next section). In this case (see figure 2(a)), sub-trees directly connected to the support a_0 are interpreted as clusters. But the tree-structured organization of the data can be useful for the interpretation of the results, which is one of the advantages of hierarchical clustering over flat clustering. As shown in figure 2(b), the interface allows the user to interactively visualize the hierarchy of data using an hyperbolic display. The user may click on data and dynamically change the display. For instance, as one goes deeper into the tree, the classification error decreases as expected. The user may detect sub-structures in the data, like for instance a class which is divided into several sub-classes. This would not be possible with flat clustering.

First, we have tried to find the best strategy for data initial sorting. The initialization step of AntTree influences the results, in particular because the first ants (i.e. first data in the database) will be in general connected first to the support. Since the data that we have used are generally sorted by their class label in the databases, this may completely bias the results. So we have sorted the data in random order, but also with more clever ordering methods: for each data, one may measure its mean similarity with other data. This mean

Table 1. Results obtained by AntTree for different initial order of data, averaged over 50 runs. Ec denotes the classification error and K' the number of classes. σ_{Ec} and $\sigma_{K'}$ are the corresponding standard deviations.

Database	AntTree (decreasing order)		AntTree (increasing order)		AntTree (random order)			
	Ec [σ_{Ec}]	K' [$\sigma_{K'}$]	Ec [σ_{Ec}]	K' [$\sigma_{K'}$]	Ec [σ_{Ec}]	K' [$\sigma_{K'}$]	K	N
Art1	0.75 [0.00]	1 [0.00]	0.17 [0.00]	4 [0.00]	0.44 [0.01]	2.36 [0.08]	4	400
Art2	0.50 [0.00]	1 [0.00]	0.18 [0.00]	3 [0.00]	0.27 [0.01]	1.94 [0.03]	2	1000
Art3	0.58 [0.00]	1 [0.00]	0.21 [0.00]	3 [0.00]	0.37 [0.01]	2.26 [0.08]	4	1100
Art4	0.43 [0.00]	3 [0.00]	0.25 [0.00]	4 [0.00]	0.27 [0.00]	3.80 [0.05]	2	200
Art5	0.36 [0.00]	2 [0.00]	0.20 [0.00]	4 [0.00]	0.36 [0.02]	3.36 [0.09]	9	900
Art6	0.53 [0.00]	1 [0.00]	0.34 [0.00]	2 [0.00]	0.53 [0.02]	1.68 [0.06]	4	400
Art7	0.54 [0.00]	4 [0.00]	0.72 [0.00]	4 [0.00]	0.66 [0.00]	3.68 [0.06]	1	100
Art8	0.61 [0.00]	5 [0.00]	0.76 [0.00]	5 [0.00]	0.66 [0.01]	3.74 [0.06]	1	1000
Iris	0.67 [0.00]	1 [0.00]	0.24 [0.00]	4 [0.00]	0.27 [0.01]	2.36 [0.08]	3	150
Wine	0.65 [0.00]	2 [0.00]	0.64 [0.00]	2 [0.00]	0.64 [0.00]	2.04 [0.04]	3	178
Glass	0.71 [0.00]	3 [0.00]	0.42 [0.00]	5 [0.00]	0.67 [0.01]	2.98 [0.07]	7	214
Pima	0.45 [0.00]	1 [0.00]	0.42 [0.00]	3 [0.00]	0.45 [0.00]	1.92 [0.11]	2	798
Soybean	0.15 [0.00]	3 [0.00]	0.08 [0.00]	4 [0.00]	0.10 [0.00]	3.90 [0.04]	4	47
Thyroid	0.38 [0.00]	3 [0.00]	0.24 [0.00]	3 [0.00]	0.36 [0.00]	2.76 [0.06]	3	215
CERIES	0.76 [0.00]	2 [0.00]	0.33 [0.00]	4 [0.00]	0.58 [0.02]	2.30 [0.11]	6	259

Table 2. Results obtained with 10-means and AntClass algorithms

database	10-Means		AntClass	
	Ec [σ_{Ec}]	K' [$\sigma_{K'}$]	Ec [σ_{Ec}]	K' [$\sigma_{K'}$]
Art1	0.18 [0.01]	8.58 [0.98]	0.15 [0.05]	4.22 [1.15]
Art2	0.38 [0.01]	8.52 [0.96]	0.41 [0.01]	12.32 [2.01]
Art3	0.31 [0.01]	8.28 [0.96]	0.35 [0.01]	14.66 [2.68]
Art4	0.32 [0.02]	6.38 [0.75]	0.29 [0.23]	1.68 [0.84]
Art5	0.08 [0.01]	8.82 [0.91]	0.08 [0.01]	11.36 [1.94]
Art6	0.10 [0.02]	8.46 [1.08]	0.11 [0.13]	3.74 [1.38]
Art7	0.87 [0.02]	7.76 [1.03]	0.17 [0.24]	1.38 [0.60]
Art8	0.88 [0.01]	8.78 [0.83]	0.92 [0.01]	13.06 [2.18]
Iris	0.18 [0.03]	7.12 [1.11]	0.19 [0.08]	3.52 [1.39]
wine	0.27 [0.01]	9.64 [0.52]	0.51 [0.11]	6.46 [2.10]
Glass	0.29 [0.02]	9.44 [0.70]	0.40 [0.06]	5.60 [2.01]
Pima	0.50 [0.01]	9.90 [0.36]	0.47 [0.02]	6.10 [1.84]
Soybean	0.13 [0.02]	8.82 [0.97]	0.54 [0.17]	1.60 [0.49]
Thyroid	0.42 [0.02]	9.56 [0.57]	0.22 [0.09]	5.84 [1.33]
CERIES	0.11 [0.01]	9.38 [0.63]	0.27 [0.15]	3.40 [1.06]

similarity can be used to sort the data either in increasing/decreasing order. With an increasing order, the first connected ants are those which are the less similar to all the others and therefore close to their cluster and far away from the others. With a decreasing order, the first ants to connect are the most similar to all the others. Thus an ant belonging to a different cluster will have more chance to be connected than in the increasing case. The results obtained are given in table 1. It is obvious that the increasing order is more interesting regarding to the clustering error. We have consequently adopted this strategy in the following (with $L_{max} = 10$: at most 10 incoming links per ant).

We have compared AntTree with other clustering algorithms: AntClass [10, 11], a clustering algorithm inspired by a colony of artificial ants, and the Kmeans algorithm initialized with 10 randomly generated initial partitions (the data used for experimentation do not contain more than 10 clusters). Table 2 shows the results obtained for the 10-means and AntClass. We can see that AntTree gives an averaged error which is lower than AntClass for Art2, Art3, Art4, Art8, pima and soybean, and almost similar for Art1, glass, thyroid. Moreover, for the majority of the databases, the number of clusters found by AntTree is closer to the number of real classes than the number found by AntClass (10 databases out of 15).

AntTree is also better than the 10-means method for Art2, Art3, Art4, Art7, Art8, pima, soybean and thyroid. Moreover, the number of classes found by AntTree is also better (14 databases out of 15) for these second results. According to the standard deviations, we can also notice that AntTree is more precise than AntClass and 10-means.

Averaged computational time for one run is between 1 ms (thyroid database) and 0.5 s (Art3 composed of 1100 data). 10-means and AntClass were programmed in C, AntTree in C++ and the tests were performed on a standard PC (PIII 700 MHz). The averaged computational times of AntTree are relatively low compared to the two other methods because when an ant is connected to the tree, it will not move anymore. AntTree benefits from the fact that tree-based method often have a low complexity (like the heap sorting algorithm for instance).

5 Conclusion

In this paper we have described a new algorithm which is directly inspired from the ants self-assembly behavior. We have shown how it can be applied to the unsupervised learning problem and how it can obtain a tree-structured organization of the data. This tree can be either visualized in a hierarchical way or it can be interpreted as a flat partition. This method has been successfully compared with the Kmeans and AntClass, both in terms of clustering errors, number of classes and computational time. Those results are extremely encouraging and the main perspective of this work is to keep on studying this promising model.

More precisely, future work will deal with the unhooking capacity of ants (like drops of ants for the *L humiles* and the disaggregate of chains for *Oecophylla longinoda*). Each ant will have the possibility to disconnect itself from its position and to move on other ants that may be more similar. We are also interested in a probabilistic approach of our algorithm: for each ant we associate a probability of connecting which depends on its similarity with its close ants. We also want to automatize the initialization and the updating of the thresholds. Finally we plan to apply this algorithm to the automatic hierarchical clustering of documents (Web pages) for the generation of portal sites.

References

1. Shin Ando and Hitoshi Iba. Ant algorithm for construction of evolutionary tree. In W. B. Langdon, editor, *GECCO 2002: Proceedings of the Genetic and Evolutionary Computation Conference*, page 131, New York, 9–13 July 2002. Morgan Kaufmann Publishers.
2. C.L. Blake and C.J. Merz. UCI repository of machine learning databases, 1998.
3. N-R Franks and A Sendova-Franks. Brood sorting by ants: distributing the workload over the work surface. *Behav. Ecol. Sociobiol*, 30:109–123, 1992.
4. S. Goss and J.-L. Deneubourg. Harvesting by a group of robots. In Varela, editor, *Proceedings of the First European Conference on Artificial Life*, pages 195–204, Sydney, Australia, 1991. Toward a Practice of Autonomous Systems.
5. C. Guinot, D. J.-M. Malvy, F. Morizot, M. Tenenhaus, J. Latreille, S. Lopez, E. Tschachler, and L. Dubertret. Classification of healthy human facial skin. Textbook of Cosmetic Dermatology Third edition (to appear), 2003.
6. A-K Jain and R-C Dubes. *Algorithms for Clustering Data.* Prentice Hall Advanced Reference Series, 1988.
7. P. Kuntz, D. Snyers, and P. Layzell. A stochastic heuristic for visualising graph clusters in a bi-dimensional space prior to partitioning. *Journal of Heuritics*, 5(3), October 1999.
8. Arnaud Lioni, Christian Sauwens, Guy Theraulaz, and J-L Deneubourg. The dynamics of chain formation in oecophylla longinoda. *Journal of Insect Behavior*, 14:679–696, 2001.
9. E.D. Lumer and B. Faieta. Diversity and adaptation in populations of clustering ants. pages 501–508, 1994.
10. N. Monmarché. On data clustering with artificial ants. In A.A. Freitas, editor, *AAAI-99 & GECCO-99 Workshop on Data Mining with Evolutionary Algorithms: Research Directions*, pages 23–26, Orlando, Florida, July 18 1999.
11. N. Monmarché, M. Slimane, and G. Venturini. On improving clustering in numerical databases with artificial ants. In D. Floreano, J.D. Nicoud, and F. Mondala, editors, *5th European Conference on Artificial Life (ECAL'99), Lecture Notes in Artificial Intelligence*, volume 1674, pages 626–635, Swiss Federal Institute of Technology, Lausanne, Switzerland, 13–17 September 1999. Springer-Verlag.
12. Guy Theraulaz, E Bonabeau, Christian Sauwens, J-L Deneubourg, Arnaud Lioni, F Libert, L Passera, and R-V Sol . Model of droplet formation and dynamics in the argentine ant (linepithema humile mayr). *Bulletin of Mathematical Biology*, 63:1079–1093, 2001.

A Multi-agent Based Approach to Modelling and Rendering of 3D Tree Bark Textures

Ban Tao, Zhang Changshui, and Shu Wei

Tsinghua University, Beijing, 100084, P.R. China
State Key Laboratory of Intelligent Technology and Systems
bantao00@hotmail.com

Abstract. Multi-Agent System (MAS) has been a wide used and effective method to solve distributed AI problems. In this paper, we simplify the biological mechanism in tree bark growth and build a MAS model to simulate the generation of tree barks. The epidermis of the bark serves as the environment of the MAS while splits and lenticels are modelled as split agents and lenticel agents. The environment records the geometrics formed by the interactions of the agents during the life cycles. Visualization of the geometrics can result in realistic 3D tree bark textures which can give much fidelity to computer graphics applications...

1 Introduction

The modelling of natural phenomena in computer graphics has been addressing phenomena from all kingdoms: mineral [1], animal [2], and vegetable [3]. Among these, creation and rendering of plants and ecosystems have long been a heated area of research for decades. Well developed modelling methods, such as modelling methods based on L-systems [3], have led to easy creation of artificial sceneries with high fidelity in the vegetable kingdoms. Mapping realistic tree bark texture onto the trunk surface of the generated plants can further improve their fidelity. Therefore, tree bark texture synthesis is involved.

Texture synthesis methods can be divided into 2D or 3D categories according to the representations of textures [4]. Compared with 2D textures, 3D textures contain information of height variation and can give better performance when changes of viewpoints take place. Ebert introduced some algorithms to synthesis 3D textures from the procedure approach [4]. Another way to synthesize image textures is to directly simulate their physical generation processes.

In this paper, we propose a new method based on Multi-Agent System (MAS)[5] to simulate the biological mechanism of tree bark growth and render 3D tree bark textures. Here, the dynamic features of MAS are taken advantage of to illustrate the evolution of the tree bark texture. The epidermis of a tree is modelled as the environment of the MAS where split agents and lenticel agents move and interact with each other. During this process, the environment records traces of the agents, and visualization of the geometrics of the environment at successive steps can result in 3D tree bark textures with high fidelity. Thus we

W. Banzhaf et al. (Eds.): ECAL 2003, LNAI 2801, pp. 572–579, 2003.

simulate the mechanism within the physical process of tree bark generation and produce texture sequences for computer graphical applications.

The remainder of this paper is organized as follows. Section 2 details the modelling of the system and the growth algorithms. The experiments results of texture rendering are given in section 3. Conclusion is drawn in section 4.

2 Multi Agent System Modelling and the Algorithms

2.1 Biometric Mechanism of Tree Bark Growth

As a nutrient and protection organism of a tree trunk, barks serve as a critical role in the growth of a tree, and it also serves as a criterion of identification of different trees. Cork, cork cambium, and phelloderm together make up the periderm, an impermeable outer layer that protects the inner stem tissues if the outer tissues split as the stem girth increases with age. It thus takes over the functions of the epidermis. Over time one cork cambium will be supplanted by another generated from parenchyma cells further inside. Water and nutrient of outer periderm are cut off by the newly create one and turn it into dead bark. The first formed periderm is usually replaced after one year's growth for most kinds of trees. The outer tissues split as the stem girth increases with age. [6]

According to the biometric mechanism, we build a MAS model, where surface of the tree trunk serves as the environment and the splits and lenticels are modelled as split agents and lenticel agents moving and interacting with each other in the environment. The evolvement of the system simulates the growth.

2.2 The Environment

Surface surrounding a segment of a tree trunk is unwrapped into a quadrate which serves as the environment. Fig. 1a and fig. 1b show a region of the environment. The environment is divided into uniform mesh grids, each of which is called a unit. A unit presents an even area where the material of the bark tissue can be deemed as identical. Each unit is associated with parameters that can affect the acts of the agents in its neighborhood. Denote a unit at the ith row and the jth column as $U(i,j), i \in (1,2,\ldots,L), j \in (1,2,\ldots,W)$, where L and W are the height and width of the environment. Attributes of an arbitrary unit $U(i,j)$ are defined as follows.

Thickness. This attribute represents the thickness of the associated area in the epidermis. It is an accumulated parameter of multiple layers and it increase as the bark grows. Environmental factors like light, water, temperature may explain the variations of this parameter. Among these factors, light is the most important one. Simplifying the variance of intensity of the light, a distribution of light is shown in fig. 1c. The thickness of unit $U(i,j)$ is calculated as

$$T_{i,j} = 1 + \alpha_{i,j} + \Gamma_{i,j}, (\alpha_{i,j} > 0, \Gamma_{i,j} \in [0.05, 0.1]) \tag{1}$$

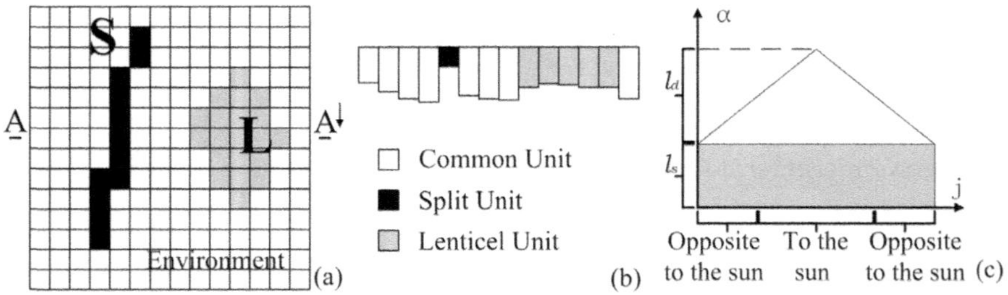

Fig. 1. (a) A local region of the MAS environment. **S** is a split agent and **L** is a lenticel agent. Other grids are common units. (b) A downward cross section at **(A-A)**. (c) Distribution of light α along a row in the environment. Light intensities in different rows may vary within a small range. Here l_d stands for the proportion of direct sun rays and l_s stands for the proportion made up of ambient light and diffuse light. j is the index of the column in the environment.

where $\alpha_{i,j}$ describes the influence of light on the thickness and $\Gamma_{i,j}$, subjected to even distribution, is a random parameter to simulate those of other factors.

Stress Intensity. The stress intensity of a unit is defined as the possibility that a split will occur at the unit. With reference to maximum stress theory [7], as the stem girth increases, the outer tissues suffer a strain perpendicular to the axis of the stem, which gives rise to splits at the frailest places. Given that the tree trunk is modelled as a regular cylinder, the stress over the bark uniform, the ultimate stress is always in direct ratio to the thickness. The stress intensity of unit $U(i,j)$ is defined as

$$I_{i,j} = C_1/T_{i,j}, \tag{2}$$

where C_1 is a constant set to 1 in this paper for computational convenience.

Age. The age of a unit $U(i,j)$, noted as $A_{i,j}$, is defined as the number of growth cycles of the outmost epidermis of the region. The age of the outmost epidermis is not level for that it includes not only the regions of unstripped epidermis but also those of inner epidermises where the outer ones are stripped off.

Categories. Each of the units in the environment falls into one of the three categories: split unit, lenticel unit, or common unit. Units passed by split agents are labelled as split units while units in lenticel agents are labelled as lenticel units. The rest are called common units.

2.3 Split Agents

The splits over the tree bark are modelled as split agents. The entity S in fig.1d gives an example of split agents. The upper and lower terminals of a split agent are defined as "head" and "tail" respectively. The prolongation of a split on the epidermis is abstracted as the extension of the terminals of the agent. A split agent has a limited eyesight, it is defined to be able to sense the status of the eight of its neighboring units and extend to one of them.

Stress Template and Direction Template. The extension of a terminal is influenced by the stress intensity in its neighborhood: the higher the stress intensity of a neighboring unit the more possible that the terminal will extend to it. Moreover, two other factors may affect their extensions: first, because of the increase in girdle, it is more possible for terminals to extend vertically than horizontally; second, a terminal tends to extend in its original direction because the tip always bears much stronger stress. We take a statistical approach to simulate the influence of the stress direction on the extension of split agents. Define a stress template as a 8-dimensional vector $P = \{p_1, p_2, ..., p_8\}^T$, where each dimension denotes the possibility that a terminal will extend to the associate direction. The subscripts here are in correspondence with the indices of the neighboring regions in fig.2a. Define a direction template to simulate the effect of the current direction of the split to its extension in the next step. The direction template is also a 8-dimensional vector noted as $Q = \{q_1, q_2, ..., q_8\}^T$, where the subscripts begin at the unit in the current direction and increase clockwise as shown in fig.2b. Denote $Q^{'} = \{q_1^{'}, q_2^{'}, ..., q_8^{'}\}^T$ as the reordered template of Q, where $q_i^{'}$ is associated with the same unit as p_i.

Stress template and reordered direction template are the main factors that define the form of the bark texture. Fig.2c and 2d give examples of the templates. For an epidermis without extraordinary distortion, the stress template and reordered direction template can be deemed as constant throughout the bark. The possibility that a terminal will extend to its kth neighbor is computed as

$$e_k = I_k \cdot p_k \cdot q_k^{'} \Big/ \sum_{l=1}^{8} I_l \cdot p_l \cdot q_l^{'}. \tag{3}$$

Active-Dormant Flag. This flag denotes the status of a terminal. "Active" means that the terminal is able to extend and "dormant" means the terminal will not prolong any more. When a split agent's head and tail are both dormant, the agent will stop acting.

Acting Rules for Split Agents. A terminal is restricted to extend to one of the 5 neighbors in or at most perpendicular to its current direction. The conjunct

(a)

p_2	p_3	p_4
p_1		p_5
p_8	p_7	p_6

(b)

q_8	q_1	q_2
q_7		q_3
q_6	q_5	q_4

(c)

0.1	0.25	0.1
0.05		0.05
0.1	0.25	0.1

(d)

0.1	0.7	0.1
0.05		0.05
0	0	0

Fig. 2. (a) Stress template in the environment. Subscript starts from the middle left neighbor of the head and increases clockwise. (b) Direction template of a head terminal. The current extending direction of the head is upward. The index of the neighbors starts from the middle up neighbor and increase clockwise. (c) An typical instance of stress template. (d) An typical reordered direction template of a head terminal.

Table 1. Algorithm Of Split Growth Cycle

Step 1	Initialization of split agents.
1.1	Evenly initialize M splits each of which only has a head and a tail;
1.2	Set up the templates and attributes of their terminals;
Step 2	Traverse all the split agents and let all of their active terminals grow one step according to the acting rule.
Step 3	For a unit passed by the split,
3.1	modify the category of the unit as a split unit; set its age to zero;
3.2	change the thickness of the unit as $T_{i,j} = T'_{i,j} \cdot (1 - I_{i,j})$, where $T'_{i,j}$ is its original thickness.
Step 4	Repeat step 2 and step 3 until all the agents are dormant.
Step 5	Stop the split growth algorithm.

Table 2. Algorithm Of Lenticel Growth Cycle

Step 1	Initialize N lenticel prototypes during environment initialization.
Step 2	For all the lenticels in the environment, carry the following operations:
2.1	traverse all of its neighbors and calculate their assimilation probabilities;
2.2	for a to be assimilated unit, label it as a lenticel unit and modify its thickness as $T_{i,j} = T'_{i,j} \cdot (1 - I_{i,j})$, where $T'_{i,j}$ is its original thickness.
Step 3	Stop the lenticel growth algorithm.

trace of head and tail forms the split on the epidermis. When a terminal meets with one of the following conditions, it stops growing and be labelled as dormant.

- Reaching the upmost or the downmost bound of the environment.
- Meeting other split agents with the same age within its neighborhood.
- Meeting a lenticel agent within its neighborhood.
- All the terminals stop with a probability of p_s, where $p_s = 0.05$ heuristically.

Algorithm Of Split Growth Cycle. The extension of splits on the bark are extracted as an absolute process called split growth cycle, when only actions of the split agent are considered. The algorithm involved is listed in Table 1.

2.4 Lenticel Agent

The lenticels on the epidermis are modelled as lenticel agents. There are a great variety of shapes for lenticels on kinds of barks. In this paper, we try to simulate the growing mechanism of typical temperate zone broad-leaved trees that are with diamond-shaped lenticels by a method base on cellular automata [8].

Acting Rules for Lenticel Agents. Because of the active and rapid mitotic division of the zone below, lenticels will expand gradually during the growth. The expansion is modelled as the assimilation of directly neighboring units under the restriction of stress distribution. A lenticel agent tends to assimilate units in

Table 3. Algorithm Of Bark Growth Cycle

Step 1	Environment initialization.
1.1	Determine the width and length of the environment;
1.2	Initialize the thickness and stress intensity of each unit based on the light distribution; set all the age attributes of the units as 0;
1.3	Initialize N lenticel agent prototypes.
Step 2	Do split growth and lenticel growth cycles according to the acting rules.
Step 3	Evolution of the environment.
3.1	Increase the width of the environment, the incremental of columns are add to the split agents.
3.2	Increase the thickness of all the units in the environment with an incremental of the thickness of a single layer.
3.3	Modify the stress intensity in the environment with respect to the thickness.
3.4	Increase the age of all the units.
Step 4	Repeat step 2 to step 3 A times and stop the growth cycle, where A is the ultimate age of the tree bark.

its 8 neighborhood. The possibility that a neighboring common unit will be assimilated is in direct ratio to its stress intensity. Note the number of all the lenticel agent's neighboring units as n, the stress intensity of the kth neighboring unit as I_k. The possibility that the kth unit will be assimilated is computed as

$$pl_k = C_2 \cdot I_k \Big/ \frac{1}{n} \sum_{i=1}^{n} I_k, \tag{4}$$

where C_2 is a constant controlling the speed of the expansion.
Algorithm Of Lenticel Growth Cycle. The expansions of lenticels on the bark are abstracted as a step called lenticel growth cycle when only expansion of lenticel agents are considered. Algorithm in the cycle is listed in Table 2.

2.5 The Tree Bark Texture Generation Algorithm

The outer layers of cells of the cork are cut off from nutrients and water that feed the plant. This outer epidermis gradually dies and scales away. As the stem increases in diameter, new splits occur. These new splits often take place within the existing splits for the stress intensity are often higher than other regions. This mechanism will lead to the gullied surface of the tree bark. To simulate this process, we abstract the evolution of the environment as life cycles, which are called bark growth cycles here. The cycles are divided into the following steps.The algorithm of the bark growth cycle is detailed in Table 3.

- Environment initialization as the very phase of primary periderm growth.
- Movements of split and lenticel agents as the corresponding phase of the formation of splits and lenticels in a growth cycle.
- Evolution of the environment as the corresponding phase of tree's growth in girth and bark's growth in thickness.
- New growth cycles of agents and the environmet.

Fig. 3. (a) Tree bark textures at different ages. (b) Tree bark textures synthesis by different parameters. (c) Texture effects in computer synthesized sceneries.

3 Post-manipulation and Rendering

During the bark growth cycle, environment records the accumulated geometrics, e.g. the thickness of the units in the environment. Visualization of the geometrics can result in a vivid structure of 3D tree bark texture. The rendering and visualization of the texture involves the following computer graphic techniques.

Visualization. Visualization of the thickness array A_{W*L} acquired from the environment, where W and L are the width and length of the environment. Here, each unit can be deemed as a vertex in the model, its abscissa i, ordinate j and thickness $T_{i,j}$ provide the 3D information.

Bump mapping. Bump mapping [9] is a technique to add more realism to synthetic images without adding a lot of geometry. It is engaged here to add per-pixel surface relief shading and increase the apparent complexity of the surface.

Erosion algorithm. To implement the effect of weathering and erosive forces, erosion algorithm [10] is adopted. Random 2×2 templates are defined to erode the common units in distinct areas respectively. Thickness of an eroded unit is modified as $T_{i,j} = T'_{i,j} \cdot (1 - I_{i,j})$, where $T'_{i,j}$ is its original thickness.

Finally, assign colors for units with different age and thickness with colors extracted from tree bark photos and render the 3D textures. Fig. 3a and 3b give some illustrative examples of the texture synthesized by our algorithm. Ultimate effects of artificial sceneries can be achieved by mapping the 3D textures onto the tree trunk surfaces. Fig. 3c gives some examples of the artificial sceneries where tree bark textures help a lot to improve the fidelity.

4 Conclusion

In this paper, a model based on MAS is acquired through analysis of the mechanism in the process of tree bark growth. Evolution of the system well simulates of the tree bark growth. Sequences of 3D textures representing tree barks of different ages can be obtained by the system. Mapping these textures onto the surfaces of tree trunks can greatly enhance the fidelity of the artificial sceneries.

The proposed model has several kinds of advantages. First of all, a sequence of synthesized tree bark textures at different ages can be used to illustrate the process of tree growth. Second, simplified mechanical model and compacted number of polygons make the algorithm and rendering doable and applicable. Finally, parameters in the model have rather obvious physical meanings.

References

[1] P. Prusinkiewicz and M. Hammel. A fractal model of mountains and rivers. In Graphics Interface 93, 174–180.
[2] Y. Chen, Y. Xu, B. Guo, and H.Y. Shum. Modeling and rendering of realistic feathers. Proc. of ACM SIGGRAPH 2002, 630–636.
[3] P. Prusinkiewicz and A. Lindenmayer. The Algorithmic Beauty of Plants. Springer-Verlag, New York, 1990.
[4] Ebert D. et al. Texturing and Modeling, A Procedural Approach, Second Edition, Cambridge 1998.
[5] G.L. Christopher. Artificial Life. The Proc. of An Interdisciplinary Workshop on the Synthesis and Simulation of Living Systems Held September, 1987
[6] Raven, H. Peter, et al. Biology of Plants, Worth Publishers, Inc., New York, 479, 1982
[7] V.L. Kolmogorov. Stresses, strains, fracture. Moscow, Metallurgy Publ., 1970.
[8] E.F. Codd, Cellular Automata. New York: Academic Press, 1968.
[9] J.F. Blinn. Simulation of Wrinkled Surfaces. Proc. of SIGGRAPH 1978, 286–292.
[10] Donald H, M.Pauline B. Computer Graphics. Prentice Hall, 417–422, 1997.

Measuring the Dynamics of Artificial Evolution[1]

Mikhail S. Burtsev

Keldysh Institute of Applied Mathematics, 4 Miusskaya sq., Moscow RU-125047, Russia
mbur@narod.ru
http://www.keldysh.ru/pages/mrbur-web/

Abstract. This paper presents results of measuring evolution in a simple ALife system. Interpretation of these results is based on the notion of dynamical systems. This approach enables the discovery of periods of high evolutionary activity, which can be treated as evolutionary transitions. Attempts were also made to locate possible cycles of trajectory in the genome phase space, and it was concluded that there were no such cycles. These results demonstrate the usefulness of a dynamical systems approach in analyzing the dynamics of artificial evolution and provide suggestions for further development.

1 Introduction

When studying artificial life models, we wish to know the key features of an evolutionary process, such as its activity and stability, diversity and complexity. Measuring these features in ALife models is often complicated due to the nature of the models and the evolutionary processes themselves.

There are no direct ways to measure the fitness of an agent, a subpopulation or a whole population, and therefore no obvious way to measure how progressive an evolutionary run was. At present, measuring adaptation and fitness is one of the main issues in evolutionary theory and artificial life.

Another problem is the genotype-phenotype mapping. The diversity and complexity of the genotype are not strictly linked to the diversity and complexity of the phenotype and its behavior. It is also difficult to link elements at the level of the genotype of an agent with the acts at the behavioral level and, finally, with the consequences for survival.

In this paper we treat a simple ALife model as a dynamical system and measure its most evident characteristics, without reference to fitness.

2 Measuring Evolution

There is no explicitly defined fitness in artificial life simulations, so adaptive fitness is estimated indirectly, using various criteria. One possible criterion is to consider larger population size as indicating higher fitness of agents in the population. But competition between agents and arms races could reduce population size while agents co-adapt and become more fit.

[1] This work was supported by the Russian Fund for Basic Research, project 02-06-80435.

W. Banzhaf et al. (Eds.): ECAL 2003, LNAI 2801, pp. 580–587, 2003.

Another indication of evolutionary activity is the speed of replication. Although a higher speed of replication for the given genotype gives it an evolutionary advantage, there is a tradeoff with the resulting disadvantages of overpopulation and overspecialization. E.g., there may be a scarcity of vital resources and decreased tolerance to environmental perturbations. Thus we must be careful of the limitations of these straightforward criteria.

Probably the most popular method for quantifying adaptive evolution in the ALife community is an evolutionary activity statistic proposed by Bedau and Packard [1,2]. The main assumption of this approach is that if a component of an evolutionary system persists, it is adaptive. One complication in applying this statistic at the genotype level is the possibility of hitchhiking. The authors suggest creating a "shadow" model to screen out neutral mutations: the shadow model is the same model, but without selection. With the aid of the shadow model, we can normalize evolutionary activities in the model and remove neutral components. However, it is not obvious for the approach, how to choose components to measure, whether they be parts of the genotype, phenotype, the agents' actions or the species. For example, if the behavior of an agent is dependant on a nonlinear interaction of genotype components, it is difficult to quantify the influence of the component on the behavior. In spite of these difficulties, the evolutionary activity statistic is a powerful method for estimating evolutionary processes, and it has been successfully applied in a number of studies [3,4].

Adami [5] proposed measuring evolution in artificial systems in terms of the complexity of an agent. In some ALife models, e.g. Avida [6], we can easily calculate the entropy per-site of an agent's genome; this measure can give some interesting insights into the dynamics of the system. Another measure based on complexity was introduced by Nehaniv [7]. This measure, called "exhibited evolvability", is proportional to the rate of increase of complexity in a population. The original measure of complexity was presented by Nehaniv and Rhodes [8] earlier.

Several additional measures were developed by Cliff and Miller [9] to evaluate adaptive progress in cases in which fitness interdependence of co-evolving species changes the fitness landscape itself over time. E.g., the evolutionary arms races (the "Red Queen effect") mentioned in the beginning of this section.

A lot of work has been done in the area of GA [10-15] to analyze the solution space as a dynamical system, but little has been done in the field of artificial life. It is difficult to find an appropriate measure of fitness in the latter case because fitness is not defined explicitly and the models are quite complex.

In this paper I use a dynamical systems approach to quantify some aspects of the evolutionary dynamics of a system, without reference to the notion of fitness.

If we consider the population in an artificial life simulation as a dynamical system we could try to answer the following questions:

1. How is the speed of the system movement in the phase space changing?
2. How stable is the trajectory of the system?
3. Does the system persist in an attractor state, or undergo transition?
4. If the system persists in an attractor state, is the attractor a stable point or a limit cycle?
5. How deep and how neutral is the basin of attraction?
6. What are the control parameters of the system? Which parameters have the greatest effect on the system dynamics?

Questions 1–4 are related to evolutionary transitions and stasis, questions 5–6 to neutrality and possibilities for diversification. This paper does not answer all these questions, but rather describes an approach that enables measurement of evolutionary activity of an Alife model as a dynamical system. The following section provides a description of the ALife model, and the subsequent section presents the results for a particular simulation.

3 The Model

The model belongs to a classic set of ALife models [16–19] with simple agents in a simple world. The model was developed to study the evolution of kin selection, but is used here as a testbed.

The world in the model is a two dimensional grid which is closed to form a torus divided into knots on the grid. The world contains two kinds of objects: agents and grass, where grass is an energy resource for agents. The number of agents in any knot is unlimited, but at most one patch of grass can exist in any knot at a given point in time. Patches of grass appear randomly and are uniformly distributed on the torus.

Each agent observes part of the local environment and performs certain actions. Specifically, each agent is oriented in space and has a field of vision. The field of vision consists of four knots: the knot the agent is in, the one in front of the agent, and the ones on the left and on the right.

Agents live in discrete time. Each agent executes one of seven actions during each time step: to rest, to eat, to turn to the left/right, to move forward to the next knot, to divide, or to fight.

When an agent rests, it changes nothing in the environment. If there is a grass patch in a knot with an agent and the agent executes the "eat" action, the patch disappears. If the agent divides, an offspring is created and placed in the knot. At any given time step, if there are other agents in the knot, the agent can interact with any of them. The agent can “fight” a randomly chosen agent in the knot.

Each agent has a limited capacity to store the energy resource internally. When an agent performs an action, its internal energy resource decreases. If the agent executes the action "eat" and there is grass in the knot, the energy resource of the agent increases. When the agent produces offspring, the parent spends some amount of energy and gives half of the rest to the newborn. If the internal energy resource goes to zero, the agent dies.

The behavior of each agent is governed by simple control system, which connects inputs to outputs: the output vector $\boldsymbol{O}$ is calculated by multiplying the input vector $\boldsymbol{I}$ by a matrix of weights $\boldsymbol{W}$. Components of $\boldsymbol{W}$ are integers in the range $[-W_{max}, W_{max}]$.

$$O_j = \sum_i w_{ij} I_i . \tag{1}$$

The number of outputs of the agent’s control system equals the number of actions an agent can perform; at each step, the agent performs the action associated with the maximum output value.

The weights of the control system are coded in the genome of the agent.

The genome of the agent $\boldsymbol{S}$ consists of three chromosomes $\boldsymbol{S} = (\boldsymbol{B}, \boldsymbol{W}, \boldsymbol{M})$. The first chromosome codes the presence or the absence of individual inputs and outputs; the second one codes the weights of the control system transformation; the third chromosome codes the marker of the agent.

If the agent executes the action "divide", an offspring appears. The genome of the offspring is a copy of its parent's genome, modified as follows:

1. for every gene corresponding to the weight of the control system, add a small random value uniformly distributed on the interval [$-p_w$, p_w], where p_w is mutation intensity;
2. with a small probability p_b , change each bit for the presence of input or output;
3. for every gene corresponding to the marker, add a small random value uniformly distributed on the interval [$-p_m$, p_m], where p_m is the mutation intensity of the marker.

4 Experiment

The simulation was run with a grid size 30 x 30 and an initial population of 200 agents. To speed up program execution, the weights of the W matrix took integer values in the range [-1000,1000] and the mutation intensity p_w was set to 50.

Each agent's generation g_i, i.e. the number of ancestors, was traced. The change of the average number of generations $\overline{g}$ in the population over time is presented in Fig. 1b. The graph can be approximated by a few lines with different slopes. If we do this, we can segment the evolution of the system into several "epochs" with different and almost constant rates of evolutionary activity within epoch. Then the rate of average generation growth $\Delta_{\overline{g}}$ (Fig. 1c) could be treated as an evolutionary (or generational) rate, i.e., the rate of generation of new solutions:

$$\Delta_{\overline{g}}^t = \overline{g}^t - \overline{g}^{t-1} . \tag{2}$$

If we use dynamical systems approach, we could represent each agent as point in a genome phase space, and the evolution of whole population could be represented as the movement of a cloud of points.

To see how the system moves in the genome phase space, one can calculate the centroid of the population $\boldsymbol{C}$ by averaging the weights over all agents and then plotting (Fig. 1d) the Euclidean distance D_C^t covered by the centroid during the given time step τ :

$$C_i = \frac{1}{N} \sum_N w_i , \tag{3}$$

$$D_C^t = \sqrt{\sum_i (C_i^t - C_i^{t-\tau})^2} , \tag{4}$$

where i = weight number and N = population size.

This measure reflects the integral displacement of the system in the genome phase space during defined time intervals. The plot shows that the speed of the system's movement was higher during epochs II, IV and V than during the relatively stable epochs I, III and VI. If we consider the relatively stable states of the second set of epochs to be attractors, then the states of the more active first set of epochs are transitions between attractors. In the context of this model, we interpret epochs II, IV and V as evolutionary transitions.

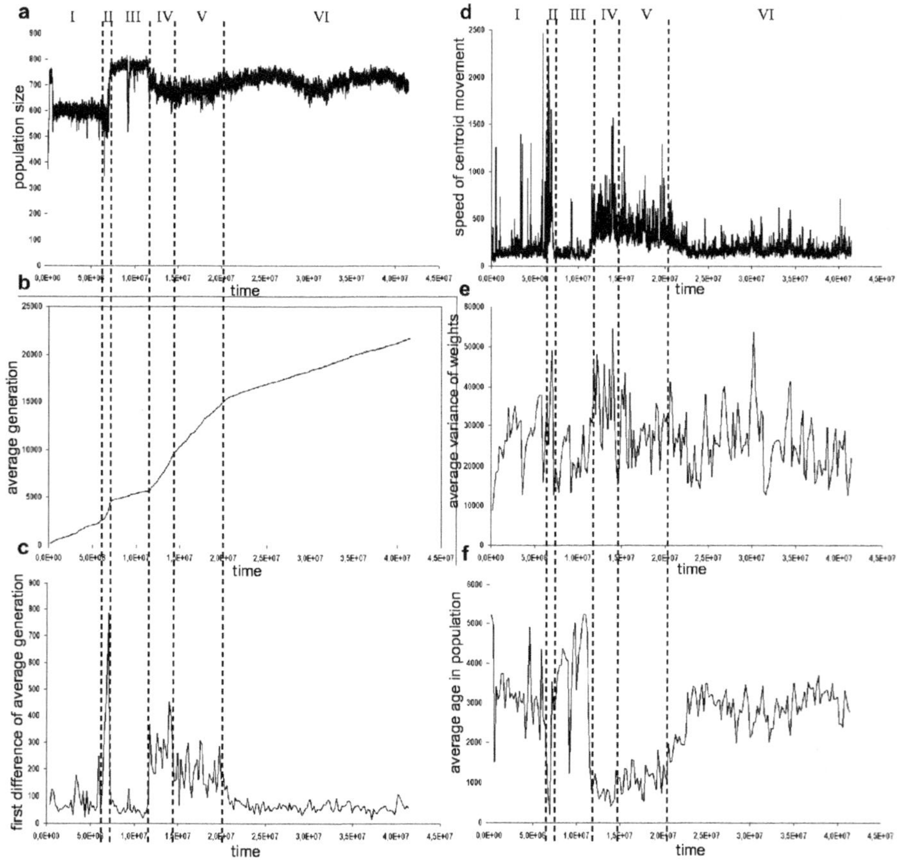

Fig. 1. Population size (a), average number of generations (b), first differences of average number of generations (c), speed of movement of the population centroid in the genome space (d), average variance of weights (e), average age of agents in the population (f).

The average variance of weights in the population (Fig. 1e) can serve as an indicator of diversity in the population. Although the average variance of genes weakly correlates with jumps in phase space (peaks in Fig. 1f), it is hard to hypothesize how this variance of genes is connected with evolutionary processes in our model.

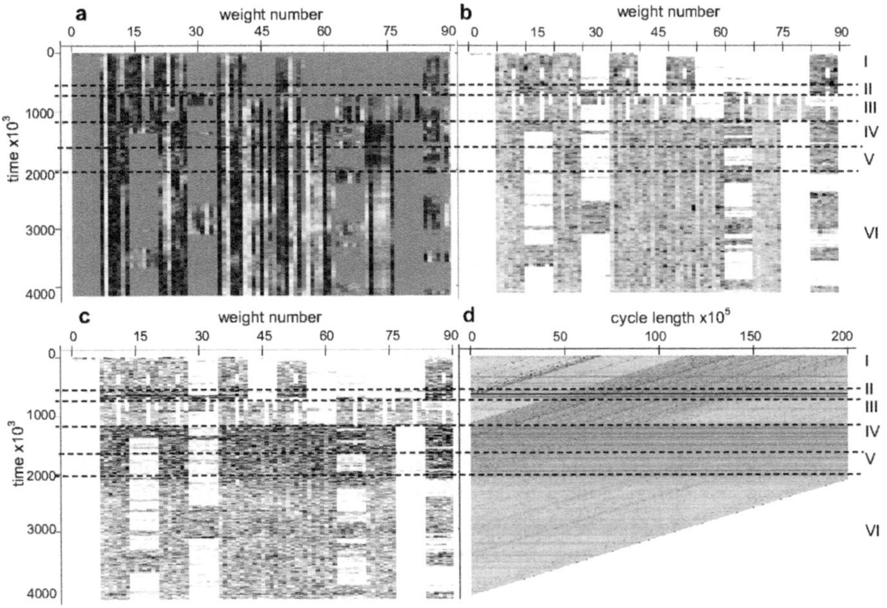

Fig. 2. Centroid weights bitmap (a) [high negative values shown in black, high positive in white, midrange in grey], weights variance bitmap (b) [here and following plots, zero shown in white and positive values in black], first differences of centroid weights bitmap (c) and distances in phase space for different cycles (d)

At the level of particular weights, evolution can be represented with the aid of bitmaps (Fig. 2). The first bitmap reflects the dynamics of the weights of the population centroid (Fig. 2a). On the bitmap, the weights are grouped by input; we see that incidents of appearance and disappearance of inputs often took place near the edges between epochs.

The complete meaning of the bitmap of the weights variance over the time (Fig. 2b) is not clear. However some vertical stripes are more uniform and brighter, i.e. have low constant variance, and perhaps the associated weights are more important for survival of agents.

The Figure 2c is a detailed analog of Figure 1e. Here the first differences of the centroid weights are plotted. The bitmap shows that during evolutionary transitions (epochs II, IV and V), the system is changing rapidly in most dimensions.

The last bitmap was created to find cycles of the population centroid trajectory in the genome phase space. (A similar approach was proposed in [9].) Such a plot can be used to help identify periodic movements of the centroid in the phase space, since the plot identifies when the system approaches a future point. The bitmap consists of vertical lines, each line for a given cycle length L. The level of gray corresponds to the Euclidean distance between the current position of the population centroid and the position after a cycle of length L:

$$d_L^t = \sqrt{\sum_i (C_i^{t+L} - C_i^t)^2} \,. \tag{5}$$

If there are cycles in the phase space, they should appear as light vertical bands on the bitmap. There are no such bands in Figure 2d, but there are dark horizontal lines concentrated inside epochs II and IV. When the system is situated at the points in its phase space corresponding to the dark lines it is equidistant from all preceding and following points. So we can infer that during evolutionary transitions, the population centroid can jump quite far from the areas where it usually persists, and never return to these points.

5 Conclusions

The evolutionary dynamics of a system can be represented by the movement of the system in a phase space, using a dynamical systems approach. Applying this approach to a simple ALife model, we identified epochs in the life of the system, which we characterized as periods of evolutionary transitions and periods of smooth development. During evolutionary transitions, the system moved rapidly in the genome phase space; we found no cycles in the trajectory of the system.

The notion of splitting the evolution of the system into epochs presented here is consistent with the notion of "epochal evolution" proposed in [20]. Epochal evolution assumes existence of subbasins of attraction in the genome phase space; these are connected by portals. The population moves from one subbasin to another by tunneling through portals. By analogy, rapid movement of the system studied here corresponds to tunneling, and slow movement corresponds to persistence in a subbasin. Further investigation might include analyses of the behavior of the system in a phenotype space, as well as analyses of relationships between the behavior at the phenotype level and at the genotype level.

Acknowledgments. Thanks to Vladimir Red'ko for long and fruitful collaboration. Without his support this work would hardly be possible. Also I would like to express appreciation to Konstantin Anokhin, Peter Turchin and Joseph Spinden for careful reviewing of this paper, and, of course, to Joseph Spinden and Gary Schiltz for correcting my English.

References

1. Bedau, M. A., and Packard, N. H.: Measurement of evolutionary activity, teleology, and life. In: C. G. Langton, C. Taylor, J. D. Farmer, and S. Rasmussen, (eds.): Proceedings of Artificial Life II, Addison-Wesley, Redwood City, CA, (1992), 431–461
2. Bedau, M. A., Snyder, E., and Packard, N. H.: A classification of long-term evolutionary dynamics. In: C. Adami, R. Belew, H. Kitano, and C. Taylor, (eds.): Proceedings of Artificial Life VI, Los Angeles, MIT Press, Cambridge, MA, (1998), 228–237

3. Channon, A. D.: Passing the ALife test: Activity statistics classify evolution in Geb as unbounded. In: J. Kelemen, P. Sosik (eds.): Advances in Artificial Life, Proceedings of 6th European Conference (ECAL 2001), Prague, Czech Republic, Springer, (2001)
4. Pachepsky, E., Taylor, T., Jones, S.: Mutualism promotes diversity and stability in a simple artificial ecosystem. Artificial Life, Vol. 8(1). MIT Press, (2002), 5–24
5. Adami, C., Ofria, C., Collier, T. C.: Evolution of biological complexity. Proc. Nat. Acad. Sci. 97, (2000), 4463–4468
6. Adami, C.: Introduction to Artificial Life. Springer, New York, (1998)
7. Nehaniv, C. L.: Measuring evolvability as the rate of complexity increase. In: C. C. Maley and E. Boudreau, (eds.): Artificial Life VII Workshop Proceedings, (2000), 55–57
8. Nehaniv, C. L. and Rhodes, J. L., The evolution and understanding of biological complexity from an algebraic perspective. Artificial Life, 6(1). MIT Press, (2000), 45–67
9. Cliff, D., Miller, G. F.: Tracking the Red Queen: Measurements of adaptive progress in co-evolutionary simulations. In: F. Moran, A. Moreno, J. J. Merelo and P. Cachon (eds): Advances in Artificial Life, Proceedings of the Third European Conference on Artificial Life (ECAL95), Lecture Notes in Artificial Intelligence 929, Springer Verlag, (1995) 200–218
10. van Nimwegen, E., Crutchfield, J. P., Mitchell, M.: Statistical Dynamics of the Royal Road Genetic Algorithm. Theoretical Computer Science, 229, (1999)
11. Nix, A. E., Vose, M. D.: Modeling genetic algorithms with Markov chains. Annals of Mathematics and Artificial Intelligence, 5, (1991)
12. Vose, M. D., Liepins, G. E.: Punctuated equilibria in genetic search. Complex Systems, 5, (1991)
13. Prugel-Bennett, A., Shapiro, J. L.: Analysis of genetic algorithms using statistical mechanics. Physical Review Letters, 72(9), (1994) 1305–1309
14. Prugel-Bennett, A., Shapiro, J. L.: The dynamics of a genetic algorithm in simple random Ising systems. Physica D, 104 (1), (1997), 75–114
15. Rattray, M., Shapiro, J. L.: The dynamics of a genetic algorithm for a simple learning problem. J. of Phys. A, 29(23), (1996), 7451–7473
16. Riziki, M.M., Conrad, M.: Computing the Theory of Evolution. Physica D, 22, (1986), 83–99
17. Packard, N.: Intrinsic adaptation in a simple model for evolution. In: C.G. Langton (ed.): Artificial life, Redwood City, Addison-Wesley, CA, (1989), 141–155
18. Ackley, D., Littman, M.: Interactions between learning and evolution. In: Langton, C. G., Taylor, C., Farmer, J. D., and Rasmussen, S. (eds.): Artificial Life II. Addison-Wesley, (1992), 487-509
19. Yaeger, L.: Computational Genetics, Physiology, Learning, Vision, and Behavior or PolyWord: Life in a New Context. In Langton, C. G. (ed.): Artificial Life III. Addison-Wesley, (1994), 263–298
20. Crutchfield, J. P.: When Evolution is Revolution – Origins of Innovation. In James P. Crutchfield, J. P. and Schuster, P. (eds.): Evolutionary Dynamics: Exploring the Interplay of Selection, Accident, Neutrality, and Function. Oxford University Press, (2002), 101–134

Pattern Recognition in a Bucket

Chrisantha Fernando and Sampsa Sojakka

School of Cognitive and Computer Sciences , University of Sussex, BRIGHTON, BN1 9QH,UK
{ctf20,sampsas}@cogs.susx.ac.uk

Abstract. This paper demonstrates that the waves produced on the surface of water can be used as the medium for a "Liquid State Machine" that pre-processes inputs so allowing a simple perceptron to solve the XOR problem and undertake speech recognition. Interference between waves allows non-linear parallel computation upon simultaneous sensory inputs. Temporal patterns of stimulation are converted to spatial patterns of water waves upon which a linear discrimination can be made. Whereas Wolfgang Maass' Liquid State Machine requires fine tuning of the spiking neural network parameters, water has inherent self-organising properties such as strong local interactions, time-dependent spread of activation to distant areas, inherent stability to a wide variety of inputs, and high complexity. Water achieves this "for free", and does so without the time-consuming computation required by realistic neural models. An analogy is made between water molecules and neurons in a recurrent neural network.

1 Introduction

Maass et al [1] have produced a model called the Liquid State Machine (LSM) which hypothesises that a cortical microcolumn consists of stereotyped recurrent circuits of integrate-and-fire neurons connected randomly according to a few parameters. The dynamics of this circuit are designed such that it can act as universal analog fading memory. Transient internal states can be read out by linear readout elements to produce stable outputs due to the high-dimensionality of the LSM. This allows robust real-time processing of time-varying inputs. All that needs to be trained is a readout module consisting of linear perceptrons.

The attraction of this model from an evolutionary robotic and a developmental neuroscience point of view is that it suggests a generic cortical architecture capable of universal real-time computation, which could potentially be specified by relatively few genetic parameters, so long as a "separation property" could be achieved[1]. Further unsupervised learning would only need to take place on a layer of linear readout elements instead of on the recurrent neural network itself.

[1] The separation property requires that separation between the trajectories of internal states of the system is roughly proportional to the distances between two different input streams that caused them.

W. Banzhaf et al. (Eds.): ECAL 2003, LNAI 2801, pp. 588–597, 2003.

Here we have taken the metaphor seriously and demonstrated that real water can be used as an LSM for solving the XOR problem and undertaking speech recognition in the Hopfield and Brody "zero-one" discrimination [2]. Doing so is computationally efficient, since water undertakes the transformation from the low-dimensional input space of motors stimulating its surface into a higher dimensional space of waves in parallel. This is achieved by placing a glass bucket of water on top of an over-head projector, and filming the interference patterns which result from vibrating the surface of the water using motors as inputs, and feeding the resulting filmed wave patterns into a perceptron. This work relates to a physical simulation of a liquid surface produced by Goldenholz [6].

Our approach differs from recent work on real non-linear media for computing. For example, Andrew Adamatzky [4] [5] has solved the XOR problem using diffusive wave front interactions in two-reactant media in a T-maze, where in the [1 1] case, interference at the junction of the T-maze cancels out the reaction so producing the necessary non-linearity. In our bucket, it is not necessary to impose task-specific structural constraints that shape the wave fronts. We will show that the perceptron is able to discover the relevant information in a remarkably large variety of wave patterns. In addition, the media does not require a chemical reaction which for example produces a precipitate that reaches a point attractor, i.e. leaves a stable pattern of sediment in some stable state that represents the input pattern. In our system the perceptron must use only real-time wave dynamics to make a stable classification.

This work relates indirectly to work on field computation which acknowledges that neural computation is a massively parallel, low speed, low-precision continuous set of analog operations [7]. Serial computations must be short as dictated by the real-time response requirements of animals. MacLennan writes "The neural mass is sufficiently large that it is useful to treat it as a continuum", and hypothesises that a similar architecture will be required to achieve artificial intelligence. He claims VLSI technology will not achieve this i.e. 15 million neurons/cm^2. The use of water-like media may serve as an alternative.

This mesoscopic approach is in contrast to mainstream neuroscience and neural networks which operates in a paradigm influenced by the work of McCulloch and Pitts [8], which views neurons as performing logical operations in circuits. Evidence for mesoscopic features relevant in conditioning tasks has been obtained by Walter Freeman [9] who describes spatial patterns of amplitude modulation in EEGs that change consistently following learning.

From an engineering perspective, the technological benefits of such an approach are within sight. Recent work by Ian Walmsley [10] in optics has demonstrated the power of exploiting interference patterns produced by the interaction of multiple sources of information to search a database consisting of 50 entries. Acoustic waves are used to encode information in the pattern of expansion and compression in tellurium dioxide. Coherent light of different frequencies (colours) is passed through the medium. Each bit of encoded information is probed by a different colour. Phase shifts due to the differences in thickness of the media (which encode the information) result in constructive or destructive interference at the other end. By examining which frequencies have passed through in-phase, the information in the database can be retrieved.

Finally, we analyse the complexity of the water using a measure of complexity devised by Sporns, Tononi and Edelman [11] and show that water satisfies an approximation to the separation property described by Maass.

Fig. 1. The Liquid Brain.

2 Methods

The "liquid brain" was constructed from a transparent water tank suspended over an overhead projector,(Figure 1). Four electric motors connected to a computer via a serial control box were attached to each side of the tank to drive a small weight into the water and thus create eight wave point sources. The interference patterns so produced were projected on an antireflective cardboard and recoded at a resolution of 320x240 pixels at 5 frames per second using a standard web-cam. Each frame obtained with the camera was convolved using a Sobel edge detection mask that was suitably thresholded to remove noise and averaged to produce a 32x24 grid. The cells of this matrix were used as the input values for 50 perceptrons in parallel which were trained using the same p-delta rule as used by Maass et al [3].

2.1 XOR Task

In order to recreate the XOR problem using the liquid, two motors were set up at opposite sides of the tank. A single motor driven at constant frequency on left and right side of the tank corresponded to the [1 0] and [0 1] cases respectively. In the [1 1] case both motors were activated simultaneously, and incoherently(i.e with no fixed phase relation). One minute footage was recorded with the video camera of each of the four cases and the frames processed as discussed above yielding a total of 3400 frames. The order of the frames in this data set was randomised and used in training the linear readout perceptron. Another set of footage was taken of the four conditions as a test set.

2.2 Speech Recognition

The speech recognition task was designed to measure the suitability of the liquid brain model for robust spatiotemporal pattern recognition in a noisy environment. Earlier experiments have shown that transient synchrony of the action potentials of a group of spiking neurons can be used for "signal" recognition of a space-time pattern across the inputs of those neurons [2]. We show that the dynamical properties of the liquid can be used to produce this synchrony and that a simple readout neuron will suffice.

Following the example of the Hopfield and Brody experiments the speech recognition task used was to distinguish a speaker saying the word "one" from the same speaker saying "zero". Twenty samples were recorded of a person saying each word with randomly varying amplitudes and intonations. The amplitude variation was mostly due to fluctuations in the distance of the speaker from the microphone (±10cm). The voice samples were recorded as 12 kHz pulse-code modulated wave files ranging from 1.5 to 2 seconds in length.

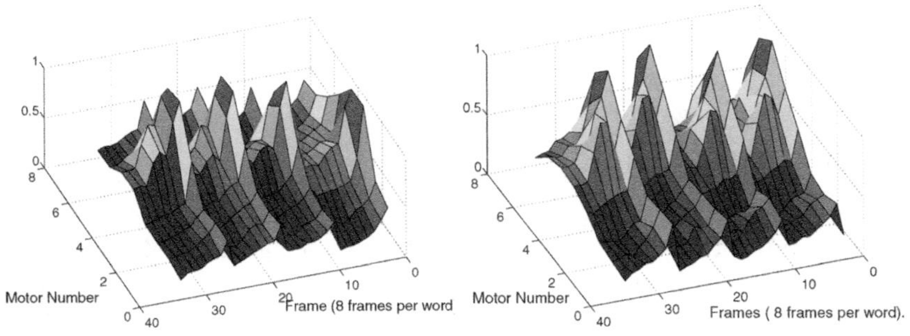

Fig. 2. Left:Four presentations of "Zero". Right: Four presentations of "One". Note the variability between and within sets.

Due to limitations in the speed of frequency response of the available motors a certain amount of pre-processing of the sound samples was necessary. Each sample was first processed using a Short-Time Fourier transform with a sliding window size set to the closest power of 2 of an eighth of the sample length thus splitting each sound sample to eight time slices. The active frequency range of 1-3000Hz of the resulting function of time and frequency was further averaged to eight frequency bands resulting in an 8x8 matrix for each sound sample, see Fig 2. The 40 matrices produced in this manner were first used to train a linear perceptron by using the p-delta rule in order to ensure that the task was not linearly separable. Each sample matrix was then used to drive the motors of the liquid so that each of the eight frequency bands controlled one motor for 4 seconds (0.5 seconds per time slice). The 20 pre-processed samples of "one" were inserted into the liquid one after another and footage recorded of the entire sequence. This procedure was then repeated for the 20 samples of the word "zero". The resulting frames were processed as above to create data sets used for the training of the readout perceptrons.

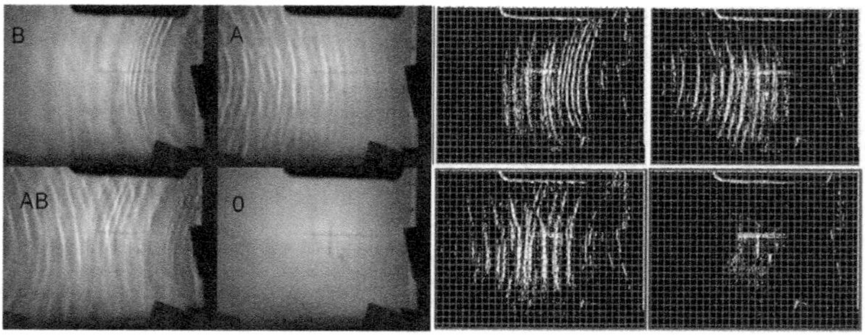

Fig. 3. Typical wave patterns for the XOR task. Top-Left: [0 1] (right motor on), Top-Right: [1 0] (left motor on), Bottom-Left: [1 1] (both motors on), Bottom-Right: [0 0] (still water). Sobel filtered and thresholded images on right.

3 Results

3.1 The XOR Problem

Figure 3 shows the waves that represent input conditions of the XOR task. Convergence to an optimal solution is obtained within 200 epochs (Figure 4) with no mistakes being made on the training set by this time. The summed outputs of the 50 perceptrons and their "correctness" for training and test sets is shown in Figure 5. The weight-matrices of these perceptrons show how the problem is being solved, see Figure 6. How can the perceptrons make the same classification in cases where the activities of inputs are radically different, i.e. both motors on versus still water? All the perceptrons learned a stereotyped weight matrix consisting of a central negative region and a lateral positive region of weights. Remarkably, this is a center-surround receptive field, with a vertical orientation bias. Thus, the solution does not in fact depend on interference patterns but rather upon mean amplitudes of wave fronts being different in different regions of the grid.

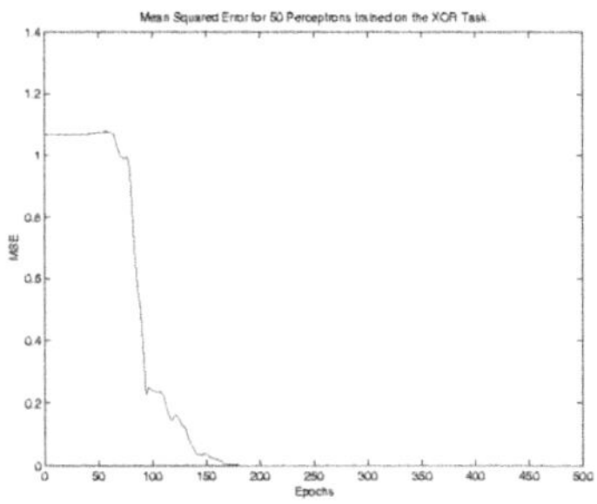

Fig. 4. Decreasing Mean Squared Error for 50 perceptrons trained on the XOR task after pre-processing by water.

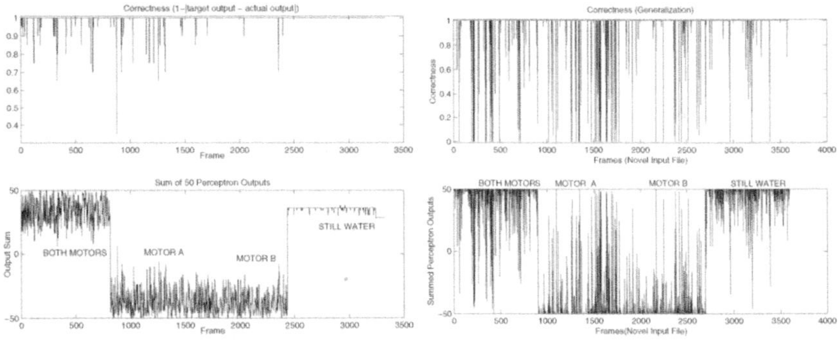

Fig. 5. Left : The summed outputs of the 50 perceptrons for the training data are shown in the bottom graph. Correctness in the top graph is defined as $(1 - \|desiredoutput - actualoutput\|)$ of the perceptrons after being passed through a stepwise squashing function. The readout module classifies the "still" and "both motor" conditions as class 1 and the "motor a" and "motor b" conditions as class -1. Although some perceptrons make mistakes, taking an average allows perfect classification. Right : Generalisation although not perfect, still results in 85% of frames being classified correctly.

3.2 Speech Recognition

The XOR problem, although non-linear is pretty easy to solve by adding just one extra dimension[2] . Our approach of adding an extra 700 dimensions is to use a machete to garnish canapes. A significantly more difficult non-linear problem is speech recognition. Speech data when fed directly into a perceptron could not achieve a classification which made any fewer than 25% mistakes, However, error decreased to 1.5% when the input was passed through water, see Figure 7. Figure

[2] This can be done in various interesting ways. For example, imagine a robot with two arms. [0 0], the robot stands still, and does not fall over, [0 1] the robot pushes against the left wall with its left arm and falls over, [1 0] the robot pushes against the right wall with its right arm and falls over, [1 1] the robot pushes against both walls with both arms and so does not fall over. Inman Harvey, Personal Communication.

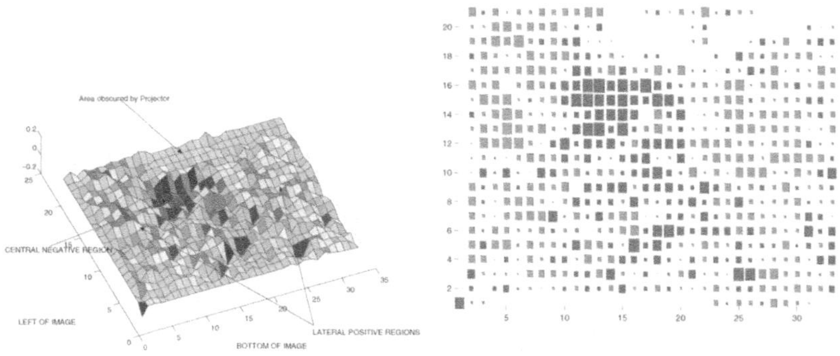

Fig. 6. The mean weight matrix of 50 perceptrons is shown as a surface plot on the left and as a Hinton diagram on the right. The center-surround receptive field works as follows. In the case of still water, positive and negative weights are activated uniformly and cancel each other out. In the case of both motors being stimulated, activity is also relatively uniform across the grid, and cancellation also occurs. However, with the [1 0] or [0 1] case, only the left or right of the field receives significant activation by wave fronts and the output of the perceptrons have a net positive value.

8 shows the summed outputs and correctness achieved by the 50 perceptrons on the training data and on the test set consisting of a novel set of footage of spoken words "zero" and "one". Generalisation is relatively poor because of the noise. In our setup the frequency response of the motors differed widely and the performance of the shafts used to drive the water deteriorated with use, the water level in the tank and hence the amplitude of the waves fluctuated due to evaporation, vibrators moved out of place, the camera and the tank itself moved, the frame rate of the camera varied and no special attempts were used to remove these sources of noise. It is remarkable any classification could be made at all. We are confident that a more pristinely engineered version of the experimental setup would eliminate most of these problems, resulting in improved performance.

The mean weight matrix for these perceptrons consist of a set of wave front detectors, positioned at the top of the image, see Figure 9. This is the region with the most distinct wave fronts as the motors are closest to the filmed area of water.

4 Discussion

4.1 Measuring the Complexity of Water

Why does using water as a pre-processor work? Here we suggest that redundancy in the representation of information in the system may allow the perceptron to solve the problem with one of many patterns of weights. An information theory measure of neural complexity by Tononi, Sporns and Edelman has been used to show that brains are high in complexity due to a pattern of intrinsic connectivity whereby groups with similar response selectivities are preferentially connected,

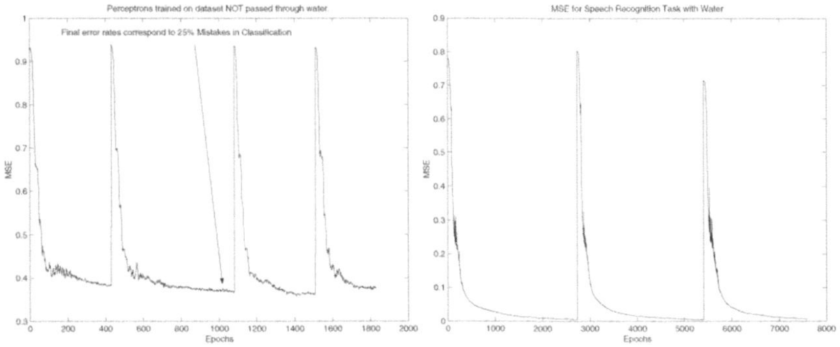

Fig. 7. Left: Error on raw data passed directly to perceptron. An Optimal classification is not possible. Right: Data passed through liquid. Optimal classification is possible.

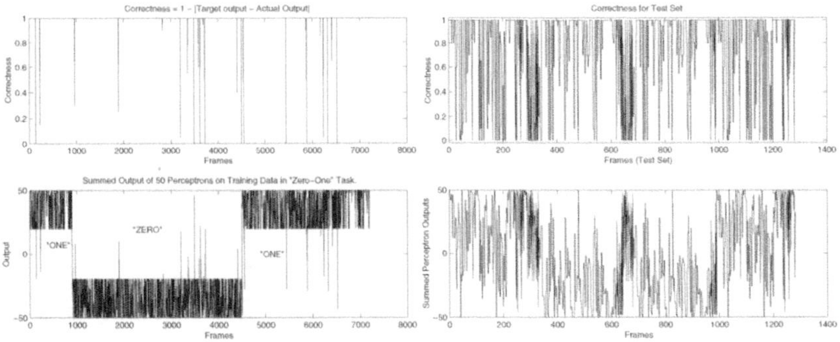

Fig. 8. Left: Perfect classification of training data. Right: Generalisation errors occur in apx 35% of frames.

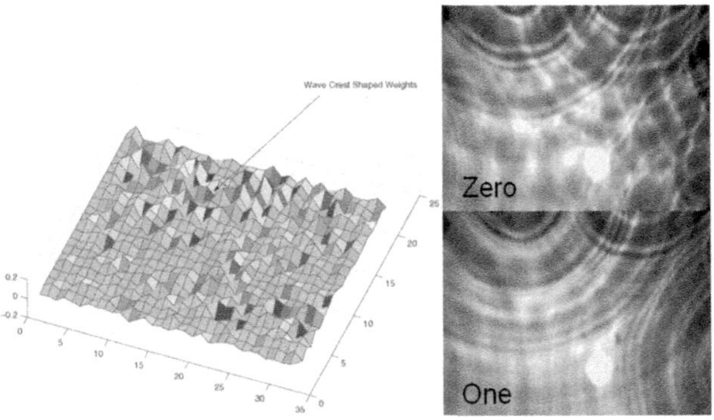

Fig. 9. Left: Mean weight matrix. Right: Typical frames for "Zero" and "One" wave patterns.

but where connectivity falls off with distance [11]. We hypothesise that water waves exhibit similar "functional connectivity". The uncertainty about the state of subset X_j of the system X, which is accounted for by the state of the rest of the system $(X - X_j)$ is the mutual information, given by

$$MI(X_j; X - X_j) = H(X_j) + H(X - X_j) - H(X) \tag{1}$$

Complexity is the area under the graph of MI for subunit sizes from $k = 1$ to $k = N/2$ where N is the size of X [11]. We measured the complexity of water under differing extents of "spontaneous activity", i.e. due to random stimulation by 0 to 8 motors. Figure 10 shows that complexity increases in a staggered fashion with increasing numbers of motors. This is due to peculiarities in the way individual motors affect the water, some making large ripples and some making relatively small ripples. An interesting feature is that the system remains complex with increasing numbers of motors. Even with 8 motors stimulating the water the system does not become chaotic, that is, by observing less than 50 randomly chosen pixels of the image of water, we gain all the information that is stored in the entire array of pixels.

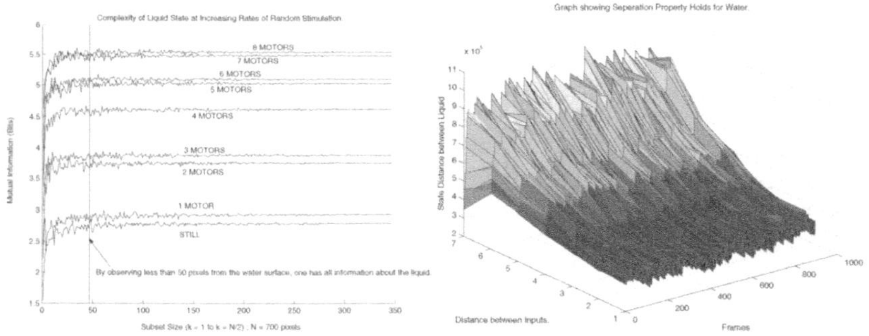

Fig. 10. Left: Complexity of water for increasing amounts of random stimulation. Right: Approximated separation property of water, where input distance is defined as the difference in number of motors stimulating water, and the state distance of the water is defined as the Euclidean distance of pixel values in the L-2 norm.

4.2 Separation Property

We approximated the separation property of water by the following method. We used the same data set as in part 4.1. The Euclidean distance between inputs was defined as the difference between numbers of motors stimulating the water between two trials. Figure 10 shows that the Euclidean distance between the state of the liquid increases linearly with Euclidean distance between inputs, just as in Maass' LSM. We did not test the separation property for more subtle measures of input distances such as the patterned motor activity used in the speech discrimination task for example. The separation property is unlikely to hold for as wide a range of measures of input distance as in Maass' LSM.

5 Conclusion

Our results show that the properties of a natural dynamical system (water) can be harnessed to solve nonlinear pattern recognition problems and that a set of simple linear readout elements will suffice to make the classification. This is possible without exhaustive tweaking of system parameters or careful efforts to decrease the level of noise in the system. Further work will use artificial evolution to create self-organising "neural networks" which are made of and exploit the epigenetic physical properties of biological materials [12].

Acknowledgements. Special thanks to Inman Harvey, Phil Husbands, Ezequiel Di Paolo, Emmet Spier, Bill Bigge, Aisha Thorn, Hanneke De Jaegher, Mike Beaton.

References

1. Maass, W., T. Natschlager, et al. (2002). "Real-Time Computing Without Stable States : A New Framework for Neural Computation Based on Perturbations." Neural Computation 14: 2531–2560.
2. Hopfield, J. J. and C. D. Brody (2000). "What is a moment? "Cortical" sensory integration over a brief interval." PNAS 97(25): 13919–13924.
3. Auer, P., H. Burgsteiner, et al. (2002). "The p-Delta Learning Rule for Parallel Perceptrons." submitted for publication. http://www.cis.tugraz.at/igi/pauer/publications.html
4. Adamatzky, A. (2001). "Computing in nonlinear media: make waves, study collisions." Lecture Notes in Artificial Intelligence 2159: 1–11.
5. Adamatzky, A., De Lacy Costello, B.P.J (2002). "Experimental logical gates in a reaction-difusion medium : The XOR gate and beyond." Physical Review E 66 046112(2002).
6. Goldenholz, D. (2002). "Liquid Computing : A Real Effect." BE707 Final Project. Boston, Boston University School of Medicine. http://www.lsm.tugraz.at/people.html.
7. MacLennan, B. (1999). "Field Computation in Natural and Artificial Intelligence". Knoxville, University of Tennessee. Report UT-CS-99-422. http://www.cs.utk.edu/ mclennan/fieldcomp.html
8. McCulloch, W. S. and W. Pitts (1943). "A logical calculus of the ideas immanent in nervous activity." Bulletin of Mathamatical Biophysics 5: 115–133.
9. Freeman, W. (2000). "Mesoscopic Neurodynamics : From neuron to brain." Journal of Physiology (Paris) 94: 303–322.
10. Walmsley, I. (2001). "Computing with interference: All-optical single-query 50-element database search." Conference on Lasers and Electro-Optics/Quantum Electronics and Laser Science, Baltimore, Maryland.
11. Tononi, E., Sporns (1998). "Complexity and Coherency: integrating information in the brain." Trends in Cognitive Sciences 2(12): 474–483.
12. Muller, G. and S. Newman (2003). Origination of Organismal Form: Beyond the Gene in Developmental and Evolutionary Biology, MIT Press.

Discovering Clusters in Spatial Data Using Swarm Intelligence

Gianluigi Folino, Agostino Forestiero, and Giandomenico Spezzano

Institute for High Performance Computing and Networking(ICAR) – CNR
Via P. Bucci 41C I-87030 – Rende (CS), Italy
{folino, forestiero, spezzano}@icar.cnr.it

Abstract. This paper presents a novel algorithm that uses techniques adapted from models originating from biological collective organisms to discover clusters of arbitrary shape, size and density in spatial data. The algorithm combines a smart exploratory strategy based on the movements of a flock of birds with a shared nearest-neighbor clustering algorithm to discover clusters in parallel. In the algorithm, birds are used as agents with an exploring behavior foraging for clusters. Moreover, this strategy can be used as a data reduction technique to perform approximate clustering efficiently. We have applied this algorithm on synthetic and real world data sets and we have measured, through computer simulation, the impact of the flocking search strategy on performance.

1 Introduction

Data mining deals with the problem of extracting interesting associations, classifiers, clusters, and other patterns from data by paying careful attention to the available computing, storage, communication, and human resources. Clustering is a data mining task concerning the process of grouping similar objects according to their distance, connectivity, or their relative density in space. In particular, spatial data mining is the discovery of interesting relationships and characteristics that may exist implicitly in spatial data. Spatial clustering has been an active area of research into data mining, with many effective and scalable clustering methods developed. These methods can be classified into partitioning methods, hierarchical methods, density-based methods, grid-based methods. Han, Kamber and Tung's paper [1] is a good introduction to this subject.

Recently, other algorithms based on biological models [2,3] have been introduced to solve the clustering problem. These algorithms are characterized by the interaction of a large number of simple agents that sense and change their environment locally. Furthermore, they exhibit complex, emergent behavior that is robust with respect to the failure of individual agents. Ants colonies, flocks of birds, termites, swarms of bees etc. are agent-based insect models that exhibit a collective intelligent behavior (*swarm intelligence*) [4] which may be used to define new algorithms of clustering.

In this paper, we present the parallel spatial clustering algorithm SPARROW-SNN (***SPA****tial Cluste****R****ing Algo****R****ithm thr****O****ugh S****W****arm Intelligence and* ***S****hared* ***N****earest-****N****eighbor Similarity*) which is based on an adaptive

W. Banzhaf et al. (Eds.): ECAL 2003, LNAI 2801, pp. 598–605, 2003.

flocking algorithm proposed by Macgill and S. Openshaw [5] as a form of effective search strategy to perform an exploratory geographical analysis. The algorithm takes advantage of the parallel search mechanism a flock implies, by which if a member of a flock finds an area of interest, the mechanics of the flock will draw other members to scan that area in more detail. SPARROW-SNN combines the flocking algorithm with a shared nearest neighbor cluster algorithm to discover clusters of arbitrary density, shape and size in spatial data. SPARROW-SNN uses the stochastic and exploratory principles of a flock of birds to detect clusters in parallel according to the shared nearest neighbor-based principles of the SNN [6] clustering algorithm and a parallel iterative procedure to merge the clusters discovered. Moreover, we have applied this strategy as a data reduction technique to perform approximate clustering efficiently [7]. We have built a SWARM [8] simulation of SPARROW-SNN to investigate the interaction of the parameters that characterize the algorithm. The first experiments show encouraging results and a better performance of SPARROW-SNN in comparison with the linear randomized search. The remainder of this paper is organized as follows. Section 2 briefly presents the heuristics of the SNN clustering. Section 3 introduces the classical flocking algorithm and presents the SPARROW-SNN algorithm. Section 4 discusses the obtained results and Section 5 draws some conclusions.

2 The SNN Clustering Algorithm

SNN is a clustering algorithm developed by Ertöz, Steinbach and Kumar [6] to discover clusters with differing sizes, shapes and densities in noisy, high dimensional data. The algorithm extends the nearest-neighbor non-hierarchical clustering technique developed by Jarvis-Patrick [9] redefining the similarity between pairs of points in terms of how many nearest neighbors the two points share. Using this new definition of similarity, the algorithm eliminates noise and outliers, identifies representative points also called *core points*, and then builds clusters around the representative points. These clusters do not contain all the points, but rather represent relatively uniform groups of points. The SNN algorithm starts performing the Jarvis-Patrick scheme. In the Jarvis-Patrick algorithm a set of objects is partitioned into clusters on the basis of the number of shared nearest-neighbors. The standard implementation is constituted by two phases. The first is a pre-processing stage which identifies the K nearest-neighbors of each object in the data set. In the subsequent clustering stage a shared nearest neighbor graph is constructed from the pre-processed data as follows. A link is created between two objects i and j if:

- i is one of the K nearest-neighbors of j;
- j is one of the K nearest-neighbors of i;
- i and j have at least K_{min} of their K-nearest-neighbors in common;

where K and K_{min} are used-defined parameters. Each link has an associate weight defined as:

$$weight(i,j) = \sum (k+1-m)(k+1-n), \text{ where } i_m = j_n \tag{1}$$

In the equation above, k is the nearest neighbor list size, m and n are the positions of a shared nearest neighbor in i and j's lists. At this point, clusters can be obtained by removing all edges with weights less than a user specified threshold and taking all the connected components as clusters. A major drawback of the Jarvis-Patrick is that, the threshold needs to be set high enough since two distinct sets of points can be merged into the same cluster even if there is only one link across them. On the other hand, if a high threshold is applied, then a natural cluster will be split into many small clusters due to the variations in the similarity in the cluster. SNN addresses these problems adding to the Jarvis-Patrick algorithm the following steps:

- for every data point in the weighted graph, calculate the total sum of weights associated with the links coming out of the point. This value is called *connectivity*;
- identify core points by choosing the point that have a value of connectivity greater than a predefined threshold (*core_threshold*);
- identify noise points by choosing the points that have a value of connectivity lower than a user specified threshold (*noise_threshold*) and remove them;
- remove all links between points with weight smaller than a threshold);
- form clusters with the connected components of points. Every point in a cluster is either a core point or is connected to a core point.

The number of clusters is not given to the algorithm as a parameter. Also note that not all the points are clustered.

3 SPARROW-SNN: A Flocking Algorithm for Spatial Clustering

In this section, we present a multi-agent clustering algorithm, called SPARROW-SNN, which combines the stochastic search of an adaptive flocking with the SNN heuristics for discovering clusters in spatial data. This approach has a number of nice properties. It has the advantages of being easily implementable on parallel computers and is robust compared to the failure of individual agents. It can also be applied to perform *approximate clustering* efficiently since the points that are, to each iteration, visited and analyzed by the agents represent a significant (in *ergodic* sense) subset of the entire data set. The subset reduces the size of the data set while keeping the loss of accuracy as small as possible. We propose to use flocking *sampling* as a data reduction technique to speed up the operations of cluster and outlier detection on large data sets collections. Our approach iteratively improves the accuracy of the clustering because, at each generation, new data points are discovered and added to each cluster with about the same increase per cent.

SPARROW-SNN uses a modified version with an *exploring* behavior of standard Reynolds' flock of birds model [10] to describe the movement rules of the agents. The behavior requires each agent to search the clusters in parallel and to signal the presence or the lack of significant patterns in the data to the other flock members, by changing its color. The entire flock then moves towards the agents (*attractors*) that have discovered interesting regions to help them, avoiding the uninteresting areas that are instead marked as obstacles. Clusters are discovered using the heuristics principles of the SNN clustering algorithm.

As first step, SPARROW-SNN computes the nearest-neighbor list for each data point using a *threshold similarity* that reduces the number of data elements to take in consideration. The introduction of the threshold similarity produces variable-length nearest-neighbor lists and therefore now i and j must have at least P_{min} of the shorter nearest-neighbor list in common; where P_{min} is a user-defined percentage. After the nearest-neighbor list is computed, SPARROW-SNN starts a fixed number of agents that will occupy a randomly generated position. The agents have an attribute that defines their color. Initially the color is the same for all. From its initial position, each agent moves around the spatial data testing the neighborhood of each location in order to verify if the point can be identified as a core point. All agents execute the same set of rules for a fixed number of times (*MaxGenerations*). When an agent falls on a data point A not yet visited, it computes the connectivity, *conn[A]*, of the point, i.e. computes the total number of strong links the points has according to the rules of the SNN algorithm. Points having a connectivity smaller than a fixed threshold (*noise_threshold*) are classified as noise and are considered to be removed from the clustering. Each agent is colored on the basis of the connectivity computed in the visited data point. The colors assigned to the agents are: **red** (*conn* > *core_threshold*), revealing core points, **green** (*noise_threshold* < *conn* <= *core_threshold*), for border points, **yellow** (*0* < *conn* < *noise_threshold*), for noise points, and **white** (*conn* = *0*), indicating an obstacle (uninteresting region).

After the coloration step, the green and yellow agents, compute their movement observing the positions of all other agents that are at some fixed distance (*dist_max*) from them, and applying the rules of Reynolds' with the following modifications:

- *Alignment* and *cohesion* do not consider yellow agents, since they move in a not very attractive zone.
- *Cohesion* is the resultant of the heading towards the average position of the green flockmates (*centroid*), of the attraction towards red agents, and of the repulsion by white agents.
- A *separation* distance is maintained from all the agents, whatever their color is.

Agents will move towards the computed destination with a speed depending from their color: green agents will move slowly than yellow agents since they will explore denser zones of clusters. Green and yellow agents have a variable speed, with a common minimum and maximum for all agents. An agent will speed up to leave an empty or uninteresting region whereas it will slow down to investigate an interesting region more carefully. The variable speed introduces an adaptive behavior in the algorithm. In fact, agents adapt their movement and change their behavior (speed) on the basis of their previous experience represented by the red and white agents. Red and white agents will stop signaling to the others the interesting and desert regions.

Note that for any agent which has become red or white, a new agent will be generated in order to maintain a constant number of agents exploring the data. In the first case, the new agent will be generated in a close random point, since the zone is considered interesting, while in the latter it will be generated in a random point over all the space.

In any case, each red agent (placed on a representative point) will run the merge procedure so that it will include, in the final cluster, the representative point discovered together with the points that share with it a significant (greater than P_{min}) number of neighbors and are not noise points. The merging phase considers two different cases: when we have never visited any of these points in the neighborhood and when we have points belonging to different clusters. In the first case, the points will be assigned the same temporary label and will constitute a new cluster; in the second case, all the points will be merged into the same cluster, i.e. they will get the label corresponding to the smallest one. So clusters will be built incrementally.

During simulations a *cage effect*, was observed; in fact, some agents could remain trapped inside regions surrounded by red or white agents and would have no way to go out, wasting useful resources for the exploration. So, a limit on their life was imposed to avoid this effect; hence, when their age exceeded a determined value (*maxLife*) they were killed and were regenerated in a new randomly chosen position of the space.

4 Experimental Results

For the experiments we used two synthetic data sets and one real life data set from a spatial database. The structure of these data sets is shown in figure 1. The first data set, called GEORGE, consists of 8000 points, characterized by a large number of noise points. The second data set, called DS1, contains 8000 points and presents different densities in the clusters. The third data set, called North-East, contains 123593 points representing postal addresses of three metropolitan areas (New York, Boston and Philadelphia) in the North East States.

We implemented our algorithm using SWARM, a multi-agent software platform for the simulation of complex adaptive systems. We first illustrate the loss of accuracy of our SPARROW-SNN algorithm in comparison with SNN algorithm when SPARROW-SNN is used as a technique for approximate clustering. To this purpose, we implemented a version of SNN and we computed the number of clusters and the number of points for cluster for the three datasets.Table 1 presents a comparison of these results with respect to the ones obtained from SPARROW-SNN when a population of 50 agents have visited respectively 7%, 12% and 22% of the entire data set.

Fig. 1. The three data sets used in our experiments.

Table 1. Number of clusters and number of points for cluster for GEORGE, DS1 and North-East data sets (percentage in comparison to the total points for cluster found by SNN) when SPARROW-SNN analyzes 7%, 12% and 22% points.

Clustering using the GEORGE data set	Per. of data points for cluster found by SPARROW-SNN		
	7%	**12%**	**22%**
G	55.8%	80.0%	88.3%
E	53.1%	69.8%	88.6%
O	62.8%	82.1%	86.2%
R	49.0%	67.5%	83.2%
G	47.9%	72.0%	78.8%
E	64.2%	77.7%	86.8%

Clustering using the DS1 data set	Per. of data points for cluster found by SPARROW-SNN		
	7%	**12%**	**22%**
1	41.35%	58.31%	70.37%
2	30.08%	53.58%	60.72%
3	29.28%	40.99%	53.02%
4	20.9%	30.5%	51.41%
5	53.38%	65.36%	76.56%
6	56.89%	69.87%	73.63%
7	33.89%	43.5%	61.58%

Clustering using the North-East data set	Per. of data points for cluster found by SPARROW-SNN		
	7%	**12%**	**22%**
Philadelfia	42.5%	65.2%	79.4%
New York	38.7%	52.3%	67.6%
Boston	46.5%	68.6%	82.3%

Note that with only 7% of points we can have a clear vision of the found clusters and with a few more points we can obtain the nearly totality of the points. This trend is well marked in GEORGE and in North-East data sets. For DS1 data set the results are not so clear because the 3 and 4 clusters have very few points, so they are very hard to discover. For the real data set we only reported the results for the three main clusters representing the towns of Boston, New York and Philadelphia. We can explain the good results through the adaptive search strategy of SPARROW-SNN that requires the individual agents to first explore the data searching for representative points whose position is not known *a priori*. Then, after the representative points are located, all the flock members are steered to move towards the representative points, that represent the interesting regions, in order to help them, avoiding the uninteresting areas that are instead marked as obstacles and adaptively changing their speed.

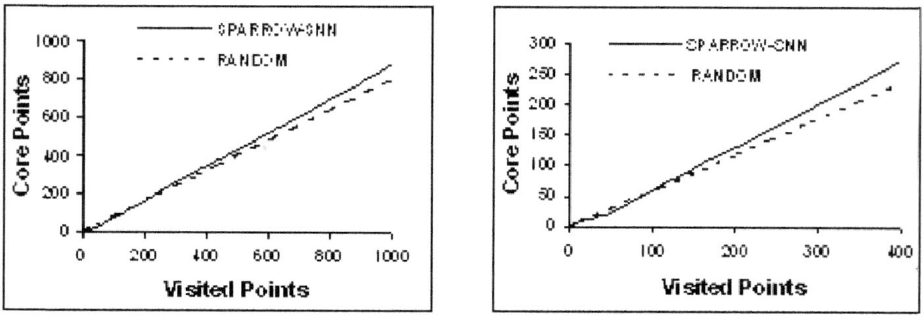

Fig. 2. Number of representative points found with SPARROW-SNN vs. total number of visited points for GEORGE and North-East datasets.

To verify the effectiveness of the search strategy we have compared SPARROW-SNN with the random-walk search strategy. Figure 2 shows, for the data sets GEORGE and North-East, the number of core points found with SPARROW-SNN and those found with the random search vs. the total number of visited points. This figure reveals that the number of core points discovered at the beginning (110 visited points for George data set and 200 for North-East data set) from the random strategy is slightly higher than the number discovered by SPARROW-SNN.

Subsequently, our strategy presents a superior behavior on the random search strategy because of the adaptive behavior of the algorithm that allows the agents to learn on their previous experience. A similar behavior has been observed for the DS1 dataset.

5 Conclusions

In this paper, we have described the parallel clustering algorithm SPARROW-SNN, which is based on the use of swarm intelligence techniques. The algorithm combines a smart exploratory strategy based on a flock of birds with a shared nearest neighbor clustering algorithm to discover clusters of arbitrary shape, size and density in spatial data. The algorithm has been implemented in SWARM and evaluated using two synthetic data sets and one real word data set. Measures of accuracy of the results show that SPARROW-SNN can be efficiently applied as a data reduction strategy to perform approximate clustering. Moreover, the adaptive search strategy of SPARROW-SNN is more efficient than that of the random-walk search strategy.

Acknowledgements. This work was supported by the CNR/MIUR project-Legge 449/97-DM 30/10/2000.

References

1. Han J., Kamber M., Tung A.K.H., Spatial Clustering Methods in Data Mining: A Survey, H. Miller and J. Han (eds.), Geographic Data Mining and Knowledge Discovery, Taylor and Francis, 2001.
2. Lumer E. D., Faieta B., Diversity and Adaptation in Populations of Clustering Ants, Proc. of the third Int. Conf. on Simulation of Adaptive Behavior: From Animals to Animats (SAB94), D. Cliff, P. Husbands, J.A. Meyer, S.W. Wilson (Eds), MIT-Press, pp. 501–508, 1994.
3. N. Monmarché, M. Slimane, and G. Venturini, "On improving clustering in numerical databases with artificial ants", in Advances in Artificial Life: 5th European Conference, ECAL 99, LNCS 1674, Springer, Berlin, pp. 626–635, 1999.
4. Bonabeau E., Dorigo M., Theraulaz G., Swarm Intelligence: From Natural to Artificial Systems, Oxford University Press, 1999.
5. James Macgill, Using Flocks to Drive a Geographical Analysis Engine, Artificial Life VII: Proceedings of the Seventh International Conference on Artificial Life, MIT Press, Reed College, Portland, Oregon, pp. 1–6, 2000.
6. Ertöz L., Steinbach M., and Kumar V., A New Shared Nearest Neighbor Clustering Algorithm and its Applications, Workshop on Clustering High Dimensional Data and its Applications, 2nd SIAM International Conference on Data Mining Arlington, VA, 2002.
7. Kollios G., Gunopoulos D., Koudas N., Berchtold S.: "Efficient Biased Sampling for Approximate Clustering and Outlier Detection in Large Datasets". IEEE Trans. on Knowledge and Data Engineering (TKDE), to appear, 2003.
8. Minar, N., Burkhart, R., Langton, C. and Askenazi, M., http://www.santafe.edu/projects/swarm, 1996.
9. 11. R. A. Jarvis and E. A. Patrick, Clustering using a similarity measure based on shared nearest neighbors, IEEE Transactions on Computers, C-22(11), 1973.
10. Reynolds C. W., Flocks, Herds, and Schools: A Distributed Behavioral Model, Computer Graphics vol. 21, n. 4, , (SIGGRAPH 87), pp. 25–34, 1987. van Leeuwen, J. (ed.): Computer Science Today. Recent Trends and Developments. Lecture Notes in Computer Science, Vol. 1000. Springer-Verlag, Berlin Heidelberg New York (1995)

When Can We Call a System Self-Organizing?

Carlos Gershenson and Francis Heylighen

Centrum Leo Apostel, Vrije Universiteit Brussel
Krijgskundestraat 33, Brussels, 1160, Belgium
{cgershen, fheyligh}@vub.ac.be
http://www.vub.ac.be/CLEA

Abstract. We do not attempt to provide yet another definition of self-organization, but explore the conditions under which we can model a system as self-organizing. These involve the dynamics of entropy, and the purpose, aspects, and description level chosen by an observer. We show how, changing the level or "graining" of description, the same system can appear self-organizing or self-disorganizing. We discuss ontological issues we face when studying self-organizing systems, and analyse when designing and controlling artificial self-organizing systems is useful. We conclude that self-organization is a way of observing systems, not an absolute class of systems.

1 Introduction

There have been many notions and definitions of self-organization, useful in different contexts (for a non-technical overview, see [1]). They have come from cybernetics [2, 3, 4, 5], thermodynamics [6], mathematics [7], information theory [8], synergetics [9], and others. Many people use the term "self-organizing", but it has no generally accepted meaning, as the abundance of definitions suggests. Also, proposing such a definition faces the philosophical problem of defining "self", the cybernetic problem of defining "system", and the universal problem of defining "organization". We will not attempt to propose yet another definition of self-organizing systems. Nevertheless, in order to try to understand these systems better, we will explore the following question: which are the necessary conditions in order to call a system "self-organizing"? We do so by combining insights from different contexts where self-organizing systems have been studied.

The understanding of self-organizing systems is essential for the artificial life community, because life obviously exhibits self-organization, and so should the systems simulating it. As a most basic example, we can look at swarming or herding behaviour, in which a disordered array of scattered agents gathers together into a tight formation. Intuitively, it seems that the system of agents has become more "organized". But precisely what does this mean?

In the following section we explore the role of dynamics in self-organizing systems. We provide examples of systems that are self-organizing at one level but not at another one. In Section 3 we note the relevance of the observer for perceiving self-organization. We discuss some deeper conceptual problems for understanding self-organizing systems in Section 4. In Section 5 we discuss applications in artificial self-

W. Banzhaf et al. (Eds.): ECAL 2003, LNAI 2801, pp. 606–614, 2003.

organizing systems, and when using this approach is appropriate. We draw concluding remarks in Section 6.

2 The Representation-Dependent Dynamics of Entropy

A property frequently used to characterize self-organization is an increase of order which is not imposed by an external agent (not excluding environmental interactions) [1]. The most common way to formalize the intuitive notion of "order" is to identify it with the negative of *entropy*. The second law of thermodynamics states that in an isolated system, entropy can only decrease, not increase. Such systems evolve to their state of maximum entropy, or thermodynamic equilibrium. Therefore, self-organizing systems cannot be isolated: they require a constant input of matter or energy with low entropy, getting rid of the internally generated entropy through the output of heat ("dissipation"). This allows them to produce "dissipative structures" which maintain far from thermodynamic equilibrium [6]. Life is a clear example of order far from thermodynamic equilibrium.

However, the thermodynamical concept of entropy as the dissipation of heat is not very useful if we want to understand information-based systems, such as those created by ALife modellers. For that, we need the more general concept of *statistical entropy* (H) which is applicable to any system for which a state space can be defined. It expresses the degree of uncertainty we have about the state s of the system, in terms of the probability distribution $P(s)$.

$$H(P) = -\sum_{s \in S} P(s) \log P(s) \tag{1}$$

In this formulation, the second law of thermodynamics can be expressed as "every system tends to its most probable state" [4]. This is in as sense a tautological law of nature, since the probabilities of the states are determined by us according to the tendencies of systems. At a molecular level, the most probable state of an isolated system is that of maximum entropy or thermodynamic equilibrium, where the molecules are distributed homogeneously, erasing any structure or differentiation. But does this apply as well to a real or artificial living organism?

We have to be aware that probabilities are relative to a level of observation, and that what is most probable at one level is not necessarily so at another. Moreover, a state is defined by an observer, being the conjunction of the values for all the variables or attributes that the observer considers relevant for the phenomenon being modelled. Therefore, we can have different degrees of order or "entropies" for different models or levels of observation of the same entity.

Let us illustrate this with the following, very simple example. Consider a system with four possible "microstates", $a1$, $a2$, $b1$, and $b2$, at the lowest, most detailed level of description. At the higher, more abstract level of description, we aggregate the microstates two by two, defining two macrostates: $A = \{a1, a2\}$ and $B = \{b1, b2\}$. This means that the system is in macrostate A if it is either in microstate $a1$ or in microstate $a2$. The probabilities of the macrostates are simply the sum of the probabilities of their sub-states. Let us suppose that we start from an initial probability distribution so that $P(a1) = P(b1) = 0.1$ and $P(a2) = P(b2) = 0.4$. This implies $P(A) =$

P(B) = 0.5. We can calculate the statistical entropy *H* using (1), getting *H*.1.72 at the lower level, and *H*=1 at the higher level.

Now consider a second distribution *P(a1)* = *P(a2)* = 0.2 while *P(b1)* = *P(b2)* = 0.3. Therefore, *P(A)* = 0.4 and *P(B)* =0.6. Now we have *H*.1.97 at the lower and *H*.0.97 at the higher levels. Subtracting the initial *H* from the second, we have M*H*/Mt.0.24 at a lower level and M*H*/Mt.!0.029 at the higher level. We have a change of distribution where entropy is *decreased* at the higher level ("self-organization"), and *increased* at the lower level ("self-disorganization"). To get the inverse change, we can just assume the final states to be the initial ones and vice versa. We would then have self-organization at the lower level and self-disorganization at the higher level.

This can be represented graphically in Figure 1, where tones of gray represent the probabilities of the states (darker colour = lower value). The macrostates provide a coarse-grained [10] representation of the system, while the microstates provide a fine-grained one. We can visualize entropy as homogeneity of colours/probabilities. At the lower level, the distribution becomes more homogeneous, and entropy increases. At the higher level, the distribution becomes more differentiated.

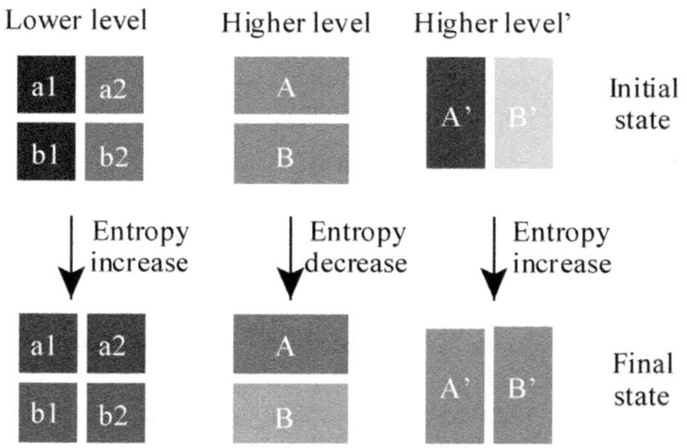

Fig. 1. Entropy, seen as homogeneity, increases at lower level, while it might increase or decrease at the higher level, depending on how we divide the states

Entropy not only depends on higher or lower levels of abstraction, but also on how we set the boundaries between states. Let us define alternative macrostates *A'*, which has *a1* and *b1* as sub-states, and *B'*, with *a2* and *b2* as sub-states. Using the same values for the probabilities, we see that for the alternative macrostates the initial *H*.0.72 and the final *H*=1. So we have that M*H*/Mt.0.27, which means that the statistical entropy increases in this macrorepresentation, while it decreases in the previous one. The system appears to be self-organizing or disorganizing, depending not only on the level at which we observe it, but also on *how* we do the "coarse-graining" of the system, that is to say, which variables we select to define the states.

The variables defined by the values (A, B), respectively (A', B'), represent two aspects of the same system, where the observer has focussed on different, independent properties. For example, a particle's state includes both its position in space and its momentum or velocity. A subsystem is defined as a physical part of a system, limited to some of its components. Similarly, an *aspect system* can be defined as a *functional part of a system*, limited to some of its properties or aspects [11].

Let us illustrate this with a typical ALife model: swarming behaviour. Groups of agents can be seen as subsystems of the swarm. The positions of all agents define one aspect system, while their velocities define another aspect system. Assume we start with non-moving agents scattered all over the simulated space. The position aspect is characterized by maximum entropy (agents can be anywhere in space), while the velocity aspect has minimum entropy (all have the same zero velocity). According to typical swarming rules, the agents will start to move with varying speeds towards the centre of the swarm while mutually adjusting their velocities so as not to bump into each other. This means that their states become more concentrated in position space, but more diffuse in velocity space. In other words, entropy decreases for the positions, while increasing for the velocities. Depending on the aspect we consider, the swarm self-organizes or self-disorganizes!

This example may appear too specific to support our argument. Let us therefore show that dynamical processes in general exhibit this kind of aspect-dependent or level-dependent behaviour, either increasing or decreasing entropy. A dynamical system, where every state is mapped deterministically onto a single next state, can be seen as special case of a Markov process, where a state s_i is mapped stochastically onto any number of other states s_j with a fixed transition probability $P(s_i \; 6 \; s_j)$. To turn a deterministic dynamics into a stochastic one, it suffices to apply coarse-graining, aggregating a number of microstates into a single macrostate. Transitions from this macrostate can now go to any other macrostate that includes one of the microstates that were the initial destinations of its microstates.

It can be proven that the statistical entropy of a distribution cannot decrease if and only if the Markov process that describes the mapping from initial to final distribution is *doubly stochastic* [12]. This means that for the matrix of transition probabilities between states, the sum over a row and the sum over a column must be one. The sum over a row (probability of a given state ending in any state of the state space) is necessarily one, by definition of probability. However, the sum over a column is not a probability but can be seen as a measure of the "attractiveness" or "fitness" F of a state s:

$$F(s_i) = \sum_j P(s_j \rightarrow s_i) \tag{2}$$

High fitness ($F > 1$) of a state s means that on average more transitions enter s than leave s (the sum of all transition probabilities leaving s can never be >1). Thus, while the process runs, high fitness states become more probable, and low fitness states less probable, increasing their differentiation, as would be expected from a process undergoing self-organization.

A doubly stochastic process is defined by the requirement that $F(s_i) = 1$ for all states s_i. This corresponds to the entropy increasing processes studied in traditional thermodynamics. In a more general process, probability eventually concentrates in the high fitness states, decreasing overall entropy if the initial distribution is more homogeneous, increasing it if it is more "peaked". Therefore, a necessary and sufficient condition for a Markov process to allow self-organization is that it has a non-trivial fitness function, *i.e.* there exist states such that $F(s)$ 1. The sum of all probabilities must be 1. Therefore, for every $F(s)>1$, there must be at least one $F(s')<1$. This condition is equivalent to saying that the dynamics has a differential "preference" for certain states *s* over other states *s'*. Any dynamics that allows attractors has such an inbuilt preference for attractor states over basin states [1]. This is the most general case, and the one typically found in ALife simulations.

For any initial and final distribution of probabilities, such as the ones in the examples we discussed, it is possible to determine a matrix of transition probabilities that maps the one onto the other. This matrix will in general not be doubly stochastic, and therefore allow self-organization as well as disorganization. Therefore, the same system, described in different aspects or levels of abstraction can be modelled as self-organizing in the one case, as self-disorganizing in another.

3 The Role of the Observer

We have to be aware that even in mathematical and physical models of self-organizing systems, it is the *observer* who ascribes properties, aspects, states, and probabilities; and therefore entropy or order to the system. But organization is more than low entropy: it is structure that has a function or *purpose* [13]. Stafford Beer [4] noted a very important issue: what under some circumstances can be seen as organization, under others can be seen as disorder, depending on the purpose of the system. He illustrates this idea with the following example: When ice cream is taken from a freezer, and put at room temperature, we can say that the ice cream disorganizes, since it loses its purpose of having an icy consistency. But from a physical point of view, it becomes more ordered by achieving equilibrium with the room, as it had done with the freezer[1]. Again, the purpose of the system is not an objective property of the system, but something set by an *observer.*

W. Ross Ashby noted decades ago the importance of the role of the observer in relation to self-organizing systems: "A substantial part of the theory of organization will be concerned with *properties that are not intrinsic to the thing but are relational between observer and thing*" ([3], p. 258, emphasis in original).

Of course there should be a correlate in the world to the observations. The question now is: How frequent is this correlate, so that we can observe self-organization? What we need is a collection of elements that interact. By generalizing the second law of thermodynamics, we can see that the system through time will reach a more "probable" or "stable" configuration. We can say that it will reach an equilibrium or attractor[2]. The observer then needs to focus his/her viewpoint, in order

---

[1] Thermodynamic entropy can be seen as order or disorder in different situations (*e.g.* [6, 4]. This also occurs with information entropy, as the debate between Wiener and Shannon showed.

[2] In some chaotic systems, this can take practically infinite time. But as systems approach an attractor, we can say that they follow this law.

to set the *purpose* of the system so that we can see the attractor as an "organized" state and to see it at the right *level* and in the right *aspect*, and then self-organization will be observed. We can see that this is much more common than what intuition tells us. Not only lasers, magnets, Bénard rolls, ant colonies, or economies can be said to be self-organizing. Even an ideal gas can be said to be self-organizing, if we say (contrary to thermodynamics) that the equilibrium state where the gas homogeneously fills its container, is "ordered" [4]. *Any* dynamical system *can* be said to be self-organizing [3]. *Self-organization is a way of modelling systems, not a class of systems.* This does not mean that there is no self-organization independently of the observer, but rather that self-organization is everywhere.

Of course, not all systems are usefully described as self-organizing. Most natural systems can be easily fit into the class "self-organizing", unlike the simple mechanisms we find in physics textbooks. Most artificial systems are hard to see as self-organizing. Many are not dynamic, others involve only one element (actually no system), and most of the rest follow sequences of rules that can be easily understood. Therefore there is no need to explain their functioning with the rather esoteric concept of "self-organization".

We have said that any dynamical system, if observed "properly", can be seen as self-organizing. But if we set a different purpose or description level, then *any* dynamical system can be self-disorganizing. An economy will not be seen as self-organizing if we look only at a short timescale, or if we look at the scale of only one small business. An ant colony will not be self-organizing if we describe only the *global* behaviour of the colony (*e.g.* as an *element* of an ecosystem), or if we only list the behaviours of individual ants. We have to remember that the description of self-organization is partially, but strongly, dependent on the observer.

4 Ontological Issues

One of the most common problems when discussing self-organizing systems is the meaning of emergence. Self-organizing systems typically have higher level properties that cannot be observed at the level of the elements, and that can be seen as a product of their interactions (more than the sum of the parts). Some people call these properties *emergent*. The problem we face is ontological. According to Aristotelean logic, a system cannot be more than one thing at a time. In this case, a system cannot be at the same time a set of elements and a whole with emergent properties. But by introducing an ontological distinction [14], we can clarify the issue.

We can distinguish two types of *being*: relative and absolute. The relative (*rel*-being) is experienced by an observer with a finite cognitive capacity. It therefore depends on her/his context, and is limited. Strictly speaking, every cognizer has a different *rel*-being of anything, since every cognizer has a different context. Theoretically, we can assume that there exists an absolute being (*abs*-being), which would be "the real thing" (Kant's *Ding-an-sich*), independent of the observer, which observers correlate to their *rel*-beings. We can observe any *abs*-being from an infinity of perspectives and describe an infinity of potential properties or aspects. Nevertheless, most *rel*-beings and contexts are similar, since they are inspired by the same *abs*-being seen by similar observers from a similar point of view. This enables us to share knowledge, but it is because of the different nuances in the different *rel*-beings and contexts that we fail to agree in every situation.

We can then say that the observation of a system at different abstraction levels or in different aspects is merely a difference in the perspective, and therefore the system *rel*-is different (only for the observers). But the system *abs*-is the same thing itself, independently of how we describe it. We can observe a cell as *rel*-being a bunch of molecules or as *rel*-being a living structure. But it *abs*-is both and even more. *Rel*-beings can be seen as different models or metaphors for describing the same thing. A change in the metaphor does not change the thing. If we define emergence as a process that requires a change of the model [15] in order to better understand and predict the system [8], then it becomes clear that there is no magic. Any dynamical system *abs*-is self-organizing and self-disorganizing at the same time, in the sense that it can potentially *rel*-be both.

Another confusion may arise when people describe systems as the lower level *causing* change in the emergent properties. Vice-versa, downward causation is the idea that higher level properties constrain or control components at the lower level [16]. Speaking about causality between abstraction levels is not accurate [14], because actually they *abs*-are the same thing. What we could say is that when we observe certain conditions in the lower level, we can expect to observe certain properties at a higher level, and vice versa. There is correlation, but not actual causation.

This leads us to what is probably the most fundamental problem. If we can describe a system using different levels, aspects, or representations, which is the one we should choose? As Prem [17] suggests, the level should be the one where the prediction of the behaviour of the system is easiest; in other words, where we need least information to make predictions[3] [8]. A possible way to formalize this requirement is by choosing the representation that minimizes the conditional entropy, *i.e.* the average uncertainty of the next state given the present state [5]. We can speculate that this is possible because of regularities in the system at that particular level, and that this is what leads people to try to describe how the simple properties cause the not so simple ones, either upward or downward.

5 Artificial Self-Organizing Systems

Independently of the definition of self-organizing systems, if we see them as a perspective for studying systems, we can use this perspective for designing, building, and controlling systems. A key characteristic of an artificial self-organizing system is that structure and function of the system "emerge" from interactions between the elements. The purpose should not be explicitly designed, programmed, or controlled. The components should *interact* freely with each other and with the environment, mutually adapting so as to reach an intrinsically "preferable" or "fit" configuration (attractor), thus defining the purpose of the system in an "emergent" way [13].

Certainly this is not the only approach for designing and controlling systems, and in many cases it is not appropriate. But it can be very useful in complex systems

[3] This argument could be also followed to decide which "graining" to choose.

where the observer cannot *a priori* conceive of all possible configurations, purposes, or problems that the system may be confronted with. Examples of these are organizations (corporations, governments, communities), traffic control, proteomics, distributed robotics, allocation of ecologic resources, self-assembling nanotubes, and complex software systems [13], such as the semantic web.

For artificial life, we believe that the perspective of self-organizing systems is essential, because we cannot explain the emergence, evolution, and development of life, whether *in vivo* or *in silico*, by restricting our models to a single level of abstraction. We need to understand how properties such as life, autocatalysis, or autopoiesis can emerge from interactions of elements without these properties, or how species or social properties can arise from individual interactions. This can only be done from a perspective of self-organizing systems. Therefore, it is important that the issues we have discussed are taken into account by artificial life researchers.

6 Conclusions

We proposed that self-organizing systems, rather than a *type* of systems, are a *perspective* for studying, understanding, designing, controlling, and building systems. This perspective has advantages and disadvantages, and there are systems that benefit from this approach, and others for which it is redundant. But even in the general case when the systems dynamics allows self-organization in the sense of entropy decrease, the crucial factor is the *observer*, who has to describe the process at an appropriate *level(s)* and *aspects*, and to define the *purpose* of the system. All these "make" the system to be self-organizing. In that sense, self-organization can be everywhere: it just needs to be observed.

We believe that this discussion on when and how to best model a system as self-organizing should be carried further in the artificial life community, since we all study and build systems from a self-organizing perspective. This would benefit not only the community, but every domain where the notion of self-organization is useful.

References

[1] Heylighen, F. (2003). The Science of Self-organization and Adaptivity, in: Knowledge Management, Organizational Intelligence and Learning, and Complexity, in: *The Encyclopedia of Life Support Systems*, EOLSS Publishers Co. Ltd.

[2] Von Foerster, H (1960). On Self-Organizing Systems and their Environments. In Yovitts, M. C. and S. Cameron (Eds.), *Self-Organizing Systems*, Pergamon, pp. 31–50.

[3] Ashby, W. R. (1962). Principles of the Self-organizing System. In von Foerster, H. and G. W. Zopf, Jr. (Eds.), *Principles of Self-organization*. Pergamon Press, pp. 255–278.

[4] Beer, S. (1966). *Decision and Control: The Meaning of Operational Research and Management Cybernetics*. John Wiley & Sons, Inc.

[5] Heylighen, F. and C. Joslyn (2001). Cybernetics and Second Order Cybernetics, in: R. A. Meyers (ed.), *Encyclopaedia of Physical Science & Technology*, Vol. 4 (3rd ed.), Academic Press, pp. 155–170.

[6] Nicolis, G. and I. Prigogine (1977). *Self-Organization in Non-Equilibrium Systems: From Dissipative Structures to Order Through Fluctuations*, Wiley.

[7] Lendaris, G. G. (1964). On the Definition of Self-Organizing Systems. *Proceedings of the IEEE*, March.

[8] Shalizi, C. R. (2001). Causal Architecture, Complexity and Self-Organization in Time Series and Cellular Automata. *Doctoral Dissertation.* University of Wisconsin at Madison.

[9] Haken, H. (1981). Synergetics and the Problem of Selforganization. In Roth, G. and H. Schwegler (eds.). *Self-organizing Systems: An Interdisciplinary Approach*, Campus Verlag, pp. 9–13.

[10] Hobbs J. (1985). 'Granularity', in : *Proceedings of the 9th International Joint Conference on Artificial Intelligence*(vol. 1), p. 423.

[11] ten Haaf W., H. Bikker, D.J. Adriaanse (2002) *Fundamentals of Business Engineering and Management, A Systems Approach to People and Organisations*, Delft University Press

[12] Koopman, B. (1978). Entropy Increase and Group Symmetry. In Tribus, M. and I. Levine (Eds.) *Maximum Entropy Formalism.* MIT Press.

[13] Heylighen, F. and C. Gershenson (2003). The Meaning of Self-organization in Computing. *IEEE Intelligent Systems*, section Trends & Controversies – Self-organization and Information Systems, May/June 2003. [in press]

[14] Gershenson, C. (2002). Complex Philosophy. *Proceedings of the 1st Biennial Seminar on Philosophical, Methodological & Epistemological Implications of Complexity Theory.* La Habana, Cuba. Also in *InterJournal of Complex Systems*, 544.

[15] Rosen, R. (1985). *Anticipatory Systems.* Pergamon.

[16] Campbell, D. T. (1974). 'Downward causation' in Hierarchically Organized Biological Systems, In Ayala, F. J. and T. Dobzhansky (Eds.), *Studies in the Philosophy of Biology*, Macmillan, pp. 179–186.

[17] Prem, E. (1993). Understanding Self-Organization: What can the speaking lion tell us? Oesterreichisches Forschungsinstitut fuer Artificial Intelligence, Wien, TR-93-14.

Contextual Random Boolean Networks

Carlos Gershenson, Jan Broekaert, and Diederik Aerts

Centrum Leo Apostel, Vrije Universiteit Brussel, Krijgskundestraat 33,
Brussels, 1160, Belgium
{cgershen, jbroekae, diraerts,}@vub.ac.be
http://www.vub.ac.be/CLEA

Abstract. We propose the use of Deterministic Generalized Asynchronous Random Boolean Networks [1] as models of contextual deterministic discrete dynamical systems. We show that changes in the context have drastic effects on the global properties of the same networks, namely the average number of attractors and the average percentage of states in attractors. We introduce the situation where we lack knowledge on the context as a more realistic model for contextual dynamical systems. We notice that this makes the network non-deterministic in a specific way, namely introducing a non-Kolmogorovian quantum-like structure for the modelling of the network [2]. In this case, for example, a state of the network has the potentiality (probability) of collapsing into different attractors, depending on the specific form of lack of knowledge on the context.

1 Introduction

Random Boolean Networks (RBNs) have been used to model a variety of phenomena and have been widely studied [3,4,5]. Especially, they were used by Stuart Kauffman [6] to study genetic regulatory networks and to correlate the number of human cell types to the attractors of a RBN with a number of nodes similar to the number of genes in the DNA. This is interesting, because one does not assume any function in the RBN, these are generated randomly. However, this approach was criticized because it assumed that the dynamics of the DNA would be discrete and synchronous in time [7,8]. The proposed alternative, non-deterministic asynchronous updating, did not give encouraging results, since the networks change drastically their properties due to the non-determinism. We proposed another type of updating: asynchronous but deterministic [1]. This is achieved by setting parameters to determine the period of the updating of each node. The problem now lies in how to find or study different updating periods.

Our approach in the present paper consists in the following: we see the parameters of the updating periods of the nodes as a *context* of the network. We then study what happens when we change this context for a given network, *i.e.* allow smaller or greater periods. We find out that the properties of the network change drastically if we change the context.

We mentioned already that the non-deterministic asynchronous updating led to effects of changes that are too drastic due to the non structured nature of

W. Banzhaf et al. (Eds.): ECAL 2003, LNAI 2801, pp. 615–624, 2003.

the non-determinism. In our approach we want to introduce non-determinism, but in a very specific and structured way: namely non-determinism related to a "lack of knowledge" on the specific context (updating period) that we consider. The idea is that we start by introducing contexts that influence the network deterministically, and call these contexts *pure contexts*. Under the effect of such a pure context the network still behaves completely deterministically, but each pure context will generate a different behaviour for the network. These pure contexts are very idealized models for context in the real world. If we want to model the effect of a real context, we have to take into account that we generally lack knowledge on how this real context interacts with the system, because of the presence of random and unpredictable fluctuations. Hence we model such a real context by introducing the idea of *mixed context*, where a mixed context is a statistical mixture of a set of pure contexts. Concretely this means that we describe a mixed context by a probability measure on the set of pure contexts. We will show that the effect of this structured type of non-determinism is that states of the network become *potentiality states*, in the sense that instead of a state evolving deterministically to one of the attractors, such a state will now be attributed a certain probability (potentiality) to *collapse* to one of the different attractors under the influence of a certain context, the probabilities being determined by the weights of the different pure contexts within the considered mixed context.

So the behaviour of the network becomes non-deterministic when we introduce mixed contexts, but it is much more structured than in asynchronous RBNs where anything can happen, and therefore it can be well studied. The structure of the probability model that arises by means of the introduction of mixed contexts has been studied in detail, and it can be proven that this structure is non-Kolmogorovian [2].

We want to study contextual Random Boolean Networks by introducing the effect of the context in the way we just explained because it is a realistic model for the effect of real contexts. There is however a second reason, namely, this type of non-deterministic contextual influence has been studied in great detail in our Brussels research group, leading to a modelling of contextuality that is quantum-mechanic-like. What we mean more specifically is that the mathematical structure that emerges from a situation where non-determinism is introduced as a consequence of the presence of mixed contexts is a quantum mechanical mathematical structure [2,9,10,11,12]. Furthermore this type of contextual models have been used in different ways, mostly to model the contextuality in situations of cognition [13,14,15,16,17]. There are reasons to believe that this type of contextuality is also present in biological systems [18,19], and that is the reason that we introduce it for Random Boolean Networks.

In the following section, we make a brief review of different types of RBNs, according to their updating scheme [1]. In Section 3, we analyse the effects of the change of pure context in the statistical properties of a RBN. In Section 4, we use a concrete example of a small contextual RBN to note the properties of a mixed context. We draw conclusions and future lines of research in Section 5.

2 Background

A Random Boolean Network [20,6] can be seen as a generalization of a boolean cellular automata. It also consists of n nodes, each with k connecting inputs, only that the connections are not restricted to neighbours, but they can be to any other node in the network. The topology is generated randomly, but it remains fixed during the dynamics of the network. These dynamics are determined by boolean rules also generated randomly in lookup tables. The state of a node will depend on the states of the nodes to which it is connected, *i.e.* its inputs.

It is interesting to study the general properties of RBNs with different number of nodes and connections, making statistical analyses, because we can obtain general properties of the dynamics of a family of networks, without assuming the rules of the dynamics. Nevertheless, it has been shown that these properties can change drastically depending on how we update the nodes [7,1,21].

We have proposed a classification of RBNs [1] according to the different updating schemes that a network might have:

Classical Random Boolean Networks (CRBNs) [20,6]. The updating is synchronous and deterministic: each node takes its value at time t+1 from the values of the nodes connected to it at time t.

Asynchronous Random Boolean Networks (ARBNs) [7]. The updating is asynchronous, but also non-deterministic. Each time step only one node is picked at random and updated.

Deterministic Asynchronous Random Boolean Networks (DARBNs) [1]. The updating is asynchronous and deterministic. To achieve this, we introduce two parameters per node: p and q, so that the node will be updated when time modulus p equals q. Therefore, p can be seen as the period and q as the translation in time of the node update. If more than one node is updated at one time step, the network is actualized in the same order after each node update.

Generalized Asynchronous Random Boolean Networks (GARBNs) [1]. The updating is non-deterministic, but semi-synchronous. We select randomly at each time step some nodes to update, and these are updated synchronously.

Deterministic Generalized Asynchronous Random Boolean Networks (DGARBNs) [1]. The updating is deterministic and semi-synchronous. We also use parameters p and q in each node to determine the period and translation of each update, but the nodes which should be updated at time t do so synchronously.

In the deterministic cases (CRBN, DARBN, DGARBN), there can be cyclic attractors (when the dynamic of the network is cycled in a subset of the state space) and point attractors (when a single state "traps" the dynamics). On the other hand, for the non-deterministic cases (ARBN, GARBN) there are point attractors, and loose attractors (a subset of the state space drags the dynamic, but this can follow several patterns inside this subset, since we do not know which node will be updated at a given time). To our knowledge loose attractors have not been studied, since it is not trivial to find them.
We will use DGARBNs to study the change of context in RBNs.

Resources: We have developed a software laboratory, "RBNLab", for studying the properties of different types of RBNs. It is available to the public, for use (via browser) and download (Java source code included) at http://student.vub.ac.be/cgershen/rbn. The results presented in this paper were obtained using RBNLab.

3 Changing Contexts in DGARBNs

We consider the set of all p's and q's as the *context* of the DGARBN. This is because in real networks some external factors, such as temperature or tension [22], can produce a change in the updating regularity of the nodes. The external factors are thus reflected in the updating periods, which are represented with the sets of parameters of p's and q's. We should note that this is a temporal context, but might reflect other types of contexts that affect a system.

After generating randomly the topology and rules of a network of given n and k, we randomly generate contexts according to a parameter maxP, which indicates the maximum allowed period for a node. The p's are random integers between 1 and maxP, while q's are also random integers, but between 0 and maxP-1. We can see that the case maxP=1 gives us CRBNs (fully synchronous).

It could be argued that it is equivalent to study CRBNs that would consider the state and context of a DGARBN as a state of a more complex CRBN, since we have shown that any deterministic asynchronous RBN can be mapped into a CRBN of higher complexity [1]. This is achieved by adding nodes (encoding the context) connected to every other node in the network. Remark that a network of given n, k, and maxP can be converted into a CRBN of, $\forall i$, $n + \text{ceiling}(\log 2(\text{LCM}(p_i)))$ and $k + \text{ceiling}(\log 2(\text{LCM}(p_i)))$. First, it is conceptually more clear to divide the network and its context, because changing the context in a DGARBN or DARBN is easy and not so in its CRBN correlate. Moreover, the family of CRBNs of extended n and k is much larger than the one of contextual RBNs, so we speculate they have different statistical properties.

About statistical properties, we should note that all types of RBNs tend to have different properties in theory (exact solutions) than in practice (statistics). This was shown already by Harvey and Bossomaier [7] for ARBNs. For these type of RBN, the exact solution (expected if one would exhaust all the RBN family, but this is unfeasible computationally) for the average number of attractors is exactly one, for any type of topology (combinations of n and k). However, the statistics give us a different result. In some networks ($k = 3$) the distribution of attractors is such that these tend to "hide" (few networks with several attractors, many with none). We are aware that in all RBNs these divergences between theory and practice are common. But we are interested more in statistical properties than in exact solutions, because we aim at finding a network of certain properties, without exhausting the network space.

Method: We generated one thousand DGARBNs for each given n and k, and explored the properties of the same networks as we varied maxP. Remark that

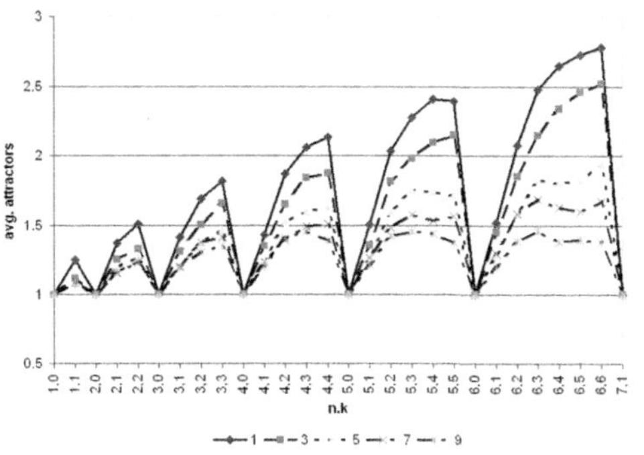

Fig. 1. Average number of attractors varying maxP

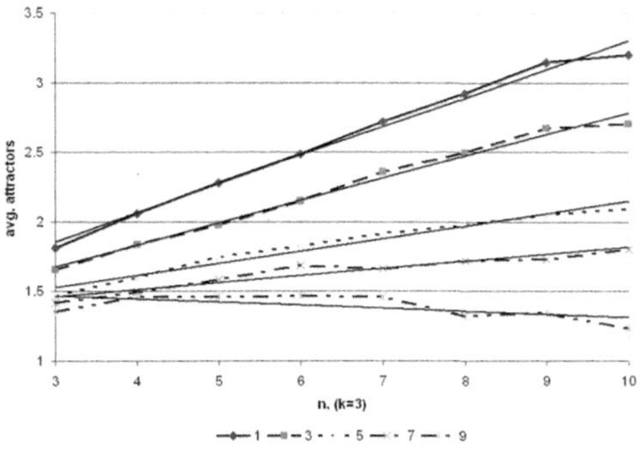

Fig. 2. Average number of attractors varying maxP

for $n > 10$, for computational resources issues, we sampled only 100 networks. Results for $n > 10$ should thus be considered illustrative. To find the properties of the networks, we explore all possible initial conditions, running the network for ten thousand steps expecting that the network will reach an attractor. The attractors take into account the state and the context, so for example a point attractor (period one in CRBN) will be considered as having a period equal to the least common multiple of all p's. This is to explore all the possible combinations of updates in the nodes.

Results: The average number of attractors of the explored networks of different n and k values can be seen in Figure 1. Figure 2 shows the results for different n's and $k = 3$.

The first thing we can see by observing these graphs is that the context change has drastic consequences on the average number of attractors. Moreover, comparing these results with the ones of Gershenson [1] contexts of small maxP clearly yield network properties very similar to synchronous RBNs (CRBNs). Still, as we increase maxP, the properties begin to look more like the ones of asynchronous non-deterministic RBNs (ARBNs and GARBNs). Another thing we can observe is that even when the average number of attractors for low maxP has a linear increment proportional to n, this decreases and becomes even decrement for high maxP. This means that for large networks, the differences in the number of attractors will grow considerably as we change the context. The attractors of the same networks collapse as we change the context and allow larger periods (increase maxP). The inverse results are obtained in CRBNs including context as part of their state, since the average number of attractors in CRBNs is increased with n and k. We also know from Gershenson [1] that the point attractors are the same for every type of RBN, so the attractors that collapse are not these ones, but the cycle attractors. The percentage of states of the networks that are part of an attractor in logarithmic scale can be seen in Figure 3 for different n's and k's and for $k = 3$ in Figure 4.

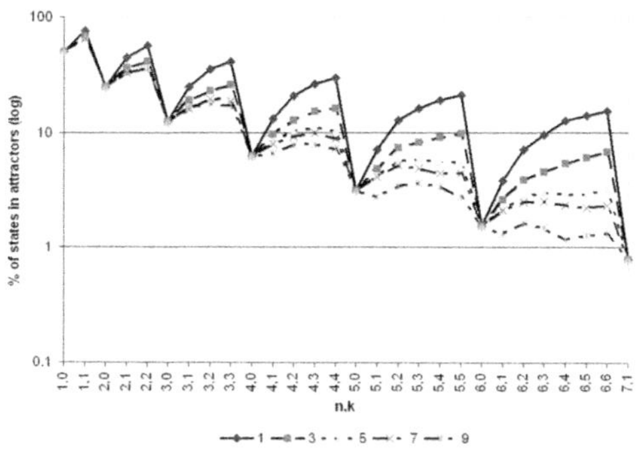

Fig. 3. Percentage of states in attractors varying maxP (log scale)

We can see that all the percentages of states in attractors decrease exponentially as we increase n. But the percentage of states in attractors decreases slower for low maxP than for a high one. This also indicates that the relative differences among contexts will be greater for larger networks. It is curious to note that for not very small n, the percentage of states in attractors for low maxP increases with k, but for a high maxP it slightly rises and falls again roughly around $k = 3$.

Even for our small statistical samples, percentages of states in attractors fit nicely an exponential curve (Figure 4), as opposed to number of attractors which roughly have linear fit (Figure 2). This suggests that there is less variance in percentages among networks than in number of attractors.

The tables with the precise data used to generate the graphics are available through the URL of RBNLab.

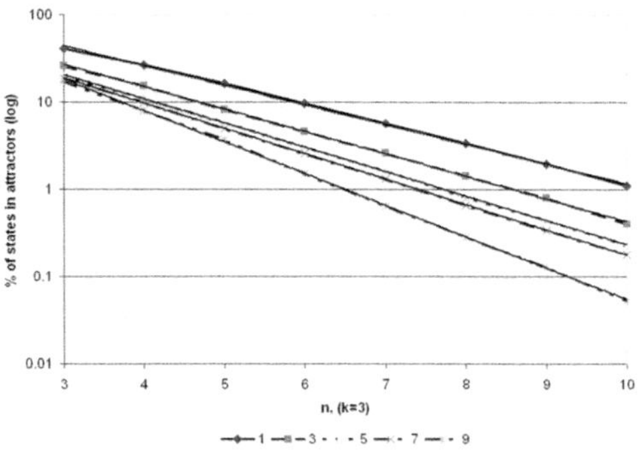

Fig. 4. Percentage of states in attractors varying maxP (log scale)

4 Lack of Knowledge about the Context

Let us now explore the situation where we do not know exactly the context in which a certain DGARBN is. We can explore a very small RBN and observe its possible pure contexts. We devised a RBN of $n = 2$, $k = 2$ that makes explicit the point we want to make. Table 1 shows its transition table.

Table 1. Transition table of a RBN $n = 2$, $k = 2$.

Net(t)	Net(t+1)
00	01
10	10
01	10
11	11

We consider now pure contexts with maxP=2. The possible dynamics can be seen in Figure 5. Other combinations of p's and q's give similar attractor basins.

All the pure contexts have at least two attractors: '11' and '10' (only D has an attractor '$01[tmod2 = 1] \rightarrow 00[tmod2 = 0] \rightarrow 01[tmod2 = 1]$'). Suppose now the network is initially in some state, e.g. '01', and suppose the uncertainty on the *mixed* context can be cast in some probabilistic measure; such that each possible *pure* context is attributed a given weight. When exposing the system to this mixed context, a probabilistic mechanism during the exposure of system to context selects one of the pure contexts, but we do not know which one exactly. Subsequently the network then proceeds in a dynamics according to the selected context. We can say that the network is in a *potentiality* state, because depending on the exact nature of the mixed context (what the probability weights on the different pure contexts are), it will go to different attractors with a probability proportional to the weights of the pure contexts in this specific mixed context. We remark that we know perfectly the dynamics of the network and the effect of the pure contexts. They are deterministic. There is no lack of knowledge about

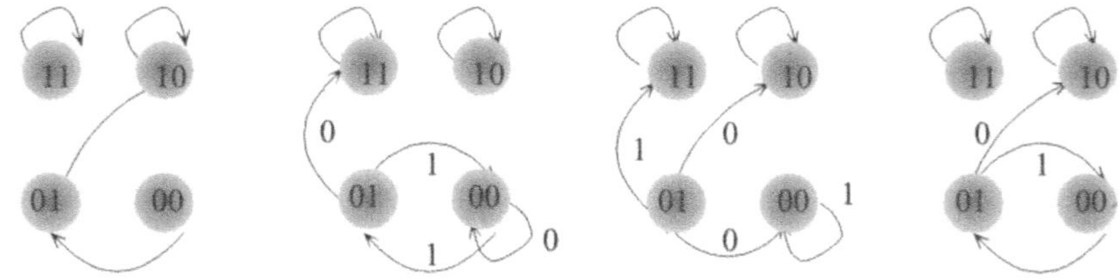

Fig. 5. Dynamics of a DGARBN $n = 2$, $k = 2$ with different pure contexts. The arrows with numbers indicate which transition will take place depending on the modulus of time over two

the functioning of the network, but there is a lack of knowledge about the pure context, expressed in the statistical mixture which is the mixed context. As it has been shown, this means that the situation we face is not a classical mechanical situation, but a quantum-mechanical-like situation [2,9,10,11,12].

Contextual RBNs of this type could be used to model, for example, how the genes and environment of a cell determine its behaviour. A stem cell has the *potentiality* of differentiating into almost any cell type. But the differentiation is not determined only by the genes of the cell, but also by the context (environment) in which it lays.

5 Conclusions and Future Work

We have used deterministic generalized asynchronous random boolean networks to model context change in discrete dynamical systems. We observed that the context changes dramatically the properties of the systems. An interesting result was noting that the properties of restricted temporal contexts (small maxP) resemble the dynamics of synchronous RBNs, while unrestricted ones (large maxP) appear more like non-deterministic asynchronous RBNs.

We introduced two types of context, pure contexts, their interaction with the network is deterministic, and mixed contexts, more precisely statistical mixtures of pure context, which have a non-deterministic interaction with the network. The presence of mixed contexts changes completely the dynamics of the network, in the sense that we get non-deterministic dynamics, generating a probability structure that is non-Kolmogorovian (quantum-like), and transforming states of the network into potentiality states that collapse with a certain probability to the attractor states.

For the pure contexts (the deterministic dynamics), it is worth mentioning that in all types of networks, the percentage of states in an attractor diminishes exponentially (see also [1]). This means that, *in theory*, independently of their connectivity (k), context, or updating scheme, large networks would always have "order for free" [6]. In other words, the wide variety of the state space, as n

increases the percentage of states which can "trap" the dynamics, *i.e.* states in an attractor, will diminish. The difference in practice lies in how long will it take to reach this order. In some chaotic networks ($k > 2$ for CRBN), this might be infinite in practice and the order will not be found. Therefore, it would be significant to study of the phase transitions between order/complexity/chaos in all types of RBNs. Another reason for studying phase transitions, specifically in contextual RBNs, is to observe if the phase transitions change with the context. In any case, we would expect to find a complexity region "at the edge of chaos", but we are interested in finding out wether this also depends on the context or not.

We also want to study the different regimes under mixed context influence. Remember that the dynamics of our network becomes non-deterministic in this case, and the probability structure is non-Kolmogorovian and quantum-like. It is well known that quantum chaos does not exist in the usual manner of classical chaos, because of the fact that in standard quantum mechanics the dynamics are always linear. The structure that we find here is however not standard quantum mechanical, it is quantum-like in the sense that the probability model is non-Kolmogorovian, but does not entail the restriction of linearity [23,24]. This means that in principle and contrary to this being the case for standard quantum mechanical systems, we must be able to find chaotic regions. In future work we want to explore the chaotic, complex, and orderly regimes and the transitions between them under the influence of non-deterministic contextuality.

The study of the effects of pure contexts in RBNs should also shed light into the debate on the feasability to correlate the number of human cell types to the attractors of a *contextual* RBN [6, 7, 8, 1].

Finally, the proposed contextual RBNs could be generalized to observe and study the dynamics of discrete n-dimensional systems interacting with m-dimensional contexts. We could study then not only the state-space of the system, but also the state-context-space.

References

1. Gershenson, C.: Classification of random boolean networks. In: Standish, R. K., Bedau, M. A. and Abbass, H. A. (eds.): Artificial Life VIII: Proceedings of the Eight International Conference on Artificial Life., MIT Press, Massachussets (2002) 1–8
2. Aerts, D.: A Possible Explanation for the Probabilities of Quantum Mechanics. J. Math. Phys. **27** (1986) 202–210
3. Wuensche, A.: Attractor Basins of Discrete Networks. PhD thesis, University of Sussex (1997). CSRP 461
4. Wuensche, A.: Attractor Basins of Discrete Networks. Technical report, SFI (1998) 11–101
5. Aldana, M. Coppersmith, S. and Kadanoff L. P.: Boolean Dynamics with Random Couplings. Accepted for publication in Springer Applied Mathematical Sciences Series, Celebratory Volume for the Occasion of the 70th birthday of Larry Sirovich (2002)
6. Kauffman, S. A.: The Origins of Order. Oxford University Press, Oxford (1993)

7. Harvey, I and Bossomaier, T.: Time Out of Joint, Attractors in Asynchronous Random Boolean Networks. In: Husbands, P. and Harvey, I. (eds): Proceedings of the Fourth European Conference on Artificial Life (ECAL97), MIT Press (1997) 67–75.
8. Di Paolo, E. A.: Rhythmic and Non-Rhythmic Attractors in Asynchronous Random Boolean Networks. Biosystems **59** (2001) 185–195
9. Aerts, D.: Quantum Structures Due To Fluctuations of the Measurement Situations. Int. J. Theor. Phys. **32** (1993) 2207–2220
10. Aerts, D.: Quantum structures: An Attempt to Explain Their Appearance In Nature. Int. J. Theor. Phys. **34** (1995) 1165–1186. Archive Ref and Link: quant-ph/0111071.
11. Aerts, D.: The Hidden Measurement Formalism: What Can Be Explained And Where Paradoxes Remain. Int. J. Theor. Phys. **37** (1998) 291–304. Archive Ref and Link: quant- ph/0105126.
12. Aerts, D.: Being and Change: Foundations of a Realistic Operational Formalism. In: Aerts, D. Czachor, M. and Durt, T. (eds): Probing the Structure of Quantum Mechanics: Nonlinearity, Nonlocality, Probability and Axiomatics, World Scientific, Singapore (2002) 71–110.
13. Aerts, D and Aerts, S.: Applications of Quantum Statistics in Psychological Studies of Decision Processes. Found. Sc. **1** (1994) 85–97
14. Aerts, D. Broekaert, J. and Smets, S.: A Quantum Structure Description of the Liar Paradox. Int. J. Theor. Phys. **38** (1999) 3231–3239. Archive Ref and Link: quant-ph/0106131
15. Aerts, D. Aerts, S., Broekaert, J. and Gabora, L.: The Violation of Bell Inequalities in the Macroworld. Found. Phys. **30** (2000) 1387–1414
16. Aerts, D., Broekaert, J. and Gabora, L.: A Case For Applying an Abstracted Quantum Formalism to Cognition. (forthcoming) (2003)
17. Gabora, L and Aerts, D.: Contextualizing Concepts Using a Mathematical Generalization of the Quantum Formalism. J. Exp. Theor. Art. Int. **14** (2002) 327–358
18. Gabora, L.: Cognitive Mechanisms Underlying the Origin and Evolution of Culture. PhD thesis, Brussels Free University (2001)
19. Aerts, D. Czachor, M. Gabora, L. Kuna, M. Posiewnik, A. Pykacz, J. Syty, M.: Quantum Morphogenesis: A Variation on a Theme by R. Thom. Accepted in Phys. Rev. E. (2003).
20. Kauffman, S. A.: Metabolic Stability and Epigenesis in Randomly Constructed Genetic Nets. J. Theor. Biol. **22** (1969) 437–467
21. Cornforth, D. Green, D. G. Newth, D. and Kirkley, M.: Do Artificial Ants March In Step? Ordered Asynchronous Processes and Modularity in Biological Systems. In: Standish, R. K. Bedau, M. A. and Abbass, H. A. (eds): Artificial Life VIII: Proceedings of the Eight International Conference on Artificial Life. MIT Press (2002) 28–32
22. Ingber, D. E.: The Architecture of Life. Scientific American **278** (1998) 48–57
23. Aerts, D. and Durt, T.: Quantum. Classical and Intermediate, An Illustrative Example. Found. Phys., **24** (1994) 1353–1369.
24. Aerts, D and Valckenborgh, F.: The Linearity of Quantum Mechanics at stake: the description of separated quantum entities. In: Aerts, D. Czachor, M. and Durt, T. (eds): Probing the Structure of Quantum Mechanics: Nonlinearity, Nonlocality, Probability and Axiomatics. World Scientific, Singapore (2002) 20–46

Representation of Genotype and Phenotype in a Coherent Framework Based on Extended L-Systems

Ole Kniemeyer[1], Gerhard H. Buck-Sorlin[12], and Winfried Kurth[1]

[1] Brandenburgische Technische Universität Cottbus, Department of Computer Science, Chair for Practical Computer Science/Graphics Systems, P.O. Box 101344, D-03013 Cottbus, Germany
[2] Institute of Plant Genetics and Crop Plant Research, Dept. Cytogenetics, Corrensstr. 3, D-06466 Gatersleben, Germany

Abstract. A formal language approach for the specification of ALife models is presented. "Relational Growth Grammars" incorporate rule-based, procedural and object-oriented concepts. By enabling parametric Lindenmayer systems to rewrite multiscaled graphs, it becomes possible to represent genes, plant organs and populations as well as developmental aspects of these entities in a common formal framework. Genetic operators (mutation, crossing-over, selection) take the form of simple graph rewrite rules. This is illustrated using Richard Dawkins' "biomorphs", whereas other applications are briefly sketched. The formalism is implemented as part of an interactive software platform.

1 Introduction

High-level, rule-based languages exhibit several features that make them suitable for biological modelling: Being both transparent in use and not requiring constant re-compilation of the code after modification (thanks to powerful string parsers), they could potentially be developed into a universal language for biology and Artificial Life—enabling easy specifications of models from the level of biochemistry and genetics up to the level of ecological interactions. We undertake one further step towards such a specialized formal language.

Since the advent of L(indenmayer)-systems in 1968 [12], these string rewriting grammars have been primarily used to model growth and architecture of individual plants [15]. By using fixed rule tables and lineage-controlled replacement, most classical models of this sort emphasize the genetically based aspects of morphogenesis (it was not until the Nineties that "globally sensitive" and "open" L-systems took the environment into account [10,13]). Nevertheless, no genetics whatsoever was usually represented in such L-systems, despite of the double-string structure of DNA which seems to predestinate it to be described in a string-rewriting formalism. The "interactive barley" model by Buck-Sorlin and Bachmann [1] is one exception, despite its not properly exploiting the string nature of the genome. Dassow and Mitrana [2] have devised a formal string

W. Banzhaf et al. (Eds.): ECAL 2003, LNAI 2801, pp. 625–634, 2003.

rewriting language called "evolutionary grammars" to represent genetic operators (e.g., mutation, recombination). However, their formalism is restricted to the genetic level and is unable to model architectural or behavioral aspects of the phenotype. Kim [9] has developed `transsys`, a software tool enabling the specification and dynamic simulation of gene expression, transcription factors and L-system-based morphogenetic processes. Though this is a step in the right direction, `transsys` treats genetic activity and macroscopic morphogenesis with different formalisms.

Godin and Caraglio [8] show how the integration of several scaling levels in one coherent framework could be done: They defined "multiple-scaled tree graphs" (MTGs) to describe branching structures simultaneously at several degrees of spatial resolution. Although their approach is basically static and does not encompass the genetic level, it served as a model for our "relational growth grammar" data structures in which binary relations play a key role.

In the following, we will sketch our improved formal language and show its applicability to ALife model specification by "reincarnating" the biomorphs from Richard Dawkins' book "The Blind Watchmaker" [3]. In the Discussion and Conclusions, our framework will be exemplified using other ALife formalisms, and the perspectives of our approach will be shown.

2 Relational Growth Grammars

2.1 The Data Structures

Typical data structures in biological models can be roughly categorized as follows:

Multisets, i.e. unordered collections where the same element can appear several times (examples: unstructured populations, molecules in a solution). L-systems operating on multisets were introduced in [11].

Strings (equivalent to 1-dimensional *arrays* or *lists*) with strict linear order amongst the elements. In our graph-related approach we will use special directed edges of type "successor" to connect elements which follow each other directly in this order relation. As classical L-systems will be embedded as a special case in our formalism, we will assume that two subsequent symbols in a grammar-generated string will automatically be connected by a successor edge. (Exceptions are the symbols "[" and "]", which will no longer be treated as parts of a string but encode branches in a more direct way than in classical L-systems, see below.) The successor relation corresponds to Godin's ">" relation [8].—Biological examples are genomes and linearly ordered architectural substructures (e.g., tree branches).

Axial trees allow to combine strings by a "lateral branch" relation, corresponding to Godin's "+" and symbolized by the bracket notation from classical L-systems [15] where the brackets are used to encode branching in a string and only become important in the (turtle) interpretation, whilst in our model new branches instantly arise at rule application. Axial trees in nature correspond to botanical trees (at the macroscopic level).

Relational structures (or edge-labelled directed graphs) generalize all these basic structures. Here, arbitrary user-defined types of edges are permitted, representing relationships not yet covered by the above basic structures—e.g., the relation between genotype and phenotype, or the "contains" relation connecting compound entities with their parts (Godin's "/"). Using the "successor" and the "contains" relation, a string can be represented as in Fig. 1.

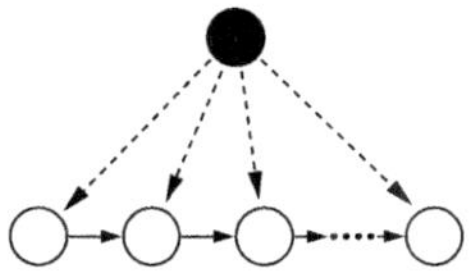

Fig. 1. A string represented by an edge-labelled graph. Continuous edges denote the successor relation, dashed edges the "contains" relation. The filled "basal" node represents an object at a lower level of resolution (multiscaled approach, cf. [8]).

The structures are implemented with Java: Each graph node is an object in the sense of object-oriented programming. Every node pair can be connected by a set of labelled edges. This flexible data structure permits communication between nodes, by sending messages along a certain edge type. It is also used in our interactive application to constitute an easy-to-use user interface, e.g. interaction controls can be connected to data nodes by edges.

2.2 The Formalism

A documentation with complete definitions of the concepts used here will be published elsewhere. Here we list only the main items characterising the Relational Growth Grammar calculus:

- *Graph grammars* are used to rewrite the data structures described above. Contrary to classical L-systems, several nodes (generalising the symbols in L-systems) can appear on the left side of a rule. In the most general case, entire subgraphs are rewritten. Our graph grammars, using both node and edge labels, were partly inspired by the PROGRES system (see [18]).
- Commands of *turtle geometry* (originally from parametric L-systems and in the systematized variant used by the Grogra software [10]) are allowed as nodes and serve to interpret the generated graphs geometrically if necessary. E.g., the command "**F**" (without parameters, with a length, or with displacement coordinates as parameters) causes the creation of a solid cylinder representing, e.g., a (botanical) internode.
- Variables, functions and relations can be defined by the user. (The declaration of several types of random variables was already implemented in Grogra [10].) Procedural programs (scripts) in a Java-like language can be inserted

into a rule and will be executed when the rule is applied (cf. [16] for C fragments in L-systems).
- We use a variant of *table L-systems* ([17]; cf. "sub-L-systems", [14]) to control directly the order in which certain groups of rules are to be applied—mainly for transparency and convenience reasons; the order of rule application could also be encoded in additional parameters and conditions.
- Essential to all rule-based languages are the processes of *matching* and *replacement*, into which the application of a rule can be decomposed. Three modes of rule application can be distinguished, according to the way these processes are evoked within a single rewriting step: *Parallel matching and parallel replacement* is the default method in classical L-systems, in other transformation systems, e.g. in CA, and also in our formalism. When using this mode one has to be aware of possible resolution conflicts with overlapping matches. The other two modes are *sequential matching and replacement immediately after each match* and *sequential replacement*, i.e. in each rewriting step only one match is exploited. For our examples only parallel matching and replacement are needed as resolution conflicts do not occur.
- *Conditions* can be specified to control the matching process, *probabilities* act upon replacement (cf. [15]).
- *Graph contexts* can be defined in a general manner, in particular they are not confined to left and right contexts in the sense of string matching.

2.3 Example: Crossing-Over

To show the potential of our new formalism, we chose the example of crossing-over (co), an important biological process essentially consisting in the exchange of aligned, homologous substrings (alleles) of matching chromosomes (cf. Fig. 2).

This process can be modelled using classical L-systems if the number of genes involved is low. With more than a few genes, the exhaustive listing of all possible multiple co events will become too long. In our new formal framework, we will represent the genomes by strings which are associated with the organisms by edges of a graph. Co will be modelled by one single graph-rewrite rule.

To encode co events, we need two additional relations (i.e., edge types): "Mating" between the two strings which are (potential) objects of co, and "alignment" between homologous gene loci. The latter can be defined on a pair j, k of nodes representing a single gene using the built-in function "index", which returns the position of an item within a string:

$$j \xrightarrow{\text{align}} k \quad := \quad \text{index}(j) == \text{index}(k)$$

Co can then be described by the graph rewrite rule shown in Figure 2a, which can be written down as follows:

$$j\,k,\ m\,n,\ j \xrightarrow{\text{align}} m,\ (\text{match}(\text{base}(j) \xrightarrow{\text{mate}} \text{base}(m)))$$
$$\longrightarrow (\text{prob}(p_r(\text{dist}(j,k))))\ ?\ j\,n,\ m\,k$$

The left hand side lists the "successor" relation between j and k, m and n, which is expressed by blanks (as in Grogra), and the alignment between j and m. The alignment between k and n, although indicated in Fig. 2a, can be omitted in the condition because it follows from $j \xrightarrow{\text{align}} m$. If a matching subgraph is found and the mating condition is met, i.e., the basal nodes are connected by a mate edge, the rule applies: With the recombination probability p_r depending on the genetic distance *dist* between two loci the "successor" relations are redirected as in Figure 2a, otherwise the rule is not applied. —Matching of the rule is checked at all possible positions, thus making double or multiple co's in extra rules (as would have to be done with classical L-systems) dispensable. In order to prevent the "mirror match" resulting from the (j, k)-(m, n)-symmetry of the left hand side no two matches may consist of the same sets of matched nodes.

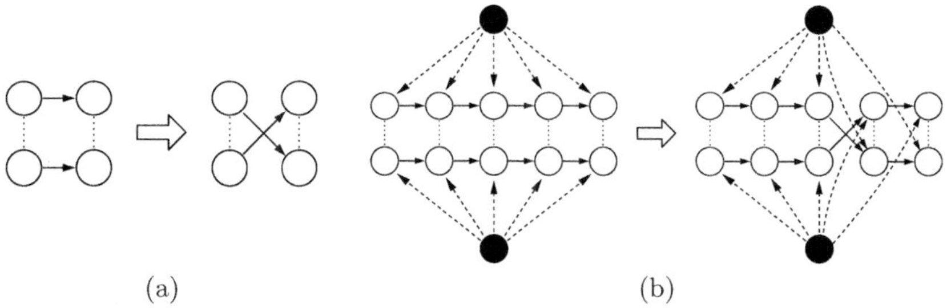

Fig. 2. A relational growth grammar rule representing co (a) and its application to a pair of strings (b). Continuous edges denote the "successor" relation, dashed edges containment, and point-edges alignment.

3 The Example of Biomorphs

We will now apply the new formalism, including the co rule from the last section, to a well-known example, the "Blind Watchmaker" programme, written by the zoologist Richard Dawkins [3] and inspired by the seemingly blind phenomenon of evolution by mutation and selection. Based on a simple genotype-phenotype model [4], it originally consists of two Pascal procedures: The first reproduces the genotype, thereby making new individuals and introducing some random chance mutation. In the second the development of the phenotype is modelled using a recursive tree-drawing routine with the genes' values as parameters (depth of recursion, direction of branches). The genotype consists of nine genes, eight of which have 19 alleles ranging from -9 over 0 to +9, the ninth a range from 0 to 9, in integral steps. The latter gene determines recursion depth. The other genes are translated into components of a matrix d which represent the horizontal, resp. vertical setoffs in a global coordinate system to be used in the development of the growing binary tree. The programme is started off with an initial set of alleles, which is then modified in the offspring by applying random mutations with given probability.

The parent and its offspring are displayed on the screen, and the user selects one of the offspring for further reproduction. By piling-up mutations in a certain direction a shortcut is taken through the multidimensional genotypic parameter space with its billions of possibilities, thereby arriving at astonishing "biomorphs". Our relational growth grammar will furthermore provide a possibility to select *two* parent individuals from which offspring is generated using the co rule shown above.

The following table of relational growth grammar rules contains both alternatives, simple vegetative reproduction (using "SelectOne") and reproduction from two parents involving co (using "SelectTwo" and "Recombine" instead).

Initialize: $\alpha \longrightarrow \mathit{population} \xrightarrow{\text{contains}} \mathit{genome}\ [1\ 1\ 1\ 1\ 1\ 0\ {-1}\ {-1}\ 5]$

Reproduce: $p{:}\mathit{population} \xrightarrow{\text{contains}} g{:}\mathit{genome}$
$\longrightarrow p\ \&(i{=}0..14,\ [\xrightarrow{\text{contains}} \mathit{germ}(200{*}i) \xleftarrow{\text{encodes}} g.\text{cloneTree}])$

Mutate: $\text{int} \longrightarrow (\text{prob}(p_m))\ ?\ \text{irandom}(-9, 9)$

Grow: $\mathit{germ}(x)\ (*\ \xleftarrow{\text{encodes}} g{:}\mathit{genome}\ *)$
$\longrightarrow \mathit{biomorph}\ \mathbf{f}(x,0,0)\ \mathit{biom}(\text{abs}(g[8])+1,\ 2);$

$b{:}\mathit{biom}(\mathit{depth}, \mathit{dir}),$
$(\text{match}(\text{root}(b) \xleftarrow{\text{encodes}} g{:}\mathit{genome})\ \&\&\ (\mathit{depth} > 0))$
$\longrightarrow \{\text{int}[\,][\,]\ d = \{\{-g[1], g[5]\}, \{-g[0], g[4]\}, \{0, g[3]\}, \{g[0], g[4]\},$
$\{g[1], g[5]\}, \{g[2], g[6]\}, \{0, g[7]\}, \{-g[2], g[6]\}\};\}$
$\mathbf{F}(\mathit{depth}{*}d[\mathit{dir} \bmod 8][0],\ 0,\ \mathit{depth}{*}d[\mathit{dir} \bmod 8][1])$
$[\mathit{biom}(\mathit{depth}-1,\ \mathit{dir}-1)]\ [\mathit{biom}(\mathit{depth}-1,\ \mathit{dir}+1)]$

SelectOne: $\mathit{population} \xrightarrow{\text{contains}} b{:}\mathit{biomorph} \xleftarrow{\text{encodes}} g{:}\mathit{genome},\ (\text{isSelected}(b))$
$\longrightarrow\ \hat{}\,\mathit{population} \xrightarrow{\text{contains}} g$

SelectTwo: $\mathit{population}\ [\xrightarrow{\text{contains}} b_1{:}\mathit{biomorph} \xleftarrow{\text{encodes}} g_1{:}\mathit{genome}]$
$[\xrightarrow{\text{contains}} b_2{:}\mathit{biomorph} \xleftarrow{\text{encodes}} g_2{:}\mathit{genome}],$
$(\text{isSelected}(b_1)\ \&\&\ \text{isSelected}(b_2))$
$\longrightarrow\ \hat{}\,\mathit{population}\ [\xrightarrow{\text{contains}} g_1]\ [\xrightarrow{\text{contains}} g_2],\ g_1 \overset{\text{mate}}{\text{———}} g_2$

Recombine: $\text{int}\ j,\ k,\ m,\ n;$
$j\ k,\ m\ n,\ j \overset{\text{align}}{\text{———}} m,\ (\text{match}(\text{base}(j) \overset{\text{mate}}{\text{———}} \text{base}(m)))$
$\longrightarrow (\text{prob}(p_r(\text{dist}(j,k))))\ ?\ j\ n,\ m\ k;$

$g{:}\mathit{genome} \overset{\text{mate}}{\text{———}} \mathit{genome} \longrightarrow g$

Rule "Initialize" replaces the axiom α by a new *population* containing a single *genome* (a string consisting of nine integer nodes). In the next step, "Reproduce", this configuration matches the left hand side. The population p is retained in the new graph, but there it no longer contains the genome g but 15 biomorph *germs* encoded by a copy ("cloneTree") of g. The "encodes" edge symbolizes the genotype-phenotype relationship. The rule makes use of the repetition operator &(range, ...).

After reproduction all genes are subject to mutation: "Mutate" replaces a gene (represented as an integer value) with a random value, uniformly distributed over the integer range $-9, \ldots, 9$. The mutation probability is p_m.

In the first rule of "Grow" germination occurs, i.e., each *germ* is replaced by a new *biomorph*. The actual geometric structure of a full-grown biomorph, its phenotype, will be represented by subordinate nodes, the first of which is the **f**(x,0,0)-node which acts as a coordinate displacement in the specified direction (thus resembling the turtle command "**f**" in classical L-systems) and ensures non-overlapping biomorphs (the parameter x was set in "Reproduce" to 200*i with the biomorph index i). The *biom*-node next to **f**(x,0,0) initiates the recursive growth process of the morphogenetic second rule of "Grow". The initial *depth* parameter of *biom* is taken from the genome g encoding the *germ*: Essentially the child node of g with index 8 is selected, the actual expression abs($g[8]$) + 1 ensures positive recursion depth values. "abs" is built in like any other usual math function. The initial *dir* parameter is set to 2. Since g is enclosed in context parentheses (*...*) it is not considered as a part of the replacement process and remains in the graph, though not listed as a node on the right hand side. However, the encoded object changes from *germ* to *biomorph*.

The morphogenetic second rule of "Grow" "grows" a biomorph according to Dawkins' model: A *biom* node splits into two *biom* branches with opposite changes of *dir*, preceded by the creation of a twig which is implemented by an **F**-node in our graph and by a line drawing instruction in Dawkins' Pascal programme. The **F**(x,y,z)-node corresponds to the **F**-command of turtle graphics, i.e., it acts as a coordinate displacement for subordinate nodes and is also a visible line segment. Its (phenotypic) vector coordinates are determined via an intermediate step (cf. the calculation of the auxiliary variable d in the Java-like script section) by the genotype g which is connected by an "encodes"-edge to the root of the growing biomorph tree, namely the *biomorph*. The growth process terminates when *depth* reaches 0.

"SelectOne" provides user interaction: If the user selects a *biomorph*, the whole *population* is replaced by a new one containing only the *genome* g of the selected *biomorph*. ^ represents the root node of the whole scene, so the new population becomes a child of this root node. Now the situation resembles that after "Initialize" but with a possibly mutated genome. Biomorph evolution proceeds with "Reproduce" as described.

So far the procedure reproduces Dawkins' programme. Upon substituting "SelectOne" by "SelectTwo" and "Recombine" the grammar lets the user select two biomorphs and performs co of their genomes before creating a new biomorph population out of the mated genome: After the user has selected two biomorphs, rule "SelectTwo" replaces the whole population with a new one containing the genomes of the two selected biomorphs, connected by a "mate"-edge indicating that co is potentially to be carried out. The first rule of "Recombine" performs this co as described in section 2.3. The second rule discards one of the two genomes, the other one remains in the graph and leads to a new biomorph population when the grammar application proceeds with "Reproduce".

This sequence of rule application can be specified formally as a table:

Initialize: 1 step
for($i = 0$; $i < nbgenerations$; i++) {
  Reproduce: 1 step
  Mutate: 1 step
  Grow: * steps
  SelectOne: 1 step /* *or* SelectTwo: 1 step; Recombine: 1 step */
}

A sample population of 15 biomorph mutants is shown in Fig. 3.

Fig. 3. Population of mutants of the genome $(1, -3, 8, -4, 5, -9, -4, 9, 8)$

4 Discussion and Conclusions

Relational growth grammars can model complex genetic operations like multiple co, while at the same time being flexible enough to model the ramified architecture of the phenotype (see biomorph example). This universality is confirmed by the successful re-implementation of other classical ALife scenarios in our formal framework. Notably, our grammars are proper extensions of L-systems, hence all plant models obtained with L-systems (e.g., [15]) can be reproduced.

Another calculus which got some attention in the ALife community is Artificial Chemistry (cf. [6]), i.e., a model of the dynamics of large numbers of artificial "molecules" in a virtual solution. A simple example is a "chemical" prime number generator [19], where the molecules are integers and their interaction in the case of collision is expressed by the following grammar rule:

$$a\text{:int},\ b\text{:int},\ (b \bmod a == 0) \longrightarrow a,\ b/a$$

Given an initial "soup" (multiset) of random integers, the iterated and nondeterministic application of the above rule leads to an increase of the "concentration" of prime numbers in the solution, finally approaching 100%.

Cellular Automata (CA) are another good example: The underlying grid of a CA could be constructed in our approach, e.g., by iterating rules associated with the symmetry group of the grid. Then the cells of the CA are the nodes of our graph, and the *context* of a cell can be defined as a multiset- or vector-valued function using the neighbourhood definition of the grid. E.g., the transition rule of Conway's famous Game of Life [7] can be expressed as follows:

$$\begin{array}{ll} x\text{:int},\ \big(x == 1\ \&\&\ x.\text{context.sum in } \{2,3\}\big) & \longrightarrow 0 \\ x\text{:int},\ \big(x == 0\ \&\&\ x.\text{context.sum} == 3\big) & \longrightarrow 1 \end{array}$$

If one were to criticize the wealth of computational tools created as an answer to the flood of biological information produced in the recent past, one should mention the lack of universality and transparency characterising many of them: They are often tailored to a specific problem and cannot be applied to a different discipline or a different organism. This rigidity is mostly due to some implicit structural rigidity "hardwired" within the source code. The extended L-system approach presented here tries to overcome this problem by providing a universal modelling language which is still transparent enough to be understood and applied by biologists or agronomists. We believe that universality and transparency are essential prerequisites to tackle future problems in biological modelling.

Acknowledgements. This research was funded by the DFG under grant Ku 847/5-1 (research group "Virtual Crops"). The second author thanks Dr. Patrick Schweizer (IPK) for providing office facilities. All support is gratefully acknowledged.

References

1. Buck-Sorlin, G.H., Bachmann, K.: Simulating the morphology of barley spike phenotypes using genotype information. Agronomie: Plant Genetics and Breeding **20** (2000) 691–702; s.a. http://taxon.ipk-gatersleben.de.
2. Dassow, J., Mitrana, V.: Evolutionary grammars: A grammatical model for genome evolution. – In: R. Hofestädt et al. (eds.): Bioinformatics. German Conference on Bioinformatics, GCB'96. September 30 - October 2, 1996, Leipzig, Germany. Springer, Berlin (1997) 199–209.
3. Dawkins, R.: The Blind Watchmaker. Longman, Harlow (1986).

4. Dawkins, R.: The Evolution of Evolvability. – In: C. Langton (ed.): Artificial Life, SFI Studies in the Sciences of Complexity, Addison-Wesley, Reading (1988) 201–220.
5. de Boer, M.J.M., Lindenmayer, A.: Map OL-Systems with Edge Label Control: Comparison of Marker and Cyclic Systems. – In: Proc. 3rd Int. Workshop on Graph-Grammars and Their Application to Computer Science, LNCS **291** (1987) 378–392.
6. Fontana, W.: Algorithmic chemistry. – In: C. G. Langton, C. Taylor, J. D. Farmer, S. Rasmussen (eds.), Artificial Life II, SFI Studies in the Sciences of Complexity, Addison-Wesley, Reading (1991) 159–209.
7. Gardner, M.: Wheels, Life, and Other Mathematical Amusements. W. H. Freeman. New York (1983).
8. Godin, C., Caraglio, Y.: A multiscale model of plant topological structures. J. Theor. Biol. **191** (1998) 1–46.
9. Kim, J.: `transsys`: A Generic Formalism for Modelling Regulatory Networks in Morphogenesis. In: J. Kelemen, P. Sosk (Eds.): Advances in Artificial Life. LNAI **2159** (2001) 242–251.
10. Kurth, W.: Growth Grammar Interpreter GROGRA 2.4: A software tool for the 3-dimensional interpretation of stochastic, sensitive growth grammars in the context of plant modelling. Introduction and Reference Manual. Berichte des Forschungszentrums Waldökosysteme der Universität Göttingen, Ser. B, **38** (1994) 190 p.
11. Lane, B., Prusinkiewicz, P.: Generating spatial distributions for multilevel models of plant communities. Proceedings of Graphics Interface 2002, Calgary (Can.), May 27–29 (2002) 69–80.
12. Lindenmayer, A.: Mathematical models for cellular interactions in development, Parts I and II. J. Theor. Biol. **18** (1968) 280–315.
13. Mech, R., Prusinkiewicz, P.: Visual models of plants interacting with their environment. ACM SIGGRAPH 1996, New Orleans, ACM (1996) 397–410.
14. Prusinkiewicz, P., Hanan, J., Mech, R.: An L-system-based plant modeling language. – In: M. Nagl, A. Schürr, M. Münch (eds.): Proceedings of the International Workshop AGTIVE'99. LNCS **1779**, Springer, Berlin (2000) 395–410.
15. Prusinkiewicz, P., Lindenmayer, A.: The Algorithmic Beauty of Plants. Springer-Verlag, New York (1990).
16. Prusinkiewicz, P., Mündermann, L., Karwowski, R., Lane, B.: The use of positional information in the modeling of plants. ACM SIGGRAPH 2001, Los Angeles, ACM (2001) 289–300.
17. Rozenberg, G.: T0L systems and languages. Information and Control **23** (1973) 357–381.
18. Schürr, A.: Programmed Graph Replacement Systems. – In: G. Rozenberg (ed.): Handbook on Graph Grammars: Vol. 1, Foundations. World Scientific, Singapore (1997) 479–546.
19. Skusa, A., Banzhaf, W., Busch, J., Dittrich, P., Ziegler, J.: Künstliche Chemie. Künstliche Intelligenz **1/00** (2000) 12–19.

Integrated-Adaptive Genetic Algorithms

Henri Luchian and Ovidiu Gheorghieş

"Al. I. Cuza" University, Iaşi, Romania
Faculty of Computer Science
{hluchian, ogh}@infoiasi.ro

Abstract. This paper proposes two general techniques for adapting operators in a genetic algorithm: one dynamically adjusts their rates, while the other customizes their specific way of operation. We show how these techniques can be integrated into a single evolutionary system, called Integrated-Adaptive Genetic Algorithm (IAGA). The IAGA exhibits fewer input parameters to adjust than the original GA, while being able to automatically adapt itself to the particularities of the optimization problem it tackles. We present a proof-of-concept implementation of this technique for royal road functions.

1 Introduction

In order to adjust a genetic algorithm [5] to the optimization problem it tackles, some kind of parameter control is usually employed [6]. Deterministic control [1] has proven useful in some particular applications, but its generalization is problematic. Adaptive control [2] uses feedback from the search to determine how the parameter values are to be altered. Self-adaptive control [9], puts additional information into the representation in order to provide some guidance to the operators.

To date, most efforts on parameter control focused on adapting only one aspect of the algorithm at a time. The most comprehensive combination of forms of control [6] was offered in [4], where the adaptation of mutation probability, crossover rate and population size were combined. Other forms of adaptive systems for genetic algorithms can be found in [10], [3], [8].

We propose here a general schema of a genetic algorithm which is able to dynamically adapt both the rate and the behavior for each of its operators. The rates of the operators are adapted in a deterministic, reinforcement-based manner. The behavior of each operator (that is, the specific way it operates) is modified by changing its parameter values. This is done by employing for each type of operator an evolutionary procedure which works in the parameters values search space.

Operators that prove themselves valuable during the optimization process are rewarded; thus, they are applied more often in subsequent generations. The behavior of each operators evolves as well during the run of the genetic algorithm, by modifying the values of the parameters which control its activity. To this end, we use for every type of operator a (mini) genetic algorithm, called

W. Banzhaf et al. (Eds.): ECAL 2003, LNAI 2801, pp. 635–642, 2003.

operator genetic algorithm (OGA). In this context, we call the genetic algorithm that works in the optimization problem's search space *main genetic algorithm* (MGA). Every OGA searches through the space of possible parameter values for its corresponding operator type. A specific set of values for the operator parameters (called an *assignment*) is considered more fit if the actual operator it induces performs better.

We exemplify our approach by implementing it for the royal road functions. We show how more effective mutation and crossover operators emerge during the optimization process and how their rates are correspondingly updated. We give a comparison of the results obtained using the standard genetic algorithm and the integrated-adaptive genetic algorithm (IAGA). We believe that the methods presented here are general enough to allow their usage with virtually no modification for genetic algorithms tackling other optimization problems.

2 Basic Constructs

In order to illustrate our ideas, we first present the basic constructs of the royal road problem [7]. Also, we propose some generalizations of the standard operators used for this problem.

Representation. A chromosome is a sequence of bits (1). All chromosomes have the same length, 2^k, where k is a constant.

$$chrom = (b_1 \ldots b_n), n = 2^k, k > 0, b_i \in \{0, 1\} \text{ for } i = \overline{1, n} \tag{1}$$

Mutation. The standard mutation operator alters genes by flipping their values. This type of operator has a parameter (*alter_rate*) which specifies the percentage of genes within a chromosome to be selected for alteration.

We propose a generalization of the mutation algorithm, called *mutation-flex*. Instead of flipping every gene selected for altering, a 1 is flipped into 0 with the probability p_{10} and a 0 is flipped into 1 with the probability p_{01}.

Crossover. We use here a generalized version of crossover, called *crossover-flex*. This operator type generates c distinct crossover points, where $1 < c < n$. One offspring inherits the odd chunks defined by these crossover points from the first parent, while the even chunks are inherited from the second parent. The second offspring is constructed dually.

Evaluation and Selection. As a quick reminder of the definition of the royal road functions we use, the genes of a chromosome are grouped into blocks. The blocks have the same size which is a power of 2 (less than k). The fitness value of a chromosome is equal to the number of blocks in the chromosome that are composed exclusively of bits set to 1.

We use a rank-based selection scheme: the highest rank receives the probability of selection $p_1 = h$; the probability of selecting the ith rank ($i > 1$) is given by $p_i = (1 - h)p_{i-1}$.

3 Integrated-Adaptive Genetic Algorithm

Real-world genetic algorithms usually define several types of operators, wether unary (mutations), binary (crossovers), or multi-ary. Each type of operator has its own set of parameters which control its behavior. When the parameters of an operator type are set to specific values, an (actual) genetic operator is obtained.

Let $\mathcal{T} = \{t^1 \ldots t^m\}$ be the types of operators employed in the genetic algorithm. For example, mutation-flex and crossover-flex are types of operators. Each type of operator $t \in \mathcal{T}$ has its own set of parameters $P(t) = (p_1^t \ldots p_l^t)$. For instance, the parameters for mutation-flex are *alter_rate*, p_{10}, p_{01}; crossover-flex has only one parameter, c, the number of crossover points.

The genetic algorithm uses actual operators to evolve new offspring. Such an operator is obtained by assigning specific values to the parameters specified by its type. For each operator type t we consider $A(t)$ the set of all such assignments. If $a \in A(t)$, and $a = (a_1 \ldots a_l)$, by $t(a)$ we denote an (actual) operator which is obtained from t by setting p_i^t to a_i, for $i = \overline{1,l}$. For example, let us consider the crossover-flex operator type. Its set of assignments is $A(\text{crossover-flex}) = \{(1) \ldots (n-1)\}$. When parameter c is set to 1, an operator which uses one crossover point is obtained—crossover-flex(1). When c is set to $n-1$, a uniform crossover is obtained—crossover-flex($n-1$).

Each operator $op = t(a)$ has a specific rate which ultimately results in the number of times the operator is applied in one generation.

Algorithm 1 presents a schematic overview of an *integrated-adaptive genetic algorithm* (IAGA).

Algorithm 1 Integrated-Adaptive Genetic Algorithm (overview)

```
initialize operators (parameters)
initialize MGA's evolution (operator rates)
while not halting condition do
  run the MGA for 1 generation
  adapt operators (parameters) using the Operator Adaptive System
  adapt operator rates using the Rates Adaptive System
```

3.1 Rates Adaptive System

The rate adaptive system (RAS) aims at increasing the rates of successful operators.

We denote by $\mathcal{O} = \{op_1 \ldots op_s\}$ the set of actual operators used by the MGA. We attach to each operator $op \in \mathcal{O}$ an information $\rho(op)$ called *reward*. The goal is to record in ρ the experience acquired so far throughout the evolution process. Recently acquired experience is treated as more relevant than the old one, whose importance is to fade away gradually.

Performance Record. The performance record on an operator indicates how well did that operator perform during the most recent generation. This is the feedback from the evolution process received by the rates adaptive system. Based on that feedback, the adaptive system makes adjustments to the rates of the operators.

Definitions. An *event* is one actual application of a specific genetic operator. An *absolute improvement* is the event in which an offspring has a better fitness than the best fitness from the previous generation. An *improvement* is the event in which an offspring has a better fitness than all of its parents, but not one which makes an absolute improvement. A *(plateau) walk* is the event in which each offspring has a fitness which is neither better, nor worse than all those of the parents. *Worsenings* are the remaining events, in which some offspring has a worse fitness than all its parents and no offspring is better than all its parents.

We define the *performance record* as a 4-tuple $\pi = (ai, i, pw, w)$ comprising the number of absolute improvements (ai), the number of improvements (i), the number of plateau walks (pw), and the number of worsenings (w). By $\sigma(\pi)$ we denote the total number of events recorded in π: $\sigma(\pi) = \pi_{ai} + \pi_i + \pi_{pw} + \pi_w$.

Comparing Performance Records. We define a (partial) strict order over the performance records in order to establish which operators performed better. Since some operators may be applied more often than others during one generation, one has to take into account relative frequencies of a specific event type, and not the absolute frequencies.

When $\sigma(\pi) > 0$, we define relative frequencies for: absolute improvements $\varphi_{ai} = \pi_{ai}/\sigma(\pi)$, improvements $\varphi_i = \pi_i/\sigma(\pi)$, plateau walks $\varphi_{pw} = \pi_{pw}/\sigma(\pi)$, worsenings $\varphi_{pw} = \pi_{pw}/\sigma(\pi)$.

If an operator whose performance record is $\sigma(\pi)$ was never applied $(\sigma(\pi) = 0)$, we consider that, if applied, it would have yielded plateau walks. That is, when $\sigma(\pi) = 0$, we define $\varphi_{ai} = 0$, $\varphi_i = 0$, $\varphi_{pw} = 1$, $\varphi_w = 0$. These definitions induce a convenient property of the partial strict order we will define: an operator which was never applied gets a chance of proving itself worthy, since it will be considered better than an operator which was actually applied and yielded only worsenings.

Any two performance records π_1 and π_2 are compared according to a 4-step strategy. Operators which yield relatively more absolute improvements are to receive a higher priority. A tie with respect to the absolute improvement rate is broken via the improvement rate: again, more is better. If we still have a draw, the operator that worsens relatively fewer candidate solutions is to be preferred. The last rule says that operators which potentially waste computational time producing relatively more plateau walks are to be discouraged.

Reinforcement-Based Rate Adaptation. The reinforcement-based rate adaptation is designed to increase the rate of operators with better performance records.

Let us consider a given moment during the run of the main genetic algorithm—namely, at the beginning of generation t. By applying the operators in use with their corresponding rates, a new population is obtained. After generation t is completed, $\pi(op)$ holds the performance record for each operator op. Let $\chi(op)$ be the rank of the operator op with respect to the relation "$>$" (note that some operators may have the same rank).

The rewards are updated at the end of each MGA generation according to the formula:

$$\rho(op) \leftarrow \delta\rho(op) + \beta + \gamma\chi(op) \tag{2}$$

Here, δ is the *diminish* rate, a constant in the interval [0..1]. Setting δ to 0 means that past experience is ignored, while setting δ to 1 means that all past experience is fully taken into account (even if it might have become irrelevant). The *gain* constant γ is used for rewarding operators that are better than others. The *base* constant β has typically a smaller value than γ and can be used to make sure that no operator will vanish altogether.

The operator rates to be used for the next generation of the MGA are set proportional with the reinforcement values.

3.2 Operator Adaptive System

The operator adaptive system aims at providing greater chances to survive and breed to those assignments which induce better actual operators.

Consider an MGA which defines the following operator types: $\mathcal{T} = \{t^1...t^m\}$. For each operator type $t \in \mathcal{T}$ we construct a population of assignments for its parameters. These assignments instantiate actual operators which may be used by the MGA. For example, for the royal road genetic algorithm we have defined two operator types: mutation-flex and crossover-flex. The MGA might use at a certain moment three instantiations of mutation-flex with $(alter\_rate, p_{10}, p_{01}) \in \{(0.1, 0.1, 0.9), (0.7, 0.8, 0), (0.4, 1, 1)\}$ and three instantiations of crossover-flex, one with one crossover point ($c = 1$), one with two crossover points ($c = 2$), and a uniform crossover ($c = n - 1$).

We propose to construct an *operator genetic algorithm* (OGA) for each operator type. The population of the OGA consists of chromosomes which represent assignments. The operators which act on such chromosomes can be defined in a straightforward manner. If the parameter values are numeric, the mutation might alter them slightly. Uniform crossover could be used. Most likely, the OGA optimizes over a relatively small search space, therefore a simplistic design would supposedly suffice.

OGA's evaluation for a chromosome (assignment) is the performance record of the induced operator during the previous generation of the MGA. Technically, it represents a side-effect of the run of the MGA. The result is that the MGA and OGA are interlaced, one generation in either of them providing data for one generation in the other.

3.3 Integration of RAS and OGAs

So far, we have defined two adaptive systems: one for adapting rates, and one for adapting operators.

When assignments for operator type parameters evolve, new operators are obtained. These operators are then involved in the reinforcement-based rate computation. To obtain a fair schema, we must provide the newcomers with a positive reward, otherwise they would not get any chance of actually doing something.

Newly obtained operators inherit their rewards form their parents. An operator which is simply copied should have the same reward as its parent. The reward of an operator obtained through a unary operator (mutation) is set to the same value as its parent. Operators which are obtained through binary operators (crossover) have their rewards set to a blend of their parents' rewards: for numeric rewards, the average could be taken.

Figure 1 shows how MGA, RAS and OGAs are integrated.

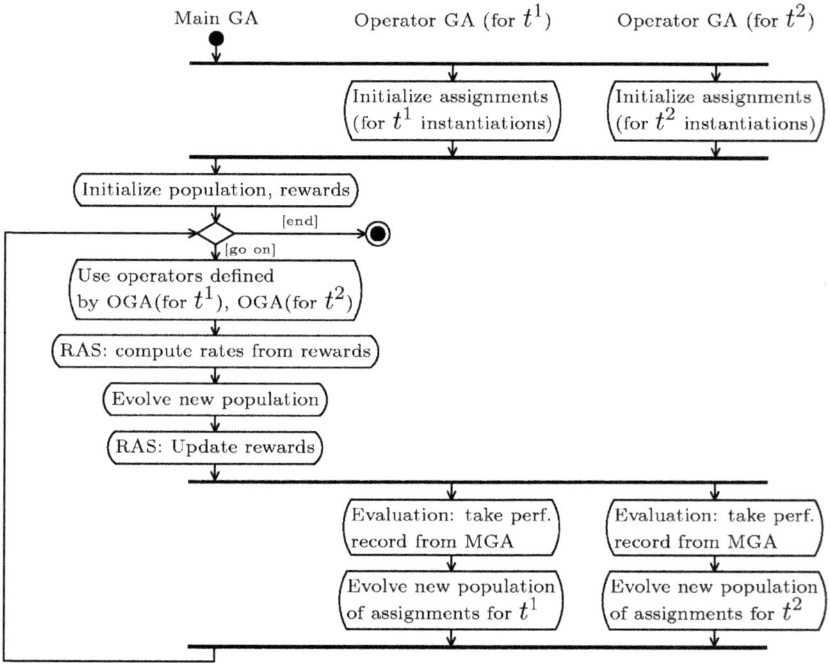

Fig. 1. Integration of MGA, RAS and two OGAs: activity diagram.

4 Experimental Results

The population of the MGA contained 100 chromosomes, each of size 256 bits. For all genetic algorithms employed, the probability of selecting the first rank was $h = 0.2$. Each run of the MGA was halted when 100000 generations were completed or when the optimal solution was found, whichever came first. The RAS parameters were set to the following values: $\eta = 0.8$, $\beta = 0.1$, $\gamma = 1.0$.

We have run experiments using the standard genetic algorithm, the genetic algorithm which makes use of the RAS only, and the IAGA. The experiments that did not employ the OGA used operators which remained unchanged during the run of the MGA. An overview of the results is depicted in table 1. There, *mf* stands for mutation-flex and *cf* for crossover-flex. For the standard genetic algorithm the settings which were used for operator rates are also indicated. The experiments that make use of the RAS started out with the same rate (20%) for each operator, since no previous knowledge on the effectiveness of operators has been assumed.

Table 1. Overview of experimental results using various degrees of adaptation. The number of generations was computed as the average over 100 different runs. For block size 16, the first setting did not lead to finding the solution before 100000 generations were completed.

				Generations to solution			
				Block size			
				2	4	8	16
Standard genetic algorithm							
mf (0.4,1,1)	mf (0.4,0.25,1)	mf (0.4,1,0.25)	cf (1)				
0.1	—	—	0.6	3757.6	6188.7	36014.8	—
0.1	0.1	0.1	0.5	259.8	760.3	1762.7	9579.6
0.2	0.2	0.2	0.2	262.5	318.3	698.0	2240.2
—	0.1	—	0.6	386.2	525.6	1190.7	2412.6
—	0.35	—	0.45	113.4	233.2	311.3	977.0
MGA with RAS only				154.3	175.2	480.0	1479.9
IAGA (MGA, RAS, mf-OGA, cf-OGA)				25.9	8.89	24.6	16.2

When RAS only was used, the mutation operator that is more likely to turn 0s into 1s gradually gained advantage over the others. Also, (1-point) crossover was constantly encouraged.

Table 1 accounts for the effectiveness of the RAS, whose performance is just behind that of the best standard GA. This comes as no surprise, since the operator rates of the best standard GA have been deliberately set to the values that were eventually discovered by the RAS.

The IAGA is a clear winner. Preliminary studies indicate that this happens because highly effective mutation operators are discovered early during the optimization process (which may not be the case for other problems); also, multi-point crossover proved very helpful. The figures for IAGA's convergence speed for different royal road instances raise some intriguing questions when compared. At this moment, it is too early for us to formulate any hypotheses to explain this behavior.

5 Conclusions and Future Work

The experiments show that we have devised a system which is able to discover in an automatic, dynamic, unsupervised manner, not only which operators better suit the optimization process, but also new, more effective, operator variants. The integrated-adaptive genetic algorithm appears to be more robust than the standard GA. The adaptive mechanism involved makes IAGA less sensitive to errors in the initial setting of its parameters.

The IAGA was very successful when applied to the royal road problem. More work has to be done to fully explain its behavior. Applying IAGAs to other optimization problems could shed some light on this issue.

Although the control of many parameters was integrated into the evolutionary system, some were still left out. Further integration of the control for population size and selection process parameters appears interesting. Evolving new operators types via genetic programming would probably enhance the power of the IAGAs.

References

1. T. Bäck. Optimal Mutation Rates in Genetic Search. In *Proceedings of the Fifth International Conference on Genetic Algorithms*, 1993.
2. B. A. Julstrom. What Have You Done for me Lately? Adapting Operator Probabilities in a Steady-State Genetic Algorithm. *Proceedings of the Sixth International Conference on Genetic Algorithms*, 1995.
3. A. Kosorukoff. Using incremental evaluation and adaptive choice of operators in a genetic algorithm. In *GECCO 2002: Proceedings of the Genetic and Evolutionary Computation Conference.*
4. J. Lis and M. Lis. Self-adapting Parallel Genetic Algorithms with the Dynamic Mutation Probability, Crossover Rate and Population Size. In *Proceedings of the 1st Polish National Conference on Evolutionary Computation*, 1996.
5. Z. Michalewich. *Genetic Algorithms + Data Structures = Evolution Programs.* Springer, Berlin, 1996.
6. Z. Michalewich and D. Fogel. *How to Solve It: Modern Heuristics.* Springer, 2002.
7. M. Mitchell, S. Forest, and J.H. Holland. The Royal Road for Genetic Algorithms: Fitness Landscapes and GA Performance. In *Proceedings of First ECAL*, 1992.
8. M. Pelikan, D.E. Goldberg, and S. Tsutsui. Combining the strengths of Bayesian optimization algorithm and adaptive evolution strategies. In *GECCO 2002: Proceedings of the Genetic and Evolutionary Computation Conference.*
9. W.M. Spears. Adapting Crossover in Evolutionary Algorithms. In *Proceedings of the 4th Annual Conference on Evolutionary Programming*, 1995.
10. D. Thierens. Adaptive mutation rate control schemes in genetic algorithms. In *Proceedings of the 2002 Congress on Evolutionary Computation CEC2002.*

Simulating the Evolution of Ant Behaviour in Evaluating Nest Sites

James A.R. Marshall[1], Tim Kovacs[2], Anna R. Dornhaus[3], and Nigel R. Franks[3]

[1]Complex Systems Modelling Group, Department of Earth Science and Engineering
Imperial College London, SW7 2AZ, UK
James.A.Marshall@imperial.ac.uk

[2]Department of Computer Science, University of Bristol, BS8 1UB, UK
kovacs@cs.bris.ac.uk

[3]Centre for Behavioural Biology, School of Biological Sciences, University of Bristol,
BS8 1UG, UK
{A.Dornhaus|Nigel.Franks}@bristol.ac.uk

Abstract. When an ant colony needs to find a new nest, scouts are sent out to evaluate the suitability of potential sites, particularly their size. It has been suggested that ant scouts of *Leptothorax albipennis* use a simple heuristic known as Buffon's needle to evaluate nest size. They do this in two stages: first laying a pheromone trail in the nest site, then, after a return to the old nest, coming back and wandering within the site assessing frequency of intersection with the pheromone trail ("two-pass" strategy). If a colony is forced to relocate from its current nest due to destruction of that nest, the time required to find a suitable new nest may be crucial. This paper details preliminary results from a computer simulation model of evaluation of nest size. The model aims to study why a "two-pass" strategy is used by ants when a "one-pass" strategy, in which the ant simultaneously lays pheromone and assesses the frequency at which it encounters its own trail, may be more time efficient. Analysis of the results indicates no clear advantage for the "two-pass" strategy, given the assumptions of the model. Possible implications of this result are discussed.

1 Introduction

Computer simulation modelling of social insects such as ants, termites and certain species of bees and wasps is an area of recent research, with models of task allocation [2], foraging in ants [3] and nest-assembly in wasps [7], among others. In this paper we consider size assessment of potential nest sites by ant scouts. Assessing the size of a potential nest site is a hard problem for a scout ant, and yet the decision of an individual scout about whether a site is suitable to house the whole colony can have a large impact on the colony's survival. Size assessment is difficult because scouts cannot simply measure diameter, since nests are often irregularly shaped. In addition, they are usually dark, and the ants' vision does not permit them to correctly estimate area. The evaluation method used by ant scouts of *Leptothorax albipennis* is therefore an

W. Banzhaf et al. (Eds.): ECAL 2003, LNAI 2801, pp. 643–650, 2003.

interesting area of study. After discounting heuristics such as perimeter length or mean free-path length, Buffon's needle has been proposed instead [4, 5]. Fundamentally, Buffon's needle works on the principle that an ant lays a pheromone trail of defined length while wandering throughout a potential nest site, then evaluates the approximate size of the nest site by wandering again within the site while assessing the frequency at which it crosses its previously laid trail. All evidence suggests that *Leptothorax* ants assess sites by working alone and that they do this by deploying individual specific trail pheromones [4].

The original *in-vivo* investigation of Buffon's needle in *L. albipennis* was extended to *in-silico* experimentation by Şahin and Franks [6]. Their model replicated nest-integrity and size-assessment behaviour, and discovered a fundamental trade-off between thigmotaxis (wall-following behaviour) and exploration of the central area of the nest site in effective assessment.

One question begs an answer: In the scouting behaviour displayed by *L. albipennis*, why is the nest evaluation process carried out in two stages, the first consisting of laying a pheromone trail in the nest site, the second in assessing frequency of intersection with that trail while wandering within the nest site? In principle it would seem more efficient to compress the two stages into one. Hence we are interested in whether the "two-pass" strategy employed by the real ants is algorithmically superior to a "one-pass" strategy in which the ant wanders within the nest site, simultaneously laying pheromone and assessing the frequency of intersection with its own trail. Is there is an algorithmic reason why a one-pass strategy is not suitable? For instance, does a one-pass strategy bias area estimation in a way which an ant cannot compensate for when classifying a nest's size? Such a bias might occur because the ant is laying pheromone and detecting it in adjacent locations, which will result in a different frequency of path crossings compared to the two unrelated random walks used with the two-pass strategy. Here we present a computer simulation model that seeks to answer this question. In order to simplify the design and analysis of that model, we do not consider perimeter evaluation behaviour as was done by Şahin and Franks [6], but focus solely on size assessment behaviour. The independent implementation of two simulation models of the same behaviour is a prime candidate for validation of the models through model docking [1]. We hope to do this in the future.

2 Methods

The computer simulation model[1] was implemented using the Swarm[2] simulation toolkit and models the process of nest site evaluation by an individual scout ant using either the one-pass or two-pass strategy. The potential nest site has a simple representation as a hollow square with one entrance in the middle of the southern wall. The ant is defined by the characteristics shown below in table 1, which are explained in the following text. The ranges of these characteristics indicate the limits within which

[1] ftp://ftp.swarm.org/pub/swarm/apps/java/contrib/Buffon-1.0-2.2.tar.gz
[2] http://www.swarm.org

they are permitted to vary by the evolutionary component of the model, described in section 2.2, for which all initial values are selected randomly.

Table 1. Ant characteristics

Characteristic	Range
scouting strategy	one-pass or two-pass
scouting time	1-1000 time units
arousal decay rate	1-10 units per time step
classification divisor	1-3000

2.1 Model of Ant Behaviour

Note that the model only simulates the size-assessing behaviour of the ant, and not the boundary-checking behaviour. The simulation of scouting behaviour proceeds as follows:

1. Place ant at entrance to nest, facing nest centre, and initialise arousal level to 0.
2. Change ant's current direction with probability:
   a. 20% turn left 45 degrees
   b. 60% maintain current direction
   c. 20% turn right 45 degrees
3. If obstacle in path along new heading repeat 2 unless all three possible headings tried in which case rotate ant 45 degrees left or right with equal probability and repeat 2.
4. Deposit pheromone at current location as dictated by strategy.
5. Move to new location indicated by current direction.
6. If pheromone detected at new location increment arousal level by 10 units. Decrement arousal level by arousal decay rate (arousal level minimum = 0).
7. Repeat from 2.

The above process repeats for the number of time steps indicated in the ant's characteristics. If the ant is using a one-pass strategy then the ant deposits pheromone continuously. If the ant is using a two-pass strategy then the ant spends the first half of its scouting time depositing pheromone and the second half not depositing pheromone but simply assessing the frequency at which it encounters its previously laid pheromone trail (the ant is reset to the entrance of the nest at the beginning of the assessment phase, just as in the case of a real ant).

The assessment behaviour of the ant just described above is based on the concept of an arousal level. This is effectively a measure of the frequency of intersections with the pheromone trail over some recent time window defined by the arousal decay rate. Note that we are not proposing here that this is how ants actually measure intersection frequency with pheromone trails, but rather we are showing how a simple and cognitively plausible mechanism is sufficient to give rise to the desired result. The arousal level scheme certainly seems cognitively simpler than the counting scheme used by Şahin and Franks [6].

The movement behaviour of the ant given above is a constrained random walk, and was designed to correspond roughly with observed movement behaviours in real ant scouts. A visualisation of the ant's simulated behaviour is shown below in figure 1, and a visualisation of actual ant behaviour is given for comparison in figure 2 below. In figure 1 dark grey represents the walls of the nest site, and light grey represents the scout ant's pheromone trail. The entrance to the simulated nests shown in figures 1 and 2 is in the middle of the right-hand wall in both cases.

At the conclusion of the scouting process the ant's assessment of nest size e is given by

$$e = (c-1) - \min\left(\operatorname{int}\left(\frac{r}{d}\right), c-1\right), \tag{1}$$

where c is the number of size categories (e.g. three categories: 0 = small, 1 = medium and 2 = large), r is the ant's arousal level at the end of the scouting process, and d is the ant's classification divisor (see table 1). This equation simply converts the ant's arousal level at the end of the scouting time, which is a measure of how often the ant encountered its own trail while scouting, into a normalised assessment of nest size.

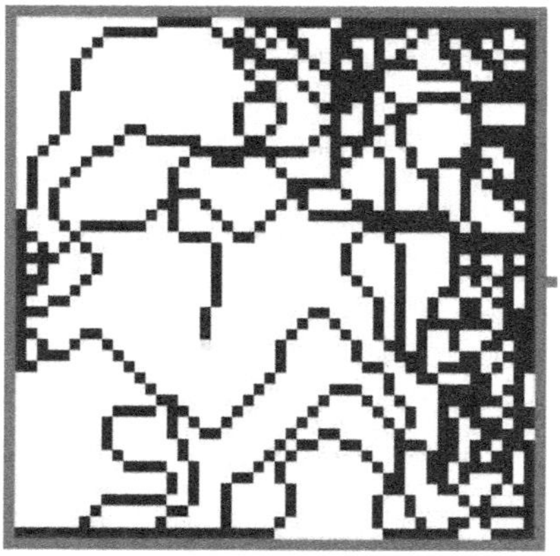

Fig. 1. Simulated ant movement in a potential nest site

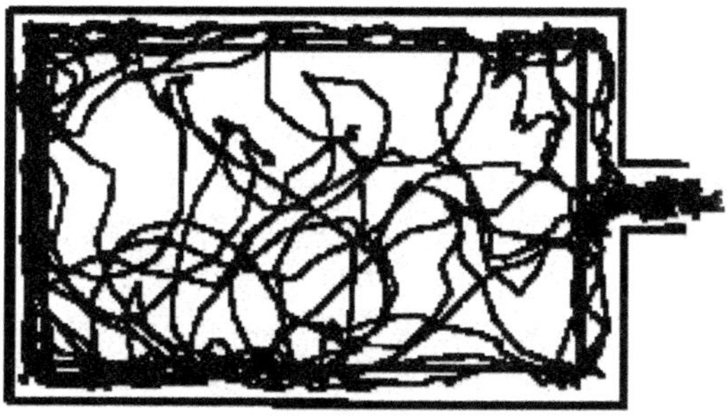

Fig. 2. Actual recorded ant movement in a potential nest site. From Mallon & Franks [4]

2.2 Model of Evolution of Ant Behaviour

A population of ants is evolved according to their average performance in assessing the sizes of potential nest sites as follows:

1. For all ants calculate average fitness based on assessment performance with three different sizes of nest (10 x 10, 30 x 30 and 50 x 50) where fitness f is given by

$$f = -q|e - s| - ta \ , \tag{2}$$

   where $-q$ (q = 1000) is the selection pressure on assessment quality, a (a = 1) is the selection pressure on assessment speed, e is the ant's estimate of nest size from equation (1), s is the actual nest size, and t is the total time the ant spends scouting.
2. Rank population according to average fitness and cull bottom 33% of population to be replaced by offspring of top 66% of population as follows:
   a. In ranked list of population mate pairs of ants with adjacent average fitness ranks to produce one offspring ant per pair, thus conserving the population size.
   b. For each mating produce offspring genotype using parameterised uniform crossover (crossover rate = 10%) and mutation operators (mutation rate = 1%, maximum value change from mutation = 10% of value range, mutations use a uniform distribution and are limited by the range of the value undergoing mutation).

The evolutionary algorithm described above is not intended realistically to simulate evolution in real ant populations, but rather to provide a simple way to apply evolutionary pressure to the behaviour under consideration.

At this point it is worth summarising the assumptions made in building our model of ant nest assessment behaviour. Table 2 below lists these assumptions.

Table 2. Model assumptions

Model Component	Assumption
ant movement	constrained random walk (see sec. 2.1)
nest shape	uniform (square)
path crossing calculation	arousal level with decay rate (see sec. 2.1)
selection pressure	relative importance of speed and accuracy is as defined in q and a in sec. 2.2

3 Results

Two different simulations were carried out: one in which all ants used the one-pass strategy, and one in which all ants used the two-pass strategy. Each simulation collected results from 40 experiments with each experiment run on a population of 21 ants over 60 generations. For each experiment, average scouting time (averaged over the fittest 66% of each generation) and number of generations required for the population to become 100% effective at assessing nest size were recorded.

The following hypotheses were tested: (1) that there is a difference in the number of generations required to establish robust nest assessing behaviour by the one-pass and two-pass strategies, and (2) that there is a difference in the time efficiency of the one-pass and two-pass strategies. Only those data where the population evolved to 100% assessment accuracy were included in the statistical analysis. In both cases the results of the analysis were non-significant, therefore the null hypotheses, that there are no differences between the one and two-pass strategies, cannot be rejected (U = 177.5 and U = 136 respectively, N_A = 20, N_B = 20; Mann-Whitney U test, Fig. 3). Furthermore for two simulations of 200 experiments each the one-pass strategy evolved to 100% accuracy in 45 cases, whereas the two-pass strategy evolved to 100% accuracy in 41 cases. Therefore the null hypothesis that there is no difference in the frequency with which one-pass and two-pass populations evolve to 100% assessment accuracy also cannot be rejected ($\chi^2 = 0.133$, $df = 1$).

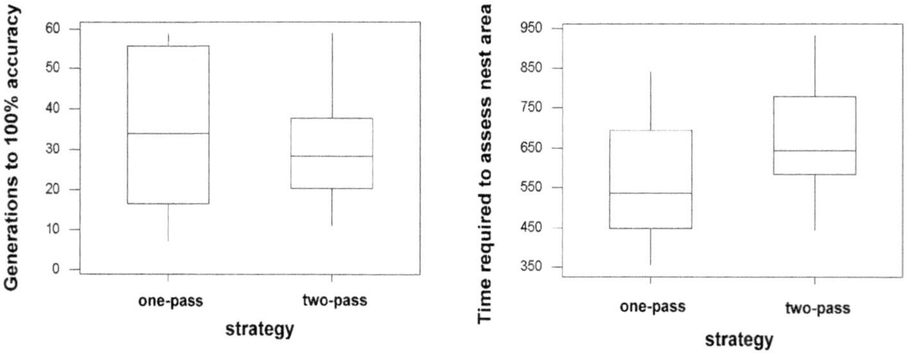

Fig. 3. Results from the simulations: neither in the number of generations to 100% accuracy (left) nor in the time then needed to perform the nest assessment (right) are the one-pass and two-pass strategies significantly different (boxes show quartiles, whiskers show range and the horizontal lines within boxes show medians; n = 20 for both strategies).

4 Discussion

Our simulation results showed no difference in effectiveness between the one-pass and two-pass strategies. That is, the evolutionary time required to establish robust nest assessing behaviour is not significantly different for the two strategies, nor is the frequency with which robust assessment behaviour is evolved using the two strategies. Furthermore, the evolved behaviours do not differ significantly in the time required to make an assessment. However, this can be qualified by two observations. Firstly as the two-pass strategy only lays pheromone for half of the scouting time, the two-pass strategy only uses half the pheromone of the one-pass strategy. So, if pheromone production is energetically costly, the two-pass strategy may be superior in energetic efficiency. On the other hand, the two-pass strategy actually takes more time outside the potential nest site than does the one-pass strategy, as the ants employing it return to their old nest between the two visits to the potential nest site. If the ant returns to the old nest site between visits, it will cover twice as much distance between nest and potential site using the two-pass strategy compared to the one-pass strategy, which costs both time and energy. These two observations each strengthen the case for a different strategy; the first indicates that the two-pass strategy may be better than the model suggests, while the second indicates that the one-pass strategy may be better than the model suggests. The fact that our model cannot distinguish the superiority of either strategy suggests that selection may not have influenced whether *L. albipennis* evolved one or two-pass strategies, and the fact that they use a two-pass strategy may be the result of a random course of evolution in that respect. Alternatively the use of a two-pass strategy by the real ants could suggest evolutionary constraint, or physical constraint based on the potential difficulty of simultaneously laying and detecting pheromone.

The main result of this paper is that the proposed one-pass strategy is able to assess nest area as quickly and as accurately as the two-pass strategy exhibited by the real ants, given the basic model presented here. That is, we have found no algorithmic reason why a one-pass strategy should not be used. To investigate the matter further, we intend to evaluate possible differences between the performance of the two strategies in more detail. In particular, the current model only utilises nest sites of differing size but uniform shape. Mallon & Franks [4] used several different shapes of nest site, including sites with a partial partition inside. We plan to use our model to assess the impact of different nest shapes on the two scouting strategies. Given more diverse nest site structures, is there any difference between the speed and effectiveness of the one-pass and two-pass strategies? Furthermore, it is possible that the arousal level profile generated by different nest shapes is a function of the movement behaviour used, therefore we plan to make this behaviour evolvable as well. Alternatively, the relative fitness of the one-pass and two-pass strategies could be evaluated if selection pressure is varied to favour speed over accuracy or vice-versa. It is also important to investigate in detail the specific effects of the assumptions used in the model. We are currently investigating the model in this manner and intend to present further results shortly.

References

1. Axtell, R., Axelrod, R., Epstein, J. M., Cohen, M. D.: Aligning Simulation Models: A Case Study and Results. Comput. Math. Organ. Theory 1 (1996) 123–141
2. Bonabeau, E., Theraulaz, G. & Deneubourg, J.-L.: Quantitative Study of the Fixed Threshold Model for the Regulation of Division of Labour in Insect Societies. Proc R. Soc. Lond. B 263 (1996) 1565–1569
3. Deneubourg, J.-L., Goss, S. & Franks, N. R.: The Blind Leading the Blind: Modelling Chemically Mediated Army Ant Raid Patterns. J. Insect. Behav. 2 (1989) 719–725

4. Mallon, E. B., Franks, N. R.: Ants Estimate Area Using Buffon's Needle. Proc. R. Soc. Lond. B 267 (2000) 765–770
5. Mugford, S. T., Mallon, E. B. Franks, N. R.: The Accuracy of Buffon's Needle: A Rule of Thumb Used by Ants to Estimate Area. Behav. Ecol. 12 (2001) 655–658
6. Şahin, E., Franks, N. R.: Simulation of Nest Assessment Behaviour by Ant Scouts. In: Dorigo, M., Di Caro, G., Sampels, M. (eds.): ANTS 2002, Lecture Notes in Computer Science, Vol. 2463. Springer-Verlag, Berlin Heidelberg (2002) 274–281
7. Théraulaz, G. & Bonabeau, E.: Modelling the Collective Building of Complex Architectures in Social Insects with Lattice Swarms. J. Theor. Biol. 177 (1995) 381–400

Evolving Evolutionary Algorithms Using Multi Expression Programming

Mihai Oltean and Crina Groşan

Department of Computer Science
Faculty of Mathematics and Computer Science
Babeş-Bolyai University, Kogălniceanu 1
Cluj-Napoca, 3400, Romania.
{moltean, cgrosan}@cs.ubbcluj.ro

Abstract. Finding the optimal parameter setting (i.e. the optimal population size, the optimal mutation probability, the optimal evolutionary model etc) for an Evolutionary Algorithm (EA) is a difficult task. Instead of evolving only the parameters of the algorithm we will evolve an entire EA capable of solving a particular problem. For this purpose the Multi Expression Programming (MEP) technique is used. Each MEP chromosome will encode multiple EAs. An nongenerational EA for function optimization is evolved in this paper. Numerical experiments show the effectiveness of this approach.

1 Introduction

Evolutionary Algorithms (EAs) [2, 5] are nonconventional tools for solving difficult real-world problems. They were developed under the pressure generated by the inability of classical (mathematical) methods to solve some real-world problems. Many of these unsolved problems are (or could be turned into) optimization problems. Solving an optimization problem means finding of solutions that maximize or minimize a criteria function [2].

Many EAs were proposed for dealing with optimization problems. Many solution representations and search operators were proposed and tested within a wide range of evolutionary models. There are several natural questions that are to be answered in all these evolutionary models: which is the optimal population size?, which is the optimal individual representation?, which are the optimal probabilities for applying specific genetic operators?, which is the optimal number of generations before halting the evolution? etc.

A breakthrough arose in 1995 when Wolpert and McReady unveiled their work on the No Free Lunch (NFL) theorems [8]. The NFL theorems state that all the black-box algorithms perform equally well over the entire set of optimization problems. (A black-box algorithm does not take into account any information about the problem or the particular instance being solved.) The magnitudes of the NFL results stroke all the efforts for developing a universal black-box optimization algorithm able to solve best all the optimization problems.

W. Banzhaf et al. (Eds.): ECAL 2003, LNAI 2801, pp. 651–658, 2003.

In their attempt to solve problems, men delegated computers to develop algorithms able to perform certain tasks. The most prominent effort in this direction is Genetic Programming (GP) [3], an evolutionary technique used for breeding a population of computer programs. Instead of evolving solutions for a particular problem instance, GP is mainly intended for discovering computer programs able to solve particular classes of problems. (This statement is only partially true, since the discovery of computer programs may be also viewed as a technique for solving a particular problem instance. The following could be an example of a problem: 'Find a computer program that calculates the sum of the elements of an array of integers.')

There are many such approaches so far in the GP literature [1]. The evolving of deterministic computer programs able to solve specific problems requires a lot of effort.

Instead of evolving deterministic computer programs we will try to evolve a full-featured evolutionary algorithm (i.e. the output of our main program will be an EA able to perform a given task). Thus we will work with EAs at two levels: the first (macro) level consists of a steady-state EA [6] which uses a fixed population size, a fixed mutation probability, a fixed crossover probability etc. The second (micro) level consists of the solution encoded in a chromosome from the GA on the first level.

Having this aim in view we use an evolutionary model similar to Multi Expression Programming (MEP) [4][1] which is very suitable for evolving computer programs that may be easily translated into an imperative language (like C or Pascal). The evolved EA is a nongenerational one (i.e. there is no cycle during evolution).

The paper is organized as follows: the MEP technique is described in section 2. The model used for evolving EAs is presented in section 3. The way in which the fitness of an MEP individual is computed is described in section 4. Several numerical experiments are performed in section 5.

2 MEP Technique

The Multi Expression Programming technique is briefly described in this section.

2.1 The MEP Algorithm

In this paper we use steady-state [6] as the underlying mechanism for MEP. The steady-state MEP algorithm starts with a randomly chosen population of individuals. The following steps are repeated until a termination condition is reached: Two parents are selected (out of 4 individuals) by using a binary tournament procedure [3] and they are recombined with a fixed crossover probability. By the recombination of two parents two, offspring are obtained. The offspring are mutated and the best of them replaces the worst individual in the current

[1] MEP source code is available at www.mep.cs.ubbcluj.ro.

population (if the offspring is better than the worst individual in the current population).

2.2 The MEP Representation

The MEP genes are (represented by) substrings of variable length. The number of genes in a chromosome is constant and it represents the chromosome length. Each gene encodes a terminal (an element in the terminal set T) or a function symbol (an element in the function set F). A gene encoding a function includes pointers towards the function arguments. Function parameters always have indices of lower values than the position of that function itself in the chromosome.

According to the proposed representation scheme, the first symbol in a chromosome must be a terminal symbol. In this way only syntactically correct programs are obtained.

Example

We use a representation where the numbers on the left positions stand for gene labels (or memory addresses). Labels do not belong to the chromosome, they are provided only for explanatory purposes. An example of a chromosome is given below (assuming that $T = \{a, b, c, d\}$ and $F = \{+, -, *, /\}$):

1: a
2: b
3: + 1, 2
4: c
5: d
6: + 4, 5
7: * 2, 6

2.3 The MEP Phenotypic Transcription

This section is devoted to describing the way in which the MEP individuals are translated into computer programs.

The MEP chromosomes are read in a top-down fashion starting with the first position. A terminal symbol specifies a simple expression. A function symbol specifies a complex expression (formed by linking the operands specified by the argument positions with the current function symbol).

For instance, genes 1, 2, 4 and 5 in the previous example encode simple expressions formed by a single terminal symbol. These expressions are: $E_1 = a$; $E_2 = b$; $E_4 = c$; $E_5 = d$.

Gene 3 indicates the operation + on the operands located at positions 1 and 2 of the chromosome. Therefore gene 3 encodes the expression: $E_3 = a + b$.

Gene 6 indicates the operation + on the operands located at positions 4 and 5. Therefore gene 6 encodes the expression: $E_6 = c + d$.

Gene 7 indicates the operation * on the operands located at positions 2 and 6. Therefore gene 7 encodes the expression: $E_7 = b * (c + d)$.

We have to choose one of these expressions ($E_1, \ldots, E_7$) to represent the chromosome. There is neither theoretical nor practical evidence that one of them is better than the others. Thus, we choose to encode multiple solutions in a single chromosome. Each MEP chromosome encodes a number of expressions equal to the chromosome length (the number of genes). The expression associated to each chromosome position is obtained by reading the chromosome bottom-up from the current position, by following the links provided by the functions pointers. The fitness of each expression encoded in a MEP chromosome is computed in a conventional manner (the fitness depends on the problem being solved). The best expression encoded in a MEP chromosome is chosen to represent the chromosome (the fitness of a MEP individual equals the fitness of the best expression encoded in that chromosome).

3 The Evolutionary Model

In order to use MEP for evolving EAs we have to define a set of terminal symbols and a set of function symbols. When we define these sets we have to keep in mind that the value stored by a terminal symbol is independent of other symbols in the chromosome and a function symbol changes the solution stored in another gene.

An EA usually has 4 types of genetic operators:

- *Initialize* - randomly initializes a solution,
- *Select* - selects the best solution among several already existing solutions
- *Crossover* - recombines two already existing solutions,
- *Mutate* - varies an already existing solution.

These operators will act as symbols that may appear into an MEP chromosome. The only operator that generates a solution independent of the already existing solutions is the *Initialize* operator. This operator will constitute the terminal set. The other operators will be considered function symbols. Thus, we have $T = \{Initialize\}$, $F = \{Select, Crossover, Mutate\}$.

A MEP chromosome C, storing an evolutionary algorithm is:

```
1: Initialize          {Randomly generates a solution.}
2: Initialize          {Randomly generates another solution.}
3: Mutate 1            {Mutates the solution stored on position 1}
4: Select 1, 3         {Selects the best solution from those}
                       {stored on positions 1 and 3}
5: Crossover 2, 4      {Recombines the solutions on positions 2 and
4}
6: Mutate 4            {Mutates the solution stored on position 4}
7: Mutate 5            {Mutates the solution stored on position 5}
8: Crossover 2, 6      {Recombines the solutions on positions 2 and
6}
```

This MEP chromosome encodes multiple evolutionary algorithms. Each EA is obtained by reading the chromosome bottom up, starting with the current gene and following the links provided by the function pointers. Thus we deal with EAs at two different levels: a micro level representing the evolutionary algorithm encoded in a MEP chromosome and a macro level GA, which evolves MEP individuals. The number of genetic operators (initializations, crossovers, mutations, selections) is not fixed and it may vary between 1 and the MEP chromosome length. These values are automatically discovered by the evolution. The macro level GA execution is bound by the known rules for GAs (see [2]).

For instance, the chromosome defined above encodes 8 EAs. They are given in Table 1.

Table 1. Evolutionary Algorithms encoded in the MEP chromosome C

EA_1	EA_2	EA_3	EA_4
i_1=*Initialize*	i_1=*Initialize*	i_1=*Initialize* i_2=*Mutate*(i_1)	i_1=*Initialize* i_2=*Mutate*(i_1) i_3=*Select*(i_1,i_2)
EA_5	**EA_6**	**EA_7**	**EA_8**
i_1=*Initialize* i_2=*Initialize* i_3=*Mutate*(i_1) i_4=*Select*(i_1, i_3) i_5=*Crossover*(i_1, i_4)	i_1=*Initialize* i_2=*Mutate*(i_1) i_3=*Select*(i_1, i_2) i_4=*Mutate*(i_3)	i_1=*Initialize* i_2=*Initialize* i_3=*Mutate*(i_1) i_4=*Select*(i_1, i_3) i_5=*Crossover*(i_2, i_4) i_6=*Mutate*(i_5)	i_1=*Initialize* i_2=*Initialize* i_3=*Mutate*(i_1) i_4=*Select*(i_1, i_3) i_5=*Mutate*(i_4) i_6=*Crossover*(i_2, i_5)

Remarks:

(*i*) In our model the *Crossover* operator always generates a single offspring from two parents. The crossover operators generating two offspring may also be designed to fit our evolutionary model.

(*ii*) The *Select* operator acts as a binary tournament selection. The best out of two individuals is always accepted as the selection result.

(*iii*) The *Initialize, Crossover* and *Mutate* operators are problem dependent.

4 Fitness Assignment

We have to compute the quality of each EA encoded in the chromosome in order to establish the fitness of a MEP individual. For this purpose each EA encoded in a MEP chromosome is run on the particular problem being solved.

Roughly speaking the fitness of a MEP individual is equal to the fitness of the best solution generated by one of the evolutionary algorithms encoded in that MEP chromosome. But, since the EAs encoded in a MEP chromosome use pseudo-random numbers it is likely that successive runs of the same EA generate completely different solutions. This stability problem is handled in the following manner: each EA encoded in a MEP chromosome is executed (run) more times and the fitness of a MEP chromosome is the average of the fitness of the best EA encoded in that chromosome over all runs. In all the experiments performed in this paper each EA encoded into a MEP chromosome was run 200 times.

5 Numerical Experiments

In this section, we evolve an EA for function optimization. For training purposes we use the Griewang's function [7].

Griewang's test function is defined by the equation (1).

$$f(x) = \frac{1}{4000} \sum_{i=1}^{n} x_i^2 - \prod_{i=1}^{n} cos\left(\frac{x_i}{\sqrt{i}}\right) + 1 \tag{1}$$

The domain of definition is $[-500, 500]^n$. We use $n = 5$ in this paper. The optimal solution is $x_0 = (0,\ldots,0)$ and $f(x_0) = 0$. Griewang's test function has many widespread local minima which are regularly distributed.

An important issue concerns the representation of the solutions evolved by the EAs encoded in an MEP chromosome and the specific genetic operators used for this purpose. The solutions evolved by the EAs encoded in MEP chromosomes are represented by using real values [2] (i.e. a chromosome of the second level EA is an array of real values). By initialization, a random point within the definition domain is generated. The convex crossover with $\alpha = \frac{1}{2}$ and the Gaussian mutation with $\sigma = 0.5$ are used.

5.1 Experiment 1

In this experiment we are interested in seeing the way in which the quality of the best evolved EA improves as the search process advances. The MEP algorithm parameters are: *Population size* = 100; *Code length* = 3000 genes; *Number of generations* = 100; *Crossover kind* = *Uniform*; *Crossover probability* = 0.7; *Mutations / chromosome* = 5; *Terminal set* = {*Initialize*}; *Function set* = {*Select*, *Crossover*, *Mutate*}. The results of this experiment are depicted in Fig.1.

Fig.1 clearly shows the effectiveness of our approach. The MEP technique is able to evolve an EA for solving optimization problems. The quality of the best evolved EA is 8.5 at generation 0. That means that the fitness of the best solution obtained by the best evolved EA is 8.5 (averaged over 200 runs). This is a good result, knowing that the worst solution over the definition domain is about 313. After 100 generations the quality of the best evolved EA is 3.36.

5.2 Experiment 2

We are also interested in seeing how the structure of the best evolved EA changed during the search process.

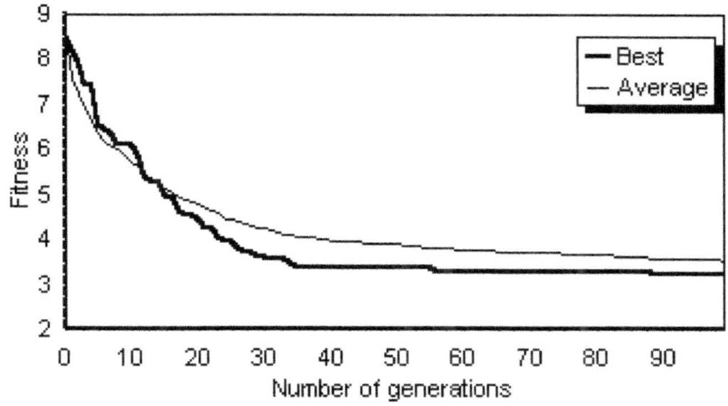

Fig. 1. The fitness of the best individual in the best run and the average (over 10 runs) of the fitness of the best individual over all runs.

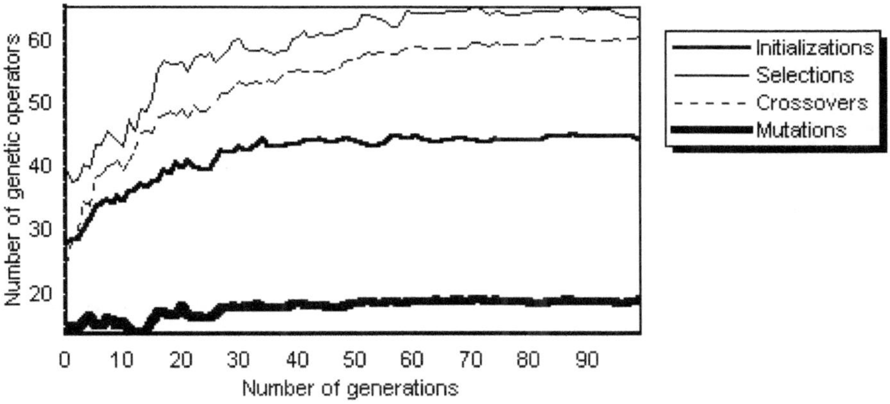

Fig. 2. The fitness of the best individual in the best run and the average (over 10 runs) of the fitness of the best individual over all runs.

The evolution of the number of the genetic operators used by the best evolved EA is depicted in Fig. 2.

From Fig. 2 it can be seen that the number of the genetic operators used by the best EA increases as the search process advances. For instance the averaged number of *Initializations* in the best EA at generation 0 is 27, while the averaged number of *Initializations* in the best evolved EA (after 100 generations) is 43. The averaged number of *Mutations* is small (less than 18) when compared to the number of occurrences of other genetic operators.

6 Conclusions and Further Work

A method for evolving evolutionary algorithms has been proposed in this paper. The numerical experiments emphasize the robustness and the effectiveness of this approach.

Further numerical experiments will analyze the relationship between the MEP parameters (such as the population size, the chromosome length, and the mutation probability) and the ability of the evolved EA to find optimal solutions. It is expected that an increased population size will bring about a substantial increase in the evolved EA performances.

The generalization ability of the evolved EA (how well it will perform on some new test problems) will also be studied. A larger set of functions should be used in order to increase the generalization ability.

An important issue is related to the amount of memory required by the evolved EA. No optimization regarding the memory used by the evolved EA was done in this paper. For instance, if the evolved EA performs 20 *Initializations*, 25 *Selections*, 50 *Crossovers* and 15 *Mutations*, the memory required by the algorithm is 110 times the memory required for storing an individual. It is obvious that this amount of memory can be reduced because some memory locations can be overridden by newly created individuals. A simple algorithm that checks whether a memory location will be accessed or not in the future can be used for this purpose.

References

1. Brameier, M., Banzhaf, W.: A Comparison of Linear Genetic Programming and Neural Networks in Medical Data Mining, IEEE Transactions on Evolutionary Computation, Vol. 5. (2001) 17–26
2. Goldberg, D.E.,: Genetic Algorithms in Search, Optimization, and Machine Learning, Addison-Wesley, Reading, MA, (1989)
3. Koza, J. R.,: Genetic Programming: On the Programming of Computers by Means of Natural Selection, MIT Press, Cambridge, MA, (1992)
4. Oltean M., Dumitrescu D.,: Multi Expression Programming, technical report, UBB-01-2002, Babeş-Bolyai University, Cluj-Napoca, Romania (available from www.mep.cs.ubbcluj.ro).
5. Holland, J.H.: Adaptation in Natural and Artificial Systems. University of Michigan Press, Ann Arbor (1975)
6. Syswerda, G.,: Uniform crossover in genetic algorithms. In Schaffer, J.D., (ed.): Proc. 3rd Int. Conf. on Genetic Algorithms, Morgan Kaufmann Publishers, San Mateo, CA, (1989) 2–9
7. Yao, X., Liu, Y., Lin, G.: Evolutionary programming made faster. IEEE Transaction on Evolutionary Computation, Vol. 3(2) (1999) 82–102
8. Wolpert, D.H., McReady W.G.,: No Free Lunch Theorems for Search, technical report SFI-TR-02-010, Santa Fe Institute (1995)

Modelling Artificial Ecosystem Selection: A Preliminary Investigation

Alexandra Penn

Centre for Computational Neuroscience and Robotics,
University of Sussex, U.K.
alexp@cogs.susx.ac.uk

Abstract. The ability of whole ecosystems to respond to selection has recently been demonstrated in artificial selection experiments, [1,2]. As well as having wide-ranging practical applications, this result significantly broadens the application of theoretical concepts of the mechanisms of heritability and variation in biological systems. Simulation models have the potential to be useful tools for the investigation of these issues. Whilst related simulation work exists [3,4], ecosystem-level selection itself has yet to be modelled. This paper presents such a model, in which ecosystems are modelled as generalised Lotka-Volterra systems and are subject to a generational selection process. A positive response to selection for diversity is demonstrated, with the only sources of variation being sampling errors arising when 'offspring' ecosystems are produced.

1 Introduction

Artificial selection at the level of the ecosystem is a new field of research, recently demonstrated in laboratory experiments by Swenson et al [1,2]. The possibility of creating 'designer ecosystems' has great potential practical usefulness. Swenson et al [2] have already demonstrated a response to selection in microbial communities evolved for improved biodegradation of the common pollutant, 3-chloroaniline. One could also envisage selection of microbial communities for biological sewage treatment, or selection of soil communities, such as mycorrhizal fungae, for increased above-ground biomass and productivity of crop plants.

Treating entire ecosystems as units of selection in their own right also raises interesting theoretical questions. Ecosystem-level selection is substantially different from individual-level selection. Firstly, selection can act on complex, 'higher-level' traits produced by the interactions between individuals and between species. Secondly, the mechanisms of variation and heritability which might allow selection at the level of the ecosystem are likely to be very different. As well as genetic variation (both within and between species), other potential sources of heritable variation include changes in the relative frequencies of different species and abiotic and stochastic factors. Swenson et al [1,2] suggest that variation is primarily introduced through the sampling process by which ecosystems are reproduced, and that the effects of this variation can be

W. Banzhaf et al. (Eds.): ECAL 2003, LNAI 2801, pp. 659–666, 2003.

significantly magnified by the dynamics of the ecosystem's subsequent development. If this is the case, then ecosystem dynamics must also play an important role in the inheritance of ecosystem-level traits.

Given the number of potential interdependent processes operating on a number of different levels and time-scales, understanding how ecosystems respond to selection presents a significant challenge. In this context, simulation may provide a useful tool for exploring intuitions and investigating hypotheses. This paper presents a preliminary investigation into modelling artificial ecosystem selection. Ecosystems, modelled using Lotka-Volterra competition equations, are successfully evolved to increase their species diversity. Two potential sources of heritable variation are modelled and compared, and the potential role of ecosystem dynamics is discussed.

2 Selection Above the Level of the Organism

The question of whether natural selection operates at a level above that of the individual organism remains a controversial one [5]. However, the requirements for selection to occur are general properties, not restricted to individual organisms [6]. Artificial selection experiments can sidestep the issues of whether selection happens in a natural setting and instead ask whether a response to selection can exist on a certain level when a particular population structure and selection pressure are imposed. A strong response to higher-level selection has been shown experimentally by a number of researchers, for example, single-species group selection with flour beetles [7], and cages of battery farm chickens [8,9], and multispecies group selection with two species of flour beetle [10]. These experiments have typically shown a strong response to group selection, and poor, or even negative, response to individual selection, and are interesting illustrations of the mechanics of higher-level selection. All have demonstrated that group-level selection can act on *interactions* between individuals or species, not directly accessible to individual-level selection [11].

The possibility of artificial selection at the ecosystem level was first suggested by Sober and Wilson, [5], and then tested experimentally by Swenson, Wilson et al [1,2]. In three separate experiments on soil and pond water microbial ecosystems, they showed statistically significant positive responses to selection for 'ecosystem-level' traits: above ground plant biomass, pH, and biodegradation of 3-chloroaniline. All experiments followed a similar procedure:

1. An initial 'population' of ecosystems was created by taking small, equally-sized, samples from the same 'source' ecosystem.
2. Samples used to inoculate a fixed amount of some sterile medium.
3. Ecosystems develop for a fixed amount of time, an 'ecosystem generation'.
4. Ecosystems evaluated for the chosen 'ecosystem-level' phenotypic trait.
5. Top scoring N_{parent} 'parent' ecosystems used to found the next generation.
6. Ecosystems reproduced 'sexually', mixing all "parents" then taking new samples, or 'asexually', taking $N_{offspring}$ samples from each 'parent'.
7. Steps then repeated from (2) for a set number of generations.

Each experiment showed wide divergence of different lines away from the value of the phenotypic trait of the original ecosystems. This was seen in both control and selected lines, however only the selected lines showed a systematic change in the direction of selection. Although the response to selection seemed somewhat erratic and variable, the difference between the first and last generations was found to be statistically significant in almost all cases; for example, a 4-fold increase in plant biomass, and 25-fold differences in H^+ concentration [1].

The idea of ecosystems as units of selection is an interesting one. The notion that selection can act at the level of ecosystems is not necessarily problematic. There are three general requirements for selection to take place amongst a population of units of any given type [6]:

1. Phenotypic variation amongst units.
2. Heritability of phenotypic differences.
3. Fitness consequences of phenotypic variation.

These requirements could, in principle, be satisfied at any level of biological organisation. In the case of artificial ecosystem selection experiments, both the creation of a population of units and the fitness consequences of phenotypic variation are imposed by the experimenters. However, for ecosystem-level selection to then occur still requires the first two conditions set out above to be met. That is, ecosystems sampled from the same original source must vary phenotypically, and that variation must be heritable. That these conditions should be fulfilled is far from intuitively obvious.

As ecosystems do not possess genomes it might be hard to see how any phenotypic variation between them could be inherited. Indeed, as all ecosystems are initially sampled from the same source ecosystem and develop in near-identical physical conditions, the source of their variation is not obvious. Swenson et al [1,2], conjecture that the main source of phenotypic variation amongst ecosystems is due to their "sensitive dependence on initial conditions". That is, that sampling results in small initial differences in species' genetic composition or population sizes, and these are magnified by ecosystem dynamics to give rise to large differences in macroscopic phenotypic traits. However, this sensitivity is a potential problem for the heritability of phenotypic traits. Heritability requires that offspring resemble their parents, whereas variation appears to arise to because the process of sampling can lead to significant differences between parent and offspring. For selection to be successful, a fine balance must be achieved between the opposing forces of variation and heritability, both of which are consequences of the underlying ecosystem dynamics. Ecosystem selection must search therefore, not only for ecosystems with the required phenotypic trait, but also for systems which will come quickly to stable local equilibria, so that their properties can reliably be transmitted to the next generation.

The differences between this approach, and individual level selection should be noted. As well as species genetic variation, an entirely different source of (potentially) heritable variation is present: the species proportions in a particular ecosystem at the end of a generation. These can differ between two ecosystems due to differences in initial species proportions, even in the absence of between-ecosystem genetic variation. Moreover, these differences may increase when the two ecosystems are themselves reproduced.

3 The Model

The model follows the same general procedure described above. Individual ecosystem dynamics were modelled using generalised Lotka-Volterra equations (described below). At the start of a run, a single source ecosystem is randomly generated and left to develop for 200 iterations. An initial population of 20 ecosystems is then created by repeatedly sampling the source ecosystem, using a fixed size 'pipette'. Although each ecosystem is generated from the same source, sampling error can introduce between-ecosystem variation in both species genetic composition, and initial species population sizes. Each ecosystem in the population is then allowed to develop for 50 iterations, and then evaluated on an ecosystem-level phenotypic trait (specified below), and the top 5 ecosystems are selected to become parents for the next generation. Offspring ecosystems are produced asexually by taking 4 fixed-size samples from each parent, thereby generating a new population of 20 individuals. As before, sampling may introduce variation between offspring of the same parent. The process of sampling, evaluation and selection is repeated until the required number of generations is reached.

Within-ecosystem dynamics are modelled using the generalised Lotka-Volterra competition equations [4,12]. For an ecosystem containing S species, the population size N_i of the ith species at time $t+1$ is given by:

$$N_{i,t+1} = N_{i,t}\left[1 + \frac{R}{K_i}\left(K_i - \sum_{j=1}^{S} N_{j,t}\alpha_{ij}\right)\right] \tag{1}$$

where K_i is the species' carrying capacity, R is the growth rate (common to all species), and α_{ij} is an interaction coefficient representing the per capita effect of species j on species i. For this model, $S = 10$ and $R = 2$ are constants. When a source ecosystem is created, each K_i is set at uniform random in the range 100:1000, and each α_{ij} is set at uniform random in the range 0:2, unless $i = j$ and then $\alpha_{ij} = 1$. Note that although all direct interactions are competitive ($a_{ij} > 0$), indirect effects may give rise to mutualisms or commensualisms.

Ecosystem reproduction involves taking a fixed-size sample from a selected parent ecosystem. In real ecosystem selection experiments, the process of sampling can introduce variation between offspring both in species genetic composition, and initial species population sizes. In the model, genetic variation due to sampling was modelled very simply. Genetic variation was assumed to affect interaction coefficients (α_{ij}) only. For each interaction coefficient, the offspring value was generated by taking a random deviate from a Gaussian distribution with a mean equal to the parent value and a standard deviation of 0.02 (i.e., 1% of the range). The initial population size for each species is calculated on the assumption that a sample contains individuals chosen at random from the parent ecosystem, thus the expected frequency of a species in a sample is equal to its frequency within the sampled ecosystem. Since species population sizes are continuous variables, sampling was modelled using the standard Gaussian approximation to a binomial distribution. Thus, N_i, the size of the species in the new sample was generated at random from a Gaussian distribution with mean, Bp_i, and standard deviation, $\sqrt{Bp_i(1-p_i)}$, where p_i is the frequency of the species in the parent ecosystem, and $B = 100$ is the mean size of the sample.

Species diversity was chosen as the ecosystem-level selection criterion, (useful in this context as calculated from the population distribution only). The degree of diversity exhibited by an ecosystem is a function of species interactions (both direct and indirect), and as such, a property of the ecosystem as whole rather than being attributable to individual species. The Shannon-Weaver Diversity Index—a commonly used ecological measure—was employed to quantify diversity [13]. It is defined as follows:

$$H = -\sum_{i}^{S} p_i \ln p_i \tag{2}$$

where S is the total number of species, and p_i is the proportion of species i in the population.

4 Results

The first question addressed was simply whether or not the model ecosystems would respond to selection. The ecosystem selection procedure was run 50 times. Each run lasted 200 generations and was started from a different randomly generated source ecosystem. Figure 1 shows the change in fitness score of best ecosystem for each generation averaged over all 50 runs. The difference between the first and last generation mean best fitness is statistically significant (Mann-Whitney U-Test $p < 0.001$). To ensure that this improvement was a result of selection, another 50 runs were carried out using the same set of source ecosystems, but parents were selected at random. This time there was no overall directional trend (see fig. 1), and no significant difference between the first and last generation means (Mann-Whitney U-Test $p > 0.96$).

An obvious question to ask is whether both sources of ecosystem variation are contributing to the successful response to selection just described. Two further sets of runs were carried out, with each using only one source of variation. The first set, employing only genetic variation, also showed a significant improvement (Mann-Whitney U-Test $p < 0.001$), achieving only a marginally lower final-generation mean than in the original experiment (0.62 compared to 0.64), which was not a significant difference (Mann-Whitney U-Test $p < 0.45$). In contrast, the set of runs employing only population size variation showed little sign of responding to selection; there was only a marginal improvement with no significant difference between a first generation mean of 0.322 and a final generation mean of 0.344 (U-Test $p < 0.75$).

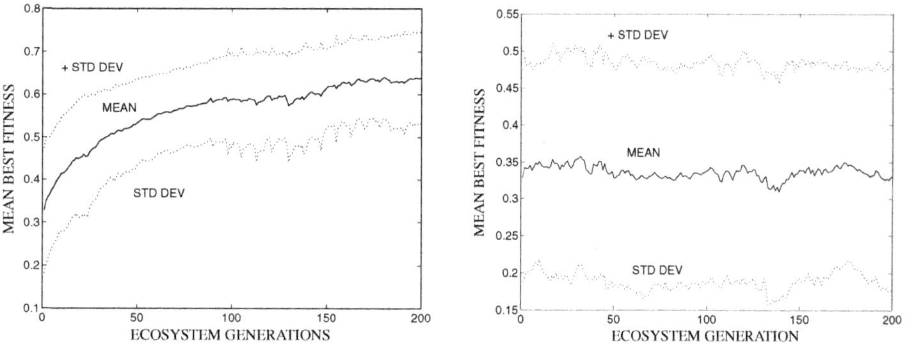

Fig. 1. Change in best ecosystem fitness over 200 generations, averaged over 50 different runs, with selection(left), random selection(right)

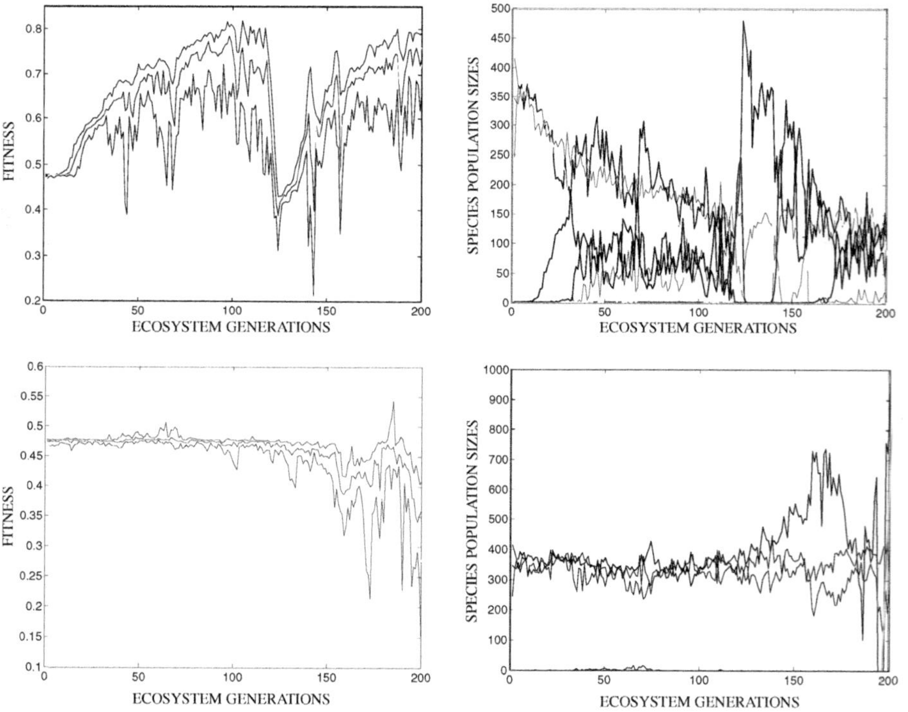

Fig. 2. Change in ecosystem population fitnesses over 200 generations for source ecosystem 4, with directed selection (top), and random selection (bottom) , showing best, worst, median, and upper and lower quartiles under random selection(left), corresponding changes in species frequencies at end of each generation for best ecosystem in population(right)

5 Discussion and Conclusions

Despite its simplicity, the model presented in this paper can illustrate some of the concepts of ecosystem-level variation and heritability introduced earlier. From the dynamical systems perspective set out in [1], phenotypic variability can arise in two ways: (i) sampling of ecosystems that are following different trajectories to the same attractor (instantaneous variation) or, (ii) ecosystems being knocked into different basins of attraction by sampling error. Figure 2 shows the dynamics of two runs starting from the same source ecosystem, one under directed and one under random selection. Phenotypic variability of both types arose under both selection procedures. This is particularly evident in the graph showing final species frequencies over 200 generations for the best ecosystem in the directed selection run (top right, fig.2). The species composition changes only slightly gen. 50-100 (accumulation of instantaneous variation), then undergoes sudden dramatic change gen. 100-150 with some species going extinct, others re-emerging from a low-level position. Finally it settles into a new 'attractor' with a totally different species distribution. These dynamics are accompanied by large changes in fitness (top left, fig.2). Fitness initially climbs, then collapses as the population dynamics suddenly change. It begins to climb again as they settle into a new equilibrium. This illustrates very well the previously discussed balance between variability and heritability. The population accumulates fitter variants produced by sampling error, but the fitness collapses as a small variation changes the underlying dynamics such that the ecosystem 'attractor' switches. This first ecosystem state is not robust enough to changes in initial conditions to be reliably heritable. Instantaneous variation is not sufficient for a reliable response to ecosystem selection. Any sustainable response to selection would require a network of ecosystem attractors (i.e. relatively stable ecosystem-level phenotypes) which could be reached via variations in species frequencies and genetic composition at the sampling phase. More heritable states would have wider basins of attraction, and come to equilibrium more quickly. In principle new ecosystem phenotypes can be reached with variation only in species proportions. However, in this model runs with population sampling error only, although displaying instantaneous variability in fitness, very rarely moved to new overall species compositions. Genetic sampling error alone did allow new species compositions to arise and fitness to increase significantly. This result may have been a consequence of the particular ecosystem dynamics used. Multispecies L-V competition systems can possess many stable attractors [13], however this does not necessarily mean that they are *sensitively* dependent on initial conditions. Ecosystems were started from small samples, and sampling error was small due to the assumption of perfect mixing. This might not have given a large enough range of possible initial species proportions for population sampling error to have played a significant role. Conversely, genetic sampling error may have been too large. This remains to be explored in future work.

Evidently the model presented here is an extreme simplification, used as a first step to explore ideas based on a dynamical systems perspective on ecosystems [1]. Real ecosystems exhibit certain patterns of connectivity, are constrained by energetic restrictions, and show higher-order interactions with the presence of some species modifying interactions between others. More realistic ecosystem dynamics may need to be incorporated, as well as stochasticity, and an abiotic environment. The abstract fitness measure chosen for these experiments satisfied the criterion of being an ecosystem-level trait, produced by the interactions of all species present. However, other fitness measures need to be investigated, including those which could be influenced by either individual or ecosystem selection. The potential for comparing an ecosystem's response to particular fitness criteria with selection at these two levels is particularly interesting. One of the next steps will be to introduce evolution within species during ecosystem generations, allowing both individual and ecosystem-level dynamics to act. Given that experiments so far have concentrated on fast-breeding, microbial ecosystems, the interplay between these dynamics is likely to be important. This model is only a first step towards the effective use of simulation tools in understanding and exploring the mechanisms of ecosystem-level selection. Ultimately it is hoped that simulations such as these can be used to help answer important practical questions on the mechanisms and conditions underlying ecosystem selection.

Acknowledgements. Thanks to Matt Quinn and Inman Harvey for constructive discussion.

References

1. Swenson, W., Wilson, D.S, Elias, R.: Artificial Ecosytem Selection, Proc. Natl. Acad. Sci. USA, 97, (2000), 9110–9114
2. Swenson, W., Arendt, J., Wilson, D.S.: Artificial Selection of Microbial Ecosystems for 3-chloroaniline Biodegradation, Environ. Microbiology, 2(5), (2000), 564–571
3. Ikegami, T, and Hashimoto, K.: Dynamical Systems Approach to Higher-level Heritability, J. Biol. Phys., 28(4),(2002), 799–804
4. Wilson, D.S.: Complex Interactions in Metacommunities, with Implications for Biodiversity and Higher Levels of Selection. Ecology, 73(6),(1997) 1984–2000
5. Sober, E., and Wilson, D.S.: Unto Others: The Evolution and Psychology of Unselfish Behaviour, Harvard University Press, Cambridge, Massachusetts (1998)
6. Lewontin, R.C.: The Units of Selection, Annu.Rev.Ecol.Syst., 1, (1970), 1–18
7. Wade, M.J.: Group Selection Among Lboratory Populations of Tribolium, Proc. Natl. Acad. Sci. USA, 73, 4604–4607
8. Craig, J.V., and Muir, W.M.: Group Selection for Adaptation to Multiple-Hen Cages: Beak-related Mortality, Feathering, and Body Weight Responses., Poultry Science 75 ,(1995), 294–302
9. Muir, W.M.: Group Selection for Adaptation to Multiple-Hen Cages: Selection Program and Direct Responses., Poultry Science 75 ,(1995), 447–458
10. Goodnight, C.J.: Experimental Studies of Community Evolution, I: The Response to Selection at the Community Level., Evolution 44,(1990), 1614–1624
11. Goodnight, C.J.: Heritability at the Ecosystem Level, Proc. Natl. Acad. Sci. USA, 97(17), (2000), 9365–9366
12. MacArthur, R.H.: Geographical Ecology, Harper and Row, N.Y., USA (1972)
13. Morin, P.J.: Community Ecology, Blackwell Science (1999)

Measuring Self-Organization via Observers

Daniel Polani

Adaptive Systems Research Group
Department of Computer Science
University of Hertfordshire
d.polani@herts.ac.uk

Abstract. We introduce *organization information*, an information-theoretic characterization for the phenomenon of self-organization. This notion, which requires the specification of an *observer*, is discussed in the paradigmatic context of the Self-Organizing Map and its behaviour is compared to that of other information-theoretic measures. We show that it is sensitive to the presence and absence of "self-organization" (in the intuitive sense) in cases where conventional measures fail.

1 Introduction

Shalizi [21] tracks back the first use of the notion of "self-organizing systems" to Ashby [21]. The bottom-up cybernetic approach of early artificial intelligence [25, 14] devoted considerable interest and attention to the area of self-organizing systems; many of the questions and methods considered relevant today have been appropriately identified almost half a century ago [e.g. 26].

The notion of *self-organization*, together with the related notion of *emergence*, are of central importance in the sciences of complexity and Artificial Life[1]. These phenomena form the backbone of those types of dynamics that lead to climbing the ladder of complexity which, as it is believed, lies ultimately behind the appearance of life-like phenomena studied in Artificial Life. Notwithstanding the importance and frequent use of these notions in the relevant literature, a both precise and useful mathematical definition remains elusive. While there is a high degree of intuitive consensus on what type of phenomena should be called "self-organizing" or "emergent", the prevailing strategy of characterization is along the line of "I know it when I see it". For a given system, the presence of self-organization or emergence is typically determined by explicit inspection. The present paper will concentrate on the discussion of how to characterize self-organization.

Specialized formal literature often does not go beyond pragmatic characterizations; e.g. Jetschke [10] defines a system as undergoing a self-organizing transition if the symmetry group of its dynamics changes to a less symmetrical one (e.g. a subgroup of the original symmetry group), typically occurring at

[1] To avoid possible misunderstandings, though they often co-occur and are mentioned together, the present paper construes "self-organization" and "emergence" to be two distinctly different phenomena, but intricately intervowen, see also Sec. 4.

W. Banzhaf et al. (Eds.): ECAL 2003, LNAI 2801, pp. 667–675, 2003.

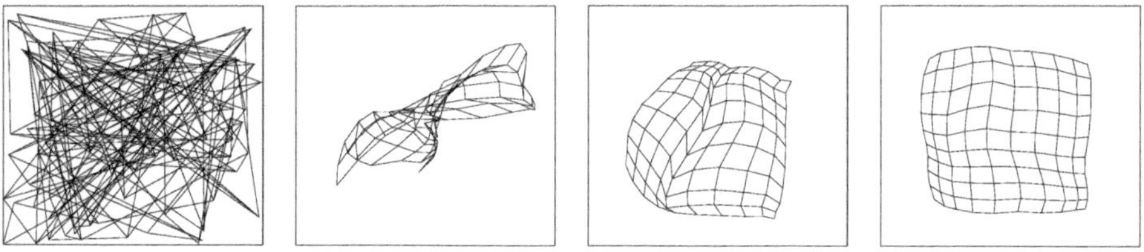

Fig. 1. An example for the training process for a Self-Organizing Map. The probability distribution used for generating the training samples is the equidistribution on the unit square $[0,1]^2 \subseteq \mathbb{R}^2$. The sequence of plots shows the SOM weights X_i where two weights X_i and X_j are connected by a line if they belong to units i and j neighbouring each other in the grid at training steps $0, 10, 100, 1000$.

phase transitions [19]. This view relates self-organization to phase transitions. However, there are several reasons to approach the definition of self-organization in a different way. The typical complex, living or artificial life system is not in thermodynamic equilibrium [see also 17]. One possible extension of the formalism is towards nonequilibrium thermodynamics, identifying phase transitions by *order parameters*. These are quantities that characterize the "deviation" of the system in a more "organized" state in the sense of Jetschke from the system in a less organized state, measured by absence or presence of symmetries. Order parameters have to be constructed by explicit inspection of the system since a generic approach is not available. Also, in Alife systems, one can not expect the a priori existence of any symmetry to act as indicators for self-organization.

2 Self-Organization

To discuss approaches to quantify self-organization, it is useful to introduce a paradigmatic system. We choose a system which is not among the typical Alife systems, but exemplifies the principles we need to develop. These principles are, however, not restricted to that model and generalize immediately to any stochastic dynamic system and will be applied to Alife systems in future. Here, our paradigm system is Kohonen's Self-Organizing Map (SOM) [11, 20, 12]

For lack of space, we do not go into details about the SOM dynamics and give only give a brief outline. A SOM is a set of units i each of which carries a weight $X_i \in \mathbb{R}^n$. The units i are located on the nodes of a (typically square) grid. The SOM is now being *trained* with samples V drawn from a probability distribution on $\mathbb{R}^n$, during which the weights X_i reorganize in such a fashion that, if possible, neighbouring features in $\mathbb{R}^n$ are represented by weights w_i and w_j belonging to units i and j which are neighbours in the grid and vice versa.

Figure 1 shows a typical training process for a SOM. The "self-organization" property is reflected by the fact that during the training the SOM representation changes from an "unorganized" to the "organized" final state. This "self-organization" property is immediately evident to a human observer, but far from

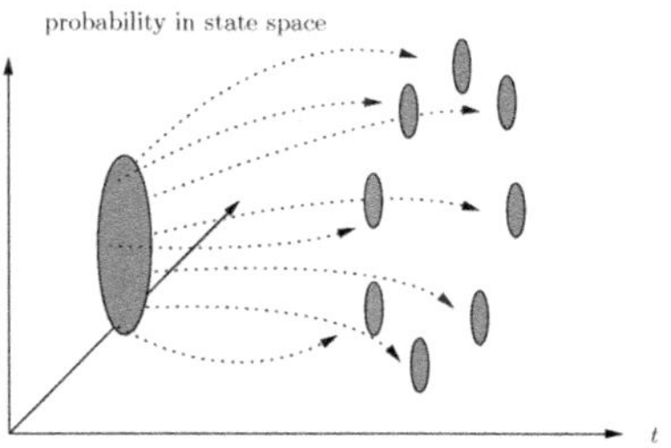

Fig. 2. Schematic representations of the evolution of probability distribution during the SOM training. The set of random initial configurations covers a larger proportion of the state space. During training the dynamics separates into distinct paths each concentrating along one of the eight probable organized solutions.

obvious to quantify [5, 24, 6, 15]. Many SOM organization measures are system-specific and therefore inadequate as universal notions of self-organization.

To alleviate this problem, Spitzner and Polani [22] attempted to apply Haken's synergetics formalism [7] to the SOM. This formalism only requires a dynamical system structure. They converted the stochastic SOM training rule to its Kushner-Clark deterministic continuous counterpart [13] and applied the synergetics formalism to it. Unfortunately, the synergetics formalism has to be applied close to fixed points of the system under which conditions it turns out to be insensitive to SOM self-organization. However, the view as a dynamical system proves fertile and opens up a selection of possible alternative characterization approaches. Figure 2 shows schematically the evolution of the state probability distribution during SOM training for the example from Fig. 1. The initial state is a random configuration of an entire vector $(X_1, X_2, \ldots, X_k)$ of initial random weights $X_i \in [0,1]^2$, $k = 1, 2, \ldots, k$ covering large parts of state space. During training, the SOM will stabilize itself along one of the 8 organized "square" solutions (rightmost plot in Fig. 1), concentrating the state probability distribution around one of the 8 organized solutions. This phenomenology will serve as a basis for the following discussions.

Due to space limitations, the formalization of the exposition is restricted to a minimum. Consider a random variable X assuming values $x \in \mathcal{X}$, $\mathcal{X}$ the set of possible values for X. For simplicity, assume that $\mathcal{X}$ is finite. Define the *entropy* of X by $H(X) := -\sum_{x\in\mathcal{X}} p(x) \log p(x)$, the *conditional entropy* of Y as $H(Y|X) := \sum_{x\in\mathcal{X}} p(x) H(Y|X = x)$ with $H(Y|X = x) := -\sum_{y\in\mathcal{Y}} p(y|x) \log p(y|x)$ for $x \in \mathcal{X}$. The *joint entropy* of X and Y is the entropy of the random variable (X, Y). The *mutual information* of random variables X and Y is $I(X;Y) := H(Y) - H(Y|X) = H(X) + H(Y) - H(X,Y)$. A generalization is the *intrinsic information*: for random variables $X_1, \ldots, X_k$, the intrinsic information is $I(X_1; \ldots; X_k) := \sum_{i=1}^{k} H(X_i) - H(X_1, \ldots, X_k)$. This notion is also known e.g. as *integration* in [23]. Similar to the mutual information, it is a measure for the degree of dependence between the different X_i.

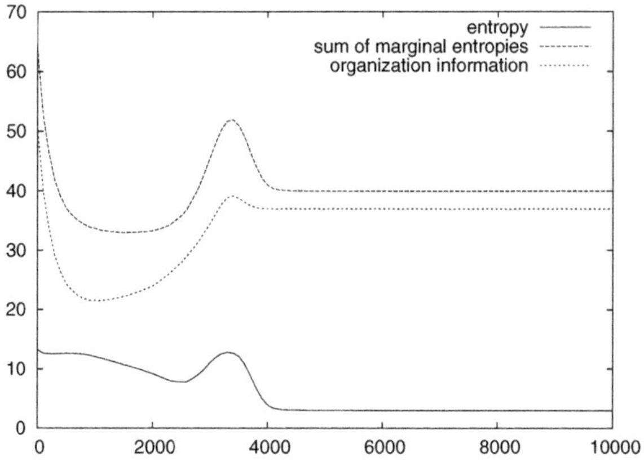

Fig. 3. Estimated entropy and information quantities for the SOM training process. For details see text (marginal entropies and organization information will be introduced in Sec. 3.)

We apply these notions to the discussion from Sec. 2. To obtain quantitative statements, we run 10000 SOM training runs of 10000 training steps each, using a SOM with units on a grid of 4×4 (as opposed to the 10×10 SOM used in Fig. 1). The entries for the total state vector variable $S = (X_{1,1}, X_{1,2}, \ldots, X_{4,3}, X_{4,4}) \in \mathbb{R}^{2\times4\times4}$ (the 2 appears because weights are two-dimensional) are boxed into 4 possible values in each component. Via frequency statistics we compute empirical estimates for the different entropies. Starting with a SOM with random initial weight vectors, one aspect of the self-organization is the concentration of the probability distribution around the organized configurations during training. If the complete state configuration of the SOM at time t is given by $S^{(t)}$, then quantify this process by the *information gain* $H(S^{(t=0)}) - H(S^{(t=t_{\text{final}})})$. The information gain is related to Ashby's redundancy measure of self-organization $(H_{\max} - H(S^{(t=t_{\text{final}})}))/H_{\max}$ where $H_{\max}$ is the maximum possible entropy of the system [9]. In our case, the initial entropy of S is given by $H(S^{(t=0)}) = 4 \times 4 \log_2 4^2 = 64\,\text{bit}$ since there are 4×4 units, the weight of each can be in one of the 4^2 quantization boxes[2] The final entropy is $H(S^{(t=t_{\text{final}})}) = 3\,\text{bit}$ matching well the empirical values from Fig. 3, reflecting the logarithm of the 8 organized states.

The information gain only takes into account the probabilities at the beginning and the end of the training. It fails to detect the symmetry-breaking whereby the initial probability distribution splits into subprobabilities as in

[2] The initial value of $\approx 13.3\,\text{bit}$ for the entropy in Fig. 3 at $t = 0$ is a strong underestimate, since the equidistribution in the high-dimensional state space is undersampled by the 10000 runs used. Only at $t = 2000 - 3000$ the probability distributions sufficiently concentrate for the empirical estimate of the entropies to become more accurate.

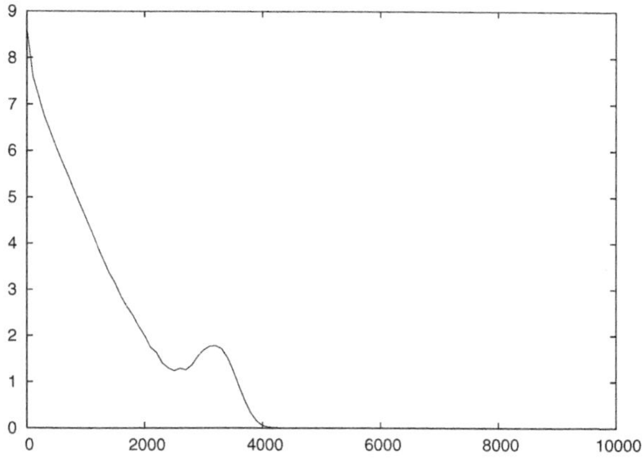

Fig. 4. Prediction entropy for the training process. For 1000 individual training runs, the single-step prediction entropy $H(S^{(t+1)}|S^{(}t))$ has been calculated for $S(t)$ at different t. For a given instance $s(t)$ of a state, 1000 individual single step probes were performed, obtaining the single-step prediction entropy for $H(S^{(t+1)}|s^{(}t))$. These were then averaged over all 1000 training runs.

Fig. 2. This split cannot be easily detected on the information-theoretic level unless trajectory history is brought into consideration[3]. If the system, after entering one of the subprobabilities regions, remains there with high probability, these can be considered stochastic attractor regions. This restriction can be detected by the *prediction entropy* $H(S^{(t+1)}|S^{(}t))$ which, for advanced t one expects to be smaller for systems with a history component than for systems whose history is quickly lost (e.g. for a system which jumps from branch to branch). Fig. 4 shows how the prediction entropy drops during training. The prediction entropy reflects the "freezing" of history. The bump at $t = 3000$ is due to the fact that in the initial part of the training process, the SOM contracts strongly, reducing the entropy, before beginning to spread its weights over $\mathcal{R} = [0, 1]^2$.

The prediction entropy is not yet sufficient to capture the whole essence of self-organization in the SOM and thus to be a promising candidate for general measures of self-organization. To see that, let us carry out a thought experiment. Modify the SOM training as to cause the SOM to always freeze in the same configuration at the end of the training. For the argument it is irrelevant whether this configuration is "organized" or a fixed random state $s^* \in [0, 1]^{2\times4\times4}$. Neither information gain nor prediction entropy are able to distinguish it from a training process generating the split probability distribution in Fig. 2.

[3] This problem is also reflected by Ashby's redundancy measure; that measure has the further problem that changed restrictions of the system induce a modification of $H_{\max}$.

3 Organization via Observers

This seems to indicate that self-organization cannot be defined via the intrinsic dynamics of the system and that one needs to assume additional structure for the system. Here, this additional structure will be an *observer*. The observer concept plays a role in obtaining entropies for physical systems [1]. It has been emphasized in the past that the observer concept may be of central importance to characterize self-organization or emergence [4, 8]. In [3, 18] the concept of emergence is being defined via observatory structures. Their approach uses the language of category theory and provides a highly general meta-mathematical framework to formulate the phenomenon of emergence. The generality of that approach, however, makes it harder to construct the pertinent notions and to evaluate them in a concrete system. It was felt that the less, but still sufficiently general information-theoretic perspective would be more directly applicable to practical purposes. Also, information theory offers a chance to provide *intrinsic* notions which do not require the introduction of additional structural language and which, at some point, might be derived as consequence directly from the system dynamics itself, which emphatically is not the case with the notions from [3, 18]. Here, we now combine the information-theoretic approach with the observer concept to characterize self-organization.

A *(perfect) observer* of a (random) variable S is a collection $S_1, S_2, \dots, S_k$ of random variables allowing full reconstruction of S, i.e. for which $H(S|S_1, S_2, \dots, S_k)$ vanishes. Let the *(observed) organization information* be the intrinsic information $I(S_1; \dots; S_k)$. Call a system *self-organizing* if the *(observed) organization information increase* is positive during the progress of its dynamics. $I(S_1; \dots; S_k)$ quantifies the dependency between the observer variables. In some respect it is related to Ashby's redundancy measure, however taking into account only the "effectively used" state space.

For the SOM, a natural choice for an observer is to set $S_1, S_2, \dots, S_k$ to $X_1, X_2, \dots, X_k$ of the k individual SOM units. Consider three cases: First, a modified dynamics that lets the training begin with a random state $S^{(t=0)}$ and end with a random state $S^{(t=t_{\text{final}})}$ independent from $S^{(t=0)}$ but with the same distribution. In this case, both the information gain and the organization information will vanish for start and end state. Both will correctly identify this system as not self-organizing. Assuming random sampling in each step, the prediction entropy will always yield the full entropy $H(S^{(t=0)})$ of the initial state with no decrease indicating an attractor; using the quantization from the example from Sec. 2 and Fig. 3 gives a prediction entropy of 64 bit. In the second case, let the dynamics start the SOM training with a random state and end up with a unique final state, as in the last example of Sec. 3. The information gain will be $H(S^{(t=0)}) - 0$ bit (64 bit in our above model). However, the organization information for both $S^{(t=0)}$ and $S^{(t=t_{\text{final}})}$ vanishes, as in the unique final state all the entropies vanish. Such a system with a single attractor point will not be regarded as self-organizing by the organization information. This is plausible since such a dynamics is doing nothing else than "freezing" the system Finally, use the original SOM dynamics. The information gain is 64 bit -3 bit $=$ 61 bit, large, but smaller than for the case of a unique final state, since there are 8 ordered final states. Thus, from the "information gain" point of view, the standard dynamics is less "self-organizing" than that freezing into a a unique state. The organization information $\sum_i H(S_i^{(t=0)}) - H(S^{(t=0)})$ vanishes in the random initial state $S^{(t=0)}$, however, since the individual (marginal) entropies $H(S_i^{(t=0)}), i = 1, \ldots, k$ are $\log 4^2 = 4$ bit each and the total entropy is $H(S^{(t=0)}) = 64$ bit. Summarizing, while the information gain "sees" only the attractor structure of the dynamics and can not distinguish between a structureless point attractor and a rich attractor landscape inducing organization in its systems, and while the prediction entropy can only quantify the "historicity" of a state class, the organization information measure is able to identify just the organizational aspect of the development of the dynamics. Final remarks on the plots: due to the underestimation of $H(S^{(t=0)})$ in the simulation runs — see footnote 2 — the value for the organization information does not seem to vanish for $t = 0$ in Fig. 3. Thus, one has to keep in mind that the value of the organization information begins with value 0 bit at $t = 0$, growing to the point where the plot becomes more accurate after $t \lessapprox 2000$.

The question how the notion of organization information depends on a given observer is still under research. At the present point a transformation formula that trans the organization information of a fine-grained observer to that of a coarse-grained observer is known. It can not be given here due to space limitations.

4 Summary and Current Work

We have introduced a notion of *organization information* to quantify a process of self-organization. It is defined for a given stochastic dynamical system for which an observer is specified. The properties of this measure have been compared with other information-theoretic measures and discussed using the Self-Organizing Map as scenario. The notion is, however, general and in its application not limited to the SOM. It is natural and appears to be versatile and sensitive to precisely the relevant self-organization processes.

On a practical level, it is envisaged to improve the numerical calculation of the quantities involved and to apply the notion to other systems of interest to validate its power. On a methodological level, the study of the influence of generalized observer change as well as the existence of "canonical", i.e. natural observers is of particular interest. This interest is fueled by strong indications that it is possible to define natural observers via recently introduced information-theoretic notions of emergence [16]. If that proves feasible, this would lead not only to an intrinsic characterization of self-organization (requiring no externally given observers), but also clarify some of the deep structural relations between the notions of self-organization and emergence.

Acknowledgements. The author would like to thank Peter Dauscher and the anonymous reviewers for helpful comments on the paper.

References

[1] Adami, C., [1998]. *Introduction to Artificial Life.* Springer.

[2] Ashby, W. R., [1947]. Principles of the self-organizing dynamic system. *J. Gen. Psychol.*, 37:125–128.

[3] Baas, N. A., Emmeche, C., [1997]. On Emergence and Explanation. *Intellectica*, 2(25):67–83.

[4] Crutchfield, J. P., [1994]. The Calculi of Emergence: Computation, Dynamics, and Induction. *Physica D*, 11–54.

[5] Erwin, E., Obermayer, K., Schulten, K., [1992]. Self-Organizing Maps: ordering, convergence properties and energy functions. *Biol. Cybern.*, 67:47–55.

[6] Goodhill, G. J., Sejnowski, T. J., [1997]. A unifying objective function for topographic mappings. *Neural Computation*, 9:1291–1304.

[7] Haken, H., [1983]. *Advanced synergetics.* Berlin: Springer-Verlag.

[8] Harvey, I., [2000]. The 3 Es of Artificial Life: Emergence, Embodiment and Evolution. Invited talk at Artificial Life VII, 1.-6. August, Portland.

[9] Heylighen, F., Joslyn, C., Turchin, V., [2003]. Principia Cybernetica Web `http://pespmc1.vub.ac.be`, March 2003

[10] Jetschke, G., [1989]. *1 Mathematik der Selbstorganisation.* Braunschweig: Vieweg.

[11] Kohonen, T., [1989]. *Self-Organization and Associative Memory*, vol. 8 of *Springer Series in Information Sciences.* Berlin, Heidelberg, New York: Springer-Verlag. Third edition.

[12] Kohonen, T., [1997]. *Self-Organizing Maps.* Springer. Second edition.

[13] Kushner, H. J., Clark, D. S., [1978]. *Stochastic Approximation for Constrained and Unconstrained Systems*, vol. 26 of *Applied Math. Science Series.* Springer.

[14] Pask, G., [1960]. The Natural History of Networks. In [26].

[15] Polani, D., [2001]. Measures for the Organization of Self-Organizing Maps. In Jain, L., Seiffert, U., editors, *Self-Organizing Neural Networks. Recent Advances and Applications.* Springer.

[16] Polani, D., [2002]. On Individuality, Emergence and Information Preservation. In Nehaniv, C. L., te Boekhorst, R., editors, *Proceedings of the Symposium on Evolvability and Individuality, 18–20 September 2002, St. Albans.* University of Hertfordshire.

[17] Prigogine, I., Nicolis, G., [1977]. *Self-Organization in Non-Equilibrium Systems: From Dissipative Structures to Order Through Fluctuations.* New York: J. Wiley & Sons.

[18] Rasmussen, S., Baas, N., Mayer, B., Nilsson, M., Olesen, M. W., [2001]. Ansatz for Dynamical Hierarchies. *Artificial Life*, 7:329–353.

[19] Reichl, L., [1980]. *A Modern Course in Statistical Physics.* Austin: University of Texas Press.

[20] Ritter, H., Martinetz, T., Schulten, K., [1994]. *Neuronale Netze.* Addison-Wesley.

[21] Shalizi, C. R., [1996]. Is the Primordial Soup Done Yet? `http://www.santafe.edu/~shalizi/Self-organization/soup-done/`, Jan 23, 2001

[22] Spitzner, A., Polani, D., [1998]. Order Parameters for Self-Organizing Maps. In Niklasson, L., Bodén, M., Ziemke, T., editors, *Proc. of the 8th Int. Conf. on Artificial Neural Networks (ICANN 98), Skövde, Sweden*, vol. 2, 517–522. Springer.

[23] Tononi, G., Sporns, O., Edelman, G. M., [1994]. A measure for brain complexity: Relating functional segregation and integration in the nervous system. *Proc. Natl. Acad. Sci. USA*, 91:5033–5037.

[24] Villmann, T., Der, R., Herrmann, M., Martinetz, T., [1997]. Topology Preservation in Self-Organizing Feature Maps: Exact Definition and Measurement. *IEEE Trans. Neural Networks*, 8(2):256–266.
[25] Walter, W. G., [1951]. A Machine that Learns. *Scientific American*, 60–63.
[26] Yovits, M. C., Cameron, S., editors, [1960]. *Self-Organizing Systems – Proceedings of an Interdisciplinary Conference, 5–6 May 1959*, Computer Science and Technology and their Application. Pergamon Press.

General Framework for Evolutionary Activity

Michael J. Raven[1] and Mark A. Bedau[2]*

[1] NYU, 503 Silver Center, 100 Washington Square East, New York NY 10003, USA
[2] Reed College, 3203 SE Woodstock Blvd., Portland OR 97202, USA,
mab@reed.edu http://www.reed.edu/~mab

Abstract. Evolutionary activity statistics have been used to visualize and quantify the adaptive evolutionary dynamics in a wide variety of artificial and natural evolving systems, but the formalism for the statistics has evolved over the years. Furthermore, the statistics can be applied to many different aspects of an evolving system, and application in any given context requires settling certain choices. In addition, the statistics involve normalization with a special-purpose "neutral" system, which requires making even more choices. So, to help make these statistics easier to use and understand, we situate them in a new and more general formal framework and then show how this framework applies to earlier work with the statistics.

1 The Need for a Framework for Evolutionary Activity

It is commonly accepted that the process of adaptation produces much of the order and functionality evident in complex systems [11,10,8], but it is often difficult to distinguish adaptive change in a system from other evolutionary phenomena, such as random genetic drift [9,12]. For natural systems the problem is often the unavailability of the relevant data. Those studying artificial evolving systems have the luxury of being able to collect virtually complete data; aside from storage space, only imagination limits what kinds of data are gathered. But this compounds rather than alleviates the problem, which is the inability to highlight the *relevant* data. The study of evolutionary dynamics in natural and artificial systems dearly needs an effective method for identifying and measuring the creation of adaptations in the course of evolution.

A decade ago Bedau and Packard devised a method for visualizing adaptive phenomena in evolving systems [2]. The method rests on the calculation of *evolutionary activity* statistics which can be applied to various kinds of components of evolving systems, including individual alleles [2], various classes of alleles [5], and whole genotypes [3,4]. A significant part of the appeal of evolutionary activity statistics is their applicability to a wide variety of evolving systems. Evolutionary activity has been measured in many artificial life systems, including Packard's Bugs [3,14],

Ray's Tierra and its derivatives [1,3,4], Lindgren's evolving iterated prisoner's dilemma [15], and Holland's Echo [4]. It has also been measured in some natural

* To whom correspondence should be addressed.

W. Banzhaf et al. (Eds.): ECAL 2003, LNAI 2801, pp. 676–685, 2003.

evolving systems, such as the biosphere as reflected in the fossil record [3,4] and the evolution of technology as reflected in the patent record [16]. Comparing evolutionary activity phenomenology within or between systems shows how evolutionary phenomena vary as a function of such factors as mutation rate and mode of selection [5], and quantifying evolutionary statistics enables adaptive evolutionary dynamics in various artificial and natural systems to be directly compared [2,3,4,14,16]. Evolutionary activity statistics have also been used to study evolutionary contingency [19], punctuated equilibria [15], mutualism [13], diversity [18], and classifications of evolutionary dynamics [4,17,6,7].

Nevertheless, there are barriers to realizing the full potential of evolutionary activity measurements. One is their very flexibility. Their applicability to an open-ended variety of different kinds of entities obscures the constraints on their proper use. Furthermore, the statistics have been defined in different ways in different publications, and this hides their underlying equivalence. It can be similarly unclear what unifies the different kinds of "neutral" systems that are used to normalize the statistics. So providing a consistent and general formal framework would help make evolutionary activity statistics more useful and more used. That is the purpose of the present paper.

2 A Formal Framework for Evolutionary Activity

The purpose of evolutionary activity is to measure the extent to which components of an evolving system are and have been resisting selection pressure. Persisting in the face of exposure to selection is the sign of an adaptation (though there are exceptions to this rule—see below). So the essential idea is to measure a component's exposure to selection, because continual exposure to selection is evidence of resistance to selection. To get started on this project, one must answer three questions about the system's components:

Question 1. What should be the components of the system?

Question 2. What should be a component's initial activity?

Question 3. What should be a component's current activity?

Choosing different answers to these questions allows one to examine evolutionary dynamics at different levels of analysis. We will consider each question in turn.

What should be the components of the system? An evolving system typically contains many different kinds of evolving components, e.g., individual genes, combinations of genes or schemata, individual phenotypic traits and clusters of them, whole genotypes, species and higher taxonomic groups. One might want to measure the adaptive evolution of any of these components. In general, there is no unique right choice of component to study, but some choices are wrong because some kinds of components in some systems are ill defined. After choosing what type of component to study, one must focus on which instances of those components are present in the system at a given time. If S is an evolving system and F is a property that identifies the components of interest, then the set of components present in S at t is:

Definition 1. $C_t = \{c : c \text{ exists in } S \text{ at } t \wedge c \text{ has property } F\}$,

while the set of all such components across T, the set of all time steps, is:

Definition 2. $C = C_{t_0} \cup C_{t_1} \cup \ldots$,

where $t_1, t_2, \ldots \in T$.[1] For example, if F is the property *being a genotype*, then C_t is the set of all genotypes extant at t and C is the set of all genotypes extant at some time or other. In most evolving systems the extant components change over time, so we have a notion of the addition ("birth" or "origination") and subtraction ("death" or "extinction") of a component, as follows:

Definition 3. c is *added to* C_t if and only if $c \notin C_{t-1} \wedge c \in C_t$.

Definition 4. c is *subtracted from* C_t if and only if $c \in C_{t-1} \wedge c \notin C_t$.

For example, instances of a particular allele at particular loci exist in a system at t if some agent has an instance of that allele at that locus at t, and a genotype exists in a system at t if some agent has that genotype at t.

What should be a component's initial activity? Computing evolutionary activity involves tracking each component's evolutionary activity over time. As a bookkeeping matter, a component's activity is stored in its *activity counter*. When a new component is added to an evolving system, its activity counter must be initialized. In some cases the proper initial value will be the same for all components at all times, but in other cases it will depend on the context. For example, if one wished to record the activity in a component's lineage, one might initialize a new component's activity with the activity of its immediate ancestor (more on this later). We can represent this formally with $\Delta^C_{init} : C \rightarrow \Re$, where the initial value of a component c's activity counter is $\Delta^C_{init}(c)$.

What should be a component's current activity? A component's activity counter is a historical record (sum) of its activity over its entire lifetime. At each moment that the component exists, its activity counter is incremented by its current activity, i.e., its current exposure to selection. Exposure to selection can be measured in various ways; some methods are easier than others and some methods reveal selection exposure more clearly. The simplest measurement of selection exposure is always a component's existence, but this method is crude. A more sensitive measure of a genotype's exposure to selection is its concentration

[1] We assume that time is discrete since artificial evolving systems usually assume discrete time. The formalism can be extended to continuous time.

in the population, and a more sensitive measure of an allele's exposure to selection is its expression or use. How best to measure a component's current activity will depend upon what one wants to learn about the target system and whether one hopes to compare this information *across* systems. A component's current activity can be represented with the partial function $\Delta^C : C \times T \rightarrow \Re$, where $\Delta^C(c)$ is c's activity at t if c is present in S at t (and is otherwise undefined).

3 Definitions of Evolutionary Activity Statistics

Evolutionary activity statistics aim to reflect how the evolutionary process is creating adaptations by observing resistance to selection pressures. A component of an evolving system can resist selection pressure only when it is "active" or exposed to selection. So, we assign each component an activity counter to record its exposure to selection pressure over its entire history.

Evolutionary activity and excess activity of a component. More precisely, we define the evolutionary activity (or activity counter) of component c at time t as c's initial activity plus the sum of c's activity increments up to t. Formally, the value of c's activity counter at t is given by $\alpha^C : C \times T \rightarrow \Re$:

Definition 5. $\alpha^C(c,t) = \Delta^C_{init}(c) + \sum_{i=Birth(c)}^{t} \Delta^C(c,i).$

where $Birth : C \rightarrow T$ gives the time step at which c is added to the system.

Every time a component is exposed to natural selection, selection can provide feedback about its adaptive value. Obviously, it will not continue to be tested by natural selection unless it has passed previous tests. So, the amount that a component has been tested by selection reflects how *successfully* it has passed those tests. If a sufficiently well-tested component persists and spreads through the population, we have positive evidence that it is persisting because of its adaptive value, i.e., that it is an adaptation. But natural selection is not instantaneous. Repeated trials might be needed to drive out maladaptive components. So persistence in the face of *some* selection is no proof of being an adaptation. Thus nonadaptive items can generate "noise" in evolutionary activity data, and to gauge resistance to selection we must filter out this noise.

One way to filter the nonadaptive noise is to determine how activity would accrue if components were persisting due solely to nonadaptive factors like random drift or architectural necessity. A general way to measure the expected evolutionary activity of such nonadaptive items is to construct a *neutral variant* of the target system, that is, a system that is similar to the target in all relevant respects except that none of its components has any adaptive significance. For example, if natural selection affects only births and deaths in a target system, then a neutral system could be just like the target system except that births and deaths are the result of random rather than natural selection. (More details about neutral systems are available elsewhere [1,3,4,14,16,5].) The accumulated activity in neutral systems provides a no-adaptation null hypothesis for the target system, which is used to screen off nonadaptive activity. If we observe significantly more evolutionary activity in the target system than in its neutral variant, we have good evidence that this "excess" activity cannot be attributed to nonadaptive factors. That is, we have good evidence that the components with excess activity are adaptations. So, we normalize target systems by subtracting the evolutionary activity accrued in the corresponding neutral system, and call the result *excess* activity. Specifically, we define the *excess activity* $\alpha^C_{\text{excess}} : C \times T \rightarrow \Re$ of a component at a time as:

Definition 6. $$\alpha^C_{\text{excess}}(c,t) = \begin{cases} \alpha^C(c,t) - \nu(c,t) & \text{if } \alpha^C(c,t) > \nu(c,t) \\ 0 & \text{otherwise} \end{cases}$$

where the function ν is determined by the specific neutral system used. As this definition indicates, the excess activity of a component is positive only if the component's raw observed activity exceeds the value of the activity observed in the neutral system.

The choice of appropriate neutral system depends on details of the target system. The ν functions implied by three recent measurements of excess evolutionary activity illustrate some possible forms of ν. Sometimes ν is a constant function which ignores c and t, in which case the significance of a component's activity depends only on its level and not on the identity of the particular component or time. But this need not be the case (see the third example below). The first example is from Bedau, Snyder, & Packard [4], who used a neutral system to determine a threshold a' above which activity could be regarded as more likely than not to be the result of an adaptation. In this case, ν is a simple constant function:

Example 1. $\nu(c,t) = a'$.

The second example is from Skusa & Bedau [16]. They determined ν by noting the maximum activity value that the neutral system produced for any component. If we let N^α be the set of values of activity counters of the neutral system at the end of the simulation, time t_{end}, then one can define ν as:

Example 2. $\nu(c,t) = \text{Max}(N^\alpha)$.

Third, Channon [7] uses a more fine grained neutral model, with a unique neutral component corresponding to each component in the target system, and he normalizes "on the fly" by taking the activity of each component target system and subtracting the activity of its corresponding component in the neutral system. If N_t is the set of neutral system components at t and $\eta : C \times T \rightarrow N_t$ is a function giving the neutral component $n \in N_t$ that corresponds to the $c \in C_t$, then one can define ν as:

Example 3. $\nu(c,t) = \alpha^N(\eta(c,t),t)$.

Extent and intensity of activity of an evolving system. Activity and excess activity are "micro" statistics defined for each component. It is possible to define various "macro" statistics that summarize the evolutionary activity in a whole system. In particular, one can measure a system's *intensity* of adaptive evolution, that is, the rate at which new adaptations are being produced by natural selection. In addition, one can measure a system's *extent* of adaptive evolution. The extent and intensity of evolutionary activity are two independently varying aspects of a systems adaptive evolution. While these statistics could be applied to either raw activity or excess activity, our concern here is with the latter.[2]

The *excess extent* of evolutionary activity is given by the function $E^C : T \rightarrow \Re$, defined simply as the sum of the excess activity of all extant components:

Definition 7. $E^C(t) = \sum_{c \in C_t} \alpha^C_{\text{excess}}(c, t)$.

This statistic measures the total continual adaptive success of all the components in the system. The *excess intensity* is given by the function $I^C : T \rightarrow \Re$, defined as the number of components that have newly given evidence of being adaptations by having positive excess activity:

Definition 8. $I^C(t) = \#\{c \in C_\tau : \alpha^C_{\text{excess}}(c, \tau - 1) = 0 \wedge \alpha^C_{\text{excess}}(c, \tau) > 0\}$.

Sometimes one is interested in the mean or median of extent and intensity statistics. It is easy to define these notions.

Note that two different kinds of time indicators appear in the definition of the excess intensity of activity. This complication arises because the intensity statistic is supposed to reflect the rate at which significant adaptations are arising, but it takes some time to determine whether a component is a significant adaptation. We let τ be the time when a component first provides evidence that it is an adaptation, i.e., the time when its excess activity first become positive. And we let t represent the time when one considers a significant component to have arisen. Now, consider some component c which eventually has positive excess activity. The choice of how to define t is the choice of when to consider c to have arisen. One option is to let t be the time when c first arises in the system, i.e., $t = Birth(c)$, in which case the intensity statistic would measure the rate of origination of new components that will *eventually* provide evidence of being an adaptation. At the other extreme, one could let t be the time when c first *does* provide significant evidence that it is an adaptation, i.e., $t = \tau$, in which case the intensity statistic would measure the rate at which components (which might have arisen some time in the past) are showing they are adaptations. The best approach will depend on the details of the system under investigation. For example, Skusa and Bedau [16] let $t = Birth(c)$ because in their system excess activity first becomes positive long after the origination of a component, but Bedau, Snyder, and Packard [4] let $t = \tau$ because this time lag in their systems was minimal. However one chooses t, it is important of course to be consistent.

[2] Here, we normalize activity before rather than after defining the macro statistics, but essentially the same result can be achieved by reversing the procedure. E.g., Rechtsteiner & Bedau [14] define excess extent after summing component activity.

4 Application to Different Kinds of Components

Different levels of activity can be measured in one and the same system at the same time. Measurements of evolutionary activity in previous work can illustrate how the present framework applies to various kinds of components. It might also help suggest how to extend the framework to new kinds of components. (In the examples below, we indicate the kind of component in question with an index on C_t and c; e.g. C_t^κ and c^κ for allele tokens, C_t^υ and c^υ for allele types, etc.

Allele tokens. An allele token is an individual allele at a locus in some particular agent. When the activity of allele tokens in asexual populations has been measured in previous work [2,5], the concern has been to see lineages of highly used alleles. Accordingly, if an allele token was inherited from an ancestor, it's activity counter was initialized to the value of the ancestral allele; otherwise, if the allele token was the product of a mutation, it was initialized to zero. So, if we let C^κ be the set of all allele tokens present at some time in the system, the activity initialization function can be defined as follows:

Example 4. $$\Delta_{init}^{C^\kappa}(c^\kappa, t) = \begin{cases} \alpha^{C^\kappa}(A(c^\kappa), t) & \text{if } A(c^\kappa) \text{ is defined} \\ 0 & \text{otherwise} \end{cases}$$

where $A : C^\kappa \to C^\kappa$ gives c^κ's immediate ancestor and is otherwise undefined.[3] Then, if we let C_t^κ be the set of allele tokens present in the system at t, the activity counter of an allele token is to be incremented just in case that allele token was used or expressed, thus:

Example 5. $$\Delta^{C^\kappa}(c^\kappa, t) = \begin{cases} 1 & \text{if } c^\kappa \in C_t^\kappa \wedge c^\kappa \text{ is used at } t \\ 0 & \text{if } c^\kappa \in C_t^\kappa \wedge c^\kappa \text{ is not used at } t \end{cases}$$

Allele types. Two agents with exactly the same kind of allele token at the same locus have the same *type* of allele at that locus. Let C_t^υ and C^υ be sets of allele types, defined analogously to the previous examples. An allele type, $c^\upsilon \in C^\upsilon$, is present in S at t just in case there is at least one allele token of type c^υ present in some agent in S at t. The lineage of a given allele is in effect an allele type,[4] so it makes most sense to initialize the activity of all allele types to zero: $\Delta_{init}^{C^\upsilon}(c^\upsilon) = 0$. When activity of allele types has been measured in previous work [5], the activity of an allele type c^υ was defined as the sum of the activity of allele tokens c^κ of that type:

Example 6. $$\Delta^{C^\upsilon}(c^\upsilon, t) = \sum_{c^\kappa \in c^\upsilon} \Delta^{C^\kappa}(c^\kappa, t)$$

[3] Example 4 can be easily modified for contexts in which reproduction is sexual.

[4] So-called "back" mutations create an exception to this rule, because our approach lumps together and ignores independent originations of the same allele. One could define a component that exactly corresponded to an individual allele lineage, if one wanted.

Phenotypic equivalence classes. Traits that are phenotypically the same can be grouped into *phenotypic equivalence classes*. One can measure the evolutionary activity of such phenotypic equivalence classes by attaching activity counters to them. A given phenotypic equivalence class is present at a moment if some member of the class is present at that moment. Let C_t^{ϕ} be the set of all phenotypic equivalence classes present at t. Phenotypic equivalence classes are like allele types in that lineages of phenotypically equivalent traits make up such classes. So it is natural to initialize the activity counters of phenotypic equivalence classes to zero: $\Delta_{init}^{C^{\phi}}(c^{\phi}) = 0$. Provided the notion of the use of a trait is well defined, it would also be natural to increment the activity of a phenotypic equivalence class c^{ϕ} by the use of its member traits:

Example 7. $\Delta^{C^{\phi}}(c^{\phi}, t) = \begin{cases} 1 & \text{if } c^{\phi} \in C_t^{\phi} \wedge c^{\phi} \text{ is used at } t \\ 0 & \text{if } c^{\phi} \in C_t^{\phi} \wedge c^{\phi} \text{ is not used at } t \end{cases}$

See [5] for one way to measure evolutionary activity of phenotypic equivalence classes.[5]

Genotypes. It is usually easy to measure the evolutionary activity of entire genotypes. Let C_t^{γ} and C^{γ} be sets of genotypes, as in previous examples. A genotype c^{γ} is present in a system at a time just in case some agent in the system has c^{γ} as its genotype. A genotype covers all the instances of the genotype in a genotype lineage, so it is natural to initialize the activity of genotypes at zero: $\Delta_{init}^{C^{\gamma}}(c^{\gamma}) = 0$. It is also natural to increment a genotype c^{γ}'s activity by its concentration in the population at t, as defined by $con : C^{\gamma} \times T \to \Re$, thus:

Example 8. $\Delta^{C^{\gamma}}(c^{\gamma}, t) = con(c^{\gamma}, t)$

where this function is defined only if $c^{\gamma} \in C_t^{\gamma}$. See [3,4,6,7,13,14,15,17,18,19] for measurements of evolutionary activity statistics of genotypes.

5 Challenges and the Future

It is straightforward to measure and interpret evolutionary activity of genotypes or phenotypic equivalence classes in almost any system. This is the most common level at which the statistics have been applied to date [3,4,6,7,13,14,15,17,18, 19]. It can be more difficult to apply the statistics at the level of genes. The evolutionary activity of individual genes is most easy to interpret when each gene has a clearly identifiable phenotypic function, because the method depends on a correlation between an component's activity and its utility for coping with selection pressure. Complications arise when genotype-phenotype mapping if many genes together affect single characters (e.g., epistasis) or if many characters are affected by single genes (e.g., pleiotropy). For instance, a gene coding for a

[5] Bedau and Raven were able to identify phenotypic equivalence classes with sets of allele tokens, because there is a one-to-one genotype-phenotype mapping in the model they studied.

given connection strength in a neural network might influence the network's behavior in different ways depending on how activation is flowing through the other nodes. In this case, it is very difficult to decompose the network's traits and behavior and assign responsibility for different pieces of it to different genes.

Nevertheless, one can usually still find a useful way to measure the evolutionary activity of genes in systems with genetic context sensitivity. For example, although a genome encoding weights a neural net typically is highly epistatic, one could still increment the activity of individual genes by their age. Long-lived weights would tend to be those that are highly adaptive in many contexts or in a few critical contexts, and these would accrue high activity. Alternatively, one could increment the activity of a network weight whenever its activation in the network exceeds some threshold. Those highly adaptive weights through which a lot of network activation consistently flows would accrue high activity.

The most significant obstacle to applying activity statistics to natural systems is collecting enough of the right kind of data, specifically, a time series of a some sort of census of components in the system. The relative ease of collecting the necessary data is the main reason why activity statistics have been applied mostly to artificial systems to date. High throughput automated methods are currently generating massive biological data bases of various kinds (genomic, proteomic, metabolomic, etc.), and electronic media are increasingly accumulating an ever growing variety of data about the evolution of various aspects of culture (patents, financial markets, internet sites, newspapers, news groups, etc.). So our ability to measure evolutionary activity in natural systems will grow significantly in the near future.

Acknowledgements Thanks to Norman Packard and our other colleagues in the exploration of evolutionary activity statistics over the years. Thanks to Seth Bullock for suggestions about how to measure activity in evolving neural nets. And thanks to the anonymous ECAL'03 reviewers for helpful comments.

References

1. Bedau, M. A., and Brown, C. T. 1999. Visualizing evolutionary activity of genotypes. *Artificial Life* 5: 17–35.
2. Bedau, M. A., and Packard, N. H. 1992. Measurement of evolutionary activity, teleology, and life. In Langton, C. G.; Taylor, C.; Farmer, J. D.; and Rasmussen, S., eds., *Artificial Life II*, 431–461. Redwood City, CA: Addison-Wesley.
3. Bedau, M. A., Snyder, E.; Brown, C. T.; and Packard, N. H. 1997. A comparison of evolutionary activity in artificial evolving systems and the biosphere. In Husbands, P., and Harvey, I., eds., *Proceedings of the Fourth European Conference on Artificial Life*, 125–134. Cambridge, MA: MIT Press.
4. Bedau, M. A.; Snyder, E.; and Packard, N. H. 1998. A classification of long-term evolutionary dynamics. In Adami, C.; Belew, R.; Kitano, H.; and Taylor, C., eds., *Artificial Life VI*, 228–237. Cambridge, MA: MIT Press.

5. Bedau, M. A. and Raven, M. J. 2002. Visualizing adaptive evolutionary activity of allele types and of phenotypic equivalence classes of alleles. In Bilotta, E.; Gross, D; Smith, T.; Lenaerts, T.; Bullock, S.; Lund, H. H.; Bird, J.; Watson, R.; Pantano, P.; Pagliarini, L.; Abbass, H.; Standish, R. ; and M. Bedau, eds., *Workshops from the Artificial Life VIII Conference*, 119–130. University of New South Wales, Sydney.
6. Channon, A. D. 2001. Passing the ALife test: Activity statistics classify evolution in Geb as unbounded. In Kelemen, J., and Sosik, P., eds., *Advances in Artificial Life*, 417–426. Heidelberg: Springer-Verlag.
7. Channon, A. D. 2002. Improving and still passing the ALife test: Component-normalised activity statistics classify evolution in Geb as unbounded. In Standish, R.; Bedau, M. A.; Abbass, H. A., eds., *Artificial Life VIII*, 173–181. Cambridge, MA: MIT Press.
8. Dawkins, R. 1987. *The Blind Watchmaker: Why the evidence of evolution reveals a universe without design.* New York: Norton.
9. Gould, S. J., and Lewontin, R. C. 1979. The spandrals of San Marco and the Panglossian paradigm: a critique of the adaptationalist programme. *Proceedings of the Royal Society B* 205: 581–598.
10. Holland, J. H. 1992. *Adaptation in Natural and Artificial Systems: An introductory analysis with applications to biology, control, and artificial intelligence*, 2nd edition. Cambridge, MA: MIT Press/Bradford Books.
11. Maynard Smith, J. 1975. *The Theory of Evolution*, 3rd edition. New York: Penguin.
12. Mayr, E. 1988. *Towards a New Philosophy of Biology*. Cambridge, MA: Harvard University Press.
13. Pachepsky, E.; Taylor, T.; and Jones, S. 2002. Mutualism promotes diversity and stability in a simple artificial ecosystem. *Artificial Life* 8: 5–24.
14. Rechtsteiner, A. and Bedau, M. A. A generic model for measuring excess evolutionary activity. In Banzhaf, W.; Daida, J.; Eiben, A. E.; Garzon, M. H.; Honavar, V.; Jakiela, M.; and Smith, R. E., eds., *GECCO-99: Proceedings of the Genetic and Evolutionary Computation Conference*, Vol. 2, 1366–1373. San Francisco, CA: Morgan Kaufmann.
15. Shannon, T. 1998. Generic behavior in the Lindgren non-spatial model of iterated two-player games. In Adami, C.; Belew, R.; Kitano, H.; and Taylor, C., eds., *Artificial Life VI*, 316–325. Cambridge, MA: MIT Press.
16. Skusa, A. and Bedau, M. A. 2002. Towards a comparison of evolutionary creativity in biological and cultural evolution. In Standish, R.; Bedau, M.; and Hussein, A., eds., *Artificial Life VIII*, 233–242. Cambridge, MA: MIT Press.
17. Standish, R. K. 2000. An Ecolab perspective on the Bedau evolutionary statistics. In Bedau, M. A.; McCaskill, J. S.; Packard, N. H.; and Rasmussen, S., eds., *Artificial Life VII*, 238–242. Cambridge, MA: MIT Press.
18. Standish, R. K. 2002. Diversity evolution. In Standish, R.; Bedau, M. A.; and Hussein, A., eds., *Artificial Life VIII*, 131–137. Cambridge, MA: MIT Press.
19. Taylor, T. and Hallam, J. 1998. Replaying the tape: An investigation into the role of contingency in evolution. In Adami, C.; Belew, R.; Kitano, H.; and Taylor, C., eds., *Artificial Life VI*, 256–265. Cambridge, MA: MIT Press.

Developing and Testing Methods for Microarray Data Analysis Using an Artificial Life Framework

Dirk Repsilber[1] and Jan T. Kim[2]

[1] Institute of Medical Biometry and Statistics
Ratzeburger Allee 160, House 4, 23538 Lübeck, Germany,
dirk.repsilber@imbs.uni-luebeck.de
[2] Institute for Neuro- and Bioinformatics,
Seelandstraße 1a, 23569 Lübeck, Germany, kim@inb.uni-luebeck.de

Abstract. Microarray technology has resulted in large sets of gene expression data. Using these data to derive knowledge about the underlying mechanisms that control gene expression dynamics has become an important challenge. Adequate models of the fundamental principles of gene regulation, such as Artificial Life models of regulatory networks, are pivotal for progress in this area.
In this contribution, we present a framework for simulating microarray gene expression experiments. Within this framework, artificial regulatory networks with a simple regulon structure are generated. Simulated expression profiles are obtained from these networks under a series of different environmental conditions. The expression profiles show a complex diversity. Consequently, success in using hierarchical clustering to detect groups of genes which form a regulon proves to depend strongly on the method which is used to quantify similarity between expression profiles. When measurements are noisy, even clusters of identically regulated genes are surprisingly difficult to detect. Finally, we suggest cluster support, a method based on overlaying multiple clustering trees, to find out which clusters in a tree are biologically significant.

1 Introduction

High throughput technology for molecular analysis of biological systems has rapidly advanced during the last decade. However, the developments in theory and modelling the fundamental dynamical principles of living systems has not kept up with the massive increase in availability of biological data. More specifically, while the amount of gene expression data has explosively grown during the last few years, an integrated theory of gene expression and regulatory networks is not yet available, even though advances in theory of transcription factor bioinformatics [1,2], modelling regulatory networks [3,4,5] and their evolution [6, 7] have been made.

As a consequence of the divergence between availability of data and theoretical foundations, the major bottleneck for making progress in understanding

W. Banzhaf et al. (Eds.): ECAL 2003, LNAI 2801, pp. 686–695, 2003.

biological systems is not due to scarcity of data, but due to lack of ways for harnessing the available data for theory and modelling. This constitutes a major challenge for Artificial Life research. One of the obstacles for linking biological data and mathematical or computer based models is the definition of adequate criteria to evaluate the degree of consistency between model and data, and for providing directions for refinements in modelling as well as in data acquisition.

Microarray data analysis has become an important tool which has been used for classification of biological systems [8], for inferring regulatory patterns and structures, such as clusters of co-regulated genes [9,10]. One of the most important long-term goals of these efforts is inferring regulatory interactions and, ultimately, entire regulatory networks [11,12,13]. In this contribution, we address this challenge by subjecting artificial regulatory networks to the same analysis procedures which are used in molecular biology and asking how much information about the underlying regulatory network can be retrieved by these approaches.

2 Systems and Methods

The transsys framework [14] provides a computer language which allows to describe regulatory gene networks by comprehensive, object-oriented *programs*. A program consists of factor (i.e. protein) and gene specifications. An *instance* represents the concentrations of the factors in a program at a specific time t. The *update method* computes the expression levels at time $t+1$ based on the levels at time t and the information provided by the program.

A set of tools for microarray simulation and analysis with transsys has been implemented using the Python programming language [15] and the `R` statistics system [16]. The software is available on the website [17].

2.1 Regulatory Network Construction

Genes can often be categorized into regulatory genes, which encode products that control gene expression (e.g. transcription factors), and structural genes, which encode products that do not function as regulators, such as proteins that form body structures or enzymes. Our method for randomly constructing transsys programs was designed to reflect this property. An example network is shown in Fig. 1.

Network construction involves two steps. Firstly, a core regulatory network with N genes is generated. Each gene encodes a unique product and has its expression controlled by K factors which are randomly chosen and randomly designated to be an activator or a repressor. The architecture of the core network is an NK network [3,6]. Secondly, structural genes are added. R genes of the core network are randomly chosen as regulators of a regulon. For each regulator, a number of structural genes, activated by the regulator, are generated. Each structural gene encodes a unique product. The entire regulon consists of the regulator and the structural genes which it controls. The structure of regulons

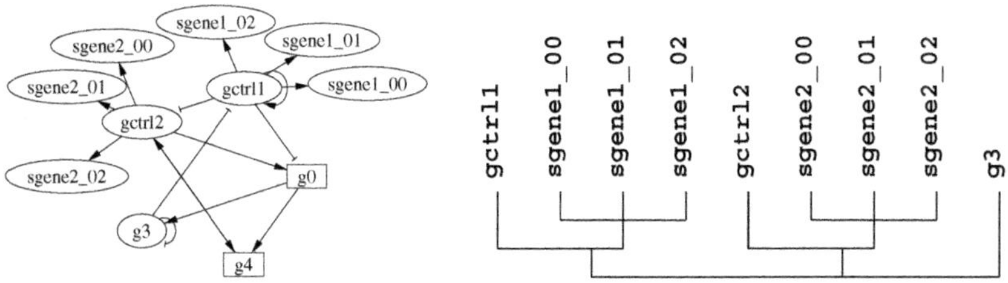

Fig. 1. Left: A regulatory network consisting of $N = 5$ regulatory genes, each of which receives $K = 2$ regulatory inputs. Edges ending in arrows depict activation, those ending with bars show repression. 2 genes, labelled `gctrl`, control a regulon. Structural genes are labelled `sgene`. Genes encoding products that are directly perturbed by the environment are boxed. Right: Regulon tree representing the regulons in the network. Genes which are directly perturbed are excluded from the regulon tree.

in a network can be represented by a hierarchical graph which we refer to as the *regulon tree*, as shown in Fig. 1.

The transsys program is constructed such that in each time step t, expression of a gene g_i results in synthesis of an amount of f_i, the product encoded by the gene, given by

$$\mathrm{synth}(f_i, t) = c + \max\{0, \sum_{k=1}^{K} \mathrm{reg}(r_k, \alpha, \beta)\}$$

where $C(f, t)$ denotes the concentration of factor f at time t and $r_1, \ldots r_K$ are the factors regulating g_i. The regulation terms are given by

$$\mathrm{reg}(r_k, \alpha, \beta) = \begin{cases} \beta C(r,t)/(\alpha + C(r,t)) & \text{if } r_k \text{ is an activator} \\ -\beta C(r,t)/(\alpha + C(r,t)) & \text{if } r_k \text{ is a repressor.} \end{cases}$$

The concentration of a gene product f at time $t + 1$ is given by $C(f, t + 1) = (1 - d)C(f, t) + \mathrm{synth}(f, t)$, where d is a decay rate.

The components of the core regulatory network are parameterized differently from those of the structural part. The parameters for the regulatory core are denoted by c_{reg}, α_{reg}, β_{reg} and d_{reg}. The parameters for the structural part are α_{struct}, β_{struct} and d_{struct}, $c = 0$ for all structural genes.

2.2 Simulation of Gene Expression Measurements

Comparative approaches have become a mainstream in gene expression analysis. In these approaches, an expression measurement performed under reference conditions is compared with measurements obtained with alternate environmental conditions (see e.g. [9,18]). Frequently, a time series of measurements is made after changing the environment. The entire set of expression changes for a gene g

is called the *expression profile* of g. In a subsequent step, distances are computed for all pairs of expression profiles. Finally, the resulting distance matrix used for hierarchical clustering. The result is a *clustering tree* in which leaves represent genes and the distance between the expression patterns of gene pairs is correlated to the length of the path connecting the genes. It is assumed that clusters in the tree (i.e. groups of genes which form subtrees) may reflect regulons. The simulation presented subsequently models this comparative approach.

For the reference measurement, a transsys instance is generated by initializing all expression levels with a random value drawn from a uniform distribution over the unit interval $[0, 1]$. Starting with this initial instance, 5000 updates are simulated to allow the system to converge into an attractor. The resulting transsys instance is used as the reference state. The concentration of factor f in the reference state is denoted by $C_{\text{ref}}(f)$.

Alterations of the environmental conditions result in changes of certain gene products, e.g. by photochemical transformations. In the simulation framework, a subset of genes (shown as boxes in Fig. 1) from the core network is designated to encode products that are subject to direct modification by the environment. An environmental impact is simulated by perturbing the concentration of each factor f encoded by this subset according to

$$C_e(f, 0) = (C_{\text{ref}}(f) + \epsilon) \cdot 2^{g\sigma_e} \tag{1}$$

where $C_e(f, 0)$ denotes the initial concentration in the new environment, g is a random value drawn from a Gaussian normal distribution, σ_e is a control parameter that allows tuning the variance and $\epsilon = 0.01$ is a small offset which allows alteration of factor concentrations which are zero in the reference state.

Starting with the transsys instance representing the initial conditions for an environment, network dynamics are simulated. Every $\Delta t = 5$ time steps, a sample is taken until 20 samples are collected. For each sample, intensity values are computed assuming the additive background model

$$a_e(f, t) = C_e(f, t) + b \cdot \exp(g\sigma_a) \tag{2}$$

where $b = 0.1$ reflects the median fluorescence of an array spot with no complementary labelled product bound to it, g is a random value from a Gaussian normal distribution and σ_a controls the variance of background fluorescence. Logarithmized intensity ratios $r_e(f, t) = \log_2(a_e(f, t)/a_{\text{ref}}(f))$ are calculated for all factors not subject to direct environmental impact. Here, $a_{\text{ref}}(f)$ denotes the intensity value calculated from $C_{\text{ref}}(f)$ according to (2). For each gene g, the expression intensities measured in environment e are aggregated into a 20-dimensional vector $X_e(g)$ with components $x_{e,i}(g) = r_e(f, \Delta t \cdot i)$. $X_e(g)$ thus represents the expression profile of gene g in environment e.

2.3 Cluster Analysis and Its Evaluation

Hierarchical clustering trees were constructed by common methods [9,10] from expression profiles of all genes that encode factors which are not directly perturbed. We used data sets of expression profiles from individual environments,

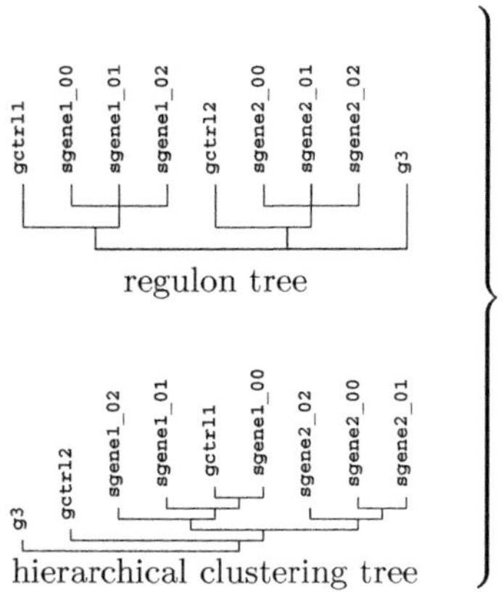

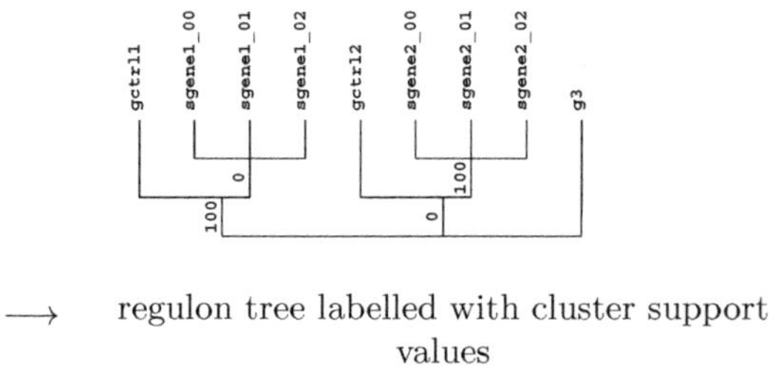

regulon	reg	str
1	100	0
2	0	100

table of cluster support values

Fig. 2. Finding consistent edges. The regulon tree and the hierarchical clustering tree both contain an edge separating regulon 1 (regulatory gene `gcntrl1` and structural genes `sgene1_01`, `sgene1_02` and `sgene1_03`) from the remaining genes. Likewise, the three structural genes of regulon 2 form a cluster in both the regulon tree and the clustering tree. Thus, the corresponding edges in the regulon tree are supported by the clustering tree, amounting to 100% because there is just one clustering tree in this example. A table shows these values more comprehensively.

and additionally a set of aggregated profiles obtained by concatenating the profiles from all environments.

For hierarchical clustering, we used Euclidean distance and the Pearson-type measure described in [9]. The latter measure $S(X,Y) \in [-1,1]$ was devised to reflect similarity of expression profile shape regardless of absolute values. As a distance measure derived from this similarity measure, we used $1 - S(X,Y)$. For clustering, the average-linkage and the single-linkage algorithm were used. For each distance measure and clustering method, a consensus tree of the clustering trees obtained for the individual environments was computed with the `consense` program [19].

Clustering trees were evaluated by comparing them to the regulon tree. For each edge in the regulon tree, we determined whether a consistent edge exists in the clustering tree. Consistency of two edges in different trees means that both edges imply the same split of the set of leaf nodes. The result can be visualized and processed into a table as illustrated in Fig. 2. We refer to the percentage of clustering trees that contain a consistent edge as the *cluster support* of an edge in the regulon tree. Cluster support values can also be computed for clustering trees instead of regulon trees. We explore using cluster support to identify biologically significant clusters in the results presented below.

Table 1. Cluster support values obtained with all combinations of Euclidean distance and Pearson-type distance, average and single linkage clustering and $\sigma_a = 0$ and $\sigma_a = 1$, as illustrated in Fig. 2.

regu-	Euclidean distance								Pearson-type distance							
lon	average linkage				single linkage				average linkage				single linkage			
	$\sigma_a = 0$		$\sigma_a = 1$		$\sigma_a = 0$		$\sigma_a = 1$		$\sigma_a = 0$		$\sigma_a = 1$		$\sigma_a = 0$		$\sigma_a = 1$	
	reg	str	reg	str	reg	str	reg	str	reg	str	reg	str	reg	str	reg	str
26	10	100	0	10	10	100	0	10	30	100	10	10	10	100	10	10
03	30	100	10	90	70	100	10	90	0	100	0	80	0	100	0	60
01	0	100	0	10	0	100	0	10	0	100	0	10	0	100	0	10
49	0	100	0	10	0	100	0	10	0	100	0	10	0	100	0	10
29	0	100	0	10	0	100	0	10	0	100	0	10	0	100	0	10
22	0	100	0	10	0	100	0	10	0	100	0	10	0	100	0	10
43	0	100	0	60	0	100	0	50	0	100	0	50	0	100	0	50
16	0	100	0	10	0	100	0	10	0	100	0	10	0	100	0	10

3 Results and Discussion

3.1 Cluster Analysis

Networks of $N = 50$ regulatory genes, each having $K = 2$ regulatory inputs, were generated with the parameters $c_{\text{reg}} = 0.2$, $\alpha_{\text{reg}} = 1$, $\beta_{\text{reg}} = 0.5$ and $d_{\text{reg}} = 0.2$. $R = 8$ regulons, each consisting of 3 structural genes with $\alpha_{\text{struct}} = 1$, $\beta_{\text{struct}} = 3$ and $d_{\text{struct}} = 0.1$, were attached to the regulatory core. 20 genes were chosen for direct perturbation. Expression dynamics were computed for 10 environments generated with $\sigma_e = 2$. Microarray simulations were performed with array measurement noise levels ranging from $\sigma_a = 0$ to $\sigma_a = 1$.

Table 1 shows the results of evaluating cluster analyses of expression profiles generated with such a network. With no noise in array measurement, all structural genes in a regulon have identical expression profiles with our current parameterization method. Thus, the groups of structural genes within all regulons receive 100% cluster support with all clustering methods. The expression profile of the regulatory gene in a regulon differs from the profiles of the structural genes. Even with no array noise, the association between the regulatory and the structural genes in a regulon is not detected by any clustering method for 6 out of 8 regulons. Only regulons 3 and 26 are recognized. Cluster support for regulon 3 is consistently higher with Euclidean distance whereas the Pearson-type distance detects regulon 26 more reliably.

When noise in array measurement is simulated, cluster support values for these groups become smaller than 100% and reveal a characteristic pattern. The cluster support for the structural gene groups in regulons 3 and 43 is at least 50% in all analyses conducted with $\sigma_a = 1$, while the structural gene groups in all other regulons receive only 10% cluster support.

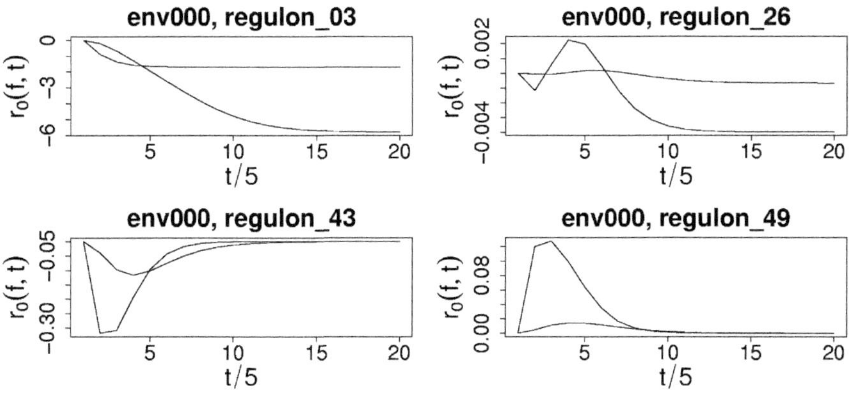

Fig. 3. Expression profiles for regulons 3, 26, 43 and 49, measured for environment 0 Graphs show the logratio profiles of the regulator and the structural genes of a regulon. The response of the structural genes lags behind the regulator, but is more pronounced.

3.2 Individual Expression Profiles

Fig. 3 shows expression profiles for regulons 3, 26, 43 and 49. Regulon 3 strongly responds to the impact of environment 0, the regulon, both the regulatory gene and the structural ones, are effecively switched off. The resulting similarity in expression profiles is captured by Euclidean distance.

Regulon 26 exhibits a complex expression profile in which an initial phase of downregulation is followed by transient upregulation until the system finally settles to a new state which is downregulated with respect to the reference state. This characteristic down-up-down shape is reflected in the expression profiles of both the regulator and the structural factors. Although the response of the structural genes lags behind that of the regulator, the Pearson-type distance proves to be suitable for detecting the characteristic similarity in shape.

Regulons 43 and 49 respond to the environmental impact by transient regulatory changes. The amplitude of the change in regulon 43 is substantially greater than in regulon 49. Therefore, the group of structural genes of regulon 43 can be detected with noisy measurements, whereas the structural gene groups of regulons 49, 26 and others with a small response amplitude become indistignuishable when noise is present.

3.3 Detecting Significant Clusters

For empirical data sets, the regulon tree is not known. A clustering tree can be constructed based on the full data set. Cluster support values for this tree can be determined by using trees computed from data subsets corresponding to individual environments. These cluster support values may be useful to indicate which edges are biologically significant in the sense that they correspond to regulons.

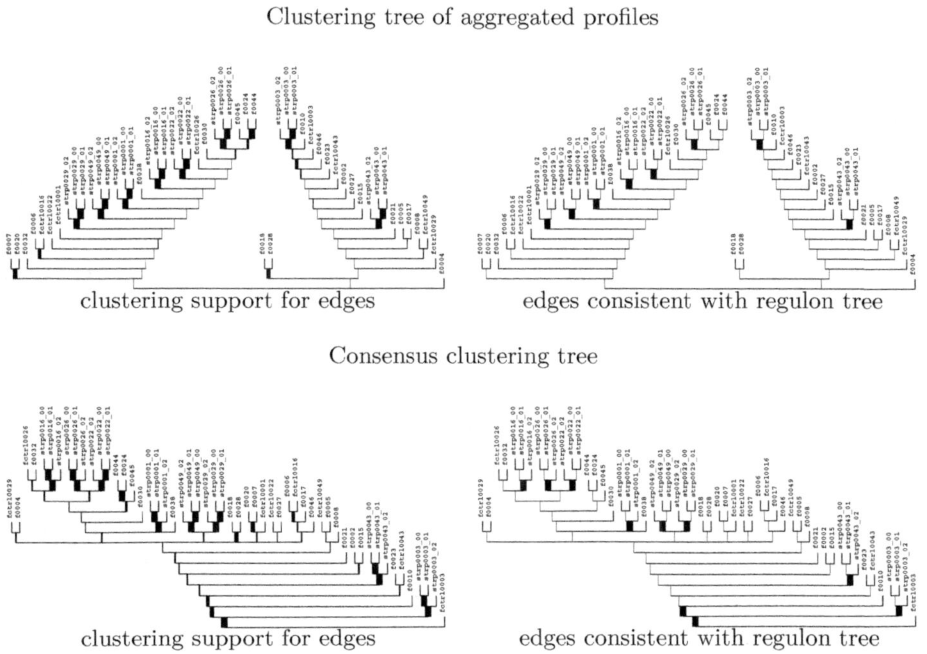

Fig. 4. Clustering tree of aggregated profiles and consensus tree, computed with $\sigma_a = 0$, Euclidean distance and single linkage clustering. Edges in trees on the left side are displayed by lines with thickness proportional to their clustering support. In trees on right side, edges which are consistent with the regulon tree are highlighted by thick lines. Clustering support and consistency with the regulon tree are clearly correlated.

To explore this approach, we have computed a clustering tree from the aggregated expression profiles and the consensus tree of the 10 clustering trees corresponding to the 10 environments, and assigned cluster support values to the edges in both trees. The results are shown in Fig. 4. In both trees, clustering support is distributed in a characteristic pattern. Elevated cluster support is strongly correlated to consistency with the regulon tree. Interestingly, the full regulon 3 is captured by the consensus tree but not by the tree of aggregated expression profiles.

4 Conclusion and Outlook

Detection of a majority of regulons on the basis of noisy expression measurements proved to be unexpectedly difficult. To a substantial extent, this difficulty is due to the fact that an environmental impact does not necessarily elicit a pronounced response of a regulon. Regulons with a small and transient response cannot be distinguished on the basis of noisy expression profiles. This finding highlights the fact that coregulation may strongly depend on the conditions under which it is observed. Genes may appear unregulated or coregulated under many conditions, and yet, under other conditions, their expression profiles may strongly differ. Even the detection of simple regulons therefore critically depends on the number and choice of conditions under which gene expression is sampled.

It is possible that small and transient responses are overrepresented in the networks which we constructed. For example, one may assume that selection induces a bias towards regulons which respond with pronounced changes to many environmental challenges. This issue is most adequately addressed by using an evolutionary algorithm for regulatory network generation, which we plan to do in future analyses.

The correlation between the expression patterns of the regulator and the structural genes in a regulon proved to be difficult to detect, even with our simplistic model in which structural genes are controlled directly and exclusively by one regulator. Depending on the particular properties of the expression profiles, different distance measures may be useful to detect the corresponding regulons. Our results indicate that, as also suggested in [20], there is no universally adequate, straightforward method to classify genes by their expression profiles. Advances in understanding the fundamental principles of regulatory networks may eventually result in the development of more generic methods of classification, and we plan to use our framework as a point of departure for research in this field. At this time, however, the choice of classification approaches should be based on the specific biological subject of research. Therefore, we also plan to extend our analysis to additional distance measures, such as mutual information [10] or jackknife correlation [21].

Cluster support can be applied to assess biological significance of clusters. We demonstrated that elevated levels of cluster support and consistency with the regulon tree are strongly correlated. This applies for clustering trees constructed from aggregated expression profiles as well as for consensus trees. It will be interesting to explore whether this correlation extends to other methods of clustering and data analysis, and whether other generic approaches such as bootstrapping [22] can be applied to assess biological significance. Performing multiple cluster analyses and selecting the most significant clusters from each analysis may provide a component of a more generic classification method.

References

1. Schneider, T.D., Stormo, G.D., Gold, L.: Information content of binding sites on nucleotide sequences. J.Mol.Biol. **188** (1986) 415–431
2. Kim, J.T., Martinetz, T., Polani, D.: Bioinformatic principles underlying the information content of transcription factor binding sites. Journal of Theoretical Biology **220** (2003) 529–544
3. Kauffman, S.A., Weinberger, E.W.: The NK model of rugged fitness landscapes and its application to maturation of the immune response. J. Theor. Biol. **141** (1989) 211–245
4. von Dassow, G., Meir, E., Munro, E.M., Odell, G.M.: The segment polarity network is a robust developmental module. Nature **406** (2000) 188–192

5. Reil, T.: Dynamics of gene expression in an artificial genome – implications for biological and artificial ontogeny. In Floreano, D., Nicoud, J.D., Mondada, F., eds.: Advances in Artificial Life. Lecture Notes in Artificial Intelligence, Berlin Heidelberg, Springer-Verlag (1999) 457–466
6. Kauffman, S.A.: Requirements for evolvability in complex systems: Orderly dynamics and frozen components. Physica D **42** (1990) 135–152
7. Bornholdt, S., Sneppen, K.: Neutral mutations and punctuated equilibrium in evolving genetic networks. Physical Review Letters **81** (1998) 236–239
8. Golub, T., Stonim, D., Tamayo, P., Huard, C., Gaasenbeek, M., Mesirov, J., Coller, H., Loh, M., Downing, J., Caligiuri, M., Bloomfield, C., Lander, E.: Molecular classification of cancer: Class discovery and class prediction by gene expression monitoring. Science **286** (1999) 531–536
9. Eisen, M.B., Spellman, P.T., Brown, P.O., Botstein, D.: Cluster analysis and display of genome-wide expression patterns. Proc. Natl. Acad. Sci. USA **95** (1998) 14863–14868
10. Michaels, G.S., Carr, D.B., Askenazi, M., Fuhrmann, S., Wen, X., Somogyi, R.: Cluster analysis and data visualization of large-scale gene expression data. In Altman, R.B., Dunker, A.K., Hunter, L., Klein, T.E., eds.: Biocomputing '98, Singapore, World Scientific (1998) 42–53
11. Akutsu, T., Miyano, S., Kuhara, S.: Inferring qualitative relations in genetic networks and metabolic pathways. Bioinformatics **16** (2000) 727–734
12. Morohashi, M., Kitano, H.: Identifying gene regulatory networks from time series expression data by *in silicio* screening and sampling. In Floreano, D., Nicoud, J.D., Mondada, F., eds.: Advances in Artificial Life. Lecture Notes in Artificial Intelligence, Berlin Heidelberg, Springer-Verlag (1999) 477–486
13. Repsilber, D., Liljenström, H., Andersson, S.G.: Reverse engineering of regulatory networks: Simulation studies on a genetic algorithm approach for ranking hypotheses. BioSystems **66** (2002) 31–41
14. Kim, J.T.: `transsys`: A generic formalism for modelling regulatory networks in morphogenesis. In Kelemen, J., Sosík, P., eds.: Advances in Artificial Life (Proceedings of the 6th European Conference on Artificial Life). Volume 2159 of Lecture Notes in Artificial Intelligence., Berlin Heidelberg, Springer Verlag (2001) 242–251
15. van Rossum, G., Drake, F.L.: Python reference manual (2002) `http://www.python.org/`.
16. Ihaka, R., Gentleman, R.: R: A language for data analysis and graphics. Journal of Computational and Graphical Statistics **5** (1996) 299–314
17. Kim, J.T.: The `transsys` home page (2003) `http://www.inb.uni-luebeck.de/transsys/`.
18. Gasch, A.P., Spellmann, P.T., Kao, C.M., Carmel-Harel, O., Eisen, M., Storz, G., Botstein, D., Brown, P.O.: Genomic expression programs in the response of yeast cells to environmental changes. Molecular Biology of the Cell **11** (2000) 4241–4257
19. Felsenstein, J.: PHYLIP – phylogeny inference package, version 3.5c. (1993) `http://evolution.genetics.washington.edu/phylip.html`.
20. Quackenbush, J.: Computational analysis of microarray data. Nature Reviews Genetics **2** (2001) 418–426
21. Heyer, L., Kruglyak, S., Yooseph, S.: Exploring expression data: identification and analysis of coexpressed genes. Genome Res. **9** (1999) 1106–1115
22. Jain, A., Moreau, J.: Bootstrap techniques in cluster analysis. Pattern Recognition **20** (1987) 547–568

Approaching Virtual Organism by PheGe

Klaus Seidl

German Research Center for Biotechnology (GBF), Braunschweig, Germany,
kse@gbf.de

Abstract. Seizing the idea of system-level thinking and to approach automated network analysis the relational platform PheGe was generated. PheGe is a pilot scheme for virtual biology on a systemic (multi-cellular organism) level. It resembles a prototype of knowledge base platforms that organize data-driven interactions in a way that attack the problem of digital recording and simulation of cell-overlapping signal cascades and cellular differentiation programs. PheGe provides static and dynamic views on the establishment and maintenance of functional units of an organism and targets the construction of a virtual organism by linking intricate systems of cellular communication to an overall response. Hereby, an implemented virtual cell studio serves as a building block for the establishment of a virtual organism where regulatory, metabolic and neuronal circuits can be analyzed and artificially induced knockouts or malfunctions can be simulated on a cellular and systemic level.

1 Introduction

The establishment of a virtual organism is one of the most fascinating visions of modern bioinformatics and implies system-level understanding of the molecular cues underlying the intrinsic wiring that define genotype-phenotype relations. Several computational approaches have been presented addressing the problem of signal interaction and signal transduction [1-4]. Although these knowledge bases target the understanding of how regulatory circuits are balanced, they lack physically incorporated spatial and temporal expression-links, which are of crucial importance for automated analysis of organism function. It becomes obvious that we need novel evaluation tools, which can elaborate the role of a single network component not only on a cellular but also on a systemic level by linkage of the individual interactions to their respective cell systems and the appropriate stage of the organism.

Presently, there are some doubts that we are able to build up a systemic view from the massive amount of information. This is still due to the lack of data that fill up the gaps between the individual network components. One of the challenges in modern bioinformatics is to provide novel concepts and tools that account for incomplete or fragmented signaling pathways and provide the appropriate knowledge necessary for successful simulations and predictions of cellular events.

Recently, PheGe has been introduced as such a concept, which organizes biological data in a way that encompasses missing actors and thus ensures the appropriate connectivity needed for bridging cellular circuits to an overall response [5]. The PheGe-system organizes data-driven interactions in a way that attack the problem of

W. Banzhaf et al. (Eds.): ECAL 2003, LNAI 2801, pp. 696–705, 2003.

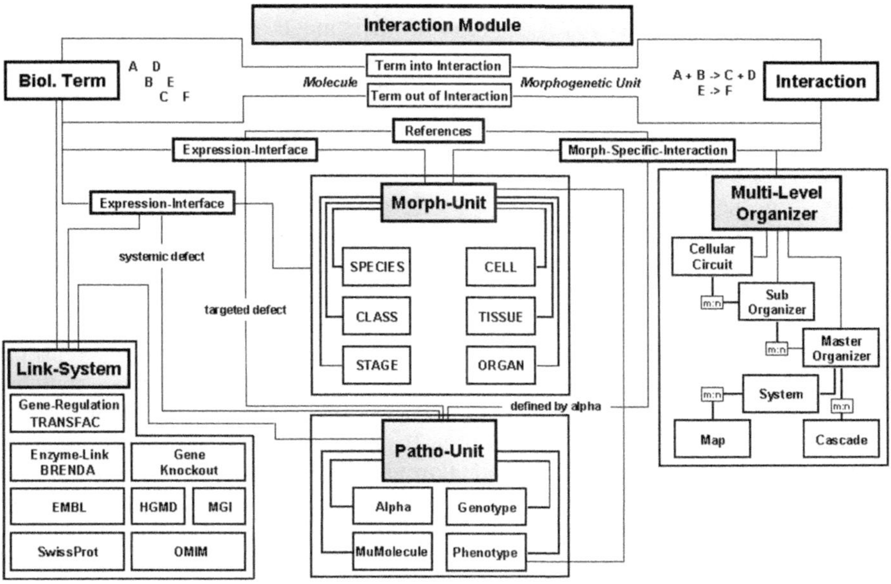

Fig. 1. Architecture of the relational PheGe-platform

digital recording and simulation of cell-overlapping signal cascades and cellular differentiation programs and enables tracing of signals through the cell, the organism and even through morphogenesis in a qualitative automated manner. Thereby, PheGe covers important areas essential for analysis of genotype-phenotype relations. The hitherto version of PheGe comprises a step-wise tracking of intra- and intercellular signaling routes. While this tool offers a first computational insight into biological strategies, it lacks essential views on the dynamics of the interplay of cellular actions towards an overall response. In the following, a new operational and dynamic part of PheGe will be presented, which includes a virtual cell studio as a basic unit for the establishment of a virtual organism. Within this virtual cell studio regulatory, metabolic and neuronal circuits can be analyzed and artificially induced knockouts (KOs) or malfunctions can be simulated on a cellular and systemic level.

2 System, Methods, Algorithm, and Biology

2. 1 PheGe, a Platform for System-Level Analysis

Visualization of signaling in PheGe is performed by directed graph technology using Visual Basic [5]. The architecture of the PheGe-platform fulfills all criteria needed for the coordination of normal and altered biological interactions to their corresponding anatomical units and the stage of the life form (Fig. 1). Furthermore, PheGe's Multi-Level Organizer (MLO) hierarchically sorts and coordinates the various interactions to their particular signaling systems and thereby guarantees the downstream cascade on a system-level even when signaling pathways are incomplete or fragmented. Interactions are sorted by the MLO according to their mode of action within a compartment. While the 'Master Organizer'-table registers interactions at the transition point between two adjacent compartments, the 'Sub Organizer'-table contains interaction shortcut-profiles that supervise intracellular pathway components defined in the 'Cellular Circuit'-table. Activation of a certain gene product in consequence of agonist receptor binding, transcription activation as well as firing of a distinct nerve as a result of mechanic or cellular stimuli can now be integrated into the signaling route, although intra- and intercellular network components are missing, obscure or just not known. Due to the physiological potential of the MLO, information can be used in a defined way to secure continuation of signaling.

Currently, PheGe comprises a static and an operational part. While the static part has been documented in detail [5], the operational part resembles a new sophisticated feature, which assists dynamical representation and analysis of PheGe's metabolic, regulatory and neuronal circuits. The data being delivered to the operational part has been pre-evaluated in the PheGe-platform. In short, all information concerning molecules, biological terms and their interaction profiles were exclusively extracted from primary literature taking advantage of the PubMed (National Library of medicine) -service, and stored in the PheGe-database. After temporal and spatial coordination the interactions were sorted in a multi-level organizer (MLO) according to their capability to ensure connectivity (for details see [5]). The implemented 'Patho-Unit', 'Link'- and 'Reference'-System provide additional and helpful information for the understanding of the molecular cues of genotype-phenotype relations. From the PheGe-knowledge base one can easily elaborate the route of signaling and mark the relevant set of components to be delivered to the operational module of the platform. Within the 'Virtual Cell Studio' all transferred components and their interactions are then automatically processed in order to provide cellular and systemic views on functional aspects of the organism and its subunits. Thereby, the data-driven interactions are evaluated by a compartment-specific management system (CSMS; see below).

2.2 The Virtual Cell Studio

The heart of the operational part is the virtual cell studio where cellular dynamics can be modulated and cellular phenotypes as well as functions can be simulated by the implemented potential of artificially induced KOs or malfunctions of individual network components. Moreover, the virtual cell studio enables bridging of the individual cellular circuits to an overall response. Accordingly, one can visualize and evaluate either a cell-specific or a systemic malfunction on an organism level. The tool allows the generation of a variety of artificially induced KOs at the same time, the possibility to rescue the activity of specific network components and even the elimination of specific cell types from the organism. In the following, establishment and potentials of the virtual cell studio are described sticking closely to the biological problem of insulin regulation and its action on glucose homeostasis.

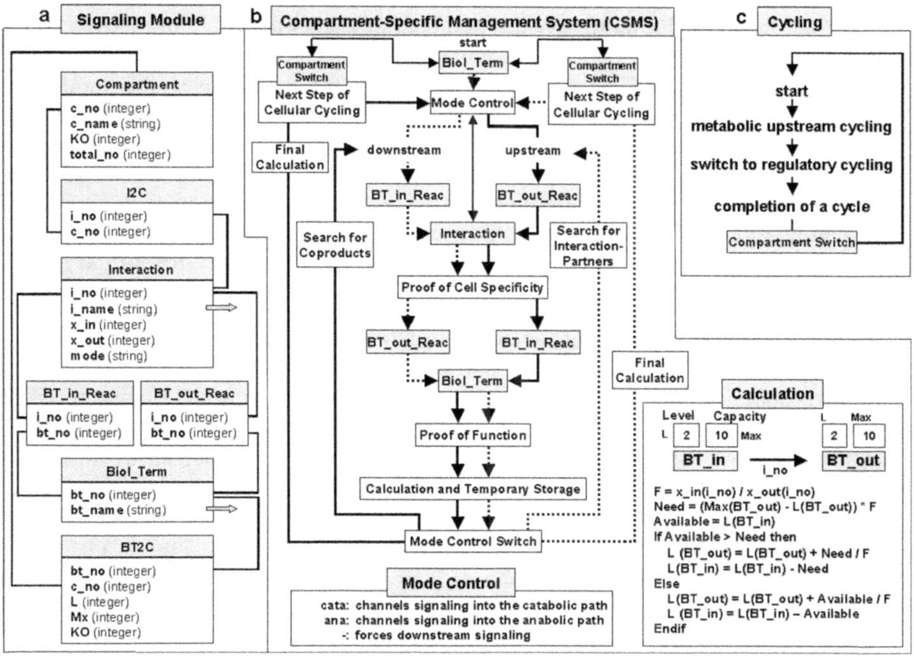

Fig. 2. Structural design of the Virtual Cell Studio. The Signaling Module (a) organizes intra- and inter-specific up- and downstream signaling and constitutes the basic platform for cycling (c), which is controlled by the CSMS (b). For details see text.

Establishment of the Model. After elaboration of the particular signaling routes within the PheGe-knowledge base the relevant data has been automatically transferred to the 'Signaling Module' of the operational part. The 'Signaling Module' comprises 7 tables, which organize intra- and interspecific up- and downstream signaling (Fig. 2a). The compartment specificity is performed by linking members of the 'Biol_Term'-table or of the 'Interaction'-table to the respective compartment via the link-tables 'BT2C' or 'I2C', respectively. In the operational part of PheGe a compartment may be defined as a cell type, the vascular system, yet also as a distinct neuronal regulation unit. Evaluation of virtual downstream signaling is achieved by linking the outcome of an interaction to the input component of the next one ('Interaction' -> 'BT_out_Reac' -> 'Biol_Term' -> 'BT_in_Reac' -> 'Interaction'; see Fig. 2a). Accordingly, upstream signaling is performed by connecting the input of an interaction to the outcome element of the upstream one ('Interaction' -> 'BT_in_Reac' -> 'Biol_Term' -> 'BT_out_Reac' -> 'Interaction'). Attributes such as actual level (L) as well as maximum level (Mx) are included into the 'BT2C'-table. This feature is an essential prerequisite for calculations of availability and turnover of signaling components within a compartment and allows visualization of cellular dynamics during simulation of cellular and systemic circuits. The 'KO' field within the 'BT2C'- and the 'Compartment'-table resembles the on/off-switch of a component's or cellular function, which on the one hand enables the creation of a knockout on cellular and systemic levels and on the other allows ablation of a total cell type while analyzing the virtual network. Furthermore, the numbers of components taking part in a given interaction as well as the numbers of molecules being produces by it are described in the 'x_in' and the 'x_out' fields of the 'Interaction'-table. Moreover, the 'mode'-field supports the 'Mode Control Switch' of the CSMS channeling signaling into the catabolic or anabolic path or just forces downstream signaling. Finally, estimations of total numbers of particular cell types in an organism can be stored into the 'total_no'-field of the 'Compartment'-table and can be used as a correction factor for whole body simulation. After assigning the dataset to the visualization platform and adjusting the attributes of the components manually the model is ready for virtual analysis and cycling.

Cycling. The basic principle of the generation of dynamic cellular views is the cyclic analysis of metabolic and regulatory pathways (Fig. 2c). At each cycle, the information flow starts at the level of energy consumption and need of ATP from the ATP-pool. The check of cellular energy status is followed by the adjustment of metabolic needs according to the physiological state, thereby moving further upstream towards fuel uptake by the cell. In detail, the amount of cellular ATP-level relative to total capacity determines the degree of ATP-formation from the responsible metabolic reactions. Following the metabolic flow within the artificial cell, the amount of pyruvate and the level of the cellular ATP-pool as well as glucose 6-phosphate level will determine the anabolic or catabolic fate of metabolic signaling within the cell. Accordingly, the amount of glucose 6-phosphate and the degree of glycolytic activity will balance processes such as glycogen synthesis, glycogenolysis and glucose capture within the cell. In each of the cycles, the metabolic flow will be terminated by the en-

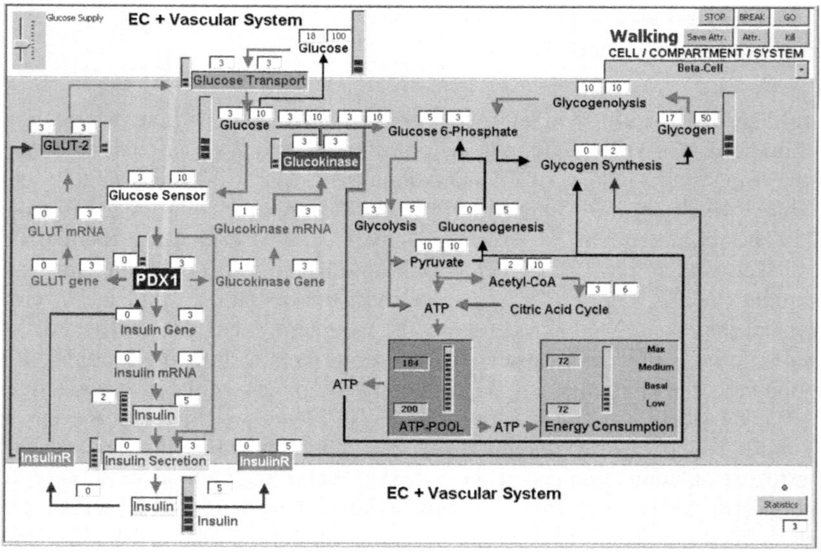

Fig. 3. The Virtual Cell Studio: Cellular view on the beta-cell. Metabolic and regulatory pathways as well as insulin-dependent reinforcement mechanisms are shown. Within the regulatory circuit PDX1 contributes to the regulation of at least three genes encoding glucokinase, glucose transporter 2 (GLUT-2) and insulin [6], thereby adapting beta-cell functions to the appropriate homeostatic conditions.

trance of glucose into the cell and the calculation of the available glucose level in the vascular system. At this stage of signaling, the compartment-specific management system will switch to regulatory cycling, search for cell-specific regulatory pathway components and proceed the regulatory information flow up to its end. Completion of a cellular cycle is determined by the failure of the CSMS to detect additional regulatory units. According to the particular situation in the cell, the regulatory run might result in a capacity or activity change of molecules or enzymes, which in turn might modulate metabolic conditions in the following cycle.

In the cellular mode of view a new cycle will be initiated within the same cell, while in the systemic approach the CSMS will start over again with the next cell type. Systemic cycling is controlled by the 'Compartment Switch', which is activated at the end of each round of cellular cycling and shifts the start of the following cycle towards the next cell type. One round of systemic cycling will be terminated after all chosen compartments have been calculated. For the purpose of systemic analysis of glucose homeostasis, energy uptake by the virtual organism is supervised by the 'Satiety Regulation System at the initial step of systemic cycling. The 'Satiety Regulation System' integrates peripheral and central energy states of the virtual organism and controls fuel uptake (see below).

Compartment-Specific Management System (CSMS). The route of signaling within the 'Virtual Cell Studio' is organized by 5 control-relays of the CSMS: the a) 'Compartment Switch', b) 'Mode Control (Switch)', c) 'Proof of Cell Specificity', d) 'Proof of Function', and e) 'Calculation' (Fig. 2b). By means of the 'Compartment'-table (Fig. 2a) the 'Compartment Switch' defines the location, where signaling takes place. It also recognizes an artificially induced ablation of a particular cell type (as marked in the 'KO'-field of the 'Compartment'-table) and blocks signaling within the respective compartment. The 'Mode Control' and the 'Mode Control Switch' retrieve information from the 'Interaction'-table of the 'Signaling Module' and trigger the fate of signaling. They decide whether to channel signaling upstream or downstream into a catabolic, anabolic, regulatory or neuronal path and control the search for interaction partners and coproducts. Proof of cell specificity is performed at any step the CSMS identifies an interaction and its associated components. A molecule, which has been virtually knocked out on a cellular or systemic level (as marked in the 'KO'-field of the 'BT2C'-table), will be detected by the 'Proof of Function'-relay and excluded from the signal path. Calculations are performed according to the procedure given in Fig. 2b. Thereby, levels and capacities (activities for enzymes) are temporary stored until the needs and availabilities of all partners and products of the particular interaction have been ascertained and corrected in a final calculation.

2.3 Biology of the System

Establishment of the Individual System Parts - Cellular View. The 'Virtual Cell Studio' enables the visualization of cellular circuits by selecting a cell type from the start menu (Fig. 3). Once the cell-specific metabolic and regulatory pathways have been built up, the Attr.- (attribute-) button can be used to organize working and maximum levels (activities for enzymes) for each of the molecules or biological terms. Activation of the GO-button will then start the simulation, which can be followed step- or round-wise. While choosing the cellular mode of view one can regulate the amount of glucose supply to the vascular system manually and pressing the 'Statistics' button will deliver additional insight into the turnover rates of the individual network components within the cell. This is of special interest, when comparing an artificially induced malfunction or KO (performed by double-click on the biological term field) with the normal situation.

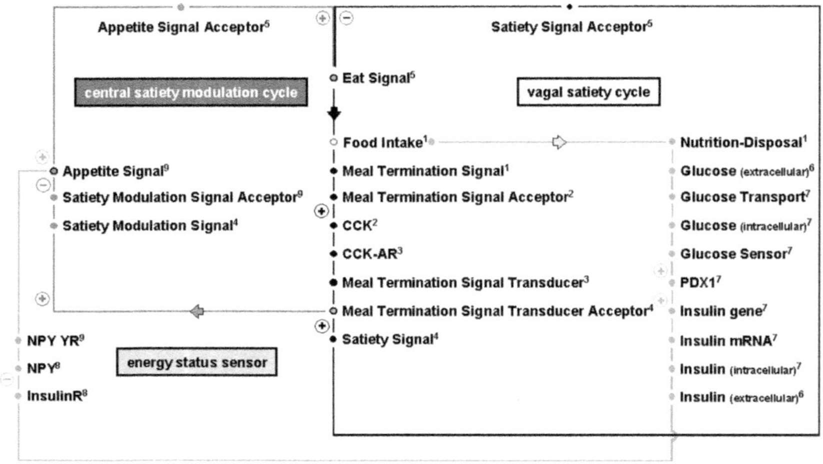

Fig. 4. The Satiety Regulation System. Numbers represent compartments: 1) gastrointestinal tract, 2) entero-endocrine I-cell, 3) vagal neuron, 4) satiety neuron of the nucleus tractus solitarii (NTS-neuron), 5) dorsomotor nucleus neuron of the vagus (DMV-neuron), 6) vascular system, 7) beta-cell, 8) adiposity neuron of the arcuate nucleus, 9) adiposity neuron of the paraventricular nucleus (lateral hypothalamic area).

Implementation of the Satiety Regulation System. Before starting to bridge the cellular arrangements to an artificial organism, special attention must be drawn to the regulation and supply of nutrition to the whole system. Fig. 4 offers a survey of PheGe's regulatory and neuronal interactions that balance eat motivation in the organism. In short, three main regulatory circuits perform the regulation of food intake at least. The first one (vagal satiety cycle) is initiated by the transformation of meal size information into a meal termination signal within the gastrointestinal tract and terminates by the concerted action of various signals that converge on the dorsal vagal complex in the caudal brainstem and trigger anorexigenic signaling through repression of the eat signal [7]. The second regulatory circuit of the food intake control is a central satiety modulation cycle that originates in the brainstem under the trigger of the vagal output and passes adiposity-neurons in the lateral hypothalamic area (LHA-adiposity-neuron) and the paraventricular nucleus (PVN-adiposity-neuron) of the hypothalamus where appetite signaling is repressed [8]. The suppression of orexigenic appetite signaling and the subsequent inactivation of the appetite signal acceptor results in a reinforcement of the dorsal vagal complex-derived satiety signaling, which finally terminates food intake motivation. The third circuit involved in the control of food intake is an energy status sensor, which connects peripheral energetic information of the organism to the central energy uptake motivation status. The point of entrance into the central regulatory pathway are the neuropeptide Y- (NPY-) positive adiposity-neurons within the arcuate nucleus of the hypothalamus, which project to the PVN- and the LHA-adiposity-neurons, where NPY is released to the synaptic cleft in order to reach its receptor NPY YR supporting appetite signaling [9]. High blood insulin levels originate from elevated glucose concentrations after a meal and resemble a high energetic status of the organism. Binding of insulin to its receptor in the arcuate nucleus will repress NPY, which consequently blocks the NPY-mediated activation of the appetite signal via the NPY YR [9]. Choosing the Satiety Regulation System from the Cell/Compartment/System-menu of the virtual cell studio activates the visualization of the interplay of these three food intake-regulating systems.

Bridging Cellular Circuits to an Overall Response - Systemic View. After establishment of the cellular views within the virtual cell studio the individual system parts were series connected and allowed to act automatically. To run the systemic simulation click on the 'Systemic View' button in the start menu, decide whether the virtual system may use food ad libitum or not and activate the 'GO' button. Each of the system parts will evaluate its own metabolic, regulatory and energetic status, compare it

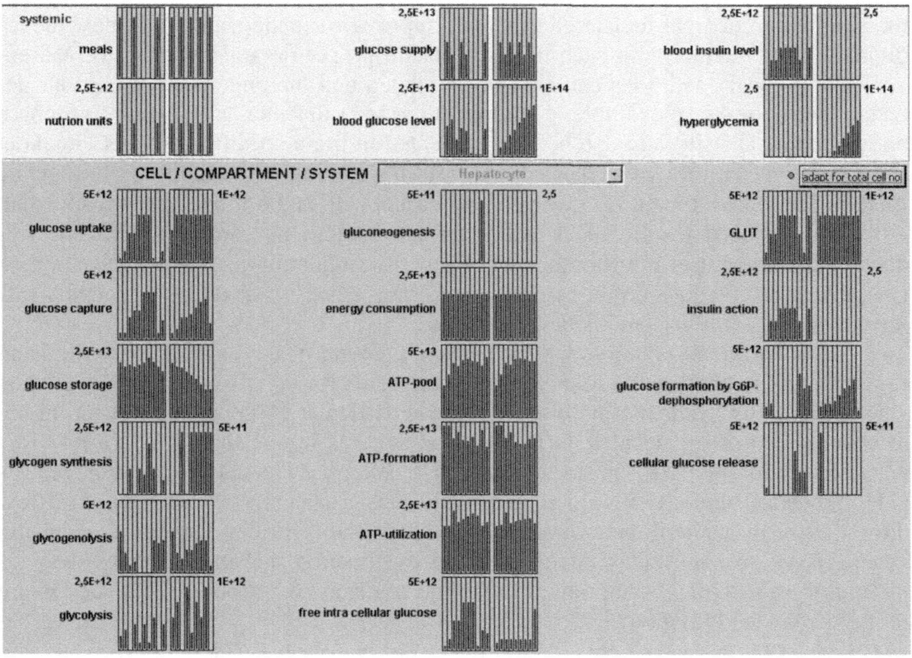

Fig. 5. Change of component levels during a period of 10 rounds of systemic cycling under normal (left) and PDX1-KO (right) conditions. Note the lack of blood insulin levels and the increase in blood glucose levels following virtual PDX1-gene deletion. For the intracellular view the hepatocyte is shown. The cell type can be changed by choosing the compartment from the control menu. Bars represent molecule numbers.

with the condition of the organism via the vascular system and respond according to the particular situation. The information flow within each cell system is recorded per cycle and the turnover of the individual network components can be retrieved after the session has terminated (Fig. 5). The user may choose to run the simulation tool with either cell-specific or systemic KOs of network components of his choice. The visualization of the systemic consequences of an artificial generation of a PDX1-KO within the beta-cell is shown in Fig. 5. Hyperglycemia appeared in consequence of PDX1-malfunction on insulin production, glucokinase activity and glucose uptake into the cell. In the liver cell maximum glucose uptake is only 1/5 in the PDX1-KO simulation as compared to the control and consequently, glycogen stores are depleted for the maintenance of the ATP-pool (Fig. 5).

3 Discussion

To the knowledge of the author, the Virtual Cell Studio is the first simulation tool, which gives insights into the dynamics of genotype-phenotype relations. While bridging cellular circuits to an overall response, metabolic, regulatory and neuronal signaling as well as the impact of a virtually induced malfunction can be viewed on a molecular level. Inactivation of the individual circuits by artificially induced KOs or the elimination of a particular cell type will improve our understanding of how the individual units interact with each other to sustain proper organism functions. Moreover, pathological conditions can now be simulated and the underlying molecular defects identified. As an example, persons with type II diabetes usually have an obese body habitus. Obesity has also been documented in brain-specific InsulinR knockout (NIRKO) mice and it was suggested that insulin signaling in the brain might regulate food intake behavior [10]. By means of the virtual cell studio it becomes obvious that artificially induced InsulinR-KO of the arcuate nucleus neurons will stimulate frequency and amount of nutrition uptake. Thus, the simulation indicates that increased food intake motivations might be one of the reasons obesity develops in patients suffering from insulin-resistant diabetes mellitus.

It is worthwhile to mention that the systemic feature of the virtual cell studio is an excellent tool to analyze the role of a single molecule for organism function even then when a naturally occurring malfunction of this particular component causes a failure to channel cell or organ fates during development, as it was shown for PDX1. Targeted disruption of PDX1 in mice results in a collapse of pancreatic morphogenesis [11]. Newborn homozygous mice selectively lack a pancreas and die within a few days after birth, while in heterozygous individuals the disturbed regulation of insulin gene expression contributes to the beta-cell dysfunction that characterizes diabetes mellitus with its typical symptom of hypergylycemia as seen from the statistic feature of the Virtual Cell Studio (Fig. 5).

Based on the reductive philosophy of PheGe, the Virtual Cell Studio serves as a sophisticated approach to bridge intricate cellular systems to an overall response. It is a pilot project that rather targets the visualization of signaling and its systemic consequences on a crude level than to account for the exact stoichiometric conditions within a living cell, which certainly remains a speculative urge for the future. General limitations of such knowledge base approaches comprise information input and data integrity checking. High expertise and persistence are essential prerequisites to expand virtual biology systems such as PheGe. Therefore, the author highly encourages experts in the particular biological fields to participate in this approach. Together with its simulation tool, PheGe is freely available upon request from the author and can be distributed within the scientific community. While the PheGe-technology will assists anyone who is interested in virtual biology, the consolidation activities of the experts will enforce the further reconstruction of a virtual organism on a high quality standard. Although a long way to go, efforts are underway to implement PheGe's virtual circuits into a more-dimensional mouse model [12], thereby closing up to the vision of time-space voyages through an organism.

References

1. Takai-Igarashi, T., Kaminuma, T.: A pathway finding system for the cell signaling networks database In Silico Biol. **1** (1999) 129–146
2. Bader, G.D., Donaldson, I., Wolting, C., Ouellette, B.F., Pawson, T., Hogue, C.W.: BIND--The Biomolecular Interaction Network Database. Nucleic Acids Res. **29** (2001) 242–245
3. Schacherer, F., Choi, C., Gotze, U., Krull, M., Pistor, S., Wingender, E.: The TRANSPATH signal transduction database: a knowledge base on signal transduction networks. Bioinformatics **17** (2001) 1053–1057
4. Xenarios, I., Salwinski, L., Duan, X.J., Higney, P.S., Kim, M., Eisenberg, D.: DIP, the Database of Interacting Proteins: a research tool for studying cellular networks of protein interactions. Nucleic Acids Res. **30** (2002) 303–305
5. Seidl, K.: PheGe, the Platform for Exploring Genotype-Phenotype Relations on Cellular and Organism Level. In: Proceedings of the IEEE Computer Society Bioinformatics Conference, CSB2002, Stanford, California (2002) 79–86
6. Hui, H., Perfetti, R.: Pancreas duodenum homeobox-1 regulates pancreas development during embryogenesis and islet cell function in adulthood. Eur. J. Endocrinol. **146** (2002) 129–141
7. Powley, T.L.: Vagal circuitry mediating cephalic-phase responses to food. Appetite **34** (2000) 184–188
8. Palkovits, M.: Interconnections between the neuroendocrine hypothalamus and the central autonomic system. Front. Neuroendocrinol. **20** (1999) 270–295
9. Baskin, D.G., Figlewicz Lattemann, D., Seeley, R.J., Woods, S.C., Porte, D. Jr., Schwartz, M.W.: Insulin and leptin: dual adiposity signals to the brain for the regulation of food intake and body weight. Brain Res **848** (1999) 114–123
10. Bruning, J.C., Gautam, D., Burks, D.J., Gillette, J., Schubert, M., Orban, P.C., Klein, R., Krone, W., Muller-Wieland, D., Kahn, C.R.: Role of brain insulin receptor in control of body weight and reproduction. Science **289** (2000) 2122–2125
11. Jonsson, J., Carlsson, L., Edlund, T., Edlund, H.: Insulin-promoter-factor 1 is required for pancreas development in mice. Nature **371** (1994) 606–609
12. Plischke,M., Wolf, K.-H., Krofczik, S., Rinkwitz, S., Grothe, C., Pretschner, D.P., Seidl, K.: 4D-Atlas der embryonalen Mausentwicklung. In: Meiler, M., Saupe,, D., Kruggel, F., Handels, H., Lehmann, T. (eds.): Informatik aktuell, Algorithmen - Systeme – Anwendungen, Proc. of the Workshop Bildverarbeitung in der Medizin 2002, Vol. 56. Springer Verlag, Heidelberg (2002) 295–298

Robustness to Damage of Biological and Synthetic Networks

Roberto Serra, Marco Villani, and Alessandro Semeria

Centro Ricerche Ambientali Montecatini via Ciro Menotti 48, I-48023 Marina di Ravenna
rserra@cramont.it

Abstract. We analyze the perturbation of the expression levels of thousands of genes when one of them is knocked-out, by measuring *avalanches*, the number of genes whose expression is affected in a knock-out experiment, and gene *susceptibilities*, which measure how often the expression of a given gene is modified. Experimental data concerning the yeast *S. cerevisiae* are available. Knock-out is simulated, using random boolean network models of gene regulation, in several experiments, using different sets of boolean functions. The major results (when only canalizing functions are allowed) are that the distributions of avalanches and susceptibilities are very similar in different synthetic networks, with very small variance, and that these two distributions closely resemble the experimental ones (a result which is even more surprising since no parameter optimization has been performed). These results strongly suggest that the distribution of avalanches and susceptibilities may be generic properties, common to many different genetic networks.

1 Introduction

Most researches in bioinformatics, in the field of genetic networks, are directed towards understanding specific genetic regulatory circuits. The importance of this line of research cannot be overestimated, since it may lead to a better understanding of the cell behaviour, as well as to the development of innovative therapies. However, as it is widely acknowledged that different gene circuits interact in a complex way in a cell, the study of specific genetic circuits may be usefully complemented by system-level analyses. In this kind of research one does not aim at determining the specific details of a given circuit or process, but at discovering possible "generic" properties, which are common to a wide set of cells and cellular processes.

The random boolean network model [1] is the best known example of such line of research. In these studies Kauffman presents an interesting coupling between topology and dynamics, finding a region where the generic dynamical properties of the whole system are determined by a limited set of parameters.

Recently, some interesting papers [2][3] focussed upon the topological properties of networks of biological interest like metabolic networks, where the nodes are small molecules involved in metabolism, and there is a link between two nodes if the corresponding molecules take part together in at least one reaction. The most interesting outcome of these studies is the discovery that several metabolic networks are of the

W. Banzhaf et al. (Eds.): ECAL 2003, LNAI 2801, pp. 706–715, 2003.

scale-free type, where the probability distribution of the number of connections per node scales as a power law.

Interesting results concerning the dynamical response of a biological system as a whole, when subject to small perturbations, have been obtained by knocking-out, one at a time, some genes in *Saccharomyces Cerevisiae*, and recording, by DNA microarray techniques, the changes in the expression levels of the other genes [4].

Gene knock-out can be simulated "in silico" using random boolean networks (briefly, RBN), as will be described in section 4. In this work we apply the random boolean network model in order to interpret the results of gene expression data in knock-out experiments.

Therefore, section 2 presents the random boolean network model; section 3 illustrates the details of the gene knock-out experiments, and section 4 describes our simulations of gene knock-out. According to the spirit of our approach, we are not interested in the detailed effects (i.e. in knowing that the silencing of a particular gene increases, or decreases, the expression of a given gene) but rather in the statistical features of the consequences of knocking-out. Comments and indications for further work are the topic of section 5.

2 Random Boolean Networks

Let us consider a network composed of N genes, or nodes, which can take either the value 0 (inactive) or 1 (active). Let $x_i(t) \in \{0,1\}$ be the activation value of node i at time t, and let $X(t)=[x_1(t), x_2(t) \ldots x_N(t)]$ be the vector of activation values of all the genes. Activation of a gene may be biologically interpreted as the presence of the enzyme corresponding to that gene.

Enzyme concentrations take continuous values, and we have introduced continuous genetic networks to take this feature into account [5][6]. However, such networks are computationally heavy, and in order to simulate large networks we adopt here the classical boolean approximation, which is often accurate (under the assumption of a sigmoidal response to solicitations of each system node, and by choosing an appropriate temporal scale [1]).

Real genes influence each other through their corresponding products and through the interaction of these products with other chemicals, by promoting or inhibiting the activation of target genes. In the corresponding artificial network these relationships are lumped in directed links (directed from the node A to the node B, if the product of gene A influences the activation of gene B) and boolean functions (which model the response of each node to the input configurations). Both the topology and the boolean function associated to each gene do not change in time.

In a classical RBN each node has the same number of incoming connections k_{in}, while the other terminals of the connections are chosen with uniform probability among the other nodes; note that self coupling is not allowed. The consequence is that the distribution of outgoing connectivities k_{out} is a Poissonian distribution [7].

The output corresponding to each input configuration is chosen at random, according to some probability distribution; let f_0 (f_1) be the number of output values equal to "0" (respectively, to "1"). Let us then define a parameter P_f for that function f as $P_f=|f_1-f_0|/2^k$. Note that $P_f \geq 0$; if both values are equally represented, $P_f=0$, while if the outputs are either all 1 or all 0, then $P_f=1$. The average of P_f on the whole network is called the internal homogeneity P of the system [1].

A boolean function is said to be a canalizing function of class C_1 if at least one value of one of its input nodes uniquely determines its output, no matter what the other input values are. A boolean function is said to be a canalizing function of class C_i if this condition holds for exactly *i* of its input nodes [1]. The higher the class of a canalizing function, the higher the regulatory effect on the system dynamical behaviour.

The network dynamics is discrete and synchronous, so all the nodes update their values at the same time. Once the connections and the boolean functions of each node have been specified, X(t) uniquely determines X(t+1). The time step must therefore be long with respect to the decay time of enzymes, so that the activations at time t+1 depend only upon those at time t, and there is no memory of the activation values at time t-1 and before.

In order to analyze the properties of an ensemble of random boolean networks, different networks are synyhesized and their dynamical properties are examined. The ensembles differ mainly in the choice of the number of nodes N, the input connectivity per node k, the internal homogeneity P and the choice of the boolean functions. While individual realizations may differ markedly from the average properties of a given class of networks, one of the major results is the discovery of the existence of two different dynamical regimes, an ordered and a disordered one, divided by a "critical zone" close to the line $1/k = 2P(1-P)$ [8].

Several observations, summarized in [1][9], indicate that biological cells tend to be found in the ordered region; this can be explained by observing that the sensitive dependence of disordered networks makes it very difficult to control the size of perturbations, and evolution might have driven the system in the ordered regime. Moreover, evolution might have driven biological systems in regions of parameter space which are not too far from the border between ordered and chaotic regimes, so to allow greater plasticity in response to perturbations or to changing environmental conditions.

3 Gene Knock-Out Experiments

Useful data from cDNA microarray measurements of gene expression profiles of *Saccharomyces cerevisiae* subject to 300 different mutations and chemical treatments are available ([4], discussed also in [10]), which refer to all characterised and uncharacterised genes (ORF). We will briefly summarize here some features of the experiments, referring the interested reader to the original references for further details; particularly, in this paper we consider only a subset of 227 experiments, where a single deletion (knock-out) has been performed.

A typical knock-out experiment compares the expression levels of all the genes in the cells with a knocked-out gene, with those in normal ("wild type") cells. The genome of *Saccharomyces Cerevisiae* includes 6312 known genes; 227 experiments have been performed. In this way, all the experimental data can be cast in matrix form E_{ij}, i=1 ... 6312, j=1... 227. E_{ij} is the ratio of the expression of gene i in experiment j to the expression of gene i in the wild type (unperturbed) cell. As it is well known, microarray techniques estimate gene expression levels by the abundance of the corresponding mRNA.

Microarray data are noisy, in part because of the experimental manipulations, in part because of intrinsic reasons, due to the difference in mRNA levels of similar cells. Therefore, in order to make precise statements about the number of genes perturbed in a given experiment, it is required that a threshold be defined, such that the difference is regarded as "meaningful" if the ratio is greater than the threshold θ (or smaller than $1/\theta$, depending upon the fact that the expression level is higher or smaller than the one in the reference state) and neglected otherwise. Here we limit our analysis to the case of a threshold equal to 4, which is the most stringent value among those examined in [10].

In order to describe the global features of these experiments, two important aggregate variables are the following.

Avalanches. Let E' be the boolean matrix which can be obtained by E by posing $E'_{ij}=1$ if $E_{ij}>\theta$, $E'_{ij}=0$ otherwise ($E'_{ij}=1$ means that the modification of the expression level of gene i in experiment j is accepted as "meaningful"). An avalanche is the size of the perturbation induced by a particular experiment. In experiment j, $A_j = \sum_i E'_{ij}$

Susceptibilities. The susceptibility of gene *n*, S_n, is equal to the number of experiments where that gene has been significantly affected: $S_n = \sum_j E'_{nj}$.

Note that the size of an avalanche depends upon the number of genes that are monitored with microarrays, and the susceptibility depends upon the number of experiments that have been performed. In our analysis these quantities are held fixed, taking the values of 6312 and 227 respectively. Of course, normalizing these data would be trivial, and it would be necessary to compare data among different organisms or different experimental sets.

Figure 1 shows the avalanches (susceptibilities) cumulative distribution, based upon the data by Hughes [4]. Note that the avalanches plot in a log-log scale is very close to a straight-line (with the exception of the initial portion): this is the characteristic signature of a power-law distribution $p(x) \sim x^{-\gamma}$, with $\gamma \cong 1.6$. The distribution of susceptibilities is plotted in a linear scale, because of its rather limited range.

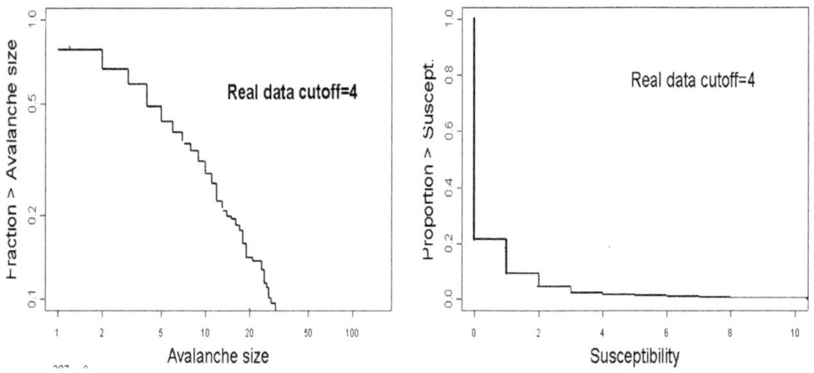

Fig. 1. Plot of avalanches cumulative distribution (a) and susceptibilities cumulative distribution (b). Note that the plot of avalanches cumulative distribution is in a log-log scale

4 The Simulation of Gene Knock-Out

In order to take into account the dynamical features of gene regulatory networks, in this preliminary work we consider computer-generated dynamical networks with a high number of genes (up to 6300, close to the number of genes of *Saccharomyces Cerevisiae* analyzed by Hughes [4] and perform simulations aimed at reproducing the experimental conditions.

We do not know the real distribution of the connectivity values for the *Saccharomyces Cerevisiae* network, nor the boolean function realised by each gene. However, theoretical reasons lead to hypothesis that:

1)real cells should lie in the ordered regime (although somehow close to the border which separates it from the chaotic regime);
2)not all the possible boolean functions have the same probabilities of occurrence.

Hypothesis (1), as it has been discussed in Section 2, is required because if the global behaviour of the biological system were too susceptible to small perturbations, its chance of surviving to environmental stress would be too low. The simplest way to implement a network in the ordered regime is that of choosing the input connectivity k_{in} =2 for every gene: in this case, for each value of the homogeneity parameter P, the network is in the ordered phase.

Hypothesis (2) involves at least a partial knowledge of the function realised by biological systems. In recent studies based upon a survey of known regulatory circuits [11], it has been observed that in most biological systems the probability of finding canalizing functions is higher than the one corresponding to a uniform choice in the ensemble of all the possible functions. On the other hand, there does not seem to be any detectable bias in favour of specific values of the homogeneity parameter P [11].

Because the importance of the choice of the set of boolean functions, we considered three alternatives:

a) an ensemble of networks with k_{in}=2 and all the possible boolean functions;
b) an ensemble of networks with k_{in}=2 and only canalizing functions (which in the k_{in}=2 case implies that XOR and NOT XOR are excluded)
c) an ensemble of networks with k_{in}=2, only canalizing functions, and the NULL function excluded (the NULL function being the one which outputs "0" for every input configuration)

Alternative (b) reflects both the fact that it is very hard to find plausible biological mechanisms able to implement such kind of functions and the observations by [11]. The exclusion of the NULL function (hypothesis (c)) comes from the consideration of the dubious detectability of genes that are always silent.

Now we have to sketch the essential steps of the experiment. Knocking-out is simulated by clamping the value of one gene to 0 in the attractor cycle. In this way, the incoming information to the gene of interest becomes useless, and the outgoing information is held fixed. Note that these experiments differ from those which have often been performed to test the stability of a given attractor, by flipping the state of a single gene and letting it evolve according to its rule [1], because here the gene is forced to take always the value 0.

A simple way to reproduce the experimental design realised on the microarray is:

1) take a specific RBN;
2) let it relax to an attractor state;
3) choose a gene (which is not always in state 0) for knock-out and clamp its value to 0 for all the following time steps;
4) let the network evolve to its new attractor;
5) compare the state of every gene in the perturbed network with the state of the corresponding gene in the original one;
6) perform the same procedure on a number of different RBN with different connections and boolean functions but belonging to the same class (like eg the class of networks with k_{in}=2 and all boolean functions allowed) and gather statistical data.

The mRNA which is used in the experiments comes from several cells which have been cultured, which are not forced to be in a peculiar phase of the cell cycle; it is therefore entirely reasonable to consider as a reference state the attractor cycle with the largest basin of attraction. This is the "least biased" choice, which is adopted here, although it implies a heavy computational weight on the simulation; we remember, nevertheless, that the different initial conditions are $2^{6300} \cong 10^{1900}$, and that under-sampling is unavoidable.

In order to compare perturbed and unperturbed samples, for each gene the ratio between the expression level in the perturbed sample and the expression level in the unperturbed sample is computed. When we are dealing with an attractor which is not a fixed point, we choose as reference, for each node, its average expression level taken over the period T of the attractor of interest. This follows from the fact that we deal with biological cells that are not synchronised with each other: therefore the temporal averages made during the simulations correspond to the ensemble average made by using many cells to extract the required mRNA.

Since boolean functions are not well suited to deal with threshold effects, in the simulation we consider as affected any gene whose expression (or average expression, if it oscillates) is different in the two cases (perturbed and unperturbed). We present here a summary of the main results, obtained by using 10 networks for each series of simulations, each network initially characterised with 630 different initial conditions. Figure 2 shows one that the plot of avalanche distributions of the first series of runs (alternative (a) – briefly, a-series), with respect to the same plot of the c-series, presents a relatively high dispersion, whereas the plots of avalanche distributions of b-series and c-series don't (the plot for the b-series is not shown).

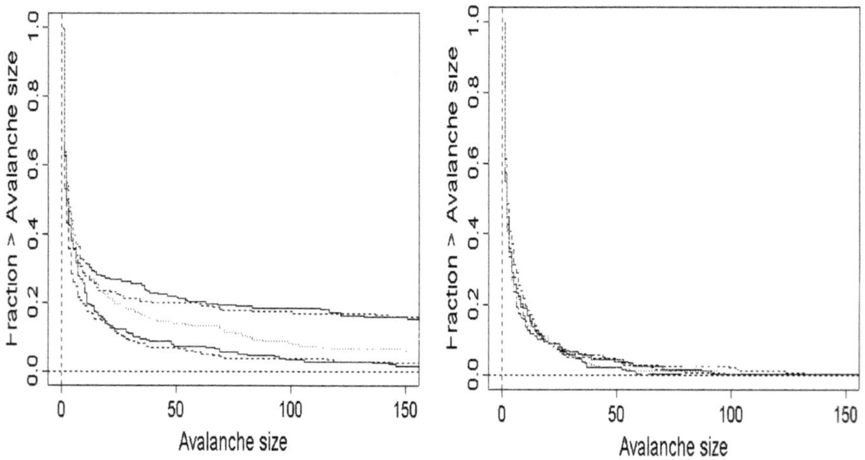

Fig. 2. The plot of avalanche cumulative distributions of the *a*-series (a) shows, with respect to the same plot of the *c*-series (b), a relatively high dispersion

Moreover, although the anaysis is complicated by the high variance, the *a*-series does not match the real data as well as the *b*-series and *c*-series; these results support the hypothesis of a low biological plausibility of an appreciable role of non canalizing functions. While it would be possible that in nature *i)* all the boloolean functions are allowed but *ii)* some peculiar networks are selected, it is simpler and biologically plausible to suppose that parity-like functions, which are difficult to implement with biological machinery, are excluded. Note that the results obtained from *b*-series and *c*-series are very similar to each other, indicating that the effect of the exclusion of the NULL function is negligible.

In figure 3 the distributions of real avalanches and susceptibilities are compared with the corresponding distributions of the *c*-series, showing a good agreement, which is even more impressive if we take into account that no tuning of the model parameters has been done (apart from the restriction to the set of canalizing boolean functions)).

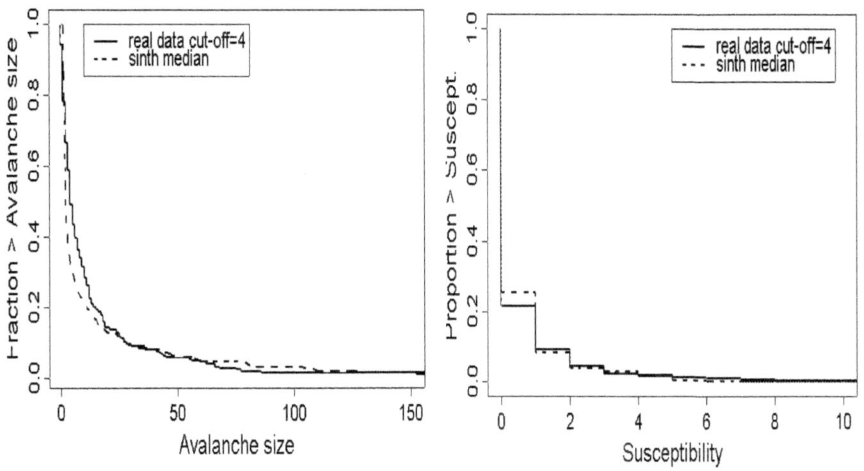

Fig. 3. Comparison between the real avalanches cumulative distribution (a) and susceptibilities cumulative distribution (b), and the corresponding distributions of the *c*-series

If we assume that the avalanche probability distribution has the form of a power-law, we can calculate the corresponding parameters, shown in tab. 1.

Table 1. The exponents of the fits done on the real (*S. Cerevisiae*) and synthetic probability density data (*a*-series, *b*-series and *c*-series), under the assumption that the avalanches probability distribution follows a power-law.

Avalanches	Exponent
S.Cerevisiae	1.6
***a*-series**	1.5
***b*-series**	1.8
***c*-series**	1.7

Further comparisons between real and synthetic data are shown in table 2, where we present some statistical indicators of the real and synthetic data. The largest avalanches of the *a*-series are much larger than those of the other synthetic networks and of the real data. The data obtained from the c-series resemble the experimental data most closely, but the differences between b- and c-series are too small to allow any definitive conclusion.

Table 2. Some features of the real and synthetic data distributions. The superscript $^{aver.}$ and ^{median} indicate respectively that the value is the average (or the median) of the corresponding values of the different synyhetic networks

	Avalanches			Susceptibilities			
	Median	Mean	Max	Median	Mean	Max	unaffected nodes
SaccCe	4.0	12.9	219	0.0	0.46	33.0	4964
***a*-series**$^{aver.}$	2.5	70.6	1565	1.3	25.41	27.4	3547
***b*-series**$^{aver.}$	2.1	7.4	113	0.0	0.27	6.8	5113
***c*-series**$^{aver.}$	2.1	12.5	266	0.0	0.45	9.6	4766
***a*-series**$^{median.}$	3.0	39.0	581	0.0	14.04	24.0	4115
***b*-series**$^{median.}$	2.0	7.3	107	0.0	0.26	6.5	5144
***c*-series**$^{median.}$	2.0	11.9	260	0.0	0.43	7.5	4736

5 Conclusions

Some care must be exercised in considering the results of the previous sections, because of the very high dimension of the network ensemble, which makes it impossible to perform an accurate sampling of the space of networks and of initial conditions.

The results of this first study require further analysis. On the experimental side, it would be interesting to consider different knock-out experiments, to check whether the generic properties which have been hypothesized here are found in a large set of organisms.

On the modelling side, it would be important to check the behaviour of networks with a larger input connectivity, and to consider also the biologically realistic case where there is a distribution of input connectivities (e.g. a scale free distribution) [2][3].

It would be interesting also to consider different updating strategies, spanning from the synchronous case considered here to the fully asynchronous one, which leads to a dynamical behaviour different from that of synchronous updating [12]. The most interesting alternative, form the viewpoint of biological plausibility, would be that of allowing each gene, at each time step, to update its state with a certain probability (but if a given gene has updated its state at time t, then there should be an interval, from t to t+δ, where no further updating of the state of that same gene is allowed).

Notwithstanding these limitations, the results achieved so far seem striking: a very simple boolean model, with almost no adjustable parameter, captures the main features of the distribution of avalanches and susceptibilities. This strongly suggests that these distributions might be generic properties, common to a wide set of models and real systems.

Acknowledgments. Several helpful discussions with David Lane, Stuart Kauffman, Annamaria Colacci, Elena Morandi and Cinzia Severini are gratefully acknowledged. This work has been supported in part by the Italian Ministry for University and Research (Miur) under grant FISR 1509 (2002).

References

1. Kauffman, S. A.: The origins of order. Oxford University Press (1993)
2. Wagner, A. and Fell, S.: The small world inside large metabolic networks. Santa Fe Institute Working Paper 00-07-041 (2000)
3. Jeong, H., et al: The large-scale organization of metabolic networks. Nature 407 (2000) 651–654
4. Hughes, T.R. et al.: Functional discovery via a compendium of expression profiles. Cell 102 (2000) 109–126
5. Serra, R. & Villani, M.: Modelling bacterial degradation of organic compounds with genetic networks. J. Theor. Biol. 189 (1) (1997) 107–119
6. Serra, R., Villani, M. & Salvemini, A.: Continuous genetic networks. Parallel Computing 27 (5) (2001) 663–683
7. Albert, R and Barabasi, A.L.: Statistical mechanics of complex systems Reviews of Modern Physics 74, 47 (2002).
8. Bastolla U., Parisi G.: The Modular Structure of Kauffman Networks. J. Phys. D 115 (1998) 219–233
9. Kauffman, S.A.: Investigations. Oxford University Press (2001)
10. Wagner, A.: Estimating coarse gene network structure from large-scale gene perturbation data. Santa Fe Institute Working Paper 01-09-051 (2001)
11. Harris, S.E., Sawhill, B.K., Wuensche, A. & Kauffman, S.A.: A model of transcriptional regulatory networks based on biases in the observed regulation rules. Complexity 7 (2002) 23–40
12. Harvey I., and Bossomaier T.R.J.: Time Out of Joint: Attractors in Asynchronous Boolean Networks. Proc. Fourth European Conference Artificial Life. Ed. P. Husbands and Harvey I. (1997) 67–75

An Agent-Based Approach to Routing in Communications Networks with Swarm Intelligence

Hengwei Shen[1] and Shoichiro Asano[2]

[1] Graduate School of Information Science and Technology,
The University of Tokyo, Tokyo, 113-8656 Japan
hwei@nii.ac.jp

[2] National Institute of Informatics, Tokyo, 101-8430 Japan
asano@nii.ac.jp
http://www.asn.r.nii.ac.jp/

Abstract. Ant colonies are distributed systems that, in spite of the simplicity of their individuals, present a highly structured social organization. When such ant agents are used to solve the adaptive routing problems in communications networks, surprisingly good properties in terms of flexibility and adaptability that are competitive with traditional routing algorithm can be achieved. In this paper, the distributed adaptive routing algorithms that explore the network and rapidly learn good routes based on simple biological ants are investigated. A forward routing algorithm we proposed will be analyzed on a simulated network and the robustness and scalability are proved on a dynamic scenario. Our algorithm can directly influence the throughput and average delays of information messages and, hence have an impact on the overall performance of the network.

1 Introduction

Swarm intelligence research originates from collective behaviors of real ants. Such systems promise to provide guiding principles for, and engineering solutions to, distributed systems management problems found, for example, in communications networks. Nowadays, the increasing capacities of computation and communication and requirements of guaranteeing different qualities of services make network management protocols, such as traffic control and routing, more necessary to be designed with new methodologies such as the agent-based routing, particularly in view of recent developments in dynamically reconfigurable networks, active networks, dynamic virtual private networks (VPNs), mobile and survivable networks, and so on.

Our goal is to devise an efficient, flexible, fault tolerance routing algorithm that discovers optimal routes expediently. Towards this goal, this paper proposes the formulation of a routing strategy that exploits the intelligent mobile agent paradigm. In comparison to ongoing research efforts that examine swarm intelligence of mobile agents as a basis for the development of a routing mechanism and compare the strategies to conventional approaches, such as distance-vector algorithm (DVA) and link-

W. Banzhaf et al. (Eds.): ECAL 2003, LNAI 2801, pp. 716–723, 2003.

state algorithm (LSA), this paper addresses the issue of improving current ant agent based routing algorithm to a more effective and efficient protocol. Specifically this paper focuses on the design and evaluation of information exchange in ant routing algorithm (ARA) that facilitates resource awareness, and fault tolerance.

This paper consists of five subsequent sections. In the next section, the essential principles of swarm intelligence are described. The paper continues with a brief overview of ant routing algorithm, and an analysis of the research effort during recent years is presented. The following section presents a generic improvement to above conventional model. Simulation issues are then briefly described. A summary of this research is then provided and the paper ends with a discussion on future work.

2 Ant Routing: Background

2.1 Routing with Swarm Intelligence

Swarm intelligence research originates from collective behaviors of real ants. It is a property of systems of unintelligent ants of limited individual capabilities exhibiting collectively intelligent behavior. The advantages of swarm intelligence are twofold. Firstly, it offers intrinsically distributed algorithms that can use parallel computation quite easily. Secondly, these algorithms show a high level of robustness to change by allowing the solution to dynamically adapt itself to global changes by letting the agents self-adapt to the associated local changes.

Since routing decisions can only be made on the basis of local and approximate information about the current and the future network states, the challenge of routing in communications networks is the need for a fully distributed algorithm that is able to find shortest paths robustly and efficiently in the face of changing network topologies as well as changing link costs. There are two major classes of adaptive and distributed packet routing algorithms: DVA and LSA. The comparison between the traditional and agent-based algorithms can be based on network routing for packet switched networks. The following aspects are here discussed.

Resource Requirements: Comparing to the ARA, the LSA pays a significant overhead in terms of state on each router and in terms of time to compute the shortest paths. The DVA has significantly lower router state, but this comes at the expense of increased routing message traffic.

Adaptiveness to Topology Changes: With the LSA, in case of topology changes, such as the corruption of router state, the recovery period tends to be quite long. For ARA, because ants are launched in the network periodically with small time interval, the algorithms can adapt to the topology changes (also traffic load) continuously. Therefore, ARA has more sophisticated mechanism to adapt to changes.

Adaptiveness to Traffic Load and Ability to Avoid Congestion: The ARA adapts better to traffic load than traditional algorithms and thus easy to route packets away from temporarily congested links or nodes. However, in traditional algorithms, because accurate packet delay is hard to measure and may change significantly with the traffic load, it is almost impossible to use the accurate packet delay as a metric.

2.2 Backward Ant Routing

There are a number of proposed agent-based routing algorithms. Ant Based Control (ABC) is a network-centric algorithm in telephone networks. The approach exploits the concept of multiple colonies of agents coexisting and in some cases coordinating with each other working towards independent goals. AntNet applies the idea of deploying agents for routing in packet switched networks. The algorithm generates mobile agents at regular intervals at different nodes in the network. These agents traverse the network to reach the destination then on their way back to the source node collect routing information. Although AntNet is an interesting approach for static networks with a good adaptive property its application in dynamic networks is yet to be explored.

Ant routing algorithms described here are all backward routing because the ants perform a form of backward learning. Backward routing is intrinsically slow since it requires the agent to reach its destination before any update to begin. This slow round trip reaction to changes in the network might induce oscillations.

For example, in figure 1, an ant agent starts from node S at the moment t_0 and moves to node j at moment t_1. It keeps track of the visited nodes in stack and of the associated distance cost d which can be the wait time and the transmission delay for each visited node, but it can not update the routing table at this time because it does not know the distance cost d_{ji}. The probability P in the routing table of node j can only be updated at moment t_5 when the agent goes backward the path with the information collected at t_2 as follows, where $P^i_{j\to D}$ is the routing table probability for node j to D via i and the denominator is defined as the sum of all the costs from node j to the destination node D.

$$P^i_{j\to D} = [1/(d_{ji}+d_{iD})] / \Sigma\,(1/d_{jD}) \qquad (1)$$

Distinctly, if distance cost d_{ji} changed at t_5, this update would induce oscillations. As an alternative, this paper introduces a more efficient asymmetric routing to overcome such oscillation problems in the next chapters: Forward Ant Routing.

3 Forward Ant Routing

Forward routing offers an alternative by removing the need of round trips. It was first introduced by Schoonderwoerd et al. in the case of virtual circuit based symmetric networks (e.g. identical costs associated with both link directions). It was then extended to packet-switching symmetric networks by Heusse et al. with CAF (Cooperative Asymmetric Forward) routing. In the CAF agent based routing algorithm, each node holds two routing tables (backward and forward routing tables) and all the routing table entries can be used at any time.

However, this paper extends above approach to asymmetric networks (e.g. packet-switching networks in which asymmetric delays occur due to different queue lengths). If the network is symmetric, the cost measured by forward agents on their path from the source to the current node can directly be used to update the estimation of the distance toward the source from the current node. For asymmetric networks, this cannot be done anymore

In the forward routing algorithm, agents cross the network to experience the state of a part of it. These agents are simple packets that use the same communication links as any ordinary data. The agents carry information of the nodes and of the distance costs visited and use it to update the routing databases stored on each node. The routing databases keep track of the state of the network and is similar to the distance costs found in the CAF routing algorithm, except that in the this case, each node holds only the forward routing tables and there are no backward routing tables anymore.

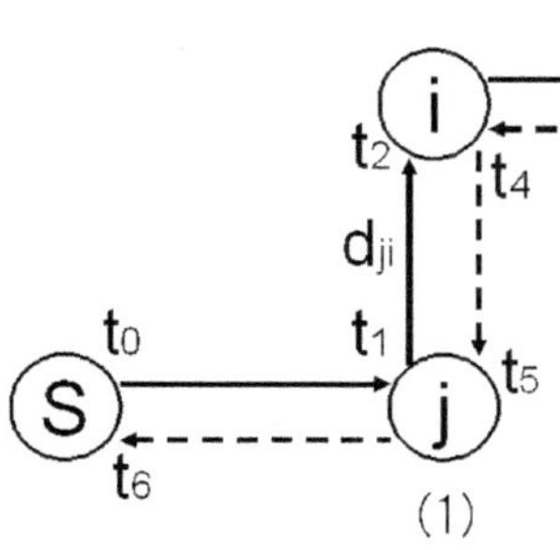

Fig. 1. Backward routing.

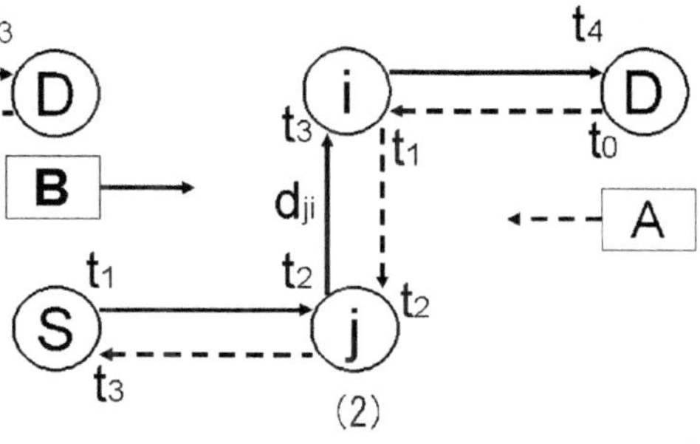

Fig. 2. Forward routing.

For example, in figure 2, an ant agent A starts from node D at moment t_0, another agent B starts from node S at t_1. When A moves to node i, it collects the information stored on that node by the agents from the opposite way and then stores the distance cost d_{ji} in its stack. When A and B move to node j at moment t_2, agent B exchanges the information with A and uses d_{ji} to update the routing table immediately. In this way, there is no round trip delay and the oscillations can be reduced.

Compared to other agent-based routing approaches, forward routing is strongly focused on ensuring good performances in terms of convergence speed and reactivity. This is achieved by assigning the exploration and update tasks to a cooperating mode. The first one, known as the agents in search mode, evaluates and tries to establish routes between the possible traffic sources and all the destinations. Agents in update mode update the routing tables according to the information that they gather from the agents on the opposite way. The two facilities are needed because a good path in one direction is not necessarily a good one in the other direction. Information exchange between the agents is an important topic that not only influences the parallel execution time but also the optimization behavior. The use and updating of the forward routing tables are more explicitly described below.

Let $F_{S \to D}$ be a routing agent emitted by a node S with a destination node D and $F_{D \to S}$ be a routing agent emitted by a node D with a destination node S. Since the network is supposed to be asymmetric, the cost encountered by $F_{S \to D}$ cannot be considered as a valuable estimation of the distance toward S for other agents from another direction. The cost used in the update is now estimated by the agent $F_{D \to S}$ which moving in the opposite direction.

To ensure a correct distance cost update in the asymmetric case however, the data packets and the routing agents have to be routed differently according to the different cost information. As a consequence, in addition to maintaining the routing table from distance costs in one direction we must now also compute the routing tables for routing agents compared with backward routing.

When the agent $F_{S\to D}$ arriving on node i, data packets from direction D→S are routed as following probabilities with costs d_{jS} and d_{ij}.

$$P^{j}_{i\to S} = [1/(d_{ij}+d_{jS})] / \Sigma(1/d_{iS}) \tag{2}$$

The routing agents from direction S→D are then routed according to the following probabilities with cost d_{Di}, where P is the routing probability information in different routing table and the denominator is the sum of the routing probabilities of all output connections (including the incoming link) to a destination.

$$P_{i\to D} = (1/d_{Di}) / \Sigma(1/d_{Di}) \tag{3}$$

Forward routing also requires a special treatment of detected loops to efficiently deal with dynamic changes in the network. In the case of a connection failure for example, the information about the open connection cannot be forwarded to the nodes on the other side, but it can at least be back propagated on the previously visited nodes.

4 Experimental Analysis

4.1 Algorithms

Suppose a network with N nodes, where S denotes a generic source node, when it generates an agent toward a destination D, Two types of agents are defined: Backward Ant $B_{S\to D}$, which will come back to S following the same path, and Forward Ant $F_{S\to D}$, which will only travel to D. Every agent transports a Stack(k) of data, where the k refers to the k-th visited node. Let N_k be set of neighboring nodes of node k and P be the probability with which the packets are routed. The following lines show the pseudo code of Backward Routing and Forward Routing used in the simulation:

Backward Routing

```
BEGIN
{
     DO (Routing Tables Set-Up)
     DO (in parallel)
     {
          At each iteration, each node S generates Bs→D to a
          randomly chosen D.
          DO (for each Bs→D)
          {
               Bs→D pushes in Stack(k) node k identifier and the
               distance associated from S to k, then will be
               routed to the next node according to P.
               DO (Avoid Loop)
          }WHILE jumping node ≠D
          When Bs→D reaches D, it will return to S following the
          backward path.
          DO (for each Bs→D)
```

```
        {
             BS→D updates the k routing table according to the
               cost carried in Stack(k,…,D), increasing prob-
               abilities associated to path used and de-
               creases other paths probabilities.
             IF k ≠S
                  BS→D will leave k and jump to a node given
                  by Stack(k-1)
             ENDIF
        }WHILE k ≠S
    }
}END
```

Forward Routing

```
BEGIN
{
    DO (Routing Tables Set-Up)
    DO (in parallel)
    {
        At each iteration, each node S generates FS→D to a
        randomly chosen D.
        DO (for each FS→D)
        {
              FS→D pushes in Stack(k) node k identifier and
              the distance associated from S to k, and then
              picks up such information stored on k by FD→S.
              FS→D updates the k routing table for data/agent
              packets separately according to the cost ex-
              changed from FD→S, increasing probabilities as-
              sociated to path used and decreases other paths
              probabilities.
              FS→D then will be routed to the next node.
              DO (Avoid Loop)
        }WHILE jumping node ≠D
    }
}END
```

4.2 Analysis

Dedicated to routing algorithm testing however, it does not need to implement a complex network environment in which some variable factors may confuse the comparisons of the algorithms. We use a topology with 9 nodes in our study. Each node has 2-4 links and all links are bi-directional and have the same transmission bandwidth according to which a given amount of waiting packets will be processed at each iteration. There is no congestion control other than the routing algorithm itself.

At each iteration, incoming packets that assumed to have the same size are routed depending on there type (routing packets or data packets), then pushed to the chosen output buffer according to the routing table. The capacities of all buffers are infinite, so packets and agents are never lost. Each outgoing link (interface) is associated with an output queue whose service rate is controlled by the transmission rate of the link.

To evaluate the result of each simulation we use the number of waiting packets as a measurement, so in our experimental system only the packet number is critical. The number of waiting packets, which is just the sum of the sizes of all the queues buffering the links, shows congestions very efficiently. In the case of infinite buffer lengths, the average packet delay is highly correlated with the overall throughput.

Figure 3 studies the algorithm responses to the bursty traffic. In the range (200, 350), the emission rate of a random node is instantly increased (from uniform 50 packets/iteration to 100 packets/iteration). The backward routing shows an oscillating response to this traffic pattern with larger amplitude. On the other hand, forward routing algorithm dynamically adapts its response to the traffic pattern with only a small transitory period.

Figure 4 shows the algorithm responses to sudden modifications in the network by a link removal of CF (at iteration 200). In such condition, routing table must be modified to achieve a new load balancing. It exhibits faster convergence and smaller waiting packets for forward routing and the advantage is distinct.

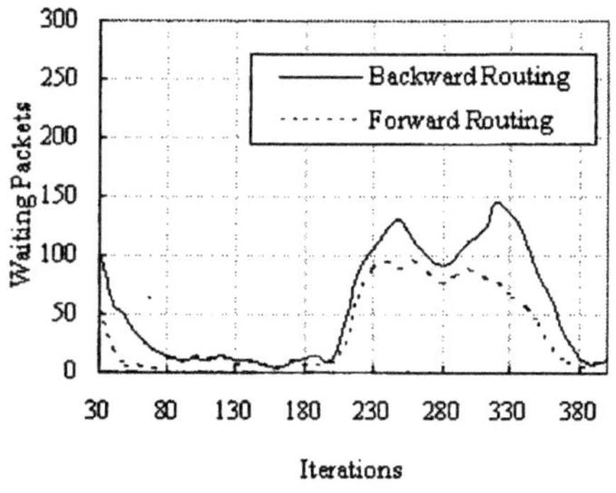

Fig. 3. Average number of waiting packets.

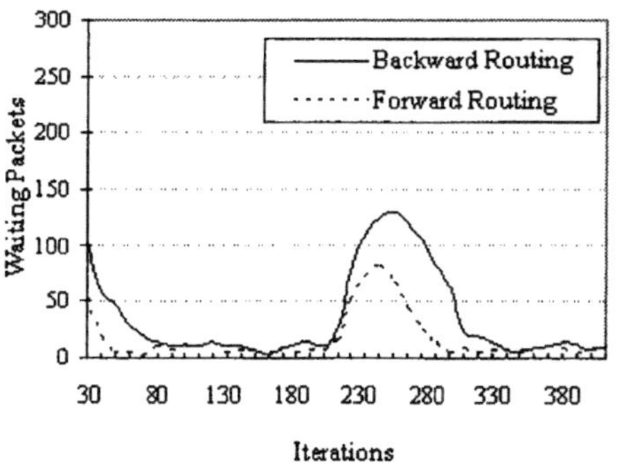

Fig. 4. Average number of waiting packets.

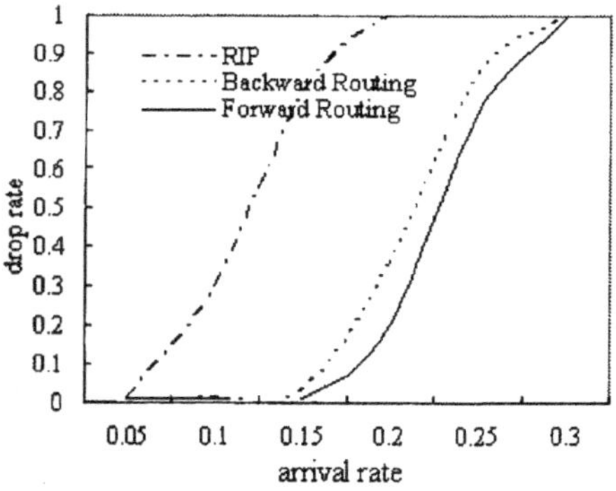

Fig. 5. Packet drop rate.

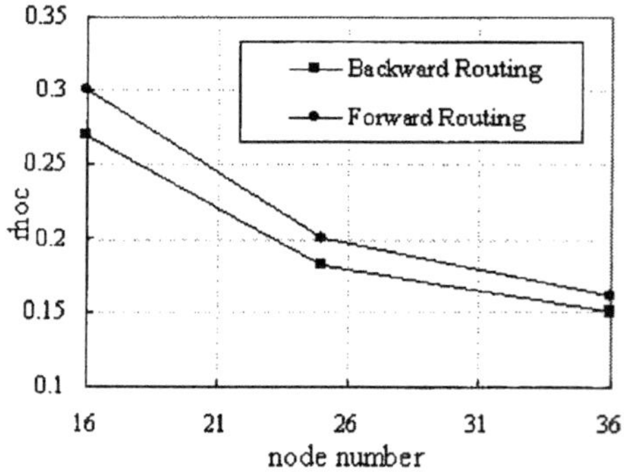

Fig. 6. Changes of network size.

4.3 Scalability

To evaluate the influence when we change the traffic load and network size, a 5×5 torus grid topology has been used. Every node is a router identical to the others and the buffer sizes were set to 100 packets. The traffic load is provided by Poissonian process. At each Poissonian event, a number of packets (10 data packets/iteration) are emitted toward a destination. There was no re-transmitting during the simulation. Figure 5 shows the change of packet drop rate when the arrival rate of traffic was changed. Packet drop rates increase rapidly when the arrival rates exceed certain threshold. These critical points are very important when we evaluate the network performance. Therefore, only the results associated with the drop rate are presented here. In this case also, forward routing performs more efficiently than the other algorithms. Figure 6 shows the influence of changes in network size (4×4, 5×5 and 6×6). The scalability of forward routing can be confirmed with this result.

5 Conclusions and Future Work

In this paper, we have presented a scheme based on agents for routing in communications networks. Our proposed algorithm based on forward ants is able to react and to deal with numerous dynamic scenarios in the communication bandwidth and the network topology. We believe that our proposed algorithm is of the highest interest for offering differentiated services, fault tolerant networks, and to deal with communication bursts in a timely manner.

The results of this paper are expected to provide alternative ways to design and implement effective and efficient routing algorithms. Particularly in view of future network situation in which logical topology layering and local traffic patterns can be dynamically constructed in real time. In our simulation, the combination of various factors was seen to play an important role in the traffic behavior and the number of agents. To reduce the number of agents in the network and the bandwidth consumed by them, more efficient techniques have to be investigated and tested in the further work although complex methods might imply increased computational complexity of the agent algorithms. A deeper investigation and trade-off analysis would be required to establish an adequate equilibrium between these important factors.

References

1. G. Theraulaz, E. Bonabeau, "A brief history of stigmergy", Artificial Life, no.5, pp.97-116, May. 1999.
2. E. Bonabeau, G. Theraulaz, "Swarm Intelligence", Oxford University Press, 2000.
3. K. Oida, M. Sekido, "An agent-based routing system for QoS guarantees", Proc. IEEE International Conf. on Systems, Man, and Cybernetics, pp.833–838, Oct. 1999.
4. G. Di Caro, M. Dorigo, "AntNet: distributed stigmergetic control for communications networks", Journal of Artifical Intelligence Research, no.9, pp.317–365, 1998.
5. T.White, B.Pagurek, "Towards multi-swarm problem solving in networks", ICMAS '98, pp 333–340, Jul. 1998.
6. D. Subramanian, P. Druschel, J. Chen, "Ants and reinforcement learning: a case study in routing in dynamic networks", Proceedings of the International Joint Conference on Artificial Intelligence, pp.832–838, Palo Alto, 1997.
7. Schoonderwoerd R., O. Holland and J. Bruten, "Ant-like Agents for Load Balancing in Telecommunications Networks". Technical Report HPL-96-35, HP Labs, Bristol, 1996.
8. M. Heusse, "Adaptive agent-driven routing and load balancing in communication network", Proc.ANTS98, Brussels, Oct.1998.

Biomorphs Implemented as a Data and Signals Cellular Automaton

André Stauffer[1] and Moshe Sipper[2]

[1] Logic Systems Laboratory, Swiss Federal Institute of Technology, Lausanne, CH-1015 Lausanne, Switzerland. andre.stauffer@epfl.ch

[2] Department of Computer Science, Ben-Gurion University, Beer-Sheva 84105, Israel. sipper@cs.bgu.ac.il

Abstract. Traditionally the cell of an automaton implements the rule table defining the state of the cell at the next time step knowing its present state and those of its neighbors. The cell deals consequently only with states. The novel cell involved here handles data and signals. It is designed as a digital system made up of a processing unit and a control unit. It allows the realization of growing structures like Biomorphs. The hardware implementation of these Biomorphs takes place in our electronic wall for bio-inspired applications, the *BioWall*.

1 Introduction: Cellular Automata

Cellular automata were originally conceived by S. M. Ulam and J. von Neumann in the 1940s to provide a formal framework for investigating the behavior of complex, extended systems. They are part of the early history of self-replicating machines and of von Neumann's thinking on the matter [1],[2]. Nowadays, they still remain the framework of less complex replicating structures: the self-replicating loops.

One of the central models used to study self-replication is that of cellular automata. These automata are dynamical systems in which space and time are discrete. A cellular automaton (CA) consists of an array of cells, each of which can be in one of a finite number of possible states, updated synchronously in discrete time steps, according to a local interaction rule. The state of a cell at the next time step is determined by the current states of a surrounding neighborhood of cells. This transition is usually specified in the form of a rule table, delineating the cell's next state for each possible neighborhood configuration. The cellular array (grid) is n-dimensional, where $n = 1, 2, 3$ is used in practice [3], [4].

Fig. 1a shows the basic automaton cell AC defined in a two-dimensional, nine-neighbor cellular space. This cell receives the states NSI, $NESI$, ESI, $SESI$, SSI, $SWSI$, WSI, and $NWSI$ from the cells to the south, southwest, west, northwest, north, northeast, east, and southeast respectively. The cell also shares its own state SO directly or indirectly with its eight neighbors. Such a cell deals consequently only with input and output states. Its implementation is a sequential machine resulting from the interconnection of the rule table TAB

W. Banzhaf et al. (Eds.): ECAL 2003, LNAI 2801, pp. 724–732, 2003.

and the state register REG (Fig. 1b). In order to simplify the design of the cell, the register REG is sometimes functionally sliced into multiple state variable groups called fields. The utilization of such fields also makes the resulting rules of the table TAB much more readable.

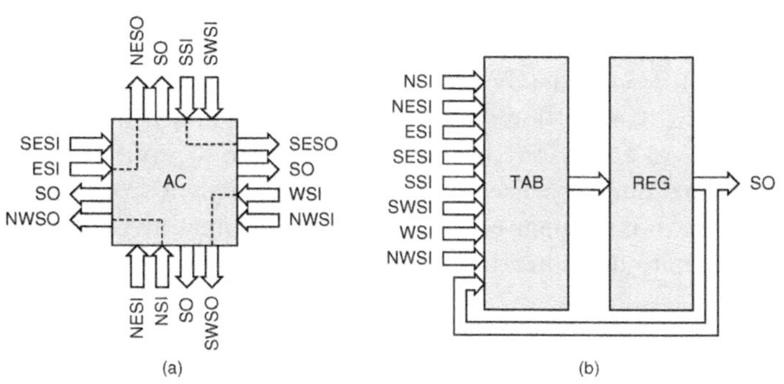

Fig. 1. Two-dimension, nine-neighbor CA. (a) Basic cell. (b) Rule table TAB and state register REG.

This paper involves a novel cellular automaton processing both data and signals. Such an automaton, whose basic cell corresponds to a digital system, is well adapted to the realization of growing structures. Section 3 introduces such growing structures: the Biomorphs. The design of the corresponding automaton cell and its hardware implementation are discussed in Section 4 and Section 5 respectively. Finally, we present concluding remarks in Section 6.

2 Data and Signals Cellular Automaton

The basic cell of our novel automaton, the data and signals cellular automaton (DSCA), works with northward (N), northeastward (NE), eastward (E), southeastward (SE), southward (S), southwestward (SW), westward (W), and northwestward (NW) directed data (D) and signals (S) (Fig. 2a). The cell computes their digital outputs O from their digital inputs I. In opposition to the state shared by the traditional cell (Fig. 1), these data and signals outputs are not necessarily identical for all the neighboring cells.

Each cell of the automaton is designed as a digital system resulting from the interconnection of a data processing unit PU and a control unit CU (Fig. 2b). In this digital system, the processing unit handles the data. It is made up of input selectors SEL, data registers REG, and output buffers BUF (Fig. 3a). The control unit of the digital system computes the signals. It combines input encoders ENC, control registers REG, and output generators GEN (Fig. 3b).

3 Biomorphs

The word *Biomorph* was devised by surrealist artist D. Morris to describe animal-like shapes in his work [5]. In the Artificial Life tradition, R. Dawkins designed a computer program to explore how complex patterns can arise from simple rules [6]. His main goal was to abstract and reduce to a minimum the amount of hand-design in order to build the Biomorphs. Simple growing rules (embryology) and guided evolution would ideally produce biologically interesting results.

In opposition to the Biomorphs defined in the literature, which implement also mutation and evolution through a combination of automatic and human-guided selection, our creatures only express a sequence of genes in order to increase. They are implemented as growing structures involving the component and data informations shown in Fig. 4.

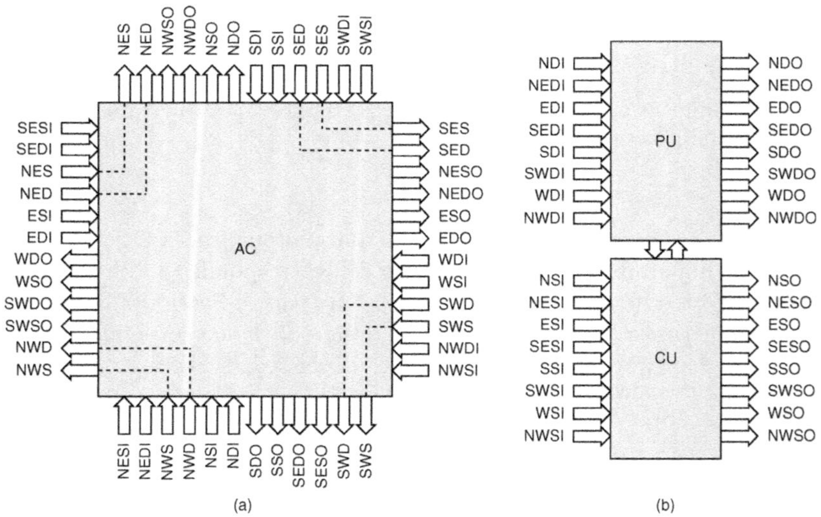

Fig. 2. Two-dimension, nine-neighbor DSCA. (a) Basic cell. (b) Processing unit PU and control unit CU.

Without any external solicitation, the central component of the initial Biomorph (Fig. 5a) presents only empty data and its structure remains unchanged (time step 0). When a physical input is provided, a gene G appears first (time step 1), propagates then along the members (time step 2), and finally expresses itself in the apex components (time step 3). Depending on the number L of apex components of its initial structure (Fig. 5b), the Biomorph will develop a creature having four, six, or eight legs.

When activated the central component produces a gene G whose value is comprised between 0 and 7. Fig. 6a shows the expression of these eight genes

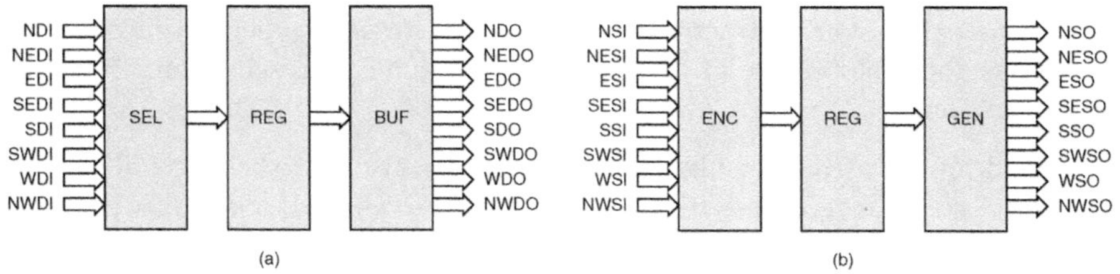

Fig. 3. Detailed architecture of the DSCA cell. (a) Processing unit. (b) Control unit.

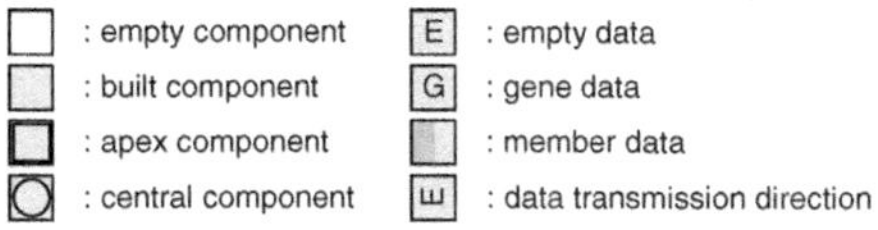

Fig. 4. Component and data informations of the Biomorphs.

for each of the eight members of the Biomorph. Starting with four-legged, six-legged, and eight-legged initial Biomorphs and applying these expressions to a sequence of four genes $G = 1, 1, 2, 2$ leads to the growth developments of Fig. 6b.

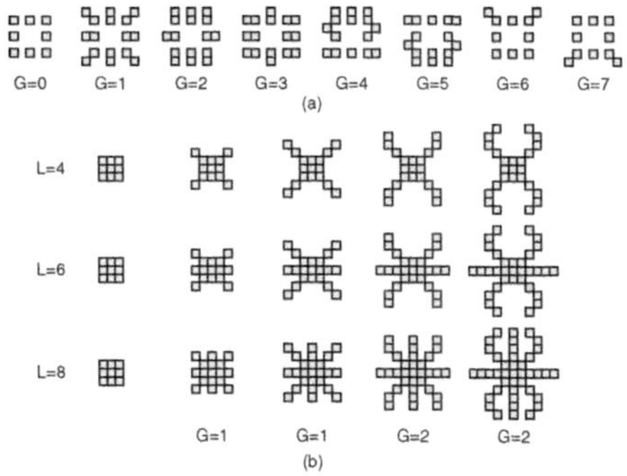

Fig. 5. Biomorph. (a) Three-time-step growth cycle. (b) Four-legged, six-legged, and eight-legged creature.

4 Cell Design

The cell is a digital system whose processing unit assures the propagation of the data and whose control unit produces the signals implied in the growth of the Biomorph. In addition to the data propagation, the digital system must perform the gene expression represented in Fig. 7. The resources needed in order to do it define the architecture of the cell (Fig. 8a). The processing unit involves the following ones:

- A 3-bit data register G2:0 for the memorization of the gene (000=empty, 001=straight growth, 010=curved in, 011=curved out, 100=curved up, 101=curved down, 110=top growth, 111=bottom growth).
- A 3-bit data register M2:0 for the memorization of the member (000=north, 001=northeast, 010=east, 011=southeast, 100=south, 101=southwest, 110=west, 111=northwest).
- A 8-input multiplexer DIMUX for the selection of one of the eight input data $NDI2:0$, $NEDI2:0$, $EDI2:0$, $SEDI2:0$, $SDI2:0$, $SWDI2:0$, $WDI2:0$ or $NWDI2:0$.
- A 2-input multiplexer DOMUX for the selection of the gene or the member as output data $DO2:0$.

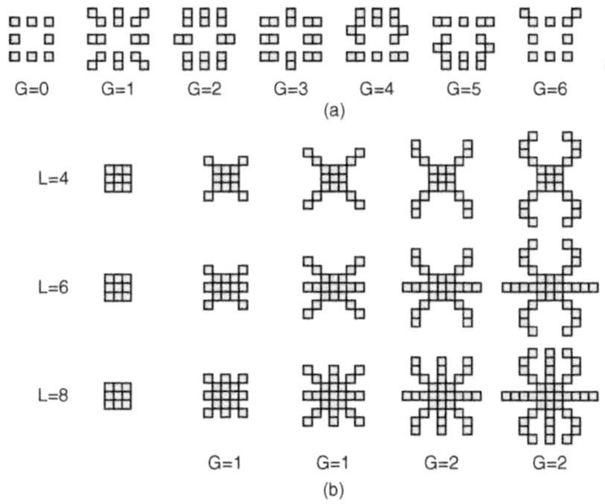

Fig. 6. Genes 0 to 7. (a) Expression for all the members of the Biomorph. (b) Developments resulting from the sequence $G = 1, 1, 2, 2$.

The control unit consists of six resources:

- A 3-bit transmission register T2:0 for the memorization of the input selection (000=northward, 001=northeastward, 010=eastward, 011=southeastward, 100=southward, 101=southwestward, 110=westward, 111=northwestward).
- A 1-bit control register B to indicate whether the component is empty ($B = 0$) or built ($B = 1$).
- A 1-bit apex register A to point out each extremity of the growing structure ($A = 1$) and perform the selection of its gene ($A = 0$) or its member ($A = 1$) as output data.

– A 1-bit control register C to define the central component of the Biomorph ($C = 1$).
– An input signals SI encoder ENC.
– An output signals SO generator GEN.

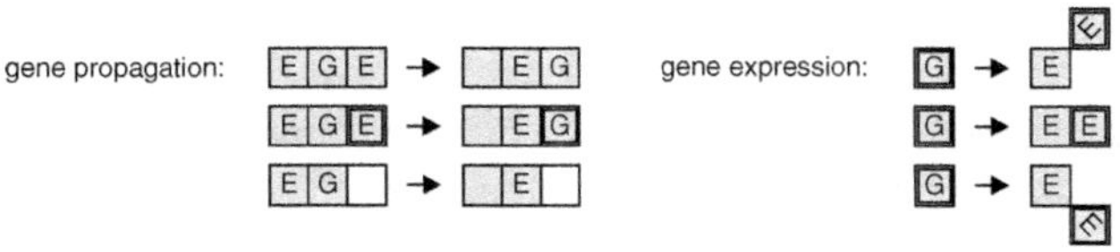

Fig. 7. Propagation and expression operations involved in the growth process.

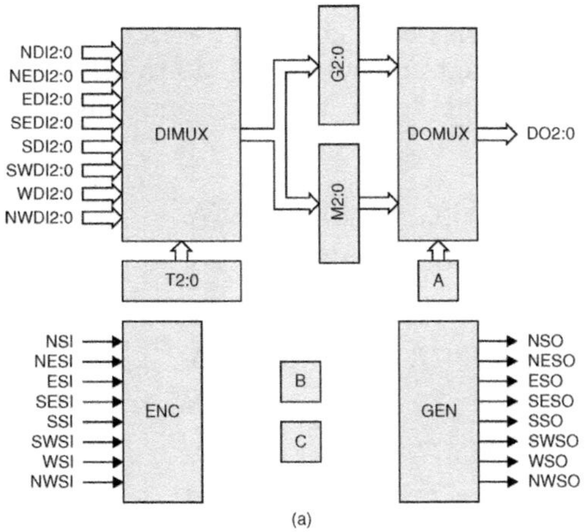

operation	description	B	A	VSI	VSO
idle cell	G <= 000	0	-	0	-
member growth	B <= 1 A <= 1 G <= 000 M <= BDI T <= TCI	0	-	1	-
gene propagation	G <= BDI	1	0	-	-
idle apex	G <= BDI	1	1	-	0
gene expression	A <= 0 G <= BDI	1	1	-	1

(b)

Fig. 8. Biomorph cell. (a) Detailed architecture. (b) Operation table.

The control variables B and A as well as the valid signal input VSI and the valid signal output VSO define the operations performed by the data and control registers (Fig. 8b). While the variables B and A proceed directly from the corresponding control registers, the input encoder ENC computes VSI and VSO according to the following equations:

$$VSI = NSI + NESI + ESI + SESI + SSI + SWSI + WSI + NWSI \quad (1)$$

$$VSO = NSO + NESO + ESO + SESO + SSO + SWSO + WSO + NWSO \quad (2)$$

In the operation table (Fig. 8b), the biomorph data information $BDI2:0$ results from the selection operated by the input multiplexer DIMUX according to its control variables $SEL2:0$:

$$SEL2 = B'.TCI2 + B.T2 \quad (3)$$

$$SEL1 = B'.TCI1 + B.T1 \quad (4)$$

$$SEL0 = B'.TCI0 + B.T0 \quad (5)$$

The transmission control information $TCI2:0$ involved in these relations realizes a priority encoding of the input signals SI:

$$TCI2 = NSI'.NESI'.ESI'.SESI'.(SSI + SWSI + WSI + NWSI) \quad (6)$$

$$\begin{aligned} TCI1 = &\ NSI'.NESI'.(ESI + SESI) \\ &+ NSI'.NESI'.ESI'.SESI'.(WSI + NWSI) \end{aligned} \quad (7)$$

$$TCI0 = NSI'.NESI \quad (8)$$

The generator GEN implements the output signals SO involved in the gene propagation and gene expression operations of the growth process (Fig. 7). The computation of these signals is based on eight memories MEM0 to MEM7, each of them defining a given signal for all the genes, members and transmission directions of the Biomorph cell:

$$NSO = C'.B.A.MEM0(G, M, T) + C.EIN \quad (9)$$

$$NESO = C'.B.A.MEM1(G, M, T) + C.EIN \quad (10)$$

$$ESO = C'.B.A.MEM2(G, M, T) + C.EIN \quad (11)$$

$$SESO = C'.B.A.MEM3(G, M, T) + C.EIN \quad (12)$$

$$SSO = C'.B.A.MEM4(G, M, T) + C.EIN \quad (13)$$

$$SWSO = C'.B.A.MEM5(G, M, T) + C.EIN \quad (14)$$

$$WSO = C'.B.A.MEM6(G, M, T) + C.EIN \quad (15)$$

$$NWSO = C'.B.A.MEM7(G, M, T) + C.EIN \quad (16)$$

According to these equations, the central component ($C = 1$) delivers an output signal to all the neighboring cells whenever the external input EIN is activated. Simultaneously, the central component produces also a random gene on its output data $DO2:0$.

5 Hardware Implementation

The hardware implementation of the nine-neighbor DSCA takes place in our two-dimensional electronic wall for bio-inspired applications, the *BioWall* (Fig. 9) [7]. In the implementation of the Biomorphs, each cell of the automaton corresponds to a unit in the wall. This unit is the combination of three elements: (1) an input device, (2) a digital circuit, and (3) an output display.

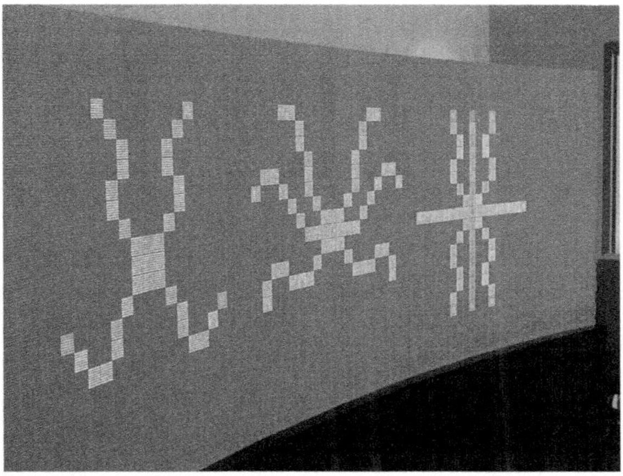

Fig. 9. The BioWall used to physically implement the Biomorphs (Photograph by A. Badertscher).

Fig. 10. Touching the central cell to physically activate the growth of the Biomorph (Photograph by A. Badertscher).

The unit's outer surface consists of a touch-sensitive panel which acts like a digital switch. This switch enables the user to click on the central cell of the Biomorph and generate a gene in order to activate the growth of its members (Fig. 10).

The unit's internal digital circuit is a field-programmable gate array (FPGA), configured so as to implement: (1) external (touch) input, (2) execution of the propagation and expression operations involved in the growth process, and (3) control of the output display. This latter is a two color light-emitting diode

(LED) display, made up of 128 diodes arranged as an 8×8 dot-matrix, each dot containing a green and a red LED. The display allows the user to view the cell's current data and to choose the gene generated by the central component.

6 Concluding Remarks

We presented a novel cellular automaton dealing with data and signals instead of states. Even though this automaton is specially well suited for the realization of growing structures, it constitutes a general model for all kind of cellular applications. It allows and simplifies in fact the design of all the cells where a great number of states leads to an explosion of the rule table.

The Biomorphs are growing structures whose complex behavior leads to the design of a data and signals automaton instead of a state based one. The data processing unit and the control unit of its basic cell introduces the characteristic resources and registers of all digital sytems. The hardware implementation of the Biomorphs takes place in the BioWall, our electronic wall for bio-inspired applications.

Acknowledgments. This work was supported in part by the Swiss National Science Foundation under grant 20-100049.1, by the Leenaards Foundation, Lausanne, Switzerland, and by the Villa Reuge, Ste-Croix, Switzerland.

References

1. J. von Neumann. Theory of Self-Reproducing Automata. University of Illinois Press, Illinois, 1966. Edited and completed by A. W. Burks.
2. M. Sipper. Machine Nature: The Coming Age of Bio-Inspired Computing. McGraw-Hill, New York, 2002.
3. M. Sipper. Evolution of Parallel Cellular Machines: The Cellular Programming Approach. Springer-Verlag, Heidelberg, 1997.
4. T. Toffoli and N. Margolus. Cellular Automata Machines. MIT Press, Cambridge MA, 1987.
5. D. Morris. The Secret Surrealist: The Paintings of Desmond Morris. Phaidon Press, Oxford, 1987.
6. R. Dawkins. The evolution of evolvability. In C. Langton (Ed), Artificial Life, 201–220, Addison-Wesley, 1988.
7. G. Tempesti, D. Mange, A. Stauffer, and C. Teuscher. The BioWall: An elctronic tissue for prototyping bio-inspired systems. In A. Stoica, J. Lohn, R. Katz, D. Keymeulen, and R. S. Zebulum (Eds.), Proceedings of the 2002 NASA/DOD Workshop Conference on Evolvable Hardware, pp.221–230, IEEE Computer Society Press, Los Alamitos CA, 2002.

Analyzing the Performance of "Winner-Take-All" and "Voting-Based" Action Selection Policies within the Two-Resource Problem

Orlando Avila-García, Lola Cañamero, and René te Boekhorst

Adaptive Systems Research Group
Department of Computer Science, University of Hertfordshire
College Lane, Hatfield, Herts AL10 9AB, UK
{O.Avila-Garcia, L.Canamero, R.TeBoekhorst}@herts.ac.uk

Abstract. The problem of action selection for an autonomous creature implies resolving conflicts between competing behavioral alternatives. These conflicts can be resolved either via competition, following a "winner-take-all" approach, or via cooperation in a "voting-based" approach. In this paper we present two robotic architectures implementing these approaches, and report on experiments we have performed to compare their underlying optimization policies. We have framed this study within the context of the "two-resource problem," as it provides a widely used standard that favors systematic experimentation, analysis, and comparison of results.

1 Introduction

The problem of behavior or action selection (AS) for an autonomous creature, animal or robot, consists in making a decision as to what behavior to execute next in order to satisfy internal needs and guarantee survival in a given environment and situation. This implies resolving conflicts between competing behavioral alternatives. Different behavior selection models have been proposed in the (robotics and ethological) literature (see [13] for a good overview), although not all of them address the same problems. Models can be classified according to various dimensions, such as their approach to behaviors (emergent functionality versus ethologically-inspired behavioral competencies), the use of external and internal factors to drive the selection (reactive versus motivated architectures), or the organization of the different architectural elements (in a "flat," parallel network versus a hierarchy of behavioral subsystems). Another dimension often referred to when comparing architectures is their arbitration policy: Conflicts among different elements in the architecture can be resolved either via competition, following a "winner-take-all" (WTA) approach, or via cooperation in a "voting-based" (VB) approach. Competition and cooperation (and therefore WTA or VB policies) can also occur at several levels[1], giving rise to different behavior selection results.

Our analysis focuses on motivated behavior-based architectures with a combination of flat and hierarchical organization, following [4]. Within this common framework, we

[1] Namely between motivational states, between behavioral subsystems, between simpler behaviors or actions, and between motor commands.

W. Banzhaf et al. (Eds.): ECAL 2003, LNAI 2801, pp. 733–742, 2003.

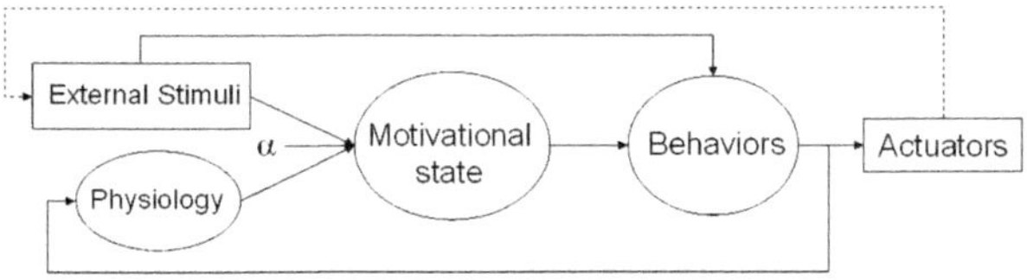

Fig. 1. Motivated behavior-based model underlying our architectures.

have defined two architectures with different arbitration policies—WTA and VB. The work presented in this paper is part of our ongoing effort to investigate the adequacy of these AS policies for different problems and environments. In previous work [5,2] we built and compared a WTA and various VB architectures in static and dynamic environments using different performance indicators drawn from the notion of viability [1]. Results obtained regarding the performance of WTA and VB architectures were complementary, and pointed to the possibility of different underlying optimization policies. To investigate this issue, we decided to place our study within the so-called "two-resource problem", in which a self-sufficient creature must continuously choose between consuming one of two resources available in the environment. This framework presents the advantages of providing a widely accepted standard that allows to compare results with those of other researchers, and of being simple enough to permit systematic experimentation and analysis. The remainder of the paper is as follows. Section 2 describes our architectures, which are framed within the two-resource problem in Section 3. Section 4 reports the robotic experiments we have performed to compare the optimization policies of our architectures. Finally, Section 5 draws some conclusions.

2 Behavior Selection Architectures

Our architectures are neither strictly flat (parallel) nor hierarchical (structured), but a combination of both. They consist of two layers—motivational and behavioral—linked through a synthetic physiology, leading to a two-step computation of intensity. This computation is parallel within each layer, but motivational intensity must be computed prior to the calculation of behavioral intensity, since the latter depends on the former. Both architectures (WTA and VB) have the same organization and elements, but they vary in the way in which these are combined (their arbitration mechanisms).

2.1 Architectural Elements

The common architectural model (Figure 1) has three main elements: a synthetic physiology, motivational states, and behaviors.

The *physiology* consists of a number of survival-related, controlled homeostatic variables—abstractions representing the level of internal resources that the agent needs in order to survive. They must be kept within a range of values for the robot to stay alive, thus defining a physiological space [8] or viability zone [1] within which survival (continued existence) is guaranteed, whereas transgression of these boundaries leads to death. These variables vary as a function of internal body dynamics and of the interactions of the robot with its environment.

Motivational states are abstractions representing tendencies to behave in particular ways as a consequence of internal and external factors. The main elements defining them are bodily needs (traditionally known as "drives") and external (incentive) stimuli. Survival-related *bodily needs* set urges to action to maintain the controlled physiological variables within a certain range. A feedback detector generates an error signal—the drive—when the value of this variable departs from its ideal value (set point), and this triggers inhibitory and excitatory controlling elements (in this case, the execution of behaviors) to adjust the variable in the adequate direction. Each motivation receiving an error signal from its feedback detector receives an intensity (activation level) proportional to the magnitude of the error. The motivational state of the robot is also influenced by the presence of *external stimuli* or environmental cues that allow to execute (consummatory) behaviors and hence to satisfy bodily needs. In addition to bodily needs and the presence of environmental cues, *other factors* can influence motivational states, such as the quality of the stimulus (e.g., palatability of food), abnormal bodily states (e.g., "illness"), etc. To account for these factors we have introduced a parameter (α), as we will see in Section 2.2. As a consequence of these combined influences, several motivations can be active at the same time (and either "competing" or "cooperating" to be satisfied), with varying degrees of intensity.

Our *behaviors* are coarse-grained subsystems (embedding simpler actions) that implement different competencies, similarly to those proposed in [7,4]. Following the classical distinction in ethology, motivated behaviors can be consummatory (goal-achieving and needing the presence of an incentive stimulus to be executed) or appetitive (goal-directed search for a particular external stimulus). In addition to modifying the external environment, the execution of a behavior has an impact on (increases or decreases) the level of specific physiological variables. Behaviors can be activated with different intensities that depend on the intensities of the motivations related to them. However, only one behavior can be executed at a time, following the assumption of the behavioral final common path [10].

2.2 Arbitration Mechanisms

This is the point in which our architectures differ. With a WTA policy the robot will execute the behavior that best satisfies its most urgent motivation, i.e., only this motivation drives behavior selection. With a VB policy, all the motivations influence the final selection since the robot will execute the behavior that best satisfies a subset. Their respective behavior selection loops are:

Behavior selection loop in WTA. Repeat forever:

1. The winner motivation j_{winner} is calculated:
   a) For each motivation j:
      i. Compute the intensity of the motivation's drive as proportional to the error[2] of its controlled variable v ($e_{v_j} \in [0, 1]$).

[2] The error is calculated as follows in both behavior selection loops:
A. Calculate the distance between the set point and the limit: $ld_v = abs(l_v - p_v)$.

   ii. Compute the effect of the presence of external stimuli k influencing the intensity of the motivation j (i.e., of incentive stimuli): $s_{k_j} \in [0, 1]$.
   iii. $m_j = e_{v_j} + (e_{v_j} \times \alpha s_{k_j})$ is the final intensity of j, where $\alpha \in [0, 1]$ is a weight controlling the influence of the incentive stimulus.
   b) The motivation with highest intensity (j_{winner}) is selected.
2. The intensity of each behavior linked (through the physiology) with the winner motivation is computed as $b_i = m_{j_{winner}} \times f_{iv}$, where $b_i, m_{j_{winner}}$ are the intensities of behavior i and the winner motivation, respectively, and f_{iv} is the effect that the execution of behavior i has on v, which is the physiological variable controlled by j_{winner}.
3. The behavior with highest intensity is selected to be executed.

Behavior selection loop in VB. Repeat forever:

1. Calculate the intensity of each motivation j:
   a) Compute the intensity of the motivation's drive as proportional to the error of its controlled variable v ($e_{v_j} \in [0, 1]$).
   b) Compute the effect of the presence of external stimuli k influencing the intensity of the motivation j (i.e., of incentive stimuli): $s_{k_j} \in [0, 1]$.
   c) $m_j = e_{v_j} + (e_{v_j} \times \alpha s_{k_j})$ is the final intensity of j, where $\alpha \in [0, 1]$ is a weight controlling the influence of the incentive stimulus.
2. The intensity of each behavior is computed as $b_i = \sum(m_j \times f_{iv})$, where b_i, m_j are the intensities of behavior i and motivation j, respectively, and f_{iv} is the effect that the execution of behavior i has on v, the physiological variable controlled by j.
3. The behavior with highest intensity is selected to be executed.

The way in which motivational and behavioral intensity are computed in the above algorithms deserves particular attention. When computing motivational intensity, the main problem is to define an appropriate combination rule between external (s_{k_j}) and internal (e_{v_j}) influences. This problem has been discussed extensively by ethologists [9]. Using an artificial creature, Spier and McFarland demonstrated how a simple multiplicative rule that they call *Cue* × *Deficit* model ($m_j = e_{v_j} \times s_{k_j}$) can show opportunism and persistence [11], and can generate optimal behavior equivalent to the quadratic utility function model [12]. One of the problems of this simple combination rule is the lack of motivational arousal (and hence, of behavioral tendency) when external incentive stimuli are absent ($s_{k_j} = 0$). This problem would be fatal in models in which the motivational state leads to the selection of the behavior to be executed and that also take into account appetitive behaviors (executed in the absence of external stimuli), as in our case. Therefore, following the formula proposed by [13] we have extended the previous model to *Deficit* + *Cue* × *Deficit*: $m_j = e_{v_j} + (e_{v_j} \times \alpha s_{k_j})$. Note that now there is motivational tendency to execute appetitive behaviors when there is no external stimulus, while we still get the benefits of the *Cue* × *Deficit* model. Another important enhancement of our

B. Calculate the distance between the actual variable value and the set point: $vd_v = abs(v_v - p_v)$
C. Calculate the normalized error:

$$e_v = \begin{cases} vd_v / ld_v & \text{if } ld_v > abs(l_v - v_v) \\ 0 & \text{otherwise} \end{cases}$$

model is the introduction of a weight (α) to control the influence of external stimuli. This weight can be taken to stand for other factors influencing motivational states[3], as explained in Section 2.1 and can have a big impact on the way optimization is achieved, as we will see in Section 4.3.

Computation of behavioral intensity is the point in which both arbitration mechanisms differ. Note that, when calculating behavioral intensity in the selection loop of WTA (step 2), only one motivation drives behavior selection; however, in VB (step 2), all the motivations influence (positively or negatively) the selection of a behavior.

3 The Two-Resource Problem

The framework we have used to compare our WTA and VB architectures and analyze their optimization policies is known as the two-resource problem. In this scenario, a self-sufficient (biological or artificial) creature must continuously decide which of its two survival-related needs to satisfy by choosing between two resources available in the environment. This framework presents the advantage of providing a standard that allows to compare results with those of other researchers, since it has been widely used to study action selection both in animals (e.g., to study feeding and drinking patterns in doves [8]) and in artificial creatures (see e.g., [3,12,6]). Also, its simplicity (although not devoid of problems) favors a systematic analysis of results, and it has actually been characterized as the minimal scenario to test action selection mechanisms [12].

Our particular implementation of the two-resource problem, detailed in Table 1, has been as follows. Like any implementation of this problem, we have used two environmental resources (external stimuli) and two physiological needs giving rise to two motivational states. However, departing from other implementations in which each motivational state can only be satisfied by one behavior, in our case two consummatory behaviors can satisfy each motivation[4] to varying degrees. The reason for this is that, with only one behavior to satisfy each motivation, a VB architecture would become a WTA one, and therefore we would not be able to compare these policies.

Consummatory behaviors can only be executed when the intensity of their related external stimulus is greater than a threshold (0.85); the robot then stops on top of the resource and "consumes" it, modifying its physiology accordingly. If a consummatory behavior cannot be executed, Search is triggered to reach the center of the resource (where >85% of stimulus is present) using both light sensors for orientation over the light gradient. Avoid is a reflex behavior that uses both bumpers to avoid the wall.

3.1 Analysis Criteria

Spier and McFarland [11] noted that, within the framework of the two-resource problem, a self-sufficient robot must perform a *basic cycle* of activities to maintain viability. The study of the physiological space should give us these cycles. Figure 2 (left) shows the basic cycle for our implementation. During phase A, our robot executes the Search

[3] In addition, this weight could be useful to introduce lifetime learning or even evolution to adapt the influence of different external stimuli to varying ecological circumstances [9].

[4] And adjust its controlled variable, also affecting the other physiological variable.

Table 1. Motivations (left) and behaviors (right) used. Controlled (physiological) variables have set points fixed at 100 and lethal boundaries at 0 ($Var_i \in [0, 100]$).

Motivation	Physiological Drive	Ext. Stim.
Cold	$\uparrow Temperature$	Heat
Fatigue	$\uparrow Energy$	Food

Behavior	Type	Stimulus	Effects on physiology
Avoid	*Reflex.*	Obstacle	$\downarrow 0.2 Temperature, \downarrow 0.2 Energy$
WarmUp1	*Consum.*	Heat	$\uparrow 0.9 Temperature, \downarrow 0.1 Energy$
WarmUp2	*Consum.*	Heat	$\uparrow 1.0 Temperature, \downarrow 0.3 Energy$
Feed1	*Consum.*	Food	$\downarrow 0.1 Temperature, \uparrow 0.9 Energy$
Feed2	*Consum.*	Food	$\downarrow 0.3 Temperature, \uparrow 1.0 Energy$
Search	*Appet.*	None	$\downarrow 0.2 Temperature, \downarrow 0.2 Energy$

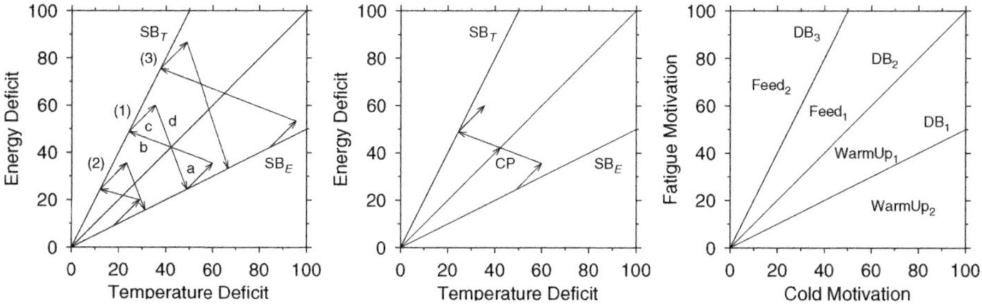

Fig. 2. Left: basic cycles in the physiological space; (1) is a stable cycle, (2) a cycle that climbs up the diagonal, (3) a cycle that descends it. Center: Optimization error as the magnitude of the vector OpE. Right: Motivational space with dominance boundaries for behavioral competition.

behavior to find the resource Heat. The phase labeled *B* corresponds to the execution of consummatory behaviors that increase Temperature (WarmUp1, WarmUp2), also decreasing Energy to a lesser extent. When this need is satisfied, another episode of Search is executed (label *C*), this time to obtain Food. This is followed by the execution of consummatory behaviors that help to correct the Energy deficit (Feed1, Feed2, that also decrement Temperature slightly) in phase *D*. Upon satiation, the same basic cycle is repeated again. Note that in our implementation the cycle is symmetric with respect to the diagonal of the physiological space because both aspects of our problem are symmetric in all respects (physiology, resources, motivations and behaviors). Figure 2 (left) shows how the shape of the basic cycle depends on its location on the diagonal with respect to the origin, and that this cycle becomes stable (i.e., its initial and final points coincide) only at a particular location.

Satiation boundaries. The point at which the satisfaction of a particular need is achieved can be termed *satiation boundary* (Figure 2, left). This limit defines the amplitude of the cycle, and can be obtained analytically. Since our two problems are symmetric, we can assert that the point at which a motivation that is being satisfied (m_{cold}) loses the competition against the other ($m_{fatigue}$) is defined by $m_{cold} = m_{fatigue}$. At this moment, the robot is consuming a resource leading to the satisfaction of m_{cold}, and hence the intensity of this external stimulus must be near the maximum ($s_{heat} \simeq 1$). In this situation, the incentive stimulus of the second motivation ($m_{fatigue}$) is disregarded

($s_{food} \simeq 0$). Therefore, using the *Deficit* + *Cue* × *Deficit* formula we have adopted to calculate motivational intensity, the final equation will be:

$$e_{energy} + (e_{energy} \times \alpha s_{food}) = e_{temper} + (e_{temper} \times \alpha s_{heat}), \quad (1)$$
$$e_{energy} \simeq (1+\alpha) e_{temper} \quad (2)$$

The line determined by this equation is the satiation boundary of m_{cold} (Figure 2, left, SB_T). Note that the parameter α, determining the influence of the external stimulus, determines also the slope of that boundary, and therefore the amplitude and location of the stable cycle; this implies that α determines the distance of the stable cycle with respect to the origin of the physiological space. This parameter thus seems to play an important role in determining how much the cycles approach the origin or ideal physiological state (Figure 2, left), and hence, how the evolution of the cycles is optimized.

Optimization criteria. Since the cycles are going to shift along the axis determined by the diagonal of the physiological space, our optimization criterion is going to be the Euclidian distance from the origin to the point where the cycle crosses the diagonal (Figure 2, center). Note that the shorter the distance, the closer the cycle gets to the origin. We have used the mean and minimum optimization error (normalized between [0,1]) obtained during the robot lifetime as performance indicators.

Dominance boundaries. Motivational states define a motivational space (Figure 2, right). Decisions made to select behaviors under different motivational states are reflected in this space, and this defines dominance boundaries [8] that divide the motivational space into regions in which a particular behavior is "dominant," i.e., is being executed. For WTA, only one dominance boundary is drawn (Dominance Boundary 2 in Figure 2, right) since in fact this policy only uses two of the four behaviors available—those that satisfy the motivations to a greater extent, Feed2 and WarmUp2. On the contrary, for VB behavioral competition defines a more complex set of dominance boundaries—three in our case, dividing the dominance regions of our four behaviors. This difference observed between WTA and VB is reflected in the different consummatory phases of basic cycles, as we will see in Section 4.3.

4 Experiments

4.1 The Robot and Its Environment

Tauron, our robot, is built using Lego Mindstorms. It has a simple "roverbot-like" design to facilitate replication of our experiments, and it is equipped with (Figure 3, left and center): two bumpers for obstacle avoidance, two frontal light sensors to detect brightness and darkness, and two motors providing differential steering to the wheels. The environment (Figure 3, right) is a $1m \times 1m$ arena surrounded by a wall. The floor is made of 9 tiles[5] (of $33cm \times 33cm$ each) of three types: "empty space" (uniform gray), two "heat sources" (white gradient), and two "food sources" (black gradient).

[5] The use of tiles does not mean that our environment is a "grid world" since our tiles have continuous physics.

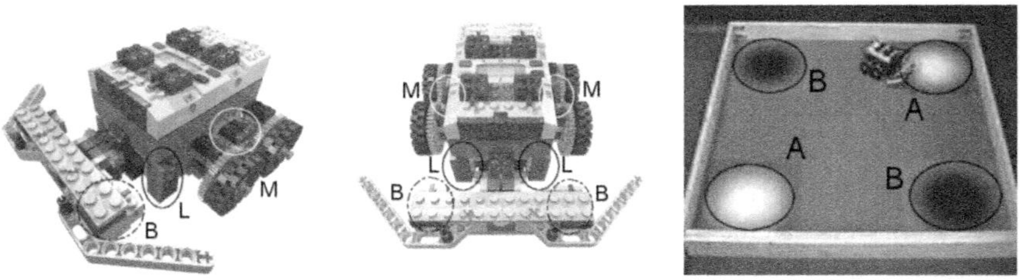

Fig. 3. Left and center: Tauron robot. Circles show the position of sensors (L: light, B: bumpers) and motors (M). Right: Environment. Circles surround resources: heat (A) and food (B).

4.2 Method

Five sets of experiments were performed to test both architectures with five different α (1, 0.7, 0.5, 0.3, 0). Each set consisted of 20 runs, 10 for each architecture, with a total of 100 runs (about 12 hours). Each run lasted 1600 steps of $260ms$ each, i.e., about 7 minutes. For each run we initialized randomly physiological variables (to values within their viability range) and location of resources in the environment.

4.3 Results

Figure 4 shows the average performance of both architectures for different α in terms of mean and minimum optimization errors. Life span is also shown as a reference indicator. We have used analysis of variance (ANOVA) to describe the results. The difference between architectures in terms of *life span* was not significant; however, there is a tendency indicating that VB performed better than WTA. For $\alpha = 0$, the life span of both architectures decreases considerably due to the lack of persistence in the execution of consummatory behaviors and therefore the dithering produced. In terms of *mean optimization error*, the difference in the performance of each architecture for different values of α is statistically highly significant (99%). The difference between architectures is also highly significant, VB outperforming WTA. In terms of *minimum optimization error*, the difference in the results obtained by VB for different values of α is statistically highly significant (99%), whereas it is not significant for WTA[6]. The difference between architectures is significant (95%), VB outperforming WTA.

Our results thus show that α strongly influences optimization—the smaller the values used for α, the bigger the optimization error, as predicted by our model (see Section 3.1). The position of the stable cycle varies between WTA and VB architectures—it gets closer to the origin in VB (Figure 5), reflecting the fact that VB outperforms WTA in terms of our two optimization indicators, as explained above. This can be illustrated taking into account the way in which both architectures consume resources as reflected in the shape of their respective cycles depicted in Figure 5. The fact that WTA only executes

[6] It should be noted the high level of variance in this case, possibly due to the big error intervals in the results obtained by this architecture for life span.

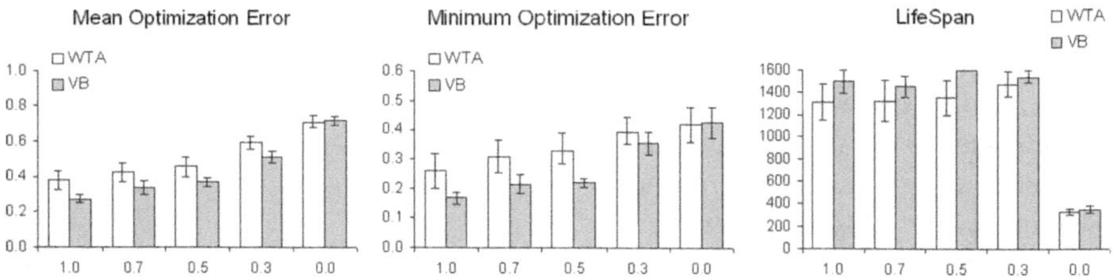

Fig. 4. Average performance of both architectures for different α (x-axis) in terms of mean (left) and minimum (center) optimization error, and life span (right). Standard error is shown.

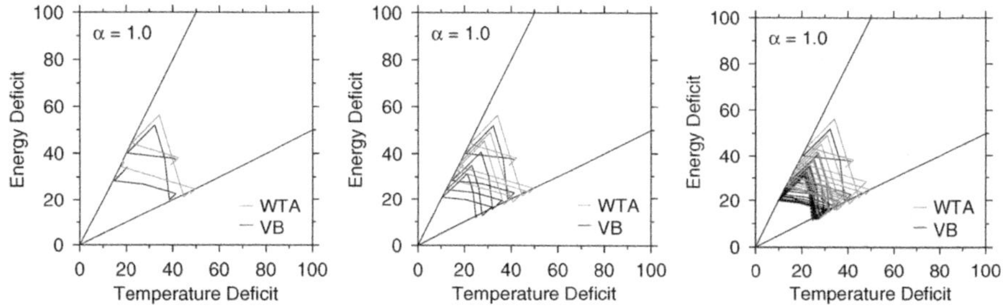

Fig. 5. Example of evolution of both architectures within the physiological space in a simulation run ((α = 1). Left: first 200 steps. Center: first 500 steps. Right: a complete run of 1600 steps.

the behavior that corrects physiological imbalance to a greater extent is reflected in the straight shape of the consummatory phases of the cycles; on the contrary, the tendency of VB to use all the available behaviors can be seen in the convex shape of its consummatory phases. Therefore, the more "cost-effective" strategy used by VB causes that its basic cycles approach the origin more than those of WTA.

5 Conclusion and Future Work

We have compared the optimization policies of simple "winner-take-all" and "voting-based" action selection mechanisms using two corresponding robotic architectures within the framework of the two-resource problem. Our results indicate that VB outperforms WTA in the sense that its basic behavioral cycles approach more the ideal physiological state of absence of deficit. The more "cost-efficient" way in which VB consumes resources accounts for this result in this two-resource problem study. However, other factors need to be considered that we could not take into account this time (but that we considered in previous work) due to the simplified architectures used to adapt them to the two-resource problem. Some of these factors are the fact that a behavior can contribute to satisfy more than one motivation, or the fact that different behaviors can satisfy the same motivation, possibly using different resources. It is also important to bear in mind that the experiments we report in this paper have been performed in static worlds only; therefore, these results cannot be generalized to dynamic environments that produce non-deterministic changes to the physiology of the robot as a result of the effects of both, behaviors and environment. This study has also shown that the weight given to external motivational factors has a big impact on the location of basic cycles in the physiological space, and hence on how well this space is managed. Our introduction of a parameter (α) in the formula used to calculate motivational intensity has allowed us to control this influence, and to determine the location of stable basic behavioral cycles within the physiological space.

To continue this study, we envisage three main directions. First, we would like to do a thorough comparison of our study with other different implementations of the two-resource problem reported in the literature. Second, we will introduce different sources of dynamism (namely mobility of resources and an "enemy" species) to investigate the anticipation and reactivity features of both AS policies. Finally, we intend to study how the dynamic variation of α (the weight given to the influence of external stimuli) during the robot's lifetime can affect on its capability to adapt to the environment.

References

1. Ashby, W.R. (1952). *Design for a Brain.* London: Chapman and Hall.
2. Avila-García, O. and Cañamero, L. (2002). A comparison of Behavior Selection Architectures Using Viability Indicators. In *Proc. International Workshop on Biologically-Inspired Robotics: The Legacy of W. Grey Walter.* 14–16 August 2002, Bristol HP Labs, UK.
3. Blumberg, B. (1994). Action Selection in Hamsterdam: Lessons from Ethology. In *Proc. Third Intl. Conf. on Simulation of Adaptive Behavior (SAB94).* Cambridge, MA: MIT Press.
4. Cañamero, L.D. (1997). Modeling Motivations and Emotions as a Basis for Intelligent Behavior. In W.L. Johnson, ed., *Proc. First Intl. Conf. on Autonomous Agents*, 148–155. New York: ACM Press.
5. Cañamero, L., Avila-García, O., Hafner, E. 2002. First Experiments Relating Behavior Selection Architectures to Environmental Complexity. In *Proc. 2002 IEEE/RSJ International Conference on Intelligent Robots and Systems* (IROS 2002), 3024–3029. IEEE Press.
6. Girard, B., Cuzin, V., Guillot, A., Gurney, K.N., and Prescott, T.J. (2002). Comparing a Brain-Inspired Robot Action Selection Mechanism with 'Winner-Takes-All'. In *Proc. Seventh Intl. Conf. on Simulation of Adaptive Behavior.* Cambridge, MA: MIT Press.
7. Maes, P. (1991). A Bottom-Up Mechanism for Behavior Selection in an Artificial Creature. In J.A. Meyer and S.W. Wilson, eds. *Proc. First Intl. Conf. on Simulation of Adaptive Behavior*, 238–246. Cambridge, MA: MIT Press.
8. McFarland, D. (1974). Experimental Investigation of Motivational State. In D. McFarland, ed., *Motivational Control Systems Analysis*, 251–282. London: Academic Press.
9. McFarland, D. (1976). Form and function in the temporal organization of behavior. *Growing Points in Ethology*, Bateson P., Hinde R. ed. Cambridge University Press.
10. McFarland, D. and Sibly, R. (1975). The behavioural final common path. *Philosophical Transactions of the Royal Society (Series B)*, 270, 265–293.
11. Spier, E. and McFarland, D. (1997). Basic Cycles, Utility and Opportunism in Self-Sufficient Robots. *Robotics and Autonomous Systems*, 20, 179–190.
12. Spier, E. and McFarland, D. (1997). Possibly Optimal Decision Making under Self-Sufficiency and Autonomy. *Journal of theoretical biology*, 189, 317–331.
13. Tyrrell, T. (1993). Computational Mechanism for Action Selection. *PhD. Thesis, Centre for Cognitive Science, University of Edinburg.*

Are There Representations in Embodied Evolved Agents? Taking Measures

Hezi Avraham[1], Gal Chechik[2], and Eytan Ruppin[1,3]

[1] School of Computer Sciences,
Tel-Aviv University, Tel-Aviv 69978, Israel,
{hezuz,ruppin}@post.tau.ac.il
[2] The Interdisciplinary Center for Neural Computation
The Hebrew University of Jerusalem, 91904, Israel,
ggal@cs.huji.ac.il
[3] School of Medicine,
Tel-Aviv University, Tel-Aviv 69978, Israel

Abstract. The question of conceptual representation has received considerable attention in philosophy, neuroscience and embodied evolved agents. Numerous theories on the interpretation of the term 'representation' exist, and many arguments have been made for and against the existence of representations in animate and animat agents. Our work studies this question in evolved artificial embodied agents in a quantitatively rigorous manner, for the first time. We develop two measures, based on information theory, to account for representations. These measures are studied by applying them to evolved agents performing a visual categorization, generalized XOR task. Our results show that having quantitative measures still leaves one with arbitrary "threshold values" decisions which permit wide freedom in determining the existence of representations. However, and more importantly, our results show that information-theoretic measures can still be used efficiently to identify discriminative neural patterns and internal structures that characterize a representation, if the latter is formed.

1 Introduction

Internal representations are thought to play a central role in our understanding of cognitive behavior and information processing of intelligent agents. Yet, the idea that representations actually exist in the brains of animate or animat agents has been seriously challenged by many. For example, Brooks [1] has proposed a bottom-up approach to building agents that are able to react in real time in their environment, while having no central model representing the world. Cliff and Noble [2] point to the lack of evidence of representations in artificial evolved systems solving simple visual tasks, where the term 'representation' denotes a sub-network's pattern of activity which marks an external object or event to the rest of the network. Rather much like Pfeifer and Scheier [3], they suggested that the workings of agents could be best understood by studying the dynamics of

W. Banzhaf et al. (Eds.): ECAL 2003, LNAI 2801, pp. 743–752, 2003.

sensory-motor coordination between the agent and the environment. The usefulness of the notion of representations has also been a focus of much debate in cognitive neuroscience, e.g., [4,5,6].

One classical way to search for representations in cognitive neuroscience research is to look for an emergence of a "correlational structure" in the patterns of neural activity studied [4,6]. Intuitively, this is tantamount to the assertion that representation of an object corresponds to neural activity that marks its existence, but is indifferent to the different instances of it. Another interesting approach, termed in information theoretic notions, has been suggested by Usher [7], where Shannon mutual information is used to characterize the information that a concept/representation carries about external items. These investigations, however, have studied representations in standard connectionist networks, or have remained on a conceptual and philosophical level. **Given this long-time and fundamental controversy in cognitive neuroscience, the present work studies, in a quantitative manner, the question of representation in embodied evolved agents**. The agents studied evolve to solve a generalized XOR task by successfully categorizing visual stimuli. This task was chosen because it is known as a non-trivial evolutionary benchmark challenge [8], and because it is characterized by vanishing correlations in the sensory input readings, on the background of which the formation of a structure in neural activity is necessary and may be readily assessed. Moreover, by this we avoid wrong conclusions using neural networks, which can exploit first-order correlations between input and output and by that ignore higher-order configurational information [5].

Our goals in this work are twofold: One is to gauge whether representations are indeed formed in our agents. The other more important one is to capitalize on the simplicity and transparency of evolved agent models and develop rigorous and quantitative measures of "representation". The latter should serve as working rulers in future studies of this fundamental issue. The paper is organized as follows. We begin by describing the model and experimental protocol by which we evolve the agents. In Sect. 3 we present two quantitative approaches for defining and measuring representations, and describe the results obtained when these methods are applied to the evolved agents. Finally, in Sect. 4, we discuss the implications of our findings for measuring and understanding representation in embodied agents.

2 Methods

The model environment (*world*) is a $14.0 \times 14.0 \times 2$ three dimensional semi continuous simulated arena, where the arena's length and width are continuous while the height is discrete. The world contains two types of *resources*, 10 *food* items and 10 *poison* items. The resources are randomly scattered in a 10.0×10.0 central *resources zone*. Each resource type has a shape comprised of two circular pellets (each pellet radius can be either 0.02 or 0.08), placed one on top of the other with a common center point (Fig. 1a,c). The agent behavioral task is eating

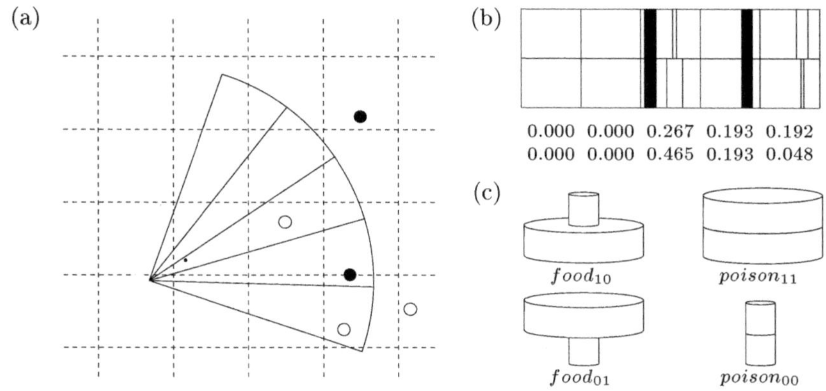

Fig. 1. The model outline. (a) A top view of the simulated world. An agent (*gray circle*) looking at two food items (*white circles*) and at two poison items (*black circles*) through a 5 sub-slices scan sensor. (b) The projected sensor retina (*box*) along with the photoreceptors values (*numbers*). (c) Resources shapes in the generalized XOR task

as much food as possible while avoiding poison. The agent is initially placed at a random location in the arena and eats a resource by simply colliding into it.

The agent consists of three main systems: a *scan sensor*, constantly giving readings of the environment; a synchronous continuous feed-forward neural network, processing the sensor input and producing two output signals; and a motor system, comprising of two independent wheels on either side of the agent's cylindrical body (0.15 radius), each driven by one output motor neuron. The sensor is mounted in front of the agent's body and processes a 3D distal resource stimuli to form a 2D retinal image. The sensor looks ahead for a distance of 3, acceptance angle of 90 ° and a height of 2. The sensor retina is realized as an array of 10 photoreceptors (Fig. 1b), each activated by the existence of a resource in the corresponding sensor's *sub-slice*. A photoreceptor value is set to the amount of total projections from its corresponding sub-slice, yielding a real number in the range of [0,1] (Fig. 1a,b).

A synchronous continuous feed-forward neural network realizes the agent's neurocontroller. We use several network architectures with 1-2 hidden layers. The network acquires its inputs from 10 dedicated sensory neurons, each transmitting the value of one photoreceptor. The output layer is composed of two output neurons, which control the agent's wheels. All network neurons set their activities according to: $V_i(t) = g(h_i(t))$, where $V_i(t)$ and $h_i(t)$ denote neuron i's activity and field (input vector sum) at network update t correspondingly, and $g(x)$ is the sigmoid function $g(x) = 1/(1 + e^{-\beta(x-\theta_i)})$, where $\beta = 4$ is the squashing factor and θ_i is a bias term, set in the genome for each neuron.

The agent motion is determined by its speed and orientation which are set by the wheels velocities and their difference, correspondingly. The agent speed is between -1 and 1 *world units* per time step, and its turn angle is in the interval of [-36 °,36 °].

An agent's lifetime lasts 50 sensory-motor cycles, followed by a normalized fitness score calculated as the number of food items it has consumed minus the number of poison items it has eaten. An evolution run lasts 10,000 generations. In each generation a population of 100 agents is evaluated, then the parents of the next generation are chosen with probability proportional to the agents fitness. At the end of an evolution session the precise fitness of an agent is assessed over 1,000 epochs. The agent's genetic encoding is a string of real valued numbers describing the neurocontroller synaptic weights and the neurons' bias terms θ_i. The initial genes values are randomly chosen in the range of -1 and 1, and have no limits during the evolution run. The genetic operators employed are uniform crossover with probability of 0.35 and mutation with probability of 0.02 and range of [-0.6,0.6].

The experiment we conducted (*generalized XOR* task) is designed to challenge the agents with a 'visual' generalized XOR problem. 5 food items have a common shape of large pellet (0.08) placed underneath a small one (0.02) ($food_{10}$). The remaining 5 food items have the opposite arrangement ($food_{01}$). Similarly, there are two types of poison: 5 poison items have large lower and upper pellets ($poison_{11}$), while the rest of the poison items have small pellets ($poison_{00}$) (Fig. 1c). Hence, the agent has to solve the generalized XOR problem in order to distinguish successfully food from poison. We evolve two kinds of neurocontrollers, one with a single hidden layer with 5 neurons (XOR-*5-2*), and another with two hidden layers, the first with 6 neurons and the second with 4 neurons (XOR-*6-4-2*).

3 Results

3.1 Performance

Direct evolution failed to come up with a good solution for the generalized XOR problem (fitness of 0.2421 and 0.2563 for XOR-*5-2* and XOR-*6-4-2*, correspondingly[1]), therefore we used incremental evolution [9]. This is executed by having large pellets with radiuses of 0.1 (instead of 0.08) and small pellets with a significantly reduced size in the initial stage of the evolution, and then gradually modifying the pellets sizes to their original values as the incremental evolution successfully progresses. Indeed, this technique succeeds and proper agents are evolved (fitness of 0.3275 and 0.3770 for XOR-*5-2* and XOR-*6-4-2*, correspondingly). Further analysis focuses on these two agents, which exhibit a well adjusted behavior, circling the arena in order to keep within the resource zone, avoiding poison, turning to food on either side and then moving with full speed ahead towards it.

[1] The maximal fitness feasible can be roughly estimated by viewing an agent evolved in a *poison-less* world with a single food type. Here, the best evolved individual attains fitness of 0.4272, having a neurocontroller of one hidden layer with 4 neurons.

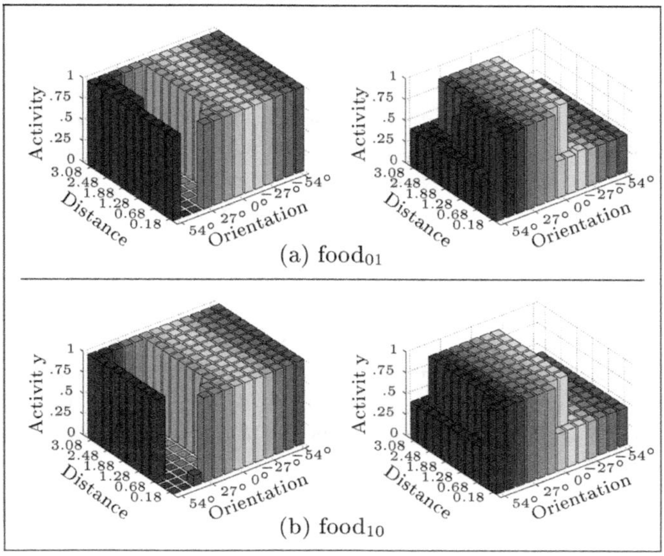

Fig. 2. Left and right output motor neurons' receptive fields in response to food stimuli, for agent XOR-*6-4-2*'s. (a) Left and right responses to food$_{01}$. (b) Left and right responses to food$_{10}$. Each bar marks the neuron's activation level in response to a food stimuli in a particular distance and orientation

3.2 Behavioral Analysis

A first clue to representations in embodied agents can be obtained from a behavioral analysis. If distinct behavioral responses are found to different types of food (or poison), the search for food representation, that is indifferent of food type, turns irrelevant since representation is not used to constitute an akin behavior. Conversely, if similar behavioral responses to both types of a resource are observed, it supports the possibility that a joint internal representation is formed.

In order to compare the agent behavior to the different resources types, we measured the output neurons' receptive fields. This was done by recording their activation levels while introducing, one at a time, all possible resource instances in the agent's field of view (110 discrete locations for each resource type). As can be clearly seen (Fig. 2), there is significant similarity in agent XOR-*6-4-2*'s output neurons receptive fields, both with regard to food types and with regard to poison types (not shown). This affirms that the agent acts similarly in response to the different types of food and similarly for the different types of poison (an analogous conclusion is made regarding agent XOR-*5-2*).

3.3 Entropy in Embedded Agents

An ideal method for studying the relationship between objects and their neural representations, in face of a stochastic noisy environment, is information theory.

It provides a rigorous and quantitative approach to measure the high order correlations that are typical to the type of non linear processing performed by neurocontrollers. Usher [7] has suggested a quantification of representation that is based on the mutual information between a coherent state of the system and the input stimuli in the environment. Following this approach, we develop a novel information related measure of representations in agents. Let us denote $S \in \{s_{\mathrm{f}}, s_{\mathrm{p}}\}$ where $S = s_{\mathrm{f}}$ if the object stimuli is food and $S = s_{\mathrm{p}}$ if it is poison. Similarly, denote $F \in \{f_{01}, f_{10}\}$ and $P \in \{p_{00}, p_{11}\}$. Let $R = \{r_1, \ldots, r_n\}$ denote the set of n coherent states of the neural system examined. To identify these states in the experiments reported below, we applied a K-means algorithm to the neural activities recorded over 2000 agent's life trials (epochs), using a Euclidean norm. The r_i's are obtained as the centroids of neural activities clusters. Each neural activity vector is tagged by the most central object in the agent's view, e.g., as food_{10} or poison_{11}. Using this notation we calculate the conditional entropy of S given R [10]

$$H(S|R) = \sum_{i=1}^{n} p(r_i) H(S|R = r_i) = -\sum_{i=1}^{n} \sum_{j=\mathrm{f,p}} p(r_i) p(s_j|r_i) \log_2 p(s_j|r_i) \ . \quad (1)$$

This measure expresses the uncertainty of knowing whether food or poison was spotted given the activity pattern of the *sub-network* examined. It obtains values in the range of 0 to 1. If food and poison yield distinct activity patterns, $H(S|R)$ will obtain vanishing values. On the other hand, similar activity patterns for both food and poison would yield high $H(S|R)$ values close to 1.

Importantly, this measure alone cannot account for a representation since low $H(S|R)$ values may be obtained when activity patterns for the two food types (or poison types) are different. Since this case does not capture the notion of representation, $H(S|R)$ must be complemented with a measure of the difference in activity patterns within different types of food or poison. Therefore, food representation should be described as the difference between $H(S|R)$ and the entropy $H(F|R)$, computed likewise using food instances only. Since the representation of food (poison) may lay in the activity of a subset of the neurons examined, we define the *Euclidean Entropy Measures* (EEM) for the subset T,

$$EEM_{\mathrm{f}}(T) = H(F|R_T) - H(S|R_T) \quad \text{and} \quad EEM_{\mathrm{p}}(T) = H(P|R_T) - H(S|R_T) \quad (2)$$

where R_T is a set of clusters obtained using K-means with Euclidean norm over the neuronal activities of all neurons in the subset T. This measure can take values in the range of $[-1, 1]$. It obtains a value of 1 when the responses for the different types of food (poison) are identical and the responses for food are completely different from those to poison, therefore denoting that a true representation of food (or poison) has been formed. A value of -1 is given when there are indistinguishable neural activities for food and for poison but distinct activities between both types of food (poison). In the case of random activity patterns, a value of zero is obtained.

Figure 3 shows the values of EEM calculated for all relevant sub-networks of the XOR-*6-4-2*'s hidden layers. It shows that EEM_{p} values are relatively constant

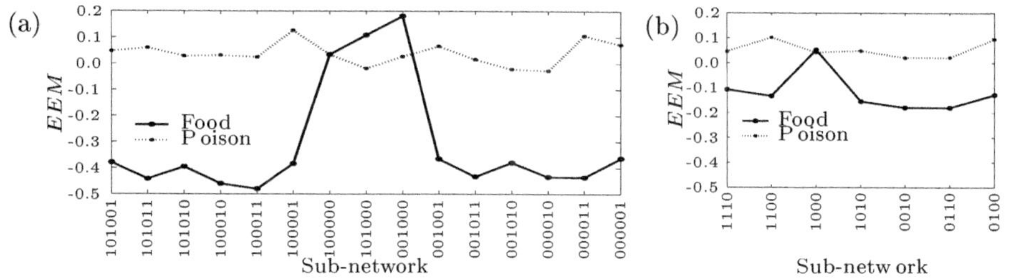

Fig. 3. Euclidean Entropy Measures for agent XOR-*6-4-2*. EEM_{f} (*solid*) and EEM_{p} (*dotted*) for (a) first hidden layer and (b) second hidden layer. A sub-network label (*x-axes*) denotes its configuration, where 1 means that the corresponding neuron exists in the sub-network. Since some neurons have a constant activity (i.e., neurons 2 and 4 in the first hidden layer and neuron 4 in the second hidden layer), only the relevant subsets for each layer are presented

across all subsets, yielding a distributed poison representation. On the other hand, EEM_{f} values are more variable and gain their maximum in few subsets, each having only few neurons, indicating a more localized food representation (qualitatively similar results were obtained for agent XOR-*5-2*).

In order to provide an index for each layer and each resource we further define

$$EEM^* = \max_T EEM(T) \ , \tag{3}$$

to measure the maximal EEM over all possible subsets. We calculated four EEM^* values for agent XOR-*6-4-2* (for food and for poison representations in each of its two hidden layers), which are generally low (0.181 for food and 0.127 for poison in the first layer, and 0.051 for food and 0.101 for poison in the second layer). This results from high $H(F|R)$ and $H(P|R)$ values but high $H(S|R)$ values (in the range [0.79,0.94]), indicating almost optimal food or poison unification but a poor food-poison separation.

The first hidden layer EEM^* values are higher than those of the second hidden layer, especially for food. One may speculate that the second hidden layer is strongly affected by the motoric constraints (e.g., same left turn for food on the left and for poison in the center), thus resource representation is more likely to form in the first hidden layer, yielding these higher EEM^* values.

3.4 Extracting Representations Using Information Bottleneck with Side Information

EEM assumes that the representation lies in a Euclidean metric on the single neuron firing patterns. **An alternative approach, is to cluster patterns of activities based on their functional relevance.** This approach was formally defined and analyzed in [11], using the Information Bottleneck method with Side Information ($IBSI$). $IBSI$ allows to search for structures in neural activities that are relevant to the discrimination between food and poison, but

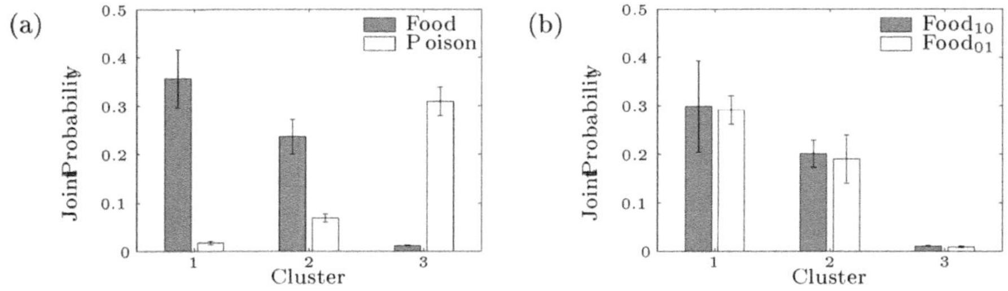

Fig. 4. *IBSI* joint probabilities. (a) $P(S, R)$ and (b) $P(F, R)$ means and *SEM* calculated on agent XOR-*6-4-2*'s first hidden layer. Clusters were obtained by applying the sequential-IBSI hard clustering algorithm with $\gamma = 1$ and *number of clusters*= 3. Similar results were obtained for wide range of γ values and number of clusters

not to the discrimination between *different types* of food or poison. IBSI can be thought of as a clustering analysis procedure that operates on the conditional distributions $P(S|R = r)$ and $P(F|R = r)$ rather than on the neural activities R. It therefore pre-assumes no particular metric between neural activity patterns at the clustering phase, and thus allows to test various kinds of metrics, rather than to use them as the basis for clustering.

Applied to the problem at hand, *IBSI* searches for clusters that compress well a discrete set of neural activities X into a set of clusters R, while maintaining information about S and removing information about F (similarly about P),

$$IBSI_{\mathrm{f}} = \min_{p(r|x)} I(X; R) - \beta[I(R; S) - \gamma I(R; F)] \ , \qquad (4)$$

where β and γ are tradeoff parameters functioning as Lagrange multipliers (see [11] for details). Inspection of the resulting set of clustered distributions $P(S, R)$ and $P(F, R)$ reveals a clear discrimination between food and poison (Fig. 4a) while hardly providing any knowledge about the different types of food (Fig. 4b).

It should be noted that both our EEM^* and $IBSI$ measures calculate very similar information theoretic measures[2]. However, EEM^* is obtained by maximizing over clusters that are formed using an Euclidean measure over neural activities, while $IBSI$ is obtained by directly maximizing the weighted difference of informations. To compare the two measures on the same scale, we computed the difference of entropies used for EEM (Eq. 2), over the clusters obtained from $IBSI$ (where T is the whole layer). $IBSI$ achieved higher values than those obtained with EEM (0.5898 for food and 0.5990 for poison in XOR-*6-4-2*'s first hidden layer, and 0.2246 for food and 0.2049 in the second hidden layer). One hypothesis regarding the large values observed in the first hidden layer is that it uses another (non-Euclidean) metric to solve the generalized XOR task while the second hidden layer metric is more similar to the Euclidean metric since

[2] In fact for $\gamma = 1$ and $\beta \to \infty$ these measures become equivalent up to a constant $H(F) - H(S)$.

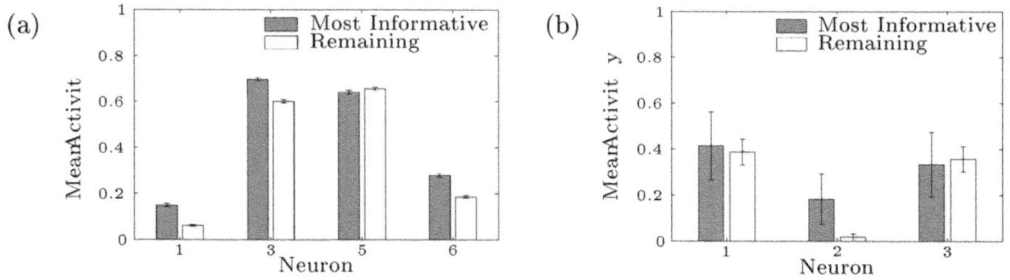

Fig. 5. *IBSI* food representations of agent XOR-*6-4-2*. Mean and *SEM* of the neural activities in the most informative cluster (*gray bars*) and in the remaining clusters (*white bars*) for (a) the first hidden layer and for (b) the second hidden layer. Only neurons with variable (and hence informative) activity are presented

is must refer to the motor actions to be executed. Another possibility is that the second hidden layer values bounds are smaller since the same motor actions should result from food and poison stimuli at different parts of the visual field.

After *IBSI* was used to extract functionally relevant clusters of neural activities, we turn to identify the nature of food and poison representations in the hidden layers neurons. To this end, we focus on the most informative cluster, defined as the cluster with majority of food stimulated responses, that maximizes the *single-symbol information* $D_{KL}[p(s|r)||p(s)] - D_{KL}[p(f|r)||p(f)]$ (and similarly for poison) where D_{KL} is the *Kullback-Liebler divergence* [10]. We first compare the mean neural activities in this cluster with the remaining activities. Figure 5 demonstrates this comparison for agent XOR-*6-4-2*, and reveals that **the mean activities of neurons 1, 3 and 6 in the first hidden layer and neuron 2 in the second hidden layer are significantly different in this cluster than in the remaining ones** (p <0.001 using t-test). This indicates that the activities of these isolated neurons is discriminative (qualitatively analogous results are obtained for agent XOR-*5-2*). Moreover, **several pairs of neurons were found to have cluster specific correlations** that were not observed in their baseline activities, which were found to be statistically significant using standard statistical test for correlation difference (for example, correlation coefficients of r_1=0.485 and r_2=-0.075 for neurons 2 and 3 in agent XOR-*6-4-2*'s second hidden layer, yielding p of 0.0156). This suggests that food representations lies not only in single neurons activities but also in function specific correlations.

4 Conclusions

This study raises two major questions: a) Are representations created in embodied evolved agents? and b) Can these representations be rigorously defined and measured? To answer these questions, we define two measures, *EEM* and *IBSI*-based, aimed to quantitatively and rigorously score inner representations. These two measures differ in their fundamental premises; the first assumes an

Euclidean metric between the neural activities, while the latter has no specific underlying metric assumption. The EEM measure produces low values for all agents and layers studied while $IBSI$ values are significantly higher, but firm conclusions about the existence or non-existence of representations remain subject to arbitrary "threshold value" decisions. However, we demonstrate that using these quantitative approaches leads to the identification of several important representational characteristics, including localized and distributed structures, discriminative neural activity patterns and cross-neuronal interactions. While much remains to be done in future studies, this work clearly shows that it's high time to corroborate the ongoing important conceptual debate about representations with a rigorous, quantitative investigation.

Acknowledgments. We acknowledge the valuable contributions made by Isaac Meilijson and Alon Keinan and the technical help provided by Oran Singer. This research has been supported by the Adams Super Center for Brain Studies in Tel Aviv University and by the Israel Science Foundation founded by the Israel Academy of Sciences and Humanities. G.C. is supported by the Israeli Ministry of Science.

References

1. Brooks, R.A.: Intelligence without representation. Artificial Intelligence **47** (1991) 139–159
2. Cliff, D., Noble, J.: Knowledge-based vision and simple visual machines. Philosophical Transactions of the Royal Society: Biological Sciences **352** (1997) 1165–1175
3. Pfeifer, R., Scheier, C.: Sensory-motor coordination: the metaphor and beyond. Robotics and Autonomous Systems, Special Issue on "Practice and Future of Autonomous Agents" **20** (1997) 157–178
4. Kosslyn, S.M., Chabris, C.F., Marsolek, C.J., Koenig, O.: Categorical versus coordinate spatial relations: computational analyses and computer simulations. Journal of Experimental Psychology: Human Perception and Performance **18** (1992) 562–577
5. Cook, N.D.: Correlations between input and output units in neural networks. Cognitive Science **19** (1995) 563–574
6. Kosslyn, S., Chabris, C., Baker, D.: Neural network models as evidence for different types of visual representations. Cognitive Science **19** (1995) 575–579
7. Usher, M.: A statistical referential theory of content: Using information theory to account for misrepresentation. Mind and Language **16** (2001) 311–334
8. Stanley, K.O., Miikkulainen, R.: Evolving neural networks through augmenting topologies. Evolutionary Computation **10** (2002) 99–127
9. Gomez, F., Miikkulainen, R.: Incremental evolution of complex general behavior. Adaptive Behavior **5** (1997) 317–342
10. Cover, T., Thomas, J.: The elements of information theory. Plenum Press, New York (1991)
11. Chechik, G., Tishby, N.: Extracting relevant structures with side information. In Becker, S., Thrun, S., Obermayer, K., eds.: Advances in Neural Information Processing Systems 15 (NIPS-2002), MIT press (2003)

Evolving Fractal Gene Regulatory Networks for Robot Control

Peter J. Bentley

Department of Computer Science, University College London, Gower Street, London.
P.Bentley@cs.ucl.ac.uk

Abstract. Fractal proteins are a new evolvable method of mapping genotype to phenotype through a developmental process, where genes are expressed into proteins comprised of subsets of the Mandelbrot Set. The resulting network of gene and protein interactions can be designed by evolution to produce specific patterns, that in turn can be used to solve problems. Here the use of fractal gene regulatory networks for learning a robot path through a series of obstacles is described. The results indicate the ability of this system to learn regularities in solutions and automatically create and use modules.

1 Introduction

Life is complex. This is true in the chemical interactions of proteins and genes within a single cell, or in the cellular interactions in a multicellular organism. While evolution is mostly to blame for this, there can be little doubt that complexity would not arise if the vast intricacies of molecular interactions and physical forces were not present. Open-ended evolution (evolution in which solutions become progressively more complex) relies on the right kind of genetic representation in the right kind of environment. In nature, this is DNA – a molecule that relies on chemical interactions in order to function.

Translating these ideas into computer science is not a simple prospect. As we know in evolutionary computation, using a fixed binary string as a genotype, prohibits complexity growth. But variable-length representations such as genetic programming do not guarantee an increase in complexity either (unless you count introns as complexity). Even if a seemingly ideal representation is found, often it is not evolvable, or it only achieves its complexity increases through the careful high-level structures (e.g. modularity) imposed on it by the developer.

This work takes a different approach. A developmental process maps variable-length genotypes to phenotypes, through the use of *fractal proteins*. Genes are expressed into complex fractal shapes (subsets of the Mandelbrot Set) that interact according to their forms. The resulting network of gene interactions can be designed by evolution to produce specific gene activation patterns, that in turn can be used to solve problems. In this paper the use of fractal gene regulatory networks for learning a robot path through a series of obstacles is described.

W. Banzhaf et al. (Eds.): ECAL 2003, LNAI 2801, pp. 753–762, 2003.

2 Background

Researchers such as Hornby [7], Bongard [5], Haddow [6] and Kumar and Bentley [10] have demonstrated that various types of development can enable smaller genotypes to represent more complex phenotypes through the ability of development to discover modularities and repetition. Other scientists in the field have been focussing on the ability of developmental methods to enable self-repairing behaviour and graceful degradation of solutions. For example, the work of Andy Tyrrell and his group create fault-tolerant hardware inspired by ideas of embryology and immune systems [8]. More recently, Julian Miller has described experiments evolving developmental programs to create "French Flag" patterns [12]. He shows that development is able to regenerate these patterns. However, work on applying developmental algorithms to robot control is less common. Most seem to rely on the development of neural networks that are then used to control motion (and in some cases generate form) [9] [7] [5].

3 Fractal Proteins

Development is the set of processes that lead from egg to embryo to adult. Instead of using a gene for a parameter value as we do in standard EC (i.e., a gene for long legs), natural development uses genes to define proteins. If expressed, every gene generates a specific protein. This protein might activate or suppress other genes, might be used for signalling amongst other cells, or might modify the function of the cell it lies within. The result is an emergent "computer program" made from dynamically forming gene regulatory networks (GRNs) that control all cell growth, position and behaviour in a developing creature [13].

Table 1. Types of objects in the representation

fractal proteins	defined as subsets of the Mandelbrot set.
Environment	contains one or more fractal proteins (expressed from the environment gene(s)), and one or more *cells*.
Cell	contains a *genome* and *cytoplasm*, and has some *behaviours*.
Cytoplasm	contains one or more fractal proteins.
Genome	comprising *structural genes* and *regulatory genes*. In this work, the structural genes are divided into different types: *cell receptor genes, environment genes* and *behavioural genes*.
regulatory gene	comprising operator (or promoter) region and coding (or output) region.
cell receptor gene	a structural gene with a coding region which acts like a mask, permitting variable portions of the environmental proteins to enter the corresponding cell cytoplasm.
environment gene	a structural gene which determines which proteins (maternal factors) will be present in the environment of the cell(s).
behavioural gene	structural gene comprising operator and cellular behaviour region.

In this work, a biologically plausible model of gene regulatory networks is constructed through the use of genes that are expressed into *fractal proteins* – subsets of the Mandelbrot set that can interact and react according to their own fractal chemistry. Further motivations and discussions on fractal proteins are provided in [1][2][3]. Table 1 describes the object types in the representation; Figure 1 illustrates the representation. Figure 2 provides an overview of the algorithm used to develop a phenotype from a genotype. Note how most of the dynamics rely on the interaction of fractal proteins. Evolution is used to design genes that are expressed into fractal proteins with specific shapes, which result in developmental processes with specific dynamics.

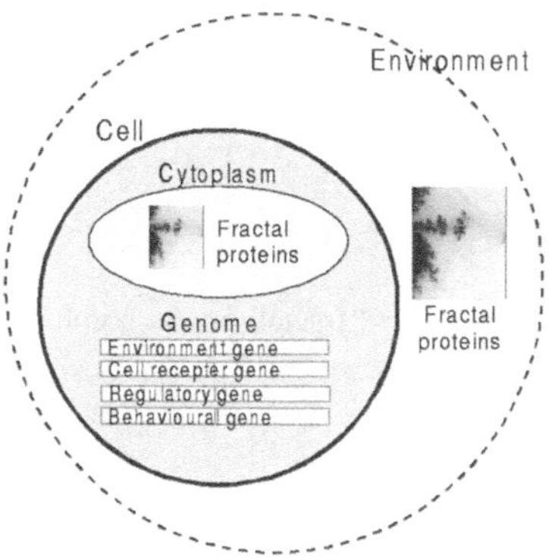

Fig. 1. Representation using fractal proteins.

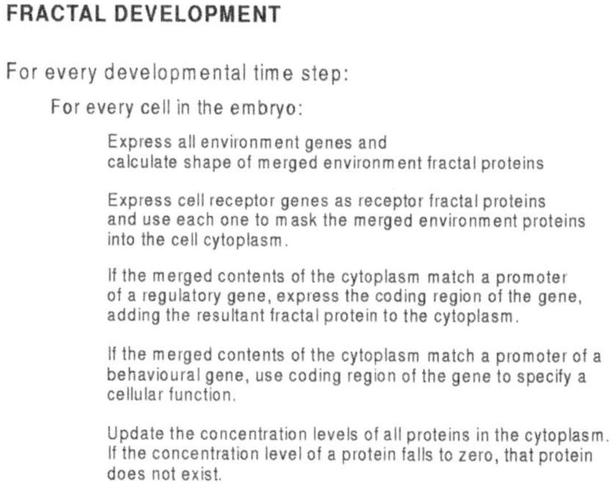

Fig. 2. The fractal development algorithm

3.1 Defining a Fractal Protein

A fractal protein is a finite square subset of the Mandelbrot set [11], defined by three codons (x,y,z) that form the coding region of a gene in the genome of a cell. Each (x, y, z) triplet is expressed as a protein by calculating the square fractal subset with centre coordinates (x,y) and sides of length z, see fig. 3 for an example. In this way, it is possible to achieve as much complexity (or more) compared to natural protein folding in nature.

Fig. 3. Example of a fractal protein defined by (x = 0.132541887, y = 0.698126164 , z = 0.468306528)

In addition to shape, each fractal protein represents a certain *concentration* of protein (from 0 meaning "does not exist" to 200 meaning "saturated"), determined by protein production and diffusion rates [1].

3.2 Fractal Chemistry

Cell cytoplasms and the environment usually contain more than one fractal protein. In an attempt to harness the complexity available from these fractals, multiple proteins are merged. The result is a product of their own "fractal chemistry" which naturally emerges through the fractal interactions.

Fractal proteins are merged (for each point sampled) by iterating through the fractal equation of all proteins in "parallel", and stopping as soon as the length of any is unbounded (i.e. greater than 2). Intuitively, this results in black regions being treated as though they are transparent, and paler regions "winning" over darker regions. See fig 4 for an example.

Fig. 4. Two fractal proteins (left and middle) and the resulting merged fractal protein combination (right).

3.3 Genes

The environment gene, cell receptor gene, regulatory genes, and behavioural genes all contain 7 real-coded values:

xp	*yp*	*zp*	*Affinity threshold*	*Concentration threshold*	*x*	*y*	*z*	*type*

where (*xp, yp, zp, Affinity threshold, Concentration threshold*) defines the promoter (operator or precondition) for the gene and (*x,y,z*) defines the coding region of the gene. The *type* value defines which type of gene is being represented, and can be one or all of the following: *environment, receptor, behavioural,* or *regulatory.* This enables the type of genes to be set independently of their position in the genome, enabling variable-length genomes. It also enables genes to be multi-functional, i.e. a gene might be expressed both as an environmental protein and a behaviour.

When *Affinity threshold* is a positive value, one or more proteins must match the promoter shape defined by (*xp,yp,zp*) with a difference equal to or lower than *Affinity threshold* for the gene to be activated. When *Affinity threshold* is a negative value, one or more proteins must match the promoter shape defined by (*xp,yp,zp*) with a difference equal to or lower than |*Affinity threshold*| for the gene to be repressed (not activated).

To calculate whether a gene should be activated, all fractal proteins in the cell cytoplasm are merged (including the masked environmental proteins) and the combined fractal mixture is compared to the promoter region of the gene. The full details of this process are lengthy; interested readers should consult [1][2][3].

Behavioural Gene. A behavioural gene is activated when other protein(s) in the cytoplasm match its promoter region (using the *affinity threshold*). For this application, a gradual activation between 'not activated' and 'activated' was required, using the x value of the coding region (x,y,z) triplet as a *fate* value to define a function, calculated as follows:

If the gene is being activated with a negative *Affinity threshold*,
$output = output - (totalconcentration - concentrationthreshold) \times fate$

If the gene is being activated with a positive *Affinity threshold*,
$output = output + (totalconcentration - concentrationthreshold) \times fate$

Note how the total concentration of proteins seen on the promoter is offset against the *Concentration Threshold* gene and scaled by the *fate* gene (x value of the coding region), allowing evolution to adjust the range of values seen on the output, and used to specify behaviours. (If there are more behavioural genes than are required, the resultant behaviour will be the sum behaviour of all genes.)

3.4 Development

As was illustrated in figure 2, an individual begins life as a single cell in a given environment. To develop the individual from this zygote into the final phenotype, fractal proteins are iteratively calculated and matched against all genes of the genome. Should any genes be activated, the result of their activation (be it a new protein, receptor or cellular behaviour) is generated at the end of the current cycle. Development continues for d cycles, where d is dependent on the problem. Note that if one of the cellular behaviours includes the creation of new cells, then development will iterate through all genes of the genome in all cells.

3.5 Evolution

The genetic algorithm used in this work has been used extensively elsewhere for other applications (including GADES [4]). A dual population structure is employed, where child solutions are maintained and evaluated, and then inserted into a larger adult population, replacing the least fit. The fittest n are randomly picked as parents from the adult population. Typically the child population size is set to 80% of the adult size and n = 40%. (For further details of this GA, refer to [4].) Because real coding was used, duplication and creep mutation is used, see [1] for complete details. Crossover is always applied; all mutations occur with probability 0.01 per gene.

4 Robot Control

Previous work has demonstrated how evolution can generate specific fractal proteins that interact with each other in order to produce desired patterns of behavioural gene activation [1][2][3]. Here the two behavioural genes are used to generate commands for a robot, with the aim of directing the robot past obstacles in its environment to reach a destination. So instead of directing cellular behaviour (i.e., cell division, differentiation or death), the fractal gene regulatory networks will direct robot behaviour. Although the tasks are clearly very different, both do rely on precise timing and accuracy, and both make use of the inherent pattern-generating properties of GRNs.

4.1 Robot

The platform used was a palm-sized six-legged robot known as a *wonderborg*™, produced by Bandai, see Fig. 5. Although not widely available outside Japan, the robot boasts several useful features: it is programmed via infrared communications, and once programmed it runs fully autonomously. It has two forward-facing infrared collision detectors, a bottom-facing infrared floor sensor, an upwards facing light sensor and two microswitch touch sensors or feelers. Its legs are driven by two motors (one for the legs on each side), operating like a tracked vehicle.

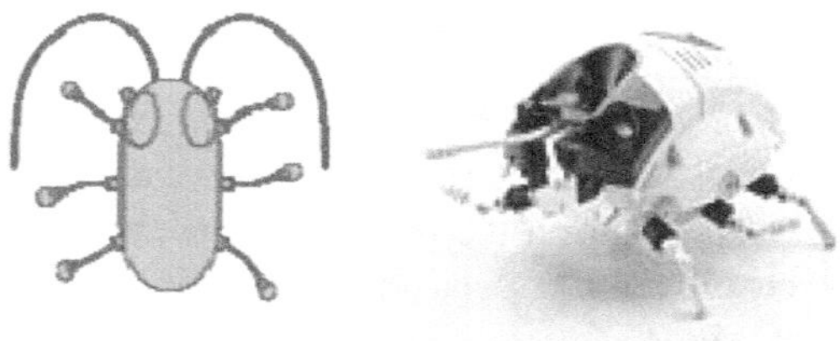

Fig. 5. The *wonderborg*™ robot (Bandai).

The *wonderborg*™ was not designed as a research platform, so unfortunately suffers from certain drawbacks. Currently the only way to send a program to the robot is through Bandai's proprietary software which forces the use of high-level commands and programming structures. Because all movement commands are high-level (e.g. "Rotate right" or "Back up left"), this prohibits the independent control of motors. To overcome the difficulties, the two behavioural genes were treated as "steering" and "acceleration," and then their values were mapped onto the appropriate high-level commands. For example, a positive value for the steering gene is taken to mean "steer right" and a value of -3 for the acceleration gene is taken to mean "move -3 backwards", see table 2. The fractal development system was then modified to convert the patterns of gene activation produced during development into a robot control file in the appropriate format for the proprietary software.

Table 2. Mapping of steering and acceleration genes to the nine predefined robot commands

Steering		Acceleration		Robot command	Symbol (n=1)
0	(None)	0	(None)	*Halt*	
-ve	(Left)	0	(None)	*Rotate left*	
+ve	(Right)	0	(None)	*Rotate right*	
0	(None)	n	(n forwards)	*Advance* by n	
-ve	(Left)	n	(n forwards)	*Turn left* by n	
+ve	(Right)	n	(n forwards)	*Turn right* by n	
0	(None)	$-n$	(n backwards)	*Back up* by n	
-ve	(Left)	$-n$	(n backwards)	*Back up left* by n	
+ve	(Right)	$-n$	(n backwards)	*Back up right* by n	

Finally, to enable high-speed evaluation of robot control programs, a *wonderborg*™ simulator was created. This reads the same file format as used by Bandai's proprietary software (and as output by the fractal developmental system) and calculates the path of the robot through an environment. The simulator was designed to be fast – on a 1 Ghz PC, approximately 40 developmental cycles and corresponding robot control simulations occur every second.

5 Experiments

Two sets of experiments were performed. The first used a simple environment with four obstacles, the second used a more difficult environment with seven obstacles, see figures 6 and 7. The robot simulator was initialised with the robot at one end of the environment. The further the robot managed to walk across the environment the higher the fitness of the corresponding controller. If the robot touched an obstacle or walked off the edges, its final position was measured as its last valid position in the environment. Note that the controllers did not use any of the robot's sensors to gauge proximity of obstacles (which might have made the task easier). Instead they specified a fixed path past the obstacles, learning the structure of the environment through generations of "hitting its head against walls." A second fitness measure (of less importance than the first) was used to provide penalties corresponding to the time taken by the robot and hence encourage efficient and fast journeys. (Without this, the robots would often rotate many times on one spot for several seconds between each movement in order to find the perfect angle.)

To evolve the controllers, the fractal development system was initialised with a single cell, 1 environment gene, 1 receptor gene, 2 behavioural genes and 6 regulatory genes. (Note that with variable length genomes, evolution was free to modify these gene numbers). The operator and coding regions of the genes were randomly initialised with the alleles that defined 10 previously evolved protein fractals [1]. 32 developmental steps were employed, and the evolutionary algorithm used a population size of 100, running for up to 500 generations in the first experiment and 1500 in the second.

6 Results and Analysis

In the first experiment, 13 out of 20 produced robot controllers able to reach the top (500 generations). In the second experiment, only 1 out of 20 produced successful robot controllers after 500 generations. When the GA was allowed to run for 1500 generations, 11 out of 20 produced robot controllers able to reach the top. The increase in success is further evidence of evolvability of this representation [1][3]; it is likely that further generations would have improved the success rate even more. Figs 6 and 7 show examples of robot paths produced by fit controllers. The final controllers have been confirmed by testing on the actual robot.
Despite the poor level of control available through the high-level commands, evolution and development manage to create remarkably precise and detailed robot paths, designed to avoid the obstacles in the environment. Significantly, these controllers often employed "modules" or "subroutines" that were repeated (sometimes with variations) to overcome the obstacles. This automatic discovery and exploitation of pattern in the solution is very important, and also very unusual – most other systems (e.g. ADFs in GP or parameterised L-Systems) must explicitly build in notions of modules. Here they emerge naturally. To illustrate a little of how this happens, figure 8 displays protein concentrations of the controller in fig 7c.

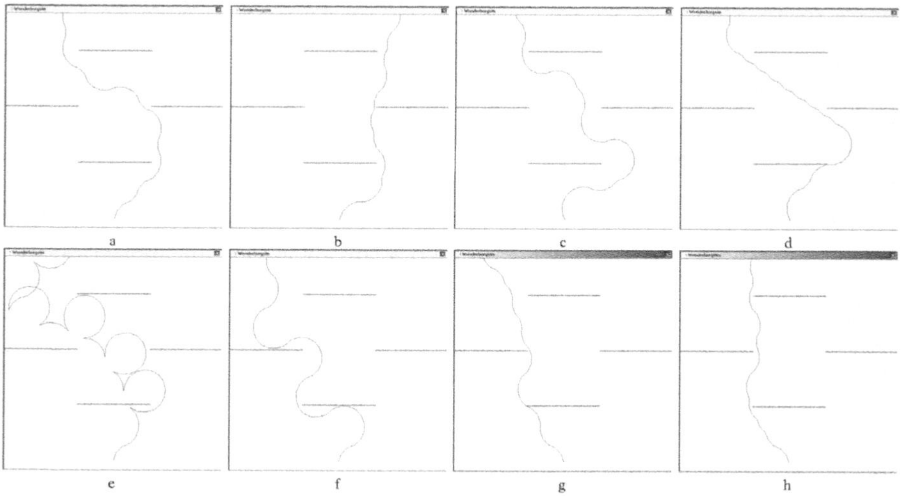

Fig. 6. Eight examples of very fit controllers for the first experiment.

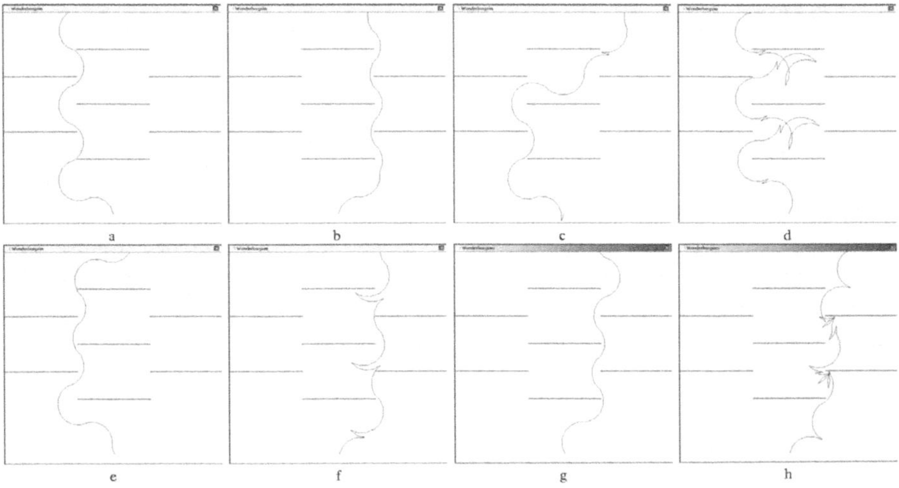

Fig. 7. Eight examples of very fit controllers for the second experiment.

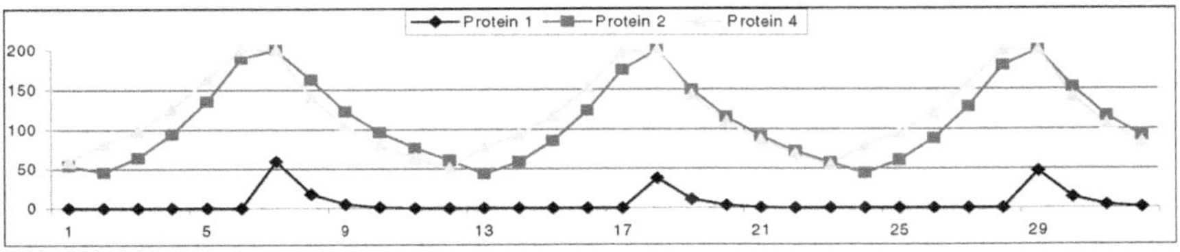

Fig. 8. Protein concentrations of the three main proteins used during development in controller shown in fig 7c. Steering and acceleration are the result of interactions between oscillating protein concentrations, resulting in very precisely controlled oscillating robot motion. The 32-step pattern is partially repeated in the final controller, producing a total of 50 commands for the robot.

7 Conclusions

It is not a trivial task to find a suitable genetic representation that enables open-ended evolution, while being evolvable, incorporating ideas of intricate chemical/physical environments, and employing developmental processes. The use of genes expressed as fractal proteins is the approach investigated in this paper. The work has shown that fractal gene regulatory networks can be successfully designed by evolution to solve problems. Here, the task of producing robot controllers capable of guiding a "bug" robot past obstacles in an environment was demonstrated. In addition to creating diverse and useful controllers, the fractal GRN also demonstrated an ability to identify patterns in solutions and create its own modules (subsolutions that were reused and used with minor modifications). This automatic learning, not just of a good solution (robot path), but of a way of *building* a good solution in an efficient manner, is seen as a significant and important property of developmental processes. Combine this with the results of previous research that shows a tendency towards efficiency and fault-tolerance of fractal GRNs [3], and the potential for this technique looks impressive.

Acknowledgments. Thanks to Sanjeev Kumar for his comments. This material is based upon work supported by the European Office of Aerospace Research and Development (EOARD), Airforce Office of Scientific Research, Airforce Research Laboratory, under Contract No. F61775-02-WE014. Any opinions, findings and conclusions or recommendations expressed in this material are those of the author and do not necessarily reflect the views of EOARD. MOBIUS is an EMBLEM project http://www.cs.ucl.ac.uk/staff/ P.Bentley/emblem.html

References

1. Bentley, P. J. Fractal Proteins. 2003a. To appear in Genetic Programming and Evolvable Machines Journal.
2. Bentley, P. J. Evolving Fractal Proteins. 2003b. In Proc. of ICES '03, the 5th International Conference on Evolvable Systems: From Biology to Hardware.
3. Bentley, P. J. Evolving Beyond Perfection: An Investigation of the Effects of Long-Term Evolution on Fractal Gene Regulatory Networks.2003c. Submitted to *Information Processing in Cells and Tissues* (IPCAT 2003).
4. Bentley, P. J. From Coffee Tables to Hospitals: Generic Evolutionary Design. 1999. Chapter 18 in Bentley, P. J. (Ed) Evolutionary Design by Computers. Morgan Kaufmann Pub. San Francisco, pp. 405–423.
5. Bongard, J. C. Evolving Modular Genetic Regulatory Networks. 2002. In *Proc.of 2002 Congress on Evolutionary Computation (CEC2002)*, IEEE Press, pp. 1872–1877.
6. P. C. Haddow, G. Tufte, and P. van Remortel. Shrinking the Genotype: L-Systems for EHW 2001. In Proc. Of 4th Int. Conf. On Evolvable Systems: From Biology to Hardware, Tokyo, Japan.
7. Hornby, G. S. Generative Representations for Evolutionary Design Automation. 2003. Brandeis University, Dept. of Computer Science, Ph.D. Dissertation.
8. A.H. Jackson, A.M. Tyrrell Implementing Asynchronous Embryonic Circuits using AARDVArc. 2002. In *Proceedings of 2002 NASA/DoD Conference on Evolvable Hardware* (EH-2002), IEEE Computing Society, Alexandria, Virginia, Pages 231–240, 15-18.
9. N. Jakobi. Harnessing Morphogenesis. 1995. *International Conference on Information Processing in Cells and Tissues*, Liverpool, UK.
10. S. Kumar and P. J. Bentley. Computational Embryology: Past, Present and Future. 2003. Invited chapter in Ghosh and Tsutsui (Eds) Theory and Application of Evolutionary Computation: Recent Trends. Springer Verlag (UK).
11. Mandelbrot, B. *The Fractal Geometry of Nature.* 1982. W.H. Freeman & Company.
12. Miller, J. and Banzhaf, W. Evolving the Program for a Cell: From French Flags to Boolean Circuits. 2003. To appear as an invited chapter in Kumar, S. and Bentley, P. J. (Eds) *On Growth, Form and Computers.* Academic Press, 2003.
13. Lewis Wolpert, Rosa Beddington, Thomas Jessell, Peter Lawrence, Elliot Meyerowitz, Jim Smith. *Principles of Development, 2nd Ed.* 2001. Oxford University Press.

Explorations of Task-Dependent Visual Morphologies in Competitive Co-evolutionary Experiments

Gunnar Buason and Tom Ziemke

Department of Computer Science, University of Skövde
Box 408, 541 28 Skövde, Sweden
{gunnar.buason, tom }@ida.his.se

Abstract. This paper presents results from a number of experiments within the area of competitive co-evolutionary robotics. The focus in these experiments has been on 'co-evolving' parts of the morphology of predator and prey robots together with their neural control system. More specifically, the results presented here focus on the evolution of the vision modules' view angle, view range and camera direction. The results have been analyzed with respect to how the evolutionary process is capable of co-adapting morphological parameters and behavioral strategies, so that task-dependent visual morphologies are evolved.

1 Introduction

It has been suggested that the use of artificial evolution is one of the most promising paths for overcoming the human designer bias in the robot design process. A number of researchers have proposed that an evolutionary approach to robot design will eventually get ahead of human design (e.g. [14]). Moreover, with the advancements in both artificial evolution, and computational power, these propositions are becoming more and more real every day. One approach that has been suggested for achieving incremental artificial evolution is competitive co-evolution (CCE). The approach is considered particularly suitable as it decreases involvement of human design in autonomous robots, and it also makes the environment more complex and dynamic in the sense that the opponent's strategies are dynamic [12].

This article focuses on CCE of two robotic species, predator and prey. Similar to their natural counterparts, these two species 'live' and 'evolve' in a delicate balance where sometimes the prey is captured by the predator, and where sometimes the prey is able to escape, bearing some similarities to nature. This has been demonstrated by a number of researchers with both simulated and real robotic experiments, e.g. Cliff and Miller [6], and Nolfi and Floreano [7, 8, 9]. Here, a step towards automatic design of autonomous robots is taken by demonstrating the possibilities of further removing the human-designer out of the loop.

This paper presents the results of ten experiments, which are a part of a series of 21 experiments documented in detail elsewhere [1, 2]. The experiments investigate the influence of successively including and removing constraints and dependencies on the evolved behavior and morphological structure of the co-evolving predator and prey

W. Banzhaf et al. (Eds.): ECAL 2003, LNAI 2801, pp. 763–770, 2003.

robots. More specifically, the camera module of the robots was in focus, and view angle, view range and camera direction were subject to the evolutionary process together with the neural control system. The first ten experiments in this series focused on validating the experimental framework and evolving the view angle and view range (c.f. [1, 3, 4]). Experiments 11-20, which are documented in this paper, use experiments 1-10 as a base and extend them by including in the evolutionary process the camera direction as an additional morphological parameter.

2 Background

The most well known scenario of CCE in Evolutionary Robotics (ER) is taken from nature, i.e. the scenario formed by predators and prey species which usually co-exist and co-adapt in a delicate balance in nature, such that neither species completely dominates over the other.

Cliff and Miller investigated a number of different aspects of CCE [6]. Their experiments considered evolution of sensor-morphology of predator and prey robots, and their results showed that "pursuers usually evolved eyes on the front of their bodies (like cheetahs), while evaders usually evolved eyes pointing sideways or even backwards (like gazelles)" [6, p. 506].

In a similar vein, Lund, Hallam and Lee [10] argued that 'hardware', such as the control system and the physical structure of a robot, should be able to change its architecture and behavior dynamically and autonomously with its environment, i.e. adapting both morphology and control mechanism, and not only the latter, fitting the robot with the task in question. Other researchers such as Pollack et al. [14] and Nolfi and Floreano [13] have also pointed out the importance of 'co-evolving' both 'brain and body' in autonomous robots.

Nolfi and Floreano [7, 8, 11] performed successful CCE experiments where they were able to transfer their simulated neural control systems onto real robots. Some of these experiments showed that the robots' sensory-motor structure, i.e. their morphology as well as their sensory and motor capacities, greatly influenced the evolution of behavioral (and learning) strategies. This work is to a large extent based on the work of Nolfi and Floreano, adopting, for example, a number of evolutionary parameters and the general experimental framework.

3 Experiments

As mentioned above, the experiments presented in this article are extensions of ten previously performed experiments documented in [1, 4]. Table 1 summarizes these earlier experiments and their results briefly.

In order to keep the number of evolved parameters small and to allow detailed analysis the previous experiments did not consider evolution of the architecture and learning rules of the neural network. Furthermore, the focus was on evolving a limited

Table 1. Overview of previous experiments

Exp	Purpose	Results
1	Replicating the predator-prey experiment in [7] where only the predator had vision.	The results from the experiments were similar to the results achieved by Floreano and Nolfi [7, 14], indicating that the experimental framework was correct. Furthermore, the results served as a baseline for comparisons with the following experiments.
2	Replicating the predator-prey experiment in [11] where both predator and prey had vision.	
3	To investigate the impact of allowing the evolutionary process to evolve view angle and view range of the predator's camera.	The results from these experiments indicate that it is possible to evolve the view angle and view range of the predator's camera while achieving a suitable pursuit behavior. Experiment 3 gave a predator with a medium view angle (187° average) with a relative long view range (356 mm average) while the introduction of constraints in experiment 4 resulted in predator with rather small view angle (63° average) and a shorter view range (289 mm average).
4	To investigate the impact of introducing a dependency between the predator's view angle and maximum speed.	
5	The focus is moved from the predator to the prey. The prey can evolve camera view angle and view range. The basic predator from experiment 1 is used.	In experiment 5 the prey was able to evolve a good overall evasion strategy 'preferring' a wider view angle and a shorter range (252° and 274 mm average respectively) than the predator in experiment 3. When a dependency between the view angle and speed was added into the behavior of the prey, as in experiment 6, the prey 'preferred' instead a smaller view angle with a relative short view range. The results of these experiments indicate that it is possible to evolve the morphology of the prey's camera, achieving a suitable evasion behavior.
6	Prey can evolve camera angle and range but has speed constraints.	
7	Predator and prey evolve camera view angle and view range.	These experiments investigated what happens when the competitive advantage is balanced out by evolving both view angle and view range in both species/robots parallel to giving the robots same maximum speed, and speed constraints. Similar results as in previous experiments were observed where the predator preferred a small view angle with a long range while the prey evolved a wide view angle with a shorter range (experiment 7 to 9). Experiment 10, on the other hand, exhibited different results. As a result of constraints on the behavior of the prey, the prey evolved a camera with a small view angle and long view range, i.e. in order to achieve minimal constraints on the speed, a tradeoff was made between speed and vision where speed was preferred.
8	Predator and prey evolve camera view angle and view range. Predator is implemented with speed constraints.	
9	Predator and prey evolve camera view angle and view range. Maximum speed of both species is set to the same value. Predator is implemented with speed constraints.	
10	Both evolve camera view angle and range, both have speed constraints, and the same maximum speed.	

number of aspects of the robot morphology, i.e. the size of the robot was kept constant, assuming a Khepera-like robot, and all the infrared sensors were used, for the sake of simplicity. The parameters that were evolved in at least some of the previous experiments were the weights of the neural control system, the view angle (0 to 360 degree) and the view range of the camera (5 to 500 mm) (cf. Fig. 1).

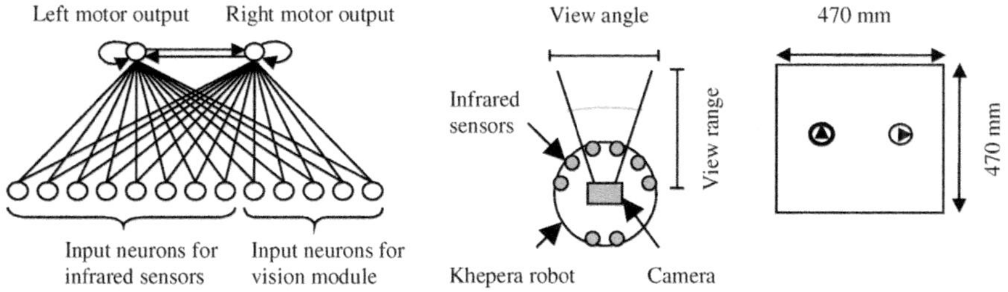

Fig. 1. Left: Neural network control architecture (adapted from [7]). Center: Khepera robot equipped with eight short-range infrared sensors and a vision module (a camera). The 'side' of the robot that has six infrared sensors is referred to as the front side. The 'side' that has two infrared sensors is referred to as the rear side. The camera is directed toward the front on the standard robot (view angle: 0 degrees). Right: Environment and starting positions. The thicker circle represents the starting position of the predator while the thinner circle represents the starting position of the prey. The triangles indicate the starting orientation of the robots, which is random for each generation.

In the new experiments documented in this paper all ten previous experiments were rerun with a crucial modification: in addition to the above the evolutionary process was allowed to evolve the direction of the camera (0 to 360 degrees). Constraints and dependencies that were introduced in the previous experiments were also maintained here, e.g. by letting the view angle constrain the maximum speed, i.e. the larger the view angle, the lower the maximum speed the robot was allowed to accelerate to. As before, each experiment was replicated three times.

3.1 Experimental Setup

The simulator that was used in this work is called YAKS, which simulates the popular Khepera robot, and is similar to the one used by [7, 8, 9, 11]. The simulation of the sensors is based on pre-recorded measurements of a real Khepera robot's infrared sensors and motor commands at different angles and distances [3].

The experimental framework that was implemented in the YAKS simulator was in many ways similar to the framework used by [7, 8, 9, 11]. What differed was that in our work we used real value encoding to represent the genotype instead of direct encoding, and the number of generations was extended from 100 to 250 generations to allow us to observe the morphological parameters over a longer period of time. Beside that, most of the evolutionary parameters were 'inherited' such as the use of *elitism* as a selection method where the 20 best individuals of a population of 100 were chosen to reproduce. In addition, a similar fitness function was used. Maximum fitness was one point while minimum fitness was zero points. The fitness was a simple time-to-contact measurement, giving the selection process finer granularity, where the prey achieved the highest fitness by avoiding the predator for as long as possible while the predator received the highest fitness by capturing the prey as soon as possible. The competition ended if the prey survived for 500 time steps or if the predator made contact with the prey before that.

For each generation the individuals were tested for ten epochs. During each epoch, the current individual was tested against one of the best competitors of the ten previous generations. At generation zero, competitors were randomly chosen within the same generation, whereas in the other nine initial generations they were randomly chosen from the pool of available best individuals of previous generations. The same environment as in [7, 8, 9, 11] was used (cf. Fig. 1).

A simple recurrent neural network architecture was used, similar to the one used in [7, 8, 9, 11] (cf. Fig. 1). The neural network had eight input nodes for the infrared sensors and five for the camera, and two sigmoid output neurons for each motor of the robot and one bias node that was connected to the output nodes (not shown in Fig. 1).

The vision module, which gave only one-dimensional input (i.e. a horizontal line), was implemented with flexible view range, view angle and camera direction, but the resolution was kept constant (five input neurons covering equal parts of the view angle).

For each experiment, the weights of the neural network were initially randomized and evolved using a Gaussian distribution with a standard deviation of 2.0. When angle, range and camera direction were evolved, the starting values were randomized using a uniform distribution function, and during evolution the values were mutated using a Gaussian distribution with a standard deviation of 5.0. Worth mentioning is that the camera used by Nolfi and Floreano had a view angle of 36 degrees and the camera direction was fixed (pointing 'forward', i.e. in the same direction as most of the infrared sensors). In our experiments the direction of the camera was in the interval of 0 to 360 degrees, where 0 respectively 360 degrees indicated that the camera was pointing in a forward direction on the robot. Both the view angle and the view range were bounded within the range of 0 to 360 degrees respectively 5 to 500 mm. If the random function generated a value that was outside the boundaries, the value was set to the last value within. For the camera this was different; if the camera direction evolved over 360 degrees it started again at 0 degrees.

Constraints, such as those used by [7, 8, 9, 11], where the maximum speed of the predator was only half the prey's, were adapted in some of the experiments here where speed was dependent on the view angle. For this, the view angle was divided up into 10 intervals of 36 degrees. The maximum speed of the robot was then reduced by 10% for each additional interval used (e.g. by 20% for an angle of 37-72 degrees).

3.2 New Experiments: Evolving Sensory Morphology

Evolving the direction of the camera can be considered a variation of the experiments performed by Cliff and Miller [6], who evolved the positions of the eye sensors on the robot. Evolving the direction of the camera is somewhat more realistic as it might be possible to perform this experiment in reality with a Khepera robot using a wide-angle camera, where the inputs of the camera are constrained to a certain specific area of the field of vision.

In order to investigate the impact of evolving the direction of the camera on the robot, all previously performed experiments (cf. Table 1) were run again with a crucial modification. This time whenever a robot had a camera in the original experiment, the direction of the camera was now also evolved.

Fig. 2 illustrates how the direction of the camera was evolved on the robots. A standard Khepera robot does not have a vision module. When the vision module is added the direction of the vision module is in a forward direction, i.e. in the same direction as the six infrared sensors. This creates a certain imbalance in the robot's sensor morphology, as there are only two infrared sensors on the rear side.

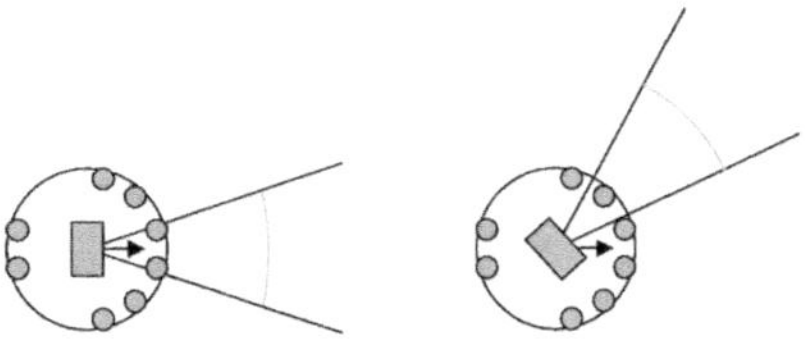

Fig. 2. Evolving the camera direction. Two robots with a vision module and a view angle of 36°. The arrow indicates the defined forward direction. The robot on the left has a camera direction of 0° while the robot on the right has evolved a camera direction of 43°, i.e. the angle is calculated in a counter-clockwise direction.

The results presented here focus on the camera direction evolved on the robots in different experiments. In sum, perhaps somewhat surprisingly, despite the additional degree of 'evolutionary freedom' the robots' overall performance degraded. In some experiments, one of the robots improved its fitness, but then again at the expense of the other robot. Usually the predator increased its performance, compared to the previous experiments (cf. Table 1), as the prey had more difficulties in finding the right balance between view angle, view range, and camera direction while avoiding the predator.

Nevertheless, the results of evolving the camera directions were interesting. Despite the variety of the experiments, i.e. different aspects being tested such as evolution of angle, range and introduction of constraints, the average camera direction evolved towards a direction of 175° with a standard deviation of 64° for the predator, and towards a direction of 178° with a standard deviation of 101° for the prey, with relatively small variations between different experiments, as illustrated in Table 2. That means, both robots 'preferred' to have the camera facing backwards, i.e. in the same direction as the two infrared sensors in the rear (cf. Fig. 1). Although the standard deviation of the prey camera is rather high, it is still possible to conclude that in most cases the prey prefers to have at least parts of the camera facing backwards.

Table 2. Evolved camera directions (averages over all replications, standard deviations in parentheses, empty fields indicate that no camera was used).

Exp	**11**	**12**	**13**	**14**	**15**	**16**	**17**	**18**	**19**	**20**	**Avg.**
Predator	179° (57°)	176° (45°)	170° (89°)	169° (57°)	185° (36°)	175° (27°)	180° (91°)	143° (91°)	226° (90°)	146° (56°)	**175° (64°)**
Prey		153° (95°)			164° (109°)	193° (110°)	170° (83°)	202° (93°)	179° (110°)	186° (109°)	**178° (101°)**

A typical scenario observed, when both robots were equipped with cameras is shown in Fig. 3. The prey, as stated earlier, prefers to have the camera facing backwards. The main reason for this can be that the camera is not able to sense the walls and is therefore not useful for obstacle avoidance. In addition, as the predator usually approaches the prey from behind, there is more use for the camera facing backwards. The prey still moves in a forward direction using the six infrared sensors for obstacle avoidance. For similar reasons the predator chooses to have the camera facing backwards and to also move in that direction. The reason behind this is to simplify the obstacle avoidance task. That is, the predator must avoid walls while chasing the prey. While doing so in most cases it rarely encounters walls, but when it catches the prey then the infrared sensors light up. The predator therefore adjusts its morphology to make it easier to capture the prey while avoiding walls. Although it is easier for the predator to learn to capture the prey, it is still left with the problem that occasionally arises, i.e. the problem of crashing into walls. Particularly surprising behaviors can occur in some of the experiments when the prey is not equipped with a camera. The prey starts to chase the predator, and the predator avoids the prey! This is caused by the prey moving in the same direction as the predator, behind it. Having so many infrared sensors in the "back", the predator believes that it is avoiding a wall and therefore moves away from the object (i.e. the prey) that in fact it should be chasing.

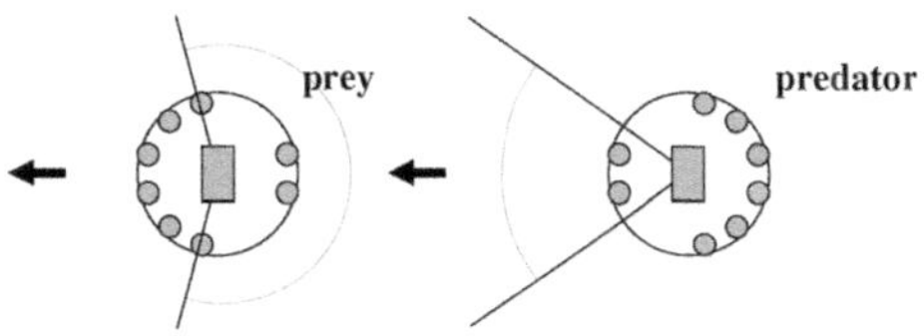

Fig. 3. Schematic illustration of the robots' 'preferred' morphologies and moving directions (indicated by the arrows). The predator (right) has the camera facing towards the 'rear side' and moves in the same direction, which allows it to keep track of the prey while chasing it. The prey also has the camera facing towards the rear, but it moves in the opposite direction, which allows it to observe the predator while running away from it [1].

4 Discussion and Conclusions

The experiments documented in this article investigated evolving the camera direction of predator-prey robots, by rerunning a number of previously performed experiments documented elsewhere [1, 4] (and summarized in Table 1). That is, robots that had cameras in previous experiments now had the ability to evolve the direction of the camera in addition to evolving view angle and view range.

The results indicate that implicitly designed constraints, which are results of human designer biases can greatly affect the behaviors evolved. In the experiments performed by Floreano and Nolfi [7], the camera was set facing forward on the robots, i.e. in the same direction as the six infrared sensors (cf. Fig. 1). By doing so, two implicit constraints were designed. First, the forward direction of the robot is assumed, and second, the robot is considered to benefit from having the camera facing in the same direction. The results of the experiments documented here, where the direction of the camera was included in the genotype, indicate two things. Firstly, both robots 'prefer' to have the camera facing in a 'backwards' direction (cf. Fig. 3), i.e. in the same direction as the two infrared sensors at the rear (cf. Fig. 1). Secondly, the predator starts to move in a backward direction while the prey moves in a forward direction. The prey benefits from this, avoiding both the predator and crashing into walls while the predator simplifies discriminating between walls, which it should avoid, and the prey, which it should capture.

In the experiments described in this article both robots had the same basic hardware structure (e.g. the camera). However, depending on the robots task different morphological structures were evolved. Hence, we conclude that by allowing the evolutionary process to 'co-evolve' both 'brain and body' of the robots, task-dependent visual morphologies and behavioral strategies are evolved, demonstrating a tight coupling between 'body and brain' in autonomous robots.

References

1. Buason, G. (2002a). *Competitive co-evolution of sensory-motor systems*. Masters Dissertation HS-IDA-MD-02-004. Dept. of Computer Science, University of Skövde.
2. Buason, G. (2002b). Competitive co-evolution of sensory-motor systems – Appendix. Technical Report HS-IDA-TR-02-004. Dept. of Computer Science, University of Skövde.
3. Buason, G. & Ziemke, T. (2003). Competitive Co-Evolution of Predator and Prey Sensory-Motor Systems. In: Cagnoni et al. (eds.) *Applications of Evolutionary Computing – EvoWorkshops 2003* (pp. 605-615). Berlin: Springer-Verlag.
4. Buason, G. & Ziemke, T. (in press) Co-Evolving Task-Dependent Visual Morphologies in Predator-Prey Experiments. *Genetic and Evolutionary Computation Conference*, to appear.
5. Carlsson, J. & Ziemke, T. (2001). YAKS – Yet Another Khepera Simulator. In: Rückert, Sitte & Witkowski (eds.), *Autonomous minirobots for research and entertainment* (pp. 235–241). Paderborn, Germany: HNI-Verlagsschriftenreihe.
6. Cliff, D. & Miller, G. F. (1996). Co-evolution of pursuit and evasion II: Simulation methods and results. In: P. Maes, M. Mataric, J.-A. Meyer, J. Pollack & , S. W. Wilson (eds.), *From animals to animats IV: Proceedings of the fourth international conference on simulation of adaptive behavior (SAB96)* (pp. 506–515). Cambridge, MA: MIT Press.
7. Floreano, D. & Nolfi, S. (1997a). God save the Red Queen! Competition in co-evolutionary robotics. In: J. R. Koza, D. Kalyanmoy, M. Dorigo, D. B. Fogel, M. Garzon, H. Iba, & R. L. Riolo (eds.), *Genetic programming 1997: Proceedings of the second annual conference*. San Francisco, CA: Morgan Kaufmann.
8. Floreano, D. & Nolfi, S. (1997b). Adaptive behavior in competing co-evolving species. In P. Husbands, & I. Harvey (eds.), *Proceedings of the fourth European Conference on Artificial Life*. Cambridge, MA: MIT Press.
9. Floreano, D., Nolfi, S. & Mondada, F. (1998). Competitive co-evolutionary robotics: From theory to practice. In: R. Pfeifer, B. Blumberg, J-A. Meyer, & S. W. Wilson (eds.), *From animals to animats V: Proceedings of the fifth international conference on simulation of adaptive behavior*. Cambridge, MA: MIT Press.
10. Lund, H., Hallam, J. & Lee, W. (1997). Evolving robot morphology. In: IEEE International Conference on Evolutionary Computation (ed.), *Proceedings of IEEE fourth international conference on evolutionary computation* (pp. 197-202). New York: IEEE Press.
11. Nolfi, S. & Floreano, D. (1998). Co-evolving predator and prey robots: Do 'arms races' arise in artificial evolution? *Artificial Life*, 4, 311–335.
12. Nolfi, S. & Floreano, D. (2000). *Evolutionary robotics: The biology, intelligence, and technology of self-organizing machines*. Cambridge, MA: MIT Press.
13. Nolfi, S. & Floreano, D. (2002). Synthesis of autonomous robots through artificial evolution. *Trends in Cognitive Sciences,* 6, 31–37.
14. Pollack, J. B., Lipson, H., Hornby, G. & Funes, P. (2001). Three generations of automatically designed robots. *Artificial Life,* 7, 215–223.

Optimal Morphology of a Biologically-Inspired Whisker Array on an Obstacle-Avoiding Robot

Miriam Fend[1], Hiroshi Yokoi[2], and Rolf Pfeifer[1]

[1] Artificial Intelligence Laboratory, Department of Information Technology, University of Zurich,
{fend,pfeifer}@ifi.unizh.ch,
[2] Research Group of Complex Systems Engineering
Graduate School of Engineering,
Hokkaido University
yokoi@complex.eng.hokudai.ac.jp

Abstract. Whiskers are versatile sensors for short-range navigation and exploration that are widespread in many animal species, especially in rodents. Their arrangement is in very precise rows and arcs on both sides of the animal's head. The controlled variations between species and the conservation within a species indicates a prominent role of their morphology for their functioning. Because of their enormous potential for robotic applications, we constructed a robot with two multi-whisker arrays, and evaluated the morphology and arrangement of the whiskers in an obstacle-avoidance task. We found that an artificial whisker array uncommon in nature performed best, and we argue that this might be explained by additional functions whiskers have in animals.

1 Introduction

Rodents and many other animals use whiskers for exploration of close objects, and for navigating in complex environments and darkness [11]. As many of them are nocturnal animals, they have to rely on other sensory information than vision. With their whiskers, they are able to discriminate textures of different roughness by actively whisking the surfaces [3] [5]. Furthermore, animals use whiskers extensively as distance and collision sensors. Fast and easy evaluation of distances to objects is crucial when moving at high speed, e.g. when fleeing from predators or when hunting. Such evaluation of sensory information can be greatly facilitated by an appropriate morphology of the sensor distribution. Thus it is not surprising that the spatial arrangement of whiskers is highly conserved within each species, where each whisker lies on a precisely defined point in the grid of rows and arcs of the whisker pad. Additionally, the length of the whiskers always increases from the snout to the back of the animal's head [2]. The use of this arrangement has not been investigated so far and is difficult to vary experimentally in animals.

Despite their enormous potential as close-distance touch sensors that do not involve heavy contact with objects [13] [12] and that are independent of light, whiskers have not received a lot of attention from roboticists. Mainly, whiskers have been used as binary touch [14] or as strain sensors [7]. It has been shown within an engineering approach

W. Banzhaf et al. (Eds.): ECAL 2003, LNAI 2801, pp. 771–780, 2003.

[6] that they can be used for fast obstacle avoidance on a robot, but so far it remains unknown, what the optimal arrangement of a whisker array on a robot is.

Since avoiding collisions is of paramount importance both in mobile robots and in animals, we used obstacle avoidance to evaluate different whisker morphologies on a robot. Obstacle avoidance is considered one of the basic behaviors in robots and has been implemented with many different methods and sensors. Many of them use vision, for example by measuring optic flow [4]. These approaches depend on an illuminated environment and often involve computationally expensive image processing. A whisker system on the other hand can work in complete darkness and will be computationally simple, if a good morphology for the sensor distribution and size is chosen.

Fig. 1. Picture of the Khepera robot with its whiskers.

In order to investigate what a suitable morphology of might be, we have built a multi-whisker array and mounted it on a robot. In this study, different whisker morphologies were compared as to how long each morphology moved through an experimental arena without getting stuck, and how well the free area was explored. We found that changing the morphology of the whisker array affected the performance of the robot, but interestingly the most successful arrangement was not the one commonly found in animals. A possible explanation for this result might be the multi-functionality of whiskers in nature. It may well be that the natural morphologies are a compromise between the use of whiskers as sensitive tactile organs and as fast collision detectors.

2 Experimental Setup

2.1 Hardware

The whisker sensor we use consists of a capacitor microphone with a natural rat whisker attached to it [8]. Physical force on the whisker hair deforms the microphone membrane and results in a voltage signal different from the resting state. This signal from the microphone is amplified on the robot and sampled on an external computer.

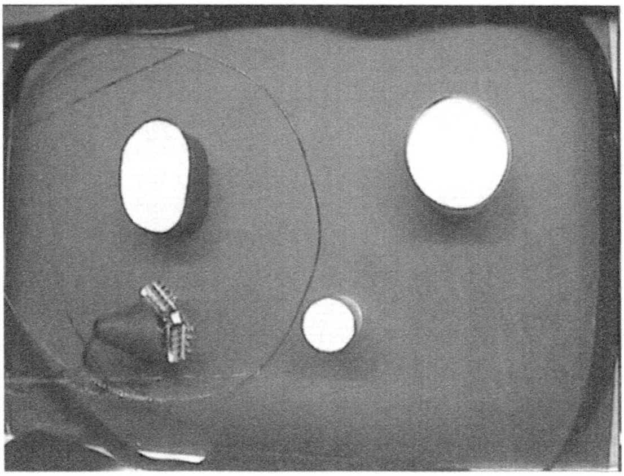

Fig. 2. Photograph of the robot arena as seen from the overhead camera. The obstacles were marked with light paper for demonstration only. During the experiment, they were black to avoid problems with the feature tracker. The robot can be seen at its starting position in the lower left corner.

One whisker array consists of 8 whiskers, which are arranged in two rows of four whiskers each. Two such arrays are mounted on a Khepera II robot [9] such that their orientation relative to the robot body can be adjusted within about 60°. In analogy to whiskers found in many different animal species, the length of the robot whiskers was increased from front to back in one condition (bottom row, figure 4), and from back to front in the other experimental condition (top row, figure 4). The latter does not correspond to a biological whisker array.

2.2 Experimental Environment

The robot environment consisted of an arena of 100x70cm. Inside were three obstacles of different size and shape, as shown in figure 2.

For every run, the robot was manually placed at the same starting position with the same orientation.

2.3 Control of the Robot

Since the focus of this work was on how a suitable sensor morphology could simplify behavioral control, the robot was equipped with a plain reflex behavior similar to the classic Braitenberg vehicle [1]. By default it moved forward with constant speed. If on one side the whiskers were stimulated above threshold, the robot turned away from the stimulated side by 45°. If both whisker arrays were stimulated above threshold, the robot drove backwards for 500 ms and then turned away from the side with the higher activity value.

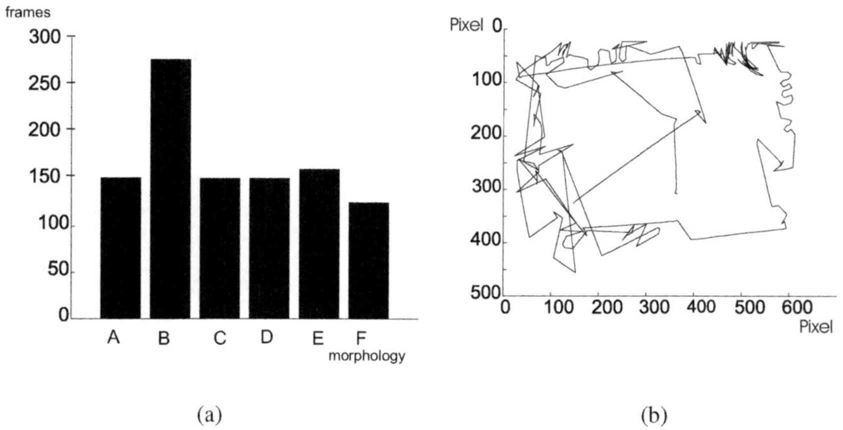

Fig. 3. a) Number of frames, in which the robot changed its position with respect to the previous frame. Three frames are approximately one second. b) Trajectory of one run of morphology B. The axes are labelled according to the pixel the robot was found at.

For the computation of the activity value, a baseline value for each whisker was determined at the beginning of each run. For every time-step of 62.5 ms (corresponds to 256 values per whisker), an activity value was computed for each whisker array by summing up all the differences from the baseline value. Sampling of the whisker signals was done at 4096 Hz per whisker. The parameters were chosen heuristically.

2.4 Tracking the Robot and Analysis

A CCD video camera was mounted above the experimental area, which was connected to a separate computer. During each run, every 300 ms a picture of 640x480 pixels was captured. The robot was marked with small, white patches that made it easily identifiable for the feature tracker. For the tracking analysis, the KLT library was used [10]. The tracking algorithm returned the (x,y) coordinates of the markers at each time step. These coordinates were used for the reconstruction of the robot trajectories, as shown in a sample trajectory in figure 3(b).

To quantify how well the robot moved through the whole environment, we computed the density of exploration for each morphology: We divided the arena in bins of 20x20 pixels and counted for all runs, how many times the robot was found in each bin. The results were plotted by assigning each bin a grayscale value corresponding to how often the robot was found in this bin (figure 5).

3 Results

For the comparison of the different morphologies (figure 4), we defined two measures: firstly, we counted the number of frames, in which the robot changed its position (figure

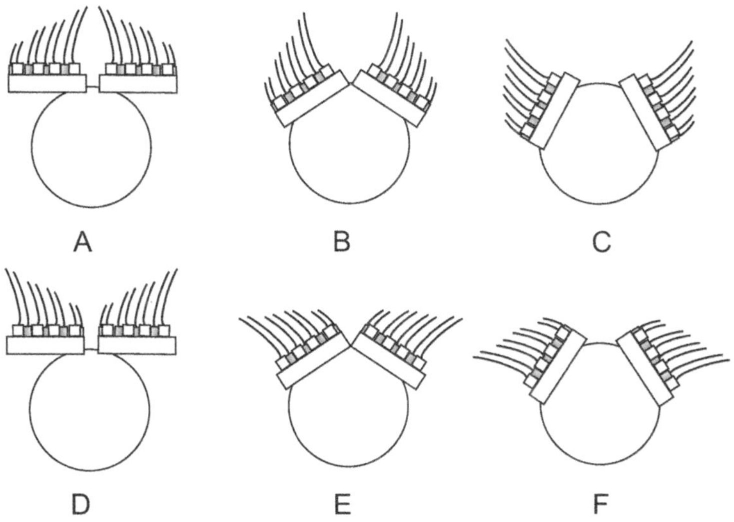

Fig. 4. Morphologies of the whisker arrays. In the top row (**A, B, C**), the long whiskers are in the center. In the bottom row (**D, E, F**), the long whiskers are on the side, corresponding to the natural arrangement. From left to right the angle between the two arrays decreases from straight (180°) to sharp (about 60°).

3(a)). This number corresponds to the amount of time, the robot moved around. Secondly, we compared how well the different morphologies were able to cover the experimental area evenly. This can be seen qualitatively in the cumulated trajectories depicted in figure 6 and more quantitatively in the densities of exploration (figure 5) and in table 1. Each morphology was tested in 20 runs, always starting at the same position and with the same orientation. If the robot got stuck, the run was interrupted, but counted. If the robot did not get stuck, the run terminated after 2500 cycles of computing the activation. In the morphologies D, E and F the natural arrangement was tested, which had the longer whiskers towards the periphery or back of the robot and short whiskers in the center. For the morphologies A, B and C, the longest whiskers were in the center and the shortest whiskers pointed outwards, in contrary to what is found in biological whisker arrays. Within each of these two configurations, the angle of the whisker array with respect to the body axis of the robot was varied from all whiskers facing to the front (A, D), a middle position with a small angle of about 20° (B, E), up to an almost parallel arrangement (C, F).

The whisker morphology B performed best on both the measures density of exploration and duration of exploration. The robot with this morphology moved around the area for about 100 s on average, whereas the robot with other morphologies moved for only $\approx$ 50 s. From the cumulated trajectories (figure 6) and the sample run in figure 3(b) of this robot it can be seen that robot B succeeded in covering the whole arena. In this arena, there are two narrow passages, namely in the upper left and especially the upper right corner of the arena (figure 2). Only robot B was able to move around the obstacle in the upper right corner. In the cumulated trajectories one can see that the morphologies E and F never even approached this narrow part, but turned away early. While all morphologies passed through the second narrow passage at least once, one can see from the trajectories that both C and F managed to do so only very few times. All figures of the trajectories (figure 6) and densities (figure 5) show a clear diagonal from the upper left to the lower right. This is the first movement of the robot at the start of each run.

Comparing the trajectories and the percentages of covered area listed in table 1 one can very well see differences in behavior between the morphologies. While robot B is clearly the best at moving through its environment, one can see that in general the arrangement with longer whiskers in the center of the robot gives better coverage of the experimental field. For example, the morphology with whiskers pointing straight to the front and long whiskers in the middle (A) covers 88% of the area that robot B covered. The more natural morphologies D, E and F perform worse, only between 72% and 82% of the performance of morphology B which corresponds to covering between 53% and 60% of the total experimental area.

Table 1. Experimental area covered by the different robot morphologies. In the second column the whole camera image is considered. In the third column, the area that was entered by robot B was taken as the reference. In the fourth column, the number of bins is listed that each robot morphology was able to enter more than five times. This shows how evenly the different robots covered the area

Morphology	% total area	% accessible area	> 5 entries
A	65	88	200
B	73	100	447
C	62	85	190
D	60	82	178
E	59	80	185
F	53	72	136

4 Discussion

During the last years, biomimetic and biologically inspired robots have become more and more attractive. Several reasons account for this changed focus: on the one hand, engineers have learned to admire the refined designs and adaptivity of biological systems. On the other hand, biologists are more and more interested in using artificial systems as a testbed for their research. For example, it is almost impossible to change the arrangement of whisker sensors in mice, even though mouse genetics are the best studied mammalian genetics. To find out, why the topology of the sensors is so well preserved, why there are longer whiskers in the back and shorter whiskers in the front of the head is an intriguing question for both biologists trying to understand this sophisticated somatosensory system

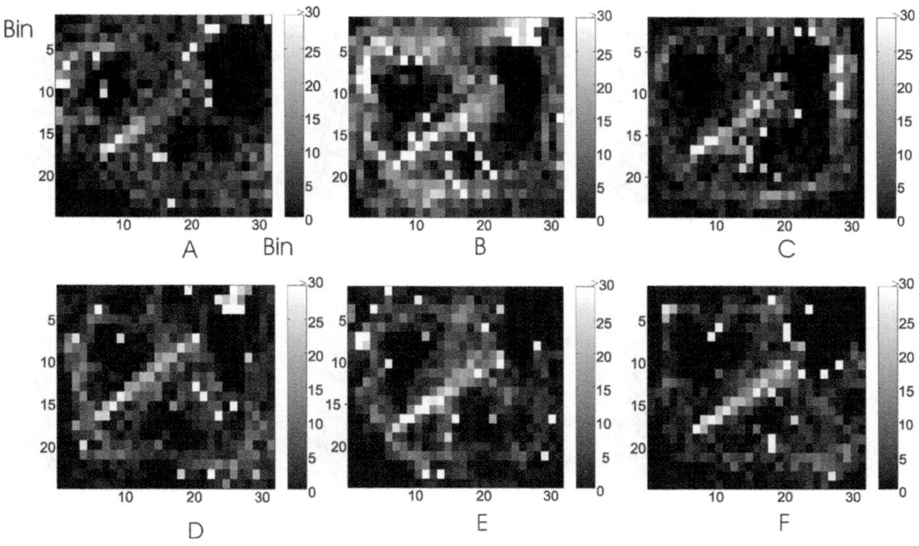

Fig. 5. Time spent in each 20x20 pixel bin. The plots A-F correspond to the morphologies A-F in figure 4. On the x and y-axis the bin numbers are marked.

and for engineers concerned with building successful robots. Our experiments show that for the simple obstacle avoidance task studied, the natural morphology is not ideal. We first discuss, which features determine the performance of the robot. Then we speculate, why the morphology we found ideal for the robot is not the solution selected for whiskers during evolution.

The relevant factors for the robot performance at the obstacle avoidance task are basically the following. How fast can the robot react after it has detected an obstacle? While animals can very well adjust their speed to how well they can perceive their environment - just as we move slower when the light is fading -, the robot moves at constant speed and turns with a constant angle. For larger obstacles in its path, it often has to turn two or three times to be sure it is avoiding. So the sooner the robot detects that it is heading towards an obstacle, the better it will avoid it. Long whiskers in the front are thus advantageous (morphologies A, B and C).

When evaluating the area covered by the robot, it is also important to consider the physical space that the robot needs. This is determined by two factors. First by the dimensions of the robot itself. The straight configurations A and D for example are the widest of the six morphologies. Thus, although it was not physically impossible for them to pass through the narrow parts of the experimental area, it was more difficult. The robot can also take up more space in its sensory dimensions. This happens mainly in the morphologies C and F. Here the long whiskers point directly to the side, so even in situations, where there is still extra space for the physical dimensions of the robot, it will turn away because in its sensory world, it is already very close to an obstacle. Another shortcoming of the mostly sideways oriented morphologies is that the front of

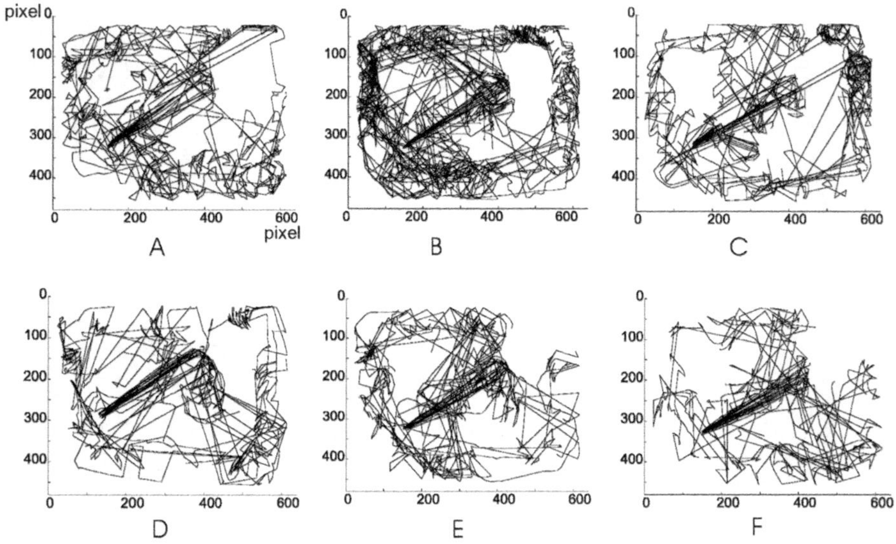

Fig. 6. Cumulated trajectories of 20 runs for every morphology A-F. The morphologies refer to the drawings A-F in figure 4. On the x and y-axis, the position of the robot is indicated as the pixels of the video image. The obstacles can be identified as area never entered by the robot.

the robot is not covered very well in the sensory space. The robot often does not detect the obstacle in time to be able to react adequately. So why is the morphology ideal in our artificial system not also found in animals? What are the differences of our task with the demands in natural systems?

First, the robot relies only on the whiskers, while animals often have visual information as well. Cats for example have a very sensitive whisker system, but they also have excellent vision. Most of the time, they do not rely solely on the tactile information. Many rodents on the other hand are night-active and have very poor vision. For these species, whiskers are of high importance for their navigation. But they also use them intensively as touch organs. A closer look at the morphologies in figure 4 shows that morphology E, which performed poorly on the obstacle avoidance, has all the whisker tips in one plane. It is thus ideally suited for palpation of objects in the front of the animal. Since whiskers are important tactile organs for these animals, evaluating the morphology of the whisker array only as a collision sensor possibly might miss features important for its other functions. Possibly, the advantages for tactile exploration outweigh the detriments of the natural whisker arrangement for obstacle avoidance. It should also be considered that most animals can move their whiskers actively. They can thus vary the exact position of their long and short whiskers much more flexibly than we can in the artificial system at the moment.

In addition to these enhanced motor capabilities, animals can also learn and adapt to different environments,while the robot controller used did not contain any possibility for learning or evolution. The aim of this sutdy was to find a morphology allowing for such a simple controller, but it cannot be excluded that the performance of different morphologies could be improved if learning was included or different environments were used. In future experiments we will look at the interplay of whisker morphology and learning of obstacle avoidance. Artificial evolution of a simulated agent with whiskers will allow us to thoroughly investigate a multitude of environments and whisker arrangements.

5 Conclusion

Overall it can be concluded that the whisker sensors are useful for obstacle avoidance even in narrow passages if an optimal morphology is chosen. Their use as touch sensors might require a different spatial arrangement than was found optimal for obstacle avoidance. To be able to study this aspect of behavior for the morphology, we will build an active artificial whisker array. With this active array we will also study the sensory processing of tactile perception and investigate, how distinction of surfaces and shapes can be achieved with just a couple of hairs.

Acknowledgements. This research has been supported by the IST-2000-28127 European project (AMOUSE). The natural rat whiskers were kindly provided by Mathew Diamond, SISSA, Cognitive Neuroscience sector, Trieste, Italy. The preamplifier board was built by Dirk Naumann, ATM Computing. Regina Mudra from the Institute of Neuroinformatics, ETH Zurich gave valuable help and advice on tracking. Thanks to Gabriel Gomez for his support in programming.

References

1. V. Braitenberg. *Vehicles. Experiments in synthetic psychology.* MIT Press, Cambridge, 1984.
2. M. Brecht, B. Preilowski, and M.M. Merzenich. Functional architecture of the mystacial vibrissae. *Behavioral Brain Research*, 84(1-2):81–97, 1997.
3. G.E. Carvell and D.J. Simons. Biometric analyses of vibrissal tactile discrimination in the rat. *Journal of Neuroscience*, 10(8):2638–2648, 1990.
4. David Coombs, Martin Herman, Tsai-Hong Hong, and Marilyn Nashman. Real-time obstacle avoidance using central flow divergence and peripheral flow. *International Conference on Computer Vision*, pages 276–283, 1995.
5. E. Guic-Robles, C. Valdivieso, and G. Guajardo. Rats can learn a roughness discrimination using only their vibrissal system. *Behavioural Brain Research*, 31(3):285–289, 1989.
6. D. Jung and A. Zelinsky. Whisker-based mobile robot navigation. *Proceedings of the IEEE/RSJ International Conference on Intelligent Robots and Systems (IROS)*, 2:497–504, 1996.
7. M. Kaneko, N. Kanayama, and T. Tsuji. Active antenna for contact sensing. *IEEE Transactions on robotics and automation*, 14(2):278–291, 1998.
8. M. Lungarella, V.V. Hafner, R. Pfeifer, and H. Yokoi. An artificial whisker sensor for robotics. *Proceedings of the IEEE/RSJ International Conference on Intelligent Robots and Systems (IROS)*, pages 2931–2936, 2002.
9. Francesco Mondada, Edoardo Franzi, and Paolo Ienne. Mobile robot miniaturisation: A tool for investigation in control algorithms. *Proceedings of the Third International Symposium on Experimental Robotics*, pages 501–513, 1993.

10. Jianbo Shi and Carlo Tomasi. Good features to track. *IEEE Conference on Computer Vision and Pattern Recognition (CVPR'94)*, pages 593–600, Jun 1994.
11. S.B. Vincent. The function of the vibrissae in the behavior of the white rat. *Behav Monographs*, 1(5), 1912.
12. S.S.M. Wang and P. Will. Sensors for computer controlled mechanical assembly. *The industrial robot*, March:9–18, 1978.
13. P. Will and D.D. Grossman. An experimental system for computer controlled mechanical assembly. *IEEE Transactions on computers*, c-24(9):879–888, 1975.
14. U. Zimmer. Self–localization in dynamic environments. *IEEE/SOFT International Workshop BIES'95, Tokyo, Japan, May 30-31 1995.*, 1995.

Phase Transitions in Self-Organising Sensor Networks

Mark Foreman[1], Mikhail Prokopenko[2], and Peter Wang[2]

Ageless Aerospace Vehicles and Smart Spaces projects
Intelligent Interactive Technology
[1] CSIRO Telecommunications and Industrial Physics
[2] CSIRO Mathematical and Information Sciences
Locked Bag 17, North Ryde, NSW 1670, Australia
{mark.foreman, mikhail.prokopenko, peter.wang}@csiro.au

Abstract. In this paper we consider a multi-cellular sensing and communication network, embedded in an *ageless aerospace vehicle*, that is expected to detect and react to impact location, intensity and damage over a wide range of impact energies. In particular, we investigate self-organisation of *impact boundaries* enclosing critically damaged areas, and measure their spatiotemporal robustness. The presented quantitative information-theoretic techniques clearly identify phase transitions, separating chaotic dynamics from ordered and robust patterns.

1 Introduction

The research results presented in this paper were obtained as part of the joint CSIRO-NASA Ageless Aerospace Vehicle (AAV) project. The aim of the AAV project is to develop and critically examine concepts for integrated smart sensing and communication networks, with the ultimate goal of developing a self-monitoring, self-repairing aerospace vehicle [1,6,11].

A modular (multi-cellular) sensing and communication network is expected to detect, report on and react to impact location, intensity and damage over a wide range of impact energies, ranging from micro-particles to meteoroids. The network is expected to self-organise in the face of damage to its parts so that robust monitoring and reporting continues as long as possible. To achieve this, we modelled and simulated a multi-agent system made of "cells", that will not only form a physical shell for an aerospace vehicle, but will also have sensors, logic, and communications. The primary principle that is followed in our work is the *emergence of global response* as a result of interactions involving transfer of information embedded locally — in order to completely avoid or reduce the number of single points-of-failure. In other words, without centralised controllers, agents (cells) are expected to self-organise and survive on the basis of local, rather than global, information (no single agent has access to what everyone else is doing). Single cells may need to make fast and automatic responses to sudden damage, while collections of cells may solve more complex tasks, for example, produce an impact boundary with desired characteristics [6] or form a spanning tree connecting cells that detected non-critical impacts [11]. Importantly, the behaviour of each cell should be as simple as possible, in terms of the internal logic and the communication policies.

W. Banzhaf et al. (Eds.): ECAL 2003, LNAI 2801, pp. 781–791, 2003.

Economy of information is directly related to the ability to manufacture new and repair or replace damaged cells: the simpler the cell, the easier it is to repair/replace it.

Recent advances in sensor networks and micro-electro-mechanical devices led to the idea of localised algorithms, in which simple local node behaviours achieve a desired global objective [3,7], while communicating only with nodes within some neighbourhood. However, despite some progress, there is a lack of a unifying methodology underlying design of localised algorithms. The main question is how to produce and retain desirable emergent behaviour while avoiding potentially adverse patterns of agents' interaction. Some promising results have been reported by Nagpal [8], in context of amorphous computing, where programmable self-assembly was demonstrated using biologically-inspired multi-agent control. Nagpal defined a small class of primitives (eg., gradients, neighbourhood query, polarity inversion, cell-to-cell contact, etc.), a set of global operations, and a translation implementing the global operations as localised algorithms using the set of primitives. However, the question of how to obtain and inter-connect global operations themselves is left unanswered. Moreover, the task of discovering the correct primitives and translating them into localised agent programs would have to be repeated for each domain.

The problem of global response engineering in multi-agent networks motivates our search for patterns and invariants in self-organisation of multi-agent systems monitoring and repairing impacts on the AAV. Our proposed methodology is based on an iterative process including the following steps: a) forward simulation leading to emergent behaviour (for a class of localised algorithms dealing with impacts of various strengths); b) quantitative measurement of bounded emergent behaviour (using information-theoretic metrics for phase transitions); c) evolutionary modelling of the desired global emergent behaviour, where the fitness functions correspond to the metrics obtained at step b). The work on impact boundaries [6] and impact networks [11] modelled the step a). This paper studies the step b).

2 Impact Boundaries

Ideally, a modular multi-cellular AAV skin should trace impact boundaries and spread despite connectivity disruptions and cell failures — analogous to the clotting of a wound on a mammal and the regeneration of neurons by re-growing severed axons within a myelin sheath. One of the immediate tasks is the formation of impact boundaries enclosing critically damaged areas. It is highly desirable that such boundaries form continuously connected closed circuits, and are robust to fluctuations caused by proximity to the impact. In short, the aim is to achieve spatiotemporal stability in impact boundaries.

Let us briefly sketch here the AAV simulator and the algorithms producing self-organising continuous and closed impact boundaries that are presented in more detail in [6]. In the AAV simulator events may occur in parallel, so the simulation is, essentially, a state machine that sweeps through the cells, updating their current state on a regular basis. Cells are represented as objects (polygons) on a two-dimensional plane, where they interact *only* with their immediate neighbours in von Neumann neighbourhood, through connected (geometrically overlapping) communication ports. Furthermore, the

AAV simulator has the ability to simulate simple environmental effects, such as the incidence of small impacts.

An important characteristic of a critical impact is that, typically, energy of the impact dissipates throughout some neighbourhood, destroying cells close to the point of the impact (epicenter) and damaging the communication links between the survived cells. The communication damage is simulated by assigning a probability of a bit error dependent on proximity of the affected communication port to the epicenter. The proximity dependency underlying the probability distribution can be modelled in many ways, and we investigated a range of functions — from linear to exponential decreases.

Our first target was formation of a boundary that is not necessarily continuously connected as a closed circuit, but is well placed in separating the cells that suffered unrecoverable communication damage (including those that were completely destroyed) from the cells that are able to communicate to their normal functional capacity. The initial algorithm enables self-organisation of both an internal "scaffolding" and the "frame" of the desired closed boundary, regardless of cells shape (triangular or square), and includes the following communication policies and behaviours:

(i) *every cycle each cell sends an "OK" message to all its neighbours;*

(ii) *upon receiving the "OK" message each cell replies with an "Acknowledgement" message;*

(iii) *if all neighbour cells failed to communicate the "OK" message, switch to the scaffolding state* S_s*;*

(iv) *if at least one neighbour cell failed to communicate the "OK" message and at least one neighbour cell did communicate the "OK" message, switch to the frame boundary state* S_f.

(v) *if all neighbour cells failed to communicate the "OK" message or there is no "Acknowledgement" message from any neighbour cell, stop sending messages.*

In order to produce *continuously* linked closed boundary, we need the following:

(vi) *if the cell state is* S_f*, and there are at least two communicating neighbours* β_1 *and* β_2*, switch to the closed state* S_c.

(vii) *if the cell state is* S_c*, 1) determine a (the) cell* α *that failed to communicate; 2) determine two communicating neighbour cells* β_1 *and* β_2 *nearest to the cell* α *on opposite sides (clockwise and counter-clockwise relative to* α*); 3) establish a boundary link between ports to* β_1 *and* β_2*; 4) map the directions to* α*,* β_1 *and* β_2 *to a desired direction* γ*; and 5) send a "Connect(*τ, γ*)" message to these neighbours with a "time to live" parameter* τ*;*

(viii) *upon receiving "Connect(*τ, γ*)" message from the cell* α*, if the cell state is not* S_c*, 1) switch to the state* S_c*; 2) determine the communicating neighbour cell* β *nearest to the cell* α *in the direction* γ*; 3) establish a boundary link between ports to* α *and* β*; 4) if* $\tau > 0$*, map the directions to* α *and* β *to a desired direction* γ' *and 5) send a "Connect(*$\tau - 1, \gamma'$*)" message to the neighbour* β.

The "time to live" parameter τ determines the maximum amount of time the message may live in the network, effectively setting a limit on the number of cells the message may pass through before being discarded. In general, the described policy achieves the desired robustness and continuity of self-organising impact boundaries for a variety of cell shapes, impact energy dissipation profiles and communication damage probability models (Figure 1). However, on some occasions the boundary is erratic and unstable — some cells frequently change their states from "frame" to "scaffolding" to normal, etc. The reason for this spatial and temporal instability is that, a simple detection of a missed acknowledgement may sometimes be insufficient — especially when the involved cells are on a periphery of the damaged area. In other words, in order to detect a persistent failure in communications with an adjacent cell, the cell needs to filter away spurious occasional miscommunications. The length ρ of such a filter (i.e., the cell-to-cell communication history, allowed to be kept by each cell), is an important parameter, directly affecting spatial and temporal stability of self-organising impact boundaries. Our experiments verified that larger values of ρ lead to stable boundaries, while $\rho = 0$ (no hysteresis at all) could not produce any boundaries (Figure 2).

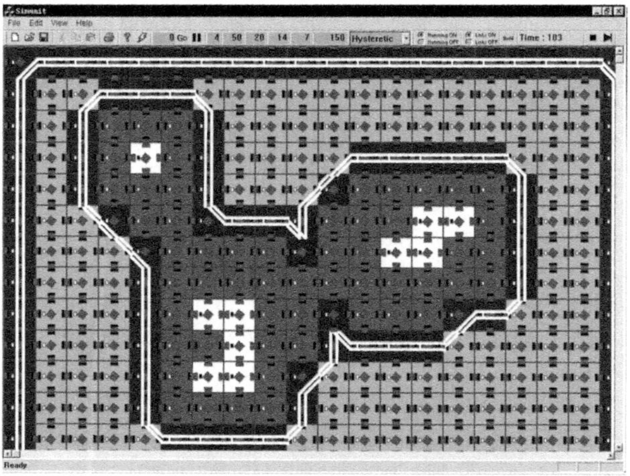

Fig. 1. A stable impact boundary: square cells ($\rho = 10$). White cells are destroyed, dark-grey cells form "scaffolding", black cells form "frame". Boundary links are shown as white double-lines.

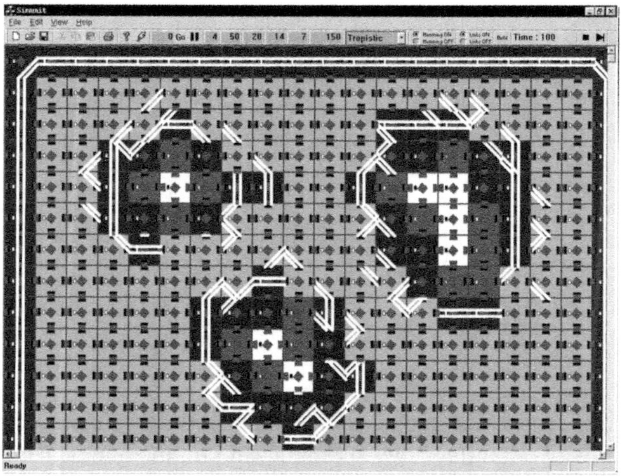

Fig. 2. An unstable impact boundary: square cells with $\rho = 0$.

3 Spatiotemporal Stability and Phase Transitions

The analysis of spatiotemporal stability of self-organising impact boundaries with respect to the parameter ρ is the primary subject of our investigation. In particular, we characterise dynamics of multi-cellular impact boundaries in terms of generic information-theoretic properties, such as the Shannon entropy, and pinpoint certain phase transitions.

Information-theoretic methods are applied in many areas exhibiting multi-agent interactions. For instance, Cellular Automata (CA) are a well-studied class of discrete dynamical systems with emergent behaviour resulting from local and short range interactions, and where information-theoretic measures of complexity (such as Shannon entropy of certain frequency distributions) were effectively used to categorise and classify distinct emergent configurations and phase transitions between them [13,5]. Langton has shown in his seminal work [5] that an increase in the mutual information (defined as a function of individual cell entropies for a particular value of the λ parameter) is an indication of a phase transition from "order" to "chaos". Wuensche [13] has used a similar quantitative metric — variance of input-entropy over time — in classifying rule-space of 1-dimensional CAs into ordered, complex and chaotic cases, related to Wolframs's qualitative classes of CA behaviour [12]. It has also been pointed out by Suzudo [10] that the entropy trajectory (plotting spatial entropy against temporal entropy) is a useful descriptor for a variety of self-organised patterns. For example, certain (non-complex) types of CA have simple entropy trajectory and do not renew their self-organised spatiotemporal pattern. Complex CA have an irregular entropy trajectory in the temporal-spatial entropy plane (similar to Figure 7), and renew the pattern.

The information-theoretic analysis of phase transitions is typically based on the notion of *mutual information*. For instance, Langton [5] investigated the mutual information of CA, defined as a simple function of the individual cell entropies, $H(A)$ and $H(B)$, and the entropy of the two cells considered as a joint process, $H(A, B)$:

$$I(A; B) = H(A) + H(B) - H(A, B)$$

and related it to phase transitions. In particular, trajectories of entropy $H(A)$ and mutual information $I(A; B)$ between a cell and itself at the next time-step were obtained while varying the parameter λ — the ratio of cells with a given property ("live" cells, for example). Interestingly, the individual cell entropy $H(A)$ was increasing with λ, and showed a discrete jump between low and high entropy values. This evidence pointed to a first-order phase transition, similar to that observed between the solid and fluid phases of matter, however the fact that the gap is not completely empty suggested the possibility of second-order transition [5]. Another intriguing feature was that the average mutual information has a distinct peak at the transition point:

> the average mutual information is essentially zero before the transition point, it jumps to a moderate value at the transition, and then decays slowly with increasing λ. The jump in the mutual information clearly indicates the onset of the chaotic regime, and the decaying tail indicates the approach to effectively random dynamics.

It is important to realise that there are two opposing forces shaping the trajectory of mutual information — individual diversity and interdependence. In the case studied by

Langton, increasing λ led to an increase in diversity and growth of individual entropy. At the same time, correlation and interdependence (reflected in the entropy of the joint process) were decreasing: at the start the cells were overly dependent, and at the end they were practically independent of each other. The "battle" between these two forces is the most intense at the transition point, or in other words, at the *edge of chaos*, where the system dynamics exhibits the most complexity.

A metric based on mutual information is suitable precisely because it incorporates these two tendencies into a single expression. Another way to achieve this combination is to use *a variance of input-entropy*. Wuensche [13] characterised rule-spaces of 1-dimensional cellular automata with the Shannon entropy of rules' frequency distribution. More precisely, given a rule-table (the rules that define a CA), the input-entropy at time step t is defined as

$$S^t = -\sum_{i=1}^{m} \frac{Q_i^t}{n} log \frac{Q_i^t}{n} ,$$

where m is the number of rules, n is the number of cells (system size), and Q_i^t is the look-up frequency of rule i at time t — the number of times this rule was used at t across the CA. The input-entropy settles to fairly *low* values for ordered dynamics, but fluctuates irregularly within a narrow *high* band for chaotic dynamics. For the complex CA, the input-entropy generally follows an attractor cycle, where order and chaos may predominate at different times causing the entropy to vary. A measure of the variability of the input-entropy curve is its variance or standard deviation, calculated over time. This variance, in a sense, captures mutual information contained in CA dynamics, combining both individual rules' diversity and their interdependence (correlation). Wuensche has convincingly demonstrated that only complex dynamics exhibits high variance of input-entropy, leading to automatic classification of the rule-space. Importantly, the peak of input-entropy variance points to a phase transition again.

We believe that a metric modelled after the input-entropy S^t may capture only the temporal stability of impact boundaries, but not the spatial stability of continuous and closed circuits. Put simply, the input-entropy S^t may be very low and may vary only slightly for temporally stable patterns that keep their disconnected shape. So it would have to be complemented with another metric designed for spatially connected patterns. The random graphs theory that dates back to seminal works of Erdos and Renyi [2] suggests to use the size of largest connected sub-graph (LCS) and its variance or standard deviation obtained across an ensemble of graphs, as an indicator of phase transitions in the graph connectivity. It is well-known that critical changes occur in connectivity of a directed graph as the number of edges increases — the size of the LCS rapidly increases as well and fills most of the graph, while the variance in the size of the LCS reaches a maximum at some critical point before decreasing [4]. A metric based on the variance in the size of the LCS may capture connectivity in impact boundaries. Again, by itself, it would be insufficient: a continuous boundary may change its shape over time, without breaking into fragments, and keeping the size of LCS almost constant. Together, temporal and spatial metrics may capture both important aspects — steadiness and continuity of impact boundaries.

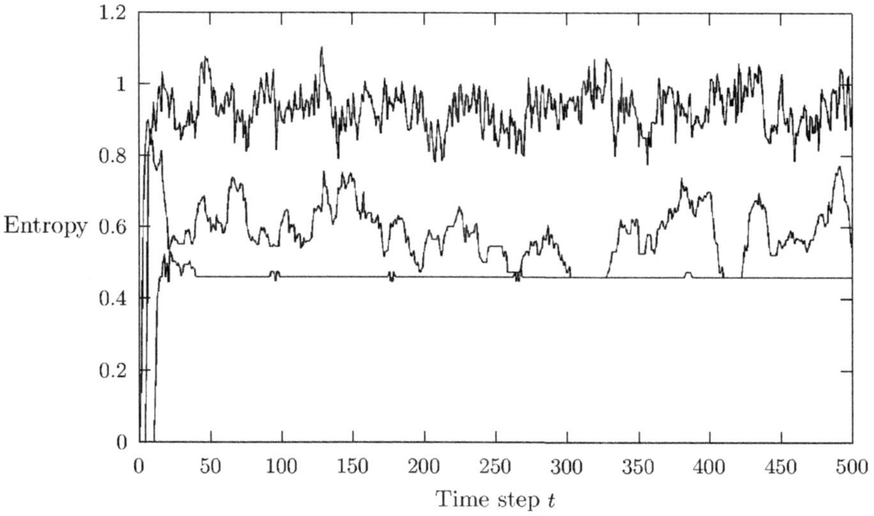

Fig. 3. Temporal entropy $H_{temp}(\rho)$ for $\rho = 0$ (top), $\rho = 4$ (middle), and $\rho = 10$ (bottom).

4 Experimental Results

In order to analyse temporal stability, we consider state changes in each cell at every time step. There are 6 symmetric boundary links possible in each square cell, connecting ports "left-right", "top-bottom", "left-top", etc. (Figure 1). Thus, there are 2^6 possible boundary states (including "no-boundary"), and $m = 2^{12}$ transitions. These transitions describe space-time dynamics inside an 1-cell neighbourhood (in CA terms), and exemplify a more general case of p-cell neighbourhoods (von Neumann neighbourhood with $p = 5$; Moore neighbourhood with $p = 9$, etc.). For each value of the parameter ρ (varying from 0 to 20), we calculate the entropy $H_{temp}(\rho)$ of a particular frequency distribution $S_i^t(\rho)$, where t is a time step, and i is a cell transition index: $1 \le i \le m$. Analogously to the analysis conducted by Wuensche [13], we define input-entropy as

$$H_{temp}(\rho) = -\sum_{i=1}^{m} \frac{S_i^t(\rho)}{n} \, log \frac{S_i^t(\rho)}{n} ,$$

where n is the system size (the total number of cells), and $S_i^t(\rho)$ is the look-up frequency of the transition i at time t. Each experiment simulates an impact at a predefined cell, and lasts 500 cycles; the first 2ρ cycles are excluded form the distribution $S_i^t(\rho)$ in order not to penalise longer history (filter) lengths. Figure 3 plots temporal entropy $H_{temp}(\rho)$ for different history lengths ρ. The $\rho = 10$ plot is almost a straight line — there is practically no variation in the entropy values, as the boundary is stable. We conducted three experiments for every value of ρ, calculating standard deviation $\sigma_{temp}(\rho)$. The plot of $\sigma_{temp}(\rho)$ is shown on Figure 4, clearly identifying a phase transition for some critical value of ρ between $\rho = 4$ and $\rho = 5$.

The analysis of spatial stability of impact boundaries is based on the concept of a connected boundary-fragment (CBF). A CBF is simply a set F of cells in the closed state

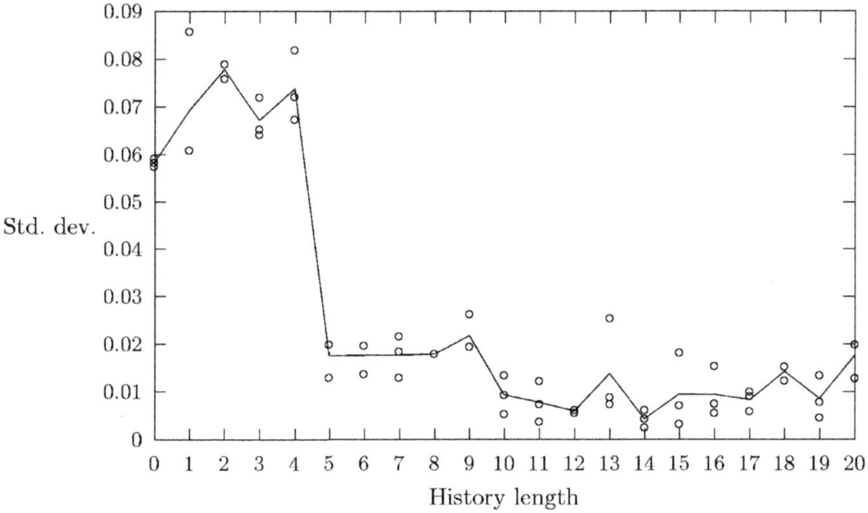

Fig. 4. Standard deviation $\sigma_{temp}(\rho)$ of temporal entropy $H_{temp}(\rho)$.

S_c such that every cell in F is connected with at least one other cell in F, and there exist no cell outside F which is connected to at least one cell in F (an analogue of a maximally connected subgraph or a graph component). We carried out 63 more experiments (3 for each value of ρ between 0 and 20), calculating the average size $H_{sp}(\rho)$ of CBF's in self-organising impact boundaries at each time-point, and its variance $\sigma_{sp}(\rho)$ over time. Figure 5 plots $H_{sp}(\rho)$ for $\rho = 0$ (bottom), $\rho = 4$ (strongly fluctuating), and $\rho = 10$ (almost straight line). It is clear that the impact boundary in this case has a size 24, and the hysteretic behaviour of length $\rho = 10$ leads to stable self-organisation. On the other hand, the cells without any hysteresis ($\rho = 0$) are not capable of self-organisation at all — an average CBF contains about 5 cells. The intermediate case ($\rho = 4$) is most unstable, however! Sometimes it creates boundaries twice as long as needed, and sometimes it collapses into a lot of small fragments. In other words, the hysteretic behaviour leads to "order", while tropistic behaviour, using a direct mapping from communication inputs to actions ($\rho = 0$), produces "chaotic" dynamics. Somewhere in the middle, we observe "complex" dynamics, and Figure 6 makes the phase transition apparent — again, between $\rho = 4$ and $\rho = 5$. The entropy trajectory in the temporal-vs-spatial plane is shown in Figure 7, plotting $\sigma_{sp}(\rho)$ against $\sigma_{temp}(\rho)$. The bottom-left "order" corner (the points after the phase transition) is clearly contrasted with the top-right "chaos" (the points before the phase transition). The small cluster in the bottom-centre region corresponds to tropistic behaviour ($\rho = 0$). Absence of any points in the middle indicates a first-order phase transition.

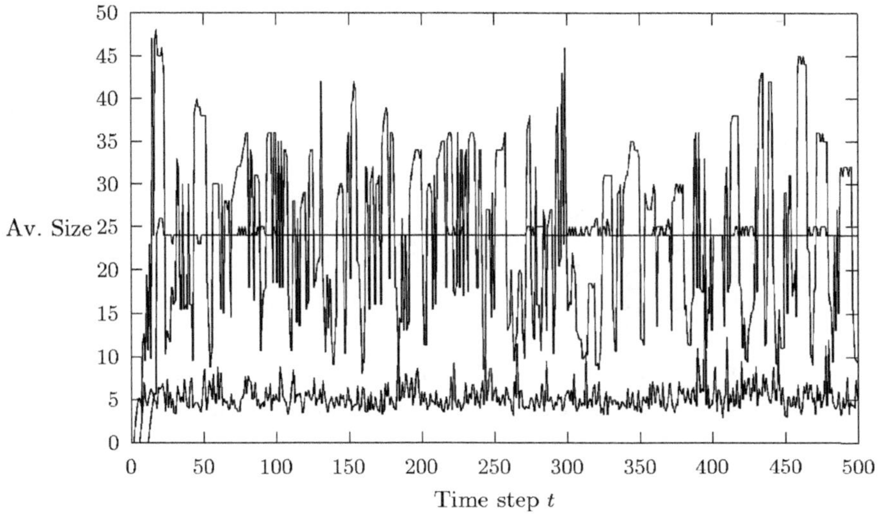

Fig. 5. Average size $H_{sp}(\rho)$ of CBF's, for $\rho = 0$ (bottom), $\rho = 4$ (strongly fluctuating), and $\rho = 10$ (almost straight line).

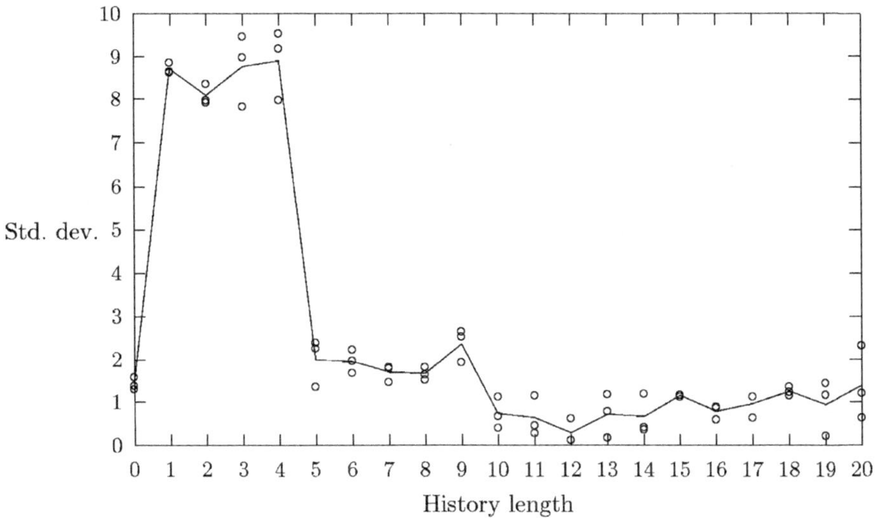

Fig. 6. Standard deviation $\sigma_{sp}(\rho)$ of average size $H_{sp}(\rho)$.

5 Conclusions

In this paper, we considered formation of impact boundaries that enclose critically damaged areas on a multi-cellular skin of ageless aerial vehicles. A number of quantitative techniques measuring spatiotemporal robustness of self-organising boundaries was pre-

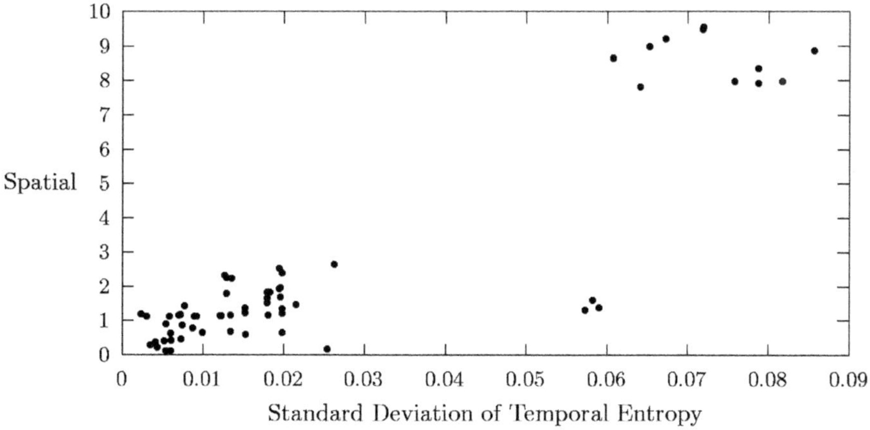

Fig. 7. The entropy trajectory: $\sigma_{sp}(\rho)$ against $\sigma_{temp}(\rho)$.

sented. In particular, we focussed on identifying the "edge of chaos", leading to discovery of evident phase transitions in sensor networks. In future, we intend to use the presented metrics as fitness functions — in *evolving the desired global behaviour*.

Acknowledgements. A part of this work was carried out under CSIRO contract to NASA. It is a pleasure to record our appreciation to Dr. Ed Generazio (NASA Langley Research Center) for his encouragement and support. The authors are grateful to other members of the AAV project, especially Philip Valencia, Don Price and Geoff Poulton. Mark Foreman's summer vacation scholarship was provided by CSIRO Telecommunication and Industrial Physics, and we would like to thank Geoff James in particular. The authors also acknowledge the helpful comments of the reviewers.

References

1. David Abbott, Briony Doyle, John Dunlop, Tony Farmer, Mark Hedley, Jan Herrmann, Geoff James, Mark Johnson, Bhautik Joshi, Geoff Poulton, Don Price, Mikhail Prokopenko, Torsten Reda, David Rees, Andrew Scott, Philip Valencia, Damon Ward, and John Winter. Development and Evaluation of Sensor Concepts for Ageless Aerospace Vehicles. Development of Concepts for an Intelligent Sensing System. NASA technical report NASA/CR-2002-211773, Langley Research Center, Hampton, Virginia.
2. Paul Erdos and Alfred Renyi. On the strength of connectedness of random graphs. Acta Mathematica Scientia Hungary 12, 261–267, 1961.
3. Deborah Estrin, Ramesh Govindan, John Heidemann, Satish Kumar. Next Century Challenges: Scalable Coordination in Sensor Networks. In Proceedings of the 5th Annual International Conference on Mobile Computing and Networks (MobiCOM '99), Seattle, 1999.
4. David Green. Emergent Behavior in Biological Systems. *Complexity International*, 1, 1994.
5. Christopher Langton. Computation at the Edge of Chaos: Phase Transitions and Emergent Computation. In S. Forest (ed.), *Emergent Computation*, MIT, 1991.

6. Howard Lovatt, Geoff Poulton, Don Price, Mikhail Prokopenko, Phil Valencia and Peter Wang. Self-organising Impact Boundaries in Ageless Aerospace Vehicles. To appear in Proc. of the 2nd Int'l Joint Conf. on Autonomous Agents and Multi-Agent Systems, 2003.
7. Nicholas J. Macias and Lisa J.K. Durbeck. Self-assembling circuits with autonomous fault handling. In Proceedings of NASA/DoD Conference on Evolvable Hardware, 2002.
8. Radhika Nagpal. Programmable self-assembly using biologically-inspired multiagent control. In the Proceedings of the First International Joint Conference on Autonomous Agents and Multi-Agent Systems (AAMAS'02), Bologna, Italy, July 2002.
9. Claude E. Shannon and Warren Weaver. *The Mathematical Theory of Communication*. University of Illinois Press, 1949.
10. Tomoaki Suzudo. The entropy trajectory: A perspective to classify complex systems. In Proceedings of the Int'l Symposium on Frontiers of Time Series Modeling, Tokyo, 2000.
11. Peter Wang, Philip Valencia, Mikhail Prokopenko, Don Price, and Geoff Poulton. Self-reconfigurable sensor networks in ageless aerospace vehicles. Submitted to the 11th International Conference on Advanced Robotics, ICAR-03, Portugal.
12. Stephen Wolfram. Universality and Complexity in Cellular Automata. *Physica D*, 10, 1984.
13. Andrew Wuensche. Classifying Cellular Automata Automatically: Finding gliders, filtering, and relating space-time patterns, attractor basins, and the Z parameter. *Complexity*, Vol. 4, no. 3, 47–66, 1999.

Artificial Metabolism: Towards True Energetic Autonomy in Artificial Life

Ioannis Ieropoulos[1], Chris Melhuish[1], and John Greenman[2]

[1] Intelligent Autonomous Systems Lab, Department of Computing, Engineering and Mathematical Sciences, The University of the West of England, Coldharbour Lane, Frenchay, Bristol, BS16 1QY, United Kingdom
{Ioannis2.Ieropoulos, Chris.Melhuish}@uwe.ac.uk
http://www.ias.uwe.ac.uk

[2] Faculty of Applied Sciences, The University of the West of England, Coldharbour Lane, Frenchay, Bristol, BS16 1QY, United Kingdom
John.Greenman@uwe.ac.uk
http://www.uwe.ac.uk/fas

Abstract. This paper reports on the proof-of-concept work to produce an energetically autonomous robot employing an artificial metabolic system using Microbial Fuel Cells. The present study compared the effects of changing a number of critical parameters, which control the fuel cell system, as a means to improve its overall performance. We demonstrate that the development of a fuel cell as an artificial metabolic system is feasible and it can provide sufficient power for a mobile robot platform to execute photo tactic 'pulsed' behaviour. The robot is code-named EcoBot I and it is the first robot in the world to be directly and entirely powered from bacterial reducing power.

1 Introduction

Autonomy comes from the Greek composite word *auto*, which means self, and *nomos*, meaning to control. Therefore, the term *autonomy* refers to and self-controlled entities that can organise and guide themselves, in a stable and reliable manner.

The term 'autonomous robot' has been ascribed to robotic systems to indicate their ability to perform tasks without human supervision. In fact, attempts to build machines that would operate without direct control date back to the ancient times. Heron of Alexandria, for example, built in 60 A.D., possibly, the first recorded example of an automaton [1]. However, the term 'autonomous' is somewhat flexible depending on the context it is used in. For example, consider the case of a robot that is released to carry out its task without external intervention but its batteries need to be charged by a human prior to its release. On completion of the task or in the event of the battery charge level getting low the robot returns to a base for recharging and/or new instructions. On one hand certain aspects of the robot's behaviour may be considered autonomous such as computational and control decisions. On the other hand, without a human in the loop, the robot would not be able to replenish its energy to accomplish the task.

W. Banzhaf et al. (Eds.): ECAL 2003, LNAI 2801, pp. 792–799, 2003.

With this in mind our long-term goal is the creation of a robot, which can generate energy for itself from its own environment. Although that energy could come from the sun or the wind, we are interested in generating energy from chemical substrates – food. We are therefore looking into a class of robot system, which demonstrates energetic autonomy by converting natural raw chemical substrate (e.g. carrots or apples) into power for essential elements of behaviour including motion, sensing and computation. This requires an artificial digestion system and concomitant artificial metabolism.

Adopting such a strategy may have an impact on the manner in which researchers and engineers incorporate their autonomous mission requirements. Four key issues are; firstly, useful energy will not (for the foreseeable future) be able to be instantly converted from raw substrate and secondly, there will be tasks (particularly those involving effectors or motion) which could not be powered continuously. The net effect is that this class of robot may have to include a 'waiting' behaviour in its repertoire in order to accumulate sufficient energy to carry out a task or sub-task. We refer to this form of behaviour as 'pulsed behaviour'. Thirdly, a robot may need to solve multi-goal action selection problems. In particular, it may be required to exhibit 'opportunistic' behaviour in terms of breaking off from its mission to forage or take advantage of energy resources such as a fallen apple. Finally, the substrate that will be consumed, which is directly related to the type(s) of microorganism(s) employed in the system will affect the robot's behaviour. In the case where the robot is designed to consume only a specific type of food, i.e. a '*specivore*' then its behaviour will be developed according to its needs [2]. On the other hand, in the case where the robot can consume a wide range of substrates, i.e. an '*omnivore*' then it could be given a more 'opportunistic' behaviour that will serve to its best interest [2]. In nature, animals, in the wild, often exhibit such behaviours and our work is obviously biologically inspired at the metabolic and behavioural levels. We argue that Energy Autonomy is a key element in future physical realisation systems of artificial life.

2 Fuel Cells

Fuel cells are an alternative to conventional energy plants or sources since they can offer a clean and environmentally friendly way of producing energy, for almost every kind of application. A fuel cell (Figure 1) is an electrochemical transducer that converts chemical energy to electrical energy from a fuel, without direct combustion. They comprise anode and cathode electrodes, electrolytes in liquid and/or solid form and catalysts.

The fuel cell principle of operation lies in the extraction of electrons as a result of the chemical reaction of separating a fuel, using a catalyst, in the anode half-cell. A catalyst is a substance, which speeds up the rate of a chemical reaction, however it remains unchanged at the end of the reaction. This allows the release of either positively or negatively charged ions that go to the cathode half-cell through the solid electrolyte, which also forms a physical barrier between the two half-cells. Catholyte and anolyte are the liquid electrolytes in the cathode and anode compartments, respectively that are necessary for electrolysis. Electrolysis defines the passage of electric current through a molten chemical or an aqueous solution. Hence, the anolyte is necessary for the transfer of electrons from the catalyst to the electrode and the catholyte is necessary for taking up incoming electrons from the electrode. In several fuel cells, where there is sufficient oxygen supply to the cathode, electrons recombine with cations and oxygen to form water.

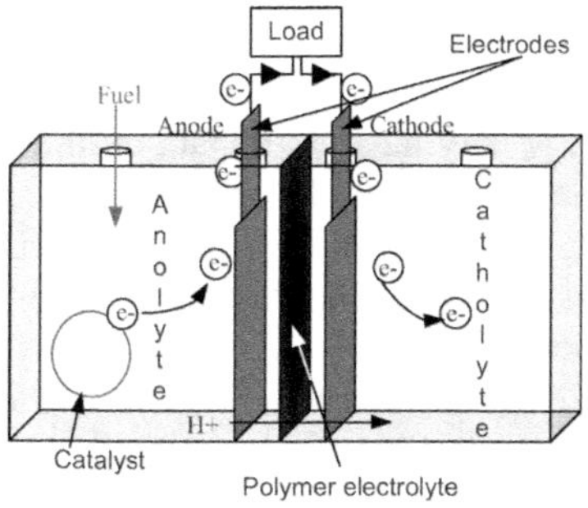

Fig.1. Schematic diagram of a general type of Fuel cell

2.1 The Microbial Fuel Cell (MFC)

Historically, MFC's date back to 1910, when Professor Michael Cresse Potter introduced the first MFC using *Escherichia coli* (*E.coli*) and platinum metal electrodes [3]. The idea of employing microorganisms to extract energy from sugars was first illustrated by Cohen at Cambridge in 1931, who revived Potter's MFC after scientists showed how enzymes in bacteria oxidise food [3].

An MFC (see Figure 2) is a *bio*-electrochemical transducer that converts *bio*-chemical energy to electrical energy, in roughly the same manner as a normal fuel cell. They fall under the Proton Exchange Membrane (PEM) fuel cell category, since that is the solid electrolyte used in the system and the fuel is *bio*catalysed by microbes. A wide range of organic substrates or nutrients of the types found in foods or food wastes can be utilised as fuel, depending on the type of microbe used.

Microorganisms metabolise the substrate, which is added to the anode half-cell, and energy (electrons) is transferred by Nicotinamide Adenine Dinucleotide (NADH) in its reduced form. This is a coenzyme, which acts as an electron carrier – a biological power source, in the metabolic pathway at the cellular level of living organisms. In its oxidised form it is symbolised as NAD^+ and as a redox couple ($NAD^+/NADH$) it can give electrons to almost every acceptor, since it is very negatively charged (-0.32V). For example, the difference in the standard redox potentials between the $NAD^+/NADH$ and ½ O_2 / H_2O is actually 1.14V, and hence the change in free energy when the two interact is actually 220kJ.

The anolyte is made up of a mediator, buffer, microbial culture and sugar solution. A mediator is a substance that penetrates the microbial cell, to a certain level and extracts the electrons from the electron transfer chain. Once the mediator taps the electron, it has become reduced and the electron carrier oxidised, i.e. NAD^+. The reduced mediator diffuses outside the microbial cell and is diverted by the electrophillic attraction of the cathode electrode. As a result the electron is released onto the anode electrode to flow through the external circuit, thus oxidizing the mediator. In the meantime, NAD^+ is re-reduced to NADH by further metabolism of substrate molecules and the mediator is re-reduced by re-oxidising NADH. The buffer keeps the anolyte pH balanced as it tends to become acidic from the bacterial acid waste production. The microorganism used in our experiments is *Escherichia coli* (*E.coli* cc17 isolate culture from UWE collection) and the biochemical reaction is as follows:

$$NAD^+ + 2H^+ + 2e^- \Leftrightarrow NADH + H^+ \quad \textbf{(1)}$$

In the cathode, incoming electrons from the external circuit reduce the buffered electrolyte in order to complete the system. In the presence of dissolved oxygen, the catholyte is barely oxidised with some water (H_2O) formation. The buffer in the solution keeps the pH balanced by consuming the hydrogen cations traveling through the PEM and tending to decrease the pH. These processes in both the anode and cathode described above are known are **red**uction-**ox**idation (redox) and are schematically shown in Figure 3.

Fig. 2. Microbial Fuel Cell in its analytical form

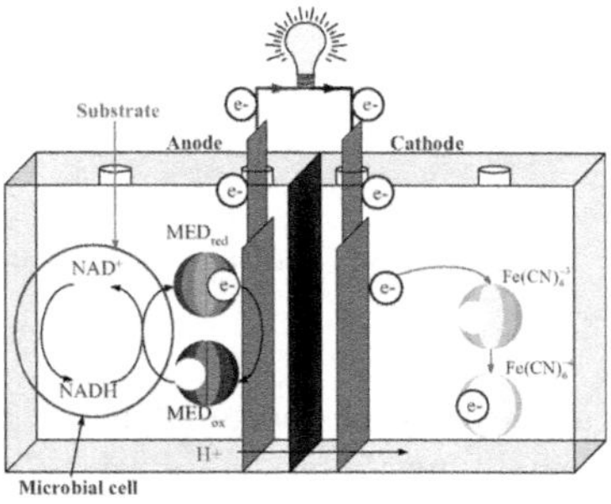

Fig. 3. Redox reactions within a Microbial Fuel Cell

3 EcoBot: An MFC Powered Robot

Using the above-mentioned technology as the sole power source, a proof-of-concept robot was built which performs photo-tactic behaviour. It was code-named EcoBot I and it was the first robot in the world to be directly and entirely powered from bacterial reducing power and the second to be running on sugar [4]. It involved no other form of conventional power source such as secondary batteries or solar panels and furthermore it was more efficient and more compact in size than Gastronome [4], which was the first such artifact to have been constructed. Inexpensive and in most cases sub-optimal substances and materials have been exploited to the maximum of their performance to give energy outputs in the range of 37.2J and achieve efficiencies of the order of 1.56%. Even though this figure is low, it is consistent with the current MFC technology. Further experiments conducted recently in our laboratory, have shown improved performance, which matches and in some cases outperforms microbial fuel cells incorporating immobilised electrodes [5]. The MFC's currently employed are of a batch (closed system / non-continuous) nature where there is no refreshment of the nutrients and vital substances. These first results imply that the longevity of such a system could be increased by a factor of 10. In nature, all living organisms operate in a continuous mode (open system) and by adopting this technique the overall efficiency should be dramatically increased. Figure 4 below is a picture of the EcoBot I prototype fully assembled.

The robot employed a bank of eight MFC's, an electronic control circuit, two photo detecting diodes and two dc geared *escap*® - 205 motors (Portescap. Switzerland). A circular piece of styrene material having a diameter of 22cm and 5cm thickness with two rectangular cuts to accommodate the MFC banks formed the main body of the robot. The overall height was 7.5cm with a total mass of 960g. The robot was balanced with the use of two caster wheels.

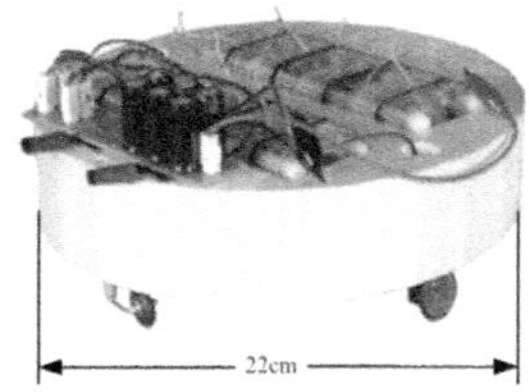

Fig. 4. UWE EcoBot I fully assembled

Energy produced by the MFC's was accumulated in a bank of six capacitors and on reaching a certain threshold, was released to fire the motors according to the photodiodes' indication. A differential drive was employed so that the robot would follow the light and when a second threshold was reached the robot would become 'idle' until enough energy was accumulated to fire the motors again. This gave the prototype a form of burst motion photo-tactic behaviour ('pulsed-behaviour'). Figure 5 below illustrates the block diagram of the robot's electronic control circuit. The robot experiment, moving from a start point to a light source, was repeated seven times. Two of these trials were video recorded, a snapshot of one of which is shown in Figure 6, and the other five were data recorded. Figure 7 illustrates the average charge/discharge cycles, in terms of average distance and time for the five trials.

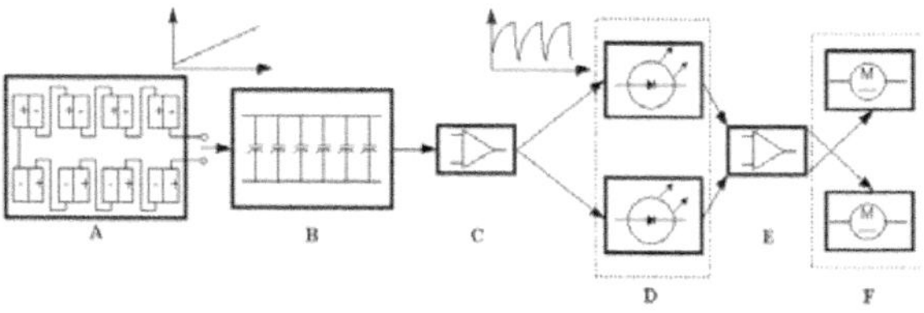

Fig. 5. Block diagram representation of the EcoBot I electronic control circuit. Block A is the MFC bank; Block B is the capacitor bank; the charge-discharge control circuit is shown as block C; block D represents the pair of photodiodes; the control circuit for differential drive is shown as block E and finally the pair of motors is shown as block F. The graph above A and B illustrates the increasing energy output from the MFCs with respect to time and the graph above C and D represents the charge-discharge cycle.

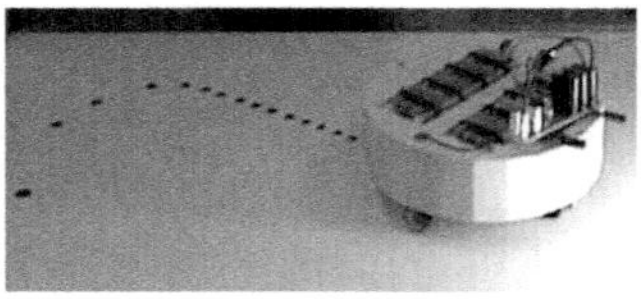

Fig. 6. EcoBot I performing photo taxis. Marks behind the robot indicate the trajectory of the photo-tactic movement and the light source is at the bottom right of the figure.

A graph of the robot's progression along a flat surface, with respect to time is shown in Figure 8. This is a graph that was produced from the recorded data of the five trials and it shows the advancements that the robot was making in burst motions. Data points show the average distance that the robot had covered for each burst motion, with the error bars indicating the maximum and minimum distance that was covered in the different trials.

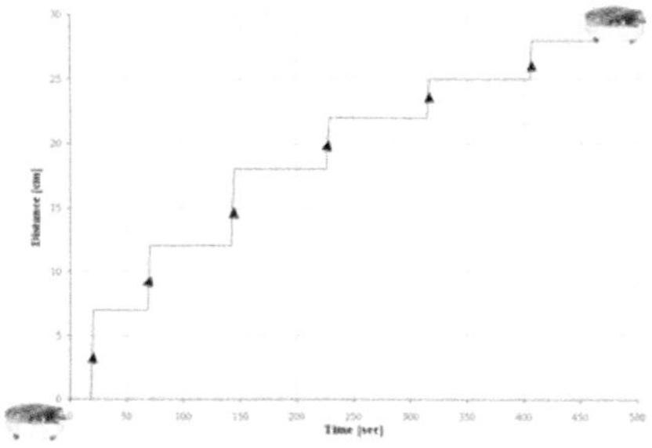

Fig. 7. Average charge/discharge cycles shown with respect to time and space

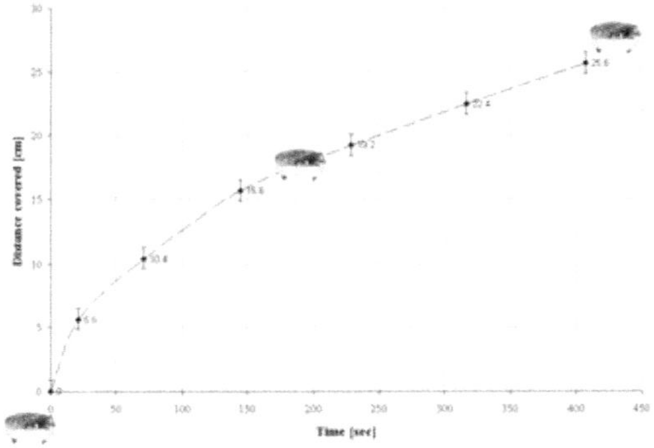

Fig. 8. Progression of EcoBot I along flat surface

As was mentioned earlier, EcoBot I was only the first step of many to produce the final envisaged system and there are a number of implications that need to be considered for any further development. These can be divided into two main categories; the hardware and the behavioural issues of the system.

In a system where continuous chemical fluid flow is employed, there will have to be a number of micro-pumps and filters to keep the system operating. Parameters such as temperature, pH and liquid level will have to be continuously monitored and controlled. As the robot will be looking for its energy source (raw food), the electronics and mechanics involved in extracting the necessary nutrients from it will have to be highly efficient and accurate. Utilisation of such devices will have an associated energy requirement, which in turn leads to the behavioural implications.

Simple "low battery" warning indications will just not be enough. The energy accumulated may be in the form of electrical charge, however this will be derived from a bio-electrochemical device. This implies that both internal sensors to provide feedback for more accurate control and external sensors to enable the agent to perform its task will have to be used. Negative feedback will be a key element in the system since in biological systems this is the basis for normal operation. It is through feedback that homeostasis is achieved at all levels of organisation in living systems – from the molecular to the social. Homeostasis means, "staying the same" and it refers to the capacity of biological systems to maintain automatic control over physio-chemical variations. Imitating this critical parameter will form a significant challenge.

4 Discussion

MFC technology is still in its infancy and the levels of power achieved are very low. It is quite clear that the power source will, for most applications, not be in a position to provide enough power for continuous operation. Therefore, the energy, as with the EcoBot I prototype, will first have to be accumulated before it is used, thus resulting in the 'pulsed' behaviour. Managing a variable energy resource is not a trivial task and therefore energy reserves must be employed to account for situations where the energy required to execute a task is more than the readily available (onboard store).

One of the near future goals is to further improve the MFC efficiency in terms of power level and longevity, as this will form the 'live engine' around which the rest of the robot will be built. Re-design and soft engineering of the MFC system is of highest priority, to achieve these goals and give the system an appropriate design configuration, in this way imitating the gut. Gas diffusion electrodes will be tested as these have the advantage of using free oxygen from the air to act as electrolytes.

Real autonomy, in the context of artificial intelligence is not only a matter of executing a task with minimum or no exogenous intervention, while having to rely on the human factor for energy requirements. Natural metabolic systems solve this problem by employing redox energy to do work. We seek to imitate nature in this respect and EcoBot I demonstrates that such an approach is feasible.

In the future, such robots could be classified according to the range and type of food that they will be consuming. Some robots may be designed to restrict themselves to one type of food (*specivores*) where others may exploit more than one type of food (*multivores / omnivores*). The costs and benefits of such artifacts merits further research.

References

1. Webb, B.: The first mobile robot. Proceedings of TIMR 99, Towards Intelligent Mobile Robots. Bristol (1999)
2. Ieropoulos, I., Melhuish, C., Greenman, J.: Imitating Metabolism: Energy Autonomy in Biologically Inspired Robots. Proceedings of the AISB'03, Second International Symposium on Imitation in Animals and Artifacts, SSAISB, Aberystwyth, Wales, pp 191-4.
3. Bennetto H.P.: Microbes come to Power. New Scientist (1987) 36–39
4. Wilkinson S.: 'Gastronome' – A Pioneering Food Powered Mobile Robot. Proceedings of the 8th IASTED, International Conference on Robotics and Applications, Paper No. 318-037. Honolulu, Hawaii, USA (2000)
5. Park, D.H., Zeikus, G.: Impact of electrode composition on electricity generation in a single-compartment fuel cell using *Shewanella putrefaciens*. Appl. Microbiol. Biotechnol., Vol. 59, (2000) 58-61

An Imitation Game for Emerging Action Categories

Bart Jansen

Vrije Universiteit Brussel, Pleinlaan 2, 1050 Brussel, Belgium,
bartj@arti.vub.ac.be
http://arti.vub.ac.be/~bartj

Abstract. An imitation game for learning action categories is proposed. It serves to invent and share a repertoire of action categories in a population of robotic agents. The agents start without any action categories built in, but they learn by imitating other agents and gradually invent categories for the actions they observe and execute. In other words, no action categories need to be defined beforehand, nor do agents have to be assigned fixed roles. If the population only consists of two agents, the repertoire of action categories can be learnt without feedback about the outcome of the game between both agents.

1 Introduction

Most of the work that has been published on learning by imitation in robotic agents focuses on the learning of action categories in a teacher - student context, either the student learns from other robots [1] or from human beings [2,3,4]. In such a set-up one agent acting as teacher already has action categories. By observing the teacher who is executing actions, the action categories can be transferred to the student: by imitating the teacher's action using for instance inverse kinematics and evaluating that action, the learner can know whether it correctly reconstructed the observed action. However, such a set-up does not explain how imitable action categories emerge and does not take into account the influence of the population dynamics of imitating agents. The emergence of categories on the population level has been studied intensively in our laboratory, for example the emergence of vowel systems [5], colour categories [6] and visual categories [7].

In this paper we propose a learning mechanism for agents such that new action categories can emerge when imitation of actions fails. This is done in a population of agents engaging in imitative interactions, called *imitation games*. The learnt action categories will be imitable. If an action is hard to observe or to imitate, it will not be learnt by other agents. The concept of imitation games used here strongly resembles the concept of imitation games used in [5] in the context of vowel systems.

Section two briefly explains the experimental set-up and the architecture of the agents. Section three describes the imitation game in detail. Section four

W. Banzhaf et al. (Eds.): ECAL 2003, LNAI 2801, pp. 800–809, 2003.

contains results showing that a repertoire of action categories can indeed emerge in a population of agents through imitation games. In section 5 a modified imitation game is presented. Playing that game, two agents can acquire a repertoire of action categories without giving feedback on each other's imitative behaviour. Section 6 discusses future work.

2 Experimental Set-up

An agent has a head with a stereo camera and a robot arm with a gripper. Using its arm, the agent can perform different kinds of gestures which it can observe using the vision system. We have built this physical set-up using a readily available commercial robot arm, called Teach-robot[1] and a vision system based on the Small Vision System (SVS)[2]. The vision system focuses on finding the coordinates of the gripper in the captured images. The resulting three-dimensional time series will be categorised during the imitation game. The vision system and the robot arm are described in more detail in [8].

The experiments described in this paper are not performed on this real physical set-up, but in simulation and serve as a proof of concept. Action execution and observation are thus not performed on real robots but are simulated using the forward and inverse kinematics of the robots. Both in the simulated execution and observation of actions, noise (e.g. 5%) is added for simulating imprecision when performed on real physical robots.

2.1 The Agent Architecture

An *action category* consists of two components: an *action* and an *observation*. This means that every category is represented by a single prototype. An action is a tuple of two 3D points, being the start- and endpoint of the gripper. Gripper points define an *action space*. The action executed by the robot will be the movement of the gripper from start point to end point. The observations are sequences of 3D points, generated by the vision system. They define an *observation space*.

Actions do not contain direct motor commands, but tuples of gripper coordinates. There is a direct mapping between both, as the forward and inverse kinematics of the agent's robot arm are known [9]. Given an action, the agent can obtain the associated observation, simply by executing the action and observing the result. Given the observation, the agent can only recover the action that was executed if the action space and the observation space are calibrated. In section 5 it is discussed whether calibration is indeed essential for the imitation game.

Agents have an *action category memory* to store the learnt action categories. At the beginning of the imitation games, these memories are empty as the action categories will emerge during the imitation games. Besides a memory for

[1] Microelectronic Kalms, http://www.teach-robot.de/
[2] http://www.ai.sri. com/konolige/svs/Papers

the learnt action categories, all agents are equipped with the same learning mechanism and have a pre-programmed drive to act and to imitate. We do not investigate how the agents can learn how to imitate nor when to imitate. We rather start with a built in learning mechanism and investigate whether this mechanism is suited for building up a shared repertoire of action categories.

3 The Imitation Game

For every game two agents are randomly selected from the population. One agent will take the role of *initiator*, the other agent will be the *imitator*. During the game, agents will observe the execution of actions and will try to categorize their observations. The category of an observation O is defined as the action category in the agents memory with an observation closest to the observation O. We use Dynamic Time Warping (DTW) [10] as distance metric on observations of actions. DTW was also used for categorising gestures in [11]. The distance metric that is used by the agents for comparing observations defines the type of categories that will emerge. If the distance metric only classifies two observations as equal when the entire trajectories match, every consistently repeatable and observable motion of the robot arm will constitute a different action category. This explains why so many categories will emerge.

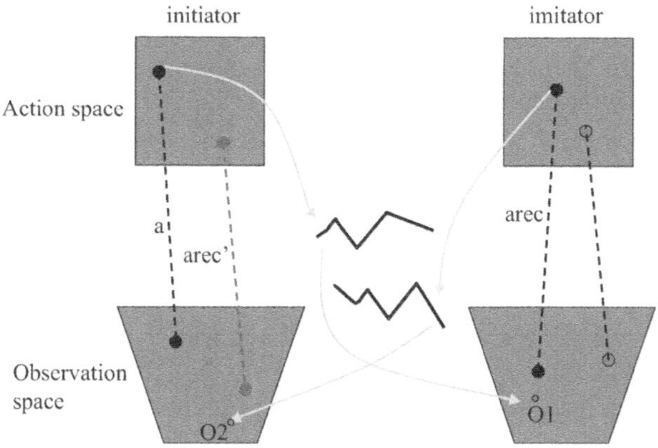

Fig. 1. The imitation game. In this example it fails, as the imitation is categorised as a'_{rec} instead of a.

In figure 1 the observation space for both the initiator and the imitator are depicted. The dashed lines represent the associations between actions and observations, which constitute categories together. Those observations and actions are indicated by large dots, the small dots indicate the actual observations.

If both the initiator and the imitator have at least one action category in their memories, they can engage in an imitation game. Otherwise, they first add a new random action category to their action category memory. The initiator then selects a random action category a from its memory and executes the action that is associated with the action category a. The imitator observes this action, denoted *O1*. It categorises the observation *O1*, denoted a_{rec}. The imitator now executes the action associated with the category a_{rec}. This action will be observed by the initiator as observation *O2* and categorized. The category will be called a'_{rec}. If the initial category and the category of the imitated action are equal ($a = a'_{rec}$), then the initiator decides that the game succeeds, otherwise that it fails. The initiator sends non-verbal feedback about the outcome of the game to the imitator. If the game succeeded, the imitator knows that the category it used is similar to the category used by the initiator. To increase similarity, the imitator shifts the category it used closer to the observation *O1* of the initiators' action. If the game fails, two different update strategies are considered. If the action category the imitator used has been successful in the past (i.e. its success-ratio is above 0.5), the failure in this game is probably caused because the initiator executed an action the imitator does not know yet. In this case a new category is constructed for the observation. If the success-ratio of the category used is below 0.5, the category itself is probably not very good. In that case, the imitator shifts the action category used towards the observation, to improve the success of the category in the future. As a last step of the game, both the initiator and the imitator update their repertoire of action categories. If an agent has an action category of which the action was not successfully imitated, the action category is removed from its repertoire (i.e. after five or more categorizations and when the success-ratio is below 0.7). Agents are forced to invent and share new action categories because with a small probability (*e.g. 0.02*) new random action categories are added to their repertoire.

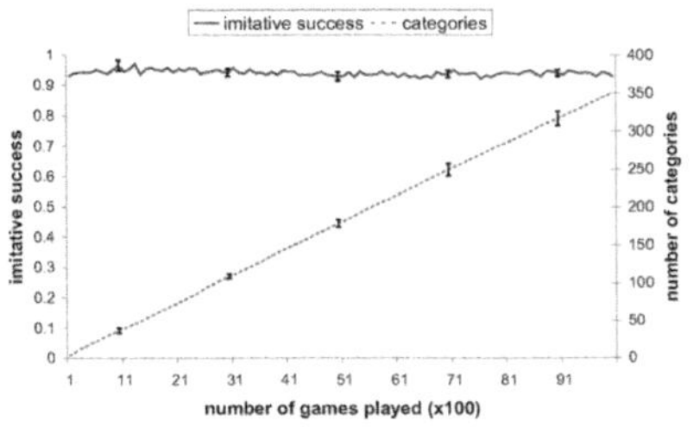

Fig. 2. Number of categories and imitative success averaged per 100 games over 10000 games, for 2 agents. Results are averaged over 10 runs.

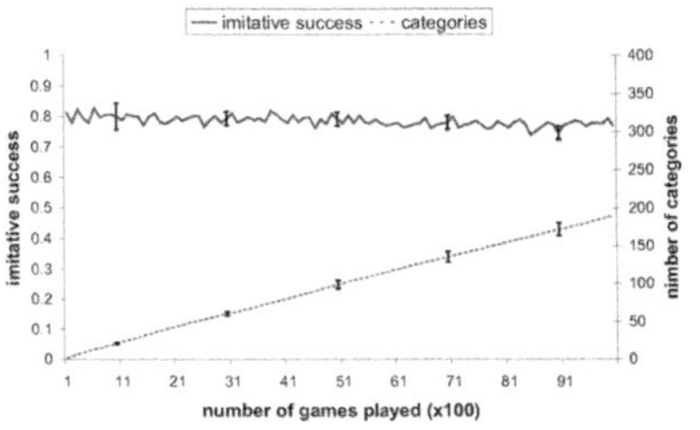

Fig. 3. Number of categories and imitative success averaged per 100 games over 10000 games, for 5 agents. Results are averaged over 10 runs.

4 Results

In the first experiment the population consists of 2 agents. In total 10000 imitation games were played. In figure 2, the imitation success is shown averaged per 100 games, averaged over the 2 agents, averaged over 10 runs. The average number of action categories per 100 games averaged over both agents and over 10 runs is shown in the same figure. The number of action categories rises steadily and reaches (345.46 ± 2.34) after 10000 games. Even while the number of action categories increases, the imitation success remains very high (0.95 ± 0.01).

These results indicate that two agents indeed build up a shared repertoire of action categories through imitation. Averaged over the 10000 games, in 95% of the cases their imitation is successful. In larger populations, agents also build a repertoire of action categories, but the imitation success is lower. In figure 3 the average number of categories and the imitation success are shown for 10000 interactions among 5 agents. The results are averaged over 10 runs. Averaged over the 10000 games, the imitation success is 0.80, which is much lower than when only two agents were used, but still higher than in the case of interactions between agents with random categories. It is not surprising that the imitative success is already very high at the beginning. As long as all agents have only one action category, there is no possibility of failure. Further experimentation is required to investigate how these results depend on the different parameters used in the game. Particularly, it needs to be shown that this game is robust enough to deal with larger amounts of noise.

Measuring category similarity. From the results shown before, it is obvious that the agents are very successful in imitating each other. Although it is easy to understand that successful imitation is not possible if the agents have very different action categories, it must be shown that the categories are shared through the population. Therefore we need to define how similar the categories of two agents are, or—in general—how similar two sets of points are in an N-dimensional space. Note that two agents can develop a different number of categories, so the sets do not contain an equal number of points. The similarity measure of the categories of two agents *Category Distance*(CD) is based on the *Weighted sum of minimum distances* metric developed in [6] and is given in equation 1. The terms A and B are agents, a and b are categories and $d(a,b)$ is the distance between the categories a and b.

$$CD(A,B) = \frac{\sum_{a\in A} \min_{b\in B} d(a,b) + \sum_{b\in B} \min_{a\in A} d(a,b)}{|A| + |B|} \tag{1}$$

We can now calculate the Category Variance (see equation 2) of the population of agents. This measure indicates how much the categories of all agents deviate from each other.

$$CV(population) = \frac{2}{N(N-1)} \sum_{i=1}^{N-1} \sum_{j=i+1}^{N} CD(A_i, A_j) \tag{2}$$

As opposed to the imitation success and the number of categories, the category variance is not averaged over 100 games. In order to reduce computation time, it was only calculated every 100 games. In figures 4 and 5, the category variances for two and five agents playing 10000 imitation games are shown. It can be seen that the category variance decreases very fast in both games, so the more imitation games the agents play, the more similar their categories become. However, the reader should notice that the CV also decreases in a population of agents with an increasing number of random categories, simply because more categories are fitted into a finite space and thus become closer to each other. In figures 4 and 5, one can see that the CV is always lower in the case of agents that learn through imitation than in the case of agents with an equally fast growing number of random categories (ECV).

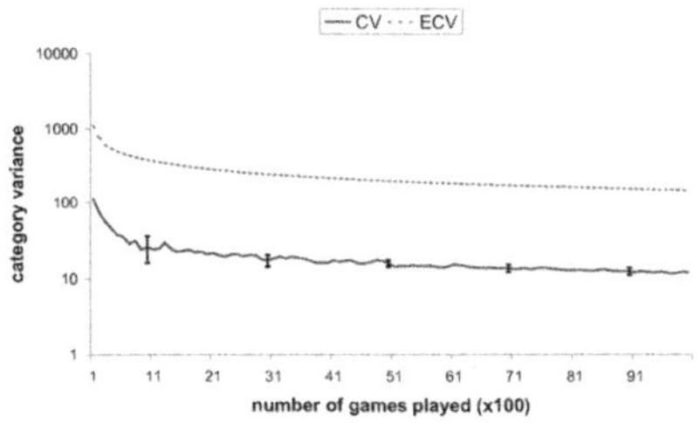

Fig. 4. The category variance for 10000 interactions among 2 agents, in a logaritmic plot. Results are averaged over 10 runs.

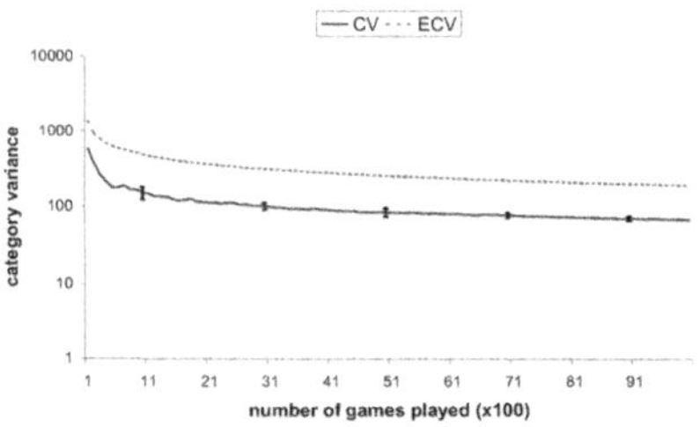

Fig. 5. The category variance for 10000 interactions among 5 agents, in a logaritmic plot. Results are averaged over 10 runs.

5 A Variation on the Game

In this section it is investigated how the paradigm of imitation games presented above can be modified such that there is no feedback required between both agents.

5.1 Who Is Learning?

In the imitation game as described above, one could ask who is learning—the initiator or the imitator? Both initiator and imitator are keeping success scores, removing bad actions and randomly adding new actions. However, most important is that the imitator considers the initiator as an agent with perfect categorization and adapts its categories by shifting to resemble the categories of the initiator more closely. As all agents take turns and take many times the role of initiator and imitator, the result is that all agents' categories become more similar. Another perspective is also possible. Suppose that the initiator considers the imitator to have full knowledge of the categories. In that case, the initiator uses the imitation to check whether its own categorization of the world is consistent with the categorization of the imitator. If the initiator's action category and the categorization of the imitation are equal, the initiator is ensured that its own categorization (for that single action) is very similar to the categorization of the imitator. So, the initiator adapts its categories to resemble the imitated action even more . If the initiators' action category and the categorization of the imitation are different, the initiator adapts its categories to decrease the probability of failure in the future. What is the difference between both perspectives? In the first game, the agent which learns by adapting its categories (the imitator) receives non-verbal feedback about the outcome of the game and adapts its categories accordingly. In the second game, there is no non-verbal feedback, as the initiator itself knows the desired and actual outcome of the game. An important difference is that in the new game, the agent that is learning—the initiator—can choose what action categories to use in the game. The initiator can investigate its own repertoire of action categories for categories that were not successfully imitable in the past. By using precisely those categories, the agent might improve its unsuccessful categories very fast. Another possibility is that the agent always selects categories with a high success ratio. Further research on this topic is required. The second game in which the initiator learns action categories is presented below.

5.2 The Learning Initiator

A modified game is proposed in which the initiator and not the imitator is learning. In this game, the initiator selects a random action category a from its memory and executes it. The imitator observes it as *O1* and categorizes it as a_{rec}. The imitator executes the associated action, which is observed by the initiator as *O2* and categorized as a'_{rec}. If a and a'_{rec} are equal, the game succeeds, otherwise it fails. So far, the game is identical to the first game. At this stage of the game the initiator does not send feedback about the outcome of the game to the imitator, as the initiator updates its action categories and the imitator does not. The initiator shifts its original action category a towards the action category a'_{rec} of the observed imitation. The shift operation does not start from the observation space but is done in the action space. In case of a failure, this same shift can be done in case of an action category a with high success. If the action category a does not have high success, it is not useful to add a new action category for the observation associated with a'_{rec}, as this will result in a new action category a''_{rec} which is very close to a'_{rec}. This addition would only confuse the agent. Therefore in this case, no new action category is added.

Shifting in the action space instead of the observation space only works if it is guaranteed that when an action is shifted towards another action, then the observation of the first action is also closer to the observation of the second one. This is the case in a real robot, as points in both spaces are related by a rotation and a translation.

The advantage of shifting directly in the action space is that the agent must never calculate what action caused a given observation. This means that the action space and the observation space need not to be calibrated anymore. This has an enormous advantage for performing experiments on real world robots, where calibration between observation space and action space would not be required anymore.

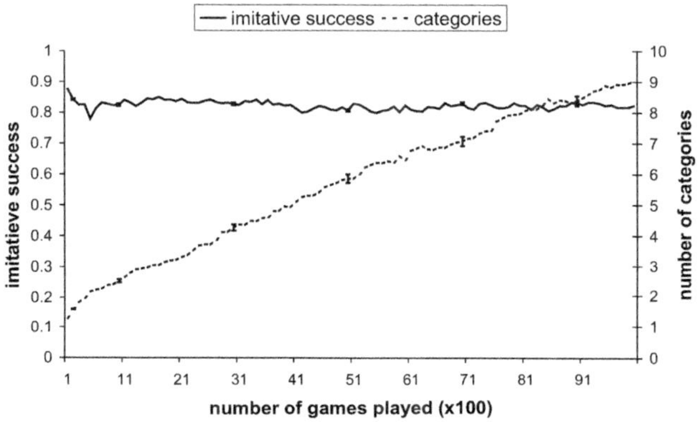

Fig. 6. Number of categories developed over 10000 games where the initiatator learns, averaged per 100 games. The initiator is adapting its repertoire. Imitative success averaged per 100 games, for the same 10000 games is shown in the same graph. Results averaged over 100 runs. The action space and the observation space are not calibrated.

Results are given in figure 6. Although convergence is slower than in the previous type of game, the agents are capable of developing a repertoire consisting of 9.04 ± 0.89 action categories, while the imitative success remains (0.82 ± 0.01). At the moment, it has not been possible to show that a population of more than 2 agents is capable of converging to a shared repertoire of action categories by playing the imitation game presented in this section. Using the same parameters as in the former imitation game, the average number of categories for a population of 5 agents after 10000 interactions was found to be 1.28 ± 0.14. Other parameter settings have not been investigated yet. The fact that the repertoires are built up much slower in this type of game and the fact that imitation games are only successful in populations of two agents is caused by the lack of feedback and by the fact that the initiator does not adapt its categories to become more similar to the categories of the imitator.

6 Future Work

This paper has shown how a repertoire of very simple action categories can be learnt and shared throughout a population of agents. However, the question remains how more complex actions can be learnt. We believe that this is possible by using the action categories as they emerge in the imitation game as building blocks for learning more complex actions. On a higher level, a new imitation game could be defined, such that complex actions can be learnt. The bigger challenge is developing an imitation game where actions involving object manipulations emerge, without the manipulation of the objects being predefined, but as an emergence of the task of imitation and a utility function.

7 Conclusion

In this paper it was shown that agents can learn action categories through imitation, even if no strict teacher-student protocol is followed where the teacher already has fully developed action categories. The agents start without any action categories and invent a repertoire of action categories while imitating other agents from the population. Due to pressure of the game to improve categorization, the action categories become shared among the population and can thus be imitated with high success (above 90%) by the other agents. This is not guaranteed if the action categories were pre-programmed. This game is also working in larger populations. The obtained results serve as a proof of concept and encourage us to repeat the same experiments on the robot set-up that we have built.

If the population is restricted to only two agents, we show that the agents can develop and share a repertoire of action categories without feedback about the outcome of the game between both agents and without the agents' action space and observation space being calibrated. This is an enormous advantage for conducting experiments on real robots. To achieve this we proposed a second imitation game in which the initiator uses the imitated action of the imitator to update its categories.

Acknowledgments. Bart Jansen is sponsored by a grant from the Institute for the Promotion of Innovation by Science and Technology in Flanders (IWT).

References

1. Vogt, P.: Grounding language about actions: Mobile robots playing follow me games. In Meyer, Bertholz, Floreano, Roitblat, Wilson, eds.: SAB2000 Proceedings Supplement Book, International Society for Adaptive Behavior (2000)
2. Billard, A., Schaal, S.: Robust learning of arm trajectories through human demonstration. In: Proceedings of the International Conference on Intelligent and Robotics Systems, IEEE Computer Society (2001) 82–90 Hawaii, November 2001.
3. Billard, A., Hayes, G.: Transmitting communication skills through imitation in autonomous robots. In Birk, A., Demiris, Y., eds.: Proceedings of EWLR97, Sixth European Workshop on Learning Robots, Lecture Notes on Artificial Intelligence. Volume 1545., Springer (1997) 79–95 Brighton, UK, July 1997.

4. Alissandrakis, A., Nehaniv, C.L., Dautenhahn, K.: Do as I do: Correspondences across different robotic embodiments. In Polani, D., Kim, J., Martinetz, T., eds.: Proceedings of the Fifth German Workshop on Artificial Life (GWAL5). (2002) 143–152 18-20 March 2002, Lübeck, Germany.
5. de Boer, B.: Self organization in vowel systems. Journal of Phonetics **28** (2000) 441–465
6. Belpaeme, T.: Factors influencing the origins of colour categories. PhD thesis, Artificial Intelligence Lab, Vrije Universiteit Brussel (2002)
7. Belpaeme, T., Steels, L., Van Looveren, J.: The construction and acquisition of visual categories. In Birk, A., Demiris, Y., eds.: Proceedings of EWLR97, Sixth European Workshop on Learning Robots, Lecture Notes on Artificial Intelligence. Volume 1545, Springer (1998) 1–12 Brighton, UK, July 1997.
8. Jansen, B., de Vylder, B., de Boer, B., Belpaeme, T.: Emerging shared action categories in robotic agents through imitation. In Dautenhahn, K., Nehaniv., C.L., eds.: Proceedings of the Second International Symposium on Imitation in Animals and Artifacts., University of Wales, Aberystwyth (2003) 145–152
9. De Vylder, B.: Forward and inverse kinematics of the teach-robot. Technical Report AI MEMO 02-02, Artificial Intelligence Lab, Vrije Universiteit Brussel, Brussels (2002)
10. Myers, C.S., Rabiner, L.R.: A comparative study of several dynamic time-warping algorithms for connected word recognition. The Bell System Technical Journal **60** (1981) 1389–1409
11. Corradini, A.: Dynamic time warping for off-line recognition of a small gesture vocabulary. In: IEEE ICCV Workshop on Recognition, Analysis, and Tracking of Faces and Gestures in Real-Time Systems (RATFG-RTS'01), IEEE Computer Society (2001) 82–90 Vancouver, B.C., Canada, July 13–August 13, 2001.

Multi-agent Model of Biological Swarming

Robert Mach[1,2] and Frank Schweitzer[1,3]

[1] Fraunhofer Institute for Autonomous Intelligent Systems, Schloss Birlinghoven, D-53757 Sankt Augustin, Germany
[2] Institute for Theoretical Physics, Cologne University, D-50923 Koeln, Germany
[3] Institute of Physics, Humboldt University Berlin, D-10099 Berlin, Germany

Abstract. An agent-based approach is used to explain the formation of vortex swarms in biological systems. The dynamics of the multiagent system is described by $3N$ coupled equations, modeling for each agent its position, its velocity and its internal energy depot. The energy depot considers the conditions for active biological motion, such as energy take-up, metabolism, and energy conversion. The equation of motion results from a superposition of *deterministic* and *stochastic* terms (random noise). The deterministic part considers indirect interactions with other agents to describe local avoidance behavior, and external influences resulting from an attractive environmental potential. Stochastic computer simulations of the multi-agent system are shown in very good agreement with the behavior observed in Daphnia swarms.

1 Introduction

One of the major conceptual challenges of system biology is the understanding of the behavior at the system level from the interactions of the entities comprising the system. To reach this goal, discrete, individual-based or *agent-based* modeling has become a very promising and powerful methodology. Recently, different computer architectures in distributed artificial intelligence have been developed to simulate the collective behavior of interacting agents. However, due to their rather complex simulation facilities many of these simulation tools lack the possibility to investigate systematically and in depth the influence of specific interactions and parameters. Instead of incorporating only as much detail as is *necessary* to produce a certain emergent behavior, they put in as much detail *as possible*, and thus reduce the chance to understand *how* emergent behavior occurs and *what* it depends on.

In our paper, we investigate a prominent example of complex behavior in biological systems, namely *swarming*. Pioneering work has been done by Reynolds[1] that is based on behavioral rules for artificial creatures (boids), while our aim is to obtain the same behavior based on interactive forces between reactive agents. Therefore, in our approach, we use *Brownian agents* [14] (see Sect. 2), that – in addition to their computational suitability – can be also investigated by means of analytical methods from statistical physics and mathematics. Swarming is

[1] http://www.red3d.com/cwr/boids/

W. Banzhaf et al. (Eds.): ECAL 2003, LNAI 2801, pp. 810–820, 2003.

a form of *collective motion* that may emerge from *local interactions* of a large number of individuals (agents). Swarms (also called herds, flocks, schools) can often be observed in certain mammals, fish, insects, and birds for various benefits such as enhaced feeding and mating as well as more successful predator avoidance. Detailed experimental investigations on swarming, however, are rare, either because of the size of the animals or because well defined conditions for experiments are difficult to realize. Fortunately, as Ordemann *et. al* have shown [9, 10], zooplankton is a suitable candidate for experiments on swarming, as briefly described in the following.

2 Modeling Swarming in Biological Systems

Daphnia (water flea) is an ideal object to study swarming behavior under well defined lab conditions, because of its intermediate size and biological complexity. The sketch of the experimental setup (Fig. 1) shows a water tank penetrated vertically from an artificial light source. While it may seem that this condition is highly artificial, we note similarities to light conditions in field observations [10]. In particular, in such a tank, Ordemann *et. al* [10] have investigated the motion of both *single* Daphnia and *many* Daphnia.

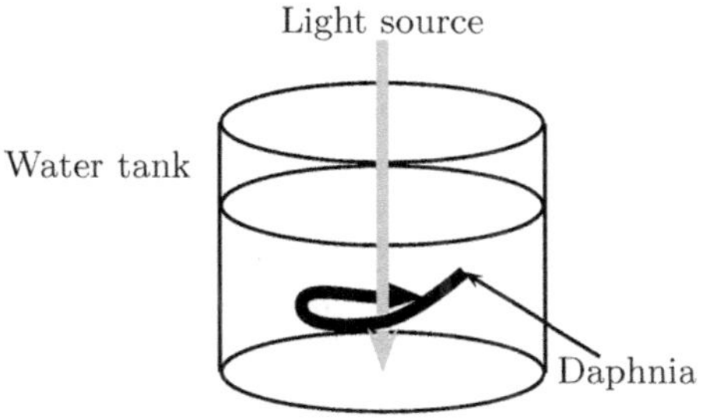

Fig. 1. Sketch of the Daphnia experiments carried out by Ordemann *et. al* [10]. The trajectory indicates the cycling motion of a single animal.[6]

In such a setup it has been found that a *single* Daphnia starts to *cycle* (i.e. rotate) around the artificial light source, keeping its cycling direction for quite a while. In repeated experiments, however, the cycling direction may change to the opposite, which lead to the conclusion that single Daphnia, while rotating around the light beam, do *not* have a preferred direction of motion.

Interestingly, the situation changes if instead of single Daphnia a larger number of animals is put in the water tank. In this case, the Daphnia start again with their cycling motion, but then *all* tend to move into the *same* direction of motion. From a physical perspective, a symmetry break is observed, i.e., the symmetry between the two possible cycling directions (left, right rotation) is clearly broken toward *one* of the possibilities (left *or* right rotation). Both of these possibilities have the same chance to occur, but only one of them is eventually realized.

It has been further observed that the Daphnia swarm, while rotating round the light beam in the same direction, mostly keeps a certain distance to the light source. This kind of swarming behavior is also called *vortex swarming.*

The vortex formation as well as the symmetry break in the cycling direction are clearly self-organized phenomena that result from the *collective* interaction of *many* animals. In order to understand this in more detail, we derive a multiagent model in the following.

Our approach is based on *Brownian agents* [14], each of them described by three state variables: spatial position $\boldsymbol{r}_i$, velocity $\boldsymbol{v}_i$ and internal energy depot e_i. The first two state variables describe the *movement* of the agent and can be observed from the outside. The agent's energy depot, however, is an *internal* variable describing its *capability* of *active movement.* Different from physical Brownian particles that move *passively* and *randomly* because of the impacts from surrounding molecules, the Brownian agent may move actively and in a preferred direction. But the term "Brownian" refers to the fact that the agent may still be subject to fluctuations that are described by a stochastic force, as explained further below.

The internal energy depot $e_i(t)$ of agent i has to consider that active motion is based on the take-up of energy from the environment, the storage of energy and conversion of stored energy into energy of motion. These three processes are summerized in the following balance equation [11]:

$$\mu \frac{d}{dt} e_i(t) = q(\boldsymbol{r}) - c\, e(t) - d_2 v^2\, e(t)\ ; \quad d_2 > 0 \tag{1}$$

$q(\boldsymbol{r})$ is the flux of energy into the internal depot. If the availability of energy is heterogeneously distributed, the energy flux may depend on the space coordinate, $\boldsymbol{r}$. In this model we assume a *homogeneous* distribution of energy, i.e. a constant influx $q(\boldsymbol{r}) = q_0$. c describes the loss of energy due to internal dissipation, which is assumed to be proportional to the internal energy. The last term in eq. (1) considers the conversion of internal energy into kinetic energy with a rate, which should be a function of the actual velocity, $\boldsymbol{v}$, of the agent. Here, $d_2 \cdot v^2$ is assumed in agreement with physical equations for the energy balance. The formal parameter μ in eq. (1) can be used to describe the time scale of relaxation of the internal energy depot, as described below.

In order to discuss the dynamics of the two external state variables, position $\boldsymbol{r}_i$ and velocity $\boldsymbol{v}_i$, we have to consider (i) the influence of the internal energy depot on the (active) motion of the agent, and (ii) the influence of the environment. The experiments described above have used a vertical beam of light that causes an attractive force on the Daphnia, which tend to cycle around it. In order to cope with this, we may choose the very simple assumption of an *external potential* of the form

$$U(\boldsymbol{r}) = \frac{a}{2}\boldsymbol{r}^2 \tag{2}$$

which generates an attractive force $\boldsymbol{F} = -\boldsymbol{\nabla}U(\boldsymbol{r}) = -a\boldsymbol{r}$ towards the center, $\boldsymbol{r} = 0$. The two effects can then be put into dynamic equations for the change of the agent's state variables, which have the following form [3, 11]:

$$\frac{d}{dt}\boldsymbol{r}_i = \boldsymbol{v}_i \; ; \quad \frac{d}{dt}\boldsymbol{v}_i = -\gamma_0\, \boldsymbol{v}_i - \left.\boldsymbol{\nabla}U(\boldsymbol{r})\right|_{\boldsymbol{r}_i} + d_2 e(t)\boldsymbol{v}_i + \sqrt{2D}\,\boldsymbol{\xi}_i(t) \tag{3}$$

Here, for the mass $m = 1$ is used. Causes for the change of the variables are summarized on the right-hand side of the equations. The change of the agent's position, $\boldsymbol{r}_i$ is caused by the movement of the agent, described by the velocity $\boldsymbol{v}_i$, that in turn can be changed by four different forces: (i) friction, with γ_0 being the friction coefficient, (ii) attraction toward the center of the light beam, expressed by the gradient of the environmental potential, (iii) active motion in forward direction driven by the energy from the internal depot, (iv) a stochastic force ξ of strength D, which describes the influence of random events on the agent's motion.

Both terms (i) and (iv) are already known from Langevin's equation to describe the motion of (passive) Brownian particles. *Without* the stochastic force, the particle would eventually come to rest because its kinetic energy would be dissipated due to friction. The stochastic force, however, keeps it (passively) moving into a random direction, where $\xi(t)$ is assumed to be Gaussian white noise with $\langle\xi_i(t)\rangle = 0$ and $\langle\xi(t)\xi(t')\rangle = \delta(t - t')$.

We conclude that in our approach of Brownian agents, the dynamics is described on the individual agent level by three coupled equations for the three state variables characterizing each agent, $\boldsymbol{r}_i$, $\boldsymbol{v}_i$, e_i, eqs. (3), (1). For theoretical investigations, these equations can be reduced to two by assuming that the internal energy depot relaxes very fast into a quasi-stationary equilibrium. This adiabatic approximation, i.e. the limit $\mu \to 0$ in eq. (1), results in

$$e_i(t) = \frac{q_0}{c + d_2 v_i^2} \tag{4}$$

and we can rewrite the equations of motion as:

$$\frac{d}{dt}\boldsymbol{r}_i = \boldsymbol{v}_i \; ; \quad \frac{d}{dt}\boldsymbol{v}_i = -\gamma(v_i^2)\, \boldsymbol{v}_i - \left.\boldsymbol{\nabla}U(\boldsymbol{r})\right|_{\boldsymbol{r}_i} + \sqrt{2D}\boldsymbol{\xi}_i(t) \tag{5}$$

$\gamma(v_i^2)$ is a *non-linear friction function*:

$$\gamma(v_i^2) = \gamma_0 - \frac{d_2\, q_0}{c + d_2 v_i^2} \tag{6}$$

which describes the active motion of the agent. It has a zero for

$$v_0^2 = \frac{q_0}{\gamma_0} - \frac{c}{d_2} \tag{7}$$

Active motion, i.e. $|v_0| > 0$ becomes possible only for a certain supercritical take-up of energy from the environment, $q_0 > c\gamma_0/d_2$. The actual motion of the agent is then a *compromise* between its active motion – which eventually would lead it everywhere, as long as internal energy is provided – and the environmental conditions which set some restrictions on this motion.

3 Simulating Swarming without Interaction

For the case of an attractive potential, eq. (2), Fig. 2 shows computer simulations for the active movement of a single agent, as described by the three eqs. (3), (1) for the state variables, $\boldsymbol{r}_i$, $\boldsymbol{v}_i$, e_i. The results clearly indicate the *cyclic motion* round the center, which has been also observed in single Daphnia motion, as explained above (*vortex motion*). Running the computer simulations for single agents with different initial conditions eventually results in the same kind of cyclic motion, but with different rotational directions, i.e. left-handed or right-handed rotations. Due to stochastic influences, also a change of the direction of motion becomes possible. Thus, we may conclude that our model of Brownian agents sufficiently describes the observed behavior of *single Daphnia*. We now turn to the case of many, i.e. $i = 1, ..., N$ Brownian agents, which is of importance for swarming. The dynamics of the multiagent system is then described by $3N$ coupled (stochastic) equations of the form (3), (1). In this case, we observe from computer simulations again the characteristic rotational motion, where, however, about half of the agents rotate clockwise, while the other half rotates counterclockwise. The two different cyclic directions can be clearly observed when looking at the *angular momentum* distribution, $\rho(L)$, where $\boldsymbol{L}$ (for $m = 1$) is defined as $\boldsymbol{L} = \boldsymbol{r} \times \boldsymbol{v}$. As Fig. 3 shows, this is a *bimodal* distribution of about equal height, indicating the both left- and righthanded rotational directions with the same probability.

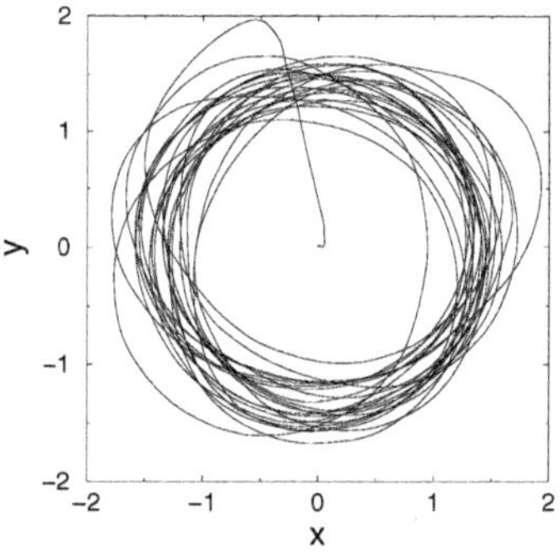

Fig. 2. Trajectory ($t = 200$) of a single Brownian agent moving in an environmental potential, eq. (2) with supercritical take-up of energy, $q_0 > c\gamma_0/d_2$. Initial conditions: $\{x(0), y(0)\} = \{0, 0\}$, $\{v_x(0), v_y(0)\} = \{0, 0\}$, $e(0) = 0$, parameters: $\gamma = 5.0$, $d_2 = 1.0$, $q_0 = 10.0$, $c = 1.0$, $D = 0.005$, $a = 0.5$.[6]

This simulation result does not quite agree with the observation of swarming in Daphnia, which apparently cycle into *one*, i.e. the same direction. The reasons for this mismatch are quite obvious: in our model, we have so far only considered "point-like" agents without any kind of mutual interaction, whereas in real biological systems the coherent motion of the swarm is certainly based

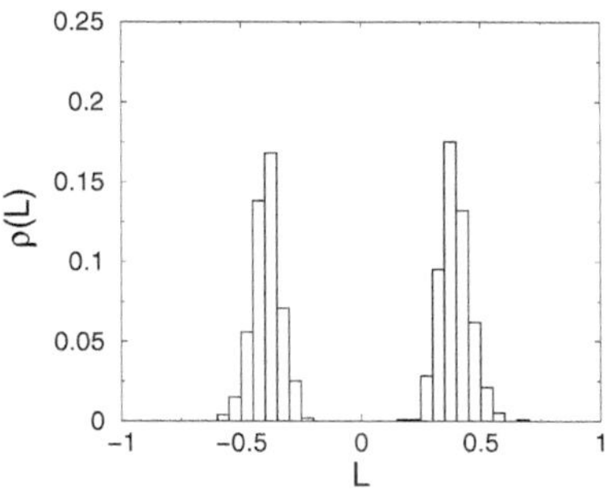

Fig. 3. Angular momentum distribution $\rho(L)$ of $N = 1000$ Brownian agents after $t = 150$. The positive or negative sign of L indicates the right- or lefthanded rotation. Parameters: $q_0 = 10.0$, $c = 1.0$ $\gamma = 20.0$, $d_2 = 10.0$, $D = 0.001$, $a = 1.0$.[6]

on interactions between the entities. Thus, the question arises, which kind of interaction may lead to the break in the rotational symmetry, as observed in the Daphnia experiments.

So far, different forms of *global* or *local* interactions have been introduced into swarming models. We mention (i) local interactions via a self-consistent field that has been created by the agents and in turn influences their further movement and/or "behavior" [5, 13] – chemotactic response is a prominent example here, (ii) local interactions based on the coupling of the agents's individual velocity to a local average velocity [2, 15, 16] (iii) global interactions, such as the coupling of the agent's individual orientation (i.e. direction of motion) to the mean orientation of the swarm [1, 2], or the coupling of the agent's individual position to the mean position (center of mass) of the swarm [7], further couplings via the mean momentum or mean angular momentum or a combined set of invariants of motion [1, 12], (iv) interactions based on hydrodynamics coupling between agents [4].

Despite the fact that some of these models simulate coherent swarm behavior or even rotation of the swarm into the same direction, there is evidence that many of the underlying assumptions for interactions can hardly be satisfied by *biological* observations, thus their biological relevance is highly questionable. Therefore, in the following section, we introduce local interactions between the agents that indeed match with biological reality.

4 Modeling Swarming with Avoidance Behavior

Experiments on Daphnia swarming has shown that these animals tend to cycle into the same direction. The simple and obvious reason for this is that animals try to avoid as much as possible collisions with other animals – which would occur much more frequently if different animals cycled into opposite directions at the same time. Thus, a biologically satisfied assumtion is to include *avoidance behavior* in our model of swarming, in order to test, whether this would lead to the observed break in the rotational symmetry described above.

Daphnia are able to visually sense their environment to a certain degree, i.e. they can detect animals approaching them from the front, and then try to avoid collisions. In our model, we account for this by assuming that there is a short-range repulsive force between agents, to prevent their collisions, which will result from a repulsive *interaction potential* $V(r_i)$ around each agent i that depends on its actual position, r_i:

$$V(r_i) = p \cdot \exp\left(-\frac{R_i}{\sigma}\right) \tag{8}$$

p denotes the strength and σ the range of the potential, the latter being a measure of the sight, i.e. the range of *visibility*. R_i is a specific function of the distance between agents, as explained in the following. Since all agents are moving, agent i needs to account for the space that will be occupied by all other agents j in the vicinity during the next time step. This space needed, depends both on the agent's positions r_j *and* their velocity of motion, v_j, so R_i is a function of these. For further specification, we introduce the unit vector in the direction of motion of agent i, $\boldsymbol{n}_i^0 = \boldsymbol{v}_i/||\boldsymbol{v}_i||$; $\boldsymbol{n}_j^0$ is defined similarly. This allows to define a new *velocity-dependent* coordinate system for agent i, namely $\boldsymbol{y}_i$ and $\boldsymbol{x}_i$ defined by:

$$\boldsymbol{y}_i = \frac{v_i {\boldsymbol{n}_i}^0 - \delta v_j {\boldsymbol{n}_j}^0}{||v_i {\boldsymbol{n}_i}^0 - \delta v_j {\boldsymbol{n}_j}^0||} \quad ; \quad \boldsymbol{x}_i \perp \boldsymbol{y}_i \quad \text{and} \quad \langle \boldsymbol{x}_i, \boldsymbol{x}_i \rangle = 1\ . \tag{9}$$

If $\delta > 0$, the direction of motion of agent j is also taken into account for agent i. The $\boldsymbol{x}_i$ can be constructed by the orthonomalization algorithm by GRAM-SCHMIDT. Using this coordinate system, the dependence of R_i on the position and velocity of agent j is now given as

$$R_i = \sqrt{\langle \boldsymbol{r}_i - \boldsymbol{r}_j, \boldsymbol{x}_i \rangle^2 + \beta^2 \langle \boldsymbol{r}_i - \boldsymbol{r}_j, \boldsymbol{y}_i \rangle^2} \tag{10}$$

with a velocity-dependent function:

$$\beta = \begin{cases} \beta' & : \langle \boldsymbol{r}_i - \boldsymbol{r}_j, \boldsymbol{y}_i \rangle \geq 0 \\ \dfrac{\beta'}{1 + \lambda \cdot v_i} & : \langle \boldsymbol{r}_i - \boldsymbol{r}_j, \boldsymbol{y}_i \rangle < 0 \end{cases} \tag{11}$$

Eventually, with the known repulsive interaction potential $V(r_i)$, the force between any two agents i and j is given as:

$$\boldsymbol{f}_{ij} = -\boldsymbol{\nabla} V(r_i) = \frac{V(r_i)}{\sigma \cdot R_i} \left(\langle \boldsymbol{r}_i - \boldsymbol{r}_j, \boldsymbol{x}_i \rangle \boldsymbol{x}_i + \langle \boldsymbol{r}_i - \boldsymbol{r}_j, \beta \boldsymbol{y}_i \rangle \beta \boldsymbol{y}_i \right) \tag{12}$$

$$= \frac{p}{\sigma \cdot R_i} \exp\left(-\frac{R_i}{\sigma}\right) (\boldsymbol{r}_i - \boldsymbol{r}_j) \tag{13}$$

The total force on agent i resulting from the assumed local interaction is then given as the sum over all 2-agent forces, $\boldsymbol{F}_i = \sum_{j \neq i} \boldsymbol{f}_{ij}$. This repulsive force of

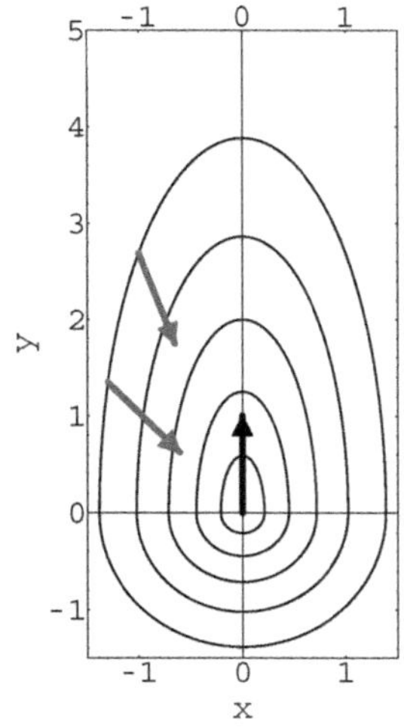

Fig. 4. Equipotential lines of the repulsive potential $V(r_i)$, eq. (8). The black arrow indicates the agent in the origin, having a velocity of $v = \{0, 1\}$. The two gray arrows representing other agents have the same absolute value of 1 and point towards the origin.[6]

course changes eq. (5) for the agents by an additional term, i.e. in the extended avoidance model, the dynamics now read:

$$\frac{d}{dt}\boldsymbol{r}_i = \boldsymbol{v}_i \; ; \quad \frac{d}{dt}\boldsymbol{v}_i = -\gamma(v^2)\,\boldsymbol{v}_i - a\boldsymbol{r}_i + \sum_{i \neq j} \boldsymbol{f}_{ij} + \sqrt{2D}\xi(t) \tag{14}$$

We note that these assumptions lead to an asymmetric repulsive potential $V(r_i)$ around each agent. Two different forms of such an asymmetrical potential have been introduced in [6]. The one used in this paper – which has been originally used to simulate the movement of pedestrians [8] – is preferred because it leads to smoother movements of the agents. The potential defined by eq. (8) with eq. (10) can be seen in Fig. 4.

Fig. 5 shows spatial snapshots of a computer simulation of the multiagent system with respect to avoidance behavior, together with the respective distribution of the angular momenta $\rho(L)$. The results can be concluded as follows: (i) On the spatial level, we observe the emergence of a *coherent motion* of the multi-agent swarm out of a random initial distribution. This collective motion is characterized by a *unique cycling direction* (either left- *or* righthanded rotation), as can be also seen from the *unimodal* distribution, $\rho(L)$. (ii) We further observe the formation of a vortex, which is rather similar to the Daphnia swarm cycling round the light beam. (iii) While in one simulation all agents cycle in the same direction, we note that in different simulations the cycling direction can be also opposite, i.e. there is *no preferred cycling direction* for the swarm, which also agrees with the observations of the Daphnia swarm. (iv) We note that for certain parameters a spontaneous change in the rotating direction can be observed. This occurs in particular if agent i takes strongly the movement of agent j into account ($\delta \approx 0.5$).

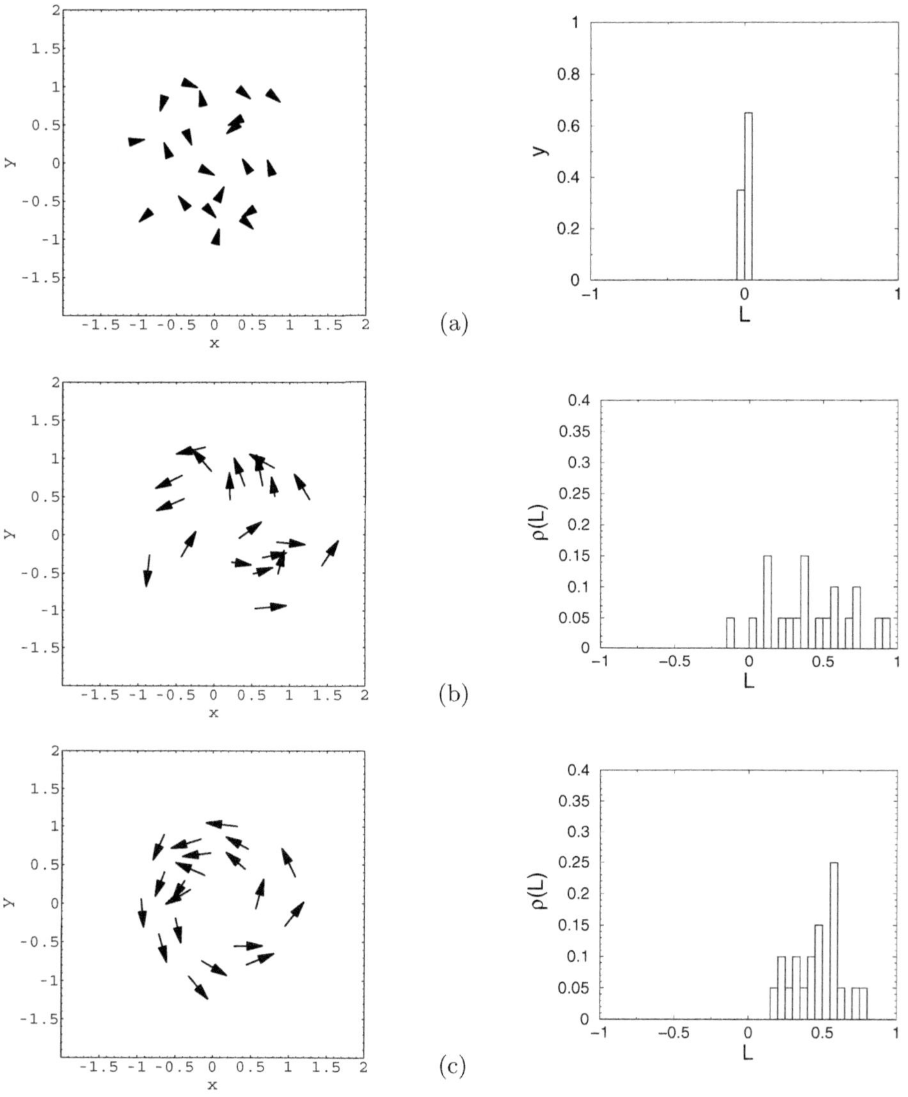

Fig. 5. Spatial snapshots (left) and distribution of angular momentum $\rho(L)$ (right) for a multiagent system ($N = 20$) at three different times: (a) $t = 0$ (b) $t = 8$ and (c) $t = 55$. The length of the arrows indicates the velocities. [6]

5 Conclusions

Our goal was to model vortex swarming behavior by means of rather *minimal* assumptions that, however, should have a clear *biological relevance.* As we have shown the model of Brownian agents introduced in this paper can indeed produce a reasonable swarming behavior that resembles real Daphnia swarms. Although reasonable assumptions about local interactions (such as local repulsion) are taken into account, a definite justification of the model is still pending. This is due to the fact that data is difficult to gather, because defined lab conditions are hard to establish. Currently, Ordemann *et al.* are testing this model on real *Daphnia* swarms.

Acknowledgment. The authors would like to thank Anke Ordemann and Frank Moss for sharing their observations on Daphnia swarms and for valuable discussion.

Bibliography

[1] A. Czirok, E. Ben-Jacob, I. Cohen, and T. Vicsek. Formation of complex bacterial colonies via self-generated vortices. *Physical Review E*, 54(2):1791–1801, 1996.

[2] A. Czirok and T. Vicsek. Collective behavior of interacting self-propelled particles. *Physica A*, 281:17–29, 2000.

[3] W. Ebeling, F. Schweitzer, and B. Tilch. Active Brownian particles with energy depots modelling animal mobility. *BioSystems*, 49:17–29, 1999.

[4] U. Erdmann and W. Ebeling. Collective motion of Brownian particles with hydrodynamic interactions. *Fluctuation and Noise Letters*, 2002 (submitted).

[5] D. Helbing, F. Schweitzer, J. Keltsch, and P. Molnár. Active walker model for the formation of human and animal trail systems. *Physical Review E*, 56(3):2527–2539, 1997.

[6] R. Mach, A. Ordemann, and F. Schweitzer. Modeling vortex swarming in Daphnia. *Journal of Theoretical Biology*, 2003 (submitted)

[7] A.S. Mikhailov and D. H. Zanette. Noise-induced breakdown of coherent collective motion in swarms. *Physical Review E*, 60:4571–4575, 1999.

[8] P. Molnár. *Modellierung und Simulation der Dynamik von Fussgängerströmen*. Shaker, Aachen 1995.

[9] A. Ordemann, G. Balazsi, and F. Moss. Motions of daphnia in a light field: Random walks with a zooplankton. *Nova Acta Leopoldina*, 2003 (in press)

[10] A. Ordemann. Vortex-swarming of the zooplankton daphnia. *The Biological Physicist*, 2(3):5–10, August 2002.

[11] F. Schweitzer, W. Ebeling, and B. Tilch. Complex motion of Brownian particles with energy depots. *Physical Review Letters*, 80(23):5044–5047, 1998.

[12] F. Schweitzer, W. Ebeling, and B. Tilch. Statistical mechanics of canonical-dissipative systems and applications to swarm dynamics. *Physical Review E*, 64(2):021110–1 – 021110–12, August 2001.

[13] F. Schweitzer, K. Lao, and F. Family. Active random walkers simulate trunk trail formation by ants. *BioSystems*, 41:153–166, 1997.

[14] F. Schweitzer. *Brownian Agents and Active Particles. Collective Dynamics in the Natural and Social Sciences*, *Springer Series in Synergetics*, Springer Berlin 2003

[15] J. Toner and Yuhai Tu. Long-range order in a two-dimensional dynamical xy model: How birds fly together. *Physical Review Letters*, 75(23):4326–4329, December 1995.

[16] T. Vicsek, A. Czirok, E. Ben-Jacob, I. Cohen, and O. Shochet. Novel type of phase transition in a system of self-driven particles. *Physical Review Letters*, 75:1226–1229, 1995.

Visually Guided Physically Simulated Agents with Evolved Morphologies

Ian Macinnes

Centre for Computational Neuroscience and Robotics (CCNR), School of Cognitive and Computing Sciences, University of Sussex, Brighton, UK

Abstract. An agent must locate and identify elements within its environment in order to manipulate them constructively. A population of physically simulated agents is produced with arbitrary evolved morphologies and neural network controllers without wheels or imposed symmetry. This is done via the use of incremental evolution. It is shown that they can locate and approach a coloured object within their environment using a light detecting sensor with a field of view. An evolved agent is examined and is shown to use a close dynamic coupling between the controller and morphology to produce a stable platform for a visual sensor, allowing it to fixate on its target.

1 Introduction

In the mid-nineties, Karl Sims produced agents with both evolved controllers and physical morphologies[1,2] in simulated worlds. Since then, evolvable morphologies have been used to explore developmental models[3], encoding schemes[5,4], and the evolution of static structures and robot bodies[6] among other things[9]. The work presented here is part of ongoing research into arbitrary evolved morphologies.

An agent must locate and identify elements within its environment in order to manipulate them constructively. A simulated agent is described with an evolved neural network controller and physical morphology that uses a light detector with a field of view to find and approach a coloured beacon.

2 Method

We use artificial evolution to synthesise the morphology and neural network controller of an agent using an incremental evolution method. Agents consist of blocks connected by powered hinge joints, each of which allow one axis of movement. This makes the strategy of using wheels or rolling parts for motion unlikely. An agent's genotype specifies the dimensions of each of its constituent blocks and joint attributes. Each agent is controlled by a genetically determined neural network. The agents exist in a simulated world with a gravity of -10 unit/s^2. The physical simulation library ODE[10] calculates the dynamics of the agents and their environment. It uses a time slice of 0.05 seconds for each physical step during fitness evaluations.

W. Banzhaf et al. (Eds.): ECAL 2003, LNAI 2801, pp. 821–828, 2003.

2.1 Neural Network Controller and Sensors

The controller is a continuous time dynamical recurrent neural network[7] with homeostatic oscillating neurons as described by Di Paolo[8]. The potential of each neuron over time is given by:

$$\tau_i \Delta y = -y_i + \sum_j w_{ji} z_j \tag{1}$$

...where τ_i is the time constant of neuron i, y_i is its potential, and w_{ji} is the weight from i to j. The activation of the neuron z_j is given by:

$$z_j = tanh(b_i - y_i) \tag{2}$$

...where b_i is the genetically determined bias value for neuron **i**. The bias for each neuron varies by the rule:

$$\Delta b_i = \frac{-b_i - y_i}{\alpha_i} \tag{3}$$

...where α_i is a bias time constant of neuron **i**.

A controller without sensors or actuators added has four fully interconnected neurons. Each actuator and sensor has its own individual set of neurons, and when added to the agent it randomly connects to existing neurons in the controller. Three types of sensor are used (the terms input and output refer to input and output to-and-from the controller):

1. A *joint* has two neurons. An input neuron specifies the angle of the joint in radians and an output neuron specifies the force the agent applies to the hinge joint, thus influencing its movement.
2. A *directional beacon* has a single input neuron whose value ranges from -1 (when the sensor points directly away from the beacon) and 1.
3. A *visual sensor* is excited when a blue object (e.g. the beacon) enters its field of view, and has an angle of projection determined by the genotype. Any body parts that get between the sensor and the beacon partially or totally block the view and hence reduces the activation. The sensor has to point at the object to be activated by it. It has an input neuron whose activation ranges from -1 (nothing blue is in the field of view) to 1 (the field of view is completely filled by something blue).

2.2 The Genetic Algorithm

The genotype divides into two sections, one for the neural network and one for the physical morphology, each containing mutable values (tables 1 and 2). Neurons map onto sensors and actuators and vice-versa via a lookup table. This table alters whenever a joint or sensor is added or removed. There are three main mutation operators. These are:

1. Adding or removing a random block with a 0.05 probability of occurring.

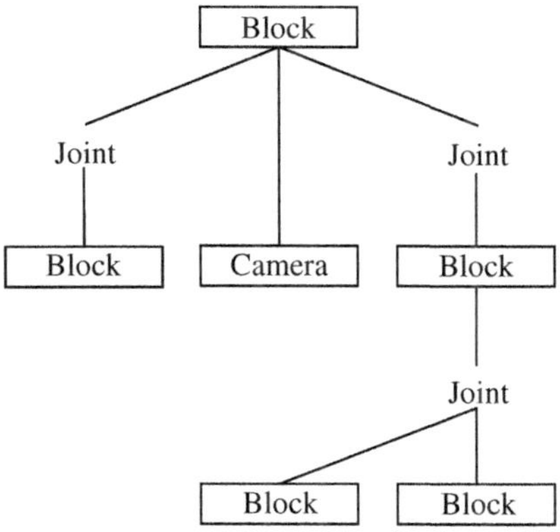

Fig. 1. Example of a physical morphology composed of five blocks and a camera node that's encoded by one section of the genotype. Each contain mutable information describing each node (table 2).

2. Altering continuous values in the genotype using one of two Gaussian distribution functions, one with a standard deviation of 0.33, and one with of 0.05, each occurring with a probability of once per reproduction event.
3. Altering discrete values in the genotype has a 0.05 probability of occurring.

A population consists of twenty-eight agents. It is seeded with randomised genotypes that produce agents constructed of three jointed blocks. The evolutionary algorithm uses rank selection with three of the highest ranking genotypes going through to the next generation.

Table 1. Each node in the neural network section of the genotype has its own set of mutable information.

Neuron Node		*Weight Node*	
attribute	*range*	*attribute*	*range*
time constant	0.4 - 10.0	weight	-20.0 - 20.0
bias	-3.0 - 3.0	from node	*node ID-not mutable*
bias time constant	0.4 - 12.0	to node	*node ID-not mutable*

We have previously found it difficult to evolve agents that use their visual sensors to approach a target. Therefore we adopted an incremental evolution strategy. Agents that successfully evolved strategies for solving one step go on to seed the population for the next step. These steps are:

1. *To move as far as possible from the initial position.*
2. *To approach a wall using directional beacons.* This takes place over a fixed period of five thousand generations. Each agent has a maximum of two sen-

Table 2. Each node in the physical morphology component of the genotype has its own set of mutable information.

Block Node		*Joint Node*		*Camera Node*	
attribute	*range*	*attribute*	*range*	*attribute*	*range*
width	1.0-10.0	x position	0.1-0.9	x position	0.0-1.0
height	1.0-10.0	y position	0.1-0.9	y position	0.0-1.0
depth	1.0-10.0	side of block	1,2,3,4,5,6	side of block	1,2,3,4,5,6
		power	0.0-500.0		
		orientation	0-2π		
		angle offset	0-2π		
		parent	*block ID*		

sors, with genetically determined positions. A wall is randomly placed on one or the other side of the agent. The reciprocal of the perpendicular distance at the end of the trial between the wall and the agent provide the fitness for that agent. The walls dimensions are 300 long by 50 units high by 3 units thick.

3. *To approach a wall using camera sensors.* This takes place over a fixed period of five thousand generations. With a population that can already approach a wall using directional beacon sensors, the directional beacon sensors steadily convert into visual sensors over evolutionary time. This happens by fractionally replacing the input of the directional beacon sensor with the input of a visual sensor. Each generation has agents with decreasing component from the directional beacon sensor and a larger component of the visual sensor, until the input from the direction beacon sensor reduces to zero. So:

$$I = \frac{(5000 - \gamma)D}{5000} + \frac{\gamma V}{5000}$$

...where I is the total input to the controller from this sensor, γ is the number of generations since the start of this evolutionary run, D is the input from the directional beacon, and V is the input from the visual sensor, both with the same position and orientation. The fitness function remains the same.
4. *Slowly reducing the size of the wall.* This occurs over five thousand generations. The wall slowly reduces to a cube with a side of three units. The fitness function remains the same.
5. *Randomly moving the beacon.* This takes place over ten thousand generations. The position of the beacon, now a cube, varies between trials. The variation increases as the generations increase, until the cube can appear anywhere within fifty units of the initial position of the agent.

3 Results

A population of agents evolved that successfully performed the task of locating and approaching a blue block using visual sensors. A typical agent consisted of

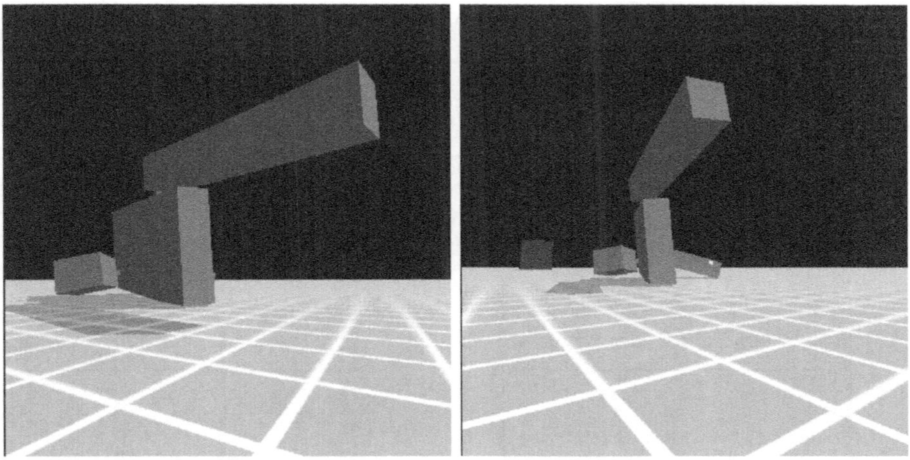

(a) beginning of trial (b) starting to turn...

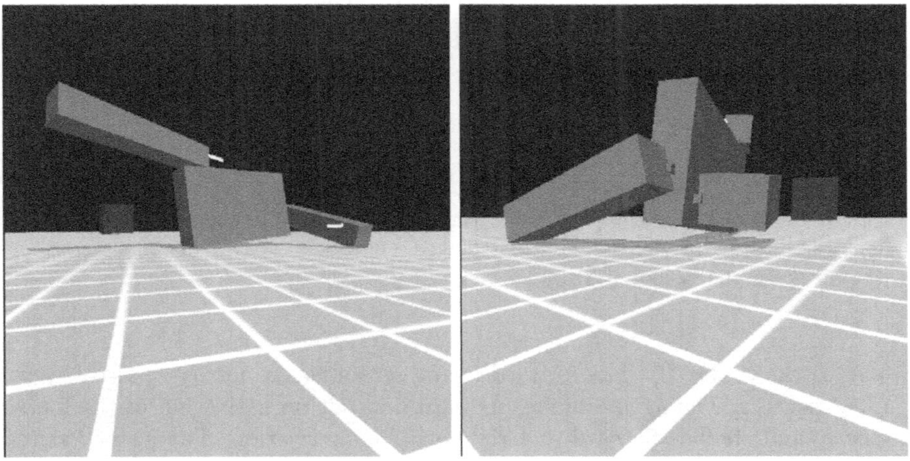

(c) ...turning further... (d) ...and moving towards the beacon

Fig. 2. (from top left to bottom right) The agent in the beginning of its trial. The white lines indicate the position and direction of the visual sensors. In (b), it is starting to turn slightly (note the beacon in the background). The agent's turn is more obviously shown by (c). By (d), the agent is moving directly towards the beacon. The agent's top block works up and down, using its momentum to edge the agent off the ground, allowing the back bottom blocks to push the agent along in small steps.

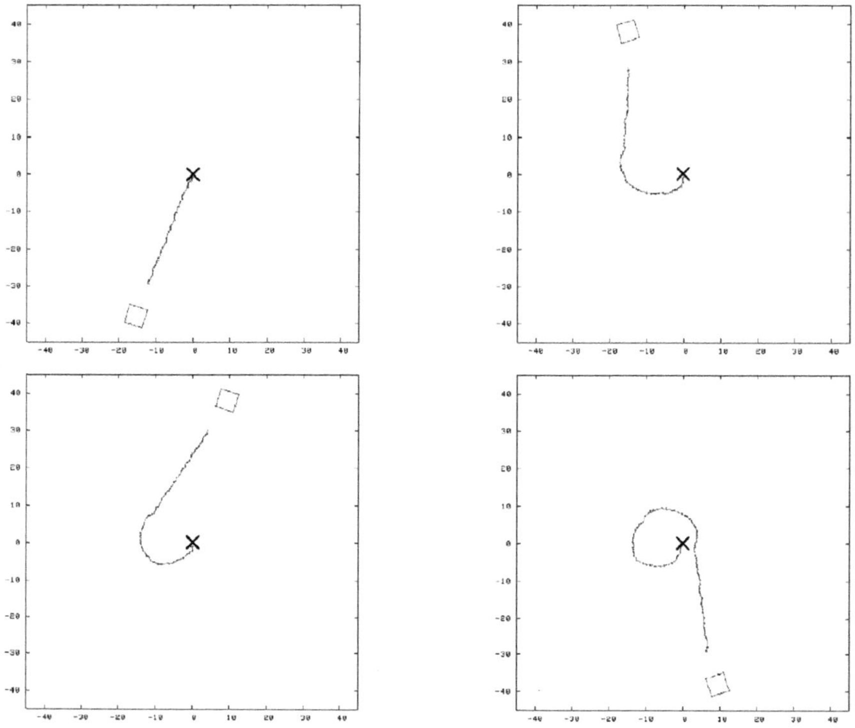

Fig. 3. Plots of the agent path when the beacon is in different positions. The cross donates the starting point. The agent makes a clockwise spiral, only moving straight when it spots the beacon in front.

four blocks (figure 2). The forward facing sensor has a narrow angle of view of 13.62 degrees. The agent shuffles forward and left on activation of the forward sensor, and shuffles forward and right with no activation. The agent locates a block by moving in a clockwise spiral until the beacon activates the front sensor. This causes the agent to move forward and very slightly to the left. The beacon slips out of view and the sensor is less activated, therefore the agent moves slightly to the right again. These two behaviours cause the beacon to be kept in or close to the field of view. If the beacon slips out of sight, these behaviours make it likely to soon be viewable again. These two behaviours can be demonstrated by showing the path of an agent with no beacon within sight (figure 4a) and that of an agent with the sensor fixed to a maximum activation (figure 4b)

The agent has two limb blocks. The limb block that pushes the agent in a clockwise direction also holds the forward looking sensor. Whether or not this sensor is stimulated alters the behaviour of the joint holding the limb in the following way:

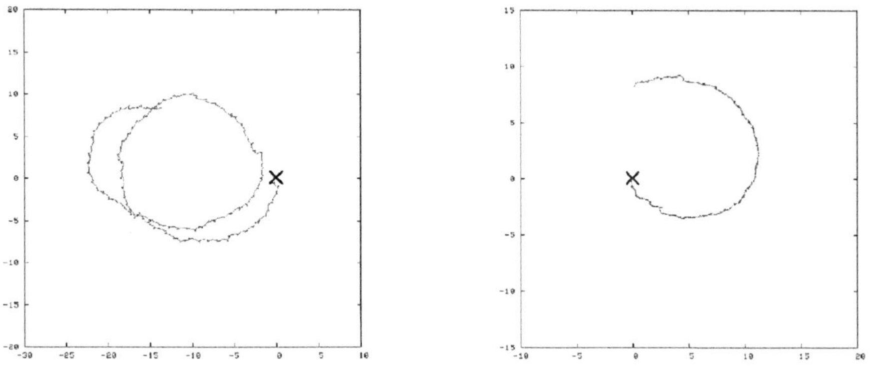

Fig. 4. The two main behaviours of the agent. The cross denotes the starting point. The left diagram shows the path of the agent when no beacon is present. It spirals to the right. The right diagram shows its path when the agent's sensor is maximumly stimulated all the time - it spirals around slowly the opposite way. Both these behaviours together allow the agent to find and approach the beacon.

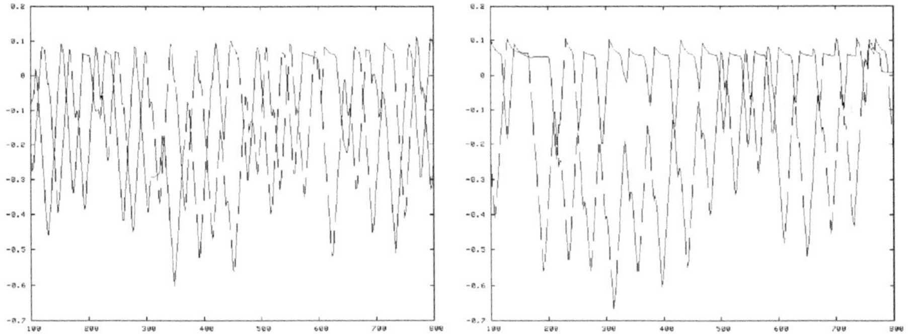

Fig. 5. The force applied to the two joints by the neural network controller. The left diagram shows the force applied when no beacon is present. The right diagram shows the force applied when the sensors are maximumly stimulated. Note that the amplitude of the force to one of the joints is less than when the sensor is not stimulated.

1. When the sensor cannot see the beacon, the neural network supplies power to this joint (figure 5a), and the agent moves in a clock wise direction.
2. When the sensor *can* see the beacon, the neural network reduces the amplitude of the power to this joint (figure 5b). This has two effects, the first of stopping the clockwise spiral of the agent, and the second of stabilising the block, and hence stabilising the visual platform.

This sensible strategy is found in animals. Eyes are usually on a stable platform, a part of the body thats jointed so that it can hold its vision steady and fixate upon a feature in the environment. In mammals, the head is very good at stabilising its orientation in respect to a focus of attention even if the body as a whole is in motion.

4 Conclusion

The adoption of an incremental strategy has allowed the evolution of the neural network controller and physical morphology of agents to find and approach a beacon. One evolved agent is examined and found to be taking full advantage of the physics and morphology of its body. It uses a close dynamic coupling of the controller and morphology to produce a stable platform for a visual sensor. This allows the agent to fixate on its target. It manages to both locate and approach a beacon using just two main behaviours - spiralling left or right, and chooses between them depending upon the activation of a visual sensor.

This work shows it is possible to evolve arbitrary morphologies to achieve behaviours that are more complex than simple locomotion and so opens the exploration of more interesting behaviours including the manipulation of the environment.

Acknowledgments. I'd like to thank Ezequiel Di Paolo, Kyran Dale, Michael Garvie, and Robert Vickerstaff for discussions and advice about this paper.

References

1. Sims, K: (1994) 'Evolving Virtual Creatures' In *Siggraph Proceedings* pp. 15–22.
2. Sims, K: (1994) 'Evolving 3D Morphology and Behaviour by Competition' In R. Brooks, P. Maes (Eds.), *Artificial Life IV* pp. 28–39. MIT Press.
3. Bongard, J.C.: (2001) 'Repeated Structure and Dissociation of Genotypic and Phenotypic Complexity in Artificial Ontogeny' In Spector, L. et al (eds), *Proceedings of the Genetic and Evolutionary Computation Conference, GECCO-2001* pp. 829–836. Morgan Kaufmann publishers.
4. Hornby, G.S., Pollack, J.B.: (2002) 'Creating High-Level Components with a Generative Representation for Body-Brain Evolution' In *Artificial Life 8* pp. 223–246. MIT Press.
5. Hornby, G.S., Pollack, J.B.: (2001) 'Evolving L-Systems to Generate Virtual Creatures' In *Computers and Graphics* 25:6 pp .1041–1048.
6. Funes, P., Pollack, J.: (1998) 'Evolutionary Body Building: Adaptive Physical Designs for Robots' In *Artificial Life Vol. 4:4* pp. 337–357. MIT Press.
7. Beer, R.D.: (1996) 'Toward the evolution of dynamical neural networks for minimally cognitive behaviour' In P. Maes, M. Mataric, J. Meyer, J. Pollack and S. Wilson (Eds.), *From animals to animats 4: Proceedings of the Fourth International Conference on Simulation of Adaptive Behaviour* pp. 421–429. MIT Press.
8. Di Paolo, E.A.: (2002) 'Evolving Robust Robots Using Homeostatic Oscillators' Cognitive Science Research Paper 526, COGS. University of Sussex.
9. Taylor, T., Massey, C.:(2001) 'Recent Developments in the Evolution of Morphologies and Controllers for Physically Simulated Creatures' In *Artificial Life 7* pp. 77–87.
10. Smith, R. et al: Open Dynamics Engine (ODE) (physical simulation library) *http://sourceforge.net/projects/opende/*

The Robot in the Swarm: An Investigation into Agent Embodiment within Virtual Robotic Swarms

Adam Lee Newton[1], Chrystopher L. Nehaniv[2], and Kerstin Dautenhahn[2]

[1] A.Newton@worc.ac.uk
[2] Adaptive Systems Research Group
Faculty of Engineering and Information Science,
University of Hertfordshire
{C.L.Nehaniv, K.Dautenhahn}@herts.ac.uk

Abstract. This paper explores the notion of 'degree of embodiment' within the context of autonomous agent research, specifically within swarms of virtual robotic agents. Swarms of virtual robots with systematically varied degrees of embodiment are designed and implemented, and a 3D world created for them. Experimental simulations are then carried out wherein groups of these robots perform swarm tasks, and levels of performance for each group are measured and analysed. Analysis of this data suggests that there is no simple linear or monotonic correlation between degree of agent embodiment and swarm performance (in this particular virtual environment), but rather that an 'ideal' degree of embodiment exists to create a superior swarm behaviour for a given task in a given environment.

1 Introduction and Background

The generally accepted non-scientific definition of embodiment is simply 'to have a physical body', and it is this concept that is extended and clarified by research in many of the Cognitive Sciences such as Philosophy, Artificial Intelligence and, perhaps more specifically, Artificial Life and Robotics. These fields share a common need to describe how an agent, whether biological, robotic, or perhaps one simulated in software, is 'embodied' or 'situated' within the surrounding environment, and how this nature of embodiment affects both agent and environment. A crucial definition of what is meant by the term *embodiment* within the context of agent research appears in (Quick et al. 1999), and it is this meaning that shall be used for the remainder of this paper:

A system X is embodied in an environment E if perturbatory channels exist between the two. That is, X is embodied in E if for every time t at which both X and E exist, some subset of E's possible states have the capacity to perturb X's state, and some subset of X's possible states have the capacity to perturb E's state.

W. Banzhaf et al. (Eds.): ECAL 2003, LNAI 2801, pp. 829–838, 2003.

Recent papers by Tom Quick of University College London and Kerstin Dautenhahn and colleagues at the University of Hertfordshire have further suggested that all bodies, (whether the body of an agent or an inanimate object) are somehow embodied and situated in their environment, but that embodiment is not a binary on-or-off attribute of an agent; rather that certain bodies may exhibit a higher 'degree of embodiment' than others (Quick et al. 1999, Nehaniv 2000, Dautenhahn et al. 2002). For instance, the greater the perturbatory 'bandwidth' connecting agent and environment, the higher the degree of embodiment. This idea is briefly summarised in Figure 1. For more information on this crucial concept, see (Quick et al. 1999, Quick and Dautenhahn 1999, Dautenhahn et al. 2002).

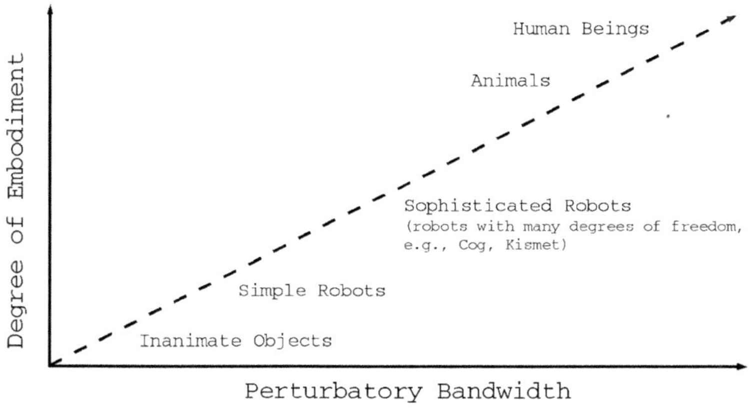

Fig. 1. A proposed correlation between degree of embodiment and perturbatory bandwidth (after Dautenhahn et al., 2002)

This theoretical correlation between the perturbatory channels and the level or degree to which an agent is embodied or situated in its environment might be experimentally investigated using virtual or physical robots. In Figure 1, the lower portion of the graph makes reference to some simplistic robotic agent having a lesser degree of embodiment than a more complex robotic agent. The investigations in this paper focus on this area, and examine whether a robot might be more or less embodied in an environment due to the perturbatory bandwidth between itself and its environment.

The number of perturbatory channels existing between a robotic agent and its environment is related to the number of sensory channels in the agent's physical construction - the number and usage of the various sensors and actuators it employs (wheels, motors, gripper arms, etc.). In the following experimental trials, groups of virtual robotic agents will be configured in a systematically differing manner, some having a higher number of sensory channels, some with a low number. Groups of these agents, placed into controlled experiments, will be given a task to perform. Performance scores, based on a simple metric, record the differences in behaviour from group to group, and this data is analysed and compared within and across the different robot groups. The central research issue this paper attempts to illuminate is how the nature of embodiment of the agents in a swarm changes the behaviour and performance of that swarm.

Though real-hardware Khepera robots were considered an option for this research, the practicalities proved prohibitive. The simulation package 'Webots' from Cyberbotics[1] was chosen due to its well-known ability to closely emulate Khepera hardware within software virtualisation. Webots provides the flexibility and rapid-prototyping facilities of other software simulation environments with the ability to adhere to Khepera reference design, allowing later cross-compilation of controller code to real-hardware robots. The GNU/Linux[2] OS platform was chosen for its reliability and scalability, and also because Linux is the most well supported OS in recent versions of Webots[3].

2 Methodology

When an agent forms part of a swarm, the 'surface' between itself and the environment, structurally connected via the aforementioned perturbatory channels, becomes even more complex. This surface now not only represents the connections between sub-states of the environment (i.e., those states that may effect change in the states of other elements connected via the perturbatory channels), but also (directly or indirectly), the states of the other agents in the swarm. In order to magnify the lower area of the graph featured in Fig. 1, and to identify how Quick et al.'s (1999) 'degrees of embodiment' theory might relate to agents that find themselves not only situated in an environment (virtualised, in this case), but also (unknowingly[4]) forming a part of a swarm, groups of systematically varied agents can be designed and developed to span different levels of sensorimotor complexity, directly relating to the amount of perturbatory bandwidth connecting each agent to its environment (and so, indirectly, to each other agent in its swarm).

The three groups of Khepera-like agents employed in these experiments vary systematically across the spectra of sensor complexity – e.g. sensory bandwidth – from simplistic to complex. Agents from group C enjoy only a low level of sensory complexity, group A enjoy a high amount, while group B is intermediate. The reference design (illustrated in Figure 1) for the Khepera robotic agent has been used as a template, and then modified to create groups A, B and C differing in

[1] http://www.cyberbotics.com

[2] The Sourcemage GNU/Linux distribution was chosen for its high performance.

[3] Version 3.2.x for these experiments.

[4] In the simplest of swarm simulations, no swarm agent explicitly 'knows' of the existence of any other agent. All communication between agents is indirect, through the environment via the process of stigmergy (see Grassé, 1959).

the number of sensory channels (infrared (IR) sensors)[5]. These groups were then populated with ten identical clones to create a uniform swarm of robotic agents. In order to explore the ramifications of design variations in both *number of* and *sensitivity of sensory channels*, each group contains three further variants - this time with the same number of 'channels' of perturbation (the same total number of sensors and actuators) but with the *sensor range* systematically altered; i.e., group A is now made up of A_i (long sensor range), A_{ii} ('normal' sensor range), A_{iii} (low sensor range). Figure 3 illustrates the swarm configurations in more detail.

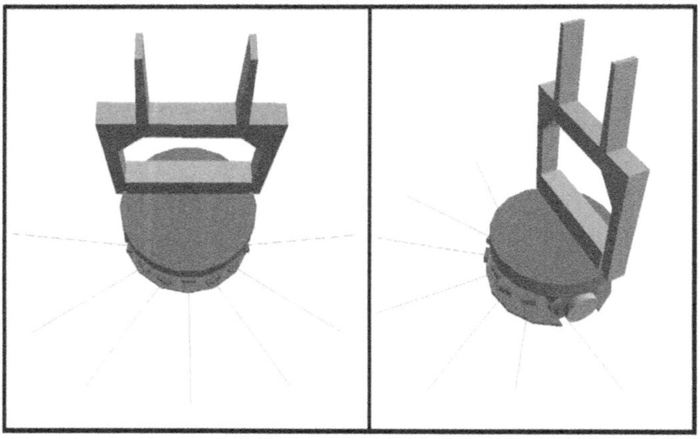

Fig. 2. A simple example of a virtual Khepera robot, modelled in Webots. (Left: top view, Right: side view)

A large arena allows the agents to roam in two dimensions, powered by their actuators. The arena is surrounded by a low wall and populated with a number of blocks, and both of these are objects that the Kheperas' IR sensors detect and respond to in a manner dictated by their controller executable. The environment shown in Figure 4 is typical of a swarm task scenario wherein a particular swarm group is challenged with roaming the arena, finding blocks and clustering them together through a simple, emergent, social-insect-like behaviour[6].

[5] The design approach taken with the 3D, VRML-based bodies and the C-based controller for each type of Khepera-like robotic agent ensured that all controllers could be later cross-compiled to real-world Khepera hardware, if available. All VRML Khepera modelling was based on the reference design supplied by Cyberbotics to ensure real-world validity at the design stage.

[6] This simplistic insect-inspired swarm task can be found in use by ants, termites, and several other species of insects. A swarm of agents roams around, picking up blocks where they find them, and putting them down near other blocks (in insects, sometimes aided by pheromones). Over time, ordered clusters of blocks emerge through a natural process of stigmergy. See Bonabeau et al. (1999) for details.

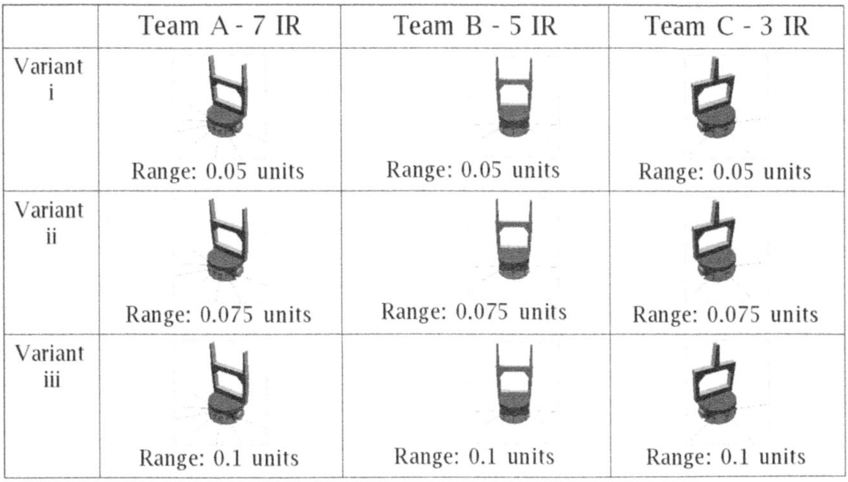

	Team A - 7 IR	Team B - 5 IR	Team C - 3 IR
Variant i	Range: 0.05 units	Range: 0.05 units	Range: 0.05 units
Variant ii	Range: 0.075 units	Range: 0.075 units	Range: 0.075 units
Variant iii	Range: 0.1 units	Range: 0.1 units	Range: 0.1 units

Fig. 3. Configuration matrix of the nine groups of robots. Each group contains ten agents to make nine swarms of ten robots each. Columns determine number of sensory channels (IR - infrared) for teams A, B, and C respectively, and rows show sensory ranges of variants i, ii, iii for each group.

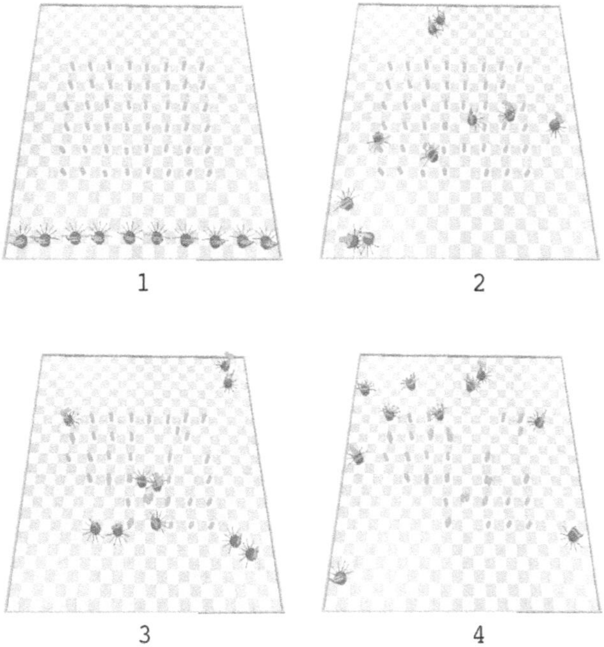

Fig. 4. An example time sequence showing a group of Kheperas at work clustering red blocks, at iteration time index 1: time $t = 0$, 2: $t = 5000$, 3: $t = 10000$, 4: $t = 15000$.

Briefly stated, the insect-like search-grab-move-drop cycle described by Bonabeau et al. (1999) breaks down into several simple behaviours. The agents must roam around, avoiding obstacles, they must be able to identify an object (in this case, the red blocks from Figure 4), pick it up, find another block, and finally put their block down next to the new one before moving away and starting the cycle all over again. Using Arkin's incremental development technique (described in (Arkin, 1998)), each of these behaviours was modelled as a simple computational routine, tested, refined, and put together to give the agents the behaviours they would need to carry out their task without being explicitly designed or programmed to do so. Figure 5 shows a simplified example of this set of behaviours in action. This code, when written in C (using the standard GNU GCC compiler) is then assigned as a controller to each agent. A cross compiler could also be used to bestow similar behaviours in physical hardware Kheperas. Each Khepera now has all the behavioural abilities it needs in order to carry out its task - it simply moves around using a random-walk[7], avoiding obstacles until it finds a block directly in front of it. The gripper arm grabs this block and the Khepera moves on until it finds another block. The block being carried is then placed adjacent to the new block, the Khepera moves away randomly to avoid picking up the same blocks, and the routine is repeated. Each successful loop of this cycle counts as one 'point' (arbitrarily chosen) in a *performance metric*, and each Khepera keeps count of its score during an experimental run of 100,000 Webots timesteps. The total from each Khepera is then collected and a swarm total score generated.

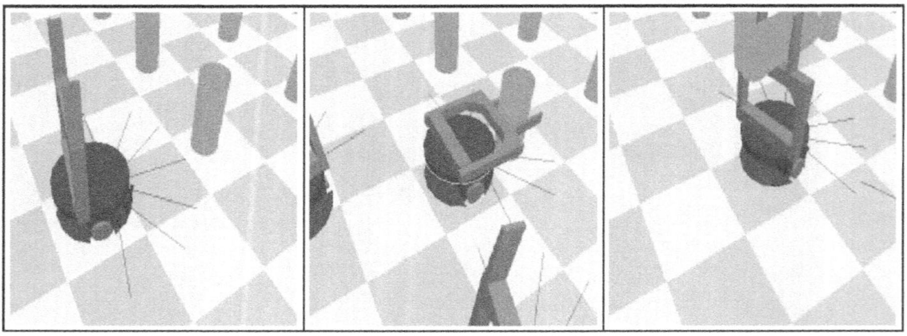

Fig. 5. A close-up on a Khepera carrying out the simple move-grab-move routine.

This process has been largely automated - each of the nine swarms (the three groups with differing number of sensors, each with their three sensor-range variants) carry out the same, standardised collective clustering task in

[7] A random-walk is an insect-like locomotion routine defined in this controller by: rotate-left(random amount), rotate-right(random amount), walk forward. This leads to a very simple 'wiggly', non-deterministic movement routine.

repeated experimental trials; simple Linux-side scripting is used to standardise the start state of the arena and the robots (in terms of their initial locations). Webots, the host software, then takes control and runs the simulation for a set number of controller iterations (common across all experiments). Further Linux-side scripting is used to gather and collate the data recorded by each agent's controller, regarding the number of successful block manipulations carried out within the time limit. These standardised results are then collated and studied.

3 Results

Each swarm repeated the simple collective clustering task fifteen times. Linux-side scripting collected the individual agent scores at the end of each run, and combined them to create a swarm performance total for that run, and an average across all fifteen runs. This data is summarised in Figure 6 showing average performance over fifteen runs for each of the nine embodiment configurations. These results show that no linear or monotonic relationship exists between sensory complexity and performance.

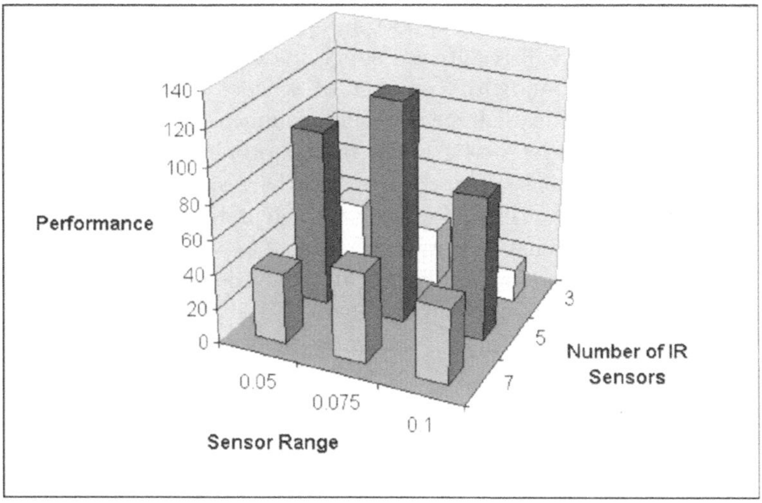

Fig. 6. Comparative Results. Average performance of the nine swarms with differing embodiments (15 runs each). The B team variants, i.e., those with an intermediate number of sensory channels, appear to be noticeably outperforming the other swarms.

Further statistical analysis was used to compare performance across sensor number and sensor range variants (as illustrated in the configuration matrix, Figure 3). Both factors, *sensor range* and *number of sensors*, showed a highly significant impact on performance with $p < .01$ for each of the two factors, although their interaction was not statistically significant (Table 1). Furthermore,

Table 1. Analysis of Variance (ANOVA) in performance for 15 experimental runs for each of the nine embodiment configurations.

Source of Variation	*SS*	*df*	*MS*	*F*	*P-value*	*F crit*
Number of Sensors	134115.6593	2	67057.8296	70.7480	2.53E-021	3.0681
Sensor Range	11114.4148	2	5557.2074	5.8630	3.68E-003	3.0681
Interaction	7718.9185	4	1929.7296	2.0359	9.33E-002	2.4436
Within	119428.0000	126	947.8413			
Total	272376.9926	134				

it is evident from Table 1 that varying the *number of sensory channels* accounted for about 16 times as much of the variance in performance in comparison to varying the sensor range.

4 Discussion

Interestingly, the results do not support an initial hypothesis that an increase in degree of embodiment will result in a superior task performance. Nevertheless, varying perturbatory bandwith (number of sensors in the experiments here) did show a highly significant impact on performance, as did varying the sensor ranges. There was no systematic linear or monotonic increase in performance with degree of embodiment (Figure 6). Rather, there appears to be an 'ideal' level or degree of embodiment that allows a given agent to most effectively traverse the terrain, avoid the arena walls and other agents, whilst successfully repeating the find-grab-drop behaviour needed to carry out the cluster task. It is possible that the variants with the high sensor ranges and larger numbers of sensors are 'too complex' for the task scenario - their movement routines are working against the functionality of the sensors. For example, the random element included in the impulse to each wheel motor at each controller iteration (each timestep) creates a random-walk, a non-deterministic (though overall forward-moving) locomotion behaviour. When sensory bandwidth is as high as found in, for example, group A_{iii}, the sensory input will become highly active as objects wiggle in and out of sensor contact as the Khepera moves. This creates a very good obstacle avoidance behaviour, but in this case means that nearly all sensor input is interpreted as indicative of an obstacle, and the blocks are avoided rather than recognised and gathered. Conversely, the variants with low sensor acuity and range will often mistake the wall or other agents for a block and waste valuable simulation cycles trying to pick it up (an agent cannot pick up a wall or another agent, so this behaviour is futile in terms of achieving the cluster task and simply wastes valuable time).

5 Conclusions

The research question, as stated previously, asked how the nature of embodiment of the agents in a swarm changes the behaviour and performance of that swarm. The experimental data drawn from this research shows that the relationship is definitely not a simple monotonic one, as the variation between scores from team to team was statistically significantly higher at an intermediate level of embodiment, rather than at high or low levels. Team B, with 5 IR sensors per Khepera, was able to out-perform both other teams, and the B_{ii} variant, with extended sensor range over both A_i, B_i, and C_i, was able to produce the overall highest average. This indicates that there is no definite correlation between either degree of embodiment, but that a certain type of embodiment (namely that found in team B_{ii}) happens to yield superior behaviour on average in each agent, and so better swarm performance overall. This is a significant result as it implies that a particular type of agent embodiment allows a swarm to operate most efficiently for a given task. The fact that the more sophisticated, more embodied agents failed to outperform the others suggests that there may be more to the nature of embodiment in terms of performance than sheer complexity (bandwidth and sensitivity) of configuration, and that further refinement of the experimental model described in this paper might shed further important light on the nature of embodiment itself.

So, in closing, the explorations that make up this paper have created questions that need to be investigated further; for example, is swarm B_{ii} the best at other tasks, too? Is the nature of embodiment (i.e., the manner in which sensors are used) more important than overall sensorimotor complexity? Further research, based on a framework similar to that described here, might well illustrate a more concrete connection between the nature of embodiment and the resultant behavioural differences expressed by an agent, and by the swarm of which that agent is a part.

Acknowledgements. The authors would like to thank the following people for their invaluable assistance with the creation and preparation of this document.

Thanks must go to Dr. René te Boekhorst (Adaptive Systems Research Group, University of Hertfordshire) for his assistance with statistical analysis of experimental data. Lastly, Adam Newton would like to thank his classmate and colleague Rehan Mushtaq for his tireless support and schooling in the finer points of C programming.

References

R. C. Arkin, *Behavior-Based Robotics.* MIT Press, Cambridge, Massachusetts, USA, 1998.

E. Bonabeau, M. Dorigo, G. Théraulaz, *Swarm intelligence: From natural to artificial systems*, Oxford University Press, 1999.

K. Dautenhahn, B. Ogden, T. Quick: "From Embodied to Socially Embedded Agents – Implications for Interaction-Aware Robots", *Cognitive Systems Research* 3(3): 397–428, 2002.

P. Grassé, "La Reconstruction du nid et les coordinations interindividuelles. La théorie de la stigmergie", Insectes Sociaux 6: 41–84, 1959.

C. L. Nehaniv, "Evolvability in Biologically Inspired-Robotics: Solutions for achieving open-ended evolution". In: G. T. Mckee and P. S. Schenker (Eds.), *Sensor Fusion and Decentralized Control in Robotic Systems III, Proc. of SPIE*, vol. 4196, pp. 13–26, 2000.

T. Quick, K. Dautenhahn, C. L. Nehaniv and G. Roberts, "On bots and bacteria: Ontology-independent embodiment", *Proc. 5th European Conference on Artificial Life ECAL'99*, Lecture Notes in Artificial Intelligence, Vol. 1674, Springer Verlag, pp. 339–343, 1999.

Quick, T., Dautenhahn, K. "Making Embodiment Measurable". *Proceedings of Fachtagung der Gesellschaft für Kognitionswissenschaft*, Bielefeld, Germany, 1999.

Building a Hybrid Society of Mind Using Components from Ten Different Authors

Ciarán O'Leary[1] and Mark Humphrys[2]

[1] Dublin Institute of Technology, School of Computing, Kevin St., Dublin 8, Ireland
Ciaran.OLeary@comp.dit.ie, www.comp.dit.ie/coleary
[2] Dublin City University, School of Computing, Glasnevin, Dublin 9, Ireland
humphrys@computing.dcu.ie, computing.dcu.ie/~humphrys

Abstract. Building large complex minds is difficult because we do not understand what the necessary components are or how they should interact. Even if the components were known it is difficult to see a situation where a single lab would have all the necessary expertise and manpower to be able to build the mind. One possible answer is to distribute the various components of the mind throughout the Internet, by letting different groups or individuals build parts of the mind. What would be needed then is a protocol that can be used to allow the components to interact, and a mechanism for managing the independent failure of components, and the switching between the required parts of the mind. This paper describes the first attempt to build such a mind. Using components that were developed independently by ten different authors, we have built a mind to solve a specific problem. Although the problem is a simple blocks world implementation, the issues dealt with, as well as the technology used, are relevant for larger and more complex minds. Robust minds require duplication and distribution. The Internet provides the ideal way to build in just such features into artificial minds.

1 Introduction

As writers such as Brooks [2] have argued, building large complex minds is difficult because we do not understand what the necessary components of the mind are or how they should interact. Even if the components (e.g. vision, navigation, planning, learning, etc.) and their interactions were known, it is difficult to see a situation where a single lab would have all the necessary expertise and manpower to be able to build the mind. A solution is to distribute the problem between all interested parties, each of whom can solve part of the problem, and then provide a way for the independently developed components to interact with each-other. The World-Wide-Mind project [8] suggests such a scheme and this paper describes the first implementation where a *Society of Mind* was built using the work of multiple (in this case ten) authors.

In other words, what we are building is a truly *hybrid* system. Many authors have argued that the mind is composed of many different types of algorithms [11], or that hybrid systems produce better solutions to problems [4], [9], [13]. The future of AI then, it seems involves hybrid symbolic/sub-symbolic systems built from quite different components.

W. Banzhaf et al. (Eds.): ECAL 2003, LNAI 2801, pp. 839–846, 2003.

1.1 Society of Mind

A Society of Mind, as described by Minsky [11] is a mind built from a multitude of agencies. Each agency is comprised of a set of agents and a master agent, which compete hierarchically for control of the body. Once a particular agency wins at one level, its sub-agents compete and so on. Duplication of functionality in multiple agents, as well as distribution of agents over the available hardware means that Societies of Mind are robust, and able to survive localised failure.

There is much agreement with Minsky that minds probably look like this, but such complex minds rarely get built. Modular, distributed minds have been built [1], [3], [6], [10], but in these cases the minds were all built by a single author/group, and all ran on localised hardware. Consequently, these minds could not incorporate the type of diversity that is necessary in a truly hybrid Society of Mind.

There are two requirements for a Society of Mind:

1. Agents: In order to build a Society of Mind, we require a diverse set of agents. To provide the necessary diversity an approach would be to allow different authors develop the components independently.
2. Arbitration: An algorithm is required for managing the competition between agents, as only one agent should have control of the body at any point in time. Several such algorithms already exist for distributed models of mind, some of which rely on the explicit design of the arbitration policy [1], [3], and others which allow this to be learned [6].

1.2 Contributions of This Paper

Requirement 1 above is met by the World-Wide-Mind. This architecture provides a simple way for an author to make an agent, or sub-mind available for others to use remotely, without having to know exactly how the sub-mind is going to be used. This is described further is section 2 below.

Once these sub-minds are online, Societies of Mind can be built by independent authors, not involved in the construction of the individual components. Authors of these sorts of minds are required to satisfy requirement 2. Although the World-Wide-Mind does not provide any instructions on how this problem should be solved, this paper describes two simple arbitration mechanisms for a specific Society of Mind.

Section 3 below introduces the problem domain for which the Society of Mind was built. Section 4 describes a hand designed Society of Mind, where the arbitration policy follows a set of heuristics developed from observing the relative performance of each of the sub-minds. Section 5 shows how such a society could be developed dynamically. Section 6 provides a set of results for the dynamically created Society of Mind, which are then discussed in full in section 7.

The existence of the World-Wide-Mind alone does not solve the problem of building large complex minds. The core part of this problem is the arbitration between the independent modules. However, as demonstrated here, the World-Wide-Mind provides a way to collect the large, diverse set of agents/sub-minds required for complex minds. This was not previously possible.

2 The World-Wide-Mind

Humphrys, [8] introduced the World-Wide-Mind project to the Artificial Life community. This is an open, online scheme where individuals, whether AI researchers, ordinary programmers or even complete novices can build components of minds and make them available for others to use.

The World-Wide-Mind provides two things; an *architecture* and a *protocol*. The architecture defines the types of entities that are required, the protocol defines how these entities interact.

2.1 Architecture

The World-Wide-Mind specification [7] defines three types of entity, a *world service*, a *mind service* and a *client*. A *world service* represents a given problem, such as a micro-world implementation of an artificial creature's habitat (for example Tyrrell's world [14]) or an embodied robot in a real world environment. The function of the world service is to provide the user with a representation of the current state of the world (for example the creature's perceptions in the micro-world, or the robot's perceptions in the real world). The *mind service* represents an implementation of (a part of) a solution to the problem faced by the creature in the world. When provided with the state of the world, the mind service will return a suggested action.

The client software is used to manage the interaction of the mind service and the world service in a *run*. A run consists of the following:

1. Using the client software, the user selects a mind service and a world service.
2. The client software contacts both services to tell them they are involved in a run.
3. The client queries the world service for its state.
4. The client forwards this state to the mind service, requesting it to suggest an action.
5. The client instructs the world service to execute this action.
6. Loop to step 3 until completion of the task at hand.

2.2 Protocol

The protocol of the World-Wide-Mind is named SOML (Society of Mind Markup Language) [12]. This protocol defines how a service is created, and once it is created, how the client (or any other entity) can communicate with it. The services built for the World-Wide-Mind are termed *Lightweight Web Services*, since they are inspired by the emerging web service standards in distributed computing [16], but lack the high level of complexity considered unnecessary for this domain. Since our services are web services, they are interacted with over HTTP, the protocol of the World-Wide-Web. In order to build a service, an author need only have access to some web server software, and have a limited knowledge of web programming using, say, CGI (a simple way of writing online programs that read HTTP messages, instead of console input). The service must support up to five SOML messages, e.g. a mind service must support a `getaction` message.

3 Blocks World

A class of final year undergraduates studying for an honours degree in Computer Science were given the task of building a solution for the well-known *blocks world* problem [15]. The world consists of a set of labelled blocks that are presented in an unordered fashion, and need to be stacked in ascending order.

This problem was delivered to the class using World-Wide-Mind technology. A world service was created that represented a randomly initialised blocks world. This world service presented the state of the world as a text string reflecting the current ordering of the blocks. Each student was then required to write a mind service that would take the state of the blocks world as input and return an action (e.g. move block X to column Y).

Each of the ten minds developed was tested ten times in the following three different worlds:

Test 1: Three blocks, three columns
Test 2: Fifteen blocks, five columns
Test 3: Five blocks, fifteen columns

The performance of the mind is measured as the number of queries (actions) that had to be sent to the world service in order to solve the problem (lower scoring minds are obviously better). A test is passed if the problem is solved.

Table 1. Performance of each mind service in problem world

	Percentage of tests passed				**Average Score**		
Mind	***Test 1***	***Test 2***	***Test 3***	***Total***	***Test 1***	***Test 2***	***Test 3***
I	80%	30%	100%	70%	7	49.33	10.6
II	90%	100%	100%	97%	5.4	57.8	8.6
III	100%	100%	90%	97%	5.8	55.2	8.4
IV	100%	90%	80%	90%	7.75	57.6	10.6
V	90%	100%	100%	97%	5.2	48.6	8.4
VI	70%	100%	100%	90%	8.29	67.4	10.8
VII	10%	0%	40%	17%	0		10
VIII	10%	0%	0%	3%	0		
IX	0%	0%	80%	27%			20.5
X	20%	0%	80%	33%	0		13
Overall	57%	52%	77%	62%	6.05	56.85	11.07

As can be seen, minds performed better when free columns were available (Test 3). Failures to pass tests were due to both coding errors by the students, and communication errors on the network.

4 Blocks World Society of Mind – Hand Designed

A Society of Mind, as discussed, is a mind constructed by arbitrating between a number of sub-minds, among which functionality is duplicated and distributed. The developer of the Society of Mind does not need to know anything about the internal operation of the sub-mind, but simply needs to be aware of its interface, in order to use it remotely.

The first Society of Mind developed for the blocks world problem was hand designed to use the best sub-mind. When this mind failed, the second best sub-mind (according to the results in table 1) was chosen and so on. The pseudocode for this simple Society of Mind is shown below:

```
order := { V, III, II, IV, VI, I, X, IX, VII, VIII }
mindIndex = 0
currentMind := order[mindIndex]
state := get state of world
while (problem not solved)
      select action from currentMind
      if (currentMind failed)
            mindIndex = mindIndex + 1 % 10;
            currentMind = order[mindIndex]
      else
            execute action in world and return state
end while
```

This Society of Mind achieved the results shown in Table 2 below.

Table 2. Results for Society of Mind – Hand designed

	Percentage of tests passed				Average Score		
Mind	***Test 1***	***Test 2***	***Test 3***	***Total***	***Test 1***	***Test 2***	***Test 3***
SOM 1	100%	100%	100%	100%	5.68	51.2	8.92

97% of the time, the best mind was used. Whenever this failed the Society of Mind was able to use an alternative mind, chosen from the ordered set of minds.

5 Blocks World Society of Mind – Dynamically Created

The mind described in section 4 above proved to be robust since it was always able to solve the problem regardless of the failure of one of the sub-minds. However, it would be preferable if we could create a Society of Mind without having to explicitly hand code which minds to call.

In order to do so, a new mind service was created, which would take as parameters the list of sub-minds that should be integrated into the Society on Mind. Once these parameters were received, each one was *automatically* subjected to the tests outlined in section 3 above. However, this time, for each test that was passed, the following information was saved:

1. Whether there was any free space at the beginning.
2. Whether there were more than three free columns at the beginning.
3. Whether all the blocks were clear at the beginning.

These were identified as three reasons why sub-minds failed to solve the problem. The information above was saved into an `Entry` object, along with the URL of the corresponding mind. Once all the tests were completed, we were left with a collection of `Entry` objects, one for each test that was passed. Obviously, a better sub-mind would have more entries than a weaker one. The mind service saved this information as a *mind configuration* and returned the ID back to the user.

In order to use the mind configuration, the user could select the mind service using the client and pass the ID of the mind configuration as a start-up parameter. Sub-minds were then given control according to the following algorithm:

1. Create a template `Entry` object, to match the current state.
2. Randomly select an `Entry` object that matches the template. The more `Entry` objects associated with a sub-mind, the more likely it is to be selected. (If one cannot be found after 20 attempts, a random `Entry` is picked.)
3. Find the sub-mind that corresponds to that particular `Entry`, and give it control.
4. Return to step 1 whenever one of the following conditions is satisfied:
   - The current sub-mind failed
   - The template for the current state changed, meaning that more/less sub-minds are now eligible for selection e.g. a sub-mind that only passed tests where all blocks were clear is now eligible.
5. Continue until the problem is solved.

In this type of Society of Mind:

1. Better minds will get picked more frequently because they will have more entries.
2. A mind will not be selected when it is unsuited to the current state.
3. A number of different sub-minds are used in each run.
4. New sub-minds can be added dynamically, since tests are performed and saved automatically.

6 Results

The dynamically created Society of Mind achieved the results shown in Table 3 below. As can clearly be seen, this Society of Mind was unable to outperform any of the individual minds, or indeed the hand designed mind. However, once again, it never failed to produce an action, and did not need to be hand designed. The better performing minds had more `Entry` objects (table 4), and consequently were chosen more frequently (table 5).

Table 3. Results for Society of Mind – dynamically created

Mind	Percentage of tests passed				Average Score		
	Test 1	*Test 2*	*Test 3*	*Total*	*Test 1*	*Test 2*	*Test 3*
SOM 2	100%	100%	100%	100%	9.84	200.3	13.23

Table 4. Percentage of entries for each sub-mind

Mind	I	II	III	IV	V	VI	VII	VIII	IX	X
Percent-age	11.3	15.6	15.6	14.5	15.6	14.5	2.7	0.5	4.3	5.4

Table 5. Percentage of times each sub-mind was called on for Society of Mind

Mind	I	II	III	IV	V	VI	VII	VIII	IX	X
Percent-age	16.5	14.0	20.1	16.8	16.8	6.7	1.4	1.4	3.3	3.0

A Society of Mind will only provide a better score to an individual mind if the individual mind is non-optimal, and its deficiencies are catered for in other minds. Clearly, this was not the case here. Several of the minds provided results that could not be improved upon by their combination with other minds. The best option for our Society of Mind would have been to pick one of the better minds e.g Mind V, and stick with it, as we did in the hand designed mind.

What has been demonstrated however, is a mechanism for handling a distributed Society of Mind, composed of independently authored components, which is robust when faced with failure of its components, and has a facility to grow with the addition of new components.

7 Discussion

This paper discusses a simple Society of Mind, developed to solve a simple blocks world problem. This particular problem has faced some criticism [5], but serves our purposes here, as our goal was to show that using independently developed components, it is possible to build robust Societies of Mind. Both of the Societies of Mind built for the purposes of this paper can be considered robust since they will always be able to produce a result, regardless of the strengths or weaknesses of the sub-minds. Each of the sub-minds failed at least one of the tests put to it, either because of coding errors, or communication errors over the Internet.

While accepting that the problem and solution were relatively simple, we expect to be able to scale up. We deliberately kept the specification for the protocol and architecture as simple as possible, and not tied to any specific platform (apart from the web), to ensure that willing authors can easily construct services. We hope to have results for more complex, multi-goal problems, such as Tyrrell's Simulated Environment [14] in the near future. We also intend to show Societies of Mind developed using authors from different institutions.

Building hybrid systems has always been difficult because of the requirement, in practice, that a single research team or single author have the necessary skills to build each of the different sub-systems. This work is a step towards a future where AI/ALife researchers will build hybrid systems out of large numbers of components they did not write.

Acknowledgements. The authors would like to thank two anonymous reviewers for their helpful comments.

References

1. Brooks, R. A.: A robust layered control system for a mobile robot. IEEE Journal of Robotics and Automation, RA-2:14–23, April (1986)
2. Brooks, R.A.: Intelligence without Reason, Proceedings of the 12th International Joint Conference on Artificial Intelligence (IJCAI-91) (1991)
3. Bryson, J.: Intelligence by Design: Principles of Modularity and Coordination for Engineering Complex Adaptive Agents, PhD Thesis, Massachusetts Institute of Technology (2001)
4. Connell, J.: SSS: A Hybrid Architecture Applied to Robot Navigation, Proceedings of the IEEE International Conference on Robotics and Automation, Nice, France, pp. 2719–24. (1992)
5. Dreyfus, H.: What Computers Still Can't Do: A Critique of Artificial Reason . Cambridge: MIT Press. (1993)
6. Humphrys, M.: Action Selection methods using Reinforcement Learning, PhD Thesis, University of Cambridge (1997)
7. Humphrys, M.: The World-Wide-Mind: Draft Proposal, Dublin City University, School of Computer Applications, Tech Report CA-0301 (2001) computing.dcu.ie/~humphrys/WWM/
8. Humphrys, M.: Distributing a Mind on the Internet: The World-Wide-Mind, Proc. 6th European Conf. on Artificial Life (ECAL-01), (2001).
9. Isla, D., Burke, R., Downie, M., and Blumberg, B.: A Layered Brain Architecture for Synthetic Creatures. Presented at The International Joint Conference on Artificial Intelligence, Seattle, USA, (2001).
10. Maes. P.: How to do the right thing. A.I. Memo 1180, MIT, Cambridge, MA, December, (1989)
11. Minsky, M.: The Society of Mind, Simon and Schuster (1985)
12. O'Leary, C., SOML (0.9) Society of Mind Markup Language 0.9, Available online at w2mind.comp.dit.ie
13. Skinner, J. and Luger, G.: Contributions of CBR to an Integrated Reasoning System, Journal of Intelligent Systems, 5:1, 19–47 (1995)
14. Tyrrell, T., Computational Mechanisms for Action Selection, PhD Thesis, University of Edinburgh, Centre for Cognitive Science (1993)
15. Winograd, T.: Understanding Natural Language. Academic Press. New York. (1972)
16. World-Wide-Web Consortium: Web Services Activity, online at w3.org/2002/ws/

Requirements for Getting a Robot to Grow up

Peter Ross[1], Emma Hart[1], Alistair Lawson[1], Andrew Webb[1], Erich Prem[2], Patrick Poelz[2], and Giovanna Morgavi[3]

[1] School of Computing, Napier University, Edinburgh EH10 5DT, UK
[2] Austrian Research Institute for AI, Schöttengasse 3. A-101 Vienna, Austria
[3] IEIIT, Via De Marini 6, 16149 Genoa, Italy

Abstract. Much of current robot research is about learning tasks in which the task to be achieved is pre-specified, a suitable technology for the task is chosen and the learning process is then experimentally investigated. In this paper we discuss a different kind of problem: how to get a robot to 'grow up' through experience of its world. We discuss what this means and what it seems to require in a robot architecture. The main contribution of this paper is to suggest a particular research agenda for the community.

1 What Does It Mean for a Robot to 'Grow up'?

A great deal of current research work in mobile robotics and autonomous systems is still focused on getting a robot to learn to do some task such as pushing an object to a known location or running as fast as possible over rough ground. The learning process may be supervised, unsupervised or a process of occasional reinforcement, but the whole aim in such work is to get the robot to achieve the task that was pre-defined by the researcher.

The next logical step along the road towards truly autonomous robots that can function productively for long periods in unpredictable environments is to investigate how one might design robots that are capable of 'growing up' through experience. By this, we mean that the robot starts with only some basic skills such as an ability to move about and an ability to sense and react to the world (such as trying to avoid obstacles), but in the course of time it develops genuinely new skills that were not entirely engineered into it at the start. In particular it should be capable of building some kind of hierarchy of skills, such that for each new skill s_{new} there is one or more sets of skills $S_1, S_2, \cdots S_n$ such that s_{new} is significantly more easily acquired if the robot has acquired all the members of some S_i than if it lacks at least one member of each of those sets. However, this is an intuitive but ill-defined notion, and not just because of the vagueness of a phrase such as "significantly more easily acquired". In human terms, skills have convenient short names such as 'wall following' or 'the ability to play the piano' but are only formally defined by performance tests such as being able to move along a wall for a given distance without straying more than another given distance away from the wall, or being able to play "Für Elise" without wrong notes. In much of robotics research, a robot trajectory of the kind shown

W. Banzhaf et al. (Eds.): ECAL 2003, LNAI 2801, pp. 847–856, 2003.

in Figure 1 (a) is readily called wall following behaviour even though, as Figure 1 (b) shows, there might not be an actual wall involved. True wall-following ought

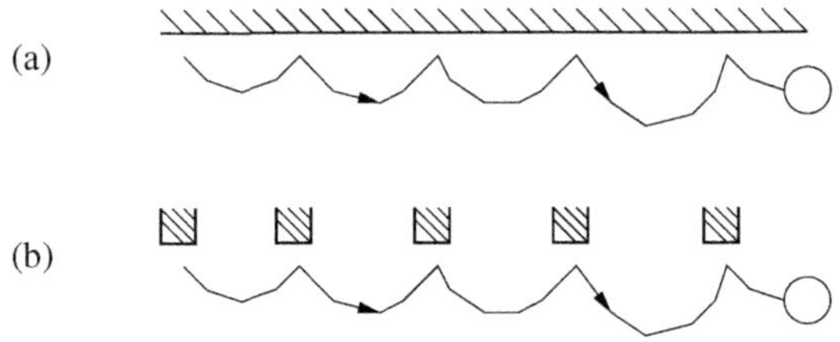

Fig. 1. What is wall-following?

to involve moving along a wall with progressively fewer 'bounces' until the robot is moving in a straight line virtually parallel to the wall so that, arguably, the robot has demonstrated at least implict knowledge of a defining property of the wall: its direction. This ability might be realised in different ways; for example, the robot might adjust some internal neural control feedback to the point where it can keep a IR sensor ray just brushing the wall, or might explicitly tune its heading so as to make the internally-recorded moving average of times between bounces tend towards zero. The point is that the skill is testable and involves the acquisition of a category of information.

There are many skills that a robot might acquire that are uninteresting or unproductive for observers or even unrecognizable, such as the ability to always take a prime number of movement steps between heading changes. With current technology, robots and humans dwell in enormously different sensory universes and although it is tempting to ask that growing up should involve an increasing competence in skills that are concisely labellable and humanly meaningful, this is perhaps asking too much. We feel that at least the levels in a hierarchy of a robot's acquired skills should be characterised by increasing temporal scales. That is, higher-level skills should be skills that are manifested on appreciably longer time-scales than lower-level ones. The ability to return to a known location, for example, involves internal and external activity over a longer time scale than the ability to avoid collisions.

Growing up should be more than a matter of prolonged supervised learning or reinforcement learning. For example, Chellapilla and Fogel [1,2] describe the evolution of a neural-net-based system that becomes very good at playing checkers. The system is only provided with the positions of pieces on the board and the piece differential; it learns simply by playing the game repeatedly, initially against itself and then against humans via the anonymity of a games website. Would this count as growing up, at least within the abstract universe of checkers-playing? We would argue that it does not, for three reasons. First, there is as yet no analysis to suggest that the route followed by this program towards expert levels of play proceeds through a hierarchy of distinct skills (such as the familiar ones of being proficient at openings or in the end-game, or being competent with kings). It is of course conceivable that an analysis would reveal some kind of developmental hierarchy. But second, and more significantly, this system does not count as an example of growing up because the sole aim throughout is to increase a numeric measure of playing success; this system is an example of externally-guided learning rather than growing up. That is, the system's *value system* does not change with experience. Third, there is no individual player in this system that develops through experience; instead, evolution transmits desirable features down through generations of individuals via mutation and recombination.

In general, the following features seem to us to be important aspects of growing up:

1. it is an individual that develops, rather than a system of individuals. And growing up is a property of the whole but not necessarily of any of the parts.
2. development involves acquiring a hierarchy of skills, in which the acquisition of new skills is facilitated by already-acquired ones;
3. although learning happens during growing up, it is not necessary to learn just one thing at a time, and learning need not happen in distinctive phases. Learning may be continual.
4. the value system – the reasons why one action is ever preferred to another – may change with experience. For example, a robot may learn the desirability of communicative dialogue as opposed to never speaking, even though dialogue has no direct relation to primitive drives such as avoiding pain or satisfying a physical need;
5. in the process of development, the individual becomes capable of purposeful action over longer timescales.

Thus, if these ideas are applied to the concept of robot growing-up, we can envisage several scenarios that could potentially be realised in practice that would illustrate that growing-up has occurred. For example, take the scenario in which a robot is told (perhaps in some stylised language) to 'find the door'. Clearly this can be realised by a robot performing random and blind search until its sensory input stream matches to something it has previously categorised as a door. This is not growing-up — however, if the same robot initially started 'life' with only a set of basic drives such as {explore, boredom, obstacle-avoidance}, and then was subject to a developmental schedule in which through experience and exposure it learned to categorise objects and sequences of actions, form internal maps, and then eventually exhibit a purposeful 'find-things' behaviour, then it can be considered to have been growing-up. Note that a developmental schedule is needed to impose structure on the whole progression; one does not learn to fly a plane by unstructured experiment with the controls (although this has been demonstrated in simulation [3], the simulator imposed some structure by interpreting commands according to context). A different scenario can be envisaged in which a robot is exposed to human speech (or structured linguistic input of any kind) but has no pre-defined competence in language or dialogue (c.f. the 'Talking Heads' project [4]). A plausible developmental route is that initially, it learns to copy patterned sounds that its hears and much later after further structured experiences, some form of dialogue emerges in which the robot listens to the human speech, pauses, then responds. The emergence of this kind of 'turn-taking' behaviour is commonly cited in the developmental psycholology literature as an example of the developing infant building on the innate babbling skills it is born with to achieve a higher cognitive function. Chipman [5] has shown that in children, good linguistic ability may appear long before any understanding of grammatical issues and there is good evidence [6] that the cognitive processes involved are not specific to linguistic mechanisms. We mention the topic of language development here because it is qualitatively different from the kind of developmental processes normally discussed in connection with robots.

Before describing the requirements of an architecture that would allow a robot to grow-up in the sense we have just described, we briefly review some of the literature currently available, which typically addresses some but not all of features listed above.

2 A Brief Survey of Work on Developmental Architectures

Weng et al [7,8] have developed both a theory and implementations of a robot developmental architecture, in particular instantiated by the SAIL-2 and Dav robots. In their approach the robot is expicitly taught by a human who issues 'good' and 'bad' reward signals. There is an 'innate behaviour' component that is responsible for generating behaviour whenever there is no clear decision, but in general action decisions are made by using the current situation (as described by a vector of both external and internal sensors that spans several consecutive timesteps) to traverse a regression tree built up through experience. The regression tree represents a form of long-term memory, and the feature combinations that are significant at any specific level of the tree represent forms of learned concept. The amount of sensory information involved is large; a specific algorithm, IHDR, has been developed to make the approach computationally tractable.

Kulakov and Stojanov [9] describe two implemented architectures for a simplistic robot capable of forward (F), left (L) and right (R) actions. In the simpler *Petitagé* model, an internal map is built of sequences of actions and perceptions. The robot may try an action sequence such as 'FFFLFFFR' suggested by a certain schema but if this is not successful enough, for example because the robot is in a corridor and so the actual percepts do not fit well with those recorded in the schema, the system will generate a modified schema and include that in the map. Links in the map record schema sequencing, that is, which schema(s) may follow the current one. The system also looks for cycles in the map, these represent a form of higher-order schema. In the more sophisticated *Vygovorotsky* model, nodes in the map represent percepts, and links between nodes are labelled with actions. Schemas represent specific sequences of actions, and each schemas has an associated reliability measure recording how well the stored percepts matched actual experience as the actions were carried out. The system has an analogy-making mechanism so that, when faced with a problem, it can retrieve a successful solution instance from its map and try to modify it to fit the current reality. There are also two schema abstraction mechanisms. One tries to generalise over similar schemas, the other tries to conjoin schemas that are consecutive and sufficiently reliable. The system also has three basic drives – hunger, seek pleasure/avoid pain, and curiosity – to fall back on for action selection.

Several other related models have been described in the literature [10,11,12, 13,14], some using an explicit dynamical systems approach and others using an explicit symbolic approach. In general all the models aim to record situations, actions and the effects of actions and then to capture sequences in some way. Conceptually they owe much to Mataric's early work on TOTO [15] and to Stein's work on MetaTOTO [16]. In particular, the latter work introduced the idea and a means whereby a robot might internally visualize the execution of an action sequence to assess it before actually doing it.

3 Growing up: Some Requirements

The previous sections have outlined what is means for a robot to grow-up. We now discuss some of the essential elements that must be incorporated in any architecture that can potentially realise 'growing-up robots'.

3.1 Sensors

An essential requirement for a robot to grow-up seems to be that it has a diverse and rich sensory world, not only because elaborate robot-human interaction seems to demand a shared universe of sensory experience but also because correlations between sensory modalities make tasks like category learning very much easier (eg two corners are more easily distinguished if they are also coloured differently). We categorise sensor information into three types. First, there are those sensors in which the stream of information gained from them is highly correlated with actions (eg in IR terms, if the robot swivels right then the sensed wall slides round to the left). Much robot research deals only with this kind of sensory information. The second type of sensor information is typified by linguistic input such as speech. This type of input occurs on an infrequent and isolated basis, need not correlate well with actions and brings severe problems of referential ambiguity, eg the robot hears *'coin'*, the French word for corner. But how does the robot know what this refers to, since there is no set of sensory snapshots temporally adjacent to the utterance that can be easily abstracted to form the concept of corner? Is such an utterance essential in the task of forming the concept?. Finally, the robot can also perceive internal sensory-information — for example an internal map of the world built up by the robot can be considered as additional sensory input. This type of sensory information is almost certainly noisy, imperfect, and unreliable but can contribute to a developing picture of the robots world at any moment in time. We also claim that it is important to mimic, somehow, the fact that a brain develops neural machinery that responds to highly specific features of the sensory stream. For example, in monkeys there are specific internal sensors for 'hand-in-view' and other such features [17]. Therefore, any architecture for a growing-up robot must be able to deal with these three types of sensory informations, occuring across differing time-frames and with differing stuctures. As far as we are aware, no work has been published concerning such a system in its full detail.

3.2 Memory

Expecting a robot to grow-up has strong implications on the kind of memory that the robot must maintain. It appears that at least three types will be required: working memory, long-term memory and episodic memory. Working memory is directly related to the incoming sensory data (from all sources) and can be thought of as a short-term queue containing recent sensory perceptions and is likely to consist of raw sensory-data, at least initially. This distantly resembles human short-term memory which experimental studies have suggested is of bounded capacity (although expandable through lengthy training) and contains things that are fairly closely linked to the sensory input – for example, when reading aloud lists of words it would appear that short-term memory encodes shape and sound information rather than more abstract semantic information. Long-term memory on the other hand must maintain some record of relationships between sensory situations, actions performed, and the effects of those actions (it becomes important in later planning stages that the effects of a given action are remembered so that the robot has some *expectation* of the consequence of performing some action). Episodic memory will be required to maintain records of sequences of events, but for a robot to perform a task such as 'find-things' in practice, it will be crucial that the episodic memory does not impose some pre-determined limit on the length of the sequences that can be remembered but can record variable-length episodes in a manner determined by the system itself. Other types of memory could also be envisaged, for example procedural memory, which contains 'how-to' knowledge, i.e. stored methods of a achieving particular goals, or internal memory, which records the system being in some particular state (which cannot necessarily be labelled by an outside observer). Preliminary work in this area using a classifier system to evolve rules to control a simulated robot through T-mazes has shown the effectiveness of such memory.

3.3 Data Abstraction

For the robot to learn online and continuously, there must be a method of abstracting and generalising data, as no system could deal with the vast quantities of raw data that will be received from the incoming sensor-streams. At the very start of the robot's 'life' only raw sensor data will be available, but we suppose that this contains far too much information to be useful. We hypothesise that a key aspect of growing up is learning what details to ignore in different situations. So, abstractions of the data must be formed – for example, natural ones to suggest at the start would contain either thresholded or thresholded-moving-average versions of raw sensory information. We envisage that therefore that a growing-up architecture must have some method of implementing *proposer processes* that suggest new data abstractions built out of all existing data items (whether raw or already abstracted) and that proposed new data items survive by being found to be useful and continue to survive only by continuing to be useful. This naturally suggests that there needs to be *critic processes* too, looking to scavenge either data items that have become useless or replace them if they have effectively been subsumed by others. However this is implemented, it is essentially that the abstraction process is continuous – over the course of time, very high level abstractions will be built, therefore we can perhaps envisage that instead of short-term memory containing raw sensory data signifying that the robot is facing a right-angled corner, the queue might contain a data-item indicating 'corner'.

In our experimental work so far, we have explored methods that do not have such processes, in particular using either multivariate signatures or multivariate scaling for categorisation and dimensionality reduction. In particular, the Isomap technique [18] is very promising. But it is not clear that sensory clustering and dimensionality reduction will meet all needs, because such techniques do not yet easily cater for radical shifts in categorisation that can arise through experience while growing-up.

3.4 Associations, Expectations, and Planning

An architecture must provide a framework for linking sensory perceptions to motor actions. However, it also seems crucial that such sensor-motor descriptions be associated with an expectation of the change in sensor values (whether raw or abstracted, internal or external) that would be caused by executing those motor actions. Without information about some (not necessarily all) expected consequences of an action, the robot would have no way other than external reinforcement to judge whether the action was appropriate. A convenient way to achieve this is to construct rule-like associations (RLAs) between the sensory context, the action taken and the perceived consequences, all of which should be available from short-term memory as described above. In fact RLAs can serve several purposes. Action selection can be based on the results of partial matching of current sensory context to stored RLAs. RLAs can be stored as nodes in a spreading-activation network of weighted links. One RLA might stimulate another if, for example, they represent temporally consecutive situations; or one might inhibit another if they represent competing accounts of very similar situations. Chains of RLAS might therefore capture a form of episodic memory. It would be interesting to see if such episodic chains had a 'natural length'; the psychological literature seems to have little to say about such matters.

In order for a robot to grow-up sufficiently to be able to perform a command such as 'find-the-door' described in section 1, the robot must eventually acquire some basic planning capabilities. The current literature reveals two possible approaches to this dilemma. The robot could develop a deliberative planner (eg

STRIPS-like) in which it essentially reasons about the expected consequences of actions, and eventually arrives at a plan by which the result can be achieved (for example, as in MetaTOTO [16]). Or the robot might produce planning-like behaviour without a phase of 'thinking about it', by cascading through a sequence of states that end with achieving the active goal; presumably the existence of the goal and details of the current sensory context trigger such a cascade. However, both styles seem to impose the same kind of memory-type requirements as discussed above. But a key question is why a robot should be motivated to develop planning or planning-like skills in the first place.

3.5 Motivation

A robot that truly grows-up must also develop the specific motivations for performing actions, categorising, learning to plan, learning to take part in dialogue and so on; these motivations would be above and beyond any original innate drives. For example, why should a 'find-things' behaviour emerge in a robot that merely has an instinct to explore at random, avoid obstacles and seek reward? As mentioned in section 1, this requires the robot to have a value system that is modified over the course of time, as the robot gains experience. For example, a learnt action associated with a particular sensory state which results in obstacle avoidance can be later overridden by a different action associated with the same sensory information to allow box-pushing. The kind of competences engineered into familiar behaviour-based systems, such as 'explore' and 'avoid obstacles', fail to address the problem of how a value system might develop and change. We see this as one of the most challenging aspects of research into growing-up.

Even the initial value system poses problems. In a behaviour-based model an 'avoid obstacles' component may be useful, but if it is never overridden then the robot can never learn to push obstacles. As an example of how this might be done we have demonstrated that a MAXSON-type second-order neural network [19] can be used to enable a simulated Khepera robot to learn satisfy the multiple goals of avoiding obstacles and exploring its environment. In such a network there are 'goal nodes' whose level of activation represent the pain of collision and the level of boredom (in the sense of familiarity with the local surroundings). In a second-order network, that contains muliplicative connections, it is possible for an active node to effectively turn off or on whole fragments of network. Crabbe et al [19] demonstrated that such a network could enable a highly-simplified robot to learn to arbitrate successfully between competing drives. We have shown that this can work for a more sophisticated robot, a simulated Khepera; but it still takes a very large number of steps to do so (50 iterations of 10,000 step sequences in which the measure of success is how much territory gets explored).

3.6 A Schedule for Development

A further important requirement for succesful growing-up is that there should be a schedule, or rather two schedules, for the whole process. First, there needs to be an *external* schedule: an ordering and duration of types of experience, representing an educational syllabus for the robot. For example: should one have pushable objects in the world in the earliest stages, before the robot has learned much about wall-following or has managed to structure its long-term memory to the point where it can in some way categorise objects as straight-sided or round or whatever? Encountering a pushable object would seem to make the business of category-learning somewhat harder. And what should be the nature of such an external schedule? Should it, for example, be like a school syllabus with exams, so that the robot does not get to see more elaborate situations until it has passed some test of competence with simpler ones? This is not a simple question; there is good evidence that 'multi-task learning', that is, learning about several related tasks in parallel, can speed learning considerably by imposing a bias towards the disovery of the common ingredients of the related tasks [20,21].

Second, there needs to be an *internal* schedule. This might control when processes first start up, and what priorities they get at different stages. For example: there is not much point awarding high priority to a process that tries to generalise sets of experiences until the robot has a reasonable set of experiences to apply it to. As with the external schedule, there is an open question about whether this schedule should be trigger-driven (by conditions becoming true, or by passing explicit tests) or should be maturational (that is, based on the passage of time or on the volume and nature of experiences).

A natural hypothesis is that these schedules matter. It will be interesting to explore how sensitive the results of growing-up are to significant alterations in either or both of these schedules.

4 Conclusions

Growing-up represents a new and wide-ranging research challenge for the ALife and autonomous systems community. In this paper we have tried to spell out what we see as some of the requirements of a system capable of growing-up. We are actively researching this, and encourage others to do so.

Acknowledgement. The authors are grateful for support from EU project IST-2000-29225 (SIGNAL) and for useful discussions with all the project partners – see http://www.ist-signal.org/

References

1. Chellapilla, K., Fogel, D.: Evolution, neural networks, games and intelligence. Proceedings of the IEEE **87** (1999) 1471–1496
2. Fogel, D.: Beyond samuel: evolving a nearly expert checkers player. In Ghosh, A., Tsutsui, S., eds.: Advances in Evolutionary Computing. Springer-Verlag (2003) 989–1004
3. Smith, R., Dike, B., Ravichandran, B., El-Fallah, A., Mehra, R.: The fighter aircraft lcs: a case of different lcs goals and techniques. In Lanzi, P.L., Stolzmann, W., Wilson, S., eds.: Learning Classifier Systems, From Foundations to Applications. Number 1813 in LNCS. Springer-Verlag (2000) 283–300

4. Labs, S.: The 'talking heads' project. (`http://talking-heads.csl.sony.fr/`)
5. Chipman, H.: Aspects of language acquisition: Developmental strategies. In Shulman, V., Restaino-Bauman, L., Butler, L., eds.: The Future of Piagetian Theory: The Neopiagetians. Plenum (1985)
6. Beilin, H.: Studies in the Cognitive Basis of Language Development. Academic Press (1975)
7. Weng, J., Hwang, W., Zhang, Y., Yang, C., Smit, R.: Developmental humanoids: Humanoids that develop skills automatically. In: Proceedings of the First IEEE-RAS International Conference on Humanoid Robots. (2000)
8. Weng, J.: A theory for mentally developing robots. In: Proceedings of the Second International Conference on Development and Learning, IEEE Computer Society Press (2002)
9. Kulakov, A., Stojanov, G.: Structures, inner values, hierarchies and stages: essentials for developmental robot architectures. In et al, C.P., ed.: Proceedings of the Second International Workshop on Epigenetic Robotics: Modeling Cognitive Development in Robotic Systems. (2002)
10. Ogawa, A., Omori, T.: Model of symbolic looking procedure acquisition process in navigation learning task. In: First International Workshop on Epigenetic Robotics: Modeling Cognitive Development in Robotic Systems. (2001)
11. Ogawa, A., Omori, T.: Looking for a suitable strategy for each problem - multiple tasks approach to navigation learning task. In et al, C.P., ed.: Proceedings of the Second International Workshop on Epigenetic Robotics: Modeling Cognitive Development in Robotic Systems. (2002)
12. Kozma, R.e.a.: Self-organizing ontogenetic development for autonomous adaptive systems (sodas). In: Proceedings of the International Joint Conference on Neural Networks (IJCNN 2001). (2001) 633–637
13. Harter, D.: Ontogenetic development of skills, strategies and goals for autonomously behaving systems. In: Proceedings of the 5th World Multi-Conference on Systemics, Cybernetics and Informatics (SCI 2001). (2001)
14. Buller, A.: Volitron: On a psychodynamic robot and its four realities. In et al, C.P., ed.: Proceedings of the Second International Workshop on Epigenetic Robotics: Modeling Cognitive Development in Robotic Systems. (2002)
15. Mataric, M.: Integration of representation into goal-driven behaviour-based robotics. IEEE Journal of Robotics Automation (1992) 304–312
16. Stein, L.: Imagination and situated cognition. Journal of Experimental and Theoretical AI **6** (1994) 393–407
17. Tanaka, K., Saito, H., Fukada, Y., Moriya, M.: Coding visual images of objects in the inferotemporal cortex of the macaque monkey. Journal of Neurophysiology **66** (1991) 170–189
18. Tenenbaum, J., de Silva, V., Langford, J.: A global geometric framework for dimensionality reduction. Science **90** (2000) 2319–2323
19. Crabbe, F., Dyer, M.: Goal-directed adaptive behaviourr in second-order neural network: learning and evolving in the maxson architecture. In Patel, M., Honavar, V., Balakrishnan, K., eds.: Advances in the Evolutionary Synthesis of Intelligent Agents. MIT Press (2001) 307–336
20. Caruana, R.: Multi-task learning. PhD thesis, School of Computer Science, Carnegie-Mellon University (1997)
21. Ghosn, J., Bengio, Y.: Bias learning, knowledge sharing. IEEE Transactions on Neural Networks (2003 (to appear))

Coevolving Communication and Cooperation for Lattice Formation Tasks

Jekanthan Thangavelautham, Timothy D. Barfoot, and Gabriele M.T. D'Eleuterio

Institute for Aerospace Studies
University of Toronto
4925 Dufferin Street
Toronto, Ontario, Canada
M3H 5T6
{thangav,tim.barfoot,gabriele.deleuterio}@utoronto.ca

Abstract. Reactive multiagent systems are shown to coevolve with explicit communication and cooperative behavior to solve lattice formation tasks. Comparable agents that lack the ability to communicate and cooperate are shown to be unsuccessful in solving the same tasks. The agents without any centralized supervision develop a communication protocol with a mutually agreed upon signaling scheme to share sensor data between a pair of individuals. The control system for these agents consists of identical cellular automata handling communication, cooperation and motion subsystems. Shannon's entropy function was used as a fitness evaluator to evolve the desired cellular automata. The results are derived from computer simulations.

1 Introduction

In nature, social insects such as bees, ants and termites collectively manage to construct hives and mounds, without any centralized supervision [1]. A decentralized approach offers some inherent advantages, including fault tolerance, parallelism, reliability, scalability and simplicity in agent design [2]. All these advantages come at a price, the need for multiagent coordination. Adapting such schemes to engineering would be useful in developing robust systems for use in nanotechnology, mining and space exploration.

In an ant colony, each individual is rarely independently working away without explicitly communicating with other individuals [1]. In fact, it is well known that ants and termites use chemicals to communicate information short distances. Cooperative effort often requires some level of communication between agents to complete a task satisfactorily [5]. The agents in our simulation can take advantage of communication and cooperation strategies to produce a desired 'swarm' behavior.

Our initial effort has been to develop a homogenous multiagent system able to construct simple lattice structures (as shown in fig. 1). The lattice formation task involves redistributing a preset number of randomly scattered objects

W. Banzhaf et al. (Eds.): ECAL 2003, LNAI 2801, pp. 857–864, 2003.

(blocks) in a 2-D grid world into a desired lattice structure. The agents move around the grid world and manipulate blocks using reactive control systems with input from simulated vision sensors, contact sensors and inter-agent communication. Genetic algorithms are used to coevolve the desired control subsystems to achieve a global consensus. A global consensus is achieved when the agents reach a consensus among the collective and arrange the blocks into one observable lattice structure. This is analogous to the heap formation task in which a global consensus is reached when the agents collect numerous piles of objects into one pile [3].

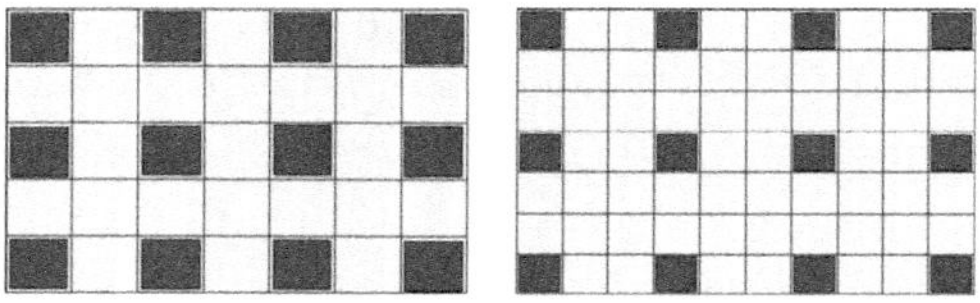

Fig. 1. The lattice structures shown include the 2×2 tiling pattern (left) and the 3×3 tiling pattern (right).

2 Related Work

The object of our study has been to determine if localized communication combined with cooperation would produce useful 'swarm' behavior to complete a predefined task. Often cooperative tasks involving coordination between numerous individuals such as table-carrying, hunting or tribal survival depend on explicit communication [5]. Communication is required for such cooperative tasks when each individual's actions depend on knowledge that is accessible to others. Like the heap forming agents, our agents can detect objects over a limited area [3]. Earlier works into communication and cooperation were based on a fixed communication language, which may be difficult to develop and may not even be an optimal solution [7,8,9].

Adaptive communication protocols have been developed combining a learning strategy such as genetic algorithms to develop a desired set of behaviors [5,10]. Maclennan and Burghardt [10] evolved a communication system in which one agent observed environmental cues and in turn 'informed' other agents. Yanco and Stein [5] used two robots, with the 'leader' robot receiving environmental cues and informing the 'follower' robot. To our advantage, coevolution has been shown in [6] to be a good strategy in incrementally evolving a solution which combines various distinct behaviors. Within a coevolutionary process competing populations (or subsystems) spur an 'arms race' where one population tries to adapt to the 'environment' created by the other and vice-versa until a stable solution is reached. The effect of this parallel evolution is a mutually beneficial end result, which is usually a desired solution [6].

3 Lattice Pattern Formation

The multiagent system discussed in this paper consists of agents on a 2-D grid world, with a discrete control system composed of cellular automata. Cellular automata (CA), it has been shown, provide a simple discrete, deterministic model of many systems including physical, biological and computational systems [13, 14].

Determining a set of local rules by hand that would exhibit a useful emergent behavior is somewhat difficult and a tedious process. By comparison, evolving such characteristics would produce desired results, provided the right fitness function is found. Using Shannon's entropy function, we devised a system able to distribute the objects (blocks) uniformly (a necessary step in forming the 3×3 tiling lattice pattern). The 2-D grid world is divided into M 3×3 cells, A_j, where the fitness value, f_i, for one set of initial condition is given as follows:

$$f_i = s \cdot \frac{\sum_{j=1}^{J} p_j \ln p_j}{\ln J} \tag{1}$$

where, $s = -100$ and is a constant scaling factor, i is an index over many sets of random initial conditions and

$$p_j = \frac{n(A_j)}{\sum_{j=1}^{J} n(A_j)} \tag{2}$$

where $n(A_j)$ is the number of blocks in cell A_j. When the blocks are uniformly distributed over J cells, we have $f_i = 100$. The total fitness, f_{total}, used to compare competing CA lookup tables is computed as follows:

$$f_{total} = \frac{\sum_{i=1}^{I} f_i}{I} \tag{3}$$

where f_i is calculated after T time steps and I is the number of simulations.

4 The Agent

Cooperative behavior between a pair of agents could be visualized as two agents being physically bolted together. To verify whether evolutionary pressure would encourage such a configuration, the agents have the option to stay paired or separate after each time step. Each agent is equipped with 3 bumper sensors, 4 spatially modal vision sensors [1,3] and 2 contact sensors wired to an accompanying trolley. The vision sensors are fitted to allow agents, blocks and empty space to be distinguished. Once the agent has chosen to 'pair up' with a neighboring agent, the physical behavior is looked up based on the input from the vision sensors and the data received from the communication session.

There are four physical behaviors which are defined as follows:

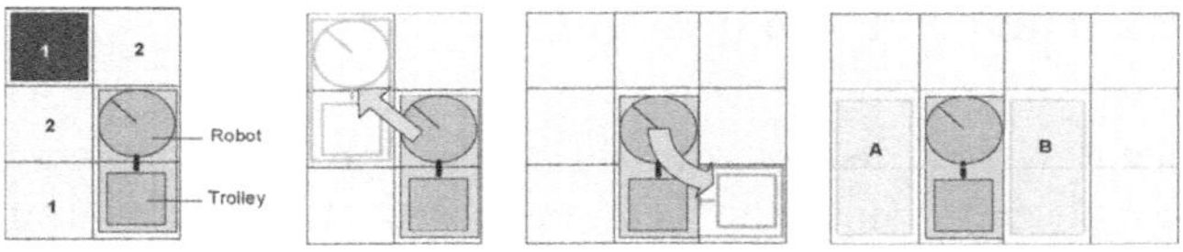

Fig. 2. (left) Each agent can detect objects (blocks,other agents or empty space) in the four surrounding squares as shown. The agent can put down a block from squares labelled (1) and pick up a block from squares labelled (2). (center left) Robot shown moving forward. (center right) Robot rotates counter-clockwise. (right) Contact sensors can detect other agents or obstructions in regions marked (A) and (B).

I Move: The agent moves diagonally to the upper left corner if the front and left-side trolley bumper detect no obstacles otherwise the agent rotates left as shown in fig. 2 (center).

II Manipulate Object: The choice of whether to put down or pick up a block is made based on whether the agent is already in possession of a block. If the agent is unable to pick up or put down a block, the 'move' command is activated.

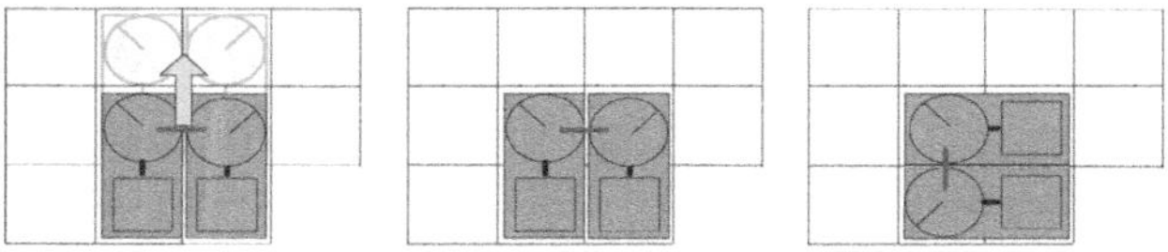

Fig. 3. (left) A pair of agents moving forward. (center),(right) Paired agents shown rotating counterclockwise.

The agents communicate one bit of information depending on what is detected using the vision sensors. The vision sensors can distinguish between an agent, a block and an empty space. The agent as shown in fig. 2 is equipped with four vision sensors with 3 possible outcomes each (block, agent, empty space), two possible outcomes for the agent (carrying a block or not) and an additional two states during a communication session (receive a 1 or 0).

When a pair of agents chooses to move forward, the collective movement is the vector sum of the diagonal movement of each individual agent (fig. 4). The contact sensors can detect a block, agent (in correct position) or empty space in two equally spaced regions next to the agent (fig. 2). The Link Lookup Table entries consist of two basis behaviors: 'Link' (paired up) and 'Unlink' (separated), which are defined as follows :

III Link: An agent will link to a neighboring agent once aligned in in one of two position (shown in fig. 4). The agents are paired only when neither agent is already paired and both have agreed to 'link'.

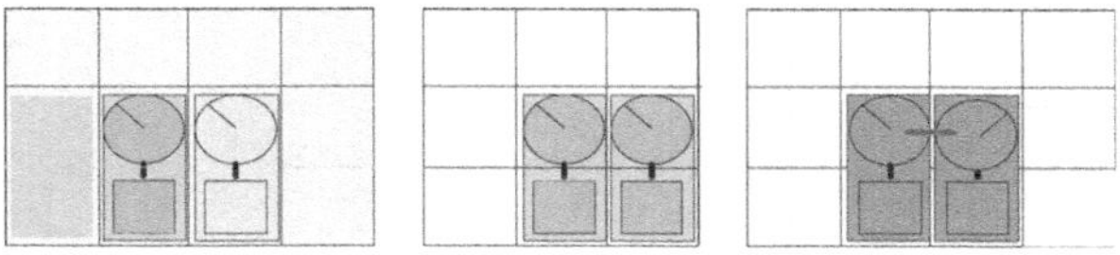

Fig. 4. (left) (1) and (2) show the two regions used to detect if a neighboring agent is in position to 'link'. (Center) Agents in position to 'link' and configuration afterwards (right).

IV Unlink: A pair of agents will 'unlink', provided the agents are already linked and either one of the agent has chosen to 'unlink'.

The total number of entries in the Physical Response Lookup Table is: $3^4 \times 2 \times 2 = 324$ entries. The Communication Lookup Table is connected to the four vision sensors, with 2 possible outcomes (block or no block), resulting in $2^4 = 16$ entries. The Link Lookup Table has two sets of sensors on each side of the agent with three possible outcomes each (obstacle, agent, empty space), which leads to $3^2 = 9$ entries. In total there are 349 lookup table entries defined for the cellular automata-based control system.

5 Simulation Results

In our simulations the GA population size was $P = 50$, number of generations $G = 300$, crossover probability $p_c = 0.7$, mutation probability $p_m = 0.005$ and tournament size of 5 (for tournament selection). For the GA run, the 2-D world size was a 16×16 grid with 24 agents, 36 blocks and a training time of 3000 time steps, where, $J = 49$ and $I = 30$ (number of initial conditions per fitness evaluation).

After 300 generations (fig. 5), the GA run converged to a reasonably high average fitness value (about 99). The agents learn to pair up and stay paired during the entire training time within the first 5-10 generations. Fig. 5 (right) shows the average similarity between the best individual and the population during each generation. The fitness time series averaged over 1000 simulations shows a smooth curve (fig. 6), which is used to calculate the emergence time. The emergence time is defined to be the number of time steps it takes for the system to have organized itself [11,3]. At a fitness value of 99, the blocks were well organized into the 3×3 tiling pattern and more importantly a global consensus (one observable lattice) was formed. For the 16×16 world, the emergence time was 2353 time steps. Fig. 7 shows some snapshots from a typical simulation at various time steps. To keep the comparison simple and meaningful, constraints had to be imposed to ensure a fair chance in arranging a perfect lattice. It was found when the ratio of agents to blocks was low, a global consensus took much longer to occur or never occurred at all. When the agents to blocks ratio is high, the collective effort is hindered by each individual (known to as *antagonism*) [12] and a global consensus is never achieved.

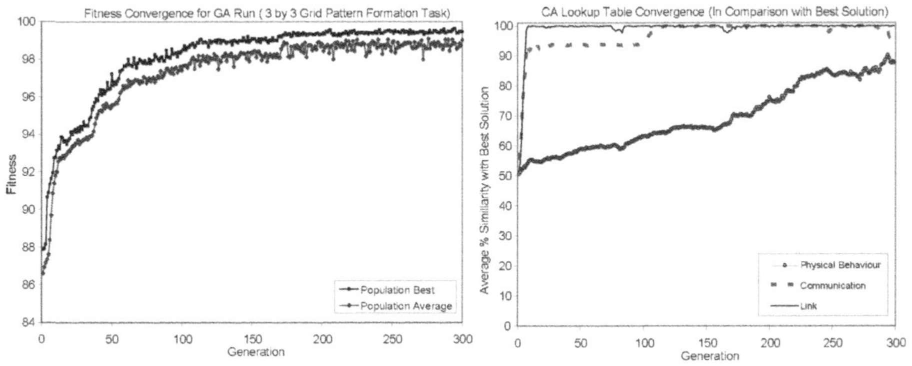

Fig. 5. (left) Convergence history for a typical GA run. (right) CA lookup table Convergence. (in comparison with Best Solution)

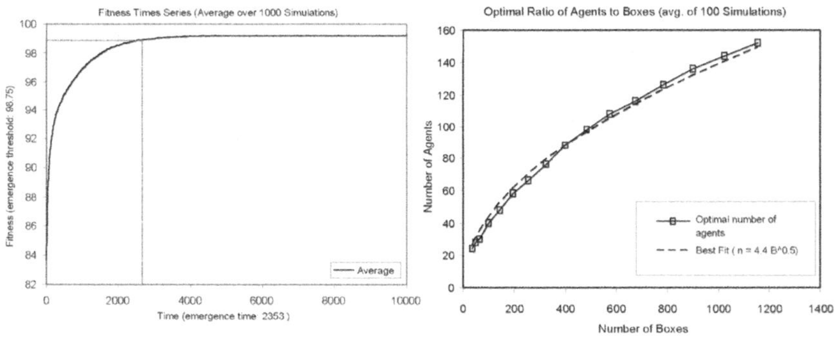

Fig. 6. (left) Average fitness time series over 1000 simulations for the 16×16 grid with 28 agents and 36 blocks. The calculated emergence time is also indicated (right) Optimal ratio of agents to blocks for problem size of up to 100×100 grid.

One of the most cited advantages of decentralized control is the ability to scale the solution to a much larger problem size. Using the optimal ratio of agents to blocks, the simulation was performed for an extended 600,000 time steps to determine the maximum fitness for various problem sizes (up to 100×100 grid). The maximum fitness value remained largely constant, as expected, due to our decentralized approach. However, further simulations will need to be conducted to confirm the scalability of the evolved solutions.

6 Discussion

It is interesting that a framework for communication between agents is evolved earlier than pair coordination. With the number of lookup table entries for the

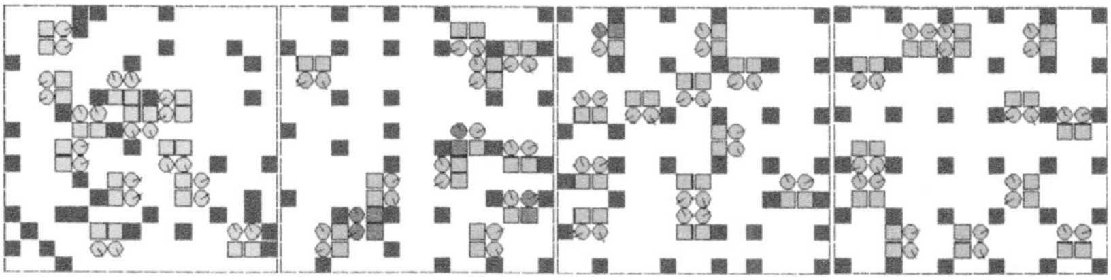

Fig. 7. Snapshot of the system taken at various time steps (0, 100, 400, 1600). The 2-D world size is a 16×16 grid with 28 agents and 36 blocks. At time step 0, neighboring agents are shown 'unlinked' (light gray) and after 100 time steps all 28 agents manage to 'link' (gray or dark gray). Agents shaded in dark gray carry a block. After 1600 time steps (far right), the agents come to a consensus and form one lattice structure.

Communication Lookup Table (CLT) being far fewer than the Physical Response Lookup Table (PRLT), it would be expected for a good solution to be found in fewer generations. Within a coevolutionary process it would be expected for competing populations or in this case subsystems to spur an 'arms race' [6]. The steady convergence in PRLT appears to exhibit this process. It was encouraging to witness our cellular automaton-based multiagent systems evolve a non-coherent communication protocol, similar to what had been observed by Yanco and Stein [5] for a completely different task. With their experiment, one of the two robots was always able to provide orders based on environmental cues to the 'follower robot'.

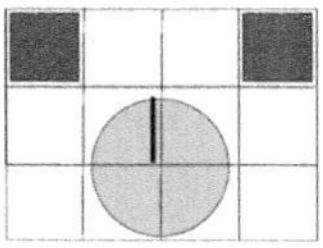

Fig. 8. The alternate configuration considered for solving the 3×3 tiling pattern formation task. The agent occupies 4 squares and can detect objects, agents and empty spaces in 7 squares as shown.

As part our effort to find optimal methods to solving the 3×3 tiling pattern formation task, a comparable agent was developed which lacked the ability to communicate and cooperate. As a result each agent had 7 vision sensors, which meant 4374 lookup table entries compared to the 349 entries for the agent discussed in the paper. After having tinkered with various genetic parameters, it was found the GA run never converged.

In this particular case, techniques employing communication and cooperation have reduced the lookup table size by a factor 12.5 and have made the GA run computational feasible. The significant factor is a correlation between the number of lookup table entries and the number of generations required to reach convergence. With the search space being too large, it is suspected the genetic algorithm was unable to find an incremental path to an optimal solution.

7 Conclusion

Our approach to designing a decentralized multiagent system uses genetic algorithms to develop a set of local behaviors to produce a desirable global consensus. The agents coevolved with localized communication and cooperative behavior can successfully form the 3×3 lattice structure. Comparable agents which have bigger lookup tables and lack the ability to communicate and cooperate are unable to perform the same tasks. Our findings show strategies employing cooperation, communication and coevolution can be used to significantly reduce the size of CA lookup tables and make a genetic search more feasible.

References

1. Kube, R., Zhang, H.: Collective Robotics Intelligence : From Social Insects to robots. In Proc. Of Simulation of Adaptive Behavior (1992) 460–468
2. Cao, Y.U., Fukunaga, A., Kahng, A. : Cooperative Mobile Robotics : Antecedents and Directions. In : Arkin, R.C., Bekey, G.A. (eds.): Autonomous Robots, Vol. 4. Kluwer Academic Publishers, Boston (1997) 1–23
3. Barfoot, T., D'Eleuterio, G.M.T.: An Evolutionary Approach to Multi-agent Heap Formation. Proc. of the Congress on Evolutionary Computation (1999)
4. Goldberg, D.: Genetic Algorithms in Search, Optimization and Machine Learning. Addison-Wesley Pub. Co., Reading, Mass., (1989)
5. Yanco, H., Stein L.: An adaptive communication protocol for cooperating mobile robots. From Animals to Animats: Proc. of the Second Int. Conference on the Simultion of Adaptive Behavior. MIT Press/Bradford Books (1993) 478–485
6. Koza, J.R.: Genetic Programming: On the Programming of Computers by Means of Natural Selection. The MIT Press. Cambridge, MA, (1992)
7. Matsumoto, A., Asama, H., Ishida, Y.: Communication in the autonomous and decentralized robot system ACTRESS. In Proc. of the IEEE international workshop on Intelligent Robots and System. (1990) 835–840
8. Shin, K., Epstein, M.: Communication Primitives for Distributed Multi-robot system. In Proc. of the IEEE Robotics and Automation Conference, (1985) 910–917
9. Fukuda, T., Kawauchi, Y.: Communication and distributed intelligence for cellular robotics system CEBOT.In Proceedings of Japan-USA symposium on Flexible Automataion, (1990) 1085–1092
10. Maclennan, B., Burghardt, G. M.: Synthetic ethology and the evolution of cooperative communication. Adaptive Behaviour, (1994) 161–188
11. Hanson, J.E, Crutchfield, J.P.: Computational mechanics of Cellular Automata : An Example. Working Paper 95-10-095, Santa Fe Institute. Submitted to Physica D, Proc. of the Int. Workshop on Lattice Dynamics. (1995)
12. Dagneff, T.,Chantemargue, F., Hirsburnner, B.: Emergence-based cooperation in a multi-agent system. Tech. report, Univ. of Fribourg. (1997)
13. von Neumann, J.: Theory of self reproducing Automata. Univ. Illinois Press, London
14. Wolfram, S.: A New Kind of Science. Wolfram Media, Champaign, IL

Evolving Aggregation Behaviors in a Swarm of Robots

Vito Trianni[1], Roderich Groß[1], Thomas H. Labella[1], Erol Şahin[2], and Marco Dorigo[1]

[1] IRIDIA – Université Libre de Bruxelles, Belgium
[2] Dept. of Computer Engineering – Middle East Technical University, Ankara, Turkey, {vtrianni,rgross,hlabella,mdorigo}@ulb.ac.be
erol@ceng.metu.edu.tr

Abstract. In this paper, we study aggregation in a swarm of simple robots, called *s-bots*, having the capability to self-organize and self-assemble to form a robotic system, called a *swarm-bot*. The aggregation process, observed in many biological systems, is of fundamental importance since it is the prerequisite for other forms of cooperation that involve self-organization and self-assembling. We consider the problem of defining the control system for the *swarm-bot* using artificial evolution. The results obtained in a simulated 3D environment are presented and analyzed. They show that artificial evolution, exploiting the complex interactions among *s-bots* and between *s-bots* and the environment, is able to produce simple but general solutions to the aggregation problem.

1 Introduction

Aggregation is a task of fundamental importance in many biological systems. It is of particular importance for the creation of functional groups of individuals, because it is at the basis of the emergence of various forms of cooperation. In fact, it can be considered a prerequisite for the accomplishment of many collective tasks. These reasons motivate our interest in the study of aggregation within the SWARM-BOTS project, whose aim is the development of a new robotic system, called a *swarm-bot* [10]. A *swarm-bot* is defined as a self-organizing, self assembling artifact composed of a swarm of *s-bots*—mobile robots with the ability to connect to/disconnect from each other. *S-bots* are equipped with simple sensors and motors and have limited computational capabilities. Their physical links are used to self-assemble into a *swarm-bot* able to solve problems that cannot be solved by a single *s-bot*.[1]

The SWARM-BOTS project takes inspiration from recent studies in *swarm intelligence*, a novel approach to the design and implementation of "intelligent" systems inspired by the effectiveness and robustness observed in social insects and in other animal societies [3]. A key role in swarm intelligence is played by the

[1] Details regarding the hardware and simulation of the *swarm-bot* are presented in [9] and on the project web-site (http://www.swarm-bots.org).

W. Banzhaf et al. (Eds.): ECAL 2003, LNAI 2801, pp. 865–874, 2003.

phenomenon of *self-organization*, whereby global level order emerges in a system from the interactions happening among the system's lower-level components. Moreover, these interactions are based only on local information, without reference to the global pattern [5]. A particular form of self-organization is observed in some social insects, which connect one to the other creating complex physical structures. This type of self-organization is referred to as *self-assembling* [1], and is one of the key elements of the SWARM-BOTS project.

Our interest in aggregation stems from the fact that, in order to be able to self-assemble, *s-bots* have to first aggregate in a common location. Additionally, we are interested in self-organized aggregation, that is, aggregation processes that are not driven by a central controller. Self-organization can lead to robust control systems that can be executed without the need of global information, but exploiting only local sensing and local interactions among *s-bots*.

We consider the problem of defining the control system for the *s-bots* using *artificial evolution*, which has gained more and more attention for robotic tasks in the last decade [8]. There are many motivations behind the use of evolutionary techniques for the design of a control system for a robot. In particular, in a multi-robot domain such as the one considered within the SWARM-BOTS project, the dynamical interactions among robots and between robots and the environment make it difficult to hand-design a control system. On the contrary, evolution can fully exploit these dynamic features, without requiring much intervention from the designer. In this paper, we show how evolution can find simple but effective behaviors, which, in some cases, scale with the number of *s-bots* involved in the experiment.

The paper is organized as follows: Section 2 presents some examples of aggregation observed in biological systems, describing the basic mechanisms that enable self-organized aggregation to emerge. Section 3 presents our experiments on evolving aggregation behavior for the *swarm-bot*. The experimental setup is described and the obtained results are analyzed. Finally, Section 4 concludes the paper.

2 Aggregation in Biological Systems

Self-organized aggregation occurs in biological systems by means of two basic mechanisms: *positive* and *negative feedback*. Positive feedback usually takes the form of attraction toward a given signal source (e.g., chemical, tactile, visual). This mechanism leads to amplifications of the source of information, which becomes more and more attractive in time. On the other hand, negative feedback serves as a regulatory mechanism, providing some form of repulsion among the system components, thus controlling the formation of the aggregate.

One of the best studied examples of self-organized aggregation is the one observed in the cellular slime mold *Dictyostelium discoideum* [4,11]. When the amoebae have enough food, they act independently, grow and reproduce rapidly. However, when they are starving, they enter a *developmental phase*: the amoebae emit a chemical attractor that diffuses in concentric waves and serves as a guide for the aggregation. When the amoebae are clustered, they form a multicellular organism called a slug, which is able to move on the substrate for some time. Eventually, the slug turns into a fruiting body that can develop and distribute spores in the environment, thus restarting the life cycle.

A similar aggregation process can be observed in many other unicellular organisms [5]. Also social and pre-social insects present multiple forms of aggregation, a particular example being the feeding clusters of larvae of the bark beetle *Dendroctonus micans* [6]. In this case, larvae emit pheromone when feeding. The pheromone diffuses in air and triggers the aggregation process. In fact, in presence of a pheromone gradient, larvae react by moving in the direction of higher concentration of pheromone, eventually forming a cluster.

Other interesting examples are given by honey bees, that cluster around the queen on a branch, while scout bees search for new nesting sites. Also birds, fish or mammals present aggregation phenomena that are self-organized, at least to some extent: for example, young penguins aggregate for warmth, many fish species create defensive or hunting schools, and some species of birds and mammals protect their cubs at the center of the group [5].

The aggregation processes described above have been shown to be self-organized. However, we know many other examples in which environmental information or other heterogeneities are used as a clustering stimulus, sometimes mixing it with self-organized behaviors. For example, the environment can offer cues for aggregation, like light or temperature for flies or humidity for sow bugs [5]. However, as mentioned before, we will focus on self-organized aggregation, showing how it can emerge as a result of artificial evolution.

3 Evolving Aggregation Behaviors

As discussed in the previous section, aggregation can be the result of a self-organizing process. After some experience in hand-coding pattern formation behaviors for a group of simulated *s-bots* placed in a grid world [10], we tried to develop *s-bots* able to aggregate by using artificial evolution. In the following, we describe the experimental setup used for the evolution of the clustering behavior. Then, the obtained controllers are analyzed and their properties and limitations are presented.

3.1 Experimental Setup

The experiments presented in this section are performed in simulation, using a software based on the rigid body dynamics simulator SDK Vortex^{TM}. This simulator reproduces the dynamics, friction and collisions between physical bodies. The *s-bots* are modeled as cylinders (radius $r = 12$ cm, height $h = 6$ cm) and are provided with two motorized wheels, a gripper that allows connections between *s-bots*, and an omni-directional speaker that continuously produces a tone that can be perceived from a distance up to 50 cm (see Fig. 1a and 1b). Each *s-bot* is also equipped with eight infrared proximity sensors, three directional microphones, three sensors for detecting established connections on the body, and a gripper sensor simulating a light barrier on the gripper, which is used to perceive the presence of a grippable object (see Fig. 1c). The environment consists of a square arena surrounded by walls. The size of the arena is chosen to be 2×2 meters and is bigger than the perceptual range of the *s-bots* to emphasize the locality of sensing.

We used a generational evolutionary algorithm for the evolution of the *s-bot* neural controller. The genotype specifies the connection weights of a simple perceptron having 17 sensory neurons that encode the state of the 16 sensors and a bias unit (i.e., a unit whose activation state is always 1.0). Each sensory neuron is directly connected to 3 motor neurons, that control the gripper and the speed of the two wheels. The transfer function of the motor neurons is a standard logistic function. Each connection weight ranges in the interval [-10, +10] and is represented in the genotype with 8 bits. Each genotype is mapped into a neural network that is cloned in every *s-bot* involved in the experiment [2]. Five *s-bots* compose the group and they are allowed to "live" for 10 "epochs" (each epoch consists of 600 cycles and each cycle simulates 100 ms of real time). At the beginning of each epoch the *s-bots* are placed in randomly selected positions and orientations within the arena.

In each epoch e, the fitness of a given genotype is estimated averaging over the last 100 cycles a measure $f_e(t)$ that describes the average distance of the group from its center of mass:

$$f_e(t) = \frac{1}{n} \sum_{i=1}^{n} \left(1 - \frac{d_i(t)}{50}\right),$$

where n is the number of *s-bots* and $d_i(t)$ is the distance of the ith *s-bot* from the center of mass, limited to 50 cm as upper bound in order to have fitness values in the interval $[0, 1]$. The final fitness is obtained by simply averaging the values obtained in each epoch.

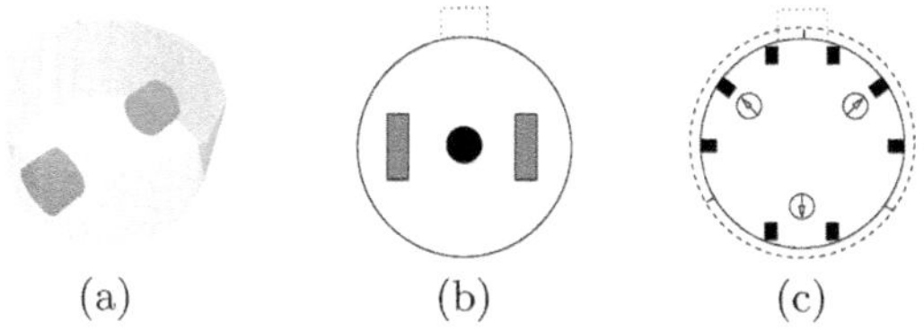

Fig. 1. The simplified *s-bot* model. (a) 3D model, where the cylindrical body is transparent to visualize the wheels (b) Actuators: motorized wheels (two gray rectangles), gripper (dotted rectangle), and omni-directional speaker (black circle). (c) Sensors: proximity sensors (black rectangles), directional microphones (white circles), connection sensors (three regions marked with a dashed line around the body), and a light barrier sensor on the gripper (dotted rectangle).

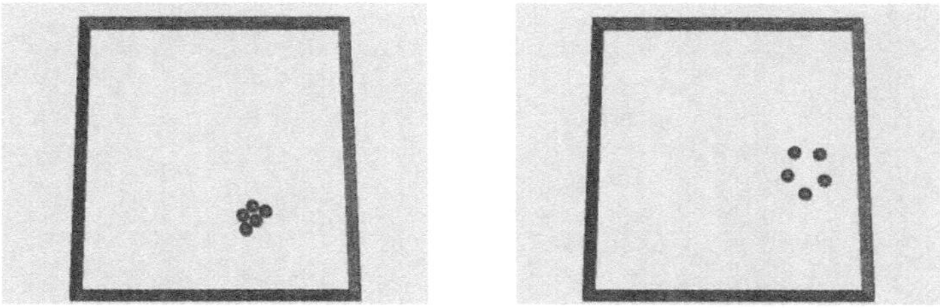

(a) (b)

Fig. 2. Snapshots of the formed aggregate. (a) Static clustering behavior. (b) Dynamic clustering behavior.

The population contains 40 genotypes. The best 8 genotypes of each generation are allowed to reproduce, each generating 5 offspring. The per-bit (flip) mutation rate is $2/L$, where L is the length of the genotype. No recombination is used. Parents are not copied to the offspring population. The evolutionary process lasts 100 generations. The experiment was replicated 10 times by starting with different randomly generated initial populations.

In all replications, good solutions are discovered quite early in the evolution, and are then slowly refined afterward, the fitness always reaching values near to the maximum. Figure 2 shows two snapshots taken at the end of the aggregation process.[2] In the following, we analyze the obtained results, classifying the evolved behaviors in two different classes.

3.2 Behavioral Analysis

By running the evolutionary experiment, we observed the emergence of two types of strategies: a *static* and a *dynamic* clustering behavior. The former creates very compact and stable aggregates in which *s-bots* do not change their relative positions (see Fig. 2a). The latter creates rather loose but moving aggregates that, as we will discuss, allow scalability of the behavior (see Fig. 2b). In the following, we analyze the most representative examples of both classes.

Static Clustering Behavior. As mentioned before, behaviors that fall into the static clustering category create very compact clusters. When far from other individuals, *s-bots* explore the arena moving backward along a circular trajectory having a diameter bigger than the arena side and avoiding walls. When two *s-bots* get close, the attraction to sound sources becomes predominant, and the trajectories change: the two *s-bots* tend to bounce against each other, due to the interplay between attraction and repulsion originating from sound and infrared sensors respectively. In fact, clusters of two *s-bots* are very unstable. However, during the time spent close to each other, the pair can attract other *s-bots*

[2] See `http://www.swarm-bots.org/index.php?main=3&sub=35&conpage=cl` for some movies of these behaviors.

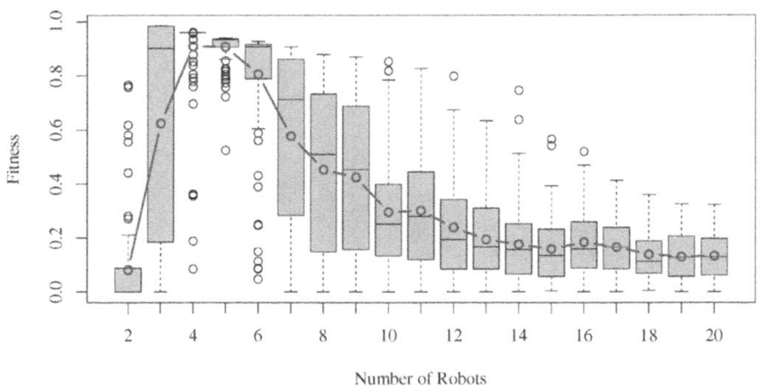

Fig. 3. Static clustering behavior: fitness evaluations obtained for different group sizes. The average fitness is drawn as a thick line. Boxes represent the interquartile range of the data, while the horizontal bars inside the boxes mark the median values. The whiskers extends to the most extreme data points within 1.5 of the interquartile range from the box. The empty circles mark the outliers.

which can join the cluster, increasing its size and stability. Finally, when the aggregate is formed, *s-bots* continuously monitor the surrounding by performing small moves or turning on the spot, in order to constantly adjust their position with respect to the other *s-bots*.

Note that the performance of the neural controller with respect to the given fitness measure is maximized by this strategy: *s-bots* are in contact when clustered, thus minimizing the distance from the center of mass. Evolution has exploited one important invariant present in the experimental setup: the number of *s-bots*. In fact, given that 5 *s-bots* are present in the environment, only clusters formed by the majority (that is, 3 *s-bots* or more) are stable, while smaller clusters (2 *s-bots*) easily disband. This suggests that when the group size is increased, it will be difficult to obtain a single cluster, but rather multiple smaller clusters will be formed.

In order to confirm this hypothesis, we analyzed the scalability of the clustering behavior with respect to the number of *s-bots* involved. We repeated the evaluation of the fitness 100 times for different group sizes[3]. The results plotted in Fig. 3 show that, as expected, the behavior does not scale well with the number of robots. In particular, the best results are achieved with group sizes around 5, showing that the evolved behavior is particularly tuned for these situations. Not surprisingly, the average fitness for a group of 2 *s-bots* is very low, as this cluster is unstable. The group of 3 *s-bots* presents an high variance: this suggests that *s-bots* were not able to form a stable cluster within the limited time, due to the lower density of *s-bots* in the arena. For group size bigger than 5 the performance quickly decreases. This can be explained by the fact that, the more

[3] Each fitness evaluation is performed over 1,000 cycles, since 600 cycles were not sufficient for the clustering of larger groups.

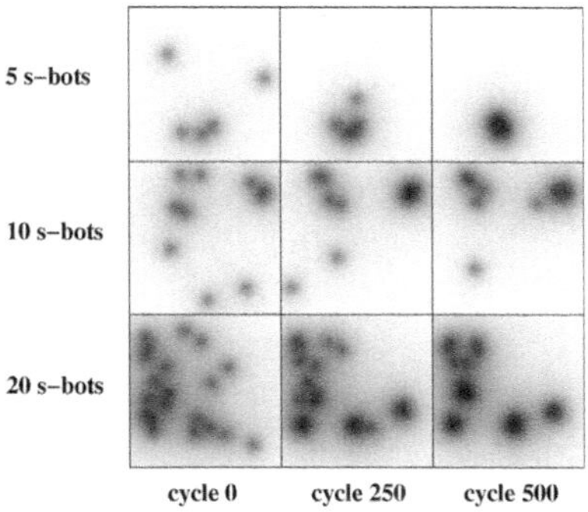

Fig. 4. Static clustering behavior: snapshots of the sound fields are shown at three instants in time. Each row corresponds to a simulation of clustering with a different group size (5, 10 and 20 *s-bots*, from top to bottom).

the *s-bots*, the higher the number of small clusters. Additionally, a high density of *s-bots* creates a high intensity of sound throughout the arena, causing all the *s-bots* to whirl in place or to join a near neighbor.

The tone continuously emitted by each *s-bot* creates a *sound field* which can be used to give an approximate indication of the sound attraction forces acting on the *s-bots*. Figure 4 plots the change in the sound field over time for groups of 5, 10 and 20 *s-bots*. In the group of five *s-bots*, a single cluster is formed, while multiple clusters appear for larger group sizes. The high intensity of sound inhibits the exploration behavior of the *s-bots* and makes them join the nearest *s-bot* or cluster, as displayed in the last row of Fig. 4.

Dynamic Clustering Behavior. The dynamic clustering behavior creates loose and moving clusters. Also in this case, a circular trajectory is observed when an *s-bot* is far from the walls and from other *s-bots*. When the *s-bots* sense each other, they aggregate and start moving together to explore the environment, in a sort of "flocking" behavior, which is the result of the interplay of attraction to sound and repulsion from too close *s-bots*. This creates moving clusters which can search and merge with other *s-bots* or clusters. When close to each other, *s-bots* continue to move and change their relative positions. In this way, small clusters can change their shape and move across the arena, having the possibility to join other clusters or attract free *s-bots* and increase the size of the aggregate. This feature makes the dynamic clustering behavior robust with respect to the formation of sub-clusters, since formed clusters can continue to explore the arena.

The scalability analysis, conducted in the same way as for the static clustering behavior, confirms the robustness of the evolved behaviors. Figure 5 shows that the performance of the group decreases almost linearly with the group size. This decrease in performance is not due to an imperfect aggregation, but to the fact that the minimum average distance to the center of the cluster grows with the

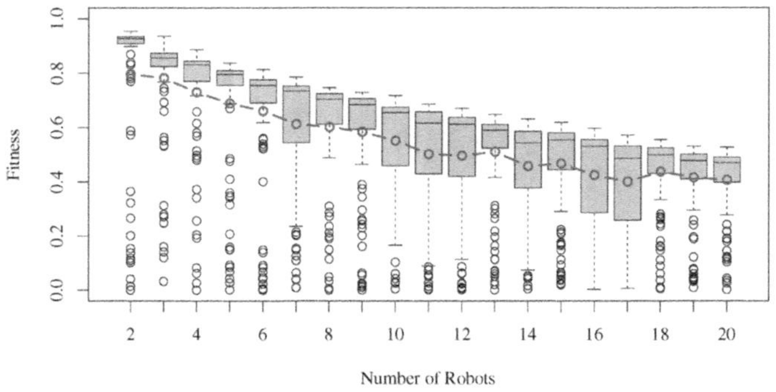

Fig. 5. Dynamic clustering behavior: fitness evaluations obtained for different group sizes. The average fitness is drawn as a thick line (see also Fig. 3 for explanation of the graph).

number of *s-bots* due to their physical embodiment. Thus, the evolved behavior is well suited for every group size, as sub-clusters, when formed, can continue the aggregation process.

Figure 6 shows the snapshots of sound fields observed for the dynamic clustering behavior. It is worth noting that, unlike the observations made on static clustering behavior, a high intensity of sound in the arena is not problematic for the movement of the *s-bots*. On the contrary, it seems to serve as a communication medium that guides the clustering.

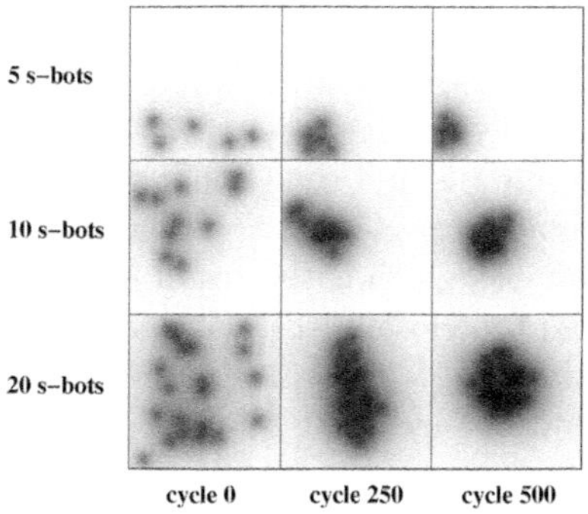

Fig. 6. Dynamic clustering behavior: snapshots of the sound fields are shown at three instants in time. Each row corresponds to a simulation of clustering with a different group size (5, 10 and 20 *s-bots*, from top to bottom).

4 Conclusions

In this paper, we have described the phenomenon of aggregation in biological systems and the evolution of controllers for a group of simulated robots in order to obtain a similar process. In our experiments, two behavioral strategies emerge from evolution. The static clustering behavior results in high fitness values, but it is tuned for a group of size 5. On the contrary, the dynamic clustering behavior obtains lower fitness values as the formed clusters are less compact. However, clusters move through the environment, leading to a scalable behavior.

The reader might have noted that, although *s-bots* can control their grippers, these are not used. In the presented work, we were mainly interested in the study of self-organized aggregation, and not in self-assembling, which requires physical connections. Accordingly, the chosen fitness function does not encourage the establishment of connections. Nevertheless, connections could have appeared because, creating a connection, two *s-bots* minimize their relative distance. What was observed, however, is that connected *s-bots* were, in most cases, unable to move in a coordinated way, making the formation of clusters difficult if possible at all.

In the literature, it is possible to find some interesting works related to the one presented in this paper. Melhuish et al. [7] studied seeded aggregation and collective movement of minimal simulated agents, using sound signals (*chorusing*) to regulate the group size. However, aggregation is not self-organized, because of the presence of an infrared beacon that serves as aggregation site. Yokoi et al. [12], taking inspiration from cellular slime molds, developed amoeba-like robots composed of connected modules. Here, the main difference with the *swarm-bot* is that, even if each module is autonomous in its control, it remains connected to other modules, lacking the full mobility of *s-bots*.

In conclusion, the obtained results show that evolution is able to find simple but effective solutions to the aggregation problem, mainly exploiting some invariants present in the environment and the complex interactions among *s-bots* and between *s-bots* and the environment. Under these conditions, effective behaviors are difficult to hand-design, as they are an "emergent property of the interaction between the robot and the environment" [8]. Furthermore, it is worth noting how the evolved aggregation mechanisms resemble the ones described in Sect. 2. In particular, the attraction to sound sources serves as a positive feedback mechanism: the higher the intensity of sound perceived, the higher the attraction toward the source, which is consequently amplified. On the other hand, the repulsion between *s-bots* constitutes the negative feedback mechanism: it makes clusters of 2 *s-bots* unstable in the static clustering behavior, and results in the movement of the clusters in the dynamic clustering behavior. The latter strategy scales with the number of *s-bots* because it does not strongly rely on environmental invariants, but is merely a result of the dynamic interaction between the *s-bots*, which makes it more similar to the processes observed in nature.

Future work will exploit the presented results to obtain more complex forms of cooperation, like aggregation around preys in order to collectively retrieve them, or aggregation for coordinated motion on rough terrain.

Acknowledgments. Marco Dorigo acknowledges support from the Belgian FNRS, of which he is a Senior Research Associate. The SWARM-BOTS project is funded by the Future and Emerging Technologies programme (IST-FET) of the European Community, under grant IST-2000-31010. The information provided is the sole responsibility of the authors and does not reflect the Community's opinion. The Community is not responsible for any use that might be made of data appearing in this publication.

References

1. C. Anderson, G. Theraulaz, and J.-L. Deneubourg. Self-assemblages in insect societies. *Insectes Sociaux*, 49:99–110, 2002.
2. G. Baldassarre, S. Nolfi, and D. Parisi. Evolving Mobile Robots Able to Display Collective Behaviours. In Hemelrijk C. K. and Bonabeau E., editors, *Proceedings of the International Workshop on Self-organisation and Evolution of Social Behaviour*, pages 11–22, Monte Verità, Ascona, Switzerland, September 8–13, 2002.
3. E. Bonabeau, M. Dorigo, and G. Theraulaz. *Swarm Intelligence: From Natural to Artificial Systems.* Oxford University Press, New York, NY, 1999.
4. J. T. Bonner. *The Cellular Slime Mold.* Princeton University Press, Princeton, NJ, 1967.
5. S. Camazine, J.-L. Deneubourg, N. Franks, J. Sneyd, G. Theraulaz, and E. Bonabeau. *Self-Organization in Biological Systems.* Princeton University Press, Princeton, NJ, 2001.
6. J.-L. Deneubourg, J. C. Gregoire, and E. Le Fort. Kinetics of the larval gregarious behaviour in the bark beetle *Dendroctonus micans*. *Journal of Insect Behavior*, 3:169–182, 1990.
7. C. Melhuish, O. Holland, and S. Hoddell. Convoying: using chorusing to form travelling groups of minimal agents. *Robotics and Autonomous Systems*, 28(2–3):207–216, 1999.
8. S. Nolfi and D. Floreano. *Evolutionary Robotics: The Biology, Intelligence, and Technology of Self-Organizing Machines.* MIT Press/Bradford Books, Cambridge, MA, 2000.
9. G. C. Pettinaro, I. W. Kwee, L. M. Gambardella, F. Mondada, D. Floreano, S. Nolfi, J.-L. Deneubourg, and M. Dorigo. Swarm robotics: A different approach to service robotics. In *Proceedings of the 33rd International Symposium on Robotics*, Stockholm, Sweden, October 7–11, 2002. International Federation of Robotics.
10. E. Şahin, T.H. Labella, V. Trianni, J.-L. Deneubourg, P. Rasse, D. Floreano, L.M. Gambardella, F. Mondada, S. Nolfi, and M. Dorigo. SWARM-BOTS: Pattern formation in a swarm of self-assembling mobile robots. In A. El Kamel, K. Mellouli, and P. Borne, editors, *Proceedings of the IEEE International Conference on Systems, Man and Cybernetics*, Hammamet, Tunisia, October 6–9, 2002. Piscataway, NJ: IEEE Press.
11. C. Van Oss, A.V. Panfilov, P. Hogeweg, F. Siegert, and C.J. Weijer. Spatial pattern formation during aggregation of the slime mould *Dictyostelium discoideum*. *Journal of Theoretical Biology*, 181:203–213, 1996.
12. H. Yokoi, W. Yu, and J. Hakura. Morpho-functional machine: Design of amoeba like robot based on vibrating potential method. *Robotics and Autonomous Systems*, 28(2–3):217–236, 1999.

Low-Level Visual Homing

Andrew Vardy and Franz Oppacher

School of Computer Science
Carleton University
Ottawa, Canada
avardy@scs.carleton.ca
http://www.scs.carleton.ca/~avardy

Abstract. We present a variant of the *snapshot model* [1] for insect visual homing. In this model a snapshot image is taken by an agent at the goal position. The disparity between current and snapshot images is subsequently used to guide the agent's return. A matrix of local low-level processing elements is applied here to compute this disparity and transform it into a motion vector. This scheme contrasts with other variants of the snapshot model which operate on one-dimensional images, generally taken as views from a synthetic or simplified real world setting. Our approach operates directly on two-dimensional images of the real world. Although this system is not a model of any known neural structure, it hopes to offer more biological plausibility than competing techniques because the processing applied is low-level, and because the information processed appears to be of the same sort of information that is processed by insects. We present a comparison of results obtained on a set of real-world images.

1 Introduction

In [1] the *snapshot model* was proposed to explain the remarkable ability of honeybees to return to a place of interest such as a nest or food source after being displaced from that place. Variants of this model have been implemented both in simulation and in physical robots [4,11,3,7,14,8,9,10]. All of these computational models operate on one-dimensional images. Even for the case of physical robots with cameras that generate two-dimensional images, the height dimension is almost immediately collapsed such that the image used for homing is effectively one-dimensional. Our approach differs from these in that we apply our processing directly on two-dimensional images. We find that the use of two-dimensional image features improves results over a similar model that operates only on one-dimensional features. Also, a number of these other models are tested either in simulated worlds or in simplified real world scenarios. We test our approach using the real world image album of [10] and compare our results with theirs.

The snapshot model was developed to match data of honeybee search patterns. A model agent is placed at the goal and allowed to capture a snapshot image. It is then displaced and allowed a return attempt. The model operates on one-dimensional panoramic binary images. Features (region centers) in the

W. Banzhaf et al. (Eds.): ECAL 2003, LNAI 2801, pp. 875–884, 2003.

current image are paired with their closest matching counterparts in the snapshot image. Each pairing generates two vectors: one correcting for bearing and one correcting for differences in apparent size. The sum of all vectors yields the motion vector. One key requirement of the snapshot model is that the agent maintains a consistent orientation. There is evidence to suggest that bees take on the same orientation when returning to a remembered place as they took when originally learning the layout of that place [2,15,6]. A robot homing via the snapshot model must employ some sort of compass system to maintain, or compensate for changes, in orientation.

In [8] a neural implementation of the snapshot model was presented. It showed that the processing necessary to implement the snapshot model could be achieved using simple neuron-like elements arranged in layers of interconnected rings. The outermost ring was exposed to a one-dimensional panoramic image of the homing agent's environment. Subsequent layers processed this image and extracted the necessary vectors to achieve homing. Our approach has been directly inspired by this work. It differs, however, in its use of two-dimensional images, and hence, two-dimensional processing layers. Also, the use of real-world test images prompted a number of modifications.

We can not state that our method is a model of any known neural circuit in an insect's brain. According to [9], *"...nothing is known so far about the neural circuits that realize visual homing in insect brains."* Still, there are three ways in which we claim that our model has some measure of biological plausibility. First, all computations are made via a matrix of locally-connected neuron-like elements. Second, this matrix has a layered structure which preserves spatial relationships from layer to layer. This is inspired by the retinotopic structure of the mammalian visual cortex where neurons on one layer seem to share receptive fields with the layers below them [5]. Our final claim to biological plausibility is in regard to the sort of information that our method employs. It operates on snapshots of a scene, stored in retinal coordinates. This is the primary contention of the snapshot model [1]. More recent work on wood ants [6] suggests the storage and use of two-dimensional templates for ant homing. We hope that our model can help to close the explanatory gap between a complete neural map of the hardware that implements visual homing in insects and a purely computational model that cares nothing for the details of implementation.

Our method operates by approximating, for each feature in the snapshot image, the direction in which that feature has moved in the current image. The features we use are image corners. The method used to determine the direction of feature movement is to examine the local gradient, created from features in the current image, at the position of the snapshot feature. Thus, we refer to our method here as the *Corner Gradient Snapshot Model*, or CGSM.

2 Processing Matrix

We will now describe the operation of the processing matrix for CGSM. The processing that is applied is depicted in figure 1. A summary of this processing follows: An input image is fed into the processing matrix. It is first smoothed and then corners are extracted as distinct features. If the agent is at the goal then the image of features is stored as the snapshot image. Gradients are formed around each feature and these gradients are locally compared with features in the snapshot image to generate vectors which indicate the direction that these features have moved in. These vectors specifying motion in the image are mapped onto vectors specifying the corresponding motion of the agent. Finally, this last set of vectors is summed to create the agent's overall motion vector. For the following layer descriptions, the image that is fed in from the previous layer is referred to as the *input image*. There is insufficient space to describe exactly how each layer's functionality can be distributed across a grid of low-level processing elements. We hope that the simplicity of each layer will provide the reader with assurance that this distribution is possible.

Gaussian Filter. Convolves the input image with a 5x5 Gaussian mask.

Corner Extraction. Extracts corners from the input image. The corner detection scheme we use is known as the Harris detector and is essentially the same as that presented in [13] (pages 82-85). The required computation is purely local in scope and requires only the application of some simple hard-coded algebra to the results of local convolutions and sums.

Local Maxima Extraction. Applies thresholding and non-maxima suppression to find peaks in the input image [13].

Gradient Formation. Forms decaying gradients around points (maxima) in the input image. These gradients allow the next layer to detect the direction in which a corner has moved from its idealized position at a stored snapshot point. Repeated gaussian filtering and diffusion are two possible ways of implementing this layer. We opt for a different approach which is quicker to compute with a serial computer. A single gradient mask is constructed with a center value of unity that decays out radially according to $1/d^{\zeta}$, where d is the distance from center and ζ is an arbitrary constant $(0 > \zeta > 1)$. Beginning with a zero image, copies of the mask are centered over each feature. The value of each pixel in the output image is taken as the maximum value from all mask copies aligned directly above it. Note that we assume that the agent moves only horizontally in the plane. Thus, corners in the top half of the image should never be paired with those in the bottom half. We implicitly prevent such pairings by forming gradients in the top and bottom halves of the image separately.

Ring Operator. Applied at every non-zero point in the snapshot image to generate a vector pointing in the direction that the feature has moved. This layer actually has two input images: the snapshot image, and the gradient image. At each non-zero point in the snapshot image, S, a ring of detector cells surrounding S is employed to detect the approximate direction of the gradient at that point. Ideally, this direction will point to the matching feature F. The vector from S to the detector cell with the highest response yields the uphill direction of the gradient. The vectors generated by this layer are *image vectors* in that they describe the motion of features within the plane of the image. Image vectors will be indicated below in lower-case ($\vec{u}$, $\vec{v}$, $\vec{i}$, and $\vec{e}$).

Vector Mapping. Maps image vectors into *agent motion vectors*. An agent motion vector describes the motion of the agent within its plane of travel that would correct (reduce to zero) the corresponding image vector. The next section provides more detail on this layer. Agent motion vectors will be indicated in upper-case ($\vec{V}$, $\vec{I}$, and $\vec{E}$).

Vector Sum. Sums the agent motion vectors in the input image to generate a single vector that will move the agent toward home.

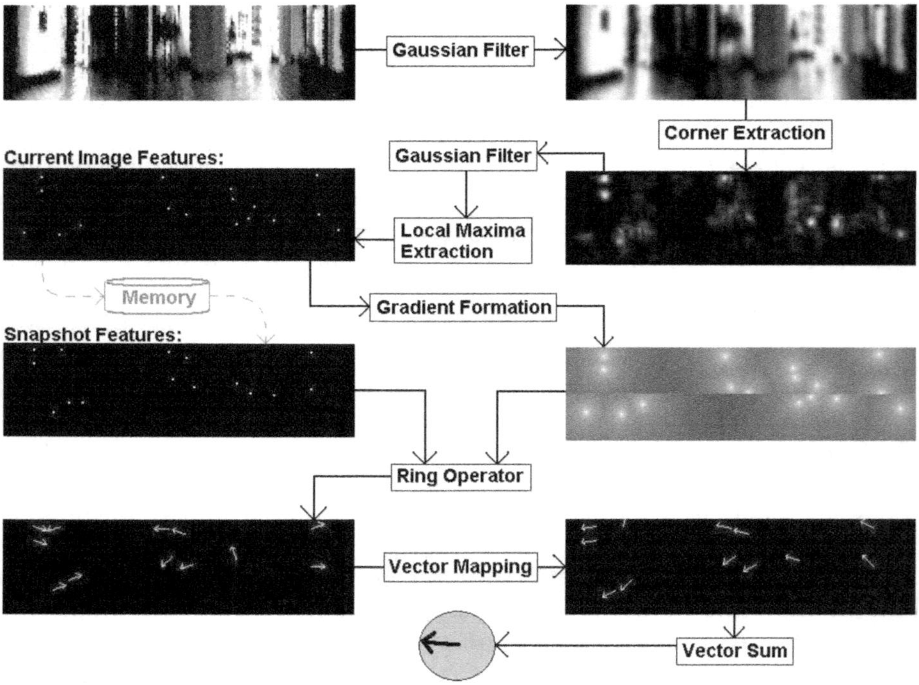

Fig. 1. Processing applied by the CGSM matrix. The input is the image in the upper-left corner. The output is the movement vector in the shaded circle at the bottom. Processing layers are shown as boxed text. Dashed arrows relate to the storage and recall of the snapshot image. The input image was taken from position (270,180) in the image album discussed in the *Results* section. This image was taken from a point directly to the right of the snapshot position (210, 180). Hence, the direction of the output movement vector is approximately correct.

2.1 Vector Mapping

Each image vector $\vec{u}$ is attached or associated with the position of a feature S in the snapshot image. $\vec{v} = -\vec{u}$ is a unit vector which points from C to S, where C is the position of a feature in the current image. If features are paired correctly then S and C will both correspond to a single *environmental feature* F. In this section we will assume that this is true. The vector mapping layer computes an agent motion vector $\vec{V}$, using $\vec{v}$ and S_x (horizontal position of snapshot feature within the image). Ideally each $\vec{V}$ would point directly home. However, there is insufficient information (e.g. distance to F) to compute $\vec{V}$ exactly, therefore we develop an approximation. First we decompose $\vec{v}$ and $\vec{V}$.

$$\vec{v} = a\vec{i} + b\vec{e} \tag{1}$$
$$\vec{V} = \alpha\vec{I} + \beta\vec{E} \tag{2}$$

$\vec{I}$ is a movement of the agent towards a feature X (the relationship of X to S and C will be made clear below). With this movement, X will move upwards in the image according to $\vec{i}$. One can imagine walking towards a light and having the image of that light travel straight upwards in one's field of vision. The walking movement is $\vec{I}$ and the corresponding movement of the light is $\vec{i}$. $\vec{E}$ is a movement of the agent 90° to the right of X. With this movement, X will move to the left according to $\vec{e}$. We can similarly imagine walking 90° to the right of a light and seeing the image of that light travel to the left in our field of vision. In this case, the walking movement is $\vec{E}$ and the movement of the light is $\vec{e}$. Figure 2 illustrates these relationships. The quantity θ_X is the horizontal angular position of X in the image, $\theta_X = \frac{2\pi X_x}{w}$, where X_x is the x−coordinate of X and w is the image width. We can now give the form of $\vec{i}$, $\vec{e}$, $\vec{I}$, and $\vec{E}$.

$$\overrightarrow{I(\theta_X)} = \begin{bmatrix} \sin\theta_X \\ \cos\theta_X \end{bmatrix} \qquad \vec{i} = \begin{bmatrix} 0 \\ 1 \end{bmatrix}$$
$$\overrightarrow{E(\theta_X)} = \begin{bmatrix} \sin(\theta_X + 90^o) \\ \cos(\theta_X + 90^o) \end{bmatrix} \qquad \vec{e} = \begin{bmatrix} -1 \\ 0 \end{bmatrix}$$

We wish to have the feature C move in the image toward S. However, the exact position of C is unknown. We make the assumption that C is not too distant from S. That is, $S \approx C$. Thus, any movement of the agent $\vec{V}$ that would make S move along $\vec{v}$ would also make C move along $\vec{v}$. Of course, S cannot move at all because it is a feature recalled from memory and is fixed in the image. The feature X is a stand-in for S intended to make it less awkward to speak about the movement of S. Summarizing: $X = S$ and $X \approx C$.

We now return to find the constants a, b, α, and β in equations 1 and 2. Given $\vec{i}$ and $\vec{e}$ we can solve equation 1 for a and b. The solution: $a = v_y$ and $b = -v_x$. We make a further approximation to find α and β. If a is much larger than b than the difference in vertical image position between S and C is greater than the difference in horizontal position. In this case, the agent's distance to the environmental feature F along $\vec{I}$ will be greater than the distance along

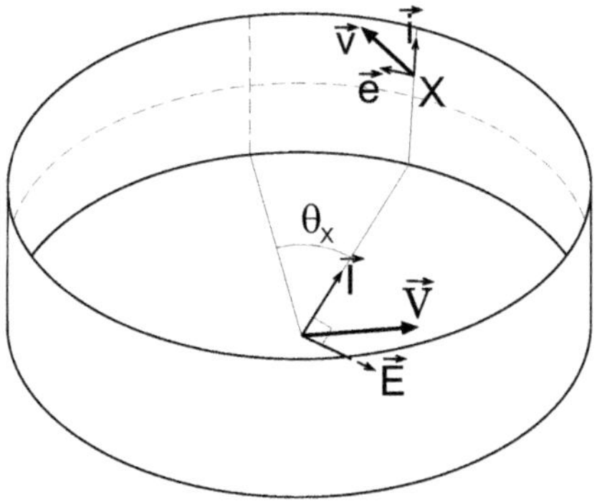

Fig. 2. Motion of the image feature X in direction $\overrightarrow{v}$ is generated by motion of the agent in direction $\overrightarrow{V}$.

$\overrightarrow{E}$. The opposite will be true if b is much larger than a. Thus, we make the approximation: $\alpha = a$ and $\beta = b$. The final mapping from $\overrightarrow{v}$ to $\overrightarrow{V}$ is as follows.

$$\overrightarrow{V} = \begin{bmatrix} v_y \sin\theta_S - v_x \sin(\theta_S + 90^o) \\ v_y \cos\theta_S - v_x \cos(\theta_S + 90^o) \end{bmatrix} \tag{3}$$

One last caveat remains. The vector $\overrightarrow{i}$ points upwards in figure 2 because movement in direction $\overrightarrow{I}$ would cause X to sweep upwards. However, if X had been a feature in the bottom half of the image then X would have swept downwards. Thus, for features in the bottom half of the image $\overrightarrow{i}$ will be $(0, -1)$.

3 Results

We compare the homing performance of CGSM against an implementation of the average landmark vector model which was designed to operate on real-world images [10,12]. The average landmark vector model can be considered a variant of the snapshot model [9]. This approach extracts features from a 1-D image that is itself extracted from a 2-D panoramic image. Once the features have been extracted the average landmark vector model is applied in its original form [7]. We shall refer to this implementation of the average landmark vector model as *XALV* (eXtracted Average Landmark Vector).

The authors of [10] have very kindly made their images available to us and we use them here to compare our results with theirs. These images were taken from a grid of positions spaced 30 cm apart within a $9m \times 3m$ area. The environment was an unmodified entrance hall of a university building. These pictures were taken by a robot-mounted camera pointed upwards at a conical mirror. The robot's orientation was kept constant throughout the image capturing process. We have taken this album of 300 images and unfolded all of them into rectangular panoramic images, 180×48 in size. Figure 3 shows an example image from the album and an unfolded version of the same image.

(a) Camera image (b) Unfolded image

Fig. 3. The image unfolding process.

Figure 4 shows the homing performance of both the XALV and CGSM models on the image dataset using the image taken at position (210,180) as the snapshot image. This image is depicted in figure 3(b). Both models calculate home vectors for all images. These home vectors are depicted in figure 4. The figure also shows the success of homing for all grid positions. A pure white cell indicates successful homing. Homing is considered successful if a path along the home vectors exists from the position in question to the home position. We calculate a quantity called the *approach index* for each position,

$$\text{approach index} = \frac{\text{starting distance} - \text{closest distance}}{\text{starting distance}}$$

The quantity *starting distance* is the euclidean distance from the starting position to the home position. *Closest distance* is the smallest distance from the current position to the home position along the path from the starting position. The approach index for a successful starting position is 1. If, however, the path from the starting position leads away or perpendicular to the home position then the approach index will be 0. If the path approaches the home position but does not reach it then the approach index will be between 0 and 1. The purpose of the approach index is to judge the success of homing even if it is not completely successful. The level of whiteness shown in figure 4 is proportional to the approach index for the corresponding position.

Figure 4 has a dashed box surrounding the left half of the dataset. This box encloses the positions used to generate the results presented in [10]. It had been our intention to replicate these results precisely, however we have not been able to reconstruct all of the necessary parameters. This being said, our re-implementation of the XALV model seems to achieve better results. The original result for positions within the dashed box was that successful homing could be achieved from 98 of the 150 grid positions. Our re-implementation achieves successful homing from 136 of these 150 positions. This improvement is eclipsed by CGSM which achieves perfect homing (150 out of 150) within the dashed box.

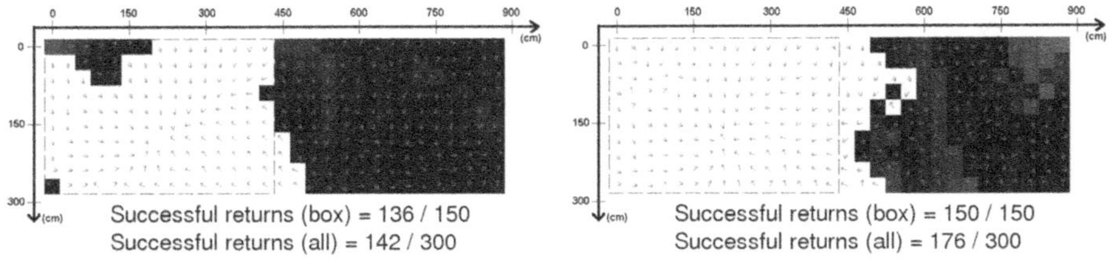

(a) XALV home vectors

(b) CGSM home vectors

Fig. 4. Homing maps of the XALV and CGSM models for goal position (210,180). Home vectors are shown for images taken at corresponding positions. The whiteness of each position is proportional to that position's approach index. Positions within the dashed box area were used for the results presented in [10].

Considering the whole of the dataset, our re-implementation of XALV achieves 142 / 300, while CGSM achieves 176 / 300. Both methods do relatively poorly in that part of the grid furthest from the home position.

The success of homing to one particular goal position does not tell the whole story. It may be that one method achieves good results for certain goal positions but not for others. We have generated 300 homing maps for the 300 image dataset using each image of the dataset, in turn, as the home position. For each homing map we count the number of successful homing positions and divide by 300. This is the *return ratio*. Figure 5 shows the return ratio for all images as the level of whiteness of each cell. Also shown is the approach index ratio which is the sum of all approach index values divided by 300.

The average and maximum return ratios are higher for CGSM than for XALV. The average and maximum approach index ratios are also higher. The standard deviation of both these ratios is lower for CGSM. For 268 of the 300 homing maps the return ratio is strictly higher for CGSM than for XALV. The approach index ratio is also strictly higher for 269 of 300 homing maps.

The area that surrounds a goal position from which successful homing can be achieved is known as the *catchment area*. We can translate the return ratio into squared metres to give an idea of the size of the catchment area for these two models. The average catchment area for CGSM is 14.6 m^2 (max. 25.3 m^2) while the average area for XALV is 6.7 m^2 (max. 21.5 m^2).

3.1 Discussion

Some caveats should first be mentioned in interpreting the results presented above. Firstly, all images were taken with constant orientation. This is an unrealistic scenario for actual robots homing in the real world. Even for robots with sophisticated odometry and compass systems, errors in orientation are inevitable.

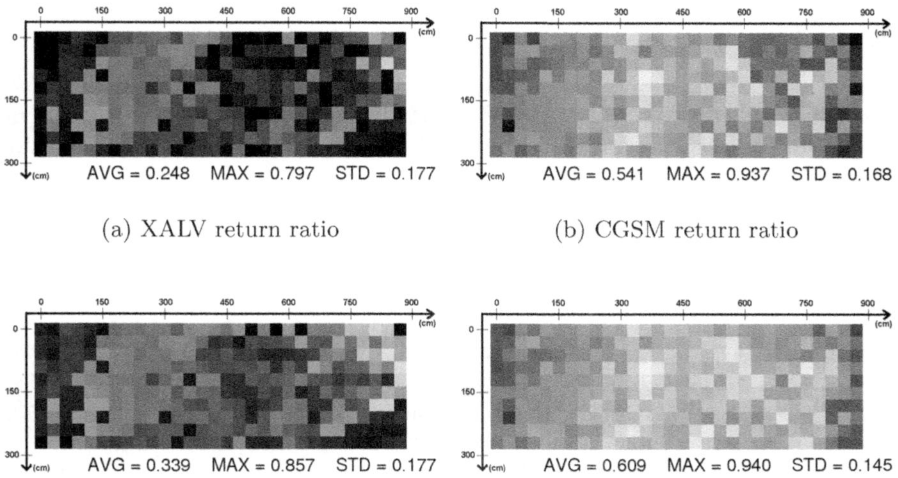

(a) XALV return ratio

(b) CGSM return ratio

(c) XALV approach index ratio

(d) CGSM approach index ratio

Fig. 5. Return ratio and approach index maps for XALV and CGSM.

Secondly, it was reported in [10] that changing lighting conditions seriously impacted on the homing performance of XALV. Thus, care must be taken in citing the catchment areas above as if they referred to actual environment-independent measures. Yet, it may also be true that the catchment area for this environment is *larger* than quoted above. If a set of images covering a larger area was employed it might be found that the catchment area would extend out further. Indeed in figure 4(b) the catchment area is prematurely bounded in places by the edges of the grid.

Our results show superior performance for CGSM over XALV. Both of these models can be considered variants of the snapshot model. We take the superior performance of CGSM to mean that snapshot-based homing models can do better by utilizing information that is implicitly encoded in both dimensions of an image seen by a homing agent. Further, we have shown that the success of a biologically plausible model tested only in simulation [8] can be transferred into a more realistic test environment where homing is based upon real-world images.

4 Conclusions

We have presented a variant of the snapshot model which is capable of successful homing using real-world test images. It exhibits improved performance over a model which did not make use of both input image dimensions. The structure and nature of our processing matrix was inspired by real biological systems. Areas for future work include free-roving robotic experiments, analysis of error conditions, and the use of alternate features and feature movement detection methods.

Acknowledgments. Thanks to Ralf Möller and Thorsten Roggendorf for making their images available, assisting with the replication of their results, and for helpful comments. This work has been partially supported by NSERC Postgraduate Scholarship B - 232621 - 2000.

References

1. B. A. Cartwright and T.S. Collett. Landmark learning in bees. *Journal of Comparative Physiology*, 151:521–543, 1983.
2. T.S. Collett and J. Baron. Biological compasses and the coordinate frame of landmark memories in honeybees. *Nature*, 368:137–140, 1994.
3. Matthias O. Franz, Bernhard Schölkopf, Hanspeter A. Mallot, and Heinrich H. Bülthoff. Where did i take that snapshot? scene-based homing by image matching. *Biological Cybernetics*, 79:191–202, 1998.
4. J. Hong, X. Tan, B. Pinette, R. Weiss, and E.M. Riseman. Image-based homing. In *Proceedings of the 1991 IEEE International Conference on Robotics and Automation, Sacremento, CA*, pages 620–625, 1991.
5. D. H. Hubel and T.N. Wiesel. Receptive fields, binocular interaction and functional architecture in the cat's visual cortex. *Journal of Physiology*, 160:106–154, 1962.
6. S.P.D. Judd and T.S. Collett. Multiple stored views and landmark guidance in ants. *Nature*, 392:710–714, 1998.
7. D. Lambrinos, R. Möller, T. Labhart, R. Pfeifer, and R. Wehner. A mobile robot employing insect strategies for navigation. *Robotics and Autonomous Systems, Special Issue: Biomimetic Robots*, 1999.
8. R. Möller, Marinus Marus, and D. Lambrinos. A neural model of landmark navigation in insects. *Neurocomputing*, 26-27:801–808, 1999.
9. Ralf Möller. Insect visual homing strategies in a robot with analog processing. *Biological Cybernetics*, 83(3):231–243, 2000.
10. Ralf Möller, Dimitrios Lambrinos, Thorsten Roggendorf, Rolf Pfeifer, and Rüdiger Wehner. Insect strategies of visual homing in mobile robots. In B. Webb and T. Consi, editors, *Biorobotics – Methods and Applications*. AAAI Press / MIT Press, 2001.
11. Thomas Röfer. Controlling a whellchair with image-based homing. In *Proceedings of AISB Workshop on Spatial Reasoning in Mobile Robots and Animals, Manchester, UK*, 1997.
12. Thorsten Roggendorf. Visuelle Landmarkennavigation in einer natürlichen, komplexen Umgebung. Master's thesis, Universität Bielefeld, 2000.
13. E. Trucco and A. Verri. *Introductory Techniques for 3-D Computer Vision*. Prentice Hall, 1998.
14. Keven Weber, Svetha Venkatesh, and Mandyam Srinivasan. Insect-inspired robotic homing. *Adaptive Behavior*, 7:65–97, 1999.
15. J. Zeil, A. Kelber, and R. Voss. Structure and function of learning flights in bees and wasps. *Journal of Experimental Biology*, 199:245–252, 1996.

Controlling a Simulated Khepera with an XCS Classifier System with Memory

Andrew Webb, Emma Hart, Peter Ross, and Alistair Lawson

Napier University, Edinburgh EH10 5DT, Scotland, UK
{a.webb,p.ross,e.hart,al.lawson}@napier.ac.uk

Abstract. Autonomous agents commonly suffer from perceptual aliasing in which differing situations are perceived as identical by the robots sensors, yet require different courses of action. One technique for addressing this problem is to use additional internal states within a reinforcement learning system, in particular a learning classifier system. Previous research has shown that adding internal memory states can allow an animat within a cellular world to successfully navigate complex mazes. However, the technique has not previously been applied to robotic environments in which sensory data is noisy and somewhat unpredictable. We present results of using XCS with additional internal memory in the simulated Khepera environment, and show that control rules can be evolved to allow the robot to navigate a variety of problems.

1 Introduction

The problem of perceptual aliasing in robotics has long been recognised. In such a problem, the robotic world contains hidden states which contain information that cannot directly be derived from the current sensory perception of the robot. Many reactive robotic systems in which agents choose their motor action using only current sensory information have been described in the literature, for example [1,2,3,4,5]. In non-Markovian environments (i.e. environments in which the choice of action depends on more than the current state) a purely reactive approach is often insufficient for dealing with hidden state problems [6]. We are interested in developing a control system for a simulated robot that enables it to find a target location, starting from a fixed initial location in an environment which contains such aliasing states, as this almost certainly needs to be addressed when producing a control system for robots in the real world.

McCallum [7] suggested using internal memory states to keep track of past states and disambiguate current perceptual information. An internal state is an intrinsic property of the system and therefore, it is unclear at the outset as to how many internal states are required by the robot. McCallum developed a reinforcement learning algorithm that utilised a window of finite size of current and past observations and actions to determine an appropriate action – a *finite-history* approach. An alternative is to use a *finite-memory* approach, in which internal register bits are used to disambiguate perceptually identical states that

W. Banzhaf et al. (Eds.): ECAL 2003, LNAI 2801, pp. 885–892, 2003.

call for different actions for ultimate success. This was first discussed in [8] and has been developed by others, eg [9].

Most work has focused on rectangular grid worlds, where animats can sense the eight surrounding cells and can move in any of eight directions: N, S, E, W, NE, etc. Kim and Hallam [10] adopted an evolutionary approach to calculate the number of states required to solve Woods problems using a Finite State Machine controller combined with a multi-objective genetic algorithm. Wilson [11] suggested that a form of internal memory could be added to his zeroth-level classifier system, ZCS. Cliff and Ross [12] implemented Wilson's proposals using ZCS and applied it to a set of Woods problems. They found that there are limitations when it becomes necessary to maintain long chains of actions before a reward is gained. Lanzi [13,14] extended Wilson's XCS [15] by adding an internal memory mechanism, (XCSM), but found that whilst XCSM could learn optimal policies in simple Woods environments, it could not find the optimal solution in more complex environments. They extended XCSM in two ways resulting in XCSMH [16]: an internal action occurs only if the external action causes a sensory change, and during the exploration cycle, the external action is chosen probabilistically, while the internal action is chosen deterministically.

The perceptual information used in animat environments is different from that gathered from real or simulated environments. In animat environments, the agent has precise information as to whether its adjacent cells are occupied, contain obstacles or contain a reward and the perceptions map naturally into a binary encoding (e.g using 2-bit codes, '00' for empty, '11' for reward, '10' for an obstacle). Furthermore, the agent always takes single steps of finite size into a new cell on the grid, therefore its position is always exactly known. The sensor information perceived by a robot however, is real-valued, noisy and may not give precise information about the surrounding environment. Motor actions are also inexact, so that the robot moves varying distances at each step. These facts present certain difficulties in mapping the classifier system approach successfully used with animats to that of a robot. Real-values must be mapped to binary values for conventional classifier system models, and it is unclear whether the system noise will prevent the system developing useful control rules. This paper presents the results of using the memory-modified XCS with a simulated Khepera robot in order to learn to find a goal location reliably.

2 Related Work

Stolzmann and Butz [17] apply the Anticipatory Classifier System (ACS) to show how a robot can learn a T-shaped maze environment without the use of reward and also learn the hand-eye coordination of a robot arm. Carse and Pipe [18] describe a fuzzy classifier system X-FCS employed on a mobile robotics problem. Hurst *et al* [19] describe the use of Wilson's Zeroth Classifier System (ZCS) for an obstacle avoidance task and also introduce TCS, a form of ZCS, on a light-seeking task, and Dorigo and Colembetti [20] have used a classifier system to control robots in real and simulated environments. There remain questions about the scalability of the approach to its use in on-line learning systems.

3 Implementing XCS for a Robotic Environment

Wilson's XCS classifier system was used in these experiments, using code freely available from [21]. XCS is not described in detail here as no modifications were made to the classifier engine, the reader should refer to [15] for details. We follow the method proposed in [14], and add an internal state register to each classifier. This extends each classifier to include both an internal *condition*, against which the current contents of the internal memory is matched, and an internal *action* which modifies the contents of the internal register. Thus, internal and external conditions/actions must be specified in the design of the classifer.

The simulated Khepera robot [22] has 6 infrared sensors at the front to detect objects, each returning a value in the range 0-1024 (where 0 means no obstacle in front), with added random noise, which were used in all experiments. The robot also has directional data available to it which was used in some experiments as extra sensor input[1]. In order to represent the real-valued sensors in binary-form suitable for the condition-action rules used by the classifier system, we use thresholded sensor values that return a 0 or 1 depending how close the sensor is to an object. If the sensor value was above a threshold of 900 (chosen through experimentation) then a 1 was returned, otherwise a 0 was returned. The list of all unique (thresholded) states that can then be experienced by the robot sensors is shown in table 1.

Table 1. Unique thresholded states that can be experienced by the robot

Ext. Condition	Interpretation
101100	object to left & front
001101	object to right & front
000000	open space
100000	object near left
000001	object near right

Table 2. Possible actions that can be taken by the robot

Ext. Action	Interpretation
00	rotate left 90°
01	rotate right 90°
10	forward far as possible
11	do nothing

The thresholded sensors corresponding to bits 2 and 5 in table 1 were always zero, so they were excluded. Also, sensors 3 and 4 are either both 1 or both 0, therefore only one of these is needed. Hence for the computational experiments, the distance sensors used were 1, 3 and 6, resulting in a 3-bit external condition. Additional sensory input in the form of the robot's current compass direction was included in some experiments. The directions north, east, south, and west were represented by the binary digits 00, 10, 01 and 11 respectively resulting

[1] The simulator also provides 8 light detectors and 2 rear infra-red detectors which were not used

in an external condition of length 5. If required, the internal condition was represented by b bits, containing symbols from the alphabet {0,1,#} and its contents are matched against the contents of the internal register. The actual number of bits b required was determined experimentally in each case.

The classifier action is a two-bit binary string that represents one of four discrete actions (Table 2). 'Move forward' means continuing to move in a straight direction until the robot is very near an obstacle. The choice of robot motor actions was influenced by the approach adopted by Hurst *et al* [19]. They used continuous actions such that the ZCS only considered a change in action at a significant event, such as a change in sensory readings. Choosing small-scale actions can overwhelm the classifier system with irrelevant information. For experiments with internal memory states, the classifier action was extended to include internal actions that change the memory register contents. A '0' or '1' in the internal action means that the corresponding memory register contents are set to '0' or '1' respectively, whilst '#' leaves the contents of a register unmodified.

4 The Test Environments

Four robot worlds were used in the computational experiments: T2, T4, T5 and T6, illustrated in figure 1. The aim is for the robot to find the target location (indicated by the shaded areas) from an initial fixed position (indicated by the unshaded robot, at the position labelled 0) in the minimum number of steps. For example, in Figure 1 (a), if the robot is at position 0 and facing north then in one step it could reach position 6. In figure 1, the shaded robots indicate the approximate locations and directions of movement the robot can take. A number of experiments were conducted with each of the problems shown in figure 1. For each problem, the size of the internal memory register was varied from 0 bits (purely reactive) to 4 bits. In problem T6, an additional experiment was performed with the addition of directional sensor information to see if this could improve performance. Progressively bigger maze problems encounter more aliasing positions, and are likely to need increasing amounts of memory and/or sensory data to solve them. At the start of each experiment, the robot was placed at the position indicated by position 0 on each of the mazes shown in figure 1. XCS is always run with a default 50/50 explore/exploit strategy, so that if it is in explore mode, it chooses with a probability 0.5 whether to select an action (both internal and external) at random or to choose the action with the highest reward. In exploit mode, the genetic algorithm part of XCS does not operate and the action which predicts the highest payoff is always selected. Each experiment is run for 10,000 steps and repeated 5 times and the best policies found presented.

5 Experimental Results

Table 3 summarises the learned behaviour obtained for Problem T2 without memory and Table 5 shows the learned behaviour obtained using one bit of internal memory. The memoryless XCS obtained a suboptimal solution to the

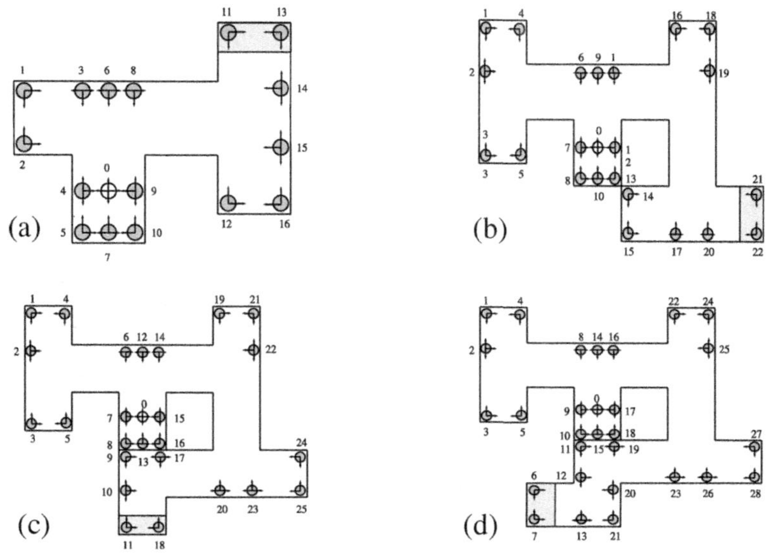

Fig. 1. Robot problem environments: (a) T2, (b) T4, (c) T5 and (d) T6

problem via continual looping movements i.e. rotating right when an obstacle is sensed at the front and moving forward otherwise. An optimal solution to the problem could be found using one bit of memory: this was found to be sufficient to disambiguate the aliasing states (sensors returning 010) encountered at locations 6 and 14 shown in Figure 1 (a).

Table 3. T2: no memory

Condition	Action	Robot location
000	f	0
010	r	6, 11, 14
100	f	6,12,14,16
110	r	12,16

Table 4. T4: no memory

Condition	Action	Robot location
000	f	0
010	r	9,19,20
100	f	9,14,15,19,20
110	r	12,14,15

Table 4 shows the result for Problem T4 using no internal memory and Table 6 shows the result obtained using two bits of internal memory. As with Problem T2, the memoryless XCS obtained a suboptimal solution to the problem via looping movements. Two bits of internal memory was the minimum required to disambiguate the three aliasing states encountered at locations 9, 19 and 20 shown in Figure 1 (b). One memory bit was insufficient to disambiguate all aliasing states and produced trajectories that were no better than the zero-memory XCS.

Table 5. T2: 1-bit memory

Condition Ext.	Int.	Action Ext.	Int.	Location
000	0	f	0	0
010	0	r	0	6
100	0	f	1	6
010	1	l	1	14
001	1	f	1	14
010	#	#	#	13

Table 6. T4: 2-bit memory

Condition Ext.	Int.	Action Ext.	Int.l	Location
000	00	f	00	0
010	00	r	00	9
100	00	f	01	9
010	01	r	01	19
100	01	f	11	19
010	11	l	11	20
001	11	f	11	20
010	##	#	##	22

Table 7 shows the trajectory obtained for for Problem T5 using three internal memory bits. The robot locations are shown in Figure 1 (c). No solution could be found for Problems T5 and T6 using the zero-memory XCS. The aliasing state '010' is encountered four times during the optimal path, which suggests that two memory bits ought to be sufficient to disambiguate them. However, in our experiments, no optimal solution was found using less than three memory bits. For Problem T6, a suboptimal solution was found using 3 memory bits plus additional 2-bit direction data. Table 8 shows the result obtained for Problem T6, the robot locations are shown in Figure 1 (d). When the simulated robot reached location 12, it achieved a left turn by performing the "rotate right 90°" action three times in succession and then moving forward – this is why the result is suboptimal.

6 Discussion

We have shown that XCS with internal memory can be used to control a simulated robot in noisy, partially observable environments and can accomplish goal finding to some extent. Four Woods-type problems have been solved using the XCS with memory, the first three problems optimally. Problem T2 which has 2 significant aliasing states required one internal state for an optimal solution. Problems T4 and T5 required two and three internal states respectively for an optimal solution to be found. An optimal policy for T6 could not be found, even using internal memory, and required the addition of directional sensor information to even find a suboptimal solution. When XCS was used without internal memory, T2 and T4 were solved but suboptimally but it was not possible to find any solution to problems T5 and T6.

The results concur with the findings of previous animat research using both XCS and ZCS, i.e. [12,13] that whilst the addition of internal memory can disambiguate between aliasing states, there are limitations, particularly when there are long sequences of actions. These problems cannot be tackled simply by increasing the number of internal states. This is a significant drawback for real problems, as the number of aliasing states encountered increases with problem size. In these experiments the bits of internal memory turn out to represent a

Table 7. T5: 3-bit memory

Condition Ext	Condition Int	Action Ext	Action Int	Location
000	000	f	001	0
010	001	r	001	12
100	001	f	000	12
010	000	r	111	22
100	111	f	011	22
010	011	l	000	23
100	000	f	101	23
010	101	l	001	10
001	001	f	000	10
010	###	#	###	11

Table 8. T6: 3-bit memory. Ext. condition has 3- bit sensor data plus 2-bit direction data

Condition External	Condition Internal	Action External	Action Internal	Location
000 00	000	f	000	0
010 00	000	r	000	14
100 01	000	f	110	14
010 01	110	r	000	25
100 10	000	f	000	25
010 10	000	r	000	26
100 11	000	f	110	26
010 11	110	r	000	12
100 00	000	r	000	12
000 10	000	r	000	12
001 10	000	f	000	13
010 10	000	r	000	13
100 11	###	f	###	7

shorthand for aspects of the history of motion. Table 8 shows that the system has learned to use only one of the three bits of memory and essentially toggles it after each T-junction in order to produce a solution that is only modestly suboptimal, whereas in T5 (see Table 7) it makes use of all three bits in its optimal answer. This work has assumed the same definition of optimal behaviour as previous work, namely that the robot should learn to get from start to goal without wrong turns. But we would question this. Ideally, a robot should learn to perform well, perhaps with some wrong turns, even in unseen environments. In the maze-like environments used to date, this would be difficult because it is the particular sequence of states that defines the successful route; the sensory information available in any state is not rich enough to provide clues of the kind that (say) hill walkers use, namely 'water flows downhill, so follow the stream'. It is perhaps time for research to move on to much richer sensory environments so that research can focus on getting the robot to learn what sensory information to ignore and on seeing if internal memory bits might correspond, to some extent, to cognitive states rather than to movement history.

Acknowledgements. The authors are grateful for support from EU project IST-2000-29225 (SIGNAL).

References

1. Maes, P., Brooks, R.: Learning to coordinate behaviors. In: Proceedings of the Eighth International Conference on Artificial Intelligence (AAAI'90). (1990) 796–802
2. Lee, W.P.e.: Applying genetic programming to evolve behaviour primitives and arbitrators for mobile robots. In: Proceedings of the IEEE International Conference on Evolutionary Computation, Indianapolis, USA (2000)

3. Sutton, R.S.: Planning by incremental dynamic programming. In: Proceedings of the Eighth International Workshop on Machine Learning. (1991) 353–357
4. Chapman, D., Kaelbling, L.P.: Learning from delayed reinforcement ina complex domain. In: Proceedings of the 12th Int. Joint Conf. on Artificial Intelligence. (1991)
5. Mahadevan, S., Connell, J.: Scaling reinforcement learning to robotics by explot-ing the subsumption architecture. In: Proceedings of the Eighth International Workshop on Machine Learning. (1991)
6. Whitehead, S., Ballad, D.H.: Learning to perceive and act by trial and error. Machine Learning **7** (1991) 45–83
7. McCallum, A.: Reinforcement Learning with Selective Perception and Hidden State. PhD thesis, University of Rochester (1996)
8. Sondik, E.: The optimal control of partially observable Markov processes. PhD thesis, Computer Science, Stanford University (1971)
9. Hansen, E.: Finite-memory control of partially observable systems. PhD thesis, Computer Science, University of Massachussetts at Amherst (1998)
10. Kim, D., Hallam, J.: An evolutionary approach to quantify internal states needed for the woods problem. In: Proceedings of the Seventh International Conference on the Simulation of Adaptive Behavior, From Animals to Animats, MIT Press (2000)
11. Wilson, S.: Zcs: a zeroth level classifier. Evolutionary Computation **2** (1994) 1–18
12. Cliff, D., Ross, S.: Adding temporary memory to zcs. Adaptive Behavior **3** (1994) 101–150
13. Lanzi, P.L.: Adding memory to xcs. In: Proceedings of the IEEE World Congress on Computational Intelligence, IEEE Press (1998) 609–61 Anchorage, Alaska.
14. Lanzi, P.L.: An analysis of the memory mechanism of XCSM. In: Genetic Programming 1998: Proceedings of the Third Annual Conference, Morgan Kaufman (1998) 643–65
15. Wilson, S.W.: Generalization in xcs. Evolutionary Computation **3** (1995) 149–175
16. Lanzi, P.: Adding memory to wilson's xcs classifier system: to learn in partially observable environments. In: Procedings of AAAI Fall Symposium on Partially-observable Markov Decision Processes, AAAI (1998) 91–98
17. Stolzmann, W., Butz, M.: Latent learning and action-planning in robots with anticipatory classifier systems. In P.L Lanzi, W.S., Wilson, S., eds.: Learning Classifier Systems: From Foundations to Application Advances in Evolutionary Computing. Springer-Verlag (2000) 301–317
18. Carse, B., Pipe, A.: X-fcs: A fuzzy classifier system using accuracy-based fitness - first results. Technical Report UWELCSG01-007, University of the West of England, Bristol (2001)
19. Hurst, J., Bull, L., Melhuish, C.: ZCS and TCS learning classifier system controllers on real robots. Technical Report UWELCSG02-002, University of the West of England, Bristol (2002)
20. Dorigo, M.: Alecsys and the autonomouse: Learning to control a real robot by distributed classifier systems. Machine Learning **19** (1995) 209–240
21. (ftp://ftp-illigal.ge.uiuc.edu/pub/src/XCS/XCS-C1.1.tar.Z)
22. (http://diwww.epfl.ch/lami/team/michel/khep-sim/SIM2.tar.gz)

Collective Decision-Making and Behaviour Transitions in Distributed Ad Hoc Wireless Networks of Mobile Robots: Target-Hunting

Jan Wessnitzer and Chris Melhuish

Intelligent Autonomous Systems Laboratory,
University of the West of England,
Coldharbour Lane,Bristol BS16 1QY, United Kingdom
jan.wessnitzer@uwe.ac.uk
http://www.ias.uwe.ac.uk

Abstract. This paper investigates coherent collective motion, collective decision-making and behaviour transitions in networks of wireless connected mobile robots. Emergent global behaviour is achieved through the multiple interactions of locally communicating agents. Collectively, the robots hunt a moving target, immobilise that target and then hunt the next target. Using explicit communication over limited local range, the robots exchange their internal state vectors with neighbouring robots and update their internal states using deterministic transition rules. Configurations of a robot's internal states and local sensor information are then translated into motion. The results demonstrate that the developed swarming algorithm is scalable and by exploiting the distributedness of the swarm shows an increasing success rate for increasing swarm sizes in domains of high noise. We have begun to validate our findings from simulation with real-robot experiments.

1 Introduction

This paper presents the preliminary results of an investigation into coherent motion, collective decision-making and behaviour transitions in swarms of mobile robots. Wireless networks of mobile robots are distributed systems where individuals require and use knowledge about their physical location to control their own motion. Since its advent, wireless local area network technology has become a viable proposition in controlling multi-robot systems. Without global coordinate system and due to the robots' mobility, it is imperative the developed algorithms only use local information and must cope with irregular and dynamically changing network topologies. Local communication over restricted range is desirable since power consumption increases with communication range. Dynamically changing network topologies may put a heavy load on the communication bandwidth. Avoiding routing tables altogether, communication between robots is directed and strictly local.

How does coherent collective action emerge from the simple local interactions between agents? Self-organisation is a term used in many disciplines, sometimes

W. Banzhaf et al. (Eds.): ECAL 2003, LNAI 2801, pp. 893–902, 2003.

under different names, such as autopoesies or autostruosis [1]. Bonabeau *et al.* identified four mechanisms making up self-organisation: positive and negative feedback, amplification of fluctuations and multiple interactions [2]. The mutual reinforcement or inhibition between the network elements, together with random fluctuations, such as noise from the environment or within the internal dynamics of the system (ie. effector noise), make up self-organisation. Swarming in mobile robots has been studied in [10], [9] and [8]. In [3], Kennedy and Eberhart discuss very simple sociocognitive principles that underlie collective intelligence and could be used to produce self-organising phenomena in collective mobile robotics. They used these principles to develop their particle swarm optimisation algorithm. They summarise three principles: *evaluation, comparison* and *imitation.* Evaluation is concerned with the sensing of environmental cues. Individuals then compare their own evaluations with those of their neighbours and imitate those neighbours who return a superior evaluation to themselves. In the framework of collective computing and robotics, this may be interpreted as (a) distributed sensing of environmental cues, (b) parallel information processing, and (c) parallel output of the results of the computation. In this paper, we present two algorithms based on the three above mentioned principles.

We present algorithms which solve the following task. Two prey-robots are randomly moving at twice the speed of the predator-robots through an enclosed environment. The predator-robots' task is to collectively hunt and immobilise one prey-robot and then move off and hunt the other prey-robot. Specifically, we identified and examine three key underpinning algorithmic mechanisms: (1) swarming through local direction control, (2) majority voting and (3) hormone-inspired decision-making. In this paper, we investigate whether such local rules can produce the desired global behaviour.

This paper proceeds as follows. In the next section, we describe, in brief, our simulation model. In the following sections, we describe the different algorithms in more detail. In sections 3 and 4 we present some results obtained from simulation and real robot experiments. Finally, we comment on the performance achieved in simulation and present some ideas for future work.

2 Agent-Based Model

Working in the same framework as developed in [4], we will demonstrate that a small and finite number of internal states, local communication and the robots' noisy sensors are sufficient for generating coherent purposeful behaviour in robot collectives.

We simulate a homogeneous, decentralised and distributed network of mobile finite automata capable of explicit communication between peer elements. A mobile automata network may be defined as a set of elementary processing units, where each unit interacts with neighbouring units and moves in a two-dimensional Euclidean space. The number of neighbours of any processor is several orders less than the total number of processors in the network. Each processor's communication graph may change over time. Each unit makes a deci-

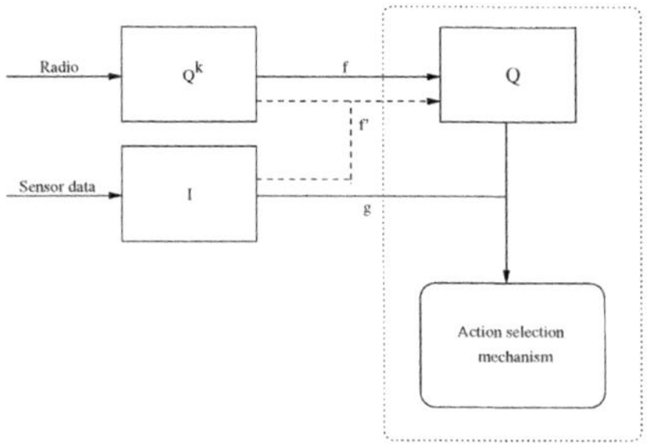

Fig. 1. The agent's action selection mechanism only takes the agent's own internal state vector Q, the internal state vectors Q^k of k neighbouring agents and local sensor information into account

sion based on its internal states and the internal states of its neighbours that are within communication range (as defined by the local communication template). Configurations of the internal states and the local sensor readings are then translated to propulsive motions of the agents. Figure 1 shows schematically how each agent updates its internal state vector using deterministic transition rules.

3 Simulation Experiments

3.1 Experimental Setup

The agents all hold an internal state vector $Q = k, v, g, s, c, t_1, t_2$. The following internal states are required by the motion (section 3.2) and majority vote algorithms (section 3.3):

- k : number of neighbours the agent is actively communicating with
- v : voting variable which simply takes two values, where each value represents a prey-robot.
- g : sensed or estimated distance towards the prey-robot. v determines which prey-robot is hunted.

The following internal states are required to achieve a behaviour switch, the algorithm is described in detail in section 3.4:

- t_1 and t_2 : these states are timers.
- s : s represents the number of agents suggesting to switch v. If s is above a certain threshold, update v.
- c : this internal state is a 'I have made a decision' flag.

At the beginning of each experiment, the agents are placed into an enclosed environment (700×700 units). The agents are able to differentiate between the two prey-robots and can sense the distances d_1 and d_2 to prey-robots 1 and 2 but cannot sense directions (The distances are in fact the Received Signal Strength

Indicator (RSSI) values cf. chemotaxis). However, the robots can sense distance and direction towards their neighbours, but we only use this information if there exists a communication link between two robots. If there is no communication link, other robots are simply seen as obstacles in space which need to be negotiated. After 2000 timesteps, the simulation runs are stopped and we measure how 'coordinated' the swarm behaviour is.

Figure 2 shows a typical simulation run where 50 agents were deployed to hunt two prey-robots. Soon after the start of the experiment, all agents hunt the same target through applying the majority vote algorithm. Consistent movement towards the target was observed. The target has been programmed with a reactive behaviour and it is simply trying to avoid nearby agents. Once the target had been immobilised, the hormone-inspired decision-making mechanism came into play: as soon as $\theta = 5$ robots sensed a stagnation in g for a pre-defined period of time, the robots began hunting the other target.

Since the task has been achieved through a combination of three algorithms, we describe the algorithms and evaluate the system dynamics further in the following subsections.

3.2 Mechanism 1: Local Direction Control

The agents constantly measure the distance towards the prey-robot and compare their measurements with those of their neighbours. In this algorithm the individuals update their direction towards the most successful agent in their topological neighbourhood. After comparing the measured distance with those of its neighbours, agent i updates its motion as follows: if it is closest to the prey-robot, it continues moving in its original direction. If, on the other hand, another agent j in its neighbourhood is closer to the prey-robot, then the agent moves towards agent j.

If no other robot occludes the direct line of sight to the target: $g = d_v$ where d_v is the sensed distance towards the target (which target is set by v). Otherwise $g = \min_{j \in u(i)}\{g_j\} + d_j$ where d_j is the sensed distance to the neighbouring agent j which supplied g_j and $u(i)$ is the neighbourhood of agent i.

Experiment 1: Swarming. In this experiment, the agents are released in one corner of the arena and are required to travel towards a goal beacon positioned in the opposite corner. Here, we measure the mean length of all the robots' trajectories while carrying out the task: all agents are required to move within a predefined radius around the goal. Figure 3 shows decreasing trajectory lengths with increasing swarm size, thereby increasing the convergence speed on the target and improving target time acquisition.

3.3 Mechanism 2: Simple Majority Vote

The purpose of this simple majority vote algorithm is to get all agents to hunt the same target. We applied this algorithm as a distributed solution to the problem

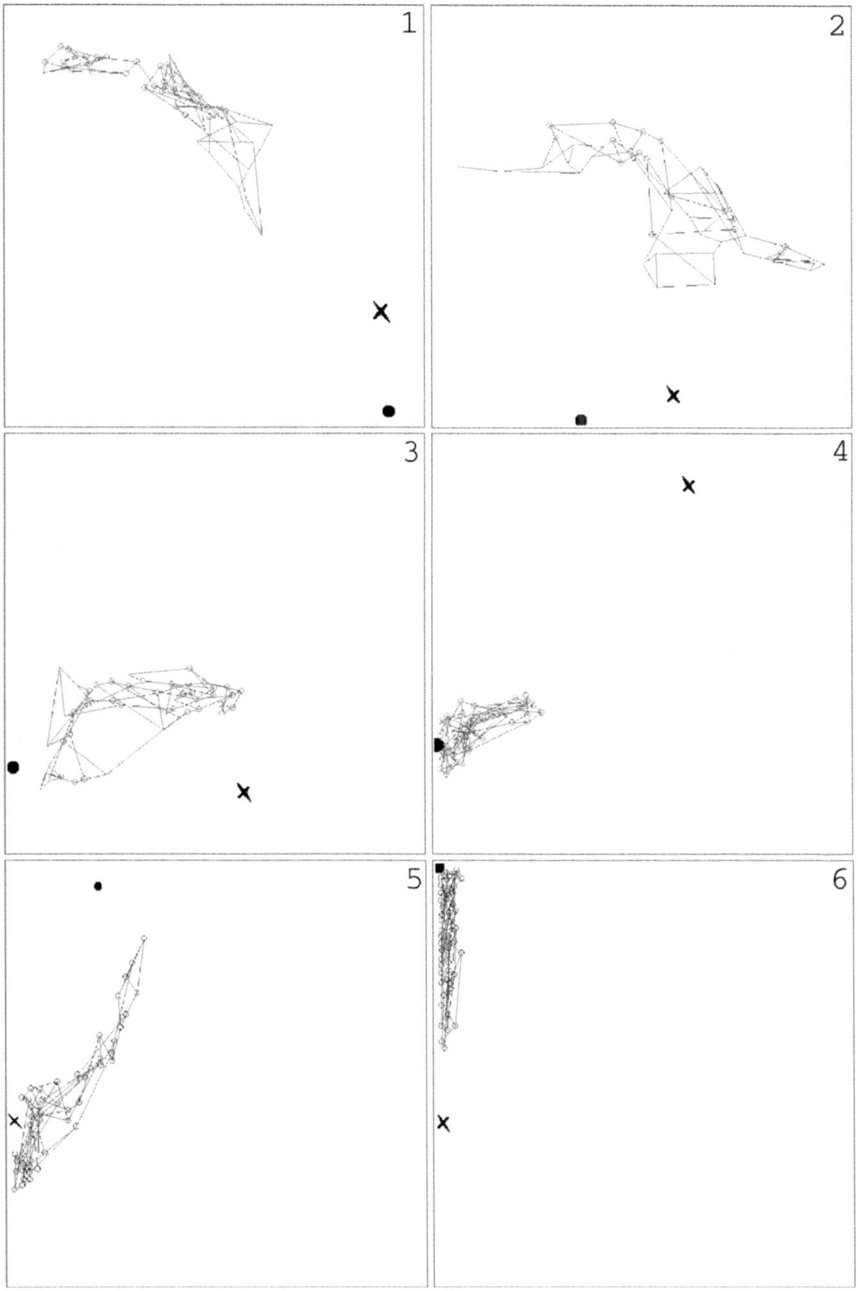

Fig. 2. Hunted prey robot is shown by the point and the ignored target is shown by the cross. 50 agents have been released. Frames 1-4 show the collective closing in on the first target. Frames 5-6 show the collective successfully hunting down the second target.

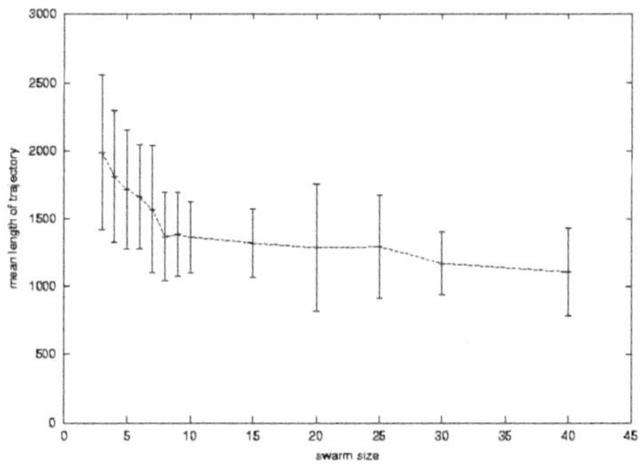

Fig. 3. Mean length of trajectories with increasing swarm size

of deciding which target should be hunted. Imagine two groups of robots, each hunting a different target. These groups converge into one large group hunting one common target. Vote v is initialised at random with equal probability. At each timestep an agent updates its vote according the votes of neighbouring agents [5]. The vote v determines which prey-robot, 1 or 2, is hunted and is in one of the two states for this study. If the majority votes 1 then the agent sets its vote v to 1. If the majority votes 2, then the agent sets its vote v to 2. If an equal number of neighbours vote 1 and 2, then vote v is decided randomly with equal probability.

Experiment 2: Voting. Table 1 shows the performance of the majority voting algorithm measured by the largest fraction of agents hunting the same target for increasing communication and degree k. As expected table 1 shows increasing performance for increasing degrees of connectivity and communication radii. However, adequate performances can still be achieved with small communication ranges and degrees of connectivity.

Table 1. Voting performance for varying degrees of connectivity k and communication radii r

	r			
k	50	100	150	200
2	0.68	0.79	0.86	0.89
3	0.80	0.89	0.94	0.94
4	0.83	0.92	0.94	0.95
5	0.81	0.91	0.93	0.94

3.4 Mechanism 3: Hormone-Inspired Decision-Making

The algorithm presented in this section addresses the problem of group behaviour transitions. When a prey-robot has been immobilised, the swarm should be able to determine this condition and start hunting the next target. An algorithm is required which doesn't solely rely on a single robot's suggestion because this could potentially result in unwanted behaviour switches due to faulty sensor equipment. Shen et al. [6] argue that, computationally speaking, a hormone signal is similar to a content-based message but has the following unique properties: (1) it has no specific destination; (2) it propagates through the network (flooding in networking terms); (3) it may have a lifetime; (4) it may trigger different actions for different receivers.

In similar fashion, robots flood the network with content-based messages to regulate the activity of the swarm. A predefined number of agents θ must sense a stagnation in g for a period of time τ. If an agent senses $|g^t - g^{t+1}| < \epsilon$ then t_1 is incremented otherwise it is decremented. If a stagnation in g has been detected for a pre-defined period of time τ, an agent proposes to switch v. It sends messages to its neighbours containing s and t_2. t_2 is the time over which agents should exchange the states s and t_2. The number of agents proposing a switch of v is stored in s. If an agent does not 'agree' to switch v, it simply forwards the message to its neighbours. In this case, s is not changed but the timer t_2 is counted down. If an agent agrees to switch v, then s is incremented and t_2 is reset allowing the information to travel further in order to reach more agents. The transition rules are defined as follows:

$$s = \max_{j \in u(i)}\{s_j\}.$$
$$t_2 = \max_{j \in u(i)}\{t_{2j}\}.$$
$$t_2 = t_2 - 1.$$

$$\text{if } t_1 > \tau \wedge c = 1 \text{ then } c = 0; s = s + 1; \text{ set } t_2.$$

$$\text{if } s > \theta \wedge t_2 > 0 \wedge c \neq \bullet \text{ then switch } v;\ c = \bullet.$$
$$\text{if } t_2 = 0 \text{ then } c = 1; s = 0.$$

Experiment 3: Connectivity. This algorithm floods the communication graph with messages and therefore the communication graph needs to be connected. Table 2 shows higher connectivity of the communication graph of the robotic network with increasing communication radius r and degree k. The interesting point here is that adequate diffusion-like exchange of information can still be achieved with low degrees of connectivity and communication range. Connectivity also indicates good swarming behaviour.

4 Real-Robot Experiment

Experiments were carried out with five LinuxBots [7]. Due to this small number of robots, the robots' task was to guide a robot towards a beacon. As in simulation, the robots can only sense the distances towards the beacon and the robot

Table 2. Global connectivity for varying degrees of connectivity k and communication radii r

	r			
k	50	100	150	200
2	0.52	0.56	0.67	0.73
3	0.56	0.86	0.94	0.96
4	0.62	0.87	0.97	0.99
5	0.64	0.90	0.95	0.99

to be guided towards that beacon. Figure 6 shows four robots (a) guiding a fifth robot (b) towards a beacon (c). Figures 4 and 5 show results obtained in simulation. Figure 4 shows the mean length of all robots' trajectories for varying noise levels and swarm sizes. Each noise level corresponds to a normally distributed error term with mean zero and standard deviation σ, $\varepsilon_i \in N(0,\sigma)$, added to the sensed distances. Figure 5 shows the number of successful runs out of 30 trials for varying noise levels and swarm sizes. It can be seen that for higher noise levels the success rate increases with swarm size whereas the success rate remains almost maximal with increasing swarm size at lower noise levels.

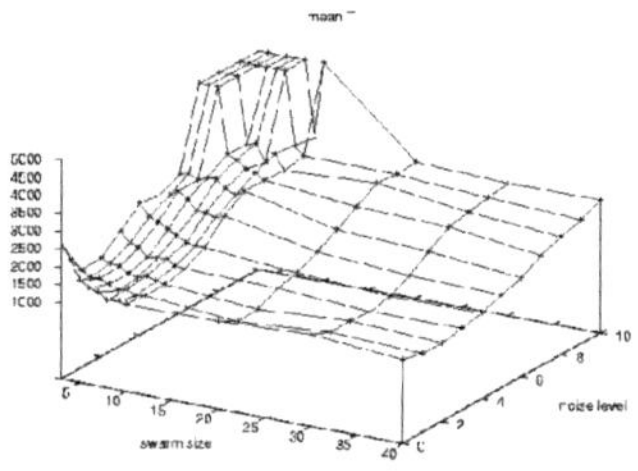

Fig. 4. Mean length of trajectories for varying noise levels and swarm sizes

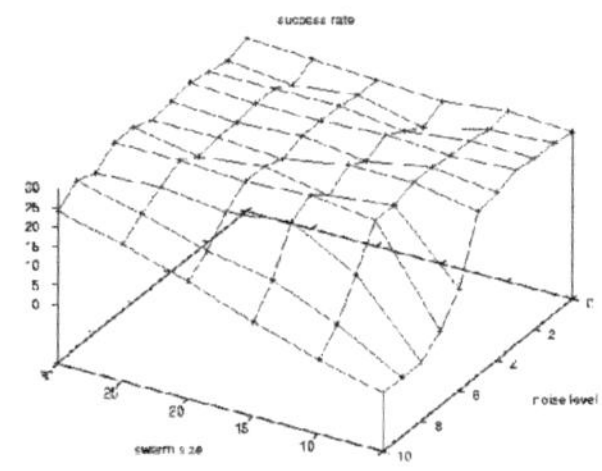

Fig. 5. Number of successful runs for varying noise levels and swarm sizes

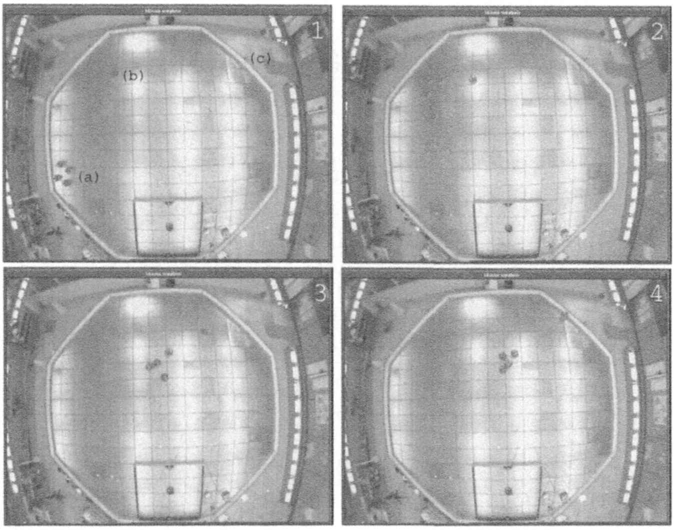

Fig. 6. Four LinuxBots guiding a fifth robot towards a beacon

As shown in figure 6, the robots successfully guide a robot towards a beacon in an arena with a diameter of roughly ten metres. When the distance between the beacon and the robot was less than a metre the experiment was successful. Four experiments have yielded mean trajectory lengths for all robots of 16.9, 18.1, 17.4 and 23.2 metres. The robots exhibited irregular motion when located in regions of low signal to noise. As soon as the robots entered regions of high signal to noise near the signal source they swiftly completed the task.

5 Discussion and Future Work

In simulation, we have demonstrated collective distributed decision-making and collective behaviour transitions in *ad hoc* networks of mobile robots. Three key mechanisms which work within the distributed framework of restricted communication range and low degrees of connectivity were identified. As expected overall performance with the task was improved with increasing connectivity and communication radius. However, it was shown that the identified mechanisms were able to produce good overall performance in the task even when the degree of connectivity and communication range were severely restricted. This has implications for the design of building large scale swarm systems in that simpler designs most likely reflect improved robustness and reduced communication ranges demand a smaller power budget and computational overhead. Coherent global behaviour has been achieved despite the limitations of local information exchange with a small number of neighbours. By exploiting the spatial distributedness of the swarm, the success rate has been shown to increase with swarm size, most noticeably in noisy environments. The real-robot experiments described were limited to five LinuxBots but future work will include studies with a larger number of robots.

Acknowledgements. We acknowledge the support of the EPSRC grant GR/R79319/01 and would like to thank Alan Winfield, Ian Horsfield and Chris Bytheway for the LinuxBot infrastructure.

References

1. Melhuish, C.: Autostruosis: Construction without explicit planning in MicroRobots – a social insect approach. Evolutionary Robotics Vol III. AAI Books Ed. Takashi Gomi, 2000.
2. Bonabeau, E., Dorigo, M., Theraulaz, G.: Swarm intelligence: from natural to artificial systems. Oxford University Press 1999.
3. Kennedy, J., Eberhart, R.: Swarm Intelligence. Academic Press 2001.
4. Wessnitzer, J., Adamatzky, A., Melhuish, C.: Towards self-organising structure formations: a decentralised approach. Proceedings of ECAL 2001.
5. Watts, D.: Small worlds – the dynamics of networks between order and randomness. Princeton University Press 1999.
6. Shen, W., Salemi, B., Will, P.: Hormone-Inspired Adaptive Communication and Distributed Control for CONRO Self-Reconfigurable Robots. IEEE Transactions on Robotics and Automation 2002.
7. Winfield, A., Holland, O.: The application of wireless local area network technology to the control of mobile robots. Microprocessors and Microsystems 23 (2000) 597–607.
8. Hayes, A., Dormiani-Tabatabaei, P.: Self-organized flocking with agent failure: Off-line optimisation and demonstration with real robots. Proceedings of the IEEE Int. Conf. on Robotics and Automation 2002.
9. Toner, J., Tu, Y.: Flocks, herds and schools: a quantitative theory of flocking. Physical review E 58 (1998) 4828–4858.
10. Reynolds, C.: Flocks, herds and schools: a distributed behavioural model. Computer graphics 21 (1987) 25–34.

Author Index

Adams, Bryant 1
Aerts, Diederik 615
Antony, Antony 387
Arita, Takaya 395
Asano, Shoichiro 716
Avila-García, Orlando 733
Avraham, Hezi 743
Azzag, H. 564

Balaam, Andy 154
Banzhaf, Wolfgang 20, 217
Barfoot, Timothy D. 857
Bedau, Mark A. 676
Benkö, Gil 10
Bentley, Peter J. 248, 753
Bersini, Hugues 164, 191
Boekhorst, René te 733
Boer, Bart de 415
Broek, Eva van den 425
Broekaert, Jan 615
Buason, Gunnar 763
Buck-Sorlin, Gerhard H. 625
Bullock, Seth 299
Burtsev, Mikhail S. 580
Busch, Jens 20
Buzing, P.C. 434

Cañamero, Lola 733
Capcarrere, Mathieu S. 278
Cartlidge, John 299
Centler, Florian 31
Changshui, Zhang 572
Chechik, Gal 743
Collier, Travis C. 525

D'Eleuterio, Gabriele M.T. 857
Dautenhahn, Kerstin 266, 829
De La Rosa, Armando 97
Dittrich, Peter 31
Dorigo, Marco 865
Dornhaus, Anna R. 643

Ebner, Marc 228
Egri-Nagy, Attila 238
Eiben, A.E. 434
Escobar, Rodrigo 97

Federici, Diego 309
Fend, Miriam 771
Fernando, Chrisantha 588
Flamm, Christoph 10
Floreano, Dario 128
Folino, Gianluigi 598
Foreman, Mark 781
Forestiero, Agostino 598
Franks, Nigel R. 643

Ganon, Zohar 319
Gershenson, Carlos 606, 615
Gheorghieş, Ovidiu 635
Glotzmann, Thomas 328
Greenman, John 792
Groşan, Crina 651
Groß, Roderich 865
Guinot, C. 564

Hart, Emma 847, 885
Hashimoto, Takashi 107
Hauhs, Michael 328
Heylighen, Francis 606
Humphrys, Mark 839
Hurford, James R. 442, 507
Hutton, Tim J. 51

Ieropoulos, Ioannis 792
Iizuka, Hiroyuki 336
Ikegami, Takashi 59, 89, 336, 482

Jansen, Bart 800
Jenkins, Tudor 452

Kaneko, Yoshiaki 146
Katai, Osamu 69
Kawakami, Hiroshi 69
Keinan, Alon 199, 319
Keller, Laurent 128
Kim, Jan T. 686
Kniemeyer, Ole 625
Kobele, Gregory M. 525
Kovacs, Tim 643
Ku, Lawrence 31
Kubík, Aleš 346

Kumagai, Yuya 107
Kurth, Winfried 625

Labella, Thomas H. 865
Lange, Holger 328
Lawson, Alistair 847, 885
Lee, Yoosook 525
Lewin, Michael 462
Lin, Ying 525
Lipson, Hod 1
Lowe, Robert 118
Luchian, Henri 635

Mach, Robert 810
Macinnes, Ian 821
Mac Dónaill, Dónall A. 41
Madina, Duraid 59
Mahdavi, Siavash Haroun 248
Maniadakis, Michail 183
Marshall, James A.R. 643
Martínez, Genaro Juárez 175
Matsumaru, Naoki 31
Matsuzaki, Shuichi 357
McIntosh, Harold V. 175
Melhuish, Chris 792, 893
Mesot, Bertrand 367
Miller, Julian F. 256
Mirolli, Marco 377
Molter, Colin 191
Monmarché, N. 564
Morgavi, Giovanna 847

Nehaniv, Chrystopher L. 238, 266, 829
Newton, Adam Lee 829

O'Leary, Ciarán 839
Oltean, Mihai 651
Ono, Naoaki 59
Oppacher, Franz 875
Osano, Minetada 357

Parisi, Domenico 377
Penn, Alexandra 659
Perez-Uribe, Andres 128
Pfaffmann, Jeffrey 31
Pfeifer, Rolf 771
Poelz, Patrick 847
Polani, Daniel 118, 667
Prem, Erich 847
Prokopenko, Mikhail 781

Quick, Tom 266

Raven, Michael J. 676
Repsilber, Dirk 686
Riggle, Jason 525
Righetti, Ludovic 278
Roberts, Graham 266
Ross, Peter 847, 885
Ruppin, Eytan 199, 319, 743

Saggie, Keren 199
Şahin, Erol 865
Salzberg, Chris 387
Sasahara, Kazutoshi 482
Sayama, Hiroki 387
Schatten, Rolf 491
Schut, M.C. 434
Schweitzer, Frank 810
Seck Tuoh Mora, Juan Carlos 175
Seidl, Klaus 696
Semeria, Alessandro 706
Serra, Roberto 706
Shen, Hengwei 716
Shimohara, Katsunori 138
Shiose, Takayuki 69
Shokur, Solaiman 278
Sipper, Moshe 724
Slimane, M. 564
Smith, Andrew D.M. 499
Smith, Kenny 507, 517
Sojakka, Sampsa 588
Spezzano, Giandomenico 598
Spier, Emmet 462
Spieth, Christian 289
Stabler, Edward P. 525
Stadler, Peter F. 10
Stauffer, André 724
Streichert, Felix 289
Sugiura, Komei 69
Suzuki, Hideaki 69, 78, 357
Suzuki, Keisuke 89
Suzuki, Reiji 395

Takahashi, Ichiro 146
Tanev, Ivan 138
Tao, Ban 572
Taylor, Charles E. 525
Teuscher, Christof 367
Thangavelautham, Jekanthan 857
Todd, Peter M. 425

Trahanias, Panos 183
Trianni, Vito 865

Ulmer, Holger 289
Unemi, Tatsuo 146

Van Looveren, Joris 472
Vardy, Andrew 875
Venturini, G. 564
Vickerstaff, Robert 209
Villani, Marco 706
Vogt, Paul 535, 545

Wang, Peter 781
Webb, Andrew 847, 885
Wei, Shu 572
Wessnitzer, Jan 893
Wilke, Claus O. 405

Yao, Yuan 525
Yokoi, Hiroshi 771

Zauner, Klaus-Peter 31
Zell, Andreas 289
Ziemke, Tom 763
Zuidema, Willem 553

Lecture Notes in Artificial Intelligence (LNAI)

Vol. 2600: S. Mendelson, A.J. Smola, Advanced Lectures on Machine Learning. Proceedings, 2002. IX, 259 pages. 2003.

Vol. 2627: B. O'Sullivan (Ed.), Recent Advances in Constraints. Proceedings, 2002. X, 201 pages. 2003.

Vol. 2631: R. Falcone, S. Barber, L. Korba, M. Singh (Eds.), Trust, Reputation, and Security: Theories and Practice. Proceedings, 2002. X, 235 pages. 2003.

Vol. 2636: E. Alonso, D, Kudenko, D. Kazakov (Eds.), Adaptive Agents and Multi-Agent Systems. XIV, 323 pages. 2003.

Vol. 2637: K.-Y. Whang, J. Jeon, K. Shim, J. Srivastava (Eds.), Advances in Knowledge Discovery and Data Mining. Proceedings, 2003. XVIII, 610 pages. 2003.

Vol. 2639: G. Wang, Q. Liu, Y. Yao, A. Skowron (Eds.), Rough Sets, Fuzzy Sets, Data Mining, and Granular Computing. Proceedings, 2003. XVII, 741 pages. 2003.

Vol. 2645: M.A. Wimmer (Ed.), Knowledge Management in Electronic Government. Proceedings, 2003. XI, 320 pages. 2003.

Vol. 2650: M.-P. Huget (Ed.), Communication in Multiagent Systems. VIII, 323 pages. 2003.

Vol. 2654: U. Schmid, Inductive Synthesis of Functional Programs. XXII, 398 pages. 2003.

Vol. 2662: H.-K. Kahng (Ed.), Information Networking. Proceedings, 2003. XVII, 1032 pages. 2003.

Vol. 2663: E. Menasalvas, J. Segovia, P.S. Szczepaniak (Eds.), Advances in Web Intelligence. Proceedings, 2003. XII, 350 pages. 2003.

Vol. 2671: Y. Xiang, B. Chaib-draa (Eds.), Advances in Artificial Intelligence. Proceedings, 2003. XIV, 642 pages. 2003.

Vol. 2680: P. Blackburn, C. Ghidini, R.M. Turner, F. Giunchiglia (Eds.), Modeling and Using Context. Proceedings, 2003. XII, 525 pages. 2003.

Vol. 2684: M.V. Butz, O. Sigaud, P. Gérard (Eds.), Anticipatory Behavior in Adaptive Learning Systems. X, 303 pages. 2003.

Vol. 2685: C. Freksa, W. Brauer, C. Habel, K.F. Wender (Eds.), Spatial Cognition III. X, 415 pages. 2003.

Vol. 2689: K.D. Ashley, D.G. Bridge (Eds.), Case-Based Reasoning Research and Development. Proceedings, 2003. XV, 734 pages. 2003.

Vol. 2691: V. Mařík, J. Müller, M. Pěchouček (Eds.), Multi-Agent Systems and Applications III. Proceedings, 2003. XIV, 660 pages. 2003.

Vol. 2699: M.G. Hinchey, J.L. Rash, W.F. Truszkowski, C. Rouff, D. Gordon-Spears (Eds.), Formal Approaches to Agent-Based Systems. Proceedings, 2002. IX, 297 pages. 2003.

Vol. 2700: M.T. Pazienza (Ed.), Information Extraction in the Web Era. XIII, 163 pages. 2003.

Vol. 2702: P. Brusilovsky, A. Corbett, F. de Rosis (Eds.), User Modeling 2003. Proceedings, 2003. XIV, 436 pages. 2003.

Vol. 2705: S. Renals, G. Grefenstette (Eds.), Text- and Speech-Triggered Information Access. Proceedings, 2000. VII, 197 pages. 2003.

Vol. 2711: T.D. Nielsen, N.L. Zhang (Eds.), Symbolic and Quantitative Approaches to Reasoning with Uncertainty. Proceedings, 2003. XII, 608 pages. 2003.

Vol. 2715: T. Bilgiç, B. De Baets, O. Kaynak (Eds.), Fuzzy Sets and Systems – IFSA 2003. Proceedings, 2003. XV, 735 pages. 2003.

Vol. 2718: P. W. H. Chung, C. Hinde, M. Ali (Eds.), Developments in Applied Artificial Intelligence. Proceedings, 2003. XIV, 817 pages. 2003.

Vol. 2721: N.J. Mamede, J. Baptista, I. Trancoso, M. das Graças Volpe Nunes (Eds.), Computational Processing of the Portuguese Language. Proceedings, 2003. XIV, 268 pages. 2003.

Vol. 2734: P. Perner, A. Rosenfeld (Eds.), Machine Learning and Data Mining in Pattern Recognition. Proceedings, 2003. XII, 440 pages. 2003.

Vol. 2741: F. Baader (Ed.), Automated Deduction – CADE-19. Proceedings, 2003. XII, 503 pages. 2003.

Vol. 2744: V. Mařík, D. McFarlane, P. Valckenaers (Eds.), Holonic and Multi-Agent Systems for Manufacturing. Proceedings, 2003. XI, 322 pages. 2003.

Vol. 2746: A. de Moor, W. Lex, B. Ganter (Eds.), Conceptual Structures for Knowledge Creation and Communication. Proceedings, 2003. XI, 405 pages. 2003.

Vol. 2752: G.A. Kaminka, P.U. Lima, R. Rojas (Eds.), RoboCup 2002: Robot Soccer World Cup VI. XVI, 498 pages. 2003.

Vol. 2773: V. Palade, R.J. Howlett, L. Jain (Eds.), Knowledge-Based Intelligent Information and Engineering Systems. Proceedings, Part I, 2003. LI, 1473 pages. 2003.

Vol. 2774: V. Palade, R.J. Howlett, L. Jain (Eds.), Knowledge-Based Intelligent Information and Engineering Systems. Proceedings, Part II, 2003. LI, 1443 pages. 2003.

Vol. 2777: B. Schölkopf, M.K. Warmuth (Eds.), Learning Theory and Kernel Machines. Proceedings, 2003. XIV, 746 pages. 2003.

Vol. 2782: M. Klusch, A. Omicini, S. Ossowski, H. Laamanen (Eds.), Cooperative Information Agents VII. Proceedings, 2003. XI, 345 pages. 2003.

Vol. 2792: T. Rist, R. Aylett, D. Ballin, J. Rickel (Eds.), Intelligent Virtual Agents. Proceedings, 2003. XV, 364 pages. 2003.

Vol. 2796: M. Cialdea Mayer, F. Pirri (Eds.), Automated Reasoning with Analytic Tableaux and Related Methods. Proceedings, 2003. X, 271 pages. 2003.

Vol. 2801: W. Banzhaf, T. Christaller, P. Dittrich, J.T. Kim, J. Ziegler (Eds.), Advances in Artificial Life. Proceedings, 2003. XVI, 905 pages. 2003.

Vol. 2807: V. Matoušek, P. Mautner (Eds.), Text, Speech and Dialogue. Proceedings, 2003. XIII, 426 pages. 2003.

Lecture Notes in Computer Science

Vol. 2758: D. Basin, B. Wolff (Eds.), Theorem Proving in Higher Order Logics. Proceedings, 2003. X, 367 pages. 2003.

Vol. 2759: O.H. Ibarra, Z. Dang (Eds.), Implementation and Application of Automata. Proceedings, 2003. XI, 312 pages. 2003.

Vol. 2761: R. Amadio, D. Lugiez (Eds.), CONCUR 2003 - Concurrency Theory. Proceedings, 2003. XI, 524 pages. 2003.

Vol. 2762: G. Dong, C. Tang, W. Wang (Eds.), Advances in Web-Age Information Management. Proceedings, 2003. XIII, 512 pages. 2003.

Vol. 2763: V. Malyshkin (Ed.), Parallel Computing Technologies. Proceedings, 2003. XIII, 570 pages. 2003.

Vol. 2764: S. Arora, K. Jansen, J.D.P. Rolim, A. Sahai (Eds.), Approximation, Randomization, and Combinatorial Optimization. Proceedings, 2003. IX, 409 pages. 2003.

Vol. 2765: R. Conradi, A.I. Wang (Eds.), Empirical Methods and Studies in Software Engineering. VIII, 279 pages. 2003.

Vol. 2766: S. Behnke, Hierarchical Neural Networks for Image Interpretation. XII, 224 pages. 2003.

Vol. 2769: T. Koch, I. T. Sølvberg (Eds.), Research and Advanced Technology for Digital Libraries. Proceedings, 2003. XV, 536 pages. 2003.

Vol. 2773: V. Palade, R.J. Howlett, L. Jain (Eds.), Knowledge-Based Intelligent Information and Engineering Systems. Proceedings, Part I, 2003. LI, 1473 pages. 2003. (Subseries LNAI).

Vol. 2774: V. Palade, R.J. Howlett, L. Jain (Eds.), Knowledge-Based Intelligent Information and Engineering Systems. Proceedings, Part II, 2003. LI, 1443 pages. 2003. (Subseries LNAI).

Vol. 2776: V. Gorodetsky, L. Popyack, V. Skormin (Eds.), Computer Network Security. Proceedings, 2003. XIV, 470 pages. 2003.

Vol. 2777: B. Schölkopf, M.K. Warmuth (Eds.), Learning Theory and Kernel Machines. Proceedings, 2003. XIV, 746 pages. 2003. (Subseries LNAI).

Vol. 2778: P.Y.K. Cheung, G.A. Constantinides, J.T. de Sousa (Eds.), Field-Programmable Logic and Applications. Proceedings, 2003. XXVI, 1179 pages. 2003.

Vol. 2779: C.D. Walter, Ç.K. Koç, C. Paar (Eds.), Cryptographic Hardware and Embedded Systems – CHES 2003. Proceedings, 2003. XIII, 441 pages. 2003.

Vol. 2781: B. Michaelis, G. Krell (Eds.), Pattern Recognition. Proceedings, 2003. XVII, 621 pages. 2003.

Vol. 2782: M. Klusch, A. Omicini, S. Ossowski, H. Laamanen (Eds.), Cooperative Information Agents VII. Proceedings, 2003. XI, 345 pages. 2003. (Subseries LNAI).

Vol. 2783: W. Zhou, P. Nicholson, B. Corbitt, J. Fong (Eds.), Advances in Web-Based Learning – ICWL 2003. Proceedings, 2003. XV, 552 pages. 2003.

Vol. 2786: F. Oquendo (Ed.), Software Process Technology. Proceedings, 2003. X, 173 pages. 2003.

Vol. 2787: J. Timmis, P. Bentley, E. Hart (Eds.), Artificial Immune Systems. Proceedings, 2003. XI, 299 pages. 2003.

Vol. 2789: L. Böszörményi, P. Schojer (Eds.), Modular Programming Languages. Proceedings, 2003. XIII, 271 pages. 2003.

Vol. 2790: H. Kosch, L. Böszörményi, H. Hellwagner (Eds.), Euro-Par 2003 Parallel Processing. Proceedings, 2003. XXXV, 1320 pages. 2003.

Vol. 2792: T. Rist, R. Aylett, D. Ballin, J. Rickel (Eds.), Intelligent Virtual Agents. Proceedings, 2003. XV, 364 pages. 2003. (Subseries LNAI).

Vol. 2794: P. Kemper, W. H. Sanders (Eds.), Computer Performance Evaluation. Proceedings, 2003. X, 309 pages. 2003.

Vol. 2795: L. Chittaro (Ed.), Human-Computer Interaction with Mobile Devices and Services. Proceedings, 2003. XV, 494 pages. 2003.

Vol. 2796: M. Cialdea Mayer, F. Pirri (Eds.), Automated Reasoning with Analytic Tableaux and Related Methods. Proceedings, 2003. X, 271 pages. 2003. (Subseries LNAI).

Vol. 2798: L. Kalinichenko, R. Manthey, B. Thalheim, U. Wloka (Eds.), Advances in Databases and Information Systems. Proceedings, 2003. XIII, 431 pages. 2003.

Vol. 2799: J.J. Chico, E. Macii (Eds.), Integrated Circuit and System Design. Proceedings, 2003. XVII, 631 pages. 2003.

Vol. 2801: W. Banzhaf, T. Christaller, P. Dittrich, J.T. Kim, J. Ziegler (Eds.), Advances in Artificial Life. Proceedings, 2003. XVI, 905 pages. 2003. (Subseries LNAI).

Vol. 2803: M. Baaz, J.A. Makowsky (Eds.), Computer Science Logic. Proceedings, 2003. XII, 589 pages. 2003.

Vol. 2805: K. Araki, S. Gnesi, D. Mandrioli (Eds.), FME 2003: Formal Methods. Proceedings, 2003. XVII, 942 pages. 2003.

Vol. 2807: V. Matoušek, P. Mautner (Eds.), Text, Speech and Dialogue. Proceedings, 2003. XIII, 426 pages. 2003. (Subseries LNAI).

Vol. 2810: M.R. Berthold, H.-J. Lenz, E. Bradley, R. Kruse, C. Borgelt (Eds.), Advances in Intelligent Data Analysis V. Proceedings, 2003. XV, 624 pages. 2003.

Vol. 2817: D. Konstantas, M. Leonard, Y. Pigneur, S. Patel (Eds.), Object-Oriented Information Systems. Proceedings, 2003. XII, 426 pages. 2003.

Vol. 2818: H. Blanken, T. Grabs, H.-J. Schek, R. Schenkel, G. Weikum (Eds.), Intelligent Search on XML Data. XVII, 319 pages. 2003.

Vol. 2832: G. Di Battista, U. Zwick (Eds.), Algorithms – ESA 2003. Proceedings, 2003. XIV, 790 pages. 2003.

Vol. 2839: A. Marshall, N. Agoulmine (Eds.), Management of Multimedia Networks and Services. Proceedings, 2003. XIV, 532 pages. 2003.

GPSR Compliance

The European Union's (EU) General Product Safety Regulation (GPSR) is a set of rules that requires consumer products to be safe and our obligations to ensure this.

If you have any concerns about our products, you can contact us on ProductSafety@springernature.com

In case Publisher is established outside the EU, the EU authorized representative is:

Springer Nature Customer Service Center GmbH
Europaplatz 3
69115 Heidelberg, Germany

Batch number: 10844661

Printed by Printforce, the Netherlands